New Directions in Civil Engineering

Series Editor
W. F. CHEN
Purdue University

Zdeněk P. Bažant and Jaime Planas
Fracture and Size Effect in Concrete and Other Quasibrittle Materials

W.F. Chen and Seung-Eock Kim
LRFD Steel Design Using Advanced Analysis

W.F. Chen and E.M. Lui
Stability Design of Steel Frames

W.F. Chen and K.H. Mossallam
Concrete Buildings: Analysis for Safe Construction

W.F. Chen and S. Toma
Advanced Analysis of Steel Frames: Theory, Software, and Applications

W.F. Chen and Shouji Toma
Analysis and Software of Cylindrical Members

Y.K. Cheung and L.G. Tham
Finite Strip Method

Hsai-Yang Fang
Introduction to Environmental Geotechnology

Yuhshi Fukumoto and George C. Lee
Stability and Ductility of Steel Structures under Cyclic Loading

Ajaya Kumar Gupta
Response Spectrum Method in Seismic Analysis and Design of Structures

C.S. Krishnamoorthy and S. Rajeev
Artificial Intelligence and Expert Systems for Engineers

Boris A. Krylov
Cold Weather Concreting

Pavel Marek, Milan Guštar and Thalia Anagnos
Simulation-Based Reliability Assessment for Structural Engineers

N.S. Trahair
Flexural-Torsional Buckling of Structures

Jan G.M. van Mier
Fracture Processes of Concrete

S. Vigneswaran and C. Visvanathan
Water Treatment Processes: Simple Options

FRACTURE AND SIZE EFFECT in Concrete and Other Quasibrittle Materials

Zdeněk P. Bažant
Walter P. Murphy Professor of Civil Engineering
and Materials Science
Northwestern University
Evanston, Illinois

Jaime Planas
Professor of Materials Science
E.T.S. Ingenieros de Caminos, Canales y Puertos
Universidad Politécnica de Madrid
Madrid, Spain

CRC Press
Boca Raton Boston London New York Washington, D.C.

Library of Congress Cataloging-in-Publication Data

Bažant, Z. P.
Fracture and size effect in concrete and other quasibrittle materials / Zdeněk P. Bažant and Jamie Planas.
p. cm.
Includes bibliographical references and index.
ISBN 0-8493-8284-X (alk. paper)
1. Concrete--Fracture,. 2. Fracture mechanics. 3. Elastic analysis (Engineering) I. Planas, Jaime. II. Title. III. Series.
TA440.B377 1997
620.1′.30426—dc21
for Library of Congress 97-26399
CIP

International Standard Book Number 0-8493-8284-X
Library of Congress Card Number 97-26399
Printed in the United States of America 1 2 3 4 5 6 7 8 9 0
Printed on acid-free paper

Preface

Our book is intended to serve as both a textbook for graduate level courses in engineering and a reference volume for engineers and scientists. We assume that the reader has the background of the B.S. level mechanics courses in the departments of civil, mechanical, or aerospace engineering. Aside from synthesizing the main results already available in the literature, our book also contains some new research results not yet published and many original derivations.

The subject of our book is important to structural, geotechnical, mechanical, aerospace, nuclear, and petroleum engineering, as well as materials science and geophysics. In our exposition of this subject, we try to proceed from simple to complex, from special to general. We try to be as concise as possible and use the lowest level of mathematics necessary to treat the subject clearly and accurately. We include the derivations or proofs of all the important results, as well as their physical justifications. We also include a large number of fully worked out examples and an abundance of exercise problems, the harder ones with hints. Our hope is that the reader will gain from the book true understanding rather than mere knowledge of the facts.

A special feature of our book is the theory of scaling of the failure loads of structures, and particularly the size effect on the strength of structures. We present a systematic exposition of this currently hot subject, which has gained prominence in current research. It has been only two decades that the classical model of size effect, based on Weibull-type statistical theory of random material strength, was found to be inadequate in the case of quasibrittle materials. Since then, a large body of results has been accumulated and is scattered throughout many periodicals and proceedings. We attempt to bring it together in a single volume. In treating the size effect, we try to be comprehensive, dealing even with aspects such as statistical and fractal, which are not normally addressed in the books on fracture mechanics.

Another special feature of our book is the emphasis on quasibrittle materials. These include concrete, which is our primary concern, as well as rocks, toughened ceramics, composites of various types, ice, and other materials. Owing to our concern with the size effect and with quasibrittle fracture, much of the treatment of fracture mechanics in our book is different from the classical treatises, which were concerned primarily with metals.

In its scope, our book is considerably larger than the subject matter of a single semester-length course. A graduate level course on fracture of concrete, with proper treatment of the size effect and coverage relevant also to other quasibrittle materials, may have the following contents: Chapter 1, highlights of Chapters 2, 3, and 4, then a thorough presentation of the main parts of Chapters 5, 6, 7, and 8, parts of Chapters 9 and 12, and closing with mere comments on Chapters 10, 11, and 13. A quarter-length course obviously requires a more reduced coverage.

The book can also serve as a text for a basic course on fracture mechanics. In that case, the course consists of a thorough coverage of Section 1.1 and Chapters 2, 3, 4, 5, and 7.

Furthermore, the book can be used as a text for a course on the scaling of fracture (i.e., the size effect), as a follow-up to the aforementioned basic course on fracture mechanics (or to courses on fracture mechanics based on other books). In that case, the coverage of this second course may be as follows: the rest of Chapters 1 and 5, a thorough exposition of Chapter 6, the rest of Chapters 7 and 8, much of Chapter 9, followed by highlights only of Chapter 10, bits of Chapter 11, and a thorough coverage of Chapter 12.

Chapters 13 and 14, the detailed coverage of which is not included in the foregoing course outlines, represent extensions important for computational modeling of fracture and size effect in structures. They alone can represent a short course, or they can be appended to the course on fracture of concrete or the course on scaling of fracture, although at the expense of the depth of coverage of the preceding chapters.

We were stimulated to write this book by our teaching of various courses on fracture mechanics, damage, localization, material instabilities, and scaling.[1] Our collaboration on this book began already in 1990, but had to proceed with many interruptions, due to extensive other commitments and duties. Most of the book was written between 1992 and 1995.

Our book draws heavily from research projects at Northwestern University funded by the Office of Naval Research, National Science Foundation, Air Force Office of Scientific Research, Waterways Experiment Station of the U.S. Army Corps of Engineers, Argonne National Laboratory, Department of Energy, and Sandia National Laboratories, as well as from research projects at the Universidad Politécnica de Madrid, funded by Dirección General de Investigación Científica y Técnica (Spain) and Comisión Interministerial de Ciencia y Tecnología (Spain). We are grateful to these agencies for their support.

The first author wishes to express his thanks to his father, Zdeněk J. Bažant, Professor Emeritus of Foundation Engineering at the Czech Technical University (ČVUT) in Prague, and to his grandfather Zdeněk Bažant, late Professor of Structural Mechanics at ČVUT, for having excited his interest in structural mechanics and engineering; to his colleagues and research assistants, for many stimulating discussions; and to Northwestern University, for providing an environment conducive to scholarly inquiry. He also wishes to thank his wife Iva for her moral support and understanding. Thanks are further due to Carol Surma, Robin Ford, Valerie Reed and Arlene Jackson, secretaries at Northwestern University, for their expert and devoted secretarial assistance.

The second author wishes to express his thanks to his mother María Rosselló, and to his sisters Joana María and María for their continuous encouragement. He also wishes to thank his wife Diana for her patience and moral support. He further expresses his thanks to Manuel Elices, professor and head of Department of Materials Science, for his continued teaching and support and for allowing the author to devote so much time to his work on this book; to assistant professor Gustavo V. Guinea for his stimulating discussions and friendly support; to Claudio Rocco, visiting scientist on leave from the Universidad de la Plata (Argentina), for providing test results and pictures for the section on the Brazilian test; to Gonzalo Ruiz, assistant professor, for providing test results and figures for the section on minimum reinforcement; and to all the colleagues, research students and personnel in the Department of Material Science, for their help in carrying out other duties which suffered from the author's withdrawal to his writing of the book.

Z.P.B. and J.P.
Evanston and Madrid
April, 1997

[1] In the case of the first author: The course on Fracture of Concrete, introduced at Northwestern University in 1988, and intensive short courses on these subjects taught at Politecnico di Milano (1981, 1993, 1997), Swiss Federal Institute of Technology, Lausanne (1987, 1989, 1994), Ecole Normale Supérieure de Cachan, France (1992), and Lulea University, Sweden (1994). In the case of the second author: The undergraduate courses on Fracture Mechanics and Continuum Mechanics and the doctoral-level courses of Physics of Continuum Media and Advanced Fracture Mechanics at the Universidad Politécnica de Madrid, and intensive short courses on Fracture Mechanics taught at Universidad Politécnica and at Universidad Carlos III in Madrid (1994, 1995), and at Universidad de la Plata, Argentina (1995).

Vector and Tensor Notation

In this book, both component and compact form are used for representation of vectors and tensors. Component notation is standard, since cartesian reference axes are used in general. For the compact notation that is used in several chapters to simplify the expressions, the following conventions are used:

1. Geometric vectors are bold faced lower case roman latin letters: e.g., $\mathbf{n}, \mathbf{t}, \mathbf{m}$.
2. Microplane or, in general, microscopic vectors are denoted by a superimposed arrow, thus $\vec{n}, \vec{\varepsilon}, \vec{\sigma}$.
3. Except for a few greek boldmath for classical stresses and strains ($\boldsymbol{\sigma}$ and $\boldsymbol{\varepsilon}$), second-order tensors are represented as bold face upper case roman latin letters, such as $\mathbf{E}, \mathbf{N}, \mathbf{M}, \mathbf{A}$, etc.
4. Fourth-order tensors are represented as bold faced upper case *italic* latin letters, such as $\boldsymbol{E}, \boldsymbol{C}, \boldsymbol{B}$, etc.
5. The transformation of a vector by a second-order tensor into another vector is represented by simple juxtaposition: $\mathbf{t} = \boldsymbol{\sigma}\mathbf{n}$ or $\mathbf{t} = \mathbf{T}\mathbf{n}$.
6. The transformation of a second-order tensor by a fourth-order tensor into another second-order tensor is represented by simple juxtaposition as well: $\boldsymbol{\sigma} = \boldsymbol{E}\boldsymbol{\varepsilon}$, $\boldsymbol{\varepsilon} = \boldsymbol{C}\boldsymbol{\sigma}$ or $\mathbf{H} = \boldsymbol{D}\mathbf{N}$, etc.
7. The inner-product of two vectors or two second-rank tensors is represented by a dot, e.g., $\vec{n} \cdot \vec{m}$, $\mathbf{n} \cdot \mathbf{m}$, $\vec{\sigma} \cdot \delta\vec{\varepsilon}$, $\boldsymbol{\sigma} \cdot \delta\boldsymbol{\varepsilon}$, $\mathbf{T} \cdot \dot{\mathbf{F}}$, etc. Accordingly, the expression $\mathbf{T} \cdot \boldsymbol{C}\mathbf{S}$ represents the inner product of the second order tensors $\mathbf{T}$ and $\mathbf{E} = \boldsymbol{C}\mathbf{S}$, the latter being the transformed by the fourth-order tensor $\boldsymbol{C}$ of the second-order tensor $\mathbf{S}$.

About the Authors

Born and educated in Prague, Dr. Zdeněk P. Bažant became in 1973 professor at Northwestern University, was named in 1990 to the distinguished W.P. Murphy Chair, and served during 1981-87 as Director (founding) of the Center for Concrete and Geomaterials. He has authored over 370 refereed journal articles and published books on Stability of Structures (1991), Concrete at High Temperatures (1996) and Creep of Concrete (1966). He served as Editor (in Chief) of ASCE Journal of Engineering Mechanics (1988-94), and is Regional Editor of the International Journal of Fracture and member of many other editorial boards. He was founding president of IA-FraMCoS, president of the Society of Engineering Science, and chairman of IA-SMiRT Division H. He has chaired many technical committees in ASCE, RILEM and ACI. He is a Registered Structural Engineer in Illinois. He has been staff consultant to Argonne National Laboratory and consulted for many firms and institutes. His awards include the Prager Medal from SES; Newmark Medal, Huber Prize and T.Y. Lin Award from ASCE; RILEM Medal; Humboldt Award; an honorary doctorate from ČVUT, Prague; Guggenheim, Kajima, JSPS, NATO and Ford Foundation Fellowships; Meritorious Paper Award from the Structural Engineering Association; Best Engineering Book of the Year Award from AAP; Medal of Merit from the Czech Society of Mechanichs; Gold Medal from the Building Research Institute of Spain; Honorary Memberships in the last two and in the Czech Society of Civil Engineers, and others. He is a Fellow of the American Academy of Mechanics, ASME, ASCE, RILEM and ACI. In 1996 he was elected to the National Academy of Engineering.

Dr. Jaime Planas received his degree in civil engineering in 1974 from the Universidad de Santander, Spain. He received his Ph.D. degree from the Universidad Politécnica de Madrid, Spain, in 1997. His doctoral thesis was awarded the Entrecanales Prize to the best thesis in soil mechanics and the Extraordinary Prize of the Polytechnical University of Madrid. He joined the Department of Materials Science of the Universidad Politécnica de Madrid, first as a lecturer (1975-77), later as assistant professor (1978-1988) and in 1989 he became professor. He teaches, or has taught, general physics, materials science, continuum mechanics, fracture mechanics and constitutive equations to students of civil engineering and of materials engineering. He conducted research on low temperature (-170 °C) mechanical properties of engineering materials and on fracture mechanics applied to materials for civil engineering. Since 1983 his main interest focused on fracture of concrete and quasibrittle materials. In this field, he actively participated in various RILEM and ESIS Technical Committees and published over 50 research papers and several reviews.

The authors in Segovia, Spain, in 1992. In the background, the aqueduct of Segovia, build by the Romans two thousand years ago.

Mému zesnulému dědečkovi Zdeňkovi, otci Zdeňkovi a matce Štěpánce,
kteří pěstovali moji touhu o poznání, a ženě Ivě, která jí podporovala.[1]
Zdeněk P. Bažant

A mon pare i ma mare, a les meves germanes, i a la meva dona,
amb el meu amor i les meves gràcies, ara i sempre.[2]
Jaime Planas

[1] To my late grandfather Zdeněk, father Zdeněk, and mother Štěpánka who nurtured my desire for knowledge, and my wife Iva who supported it.

[2] To my father and mother, to my sisters, and to my wife, with my love and my thanks, now and ever.

Dr. Zdeněk Bažant, one of the persons to whom this book is dedicated, was born on November 25, 1879, in Prostějov and died on September 1, 1954, in Nové Město na Moravě (both now in the Czech Republic). He was professor of structural mechanics and strength of materials at the Czech Technical University in Prague (ČVUT) and a member of the Czechoslovak Academy of Sciences. He was twice elected the rector (president) of ČVUT. He authored several books and many research articles in Czech, English, French, and German.

Contents

1

Why Fracture Mechanics?

Fracture mechanics is a failure theory that

1. determines material failure by *energy* criteria, possibly in conjunction with strength (or yield) criteria, and

2. considers failure to be *propagating* throughout the structure rather than simultaneous throughout the entire failure zone or surface.

While fracture mechanics has already been generally accepted in failure analysis of metal structures, especially in aerospace, naval, and nuclear engineering, its advent in the field of concrete structures is new. Therefore, after briefly outlining the history of this discipline, we will attempt in this introductory chapter to spell out the reasons for adopting the fracture mechanics approach and will focus especially on the structural size effect – the main reason for introducing fracture mechanics into the design of concrete structures.

1.1 Historical Perspective

Concrete structures are, of course, full of cracks. Failure of concrete structures typically involves stable growth of large cracking zones and the formation of large fractures before the maximum load is reached. So why has not the design of concrete structures been based on fracture mechanics, a theory whose principles have been available since the 1950s? Have concrete engineers been guilty of ignorance?

Not really. The forms of fracture mechanics that were available until recently were applicable only to homogeneous brittle materials, such as glass or to homogeneous typical structural metals. The question of applicability of these classical theories to concrete was explored long ago, beginning with Kaplan (1961) and others, but the answer was negative (e.g., Kesler, Naus and Lott 1972). Now, we understand that the reason for the negative answer was that the physical processes occurring in concrete fracture are very different from those taking place in the fracture of the aforementioned materials and, especially, that the material internal length scale for these fracture processes is much larger for concrete than for most materials. A form of fracture mechanics that can be applied to this kind of fracture has appeared only during the late 1970s and the 1980s.

Concrete design has already seen two revolutions. The first, which made the technology of concrete structures possible, was the development of the elastic no-tension analysis during 1900-1930. The second revolution, based on a theory conceived chiefly during the 1930s, was the introduction of plastic limit analysis during 1940-1970. There are now good reasons to believe that introduction of fracture mechanics into the design of concrete structures might be the third revolution. The theory, formulated mostly after 1980, finally appears to be ripe.

1.1.1 Classical Linear Theory

The stimulus for fracture mechanics was provided by a classical paper of Inglis (1913), who obtained the elastic solution for stresses at the vertex of an ellipsoidal cavity in an infinite solid and observed that, as the ellipse approaches a line crack (i.e., as the shorter axis tends to zero), the stress at the vertex of the ellipse tends to infinity (Fig. 1.1.1). Noting this fact, Griffith (1921, 1924) concluded that, in presence of a crack, the stress value cannot be used as a criterion of failure since the stress at the tip of a sharp crack in an elastic continuum is infinite no matter how small the applied load (Fig. 1.1.1b).

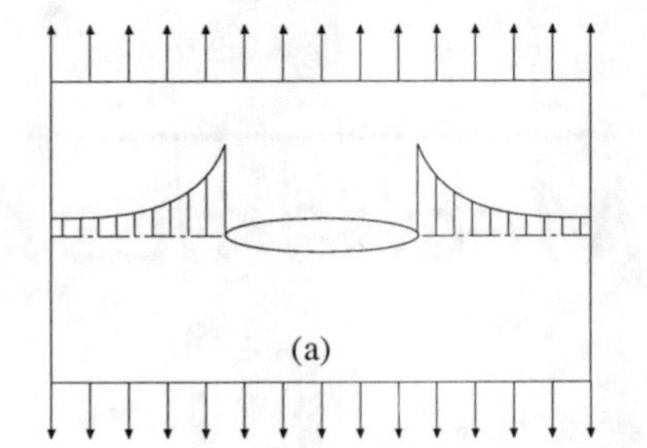

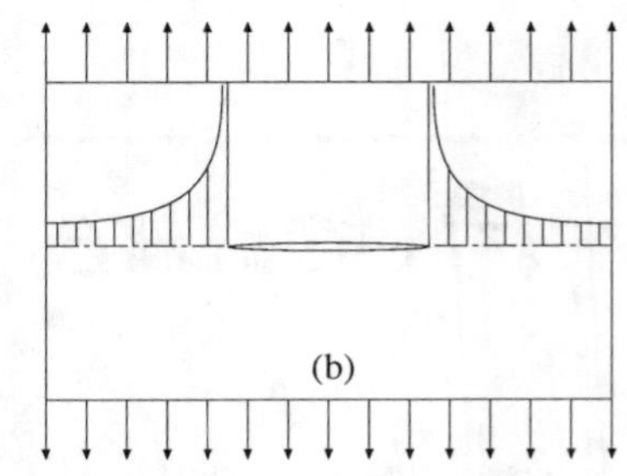

Figure 1.1.1 The stress at the ellipse vertices is finite in an elastic plate with an elliptical hole (a); but the stress concentration tends to infinity as the ellipse shrinks to a crack (b).

This led him to propose an energy criterion of failure, which serves as the basis of the classical linear elastic fracture mechanics (LEFM) or of the more general elastic fracture mechanics (EFM, in which linearity is not required). According to this criterion, which may be viewed as a statement of the principle of balance of energy, the crack will propagate if the energy *available* to extend the crack by a unit surface area equals the energy *required* to do so. Griffith took this energy to be equal to $2\gamma_s$, where γ_s is the specific surface energy of the elastic solid, representing the energy that must be supplied to break the bonds in the material microstructure and, thus, create a unit area of new surface.

Soon, however, it was realized that the energy actually required for unit crack propagation is much larger than this value, due to the fact that cracks in most materials are not smooth and straight but rough and tortuous, and are accompanied by microcracking, frictional slip, and plasticity in a sizable zone around the fracture tip. For this reason, the solid state specific surface energy $2\gamma_s$ was replaced by a more general *crack growth resistance,* $\mathcal{R}$, which, in the simplest approximation, is a constant. The determination of $\mathcal{R}$ has been, and still is, a basic problem in experimental fracture mechanics. The other essential problem of LEFM is the determination, for a given structure, of the energy available to advance the crack by a unit area. Today, this magnitude is called the *energy release rate,* and is usually called $\mathcal{G}$ (note that the rate is with respect to crack length, not time).

The early Griffith work was considered of a rather academic nature because it could only explain the failure of very brittle materials such as glass. Research in this field was not intensely pursued until the 1940s. The development of elastic fracture mechanics essentially occurred during 1940-1970, stimulated by some perplexing failures of metal structures (e.g., the fracture splitting of the hulls of the "Liberty" ships in the U.S. Navy during World War II). During this period, a good deal of theoretical, numerical, and experimental work was accomplished to bring LEFM to its present state of mature scientific discipline.

In a highly schematic vision, the essence of the theoretical work consisted in generalizing Griffith's ideas, which he had worked out only for a particular case, to any situation of geometry and loading, and to link the energy release rate $\mathcal{G}$ (a structural, or global, quantity) to the elastic stress and strain fields. The essence of the experimental work consisted in setting up test methods to measure the crack growth resistance $\mathcal{R}$. In the energetic approach, the last theoretical step was the discovery of the J-integral by Rice (1968a,b). It gave a key that closed, on very general grounds, the circle relating the energy release rate to the stress and strain fields close to the crack tip for any elastic material, linear or not, and supplied a logical tool to analyze fracture for more general nonlinear behaviors. Today, it is one of the cornerstones of elastoplastic fracture mechanics, the branch of fracture mechanics dealing with fracture of ductile materials.

The second major achievement in the theoretical foundation of LEFM was due to Irwin (1957), who introduced the concept of the stress intensity factor K as a parameter for the intensity of stresses close to the crack tip and related it to the energy release rate. Irwin's approach had the enormous advantage that the stress intensity factors are additive, while Griffith's energy release rates were not. However, his approach was limited to linear elasticity, while the concept of energy release rate was not.

1.1.2 Classical Nonlinear Theories

LEFM, which is expounded in Chapters 2–4, provides the basic tool today for the analysis of many structural problems dealing with crack growth, such as safety in presence of flaws, fatigue crack growth, stress corrosion cracking, and so on. However, soon after the introduction of the fracture mechanics concepts, it became evident that LEFM yielded good predictions only when fracture was very brittle, which meant that most of the structure had to remain elastic up to the initiation of fracture. This was not the case for many practical situations, in particular, for tough steels which were able to develop large plastic zones near the crack tip before tearing off. The studies of Irwin, Kies and Smith (1958) identified the size of the yielding zone at the crack tip as the source of the misfit. Then, various nonlinear fracture mechanics theories were developed, more or less in parallel. Apart from elastoplastic fracture mechanics (essentially based on extensions of the J-integral concept, and outside the scope of this book), two major descriptions were developed: equivalent elastic crack models and cohesive crack models.

In the equivalent crack models, which will be presented in detail in Chapters 5 and6, the nonlinear zone is approximately simulated by stating that its effect is to decrease the stiffness of the body, which is approximately the same as increasing the crack length while keeping everything else elastic. This longer crack is called the effective or equivalent crack. Its treatment is similar to LEFM except that some rules have to be added to express how the equivalent crack extends under increased forces. In this context, Irwin (1958) in general terms, and more clearly Krafft, Sullivan and Boyle (1961), proposed the so-called R-curve (resistance curve) concept, in which the crack growth resistance $\mathcal{R}$ is not constant but varies with the crack length in a manner empirically determined in advance. This simple concept still remains a valuable tool provided that the shape of the R-curve is correctly estimated, taking the structure geometry into account.

For concrete, the equivalent crack models proposed by Jenq and Shah (1985a,b) and Bažant and co-workers are among the most extended and have led to test recommendations for fracture properties of concrete (see Chapters 5 and 6). The 1980s have also witnessed a rise of interest in the size effect, as one principal consequence of fracture mechanics. A simple approximate formula for the effect of structure size on the nominal strength of structures has been developed (Bažant 1984a) and later exploited, not only for the predictions of failures of structures, but also as the basis of test recommendations for the determination of nonlinear fracture properties, including the fracture energy, the length of the fracture process zone, and the R-curve. This R-curve concept has also been applied to ceramics and rocks with some success, although until recently it has not been recognized that the R-curves are not a true material property but depend on geometry.

The cohesive crack models, which are discussed in detail in Chapter 7, were developed to simulate the nonlinear material behavior near the crack tip. In these models, the crack is assumed to extend and to open while still transferring stress from one face to the other. The first cohesive model was proposed by Barenblatt (1959, 1962) with the aim to relate the macroscopic crack growth resistance to the atomic binding energy, while relieving the stress singularity (infinite stress was hard to accept for many scientists). Barenblatt simulated the interatomic forces by introducing distributed cohesive stresses on the newly formed crack surfaces, depending on the separation between the crack faces. The distribution of these cohesive atomic forces was to be calculated so that the stress singularity would disappear and the stresses would remain bounded everywhere. Barenblatt postulated that the cohesive forces were operative on only a small region near the crack tip, and assumed that the shape of the crack profile in this zone was independent of the body size and shape. Balancing the external work supplied to the crack tip zone —which he showed to coincide with Griffith's $\mathcal{G}$— against the work of the cohesive forces —which was $2\gamma_s$ by definition— he was able to recover the Griffith's results while eliminating the uncomfortable stress singularity.

Dugdale (1960) formulated a model of a line crack with a cohesive zone with constant cohesive stress (yield stress). Although formally close to Barenblatt's, this model was intended to represent a completely different physical situation: macroscopic plasticity rather than microscopic atomic interactions. Both models share a convenient feature: the stress singularity is removed. Although very simplified, Dugdale's approach to plasticity gave a good description of ductile fracture for not too large plastic zone sizes. However, it was not intended to describe fracture itself and, in Dugdale's formulation, the plastic zone extended forever without any actual crack extension.

More elaborate cohesive crack models have been proposed with various names (Dugdale-Barenblatt models, fictitious crack models, bridged crack models, cracks with closing pressures, etc.). Such models include specific stress-crack opening relations simulating complete fracture (with a vanishing transferred stress for large enough crack openings) to simulate various physically different fracture mechanisms: crazing in polymers (which must take viscoelastic strains into account, see Chapter 11), fiber and crack bridging in ceramics, and frictional aggregate interlock and crack overlapping in concrete. All these models share common features; in particular, a generic model can be formulated such that all of them become particular cases, and the mathematical and numerical tools are the same (Elices and Planas 1989).

However, the fictitious crack model proposed by Hillerborg for concrete (Hillerborg, Modéer and Petersson 1976) merits special comment . In general, all the foregoing fracture mechanics theories require a preexisting crack to analyze the failure of a structure or component. If there is no crack, neither LEFM nor EFM, equivalent crack models or classical cohesive crack models, can be applied. This is not so with Hillerborg's fictitious crack model. It is a cohesive crack in the classical sense described above, but it is more than that because it includes crack initiation rules for any situation (even if there is no precrack). This means that it can be applied to initially uncracked concrete structures and describe all the fracture processes from no crack at all to complete structural breakage. It provides a continuous link between the classical strength-based analysis of structures and the energy-based classical fracture mechanics: cohesive cracks start to open as dictated by a strength criterion that naturally and smoothly evolves towards an energetic criterion for large cracks. We will discuss this model in detail in Chapter 7.

1.1.3 Continuum-Based Theories

The foregoing description is, at least for concrete, only half of the story: the half dealing with researchers interested in discovering when and how a preexisting crack-like flaw or defect would grow. The other half deals with structural engineers wanting to describe the crack formation and growth from an initially flaw-free structure (in a macroscopic sense). The first finite element approaches to that problem consisted in reducing to zero the stiffness of the elements in which the tensile strength was reached (Rashid 1968). Later, more sophisticated models were used with progressive failure of the elements (progressive softening) and, starting with the work of Kachanov (1958), there was a great proliferation of continuum damage mechanics models with internal variables describing softening.

However, even though some results were very promising, it later became apparent that numerical analysis using these continuum models with softening yielded results strongly dependent on the size of the elements of the finite element mesh (see the next section for details). To overcome this difficulty while keeping the continuum mechanics formulation —which seems more convenient for structural analysis— Bažant developed the crack band model in which the crack was simulated by a fracture band of a fixed thickness (a material property) and the strain was uniformly distributed across the band (Bažant 1976, 1982; Bažant and Cedolin 1979, 1980; Bažant and Oh 1983a; Rots et al. 1985). This approximation, analyzed in depth in Chapter 8, was initially rivalling Hillerborg's model, but it soon became apparent that they were numerically equivalent (Elices and Planas 1989).

Since the 1980s, a great effort, initiated by Bažant (1984b) with the imbricate continuum, was devoted to develop softening continuum models that can give a consistent general description of fracture processes without further particular hypotheses regarding when and how the fracture starts and develops. In the nonlocal continuum approach, discussed in Chapter 13, the nonlinear response at a point is governed not only by the evolution of the strain at that point but also by the evolution of the strains at other points in the neighborhood of that point. These models, which probably constitute the most general approach to fracture, evolved from the early nonlocal elastic continua (Eringen 1965, 1966; Kröner 1967) to nonlocal continua in which the nonlocal variables are internal irreversible variables such as damage or inelastic strain (Pijaudier-Cabot and Bažant 1987; Bažant and Lin 1988a,b). Higher-order continuum models, in which the response at a point depends on the strain tensor and on higher order gradients (which include Cosserat continua) are related to the nonlocal model and are also intended to handle fracture in a continuum framework (e.g., de Borst and Mühlhaus 1991). However, the numerical difficulties associated with using generalized continuum models make these models available for practical use to only a few research groups. Moreover, sound theoretical analysis concerning convergence and uniqueness is still lacking, which keeps these models somewhat provisional. Nevertheless, the generalizing power of these models is undeniable and they can provide a firm basis to extend some simpler and well accepted models. It has been recently shown, for example, that the cohesive crack models arise as rigorous solutions of a certain class of nonlocal models (Planas, Elices and Guinea 1993).

1.1.4 Trends in Fracture of Quasibrittle Materials

The research activity in fracture mechanics of quasibrittle materials —concrete, rocks, ceramics, composites, ice, and some polymers— experienced a burst of activity during the 1980s. Much research effort was —and still is— devoted to refine the foregoing models, to improve the analytical and numerical tools required to handle the models, to develop experimental methods to measure the parameters entering the various theories, and to relate the macroscopic fracture behavior to the microstructural features of the materials. In this respect, idealized models reflecting the heterogeneous nature of concrete have been developed to help understanding of the macroscopic behavior (see Chapter 14 for details). Extensive bibliographies and historical reviews of concrete fracture mechanics have recently appeared in the reports of various committees (Wittmann 1983; Elfgren 1989; ACI Committee 446 1992).

Recently, it is being recognized that fractures of concrete and of modern toughened ceramics exhibit strong similarities. Their exploitation should benefit both disciplines. In fact, the way to toughen ceramics is to make them behave more like concrete, especially reinforced concrete.

At present, we are entering a period in which introduction of fracture mechanics into concrete design is becoming possible (see Chapter 10). This will help achieve more uniform safety margins, especially for structures of different sizes. This, in turn, will improve economy as well as structure reliability. It will make it possible to introduce new designs and utilize new concrete materials. Fracture mechanics will no doubt be especially important for high-strength concrete structures, fiber-reinforced concrete structures, concrete structures of unusually large sizes, and other novel structures. Applications of fracture mechanics are most urgent for structures such as concrete dams and nuclear reactor vessels or containments, for which the safety concerns are particularly high and the consequences of a potential disaster enormous.

One of the simplest ways to incorporate fracture mechanics into design practice is through the size effect, or modification of structural strength with the size of the structure. The analysis of size effect starts later in this chapter and permeates most of the book.

1.2 Reasons for Fracture Mechanics Approach

Since concrete structures have been designed and successfully built according to codes that totally ignore fracture mechanics theory, it might seem unnecessary to change the current practice. Nevertheless, there are five compelling reasons for doing so.

Reason 1: Energy required for crack formation must be taken into account.

Reason 2: The results of the structural analysis must be objective.

Reason 3: The structural analysis must agree with the absence of yield plateau from the load-deflection diagram.

Reason 4: The structural analysis must adequately compute the energy absorption capability and ductility.

Reason 5: The structural analysis must capture the size effect.

Let us examine these reasons in more detail.

1.2.1 Energy Required for Crack Formation

From the strictly physical viewpoint, it must be recognized that while crack initiation may depend on stress, the actual formation of cracks requires a certain energy —the fracture energy. Hence, energy criteria should be used. This reason might suffice to a physicist, but not to a designer, at least at a first glance. There are, however, more practical reasons for taking the fracture mechanics approach.

1.2.2 Objectivity of Analysis

A physical theory must be objective, in the sense that the results of calculations made with it must not depend on subjective aspects such as the choice of coordinates, the choice of mesh, etc. If a theory is found

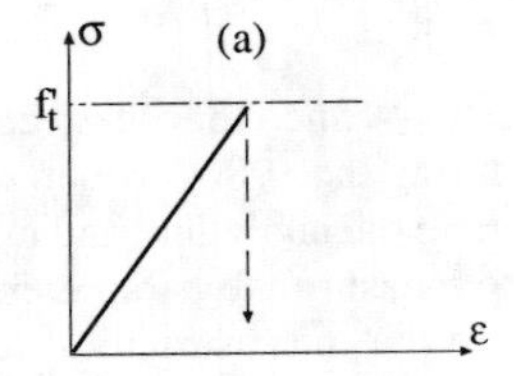

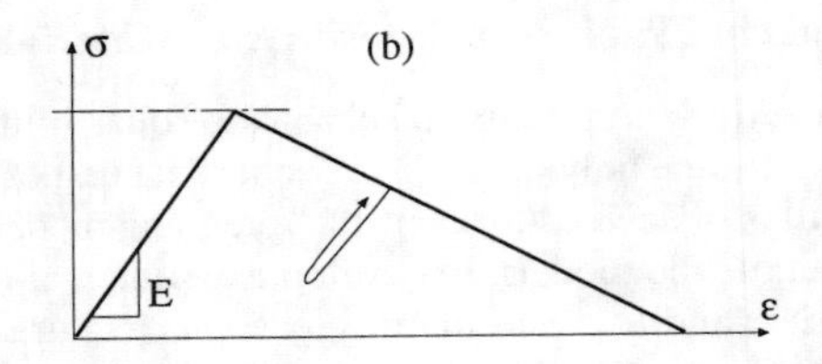

Figure 1.2.1 Softening stress-strain curves in smeared cracking models: (a) step softening; (b) progressive softening (from ACI Committee 446 1992).

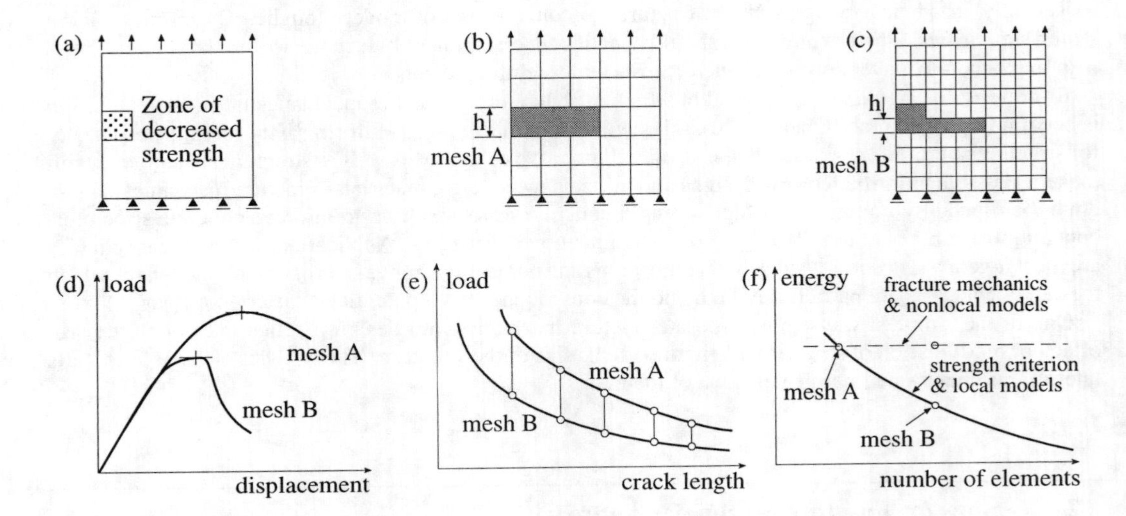

Figure 1.2.2 Illustration of lack of mesh-objectivity in classical smeared crack models (adapted from ACI Committee 446 1992).

to be unobjective, it must be rejected. There is no need to even compare it to experiments. Objectivity comes ahead of experimental verification.

A powerful, widely used approach to finite element analysis of concrete cracking is the concept of smeared cracking, introduced by Rashid (1968), which does not utilize fracture mechanics. According to this concept, the stress in a finite element is limited by the tensile strength of the material, f_t'. After the strength limit is reached, the stress in the finite element must decrease. In the initial practice, the stress was assumed to drop suddenly to zero, but it was soon realized that better and more realistic results are usually obtained if the stress is reduced gradually, i.e., the material is assumed to exhibit gradual strain softening (Scanlon 1971; Lin and Scordelis 1975); see Fig. 1.2.1. The concept of sudden or gradual strain-softening, though, proved to be a mixed blessing. After this concept had been implemented in large finite element codes and widely applied, it was discovered that the convergence properties are incorrect and the calculation results are unobjective as they significantly depend on the analyst's choice of the mesh (Bažant 1976, 1983; Bažant and Cedolin 1979, 1980, 1983; Bažant and Oh 1983a; Darwin 1985; Rots et al. 1985).

This problem, known as spurious mesh sensitivity, can be illustrated, for example, by the rectangular panel in Fig. 1.2.2a, which is subjected to a uniform vertical displacement at the top boundary. A small region near the center of the left side is assumed to have a slightly smaller strength than the rest of the panel, and consequently a smeared crack band starts growing from left to right. The solution is obtained by incremental loading with two finite element meshes of very different mesh sizes, as shown (Fig 1.2.2b,c). Stability check indicates that cracking must always localize in this problem into a band of single-element width at the cracking front. Typical numerical results for this as well as other similar problems are illustrated in Fig. 1.2.2d–f. In the load-deflection diagram (Fig. 1.2.2d), one can see that the

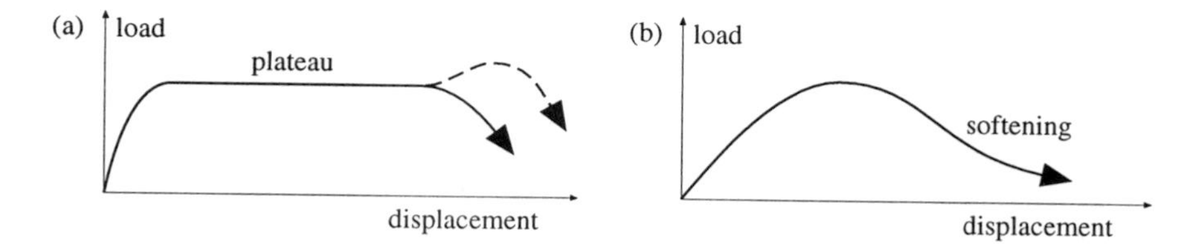

Figure 1.2.3 Load-deflection curves with and without yielding plateau (adapted from ACI Committee 446 1992).

peak load as well as the post-peak softening strongly depends on the mesh size, the peak load being roughly proportional to $h^{-1/2}$ where h is the element size. Plotting the load vs. the length of the crack band, one again finds large differences (Fig. 1.2.2e). The energy that is dissipated due to cracking decreases with the refinement of the mesh (Fig. 1.2.2f), and converges to zero as $h \rightarrow 0$, which is, of course, physically unacceptable.

The only way to avoid the foregoing manifestations of unobjectivity is some form of fracture mechanics or nonlocal model. By specifying the energy dissipated by cracking per unit length of the crack or the crack band, the overall energy dissipation is forced to be independent of the element subdivision (see the horizontal dashed line in Fig. 1.2.2f), and so is the maximum load.

1.2.3 Lack of Yield Plateau

Based on load-deflection diagrams, one may distinguish two basic types of structural failure: plastic and brittle. The typical characteristic of plastic failure is that the structure develops a single-degree-of-freedom mechanism such that the failure in various parts of the structure proceeds simultaneously, in proportion to a single parameter. Such failures are manifested by the existence of a long yield plateau on the load-deflection diagram (Fig. 1.2.3a). If the load-deflection diagram lacks such a plateau, the failure is not plastic but brittle (Fig. 1.2.3b) . When there are nosignificant geometric effects (such as the P–Δ effect in buckling), the absence of a plateau implies the existence of softening in the material due to fracture, cracking, or other damage. This further implies that the failure process cannot develop a single-degree-of-freedom mechanism but consists of propagation of the failure zones throughout the structure. The failure is nonsimultaneous and propagating.

To illustrate such behavior, consider the punching shear failure of a slab (Fig. 1.2.4). The typical (approximate) distributions of tensile stress σ along the failure surface are drawn in the figure. If the material is plastic, the cross-section gradually plasticizes until all its points are at the yield limit (Fig. 1.2.4b). However, if the material exhibits strain softening, then the stress peak moves across the failure zone, leaving a reduced stress (strain softening) in its wake (Fig. 1.2.4c,d). The stress reduction in the wake is mild if the structure is small, in which case the plastic limit analysis is not too far off (Fig. 1.2.4c). If the structure is large, however, the stress profile develops a steep stress drop behind the peak-stress point, and then the limit analysis solutions grossly overestimate the failure load (Fig. 1.2.4d).

1.2.4 Energy Absorption Capability and Ductility

The area under the entire load-deflection diagram represents the energy that the structure will absorb during failure. Consideration of this energy is important, especially for dynamic loading, and determines the ductility of the structure. Plastic limit analysis can give no information on the post-peak decline of the load and the energy dissipated in this process. According to plasticity, the load is constant after the peak, and the energy absorption theoretically unlimited. So some form of fracture mechanics is inevitable.

1.2.5 Size Effect

The size effect is, for design engineers, the most compelling reason for adopting fracture mechanics. Therefore, we discuss it more thoroughly now, and we will return to it in considerable detail in future

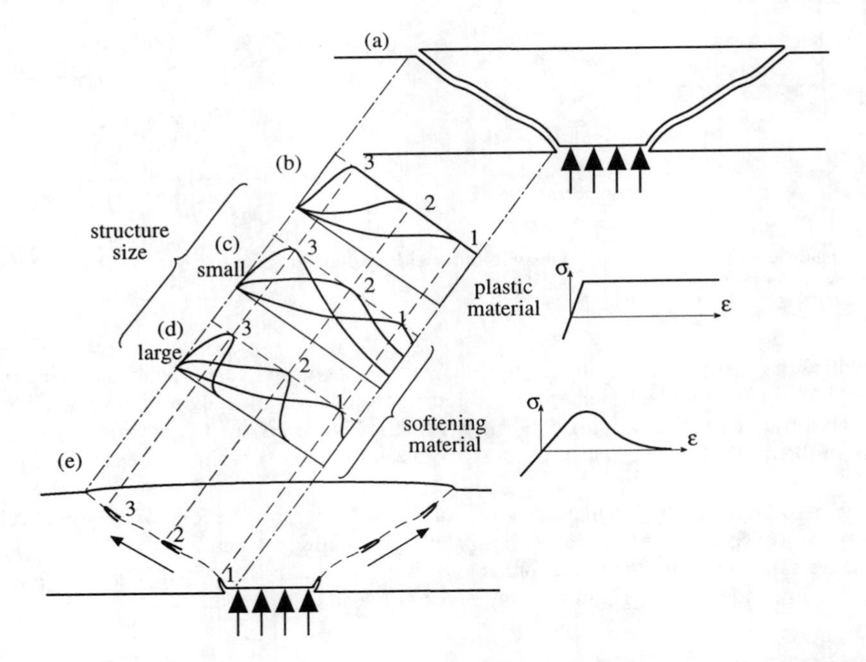

Figure 1.2.4 Influence of the structure size on the length of the yielding plateau in a punched slab (from ACI Committee 446 1992).

chapters. By general convention, the load capacity predicted by plastic limit analysis or any (deterministic) theory in which the material failure criterion is expressed in terms of stress or strain (or both) are said to exhibit no size effect. The size effect represents the deviation from such a prediction, i.e., the size effect on the structural strength is the deviation, engendered by a change of structure size, of the actual load capacity of a structure from the load capacity predicted by plastic limit analysis (or any theory based on critical stresses or strains).

The size effect is rigorously defined through a comparison of geometrically similar structures of different sizes. It is conveniently characterized in terms of the nominal strength, σ_{Nu}, representing the value of the nominal stress, σ_N, at maximum (ultimate) load, P_u. The nominal stress, which serves as a load parameter, may, but need not, represent any actual stress in the structure and may be defined simply as $\sigma_N = P/bD$ when the similarity is two-dimensional or as P/D^2 when the similarity is three-dimensional; b = thickness of a two-dimensional structure, and D = characteristic dimension of the structure, which may be chosen as any dimension, e.g., the depth of the beam, or the span, or half of the span, since only the relative values of σ_N matter. The nominal strength is then $\sigma_{Nu} = P_u/bD$ or P_u/D^2 (see Section 1.4.1 for more details).

According to the classical failure theories, such as the elastic analysis with allowable stress, plastic limit analysis, or any other theory that uses some type of a strength limit or failure surface in terms of stress or strain (e.g., viscoelasticity, viscoplasticity), σ_{Nu} is constant, i.e., independent of the structure size, for any given geometry, *notched or not.* We can, for example, illustrate it by considering the elastic and plastic formulas for the strength of beams in bending shear and torsion. These formulas are found to be of the same form except for a multiplicative factor. Thus, if we plot $\log \sigma_{Nu}$ vs. $\log D$, we find the failure states, according to strength or yield criteria, to be always given by a horizontal line (dashed line in Fig. 1.2.5). So, the failures according to the strength or yield criteria exhibit no size effect.

By contrast, failures governed by linear elastic fracture mechanics exhibit a rather strong size effect, which in Fig. 1.2.5 is described by the inclined dashed line of slope $-1/2$, as we shall justify in Chapter 2. The reality for concrete structures is a transitional behavior illustrated by the solid curve in Fig. 1.2.5. This curve approaches a horizontal line for the strength criterion if the structure is very small, and an inclined straight line of slope $-1/2$ if the structure is very large.

There is another size effect that calls for the use of fracture mechanics. It is the size effect on ductility

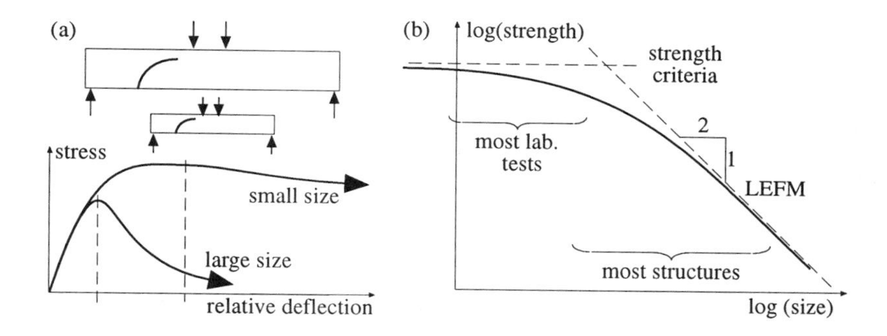

Figure 1.2.5 Size effects: (a) on the curves of nominal stress vs. relative deflection, and (b) on the strength in a bilogarithmic plot (adapted from ACI Committee 446 1992).

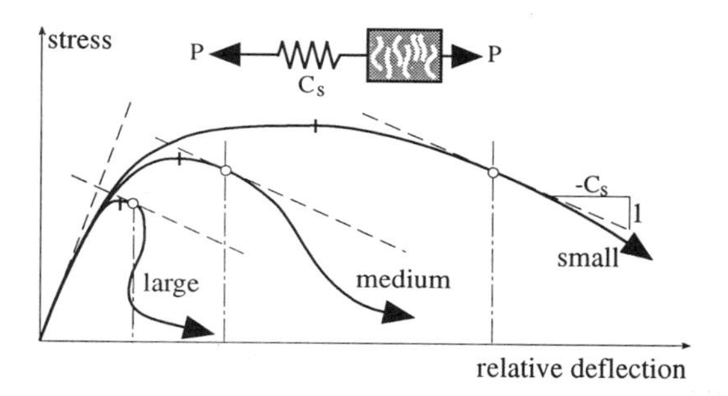

Figure 1.2.6 Size effect on the structural ductility (adapted from ACI Committee 446 1992).

of the structure, which is the opposite of brittleness, and may be characterized by the deformation at which the structure fails under a given type of loading. For loading in which the load is controlled, structures fail (i.e., become unstable) at their maximum load, while for loading in which the displacement is controlled, structures fail in their post-peak, strain-softening range. In a plot of σ_N vs. the deflection, the failure point is characterized by a tangent (dashed line in Fig. 1.2.6) of a certain constant inverse slope $-C_s$ where C_s is the compliance of the loading device (see e.g., Bažant and Cedolin 1991, Sec. 13.2). Geometrically similar structures of different sizes typically yield load-deflection curves of the type shown in Fig. 1.2.6. As illustrated, failure occurs closer to the peak as the size increases. This effect is again generally predicted by fracture mechanics, and is due to the fact that in a larger structure more strain energy is available to drive the propagation of the failure zone.

The well-known effect of structure size or member size on crack spacing and crack width is, to a large extent, also explicable by fracture mechanics. It may also be noted that the spurious effect of mesh size (Reason 2, Section 1.2.2) can be regarded as a consequence of the structural size effect.

1.3 Sources of Size Effect on Structural Strength

There are six different size effects that may cause the nominal strength to depend on structure size:

1. *Boundary layer effect*, also known as the wall effect. This effect is due to the fact that the concrete layer adjacent to the walls of the formwork has inevitably a smaller relative content of large aggregate pieces and a larger relative content of cement and mortar than the interior of the member. Therefore, the surface layer, whose thickness is independent of the structure size and is of the same order of magnitude as the maximum aggregate size, has different properties. The size effect is due to the fact that in a smaller member, the surface layer occupies a large portion of

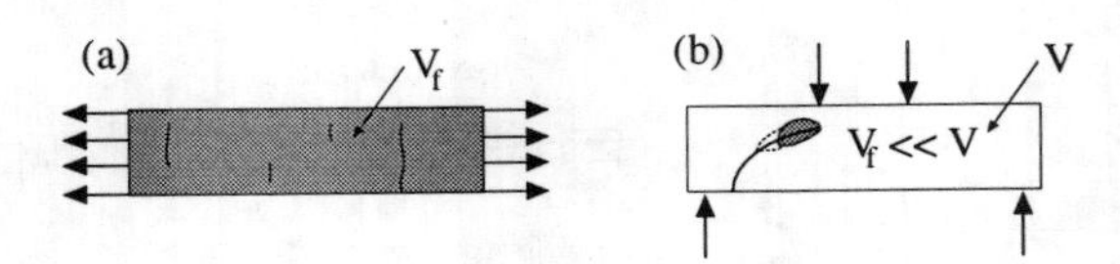

Figure 1.3.1 The essence of the difference between statistical and fracture size effect (adapted from ACI Committee 446 1992).

the cross-section, while in a large member, it occupies a small part of the cross-section. In most situations, this type of size effect does not seem to be very strong. A second type of boundary layer effect arises because, under normal stress parallel to the surface, the mismatch between the elastic properties of aggregate and mortar matrix causes transverse stresses in the interior, while at the surface these stresses are zero. A third type of boundary layer size effect arises from the Poisson effect (lateral expansion) causing the surface layer to nearly be in plane stress, while the interior is nearly in plane strain. This causes the singular stress field at the termination of the crack front edge at the surface to be different from that at the interior points of the crack front edge (Bažant and Estenssoro 1979). A direct consequence of this, easily observable in fatigue crack growth in metals, is that the termination of the front edge of a propagating crack cannot be orthogonal to the surface. The second and third types exist even if the composition of the boundary layer and the interior is the same.

2. *Diffusion phenomena*, such as heat conduction or pore water transfer. Their size effect is due to the fact that the diffusion half-times (i.e., half-times of cooling, heating, drying, etc.) are proportional to the square of the size of the structure. At the same time, the diffusion process changes the material properties and produces residual stresses which in turn produce inelastic strains and cracking. For example, drying may produce tensile cracking in the surface layer of the concrete member. Due to different drying times and different stored energies, the extent and density of cracking may be rather different in small and large members, thus engendering a different response. For long-time failures, it is important that drying causes a change in concrete creep properties, that creep relaxes these stresses, and that in thick members the drying happens much slower than in thin members.
3. *Hydration heat* or other phenomena associated with chemical reactions. This effect is related to the previous one in that the half-time of dissipation of the hydration heat produced in a concrete member is proportional to the square or the thickness (size) of the member. Therefore, thicker members heat to higher temperatures, a well-known problem in concrete construction. Again, the nonuniform temperature rise may cause cracking, induce drying, and significantly alter the material properties.
4. *Statistical size effect,* which is caused by the randomness of material strength and has traditionally been believed to explain most size effects in concrete structures. The theory of this size effect, originated by Weibull (1939), is based on the model of a chain. The failure load of a chain is determined by the minimum value of the strength of the links in the chain, and the statistical size effect is due to the fact that the longer the chain, the smaller is the strength value that is likely to be encountered in the chain. This explanation, which certainly applies to the size effect observed in the failure of a long concrete bar under tension (Fig. 1.3.1), is described by Weibull's weakest-link statistics. However, as we will see in Chapter 12, on closer scrutiny, this explanation is found to be inapplicable to most types of failures of reinforced concrete structures. In contrast to metallic and other structures, which fail at the initiation of a macroscopic crack (i.e., as soon as a microscopic flaw or crack reaches macroscopic dimensions), concrete structures fail only after a large stable growth of cracking zones or fractures. The stable crack growth causes large stress redistributions and a release of stored energy, which, in turn, causes a much stronger size effect, dominating over any possible statistical size effect. At the same time, the mechanics of failure restricts the possible locations of the decisive crack growth at the moment of failure to a very small zone. This causes the random strength values outside this zone to become irrelevant, thus suppressing the statistical size effect. We will also see that some recent experiments on diagonal shear failure of reinforced concrete beams contradict the prediction of the statistical theory.

5. *Fracture mechanics size effect,* due to the release of stored energy of the structure into the fracture front. This is the most important source of size-effect, and will be examined in more detail in the next section and thoroughly in the remainder of the book.
6. *Fractal nature of crack surfaces.* If fractality played a significant role in the process of formation of new crack surface, it would modify the fracture mechanics size effect. However, such a role is not indicated by recent studies (Chapter 12 and Bažant 1997d). Probably this size effect is only a hypothetical conjecture.

In practical testing, the first 3 sources of size effect can be, for the most part, eliminated if the structures of different sizes are geometrically similar in two rather than three dimensions, with the same thickness for all the sizes. Source 1 becomes negligible for sufficiently thick structures. Source 2 is negligible if the specimen is sealed and is at constant temperature. Source 3 is significant only for very massive structures. The statistical size effect is always present, but its effect is relatively unimportant when the fracture size effect is important. Let us now give a simple explanation of this last and dominant size effect.

1.4 Quantification of Fracture Mechanics Size Effect

In the classical theories based on plasticity or limit analysis, the strength of geometrically similar structures is independent of the structure size. As already pointed out, however, concrete structures and, in general, structures made of brittle or quasibrittle materials, do not follow this trend. In this section we first define what is understood by strength and size of a structural element and then examine how the strength depends on the size. We finally give a simple justification of Bažant's size effect law. The experimental evidence supporting the existence of size effect will be presented in the next section.

1.4.1 Nominal Stress and Nominal Strength

The size effect is understood as the dependence of the structure strength on the structure size. The strength is conventionally defined as the value of the so-called *nominal stress* at the peak load. The nominal stress is a load parameter defined as proportional to the load divided by a typical cross-sectional area:

$$\sigma_N = c_N \frac{P}{bD} \quad \text{for } 2D \text{ similarity}, \qquad \sigma_N = c_N \frac{P}{D^2} \quad \text{for } 3D \text{ similarity} \tag{1.4.1}$$

in which P = applied load, b = thickness of a two-dimensional structure (which, for certain reasons, in experiments should be preferably chosen the same for all structure sizes); D = characteristic dimension of the structure or specimen; and c_N = coefficient introduced for convenience, which can be chosen as $c_N = 1$, if desired. For $P = P_u$ = maximum load, Eq. (1.4.1) gives the nominal strength, σ_{Nu}.

Coefficient c_N can be chosen to make Eq. (1.4.1) coincide with the formula for the stress in a certain particular point of a structure, calculated according to a certain particular theory. For example, consider the simply supported beam of span S and depth h, loaded at mid-span by load P, as shown in Fig. 1.4.1a. Now we may choose, for example, σ_N to coincide with the elastic bending formula for the maximum normal stress in the beam (Fig. 1.4.1b), and the beam depth as the characteristic dimension ($D = h$), in which case we have

$$\sigma_N = \frac{3PS}{2bh^2} = c_N \frac{P}{bD}\,, \quad \text{with } c_N = 1.5\frac{S}{h} \tag{1.4.2}$$

It appears that c_N depends on the span-to-depth ratio which can vary for various beams. It is thus important to note that the size effect may be consistently defined only by considering geometrically similar specimens or structures of different sizes, with geometrically similar notches or initial cracks. Without geometric similarity, the size effect would be contaminated by the effects of varying structure shape. With this restriction (most often implicitly assumed), coefficient c_N is constant because, for geometrically similar structures, S/h is constant by definition.

The foregoing definition of σ_N is not the only one possible. Alternatively, we may choose σ_N to

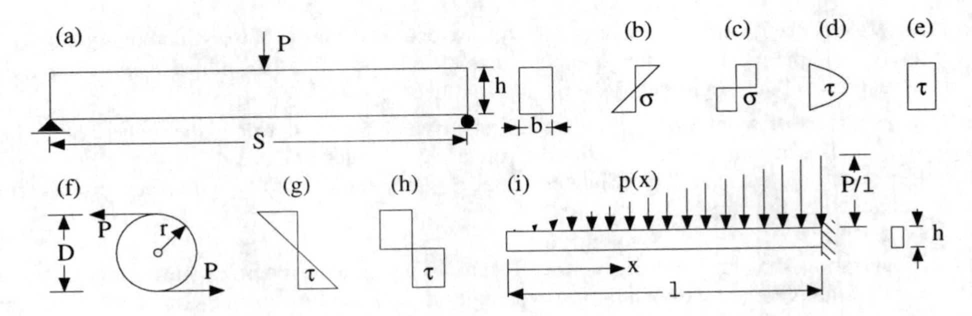

Figure 1.4.1 (a) Three-point bent beam. (b) Elastic stress distribution. (c) Plastic stress distribution. (d) Elastic shear stress distribution. (e) Plastic shear stress distribution. (f) Shaft subjected to torsion. (g) Elastic shear stress distribution. (h) Plastic shear stress distribution. (i) Cantilever beam with linearly distributed load.

coincide with the plastic bending formula for the maximum stress (Fig. 1.4.1c), in which case we have

$$\sigma_N = \frac{PS}{bh^2} = c_N \frac{P}{bD}, \text{ with } c_N = \frac{S}{h} \ (= \text{constant}) \tag{1.4.3}$$

Alternatively, we may choose as the characteristic dimension the beam span instead of the beam depth ($D = S$), in which case we have

$$\sigma_N = \frac{3PS}{2bh^2} = c_N \frac{P}{bD}, \text{ with } c_N = 1.5\frac{S^2}{h^2} \ (= \text{constant}) \tag{1.4.4}$$

We may also choose σ_N to coincide with the formula for the maximum shear stress near the support according to the elastic bending theory (Fig. 1.4.1d), in which case we have, with $D = h$,

$$\sigma_N = \frac{3P}{4bh} = c_N \frac{P}{bD}, \text{ with } c_N = 0.75 \ (= \text{constant}) \tag{1.4.5}$$

Alternatively, using the span as the characteristic dimension ($D = S$), we may write

$$\sigma_N = \frac{3P}{4bh} = c_N \frac{P}{bD}, \text{ with } c_N = \frac{3S}{4h} \ (= \text{constant}) \tag{1.4.6}$$

All the above formulae are valid definitions of the nominal strength for three-point bent beams, although the first one (1.4.2) is the most generally used (and that used throughout this book). Other examples are given next.

Example 1.4.1 Consider torsion of a circular shaft of radius r, loaded by torque $T = 2Pr$ where P is the force couple shown in Fig. 1.4.1f. Using σ_N to coincide with the elastic formula for the maximum shear stress, we may write, taking $D = 2r$ = diameter,

$$\sigma_N = \frac{4P}{\pi r^2} = \frac{16P}{\pi D^2} = c_N \frac{P}{D^2}, \text{ with } c_N = \frac{16}{\pi} \ (= \text{constant}) \tag{1.4.7}$$

If, instead, we chose the radius as the characteristic dimension ($D = r$), we may write

$$\sigma_N = \frac{4P}{\pi r^2} = c_N \frac{P}{D^2}, \text{ with } c_N = \frac{4}{\pi} \ (= \text{constant}) \tag{1.4.8}$$

Note that in this case we have a three-dimensional similarity. □

Example 1.4.2 Consider the cantilever of span ℓ and cross-section depth h shown in Fig. 1.4.1i, which is loaded by distributed load $p(x)$ increasing linearly from the cantilever end. We chose the value of the distributed load at the fixed end to be denoted as P/ℓ (ℓ is used to achieve the correct dimension). Now, choosing σ_N to coincide with the elastic bending formula for the maximum stress, and the characteristic dimension to coincide with the beam depth ($D = h$), we may write

$$\sigma_N = \frac{P\ell}{bh^2} = c_N \frac{P}{bD}, \text{ with } c_N = \frac{\ell}{h} \text{ (=constant)} \tag{1.4.9}$$

which is again of the same form. ▯

To sum up, the nominal stress can be defined by the simple equation (1.4.1) regardless of the complexity of structure shape and material behavior, and can be used as a load parameter having the dimension of stress.

1.4.2 Size Effect Equations

With the foregoing definitions, the size effect consists in the variation of the nominal strength σ_{Nu} with size D. There are various possible plots showing special aspects of the size effect, but the most widely used is the bilogarithmic plot already shown in Fig. 1.2.5 in which $\log \sigma_{Nu}$ is plotted vs. $\log D$. As previously discussed in Section 1.2.5, the strength theory (based on yield or strength criteria) predicts no size effect (horizontal dashed line in Fig. 1.2.5b); this is the kind of response assumed in most engineering approaches and codes (see Chapter 10 for a detailed discussion about the need of including the size effect in the codes.) On the other extreme, we have the purely brittle behavior of structures that fail by crack instability at a fixed crack-to-size ratio (relative crack length). In the next chapter, after presenting the essentials of linear elastic fracture mechanics (LEFM) we will see that the size effect in such a case is shown in the plot of Fig. 1.2.5b as an inclined line of slope -1/2. The actual size effect behavior is best described by a transitional curve having the two straight lines as asymptotes, as sketched in Fig. 1.2.5b.

The simplest size effect law satisfying this condition was derived by Bažant (1984a) under very mild assumptions which apply, approximately, to a large number of practical cases. Bažant's size effect equation brings into play the energy required for crack growth as shown in the next paragraph, where a short derivation is presented. However, the final expression can be written (without explicitly showing the fracture energy term) as a function depending on only two parameters as

$$\sigma_{Nu} = \frac{Bf'_t}{\sqrt{1 + D/D_0}} \tag{1.4.10}$$

where f'_t is the tensile strength of the material, introduced only for dimensional purposes, B is a dimensionless constant, and D_0 is a constant with the dimension of length. Both B and D_0 depend on the fracture properties of the material and on the geometry (shape) of the structure, but not on the structure size. Simple derivations of this size effect law are given next.

1.4.3 Simple Explanation of Fracture Mechanics Size Effect

Consider a uniformly stressed panel as shown in Fig. 1.4.2. Imagine first that fracture proceeds as the formation of a crack band (or fracture band) of thickness h_f across the central section of the panel. Now, the extension of the crack band by a unit length will require a certain amount of energy that, per unit thickness of the specimen, is called *fracture energy* and is denoted as G_f. The value of G_f may be considered, for the present purposes, approximately a material constant. To determine the load required to propagate the band, an energy balance condition must be imposed by writing that the energy available is equal to the energy required for band extension.

To do so, one writes that the strain energy released from the structure at constant σ_N (which is the condition of maximum load) is used to further propagate the crack band. As an approximation, we may assume that the presence of a crack band of thickness h_f reduces the strain energy density in the band and cross-hatched area from $\sigma_N^2/2E$ (for the intact panel) to zero (E = elastic modulus of material). The cross-hatched area is limited by two lines of some empirical slope k. When the crack band extends by Δa at no boundary displacements, the additional strain energy that is released comes from the densely cross-hatched strip of horizontal dimension Δa (Fig. 1.4.2a). If the failure modes are geometrically similar, as

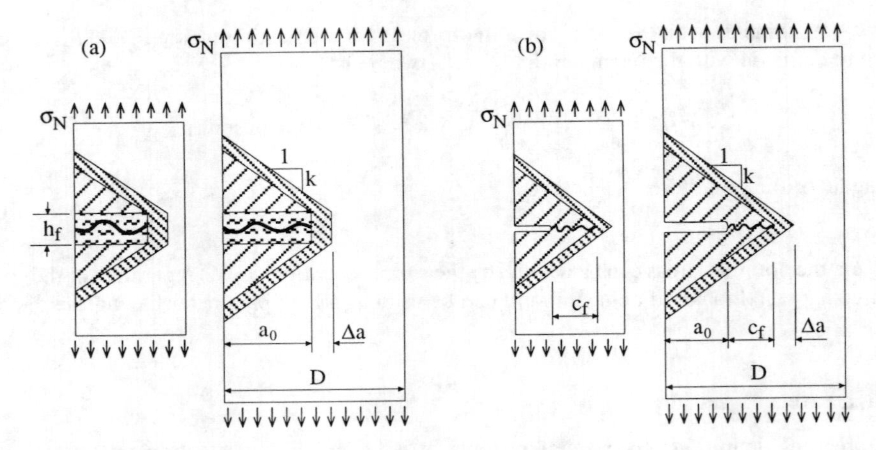

Figure 1.4.2 Sketches for explaining size effect: (a) blunt crack band, (b) slit-like process zone (adapted from ACI Committee 446 1992).

is usually the case, then the larger the panel, the longer is the crack band at failure. Consequently, the area of the densely cross-hatched strip for a larger panel is also larger. Therefore, in a larger structure, more energy is released from the strip by the same extension of the crack band. This is the source of size effect.

Quantitatively, the energy released per unit panel thickness is given by the area of the densely cross-hatched region $h_f\Delta a+2ka_0\Delta a$ times the thickness, times the energy density of the intact panel $\sigma_N^2/2E$. Therefore, the release of energy from the aforementioned strip (at constant boundary displacement) is $b(h_f\Delta a + 2ka_0\Delta a)\sigma_N^2/2E$, where b is the panel thickness. This must be equal to the energy required to create the fracture, which is $G_f b\Delta a$.Therefore,

$$b(h_f\Delta a + 2ka_0\,\Delta a)\frac{\sigma_N^2}{2E} = G_f\,b\Delta a \tag{1.4.11}$$

Solving for the nominal stress, one obtains the size effect law (1.4.10) in which

$$Bf_t' = \sqrt{\frac{2G_fE}{h_f}} = \text{constant,} \quad \text{and} \quad D_0 = \frac{h_fD}{2ka_0} = \text{constant.} \tag{1.4.12}$$

Note that D_0 depends on the structure shape through the constant k but is independent of the structure size if the structures are geometrically similar (D/a_0 = constant); f_t' = tensile strength, introduced for convenience; and h_f = width of the fracture band front, which is treated here approximately as a constant, independent of structure size.

Lest one might get the impression that this explanation of size effect works only for a crack band but not for a sharp line crack, consider the similar panels of different sizes with line cracks as shown in Fig. 1.4.2b. In concrete, there is always a sizable fracture process zone ahead of the tip of a continuous crack, of some finite length which may, in the crudest approximation, be considered constant. Over the length of this zone, the transverse normal stress gradually drops from f_t' to 0. Because of the presence of this zone, the elastically equivalent crack length that causes the release of strain energy from the adjacent material is longer than the continuous crack length, a_0, by a distance c_f which can be assumed to be approximately a material constant.

When the crack extends by length Δa, the fracture process zone travels with the crack tip, and the area from which additional strain energy is released consists of the strips of horizontal dimension Δa that are densely cross-hatched in Fig. 1.4.2b. Following the same procedure as before for the crack band, we see that the area of the zone from which energy is released is $2k(a_0 + c_f)\Delta a$. So the total energy release is $b2k(a_0 + c_f)\Delta a\sigma_N^2/2E$, which must be equated to the energy required for crack extension, $bG_f\Delta a$, thus delivering the equation

$$b2k(a_0 + c_f)\Delta a\frac{\sigma_N^2}{2E} = G_f b\Delta a \tag{1.4.13}$$

Solving for σ_N, one again obtains the size effect law in (1.4.10) in which now

$$Bf'_t = \sqrt{\frac{G_f E}{k c_f}} = \text{constant}, \quad \text{and} \quad D_0 = c_f \frac{D}{a_0} = \text{constant}. \tag{1.4.14}$$

The foregoing equations are only approximate in their details, because of the simplifying assumptions in determining the structural energy release. However, their structure is correct. The same form is obtained using simplified theories for other geometries (e.g., bending). The fine-tuned equations require the use of more sophisticated fracture mechanics concepts, and their presentation will be deferred until Chapter 6.

As will be shown in Chapter 9, Eq. (1.4.10) can also be derived, in a completely general way, by dimensional analysis and similitude arguments (Bažant 1984a). This general derivation rests on two basic hypotheses: (1) the propagation of a fracture or crack band requires an approximately constant energy supply (the fracture energy, G_f) per unit area of fracture plane , and (2) the energy released by the structure due to the propagation of the fracture or crack band is a function of both the fracture length and the size of the fracture process zone at the fracture front.

Applications of Eq. (1.4.10) to brittle failures of concrete structures rest on two additional hypotheses: (3) the failure modes of geometrically similar structures of different sizes are also geometrically similar (e.g., a diagonal shear crack has at failure about the same slope and the same relative length), and (4) the structure does not fail at crack initiation (which is really a requirement of good design).

These hypotheses are never perfectly fulfilled, so it must be kept in mind that Eq. (1.4.10) is approximate, valid only within a size range of about 1:20 for most structures (for a broader size range, a more complicated formula would be required). This size range is sufficient for most practical purposes, but for some structures the range of interest extends beyond the applicability range. This is so because a sufficiently large change of structure size may alter the failure mode and thus render Eq. (1.4.10) inapplicable beyond that size; this happens, for example, for the brazilian split-cylinder tests. The analysis of such 'anomalous' size effect will be deferred until Chapter 9.

Exercises

1.1 In fracture mechanics manuals, it is customary to use for σ_N the maximum tensile stress computed elastically for an unnotched specimen. Express σ_N for a beam in terms of the maximum bending moment M, the beam depth D, and the central moment of inertia of the cross-section I. (Answer: $\sigma_N = MD/2I$)

1.2 Determine σ_N as in the previous exercise for a hollow cylindrical bar of outer diameter D and inner diameter αD ($\alpha < 1$) subjected to a torsional moment M_T. (Answer: $\sigma_N = 16M_T/[\pi D^3(1 - \alpha^4)]$)

1.3 With the same criteria as in the previous exercises, determine σ_N for a circular bar of diameter D subjected to simultaneous tension and torsion; let P be the tensile force and $M_T = \beta P D$ the torque, where β is some dimensionless constant. Give the coefficient c_N corresponding to Eq. (1.4.1). Hint: use Mohr's circle to find the maximum tensile stress.

1.4 Results from the literature were analyzed by the authors using a characteristic specimen dimension D and a nominal stress σ_N defined with $c_N = 1$. The results for the best fit of Bf'_t and D_0 were 1.15 MPa and 322 mm, respectively. To compare with other results, you want to use a nominal stress defined using the same characteristic size D, but a constant $\hat{c}_N = 2.5$. What would the values of the best-fit constants (say $\hat{B}f'_t$ and $\hat{D}_0$) in this case. (Answer: $\hat{B}f'_t = 2.88$ MPa, $\hat{D}_0 = 322$ mm.)

1.5 Generalize the previous exercise and prove that if, for a particular selection of c_N and D, the size effect parameters are Bf_t and D_0, for a different selection $\hat{c}_N$ and $\hat{D}$ (where $\hat{D}/D$ = constant), their value is $\hat{B}f'_t = (\hat{c}_N/c_N)Bf'_t$ and $\hat{D}_0 = (\hat{D}/D)D_0$.

1.6 Find the relationship between h_f and c_f that make identical Eqs. (1.4.12) and (1.4.14).

Table 1.5.1 Summary of size effect test series.

Series	Material	Specimen type[a]	a_0/D	S/D	b (mm)	c_N	Reference
A1–A6	concrete	SEN-TPB	1/3	4	76.2	6	Walsh 1972
B1	concrete	SEN-TPB	1/6	2.5	38.1	3.75	Bažant and Pfeiffer 1987
B2	concrete	DEN-EC	1/6	—	38.1	1	Ibid.
B3	concrete	DEN-T	1/6	—	19.1	1	Ibid.
B4	concrete	DEN-S	1/6	—	38.1	1	Bažant and Pfeiffer 1986
C1	mortar	SEN-TPB	1/6	2.5	38.1	3.75	Bažant and Pfeiffer 1987
C2	mortar	DEN-EC	1/6	—	38.1	1	Ibid.
C3	mortar	DEN-T	1/6	—	19.1	1	Ibid.
C4	mortar	DEN-S	1/6	—	38.1	1	Bažant and Pfeiffer 1986
D1	HSC	SEN-TPB	1/3	2.5	38.1	3.75	Gettu, Bažant and Karr 1990
E1	marble	SEN-TPB	0.5	4	30	6	Fathy 1992
E2	granite	SEN-TPB	0.5	4	30	6	Ibid.
F1	limestone	SEN-TPB	0.4	4	13	6	Bažant, Gettu and Kazemi 1991
G1	SiO_2	SEN-TPB	0.2[b]	4	$= D$	6	McKinney and Rice 1981[c]
G2	SiC CN-137	SEN-TPB	0.2	4	$= D$	6	Ibid.
G3	SiC CN-163	SEN-TPB	0.2	4	$= D$	6	Ibid.
H1–2	concrete	DP	—	—	—	0.4	Marti 1989
I1	microcon.	BPO	—	—	—	$4/\pi$	Bažant and Şener 1988
J1	mortar	UPT	—	—	—	0.75	Bažant, Şener and Prat 1988
J2	R. mortar	RPT	—	—	—	0.75	Ibid.
K1	R. mortar	LRB-UB	—	—	—	0.5	Bažant and Kazemi 1991
K2	R. mortar	LRB-AB	—	—	—	0.5	Ibid.
L1	microcon.	PS	—	—	—	$1/\pi$	Bažant and Cao 1987

[a] See specimen types in Fig. 1.5.1.

[b] Variable. Results have been reduced to a fixed $a_0/D = 0.2$.

[c] See also Bažant and Kazemi 1990b.

1.5 Experimental Evidence for Size Effect

The size effect law proposed by Bažant (Eq. (1.4.10)) has been verified by a large number of experimental data, for both notched fracture specimens and unnotched structures. We will postpone a full discussion of the implications of size effect for structural analysis until Chapter 10, but we will review the evidence now.

Results from 23 test series, among those available, are briefly examined in the following, corresponding to various authors, materials, and specimens. The essentials of the specimen characteristics are summarized in Table 1.5.1 and Fig. 1.5.1. In particular, the coefficient c_N corresponding to Eq. (1.4.1) to each series of experiments is included in Table 1.5.1. Some characteristics of the materials, together with the size range and the best fits for Bf'_t and D_0, are summarized in Table 1.5.2, where the sources of the data are also displayed. Further details of the tests are given in the following; however, detailed descriptions are not given; they can be obtained from the referenced sources.

Since various fitting procedures have been used in the literature to get the optimal fit of the size effect law to the experimental data, a unified procedure has been used in this book. In this procedure the nonlinear regression is directly performed on the data presented as a bilogarithmic diagram (see Chapter 6 for a discussion of the fitting procedures). To do so, we call $v = \ln \sigma_{Nu}$, and take v to be the variable whose quadratic error is to be minimized with the following approximating equation —equivalent to (1.4.10):

$$v = \ln(Bf'_t) - 0.5 \ln\left(1 + \frac{D}{D_0}\right) \tag{1.5.1}$$

The optimal fits for Bf'_t and D_0 were calculated using a standard Levenberg-Marquardt algorithm. The values of Bf'_t and D_0 with their standard errors are given in Table 1.5.2.

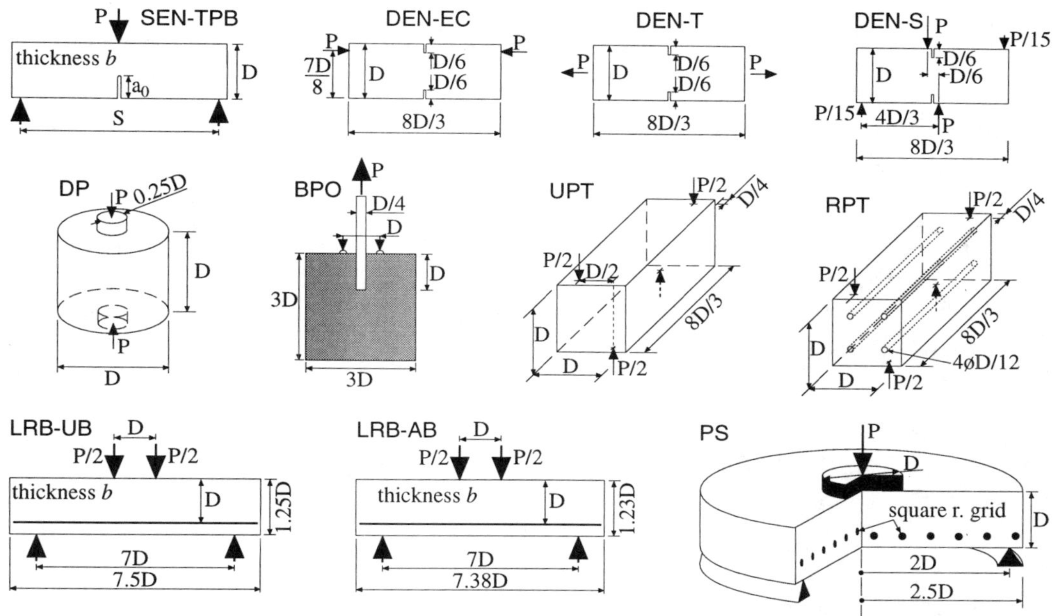

Figure 1.5.1 Summary of specimens used for size effect verification.

Table 1.5.2 Essential data of size effect tests

Series	Material	d_a [a] (mm)	Strength [b] (MPa)	Elastic Modulus (GPa)	Size Range (mm)	Bf_t' (MPa)	D_0 (mm)
A1	concrete	12.7	23.1	n.a.	76–381	4.5 ± 0.9	36 ± 17
A2	concrete	12.7	35.4	n.a.	76–381	2.8 ± 0.5	157 ± 99
A3	concrete	12.7	14.3	n.a.	76–381	3.2 ± 2.1	34 ± 52
A4	concrete	12.7	15.6	n.a.	76–381	1.7 ± 0.3	126 ± 78
A5	concrete	12.7	46.8	n.a.	76–381	2.9 ± 0.4	212 ± 114
A6	concrete	12.7	32.7	n.a.	76–381	4.1 ± 0.7	55 ± 23
B1	concrete	12.7	34.1	n.a.	76–305	6.0 ± 0.3	60 ± 10
B2	concrete	12.7	37.4	n.a.	76–305	3.9 ± 0.2	54 ± 10
B3	concrete	12.7	29.1	n.a.	76–152	2.7 ± 0.1	184 ± 38
B4	concrete	12.7	39.7	n.a.	76–305	4.6 ± 0.1	719 ± 130
C1	mortar	4.83	48.4	n.a.	76–305	14 ± 4	7.7 ± 4.4
C2	mortar	4.83	48.1	n.a.	76–305	6.6 ± 1.3	10 ± 5
C3	mortar	4.83	46.4	n.a.	76–152	3.3 ± 0.1	95 ± 17
C4	mortar	4.83	49.0	n.a.	76–305	5.9 ± 0.2	190 ± 25
D1	HSC	9.5	96.0	n.a.	38–152	32 ± 8	19 ± 11
E1	marble	4	7.7 [c]	36	12.5–100	3.7 ± 0.2	47 ± 10
E2	granite	2	12.3 [c]	39	12.5–100	5.1 ± 0.2	35 ± 5
F1	limestone	1.5	3.45 [c]	30.5	13–102	3.3 ± 0.1	45 ± 6
G1	SiO_2	0.02	—	58	5–32	17 ± 3	2.6 ± 1.2
G2	SiC CN-137	2	—	130	7–37	36 ± 4	101 ± 157
G3	SiC CN-163	2	—	140	7–37	35 ± 3	7.3 ± 2.2
H1	concrete	9.5	33.3	n.a.	76–610	2.7 ± 0.1	352 ± 63
H2	concrete	9.5	23.6	n.a.	76–1219	1.95 ± 0.05	601 ± 83
I1	microcon.	6.35	45.8	13–51	45.9	36 ± 14	6 ± 6
J1	mortar	4.83	43.7	n.a.	38–152	5.4 ± 0.3	37 ± 6
J2	mortar	4.83	43.6	n.a.	38–152	6.2 ± 2.1	23 ± 20
K1	mortar	4.83	46.8	n.a.	41–163	2.2 ± 0.1	151 ± 40
K2	mortar	4.83	46.2	n.a.	21–330	4.6 ± 1	11 ± 6
L1	microcon.	6.35	na	48–53	25–102	10.3 ± 1.3	178 ± 178

[a] Maximum aggregate or grain size.

[b] Compressive unless otherwise stated.

[c] Splitting tensile strength.

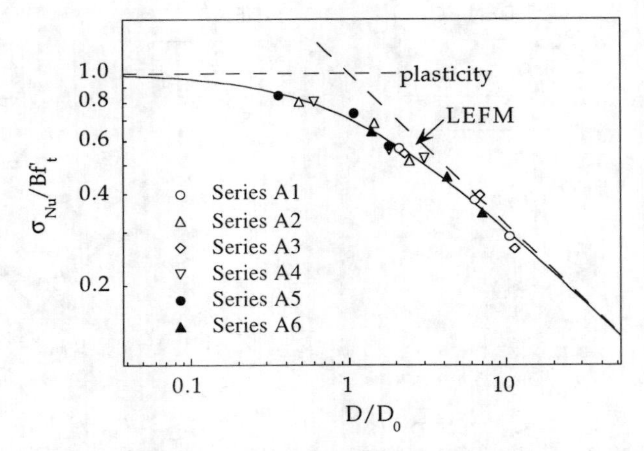

Figure 1.5.2 Size effect results of Walsh (1972).

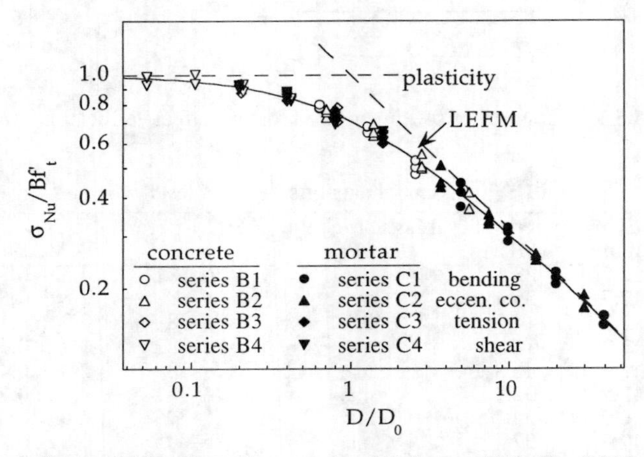

Figure 1.5.3 Size effect results of Bažant and Pfeiffer (1986, 1987).

1.5.1 Structures with Notches or Cracks

The size effect law was originally verified (Bažant 1983, 1984a) by comparisons with the tests of Walsh (1972), whose results are summarized in Fig.1.5.2. Walsh used single-edge-notched beams in three-point bending (SEN-TPB, see Table 1.5.1 and Fig. 1.5.1, test series A1–6). Walsh was apparently first to plot the test results as $\log \sigma_{Nu}$ vs. $\log D$, but did not try to describe this plot mathematically or generalize it. Walsh's classical tests, however, were of limited range, too short for the scatter obtained, and included only one type of fracture specimen, and so the comparisons were not completely conclusive.

A stronger experimental verification was presented by Bažant and Pfeiffer (1986, 1987), covering a broader size range and four very different types of specimens : SEN-TPB, double-edge-notched in eccentric compression (DEN-EC), double-edge notch in eccentric compression (DEN-EC), and double-edge notch in shear (DEN-S); see Table 1.5.1 and Fig. 1.5.1, test series B1–4 and C1–4. The research included tests on concrete (series B1–4) and on mortar (series C1–4). The test results and their optimum fits by the size effect law are shown in Fig. 1.5.3. Obviously, the comparison provides a strong justification for the size effect law (for statistical comparisons, see the original paper). The data points in Fig. 1.5.3 refer to individual tests.

An experimental size effect study of high strength concrete has been conducted by Gettu, Bažant and Karr (1990), using SEN-TPB. The results are shown in Fig. 1.5.4, again in comparison with the optimum fits by the size effect law and its asymptotes.

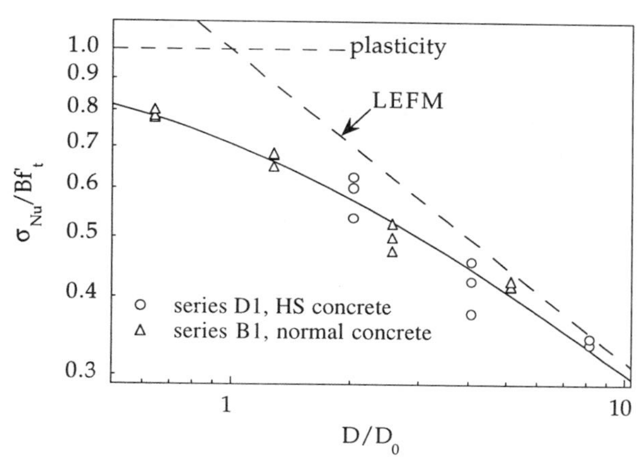

Figure 1.5.4 Size effect results of Gettu, Bažant and Karr (1990). Results of Bažant and Pfeiffer (1987) for ordinary concrete are also included for comparison.

For concrete, the foregoing results have been complemented by the extensive size effect data for SEN-TPB specimens published by Bažant and Gettu (1992). These tests (as well as similar tests of fracture of limestone by Bažant, Bai and Gettu 1993), however, also included a systematic investigation of the effect of the loading rate, and, therefore, the presentation of these data is better postponed to Chapter 11. Further, it has been shown (Bažant, Kim and Pfeiffer 1986) that the size effect law agrees well with the results of the size effect tests by Jenq and Shah (1985a,b), although a good fitting was not possible because the size range was too limited compared to the scatter obtained.

The ability of the size effect law (1.4.10) to describe the strength variation of notched specimens in materials other than concrete has been investigated, too, particularly for rocks and ceramics. Moving from coarser grained rocks to ceramics, Fathy (1992) tested marble and granite; Bažant, Gettu and Kazemi (1991) limestone; and McKinney and Rice (1981) slip-cast fused silica (SiO_2) and nitridized silicon carbide (SiC CN-137 and SiC CN-137). Fig. 1.5.5 summarizes their test results. It can be seen that the size effect law describes the various results acceptably well.

Recent experimental data indicate a similar degree of agreement for composites and for ice. Bažant, Daniel and Li (1996) tested in tension both single- and double-edge notched specimens of highly orthotropic carbon-epoxy fiber laminates and they found good agreement with the size effect law for a 1:8 range of sizes. Adamson et al. (1995) and Mulmule, Dempsey and Adamson (1995) performed various series of tests on sea ice using various specimen geometries. In one of the series, square plates with a notch subjected to opening forces at the notch mouth were tested with a size ratio of 1:160 (the specimens ranged from 0.5 to 80 m in size). This is the test series of wider size range known to date. The results showed a very good agreement with the size effect law.

1.5.2 Structures Without Notches or Cracks

Extensive tests have been carried out to verify (1.4.10) for various types of failure of unnotched concrete structures. Good agreement of (1.4.10) with test results has been demonstrated for:

1. Double-punch tests of cylinders (Marti 1989).

2. Pullout failure of bars (Bažant and Şener 1988), pullout of studded anchors (Eligehausen and Ožbolt 1990), and bond splices (Şener 1992).

3. Failures of unreinforced pipes (Gustafsson and Hillerborg 1985; Bažant and Cao 1986).

4. Torsional failure of beams (Bažant, Şener and Prat 1988).

5. Diagonal shear failure of longitudinally reinforced beams without or with stirrups, unprestressed or prestressed (Bažant and Kim 1984; Bažant and Sun 1987; Bažant and Kazemi 1991).

6. Punching shear failure of slabs (Bažant and Cao 1987).

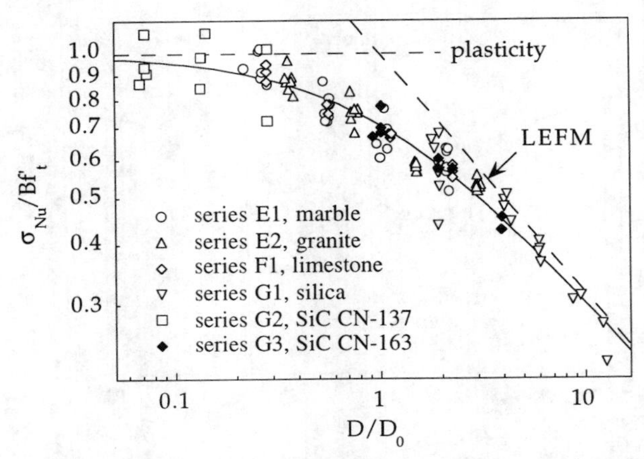

Figure 1.5.5 Size effect results for various kinds of rocks and ceramics (Fathy 1992; Bažant, Gettu and Kazemi 1991; McKinney and Rice 1981).

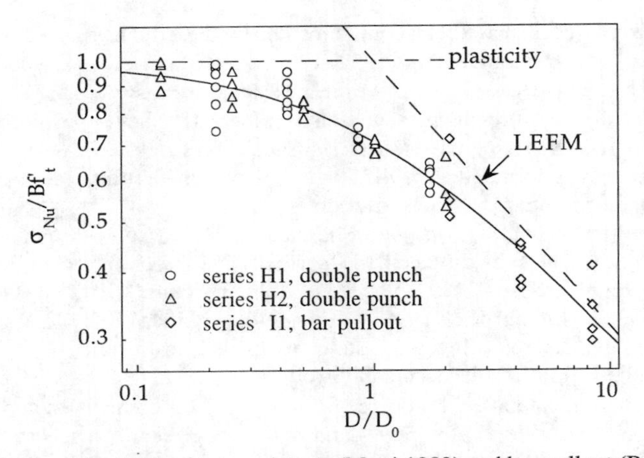

Figure 1.5.6 Size effect in double punch tests (Marti 1989) and bar pullout (Bažant and Şener 1988).

A sample of the results, which can be regarded as an additional verification of applicability of fracture mechanics to brittle failures of concrete structures, are shown in Figs. 1.5.6 and 1.5.7.

As further evidence of applicability of fracture mechanics, Fig. 1.5.8 shows, for the punching shear failure, that the post-peak load drop becomes steeper and larger as the size increases. This is because, in a larger specimen, there is (for the same σ_N) more energy to be released into a unit crack extension. The load must be reduced since the fracture extension dissipates the same amount of energy.

The existing test data on concrete specimens with regular-size aggregate reported in the literature also offer evidence of size effect, and the need for a fracture mechanics based explanation has been pointed out by various researchers, beginning with Reinhardt (1981a, 1981b). The data from the literature are generally found to agree with Fig. 1.2.5b although often the evidence is not very strong because the data exhibit very large statistical scatter and the size range is insufficient.

Due to large scatter and size range limitation, about equally good fits can often be obtained with other theories of size effect, e.g., Weibull's statistical theory. However, the measured size effect curves in the Figs. 1.5.2–1.5.7 do not agree with the Weibull-type statistical theory. This theory gives a straight line of slope –1/6 for two-dimensional similarities and –1/4 for two-dimensional similarity, which are significantly smaller than seen in the figures.

Much of the scatter probably stems from the fact that the test specimens of various sizes were not geometrically similar. Theoretical adjustments must, therefore, be made for the factors of shape before

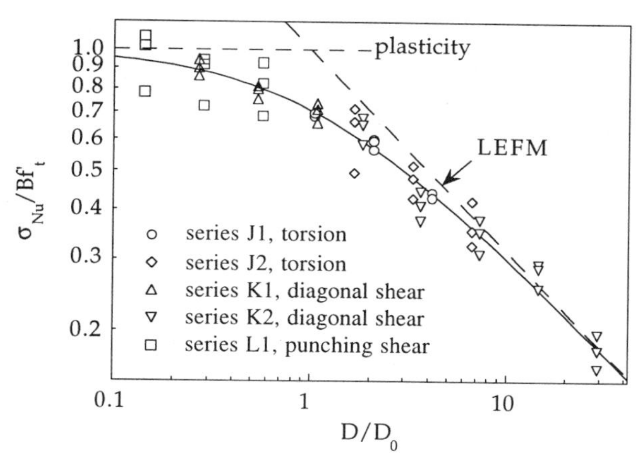

Figure 1.5.7 Size effect in torsion for plain (J1) and reinforced (J2) concrete prisms (Bažant, Şener and Prat 1988); diagonal shear failure of longitudinally reinforced beams for unanchored (K1) and anchored (K2) bars (Bažant and Kazemi 1991); and punching shear of slabs (Bažant and Cao 1987).

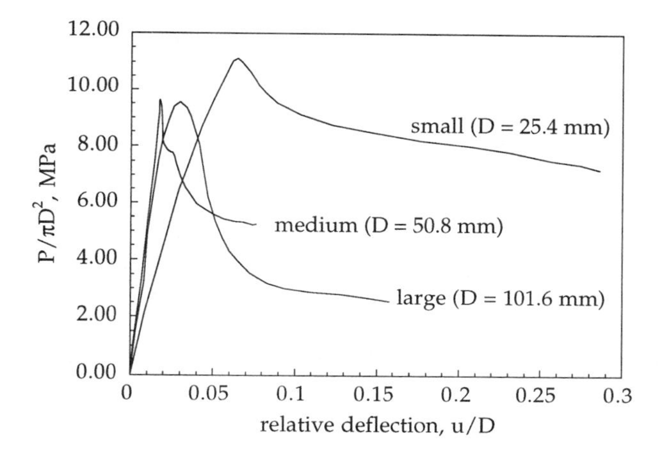

Figure 1.5.8 Size effect on the load-deflection curve in punching of slabs (adapted from Bažant and Cao 1987; series L1 in Tables 1.5.1 and 1.5.2).

the comparison with (1.4.10) can be made and, since the exact theory is not known, such adjustments introduce additional errors, manifested as scatter.

Exercises

1.7 In the plastic limit ($D/D_0 \ll 1$), the nominal strength is given by Bf'_t. Determine the largest size for which Bažant's size effect law differs from the plastic limit less than 5%. Hint: use the approximation $(1+x)^{-1/2} \approx 1 - x/2$ for $x \ll 1$.

1.8 Apply the result of the preceding exercise to the tests in Tables 1.5.1 and 1.5.2. Decide, for each test series, whether specimens of such size can be representative of the material. Hint: compare the specimen size with the maximum aggregate or grain size.

1.9 In the LEFM limit ($D/D_0 \gg 1$), the nominal strength is given by $Bf'_t\sqrt{D_0/D}$. Determine the smallest size for which Bažant's size effect law differs from the LEFM limit less than 5%. Hint: use the approximation $(1+x)^{-1/2} \approx 1 - x/2$ for $x \ll 1$.

1.10 Apply the result of the preceding exercise to the tests in Tables 1.5.1 and 1.5.2. Decide, for each test series, whether specimens of such size can be manufactured for laboratory testing.

1.11 We say that one structure is more brittle than another when its situation on the log σ_{Nu} vs. log D size effect plot is closer to LEFM limit. If (and only if) two structures are geometrically identical but made of different materials, the difference in the brittleness is entirely due to the difference in material brittleness. Decide which are the more brittle materials in the following tests (defined in Tables 1.5.1 and 1.5.2 and Figs. 1.5.1–1.5.7): (a) Concrete in series B1–4 and mortar in series C1–4; (b) Marble in series E1 and granite in series E2; (c) Concrete in series H1 and concrete in series H2; (c) ceramic materials in series G1, G2, and G3.

2

Essentials of LEFM

Linear elastic fracture mechanics (LEFM) is the basic theory of fracture, originated by Griffith (1921, 1924) and completed in its essential aspects by Irwin (1957, 1958) and Rice (1968a,b).

LEFM is a highly simplified, yet sophisticated, theory, that deals with sharp cracks in elastic bodies. As we shall see, LEFM is applicable to any material as long as certain conditions are met. These conditions are related to the basic ideal situation analyzed in LEFM in which all the material is elastic except in a vanishingly small region (a point) at the crack tip. In fact, the stresses near the crack tip are so high that some kind of inelasticity must take place in the immediate neighborhood of the crack tip; however, if the size of the inelastic zone is small relative to linear the dimensions of the body (including the size of the crack itself), the disturbance introduced by this small inelastic region is also small and, in the limit, LEFM is verified exactly.

Thus, LEFM is the basic theoretical reference to describe the behavior of any material with cracks, even if, as it happens for concrete, the geometry and dimensions of structures built in practice do not allow direct use of LEFM.

This and the next two chapters give an overall view of the mathematical theory of LEFM with some straightforward applications to idealized cases. They are not intended as a substitute for handbooks or treatises on LEFM. Their objective is to simplify the access of the reader to the concepts required in the remaining chapters and, at the same time, to provide in this book a self-contained presentation, so that recourse to external references be minimized.

This introductory chapter gives a short account of the most essential concepts in LEFM. Section 2.1 develops the energetic approach to fracture —the Griffith approach. It introduces the concept of energy release rate $\mathcal{G}$, representing the energy available for fracture, and the fracture energy or crack resisting force $\mathcal{R}$, representing the energy required for fracture. Also included in this section are the basic expressions for $\mathcal{G}$, and some techniques to describe fracture processes (Section 2.1). The systematic analysis of the techniques available to compute $\mathcal{G}$ and to measure $\mathcal{R}$ is deferred until the next chapter.

Section 2.2 introduces the concept of stress intensity factor K_I based on a simple example and describes the general properties of the stresses and displacements near the crack tip (the formal derivation of such properties is skipped in this introductory chapter and postponed until Chapter 4). It also shows that the energetic approach and the approach based on K_I —Irwin's approach— are equivalent, and rewrites the crack growth criterion in terms of the stress intensity factor. The presentation of the methods to compute stress intensity factors and other related quantities is postponed until the next chapter.

The final Section 2.3 deduces the size effect laws for classical plasticity and for LEFM. As explained in the previous Chapter, these are the reference laws for any nonlinear fracture model, and are extensively used for comparison with experimental as well as theoretical nonlinear size effects in the remainder of the book.

2.1 Energy Release Rate and Fracture Energy

Formation of a crack in an elastic solid initially subjected to uniform uniaxial tension disrupts the trajectories of the maximum principal stress in the manner depicted in Fig. 2.1.1a. This indicates that stress concentrations must arise near the crack tip. They were calculated by Inglis (1913) as the limit case of his solution for an elliptical hole.

From Inglis's solution, Griffith (1921, 1924) noted that the strength criterion cannot be applied because the stress at the tip of a sharp crack is infinite no matter how small the load is (Fig. 2.1.1b). He further concluded that the formation of a crack necessitates a certain energy per unit area of the crack plane,

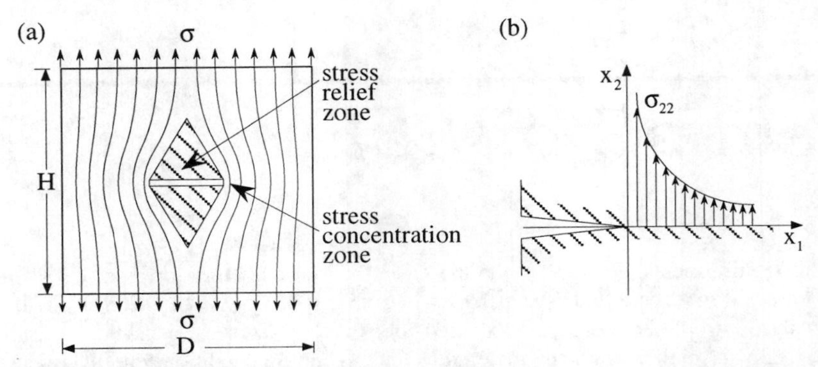

Figure 2.1.1 (a) Disruption of the trajectories of maximum principal stress by a crack; (b) singular distribution of normal stress ahead of the crack tip (adapted from ACI Committee 446 1992).

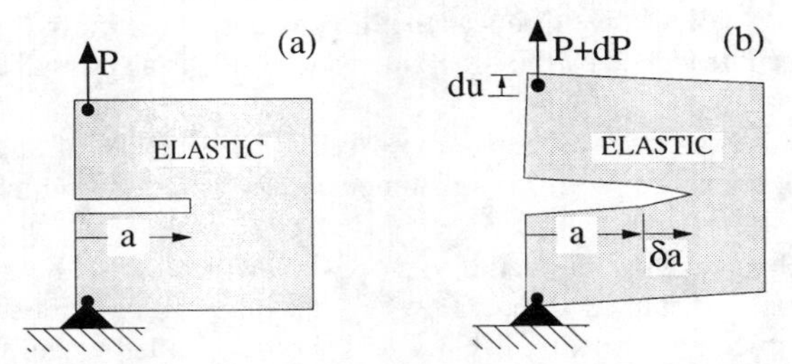

Figure 2.1.2 Crack growth in a cracked specimen: (a) initial situation; (b) co-planar crack growth upon further loading.

which is a material property, provided the structure is so large that the crack tip region in which the fracture process takes place is negligible. However, more general approaches accept that the specific energy required for crack growth may depend on the cracking history instead of being a constant. In such cases, the energy required for a unit advance of the crack is called the crack growth resistance, $\mathcal{R}$.

The basic problem in fracture mechanics is to find the amount of energy available for crack growth and to compare it to the energy required to extend the crack. Although conceptually simple, the problem is far from trivial and deserves a detailed analysis.

2.1.1 The General Energy Balance

Consider a plane structure of thickness b in which a preexisting straight crack of length a is present (Fig. 2.1.2). Assume that upon quasi-static loading, a certain load level is attained at which the crack advances an elemental length δa in its own plane, sweeping an element of area $\delta A = b\,\delta a$. The energy $\delta\mathcal{W}^F$ required to do so is the increment of area times the crack growth resistance:

$$\delta\mathcal{W}^F = \mathcal{R}\, b\, \delta a \tag{2.1.1}$$

Alternative notations found in the literature for the crack growth resistance $\mathcal{R}$ are G_R when it is history-dependent, and G_c and G_{Ic} (critical energy release rate) when it is a material property not dependent on the cracking history. In this book, we use G_f for the latter case because this is the most widespread notation in the field of concrete fracture. G_f is called the *specific fracture energy*, or fracture energy, for short. In Section 2.1.5 we will justify that the only case consistent with the hypotheses of elastic fracture, where the inelastic zone is negligibly small, is that of $\mathcal{R}$ = constant = G_f.

The total energy supply to the structure is the external work, which, in the infinitesimal process under consideration, is denoted as $\delta\mathcal{W}$. From this total supply, a part is stored in the structure as elastic energy, $\delta\mathcal{U}$. The remainder is left to drive other processes and to generate kinetic energy $\delta\mathcal{K}$. When the only energy-consuming process is fracture, and the process is quasi-static ($\delta\mathcal{K} = 0$), this remainder is the available energy for fracture, or elemental energy release $\delta\mathcal{W}^R$:

$$\delta\mathcal{W}^R = \delta\mathcal{W} - \delta\mathcal{U} \tag{2.1.2}$$

Although Eq. (2.1.2) may be directly handled in many cases (as done by Griffith 1921), it is often more convenient to work with specific energies (energies per unit area of crack growth). The specific available energy, usually called the energy release rate, $\mathcal{G}$, is thus defined so that

$$\mathcal{G}\, b\, \delta a = \delta\mathcal{W}^R = \delta\mathcal{W} - \delta\mathcal{U} \tag{2.1.3}$$

The essential advantage of using $\mathcal{G}$ is that, as it will turn out in the next paragraph, $\mathcal{G}$ is a *state function.* This means that $\mathcal{G}$ depends on the instantaneous geometry and boundary conditions, but not on how they vary in the actual fracture process, or on how they have been attained.

The balance of energy requires that, in a quasi-static process,

$$\mathcal{G}\, \delta a = \mathcal{R}\, \delta a \qquad \text{for quasi-static growth} \tag{2.1.4}$$

and in a more general incipiently dynamic situation (initial kinetic energy $\mathcal{K} = 0$, kinetic energy increase $\delta\mathcal{K} \geq 0$),

$$\mathcal{G}\, \delta a = \mathcal{R}\, \delta a + \delta\mathcal{K}/b \tag{2.1.5}$$

Since $\delta\mathcal{K} \geq 0$ (because initially $\mathcal{K} = 0$ and always $\mathcal{K} \geq 0$), the equations may be made to hold in any circumstance (as they should, the balance of energy being a universal law, the first law of thermodynamics) if the following fracture criterion is met:

$$\text{if} \quad \mathcal{G} < \mathcal{R} \quad \text{then} \quad \delta a = 0 \quad \text{and} \quad \delta\mathcal{K} = 0 \quad \text{No crack growth (stable)} \tag{2.1.6}$$

$$\text{if} \quad \mathcal{G} = \mathcal{R} \quad \text{then} \quad \delta a \geq 0 \quad \text{and} \quad \delta\mathcal{K} = 0 \quad \text{Quasi-static growth possible} \tag{2.1.7}$$

$$\text{if} \quad \mathcal{G} > \mathcal{R} \quad \text{then} \quad \delta a > 0 \quad \text{and} \quad \delta\mathcal{K} > 0 \quad \text{Dynamic growth (unstable)} \tag{2.1.8}$$

This system of conditions summarizes what seems obvious: If the energy available is less than that required, then the crack cannot grow (and the structure is stable). If the energy available equals the required energy, then the crack can grow statically, i.e., with negligible inertia forces (and the structure can be stable or unstable depending on the variation of $\mathcal{G} - \mathcal{R}$ with displacements). If the energy available exceeds that required, then the structure is unstable and the crack will run dynamically (the excess energy being turned into kinetic energy).

The central problems of elastic fracture mechanics are to measure the crack growth resistance, $\mathcal{R}$, for particular materials and situations, on one hand, and to calculate the energy release rate $\mathcal{G}$, on the other. This latter problem may be handled in various equivalent ways, the bases of which are explored next.

2.1.2 Elastic Potentials and Energy Release Rate

Consider the plane elastic specimen in Fig. 2.1.2, in which the crack length a can take any value. Let P be the load and u the load-point displacement. By definition, the elementary work is

$$\delta\mathcal{W} = P\delta u \tag{2.1.9}$$

for any incremental process. For an equilibrium situation and given any crack length a, there is a unique relationship between the equilibrium force and the displacement (which can be calculated by solving the elastic problem). So we can write

$$P = P(u, a) \tag{2.1.10}$$

where $P(u, a)$ can be determined by elastic equilibrium analysis of the structure. Based on the corresponding elastic solution, the stored (elastic) strain energy can also be calculated for any u and a:

$$\mathcal{U} = \mathcal{U}(u, a) \tag{2.1.11}$$

Consider, then, that the elastic body with a crack of length a is subjected, in a static manner, to displacement δu, with *no crack growth.* In this situation, all the work is stored as strain energy, so that

$$\delta\mathcal{W} - [\delta\mathcal{U}]_a = 0 \tag{2.1.12}$$

where subscript a indicates that the crack length remains constant in this process.

Consider now a general process where both u and a are allowed to vary. Then, Eq. (2.1.3), which is the definition of $\mathcal{G}$, can be written as

$$\mathcal{G}\, b\,\delta a = P(u,a)\delta u - \left\{\left[\frac{\partial\mathcal{U}(u,a)}{\partial u}\right]_a \delta u + \left[\frac{\partial\mathcal{U}(u,a)}{\partial a}\right]_u \delta a\right\} \tag{2.1.13}$$

Considering equilibrium variation at $\delta a = 0$, one obtains the well-known second Castigliano's theorem:

$$P(u,a) = \left[\frac{\partial\mathcal{U}(u,a)}{\partial u}\right]_a \tag{2.1.14}$$

So the first two terms in (2.1.13) in an equilibrium process cancel and we get

$$\mathcal{G} = \mathcal{G}(u,a) = -\frac{1}{b}\left[\frac{\partial\mathcal{U}(u,a)}{\partial a}\right]_u \tag{2.1.15}$$

This basic result shows that the energy release rate $\mathcal{G}$ is indeed a state function, because it depends only on the instantaneous boundary conditions and geometry (in this case uniquely defined by u and a).

Sometimes one may prefer to use the equilibrium load P rather than the equilibrium displacement u as an independent variable. In such case, it is preferable to introduce a dual elastic potential, the complementary energy $\mathcal{U}^*$, defined as

$$\mathcal{U}^* = Pu - \mathcal{U} \tag{2.1.16}$$

Substituting $\mathcal{U}$ from this equation in the expression for the available energy (2.1.2), together with the expression (2.1.9) for the elemental work, one gets for the elemental energy release

$$\delta\mathcal{W}^R = \delta\mathcal{W} - \delta\mathcal{U} = P\delta u - \delta(Pu - \mathcal{U}^*) = \delta\mathcal{U}^* - u\,\delta P \tag{2.1.17}$$

Writing the complementary energy and the displacement as functions of the applied load and the crack length

$$u = u(P,a)\,, \qquad \mathcal{U}^* = \mathcal{U}^*(P,a) \tag{2.1.18}$$

and considering an equilibrium process in which both P and a are allowed to vary, Eq. (2.1.17) yields

$$\mathcal{G}\, b\,\delta a = -u(P,a)\delta P + \left\{\left[\frac{\partial\mathcal{U}^*(P,a)}{\partial P}\right]_a \delta P + \left[\frac{\partial\mathcal{U}^*(P,a)}{\partial a}\right]_P \delta a\right\} \tag{2.1.19}$$

Considering equilibrium variation at $\delta a = 0$, one gets the well-known first Castigliano's theorem

$$u(P,a) = \left[\frac{\partial\mathcal{U}^*(P,a)}{\partial P}\right]_a \tag{2.1.20}$$

So the first two terms in (2.1.19) cancel and we have

$$\mathcal{G} = \mathcal{G}(P,a) = \frac{1}{b}\left[\frac{\partial\mathcal{U}^*(P,a)}{\partial a}\right]_P \tag{2.1.21}$$

The couple of equations (2.1.20) and (2.1.21) are strictly equivalent to the couple (2.1.14) and (2.1.15). Indeed, in this single-point-load problem there are 4 mechanical variables, namely, P, u, a, and $\mathcal{G}$, but only two of them are independent variables (Elices 1987). The choice of the independent variables is arbitrary, and is usually done depending on the boundary conditions and the available data.

Remark: Under isothermal conditions (slow loading, slow crack growth), $\mathcal{U}$ represents the Helmholtz's free energy of the structure and $\mathcal{U}^*$ its Gibbs' free energy. Under isentropic (or adiabatic) conditions (rapid loading, rapid growth), $\mathcal{U}$ represents the internal (total) energy of the structure and $\mathcal{U}^*$ represents the enthalpy (see e.g., Bažant and Cedolin 1991, Sec. 10.1).

Other potentials can be used to perform the foregoing analysis. For example, the potential energy Π of the structure-load system, defined as $\Pi = \mathcal{U} - W_a(u)$, where $W_a(u) = \int_0^u P_a(u')\,du'$ is the work of the applied load $P_a(u)$, which is assumed to be defined independently of the structure. The energy release rate is easily expressed in terms of the potential energy as

$$\mathcal{G} = \mathcal{G}(u, a) = -\frac{1}{b}\left[\frac{\partial \Pi(u, a)}{\partial a}\right]_u \tag{2.1.22}$$

Same as for the strain energy, a dual potential can be defined for the potential energy, the complementary potential energy Π^* of the structure-load system, defined as $\Pi^* = \mathcal{U}^* - W_a^*(P)$, where $W_a(P) = \int_0^P u(P')\,dP'$ is the complementary work of the applied load (which is 0 for dead loads). The energy release rate is easily expressed in terms of the complementary potential energy as

$$\mathcal{G} = \mathcal{G}(P, a) = \frac{1}{b}\left[\frac{\partial \Pi^*(P, a)}{\partial a}\right]_P \tag{2.1.23}$$

In this book, the potential energy and complementary potential energy of the structure-load system will not be used. △

The foregoing results may seem too particular because no distributed loads were considered. This limitation may be overcome in most practical cases by defining *generalized forces and displacements.* A generalized force Q and its associated generalized displacement q are defined in such a way that the external work $\delta\mathcal{W}$ may be written as

$$\delta\mathcal{W} = Q\,\delta q \tag{2.1.24}$$

With this definition, all the foregoing expressions hold as long as one interprets P as a generalized force and u as its generalized displacement.

There are many well known cases of generalized forces used in engineering. For example: the generalized displacement associated with a torque is the angular rotation; the generalized displacement associated with a pressure acting inside a cavity is the volume variation of the cavity.

Example 2.1.1 To illustrate the application of the above equations, consider a long-arm double cantilever beam (DCB) specimen subjected to constant moments M as depicted in Fig. 2.1.3a. Assume further that the material is linear elastic, and that the arms are slender enough for the classical theory of bending to apply. With these hypotheses, the elastic or complementary energy per unit length of the beam is known from the theory of strength of materials (e.g., Timoshenko 1956):

$$\frac{d\mathcal{U}}{dx} = \frac{d\mathcal{U}^*}{dx} = \frac{M^2}{2EI} \tag{2.1.25}$$

where x is the coordinate of a cross-section along the beam axis, E the elastic modulus, and I the inertia moment of the cross-section of the beam. We thus compute the elastic or complementary energy of the specimen as the energy of two pure bent cantilever beams of length a:

$$\mathcal{U} = \mathcal{U}^* = 2a\frac{M^2}{2EI}, \tag{2.1.26}$$

and compute the energy release rate by direct application of (2.1.21)

$$\mathcal{G} = \frac{M^2}{bEI}, \tag{2.1.27}$$

with M taking the place of P. ▯

Example 2.1.2 As another example, consider the double cantilever specimen in Fig. 2.1.3b. The bending moment distribution for the upper arm, $M = Px$, is also shown in this figure. Within the

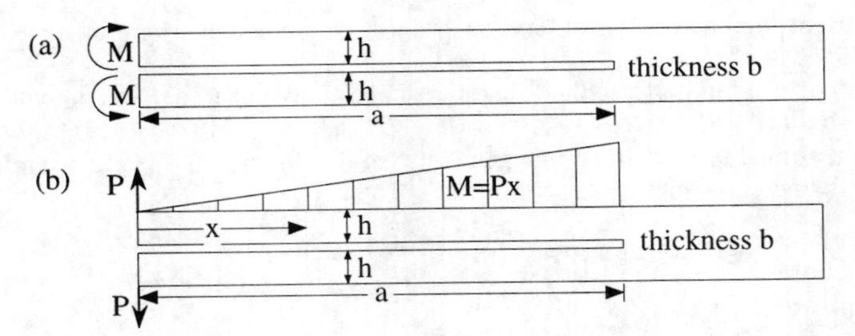

Figure 2.1.3 Long double cantilever beam specimen subjected to (a) pure bending, and (b) opening end forces.

classical beam theory (neglecting shear), the corresponding complementary energy per unit length of one arm is given now, according to (2.1.25), by $d\mathcal{U}^*/dx = P^2x^2/2EI$. The total complementary energy is obtained by integration along both arms of the specimen

$$\mathcal{U}^* = 2\frac{P^2}{2EI}\int_0^a x^2\,dx = \frac{P^2a^3}{3EI} \tag{2.1.28}$$

With this, Equations (2.1.20) and (2.1.21) provide expressions for the relative displacement between the load points, u, and for the energy release rate $\mathcal{G}$

$$u = \frac{2Pa^3}{3EI} = \frac{8Pa^3}{Ebh^3}, \qquad \mathcal{G} = \frac{P^2a^2}{bEI} = \frac{12P^2a^2}{Eb^2h^3} \tag{2.1.29}$$

in which we set $I = bh^3/12$. Except for a factor 2, the first expression for u in the previous equation is very well known in the field of strength of materials. The factor 2 comes from the relative displacement of the forces (working displacement) being twice the deflection of one beam. □

2.1.3 The Linear Elastic Case and the Compliance Variation

The foregoing general results are greatly simplified in the particular, yet essential, case of linear elasticity, because of the linear relationship between u and P at constant a. This may be written as

$$u = C(a)\,P \tag{2.1.30}$$

where $C(a)$ is the (secant) compliance for a crack length a. After substituting u from Eq. (2.1.30) into Eq. (2.1.20), it immediately follows by integration that the complementary energy must be

$$\mathcal{U}^* = \frac{1}{2}C(a)P^2. \tag{2.1.31}$$

Substitution of this expression into Eq. (2.1.21) gives the following result for the energy release rate:

$$\mathcal{G}(P,a) = \frac{P^2}{2b}\frac{dC(a)}{da} \equiv \frac{P^2}{2b}C'(a), \tag{2.1.32}$$

where, in the second expression, the first derivative of the compliance has been briefly denoted as $C'(a)$.

In the foregoing derivation, (P,a) were taken as independent variables. But one can equally well use (u,a) as independent variables. Substituting P from Eq. (2.1.32) into Eq. (2.1.14), it follows by immediate integration that the elastic energy must be

$$\mathcal{U}(u,a) = \frac{u^2}{2C(a)} \tag{2.1.33}$$

Henceforth, from Eq. (2.1.21), the energy release rate is found to be

$$\mathcal{G}(u,a) = \frac{u^2}{2bC^2(a)}\frac{dC(a)}{da} \equiv \frac{u^2}{2bC^2(a)}C'(a) \tag{2.1.34}$$

which, in view of (2.1.30), turns out to be identical to the previous Eq. (2.1.32), as it must.

At this point it is worth to recall the well-known fact that, in linear elasticity, the elastic energy and the complementary energy always take the same value (although they are conceptually different, as graphically shown in the next subsection). In the case of a single point load, it is sometimes useful to rewrite Eqs. (2.1.31) and (2.1.33) in the form

$$\mathcal{U} = \mathcal{U}^* = \frac{1}{2}Pu \tag{2.1.35}$$

Example 2.1.3 Consider again the pure bent DCB in Fig. 2.1.3a with the same hypotheses as stated in the previous section. Taking M as the generalized force, the relative rotation θ of the arm ends is the corresponding generalized displacement. Since the rotation of each beam end is $\theta/2$, and such rotation has an expression well known from the strength of materials: $\theta/2 = aM/EI$. Therefore, the generalized compliance is

$$C = \frac{\theta}{M} = \frac{2a}{EI} \tag{2.1.36}$$

The use of (2.1.32) leads again to Eq. (2.1.27) for $\mathcal{G}$. ▯

Example 2.1.4 Consider again the long-arm DCB specimen of Fig. 2.1.3b subjected to loads P at the arm tips. In this case, the deflection of each arm is well known to be $u_1 = Pa^3/3EI$. Thus the displacement over which the loads P work is $u = 2u_1 = 2Pa^3/3EI$, from which it follows that

$$C = \frac{2a^3}{3EI} \tag{2.1.37}$$

Using (2.1.32) again yields the result (2.1.29). ▯

Example 2.1.5 Consider the center cracked panel depicted in Fig. 2.1.1a. Let the dimensions of the panel —width, height, thickness— be, respectively, D, H, and b; and assume a central crack of total length $2a$. (Note: it is customary to use $2a$ instead of a for the crack length for this kind of internal cracks; this requires special care when differentiating with respect to crack length, see below.) A detailed elastic analysis (Chapter 4) delivers the relative displacement of the upper and lower edges of the panel as a function of the crack size. For small cracks ($2a \ll D, 2a \ll H$) and plane stress, this displacement turns out to be

$$u = \frac{\sigma H}{E}\left(1 + \frac{2\pi a^2}{DH}\right) = \frac{PH}{BDE}\left(1 + \frac{2\pi a^2}{DH}\right) \tag{2.1.38}$$

where we wrote that the resultant load is $P = \sigma BD$. From the last expression we get $C = u/P$ and using (2.1.32) with a replaced by $2a$, we get

$$\mathcal{G} = \frac{P^2}{2B}\frac{dC}{d(2a)} = \frac{P^2\pi a}{B^2D^2E} = \frac{\sigma^2}{E}\pi a \tag{2.1.39}$$

This is one of the most celebrated Griffith's results (although Griffith, obtained it in a different way). ▯

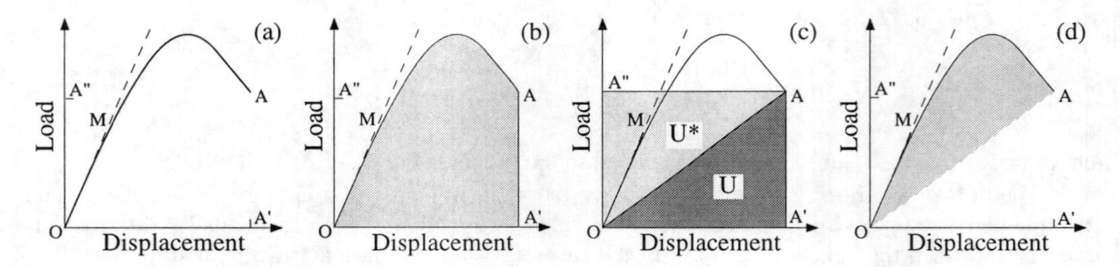

Figure 2.1.4 (a) Quasi-static load-displacement curve. (b) Area representing the total work supply. (c) Areas representing elastic strain energy and complementary energy. (d) Area representing energy supply for fracture and energy dissipated in fracture.

2.1.4 Graphical Representation of Fracture Processes

The energetic equations allow graphical interpretation which, in many instances, supply vivid pictures helpful for problem-solving and explaining. A loading process of a specimen or structure is sometimes best followed on a load-displacement plot. In the case of a single load P, the displacement to consider is the load-point displacement u. Let the plot of a *quasi-static* P-u curve for a given specimen be as shown in Fig. 2.1.4a. The work supplied to the specimen from the beginning of the loading, point O, up to point A, is the integral of $P\,du$, which is equal to the area $OMAA'O$ shown in Fig. 2.1.4b. The area that complements this to rectangle $OA''AA'$ is the integral of $u\,dP$, and is called the complementary work.

If all of the material remains linear elastic except for a zone of negligible volume along the crack path, the (elastic) strain energy $\mathcal{U}$ is represented by the area of the triangle OAA' in Fig. 2.1.4c. The complementary energy $\mathcal{U}^*$ is the area of the triangle OAA'' which complements OAA' to the rectangle $OA'AA''$, of area Pu.

The energy supplied for fracture is the difference between the work and the fracture energy, hence, the area $OMAO$ in Fig. 2.1.4d (it is also the difference between the complementary energy and the complementary work). If the curve shown corresponds to an actual quasi-static fracture process, then the energy supplied for fracture must coincide with the energy consumed by fracture; hence, the area $OMAO$ in Fig. 2.1.4d is also the energy consumed in fracture.

An equilibrium fracture process from point A, where the crack length was a, to a nearby point B, where the crack length has increased by Δa, may be represented as shown in Fig. 2.1.5a. The energy release available for fracture is the area of triangle OAB. This area coincides, except for second-order small terms, with those corresponding to the *virtual* (nonequilibrium) processes represented in Figs. 2.1.5b, c, and d, respectively, corresponding to constant displacement, constant load, and arbitrary $\Delta P/\Delta u$. The energy release rate $\mathcal{G}$ is the limit of the ratio of the area of *any* of the shaded triangles to the crack extension. This shows, again, that $\mathcal{G}$ is path independent, hence, a state function. When Fig. 2.1.5b is used, Eq. (2.1.15) is obtained. When Fig. 2.1.5c is used, Eq. (2.1.21) is obtained. Both turn out to be identical to Eq.(2.1.32) as the reader may easily check. For example, taking the shaded triangle in Fig. 2.1.5c, we express the fracture energy as

$$\mathcal{G}\, b\,\Delta a = \text{area}(OAB\text{"}) = \frac{1}{2}P\,(\overline{AB\text{"}}) = \frac{1}{2}P[PC(a+\Delta a) - PC(a)] = \frac{1}{2}P^2C'(a)\Delta a \quad (2.1.40)$$

from which Eq. (2.1.32) immediately follows.

As previously stated, only two of the four variables P, u, a, and $\mathcal{G}$ can be taken as independent variables. Any pair of them may be used to define the entire fracture process. However, it is useful to take conjugate variable pairs, as is customary in thermodynamics, because then the areas in the graphical representation have direct energetic interpretations. One of such pair is the P-u representation just analyzed. The other is the $\mathcal{G}$-a representation. This representation has the advantage that $\mathcal{G}$ is the "driving force" for crack growth which is directly related to the material property $\mathcal{R}$, the "resisting force".

If one then imagines a plot of a loading process in a $\mathcal{G}$-a plane, such as that in Fig. 2.1.6a, one finds that OM is a loading at constant crack length under increasing $\mathcal{G}$. At point M, the crack starts to increase under increasing $\mathcal{G}$ up to point A. The total energy released is the integral of $\mathcal{G}\,da$, equal, henceforth,

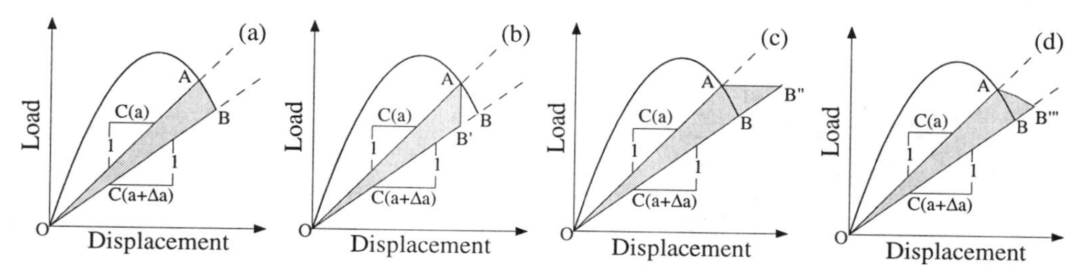

Figure 2.1.5 (a) Actual (equilibrium) incremental fracture process. (b-d) Virtual (nonequilibrium) incremental fracture processes: (b) At constant displacement; (c) at constant load; and (d) at arbitrary $\Delta P/\Delta u$.

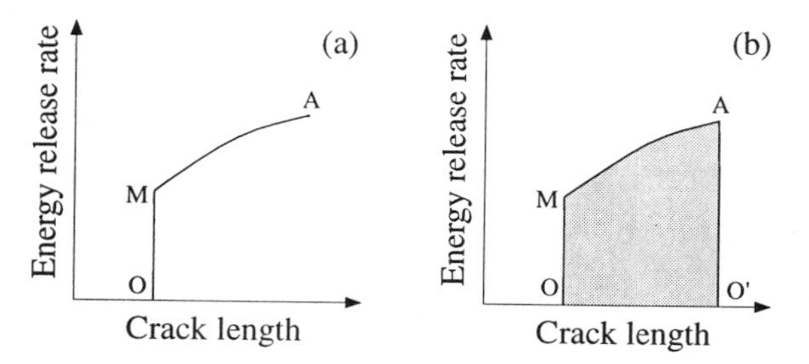

Figure 2.1.6 (a) Loading path in a $\mathcal{G} - a$ plot. (b) Area representing the total energy supply for fracture and the energy dissipated in fracture.

to the area $OMAO'O$ in Fig. 2.1.6b. Moreover, if the process is an actual quasi-static (equilibrium) process, this coincides with the total fracture energy.

Furthermore, the instantaneous $\mathcal{G}$ on the MA portion, where the crack is actually growing, must coincide with the instantaneous fracture energy which, in this example, is not constant. This is an example of the so-called R-curve behavior, a short for resistance curve behavior, in which the crack growth resistance increases with the crack extension (Chapter 5). In Section 2.1.6 we argue that this is not a kind of behavior consistent with the hypotheses of LEFM, which really imply that the crack growth resistance must be a constant.

2.1.5 Rice's J-Integral

One of the most famous equations in fracture mechanics is the J-integral, due to J. Rice (1968a). Although in its original derivation, the J-integral was not directly related to $\mathcal{G}$, soon after that Rice (1968b) realized that J was equal to $\mathcal{G}$. In the following paragraphs we introduce the J-integral as a particular form of expressing the energy release rate.

In deriving the J integral, the general expression (2.1.3) is used, together with a particular way of expressing the work and the elastic energy and a particular virtual process. To start fixing the main concepts, we first notice that although we have been continuously referring to a given body, any part of a body is another body in mechanical terms. Henceforth, all the equations used so far may be used for any subbody. In a plane case, we can take any contour Γ surrounding the crack tip to define a subbody and apply to it the energy balance equations —in particular Eq. (2.1.3)— to find $\mathcal{G}$. This is what is done in the derivation that follows.

The derivation of the J-integral is done in many books in an Eulerian framework, where the axes and the contour Γ move with the crack tip. It is important to realize that ours use the Lagrangian coordinates (coordinates of material points in the initial state), and then the reference subbody defines a closed system. If that were not the case, the flow (transport) terms would have to be included in the energy balance equations.

In the plane case, the elemental external work $\delta\mathcal{W}$ is (for stress-free crack surfaces) just the work done

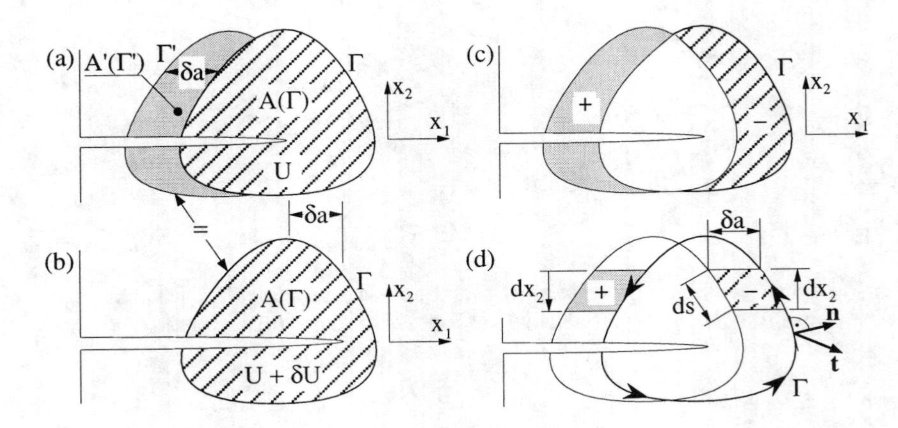

Figure 2.1.7 Determination of the variation of elastic energy by the J-integral.

by the surface tractions on the boundary Γ:

$$\delta \mathcal{W} = b \int_{\Gamma} t_i \delta u_i \, ds \tag{2.1.41}$$

where ds is the differential of arc-length along the contour Γ, t_i the components of the surface traction vectors acting on this boundary, and u_i the components of the displacement vector (summation is implied by repeated indices).

We write the elastic energy $\mathcal{U}$ as the integral of the elastic (or strain) energy density, $\overline{\mathcal{U}}$, throughout the volume of the subbody defined by Γ, which, for a plane case reads

$$\mathcal{U} = b \int_{\mathcal{A}(\Gamma)} \overline{\mathcal{U}} \, dA \tag{2.1.42}$$

where $\mathcal{A}(\Gamma)$ stands for the plane area of the subbody.

Substitution of the above equations (2.1.41) and (2.1.42) into Eq. (2.1.3) leads to:

$$\mathcal{G} \, \delta a = \int_{\Gamma} t_i \delta u_i \, ds - \delta \left[\int_{\mathcal{A}(\Gamma)} \overline{\mathcal{U}} \, dA \right] \tag{2.1.43}$$

The next step is to evaluate the variations δ in any virtual elemental process. To obtain the J-integral expression, we select a virtual process in which we translate all the fields (displacement, stress, energy density) a distance δa parallel to the crack, while extending the crack by this same amount. The variation of the displacement of a given material point, situated at (x_1, x_2) due to the translation is easily obtained:

$$\delta u_i(x_1, x_2) = u_i(x_1 - \delta a, x_2) - u_i(x_1, x_2) = -u_{i,1}(x_1, x_2) \delta a \tag{2.1.44}$$

where $u_{i,1}$ stands for $\partial u_i / \partial x_1$.

The other variation to be computed is that of the elastic energy — the integral in Eq. (2.1.43). It may be evaluated in various ways. Direct analytical treatment using an expression for the elastic energy density similar to the previous equation (2.1.44) is straightforward. However, the solution may be obtained in a much more physical (and graphical) way as follows: Let the cross-hatched area shown in Fig. 2.1.7a be the subbody $\mathcal{A}(\Gamma)$ defined by the contour Γ in its initial situation. When the crack is extended and the fields are translated by δa, the subbody reaches a state as defined in Fig. 2.1.7b. Because, by construction, the fields in part (b) of the figure are those in part (a) translated, the final energy of the subbody $\mathcal{A}(\Gamma)$ coincides with the initial energy of a subbody $\mathcal{A}'(\Gamma')$ defined by a contour Γ' obtained by displacing Γ a distance δa towards the left, as shown in Fig. 2.1.7a by a lightly shaded area partially hidden by $\mathcal{A}(\Gamma)$. Therefore, the *variation* of $\mathcal{U}$ between the initial and final states, is equal to the *difference* of initial energies between the two bodies $\mathcal{A}'(\Gamma')$ and $\mathcal{A}(\Gamma)$:

$$\delta\mathcal{U} = \mathcal{U}[\mathcal{A}'(\Gamma')] - \mathcal{U}[\mathcal{A}(\Gamma)] \tag{2.1.45}$$

and, graphically, this energy reduces to the energy contained in the two crescents shown in Fig. 2.1.7c, positive for the lightly shaded part (on the left) and negative for the cross-hatched part (on the right). The result may be expressed as a contour integral using the infinitesimal surface elements and the contour orientation depicted in Fig. 2.1.7d:

$$\delta\mathcal{U} = b\delta\left[\int_{\mathcal{A}(\Gamma)} \overline{\mathcal{U}}\, dA\right] = -b\int_\Gamma \overline{\mathcal{U}}\, dx_2\ \delta a \tag{2.1.46}$$

Substitution of this result and that in Eq. (2.1.44) for δu_i in Eq. (2.1.44), finally leads to the following expression for the energy release rate:

$$\mathcal{G} = \int_\Gamma (\overline{\mathcal{U}}\, dx_2 - t_i\, u_{i,1} ds) \tag{2.1.47}$$

The integral expression in the right hand member is Rice's J-integral. This integral can be computed (i.e., it is defined) whenever all points on contour Γ are elastic, even in situations where elastic fracture mechanics does not apply. However, the J-integral is equal to the energy release rate, $\mathcal{G}$, only if (1) the nonelastic zone reduces to a point in the interior of Γ, (2) the crack faces are traction-free, and (3) the crack is plane and extends in its own plane.

The J-integral as written in 2.1.47 is —because the factor dx_2 must take the proper sign— an *oriented* line integral. It must be performed anti-clockwise, from the lower to the upper face of the crack, to give the correct result. This need may be avoided by realizing that the sign is correctly captured if one writes $dx_2 = n_1 ds$ where n_1 is the component along the crack line of the unit outward normal to the contour, and ds is the unoriented arc-length differential. The J-integral may then be written as an unoriented contour integral with positive arc-length differential. (This is, in fact, equivalent to considering the line integral as a surface integral per unit thickness, where $ds = dA/b$ is the lateral area per unit thickness.) The resulting expression is

$$J = \int_\Gamma (\overline{\mathcal{U}}\, n_1 - t_i\, u_{i,1}) ds \tag{2.1.48}$$

Although the J-integral can be analytically evaluated in only very few cases (one of which is shown in the forthcoming example), it is a very powerful theoretical tool.

Example 2.1.6 Consider again the pure bent DCB in Fig. 2.1.3a with the same hypotheses as stated in the preceding section. Let us compute the J integral following the path $ABCDEF$ shown in Fig. 2.1.8. Since $n_1 = 0$ and $t_i = 0$ along BC and DE, the contribution of these two segments to the J integral is zero. So is the contribution from CD if one assumes that this segment is far enough from the crack tip to be stress free. Therefore, only the segments AB and AF contribute to the integral. Moreover, their contribution is identical, by symmetry arguments. If we call σ the bending stress, which is the normal stress σ_{11} in the direction of the arm axis and the only nonzero component of the stress tensor, we have (along FE, for example): (1) $n_1 = -1$; (2) $\overline{\mathcal{U}} = \sigma^2/2E$; (3) $t_1 = -\sigma$ (note the sign); and (4) $u_{1,1} = \varepsilon_{11} = \sigma/E$. Thus

$$J = \frac{1}{2E}\int_E^F \sigma^2 ds$$

Using $\sigma = Mz/2I$ where z is measured from the center line and $ds = dz$, Eq. (2.1.27) for $\mathcal{G}$ is readily recovered. ▯

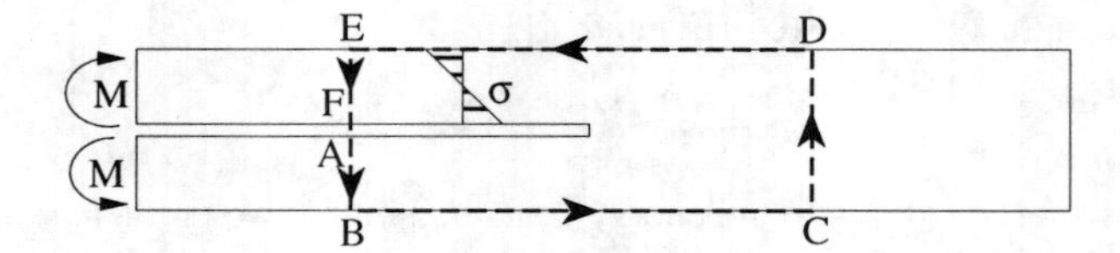

Figure 2.1.8 Integration path (dashed line) used to compute J in the pure bending DCB.

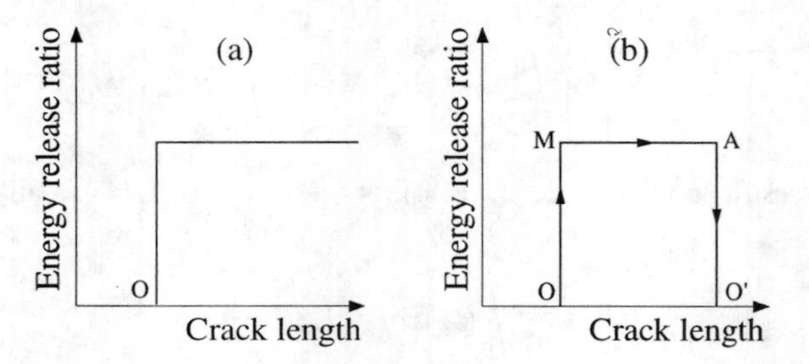

Figure 2.1.9 (a) In true LEFM, crack growth must take place under constant $\mathcal{G}$. (b) A loading-cracking-unloading process in a LEFM specimen.

2.1.6 Fracture Criterion and Fracture Energy

In true LEFM, in which nonlinear behavior and fracture occur at a single mathematical point at the crack tip, the crack must grow statically under constant $\mathcal{G}$, because the material has no memory of the previous loading. This implies that the crack growth resistance is a constant: $\mathcal{R} = G_f$. So the quasi-static loading path in a $\mathcal{G}$-a plot is a step function as depicted in Fig. 2.1.9a.

When the LEFM limit is applicable, a loading-cracking-unloading path in the $\mathcal{G}$-a plot looks as shown in Fig. 2.1.9b. Along the segment OM, $\mathcal{G}$ increases while the crack retains its initial length a_0. Along MA, the crack grows under constant $\mathcal{G} = G_f$. If at point A the specimen is unloaded, the crack will not heal, and so the unloading AO' will take place at constant crack length $a = a_1$ down to zero load.

This process may be also plotted in a P-u plot, a much more usual way of plotting experimental results. In such a plot, the constant crack length segments OM and AO' become constant compliance lines, i.e., straight lines through the origin. The MA segment is an iso-$\mathcal{G}$ curve corresponding to $\mathcal{G} = G_f$, the equation of which is obtained by eliminating the crack length a from equations 2.1.21 and 2.1.22:

$$u = C(a)\, P \tag{2.1.49}$$

$$G_f = \frac{P^2}{2b} C'(a) \tag{2.1.50}$$

The resulting P-u plot for the process shown in Fig. 2.1.9b typically looks as shown in Fig. 2.1.10a, with negative slope for the iso-$\mathcal{G}$ curve. However, there exist certain geometries where the iso-$\mathcal{G}$ curves display positive slope as depicted in Fig. 2.1.10b.

Example 2.1.7 A DCB specimen similar to that in Fig. 2.1.3b has a thickness $b = 10$ mm, width $2h = 20$ mm, and initial crack length $a_0 = 80$ mm, with $E = 300$ GPa and $G_f = 100$ N/m. We want to describe the evolution of the crack by means of the $\mathcal{G}(a)$ and $P(u)$ curves in a quasi-static test in which the displacement is monotonically increased until the crack doubles its initial length and then is decreased until complete unloading.

The $\mathcal{G}(a)$ curve is immediate (Fig. 2.1.11a): the crack length is constant and equal to $a_0 = 80$ mm while $\mathcal{G}$ is increasing up to $G_f = 100$ N/m (segment OM); from this point on, the fracture energy is kept constant at 100 N/m until the crack reaches 160 mm (point A) where unloading begins at constant crack length down to O'.

To follow the $P(u)$ process we use the approximate results for u and $\mathcal{G}$ from Examples 2.1.2 and 2.1.4.

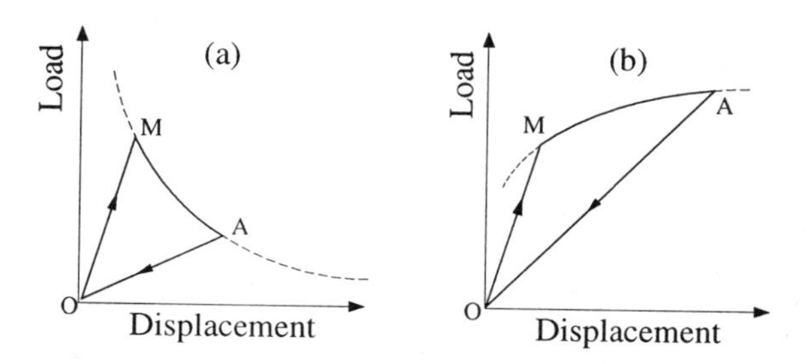

Figure 2.1.10 P-u plot of a loading-cracking-unloading in LEFM: (a) For a typical structure where iso-$\mathcal{G}$ curves display downwards slope; (b) for more exotic structures or loadings the slope of the iso-$\mathcal{G}$ curves may be positive.

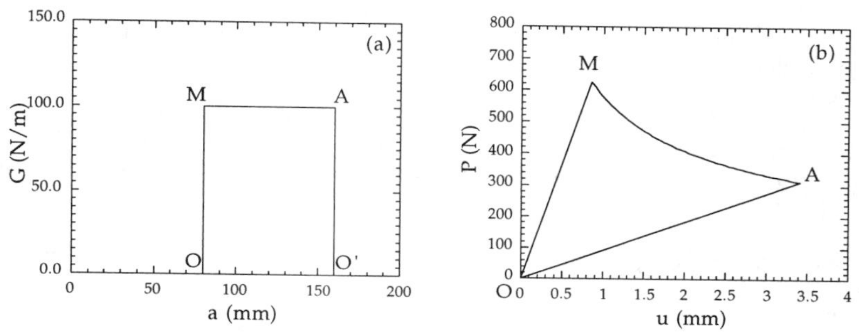

Figure 2.1.11 Process experienced by the DCB specimen in Example 2.1.7; (a) $\mathcal{G}$–a; (b) P–u diagram.

Initially, $a = a_0 = 80$ mm, and the load grows proportionally to the displacement following the first of (2.1.29) with $a = a_0$:

$$u = \frac{8Pa_0^3}{Ebh^3} \qquad \Rightarrow \qquad P(\text{N}) = 732.4\ u(\text{mm}) \tag{2.1.51}$$

where the dimensions in parentheses indicate the units for the load and the displacement.

Point M corresponds to $\mathcal{G} = G_f$ with $a = a_0$; the load P_M at which this condition is met is obtained from the second of (2.1.29):

$$G_f = \frac{12P_M^2 a_0^2}{Eb^2h^3} \qquad \Rightarrow \qquad P_M = 625\ \text{N} \tag{2.1.52}$$

corresponding to a displacement u_M that, according to (2.1.51), is given by $u_M = 0.8533$ mm. From this point on, the crack will grow and the displacement and load will evolve following Eqs. (2.1.29) with $G = G_f$. The results are:

$$P(\text{N}) = 625\ \frac{a_0}{a} \qquad \text{and} \qquad u(\text{mm}) = 0.8533\ \frac{a^2}{a_0^2} \tag{2.1.53}$$

which are the parametric equations of the P–u curve during the process MA in Fig. 2.1.11a. The cartesian equation is obtained by eliminating a from the foregoing equations. The result is $P = P_M\sqrt{u_M/u}$ or $P(\text{N}) = 577.4\ u^{-1/2}(\text{mm}^{-1/2})$.

Values P_A and u_A at the unloading point A are obtained by setting $a = a_A = 2a_0$ in the parametric equations (2.1.53) with the result $P_A = 312.5$ N and $u_A = 3.413$ mm. The unloading branch then follows as a linear equation down to the origin, i. e., $P = P_A u/u_A$ or $P(\text{N}) = 91.55\ u(\text{mm})$. Figure 2.1.11b shows the curve followed in the P–u diagram. □

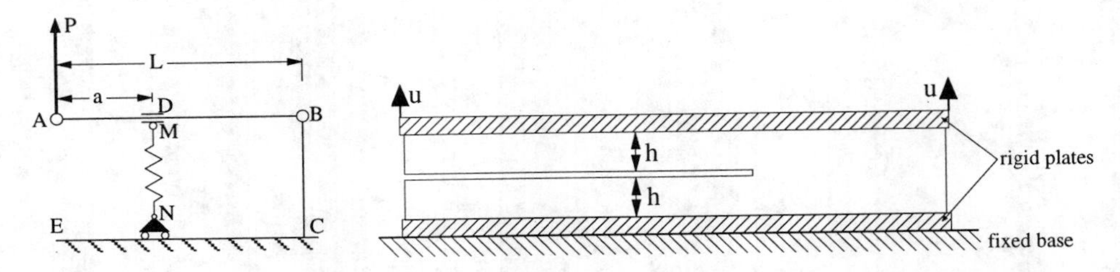

Figure 2.1.12 Illustrations for exercises 2.1 (left) and 2.8 (right).

Exercises

2.1 Consider the mechanism in Fig. 2.1.12 consisting of two rigid bars AB and BC joined by a hinge at B, and a spring MN of constant k. The spring is connected to bar AB by a frictional slider D and to the rigid support EC by a rolling support. Let L be the known length of bar AB, a the position of the slider on this bar, P the vertical load applied at point A, and u the vertical displacement of point A. (a) Find the elastic energy of the system as a function of u and a. (b) Find the load as a function of u and a. (c) Find the energy available to displace the slider a unit length, called $\mathcal{G}$ by analogy with crack growth. (d) Show that $\mathcal{G}$ is no more than the component along AB of the force that the spring exerts on the slider. (e) Show that the slider tends to move towards point B whatever the direction of the load P.

2.2 Give a detailed proof of Eqs. (2.1.20) and (2.1.21).

2.3 Derive the equivalent of equation (2.1.3) for the energy release rate in circular cracks of growing radius a in an axisymmetric stress field. (Answer: $2\pi a \mathcal{G}\, \delta a = \delta \mathcal{W}^R$.)

2.4 Show that the generalized displacement associated with the resultant force of a uniform traction distribution is the average displacement in the direction of the tractions over the area of their application. (Hint: write $t_i = \sigma e_i$, where σ is a variable stress and e_i is a fixed unit vector.)

2.5 To simulate rock fracture in the laboratory, a very large panel of thickness b with a relatively small crack of length $2a$ is tested by injecting a fluid into the crack. From Inglis (1913) results it is known that under pressure p the straight crack adopts an elliptical shape, with minor axis $c = 2pa/E'$, where the effective modulus E' is equal to the Young modulus E for generalized plane stress and equal to $E/(1-\nu^2)$ for plane strain, with $\nu =$ Poisson's ratio. (a) Find the complementary energy as a function of p and a. (b) Find the energy release rate for this case (note that a is the half crack length, not the crack length).

2.6 To test the fracture behavior of rock, a large 50-mm-thick slab will be tested in a laboratory by injecting a fluid into a central crack of initial length $2a_0 = 100$mm. Let p be the fluid pressure and V the volume expansion of the crack. In the assumption of full linear elastic behavior with $G_f = 20$ N/m, find and plot the p-V and $\mathcal{G}$-a curves the panel experiences when it is subjected to a controlled-volume injection until the crack grows up to 1000 mm, after which it is unloaded to zero pressure. Use effective elastic modulus $E' = 60$ MPa in the expression for $\mathcal{G}$ obtained in exercise 2.5.

2.7 For the panel of exercise 2.6, find and plot the p-V and $\mathcal{G}$-a curves for a test in which the crack extends from 100 mm to 1000 mm under volume expansion control and then the panel is unloaded. Assume that resistance to crack growth varies with crack extension Δa in the form

$$\mathcal{R} = 2G_f \left[1 - \frac{\Delta a}{2\lambda}\right] \qquad \text{for} \qquad 0 \le \Delta a \le \lambda \tag{2.1.54}$$

$$\mathcal{R} = G_f \qquad \text{for} \qquad \Delta a > \lambda \tag{2.1.55}$$

where $G_f = 100$ N/m and $\lambda = 276$ mm. Find the peak pressure and the maximum increase in volume. Use an effective elastic modulus $E' = 60$ GPa.

2.8 Find the J integral for an infinite strip, of thickness b and width $2h$, with a symmetric semi-infinite crack subjected to imposed zero displacements on its lower face and constant vertical displacement u on its upper face (Fig. 2.1.12; Rice 1968a). Assume linear elasticity and plane strain with known elastic modulus E and Poisson's ratio ν.

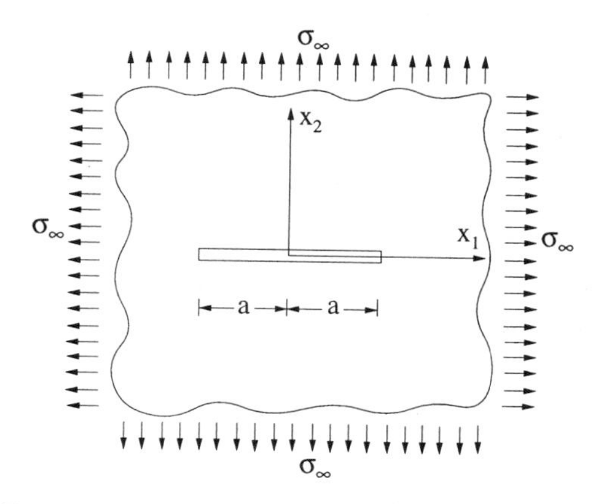

Figure 2.2.1 Center cracked infinite panel subjected to remote equiaxial tension.

2.9 A double cantilever beam specimen with arm depths h = 10 mm, thickness b = 10 mm, and initial crack length a_0 = 50 mm, is made of a material with a fracture energy G_f = 180 J/m^2 and an elastic modulus E = 250 GPa. The specimen is tested at a controlled displacement rate so that the load goes through the maximum and then decreases, at still increasing displacement, down to 25% of the peak load. When this point is reached, the specimen is completely unloaded. Assuming that LEFM and the beam theory apply, find the $P(u)$ and $G(a)$ curves. Give the equations of the different arcs and the coordinates of the characteristic points.

2.2 LEFM and Stress Intensity Factor

It was a great achievement of Irwin to reformulate LEFM problem in terms of the stress states in the material close to the crack tip rather than energetically and prove that this, so-called local, approach was essentially equivalent to the Griffith energetic (or global) approach.

The essential fact is that when a body contains a crack, a strong stress concentration develops around the crack tip. If the behavior of the material is isotropic and linear elastic except in a vanishingly small fracture process zone, it happens that this stress concentration has the same distribution close to the crack tip whatever the size, shape, and specific boundary conditions of the body. Only the intensity of the stress concentration varies. For the same intensity, the stresses around and close to the crack tip are identical.

To prove this assertion and to be able to solve problems for cracked structures of interest in engineering, mathematical tools specially suited to handle problems of elastic bodies with cracks were developed in the theory of elasticity. However, a user of LEFM (even a proficient one) does not need to make use of the sophisticated mathematical formalisms required to prove the most general properties of the elastic fields in cracked bodies. Therefore, in this section we present the most important results regarding linear elastic bodies with cracks. Chapter 4 gives the mathematical treatment and derivation of these results, intended only for those readers who wish to understand the sources of LEFM in greater depth.

We also restrict the analysis to the so-called *mode I*, by far the most often encountered mode in engineering practice. This is the mode where the crack lies in a plane of geometrical and loading symmetry of the structure and, therefore, no shear stresses act on the crack plane. The shear loading modes (II and III) and the fracture criteria associated with them will be analyzed in Chapter 4.

2.2.1 The Center Cracked Infinite Panel and the Near-Tip Fields

Consider a crack of length $2a$ in a two-dimensional infinite linear elastic isotropic solid, subjected to uniform normal stress σ_∞ at infinity in all directions (Fig. 2.2.1). The solution of this problem was obtained by Griffith (1921), as a particular solution of the panel with an elliptical hole obtained by Inglis (1913), and is derived in full in Chapter 4. Using the central axes shown in Fig. 2.2.1, the normal stresses

σ_{22} along the uncracked part of the crack plane ($x_2 = 0, x_1^2 - a^2 > 0$) are expressed as:

$$\sigma_{22} = \sigma_\infty \frac{|x_1|}{\sqrt{x_1^2 - a^2}} \tag{2.2.1}$$

This result shows that the stresses tend to infinity when the crack tips are approached from the solid. So the stress field has a singularity at the crack tip. In order to determine the asymptotic near-tip stress field, we write the stresses as a function of the distance r to the right crack tip, that is, replacing $x_1 - a$ with r (and x_1 with $r + a$, and $x_1 + a$ with $r + 2a$). Then, setting $x_1^2 - a^2 = (x_1 + a)(x_1 - a)$, we get for σ_{22} the following asymptotic approximation:

$$\sigma_{22} = \frac{\sigma_\infty \sqrt{a}}{\sqrt{2r}} \left[1 + \frac{3r}{4a} - \frac{5r^2}{32a^2} + ... \right] \tag{2.2.2}$$

where the factor in square brackets shows the first three terms of the Taylor series expansion of $(1 + r/a)/\sqrt{1 + r/2a}$. This factor obviously tends to 1 for $r \ll a$. It is now customary to denote

$$K_I = \sigma_\infty \sqrt{\pi a} \tag{2.2.3}$$

and call it the stress intensity factor (Subscript I refers to the opening mode of fracture, or mode I, to be distinguished from the shear modes II and III whose discussion is deferred to Chapter 4). The near-tip ($r/a \to 0$) expression for σ_{22} now becomes

$$\sigma_{22} = \frac{K_I}{\sqrt{2\pi r}} \tag{2.2.4}$$

which shows that the stress displays a singularity of order $r^{-1/2}$ at the crack tip.

For the normal displacements u_2 along the crack faces ($x_1^2 - a^2 < 0$), the elastic solution delivers

$$u_2^\pm = \pm \frac{2\sigma_\infty}{E'} \sqrt{a^2 - x_1^2} \tag{2.2.5}$$

where u_2^+ and u_2^- are, respectively, the vertical displacement of the upper and lower faces of the crack, and E' is the effective elastic modulus defined as

$$\begin{aligned} E' &= E && \text{for plane stress} \\ E' &= E/(1-\nu^2) && \text{for plane strain} \end{aligned} \tag{2.2.6}$$

The crack opening w is the jump in displacement between the faces of the crack, $w = u_2^+ - u_2^-$, and is obtained from Eq. (2.2.5) as

$$w = \frac{4\sigma_\infty}{E'} \sqrt{a^2 - x_1^2} \tag{2.2.7}$$

To see how the crack opening behaves in the neighborhood of the crack tip, we rewrite the last equation as a function of the distance r from the right crack tip (now $r = a - x_1$) and substitute $\sigma_\infty = K_I/\sqrt{\pi a}$ to get

$$w = \frac{8}{\sqrt{2\pi} E'} K_I \sqrt{r} \left[1 - \frac{r}{4a} - \frac{r^2}{32a^2} + ... \right] \tag{2.2.8}$$

where the factor in the square brackets shows the first three terms of the Taylor series expansion of $\sqrt{1 - r/2a}$. Thus, the near-tip ($r \ll a$) expression for the crack opening w is

$$w = \frac{8}{\sqrt{2\pi} E'} K_I \sqrt{r} \tag{2.2.9}$$

which shows that the profile of the deformed crack is parabolic (more precisely, a parabola of the second degree with its axis coincident with the crack line.)

Although the above near-tip results made use of a quite particular case, they remain valid for all the mode I loading cases. This will be further discussed next.

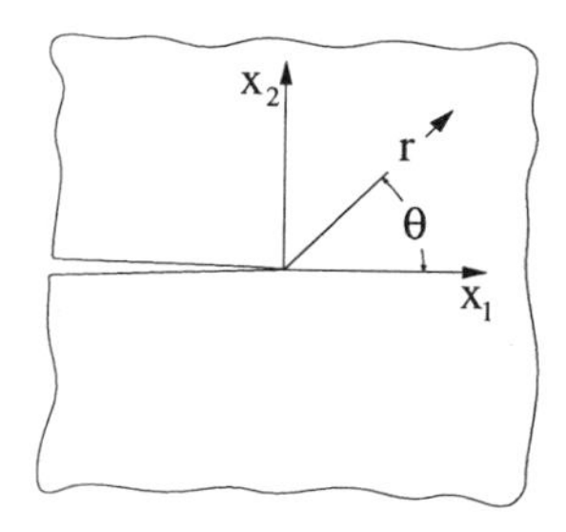

Figure 2.2.2 Axes for near crack tip field description.

2.2.2 The General Near-Tip Fields and Stress Intensity Factors

The results of the previous section may be generalized to any mode I loaded crack using various mathematical formulations described in Chapter 4 to come. The analysis shows that the stresses $\sigma_{ij}(r,\theta)$ at any distance r of the crack tip may be written as are given by

$$\sigma_{ij}(r,\theta) = \frac{K_I}{\sqrt{2\pi r}}\, S_{ij}(r,\theta) \tag{2.2.10}$$

where K_I is the mode I stress intensity factor proportional to the load, and functions $S_{ij}(r,\theta)$ of polar coordinates (r,θ) —Fig. 2.2.2— are regular everywhere, except at load points, other crack tips and reentrant corners. These functions are dimensionless, and thus independent of structure size and load, but they depend on the geometry of the structure and of the loading . When the crack tip is approached $(r \to 0)$, the general near-tip expression may be written as

$$\sigma_{ij}(r,\theta) = \frac{K_I}{\sqrt{2\pi r}}\, S^I_{ij}(\theta) \tag{2.2.11}$$

where K_I is proportional to the load and the dimensionless functions $S^I_{ij}(\theta) = S_{ij}(0,\theta)$ are independent of geometry and the same for all mode I situations. They are given in Section 4.3.2, Eqs. (4.3.18)–(4.3.19).

This result means that two different linear elastic cracked bodies (different sizes, shapes, and material) subjected to mode I loading have identical stress distribution close and around the crack tip if the values of the stress intensity factors K_I are the same for both of them.

When r is not very small, the expression (2.2.11) represents the first term (the dominant one) of the series expansion in powers of r of the general expression (2.2.10). To illustrate this, the general power series expansion for the σ_{22} stress component along the crack plane, which is analogous to that for the center cracked panel, Eq. (2.2.2), is

$$\sigma_{22} = \frac{K_I}{\sqrt{2\pi r}} \left[1 + \beta_1 \frac{r}{D} + \beta_2 \frac{r^2}{D^2} + \ldots + \beta_m \frac{r^m}{D^m} + \ldots\right] \tag{2.2.12}$$

where D is a characteristic dimension of the structure (which may be, but in general need not be, chosen as the crack length a, as it was for the center-cracked panel). The dimensionless coefficients β_m, depend on the details of geometry and loading.

In the case of the center-cracked panel subjected to equiaxial remote stress, the series expansion is given by Eq. (2.2.2). For this geometry the first term is dominant, with error under within 3%, at distances $r \leq 0.04a$. For other geometries, the first term of the above series is identical, but the subsequent terms may differ appreciably (Wilson 1966; Knott 1973). However, if the size of the fracture process zone is much less than the K_I-dominated zone (a few percent of the crack size, in general) the remaining terms can be neglected and LEFM holds. If the fracture zone is too large, some inelastic fracture theory must be introduced. This is the case for concrete in most practical situations, and the main concern of this book.

Similar conclusions are reached if the displacement field around the crack tip is analyzed. The general

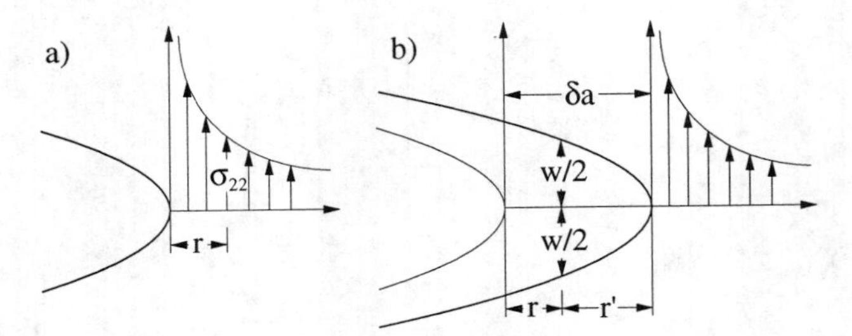

Figure 2.2.3 Near-tip stresses and crack openings in crack propagation: (a) initial situation, (b) after crack advanced an amount δa.

solution using polar coordinates at the crack tip is of the form

$$u_i(r,\theta) = \frac{K_I}{E'}\sqrt{\frac{8r}{\pi}}\,D_i(r,\theta) \tag{2.2.13}$$

where the dimensionless functions $D_i(r,\theta)$ are regular and depend on the geometry and loading. The near-tip distribution for $r \longrightarrow 0$, however, is geometry independent:

$$u_i = \frac{K_I}{E'}\sqrt{\frac{8r}{\pi}}\,D_i^I(\theta) \tag{2.2.14}$$

The dimensionless functions $D_i^I(\theta) = D_i(0,\theta)$ are given in Section 4.3.2, Eq. (4.3.20).

An important consequence of Eq. (2.2.13) is the expression of the crack opening profile. The upper crack face corresponds to $\theta = \pi$ and the lower crack face to $\theta = -\pi$, so we can write $w = u_2(r,\pi) - u_2(r,-\pi)$. Thus, using (2.2.13) and expanding the resulting expression in power series of r we get an expression similar to Eq. (2.2.8) for the center cracked panel:

$$w = \frac{8}{\sqrt{2\pi}E'}K_I\sqrt{r}\left[1 + \gamma_1\frac{r}{D} + \gamma_2\frac{r^2}{D^2} + ... + \gamma_m\frac{r^m}{D^m} + ...\right] \tag{2.2.15}$$

where D is the characteristic dimension of the structure previously introduced in Eq. (2.2.12), and the dimensionless coefficients γ_m depend again on the geometry of structure and loading. It can be proved (see Chapter 4) that they are related to coefficients β_m of the stress expansion as follows:

$$\gamma_m = \frac{(-1)^m}{2m+1}\beta_m \tag{2.2.16}$$

2.2.3 Relationship Between K_I and $\mathcal{G}$

Since the asymptotic near-tip stress field is unique, and since the rate of energy flow into the crack tip must depend only on this field, there must exist a unique relationship between the energy release rate $\mathcal{G}$, and the stress intensity factor K_I. There are various ways to derive it. The simplest is to calculate the work of stress during the opening of a short slit ahead of the crack. We consider Mode I and imagine the crack tip to be advanced by an infinitesimal distance δa in the direction of axis x_1 (Fig. 2.2.3a,b). Let A and B be the initial and final state. We use the procedure illustrated in Fig. 2.2.4, where the initial stress state A has been preserved by introducing a line slit of length δa ahead of the preexisting crack, but keeping it closed by means of external surface tractions equal to the stresses existing in the actual state A (Fig. 2.2.4a). The final state B is then reached by reducing these stresses proportionally down to zero. In doing so, the intermediate states are such as the one depicted in Fig. 2.2.4b, in which the surface traction (closing tractions) are reduced to $\tau\sigma_{22}^A$, where τ is a scalar load parameter varying from 1 in the initial state A to 0 in the final state B (Fig. 2.2.4d). The crack openings in the intermediate states must vary linearly with τ because the structure is elastic, and so they must be proportional to $(1-\tau)$ and,

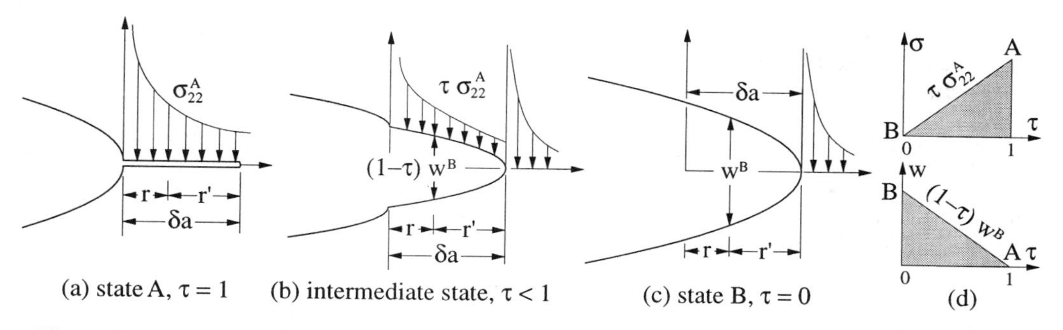

(a) state A, τ = 1 (b) intermediate state, τ < 1 (c) state B, τ = 0 (d)

Figure 2.2.4 Proportional release of the closing tractions are used to compute the energy release rate.

therefore, equal to $(1-\tau)w^B$ (Fig. 2.2.4d). Since the remote boundaries are assumed clamped, the only energy supply in going from the initial to the final state is the work done by the closing surface tractions. The elemental work per unit area at a given location on the crack surface when the crack opens dw under tractions σ_{22} is $-\sigma_{22}dw$, the minus sign coming from the different orientation of σ_{22} (closing) and dw (opening). Using this for an elemental intermediate step in which the load factor τ varies by $d\tau$, the work (external energy supply) per unit surface of the slit turns out to be

$$d\left(\frac{d\mathcal{W}}{b\,dr}\right) = -(\tau\sigma_{22}^A)d[(1-\tau)\;w^B] = -\tau\sigma_{22}^A(-d\tau\;w^B) \tag{2.2.17}$$

where b is the thickness of the body. Integration yields the total work per unit surface done by the surface tractions in passing from state A to B:

$$\frac{d\mathcal{W}_{A-B}}{b\,dr} = \sigma_{22}^A\;w^B\int_1^0 \tau d\tau = -\frac{1}{2}\,\sigma_{22}^A\;w^B \tag{2.2.18}$$

Therefore, the total external work supply — thus also the elastic energy variation at clamped boundaries — is obtained by integration of the previous equation with respect to r:

$$\mathcal{U}_B - \mathcal{U}_A = \mathcal{W}_{A-B} = -\frac{1}{2}\,b\int_0^{\delta a} \sigma_{22}^A\,w^B\;dr \tag{2.2.19}$$

Since δa is vanishingly small, one may now use the near-tip field expressions (2.2.4) and (2.2.9),

$$\sigma_{22}^A = \frac{K_I^A}{\sqrt{2\pi r}} \quad ; \quad w^B = \sqrt{\frac{32}{\pi}}\frac{K_I^B}{E'}\sqrt{\delta a - r} \tag{2.2.20}$$

After substitution into Eq. (2.2.19), this leads to

$$\delta\mathcal{U} = \mathcal{U}_B - \mathcal{U}_A = -b\frac{2K_I^A K_I^B}{\pi E'}\int_0^{\delta a}\sqrt{\frac{\delta a - r}{r}}dr = -b\frac{K_I^A K_I^B}{E'}\;\delta a \tag{2.2.21}$$

in which the integration has been performed by means of the substitution $r = \delta a\;\sin^2 t$. Noting that under fixed boundary conditions $\mathcal{G}b\delta a = -\delta\mathcal{U}$, and that $K_I^B \to K_I^A = K_I$ for $\delta a \to 0$, we obtain the celebrated Irwin's result:

$$\mathcal{G} = \frac{K_I^2}{E'} \tag{2.2.22}$$

This shows that Griffith's and Irwin's approaches are equivalent, and allows us to discuss fracture criteria.

2.2.4 Local Fracture Criterion for Mode I: K_{Ic}.

Mode I is quite simple. Since the stress state of the material surrounding a very small fracture process zone —the crack tip— is uniquely determined by K_I, the crack will propagate when this stress intensity factor reaches a certain critical value K_{Ic}, called fracture toughness. K_{Ic} for the given material may be determined performing a fracture test and determining the K_I value that provoked failure. Because the energy fracture criterion must also hold, and indeed does according to the fundamental relationship (2.2.22), K_{Ic} is related to the fracture energy G_f by

$$K_{Ic} = \sqrt{E' G_f} \tag{2.2.23}$$

With this definition, the local fracture criterion for pure mode I may be stated in analogy to the energy criterion: indexfracture criterion!in local approach

$$\text{if} \quad K_I < K_{Ic} \quad \text{then:} \quad \text{No crack growth (stable)} \tag{2.2.24}$$

$$\text{if} \quad K_I = K_{Ic} \quad \text{then:} \quad \text{Quasi-static growth possible} \tag{2.2.25}$$

$$\text{if} \quad K_I > K_{Ic} \quad \text{then:} \quad \text{Dynamic growth (unstable)} \tag{2.2.26}$$

For loadings that are not pure mode I, the problem becomes more difficult because, in general, an initially straight crack kinks upon fracture and the criteria must give not only the loading combination that produces the fracture, but also the kink direction. This is still an open problem today, and the interested reader may find a summary of the most widely used criteria in Chapter 4. For most of the discussions in this book, LEFM mode I fracture is all that is needed.

Exercises

2.10 Estimate the strength of a large plate under uniaxial tensile stress if it contains through cracks of up to 10 mm. The plate is made of a brittle steel with $K_{Ic} = 60$ MPa$\sqrt{\text{m}}$.

2.11 Assuming plane stress and applicability of beam theory, find the expression for the stress intensity factor (mode I) of a double cantilever beam specimen of thickness b, arm length a, and arm depth h subjected to two opposite bending moments M (see Fig. 2.1.3).

2.12 Determine the stress intensity factor of the center-cracked panel subjected to internal pressure, described in exercise 2.5 (a) Use Irwin's relationship, (b) use the near-tip expansion for the crack opening.

2.13 To test the fracture behavior of rock, a large 50-mm-thick panel of this material will be tested in a laboratory by injecting a fluid into a central crack of initial length $2a_0 = 100$mm. Let p be the fluid pressure and V the volume expansion of the crack. Assuming linear elastic behavior with $K_{Ic} = 35$ kPa$\sqrt{\text{m}}$, find and plot the p-V and K-a curves the panel would experience if it were subjected to a controlled volume injection until the crack grew up to 1000 mm, and was then unloaded to zero pressure. Use Inglis result given in exercise 2.5 and an effective elastic modulus $E' = 60$ MPa in the determination of the volume increase.

2.14 Find the stress intensity factor for the infinite strip of exercise 2.8.

2.15 Check that the coefficients of the near-tip power expansions for σ_{22} and w for the center cracked panel subjected to remote equiaxial stress satisfy the relationship (2.2.16).

2.3 Size Effect in Plasticity and in LEFM

Scaling laws are the most fundamental aspect of every physical theory. As discussed in Chapter 1, the problem of scaling law and size effect in the theories of structural failure has received considerable attention, particularly with regard to distributed damage and nonlinear fracture behavior. The necessity of using theories that correctly predict the size effect was emphasized in Section 1.2.5 and the basis for the general analysis together with some simple size effect derivations and examples was presented in Section 1.4. It was shown that for plasticity theory (and for allowable-stress analysis, too) no size effect was predicted, but that for LEFM the nominal structural strength decreased with increasing size as $D^{-1/2}$. The main objective of the present section is to derive these properties from the basic theories.

To do so, we first study the implications of limit analysis for size effect. Then we set up the general forms that the expressions for K_I and $\mathcal{G}$ must take, and derive the size effect for LEFM. We end the section with a brief discussion of the effect of structure size on the strength of structures containing relatively very small flaws whose size is independent of the size of the structure.

We remember from Section 1.4 that, unless otherwise stated, the size effect is defined by comparing geometrically similar structures of different sizes (in the case of notched or fractured structures, the geometric similarity, of course, means that the notches or initial traction-free cracks are also geometrically similar). In this section, our interest is in analyzing the effect of the size on the nominal strength σ_{Nu} which was defined in Section 1.4.1. We recall that its general form for plane problems is

$$\sigma_{Nu} = c_N \frac{P_u}{bD} \tag{2.3.1}$$

where c_N depends only on geometrical ratios and thus is a constant for geometrically similar structures.

We also recall from Section 1.4 that, since any two consistently defined nominal stresses are related by a constant factor, the general trend of the size effect is independent of the exact choices for c_N and D. However, for quantitative analysis and, specially, for comparison of results from various sources, it is useful to make a consistent choice throughout.

2.3.1 Size Effect for Failures Characterized by Plasticity, Strength, or Allowable Stress

Consider a family of geometrically similar structures subjected to proportional loading characterized by the nominal stress σ_N. Assume that the response of the material can be fully described by a certain constitutive equation relating the stress and strain tensors. No restrictions other than rate independence are imposed on this relation. It may be linear or nonlinear elastic, elasto-plastic, or of some other more sophisticated kind. The point is that the constitutive equation and fracture criterion contain parameters which either are dimensionless (hardening exponents, strain thresholds, etc.) or have the dimension of stress (elastic moduli, yield stresses), but do not contain any parameter with dimension of length. In other words, no characteristic size exists.

Consider a reference structure of size D and a scaled geometrically similar structure of size $D' = \lambda D$. Assume that for the reference structure (that of size D) the stress at an arbitrary point of coordinates (x_1, x_2) for a given load characterized by σ_N is given by $\sigma_{ij}(\sigma_N, x_1, x_2)$. This stress distribution satisfies the equilibrium equations and the traction-imposed boundary conditions. Considering the scaled structure, it turns out that if the stresses at homologous points of coordinates $x_1' = \lambda x_1$ and $x_2' = \lambda x_2$ is taken to be identical to those for the reference structure, then this state corresponds to the actual solution for the second structure. This correspondence may be analytically stated as:

$$\sigma'_{ij}(\sigma_N, x'_1, x'_2) = \sigma_{ij}(\sigma_N, x_1, x_2) \quad \text{with } x'_1 = \lambda x_1 \text{ and } x'_2 = \lambda x_2 \tag{2.3.2}$$

Then, it is easy to prove that, with this condition, (1) the traction-imposed boundary conditions are automatically satisfied because of the similitude (the imposed surface tractions at the boundary are identical at homologous points); (2) the equilibrium conditions $\sigma_{ij,j} = 0$ are also trivially verified; (3) the constitutive equation is also satisfied if the strain fields are related by an equation similar to (2.3.2):

$$\varepsilon'_{ij}(\sigma_N, x'_1, x'_2) = \varepsilon_{ij}(\sigma_N, x_1, x_2) \quad \text{with } x'_1 = \lambda x_1 \text{ and } x'_2 = \lambda x_2 \tag{2.3.3}$$

where ε_{ij} and ε'_{ij} are the strain tensors for the structures of sizes D and D', respectively; and (4) the last equation is verified if the similitude law for the displacements is given by

$$u'_i(\sigma_N, x'_1, x'_2) = \lambda u_i(\sigma_N, x_1, x_2) \quad \text{with } x'_1 = \lambda x_1 \text{ and } x'_2 = \lambda x_2 \tag{2.3.4}$$

which is proven as follows:

$$\varepsilon'_{ij}(\sigma_N, x'_1, x'_2) = \frac{1}{2}\left(\frac{\partial u'_i}{\partial x'_j} + \frac{\partial u'_j}{\partial x'_i}\right) = \frac{1}{2}\left(\frac{\lambda \partial u_i}{\partial(\lambda x_j)} + \frac{\lambda \partial u_j}{\partial(\lambda x_i)}\right) =$$

$$= \varepsilon_{ij}(\sigma_N, x_1, x_2) \quad \text{with } x'_1 = \lambda x_1 \text{ and } x'_2 = \lambda x_2 \tag{2.3.5}$$

Therefore, the laws of similitude (2.3.2) and (2.3.3) just state that for a given nominal stress σ_N, the stresses and strains at homologous points of two geometrically similar structures are identical. This implies, in particular, that the stress and strain maxima and minima also occur at homologous points.

If failure is assumed to occur when the stress, strain, or, in general, a certain function of the two $\Phi(\sigma_{ij}, \varepsilon_{ij})$ reaches a critical value, i.e., when:

$$\Phi(\sigma_{ij}, \varepsilon_{ij}) = \Phi_c \tag{2.3.6}$$

where Φ_c is a given critical or allowable value, then the two similar structures will fail at the same nominal stress. Thus, for theories such as plasticity or elasticity with strength limit or allowable stress, the nominal strength of two similar structures of sizes D and D' is identical:

$$\sigma'_{Nu} = \sigma_{Nu} \tag{2.3.7}$$

In such a case we say that there is no size effect.

The foregoing result may be also obtained from dimensional analysis. Indeed in this kind of problem, the external load at fracture is completely determined by σ_{Nu}, D and a number of geometrical ratios γ_i (those defining the shape, for example, the span-to-depth ratio). The material response is determined by a certain critical or allowable stress σ_c and a number of dimensionless ratios μ_i (the hardening exponent, strain parameters, and ratios of elastic modulus to the allowable stress, E/σ_c, for example). With these variables, the only dimensionally correct expression for the nominal strength is

$$\sigma_{Nu} = \sigma_c \phi(\gamma_i, \mu_i) \tag{2.3.8}$$

where $\phi(\gamma_i, \mu_i)$ is a dimensionless function. Since the arguments γ_i are geometrical ratios that remain constant for geometrically similar structures, and since the μ_i are constants for a given material, (2.3.7) follows.

2.3.2 General Forms of the Expressions for K_I and $\mathcal{G}$

Since crack growth in LEFM is defined by the condition $\mathcal{G} = G_f$ or $K_I = K_{Ic}$, we need to know the structure of the equations for K_I and $\mathcal{G}$ if we want to investigate the influence of the size. And since size effect is one of the main topics of this book, it is also convenient to define the conventional forms of the equations we are going to use so that the size D is made explicit. Systematization of the presentation of the existing results also requires using general mathematical forms of the equations for K_I and $\mathcal{G}$, so that a single experimental or numerical result may be used for any similar specimen or structure.

To determine the general form, we consider a family of geometrically similar structures subjected to the same type of loading (for example, the center cracked panel in Fig. 2.1.1 or the DCB specimens in Fig. 2.1.3). Let D be a characteristic dimension (for example, the panel width or the arm depth in the DCB specimen), all the remaining dimensions being proportional (for example, the height-to-width ratio of the panel and the total length-to-depth ratio for the DCB), except for the crack-to-depth ratio a/D, which is free to vary. The purpose of the analysis is to obtain the general expression for $\mathcal{G}$ and K_I showing explicitly the dependence on the variables P (or σ_N), D and $\alpha = a/D$. Let us first elaborate the examples:

Example 2.3.1 For the center cracked panel with a short crack ($\alpha = a/D \ll 1$) the expression for the stress intensity factor (2.2.3) can be written in either of the two following forms:

$$K_I = \sigma_N \sqrt{D}\sqrt{\pi\alpha} = \frac{P}{b\sqrt{D}}\sqrt{\pi\alpha} \tag{2.3.9}$$

where we set $\sigma_N = \sigma = P/bD$, and b is the panel thickness. ▯

Example 2.3.2 For the DCB specimen of Fig. 2.1.3b, we take the approximate expression (2.1.29), substitute it into Irwin's relationship (2.2.22) (assuming plane stress, $E' = E$), and then we get

$$K_I = \sigma_N \sqrt{D} 2\alpha\sqrt{3} = \frac{P}{b\sqrt{D}} 2\alpha\sqrt{3} \tag{2.3.10}$$

where we substituted $D = h$, $\sigma_N = P/bD$. ▯

The resemblance of the expressions for the these two simple cases is evident. They only differ in the factors depending on α ($\sqrt{\pi\alpha}$ for the panel and $2\alpha\sqrt{3}$ for the DCB specimen). This result is general. Indeed, because the response is linear elastic, the stress intensity factor must be proportional to the force per unit thickness P/b. Since the stress intensity factor must have dimensions of Force×Length$^{-3/2}$ and must depend on the relative crack depth α, the only possible forms of the expression based on P and σ_N are:

$$K_I = \frac{P}{b\sqrt{D}}\,\hat{k}(\alpha) = \sigma_N\sqrt{D}\,k(\alpha) \tag{2.3.11}$$

where $\hat{k}(\alpha)$ and $k(\alpha)$ are dimensionless functions, $\alpha = a/D$ is the relative crack depth, and $k(\alpha) = \hat{k}(\alpha)/c_N$. The convention of using 'hatted' k for expressions based on P and plain k for expressions based on σ_N will be retained throughout the book.

The general forms for the energy release rate $\mathcal{G}$ may be obtained directly from the foregoing by using Irwin's relationship (2.2.22). They are

$$\mathcal{G} = \frac{P^2}{b^2DE'}\,\hat{g}(\alpha) \qquad \text{or} \qquad \mathcal{G} = \frac{\sigma_N^2}{E'}D\,g(\alpha) \tag{2.3.12}$$

where

$$\hat{g}(\alpha) = \hat{k}^2(\alpha) \qquad \text{and} \qquad g(\alpha) = \frac{1}{c_N^2}\hat{g}(\alpha) = k^2(\alpha) \tag{2.3.13}$$

In what follows, we systematically use the forms (2.3.11) and (2.3.12). Other equivalent forms can be found in the literature, as discussed latter.

Another simple argument leading to (2.3.12) is to note that the complementary energy of the structure must be expressible as $\mathcal{U}^* = 2\overline{U}V f(\alpha)$, where $\overline{U} = \sigma_N^2/2E'$ is a nominal energy density, $V = bD^2$ is the volume of the structure and $f(alpha)$ is a dimensionless function. Then $\mathcal{G} = \partial\mathcal{U}^*/b\partial a = (\partial\mathcal{U}^*/\partial\alpha)/bD = (\sigma_N^2/E')Df'(\alpha)$. Setting $g(\alpha) = f'(\alpha)$, one gets Eq. (2.3.12).

2.3.3 Size Effect in LEFM

Consider now a family of geometrically similar plane cracked structures loaded in mode I. Let a_0 be the initial crack length and $\alpha_0 = 0/D$ the initial relative crack length. From (2.3.11), the crack growth condition, $K = K_{Ic}$ is fulfilled when σ_N reaches a value σ_{Ni} (initiation stress) given by

$$\sigma_{Ni} = \frac{K_{Ic}}{\sqrt{D}k(\alpha_0)} \tag{2.3.14}$$

After reaching this point, the crack will grow and the nominal stress will vary to keep $K_I = K_{Ic}$, i.e.,

$$\sigma_N = \frac{K_{Ic}}{\sqrt{D}k(\alpha)} \qquad \text{for} \quad \alpha > \alpha_0 \tag{2.3.15}$$

Obviously, if $k(\alpha)$ increases with α, then σ_N decreases after the crack starts to grow and the peak load coincides with the onset of crack growth. If, on the other hand, $k(\alpha)$ decreases with α, then σ_N increases after the crack starts to grow and, eventually, reaches a maximum when $k(\alpha)$ reaches a minimum. The first case corresponds to the so-called *positive geometries* (Planas and Elices 1989a) and for them

$$\sigma_{Nu} = \sigma_{Ni} = \frac{K_{Ic}}{\sqrt{D}k(\alpha_0)} \qquad \text{if} \quad k'(\alpha_0) > 0 \tag{2.3.16}$$

where $k'(\alpha_0)$ stands for the derivative of $k(\alpha)$ at $\alpha = \alpha_0$. For negative geometries, the peak load occurs when the crack length reaches a value α_M for which $k(\alpha)$ goes through a minimum, thus,

$$\sigma_{Nu} = \frac{K_{Ic}}{\sqrt{D}k(\alpha_M)} \qquad \text{if} \quad k'(\alpha_0) > 0 \quad k(\alpha_M) = \text{minimum} \tag{2.3.17}$$

In any case, since both α_0 and α_M are constant for geometrically similar structures, it turns out that the nominal strength is always inversely proportional to the square root of the size. Hence, for similar precracked structures of sizes D_1 and $D = \lambda D_1$, the nominal strengths are related by

$$\sigma_{Nu} = \sigma_{Nu_1}\sqrt{\frac{D_1}{D}} = \lambda^{-1/2}\sigma_{Nu_1} \propto D^{-1/2} \tag{2.3.18}$$

Thus, it has been generally proved that geometrically similar structures following LEFM exhibit the inverse square root size effect.

2.3.4 Structures Failing at Very Small Cracks Whose Size is a Material Property

The foregoing size-effect analysis applies always to structures in which the crack length at maximum load is proportional to the size of the structure. This kind of size effect, however, differs from that found in normal kinds of metallic and other structures which become unstable and fail (or must be assumed to fail) before a small flaw, represented by a microcrack, can become a macrocrack of significant length compared to the structure size.

If the crack is small compared to the distances over which the stresses vary appreciably (let's call them microcracks, for short), it is easy to show (see the superposition method to compute K_I in the next chapter) that the stress intensity factor always takes the form

$$K_I = k_0 \sigma \sqrt{a} \tag{2.3.19}$$

where σ is the stress normal to the microcrack plane at the microcrack location computed as if no microcrack existed in the structure; a is the half-length or radius (for a pennyshape) of the microcrack and k_0 a constant depending on the exact shape and location of the crack (but not on microcrack or structure size). Consider now two structures that are similar (which means macro-geometrically similar) and contain identical microcracks in homologous positions. Since we have proved in Section 2.3.1 that the stresses at homologous positions are identical, it follows that the microcrack at a specific site starts to grow in both structures at the same nominal stress level. If one further assumes that *global failure follows immediately after one of the cracks starts to grow,* it turns out that the two structures will fail at the same nominal stress level. Hence, in this type of similitude no size effect is present because, in fact, it is equivalent to analysis based on allowable-stress or strength criterion (in which the allowable stress or strength varies from microcrack site to microcrack site).

However, getting similar structures with identical distribution of microcracks is practically impossible, so actual structures are macroscopically similar but microscopically random. Then the strength of the structure can be defined only on statistical (probabilistic) grounds, and a size effect appears because the probability of getting larger flaws in the more highly stressed regions of the structure increases with the structure size. The analysis of this kind of size effect will be deferred until Chapter 12, where we prove that the statistical size effect vanishes asymptotically when macrocracks or notches exist at the start of failure in the body (see also the short discussion in Section 1.3).

With respect to quasibrittle materials, and particularly concrete, it is important to note that they contain plenty of microcracks, but failure does not happen as soon as one of these microcracks starts to grow. It only occurs after many microcracks have grown and coalesced to form a macroscopic fracture process zone. This is a feature that makes the classical statistical theory of strength inapplicable to these materials.

Exercises

2.16 Show that the stress intensity factor for a penny shaped crack of radius a coaxial to a cylindrical bar of diameter D subjected to uniaxial stress σ must take the form (2.3.11). What is the form if K_I is written in terms of the axial load $P = \sigma\pi D^2/4$?

2.17 Find the general forms for the energy release rate of a penny shaped crack of radius a coaxial to a cylindrical bar of diameter D subjected to uniaxial stress σ.

2.18 In most handbooks on stress intensity factors, K_I is written in the form $K_I = Y\sigma\sqrt{\pi a}$ where a is the crack length, σ a characteristic stress, and Y a dimensionless factor depending only on geometrical ratios, in particular on the relative notch depth a/D where D is a characteristic linear dimension of the body. Rewrite this in the form (2.3.11) and find the relationship between Y and $k(\alpha)$.

2.19 Write the stress intensity factor of the DCB specimen of Fig. 2.1.3b in terms of the relative displacement u of the load-points. Show that the general form of K_I for imposed displacement must be $K_I = (Eu/\sqrt{D})\tilde{k}(\alpha)$ where E is the elastic modulus, u the displacement, D a characteristic dimension of the body, $\alpha = a/D$ the relative crack depth, and $\tilde{k}(\alpha)$ a nondimensional function.

2.20 To analyze the behavior of a large structure with cracks, which is assumed to behave in an essentially linear elastic manner, a reduced scale model is built at a 1/10 scale. The model is tested so that the stresses at homologous points are identical in both model and reality, and we require the full scale and reduced models to break at the same stress level. Determine the scale factors for (a) loads, (b) toughness, and (c) fracture energy.

3

Determination of LEFM Parameters

The application of LEFM to practical problems requires knowledge of the stress intensity factors or the energy release rates for the actual geometry and type of loading. In many cases one further needs the evolution of K_I with crack length. The selection of the method adequate to treat a particular problem depends very much on external inputs: economical importance of the problem; time available for analysis; bibliographical, analytical, numerical, and experimental facilities available to the analyst.

Fracture mechanics literature contains a vast number of closed form solutions of various elastic bodies with cracks. If the problem at hand can be approximated by one of the cases in the handbooks or papers, the problem is solved with ease. Section 3.1.1 briefly shows the use of closed form solutions from the handbooks. Quite often the problem does not coincide with any of the cases of the handbook, but can be obtained by superposition of other cases. The superposition may range from simple two-state cases to continuous weighted superposition in the sense of Green's function. Section 3.1.2 illustrates by some examples the use of the superposition method.

Close to the handbook cases solved with ease are the cases where the elastic energy release rate can be analytically calculated using the energetic approaches of Section 2.1, together with adequate simplifying assumptions. The stress intensity factor, usually mode I intensity factor, then follows from Irwin's relationship $K_I = \sqrt{E'\mathcal{G}}$. Section 3.2 illustrates some of the available approximations.

When no closed-form solutions are available, other strategies are at hand for the analyst to choose. The first one is to try to find an analytical solution. This is a highly specialized mathematical task out of the reach of most engineering practitioners and researchers. It can be accomplished by one of the formal approaches described in Chapter 4, and is outside the scope of this book.

When all the analytical treatments fail — because Green's functions are not available for the geometry of interest, for example — one may resort to numerical methods, an expedient that is getting increasingly easy to handle, increasingly reliable, and becoming readily accessible to engineers (Section 3.3.1). Alternatively, the stress intensity factor of reduced-scale elastic specimens can be experimentally measured in various ways (Section 3.3.2).

Application of LEFM to practical cases requires the fracture parameters of the given materials to be known, too. The main aspects of the determination of K_{Ic} and G_f are presented in Section 3.4.

An aspect often involved in fracture problems is the determination of displacements and similar variables such as crack volume or crack opening profile. Section 3.5 shows how these displacements can be calculated when closed form expressions for the stress intensity factor as a function of crack length are known. As a corollary, the stress intensity due to a point load on the crack faces is determined from the expressions of the crack opening profile and stress intensity factor for another arbitrary loading —an expression known as Bueckner's (1970) weight function.

3.1 Setting up Solutions from Closed-Form Expressions

3.1.1 Closed-Form Solutions from Handbooks

A large number of solutions for stress intensity factors have been collected in handbooks (Sih 1973; Rooke and Cartwright 1976; Tada, Paris and Irwin 1985; Murakami 1987). The energy release rates are not included in these handbooks because the expressions for K_I are simpler, and the expressions for $\mathcal{G}$ are very easily obtained from those for K_I using Irwin's equation.

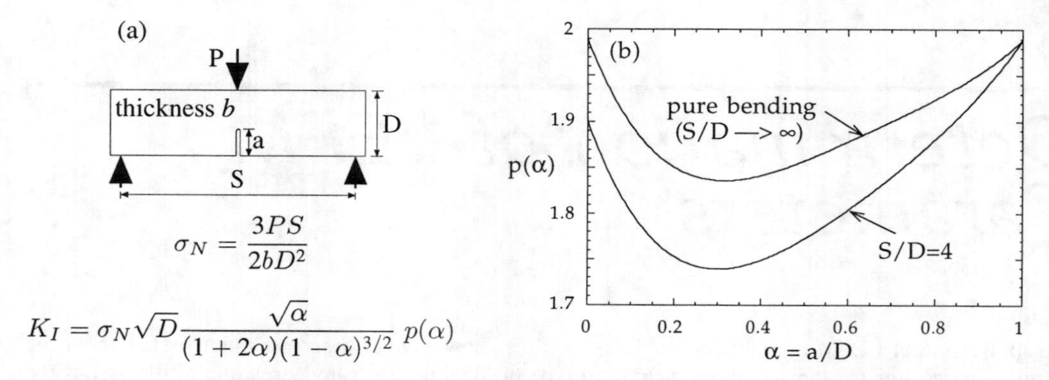

Figure 3.1.1 Notched beam in three-point bending. (a) Definition of geometry and expression for K_I; (b) variation of the shape factor $p(\alpha)$.

A few of the collected solutions are exact. Most of them are empirical fits of approximate but accurate numerical results. In a few cases no analytical expressions are given, but a graphical representation of the results is provided. In most cases of complex analytical expressions, a graphical representation is provided as well as the closed-form expression. Different fits, with different ranges of applicability and different accuracies, may be available for a given case, a point that must never be overlooked.

In this book, we write the expressions for K_I in the form (2.3.11) because of our interest in the size effect. This is to be taken into account when comparisons are made with handbooks in which the prototype expression for a stress intensity factor is taken to be that for a center cracked infinite panel, so most handbooks use the form $K_I = \sigma_N \sqrt{\pi a} F(\alpha)$, where $F(\alpha)$ is a dimensionless function of the relative crack length. Comparing this with (2.3.11), it turns out that the relationship between $k(\alpha)$ and $F(\alpha)$ is $k(\alpha) = \sqrt{\pi\alpha} F(\alpha)$.

Example 3.1.1 For a single-edge cracked beam subjected to three-point bending (Fig. 3.1.1a), the expression for K_I depends mildly on the shear force magnitude near the central cross-section, i.e., on the span-to-depth ratio. Fig. 3.1.1b shows a plot of $k(\alpha)$ for the limiting case of pure bending (formally equivalent to $S/D \to \infty$) and for the case $S/D = 4$ (a standard ASTM testing geometry). Analytical approximate expressions for these two cases were produced by Tada, Paris and Irwin (1973), for pure bending, and by Srawley (1976), for $S/D = 4$. Recently, Pastor et al. (1995) produced expressions accurate within 0.5% for any a/D. The latter expressions have the advantage over the former that their structure is identical (additionally, they correct a 4% error that crept in the Srawley formula in the limit of short cracks). With the definition of σ_N shown in Fig. 3.1.1, the shape factor takes the form

$$k_{S/D} = \sqrt{\alpha}\, \frac{p_{S/D}(\alpha)}{(1+2\alpha)(1-\alpha)^{3/2}} \tag{3.1.1}$$

where $p_r(\alpha)$ is a fourth degree polynomial in α. The expression of the polynomials for $S/D = 4$ and ∞ (pure bending) are

$$p_4(\alpha) = 1.900 - \alpha\left[-0.089 + 0.603(1-\alpha) - 0.441(1-\alpha)^2 + 1.223(1-\alpha)^3\right] \tag{3.1.2}$$

$$p_\infty(\alpha) = 1.989 - \alpha(1-\alpha)\left[0.448 - 0.458(1-\alpha) + 1.226(1-\alpha)^2\right] \tag{3.1.3}$$

Note that for very short cracks ($\alpha \to 0$) the shape factors $k(\alpha)$ behave as $c_0\sqrt{\pi\alpha}$, where c_0 is a constant close to 1.12. For very deep cracks ($\alpha \to 1$), $k(\alpha) \propto c_1(1-\alpha)^{-3/2}$, where c_1 is a constant close to 2/3. This is the general trend for specimens in which the resultant force over the crack plane is zero.
□

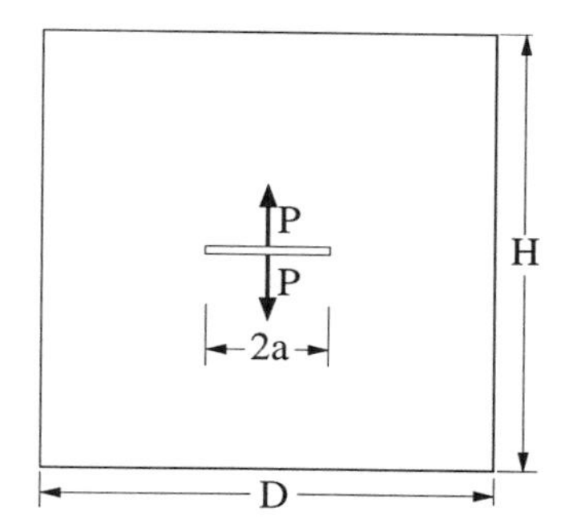

Figure 3.1.2 Center crack loaded by symmetric concentrated forces.

Example 3.1.2 The stress intensity factor for the center cracked panel of Fig. 2.1.1 was obtained numerically by Isida (1973) with very high accuracy for $H \gg D$. These results may be approximated within 0.1% by the Feddersen-Tada expression (Feddersen 1966; Tada, Paris and Irwin (1973))

$$K_I = \sigma\sqrt{D}k(\alpha), \qquad k(\alpha) = \sqrt{\frac{\pi\alpha}{\cos\pi\alpha}}\,(1 - 0.1\alpha^2 + 0.96\alpha^4) \tag{3.1.4}$$

In this case, the behavior for short cracks coincides with that for an infinite panel, $k(\alpha) \to \sqrt{\pi\alpha}$; for long cracks ($\alpha \to 0.5$), $k(\alpha) \to c_2(1-2\alpha)^{-1/2}$ where c_2 is a constant very close to 1. □

Example 3.1.3 The stress intensity factors for cracks with concentrated loads applied on the crack faces display a completely different type of dependence on the crack length. The simplest case is that of a center-cracked infinite panel loaded with two equal and opposite forces at the centers of the crack faces (Fig. 3.1.2, with a/H and $a/D \ll 1$). The stress intensity factor is then written as

$$K_I = \frac{P}{b\sqrt{\pi a}} \tag{3.1.5}$$

which shows that for a given load P the stress intensity factor *decreases* as the crack length increases. It is obvious, however, that this decrease cannot be indefinite for a real (finite) plate. Indeed, based on numerical results by Newman (1971), Tada, Paris and Irwin (1973) proposed the following modified formula for a finite panel of width D and height $H = 2D$:

$$K_I = \frac{P}{b\sqrt{D}}\hat{k}(a/D), \qquad \hat{k}(\alpha) = \frac{1 - 0.5\alpha + 0.957\alpha^2 - 0.16\alpha^3}{\sqrt{\pi\alpha(1-2\alpha)}} \tag{3.1.6}$$

with error less than 0.3% for any a/D. Note that for a relatively small crack length ($a/D \to 0$), this expression coincides with that for the infinite panel. On the other hand, for large cracks ($a/D \to 0.5$), $\hat{k}(\alpha) \to c_2(1-2\alpha)^{-1/2}$, which coincides with the previous example if one sets $\sigma_N = P/bD$. □

3.1.2 Superposition Methods

One of the advantages of LEFM is that the solutions for different loading cases are additive in stresses, strains, and displacements. Since stress intensity factors are nothing but parameters of the stress field, the stress intensity factors are also additive. This explains why Irwin's local approach is more popular than Griffith's global approach, even though they are generally equivalent. The energy release rates are not additive (although they are square-root additive); however, special care must be taken in problems in which geometrical nonlinearities arise. Such is generally the case when the resultant mode I stress intensity factor at any of the crack tips becomes negative. This implies interpenetration of the faces of the crack which is in reality impossible; instead, there is partial closing of the crack with face-to-face compression, which is a nonlinear phenomenon.

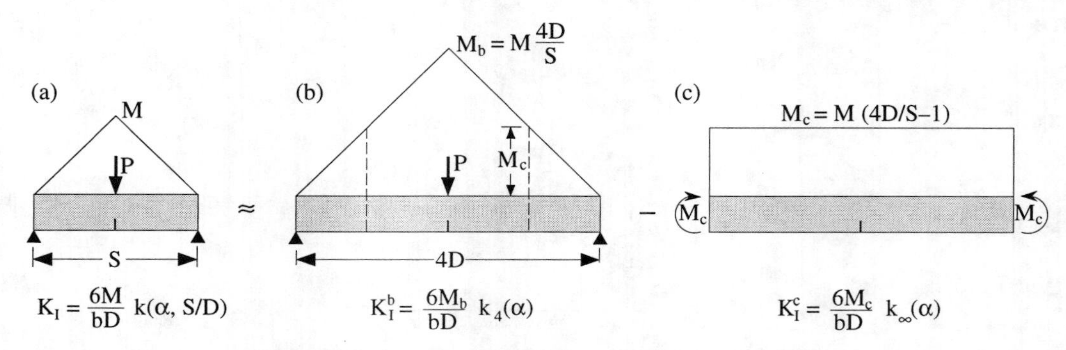

Figure 3.1.3 Load decomposition in three-point bent beams.

The literature is replete with problems solvable by superposition. Many cases are quite obvious: the actual loading often is just a superposition of various simpler loadings for which the stress intensity factors are known. This is the case for eccentric tension or compression of cracked strips which are solved by superposition of a pure tension or compression state and a pure bending state. Sometimes the simple states are not so obvious, as the following example shows.

Example 3.1.4 To obtain the stress intensity factor of a single-edge cracked specimen subjected to three-point bending for arbitrary span-to-depth ratio, Guinea (1990) and Pastor et al. (1995) used the approximate superposition illustrated in Fig. 3.1.3. In this approximation, the solutions for $S/D = 4$ (Fig. 3.1.3b) and for pure bending (Fig. 3.1.3c) are superposed so that the resulting bending moment distribution over the central part is the same as that for the actual beam (Fig. 3.1.3a). The result may be written as:

$$k_{S/D}(\alpha) = k_\infty(\alpha) + \frac{4D}{S}[k_4(\alpha) - k_\infty(\alpha)] \tag{3.1.7}$$

where $k_4(\alpha)$ and $k_\infty(\alpha)$ are the solutions for $S/D = 4$ and ∞ given in the example 3.1.1. Rearranging, the final expression turns out to be of the form (3.1.1) with $p_{S/D}(\alpha)$ given by

$$p_{S/D}(\alpha) = p_\infty(\alpha) + \frac{4D}{S}[p_4(\alpha) - p_\infty(\alpha)] \tag{3.1.8}$$

with $p_4(\alpha)$ and $p_\infty(\alpha)$ given by (3.1.2) and (3.1.3). This solution was checked against existing results in the literature for $S/D = 8$ (Brown and Srawley 1966) and finite element results using very small singular quarter-node elements for $S/D = 8$ and 2.5 (Pastor et al. 1995). The results coincided within 1%. ☐

A particularly important class of superposition is that in which the effect of remotely applied stresses is first reduced to the effect of a stress distribution over the crack faces and then the stress intensity factor, due to this stress distribution, is obtained by integration of the stress intensity factor due to a point load at an arbitrary location on the crack faces. Let us illustrate these steps by examples; first the reduction to a case with stresses on the crack faces.

Example 3.1.5 Consider a center-cracked panel subjected to remote uniaxial stress σ. We may decompose the whole elastic solution (Fig. 3.1.4a) as the solution for an *uncracked* panel (Fig. 3.1.4b) and the solution for a cracked panel with the faces of the cracks subjected to stresses identical but opposite to those in the uncracked panel (Fig. 3.1.4c). In this way, the remote boundary conditions are satisfied, as well as the boundary conditions on the crack face. Of course, the stress intensity factor for load case (b) is zero (there is no singularity in an uncracked panel) so that one finds that the stress intensity factors of cases (a) and (c) are identical. This particular example proves that the stress intensity factor for a center-cracked panel subjected to remote uniaxial stress σ is identical to that corresponding to a center-cracked panel with the crack subjected to internal pressure $p = \sigma$. ☐

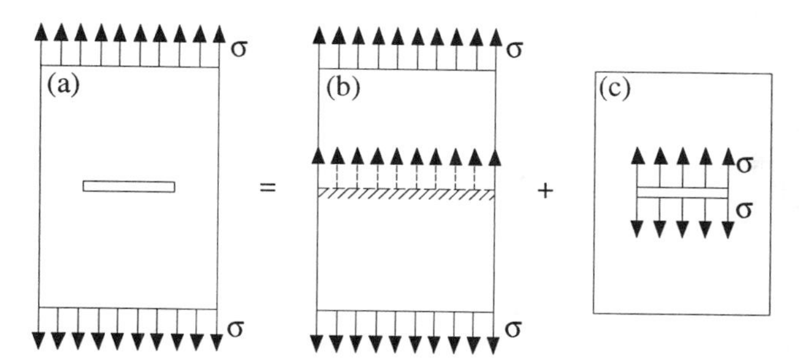

Figure 3.1.4 Solution for a cracked panel expressed as superposition of the solution for an uncracked panel and the solution for a loaded crack with no remote stresses.

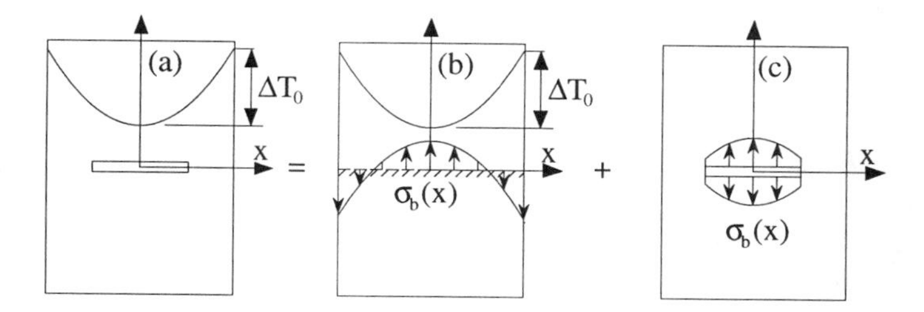

Figure 3.1.5 Superposition used to analyze thermally induced stress intensity factors.

Even if the foregoing examples may seem very simple, the basic procedure is always the same. In particular, this is the most usual method when internal stresses build up due to thermal or moisture gradients. In those cases, the internal stress distribution is first computed for the body without crack (but with non-uniform temperature or moisture distribution), and then equal and opposite stresses are applied on the crack faces (while keeping uniform temperature).

Example 3.1.6 Consider a long center-cracked panel with free ends subjected to heating on both its sides. Assume that at a given instant the temperature profile is parabolic across the section: $\Delta T = \Delta T_0(2x/D)^2$ (Fig. 3.1.5a). We decompose this state in the state shown in Fig. 3.1.5b, with thermal gradient and no crack, and the state shown in Fig. 3.1.5c, with no thermal gradient and stresses on the crack faces equal and opposite to those in case (b). The stresses in case (b) may be estimated in the classical way by assuming that initially plane sections remain plane. If we call β the coefficient of linear thermal expansion, the result for the stress distribution along the crack plane is

$$\sigma^b = \frac{1}{3}\beta E \Delta T_0 \left[1 - 2\left(\frac{2x}{D}\right)^2\right] \tag{3.1.9}$$

Except for a change in sign, this is the stress distribution to be applied on the crack faces in state (c). As in the previous example, the stress intensity factor for the original state (a) is equal to that for state (c) because in state (b) there is no crack and thus there is no stress singularity. ▯

Once the equivalent problem with stresses on the crack faces has been obtained, the problem remains of finding the stress intensity factor for this case. This may be done directly if the so-called *Green function* for the problem is known. The Green function is no more than the expression for the stress intensity factor engendered by a unit point load applied at any location on the crack face.

To be systematic, we write the stress intensity factor generated by a load-pair P_x located at point x on

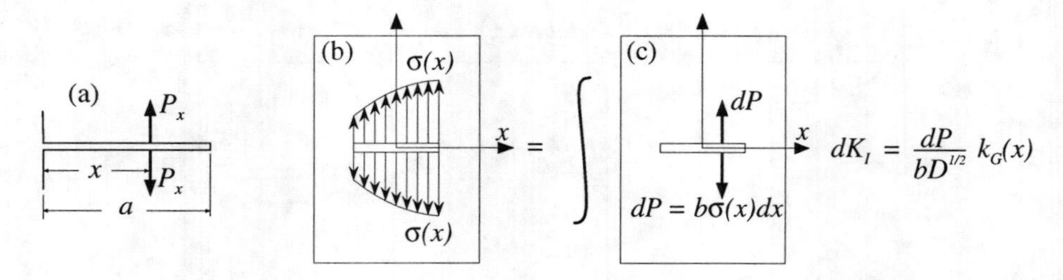

Figure 3.1.6 Superposition based on Green's function: (a) Base problem; (b) general problem for a center cracked panel; (c) Superposition corresponding to the general problem.

the crack faces (Fig. 3.1.6a) in the form

$$K_I = \frac{P_x}{b\sqrt{D}} k_G(\alpha, x/D) \tag{3.1.10}$$

where $k_G(\alpha, x/D)$ is the dimensionless Green function (the Green's function with dimensions includes the factor $1/b\sqrt{D}$). If k_G is known, the stress intensity factor generated by an arbitrary stress distribution over the crack faces is easily obtained by integration, as shown in the following example (which can easily be generalized to other more complex situations).

Example 3.1.7 Consider the center-cracked panel subjected to a known normal stress distribution over the crack faces, symmetric with respect to the crack plane (Fig. 3.1.6b). Let the stress at relative distance x/D from the crack center be $\sigma(x/D)$. If we subdivide the crack into infinitesimal length elements dx, the element at x contributes with an elemental concentrated load $dP_x = b\,\sigma(x/D)$. According to (3.1.10) this produces an infinitesimal stress intensity factor $dK_I = (\sigma(x/D)/\sqrt{D})k_G(\alpha, x/D)\ dx$. Now, adding up the contributions from all the elements we get

$$K_I = \int dK_I = \frac{1}{\sqrt{D}} \int_{-a}^{a} \sigma(x/D) k_G(\alpha, x/D)\ dx \tag{3.1.11}$$

or, with the change $x/D = u$

$$K_I = \sqrt{D} \int_{-\alpha}^{\alpha} \sigma(u)\ k_G(\alpha, u)\ du \tag{3.1.12}$$

To check this approximation, let us consider the previous Examples 3.1.5 and 3.1.6 for the limiting case $\alpha = a/D \ll 1$. In such a limit, the plate may be considered to be infinite, and the function $k_G(\alpha, u)$, with $u = x/D$, simplifies to

$$k_G(\alpha, u) = \frac{1}{\sqrt{\pi\alpha}} \frac{\alpha + u}{\sqrt{\alpha^2 - u^2}} \tag{3.1.13}$$

Substitution of this expression into (3.1.12) gives

$$K_I = \sqrt{D}\ \frac{1}{\sqrt{\pi\alpha}} \int_{-\alpha}^{\alpha} \sigma(u)\ \frac{\alpha + u}{\sqrt{\alpha^2 - u^2}}\ du \tag{3.1.14}$$

For the case of uniform tension $\sigma(u) = \sigma =$ constant, the integration readily delivers the well-known result $K_I = \sigma\sqrt{\pi a}$ (in the integral, set $u = \alpha \sin t$, $du = \alpha \cos t\ dt$). For the case of the parabolic distribution of temperature in Example 3.1.6 we notice that since we are considering $a/D \ll 1$, and since $x \leq a$, then the term $2(2x/D)^2$ in (3.1.9) is also negligible. Thus, the stress intensity factor for this case is just $K_I = \beta E \Delta T_0 \sqrt{\pi a}/3$. ☐

The function $m_G = K_I/P_x = k_G(\alpha, x/D)/\sqrt{D}$ is called the weight function, of which k_G is a dimensionless version. Finding the weight function for a particular case is a difficult problem of elasticity theory, which will be briefly outlined in Section 3.5.5. For a systematic approach to the weight function method, see Wu and Carlsson (1991).

Exercises

3.1 A long strip of width $D = 400$ mm is subjected to variable uniaxial stress with peaks of 14 MPa. In these conditions, edge cracks may be assumed to grow due to fatigue, and the designer wants the strip not to fail before the fatigue cracks are clearly visible. Determine the required toughness (K_{Ic}) if the crack length at which the strip fails must be (a) at least 10 mm, (b) at least 50 mm. Give also the values of the fracture energy if the material is a steel ($E' \approx 200$ GPa).

3.2 In a long strip of width $D = 300$ mm, the expected peak stress (uniaxial) is 30 MPa. If there exist welding flaws which resemble a center crack, determine the maximum flaw size allowable if the fracture toughness is 96 MPa$\sqrt{\text{m}}$ and (a) the strength safety factor is 1; (b) the strength safety factor is 2. (Hint: make a first estimate of a assuming $a/D \ll 1$, and then iterate until 1% accuracy of the result.)

3.3 Estimate the stress intensity factor for eccentric tension in a single-edge cracked strip. Let P be the load, D and b the strip width and thickness, and e the eccentricity (positive towards the cracked side). Write the results as (a) $(P/b\sqrt{D})k(\alpha)$; (b) $\sigma_N\sqrt{D}k(\alpha)$, with σ_N equal to the mean remote tensile stress; (c) same, but with σ_N equal to the maximum remote tensile stress; show that in this latter case, the shape function $k(\alpha)$ for very short cracks tends to the same value as that for a semi-infinite plate with an edge crack.

3.4 A thin slit of length $2c$ is machined in a large panel of thickness b made of a brittle material. A flat jack of identical length is inserted into the slit and pressure is applied to it until a crack propagates symmetrically. Determine the stress intensity factor at the crack tips for arbitrary crack length $2a$ and jack pressure p. Show that when $a \gg c$, the stress intensity factor approaches that corresponding to a center crack loaded at its center by a pair of forces equal to the jack force.

3.5 In the pressurized panel of the previous exercise, plot the evolution of the pressure in the jack vs. the crack length for quasi-static crack growth. Assume that the initial slit behaves as a crack, and that the fracture toughness K_{Ic} is known. (Hint: plot $p\sqrt{c}/K_{Ic}$ vs a/c.)

3.6 A large panel has a center crack of length $2a$ subjected to a symmetric internal pressure distribution which takes the value p_0 at the crack center and decreases linearly to zero at the crack tips. Determine the stress intensity factor.

3.7 Show that if a center crack in a large panel is subjected to an arbitrary symmetric pressure distribution of the type $p = p_0\phi(x/a)$, where $\phi(x/a)$ is a dimensionless function of the relative coordinate x/a along the crack, the resulting stress intensity factor is always of the form $K_I = kp_0\sqrt{\pi a}$ where k is a constant.

3.2 Approximate Energy-Based Methods

Building on the results of Section 2.1, let us now review the techniques that provide approximate solutions for $\mathcal{G}$ in a number of cases. The presentation is not intended to be exhaustive. Its purpose is twofold: to give at least a minimum of insight on how to obtain the energy release rate, and to provide a number of simple equations for $\mathcal{G}$ which can be used in solving typical fracture problems.

The section covers first the simplified approaches to truly linear problems, relying only on simple mechanics of materials. After that, a certain geometrically nonlinear case is briefly analyzed: the extension of a surface delamination crack due to buckling under compression.

3.2.1 Examples Approximately Solvable by Bending Theory

In Section 2.1 we already presented the DCB specimen as an example of structures in which the energy release rates may be approximately calculated by bending theory (Fig. 2.1.3). The energy release rates of

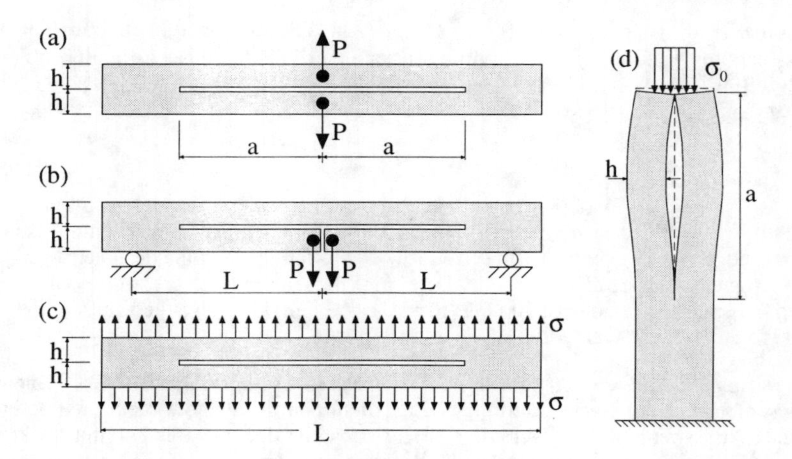

Figure 3.2.1 Structures with energy release rate approximately solvable by beam theory.

the beams or structures shown in Fig. 3.2.1 can also be solved in this way if they are slender and the cracks are assumed to grow straight ahead. The solutions are asymptotically exact as the slenderness tends to infinity.

The approximation, which has already been used in Examples 2.1.1–2.1.4, consists of using the classical beam theory to determine the load-point displacement and the elastic energy for the arms at both sides of the crack, assuming fixed ends at the crack root sections. Two basic approaches may be used. In the first, the bending moment distribution is computed; then the energy per unit length of beam, $M^2/2EI$, is integrated to find the total elastic energy or the complementary energy and the energy release rate is determined by differentiation with respect to crack length according to Eqs. (2.1.15) or (2.1.21). This procedure was illustrated in Examples 2.1.1 and 2.1.2. The second approach is to compute first the compliance and use Eq. (2.1.32) to determine the energy release rate in the manner in Examples 2.1.3 and 2.1.4.

All the structures shown in Fig. 3.2.1 can be solved in either way, although some structures are statically indeterminate and then the redundant forces must be solved first. Care should be taken regarding the value of the crack length in these structures, which is a when a single crack tip exists, but Na when N crack tips are present. Hence, partial derivatives must be with respect to Na to obtain the energy release rate per crack tip.

3.2.2 Approximation by Stress Relief Zone

Consider the center cracked panel of Fig. 3.2.2a subjected to remote stress σ_∞ perpendicular to the crack plane, and assume that the crack length is much less than the remaining dimensions of the plate. The principal stress trajectories in Fig. 2.1.1a reveal that the formation of a crack causes stress relief in the shaded, approximately triangular regions next to the crack. As an approximation, one may suppose the stress relief region to be limited by lines of some constant slope k (Fig. 3.2.2a), called the "stress diffusion" lines, and further assume that under constant boundary displacements the stresses inside the stress relief region drop to zero while *remaining unchanged outside*. Based on this assumption, the total loss of strain energy due to the formation of a crack of length $2a$ at controlled (fixed) boundary displacements is $\Delta\mathcal{U} = -2ka^2b(\sigma_\infty^2/2E)$ where $\sigma_\infty^2/2E$ is the initial strain energy density. Writing that $\sigma_\infty = Eu/L$ where u and L are the panel elongation and length, we can rewrite the loss of strain energy as $\Delta\mathcal{U} = -2ka^2bE(u^2/2L^2)$ and, therefore, the energy release rate per crack tip is

$$\mathcal{G} = -\frac{1}{b}\left[\frac{\partial(\Delta\mathcal{U})}{2\partial a}\right]_u = 2kaE\frac{u^2}{L^2} = 2ka\frac{\sigma_\infty^2}{E} \tag{3.2.1}$$

where in reaching the last equality we assume that after crack formation the relationship $\sigma_\infty = Eu/L$ is still approximately valid, an assumption that we will show to hold later.

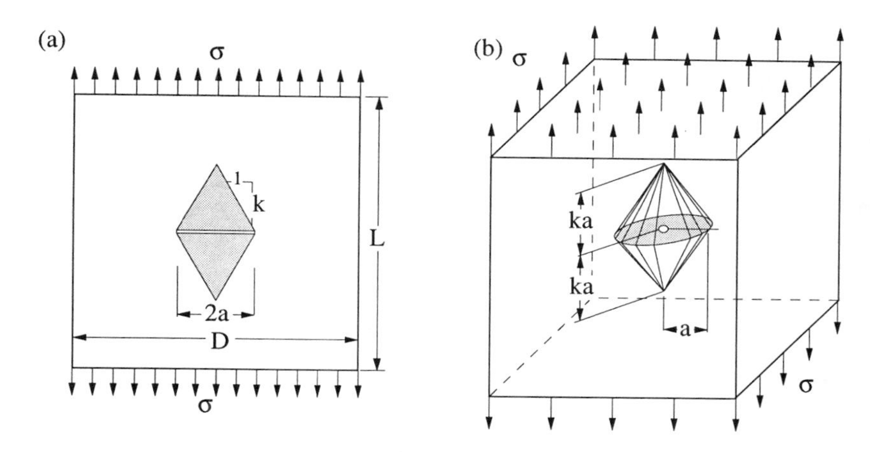

Figure 3.2.2 Approximate zones of stress relief: (a) for a center cracked panel, (b) for a penny-shaped crack.

The foregoing approximate result is in exact agreement with Griffith solution (1924) if one assumes that $k = \pi/2 = 1.571$. Even if k is unknown, the form of this equation obtained by the stress relief argument is correct. If the stress relief zone is assumed to be a circle of radius a passing through the crack tips, the result also happens to be exact. The same is true when the zone is taken as a rectangle of width $2a$ and height $\pi a/2$, or any geometrical figure whose area is πa^2.

For the penny-shaped crack in an infinite elastic space subject to remote tension σ_∞ (Fig. 3.2.2b), the stress relief region may be taken to consist of two cones of base πa^2 and height ka. Therefore, $\Delta\mathcal{U} = -2\frac{1}{3}\pi a^2 ka(\sigma_\infty^2/2E) = -\frac{12}{3}\pi k a^3 E u^2/L^2)$. Also, $\mathcal{G} = -[\partial\Delta\mathcal{U}/\partial(\pi a^2)]_u = -[\partial\Delta\mathcal{U}/\partial a]_u/2\pi a$, i.e.,

$$\mathcal{G} = \frac{k\sigma_\infty^2 a}{2E} = \frac{k\sigma_\infty^2 a}{2E} \tag{3.2.2}$$

Again, this equation is of the correct form and is in exact agreement with the analytical result (Sneddon 1946) if one assumes that $k = 8/\pi$. The exact value also results if the stress relief zone is assumed to be a rotational ellipsoid of minor semiaxes a, a and $8a/\pi$ or any geometrical figure whose volume is $16a^3/3$.

The approximate method of stress relief zones can be applied in diverse situations for a quick estimate of $\mathcal{G}$. The value of k depends on geometry and its order of magnitude is 1 (except in the case of high orthotropy). The error in intuitive estimations of k can be substantial; however, the form of the equation obtained for $\mathcal{G}$ is correct.

There is a dichotomy in the method of stress relief zone which one must be aware of. Since the stress relief zone in Fig. 3.2.2 does not reach the top and bottom boundaries, the stress σ at top and bottom can remain constant and equal to σ_∞. Since there is a continuous zone of constant stress $\sigma = \sigma_\infty$ connecting the top and bottom boundaries, the displacements at top and bottom also remain constant during the crack extension at constant σ_∞.

However, from Eq. (2.1.32), it follows that a non-zero $\mathcal{G}$ implies an increase of the compliance due to the presence of the crack. This, in turn, implies that the stress cannot remain constant while the crack extends at constant displacement. The variation of compliance due to crack extension may be obtained from Eq. (3.2.1) and from this the stress variations at constant displacement may be inferred. Let D be the width of the panel, L its length, u the relative displacement between the top and bottom boundaries. The resultant load and initial (uncracked) compliance then are:

$$P = \sigma_\infty bD \tag{3.2.3}$$

$$C_0 = \frac{L}{bDE} \tag{3.2.4}$$

Inserting the foregoing expression for P into Eq. (2.1.32) (taking care to change the total crack length

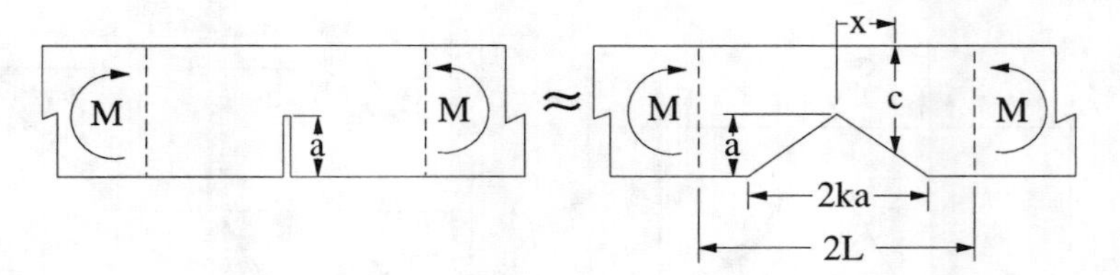

Figure 3.2.3 Herrmann's approximation.

a to $2a$) and equating the result to Eq. (3.2.1), one easily finds

$$\frac{dC(a)}{da} = \frac{8ka}{bD^2E} \tag{3.2.5}$$

which integrates to

$$C(a) = C_0 + \frac{4ka^2}{bD^2E} \tag{3.2.6}$$

From this, one sees that the variation of compliance contains the factor $(a/D)^2$ whose absolute value tends to vanish when the size of the panel is much larger than the crack length. The remote stress drop due to crack extension at constant displacement is also shown to vanish as $(a/D)^2$. Indeed, by differentiating the relation $u = C(a)P$ one finds the first-order approximation for the remote stress drop as

$$\Delta\sigma_\infty(u, a) \approx -\sigma_\infty(u, 0)\frac{4ka^2}{LD} \tag{3.2.7}$$

where the higher order terms in $4ka^2/LD$ have been neglected.

Henceforth, the initial contradiction between the hypotheses of both constant remote stress and displacement exists at the theoretical level, but is resolved at the approximation level because it has been proved *a posteriori* that, in this case, the stress drop is vanishingly small when the crack is small relative to the size of the panel.

3.2.3 Herrmann's Approximate Method to Obtain $\mathcal{G}$ by Beam Theory

A remarkably simple method for close approximation of $\mathcal{G}$ in notched beams was discovered by Kienzler and Herrmann (1986) and Herrmann and Sosa (1986). The method was derived from a certain unproven hypothesis (postulate) regarding the energy release when the thickness of the fracture band is increased. Bažant found a different derivation of this method (Bažant 1990a) which is simpler and at the same time indicates that the hypothesis used by Herrmann et al. might not be exact but merely a good approximation. Also, Herrmann's method relies on more sophisticated concepts (material forces) which are elegant but seem more complicated than necessary to obtain the result. An even a simpler method of deriving Bažant's and Herrmann's results has been recently developed by Planas and Elices (1991d). This last will be presented now.

The method consists of approximating the cracked beam by a triangularly notched beam as shown in Fig. 3.2.3, and calculating its energy in the frame of the strength of materials theory (bent beam of variable inertia.)

Let k be the slope of the sides of the triangular notch, to be determined empirically. Let b and D be, respectively, the thickness, and depth of the unnotched beam, and let M be the constant bending moment over a central portion of the beam of length $2L \geq 2ka$. With the axis shown in Fig. 3.2.3, the complementary energy of the central portion is

$$\mathcal{U}^* = M^2 \int_0^{ka} \frac{1}{EI(c)}\, dx + (L - ka)\frac{M^2}{EI_1} \tag{3.2.8}$$

where EI_1 is the bending stiffness of the unnotched beam and $EI(c)$ is the bending stiffness when the

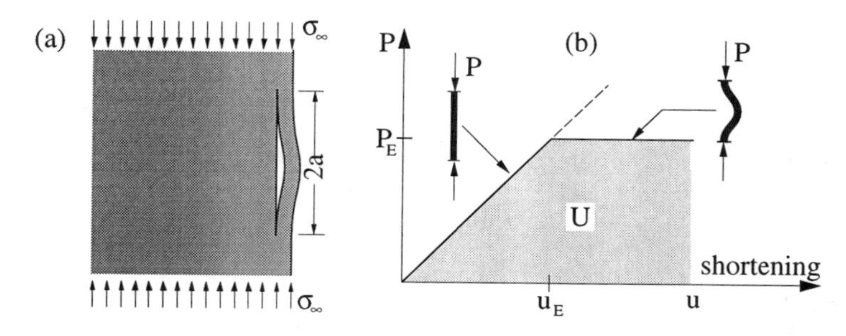

Figure 3.2.4 (a) Buckling of a subsurface layer. (b) Load-displacement curve for a buckling beam.

depth has been reduced to c,

$$c = D - a + \frac{x}{k} \tag{3.2.9}$$

Applying Eq. (2.1.21) to Eq. (3.2.8), one finds:

$$b\mathcal{G} = M^2 \int_0^{ka} \frac{\partial}{\partial c}\left(\frac{1}{EI(c)}\right)\frac{\partial c}{\partial a}\,dx + M^2\frac{1}{EI(D)}k - k\frac{M^2}{EI_1} \tag{3.2.10}$$

where, since $I(D) = I_1$, the second and third terms cancel out. Moreover, from Eq. (3.2.9), $\partial c/\partial a = -k\,\partial c/\partial x$. Integration now delivers Herrmann's result

$$\mathcal{G} = \frac{k}{b}\left(\frac{1}{EI_2} - \frac{1}{EI_1}\right)M^2 \tag{3.2.11}$$

where EI_2 is the bending stiffness of the central (notched) section.

When the remote flexural stress $\sigma_f = MD/2I_1$ is used as a measure of the load, and the expressions of the inertia moments for rectangular cross-section are substituted, the previous equation reads

$$\mathcal{G} = \frac{kD\sigma_f^2}{3E}\left[\left(\frac{D}{D-a}\right)^3 - 1\right] \tag{3.2.12}$$

According to Kienzler and Herrmann (1986, Figs. 3 and 4), this compares (for $k = 1$) very well with accurate solutions from handbooks. However, it appears that the agreement would be even better for some value $k \neq 1$ (Bažant 1990a). For very shallow notches, this method requires rather large k-value (about 4) to accurately fit the results. But in this case, the approximation by a beam of variable thickness is poor.

The results of Herrmann and Sosa (1986) for double-edge-notched and center-cracked specimens may also be obtained using the Planas and Elices expedient of approximating the crack by a triangular notch (coinciding with the shaded areas in Fig. 3.2.2a for the center cracked panel), and performing a classical analysis with the assumption that the cross-sections remain plane.

3.2.4 Subsurface Cracking in Compression by Buckling

A slightly exceptional case in LEFM is that shown in Fig. 3.2.4a, where a subsurface (delamination) crack may grow due to buckling of the layer above it, induced by a remotely applied compressive stress σ_∞. In this case the computation of $\mathcal{G}$ must take into account the geometrical nonlinearity implied in buckling.

To do this, one computes the elastic energy of a buckled beam, $\mathcal{U}_b$, as the work supply when no dissipative processes take place. This coincides with the area swept by the P-u curve, where P is the compressive load applied to the beam and u is the beam shortening. When h, b, and $2a$ are, respectively, the depth, thickness, and length of the beam, the P-u curve coincides with the straight line $P = bhEu/2a$ for loads below Euler's buckling load P_E and is horizontal ($P = P_E$) for further displacements (Fig. 3.2.4). The area under the curve $\mathcal{U}_b$ is

$$\mathcal{U}_b = P_E\, u - \frac{1}{2} P_E\, u_E = 2abh\sigma_E \left(\frac{u}{2a} - \frac{\sigma_E}{2E} \right) \tag{3.2.13}$$

where $\sigma_E = P_E/bh$ is the buckling stress, which, for fixed ends and rectangular cross-section, is

$$\sigma_E = \frac{\pi^2 h^2}{12a^2} E \tag{3.2.14}$$

The fundamental simplification in our problem consists of assuming that the displacement of the ends of the buckling layer of length $2a$ is imposed by the deformation of the surrounding material which stays at stress level σ_∞. So we have

$$u = 2a \frac{\sigma_\infty}{E} \tag{3.2.15}$$

and, after substitution into Eq. (3.2.13), the strain energy of the buckled layer is

$$\mathcal{U}_b = abh \frac{\sigma_E}{E} (2\sigma_\infty - \sigma_E) \tag{3.2.16}$$

The energy $\mathcal{U}$ of the whole system is $\mathcal{U}_b$ plus the strain energy of the surrounding material, which is the strain energy density $\sigma_\infty^2/2E$ times the surrounding volume, equal, in turn, to the total (constant) volume of the body V minus the volume of the buckling layer $2abh$. The resulting expression is

$$\mathcal{U} = (V - 2abh)\frac{\sigma_\infty^2}{2E} + abh\frac{\sigma_E}{E}(2\sigma_\infty - \sigma_E) = V\frac{\sigma_\infty^2}{2E} - 2abh\frac{(\sigma_\infty - \sigma_E)^2}{2E} \tag{3.2.17}$$

From this, the expression for $\mathcal{G}$ follows at once using Eq. (2.1.15) with the condition, following from the simplifications used in the derivation, that σ_∞ remains constant at constant displacement. After inserting Eq. (3.2.14) and differentiating with respect to a, one gets the following expression for the energy release rate:

$$\mathcal{G} = -\frac{1}{b}\left[\frac{\partial \mathcal{U})}{2\partial a}\right]_u = \frac{h}{E}\left[\sigma_\infty^2 + \frac{\pi^2 h^2}{6a^2}\sigma_\infty E - \frac{\pi^4 h^4}{48a^4}E^2\right] \tag{3.2.18}$$

This result captures some, but not all, of the important aspects of the problem of delamination in layered composites (Sallam and Simitses 1985, 1987; Yin, Sallam, and Simitses 1986).

3.3 Numerical and Experimental Procedures to Obtain K_I and $\mathcal{G}$

In many practical problems there is no analytical solution for the energy release rate, and one must resort to experimental or numerical approximations. While experimental procedures have been extensively used in the past, modern computers have made the numerical procedures relatively easy. In this section we give only a brief sketch of the available methods for the case in which the propagation direction is known.

3.3.1 Numerical Procedures

There are various numerical approaches to solving linear elastic fracture problems. For our purposes, the best classification is based on commercial availability. Special purpose computer programs, which make use of special properties of the fields in plane elasticity (see Chapter 4) and are usually usually not available commercially, may be very accurate, but are generally restricted to research by specialists. This is the case of the so-called boundary collocation, in which special power expansions of the unknowns are used for a particular problem and the coefficients of the expansion are determined so that the boundary conditions be satisfied only in some average sense.

Restricting attention to commercial programs, we may distinguish between special purpose programs and general purpose programs. Special purpose programs are specifically designed to deal with cracks and determine stress intensity factors, so that the user may access to post-process routines that will readily compute the stress intensity factor from a basic numerical solution. General purpose programs are those available to solve general elasticity problems, which can be used, with special strategies, to solve fracture problems. With the general purpose programs, two basic strategies may be used: (1) incremental stiffness method, and (2) near-tip field fitting.

The incremental stiffness method essentially consists of determining the compliance for two different, but close, crack lengths $a - \Delta a$ and $a + \Delta a$. Then the energy release rate G is estimated as

$$\mathcal{G} \approx \frac{P^2}{2b} \frac{C(a+\Delta a) - C(a-\Delta a)}{2\Delta a} \tag{3.3.1}$$

Any finite element, finite difference, or boundary element code may be used to produce the two compliance values for two close crack lengths. Numerical resolution and mesh refinement limit the accuracy of this procedure. In very general terms, there are two possible approaches: (1) use only one mesh and simulate crack extension by freeing one node, so that the crack extends by one element, or (2) modify the mesh for the second calculation in which the node at the crack tip is displaced by Δa. The first method is easy to use and does not require modification of the global stiffness matrix, but requires a fine mesh so that the numerical differentiation in (3.3.1) gives accurate results. The second method decouples the crack extension from the mesh size, but requires partial recalculation of the stiffness matrix.

Example 3.3.1 A commercial finite element code was used to analyze a single-edge cracked beam in three-point bending, with a span-to-depth ratio $S/D = 4$ (Guinea 1990). Half the beam was discretized so that 100 equally sized elements were placed along the crack plane. The crack length was varied by changing the boundary conditions along the nodes in the crack plane, from opening displacement prevented (no crack at this node) to load free (crack at this node). Computations were performed in plane stress with $D = 1$, $b = 1$, and $E = 1$, for a load $P = 1$ (in arbitrary, but consistent, units). Although, the purpose of the computations was other than determining K_I, the results can be used to examine the accuracy of the differential stiffness method.

Consider, for example, the case $a = 0.5D$. The displacements computed for crack lengths $a_1 = 0.49D$ $a_2 = 0.51D$ were, respectively, 57.261 and 61.663, numerically identical to the compliance values (because $P = 1$). The energy release rate is then evaluated from (3.3.1) as $\mathcal{G} \approx (1^2/2)(61.663 - 57.261)/0.02 = 110.1$ in appropriate units. Now, since we always write $\mathcal{G} = (P/bD)\hat{g}(\alpha)$, it turns out that in our calculation the numerical value of $\mathcal{G}$ (with its arbitrary units) coincides with the dimensionless value of $\hat{g}(0.5)$. Therefore, $\hat{g}(0.5) \approx 110$. The stress intensity factor follows from Irwin's equation as $K_I = \sqrt{E'\mathcal{G}} \approx \sqrt{110} = 10.5$. Now, since we write $K_I = \sigma_N \sqrt{D} k(\alpha)$, we may easily find $k(0.5)$ upon noting that for a span-to-depth ratio of 4, $\sigma_N = 6P/bD$. The result is $k(0.5) \approx 10.5/6 = 1.75$. This value is to be compared with that given by equations (3.1.1) and (3.1.2) (or Fig. 3.1.1) which give $k(0.5) = 1.77$. Thus, the numerical estimate turns out to be about 1.2% lower than the more accurate value. □

The near-tip field fitting consists of making use of the known near-tip behavior of the stress and displacement or crack opening fields to make an estimate for K_I. It can make use, for example, of the stress distribution ahead of the crack tip, which is known to behave as $\sigma_{22} = K_I/\sqrt{2\pi r}$, where σ_{22} is the stress normal to the crack plane, and r is the distance to the crack tip. This means that a plot of $\sigma_{22}\sqrt{2\pi r}$ vs. r should tend to K_I as r approaches zero. It is also possible to use the displacement field, particularly the crack opening, which is known to behave as $w = 8K_I\sqrt{r}/E'\sqrt{2\pi}$. Therefore, the limit of $wE'\sqrt{2\pi}/8\sqrt{r}$ as $r \to 0$ is also K_I.

Example 3.3.2 The results of the nodal reactions along the uncracked ligament, or the crack opening distribution, may be used to make a near-tip field fit. We use $K_I = \lim_{r\to 0} \sigma_{22}\sqrt{2\pi r}$, and agree to write $K_I = \sigma_N \sqrt{D} k(\alpha)$ and for a given value of α, we define

$$\kappa(r) = \frac{\sigma_{22}}{\sigma_N}\sqrt{2\pi \frac{r}{D}} \tag{3.3.2}$$

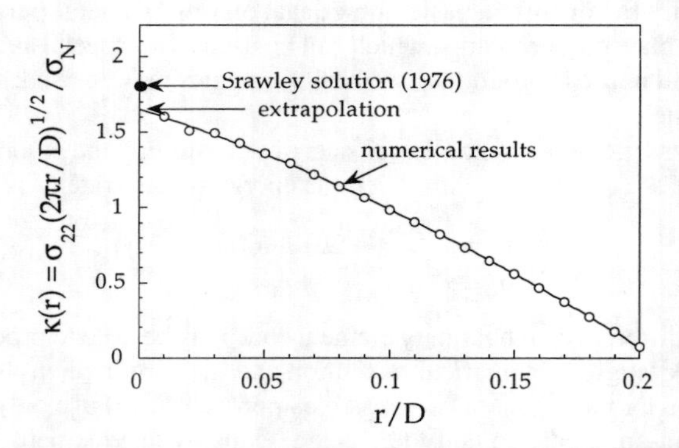

Figure 3.3.1 Plot of normal stresses times $\sqrt{r}$ vs. r (r =distance to crack tip). Extrapolation to zero gives the dimensionless stress intensity factor.

then $k(\alpha) = \lim_{r\to 0} \kappa(r)$. Plotting the nodal values of κ_n vs. the distance to the crack tip, and extrapolating to zero, we get an estimate of $k(\alpha)$. This was used in the finite element computations described in the previous example (now for $a/D = 0.5$). The nodal normal stresses were obtained as $\sigma_{22n} \approx R_n/bh$, where R_n is the nodal reaction and h the width of the elements. From this, the nodal values of κ were obtained and plotted as shown in Fig. 3.3.1. The extrapolated value gives $k(0.5) \approx 1.65$. This value differs by 7% from the more accurate value $k(0.5) = 1.77$ obtained from equations (3.1.1) and (3.1.2) (or Fig. 3.1.1). □

The foregoing examples show two of the ways to determine K_I and $\mathcal{G}$ from numerical results. The determination of K_I from the crack opening profile is left as one of the exercises. The general experience is that the differential stiffness method is more accurate for a given mesh size. This is probably due to the cancellation of constant errors in the differentiation process. However this method requires two computations, while the near-tip field fitting requires only one, although this is really not a problem with the kind of computers available today.

Getting good results (less than 5% error) with near-tip analysis requires extremely fine meshes, because of the difficulty in representing the crack tip singularity with ordinary finite elements. Indeed, careful studies of convergence by Wilson (1971) and Oglesby and Lamackey (1972) showed that the near-tip approximate solution may not converge to the analytical solution whatever the mesh refinement. To solve this problem, one needs special elements whose shape functions include a $r^{-1/2}$ singular term.

Various singular finite elements have been developed (see, e.g., Aliabadi and Rooke 1991), but most of them incorporate special shape function and require specially designed finite element codes. A remarkable exception is the so-called quarter-node isoparametric element (Barsoum 1975, 1976; Henshell and Shaw 1975). In this formulation, a standard 8-node isoparametric quadrilateral element is collapsed, as shown in Fig. 3.3.2a, to a triangular quarter-point element. The vertex corresponding to the collapsed nodes 1–7–8 becomes the singular point, and a $r^{-1/2}$ singularity is achieved by placing nodes 2 and 6 at a quarter (from the singular vertex) of the radial sides of the triangle. These elements are placed in a rosette around the crack tip as shown in Fig. 3.3.2b.

The stress intensity factor may be evaluated from the displacement fields of any of the elements, but most usually K_I is obtained from the values of the crack opening evaluated at the two nodes along the crack faces. With this method, values of K_I accurate within a few percent may be obtained without much mesh refinement. However, recent recommendations by ESIS Technical Committee 8 (1991) suggest, again, that best results for stress intensity factors are obtained if energetic approaches based on the determination of $\mathcal{G}$ are used instead of near-tip fields.

The differential stiffness method is not the only way to determine $\mathcal{G}$. The J-integral and other path-independent integral expressions may be (and have been) used to determine the energy release rate. This has the advantage that the evaluation of J is made using values of the fields at points far from the crack tip, where the errors are expected to be smaller. It also avoids numerical differentiation, and a single

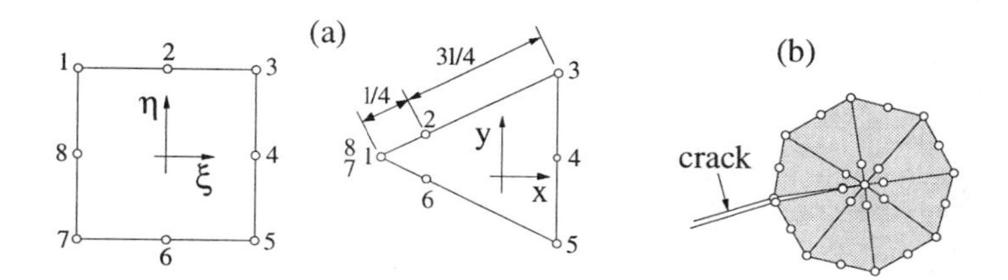

Figure 3.3.2 (a) Collapsing of an 8-node quadratic isoparametric element into a singular quarter-node element. (b) Rosette of singular finite elements at a crack tip.

computation is enough. It requires, however, special postprocessing routines, both in finite elements codes and in boundary element codes.

Although finite element codes dominate the market, commercial codes based on boundary elements have recently become available. They yield K_I-values of much higher accuracy than finite element codes. The main advantage of this kind of formulation is that only the boundary of the elements must be modeled, so that the number of degrees of freedom is greatly reduced. This is, of course, achieved at the cost of larger complexity of the code, especially the postprocessing. In particular, handling cracks may require special formulations and special postprocessing which are outside the scope of this book (for details, see Aliabadi and Rooke 1991).

3.3.2 Experimental Procedures

Experimental procedures to determine the stress intensity factor were often used with some intensity in the past, when numerical calculations were of limited availability. All the methods relied on measuring some features of the displacement fields of elastic specimens, and relating them to the energy release rate or to the stress intensity factor.

The simplest method, which is generally used in laboratory environments, uses the experimental version of the differential stiffness method. It is implemented by measuring the compliance of a specimen for various crack depths and determining $\mathcal{G}$ from $\mathcal{G} = (P^2/2b)dC/da$. In principle, two tests with two slightly different crack lengths are enough to get a result for a given crack length. However, experimental accuracies being always very limited, it is usually better to make a larger number of tests over a finite range, fit a smooth curve to these results, and then perform the differentiation. Because the experimental accuracy in obtaining the compliance is rarely better than 1%, this method is not very reliable unless the compliance variation due to the growth of the crack is a sufficiently large fraction of the total compliance. This excludes large specimens or structures with tiny cracks (or, generally speaking, small relative crack depths). In some test setups, a further source of error is that, in order to have a good control of the geometry, cracks are substituted by cut slits (notches). In this case, the notch width must be much less than any relevant dimension of the specimen (crack length, remaining ligament length, distance of applied loads from the crack tip, etc.).

Other methods rely on the analysis of the properties of the strain or displacement fields close to the crack tip. These include: strain gauge techniques, photoelastic techniques, interferometric techniques, and the caustics method.

The strain gauge technique measures the strain and stress at a set of points around the crack tip by means of bonded electrical strain gauges. In the photoelastic technique, the shear strain field around the crack is measured in a specimen made of a photoelastic polymer. In the interferometric techniques, the displacement field (usually the component normal to the crack plane) is mapped by interferometry. From the experimental results of stress, shear-strain, or displacement vs. the distance to the crack tip, near-tip fitting techniques similar to those sketched for numerical methods are used to infer the value of the stress intensity factor.

The principle of the caustics method is different of the former in that it uses the out-of-plane displacements to find the stress intensity factor. Due to Poisson effect, a depression of the surface of the specimen is produced around the crack tip. If the surface is polished, a mirror with a profile determined by the elastic field is produced. When a beam of light impinges normally over this mirror, the reflected rays produce a bright kidney-shaped spot whose size is related to the stress intensity factor. If transparent specimens are used, transmitted light can be used and then the specimen acts as a lens with a profile determined by the elastic field.

In all these techniques, it is essential to guarantee that the plastic zone is small compared to the size of the region over which the stresses, strains, or displacements are measured. If notches, instead of cracks, are used (which is usual in photoelastic techniques), corrections are required to take into account the finite radius at the tip.

For details of the experimental techniques, see Smith and Kobayashi 1993.

3.4 Experimental determination of K_{Ic} and G_f

When a crack in a laboratory specimen may be guaranteed to behave in a linear elastic way, the experimental determination of K_{Ic} or G_f is conceptually easy. The simplest way is to use a specimen in which the crack growth initiation coincides with the peak load (all the standard specimens belong to this category). In this case, one simply loads the specimen up to failure and records the peak load P_u. If LEFM conditions are fulfilled, the value of the stress intensity factor for this load coincides with K_{Ic}:

$$K_{Ic} = K_{Iu} = \frac{P_u}{b\sqrt{D}} k(\alpha) \tag{3.4.1}$$

where $\alpha = a/D$ is the relative crack length at the beginning of the test.

The difficulties in this kind of testing arise at two different levels: (1) Specimen preparation (precracking), and (2) verification of LEFM conditions. These aspects are well defined for metals in most national standards, particularly in ASTM E 399. The crack is grown from a normalized starter notch by fatigue under controlled conditions. The LEFM conditions are verified in two ways. First, the nonlinearity of the load displacement curve before peak is limited (an ideally brittle material is completely linear up to failure). This is done as shown in Fig. 3.4.1a by defining a kind of conventional (load) elastic limit P_5 for which the secant stiffness is 95% of the initial tangent stiffness. Deviation from linearity is acceptable if either the peak load occurs before the elastic limit or the ratio P_u/P_5 is less than 1.1 (see the standards for details).

Apart from this direct verification of linearity, there is a further condition which verifies that the specimen thickness and size are large enough for the nonlinear zone at fracture to be negligible (for engineering purposes). Since the standard specimens are designed so that their thickness is one-half of their width or depth ($b = 0.5D$) and the crack length is close to half the depth ($\alpha \approx 0.5D$), the thickness and size conditions are expressed in a single condition:

$$b \geq 2.5 \left(\frac{K_{Ic}}{\sigma_c}\right)^2 \tag{3.4.2}$$

where σ_c is a conventional flow stress (usually a value between the conventional 0.2% proof stress and the tensile strength). The origin of the foregoing equation is discussed in detail in the next chapter. Here, it is enough to say that the factor $(K_{Ic}/\sigma_c)^2$ is proportional to the size of the plastic zone, so the equation really places a limit on the extent of the plastic zone relative to the specimen size.

For materials other than metals, the situation is more complex. Cracks in polymers and structural ceramics cannot easily be grown using cyclic loading. For polymers, cracking by forcing a razor blade into the notch root has been chosen by ASTM standards (ASTM 1991). For fine ceramics, no standards are yet available, and round robins are being performed to compare toughness test results on specimens with different kinds of notches and cracks, as that promoted by ESIS TC 6 (Pastor 1993; Primas and Gstrein 1994). Specifications for the minimum size required for LEFM to apply have been set for polymers, and are similar to those previously stated for metals. No agreed limitations have been set yet for ceramics.

For concrete, it is generally accepted today that the sizes required for LEFM to apply are really huge (several meters or even tens of meter). Therefore, special purpose tests taking into account the nonlinear

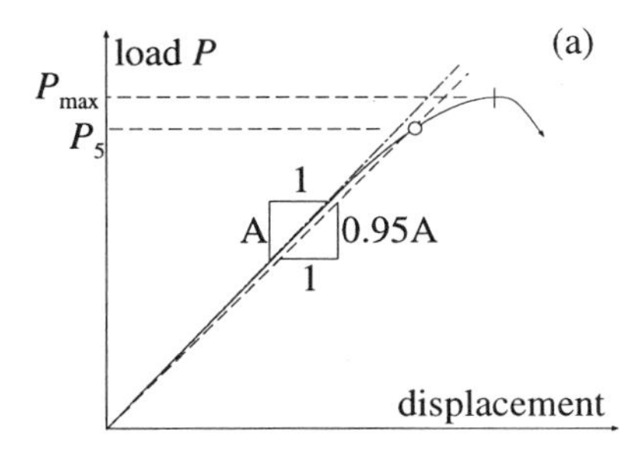

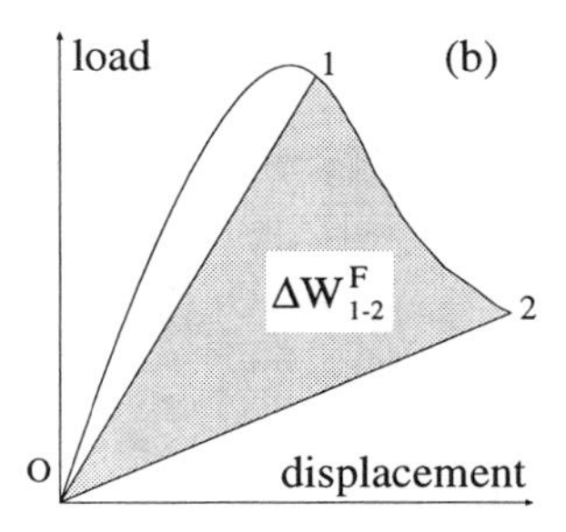

Figure 3.4.1 Experimental determination of fracture properties: (a) Load-displacement curves and definition of the conventional limit P_5 (After ASTM E 399, simplified); (b) determination of G_f from experiment.

fracture behavior of concrete have been set. They will be analyzed in the following chapters, where nonlinear models are introduced.

For any material, nonstandard tests may also be used to determine the fracture properties of the material, whenever the size of the specimen is large enough for LEFM to apply. One such method, based on an energetic analysis, consists of performing a stable test (controlling the displacement rather than the applied load) and simultaneously measuring the load, P, the load point displacement, u, and the crack length, a. Let the $P-u$ curve be known between the points 1 and 2 at which the crack lengths were measured to be, respectively, a_1 and a_2 (Fig. 3.4.1b). Then, according to Section 2.1.4, the energy consumed in fracture between points 1 and 2, $\Delta\mathcal{W}^F_{1-2}$, is the area of the curvilinear triangle $O12$, while the area of the newly formed crack is $b(a_2 - a_1)$; hence,

$$G_f = \frac{\Delta\mathcal{W}^F_{1-2}}{b(a_2 - a_1)} \tag{3.4.3}$$

The accuracy of the result depends on the accuracy of the individual measurements, which may be controlled to some extent by adequate experimental design, but it also depends on the degree of accuracy of the hypotheses underlying the equation above. The method becomes inaccurate, even invalid, if the inelastic zone ahead of the crack tip is so large that the hypothesis of negligible inelastic zone is no longer acceptable.

Determination of how large the inelastic zone is, relative to the specimen dimensions, and how large its size must be to stay reasonably close to LEFM is, to a great extent, one of the objectives of the various inelastic fracture mechanics approaches that will be analyzed in the remaining chapters. At this stage, we only list the most obvious conditions that the experimental outputs should fulfill:

1. Deviation from linearity prior to the peak load should be small. This applies to specimens where the iso-$\mathcal{G}$ curves are monotonically decreasing as in Fig. 2.1.10a. The more rounded the peak, the farther the behavior is from LEFM. Quantitative criteria to ensure prepeak linearity can be formulated, similar to those previously given by ASTM E 399.
2. The $P-u$ curve after the peak should be an iso-$\mathcal{G}$ curve. The most direct way to check this point is to take various arcs 1-2, 2-3, 3-4, and so on, and calculate a value of G_f for each of these arcs. They should be equal if LEFM applies.
3. When unloading is performed, the unloading curve should be straight and unload to the origin. Deviation from this behavior indicates deviation from LEFM.

Exercises

3.8 Find the expression for the energy release rate of the structure in (a) Fig. 3.2.1a, (b) Fig. 3.2.1c.

3.9 A brittle material may contain planar voids. If these voids are similar to penny-shaped cracks, determine the maximum diameter of the voids which allow the material to be used up to 90 percent of its elastic limit. Complementary tests delivered values of 55 MPa for the yield strength and 16 kJ/m^2 for the fracture energy.

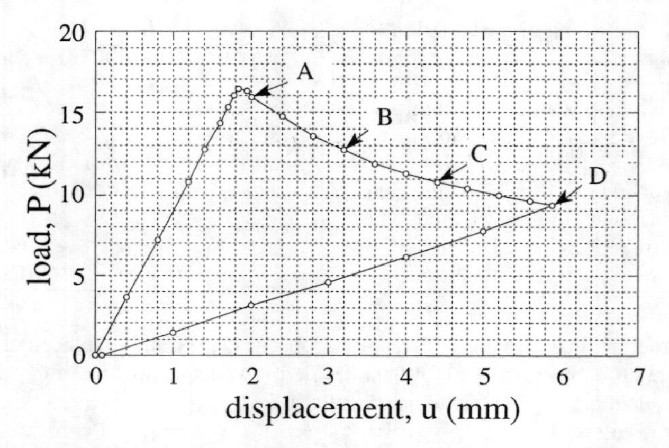

Figure 3.4.2 Test output in an experimental determination of G_f.

3.10 Use the triangular notch approximation of Planas and Elices to get the energy release rate of a center-cracked panel such as that depicted in Fig. 3.2.2. Show that for the limiting case $a/D \to 0$ the expression coincides with that obtained from the stress relief zone approximation in Section 3.2.2.

3.11 The results of the finite element calculation of Guinea (described in example 3.3.1) gave, for $a/D = 0.5$, the crack opening profile near the crack tip included in the table below, where r and w are, respectively, the distance to the crack tip and the crack opening (in appropriate length units; remember that in the computation $b = D = 1$, $P = 1$, $E = 1$ and plane stress was used). (a) Show that $K_I = \lim_{r \to 0} wE'\sqrt{\pi/32r}$. (b) Plot $wE\sqrt{\pi/32r}$ vs. r and get an estimate of K_I for the computed case. (c) For the usual definition of σ_N shown in Fig. 3.1.1, determine from the previous result an estimate of the shape factor $k(0.5)$. (d) Evaluate the error of the estimate in comparison with the more precise equations of example 3.1.1.

r	0.01	0.02	0.03	0.04	0.05	0.06	0.07	0.08	0.09
w	3.1669	4.5816	5.7928	6.8123	7.7424	8.6050	9.4201	10.198	10.947

3.12 Compliance tests have been performed on single-edge crack specimens of 25 mm thickness and 50 mm depth made of a material with an elastic modulus of $E = 3$ GPa. Nine specimens with crack lengths ranging between 23 and 27 mm were subjected to the same load, $P = 500$ N, and their displacement was measured. The results are shown in the following table. Give an estimate of the energy release rate, in J/m^2, for a specimen of this particular shape, size, and material with a crack length of 25 mm, for any load, P expressed in N.

a (mm)	23.1	23.6	23.9	24.4	24.9	25.6	25.8	26.5	27.0
u (μm)	270	291	289	310	317	347	348	369	397

3.13 A specimen geometrically similar to that in the preceding exercise has been loaded up to crack initiation. The specimen had a thickness of 10 mm and a depth of 100 mm, with an initial crack of 50 mm. The load at which the crack started to grow was 17.5 kN. Additional testing provided for the elastic modulus of that material, the value $E = 100$ GPa. Estimate the fracture energy under the assumption that LEFM applies.

3.14 A double-cantilever beam specimen subjected to opposite point loads at the ends of its arms has been tested under displacement control. The specimen thickness was 50 mm and the arm depth 30 mm. The resulting load-displacement curve is shown in Fig. 3.4.2b. If the initial crack length was 150 mm, (a) give an estimate of the elastic modulus of the material. (b) Give an estimate of G_f using the peak load value and the initial crack length. (c) Give estimates of the crack length at points A, B, C, and D. (d) Give average values of the crack growth resistance over arcs AB, BC, and DE. (e) Decide whether LEFM is a reasonably good approximation, and, if it is, give a final estimate for G_f.

3.5 Calculation of Displacements from K_I-Expressions

3.5.1 Calculation of the Displacement

The procedure to obtain the displacement from the expression of K_I as a function of the crack depth is, in fact, the reverse of that used in Section 2.1 to obtain $\mathcal{G}$ from the expression of the compliance. To obtain the compliance (from which the displacement follows trivially), we couple (2.1.32) with Irwin's relationship (2.2.22) to get

$$\frac{1}{2b}P^2\frac{dC(a)}{da} = \frac{K_I^2}{E'} \tag{3.5.1}$$

This provides the basic equation to solve for the compliance, which may be simplified by using the general expression (2.3.11) for K_I:

$$\frac{dC(a)}{da} = \frac{2}{bDE'}\hat{k}^2(\alpha) \tag{3.5.2}$$

This equation may be integrated between the limits for no crack, for which the compliance is C_0, and an arbitrary crack length a:

$$C(a) = C_0 + \frac{2}{bDE'}\int_0^a \hat{k}^2(\alpha)da = C_0 + \frac{2}{bE'}\int_0^{a/D} \hat{k}^2(\alpha)\, d\alpha \tag{3.5.3}$$

where the second expression follows by setting $da = D\, d\alpha$.

Thus, setting $u = CP$, the displacement can be written as

$$u = \frac{P}{bE'}\hat{v}(\alpha), \qquad \alpha = \frac{a}{D} \tag{3.5.4}$$

The dimensionless function $\hat{v}(\alpha)$ is given by

$$\hat{v}(\alpha) = \hat{v}_0 + \hat{v}^c(\alpha), \qquad \hat{v}_0 = u_0\frac{bE'}{P}, \qquad \hat{v}^c(\alpha) = 2\int_0^\alpha \hat{k}^2(\alpha')\, d\alpha' \tag{3.5.5}$$

in which u_0 is the elastic displacement of the structure in the absence of crack and $\hat{v}^c(\alpha)$ is the additional displacement due to the crack.

Often the displacement is expressed in terms of the nominal strength $\sigma_N = c_N P/bD$ instead of the load P. Making the substitution and taking into account the definitions of functions $\hat{k}(\alpha)$ and $k(\alpha)$ given in and below (2.3.11), one gets

$$u = \frac{\sigma_N}{E'}Dv(\alpha), \qquad \alpha = \frac{a}{D} \tag{3.5.6}$$

The dimensionless function $v(\alpha)$ is given by

$$v(\alpha) = v_0 + v^c(\alpha), \qquad v_0 = u_0\frac{E'}{\sigma_N D}, \qquad v^c(\alpha) = 2\int_0^\alpha k^2(\alpha')\, d\alpha' \tag{3.5.7}$$

in which, again, u_0 is the elastic displacement the structure would experiment if uncracked. Note that the equations are formally the same as before, except that the functions labeled by hats (which we always use with P) are replaced by functions without hats, representing variables expressed in terms of σ_N.

Example 3.5.1 Consider a large plate with a very short edge crack ($a \ll D$) subjected to remote stress σ (Fig. 3.5.1). The approximate expression of the stress intensity factor is $K_I = 1.1215\sigma\sqrt{\pi a}$, which may be rewritten as $K_I = \sigma\sqrt{D}1.1215\sqrt{\pi\alpha}$. Therefore, taking $\sigma_N = \sigma$ we have

$$k(\alpha) = 1.1215\sqrt{\pi\alpha} \tag{3.5.8}$$

Thus, from the last equation of (3.5.7), we get

$$v^c(\alpha) = 2 \times 1.258\pi\int_0^\alpha \alpha'\, d\alpha' = 1.258\pi\alpha^2 \tag{3.5.9}$$

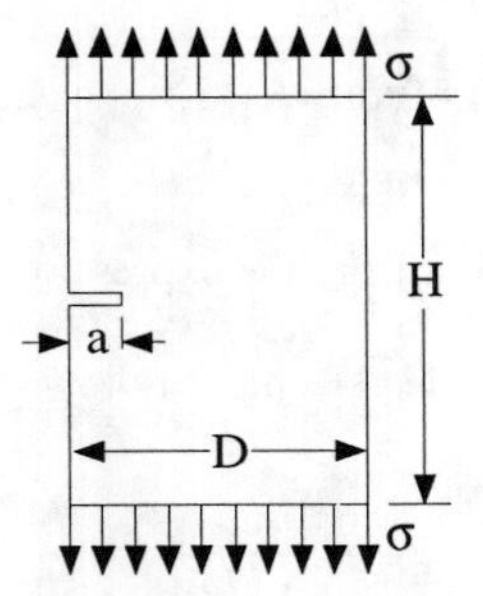

Figure 3.5.1 Single-edge cracked panel subjected to remote uniaxial stress.

Since the displacement for an uncracked panel is $u_0 = \sigma H/E$, the second equation of (3.5.7) gives $v_0 = H/D$. Therefore, the total displacement is

$$u = \frac{\sigma_N}{E'} D \left(\frac{H}{D} + 1.258 \pi \alpha^2 \right) \tag{3.5.10}$$

Identical results are obtained if calculations are done in terms of the resultant load $P = \sigma b D$. ▯

The foregoing expressions hold for a single-tipped crack, and a stands for the total crack length. For the center cracked panel, or more generally for internal cracks, where the total crack length is customarily represented by $2a$, the matters become a little more complex in the case of loadings that are not symmetric with respect to the axis normal to the crack. In such a case, the energy release rates are different at one and the other tip. This means that in reality $\partial \mathcal{U}^*/\partial a = \mathcal{G}^+ + \mathcal{G}^-$ where the superscripts $+$ and $-$ refer to the right and left tips, respectively. Therefore, we must also distinguish the stress intensity factors K_I^+ and K_I^- and their associated shape factors $k^+(\alpha)$ and $k^-(\alpha)$. A development strictly parallel to the previous one for a single-tipped crack leads to identical expressions for the displacement, except that function $v^c(\alpha)$ is now given by

$$v^c(\alpha) = 2 \int_0^\alpha \left[k^{+^2}(\alpha') + k^{-^2}(\alpha') \right] d\alpha' \tag{3.5.11}$$

and similarly for the functions with hats. This shows that, for symmetric loadings ($k^+ = k^- = k$), the double-tip case reduces to the single-tip one just by replacing $d\alpha$ by $2d\alpha$. But for nonsymmetric loading, the expression is quite different.

3.5.2 Compliances, Energy Release Rate, and Stress Intensity Factor for a System of Loads

To calculate arbitrary displacements from K_I expressions, it is convenient to establish a more general framework in which a system of independent forces P_i ($i = 1, \ldots, n$) is assumed to act on a cracked elastic body. Let the displacement of the load-point i in the direction of P_i be u_i. The displacements may be written as linear functions of the loads:

$$u_i = \sum_{j=1}^{n} C_{ij}(a) P_j \qquad i = 1, \ldots, n \tag{3.5.12}$$

where $C_{ij}(a)$ are the elements of the compliance matrix (which depend upon the crack length a). As a consequence of the reciprocity theorem, the compliance matrix is symmetric: $C_{ij} = C_{ji}$.

The elastic and complementary energies are equal, and are given by $\mathcal{U} = \mathcal{U}^* = \sum_{i=1}^{n} P_i u_i/2$. Substituting (3.5.12), the complementary energy is obtained as

$$\mathcal{U}^* = \frac{1}{2} \sum_{i=1}^{n} \sum_{j=1}^{n} C_{ij}(a) P_i P_j \tag{3.5.13}$$

From this expression and (2.1.21), we obtain the following expression for the energy release rate:

$$\mathcal{G} = \frac{1}{2b}\sum_{i=1}^{n}\sum_{j=1}^{n} P_i P_j \frac{dC_{ij}(a)}{da} \tag{3.5.14}$$

For a general system of loads, this equation is the equivalent of (2.1.34) .

For the stress intensity factor, we may apply the superposition principle and write

$$K_I = \sum_{i=1}^{n} K_{Ii} = \frac{1}{b\sqrt{D}}\sum_{i=1}^{n} P_i\, \hat{k}_i(\alpha) \tag{3.5.15}$$

where $K_{Ii} = (P_i/b\sqrt{D})\hat{k}_i(\alpha)$ is the stress intensity factor due to P_i alone. Now, according to Irwin's equation, $\mathcal{G} = K_I^2/E'$, and so

$$\frac{1}{2b}\sum_{i=1}^{n}\sum_{j=1}^{n}\frac{dC_{ij}(a)}{da}P_iP_j = \frac{1}{b^2DE'}\sum_{i=1}^{n}\sum_{j=1}^{n}\hat{k}_i(\alpha)\hat{k}_j(\alpha)P_iP_j \tag{3.5.16}$$

The values of the P_is are arbitrary. So for the equality to hold for any P_i, the coefficients of the products P_iP_j on both sides of the equation must be identical. We thus find that

$$\frac{dC_{ij}(a)}{da} = \frac{2}{bDE'}\hat{k}_i(\alpha)\hat{k}_j(\alpha) \qquad (i,\, j = 1,\, \ldots, n) \tag{3.5.17}$$

which is the generalization of (3.5.2) to any system of forces. This equation can be integrated in the same way as before, to obtain

$$C_{ij}(a) = C_{ij0} + \frac{2}{bE'}\int_0^{a/D}\hat{k}_i(\alpha)\hat{k}_j(\alpha)d\alpha \qquad (i,\, j = 1,\, \ldots, n) \tag{3.5.18}$$

where C_{ij0} is the component of the compliance matrix for the uncracked body ($a = 0$). This equation provides the means of obtaining the displacements u_j at various points caused by only one force, say P_1 (all the remaining forces being zero). The result is, obviously, $u_j = C_{j1}P_1$.

The foregoing equation can be recast in terms of the full expression of the stress intensity factor, by setting that $\hat{k}(\alpha) = b\sqrt{D}\,K_{Ii}/P_i$ and thus

$$C_{ij}(a) = C_{ij0} + \frac{2b}{E'}\int_0^{a}\frac{K_{Ii}K_{Ij}}{P_iP_j}da \qquad (i,\, j = 1,\, \ldots, n) \tag{3.5.19}$$

which does not require a particular form of expressing the stress intensity factor, and can be directly used when P_i are generalized forces rather than point loads.

Note, again, that in the foregoing expressions a stands for the total crack length, and the energy release rates correspond to a single crack tip. For the center cracked panel, the two crack tips must be made explicit, as previously done for the single force loading, Eq. (3.5.11). The general expression for multiple loading is:

$$C_{ij}(a) = C_{ij0} + \frac{2}{bE'}\int_0^{a/D}\left[\hat{k}_i^+(\alpha)\hat{k}_j^+(\alpha) + \hat{k}_i^-(\alpha)\hat{k}_j^-(\alpha)\right]d\alpha \qquad i,\, j = 1,\, \ldots, n \tag{3.5.20}$$

where $\hat{k}_i^+(\alpha)$ and $\hat{k}_i^-(\alpha)$ are the shape factors for the stress intensity factor created by load P_i at the right and at the left crack tips, respectively.

3.5.3 Calculation of the Crack Mouth Opening Displacement

As an important application of the foregoing general result, let us calculate the expression for the crack mouth opening displacement (CMOD), which we denote as w_M (Fig. 3.5.2). Aside from the actual load P, we also consider a virtual loading P_M consisting in a pair of forces at the crack mouth working

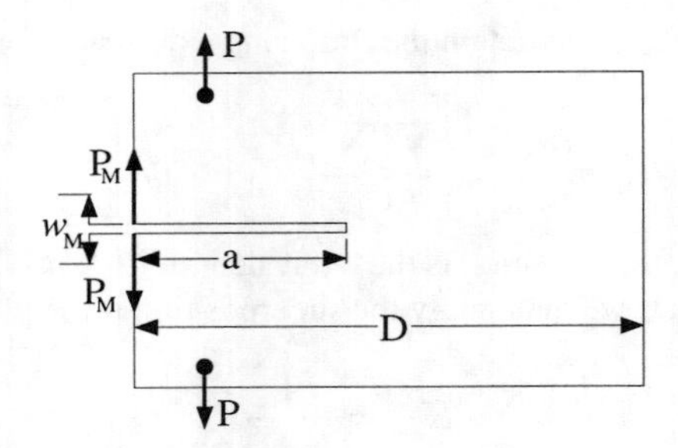

Figure 3.5.2 Crack mouth opening w_M, applied load P, and crack mouth load P_M.

through w_M. In the previous expressions we now have $n = 2$, and we set $P_1 \equiv P$, $P_2 \equiv P_M$, $u_1 \equiv u$, $u_2 \equiv w_M$, $C_{11}(a) \equiv C(a)$, and $C_{12}(a) \equiv C_M(a)$ so that we write the displacements as

$$u = C(a)\,P + C_M(a)\,P_M \tag{3.5.21}$$

$$w_M = C_M(a)\,P + C_{MM}(a)\,P_M \tag{3.5.22}$$

We also set $\hat{k}_1(\alpha) \equiv \hat{k}(\alpha)$ and $\hat{k}_2(\alpha) \equiv \hat{k}_M(\alpha)$ for the shape factors corresponding to forces P and P_M. Noting that $C_{M0} \equiv C_{120} = 0$ (because when the crack length is zero, the crack opening is also zero), the cross-compliance for the CMOD, $C_M(a)$, follows from (3.5.18):

$$C_M(a) = \frac{2}{bE'} \int_0^{a/D} \hat{k}(\alpha)\hat{k}_M(\alpha) d\alpha \tag{3.5.23}$$

Thus, according to (3.5.22), the crack mouth opening displacement when the structure is loaded by P alone is

$$w_M = \frac{P}{bE'}\hat{v}_M(\alpha)\,, \quad \hat{v}_M(\alpha) = 2\int_0^{\alpha} \hat{k}(\alpha')\hat{k}_M(\alpha')d\alpha' \tag{3.5.24}$$

Again, this can be expressed in terms of σ_N instead of P; the result is

$$w_M = \frac{\sigma_N}{E'} D v_M(\alpha)\,, \quad v_M(\alpha) = 2\int_0^{\alpha} k(\alpha')\hat{k}_M(\alpha')d\alpha' \tag{3.5.25}$$

where we notice again that the expression is identical to the previous one except that the hat is removed for $k(\alpha)$ (but not for $\hat{k}_M(\alpha)$).

Example 3.5.2 Consider again the plate of Fig. 3.5.1. When a pair of loads P_M is applied to the crack mouth (as shown in Fig. 3.5.2), the corresponding stress intensity factor is expressed as $K_I = 2.594 P_M / b\sqrt{\pi a}$ (Ouchterlony 1975; also Tada, Paris and Irwin 1985). This can be rewritten as $K_I = (P_M / b\sqrt{D}) 2.594/\sqrt{\pi\alpha}$. Therefore, the shape function $\hat{k}_M(\alpha)$ is

$$\hat{k}_M(\alpha) = \frac{2.594}{\sqrt{\pi\alpha}} \tag{3.5.26}$$

Substituting this and (3.5.8) into (3.5.25) we get the CMOD:

$$w_M = \frac{\sigma}{E'} D 2 \int_0^{\alpha} 2.909\, d\alpha' = 5.818 \frac{\sigma}{E'} D\alpha = 5.818 \frac{\sigma}{E'} a \tag{3.5.27}$$

which is the expression found, for example, in Tada, Paris and Irwin (1985). □

3.5.4 Calculation of the Volume of the Crack

A further interesting application of the general relation (3.5.18) is the determination of the volume of the opened crack. To this end, we must consider a loading which is work-conjugate to volume, i.e., such that the work is expressible as the product of the conjugate generalized force with the variation of crack volume dV. Such a loading is a uniform internal pressure p over the crack faces; then p and V are conjugate variables and can be used directly in the energetic expressions. It is now necessary to adjust (3.5.18), because the dimensions of the variables are different. This is fairly easy and is left to the reader as an exercise. Here we execute a simple trick to be able to use (3.5.18) as it is: we define a generalized force P_V from p, and its associated displacement $u_v \propto V$ so that $P_V u_V = pV$ and the dimensions be those of force and length:

$$P_V = pbD \qquad \text{and} \qquad u_V = \frac{V}{bD} \tag{3.5.28}$$

(Note that the crack length a must *not* appear in the definition of P_v and u_V.) Working now as in our calculation of the CMOD, we have $n = 2$, and set $P_1 \equiv P$, $P_2 \equiv P_V$, $u_1 \equiv u$, $u_2 \equiv u_V$, $C_{11}(a) \equiv C(a)$, and $C_{12}(a) \equiv C_V(a)$. So we may write the displacements as

$$u = C(a)\, P + C_V(a)\, P_V \tag{3.5.29}$$

$$u_v = C_V(a)\, P + C_{VV}(a)\, P_V \tag{3.5.30}$$

We also set $\hat{k}_1(\alpha) \equiv \hat{k}(\alpha)$ for the shape factor corresponding to force P, and $\hat{k}_2(\alpha) \equiv \hat{k}_V(\alpha)$ for the one corresponding to the internal pressure — in which the stress intensity factor must be written in the form $K_I = (P_V/b\sqrt{D})\hat{k}_V(\alpha)$. Thus, the cross-compliance for u_V, $C_V(a)$, follows from (3.5.18) with $C_{V0} \equiv C_{120} = 0$ (because when the crack length is zero, the crack volume is also zero):

$$C_V(a) = \frac{2}{bE'} \int_0^{a/D} \hat{k}(\alpha)\hat{k}_V(\alpha)d\alpha \tag{3.5.31}$$

The crack volume follows from (3.5.30) and (3.5.28):

$$V = bDu_V = bDC_V P = \frac{PD}{E'}\hat{v}_V(\alpha) \tag{3.5.32}$$

where

$$\hat{v}_V(\alpha) = 2\int_0^{\alpha} \hat{k}(\alpha')\hat{k}_V(\alpha')d\alpha' \tag{3.5.33}$$

If we write V in terms of σ_N instead of P, we get

$$V = \frac{\sigma_N}{E'} bD^2 v_V(\alpha)\,, \qquad v_V(\alpha) = 2\int_0^{\alpha} k(\alpha')k_V(\alpha')d\alpha' \tag{3.5.34}$$

where $k(\alpha')$ is the shape factor defined in (2.3.11) and $k_V(\alpha')$ is defined so that the stress intensity factor created by a uniform pressure inside the crack is written as $K_I = p\sqrt{D}k_V(\alpha)$.

Example 3.5.3 Consider again the plate of Fig. 3.5.1. When a uniform pressure p is applied to the crack faces, the superposition sketched in Fig. 3.5.3 shows that the stress intensity factor is identical to that corresponding to a remote uniaxial stress $\sigma = p$. The corresponding stress intensity factor is $K_I = 1.1215p\sqrt{\pi a}$, and so the shape function $k_V(\alpha)$ is

$$k_V(\alpha) = 1.1215\sqrt{\pi\alpha} \tag{3.5.35}$$

Substituting this and (3.5.8) in (3.5.34), we get the crack volume:

$$V = \frac{\sigma}{E'} bD^2 2\int_0^{\alpha} 1.258\pi\alpha'\, d\alpha' = \frac{\sigma}{E'} bD^2\alpha^2 = \frac{\sigma}{E'} 1.258\pi ba^2 \tag{3.5.36}$$

where we wrote $D\alpha = a$. □

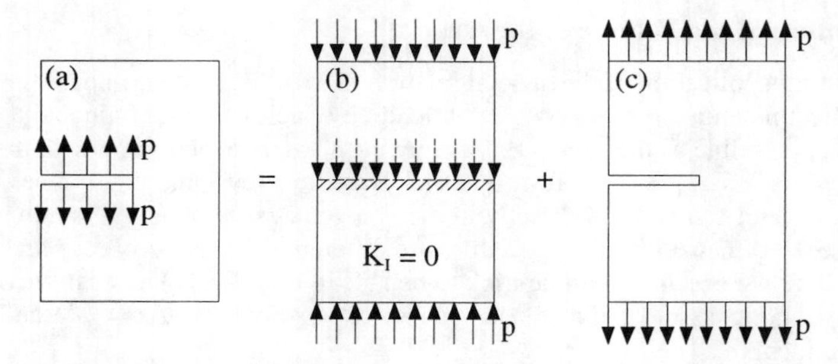

Figure 3.5.3 Stress intensity factors for internal pressure p and remote uniaxial stress p are identical.

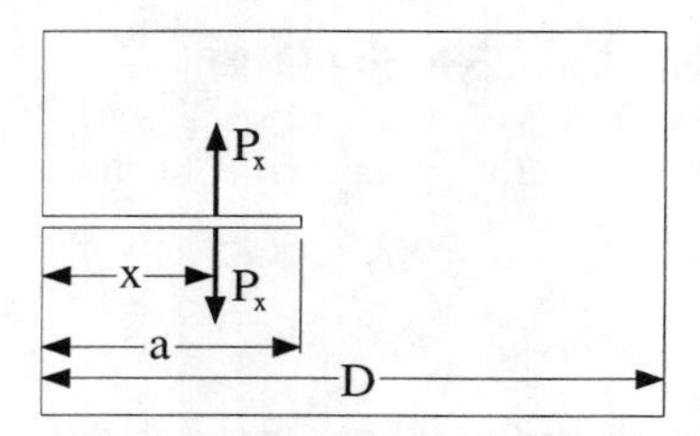

Figure 3.5.4 Virtual loading used in the computation of the crack opening profile.

For an internal crack of length $2a$, the results are similar, except that the shape factors at both tips of the crack appear explicitly, as in previous sections. Then the expression for $v_V(\alpha)$ in (3.5.34) must be replaced by

$$v_V(\alpha) = 2\int_0^\alpha \left[k^+(\alpha')k_V^+(\alpha') + k^+(\alpha')k_V^-(\alpha')\right] d\alpha' \tag{3.5.37}$$

3.5.5 Calculation of the Crack Opening Profile

To obtain the crack opening profile from K_I expressions, we need to know the stress intensity factor produced by a pair of point loads at an arbitrary position x along the crack. Let P_x be the magnitude of the loads located at x (Fig 3.5.4). The displacement conjugate of the pair P_x is the crack opening at point x, $w(x)$. The stress intensity factor may be written as in (3.1.10), where now it is essential to remark that when the point loads are applied on the uncracked part of the crack plane (i.e., when $x > a$) they do not generate any stress intensity factor at the crack tip, so that we can write

$$k_G(\alpha, x/D) = 0 \quad \text{for} \quad x > a \ (\text{or } x/D > \alpha) \tag{3.5.38}$$

Then we proceed as we did in calculating the CMOD. We first write the displacements as

$$u = C(a)\,P + C_x(a)\,P_x \tag{3.5.39}$$

$$w(x) = C_x(a)\,P + C_{xx}(a)\,P_x \tag{3.5.40}$$

The compliance $C_x \equiv C_{12}$ is obtained from (3.5.18) with $\hat{k}_1(\alpha) = \hat{k}(\alpha)$, $\hat{k}_2(\alpha) = k_G(\alpha, x/D)$, and $C_{120} = 0$:

$$C_x(a) = \frac{2}{bE'}\int_0^{a/D} \hat{k}(\alpha)k_G(\alpha, x/D)d\alpha \tag{3.5.41}$$

Now, taking into account (3.5.38) and (3.5.40), the crack opening profile is obtained as

$$w(x,a) = \frac{P}{bE'}\hat{v}_x(\alpha)\,, \qquad \hat{v}_x(\alpha) = 2\int_{x/D}^{\alpha} \hat{k}(\alpha')k_G(\alpha', x/D)d\alpha' \tag{3.5.42}$$

where it is understood that this equation is valid for $x \le a$, and that, obviously, $w(x,a) = 0$ for $x \ge a$. The foregoing equation can be rewritten in terms of σ_N as

$$w(x,a) = \frac{\sigma_N}{E'} D v_x(\alpha)\,, \qquad v_x(\alpha) = 2\int_{x/D}^{\alpha} k(\alpha')k_G(\alpha', x/D)d\alpha' \tag{3.5.43}$$

where, again, $k(\alpha)$ is defined in (2.3.11).

For an internal crack for which a is the half crack length, the previous adjustment for two crack tips must be performed again. The result is identical to (3.5.43) except that $v(\alpha)$ is now replaced by

$$v_x(\alpha) = 2\int_{x/D}^{\alpha} \left[k^+(\alpha')k_G^+(\alpha', x/D) + k^-(\alpha')k_G^-(\alpha', x/D)\right] d\alpha' \tag{3.5.44}$$

Example 3.5.4 Consider the center cracked panel of Fig. 2.1.1 subjected to remote uniaxial stress σ, and assume the crack to be very small relative to the dimensions of the panel. To obtain the crack opening profile we use the stress intensity factor for a pair of point loads on the crack faces introduced in Example 3.1.7 (Fig. 3.1.6c), with $k_G^+(\alpha, x/D)$ given Eq. (3.1.13). Because of the symmetry, the shape factor for the tip on the left is $k_G^-(\alpha, x/D) = k_G^+(\alpha, -x/D)$. Introducing this and $k^+(\alpha) = k^-(\alpha) = \sqrt{\pi\alpha}$ into (3.5.44), and substituting the result in the first of (3.5.43), we get

$$w(x,a) = \frac{\sigma}{E'} D \int_{x/D}^{\alpha} \frac{4\alpha'}{\sqrt{\alpha'^2 - (x/D)^2}} d\alpha' = \frac{4\sigma}{E'}\sqrt{a^2 - x^2} \tag{3.5.45}$$

which does coincide with the solution obtained by the complete elastic analysis of the problem; see Section 2.2.1. ▯

3.5.6 Bueckner's Expression for the Weight Function

Bueckner (1970) devised a procedure to obtain the weight function $k_G(\alpha, x/D)$ from the solution for the stress intensity factor and crack opening profile for arbitrary loading. This is the method exploited in the book by Wu and Carlsson (1991). Here it suffices to exploit (3.5.42) for demonstrating the simplest version of Bueckner's result. Differentiating that equation with respect to a (and keeping in mind that $\alpha = a/D$), we get:

$$\frac{\partial w(x,a)}{\partial a} = \frac{\sigma_N}{E'} D \frac{\partial \hat{v}_x(\alpha)}{\partial \alpha}\frac{1}{D} = \frac{\sigma_N}{E'} 2k(\alpha)k_G(\alpha, x/D) = \frac{2K_I}{E'\sqrt{D}} k_G(\alpha, x/D) \tag{3.5.46}$$

where in the last expression we substituted (2.3.11), i.e., $K_I = \sigma_N \sqrt{D} k(\alpha)$. Solving for $k_G(\alpha, x/D)$, we get

$$k_G(\alpha, x/D) = \frac{E'\sqrt{D}}{K_I(a)} \frac{\partial w(x,a)}{2\partial a} \tag{3.5.47}$$

The arguments a and x have been made explicit for clarity. Note that, compared to the expressions in other texts, our analysis is limited to pure mode I (structures and loadings symmetric with respect to the crack plane), and so the half crack opening $w/2$ is equal to half the displacement of the upper face, which is the usual variable included in the weight function expressions. We use the crack opening rather than the displacement of one crack face because in the following chapters the crack opening is the essential variable.

Consider now the center-cracked panel. Since it has two crack tips, we cannot get both weight functions from a single loading. This is clear from (3.5.44) where two unknowns are present, namely $k_G^+(\alpha, x/D)$ and $k_G^-(\alpha, x/D)$. In particular, from the solution for a symmetric loading for which $\hat{k}_s^+(\alpha) = \hat{k}_s^-(\alpha) =$

$\hat{k}_s(\alpha)$, we can find only the symmetric part of the weight function. Indeed, proceeding as before, we get

$$k_G^+(a/D, x/D) + k_G^-(a/D, x/D) = \frac{E'\sqrt{D}}{K_{Is}(a)} \frac{\partial w_s(x,a)}{2\partial a} \tag{3.5.48}$$

where subscript s indicates that the loading corresponding to this solution must be symmetric. The symmetric part of k_G is all that is needed to obtain further crack opening profiles for other symmetric loadings. To obtain the complete expression for the right and left weight functions, we also need to solve the antisymmetric case, for which $\hat{k}_a^+(\alpha) = -\hat{k}_a^-(\alpha)$. If this solution is available, it is easy to find the antisymmetric part as

$$k_G^+(a/D, x/D) - k_G^-(a/D, x/D) = \frac{E'bD}{P_a \hat{k}_a^+(a/D)} \frac{\partial w_a(x,a)}{2\partial a} = \frac{E'\sqrt{D}}{K_{Ia}^+(a)} \frac{\partial w_a(x,a)}{2\partial a} \tag{3.5.49}$$

where subscript a refers to antisymmetric loading. Combining (3.5.49) and (3.5.48), one can easily obtain the expression for the weight functions corresponding to both crack tips.

Exercises

3.15 The stress intensity factor for a center cracked strip of width D, with a crack of length $2a$ subjected to remote uniaxial stress, may be approximated by the Feddersen-Tada expression (3.1.4) within 0.1%. Write the equation for the additional compliance of the strip due to the crack. Take two terms of the series expansion of the integrand in powers of a/D, calculate the additional compliance, and estimate the values a/D for which this result is accurate within 2%.

3.16 For the panel in Exercise 3.4, find the volume of hydraulic fluid injected into the jack for given crack length $2a$ and pressure p, assuming the fluid to be incompressible. Hints: Define $P = 2pbc$ and $u = V/2bc$. Watch the integration limits. Integrate by parts twice.

3.17 Find the volume of a centrally located crack in a large panel subjected to equal and opposite normal forces at the crack center.

3.18 Find the crack opening profile of a centrally located crack in a large panel subjected to equal and opposite normal forces at the crack center. Note the logarithmic singularity at $x = 0$.

4

Advanced Aspects of LEFM

In this chapter, we summarize some advanced topics in LEFM that were not covered in depth in the preceding chapters. First, we present the theoretical framework to analytically handle plane elasticity problems with cracks. Emphasis is put on the understanding of various methods of solution, such as the complex potentials (expounded in Section 4.1), Westergaard stress functions (presented in Section 4.2), and Airy stress functions (developed as exercises). The presentation does not aim at complete, formal presentations (for this purpose, see, e.g., England 1971). Neither it aims at teaching the skills to obtain the solution from scratch. It only aims at facilitating insight into the use of complex potentials and Westergaard stress functions to obtain stress and displacement fields. As a basic example, these methods are applied to the analysis of the infinite center-cracked panel.

The complex potentials are next used to analyze the near-tip fields (Section 4.3). The in-plane case, involving fracture modes I (pure opening) and II (in-plane shear), is discussed first. Then the formalism to handle the antiplane case of mode III (antiplane shear) is introduced and the general antiplane stress and displacement near-tip fields are obtained.

The next topic covered is that of the path-independent integrals, of which the J–integral is the most important. Section 4.4 shows formally that Rice's J–integral is path-independent under certain assumptions; it introduces a further path-independent integral for the LEFM case which is based on the reciprocity theorem and is used to provide another derivation of Irwin's relation. Finally, other path independent integrals are briefly discussed (I–, J_k–, L–, and M–integrals).

The last section deals with the topic of mixed mode fracture in LEFM. The existing fracture criteria are briefly described, with emphasis put on the single-parameter models, especially the maximum principal stress criterion (Erdogan and Sih 1963).

4.1 Complex Variable Formulation of Plane Elasticity Problems

4.1.1 Navier's Equations for the Plane Elastic Problem

We take axes x_1, x_2 lying in the plane of the structure, and axis x_3 perpendicular to it. Plane states always require $\sigma_{13} = \sigma_{23} = 0$, while $\sigma_{33} = 0$ in generalized plane stress, and $\varepsilon_{33} = 0$ in plane strain.

Restricting attention to the in-plane components of vectors and tensors (i.e., restricting indices to values 1 and 2), the equilibrium equation for negligible body forces are reduced to state that the 2D divergence of the stress tensor must vanish:

$$\sigma_{ij,j} = 0 \tag{4.1.1}$$

where subscript $_{,j}$ implies partial derivative with respect to the corresponding cartesian coordinate (i.e., $f_{,j} = \partial f / \partial x_j$). Repeated indices imply summation over $i = 1, 2$.

The plane version of Hooke's law may be reduced to (see, for example, Malvern 1969):

$$\sigma_{ij} = \lambda' \varepsilon_{kk} \delta_{ij} + 2G\varepsilon_{ij} \tag{4.1.2}$$

where G is the shear modulus and λ' is an effective plane Lamé constant. These elastic constants can be written as

$$G = \frac{E}{2(1+\nu)}, \qquad \lambda' = \frac{E'\nu'}{(1-\nu'^2)} \tag{4.1.3}$$

where E' and ν' are, respectively, the effective elastic modulus and the effective Poisson's ratio, defined in terms of the elastic modulus E and Poisson's ratio ν as

$$E' = E\,, \qquad \nu' = \nu \qquad \text{for plane stress} \tag{4.1.4}$$

$$E' = \frac{E}{(1-\nu^2)}\,, \qquad \nu' = \frac{\nu}{(1-\nu)} \qquad \text{for plane strain} \tag{4.1.5}$$

Writing the strain tensor as the symmetric part of the gradient of the displacement vector, $\varepsilon_{ij} = \frac{1}{2}(u_{i,j} + u_{j,i})$, and substituting it into the stress-strain law (4.1.2) and the result in the equilibrium equation (4.1.1), the Navier's equations for plane elasticity are found:

$$(\lambda' + G)u_{k,ki} + Gu_{i,kk} = 0 \tag{4.1.6}$$

which may be found in the literature in the compact notation forms

$$(\lambda' + G)\text{grad}(\text{div}\,\mathbf{u}) + G\text{div}(\text{grad}\,\mathbf{u}) = 0 \tag{4.1.7}$$

or

$$(\lambda' + G)\nabla(\nabla\cdot\mathbf{u}) + G\,\nabla^2\mathbf{u} = 0 \tag{4.1.8}$$

where ∇ is the gradient operator and $\nabla^2 = \nabla\cdot\nabla$ is the Laplace operator.

4.1.2 Complex Functions

One effective way to solve plane elasticity problems is to use complex variables. The (x_1, x_2) plane is projected into the complex plane by defining the complex variable z as $z = x_1 + ix_2$ where i is the imaginary unit, $i^2 = -1$. A complex displacement, υ, is similarly defined as $\upsilon = u_1 + iu_2$.

In general, the displacement field is given by expressing u_1 and u_2, hence, the complex variable υ, as a function of the real variables x_1 and x_2. To write them in terms of the complex variable z, it is convenient to introduce z^*, the complex conjugate of z: $z^* = x_1 - ix_2$, so that $x_1 = (z + z^*)2$ and $x_2 = (z - z^*)/2i$. In this way, an arbitrary complex function of (x_1, x_2) is written as an arbitrary complex function of (z, z^*), and to find the complete displacement field, we are bound to look for a complex displacement function $\upsilon(z, z^*)$.

Let $f(z, z^*)$ be an arbitrary complex function. The partial derivatives with respect to the real variables (x_1, x_2) are obtained following standard differentiation rules as

$$f_{,k} = \frac{\partial f(z,z^*)}{\partial z} z_{,k} + \frac{\partial f(z,z^*)}{\partial z^*} z^*_{,k} \tag{4.1.9}$$

and since, trivially, $z_{,1} = 1$, $z_{,2} = i$, $z^*_{,1} = 1$ and $z^*_{,1} = -i$, it follows that

$$f_{,1} = \frac{\partial f(z,z^*)}{\partial z} + \frac{\partial f(z,z^*)}{\partial z^*} \tag{4.1.10}$$

$$f_{,2} = i\left[\frac{\partial f(z,z^*)}{\partial z} - \frac{\partial f(z,z^*)}{\partial z^*}\right] \tag{4.1.11}$$

These equations may obviously be solved for $\partial f/\partial z$ and $\partial f/\partial z^*$ to get

$$\frac{\partial f}{\partial z} = \frac{f_{,1} - if_{,2}}{2} \tag{4.1.12}$$

$$\frac{\partial f}{\partial z^*} = \frac{f_{,1} + if_{,2}}{2} \tag{4.1.13}$$

4.1.3 Complex Form of Hooke's and Navier's Equations

We will need the complex counterpart of Hooke's and Navier's equations in terms of the complex displacement v. To this end, after taking the real part of (4.1.12, we apply Eqs. (4.1.12) and (4.1.13) with $f(z, z^*) = v(z, z^*)$ to get the complex equations for the strain components:

$$u_{1,1} + u_{2,2} = u_{k,k} = \varepsilon_{kk} = \frac{\partial v}{\partial z} + \frac{\partial v^*}{\partial z^*} \tag{4.1.14}$$

$$u_{1,1} - u_{2,2} + i(u_{2,1} + u_{1,2}) = \varepsilon_{11} - \varepsilon_{22} + 2i\varepsilon_{12} = 2\frac{\partial v}{\partial z^*} \tag{4.1.15}$$

With this, Hooke's equations (4.1.2) are written in terms of the complex displacement as

$$\sigma_{11} + \sigma_{22} = 2(\lambda' + G)\left[\frac{\partial v}{\partial z} + \frac{\partial v^*}{\partial z^*}\right] \tag{4.1.16}$$

$$\sigma_{11} - \sigma_{22} + 2i\sigma_{12} = 4G\frac{\partial v}{\partial z^*} \tag{4.1.17}$$

The two component equations implied in Navier's (4.1.6) for $i = 1$ and 2 are reduced to a single complex equation by multiplying the second equation by i and adding it to the first:

$$(\lambda' + G)(u_{k,k1} + iu_{k,k2}) + Gv_{,kk} = 0 \tag{4.1.18}$$

The first term of this equation is transformed by setting $f = u_{k,k}$ in Eqs. (4.1.10) and (4.1.11), multiplying the second by i, and adding the two equations. The $v_{,kk}$ is expressed in terms of the partial derivatives with respect to z and z^* by applying (4.1.12) with $f = v$ and then (4.1.13) to the result. The ensuing complex variable version of the Navier's equation is

$$(\lambda' + G)\frac{\partial}{\partial z^*}\left[\frac{\partial v}{\partial z} + \frac{\partial v^*}{\partial z^*}\right] + 2G\frac{\partial^2 v}{\partial z^* \partial z} = 0 \tag{4.1.19}$$

or

$$\frac{\partial}{\partial z^*}\left[(\lambda' + 3G)\frac{\partial v}{\partial z} + (\lambda' + G)\frac{\partial v^*}{\partial z^*}\right] = 0 \tag{4.1.20}$$

4.1.4 Integration of Navier's Equation: Complex Potentials

It is possible to obtain a closed-form general solution of the foregoing Navier equation in terms of two arbitrary holomorphic functions called the complex potentials. A complex function $M(z)$ is called holomorphic if, in addition to being single valued and differentiable over the region of interest, it verifies

$$\partial M/\partial z^* = 0, \tag{4.1.21}$$

i.e., it is independent of z^*, so it can be expressed in terms of z alone. For holomorphic functions the partial derivative with respect to z becomes, then, a total derivative, which is denoted as

$$\frac{\partial M(z)}{\partial z} = \frac{dM(z)}{dz} = M'(z) \tag{4.1.22}$$

Hence, (4.1.20) simply states that the function in the square brackets is holomorphic. We thus write:

$$(\lambda' + 3G)\frac{\partial v}{\partial z} + (\lambda' + G)\frac{\partial v^*}{\partial z^*} = M(z) \tag{4.1.23}$$

where $M(z)$ is an arbitrary holomorphic function.

Next, the term $\partial v^*/\partial z^*$ may be eliminated from Eq. (4.1.23) using its complex conjugate as a second equation; this provides:

$$4G(\lambda' + 2G)\frac{\partial v}{\partial z} = (\lambda' + 3G)M(z) - (\lambda' + G)[M(z)]^* \tag{4.1.24}$$

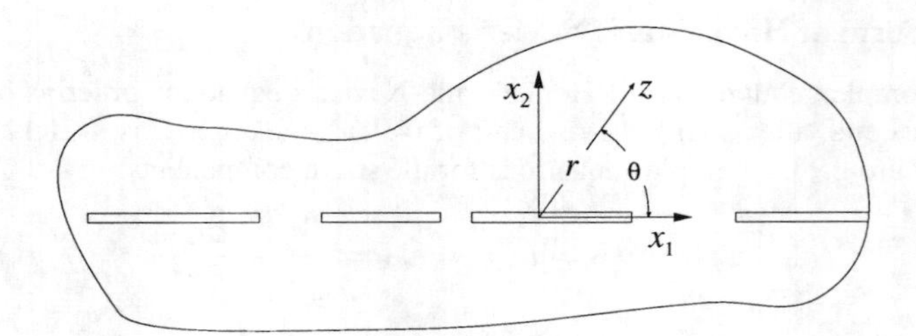

Figure 4.1.1 Coordinate axes for collinear crack problems.

This equation may be integrated directly taking into account that $[M(z)]^* = M^*(z^*)$[1] is a function depending only on z^*, and not on z, so $\int [M(z)]^* dz = [M(z)]^* \int dz$, and we can write

$$4G(\lambda' + 2G)v = (\lambda' + 3G) \int M(z)dz - (\lambda' + G)[M(z)]^* \, z + [N(z)]^* \tag{4.1.25}$$

where, again, $[N(z)]^* = N^*(z^*)$ is a function depending only on z^*, but otherwise arbitrary.

It is customary to use special forms for the arbitrary holomorphic functions $M(z)$ and $N(z)$ to simplify the expressions for the stresses. We rewrite the equations in terms of two arbitrary holomorphic functions $\phi(z)$ and $\chi(z)$ defined through

$$M(z) = \frac{4(\lambda' + 2G)}{\lambda' + G} \phi'(z) \, , \qquad N(z) = -4(\lambda' + 2G)\chi(z) \tag{4.1.26}$$

from which the general solution for the complex displacement field becomes

$$2G \, v(z, z^*) = \kappa \phi(z) - z[\phi'(z)]^* - [\chi(z)]^* \, , \quad \kappa = \frac{3 - \nu'}{1 + \nu'} \tag{4.1.27}$$

The expressions for the stress tensor components follow by direct substitution into Eqs. (4.1.16) and (4.1.17):

$$\sigma_{11} + \sigma_{22} = 2 \left[\phi'(z) + [\phi'(z)]^* \right] \tag{4.1.28}$$

$$\sigma_{22} - \sigma_{11} + 2i\sigma_{12} = 2 \left[z^* \phi''(z) + \chi'(z) \right] \tag{4.1.29}$$

The problem of finding displacements and stresses has thus been reduced to determining two holomorphic functions $\phi(z)$ and $\chi(z)$, which must be chosen so as to satisfy the boundary conditions.

When particular cases involving collinear straight cracks are considered, as in many practical situations, it is convenient to fix the axes in order to simplify the equations. In the following, the x_1 axis is always taken to lie along the crack as shown in Fig. 4.1.1. In this way, the components of the tractions across the crack plane are defined by σ_{22} (normal component) and σ_{12} (shear component). This means that the complex form of the tractions across the crack line is just $\sigma_{22} + i\sigma_{12}$, which can be obtained by adding together Eqs. (4.1.16) and (4.1.17). In doing so, it is convenient to replace $\chi(z)$ by introducing a new complex function, $\psi(z)$:

$$\chi(z) = \psi(z) - z\phi(z) \tag{4.1.30}$$

The resulting equations for stresses and displacements are

$$\sigma_{11} + \sigma_{22} = 2 \left[\phi'(z) + [\phi'(z)]^* \right] \tag{4.1.31}$$

$$\sigma_{22} + i\sigma_{12} = [\phi'(z)]^* + \psi'(z) - (z - z^*)\phi''(z) \tag{4.1.32}$$

$$2G \, v(z, z^*) = \kappa \phi(z) - [\psi(z)]^* - (z - z^*)[\phi'(z)]^* \tag{4.1.33}$$

---

[1] A holomorphic function can be represented (except in the neighborhood of singular points) by a power series expansion in z: $M(z) = \sum m_n \, z^n$, where m_n are complex coefficients. Its complex conjugate is $[M(z)]^* = \sum m_n^* \, z^{*n} = M^*(z^*)$, where $M^*(t) = \sum m_n^* \, t^n$.

With this form, the equations greatly simplify for points lying on the crack line ($x_2 = \pm 0$), because for them $z - z^*$ vanishes;

$$\sigma_{22} + i\sigma_{12} = [\phi'(z)]^* + \psi'(z) \quad \text{for} \quad z = x_1 \pm 0i \tag{4.1.34}$$

$$2G\, v(z, z^*) = \kappa\phi(z) - [\psi(z)]^* \quad \text{for} \quad z = x_1 \pm 0i \tag{4.1.35}$$

We write $\pm 0i$ because, in general, the complex potentials can be discontinuous across the crack line, such that their values for $z = x_1 + 0i$ (upper face) and for $z = x_1 - 0i$ (lower face) may differ.

In some special cases, a single holomorphic function is enough to find a solution. These cases are often found in practice and many handbooks use the simplified formalism involving a single complex function: the Westergaard function. This is described in the next section.

Exercises

4.1 Show that a holomorphic function $M(z)$ cannot be everywhere real unless it is a real constant. Similarly, show that it cannot be everywhere imaginary unless it is an imaginary constant.

4.2 Let $f(z, z^*)$ be a general complex function. Show that its Laplacian $\nabla^2 f$ can be expressed as

$$\nabla^2 f = 4\frac{\partial^2 f}{\partial z \partial z^*} \tag{4.1.36}$$

4.3 Show that a general complex function $f(z, z^*)$ is harmonic (i.e., $\nabla^2 f = 0$) if and only if it can be expressed as $f(z, z^*) = M(z) + [N(z)]^*$, where $M(z)$ and $N(z)$ are holomorphic functions. Show that if $N(z) = M(z)$, then $f(z, z^*)$ is real.

4.4 Show that the complex potentials $\phi(z) = \phi_1 z$ and $\psi(z) = \psi_1 z$, where ϕ_1 and ψ_1 are complex constants, correspond to states of uniform stress and that, furthermore, the displacement at the origin of coordinates is zero.

4.5 Show that the complex potentials $\phi(z) = \sigma z/4$ and $\psi(z) = -\sigma z/4$ provide the solution for an infinite panel subjected to uniform uniaxial stress σ parallel to the x_1 axis.

4.6 Show that the complex potentials $\phi(z) = \sigma z/4$ and $\psi(z) = 3\sigma z/4$ are the solution for an infinite panel subjected to uniform uniaxial stress σ parallel to the x_2 axis.

4.7 Show that the complex potentials $\phi(z) = i\tau z/2$ and $\psi(z) = -i\tau z/2$ give the solution for an infinite panel subjected to uniform shear τ.

4.8 Use Mohr's circle expressions to obtain the stress components in cylindrical coordinates. In particular, show that the cylindrical counterparts of Eqs. (4.1.31) and (4.1.32) are

$$\sigma_{rr} + \sigma_{\theta\theta} = 2\left[\phi'(z) + [\phi'(z)]^*\right] \tag{4.1.37}$$

$$\sigma_{\theta\theta} + i\sigma_{\theta r} = \phi'(z) + [\phi'(z)]^* + \left[-\phi'(z) + \psi'(z) - (z - z^*)\phi''(z)\right] e^{2\theta i} \tag{4.1.38}$$

4.9 Show that a general complex function $f(z, z^*)$ is biharmonic (i.e., $\nabla^2(\nabla^2 f) = 0$) if and only if it can be expressed as $f(z, z^*) = M(z)z^* + [N(z)]^* z + P(z) + [Q(z)]^*$, where $M(z)$, $N(z)$, $P(z)$ and $Q(z)$ are arbitrary holomorphic functions. Show that $f(z, z^*)$ is real if $N(z) = M(z)$ and $P(z) = Q(z)$, in which case $f(z, z^*) = \text{Re}[H(z)z^* + L(z)]$ where Re ω indicates the real part of complex ω; $H(z)$ and $L(z)$ are holomorphic functions.

4.10 An alternative approach to plane elasticity problems is to use the Airy stress function Φ (see e.g., Timoshenko and Goodier 1951). In that approach

$$\sigma_{11} = \Phi_{,22}\,, \quad \sigma_{22} = \Phi_{,11}\,, \quad \sigma_{12} = -\Phi_{,12} \tag{4.1.39}$$

and Φ is a biharmonic function. Consider Φ as a complex function of z and z^* and express the components σ_{ij} as a function of the partial derivatives of Φ with respect to z and z^*. In particular, show that

$$\sigma_{11} + \sigma_{22} = 4\frac{\partial^2\Phi}{\partial z \partial z^*}\,, \quad \text{and} \quad \sigma_{11} + i\sigma_{12} = 2\frac{\partial^2\Phi}{\partial z^2} + 2\frac{\partial^2\Phi}{\partial z \partial z^*} \tag{4.1.40}$$

4.11 Show that the Airy stress function Φ is related to the complex potentials $\phi(z)$ and $\psi(z)$ in (4.1.31) and (4.1.32) by

$$\Phi = \text{Re}[\overline{\phi}(z) + \overline{\psi}(z) - (z\text{-}z^*)\phi(z)] \tag{4.1.41}$$

where $\overline{\phi}(z) = \int \phi(z)dz$ and $\overline{\psi}(z) = \int \psi(z)dz$. Check the biharmonicity condition against the result in exercise 4.9.

4.2 Plane Crack Problems and Westergaard's Stress Function

4.2.1 Westergaard Stress Function

For various special cases, it suffices to use a single complex function to solve the problem. Westergaard found a number of them. The corresponding functions are called Westergaard stress functions and are denoted as $Z(z)$. Although Westergaard derived them using a different reasoning, his findings may be easily understood in the frame of our previous development.

Consider, first, the case of a loading symmetric with respect to the crack line, with no shear tractions applied on the crack faces. In such a case, the shear stress σ_{12} along the x_1 axis must vanish identically. A look at Eq.(4.1.34) shows that a sufficient condition is $\phi'(z) = \psi'(z)$. Westergaard's formulation is obtained by setting

$$\phi'(z) = \psi'(z) = \frac{1}{2}Z(z) \tag{4.2.1}$$

from which the stresses and displacements are found to be

$$\sigma_{11} + \sigma_{22} = Z(z) + [Z(z)]^* \tag{4.2.2}$$

$$2\sigma_{22} + 2i\sigma_{12} = [Z(z)]^* + Z(z) - (z - z^*)Z'(z) \tag{4.2.3}$$

$$4G\, v(z, z^*) = \kappa\overline{Z}(z) - [\overline{Z}(z)]^* - (z - z^*)[Z(z)]^* \tag{4.2.4}$$

where $\overline{Z}(z)$ stands for the primitive function of $Z(z)$, i.e., $\overline{Z}(z) = \int Z(z)dz$.

When the loading is antisymmetric about the crack plane, $\sigma_{22} = 0$ along the crack line, and Eq. (4.1.34) shows a sufficient condition to be $\phi'(z) = -\psi'(z)$. Westergaard's formulation for this case follows by setting

$$\phi'(z) = -\psi'(z) = \frac{1}{2}Z(z) \tag{4.2.5}$$

The corresponding stresses and displacements are

$$\sigma_{11} + \sigma_{22} = Z(z) + [Z(z)]^* \tag{4.2.6}$$

$$2\sigma_{22} + 2i\sigma_{12} = [Z(z)]^* - Z(z) - (z - z^*)Z'(z) \tag{4.2.7}$$

$$4G\, v(z, z^*) = \kappa\overline{Z}(z) + [\overline{Z}(z)]^* - (z - z^*)[Z(z)]^* \tag{4.2.8}$$

The Westergaard approach will next be applied to a classical problem: the center-cracked infinite panel.

4.2.2 Westergaard's Solution of Center-Cracked Infinite Panel

Let us now consider a crack of length $2a$ in a two-dimensional infinite solid, subjected to uniform normal stress σ_∞ at infinity in all directions (Fig. 4.2.1). For this case, Westergaard's stress function may be assumed in the form

$$Z = \frac{\sigma_\infty z}{\sqrt{z^2 - a^2}} \tag{4.2.9}$$

of which the primitive and the derivative are

$$\overline{Z} = \sigma_\infty\sqrt{z^2 - a^2}\,; \qquad Z' = -\sigma_\infty\frac{a^2}{(z^2 - a^2)\sqrt{z^2 - a^2}} \tag{4.2.10}$$

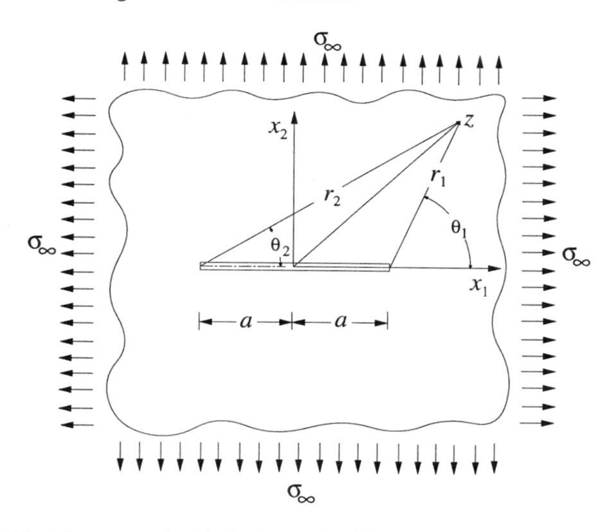

Figure 4.2.1 Center-cracked infinite panel subjected to remote equiaxial tension.

Table 4.2.1 Values of variables that can be discontinuous across the crack line

Variable	Crack side	x_1 Interval		
		$x_1 < -a$	$-a < x_1 < a$	$a < x_1$
θ_1	upper	$+\pi$	$+\pi$	0
	lower	$-\pi$	$-\pi$	0
θ_2	upper	$+\pi$	0	0
	lower	$-\pi$	0	0
$\sqrt{z^2 - a^2}$	upper	$-\sqrt{x_1^2 - a^2}$	$+i\sqrt{a^2 - x_1^2}$	$\sqrt{x_1^2 - a^2}$
	lower	$-\sqrt{x_1^2 - a^2}$	$-i\sqrt{a^2 - x_1^2}$	$\sqrt{x_1^2 - a^2}$

These functions are holomorphic except at the crack surface ($x_2 = 0, -a \leq x_1 \leq a$). The field differential equations are thus satisfied automatically, and the boundary condition at infinity is also satisfied because, for $|z| \to \infty$, $\lim Z = \lim(\sigma_\infty z/\sqrt{z^2}) = \sigma_\infty$.

At the crack line, the complex plane is said to have a cut, which implies a displacement discontinuity. This discontinuity is contained in the square root appearing in Z, and considerable care must be exerted to get consistent results. One of the systematic ways of doing so is to write the complex expressions in the exponential form, i.e., $c = |c|e^{arg(c)i}$ where c is any complex number or expression, and to keep the argument in the $[-\pi, +\pi]$ interval.

For our particular case, it is useful to write

$$z - a = r_1 e^{\theta_1 i}; \qquad z + a = r_2 e^{\theta_2 i} \tag{4.2.11}$$

where r_1, θ_1, r_2 and θ_2 are depicted in Fig. 4.2.1. Then

$$\sqrt{z^2 - a^2} = \sqrt{r_1 r_2} e^{i(\theta_1 + \theta_2)/2} \tag{4.2.12}$$

The values of θ_1 and θ_2 and of $\sqrt{z^2 - a^2}$ along the two sides of the crack plane are given in Table 4.2.1. With this and the introduction of the Heaviside function, $H(x)$ — equal to 1 for positive x and to 0 for negative x — the following compact expressions for Z and $\overline{Z}$ along the crack plane may be written:

$$Z(x_1, \pm 0) = \sigma_\infty \frac{|x_1| H(x_1^2 - a^2) \mp i x_1 H(a^2 - x_1^2)}{\sqrt{|a^2 - x_1^2|}} \tag{4.2.13}$$

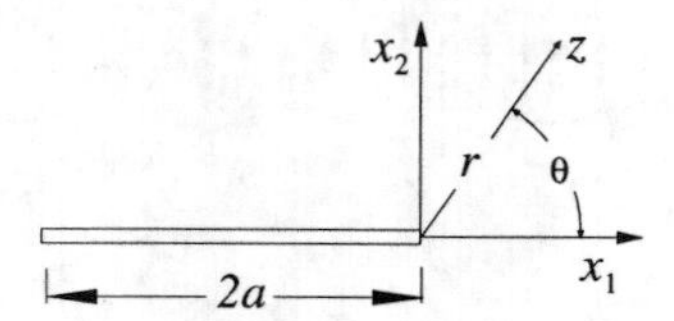

Figure 4.2.2 Axes used to analyze the near-tip solution.

$$\overline{Z}(x_1, \pm 0) = \sigma_\infty \left[H(x_1^2 - a^2) \operatorname{sign} x_1 \pm iH(a^2 - x_1^2) \right] \sqrt{|a^2 - x_1^2|} \tag{4.2.14}$$

where $\operatorname{sign} x_1 = 1$ if $x_1 > 0$ and-1 if $x_1 < 0$, and the upper and lower signs of the imaginary parts correspond, respectively, to the upper and lower sides of the crack plane. Substitution of the foregoing expressions into Eqs. (4.2.2) to (4.2.4) delivers the following expressions for the stresses and displacements along the crack plane:

$$\sigma_{11}(x_1, \pm 0) = \sigma_{22}(x_1, \pm 0) = \sigma_\infty \frac{|x_1|}{\sqrt{x_1^2 - a^2}} H(x_1^2 - a^2) \tag{4.2.15}$$

$$\sigma_{12}(x_1, \pm 0) = 0 \tag{4.2.16}$$

$$u_1(x_1, \pm 0) = \frac{(1 - \nu')\sigma_\infty}{E'} \sqrt{x_1^2 - a^2}\, H(x_1^2 - a^2) \operatorname{sign} x_1 \tag{4.2.17}$$

$$u_2(x_1, \pm 0) = \pm \frac{2\sigma_\infty}{E'} \sqrt{a^2 - x_1^2}\, H(a^2 - x_1^2) \tag{4.2.18}$$

The two first equations above show that the assumed Westergaard function does satisfy, in addition to boundary conditions at infinity, also the boundary conditions on the crack faces, namely, $\sigma_{22} = \sigma_{12} = 0$ for $|x_1| < a$. Moreover, the only function displaying a discontinuity on going across the crack is u_2. The jump in displacement is the crack opening, $w(x_1)$, which, in general, is written in one of the equivalent forms:

$$w(x_1) = [u_2](x_1) = u_2^+(x_1) - u_2^-(x_1) = u_2(x_1, +0) - u_2(x_1, -0) \tag{4.2.19}$$

In this particular case, Eq. (4.2.18) gives for w

$$w(x_1) = \frac{4\sigma_\infty}{E'} \sqrt{a^2 - x_1^2}\, H(a^2 - x_1^2) \tag{4.2.20}$$

4.2.3 Near-Tip Expansion for the Center-Cracked Panel

The result in Eq. (4.2.15) shows that the stresses tend to infinity when the crack tips are approached from the solid, so the stress field has a singularity at the crack tip. In order to determine the asymptotic near-tip stress field, we transform coordinates by translating the coordinate center into the crack tip, that is, replacing $z - a$ with z (and z with $z + a$, and $z + a$ with $z + 2a$; Fig. 4.2.2). Then, setting $z^2 - a^2 = (z + a)(z - a)$, we get for Z the following asymptotic approximation from Eq. (4.2.9):

$$Z = \frac{\sigma_\infty (z + a)}{\sqrt{z(z + 2a)}} = \sigma_\infty \sqrt{\frac{a}{2z}} \left[1 + \frac{3z}{4a} - \frac{5z^2}{32a^2} + \ldots \right] \tag{4.2.21}$$

where the factor in the square brackets shows the first three terms of the MacLaurin series expansion of $(1 + z/a)/\sqrt{1 + z/2a}$ which obviously tends to 1 for $|z| \ll a$. It is now customary to denote

$$K_I = \sigma_\infty \sqrt{\pi a} \tag{4.2.22}$$

and call it the stress intensity factor. (Subscript I refers to the opening mode of fracture, to be distinguished from the shear modes II and III to be discussed later). The near-tip expression for Z now becomes

$$Z = \frac{K_I}{\sqrt{2\pi z}} \qquad \text{for} \qquad |z| \ll a \tag{4.2.23}$$

The stresses and crack openings along the crack plane and near the tip may be directly obtained from the foregoing expression or may be derived form Eqs. (4.2.15) to (4.2.18) replacing $x_1 - a$ with x_1, x_1 with $x_1 + a \simeq a$, and $x_1 + a$ with $x_1 + 2a \simeq 2a$. In particular, the normal stress distribution and openings close to the crack tip (i.e., for $|x_1| \ll a$) are

$$\sigma_{22}(x_1, \pm 0) = \frac{K_I}{\sqrt{2\pi x_1}}\; H(x_1) \tag{4.2.24}$$

$$w(x_1) = \frac{8K_I}{\sqrt{2\pi}E'}\sqrt{|x_1|}\; H(-x_1) \tag{4.2.25}$$

These equations show that the singularity in stresses is of the order of $x_1^{-1/2}$ and that the profile of the deformed crack is parabolic (more precisely, a parabola of the second degree, with its axis coincident with the crack line).

Although the near-tip fields have been derived for quite a particular case, they remain valid for all the symmetric loading cases. This, and the homologous result for the antisymmetric case, will be shown in the forthcoming section.

Exercises

4.12 A straight crack does not disturb the stress field of a uniform uniaxial stress parallel to crack. Using this fact, show that the complex potentials corresponding to a center-cracked panel subjected to remote uniaxial stress normal to the crack are

$$\phi(z) = \sigma_\infty\left(\sqrt{z^2 - a^2} - \frac{1}{4}z\right), \quad \psi(z) = \sigma_\infty\left(\sqrt{z^2 - a^2} + \frac{1}{4}z\right) \tag{4.2.26}$$

where a is the half-length of the crack. (Hint: superpose the Westergaard solution (4.2.9) on the solution of a compressive uniaxial stress parallel to the crack.)

4.13 By analysis analogous to that in Section 4.2.2, show that for a center-cracked panel subjected to remote constant shear stress τ_∞, the Westergaard function to be substituted into Eqs. (4.2.5)–(4.2.8) is

$$Z(z) = \frac{\tau_\infty i z}{\sqrt{z^2 - a^2}} \tag{4.2.27}$$

Also prove that the stress component σ_{12} along the crack plane takes exactly the same form as σ_{22} for the symmetric loading case, i.e. Eq. (4.2.15), with τ_∞ in place of σ_∞.

4.14 Find the near-tip one-term expansion for stresses and crack-sliding displacements for a center-cracked infinite panel subjected to remote shear stress τ_∞.

4.3 The General Near-Tip Fields

4.3.1 In-Plane Near-Tip Asymptotic Series Expansion

There exists an important result concerning the near-tip stress singularity found in the previous section: it is general, applicable to isotropic bodies of any size and geometry, loaded in an arbitrary manner. This result can be proven in various ways; we use the general complex function formulation presented in the first section of this chapter, with the axes shown in Fig. 4.3.1.

In the context of the general complex formulation in Eqs. (4.1.31)–(4.1.33), the fact that the Westergaard function for the center-cracked panel behaves as $z^{-1/2}$ in the neighborhood of the crack tip suggests — in the frame — looking for holomorphic functions $\phi'(z)$ and $\psi'(z)$ of the form

$$\phi'(z) = Az^\lambda \qquad \text{and} \qquad \psi'(z) = Bz^\lambda \tag{4.3.1}$$

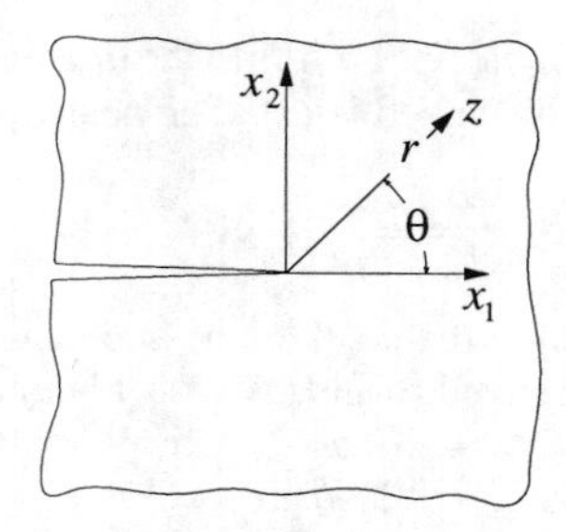

Figure 4.3.1 Axes for near crack tip field description.

where A and B are arbitrary complex constants and λ is a real exponent. The problem is to determine the set of exponents λ that satisfy the near-tip boundary conditions, namely, that the faces of the cracks are stress free. With the coordinate axes shown in Fig. 4.3.1, this requires the complex quantity $\sigma_{22} + i\sigma_{12}$ to vanish for $z = re^{\pi i}$ and for $z = re^{-\pi i}$. If the expressions (4.3.1) are substituted into Eq. (4.1.32), one obtains for the upper and lower surfaces the conditions:

$$A^* r^\lambda e^{-\lambda\pi i} + B r^\lambda e^{\lambda\pi i} = 0 \tag{4.3.2}$$

$$A^* r^\lambda e^{\lambda\pi i} + B r^\lambda e^{-\lambda\pi i} = 0 \tag{4.3.3}$$

This is a set of two homogeneous (complex) equations for the unknowns A^* and B which has a nontrivial solution only if the determinant of the matrix of coefficients vanishes. Imposing this condition, one easily gets the restriction

$$r^{2\lambda}\left(e^{-2\lambda\pi i} - e^{2\lambda\pi i}\right) = 0 \tag{4.3.4}$$

After setting $e^{2\lambda\pi i} - e^{-2\lambda\pi i} = 2i\,\sin(2\lambda\pi)$, one gets the final condition $\sin(2\lambda\pi) = 0$ or

$$\lambda = \frac{n}{2} \quad \text{for} \quad n = ..., -2, -1, 0, 1, 2, ... \tag{4.3.5}$$

However, not all these integer values of n correspond to physically admissible fields because some of them lead to infinite displacements at the crack tip, which are impossible. A single look to Eq. (4.1.33) shows that this is the case for $n < -1$. Therefore, the admissible eigenvalues are

$$\lambda = \frac{n}{2} \quad \text{for} \quad n = -1, 0, 1, 2, ... \tag{4.3.6}$$

With this, Eq. (4.3.2) leads to the following relation between A and B for a given exponent order, n:

$$A^* = -B\cos(n\pi) = (-1)^{n+1} B \tag{4.3.7}$$

Since any solution of the form (4.3.1) (with λ from the set defined in (4.3.6) and A and B satisfying (4.3.7)) complies with the boundary conditions at the crack surface, one may expect that a linear superposition of such terms, i.e., a series, will be a general solution. Thus, we write the general near-tip solution for the complex potentials as the series (in which we set $m = n + 1$)

$$\phi'(z) = \sum_{m=0}^{\infty} A_m\, z^{(m-1)/2} \tag{4.3.8}$$

$$\psi'(z) = \sum_{m=0}^{\infty} (-1)^m A_m^*\, z^{(m-1)/2} \tag{4.3.9}$$

where the complex coefficients A_m have to be determined so as to satisfy the remaining boundary conditions.

When these coefficients are known, the full stress and displacement fields can be determined from Eqs. (4.1.31)–(4.1.33).

Before proceeding with the dominant term of the solution, let us calculate the stresses along the crack line $x_2 = 0$, to get a feeling of how the coefficients A_m influence the solution. The result is obtained in complex form from (4.1.34):

$$\sigma_{22}(x_1, 0) + i\sigma_{12}(x_1, 0) = |x_1|^{-1/2} \sum_{p=0}^{\infty} 2A_{2p}^* x_1^p \; H(x_1) \tag{4.3.10}$$

where $H(x_1)$ is again the Heaviside step function. This equation shows a number of features. First, it depends only on the coefficients with even subscripts; this is extensible to crack openings and crack slidings (relative motions of crack faces). Furthermore, if the coefficients A_{2p} are all real, then $\sigma_{12} = 0$ which corresponds to a loading symmetric with respect to the crack plane (called mode I loading). Conversely, if they are all imaginary, then $\sigma_{22} = 0$, which corresponds to an antisymmetric loading (called mode II loading). This result may be shown to be general: the real parts of coefficients A_m (even and odd) contribute to the symmetric (mode I) part of the loading, and their imaginary parts contribute to the antisymmetric (mode II) part of the loading. We can thus separate both kinds of loading by separating the real and imaginary parts of the coefficients A_m, writing

$$A_m = A_m^I + iA_m^{II} \tag{4.3.11}$$

where A_m^I and A_m^{II} are real.

There is no point in performing the algebra leading to the separate general expressions for the stresses and displacements, although we will give some of these expressions for mode I after performing the analysis of the dominant term of the series.

4.3.2 The Stress Intensity Factors

From the solution of stresses in the previous equation, it is obvious that the stresses diverge when approaching the crack tip, and do so as the inverse of the square root of the distance. For points close enough to the crack tip, all the terms in the series will be negligible compared to the first. The stresses will then behave as

$$\sigma_{22}(x_1, 0) + i\sigma_{12}(x_1, 0) \approx 2A_0^* \; |x_1|^{-1/2} \; H(x_1) \tag{4.3.12}$$

and the usual expressions for the singular terms are obtained by setting

$$2A_0^* = \frac{K_I + iK_{II}}{\sqrt{2\pi}} \tag{4.3.13}$$

where K_I and K_{II} are called the mode I stress intensity factor and the mode II stress intensity factor, respectively.

With this notation, the dominant (singular) terms of ϕ' and ψ', and their integrals are

$$\phi'(z) \approx \frac{K_I - iK_{II}}{2\sqrt{2\pi z}}, \qquad \psi'(z) \approx \frac{K_I + iK_{II}}{2\sqrt{2\pi z}} \tag{4.3.14}$$

$$\phi(z) \approx (K_I - iK_{II})\sqrt{\frac{z}{2\pi}} \quad \text{and} \quad \psi(z) \approx (K_I + iK_{II})\sqrt{\frac{z}{2\pi}} \tag{4.3.15}$$

From the foregoing expressions, the dominant term for stresses and displacements are easily found from Eqs. (4.1.31) to (4.1.33) after setting $z = re^{\theta i}$:

$$\sigma_{ij} = \left[K_I \, S_{ij}^I(\theta) + K_{II} \, S_{ij}^{II}(\theta)\right] \frac{1}{\sqrt{2\pi r}} \tag{4.3.16}$$

$$u_i = \left[\frac{K_I}{E'} \, D_i^I(\theta) + \frac{K_{II}}{E'} \, D_i^{II}(\theta)\right] \sqrt{\frac{8r}{\pi}} \tag{4.3.17}$$

where $S_{ij}^I(\theta)$, $S_{ij}^{II}(\theta)$, $D_i^I(\theta)$, and $D_i^{II}(\theta)$ are functions depending only on the polar angle θ (except the D_is that depend also on ν'). The angular functions S_{ij} and D_i for in-plane modes are listed next.

Note that the expressions for displacement functions differ from the usual ones found in most textbooks because we have chosen E' and ν' as the explicit elastic constants instead of G and κ.

Mode I Angular Functions

$$S_{11}^{I} = 2\cos\frac{\theta}{2} - S_{22}^{I}\,,\quad S_{22}^{I} = \cos\frac{\theta}{2}\left(1+\sin\frac{\theta}{2}\sin\frac{3\theta}{2}\right)\,,\quad S_{12}^{I} = \cos\frac{\theta}{2}\sin\frac{\theta}{2}\cos\frac{3\theta}{2} \tag{4.3.18}$$

$$S_{23}^{I} = S_{23}^{I} = 0\,,\quad S_{33}^{I} = \begin{cases} 0 & \text{for plane stress} \\ 2\nu\cos(\theta/2) & \text{for plane strain} \end{cases} \tag{4.3.19}$$

$$D_{1}^{I} = \frac{5-3\nu'}{8}\cos\frac{\theta}{2} - \frac{1+\nu'}{8}\cos\frac{3\theta}{2}\,,\quad D_{2}^{I} = \frac{7-\nu'}{8}\sin\frac{\theta}{2} - \frac{1+\nu'}{8}\sin\frac{3\theta}{2} \tag{4.3.20}$$

Mode II Angular Functions

$$S_{11}^{II} = -2\sin\frac{\theta}{2} - S_{22}^{II}\,,\quad S_{22}^{II} = \sin\frac{\theta}{2}\cos\frac{\theta}{2}\cos\frac{3\theta}{2}\,,\quad S_{12}^{II} = \cos\frac{\theta}{2}\left(1-\sin\frac{\theta}{2}\sin\frac{3\theta}{2}\right) \tag{4.3.21}$$

$$S_{23}^{II} = S_{23}^{II} = 0\,,\quad S_{33}^{II} = \begin{cases} 0 & \text{for plane stress} \\ -2\nu\sin(\theta/2) & \text{for plane strain} \end{cases} \tag{4.3.22}$$

$$D_{1}^{II} = \frac{9+\nu'}{8}\sin\frac{\theta}{2} + \frac{1+\nu'}{8}\sin\frac{3\theta}{2}\,,\quad D_{2}^{II} = -\frac{3-5\nu'}{8}\cos\frac{\theta}{2} - \frac{1+\nu'}{8}\cos\frac{3\theta}{2} \tag{4.3.23}$$

The foregoing analysis has the interesting property of being independent of the particular geometry and loading at hand. Only the values of the A_m coefficients, hence also of the stress intensity factors K_I and K_{II} depend of the particular case envisaged.

A similar solution of the near-tip asymptotic series stress field by separation of variables was presented by Knein (1927), in a program directed by J. von Kármán in Aachen. Knein, however, after getting the correct expression, incorrectly rejected the term with singular stress as meaningless.[2] Later, the full, correct expansion was obtained, on the basis of Airy stress function, by Williams (1952). The method can be generalized to bodies with sharp corners and to various three-dimensional situations (e.g., Bažant 1974; Bažant and Keer 1974; Bažant and Estenssoro 1979), to dynamic crack propagation (Achenbach and Bažant 1975; Bažant, Glazik and Achenbach 1976; Achenbach, Bažant and Khetan 1976a,b)) and to cracks at the interface between two different elastic solids. For anisotropic elastic materials, the near-tip solution has been presented by Karp and Karal (1962). For corners, one finds that $\lambda > -1/2$, which means that the singularity is weaker than for a crack.

There have been repeated attempts to use corner singularities in fracture criteria. However, such interpretations are dubious since for $\lambda > 1/2$ the flux of energy into the corner tip is zero so there is no energy available for fracture. For interface cracks between two dissimilar solids, there is another difficulty: the exponent λ is complex, which implies overlapping of the opposite stress surfaces — a physically impossible situation.

4.3.3 Closer View of the Near-Tip Asymptotic Expansion for Mode I

Consider now a purely symmetric loading in which the A_m coefficients are all real, such that $A_m = A_m^* = A_m^I$. Substituting these values into (4.3.8) and (4.3.9) and the results into the expressions (4.1.31)–(4.1.33),

---

[2]To accept singular stresses was not an easy step to make in early history. The fact that the singular stresses cannot be dismissed became clear from Griffith (1921) analysis of a crack as the limit case of elliptical hole (Inglis 1913); indeed, as the minor axis b of the ellipse approaches 0, the stress field must change continuously but, if the singular stresses were dismissed, there would be a discontinuity for $b \to 0$.

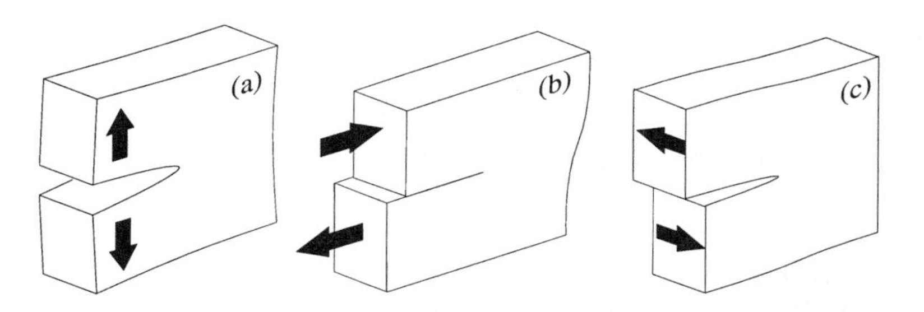

Figure 4.3.2 Sketches of the three pure modes: (a) Mode I or pure opening mode; (b) mode II or in-plane shear mode; and (c) mode III or antiplane shear mode.

one can find the expressions for the stress and displacement fields in polar coordinates. They take the form

$$\sigma_{ij} = \frac{K_I}{\sqrt{2\pi r}} S_{ij}(r,\theta) \,, \quad u_i = \frac{4K_I}{\sqrt{2\pi}E'}\sqrt{r}D_i(r,\theta) \tag{4.3.24}$$

where $S_{ij}(r,\theta)$ and $D_i(r,\theta)$ are regular everywhere (except at load points and reentrant corners other than the crack tip). There is no point in writing the full expressions for these functions as a series expansion (of positive, half-integer powers of r). Just note that their expressions for $r = 0$ give the singular terms listed in the previous section: $S_{ij}(0,\theta) = S^I_{ij}(\theta)$ and $D_i(0,\theta) = D^I_i(\theta)$.

The expressions for stresses and crack openings along the crack plane ($x_2 = \pm 0$) follow by substitution of (4.3.8) and (4.3.9), with $A_m = A^*_m = A^I_m$, into (4.1.34) and (4.1.35):

$$\sigma_{22}(x_1, \pm 0) = \frac{K_I}{\sqrt{2\pi|x_1|}}\left[1 + \sum_{p=1}^{\infty}\beta_p x_1^p\right] H(x_1) \tag{4.3.25}$$

$$w(x_1) = 8\frac{K_I}{\sqrt{2\pi}E'}\sqrt{|x_1|}\left[1 + \sum_{p=1}^{\infty}\frac{(-1)^p}{2p+1}\beta_p x_1^p\right] H(-x_1) \tag{4.3.26}$$

in which $\beta_p = A^I_{2p}/A^I_0$. Note, again, that only the coefficients A_m with even indices are involved in these expressions. This does not mean that the remaining coefficients are zero or that they do not influence other aspects of the fields or their values at points other than those on the crack plane.

4.3.4 The Antiplane Shear Mode

Up to now, only the plane fracture cases have been analyzed. It has been proved that the dominant term of the solution may be decomposed into a symmetric, pure opening mode (mode I) and an antisymmetric, in-plane shear mode (mode II). These modes are illustrated in Fig. 4.3.2. Furthermore, in three dimensions there exists another, antisymmetric mode, mode III, representing the antiplane shear mode (Fig. 4.3.2), with stress intensity factor K_{III}.

A detailed analysis of mode III can be easily performed using a complex variable formulation similar, but simpler, than that used for modes I and II. Indeed, the pure mode III is characterized by a very simple displacement field:

$$u_1 = u_2 = 0 \,, \qquad u_3 = u_3(x_1, x_2) \tag{4.3.27}$$

which implies that all strain tensor components are zero except ε_{13} and ε_{23}, which are

$$\varepsilon_{13} = \frac{1}{2}u_{3,1} \,, \qquad \varepsilon_{23} = \frac{1}{2}u_{3,2} \tag{4.3.28}$$

Accordingly, the only nonzero stress components are σ_{13} and σ_{23}, given by Hooke's law as

$$\sigma_{13} = Gu_{3,1}\ , \qquad \sigma_{23} = Gu_{3,2} \tag{4.3.29}$$

The static equilibrium equations $\sigma_{ij,j} = 0$ are reduced to

$$u_{3,11} + u_{3,22} = 0 \tag{4.3.30}$$

which states that the bidimensional Laplacian of the function u_3 must vanish.

The solution of this general problem may take advantage of the complex function formalism by letting u_3 to be expressed as the real part of a harmonic complex variable function $f(z, z^*)$:

$$u_3 = \mathrm{Re} f(z, z^*)\ , \qquad f_{,kk} = 0 \tag{4.3.31}$$

The use of relations (4.1.10) and (4.1.11) transforms the harmonicity condition $f_{,kk} = 0$ into

$$\frac{\partial^2 f}{\partial z^* \partial z} = 0 \tag{4.3.32}$$

This is readily integrated twice to furnish

$$f(z, z^*) = M(z) + [N(z)]^* \tag{4.3.33}$$

where $M(z)$ and $N(z)$ are arbitrary holomorphic functions (see exercises 4.2 and 4.3). The displacement is obtained by taking the real part, with the result

$$u_3 = \frac{1}{2G}\left[\zeta(z) + [\zeta(z)]^*\right] \tag{4.3.34}$$

The arbitrary holomorphic function $\zeta(z) = G[M(z) + N(z)]$ is introduced for convenience. It must be determined by imposing the boundary conditions. The stresses are

$$\sigma_{31} + i\sigma_{32} = [\zeta'(z)]^* \tag{4.3.35}$$

4.3.5 Antiplane Near-Tip Asymptotic Series Expansion

The near-tip analysis is performed using again the axes shown in Fig. 4.3.1 and assuming power-law expressions for the holomorphic function $\zeta(z)$:

$$\zeta'(z) = Az^\lambda \tag{4.3.36}$$

The eigenvalues are obtained by expressing the conditions that $\sigma_{32} = 0$ on the crack faces, i.e., for $\theta = \pi$ and $\theta = -\pi$. Setting $A = a + bi$, with a and b real, the resulting equations are:

$$a\sin(\lambda\pi) + b\cos(\lambda\pi) = 0 \tag{4.3.37}$$

$$-a\sin(\lambda\pi) + b\cos(\lambda\pi) = 0 \tag{4.3.38}$$

This set of homogeneous equations has a nontrivial solution only if the determinant of the coefficients vanishes, i.e., for $\sin(2\lambda\pi) = 0$. When, furthermore, the condition of bounded displacements at the crack tip is imposed, the set of admissible values of λ is

$$\lambda = \frac{n}{2} \qquad \text{for} \qquad n = -1, 0, 1, 2, ... \tag{4.3.39}$$

The corresponding solutions for a and b are, from (4.3.37)

$$a = 0, \quad b = \text{arbitrary} \quad \text{for} \ \ n = 2p - 1\ , \ p = 0, 1, 2, ... \tag{4.3.40}$$

$$a = \text{arbitrary}, \quad b = 0 \quad \text{for} \ \ n = 2q\ , \ q = 0, 1, 2, ... \tag{4.3.41}$$

Hence, the solution is conveniently expressed as a sum of a half-integer power series and an integer power series:

$$\zeta'(z) = z^{-1/2} \sum_{p=0}^{\infty} a_p\, z^p + i \sum_{q=0}^{\infty} b_q\, z^q \tag{4.3.42}$$

where the real coefficients a_p and b_q must be determined for each particular geometry and remote boundary conditions.

At short distances from the crack tip, the first term of the first series is dominant. Traditionally one sets

$$a_0 = \frac{K_{III}}{\sqrt{2\pi}} \tag{4.3.43}$$

and so the near-tip approximation reads

$$\zeta'(z) \approx \frac{K_{III}}{\sqrt{2\pi z}} \tag{4.3.44}$$

The near-tip stresses and displacements are then obtained from Eqs. (4.3.35) and (4.3.34) as

$$\sigma_{ij} = \frac{K_{III}}{\sqrt{2\pi r}} S_{ij}^{III}(\theta) \tag{4.3.45}$$

$$u_i = \frac{4K_{III}}{2G\sqrt{2\pi}} \sqrt{r} D_i^{III}(\theta) \tag{4.3.46}$$

in which the angular functions for the antiplane mode are given by

$$S_{11}^{III} = S_{22}^{III} = S_{33}^{III} = S_{12}^{III} = 0\,, \quad S_{13}^{III} = \cos\frac{\theta}{2}\,, \quad S_{23}^{III} = \sin\frac{\theta}{2} \tag{4.3.47}$$

$$D_1^{III} = D_2^{III} = 0\,, \quad D_3^{III} = \cos\frac{\theta}{2} \tag{4.3.48}$$

4.3.6 Summary: The General Singular Near-Tip Fields

To sum up, for a general loading, the stresses near the crack tip display a singularity. The dominant term of the series expansion of the fields in powers of r depend only on the stress intensity factors K_I, K_{II}, and K_{III}. The general expressions for the dominant terms in stresses and displacements are

$$\sigma_{ij} = \frac{1}{\sqrt{2\pi r}} [K_I S_{ij}^I(\theta) + K_{II} S_{ij}^{II}(\theta) + K_{III} S_{ij}^{III}(\theta)] \tag{4.3.49}$$

$$u_i = \sqrt{\frac{8r}{\pi}} \left[\frac{K_I}{E'} D_i^I(\theta) + \frac{K_{II}}{E'} D_i^{II}(\theta) + \frac{K_{III}}{2G} D_i^{III}(\theta) \right] \tag{4.3.50}$$

in which the angular functions S_{ij} and D_i are given by (4.3.18)–(4.3.23) and (4.3.47)–(4.3.48).

Exercises

4.15 Derive in detail conditions (4.3.2) and (4.3.2).

4.16 Carry out the algebra leading to (4.3.7).

4.17 Carry out the calculations leading to (4.3.10).

4.18 Take two terms of the series in (4.3.8) and (4.3.9) and find σ_{11} and σ_{22} along the crack plane. Which of these variables is influenced by coefficient A_1?

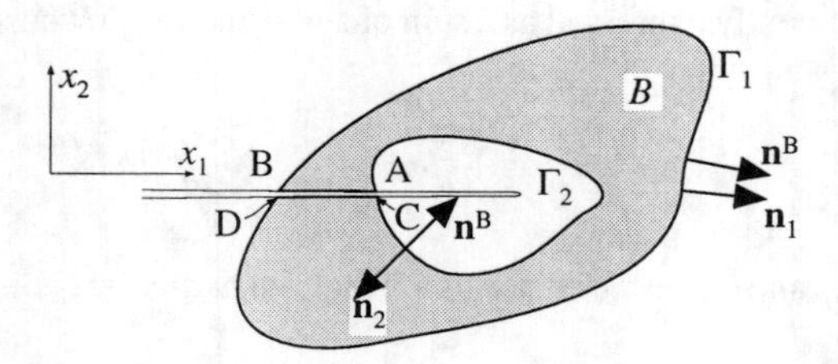

Figure 4.4.1 Contours Γ_1 and Γ_2 define a subbody B (dashed) that does not include the crack tip.

4.4 Path-Independent Contour Integrals

In this section we deepen the analysis of the J–integral, introduce further path-independent expressions to express $\mathcal{G}$, give a further proof of Irwin's relationship between $\mathcal{G}$ and the stress intensity factor, generalizing it to mixed mode situations, and give a short account of other path-independent integrals.

4.4.1 Path Independence of the J-Integral

That Rice's J-integral is independent of the contour over which the integration is performed is clear from the derivation given in Section 2.1.5, in which we proved that $J = \mathcal{G}$. Since $\mathcal{G}$ is a state function, and thus unique for a given state, the value of J must be unique whatever the contour actually used. However, it is instructive to give an independent proof, based on purely mathematical grounds, because then the hypotheses required to fulfill path-independence emerge with clarity.

The key to obtain the proof is to use two arbitrary contours Γ_1 and Γ_2, as shown in Fig. 4.4.1, to define a body $\mathcal{B}$ (shaded in the figure) and to extend to its boundary a surface integral with an integrand similar to that for the J-integral. Let $S_{\mathcal{B}}$ be the surface of the body B defined by the two contours Γ_1 and Γ_2 and the two segments AB and CD. Let J_ϕ be the surface integral defined as:

$$J_\phi = \oint_{S_{\mathcal{B}}} (\overline{\mathcal{U}} n_1^{\mathcal{B}} - t_i^{\mathcal{B}} u_{i,1}) dA \tag{4.4.1}$$

where as in (2.1.48) $\overline{\mathcal{U}}$ is the elastic energy density, $n_1^{\mathcal{B}}$is the component along the x_1 axis of the unit exterior normal to $\mathcal{B}$, $t_i^{\mathcal{B}}$ is the traction vector acting on the boundary of B, and $u_{i,1} = \partial u_i/\partial x_1$; dA is the element of area. Superscripts $\mathcal{B}$ are used to distinguish the normal and traction vectors defined with respect to the body $\mathcal{B}$ and with respect to the Γ-contours.

First, we transform this integral to show that it is proportional to the difference of the J-integrals performed along Γ_1 and Γ_2; next, we prove that this integral must be zero if no singular points are included in $\mathcal{B}$. To prove the first part we realize that, for the bases of the prism (shaded area), $n_1 = 0$ and either $t_i = 0$ (for plane stress) or $u_i = 0$ (for plane strain). So these areas do not contribute to the integral. For the lateral area, we have that for segments AB and CD both $n_1 = 0$ and $t_i = 0$, and so these areas do not contribute to the integral either. The only contributing parts are the lateral areas defined by contours Γ_1 and Γ_2. Since the state is uniform across the thickness b, we set $dA = b\, ds$, where ds is the arc length along the contour, and write

$$J_\phi = b \int_{\Gamma_1} (\overline{\mathcal{U}} n_1^{\mathcal{B}} - t_i^{\mathcal{B}} u_{i,1}) ds + b \int_{\Gamma_2} (\overline{\mathcal{U}} n_1^{\mathcal{B}} - t_i^{\mathcal{B}} u_{i,1}) ds \tag{4.4.2}$$

Now, the integrals appearing in this expressions are much like the J-integral expression in (2.1.48). Recalling, however, the convention of signs for the normal and traction vectors in the J-integral expression, it turns out that $n_1^{\mathcal{B}} = n_1$ and $t_i^{\mathcal{B}} = t_i$ for the outermost contour (Γ_1 in the figure), and $n_1^{\mathcal{B}} = -n_1$ and $t_i^{\mathcal{B}} = -t_i$ for the innermost contour (Γ_1 in the figure). Thus, we finally get

$$J_\phi = b[J(\Gamma_1) - J(\Gamma_2)] \tag{4.4.3}$$

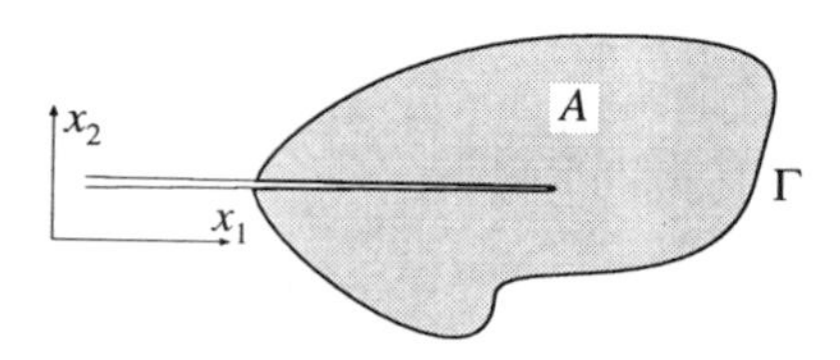

Figure 4.4.2 A contour Γ surrounding the crack tip defines a subbody that includes the crack tip.

which is half of the proof. The other half is obtained by applying the divergence theorem to transform J_ϕ to a volume integral; dropping the superscripts $\mathcal{B}$ which are no longer necessary, setting $t_i = \sigma_{ij}n_j$ and applying the divergence theorem, we get:

$$J_\phi = \oint_{S_\mathcal{B}} (\overline{\mathcal{U}}n_1 - u_{i,1}\sigma_{ij}n_j)dA = \int_\mathcal{B} [\overline{\mathcal{U}}_{,1} - (u_{i,1}\sigma_{ij})_{,j}]dV \tag{4.4.4}$$

Now, the elastic energy density $\overline{\mathcal{U}}$ can be always written as a function of the deformation gradient $u_{i,j}$ (in fact, we are dealing with hyperelastic materials, see e.g., Malvern 1969). So the first term in the integral can be written as:

$$\overline{\mathcal{U}}_{,1} = \frac{\partial \overline{\mathcal{U}}(u_{i,j})}{\partial x_1} = \frac{\partial \overline{\mathcal{U}}(u_{i,j})}{\partial u_{i,j}} u_{i,j1} = \sigma_{ij}u_{i,j1} \tag{4.4.5}$$

in which the last equality comes from the well-known property of the hyperelastic materials that the strain energy density acts as a potential for the stresses (see, e.g., Malvern 1969). Substituting this result into (4.4.4), expanding the second term in the integral, and setting $u_{i,j1} = u_{i,1j}$, we get

$$J_\phi = -\int_\mathcal{B} u_{i,1}\sigma_{ij,j}dV \tag{4.4.6}$$

Since the equilibrium equation in absence of body forces requires $\sigma_{ij,j} = 0$, it turns out that the integral $J_\phi = 0$ whatever the body actually defined by the contours, and from (4.4.3) we conclude that $J(\Gamma_1) = J(\Gamma_2) = J$, whatever Γ_1 and Γ_2.

Now we are ready to summarize the conditions that must be met for the J–integral to be path-independent:

1. The crack faces must be stress-free (this makes zero the contribution from segments AB and BC).
2. The body defined by the two contours must be elastic (hyperelastic). If the integral should be path independent no matter how small the innermost contour, the whole body must be elastic except at the very crack tip (or in a vanishingly small area around it).
3. The body $\mathcal{B}$ defined by the two contours must not contain any singularity (this is required for the divergence theorem to be applicable).
4. No body forces must act (this is required for the equilibrium equation to reduce to $\sigma_{ij,j} = 0$).

4.4.2 Further Contour Integral Expressions for $\mathcal{G}$ in LEFM

Other expressions for $\mathcal{G}$ taking the form of contour integrals can be obtained. Consider the simplest cases of plane bodies in absence of volume forces, with stress-free crack faces and linear elasticity valid everywhere. Then we can consider a subbody $\mathcal{A}$ defined by the contour Γ as shown in Fig. 4.4.2. The external incremental work performed on this subbody during an arbitrary incremental process, in which the crack length grows by δa and the displacements increase by δu_i, is just the work performed by the surface tractions t_i acting on the contour Γ. So:

$$\delta\mathcal{W} = b\int_\Gamma t_i \delta u_i \, ds \tag{4.4.7}$$

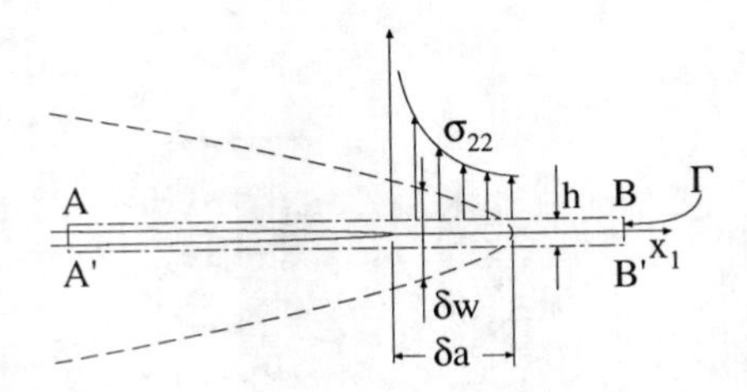

Figure 4.4.3 Contour used to deduce Irwin's relationship.

On the other hand, the strain energy $\mathcal{U}$ of a linear elastic body loaded only by surface tractions is known to be expressible as the surface integral of $t_i u_i/2$. Thus, we have

$$\mathcal{U} = b \int_\Gamma \frac{1}{2} t_i u_i \, ds \tag{4.4.8}$$

from which it follows that the variation of strain energy is expressible as

$$\delta\mathcal{U} = b \int_\Gamma \frac{1}{2} (\delta t_i \; u_i + t_i \; \delta u_i) \, ds \tag{4.4.9}$$

Substituting (4.4.7) and (4.4.9) into (2.1.3) and rearranging, we get the general expression

$$\mathcal{G}\delta a = \frac{1}{2} \int_\Gamma (t_i \; \delta u_i - \delta t_i \; u_i) \, ds \tag{4.4.10}$$

As already pointed out for the J-integral, the foregoing contour integral must be path independent because of its physical derivation. However, it can be shown purely mathematically to be independent of the contour Γ (as long as it contains the crack tip) using an approach similar to that used for the J-integral, except that the use of the divergence theorem is replaced by the application of the reciprocity theorem.

The virtual variations δ in the previous equations correspond to any arbitrary process as long as the equilibrium equations and Hooke's law are satisfied and the crack faces remain traction-free. One such virtual process consists in a translation of the fields by δa parallel to the crack line, i.e., along the x_1 axis. In such a case, the displacements and tractions vary as

$$\delta u_i = u_i(x_1 - \delta a, x_2) - u_i(x_1, x_2) = -u_{i,1} \; \delta a \tag{4.4.11}$$

$$\delta t_i = \sigma_{ij}(x_1 - \delta a, x_2) n_j - \sigma_{ij}(x_1, x_2) n_j = -\sigma_{ij,1} n_j \; \delta a \tag{4.4.12}$$

Substitution of these expressions into (4.4.10) leads to the following contour integral expression for $\mathcal{G}$:

$$\mathcal{G} = \frac{1}{2} \int_\Gamma (\sigma_{ij,1} \; u_i - \sigma_{ij} \; u_{i,1}) n_j \, ds \tag{4.4.13}$$

4.4.3 Further Proof of the Irwin Relationship

A simple proof of the Irwin relationship $\mathcal{G} = K_I^2/E'$ was presented in Chapter 2. It is, of course, possible to prove it using the J-integral performed along a contour of very small size around the crack tip, so that only the singular term of the near-tip expansion has to be considered (it can further be proved that the contribution of the other terms to the J-integral are identically zero for any contour). However, the resulting algebra is quite lengthy, so we prefer to develop an alternative proof using the contour integral expression in (4.4.10).

To facilitate the computation of the integral, we take the contour Γ to be a very narrow rectangle that encloses the initial and final crack tips and has a vanishingly small depth h (Fig. 4.4.3). We consider pure Mode I. The only contribution to the integral is that corresponding to the upper and lower faces of the rectangle, which correspond, in the limit, to the planes $x_2 = 0^+$ and $x_2 = 0^-$. In this situation, the only nonvanishing component of the traction vector is t_2, and the only relevant component of the displacement is u_2. Also taking into account that, for the upper face, $t_2 = t_2^+ = \sigma_{22}$ and that, for the lower face, $t_2 = t_2^- = -\sigma_{22}$, Eq. (4.4.10) can be rewritten as

$$\mathcal{G}\,\delta a = \frac{1}{2}\int_{x_A}^{x_B}\left[\sigma_{22}(\delta u_2^+ - \delta u_2^-) - \delta\sigma_{22}(u_2^+ - u_2^-)\right]dx_1 = \frac{1}{2}\int_{x_A}^{x_B}\left[\sigma_{22}\delta w - \delta\sigma_{22} w\right]dx_1 \tag{4.4.14}$$

where w is the crack opening. Since obviously

$$\begin{aligned} \sigma_{22} = \delta\sigma_{22} &= 0 \quad \text{for} \quad x_1 < 0 \\ w &= 0 \quad \text{for} \quad x_1 > 0 \\ \delta w &= 0 \quad \text{for} \quad x_1 > \delta a \end{aligned} \tag{4.4.15}$$

the expression is reduced to

$$\mathcal{G}\,\delta a = \frac{1}{2}\int_0^{\delta a}\sigma_{22}\,\delta w\,dx_1 \tag{4.4.16}$$

We now use the near-tip field expressions (4.2.24) and (4.2.25) to find that, over the integration interval and at a constant K_I

$$\sigma_{22} = \frac{K_I}{\sqrt{2\pi x_1}} \quad , \quad \delta w = \frac{8K_I}{\sqrt{2\pi}E'}\sqrt{\delta a - x_1} \tag{4.4.17}$$

After substitution into Eq. (4.4.16), this finally leads to

$$\mathcal{G}\,\delta a = \frac{2K_I^2}{\pi E'}\int_0^{\delta a}\sqrt{\frac{\delta a - x_1}{x_1}}dx_1 dx_1 = \frac{K_I^2}{E'}\,\delta a \tag{4.4.18}$$

in which the integration has been performed by means of the substitution $x_1 = \delta a\ \sin^2 t$. This ends the proof of Irwin's relationship for mode I. The proof is easily extended to a general loading. One can show that the three modes contribute with independent terms to the expression for $\mathcal{G}$. The contribution of the in-plane shear mode II is obtained by replacing σ_{22}, u_2, and K_I in the foregoing equations by σ_{21}, u_1, and K_{II}, respectively. In a similar way, the contribution of the antiplane shear mode III is obtained by replacing σ_{22}, u_2, K_I, and E' by σ_{23}, u_3, K_{III}, and $2G$, respectively. Thus, the general Irwin relationship (valid for coplanar crack growth only) is

$$\mathcal{G} = \frac{K_I^2}{E'} + \frac{K_{II}^2}{E'} + \frac{K_{III}^2}{2G}. \tag{4.4.19}$$

4.4.4 Other Path-Independent Integrals

As just seen, there are various contour integral expressions which are path-independent when the path encloses a crack tip. They usually correspond to contour integrals that vanish identically when the contour does not contain any crack tip. This means that the value of the path-independent integral is related to the crack-tip singularity and provides a measure of its intensity. Some of the integrals, such as the J-integral and the integrals in (4.4.10) and (4.4.13), give $\mathcal{G}$ for self-similar (coplanar) crack growth. Others are related to different kinds of kinetics of the fracture or singular zone; still others are extensions of the classical path-independent integrals to take into account situations in which some of the hypotheses listed at the end of Section 4.4.1 no longer hold (body forces or loading at the crack faces being the most usual modifications). For an account of the most important path-independent integrals, the reader is referred to the book by Kanninen and Popelar (1985). Here we only mention four path-independent integrals: I, J_α, L_3, and M.

The I–integral was proposed by Bui (1978) based on complementary energy analysis; it is the dual of the J–integral and is written as:

$$I = -\int_\Gamma\left(\overline{\mathcal{U}}^* n_1 - \sigma_{ij,1}n_j u_i\right)ds \tag{4.4.20}$$

where $\overline{\mathcal{U}}^*$ is the complementary energy density.

The J_k integrals (component of the vectorial $\boldsymbol{J}$–integral) and the L and M integrals established by Knowles and Sternberg (1972), can be understood as generalizations of the J–integral by appropriate changes of its integrand. Their definition is as follows:

$$J_k = \int_\Gamma \left(\overline{\mathcal{U}} n_k - t_i u_{i,k}\right)\, ds \tag{4.4.21}$$

$$L = \int_\Gamma e_{3kl}\left(\overline{\mathcal{U}} n_k x_l + t_k u_l - t_i u_{i,k} x_l\right)\, ds \tag{4.4.22}$$

$$M = \int_\Gamma \left(\overline{\mathcal{U}} n_k x_k - t_i u_{i,k} x_k\right)\, ds \tag{4.4.23}$$

where e_{mkl} is the permutation symbol, such that the vector cross product of two vectors $\boldsymbol{c} = \boldsymbol{a} \times \boldsymbol{b}$ is written in component form as $c_m = e_{mkl} a_k b_l$. The J_k, L, and M integrals vanish if performed along a closed contour when hypotheses 1–4 listed at the end of Section 4.4.1 hold; M further requires the material to be linear elastic. When the contour encloses a singularity, these integrals give, in general, a nonzero but path-independent value; however, to ensure that J_2 be path-independent, the crack faces up to the crack tip must be included as part of the contour.

4.4.5 Exercises

4.19 Show that the integral $\int_\Gamma d(u_i \sigma_{i2})$ is path-independent when Γ is a contour such as that in Fig. 4.4.2. Show, moreover, that its value is zero if the crack faces are stress free.

4.20 Show that $I = J - \int_\Gamma d(u_i \sigma_{i2})$ and demonstrate that $I = J = \mathcal{G}$.

4.21 Show that $M = 0$ for a closed contour that includes no singularity. (Hint: write the elastic energy density $\overline{\mathcal{U}}$ for a linear elastic material as $\sigma_{ij} u_{i,j}$.)

4.22 Show that if the contour includes a singularity, the M–integral is sensitive to the location of the origin of coordinates. In particular, if $\{O, x_1, x_2\}$ is a coordinate system with origin at O and $\{O', x_1', x_2'\}$, a second system parallel to the first, with O' defined by its coordinates (x_1^O, x_2^O) with respect to O, show that $M_O = M_{O'} + J_k x_k^O$.

4.23 Show that if the origin of coordinates is taken to lie at the tip of a sharp crack in an elastic body, with x_1 parallel to the crack line, then $M = 0$. (Hint: integrate along a circle of a small radius r with center at the crack tip, and use the near-tip expansion to show that $M = r \int_{-\pi}^{\pi} f(\theta) d\theta$ where $f(\theta)$ is regular. Then take the limit for $r \to 0$.)

4.24 Show that if O is the position of the initial crack tip and the crack growth is coplanar, the value of the M–integral with origin at O is $M = J\Delta a = \mathcal{G}\Delta a$, where J and $\mathcal{G}$ are the instantaneous values of the J–integral and of the energy release rate. Show that if the crack growth is quasi-static, then $M = \mathcal{R}\Delta a$, where $\mathcal{R}$ is the crack growth resistance. It is understood that the integration contours contain the crack tip at any instant over the interval of interest.

4.5 Mixed Mode Fracture Criteria.

In Chapter 2 we derived that, in LEFM, mode I fracture should take place as the energy release rate $\mathcal{G}$ reaches a critical constant value G_f. Based on the near-tip field and Irwin's relationship, this was shown to be equivalent to stating that fracture happens when $K_I = K_{Ic}$, with $K_{Ic} = \sqrt{E' G_f}$ = fracture toughness.

This order of reasoning is inverted in many textbooks, which start with the evaluation of the stress field around the crack tip to show that, for pure opening mode and in the limit of LEFM, the vanishingly small fracture process zone is surrounded by a linear elastic material with stress and strain fields uniquely determined, for any type of loading, geometry, or structure size, by the stress intensity factor K_I. It follows that a critical value K_{Ic} must exist so that when the actual K_I is lower, no crack growth can take

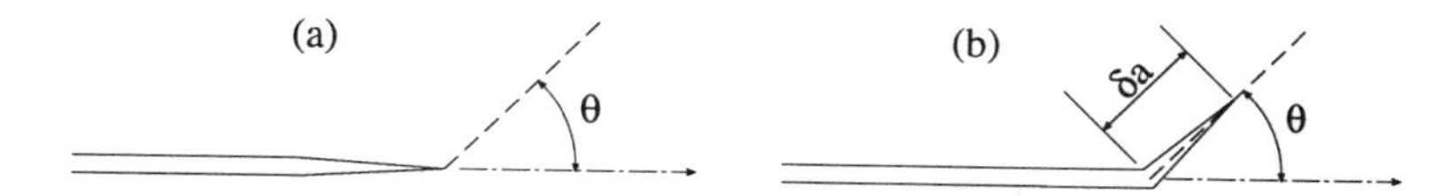

Figure 4.5.1 Crack kinking in mixed mode crack growth: (a) initial situation; (b) after crack advanced by δa.

place. Obviously, quasi-static mode I crack growth requires $K_I = K_{Ic}$, and exceeding the critical value will lead to dynamic crack growth. (With the static analysis performed in this book, we can fully describe quasi-static processes and detect the onset of dynamic cracking. To describe a dynamic crack growth, the analysis must take into account the inertia terms in the equilibrium equation. This is the subject of dynamic linear elastic fracture mechanics, a topic outside the scope of our book. See, for example, Kanninen and Popelar 1985.)

This reasoning may be extended to other fracture modes to obtain phenomenological fracture criteria. Hence, for pure modes II and III, critical stress intensity factors K_{IIc} and K_{IIIc} may be defined such that the crack growth may occur when the critical values are reached. But these parameters give only information for pure mode loadings, and do not allow following the cracking process, which will, in general, involve change from pure to mixed modes.

For mixed modes, the straight phenomenological approach consists in postulating that fracture may initiate when the values of K_I, K_{II}, and K_{III} verify a critical condition. This may be written in the form

$$f(K_I, K_{II}, K_{III}) = 0 \tag{4.5.1}$$

where f is an empirical function. An example is the ellipsoidal failure locus

$$\frac{K_I^2}{K_{Ic}^2} + \frac{K_{II}^2}{K_{IIc}^2} + \frac{K_{III}^2}{K_{IIIc}^2} - 1 = 0 \tag{4.5.2}$$

where K_{Ic}, K_{IIc}, and K_{IIIc} are material parameters to be determined by experiment.

This pure phenomenological approach may prove practically useful, but gives no insight into the way fracture proceeds. This is why there have been various attempts to extend to mixed modes the single-parameter coming from mode I analysis.

The simplest approach is to take advantage of the general expression (4.4.19) for the energy release rate and assume that the energy required to fracture an isotropic material is independent of the mode of fracture. Then the fracture criterion may be written as failure locus

$$\frac{K_I^2}{E'} + \frac{K_{II}^2}{E'} + \frac{K_{III}^2}{2G} - G_f = 0 \tag{4.5.3}$$

or

$$K_I^2 + K_{II}^2 + (1+\nu')K_{III}^2 - K_{Ic}^2 = 0 \tag{4.5.4}$$

However, these equations make sense only when the crack grows straight ahead, in its own plane. For isotropic bodies, this is observed to happen only for symmetric loading (pure mode I).

Restricting attention to in-plane loading, experiments show that an initially straight crack will kink when loaded in mixed mode, as shown in Fig. 4.5.1. To describe this behavior, two basic approaches have been used. The first is a direct extension of the energetic balance condition to crack kinking and is called the *maximum energy release rate criterion*. The second is based on the near-tip stress distribution and is called the *maximum principal stress criterion*. These two approaches will be examined next[3].

4.5.1 Maximum Energy Release Rate Criterion

This criterion may be set based on the following reasoning: for an isotropic body, for which all directions are materially identical, one may assume that the energy required to crack a unit surface, G_f, is independent

[3]A third approach, called the *strain energy density criterion*, may rise some theoretical objections and is given as an appendix.

of the orientation of this surface; however, the energy release rate does depend on the orientation θ of the newly cracked surface, i.e., $\mathcal{G} = \mathcal{G}(\theta)$, and it appears logical to assume that cracking will happen along the direction for which the condition $\mathcal{G}(\theta) = G_f$ is first met.

To be more precise, one has to express the energy release rate as a function of the direction of propagation and the loading. For infinitesimal crack propagation, the loading is completely defined by the near-tip fields of the initial crack, that is, by K_I and K_{II}. So,

$$\mathcal{G} = \mathcal{G}(K_I, K_{II}, \theta) \tag{4.5.5}$$

Assuming, as it appears logical, that the above function is monotonic both in K_I and K_{II}, the cracking direction, θ_c, and the loading for which cracking occurs are determined from the system of equations:

$$\mathcal{G}(K_I, K_{II}, \theta)\,|_{\theta=\theta_c} = \text{maximum} \tag{4.5.6}$$

$$\mathcal{G}(K_I, K_{II}, \theta_c) = G_f \tag{4.5.7}$$

The explicit form of these equations may be obtained by elastic analysis of a semi-infinite crack with a kink at its tip, a subject outside the scope of this book. The analysis shows that the result is essentially identical (Nuismer 1975) to that of the maximum principal stress criterion, which is conceptually different and is analyzed next.

4.5.2 Maximum Principal Stress Criterion

Erdogan and Sih (1963) proposed this criterion, which states that, for in-plane mixed mode, crack growth will happen along the direction for which the initial normal stress across the possible crack path (prior to crack growth) is tensile, principal, and maximum.

The tractions over a potential fracture plane (Fig. 4.5.1a) inclined at θ are $\sigma_{\theta\theta}$ (normal) and $\sigma_{r\theta}$ (tangential). The cartesian expressions for the stress components given by Eqs. (4.3.16) and (4.3.18) are easily transformed to cylindrical components:

$$\sigma_{\theta\theta} = \frac{1}{\sqrt{2\pi r}}\left[K_I \cos^3\frac{\theta}{2} - 3K_{II}\cos^2\frac{\theta}{2}\sin\frac{\theta}{2}\right] \tag{4.5.8}$$

$$\sigma_{r\theta} = \frac{1}{\sqrt{2\pi r}}\left[K_I \cos^2\frac{\theta}{2}\sin\frac{\theta}{2} + K_{II}\cos\frac{\theta}{2}\left(\cos^2\frac{\theta}{2} - 2\sin^2\frac{\theta}{2}\right)\right] \tag{4.5.9}$$

It may be noticed that, remarkably enough,

$$\frac{\partial\sigma_{\theta\theta}}{\partial\theta} = -\frac{3}{2}\sigma_{r\theta} \tag{4.5.10}$$

which implies that the following criteria are equivalent:

a. The circumferential stress is principal ($\sigma_{r\theta} = 0$).
b. The circumferential stress $\sigma_{\theta\theta}$ is maximum with respect to θ.
c. The magnitude $\sqrt{\sigma_{\theta\theta}^2 + \sigma_{r\theta}^2}$ of the traction vector is maximum with respect to θ.

any of which may be taken as the criterion for the direction of fracture. From the foregoing statement (a) and (4.5.9)

$$K_I \cos^2\frac{\theta_c}{2}\sin\frac{\theta_c}{2} + K_{II}\cos\frac{\theta_c}{2}\left(\cos^2\frac{\theta_c}{2} - 2\sin^2\frac{\theta_c}{2}\right) = 0 \tag{4.5.11}$$

This equation has one solution for $\cos(\theta_c/2) = 0$ ($\theta_c = \pm\pi$), which obviously corresponds to the preexisting crack surfaces and to a minimum of circumferential stress. Dividing the above equation by $\cos^3(\theta_c/2)$ one gets a second degree equation in $\tan(\theta_c/2)$ which gives the cracking directions as a function of the ratio K_I/K_{II}:

$$\tan\frac{\theta_c}{2} = \frac{K_I}{4K_{II}}\left(1 \pm \sqrt{1 + 8\frac{K_{II}^2}{K_I^2}}\right) \tag{4.5.12}$$

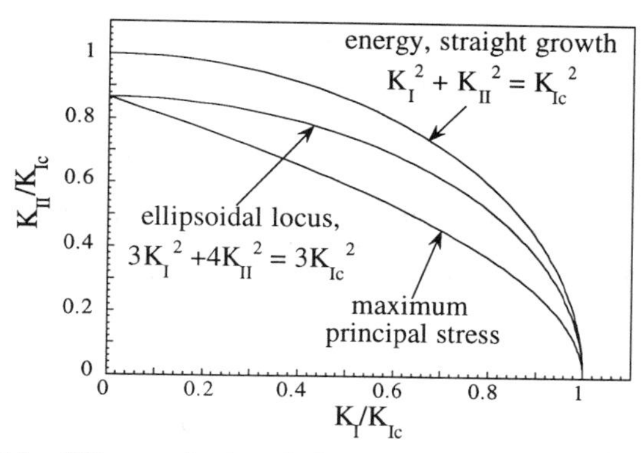

Figure 4.5.2 Failure loci for different mixed mode fracture criteria (drawn for positive K_{II} only; the loci are symmetric with respect to the K_I axis).

To find the load level at which crack growth initiates, we notice that, for a line oriented at θ_c, the traction vector is normal and takes the form $\sigma_{\theta\theta} = K(\theta_c)/\sqrt{2\pi r}$, where $K(\theta)$ is the function appearing in square brackets in Eq. (4.5.8). Hence, the traction distribution over the cracking plane is identical in shape to that found for pure mode I. It is logical, then, to assume that cracking will happen when the intensity $K(\theta_c)$ reaches the critical value K_{Ic}:

$$K_I \cos^3 \frac{\theta_c}{2} - 3K_{II} \cos^2 \frac{\theta_c}{2} \sin \frac{\theta_c}{2} = K_{Ic} \tag{4.5.13}$$

The loading states that produce crack initiation, the *failure locus*, are obtained by substituting the solution for the cracking direction given in Eq. (4.5.12) into Eq. (4.5.13). However, this procedure is dull, tedious work. A better way is to first find an explicit parametric representation of the failure locus. To this end, we solve the system of Eqs.(4.5.11) and (4.5.13) for K_I and K_{II} and get

$$K_I = K_{Ic} \frac{\cos^2(\theta_c/2) - 2\sin^2(\theta_c/2)}{\cos^3(\theta_c/2)} \tag{4.5.14}$$

$$K_{II} = -K_{Ic} \frac{\sin(\theta_c/2)}{\cos^2(\theta_c/2)} \tag{4.5.15}$$

These parametric equations show that K_I at failure is an even function of the kink angle θ_c, while K_{II} is an odd function of it. Moreover, when θ_c spans the interval $[-\pi, \pi]$, K_I/K_{Ic} spans the interval $[-\infty, 1]$, while K_{II}/K_{Ic} spans the whole real line. However, negative values of K_I imply interpenetration of the preexisting crack faces, which is, in general, forbidden and implies contact stresses (normal and frictional), which would require a nonlinear analysis. Therefore, we restrict our attention to the interval for which $K_I > 0$, which turns out to correspond to $|\theta_c| < 2\arcsin(1/\sqrt{3}) \approx 70.5°$. Fig. (4.5.2) shows the failure locus for this interval. It also shows two other failure loci; one corresponds to the coplanar crack growth, Eq. (4.5.4); the other to a more general criterion such as the ellipsoidal Eq. (4.5.2), with $K_{IIc} = \sqrt{3}K_{Ic}/2$; in both cases the loci have been plotted for $K_{III} = 0$.

Exercises

4.25 A large panel subjected to tension has a crack of 40 mm in length lying at an angle of 45° with the loading direction. Assuming that the material behaves as dictated by LEFM with $K_{Ic} = 40$ MPa m$^{1/2}$, determine the failure stress and the crack propagation direction for (a) straight propagation with $\mathcal{G} = K_{Ic}^2/E'$ criterion, and (b) maximum principal stress criterion.

4.26 A pipe 500 mm in diameter and 5 mm in thickness is subjected to internal pressure of 10 MPa. Joints ensure that the pipe is nearly unstressed in the axial direction. Flaws may grow along the helical weld at an angle of 60° to the axis of the pipe. If the material shows a brittle behavior with $K_{Ic} = 50$ MPam$^{1/2}$, find the maximum size of the through-crack that the pipe can withstand without catastrophic failure assuming:

(a) straight propagation with $\mathcal{G} = K_{Ic}^2/E'$ criterion, and (b) maximum principal stress criterion. Compare the results with the maximum size of cracks lying parallel to the pipe axis.

4.27 A cylindrical pressure vessel 1 m in diameter and 20 mm in thickness is ended by semi-spherical caps, and is subjected to internal pressure of 4 MPa. Flaws may grow along the helical weld of the cylindrical wall at an angle of 45° with the axis of the cylinder. If the material shows a brittle behavior with $K_{Ic} = 45$ MPam$^{1/2}$, find the maximum size of the through-cracks that the vessel can withstand without catastrophic failure assuming: (a) straight propagation with the criterion $\mathcal{G} = K_{Ic}^2/E'$, and (b) maximum principal stress criterion. Compare the result with the maximum size of cracks lying parallel to the pipe axis and normal to the pipe axis.

Appendix: Strain Energy Density Criterion

Proposed by Sih (1974), this criterion postulates that the crack growth will take place in the direction in which the strain energy density $\overline{\mathcal{U}} = \sigma_{ij}\varepsilon_{ij}/2$ is minimum. This criterion, however, does not appear to rest on a sound foundation in the thermomechanics of solids. It has, nevertheless, been used to describe mixed mode fracture direction, with reasonable results, and so we present it briefly.

The equations defining the failure locus and the cracking angle when LEFM conditions hold may be obtained with no special difficulties for the case of in-plane loading. They ensue from the near-tip expression (4.3.16) by using the expression $\overline{\mathcal{U}} = [(1+\nu')\sigma_{ij}\sigma_{ij} - \nu'(\sigma_{kk})^2]/2E'$ for the strain energy density. The details of the algebra will not be given here (for further details and problem solving using this method, see Gdoutos 1989). The essential result is that the near-tip strain energy density distribution takes the form

$$\overline{\mathcal{U}} = \frac{1}{r} S(\theta, K_I, K_{II}) \tag{4.5.16}$$

where $S(\theta, K_I, K_{II})$ is the *strain energy density factor* given by

$$S(\theta, K_I, K_{II}) = \frac{1+\nu'}{8\pi E'}[s_{11}(\theta)K_I^2 + 2s_{12}(\theta)K_I K_{II} + s_{22}(\theta)K_{II}^2] \tag{4.5.17}$$

in which

$$s_{11}(\theta) = (1+\cos\theta)(\kappa - \cos\theta) \tag{4.5.18}$$

$$s_{12}(\theta) = \sin\theta[2\cos\theta - (\kappa - 1)] \tag{4.5.19}$$

$$s_{22}(\theta) = (\kappa+1)(1-\cos\theta) + (1+\cos\theta)(3\cos\theta - 1) \tag{4.5.20}$$

with κ given in (4.1.27).

We first note from (4.5.16) that the energy density is singular at the origin, so that the extremes are not well defined. Sih (1974) postulated that the energy density at a fixed distance r_0 from the crack tip must be minimum with respect to the orientation and equal to a critical value. This is equivalent to minimizing the strain energy density factor $S(\theta, K_I, K_{II})$. Thus, θ_c and the failure locus are defined by the equations:

$$\left.\frac{\partial S(\theta, K_I, K_{II})}{\partial\theta}\right|_{\theta=\theta_c} = 0, \qquad \left.\frac{\partial^2 S(\theta, K_I, K_{II})}{\partial\theta^2}\right|_{\theta=\theta_c} > 0 \tag{4.5.21}$$

$$S(\theta_c, K_I, K_{II}) = S_C \tag{4.5.22}$$

where S_C is the critical strain energy density factor, a material property.

For mode I this criterion is strictly equivalent to the other criteria, and the following relationship may be shown to exist between S_C and the critical stress intensity factor K_{Ic} and the fracture energy G_f:

$$S_C = \frac{1-\nu'}{2\pi}\frac{K_{Ic}^2}{E'} = \frac{1-\nu'}{2\pi} G_f \tag{4.5.23}$$

With this relation, the equations (4.5.21) and (4.5.22) may be reduced, after lengthy algebra, to the following complicated equation system:

$$\sin\theta_c(\cos\theta_c - m)K_I^2 + 2(\cos 2\theta_c - m\cos\theta_c)K_I K_{II} + \sin\theta_c(m - 3\cos\theta_c)K_{II}^2 = 0 \tag{4.5.24}$$

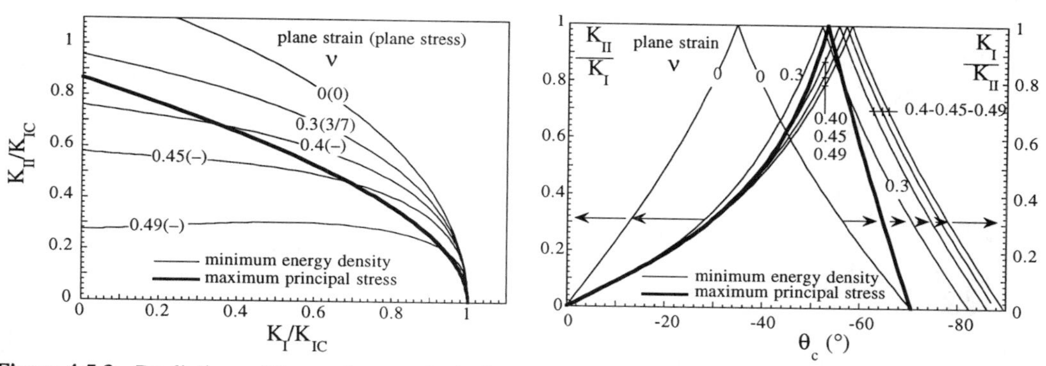

Figure 4.5.3 Predictions of the maximum principal stress criterion compared to the predictions of the minimum strain energy density criterion for various Poisson ratios: failure loci (left), and crack growth angle θ_c vs. mixed mode proportion (right).

$$(1+\cos\theta_c)\,(2m+1-\cos\theta_c)\,K_I^2+4\sin\theta_c\,(\cos\theta_c-m)\,K_I K_{II}+$$
$$+\left[2(m+1)(1-\cos\theta_c)+(1+\cos\theta_c)(3\cos\theta_c-1)\right]K_{II}^2=4mK_{Ic}^2 \qquad (4.5.25)$$

where $m=(1-\nu')/(1+\nu')$.

The best way to produce the failure locus is to handle the foregoing system parametrically, as follows: For a given cracking direction θ_c, first solve for the ratio K_I/K_{II} from the first equation; then use the second equation to solve for K_{II}/K_{Ic}; finally, substitute back into the K_I/K_{II} ratio previously obtained to get K_I/K_{Ic}. In solving for K_I/K_{II} from the first equation, care must be taken to select the appropriate root of the quadratic equation. This is best done by imposing the condition of minimum in (4.5.21), which is obtained by differentiating (4.5.24) with respect to θ_c:

$$(\cos 2\theta_c-m\cos\theta_c)\left(\frac{K_I}{K_{II}}\right)^2+2(m\sin\theta_c-2\sin 2\theta_c)\frac{K_I}{K_{II}}+(m\cos\theta_c-3\cos 2\theta_c)>0 \quad (4.5.26)$$

The foregoing procedure has been used to determine the failure loci and crack direction curves for various values of ν'. Fig. 4.5.3 shows the results, together with the predictions of the maximum circumferential stress criterion. It is important to note that the curves corresponding to the minimum energy density criterion show a strong dependence on Poisson's ratio and on deformation mode (plane strain or plane stress). In particular, for plane strain and an incompressible material ($\nu=0.5,\nu'=1$), the equations degenerate and pure mode II failure may take place under a vanishingly small load. This is a serious questionable aspect of this criterion. For the usual value of Poisson ratio for metals (around 0.3), the predictions lie not too far from those of the maximum principal stress criterion.

5

Equivalent Elastic Cracks and R-curves

As already mentioned in the introductory chapter, the fracture behavior of concrete deviates significantly from LEFM, and various nonlinear theories have been proposed to deal with this problem. In this chapter, we first review experimental evidence concerning the inability of LEFM to describe the fracture of concrete. Then we examine the simplest nonlinear theories, based on equivalent elastic cracks.

All the problems lie in the fact that, for concrete as well as other (apparently brittle) materials, the nonlinear zone ahead of the crack tip is relatively large *compared to* structure dimensions and, therefore, its effect cannot be neglected as in LEFM. The reason that this zone is so large, as compared to ductile material with homologous strength and toughness, will be discussed in Section 5.2. It will be seen that progressive microcracking and softening is at the root of the problem.

After that, the simplest nonlinear approximation, the effective crack concept, will be presented and analyzed (Section 5.3). This is a simple approach, no more than a first-order correction to LEFM which, however, embodies important concepts because the fracture behavior of *all* quasibrittle materials can be described in this way for large enough cracks in large enough structures. The intrinsic structure size—a concept that allows consistent comparison of the response of structures of different geometries, and the analysis of how large a test specimen should be for its fracture to be describable by LEFM within a given maximum error, can be based on this simple model (although originally more sophisticated models were used).

The concept of effective elastic crack has been used to determine the fracture toughness in concrete. Section 5.4 briefly describes two of these approaches, one used by Swartz and co-workers at the Kansas State University and the other by Nallathambi and Karihaloo at the University of Sydney. These models require measuring the effective crack extension at peak load and then computing the toughness using LEFM equations. They differ only in how the equivalent crack extension is measured (or computed from the measured variables).

The foregoing approximations cannot be used to predict the behavior of a general structure, unless the structure as well as the crack is so large that LEFM applies. This is so because these approximations do not include a predictive equation for the effective crack extension at peak load. The simplest of the models used for concrete that can predict the peak load of virtually any structure (including uncracked ones, if some simple extra hypotheses are added) is the Jenq-Shah two-parameter model which is analyzed in Section 5.5. A brief extract of the RILEM recommendation for the determination of the two parameters of the Jenq-Shah model is also included.

Next in complexity are the R-curve models in which the nonlinearity is included in a crack growth rule in which the crack growth resistance $\mathcal{R}$ is assumed to evolve as the crack extends. This represents an entire family of complete mechanical models, from which, given the R-curve, the full load-displacement response of any precracked structure can be predicted. These models and their use are examined in some detail in Section 5.6, together with a brief reference of the methods to determine them. Section 5.7 discusses the structural stability based on R-curves and closes this chapter. The determination of R-curves from size effect is deferred until the next chapter.

5.1 Variability of Apparent Fracture Toughness for Concrete

The early researchers on concrete fracture often attempted to apply LEFM directly to evaluate the fracture tests. In this case, the fracture toughness K_{Ic} is calculated from the peak load P_u (or the nominal strength

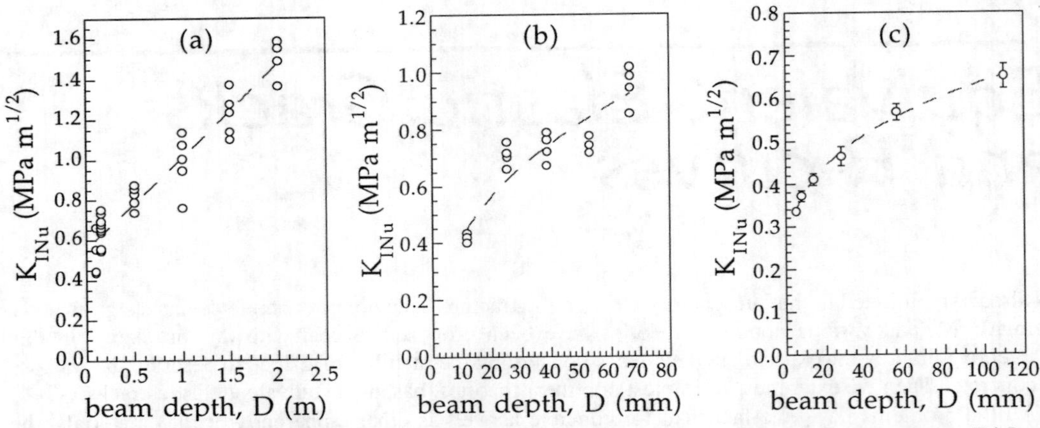

Figure 5.1.1 Dependence of the apparent fracture toughness on the specimen size: (a) concrete (Di Leo's data, after Hillerborg 1984); (b) mortar (after Ohgishi et al. 1986); (c) hardened cement paste (after Higgins and Bailey 1976).

σ_{Nu}) using the LEFM formula for a propagating crack in a specimen. If the geometry is so-called *positive* (K_I at constant load increasing with the crack length), the onset of crack extension coincides with the peak load and (2.3.11) yields:

$$K_{Ic} = \sigma_{Nu}\sqrt{D}\,k(\alpha_0) \tag{5.1.1}$$

where α_0 is the relative initial crack depth (usually the notch depth in concrete). The condition of positiveness of the geometry is the same as $k'(\alpha_0) > 0$, where the prime indicates the derivative. Most of the specimens used in fracture testing have positive geometry.

It turned out, however, that the toughness values thus determined were not invariant with respect to the beam size, shape, or notch depth. This means that (5.1.1) did not supply the true fracture toughness, but an apparent fracture toughness that varied with testing conditions. To emphasize this fact, we use in this book a special notation for the apparent fracture toughness: first, we define the nominal stress intensity factor K_{IN} as the stress intensity factor computed for the actual load and the initial crack length, i.e.,

$$K_{IN} = \sigma_N\sqrt{D}\,k(\alpha_0) \tag{5.1.2}$$

Then, the apparent fracture toughness K_{INu} is defined as the value of K_{IN} at peak load:

$$K_{INu} = \sigma_{Nu}\sqrt{D}\,k(\alpha_0) \tag{5.1.3}$$

This equation is nearly identical to (5.1.1), but the conceptual difference is huge: K_{Ic} in (5.1.1) must be a material property, independent of testing details; K_{INu} need not. Of course, for positive geometries,

$$K_{INu} = K_{Ic} \tag{5.1.4}$$

if LEFM applies to the particular experimental conditions. The discussion of the conditions for which this is satisfied is one of the main concerns of this chapter.

That the apparent fracture toughness was dependent on various experimental conditions was the subject of many studies during the 1970s. The initial results were somewhat contradictory, as shown in the historical review by Mindess (1983). However, the dependence of the apparent toughness of concrete on the size of the specimen was already documented by Walsh (1972, 1976) and today it is widely accepted. This effect (the size effect) is perfectly clear in the examples in Chapter 1 where the experimental evidence of transitional size effect was presented: if LEFM applied, all the results in Figs. 1.5.2–1.5.7 would have to lie on the LEFM straight line of slope $-1/2$, and they clearly do not . Further evidence is furnished by Di Leo's data cited by Hillerborg (1984), by the work of Ohgishi et al. (1986), for mortar, and by the results of Higgins and Bailey (1976) for hardened cement paste (Figs. 5.1.1a–c).

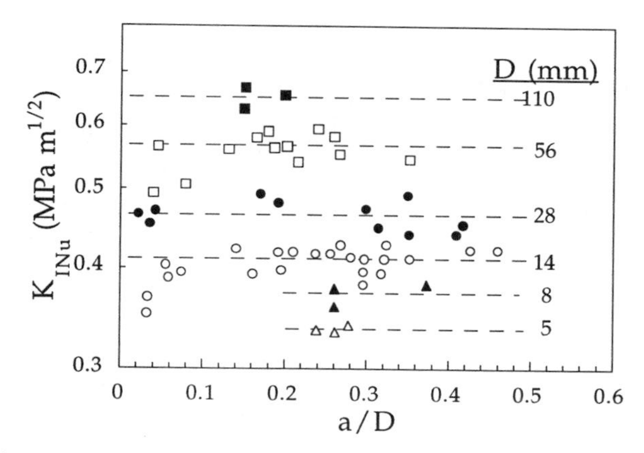

Figure 5.1.2 Dependence of apparent fracture toughness on specimen size and initial notch depth for hardened cement paste (after Higgins and Bailey 1976).

The dependence of the apparent toughness on the notch or crack length is not so clear, and one can find in the literature experimental results supporting any possibility: apparent toughness increasing, decreasing, or nearly constant with crack length. However, one must keep in mind that constancy of apparent fracture toughness with crack length, although necessary, is not sufficient to guarantee that LEFM holds. To illustrate this, Fig. 5.1.2 shows experimental results of Higgins and Bailey (1976) on hardened cement paste, from which it is clear that the apparent toughness depends on the size while being nearly constant with the initial notch depth, except for notches shorter than 0.1 of the beam depth.

Now, the question is why apparently brittle materials such as concrete, mortar, or hardened cement paste do not follow LEFM. An early explanation of the size and geometry dependence of K_{INu}, offered by Shah and McGarry (1971), was that the microcracks were arrested by the relatively large aggregate particles; this indeed does occur, but it cannot be the fundamental reason, because the same kind of deviation is observed in tough ceramics which have no large particles, and in polymers — the so-called crazing— which have no particles at all. Another reason for the variation of K_{INu} was suggested by Swartz, Hu et al. (1982), Swartz, Hu and Jones (1978) and Swartz and Go (1984), who noted that the data evaluations in the early tests (Naus and Lott 1969; Walsh 1972, 1976) used the notched depth a_0 to substitute for the crack length a, which violated one of the requirements of the ASTM method for metals, namely, that a true crack, not a notch, should be used. However, other experimental results on cement paste and mortar show that the values of K_{INu} based on notched specimens are essentially independent of the notch width (Ohgishi et al. 1986). In fact, a notch with a width smaller than the major inhomogeneities in the material (the aggregate or grain size) should be approximately equivalent to a crack from a macroscopical point of view.

Today it is clear that the fundamental reason for the deviation of the fracture behavior of concrete and other quasibrittle materials from LEFM is the existence of fracture process zone that is not small enough compared to the structure dimensions. This is the consequence of the progressive softening of the material. The next section gives a crude approach to estimating the size of the fracture process zone and the effect of progressive softening, which is the key to explaining the observed deviations from LEFM.

5.2 Types of Fracture Behavior and Nonlinear Zone

As just explained, the deviation from LEFM displayed by concrete and other quasibrittle materials is mainly related to the development of a large inelastic zone ahead of the crack tip. In this section, we investigate the differences in fracture behavior between ductile and quasibrittle materials and make some rough estimates of the size of the nonlinear zone. In particular, we show that the fracture process zone of a quasibrittle material can be one order of magnitude larger than that for a ductile material of similar macroscopic strength and toughness.

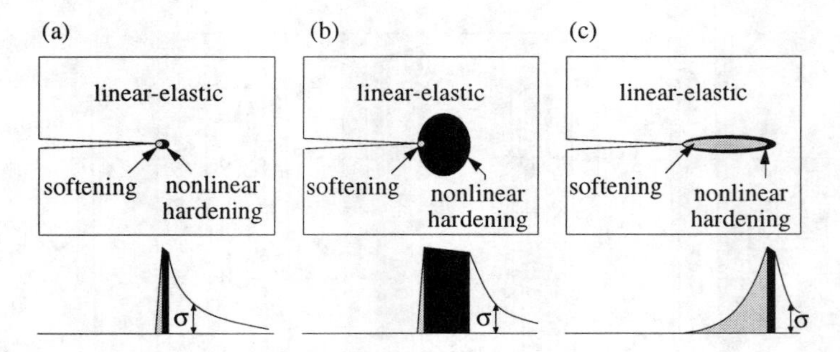

Figure 5.2.1 Types of fracture process zone. Diagrams at the bottom show the trends of the stress distributions along the crack line. Adapted from Bažant (1985a).

5.2.1 Brittle, Ductile, and Quasibrittle Behavior

The fracture process zone is a nonlinear zone characterized by progressive softening, for which the stress decreases at increasing deformation. This zone is surrounded by a nonsoftening nonlinear zone characterized by hardening plasticity or perfect plasticity, for which the stress increases at increasing deformation or remains constant. Together, these two zones form a nonlinear zone. Depending on the relative sizes of these two zones and of the structure, one may distinguish three types of fracture behavior (Fig. 5.2.1). In the first type of behavior (Fig. 5.2.1a), the whole nonlinear zone (and thus also the fracture process zone) is small compared to the structure size. Then, the entire fracture process takes place almost at one point—the crack tip, the whole body is elastic, and linear elastic fracture mechanics (LEFM), which was described in the preceding three chapters, can be used. This type of model is a good approximation for certain very brittle materials such as plexiglass, glass, brittle ceramics, and brittle metals. It must be emphasized that applicability of LEFM is relative — the structure must be sufficiently large compared with the fracture process zone. Thus, the overall behavior of extremely large concrete structures, such as gravity dams, can be described by LEFM, while the behavior of a very small part made of a brittle fine-grained ceramic cannot (and is of the third type as described below).

In the second and third types of behavior (Fig. 5.2.1b,c) the ratio of the nonlinear zone size to the structure size is not sufficiently small, and then LEFM is inapplicable, although it can still be applied in a certain equivalent sense if the nonlinear zone is not very large. In the second type of behavior (Fig. 5.2.1b), we include situations where most of the nonlinear zone consists of elastoplastic hardening or perfect yielding, and the size of the actual fracture process zone in which the breaking of the material takes place is still small. Many ductile metals, especially various tough alloys, fall into this category, and its general type may defined as *ductile.* This kind of behavior is best treated by a specialized branch of fracture mechanics—the elasto-plastic fracture mechanics—which is outside the scope of this book.

The third type of behavior (Fig. 5.2.1c), which is of main interest in this book, includes situations in which a major part of the nonlinear zone undergoes progressive damage with material softening, due to microcracking, void formation, interface breakages and frictional slips, and other similar phenomena (e.g., crystallographic transformation in certain ceramics). The zone of plastic hardening or perfect yielding in this type of behavior is often negligible, i.e., there is a rather abrupt transition from elastic response to damage. This happens for concrete, rock, ice, cemented sands, stiff clays, various toughening ceramics and fiber composites, wood particle board, paper, coal, etc. We call these materials *quasibrittle* because even if no appreciable plastic deformation takes place, the size of the fracture process zone is large enough to have to be taken into account in calculations, in contrast with the true brittle behavior in which LEFM applies. The distinction from the second type of behavior, however, may be blurred. For example, when concrete is subjected to a large compression parallel to the crack plane (as in splitting fracture), both the hardening nonlinear zone and the fracture process zone may be large.

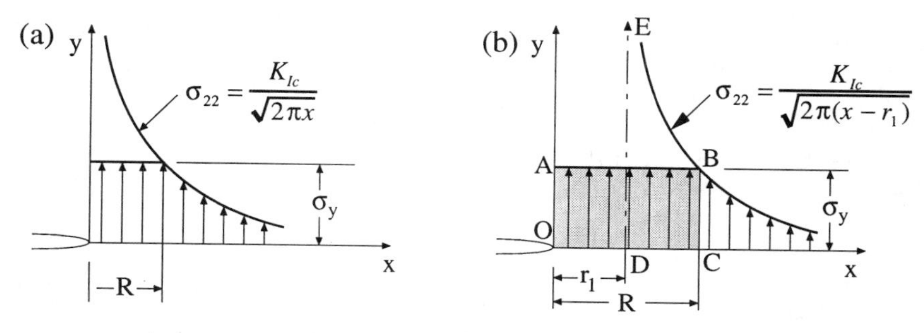

Figure 5.2.2 Irwin's estimate of effective crack extension.

5.2.2 Irwin's Estimate of the Size of the Inelastic Zone

Although linear elastic fracture mechanics assumes the size of the inelastic zone at the crack tip to be zero, in reality, this zone must have some finite size, R (Fig. 5.2.2a). This size was estimated by Irwin (1958) for plastic (ductile) materials. For a crude estimate, one may simply retain the singular term of the elastic stress distribution along the crack line, and cut the singularity off by putting a limit to the admissible stress equal to the yield strength σ_y, as Fig. 5.2.2a shows. The limit of the zone over which the tensile strength has been exceeded lies at the distance $x = R$ from the crack tip such that $\sigma_{22} = \sigma_y$ for $x = R$. For mode I, and for x small relative to the size of the structure, the singular term of the near-tip field is dominant and substituting x instead of r in (2.2.4) $\sigma_{22} = K_I(2\pi x)^{-1/2}$. The solution for R is immediately found:

$$R = \frac{1}{2\pi}\left(\frac{K_I}{\sigma_y}\right)^2 \tag{5.2.1}$$

In reality, the nonlinear zone must be larger than this because cutting off the singular stress peak (in Fig. 5.2.2a) decreases the stress resultant. If the stress resultant from an inelastic zone of size R is made equal to the stress resultant of the elastically calculated stresses $\sigma_{22} = K_I(2\pi x)^{-1/2}$, then the stress field farther from the nonlinear process zone would be unaffected, due to Saint-Venant's principle. Equal magnitudes of these resultants may be achieved by assuming the tip of an equivalent elastic crack to be shifted forward by distance r_1 ahead of the true crack tip. Graphically, this means that the area under the elastic stress solution in Fig. 5.2.2 must be equal to the area under the modified "plastic" stress distribution. Thus, the area of the region $BCDE$ must be equal to the area of the shaded rectangle $OABCO$:

$$\int_{r_1}^{R} K_I[2\pi(x - r_1)]^{-1/2}dx = R\sigma_y \tag{5.2.2}$$

The condition $\sigma_{22} = \sigma_y$ for $x = R$, now with $\sigma_{22} = K_I[2\pi(x - r_1)]^{-1/2}$, immediately leads to

$$R - r_1 = \frac{1}{2\pi}\left(\frac{K_I}{\sigma_y}\right)^2 \tag{5.2.3}$$

Solving R from the above two equations, Irwin (1958) obtained, for the length of the yielding zone, the improved estimate

$$R = \frac{1}{\pi}\left(\frac{K_I}{\sigma_y}\right)^2 \tag{5.2.4}$$

Even this improved estimate is very rough. Indeed, the stress distribution used to find it is mostly a guess, and probably cannot be a correct solution for the elasto-perfectly plastic behavior. In fact, it is not. However, the solution still captures the right structure for the size of the process zone. It has to be proportional to the square of the ratio between the applied stress intensity factor and the strength of the material. Only the value of the proportionality factor will depend on the detailed features of the problem: the type of material and the exact loading situation.

From a practical point of view, the foregoing estimate may be useful to decide whether a particular case may be described by LEFM. Assume that this is the case, so that upon loading the specimen, the crack growth would take place when $K_I = K_{Ic}$ and that subsequently the crack would grow in a self-similar way, with constant K_I. In this case, the maximum size of the process zone is that corresponding to the critical situation. Its value R_c is then given by Eq. (5.2.4) with $K_I = K_{Ic}$:

$$R_c = \frac{1}{\pi}\left(\frac{K_{I_c}}{\sigma_y}\right)^2 \tag{5.2.5}$$

Inverting the reasoning, if LEFM is to be applicable, then R_c must be small in comparison with the specimen or structure dimensions. The result is that the size of the specimen must be larger than a certain size proportional to R_c. How large, depends on the specific nature of the material and on the type of the specimen itself. For plane specimens of metals, the criterion usually accepted is that the minimum distance from the crack tip to any surface or load point of the specimen, D_s, must verify the condition (ASTM E 399 1983)

$$D_s > 2.5\left(\frac{K_{Ic}}{\sigma_y}\right)^2 \tag{5.2.6}$$

where for metals σ_y is taken to be the 0.2% conventional yield strength. This assures that LEFM is applicable within an error around 5%. Eliminating K_{Ic}/σ_y from this equation and (5.2.5), one obtains the relationship between the D_s and R_c:

$$D_s > 7.9R_c \tag{5.2.7}$$

which shows that, for LEFM to apply, R_c must be less than about 13% of the distance from the crack tip to the nearest point of the specimen or structure boundary.

One could try to extrapolate the preceding analysis, particularly Eq. (5.2.5), to concrete and other quasibrittle materials by substituting the tensile strength f'_t instead of σ_y. However, this does not work; as we shall see, the predicted R_c value is much less than the actual value. The reason is that concrete does not yield, it softens and microcracks. Indeed, all the foregoing derivation assumes that the material at the crack tip initially loads elastically, then starts to yield at constant stress and, after some yielding, it suddenly breaks off. Concrete behaves differently. When the tensile strength is reached (actually before, but this is not important for our crude analysis), internal bonds start to break, broken bonds coalesce into microcrack and slips, and these eventually into a full crack. The result is that, after peak stress, concrete softens progressively or undergoes damage. This makes the fracture process zone larger as we show next.

5.2.3 Estimate of the Fracture Zone Size for quasibrittle Materials

To see how the softening influences the size of the fracture process zone, consider the approximation depicted in Fig. 5.2.3. It is similar to Irwin's, but it assumes that when the critical situation $K_I = K_{Ic}$ is reached, the point at the crack tip has already completely softened and the points ahead of it are in intermediate states of fracture, in such a way that the stress distribution over the fracture process zone is linear. Imposing the same conditions as before, it turns out that the only modification is that the area of the region $BCDEB$ must now be equal to that of the shaded triangle $OBCO$. The result is that the estimate of the fracture process zone size now *doubles* Irwin's estimate.

Further analysis shows that if the progressive softening is concave upwards, instead of linear, one can get still much larger sizes of the fracture process zone for quasibrittle materials.

To prove this result, take the case shown in Fig. 5.2.4, in which the stress profile along the fracture zone is parabolic of degree n. We again force the areas $OBCDO$ and $DEBCD$ to be equal. This leads to an equation similar to (5.2.2) except that now K_{Ic} replaces K_I on the left-hand side while the right-hand side is replaced by the shaded area under the parabola:

$$\int_{r_1}^{R_c} K_{Ic}[2\pi(x - r_1)]^{-1/2}dx = \int_0^{R_c} f'_t\left(\frac{x}{R_c}\right)^n dx = \frac{1}{n+1}R_c f'_t \tag{5.2.8}$$

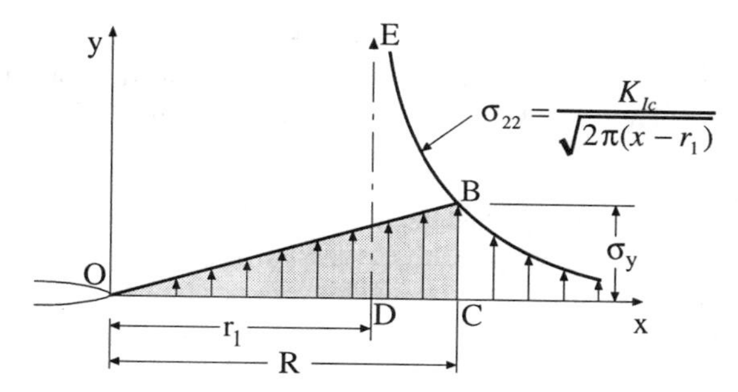

Figure 5.2.3 Estimate of the fracture process zone size for a progressively softening material: linear distribution.

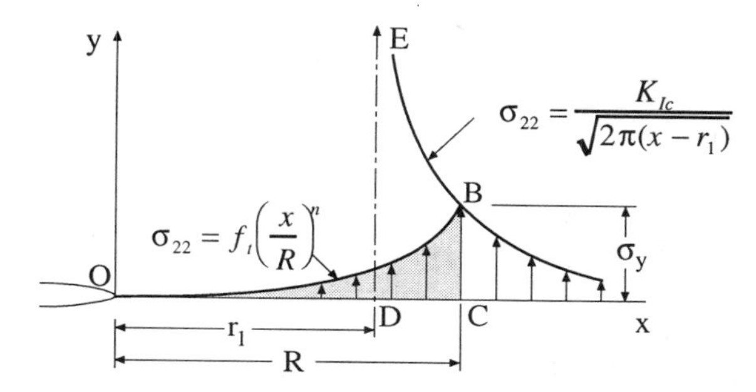

Figure 5.2.4 Estimate of the fracture process zone size for a progressively softening material: parabolic distribution of degree n.

This equation together with (5.2.3), which does not change, leads to the solution

$$R_c = \frac{n+1}{\pi}\left(\frac{K_{Ic}}{f'_t}\right)^2 \tag{5.2.9}$$

This shows that the nonlinear zone size may be many times Irwin's estimate for metals. Estimates using the more sophisticated models described in Chapter 7 indicate that values of n for concrete may be of the order of 7 and as large as 14.

To sum up, one may state that the nonlinear zone ahead of the crack tip at the critical state for a very large structure may be written as

$$R_c = \eta\left(\frac{K_{Ic}}{f'_t}\right)^2 = \eta\,\frac{E'G_f}{f'^2_t} \tag{5.2.10}$$

where the second equality stems from Irwin's relationship $K_{Ic} = \sqrt{E'G_f}$ (Chapter 2), and η is some dimensionless constant taking the value $1/\pi$ for Irwin's estimate, and some possibly much larger value for quasibrittle materials, roughly between 2 and 5 for concrete.

In the context of cohesive crack models, to be described in Chapter 7, the fraction in the right-hand side of the previous equation, which has dimensions of length, is called the *characteristic size*:

$$\ell_{ch} = \left(\frac{K_{Ic}}{f'_t}\right)^2 = \frac{E'G_f}{f'^2_t} \tag{5.2.11}$$

(Hillerborg, Modéer and Petersson 1976 called it the characteristic length, but this term conflicts with the previously established meaning of characteristic length of a nonlocal continuum.) So the final result may be written as

$$R_c = \eta\ell_{ch} \tag{5.2.12}$$

Taking into account that the reported values of ℓ_{ch} for concrete fall in the range 0.15 m–0.40 m, the limiting values for the size of a fully developed fracture process zone (i.e., one in an infinite body) range from 0.3 m to 2 m. This explains why there is no way to approximate the behavior of regular laboratory specimens by LEFM.

The foregoing results, even though only estimates, show that the size of the fracture process zone strongly depends on the precise way in which the material softens. Henceforth, describing the softening behavior in the fracture process zone is a real necessity when dealing with quasibrittle materials. This will be done in Chapters 7–9 and 13–14. Now we will investigate how to characterize the fracture process zone in a simplified way.

Exercises

5.1 Work out the algebra leading from (5.2.2) and (5.2.3) to (5.2.4)

5.2 For the metallic materials whose properties are given in the table below (extracted from Broek 1986), determine: (a) the size of the critical inelastic zone using Irwin's estimate; (b) the minimum distance from the crack tip to the specimen surface, D_s, according to ASTM E 399; (c) the characteristic size ℓ_{ch} obtained by setting σ_y instead of f'_t in (5.2.11).

Material	σ_y (MPa)	K_{Ic} (MPa$\sqrt{\text{m}}$)
Maraging steel 300	1670	93
D 6 AC steel (heat treated)	1470	96
A 533 B (reactor steel)	343	195
Low strength carbon steel	235	>220
Titanium, 6Al-4V	1100	38
Titanium, 4Al-4Mo-2Sn-05Si	940	70
Aluminum 7075, T651	540	29
Aluminum 7079, T651	460	33
Aluminum 2024, T3	390	34

5.3 What should be the minimum distance D_s from the crack tip to the surface for a quasibrittle material? Express D_s as a function of coefficient η in equation (5.2.12) and assume the same ratio D_s/R_c as for metals. Give the value of D_s for a concrete with $\ell_{ch} = 0.25$ m and $\eta = 4$. Find the minimum depth of single-edge notch beams with a notch-to-depth ratio of 0.5 that is needed to make LEFM applicable.

5.3 The Equivalent Elastic Crack Concept

The problem of fracture of a quasibrittle material, in which the fracture process zone is not negligible relative to the dimensions of the specimen, is a nonlinear problem. If the fracture process zone is very large, so that it extends over a significant portion of the unbroken ligament, relatively complex models are required for describing the material behavior in the process zone. However, if only a small (but non-negligible) part of the ligament is fracturing, then one can use simplified, partly linearized models which simulate the response of the specimen far from the crack tip by an equivalent elastic crack with the tip somewhere in the fracture zone, as was implicitly done in the previous section (Figs. 5.2.2–5.2.4). Recent analysis by Elices and Planas (1993) and Planas, Elices and Ruiz (1993) showed that for arbitrary sizes there are many ways of defining the equivalence between the actual specimen and the effective elastic one. However, for the case of relatively small fracture process zones, all these types of equivalence merge in a single one called the *far field equivalence*. The reason is that in that case, the fields of stresses, strains, displacements, etc., far from the crack tip are identical within terms of the order of $(R_c/D)^2$ where R_c is the size of the process zone and D the size of the specimen.

The problem in defining this equivalence is to establish where the tip of the equivalent crack is located for a particular load level, i.e., what is the equivalent crack extension

$$\Delta a_e = a_e - a_0 \tag{5.3.1}$$

in which a_e is the equivalent crack length and a_0 the initial crack length.

In general, this can be done by stating a crack growth rule relating the effective (or equivalent) crack extension Δa_e to the crack driving force $\mathcal{R}$. This is the so-called R-curve approach, which can approximately describe the complete crack growth process and will be analyzed in Section 5.6. But, before describing those models, we are going to analyze more simplified models that deal with the critical (peak load) condition only.

We start, in this section, with a crude approximate analysis of the equivalent crack extension based on the general and simple concepts presented in the preceding section. More refined models will be presented in subsequent sections, but the essential facts and order of magnitudes estimates will remain the same.

5.3.1 Estimate of the Equivalent LEFM Crack Extension

According to the preceding discussion, the equivalent elastic crack can be determined by forcing the remote fields of the actual structure to coincide with the remote fields of the equivalent crack if subjected to the same load. Focusing on the σ_{22} stress component along the crack line, and on very large specimens such as those examined in the previous section, the graphical constructions of Figs. 5.2.2 to 5.2.4 clearly show that far from the crack tip these stresses are identical to those for a sharp crack with the tip located at point D. Therefore, the effective crack extension Δa_e is, in this approximation, identical to the value r_1 given in the figures. So from (5.2.3),

$$\Delta a_e = r_1 = R - \frac{1}{2\pi}\left(\frac{K_I}{f'_t}\right)^2 \tag{5.3.2}$$

Now, applying this to the critical situation and using (5.2.11) and (5.2.12), one obtains the critical effective crack extension as

$$\Delta a_{ec} = \beta \ell_{ch} \quad \text{with} \quad \beta = \eta - \frac{1}{2\pi} \tag{5.3.3}$$

where, according to the values of η discussed in the previous section, β is equal to $1/2\pi$ for the Irwin estimate and may range roughly from 2 to 5 for concrete. What is essential is that β as well as Δa_{ec} are constants for a given material.

Now, we must realize that in the previous analysis the specimen is considered to be virtually infinite (because the stress distribution is completely dominated by the singular term). Since, in the forthcoming sections, we are going to consider other less restrictive situations, it is convenient to adopt a notation clearly indicating the value for virtually infinite size D. We thus define the critical effective crack extension for a semi-infinite crack in an infinite body as c_f such that

$$c_f = \lim_{D\to\infty} \Delta a_{ec} \tag{5.3.4}$$

Thus, Δa_{ec} in (5.3.3) is the value for infinite size and, therefore, coincides with c_f. The estimate for c_f is thus

$$c_f = \beta \ell_{ch} \quad \text{with} \quad \beta = \eta - \frac{1}{2\pi} \tag{5.3.5}$$

so c_f is a material property.

5.3.2 Deviation from LEFM

Let us now examine how the apparent toughness K_{INu} defined in Section 5.1 deviates from the LEFM prediction. For a large, but not infinitely large specimen, we use the definition of critical equivalent crack and write the condition that the stress intensity at peak load for the equivalent crack must be the true fracture toughness; that is

$$K_{Ic} = \sigma_{Nu}\sqrt{D}\, k(\alpha_{ec}) \tag{5.3.6}$$

Eliminating σ_{Nu} from this equation and Eq. (5.1.3), the apparent fracture toughness is obtained as

$$K_{INu} = K_{Ic}\frac{k(\alpha_0)}{k(\alpha_{ec})} \tag{5.3.7}$$

Now, recalling that all our reasoning in the preceding sections refers to a specimen much larger than the fracture process zone, we see that $\Delta a_{ec} \approx c_f \ll D$. Because $\alpha = \alpha_0 + \Delta a_{ec}/D$, we may use a two-term Taylor expansion for $k(\alpha)$

$$k(\alpha_{ec}) \approx k(\alpha_0) + k'(\alpha_0)\frac{c_f}{D} \tag{5.3.8}$$

Substitution of (5.3.8) into (5.3.7), followed by series expansion of the result, leads to the asymptotic two-term expression

$$K_{INu} = K_{Ic}\left(1 - \frac{k'(\alpha_0)}{k(\alpha_0)}\frac{c_f}{D}\right) \quad \text{as } \frac{c_f}{D} \to 0 \tag{5.3.9}$$

This shows that the relative deviation of the apparent toughness from the true one is proportional to c_f.

Example 5.3.1 Consider an infinite panel with a central crack of length $2a_0$, subjected to remote tensile stress σ_∞ normal to the crack. The well-known expression for the stress intensity factor $K_I = \sigma_\infty\sqrt{\pi a}$ does follow the form of (5.3.6) if we identify σ_{Nu} with $\sigma_{\infty u}$, D with a_0 (the only relevant specimen dimension), and take $k(\alpha) = \sqrt{\pi\alpha}$. Therefore, we have $k'(\alpha_0)/k(\alpha_0) = 1/2$ which substituted into (5.3.9), leads to

$$K_{INu} = K_{Ic}\left(1 - \frac{c_f}{2a_0}\right) \tag{5.3.10}$$

This, indeed, has the expected form. ▯

5.3.3 Intrinsic Size

The relationship in the previous section provides a means of defining in a precise and objective way what we mean by size. Up to now, the size was some user-selected specimen dimension, one among various choices: beam depth, loading span, initial crack length, initial ligament length are some of the possibilities among many others. This choice is not important as long as geometrically similar structures are referred. But the arbitrariness of the choice makes it difficult to compare different kinds of specimens. This may be solved if we define an *intrinsic size* $\overline{D}$ in such a way that *any two specimens of the same intrinsic size experience the same asymptotic deviation from LEFM*, as expressed by (5.3.9).

To do so, we note that the geometry and size enter (5.3.9) only through the factor $k'(\alpha_0)/k(\alpha_0)D$, which has dimensions of inverse of length. Therefore, any two specimens with this same factor will deviate the same amount from LEFM. Therefore, we may take the intrinsic size to be proportional to the inverse of this factor, as done independently by Planas and Elices (1989a), and Bažant, Kazemi and Gettu (1989), and define $\overline{D}$ by the relation

$$\frac{1}{\overline{D}} = 2\frac{k'(\alpha_0)}{k(\alpha_0)}\frac{1}{D} \tag{5.3.11}$$

where D is any arbitrarily selected specimen dimension (the reason for the factor 2 will become clear in Section 6.1). Because $\alpha = a/D$, this definition is equivalent to any of the following:

$$\frac{1}{\overline{D}} = \left.\frac{\partial \ln k^2(\alpha)}{\partial a}\right|_{a=a_0} = \left.\frac{\partial \ln K_I^2(\alpha)}{\partial a}\right|_{a=a_0} = \left.\frac{\partial \ln \mathcal{G}(\alpha)}{\partial a}\right|_{a=a_0} \tag{5.3.12}$$

These definitions must be restricted to positive geometries. For negative geometries, the intrinsic size becomes negative because $k'(\alpha_0) < 0$, which is an unacceptable property and plainly means that the equation upon which the definition is built is not valid for such a case. (Planas and Elices 1990b and Elices and Planas 1992 have extended the definition of intrinsic size to negative geometries based on the R-curve concept; it turns out that, in this case, the asymptotic deviation from LEFM vanishes to first order, so that $\overline{D} \to \infty$.)

With the definition of intrinsic size, the asymptotic deviation from LEFM (5.3.9) reduces to

$$K_{INu} = K_{Ic}\left(1 - \frac{c_f}{2\overline{D}}\right) \tag{5.3.13}$$

Example 5.3.2 For a small surface crack of length a in a body subjected to constant stress σ perpendicular to the crack, the stress intensity factor is known to be given by $K_I = 1.12\sigma\sqrt{\pi a}$. Using the second of the equations (5.3.12), one easily finds

$$\frac{1}{\overline{D}} = \left.\frac{\partial \ln K_I^2(\alpha)}{\partial a}\right|_{a=a_0} = \left.\frac{\partial \ln a}{\partial a}\right|_{a=a_0} = \frac{1}{a_0} \tag{5.3.14}$$

Thus, for a small surface crack, the intrinsic size is the initial crack length, independently of the remaining in-plane dimensions of the structure (which may be very large). This is consistent with the usual statement that for LEFM to apply, the distance from the crack tip to any outer boundary must be large; in this case, the minimum distance to the boundary is the initial crack length. From (5.3.13) one finds:

$$K_{INu} = K_{Ic}\left(1 - \frac{c_f}{2a_0}\right) \tag{5.3.15}$$

which turns out to coincide with Eq. (5.3.10) for the center-cracked infinite panel of the preceding example. The reader may easily check that the intrinsic size of the tensioned infinite panel with a center crack is indeed a_0, the half-length of the crack. ▯

As a corollary of the results of this simple example, one finds that the intrinsic size of a structure with no precrack ($a_0 \to 0$) is zero, so that the deviation from LEFM is infinity: this is consistent with the well-known fact that LEFM cannot be applied to uncracked structures (modified LEFM models can, however).

5.3.4 How Large the Size Must Be for LEFM to Apply?

Based on some comparisons of notched beam tests with the modulus of rupture, Walsh (1972, 1976) stated that the beam depth D should be at least 230 mm (9 in.). However, in the light of subsequent studies of size effect, this now appears to be a gross underestimation. According to the size effect law of Bažant, the ratio $\beta = D/D_0$ in (1.4.10) must be at least 25 if the deviation from the straight-line asymptote for LEFM should be less than 2%. Bažant and Pfeiffer's (1987) tests of specimens of different types show that: (1) for eccentric compression fracture specimens $D_0 = 1.85d_a$, (2) for three-point bend beams, $D_0 = 5.4d_a$, and (3) for centric tension specimens, $D_0 = 16.8d_a$. Consequently, for the theoretical errors to be under 2%, the minimum cross-section depths of these specimens that are required for direct applicability of LEFM are, as follows:

$$\begin{aligned} \text{For eccentric compression specimens: } & D \geq 46d_a \\ \text{For three-point bend specimens: } & D \geq 135d_a \\ \text{For concentric tension specimens: } & D \geq 420d_a \end{aligned} \tag{5.3.16}$$

These dimensions are too large for laboratory tests. Their use would be financially wasteful, since other methods can provide adequate information on the material fracture parameters using much smaller specimens. This is one of the central topics of this book.

In order to give a general expression of the required size, valid in principle for any material and specimen geometry, we can use the foregoing results to define the intrinsic size $\overline{D}_\epsilon$ required for the apparent toughness to have a maximum relative deviation ϵ from the true fracture toughness. From (5.3.13), we immediately get

$$\overline{D}_\epsilon = \frac{1}{2\epsilon} c_f \tag{5.3.17}$$

The prediction of this equation may be matched to the previous results of Bažant and Pfeiffer if the critical effective crack extension is taken to be $c_f \approx d_a$. However, this appears to be a very low value for many ordinary concrete mixes. Following an interpretation of the experimental results based on cohesive crack models (Chapter 7), Planas and Elices (1989a, 1991a, 1992a, 1993a) found that effective crack extensions larger than that by one order of magnitude may be expected for ordinary concrete.

Unfortunately, the theories used to interpret the small size results and to extrapolate to large sizes do influence the results for the large size estimates. Nevertheless, even the most optimistic estimates indicate that the specimen dimensions required for LEFM to apply within a few percent are too large for any practical purpose.

Exercises

5.4 Use the approximate beam theory to find the intrinsic size of a double-cantilever beam specimen subjected to concentrated forces at the cantilever end. Give the result in terms of the initial crack length a_0.

5.5 Use the closed form solutions of example 3.1.1 to find the intrinsic size of a notched three-point-bend beam in terms of the beam depth D; consider a span-to-depth ratio of 4 and a notch-to-depth ratio of (a) 1/3, (b) 0.5 . Hint: use numerical rather than analytical differentiation.

5.6 Use the closed form solutions of example 3.1.1 to prove that the intrinsic size of a notched three-point-bend beam tends toward 1/3 of the initial ligament size $b_0 = D - a_0$ for very deep notches ($b_0/D \ll 1$). Hint: before differentiation, expand the K_I expression in powers of b/D and take the most significant term.

5.7 Consider a three-point-bend notched specimen of depth D and notch-to-depth ratio of 0.5. Find the minimum beam depth (in terms of ℓ_{ch}) required for LEFM to apply within 2% for: (a) a "ductile" material with R_c approximately given by Irwin's estimate, (b) a quasibrittle material with a power-law stress distribution ahead of the crack tip, of exponent n= 20. In the latter case, find the depth in meters for an ordinary concrete for which $\ell_{ch} = 250$ mm.

5.8 Consider a double-cantilever beam specimen with concentrated loads at the cantilever ends. Let D be the depth of the arms (half the total specimen depth) and $a_0 = 6D$ the initial crack length. Find the minimum size D and the minimum crack length a_0 (in terms of ℓ_{ch}) required for LEFM to apply within 2% for: (a) a "ductile" material, with R_c approximately given by Irwin's estimate, (b) a quasibrittle material with a power-law stress distribution ahead of the crack tip, of exponent n= 20. In the latter case, find the depth and crack length (in meters) for an ordinary concrete for which $\ell_{ch} = 250$ mm.

5.4 Fracture Toughness Determinations Based on Equivalent Crack Concepts

As we have seen, LEFM cannot be directly applied for most practical and laboratory situations because the size of concrete structures is too small. In this section we discuss the simplest modifications of LEFM introduced to obtain a size independent fracture toughness by the research groups of Swartz at Kansas University and of Karihaloo at Sydney University. These approaches are based on the equivalent crack concept, but they are experimental methods rather than complete models. In particular, as it will be shown, they are not predictive. They rely on the measurement of the equivalent crack extension, but they cannot predict it for a general situation. The value of the toughness so measured is of practical use, in principle, only for predicting the behavior of very large structures where LEFM does apply.

The basis of all these approaches is the same: they assume that when the peak load is reached, the far fields may be described by an equivalent elastic crack such that $a_e = a_0 + \Delta a_{ec}$, and $K_I = K_{Ic}$ (fracture toughness). This is related to the peak load or nominal strength σ_{Nu} by an equation identical to (5.3.6), except that in $\alpha_e = (a_0 + \Delta a_{ec})/D$, Δa_{ec} is no longer equal to c_f because the specimens are not necessarily quasi-infinite. This means that further equations are required for being able to solve for K_{Ic} from (5.3.6). In particular, the critical effective crack extension δa_{ec} is required. In the remainder of this section, we summarize the methods used by Swartz et al. and by Nallathambi and Karihaloo to determine Δa_{ec} (or a_{ec}) from experiments.

5.4.1 Compliance Calibration of Equivalent Crack Length

To estimate the critical equivalent crack length, Swartz, Hu and Jones (1978) used the compliance calibration method. It is based on the property that in LEFM the displacements are proportional to the load, and the associate compliance is uniquely related to the crack length. Although there is a theoretical relationship between the compliance and the crack length (see Section 3.5), Swartz, Hu and Jones used

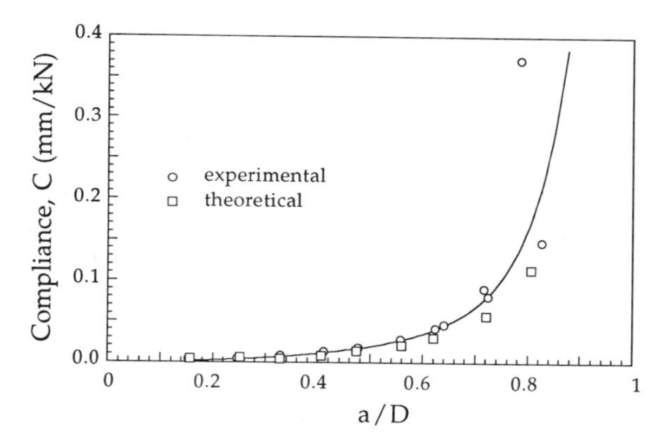

Figure 5.4.1 Compliance calibration curve for notched beams (after Swartz, Hu and Jones 1978).

a direct calibration method, which uses a calibration curve for each given geometry and each material (concrete mix). The calibration curve is measured using a beam which is saw-notched at progressively increasing notch depths. At a given notch depth a_0, the beam is loaded to less than $1/3$ of the maximum load expected for this notch depth, and the load is then cycled three times while measuring the crack mouth opening displacement (CMOD, denoted as w_M). The corresponding compliance for this notch depth is then determined as $C_M = \Delta w_M/\Delta P$. This is then repeated for all the notch depths, which provides the C_M–a_0 plot exemplified in Fig. 5.4.1 (from Swartz, Hu and Jones 1978). This plot is then used to estimate the actual crack length of the beams at maximum load from the measured unloading-reloading compliance.

Swartz, Hu and Jones (1978) used pre-cracked beams which had only a short notch in actual testing. The beam was first preloaded beyond the maximum load point so that a stable crack growth was produced until the unloading or reloading compliance matched the value for the desired crack length, as given by the calibration curve, which was called a_i (for initial crack length). Then the fracture test was carried out to measure the maximum load P_u, from which K_{Ic} was then calculated using (5.3.6) (with a_{ec} replaced by a_i). To achieve stable crack growth in this type of testing, it may be necessary to use a closed-loop control system which can perform tests at a controlled CMOD.

Swartz, Hu et al. (1982) found that the computed K_{Ic} for the precracked beams was always higher than for the notched beams. However, this approach was criticized in two respects: (1) notched, rather than precracked, beams were used to construct the compliance calibration curves, and (2) the initial crack length a_i determined as described above is not a realistic value because it does not include further slow crack growth up to the peak. Therefore, the compliance calibration method was modified.

5.4.2 Modified Compliance Calibration Method

The modified procedure used precracked beams that were impregnated with a dye to reveal the profile of the crack front throughout the specimen thickness. The extended crack length a_c was then measured to determine K_{Ic} (Swartz and Go 1984; Go and Swartz 1986; Swartz and Refai 1989). Precracked beams were used both for determining the compliance calibration curves and for the fracture tests. The test method was very time consuming since a number of beams were required to obtain the compliance calibration curve. In these tests, the beam was loaded beyond the maximum load point, controlling the CMOD, and then unloaded. Then the compliance of P vs. w_M was used by loading up to 1/3 of P_u. A dye was introduced and the load was cycled to force the dye to better penetrate the crack (the beam was positioned so that the crack would grow downward). When the dyed surface was dried, the beam was loaded to failure and P vs. w_M plotted. The initial slope of this plot was taken as the initial compliance C_i. After failure, the dyed surface area was measured, and the initial crack was obtained as a_i =(dyed surface area)$/b$; the dyed surface front was very irregular, and so a direct measurement of the crack length was not possible; further details are given in Swartz and Refai (1989). The foregoing steps were repeated for different crack length to establish the entire curves relating C_i to a_i. At the point on the unloading diagram corresponding to the onset of unstable crack growth, taken as $0.95P_u$, the extended crack length a_c was determined from the $P_u(a_i)$ plot. The value of K_{Ic} was then computed from load 0.95 P_u and crack length a_c.

The results obtained by this method indicated K_{Ic} to be approximately invariant with respect to crack length as well as beam size, although only beams of two sizes (depth 8 in. and 12 in.), too close to each other, were used. It is doubtful that this method can give invariant K_{Ic} for a broad range of sizes, because the fracture process zones were not negligible compared to the cross-section dimensions and the proximity of the boundaries must have influenced the process zone size, causing it to be different from that for a very large beam.

5.4.3 Nallathambi-Karihaloo Method

Another variant of adapting LEFM to concrete was proposed by Nallathambi and Karihaloo (1986a, 1986b). The method was developed on the basis of standard three-point-bend beam tests, and uses the secant compliance to determine the effective crack extension.

The basic idea consists in determining the critical effective crack length from the deformability of the specimen. The basic equation relates the mid-span total deflection u to the load P (Section 3.5.1). Taking into account that the weight mg of the specimen is also acting, one writes

$$u = \frac{P}{bE'}\hat{v}_1(\alpha) + \frac{mg}{bE'}\hat{v}_2(\alpha) \tag{5.4.1}$$

where α is the relative crack depth a/D, a the crack length, D the beam depth, and $\hat{v}_1(\alpha)$ and $\hat{v}_2(\alpha)$ dimensionless shape functions.

Functions $\hat{v}_i(\alpha)$, $i = 1, 2$, can always be written in the form

$$\hat{v}_i(\alpha) = \hat{v}_{i0} + \hat{v}_i^c(\alpha) \tag{5.4.2}$$

where $\hat{v}_{i0}$ takes into account the deformability of the beam without any crack, and $\hat{v}_i^c(\alpha)$ the extra deformability due to the crack (see Section 3.5.1). The Nallathambi-Karihaloo basic approach has evolved along the years and one may find various expressions for the foregoing equations, which differ in how the dead-weight is taken into account, whether or not the shear deformation is included, and how the compliance due to the crack is evaluated. In the following, we give expressions that are equivalent to those proposed in the State of Art Report of RILEM Technical Committee 89-FMT (Karihaloo and Nallathambi 1991).

The components $\hat{v}_{i0}$ are computed using beam theory with the following results

$$\hat{v}_{10} = \frac{1}{4}\left(\frac{S}{D}\right)^3 \left[1 + 2.70\left(\frac{D}{S}\right)^2 - 0.84\left(\frac{D}{S}\right)^3\right] \tag{5.4.3}$$

$$\hat{v}_{20} = \frac{5}{32}\left(\frac{S}{D}\right)^3 \left[1 + 2.16\left(\frac{D}{S}\right)^2\right] \tag{5.4.4}$$

where S is the beam loading span.

The contribution of the crack to the deformability is obtained by Karihaloo and Nallathambi as described in Bažant (1987b) and Section 3.5.1. The weight of the beam is taken into account by taking its effect to be identical to that for a central point-load giving the same bending moment at the central section: $P_{mg} = mg/2$. The result is:

$$\hat{v}_1^c(\alpha) = \frac{9}{2}\left(\frac{S}{D}\right)^2 F(\alpha)\,, \quad \hat{v}_2^c(\alpha) = \frac{1}{2}\hat{v}_1^c(\alpha) \tag{5.4.5}$$

where

$$F(\alpha) = \int_0^\alpha k^2(\alpha')\,d\alpha' \tag{5.4.6}$$

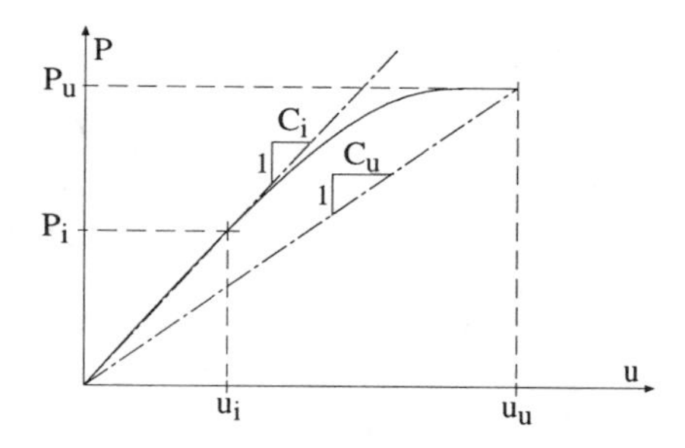

Figure 5.4.2 Load-deflection curve and initial and secant compliance.

and $k(\alpha)$ is the shape function for the stress intensity factor as expressed in (2.3.11), with the nominal stress given by (see Fig. 3.1.1a)

$$\sigma_N = \frac{3PS}{2bD^2} \tag{5.4.7}$$

The expression for $k(\alpha)$ used by Karihaloo and Nallathambi is that due to Srawley (1976):

$$k(\alpha) = \sqrt{\alpha}\, \frac{1.99 - \alpha(1-\alpha)(2.15 - 3.93\alpha + 2.7\alpha^2)}{(1+2\alpha)(1-\alpha)^{3/2}} \tag{5.4.8}$$

Nallathambi and Karihaloo use a first experimental point on the initial linear part of the P–u plot (Fig. 5.4.2) to get the effective elastic modulus E'. If the coordinates of this point are (u_i, P_i), E' is obtained from (5.4.1) substituting these values and setting that along the linear portion of the curve $\alpha = \alpha_0$:

$$E' = \frac{P_i}{bu_i}\left[\hat{v}_1(\alpha_0) + \frac{mg}{P_i}\hat{v}_2(\alpha_0)\right] \tag{5.4.9}$$

The relative critical effective crack depth α_{ec} is sought as the solution of Eq. (5.4.1) written for the peak:

$$E' = \frac{P_u}{bu_u}\left[\hat{v}_1(\alpha_{ec}) + \frac{mg}{P_u}\hat{v}_2(\alpha_{ec})\right] \tag{5.4.10}$$

where P_u is the peak load and u_u the deflection at the peak. This equation must be solved numerically. Karihaloo and Nallathambi suggest using an incremental procedure, so that the right-hand side of this equation is evaluated at α increments $\Delta\alpha = 0.001$ until its value differs from the previously found value of E' by less than 0.5%. The procedure includes a numerical integration to get function $F(\alpha)$ (although analytical integration is possible, it is practical only if one of the modern computer programs allowing symbolic analysis is available).

To avoid the necessity of making this calculation, Nallathambi and Karihaloo computed α_{ec} for a large number of experimental results, and found a regression formula from which the value of the effective crack length may be estimated. The formula is:

$$\alpha_{ec} = \gamma_1 \left(\frac{\sigma_{Nu}}{E'}\right)^{\gamma_2} \alpha_0^{\gamma_3} \left(1 + \frac{d_a}{D}\right)^{\gamma_4} \tag{5.4.11}$$

where σ_{Nu} follows from (5.4.7) with $P = P_u + mg/2$ (to take into account the weight); d_a is the maximum aggregate size, and E' the effective elastic modulus obtained from (5.4.9). The set of constants is then: $\gamma_1 = 0.088 \pm 0.004$, $\gamma_2 = -0.208 \pm 0.010$,$\gamma_3 = 0.451 \pm 0.013$, and $\gamma_4 = 1.653 \pm 0.109$. If the effective modulus E' is approximated by the standard modulus obtained on cylindrical specimens, a correlation still exists, but the constants take different values: $\gamma_1 = 0.198 \pm 0.015$, $\gamma_2 = -0.131 \pm 0.011$, $\gamma_3 = 0.394 \pm 0.013$, and $\gamma_4 = 0.600 \pm 0.092$. The great variation of all these constants depending on the method used to measure E appears hard to explain.

Once the critical equivalent crack extension has been found, the critical stress intensity factor is obtained from (5.3.6) in which, again, σ_{Nu} is given by (5.4.7) with $P = P_u + mg/2$ (to take into account the weight).

This approach was proposed for concrete mixes with maximum aggregate sizes d_a in the range 5–24 mm, and for the following ranges of specimen dimensions: b = 40–80 mm ($\geq 3d_a$); D = 50–300 mm ($\geq 3d_a$); S = 200–1800 mm ($\geq 3d_a$); and for $4 \leq S/D \leq 8$; a_0/D = 0.2–0.6 (preferably 0.3–0.4). The effective crack length formula was calibrated by an extensive test program including more than 950 beams from various investigators (Karihaloo and Nallathambi 1991; Nallathambi and Karihaloo 1986a), in which various maximum aggregate sizes, beam sizes, and relative notch depths were used. The coefficient of variation of these test was less than 10%. The results for K_{Ic} were shown to be in good agreement with Bažant's size effect method as well as Jenq and Shah's method for K_{Ic} (Section 5.5). However, it is not clear how this method could be applied to problems other than notched three-point-bend beams, and neither is how the value of K_{Ic} so obtained should be used to make practical predictions of structural strength.

Apart from the aforementioned objection, it seems that the way the mid-span displacements are interpreted in the foregoing equations may differ from what is actually measured in most laboratories (Planas 1993). Indeed, in the foregoing u is the mid-span deflection measured from the stress-free configuration. However, in most experimental records, the displacement is measured with respect to an initial configuration in which the self weight is already acting. This means that, in reality, the P–u plot shown in Fig. 5.4.2 ought to be the P–Δu plot, where

$$\Delta u = u - u_0 \tag{5.4.12}$$

in which u_0 is the initial deflection produced by the beam weight alone, so that according to (5.4.1)

$$u_0 = \frac{mg}{bE'}\hat{v}_2(\alpha_0) \tag{5.4.13}$$

Taking into account (5.4.1)–(5.4.5), one finds for Δu the expression:

$$\Delta u = \frac{P}{bE'}\hat{v}_1(\alpha) + \frac{mg}{2bE'}\left[\hat{v}_1^c(\alpha) - \hat{v}_1^c(\alpha_0)\right] \tag{5.4.14}$$

Therefore, if Δu_i and Δu_u are what is actually measured (rather than u_i and u_u), the previous equations (5.4.9) and (5.4.10) ought to be replaced by

$$E' = \frac{P_i}{b\Delta u_i}\hat{v}_1(\alpha_0) \tag{5.4.15}$$

$$E' = \frac{P_u}{b\Delta u_u}\left[\hat{v}_{10} + \left(1 + \frac{mg}{2P_u}\right)\hat{v}_1^c(\alpha_{ec}) - \frac{mg}{2P_u}\hat{v}_1^c(\alpha_0)\right] \tag{5.4.16}$$

This discrepancy might contribute to the differences in the correlation values obtained by Nallathambi and Karihaloo for the γ_is depending on whether E' is evaluated from the initial compliance or from standard compression tests.

5.5 Two-Parameter Model of Jenq and Shah

Jenq and Shah (1985a,b) proposed an equivalent crack model which may predict, in principle, peak loads of precracked specimens of any geometry and size. However, it cannot, in general, predict the complete structural behavior (load-displacement curves and crack evolution).

Although the underlying physics is different, the Jenq-Shah model exhibits a close analogy with that proposed by Wells (1963) and Cottrell (1961) which has found wide application for the ductile fracture of metals and has been adopted for the standards of many European countries, Japan, China, and Australia (see also Atkins and Mai 1985, and Knott 1973). In both models, failure is assumed to happen when the crack opening at the initial crack tip reaches a certain critical value.

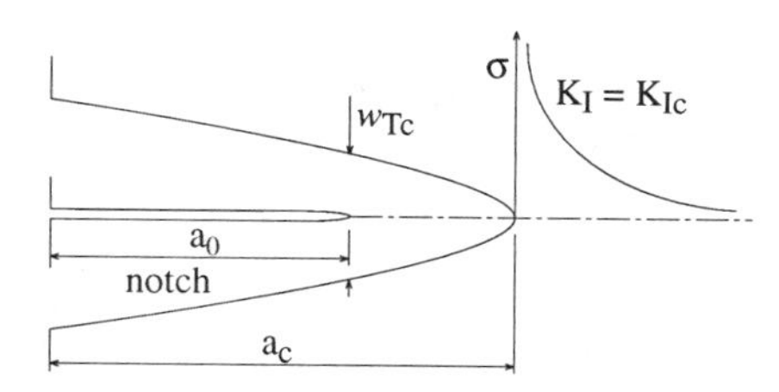

Figure 5.5.1 Equivalent elastic crack profile and definition of crack-tip opening displacement in Jenq-Shah model.

In the Jenq-Shah two-parameter model, the actual crack is replaced by an equivalent fictitious LEFM crack whose length a_{ec} is determined from the condition that the so-called crack tip opening displacement (CTOD, denoted as w_T) be equal to a certain critical value w_{Tc}. This displacement is defined as the opening displacement of the crack at the tip a_0 of the traction-free crack, that is, at the beginning of the inelastic zone (Fig. 5.5.1). For notched specimens of positive geometry, w_T is the opening displacement at the notch tip. The criterion for crack propagation is expressed by means of two parameters: (1) the critical value of the stress intensity factor of the equivalent LEFM crack tip, K_{Ic}, and (2) w_{Tc}.

The value of K_{Ic} from this model has been shown to be essentially independent of the geometry of the specimens. The results obtained from compact tension tests, wedge-splitting cube tests and large, tapered, double-cantilever beams showed that the two fracture parameters (K_{Ic} and w_{Tc}) might be considered as geometry-independent material parameters (Jenq and Shah 1988a), although a study by Jenq and Shah (1988b) indicated a significant influence of geometry. Other researchers reported substantial size effect both on K_{Ic} and on w_{Tc} (Elices and Planas 1991).

Tang, Shah and Ouyang (1992) showed that their two-parameter model may be applied to unnotched (uncracked) specimens. In particular, the model gives realistic predictions of the uniaxial tensile strength, the split-cylinder strength, the size effect on conventional (apparent) K_{Ic}, the size effect on the modulus of rupture, and the size effect in the split-cylinder (Brazilian) test. Extensions of this model to mixed-mode loading and to impact load have been proposed (Shah and John 1986; John and Shah 1986, 1990).

5.5.1 The Basic Equations of Jenq-Shah Model

The two-parameter model of Jenq and Shah can be characterized by two LEFM relations. The first one just states that, at peak load, the stress intensity factor takes its critical value K_{Ic}, the crack length being equal to the (unknown) critical effective crack length a_{ec}. So, from (5.3.6),

$$\sigma_{Nu}\sqrt{D}k(\alpha_{ec}) = K_{Ic} \tag{5.5.1}$$

The second equation states that, at peak load, w_T (i.e., the CTOD) takes its critical value w_{Tc}. This requires the knowledge of the crack opening profile for any crack length, a function that is not known in closed form except for a few cases. However, it follows from the general elastic solution (3.5.43) for the crack opening profile, that the crack-tip opening displacement can be written in the form

$$w_T = \frac{\sigma_N}{E'} D v_T(\alpha, \alpha_0) \tag{5.5.2}$$

where $v_T(\alpha, \alpha_0)$ is the value of the shape function $v_x(\alpha)$ in (3.5.43) for $x = a_0$.

From this and the basic postulate that $w_T = w_{Tc}$ at peak load, one gets the condition

$$\frac{\sigma_{Nu}}{E'} D v_T(\alpha_{ec}, \alpha_0) = w_{Tc} \tag{5.5.3}$$

Given the initial crack length, the specimen dimensions, and the material properties K_{Ic} and w_{Tc}, the two foregoing equations (5.5.1) and (5.5.3) can be solved for a_{ec} and σ_{Nu}, thus giving the peak load and the critical equivalent crack extension. This is illustrated in the following example.

Example 5.5.1 To determine the strength predicted by the Jenq-Shah model for an infinite panel with a central crack of length $2a_0$, subjected to remote tensile stress σ_∞ normal to the crack, we use the well-known solution for the stress intensity factor $K_I = \sigma_\infty\sqrt{\pi a}$ to write (5.5.1) as

$$\sigma_{\infty u}\sqrt{\pi a_{ec}} = K_{Ic} \tag{5.5.4}$$

This takes the form of (5.5.1) if we identify σ_{Nu} with $\sigma_{\infty u}$, D with a_0 (the only relevant specimen dimension), and take $k(\alpha) = \sqrt{\pi\alpha}$. The second equation follows from the crack opening profile (2.2.7), $w = (4\sigma_\infty/E')\sqrt{a^2 - x^2}$, where x is the distance measured from the center of the crack. Therefore, putting $x = a_0$ and $a = a_{ec}$, we get w_T (i.e., the CTOD). The resulting equation for the peak load state thus is

$$\frac{4\sigma_{\infty u}}{E'}\sqrt{a_{ec}^2 - a_0^2} = w_{Tc} \tag{5.5.5}$$

This also coincides with the form (5.5.3) with the same identifications as before and with $v_T(\alpha, \alpha_0) = 4\sqrt{\alpha^2 - 1}$ (in this particular case, $\alpha_0 = 1$, and v does not depend on α_0). Now, solving for a_{ec} from (5.5.4), substituting in (5.5.5) and solving for $\sigma_{\infty u}$, we get the solution

$$\sigma_{\infty u} = \frac{K_{Ic}}{\sqrt{\pi a_0}}\left[\sqrt{1 - \left(\frac{c}{a_0}\right)^2} - \frac{c}{a_0}\right]^{1/2} \quad \text{with} \quad c = \frac{\pi}{32}\left(\frac{w_{Tc}E'}{K_{Ic}}\right)^2 \tag{5.5.6}$$

For $a_0 \to \infty$, this solution converges to the LEFM solution $\sigma_{\infty u} = K_{Ic}/\sqrt{\pi a_0}$. Solving now for a_{ec}, we find

$$a_{ec} = a_0 + a_0\frac{1 + c/a_0 - \sqrt{1 - (c/a_0)^2}}{\sqrt{1 - (c/a_0)^2} - c/a_0} \tag{5.5.7}$$

Taking the limit for $a_0 \to \infty$ for the crack extension $\Delta a_{ec} = a_{ec} - a_0$, we get

$$\lim_{a_0\to\infty} \Delta a_{ec} = c_f = c = \frac{\pi}{32}\left(\frac{w_{Tc}E'}{K_{Ic}}\right)^2 \tag{5.5.8}$$

which gives the limit of the effective crack extension for very large cracks. □

This example shows that, when the size of the specimen (represented in this particular case by a_0) tends to infinity, the solution for the strength tends to LEFM, while the critical crack extension tends to a constant. This is indeed generally true. To prove it, it is better to recast equation (5.5.3) so that the shape of the crack profile near the crack tip would be explicit; based on the general form (2.2.16), we can write $w = (8K_I/\sqrt{2\pi}\,E')\sqrt{r}\,L(r/D)$ where r is the distance to the crack tip and function $L(r/D)$ is a dimensionless regular function which depends implicitly on the crack length and satisfies the condition $L(0) = 1$ (Planas and Elices 1989a). Writing this equation for the initial crack tip and peak load condition ($r = \Delta a_{ec}$, $K_I = K_{Ic}$), we get

$$\frac{8}{\sqrt{2\pi}\,E'}K_{Ic}\sqrt{\Delta a_{ec}}\,L(\Delta\alpha_{ec}, \alpha_0) = w_{Tc} \tag{5.5.9}$$

where the dependence of L on the initial crack length has been made explicit, and $L(0, \alpha_0) = 1$.

This equation has the advantage of having a single unknown, Δa_{ec}. Upon squaring, it can be rearranged to read

$$\Delta\alpha_{ec}L^2(\Delta\alpha_{ec}, \alpha_0) = \frac{c_f}{D} \tag{5.5.10}$$

where, again,

$$c_f = \frac{\pi}{32}\left(\frac{w_{Tc}E'}{K_{Ic}}\right)^2 \tag{5.5.11}$$

We clearly see that $D \to \infty$ implies $\Delta\alpha_{ec} \to 0$, and, thus, $L(\Delta\alpha_{ec}, \alpha_0) \to 1$. This proves the result (5.5.8) to be completely general (geometry independent).

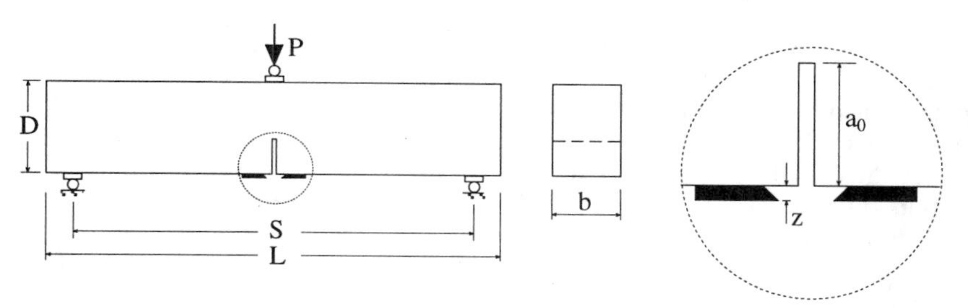

Figure 5.5.2 Testing configuration and geometry of specimen for the determination of Jenq-Shah fracture parameters (adapted from RILEM 1990a).

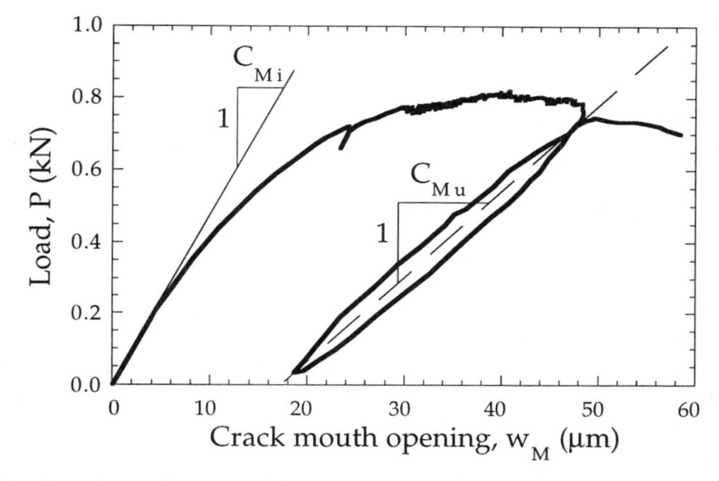

Figure 5.5.3 Experimental Load-CMOD curve in Jenq-Shah method (adapted from RILEM 1990a).

5.5.2 Experimental Determination of Jenq-Shah Parameters

In the previous section, it was seen how the peak load and the critical equivalent crack length may be obtained when the model parameters K_{Ic} and w_{Tc} are given. To obtain the parameters from an experiment, the same two equations are used, but now the peak load and the critical equivalent crack length must be measured. The measurement of the effective crack length cannot be done by direct methods, because it corresponds to a visually ill-defined point somewhere in the fracture process zone. Therefore, Jenq and Shah (1985a, b) devised a method based on the measurement of the compliance. To do so, they use the LEFM basic equation (3.5.24) relating the crack mouth opening displacement (CMOD) w_M to the applied load P and to the relative crack length $\alpha = a/D$. In the RILEM (1990a) Recommendation, a notched three-point-bend beam with a span-to-depth ratio of 4 and a notch-to-depth ratio of 1/3 is proposed (Fig 5.5.2). For this specimen, D is the beam depth and the shape function $\hat{v}_M(\alpha)$ in (3.5.24) is approximately given by (Tada, Paris and Irwin 1985):

$$\hat{v}_M(\alpha) = 24\alpha p_M(\alpha)\,, \quad p_M(\alpha) = 0.76 - 2.28\alpha + 3.87\alpha^2 - 2.04\alpha^3 + 0.66(1-\alpha)^{-2} \qquad (5.5.12)$$

From this, the elastic compliance $C_M = w_M/P$ is given by

$$C_M = \frac{24}{bE'}\alpha p_M(\alpha) \qquad (5.5.13)$$

which is the relationship relating the compliance to the crack length.

With these equations as the basis, the test is performed by loading the specimen under CMOD control up to the point beyond the peak at which the load has been reduced approximately to 95% of the peak (Fig. 5.5.3) and then it is unloaded and reloaded. The initial compliance C_{Mi} is determined from the initial tangent, and the mean unloading compliance C_{Mu} is calculated from a linear fit to the unloading-reloading loop (Fig. 5.5.3). Equation (5.5.13) is then used to obtain the elastic modulus E' from the initial compliances and crack lengths, and then used again to find the equivalent elastic crack length at the unloading-reloading cycle (which is taken to coincide with the critical equivalent crack length). However, since, in general, the CMOD is not measured exactly at the crack mouth because of the thickness of the knife edges supporting the clip gauge (Fig 5.5.2), a correction is introduced such that α in $p_M(\alpha)$ is replaced by $\alpha' = (a+z)/(D+z)$, and then one has, for the elastic modulus,

$$E' = \frac{24}{bC_{Mi}}\alpha_0 p_M(\alpha_0'), \quad \alpha_0' = \frac{a_0+z}{D+z} \tag{5.5.14}$$

The critical effective crack length is obtained by substituting the value of E' and of the unloading compliance into (5.5.13):

$$\alpha_{ec}\, p_M(\alpha_{ec}') = \frac{bC_{Mu}E'}{24}, \quad \alpha_{ec}' = \frac{a_{ec}+z}{D+z} \tag{5.5.15}$$

which must be solved iteratively.

Direct iteration solving for the new value in the form $\alpha_{ec} = bC_{Mu}E'/[24\, p_M(\alpha_{ec}'^{pre})]$ may easily diverge (superscript ^{pre} indicates the value from the previous iteration). A modified weighted direct iteration scheme of the form

$$\alpha_{ec} = (1-\omega)\frac{bC_{Mu}E'}{24\, p_M(\alpha_{ec}'^{pre})} + \omega\, \alpha_{ec}^{pre} \tag{5.5.16}$$

where ω is some appropriately chosen weight, may be useful for a fast and systematic solution of the equation (see the forthcoming example).

Once α_{ec} is known, K_{Ic} follows from (5.5.1), with $k(\alpha)$ given by Srawley formula (5.4.8). Note that in the computation of the nominal stress σ_{Nu}, the ultimate central bending moment M_u must include all the contributions, including those from the weight of the specimen and the weight of all the ancillary equipment hanging on the specimen; thus, we write

$$\sigma_{Nu} = \frac{6M_u}{bD^2}, \quad \text{with} \quad M_u = \frac{P_u S}{4} + M_{weight} \tag{5.5.17}$$

where P_u is the ultimate central load and M_{weight} is the bending moment at the central cross-section caused by the dead weights.

Similarly, w_{Tc} is obtained from (5.5.3), with $v_T(\alpha, \alpha_0)$ approximately given for the recommended geometry by (RILEM Recommendation 1990a)

$$v_T(\alpha,\alpha_0) = 4p_M(\alpha)\left[(\alpha-\alpha_0)^2 + (1.081 - 1.149\alpha)\alpha_0(\alpha-\alpha_0)\right]^{1/2} \tag{5.5.18}$$

in which $p_M(\alpha)$ is given in (5.5.12).

Example 5.5.2 The results of Fig. 5.5.3 were obtained by Jenq and Shah for a mortar beam of $b \times D \times L = 28.6 \times 76 \times 404$ mm, with a central notch of 22.4 mm and a loading span $S = 4D = 304$ mm. The distance from the measuring clip gauge to the surface of the specimen was $z = 5$ mm, the density of the mortar 2350 kg/m^3, and the weight of the equipment resting on the specimen was negligible. To calculate the Jenq-Shah fracture parameters, we first estimate $C_{Mi} \approx 2.1\ 10^{-8}$ m/N from the initial tangent of the graph. Next, from the mean of the unloading-reloading loop, we get $C_{Mu} \approx 4.2\ 10^{-8}$ m/N. From (5.5.14) we then find $E' \approx 21.9$ GPa, and after iterating 4 times (5.5.16) with $\omega = 0.5$, we get $\alpha_{ec} \approx 0.4235$ (roughly, one iteration per digit but the accuracy of the coefficients in (5.5.18) does not justify more than 3-digit accuracy). Now, from the graph we find the maximum applied load to be $P_u \approx 0.81kN$, and we get the ultimate bending moment as $M_u = P_u S/4 + (mg/2)(S/2 - L/4)$, where the second term is the bending moment due to the weight of the beam (m is the mass of the beam and g the acceleration of gravity, 9.8 m/s^2). Working out the computations, with $m = 2350 \times 0.404 \times 0.076 \times 0.0286 = 2.06$ kg, we get $M_u = 62.1$ Nm, which gives an ultimate nominal stress $\sigma_{Nu} = 2.25$ MPa. Using for K_{Ic} the expression (5.5.1) (together with (5.4.7) and (5.4.8)), and Eqs. (5.5.3) and (5.5.18) for w_{Tc}, we get $K_{Ic} = 0.87$ MPa m$^{1/2}$ and $w_{Tc} = 14.2\ \mu$m. □

The parameters of the Jenq and Shah's model can also be determined by Bažant's size effect method from G_f and c_f, which has some advantages; see Section 6.1.

Exercises

5.9 Determine the critical effective crack extension for infinitely large specimens (and cracks) corresponding to the Jenq-Shah model with $K_{Ic} = 0.3$ MPa m$^{1/2}$ and w_{Tc} =10 μm.

5.10 For the test described in example 5.5.2, compare the apparent fracture toughness (obtained from peak load and initial crack length) with the Jenq-Shah toughness.

5.11 Compare the critical effective crack extension for the test described in example 5.5.2 with the equivalent crack extension for an infinitely large specimen.

5.12 Determine —according to Jenq-Shah model—the failure load of a three-point-bend beam if the specimen dimensions and material properties are as follows: thickness $b = 6$ cm, span $S = 60$ cm, beam depth $D = 15$ cm, initial notch length $a_0 = 6$ cm, $K_{Ic} = 0.3$ MPa m$^{1/2}$, and w_{Tc} =10 μm. Compare the critical effective crack extension with that corresponding to a very large specimen. Determine the apparent fracture toughness.

5.13 Determine —according to Jenq-Shah model—the rupture modulus of a three-point-bend beam (unnotched) if the specimen dimensions and material properties are as follows: thickness $b = 6$ cm, span $S = 60$ cm, beam depth $D = 15$ cm, $K_{Ic} = 0.3$ MPa m$^{1/2}$, and w_{Tc} =10 μm. Assume that a crack starts to grow from the tensile surface along the middle cross-section. What is the critical effective crack extension? And the apparent fracture toughness?

5.14 A prism of a quasibrittle material is tested in pure tension. Assume that the failure mechanism is such that a surface crack develops starting from some microscopic defect and that the peak load can be calculated in accordance with the Jenq-Shah model. Determine the relationship between the Jenq-Shah parameters K_{Ic} and w_{Tc} and the tensile strength f'_t measured in this test. The crack opening at the surface is approximately given by $w_T = 5.816 a\sigma/E'$ where σ is the applied tensile stress and a is the elastic crack (Tada, Paris and Irwin 1985).

5.6 R-Curves

The models expounded in the preceding sections are applicable only to the critical situation (peak load). The R-curve models are generalizations of such models that keep the concept of equivalent elastic crack but allow full mechanical analysis, i.e., description of the complete crack growth process and the determination of load-displacement curves. In such models, the basic LEFM formalism is preserved, but instead of a fracture criterion $\mathcal{G} = G_f$ or $K_I = K_{Ic}$, a crack growth rule relating a variable crack growth resistance $\mathcal{R}$ to the crack evolution is used.

The most prevalent kind of R-curve model assumes the resistance to depend only on the equivalent crack extension Δa_e. This is the default assumption in most papers dealing with this topic. But, in principle, other possibilities exist, in which the resistance does not depend explicitly on the effective crack extension, but on some other variable such as the crack tip opening displacement (CTOD), a case that will be discussed at the end of this section. First we will discuss the R-Δa formulation.

5.6.1 Definition of an R-Δa Curve

The formal definition of an R-curve model is very simple. One needs only to express the crack growth resistance $\mathcal{R}$ as a function of the effective crack extension Δa_e. With the understanding that from now on we always refer to effective or equivalent cracks, we drop subscript $_e$ and write simply Δa for the equivalent crack extension. With this convention, the R-curve as a function R of Δa reads

$$\mathcal{R} = R(\Delta a) \tag{5.6.1}$$

To some extent, function $R(\Delta a)$ may be approximately considered to be a fixed material property, as proposed by Irwin (1960) and in more detail by Krafft, Sullivan and Boyle (1961). Later it was found,

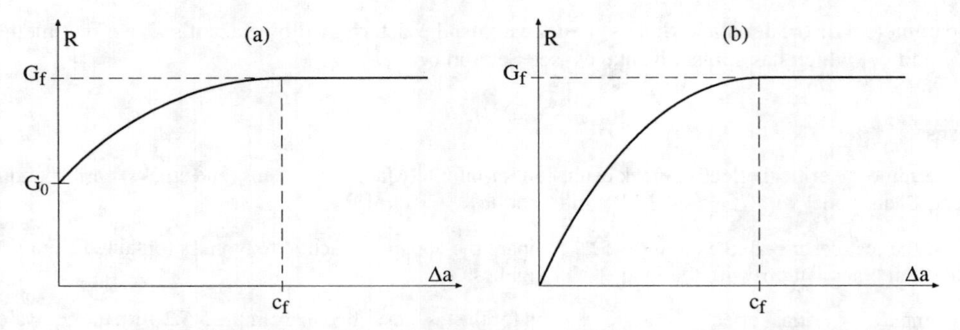

Figure 5.6.1 Two typical R-curves: (a) with threshold, and (b) without threshold.

however, that the shape of the R-curve depends considerably on the shape of the specimen or structure and, according to the older definitions of the R-curve, also on the structure size. The R-curve can be assumed to be unique only for a narrow range of specimen or structure geometries. Thus, it is necessary to determine the R-curve for the given geometry prior to undertaking fracture analysis. Let us now assume that the R-curve for the given structure geometry is already known.

Although the only structural variable that explicitly appears in the equation is the effective crack extension, other parameters depending on the material must also enter the equation. There are a number of parameters which have a physical meaning. Consider, for example, the R-curves depicted in Fig. 5.6.1. In the curve corresponding to Fig. 5.6.1a, the crack driving force $\mathcal{G}$ must reach a threshold value G_0 before the crack starts to grow, while in that in Fig. 5.6.1a, the crack starts to grow from the very beginning. In both cases, the resistance increases with crack growth until a plateau is reached. This happens when the fracture process zone is fully developed and starts to travel in a self similar way. Let us call c_f the value of the crack extension at which the R-curve reaches the plateau, and G_f the value of the resistance at the plateau, then, c_f and G_f depend on the material. If the assumption is valid that the R-curve is geometry and size independent, then they depend *only* on the material and, thus, are material properties. Because of the required dimensional compatibility of (5.6.1), this equation must accept the form

$$\mathcal{R} = G_f\, R\!\left(\frac{\Delta a}{c_f}\right) \tag{5.6.2}$$

where now $R(\Delta a/c_f)$ is a nondimensional function.

Example 5.6.1 As a simple example, consider the R-curve given by

$$\mathcal{R} = \begin{cases} G_f\left[1-\left(1-\Delta a/c_f\right)^n\right] & \text{for } \Delta a \le c_f \\ G_f & \text{for } \Delta a > c_f \end{cases} \tag{5.6.3}$$

This is indeed a family of R-curves depending on three parameters, namely, G_f, c_f, and n. To specify a particular material of this family, one has to give particular values of the three parameters. □

As explained in Section 2.1.4, the energy that must be supplied per unit crack front to produce a crack extension δa is $\mathcal{R}\delta a$. So, in a plane specimen, the fracture work per unit thickness, $\mathcal{W}_F/b$, required to extend the initial crack by Δa is given by the area under the corresponding arc of the R-curve:

$$\frac{\mathcal{W}_F}{b} = \int_0^{\Delta a} R(\Delta a)\, d\Delta a \tag{5.6.4}$$

This expression can be integrated in a particular case to give a function of the crack extension depending on the same material constants as the R-curve itself.

Example 5.6.2 Taking the R-curve given in the previous example, it is an easy matter to find the energy supply required to extend the crack by Δa from (5.6.4). The result is

$$\frac{\mathcal{W}_F}{b} = \begin{cases} \frac{1}{n+1} G_f c_f \left[1 - (1 - \Delta a/c_f)^{n+1}\right] & \text{for } \Delta a \leq c_f \\ G_f \Delta a - \frac{n}{n+1} G_f c_f & \text{for } \Delta a > c_f \end{cases} \tag{5.6.5}$$

Note that $\mathcal{W}_F/b\Delta a \rightarrow G_f$ for $\Delta a \gg c_f$. ▯

Some authors prefer to present the R-curves in a K_I vs. Δa plot, rather than in an $\mathcal{G}$–Δa graph. This is perfectly equivalent if one defines a resistant stress intensity factor K_R related to $\mathcal{R}$ by the Irwin relationship between K_I and $\mathcal{G}$:

$$K_R = \sqrt{E'\mathcal{R}} \tag{5.6.6}$$

Although minor mathematical transformations are needed, the two formulations are equivalent. But the direct relationship between the work of fracture and the area under the R-curve is lost in the K_R formulation. That is why we will adhere to the energetic representation.

5.6.2 Description of the Fracture Process

In a quasi-static fracture process, the equality $\mathcal{G} = \mathcal{R}$ is satisfied whenever the crack is growing. Therefore, in a $\mathcal{G}$–a plot the process must follow the R-curve. For a structure with a single load P (or a system of proportional loads in which the loading parameter P is taken to be a generalized force), one can build the $P(u)$ curve from the R-curve just by writing that $\mathcal{G} = \mathcal{R}$ and using the expressions for $\mathcal{G}$ and u derived from LEFM. Considering a pure mode I loading, we express $\mathcal{G}$ and u as a function of the load P and $a = a_0 + \Delta a$. The resulting system of two equations is:

$$\mathcal{G}(P, a) = R(a - a_0) \tag{5.6.7}$$

$$u = C(a)P \tag{5.6.8}$$

which furnishes a parametric representation of the $P(u)$ curve: for each value of the parameter a, one solves the first equation for P and then uses the second equation to get the displacement.

Example 5.6.3 Consider a material characterized by the R-curve (5.6.3) with $n = 3$, as shown in Fig. 5.6.2a. Consider a DCB specimen (double-cantilever beam specimen, Fig. 2.1.3, and sketch in Fig. 5.6.2b), with an initial crack length $a_0 = 8h$ and a thickness b. The expressions for $\mathcal{G}$ and u were obtained in Section 2.1.2, Eq. (2.1.29). The parametric equation for P, obtained from (5.6.7) and the second of (2.1.29), is

$$P = \begin{cases} (P_0 a_0/a)\left[1 - \left(1 - (a - a_0)/c_f\right)^3\right]^{1/2} & \text{for } a - a_0 \leq c_f \\ P_0 a_0/a & \text{for } a - a_0 > c_f \end{cases} \tag{5.6.9}$$

where

$$P_0 = \frac{bh}{6a_0}\sqrt{3EG_f h} \tag{5.6.10}$$

which is the load required to ensure that $\mathcal{G} = G_f$ at $a = a_0$. Now the expression for u follows from the first of (2.1.29):

$$u = \begin{cases} u_0 (a/a_0)^2 \left[1 - \left(1 - (a - a_0)/c_f\right)^3\right]^{1/2} & \text{for } a - a_0 \leq c_f \\ u_0 (a/a_0)^2 & \text{for } a - a_0 > c_f \end{cases} \tag{5.6.11}$$

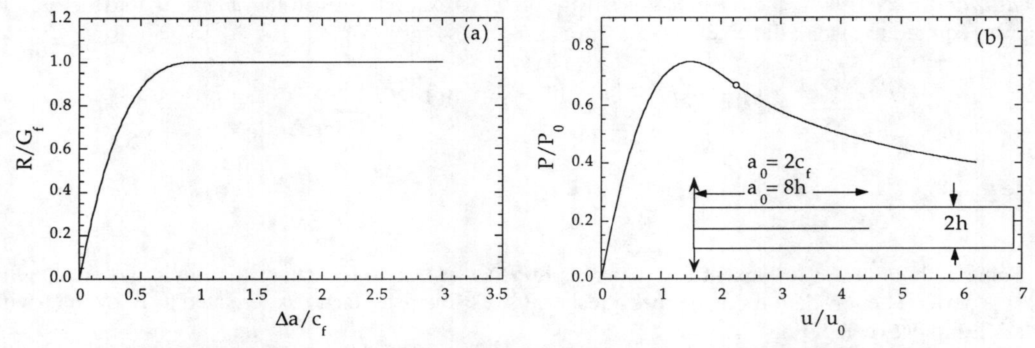

Figure 5.6.2 (a) Example of R-curve. (b) P–u curve for a double-cantilever beam.

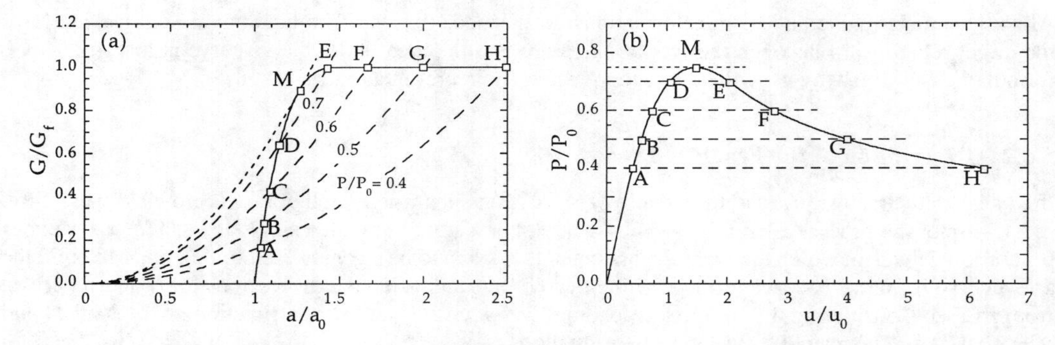

Figure 5.6.3 Representation of fracture processes in the DCB specimen of Fig. 5.6.2b. Dashed lines are curves of constant load. Full line is the fracture path. (a) Appearance in the $\mathcal{G}$–a plot. (b) Appearance in the P–u plot.

where u_0 is the displacement one would have for $P = P_0$ and $a = a_0$:

$$u_0 = \frac{4a_0^2}{3h^2}\sqrt{\frac{3G_f h}{E}} \tag{5.6.12}$$

The plot of the $P(u)$ curve is obtained by giving values to a in (5.6.9) and (5.6.11) and plotting the resulting (P, u) pairs. Fig. 5.6.2b shows such a plot for a specimen with $h = c_f/2$. The open circle marks the point where the plateau is reached ($\Delta a = c_f$). ▯

5.6.3 The Peak Load Condition

Eq. (2.3.12) shows that, according to LEFM, $\mathcal{G}(P, a) = P^2\hat{g}(\alpha)/Eb^2D$, where α is the relative crack depth a/D. So $\mathcal{G}(P, a)$ is proportional to P^2. For most structures and fracture specimens, $\hat{g}(\alpha)$, and thus also $\mathcal{G}(P, a)$, are increasing functions of the crack depth (in this case, we speak of positive geometry). The plots of functions $\mathcal{G}(P, a)$ for a succession of increasing values $P = P_1, P_2, P_3, \ldots$ then look as shown by the dashed curves in Fig. 5.6.3a, which corresponds to the example in Fig. 5.6.2. In the P–u plot, these lines are horizontal, as shown in Fig. 5.6.3b. Note that, in principle, each dashed curve cuts the curve describing the fracture process (full lines) twice, once before reaching the peak load and once after that. As the crack grows, the loading follows the sequence of points A–B–C–D–M–E–F–G–H. Along the arc A–B–C–D–M, the load is increasing, i.e., the structure is hardening. Along M–E–F–G–H the load is decreasing, i.e., the structure is softening.

To find the peak load, it is not necessary to plot the $P(u)$ curve. Indeed, it is quite clear that the peak is defined in the $\mathcal{G}$–a plot by the condition that the R-curve be tangent to a constant-load curve (point M

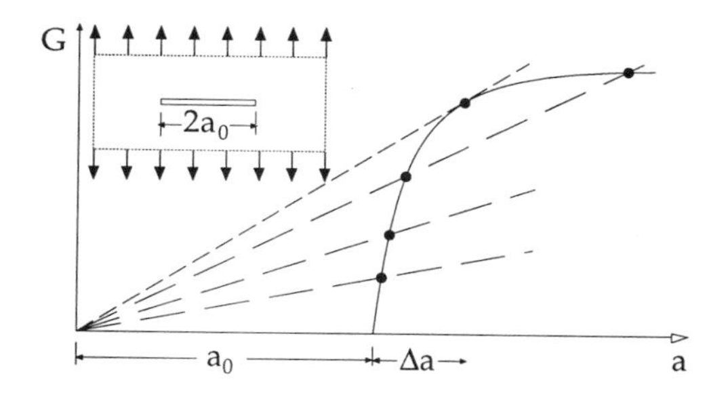

Figure 5.6.4 $\mathcal{G}$–a curves at constant load (dashed lines), for a center-cracked panel subjected to remote uniform tension. The peak load corresponds to the finely dashed line which is tangent to the R-curve, shown by a full line.

in Fig. 5.6.3b). This condition is very easy to impose graphically, although it may not always be easy to solve analytically or numerically.

Example 5.6.4 Consider a very large panel (infinite for any practical purpose) with a central crack of length $2a_0$ subjected to a remote uniaxial stress σ normal to the crack plane. Assume that the material fracture is described by an R-curve of the type (5.6.3) with $n = 2$. To find the peak stress σ_u, we first recall that, for the center-cracked panel, $\mathcal{G} = \sigma^2\pi a/E'$ which means that the constant-load $\mathcal{G}$–a curves are straight lines. Fig. 5.6.4 shows the peak load tangency condition. The analytical solution is obtained by writing that the straight line and the parabola are tangent: they have same ordinate and same slope at the contact point. Imposing the first condition, we have

$$\frac{\sigma^2\pi a}{E'} = G_f\left[1 - \left(1 - \frac{a - a_0}{c_f}\right)^2\right] \tag{5.6.13}$$

where we wrote $a - a_0$ for Δa in the expression of the R-curve. Differentiating this equation with respect to a, we get the condition for equal slopes:

$$\frac{\sigma^2\pi}{E'} = \frac{2G_f}{c_f}\left(1 - \frac{a - a_0}{c_f}\right) \tag{5.6.14}$$

These two equations form a system whose solution for σ and a are the values at the peak load point. In this case, the solution may be obtained analytically solving (5.6.14) for a, substituting this in (5.6.13) and solving for σ. The result is

$$\sigma_u = \sigma_0\,\gamma\left(\frac{c_f}{a_0}\right) \tag{5.6.15}$$

where σ_0 and γ are given by

$$\sigma_0 = \sqrt{\frac{E'G_F}{\pi a_0}}\,, \qquad \gamma(\rho) = \left[\frac{\rho^2}{2\left(1 + \rho - \sqrt{1 + 2\rho}\right)}\right]^{-1/2} \tag{5.6.16}$$

Note that σ_0 is the peak load one would have for this panel if LEFM were applicable with a constant $\mathcal{R} = G_f$. □

This example illustrates the general approach to finding the peak load: graphically, one must find the $\mathcal{G}$–a curve that is tangent to the R-curve. Analytically, the peak load P_u is the solution for P of the system of equations

$$\mathcal{G}(P, a) = R(a - a_0) \tag{5.6.17}$$

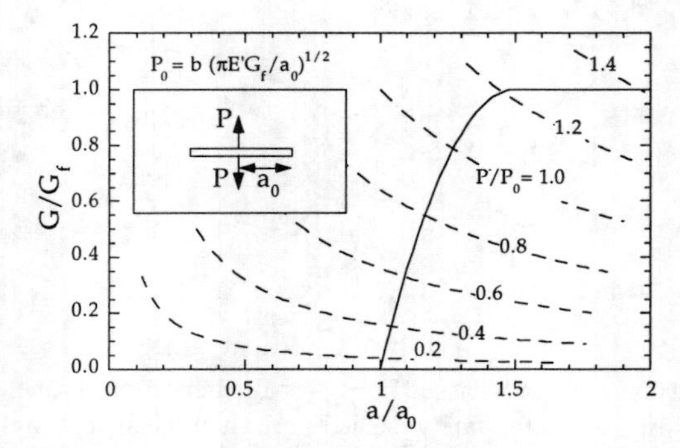

Figure 5.6.5 $\mathcal{G}$–a curves at constant load (dashed lines), for a center-cracked panel subjected to two central opening point loads. None of these curves ever becomes tangent to the R-curve.

$$\frac{\partial \mathcal{G}(P,a)}{\partial a} = R'(a - a_0) \tag{5.6.18}$$

where $R'(\Delta a)$ stands for the derivative of $R(\Delta a)$ with respect to its argument.

5.6.4 Positive and Negative Geometries

The examples in the previous section correspond to positive geometries, for which $\mathcal{G}$ increases with a at constant load. This is, as already pointed out, the most common type of geometry. However, negative geometries may be encountered and one needs to now how to handle them. The graphical analysis may proceed as in the previous examples, except that the $\mathcal{G}$–a curves will not be always increasing. In a particular special case, negative geometry can lead to ever increasing load required for propagating the crack, as shown in the following example.

Example 5.6.5 Consider an infinite panel of thickness b with a center crack of length $2a$ subjected to two equal opening opposite forces P applied at the center of its faces (example 3.1.2). The stress intensity factor is given by $K_I = P/b\sqrt{\pi a}$. Thus, according to Irwin's relationship, the energy release rate is given by $\mathcal{G}(P,a) = P^2/bE'\pi a$. The curves $\mathcal{G}$–a at constant load are hyperbolas which never become tangent to the R-curve, for well-behaved (nondecreasing) R-curves. This is illustrated in Fig. 5.6.5 and shows that there is no peak load: the load must keep increasing in order to keep the crack growing. Analytically, one may show it by writing $\mathcal{G}(P,a) = R(a - a_0)$, from which the evolution of P with the crack growth is:

$$P = \sqrt{bE'\pi\, aR(a - a_0)} = \sqrt{bE'\pi\, aR(\Delta a)} \tag{5.6.19}$$

Since $R(\Delta a)$ is positive and nondecreasing, $aR(\Delta a)$ is monotonically increasing, and so is P. □

This example corresponds to an unbounded specimen, a limiting case. In actual panels, having a finite width, the geometry is negative while the crack is small, but shifts to positive when the crack tip approaches the boundaries. The $\mathcal{G}$–a curves at constant load go through a minimum as in Fig. 5.6.6. In these cases, there is a peak load whose value may be determined using a general method to find the tangency condition, as shown by the finely dashed curve in Fig. 5.6.6.

5.6.5 *R*-Curve Determination from Tests

To determine the R-curve, it is necessary to obtain a set of pairs of values $(\mathcal{G}, \Delta a)$ at various quasi-static crack growth instants. Since in quasi-static crack growth $\mathcal{G} = \mathcal{R}$, these pairs define points of the R–Δa curve. There are many ways of determining the R-curve (for a complete discussion, see Elices and Planas 1993, and Planas, Elices and Ruiz 1993). However, in practice, the load is directly measured, and the

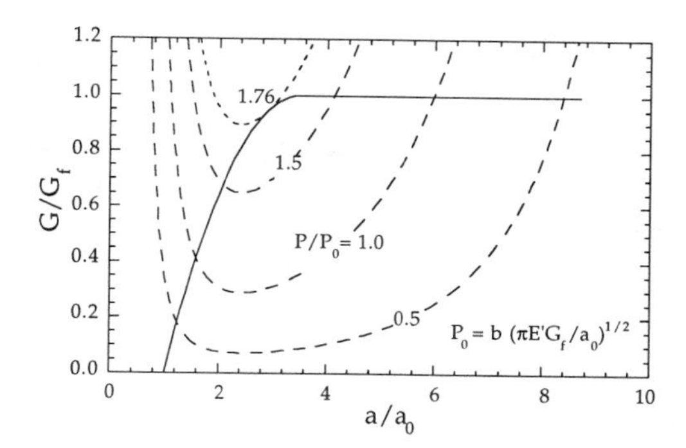

Figure 5.6.6 $\mathcal{G}$–a curves at constant load (dashed lines), for an initially negative geometry. The peak load corresponds to the finely dashed curve which is tangent to the R-curve, shown as a full line.

equivalent elastic crack length is inferred from a second measurement. Since measuring the load is a standard procedure, the various methods differ only on how the equivalent crack extension is determined. There are basically five methods:

1. *Direct determination of a.* In this method, the crack length along a slow fracture test is measured by optical means. For concrete and other highly heterogeneous materials, there are two problems: (1) the measured crack length is not clearly defined and depends strongly on the resolution of the optical set-up; and (2) the crack can be measured only at the surface of the specimen, while the crack front can be deeper (or shallower) in the core of the specimen. In general, other kinds of measurement giving an average crack length are more reliable.
2. *Unloading compliance measurement of a.* This approach includes, as particular cases, the techniques, previously described, of Swartz (Section 5.4.1) and of Jenq and Shah (Section 5.5). The procedure is always the same: the specimen is subjected to stable crack growth up to the required point, and then one or various unloading-reloading cycles are performed. The effective crack length is computed from the measured unloading-reloading compliance, assuming that LEFM applies in such a process (for three-point-bend specimens, see the details in Section 5.5.2). In principle, any displacement can be used, but the crack-mouth-opening displacement is the most frequent. The main drawback of this method is that, due to crack closure and other stiffening effects, the measured equivalent crack length can be too small. By crack closure, one usually means the crack lips come into contact upon load release. This is usually due to roughness mismatch or debris. Other stiffening effects stem from the fact that the fracture process zone may contribute a larger deformation during virgin loading than during unloading and reloading, the effect being similar to that stemming from crack closure.
3. *Secant compliance measurement of a.* In the secant compliance method, the equivalent elastic crack is defined as that giving the same displacement as observed in virgin loading. The Nallathambi-Karihaloo method described in Section 5.4.3 is of this type (although it is applied only to the peak load; application to other arbitrary points along the load-displacement curve follows the same trend). The compliance may refer, as before, to any displacement: load-point displacement —as in the Nallathambi-Karihaloo method— or CMOD. Although it has been shown that the two possibilities are not strictly equivalent, the R-curves so obtained do not differ too much (Elices and Planas 1993).
4. *R-curve determination based on size effect.* The R-curve can also be determined from the property, described in Section 5.6.3, that at peak load the $\mathcal{G}$–a curve is tangent to the R curve. Then, if one performs tests on specimens of various sizes, records the peak loads P_u and draws the curve $\mathcal{G}(P_u, a)$ for each specimen size, these curves must be tangent to the R-curve at different points (because the specimens have different sizes). Thus, the R-curve is the envelope of all the $\mathcal{G}(P_u, a)$ curves for all sizes. Bažant, Kim and Pfeiffer (1986) put forth an analytical-numerical method that uses tests with a few sizes and extends them to other sizes by means of Bažant's size effect law (1.4.10); this and the more recent fully analytical method devised by Bažant and Kazemi (1990a) will be the object of a detailed presentation in next chapter. Although the method can, in

principle, be applied on the basis of purely experimental results, with no recourse to the size effect law, using the size effect law eliminates the difficulty of getting the envelope of curves that have a large experimental scatter.

5. *R-curve determination based on shape effect.* In this method, the shape of the specimen, e.g., the relative notch length $\alpha_0 = a_0/D$, is varied. The R-curve is the envelope of the the $\mathcal{G}(P_u, \alpha_0)$ curve for all α_0. This method, however, cannot determine the full, or even a broad part of the range of the R-curve, because variation of α_0 cannot change the brittleness of single-size fracture specimens by more than 1:3 (Bažant and Li 1996). To cover a broader range, the size D must be varied also.

5.6.6 R-CTOD Curves

In the preceding analysis of the R-curves we have assumed that the resistance curve depends, aside from some material parameters, only on the crack extension; in particular, the R–Δa curve is usually assumed to be independent of specimen size and geometry. This is really not so. In general, the R-curves depend on geometry and size, and also on the method used to determine them. The only method that leads to size-independent R–Δa curves is the size effect method, although the R-curves so obtained still are geometry-dependent. All the other methods lead to R-curves that depend on geometry and size, except for the limit of large sizes in which the various R-curves merge into a unique curve (Planas and Elices 1991b, Elices and Planas 1993; Planas, Elices and Ruiz 1993).

One formulation that leads to equations that are relatively less dependent on the size and geometry is the so-called R-CTOD curve (Planas, Elices and Toribio 1989; Elices and Planas 1992; Planas, Elices and Ruiz 1993). In this formulation, it is assumed that the crack growth resistance $\mathcal{R}$ depends explicitly only on the value w_T of the crack-tip-opening displacement, defined (as in the Jenq-Shah model) as the crack opening at the initial crack (or notch) tip; Fig. 5.5.1. Thus, Eq. (5.5.1) for the R–Δa curve is replaced in the R-CTOD formulation by

$$\mathcal{R} = R(w_T) \tag{5.6.20}$$

This curve, together with the equations of LEFM for the stress intensity factor and for the CTOD, determines crack growth completely . This is so because Irwin's relationship $\mathcal{G} = K_I^2/E'$ and the general expression (2.3.11) for K_I yield

$$\mathcal{G} = \frac{\sigma_N^2}{E'} D k^2(\alpha) \tag{5.6.21}$$

Furthermore, w_T is given by (5.5.2) which we repeat here for convenience:

$$w_T = \frac{\sigma_N}{E'} D v_T(\alpha, \alpha_0) \tag{5.6.22}$$

The analysis of the fracture process takes a form completely analogous to the $\mathcal{R}$–Δa formulation. One simply eliminates α from equations (5.6.21) and (5.6.22) to get an expression for $\mathcal{G}$ as a function of σ_N and w_T:

$$\mathcal{G} = \mathcal{G}(\sigma_N, w_T) \tag{5.6.23}$$

Once this equation is obtained, the analysis follows exactly as for the $\mathcal{R}$–Δa curves, except that w_T replaces Δa. The plots are similar and the peak load condition is again that the $\mathcal{G}$–w_T curve at constant load obtained form (5.6.23) be tangent to the $\mathcal{R}$–w_T curve (5.6.20). If elimination of α is analytically impossible, (5.6.21) and (5.6.22) serve as a parametric representation of (5.6.23), with parameter α.

Example 5.6.6 Consider determining equation (5.6.23) for an infinite panel with a central crack of initial half-length a_0, subjected to a remote tensile stress σ. Since, for this geometry, $K_I = \sigma\sqrt{\pi a}$, the energy release rate is

$$\mathcal{G} = \frac{\sigma^2}{E'} \pi a \tag{5.6.24}$$

The crack-tip opening is (see example 5.5.1)

$$w_T = \frac{4\sigma}{E'}\sqrt{a^2 - a_0^2} \tag{5.6.25}$$

Solving for a from this last equation we get $a = \sqrt{a_0^2 + w_T^2 E'^2/(16\sigma^2)}$; substitution of this result into (5.6.24) leads to

$$\mathcal{G} = \frac{\sigma^2}{E'}\pi a_0 \sqrt{1 + \frac{w_T^2 E'^2}{16 a_0^2 \sigma^2}} \tag{5.6.26}$$

which is the form of (5.6.23) for this particular case. □

Exercises

5.15 Consider a very large panel with a short crack of length a_0 at its edge, subjected to a tensile stress σ parallel to the cracked edge. Assume that the crack growth can be approximated by a R-curve with a threshold, such as the one in Fig. 5.6.1a. Draw a plot similar to that in Fig. 5.6.4 and find the peak stress for very small a_0. Show that the fracture strength of the panel becomes infinitely large as a_0 becomes infinitely small.

5.16 Consider a very large panel with a short crack of length a_0 at its edge, subjected to a tensile stress σ parallel to its cracked edge. Assume that the crack growth can be approximated by a convex R-curve without threshold, such as the one in Fig. 5.6.1b. Draw a plot similar to that in Fig. 5.6.4 and show graphically that when $a_0 \to 0$ the fracture strength of the panel tends to a finite limit. Find the value of this limit if the initial slope of the R-curve is R'_0.

5.17 A panel subjected to tensile stress parallel to one of its edges is made of a material with an R-curve given by (5.6.4). Determine the fracture strength of the panel for vanishingly small edge cracks as a function of G_f, E', c_f, and n.

5.18 (a) Find the external work required to completely break a three-point-bend notched beam of thickness b, depth D, and notch-to-depth ratio α_0, under the assumption that the fracture of the material can be described by a R-curve given by $\mathcal{R} = G_f[1 - \exp(-\Delta a/c_0)]$ where G_f and c_0 are constants. (b) Find the dependence of the mean fracture energy on the beam depth; the mean fracture energy is the total work supplied, divided by the initial ligament area.

5.19 Consider a very large panel with a center crack of length $2a_0 = 100$ mm, subjected to remote tension normal to the crack plane. Assuming that the material follows a R-curve defined by (5.6.4) with $n = 2$, $G_f = 120$ N/m, and $c_f = 50$ mm, and that $E' = 600 GPa$, determine (a) the peak stress; (b) the apparent fracture toughness K_{INu}; (c) the ultimate fracture toughness K_{Iu} (value of K_I at the equivalent crack tip at peak load); and (d) the true fracture toughness K_{Ic} (value of K_I at peak for an infinitely long crack).

5.20 Repeat the previous exercise for an initial crack length 10 times longer. What is now the difference between K_{INu}, K_{Iu}, and K_{Ic}?

5.21 Consider a test carried out on a DCB specimen of thickness b, total depth $2h$, and initial crack length $a_0 = 10h$. The output of the test is a load–displacement curve digitally recorded as a number of points (P_i, u_i) $(i = 1, 2, \cdots)$. Write the equations that should be included in a subroutine to compute: (a) the effective elastic modulus, given that, from visual inspection, the first m points are on the linear portion of the P–u plot; (b) the value of the equivalent crack extension Δa_i corresponding to each experimental point (P_i, u_i); (c) the value of the crack growth resistance $\mathcal{R}_i$ for each experimental point. Which kind of method are you using to determine the R-curve?

5.22 Consider a very large panel with a center crack of length $2a_0 = 100$ mm, subjected to remote tension normal to the crack plane. Assuming that the material follows a R–w_T curve that rises linearly up to $\mathcal{R} = G_f = 120$ N/m when $w_T = w_{Tc} = 20$ μm and is thereafter constant, and that $E' = 600 GPa$, determine (a) the peak stress; (b) the apparent fracture toughness K_{INu}; (c) the ultimate fracture toughness K_{Iu} (value of K_I at the equivalent crack tip at peak load); (d) the true fracture toughness K_{Ic} (value of K_I at peak load for an infinitely long crack).

5.23 Repeat the previous exercise for a 10-times longer initial crack. How do K_{INu} and K_{Iu} evolve with respect to K_{Ic}?

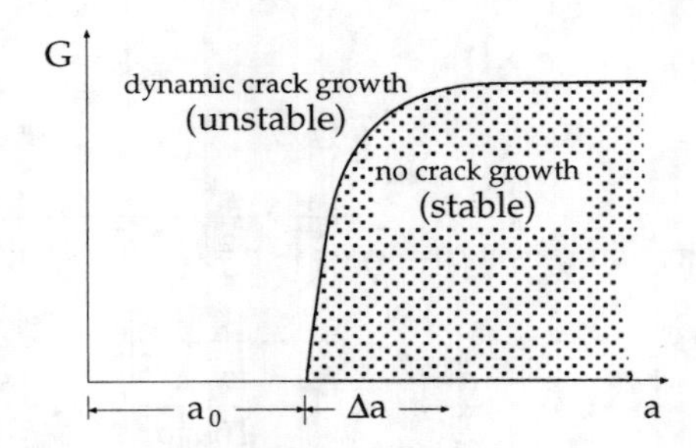

Figure 5.7.1 The R-curve divides the space into a stability region (region where no crack growth can take place, shaded area), and an instability region (or region of dynamic crack growth).

5.24 Show that, for positive geometries, the Jenq–Shah model can be considered a special case of R-CTOD model in which the $\mathcal{R}$–w_T curve is concave upward up to a cusp point for $w_T = w_{Tc}$ and $\mathcal{R} = K_{Ic}^2/E'$, followed by a plateau. Show that, for positive geometries, the only condition is that the concavity of the initial branch be strong enough for the usual $\mathcal{G}$–w_T curves to be always tangent to the R-curve at the cusp point.

5.7 Stability Analysis in the R-Curve Approach

Stability is an important issue both in experimental and practical situations. The stability analysis is easy to perform if the quasi-static P–u response of the structure is known (see, e.g., Bažant and Cedolin 1991).We saw in the previous section how to obtain this response from the R-curve, but this was not always simple. Therefore, in the following we are going to base the analysis of stability on a $\mathcal{G}$–a plot rather than on a P–u plot. The load-controlled case will be analyzed first on the basis of the examples in the previous section.

5.7.1 Stability under Load-Control Conditions

The most common case of structural boundary condition is the load-control, in which the forces acting on the structure are prescribed while the structure is free to deform. It is the case of loading by dead weights, wind loads, most cases of water pressure, and so on. The stability analysis is very simple: assuming that we start at an equilibrium state with $\mathcal{G} = \mathcal{R}$, the question is whether the crack can further propagate if we keep the load constant. Now, we remember from Section 2.1.1, that the crack cannot propagate if $\mathcal{G} < \mathcal{R}$, and that the crack will propagate dynamically (i.e., with excess energy turning into kinetic energy) whenever $\mathcal{G} > \mathcal{R}$. The equilibrium is unstable at constant load if the crack can go into the dynamic regime. In the $\mathcal{G}$–a plot, the limit between the no-crack-growth states and the dynamic-crack-growth states is the R-curve, as sketched in Fig. 5.7.1. To know whether the equilibrium at a point is stable, we must know whether a virtual crack growth along the constant-load $\mathcal{G}$–a curves would drive the crack into the stability region or the instability region.

Consider then the case of the DCB specimen shown in Figs. 5.6.2 and 5.6.3, and focus on the $\mathcal{G}$–a plot in Fig. 5.6.3a. Pick an equilibrium point such as point C. If the crack is assumed to grow a small distance, while keeping the load constant, the point will move along the dashed line CF into the stability region. Thus, the crack must stop, and so the equilibrium is stable. If we now pick an equilibrium point such as point G and repeat the foregoing mental exercise, we see that now a small crack growth at constant load drives the point along BG into the instability region. Thus, point G is an unstable equilibrium point.

This analysis shows then that equilibrium points on the branch A–M are stable at constant load, while those on the branch M–H are unstable. It is obvious that M itself, which corresponds to the peak load, is a point of unstable equilibrium. Therefore, if we load the structure progressively (for example, by adding dead weights), it will fail catastrophically upon reaching point M.

Of course, structures such as the that analyzed in Example 5.6.5, which have monotonically decreasing $\mathcal{G}$–a curves at constant load (Fig. 5.6.5), are always stable. For more general structures such as that one

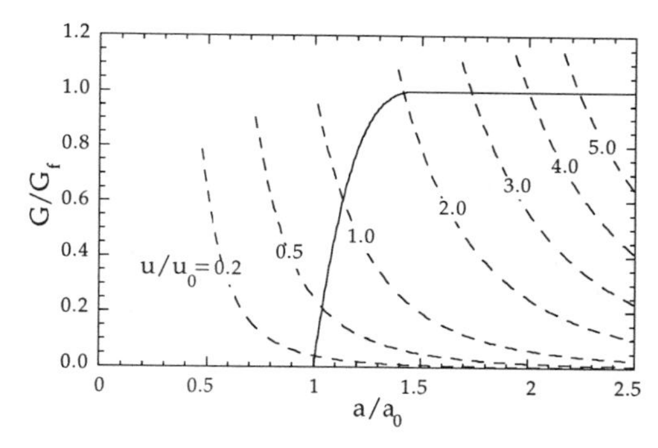

Figure 5.7.2 $\mathcal{G}$–a curves at constant displacement (dashed lines), for the DCB specimen of Fig. 5.6.2b. All the equilibrium points are stable, and u_0 is given by (5.6.12).

having the $\mathcal{G}$–a diagram shown in Fig. 5.6.6, the instability at constant load occurs again at the peak load, for which the $\mathcal{G}$–a curve at constant load is tangent to the R-curve.

5.7.2 Stability under Displacement-Control Conditions

In some practical cases and in many experimental situations, the structure is loaded in such a way that the displacement, rather than the load, is imposed. This is the case of some thermal and shrinkage loadings, and loading by very stiff devices such as wedges or screws. For example, loading through a bolt with a pitch of 1 mm will produce a displacement of approximately 0.125 mm when we turn the bolt by 1/8 of a revolution.

For such a kind of loading devices, the stability can be analyzed the same way as in the load-controlled case, except that the displacement is kept constant instead of the load. Therefore, in the G–a plot, we need to draw the G–a curves at constant displacement. This is done by eliminating P from the equations giving $\mathcal{G} = \mathcal{G}(P, u)$ and $u = C(a)P$, and plotting the resulting equation $\mathcal{G} = \mathcal{G}(u, a)$. This is best shown through an example.

Example 5.7.1 Analyze stability under controlled displacement of the DCB beam in example 5.6.3. The problem is solved by eliminating P from the equations (2.1.29). The result is

$$\mathcal{G} = \frac{3Eh^3u^2}{16a^4} = G_f \frac{u^2 a_0^4}{u_0^2 a^4} \tag{5.7.1}$$

with u_0 given by (5.6.12). Fig. 5.7.2 shows a number of constant displacement $\mathcal{G}$–a curves (dashed lines) and the R-curve. Following the discussion in the previous section, it is obvious that all the equilibrium states are stable since further crack growth at constant displacement drives the state point into the zone of stability according to Fig. 5.7.1.

It turns out that, for this geometry, the displacement controlled tests are always stable. However, this is not a general property. Some geometries may display instabilities at controlled displacement (this is usually called a *snap-back* instability, or snap-back, for short; see the next section). ▯

5.7.3 Stability under Mixed-Control Conditions

In many circumstances, the structure of interest is connected to a surrounding, and together they make a larger structure. In such cases, the load applied to the whole structure is transmitted to our substructure in a more complex manner which implies that at the end, our structure is subjected to a mixed control, with a prescribed relation of load to displacement. The simplest case is the series coupling shown in Fig. 5.7.3a. This corresponds to a common situation in testing practice, in which a specimen is loaded using a displacement-controlled machine with a finite stiffness. In most cases the deformation of the machine

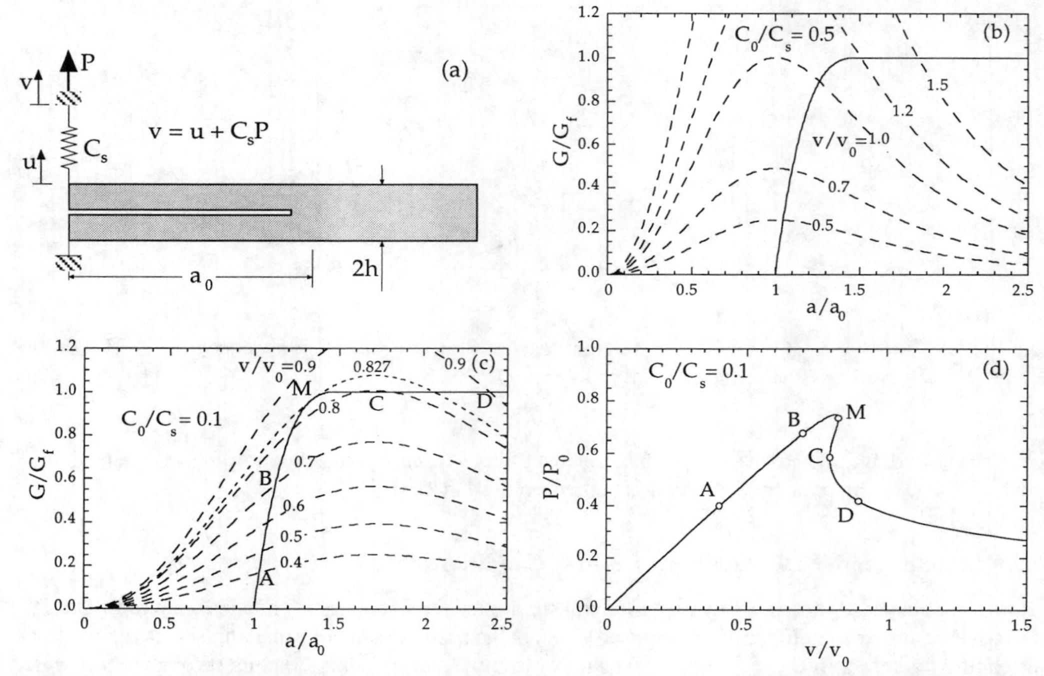

Figure 5.7.3 (a) DCB specimen coupled in series with a spring. (b) $\mathcal{G}$–a plot for a relatively stiff spring. (c) $\mathcal{G}$–a plot for a relatively soft spring. (d) P–v plot for the soft spring showing snap-back along segment M–C. v_0 and C_0 are given in (5.7.4), P_0 in (5.6.10).

may be taken to be linear, and thus equivalent to a spring of compliance C_s, as shown ($C_s = 1/k$ where k is the spring constant). The machine driving system controls the displacement v which is the sum of the displacement u of the specimen, plus the elongation C_sP of the spring:

$$v = u + C_sP \tag{5.7.2}$$

Since the control variable is v, the $\mathcal{G}$–a curves to be plotted are constant-v curves. To obtain them, it is necessary to eliminate P and u from (5.7.2) and the equations $\mathcal{G} = \mathcal{G}(P, a)$, and $u = C(a)P$, as done in the following example.

Example 5.7.2 Consider the DCB specimen of example 5.6.3 loaded as shown in Fig. 5.7.3a. We eliminate P and u from (5.7.2) and the two equations (2.1.29). The result is

$$\mathcal{G} = G_f \frac{v^2}{v_0^2} \left[\frac{(C_0 + C_s)a_0^2 a}{C_0 a^3 + C_s a_0^3} \right]^2 \tag{5.7.3}$$

where C_0 is the initial compliance of the specimen and v_0 is the total displacement required to give $\mathcal{G} = G_f$ for the initial notch depth:

$$C_0 = \frac{8a_0^3}{Ebh^3} \quad , \quad v_0 = u_0 \frac{C_0 + C_s}{C_0} \tag{5.7.4}$$

u_0 is given by (5.6.12). The $\mathcal{G}$–a curves at constant v have asymmetric bell shapes as depicted in Figs. 5.7.3b,c. The position of the maximum depends on the value of the parameter C_0/C_s. It is easy to verify that, for $C_0/C_s = 0.5$, the maximum is located precisely at $a = a_0$, so that at the intersection with the R-curve, the slope of the $\mathcal{G}$–a curve is always negative whatever the R-curve, and the fracture is stable. For $C_0/C_s > 0.5$ the maximum is located at the left of $a = a_0$ and the fracture is also stable. For $C_0/C_s < 0.5$ the stability depends on the R-curve, but for low enough values of C_0/C_s, instability is always reached.

As an illustration, consider the R-curve of example 5.6.3 and assume $C_0/C_s = 0.1$. Carrying out the computations, we obtain the curves in Fig. 5.7.3c, which shows that the branch ABM is stable, the arc MC is unstable, and the curve from C on is stable again. Point M is a maximum *relative to* v, and a snap-back is initiated at M.

The snap-back is best seen on a P–v plot. This plot is constructed writing $\mathcal{G} = \mathcal{R}$. The result for the parametric equation for P is identical to that in example 5.6.3, and the result for v follows by setting $\mathcal{G} = \mathcal{R}$ in (5.7.3):

$$v = \begin{cases} v_0 \dfrac{C_0 a^3 + C_s a_0^3}{(C_0 + C_s) a_0^2 a} \left[1 - \left(1 - \dfrac{a - a_0}{c_f} \right)^3 \right]^{1/2} & \text{for } a - a_0 \leq c_f \\ v_0 \dfrac{C_0 a^3 + C_s a_0^3}{(C_0 + C_s) a_0^2 a} & \text{for } a - a_0 > c_f \end{cases} \tag{5.7.5}$$

Fig. 5.7.3d shows the P–v plot for quasi-static crack growth. It is clear that, in order to keep static equilibrium, v must be reduced upon reaching point M and must move back and down to point C: this is a snap-back. From C on, the test becomes stable again. If an actual test is conducted at controlled v, the test will become unstable at point M and dynamic failure will happen there. This does not mean that the specimen will completely break at M. Rather, it means that the excess of energy of the system will be transformed into kinetic energy after that point. This energy gets partly emitted in sound waves and partly it goes into the kinematic energy of the dynamic near-tip field (Achenbach and Bažant 1975; Bažant, Glazik and Achenbach 1976; Achenbach, Bažant and Khetan 1976a,b) transformed by damping and friction into heat or is emanated far away, in further crack propagation. After the kinetic energy is transformed by damping and friction into heat or is emanated far away, the crack will be arrested and fracture will then progress again quasi-statically.

It is obvious that the intensity of the snap-back will depend on the relative stiffness of the spring: the lesser the parameter C_0/C_s, the worse the snap-back.

As a final remark, note that when $C_s \to 0$, $v \to u$, and we approach the displacement-controlled case; when $C_s \to \infty$, then $v/C_s \to P$ and the situation tends towards the load-controlled situation. □

The foregoing example illustrates one of the many possible couplings between the controlled variable v and the basic mechanical variables P and u. In general, one may have any type of coupling, represented by a relation of the type $f(v, P, u) = 0$, which may be used together with the relation $u = C(a)P$ to eliminate P and u from the expression of $\mathcal{G}$.

Modern experimental equipment has sophisticated electronics which is able to routinely control tests in load, displacement, or any other direct experimental variable (measured by means of a transducer). A typical example is the test controlled by crack-mouth opening displacement (CMOD), but there are more sophisticated examples, such as those using compound signals, as an average (or sum) of several transducer signals. This is done without much trouble because electronic (even electric) addition of transducer signals is easy. In principle, it is also possible to control the test using a computed variable: a variable that is calculated from the values read from a number of transducers and then used as a feed-back signal. However, this is not a routine task. By far, the most common of the modern control variables is the CMOD, w_M, because it usually leads to completely stable tests. The analysis of one such a test is similar to the analysis of the displacement-controlled test, except that u is replaced by w_M, and the equation to use for eliminating P is $w_M = C_M(a)P$, where $C_M(a)$ is the cross-compliance associated to CMOD. This compliance must be computed using classical techniques in LEFM, such as those described in Section 3.5.

Exercises

5.25 Show, based on the foregoing concepts of R-curve analysis, that a positive geometry becomes unstable at controlled load as soon as the peak load is reached.

5.26 Consider a large panel with a center slit of length 100 mm on which a hydraulic-fracturing simulation is being performed by injecting a liquid inside the crack. Assuming that LEFM applies, with $G_f = 140\ N/\text{m}$ and $E' = 500$ MPa, and that the liquid is incompressible, analyze the test stability if it is carried out (a) at controlled pressure; (b) at controlled volume of injected liquid.

5.27 Carry out the calculations required to plot the P-v curve in Fig. 5.7.3d.

5.28 Consider the panel in the previous exercise. Analyze the stability of the tests assuming that they are performed controlling the mass of injected fluid. Consider that the volume is measured by the displacement of the injection piston, which measures the volume of the crack V, plus an extra volume V' due to the elastic compression of the fluid and the dilation of the piston itself and the piping. Calibration tests indicate that V' is proportional to the pressure so that $V' = C_s p$ with $C_s = 10^{-6}$ m^3/MPa.

5.29 Consider the panel in the previous exercise. What is the maximum value of C_s below which one attains a completely stable test?

6

Determination of Fracture Properties From Size Effect

As pointed out in the first chapter, the structural size effect is probably the most important manifestation of fracture phenomena, at least from the engineering point of view. Therefore, it is important to relate the size effect behavior to the fracture properties of the material, such as K_{Ic} and G_f. This is the objective of this chapter.

In Section 6.1, we begin by analyzing the size effect predicted by equivalent elastic crack models. It turns out that all such models, including Jenq-Shah two parameter model, lead to an asymptotic size effect (asymptotically approached for large sizes) which is identical to Bažant's size effect law (1983, 1984a), derived in a simple way in Section 1.4. This fact not only provides a further justification of Bažant's size effect law, but also the basic equations relating the parameters of this law to the fundamental fracture parameters. These equations are used in Section 6.2 to find the basic relations permitting identification of the fracture properties from the observed size effect behavior. This section also introduces an intrinsic representation of the size effect law which exploits the concept of intrinsic size introduced in the previous chapter to formulate the size effect in a way that is independent of the specimen geometry. Such a formulation allows comparing size effect results from various geometries on a common basis and describing combined size and shape effects.

In view of the fact that the size effect law (in the form proposed by Bažant and Kazemi 1990a) describes not only the effect of size, but also the effect of geometry, a more accurate term would be the "size-shape effect law". But the shorter term seems to have already gotten stuck.

With the theoretical background of the first two sections, Section 6.3 describes the experimental determination of fracture properties based on size effect, which includes a discussion of the basic regression relations, the experimental and calculation procedures recommended by the existing RILEM Recommendation, and some recent improvements in the regression relations incorporating weighted least-square fitting.

Finally, Section 6.4 shows how to incorporate the fracture parameters determined by the size effect method into an R-curve formulation, from which one can determine the structural response of any precracked specimen. The techniques required to compute the R-curve from the size effect law, and to use it to predict the load-displacement response of a cracked specimen or structure is described in detail.

6.1 Size Effect in Equivalent Elastic Crack Approximations

6.1.1 Size Effect in the Large Size Range

We have seen in the preceding chapter that, at peak load, all the equivalent crack models satisfy (5.3.6). Moreover, for large specimen sizes, the critical equivalent crack extension tends to a limiting constant value c_f; $\Delta a_{ec} \to c_f$ for $D \to \infty$; see (5.3.4).

To find the size effect implied by (5.3.6) in the large size range, we first set in it $\alpha_{ec} = \alpha_0 + c_f/D$ and we solve for σ_{Nu} as

$$\sigma_{Nu} = \frac{K_{Ic}}{\sqrt{D\, k^2(\alpha_0 + c_f/D)}} \tag{6.1.1}$$

We next approximate $k^2(\alpha_0 + c_f/D)$ by its two-term Taylor series expansion at α_0:

$$k^2(\alpha_0 + c_f/D) \approx k_0^2 + 2k_0k_0'\frac{c_f}{D} \tag{6.1.2}$$

where $k_0 = k(\alpha_0)$ and $k_0' = k'(\alpha_0)$ stand for the values of $k(\alpha)$ and its first derivative for the initial crack length. Inserting this approximation into (6.1.1), we get the following result (shown by other means by Bažant and Kazemi (1990a); see also Bažant 1989b):

$$\sigma_{Nu} = \frac{K_{Ic}}{\sqrt{k_0^2 D + 2k_0k_0'c_f}} = \frac{K_{Ic}}{\sqrt{2k_0k_0'c_f}\,\sqrt{1 + D/(2k_0'c_f/k_0)}} = \sqrt{\frac{E'G_f}{g_0'c_f + g_0 D}} \tag{6.1.3}$$

in which $g_0 = g(\alpha_0) = k_0^2$ and $g_0' = g_0'(\alpha_0) = 2k_0k_0'$. If we set:

$$Bf_t' = \frac{K_{Ic}}{\sqrt{2k_0k_0'c_f}} = \frac{K_{Ic}}{\sqrt{g_0'c_f}}, \quad \text{and} \quad D_0 = \frac{2k_0'}{k_0}c_f = \frac{g_0'}{g_0}c_f \tag{6.1.4}$$

we obviously get the classical form (1.4.10) of Bažant's size effect law:

$$\sigma_{Nu} = \frac{Bf_t'}{\sqrt{1 + D/D_0}} \tag{6.1.5}$$

The form (6.1.3) might be more appropriately called the *size-shape effect law,* because the function $k(\alpha)$ introduces the effect of geometry (shape). This equation, unlike Eq. (1.4.10) or (6.1.5), can be applied to dissimilar structures.

With the foregoing derivation, we have achieved another independent justification for Bažant's size effect law, and have shown that all the equivalent crack models asymptotically converge to it for large sizes. At the same time we obtained an interpretation of its constants Bf_t' and D_0, in terms of the more fundamental fracture parameters K_{Ic} and c_f. The relations (6.1.4) are at the base of the experimental determination of the fracture properties of concrete based on size effect, the main topic in the next sections. Here we use them to explore the structure of the size effect parameters D_0 and B.

The second expression in (6.1.4) reveals the basic characteristics of the transitional size D_0. First, it is proportional to the effective length of the fracture process zone c_f, which, in turn, is approximately proportional to the inhomogeneity size of the material (dictated by microstructural features), and also to the characteristic size $\ell_{ch} = K_{Ic}^2/f_t'^2$, as discussed in Section 5.3.1. Second, it is proportional to the ratio $2k'(\alpha_0)/k(\alpha_0)$, or , equivalently, $g'(\alpha_0)/g(\alpha_0)$, which is independent of material properties and introduces the effect of structure geometry (shape). It is interesting to note that the same ratio appears in the definition of the intrinsic size $\overline{D}$ previously introduced in Section 5.3.3 for the fracture process zone of concrete, and appears in some other works dealing with nonlinear fracture models, particularly those of Horii, Planas and Elices, and Bažant et al. (Horii 1989; Horii, Hasegawa and Nishino 1989; Horii, Zihai and Gong 1989; Planas and Elices 1989a, 1990b, 1991a, 1992a; Bažant, Kazemi and Gettu 1989; Bažant and Kazemi 1990a,b; Bažant 1989b; Llorca, Planas and Elices 1989; Elices and Planas 1991, 1992). Though certainly not an exact characterization of structure geometry, this ratio apparently captures its main effect on fracture.

The first of (6.1.4), in turn, reveals the basic structure of B: it can be rewritten in terms of ℓ_{ch} as

$$B = \frac{1}{\sqrt{2k_0k_0'}}\sqrt{\frac{\ell_{ch}}{c_f}} = \frac{1}{\sqrt{g_0'}}\sqrt{\frac{\ell_{ch}}{c_f}} \tag{6.1.6}$$

which shows that B also consists of a product of a geometrical function times a material parameter. This material parameter was called $\beta = c_f/\ell_{ch}$ in Section 5.3.1— Eqs. (5.3.3) and (5.3.5)— and is related to the softening behavior of the material; for concrete, its value can be estimated in the range 2 to 5.

6.1.2 Size Effect in the Jenq-Shah Model

To analyze the size effect implied by the Jenq-Shah model (Bažant 1993d), we consider geometrically similar structures (which also implies similar notches) and rewrite the governing equations (5.5.1) and

(5.5.10) in parametric form. To simplify the expressions, we drop the subscript from $\Delta\alpha_{ec}$ and set $\alpha_{ec} = \alpha_0 + \Delta\alpha$. Then we solve for σ_{Nu} from (5.5.1):

$$\sigma_{Nu} = \frac{K_{Ic}}{\sqrt{Dk^2(\alpha_0 + \Delta\alpha)}}, \tag{6.1.7}$$

Note that this is identical to (6.1.1), except that $\Delta\alpha$ is not given but must be obtained from (5.5.10), which we leave as it was (except for the subscript):

$$\Delta\alpha\, L^2(\Delta\alpha, \alpha_0) = \frac{c_f}{D} \tag{6.1.8}$$

where c_f is given in terms of the material properties E', K_{Ic}, and w_{Tc} by (5.5.11). The two foregoing equations are the parametric equations of the size effect curve for the Jenq-Shah model. Elimination of parameter $\Delta\alpha$ yields the size effect curve of σ_{Nu} vs. D.

This elimination is not feasible analytically, in general. However, the solution for relatively large sizes can be found by expanding the equations in series of powers of $\Delta\alpha$ and c_f/D. In this regard, we remark that $L(c_f/D, \alpha_0)$ is regular in c_f/D and accepts a power series expansion. So does $L^2(c_f/D, \alpha_0)$, and so, recalling that $L(0, \alpha_0) = 1$, one can write

$$L^2(\Delta\alpha, \alpha_0) = 1 + \sum_{n=1}^{\infty} a_n\, (\Delta\alpha)^n \tag{6.1.9}$$

where, for brevity, we do not show explicitly that the coefficients a_n depend on α_0.

Now, one may seek a power expansion of the solution of (6.1.8) in the form

$$\Delta\alpha = \frac{c_f}{D} + \sum_{n=1}^{\infty} b_n \left(\frac{c_f}{D}\right)^{n+1} \tag{6.1.10}$$

The b_ns are determined by substituting (6.1.9) and (6.1.10) into (6.1.8) and comparing the coefficients of like powers of c_f/D on both sides of the resulting equation. One easily finds the first few terms of the expansion: $b_1 = -2a_1$, $b_2 = 2a_1^2 - 3a_2$, ... Now we can go back to (6.1.7) and expand $k^2(\alpha)$ in Taylor series at $\alpha = \alpha_0$;

$$k^2(\alpha_0 + \Delta\alpha) = k_0^2\left[1 + c_1\Delta\alpha + c_2\,(\Delta\alpha)^2 + \ldots\right] \quad , \text{ where } \quad c_n = \frac{1}{k_0^2 n!}\left.\frac{d^n k^2(\alpha)}{d\alpha}\right|_{\alpha=\alpha_0} \tag{6.1.11}$$

Finally, we substitute this and $\Delta\alpha$ given by (6.1.10) into (6.1.7), and obtain the following expression for the size effect corresponding to the Jenq-Shah model:

$$\sigma_{Nu} = Bf_t'\left[1 + \frac{D}{D_0} + d_1\frac{c_f}{D} + d_2\left(\frac{c_f}{D}\right)^2 + \ldots\right]^{-1/2} \tag{6.1.12}$$

where B and D_0 are the same as defined in (6.1.4), and d_n are coefficients that may be obtained in terms of the coefficients b_n and c_n. The first two coefficients are, for example, $d_1 = (c_1 b_1 + c_2)/c_1$, and $d_2 = (c_1 b_3 + 2c_2 b_1 + c_3)/c_1$.

Equation (6.1.12) represents a general asymptotic description of the size effect of the Jenq-Shah model. This infinite asymptotic series is similar to that derived in Bažant (1985b, 1986a) by another argument (see Chapter 9). For not-too-small size D, the terms with c_f/D, $(c_f/D)^2$, etc. are negligible compared to 1. Dropping them, we see that (6.1.12) reduces to the size effect law in (6.1.5).

So we may conclude that the equivalent LEFM model based on the critical crack-tip opening displacement, including Jenq and Shah's two-parameter model for concrete, gives a size effect that is asymptotically equivalent to the size effect law, Eq. (1.4.10) or (6.1.5). This conclusion implies that Jenq and Shah's model must give overall similar results as the R-curve model based on the size effect law. This has been confirmed experimentally (e.g., Karihaloo and Nallathambi 1991). Furthermore, it follows that the material parameters of Jenq and Shah's model can be determined from size effect measurements.

6.2 Size Effect Law in Relation to Fracture Characteristics

6.2.1 Defining Objective Fracture Properties

Experience shows that different experimental techniques or different analysis (i.e., using different models for the interpretation of tests results) may lead to different values of theoretically identical fracture parameters: fracture toughness, fracture energy, size of the fracture process zone, etc. There is, however, a way to uniquely define such parameters: use the values of the parameters corresponding to the extrapolation of specimen size to infinity. The reason: as discussed in the preceding chapter, in an infinitely large specimen, the fracture process zone occupies an infinitely small fraction of the specimen volume. Therefore, in the limit, all of the specimen volume is in an elastic state; and since from linear elastic fracture mechanics it is known that the near-tip asymptotic field of displacements and stresses is the same regardless of the shape and size of the specimen or structure, it turns out that the fracture process zone in an infinitely large specimen is exposed along its boundary to the same stress or displacement field, regardless of the specimen shape, and so it must behave in the same manner. In particular (in the sense of statistical continuum smoothing of the random micro scatter), it must have the same distribution of strains and microcracks, the same length and width, and the same energy dissipation. Consequently, an unambiguous definition (proposed by Bažant 1987a, 1989a) is as follows:

The fracture energy G_f and the effective fracture process zone length c_f are, respectively, the energy release rate required for crack growth and the distance from the notch tip to the tip of the equivalent LEFM crack in an infinitely large specimen of any shape (provided it has positive geometry).

Without the foregoing asymptotic definition, the problem of defining and determining the material fracture characteristics becomes ambiguous and more difficult. The fracture process zone length and width depends on the specimen shape because it is influenced by the proximity of the specimen boundary. It appears impossible to eliminate from measurements these parasitic influences with high accuracy, without making an extrapolation to infinite size.

Because the failure of specimen (peak load) is dictated by the material characteristics, it must be possible to determine these characteristics from size effect measurements. In fact, by virtue of the foregoing asymptotic definition, the determination of fracture characteristics is reduced to the calibration of the size effect law. If we knew this law exactly, we would get exact results, but the exact size effect law, applicable up to arbitrarily large sizes, is not known. Therefore, the size effect method, like others, yields only approximate results in practice. Nevertheless, the validity of Bažant's size effect law (6.1.5) is rather broad, covering a size range of up to about 1:20, which suffices for most practical purposes. We shall see in the next chapter that other extrapolations lead to different values for parameters named identically; however, if the values are used consistently (i.e., within the models on which their determination has been based), then realistic predictions are obtained for typical aspects of the structural response (peak load, load-displacement curves, or the like). Next we determine the relation of the parameters in Bažant's size effect law to the more fundamental fracture parameters.

6.2.2 Determination of Fracture Parameters from Size Effect

In the previous section we showed that the size effect predicted by the effective crack models converges to Bažant's size effect law for large sizes, and we determined the relation of the size effect parameters D_0 and Bf'_t to the asymptotic fracture parameters, Eqs. (6.1.4).

Thus, if we assume that D_0 and Bf'_t are known from experiment (see the next section for the detailed experimental procedures), we can solve for K_{Ic} and c_f from (6.1.4) and get

$$K_{Ic} = Bf'_t\sqrt{D_0}\,k_0 \tag{6.2.1}$$

$$c_f = \frac{k_0}{2k'_0}D_0 \tag{6.2.2}$$

For infinite size, LEFM must hold, and according to Irwin's relationship, $G_f = K_{Ic}^2/E'$, so from

(6.2.1), we get

$$G_f = \frac{(Bf_t')^2 D_0 k_0^2}{E'} \tag{6.2.3}$$

6.2.3 Determination of Fracture Parameters from Size and Shape Effects and Zero Brittleness Method

The shape (or geometry) effect, e.g., the effect of relative notch length, is embedded in Eqs. (6.2.1) and (6.2.2) through $k(\alpha_0)$ and $k'(\alpha_0)$. Thus, one might think that G_f and c_f could be determined by testing specimens of the same size but with variable notch length. However, this is not feasible in practice because the range of brittleness number of the fracture specimens that can be achieved in this manner does not exceed about 1:3, which is insufficient (Bažant and Li 1996). A sufficient range (about 1:6) can be achieved by varying the relative notch length and using also two specimen sizes in the ratio 1:2.

There may, nevertheless, be a way to use fracture specimens of one size: vary $\alpha_0 = a_0/D$ all the way to 0, which means that tests of unnotched (in addition to notched) fracture specimens must be included. The evaluation of such test results, called the zero brittleness method, requires fitting of the data by the universal size effect law (9.1.48) discussed latter (in Section 9.1).The fitting is nonlinear but can be accomplished by a sequence of linear regressions. See Bažant and Li (1996) and also Section 9.3.4.

6.2.4 Intrinsic Representation of the Size Effect Law

In the foregoing, the size of the specimen, D, is just a 'user-selected' (i.e., arbitrary) linear dimension. It turns out, then, that different users can obtain different size effect parameters. Thus, these parameters are not intrinsic and cannot be directly compared if different definitions of D are used or, more important, different geometries are involved.

The concept of intrinsic size, introduced in Section 5.3.3, allows rewriting Bažant's size effect law (6.1.5) in an intrinsic form in which the geometrical and material properties are decoupled (Bažant, Kazemi and Gettu 1989; Bažant and Kazemi 1990a). So, in (6.1.3), we set $D/(2k_0'/k_0) = \overline{D}$ and, in accordance with the definition (5.3.11) of the intrinsic size $\overline{D}$, we get

$$\sigma_{Nu} = \frac{K_{Ic}}{\sqrt{2k_0 k_0'(c_f + \overline{D})}} \tag{6.2.4}$$

Now, if we introduce the intrinsic nominal strength $\overline{\sigma}_{Nu}$ defined as

$$\overline{\sigma}_{Nu} = \sigma_{Nu}\sqrt{2k_0 k_0'} \tag{6.2.5}$$

we get a modified size effect law in which the parameters depend only on the material properties:

$$\overline{\sigma}_{Nu} = \frac{\overline{B}f_t'}{\sqrt{1 + \overline{D}/\overline{D}_0}} \tag{6.2.6}$$

in which

$$\overline{B}f_t' = \frac{K_{Ic}}{\sqrt{c_f}}, \quad \overline{D}_0 = c_f \tag{6.2.7}$$

Thus, if the size effect is expressed in terms of the intrinsic strength $\overline{\sigma}_{Nu}$ and of the intrinsic size $\overline{D}$, the structure of the equation is retained, but the dependence on the geometry (shape) of the specimen is eliminated. This allows using data from different specimens shapes and sizes in a single plot, a fact that can be useful to compare apparently dissimilar results.

Exercises

6.1 Determine K_{Ic}, G_f, and c_f for test series B1 and C1 in Section 1.5. Sort the materials by (1) strength and (2) toughness. Note: Estimate the elastic modulus from ACI formula $E = 4735\sqrt{f_c'}$ (both E and f_c' in MPa). Take $k_0 = 0.687$ and $k_0' = 2.10$ (the method to obtain them will be explained in Section 6.3.4).

6.2 According to (6.2.6), for a given intrinsic size $\overline{D}$, the behavior is fully brittle (i.e., LEFM applies exactly) when $c_f = 0$, and the behavior is fully ductile when $c_f \to \infty$. We may define a material brittleness γ as the inverse of c_f: $\gamma = 1/c_f$ (with unit m^{-1} or mm^{-1}). Compute the material brittleness for series B1 and C1.

6.3 Size Effect Method: Detailed Experimental Procedures

6.3.1 Outline of the Method

In the size effect method, a number of geometrically similar notched specimens of various sizes are tested for the peak load. The nominal strength is then computed, its plot vs. the size is considered, and the values of the parameters Bf_t' and D_0 are obtained by optimal least-square fitting of the size effect law to the experimental results. Finally, K_{Ic}, G_f, and c_f are obtained from Eqs. (6.2.1)–(6.2.3).

It should be noted that the size effect method is applicable only for specimens of positive geometry, i.e., those for which $k'(\alpha_0)$ is positive (and not very small). When $k'(\alpha_0)$ is approximately zero or negative, the method is inapplicable. This happens, for example, during the initial crack growth in a center-cracked specimen loaded on the crack. One reason for the failure of the size effect method is that in such specimens the crack length at maximum load can be much longer than the notch, due to stable crack growth, and thus the crack lengths at maximum loads of specimens of different sizes are not similar. Another reason is that Eq. (6.2.2) gives, in this case, either infinite or negative c_f, which is impossible.

Eqs. (6.2.1)–(6.2.3) are used for regressions based on the size effect law in its ordinary form (6.1.5). One can, however, also use the size effect law in the intrinsic form (6.2.6), which directly involves material parameters G_f and c_f. In that case, these parameters can be obtained directly by optimal fitting of Eq. (6.2.6) to the measured values of $\overline{\sigma}_{Nu}$ for various values of $\overline{D}$. When such a method is used, the specimen shapes do not necessarily have to be geometrically similar, and test results for different specimen geometries can be mixed in one and the same regression. However, the parameter that takes into account the specimen shape, namely, the ratio $k'(\alpha_0)/k(\alpha_0)$, is only approximately known and thus it introduces some additional error. To avoid this error, it is preferable to use specimens that are geometrically similar.

In either case, the fitting can be accomplished easily by nonlinear regression using some standard nonlinear optimization subroutine such as the Levenberg–Marquardt algorithm. One advantage of this procedure is that this subroutine also directly gives the coefficients of variation of G_f and c_f. However, other simpler methods involving linear regression can be used, too. In the next section, we examine the simplest regression procedures.

6.3.2 Regression Relations

In a subsequent section, we will justify that the best approach to identifying D_0 and Bf_t' from experiments is nonlinear optimization, provided that a computer subroutine such as Levenberg-Marquardt algorithm is available. From measurements, one gets a series of nominal strength values σ_{Nu_k} corresponding to sizes $D_k (k = 1, 2, \ldots)$. Values of D_0 and Bf_t' are sought such that the sum of squared deviations in a log–log plot be minimum (Fig. 6.3.1a). This is equivalent to a fit in classical coordinates (x, y) in which the curve to fit is written as

$$y = \ln \frac{M}{\sqrt{N + e^x}} \tag{6.3.1}$$

where

$$x = \ln D, \;\; y = \ln \sigma_{Nu}, \;\; Bf_t'\sqrt{D_0} = M, \;\; D_0 = N \tag{6.3.2}$$

One readily gets K_{Ic}, G_f and c_f in terms of the best fit parameters M and N from Eqs. (6.2.1)–(6.2.3):

$$K_{Ic} = k_0 M \;, \quad G_f = \frac{k_0^2}{E'} M^2 \;, \quad c_f = \frac{k_0}{2k_0'} N \tag{6.3.3}$$

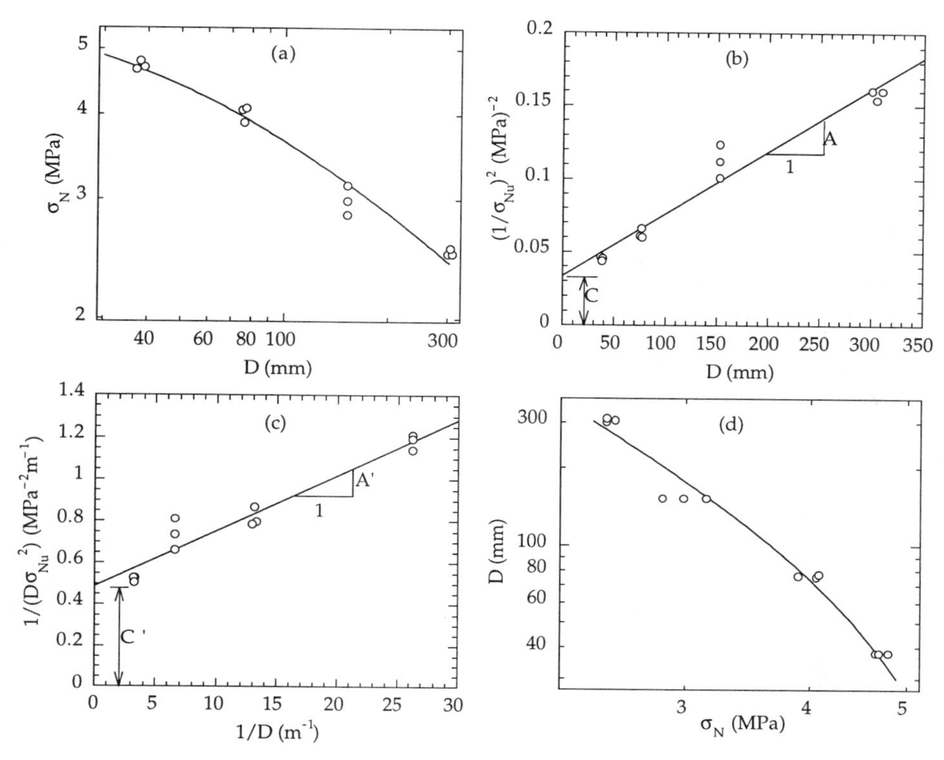

Figure 6.3.1 Regression plots: (a) bilogarithmic; (b) linear regression I; (c) linear regression II; (d) inverse bilogarithmic plot. The experimental data correspond to tests by Bažant and Pfeiffer (1987) reported as Series B1 in Chapter 1 (see Tables 1.5.1 and 1.5.2).

Note that M and N have been chosen so that G_f depends only on M and c_f only on N. In this way, the errors for the fit parameters, computed automatically by the optimization routine, can be directly used to estimate the errors for G_f and c_f. If other parameters are used, their correlation coefficient must also be calculated, which is not a standard feature in many commercial optimization routines.

If a nonlinear optimization program is unavailable, one can exploit the fact that the size effect in Eq. (1.4.10) can be algebraically rearranged to a linear regression plot (Bažant 1984a; Fig. 6.3.1b):

$$Y = AX + C \qquad \text{(Linear Regression I)} \tag{6.3.4}$$

in which

$$X = D,\ Y = \left(\frac{1}{\sigma_{Nu}^2}\right),\quad Bf_t' = \frac{1}{\sqrt{C}},\ D_0 = \frac{C}{A} \tag{6.3.5}$$

As before, K_{Ic}, G_f, and c_f follow from Eqs. (6.2.1)–(6.2.3):

$$K_{Ic} = k_0 \frac{1}{\sqrt{A}}, \qquad G_f = \frac{k_0^2}{E'}\frac{1}{A}, \qquad c_f = \frac{k_0}{2k_0'}\frac{C}{A} \tag{6.3.6}$$

Another algebraic rearrangement of Eq. (1.4.10) yields an alternative linear regression plot (Planas and Elices 1989a; Fig. 6.3.1c):

$$Y' = A'X' + C' \qquad \text{(Linear Regression II)} \tag{6.3.7}$$

in which

$$X' = \frac{1}{D},\ Y' = \frac{1}{\sigma_{Nu}^2 D},\ Bf_t' = \frac{1}{\sqrt{A'}},\ D_0 = \frac{A'}{C'} \tag{6.3.8}$$

Now the expressions for K_{Ic}, G_f, and c_f are:

$$K_{Ic} = k_0 \frac{1}{\sqrt{C'}}, \qquad G_f = \frac{k_0^2}{E'} \frac{1}{C'}, \qquad c_f = \frac{k_0}{2k_0'} \frac{A'}{C'} \tag{6.3.9}$$

Note that, in the first type of linear regression, the fracture energy is inversely proportional to the regression slope, and in the second type of regression, it is inversely proportional to the intercept. The linear regression plot of type I gives a better visual display of the test data for smaller specimen sizes while the regression plot of the type II gives a better display of the extrapolation to infinite size. LEFM corresponds to $c_f \to 0$, that is, to $C = 0$ in the first type of regression (line through the origin), and to $A' = 0$ in the second type of regression (a horizontal line). The strength theory corresponds to $D_0 \to \infty$, that is to $A/C = 0$ in the first type of regression (a regression line with a negligible slope), and to $C'/A' = 0$ in the second type of regression (a regression line through the origin).

The regressions in Eqs. (6.3.1), (6.3.4), and (6.3.7) are not completely equivalent and do not yield exactly the same results. The reason is that these regressions imply different weighing of the data points. An improved regression method taking weights into account will be introduced in Section 6.3.6. Before we examine the RILEM recommendation, which uses the linear regression of type I with equal weights, we apply the foregoing regressions to an example.

Example 6.3.1 The following table summarizes the results of the tests on notched three-point bend concrete beams by Bažant and Pfeiffer (1987); they correspond to Series B1 in Tables 1.5.1 and 1.5.2. From the raw data in the first two rows, the coordinates (x, y), (X, Y), and (X', Y') have been computed according to Eqs. (6.3.2), (6.3.5), and (6.3.8) and are included in the following rows of the table.

		Specimen No.											
Var.	Units	#1	#2	#3	#4	#5	#6	#7	#8	#9	#10	#11	#12
D	mm	38	38	38	76	76	76	152	152	152	305	305	305
σ_{Nu}	MPa	4.65	4.69	4.79	3.89	4.06	4.08	2.84	2.99	3.15	2.50	2.50	2.55
x	ln(mm)	3.64	3.64	3.64	4.33	4.33	4.33	5.03	5.03	5.03	5.72	5.72	5.72
y	ln(MPa)	1.54	1.55	1.57	1.36	1.40	1.41	1.05	1.09	1.15	.915	.915	.935
$10^3 X$	m	38	38	38	76	76	76	152	152	152	305	305	305
$10^3 Y$	MPa^{-2}	46.2	45.5	43.6	66.1	60.8	59.9	124	112	101	160	160	154
X'	m^{-1}	26.2	26.2	26.2	13.1	13.3	13.0	6.56	6.56	6.56	3.33	3.23	3.28
Y'	(MPa$\sqrt{\text{m}}$)$^{-2}$	1.21	1.19	1.14	.867	.798	.786	.811	.735	.662	.526	.526	.506

The foregoing values were fed to a commercial data analysis program, which drew the graphics in Fig. 6.3.1 and computed the best fit values for (M, N), (A, C), and (A', C') according to the nonlinear regression (6.3.1), the linear regression I (6.3.4), and the linear regression II (6.3.7), respectively. With units MPa and m, the results were: $M = 1.46, N = 0.0596, A = 0.427, C = 0.0334, A' = 0.0267$, and $C' = 0.486$. From these, values of K_{Ic}, G_f, and c_f were computed using (6.3.3), (6.3.6), and (6.3.9); the following table summarizes the results; the values in parentheses are the coefficient of variation in percent, as furnished by the program.

		Regression		
Parameter	Units	Nonlinear	Linear I	Linear II
$(1/k_0)\, K_{Ic}$	MPa m$^{1/2}$	1.46 (± 3.4%)	1.53 (± 3.4%)	1.43 (± 3.4%)
$(E'/k_0^2)\, G_f$	(MPa)^2 m	2.13(± 6.8%)	2.34(± 6.8%)	2.06 (± 6.8%)
$(2k_0'/k_0)\, c_f$	mm	60 (± 16%)	78 (± 22%)	55 (± 14%)

We see that the values for K_{Ic} have little statistical error (less than 4%), but the mean values provided by the three methods differ slightly, the maximum difference relative to the nonlinear fit being about 5%. This difference is twice as large for G_f (because of the squaring) and becomes as large as 30% for c_f. Note, however, that the statistical error is also very large for c_f and that the relative difference between any two results is always lower than twice the coefficient of variation. ▯

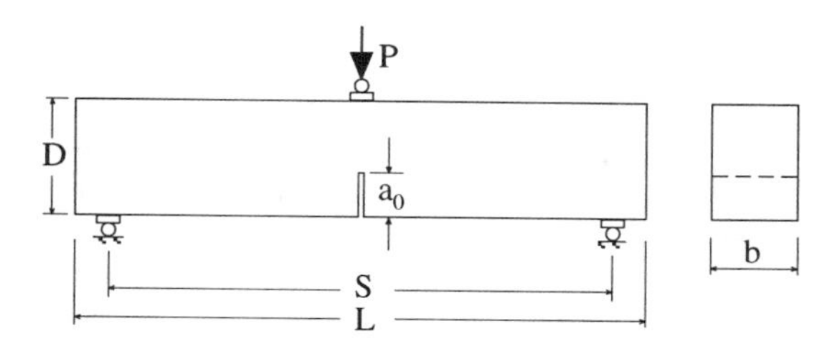

Figure 6.3.2 Suggested specimen geometry of the RILEM draft recommendation.

6.3.3 RILEM Recommendation Using the Size Effect Method: Experimental Procedure

One of the RILEM Recommendations on concrete fracture recommends determining the material fracture characteristics from the size effect law (RILEM 1990b). The idea of this method, proposed by Bažant (1987a) and Bažant and Pfeiffer (1987), is that one first determines the parameters of the size effect law by linear regression I and then the material fracture parameters ensue from Eqs. (6.3.6).

The method works equally well for a number of geometries (provided they are positive geometries), but three-point-bend beams are recommended for the purpose of standardization (Fig. 6.3.2). The loads and reactions are applied through one hinge and two rollers with a minimum possible rolling friction, and through stiff bearing plates (of such a thickness that they could be considered as rigid). The bearing plates are either glued with epoxy or are set in wet cement. The distance from the end of the beam to the end support must be sufficient to prevent spalling and cracking at beam ends. The span-to-depth ratio of the specimen, S/D, should be at least 2.5 (this has been set only for the purpose of standardization; the theory does not prevent smaller values). The ratio of the notch depth to the beam depth, a_0/D, should be between 0.15 and 0.5. The notch width at the tip should be as small as possible and must not exceed 0.5 d_a where d_a = maximum aggregate size. The width b of the beam and the depth d should not be less than $3d_a$.

An important point is the choice of specimen sizes $D_k(i = 1, 2, \ldots n)$ for which the tests should be carried out. If only two sizes were used, the regression line could be passed through the average of the σ_{Nu}-values for each size exactly. Thus, even though the coefficients of variation could be calculated, one would have no idea how well the observed dependence on D agrees with the size effect law. Therefore, at least three different sizes should be used. Their values are best chosen as a geometric progression, i.e., $D_1/D_2 = D_2/D_3 =$ constant.[1]

In summary, the Recommendation indicates that specimens of at least three different sizes, characterized by beam depths $D = D_1, \ldots D_n$ should be tested. The smallest depth D_1 must not exceed $5d_a$ and the largest depth D_n must not be smaller than $10d_a$. Because of the magnitude of random scatter exhibited by concrete, the ratio D_n/D_1 must be at least 4. The ratios of the adjacent sizes, D_{i+1}/D_k, should be approximately constant. Optimally, the size range should be as broad as feasible. Thus, for instance, the choice $D/d_a = 4, 8, 16$ is usually acceptable, but the choices $D/d_a = 3, 6, 12, 24$ or $3, 9, 27$ are preferable.

For statistical reasons, at least three identical specimens should be tested for each specimen size. All the specimens of all the sizes must be cast from the same batch of concrete, and the quality of concrete must be as uniform as possible. The curing procedure and the environments to which the specimens are exposed, including their histories, must be the same for all the specimens. To avoid differences in the hydration heat effects and to minimize other types of size effect (see Section 1.3), all the specimens should be geometrically similar in two dimensions, that is, the third dimension (thickness b) should be the same for all the specimens.

---

[1] The reason becomes clear by inverting the coordinates (Fig. 6.3.1d). One may regard σ_{Nu_k} as the given coordinates, and the corresponding D_k as the size values for which each σ_{Nu_k} would be obtained. These D_k values will, in general, differ from the $D-$value corresponding to the size effect law. Therefore, $\Delta D_k = D_k - D$ are the errors (Fig. 6.3.1d). Now, for the same reasons as before, it seems reasonable to consider that the inverse of the size effect law has roughly constant relative errors $\Delta D_k/D_k$ rather than constant absolute errors, ΔD_k. Consequently, since $\Delta D_k/D_k = \Delta(\ln D_k)$, a logarithmic size scale should be used. This means that the sizes D_k of the specimens to be tested should be chosen uniformly spaced in the logarithmic scale. Note that if, on the other hand, the chosen sizes are crowded in one portion of the ln D scale, one, in fact, imposes a bias for that portion of the size range.

It is sufficient to use an ordinary uniaxial testing machine without high stiffness. However, closed-loop control and high stiffness of the loading frame lead to more consistent results, and they also permit determining the postpeak response, which is useful for calibration of more sophisticated fracture models. The same machine ought to be used for testing all of the specimens. The specimens should be loaded at constant (or almost constant) displacement rates (this could be the load-point displacement, but better the crack mouth opening displacement). Although the size effect law is applicable over a very broad range of loading rates (its applicability has been demonstrated for tests at which the maximum loads are reached in times ranging from 1 s to 10^6 s; see Chapter 11), for the purpose of standardization it is desirable that the maximum load be reached in about 5 minutes.

Aside from the aforementioned maximum load values $P_1, \ldots P_n$ for specimens of sizes $D_1, \ldots D_n$, the following data should also be obtained and reported: Young's modulus E_c, standard compression strength f'_c (and preferably also the modulus of rupture, f_r); all the dimensions of the beam and bearing plates; the maximum aggregate size; the ratios (by weight) of water:cement:sand and gravel in the mix; the type of cement, its fineness, and admixtures; the mineralogical type of aggregate; the curing and storing conditions; temperature and relative humidity during the test; and the mean mass density of the concrete.

6.3.4 RILEM Recommendation Using the Size Effect Method: Calculation Procedure

The raw data for the calculations are the specimen dimensions, particularly the beam depths D_k and the measured maximum loads $P^0_{u_k}$ ($k = 1, 2, \ldots, n; n$ = number of specimens). The first step is to compute σ_{Nu} for all the specimens. For heavy specimens, the own weight of the specimens may have to be taken into account. To this end, the measured maximum loads $P^0_{u_1}, \ldots, P^0_{u_n}$ should best be corrected in such a manner that the corrected loads $P_{u_1}, \ldots, P_{u_n}$ without own weight would produce the same stress intensity factor according to LEFM. This is approximately equivalent to requiring that the bending moments at the notch section produced by loads $P^0_{u_k}$ plus the own weight be equal to the bending moment due to P_{u_k} alone. If a normal test configuration in which the notch is located at the bottom of the beam is used, if the specimen length L_k is almost the same as the span S_k, and no ancillary equipment (loading plates or rods, e.g.) is resting on the specimen, this is approximately achieved by taking $P_k = P^0_k + (m_k g/2)$ $(i = 1, \ldots, n)$, in which g is the acceleration of gravity and $m_k g$ are the weights of all the individual specimens. If L_k differs from S_k substantially, then

$$P_{u_k} = P^0_{u_k} + m_k g(2S_k - L_k)/2S_k \tag{6.3.10}$$

Based on these corrected ultimate loads, the nominal strength of each specimen is computed from the first of (1.4.1). In the original recommendation, $c_N = 1$ is used, and the shape factors for the stress intensity factor are modified accordingly. Here we use a different value of c_N so that the expressions in chapters 2–5 could be directly used. Specifically, the nominal strength is computed for each specimen as

$$\sigma_{Nu_k} = \frac{3P_{u_k} S_k}{2b_k D_k^2} \qquad k = 1, 2, \ldots, n \tag{6.3.11}$$

Finally, one calculates the coordinates of the data points in the linear regression of type I, Eq. (6.3.4) $X_k = D_k$ and $Y_k = (1/\sigma_{Nu_k})^2$ where D_k are the sizes corresponding to P_{u_k}. An example of such correlation is shown in Fig. 6.3.1b.

The slope A and intercept C of the regression line may now be calculated from the well-known linear regression equations (e.g., Pugh and Winslow 1966, Section 11.6; Press et al. 1992, Section 15.2). For future convenience, we present here the equations in a form slightly different from that used in the RILEM (1990b) Recommendation. In this presentation, it is useful to define the following sums (which are automatically performed by most hand calculators):

$$\Sigma = \sum_{k=1}^{n} 1^k = n\,, \quad \Sigma_x = \sum_{k=1}^{n} X_k\,, \quad \Sigma_y = \sum_{k=1}^{n} Y_k$$

$$\Sigma_{xx} = \sum_{k=1}^{n} (X_k)^2\,, \quad \Sigma_{xy} = \sum_{k=1}^{n} Y_k X_k\,, \quad \Sigma_{yy} = \sum_{k=1}^{n} (Y_k)^2 \tag{6.3.12}$$

Then the regression coefficients A and C are obtained as

$$A = \frac{\Sigma\Sigma_{xy} - \Sigma_x\Sigma_y}{\Delta}, \qquad C = \frac{\Sigma_{xx}\Sigma_y - \Sigma_x\Sigma_{xy}}{\Delta} \tag{6.3.13}$$

where

$$\Delta = \Sigma\Sigma_{xx} - (\Sigma_x)^2 \tag{6.3.14}$$

Note that in the foregoing Y_k are the measured individual data (not the averages of the data for each size!). It should be checked whether the plot of all the data points is approximately linear (if not, the test procedure was probably tarnished by some errors or inadequate control of test conditions).

After computing A and C, one calculates the geometrical factors $k_0 = k(\alpha_0)$ and $k'_0 = k'(\alpha_0)$ using the shape factor function $k(\alpha)$ given in the recommendation. Three shape functions are given, for span-to-depth ratios $S/D = 2.5$, 4 and 8. Although interpolation between them is deemed acceptable, the recommendation suggested sticking to these values to avoid introducing additional errors. Today, however, these expressions can be advantageously replaced by the general formulas of Pastor et al. (1995) given in examples 3.1.1 and 3.1.4; see Eqs. (3.1.1)–(3.1.3) and (3.1.8). We summarize them here for the reader's convenience: The shape factor for a span to depth ratio S/D takes the form

$$k_{S/D}(\alpha) = \sqrt{\alpha}\,\frac{p_{S/D}(\alpha)}{(1+2\alpha)(1-\alpha)^{3/2}}\,, \qquad p_{S/D}(\alpha) = p_\infty(\alpha) + \frac{4D}{S}[p_4(\alpha) - p_\infty(\alpha)] \tag{6.3.15}$$

where the polynomials $p_4(\alpha)$ and $p_\infty(\alpha)$ are given by Eqs. (3.1.2) and (3.1.3). With the values of k_0 and k'_0 computed from these formulae, the fracture parameters are calculated from (6.3.6). Note that numerical differentiation is the faster method to obtain k'_0; analytic differentiation is feasible, but the resulting formula is too long to be practical.

It is also desirable to calculate statistics of the results. If the standard deviation of the data population is not known, an estimate of it is obtained from the quadratic deviation from the regression line, χ^2:

$$\chi^2 = \sum_{k=1}^{n}(Y_k - \hat{Y}_k)^2 = \sum_{k=1}^{n}(Y_k - AX_k - C)^2 = \Sigma_{yy} - A\Sigma_{xy} - C\Sigma_y \tag{6.3.16}$$

Then, the coefficients of variation of A and C, i.e., ω_A and ω_C, and the relative width of the scatter band m are determined as

$$\omega_A^2 = \frac{1}{A^2}\,\frac{\chi^2\Sigma}{(n-2)\Delta}\,, \qquad \omega_C^2 = \frac{1}{C^2}\,\frac{\chi^2\Sigma_{xx}}{(n-2)\Delta}\,, \qquad m^2 = \frac{(n-1)\chi^2\Sigma(\Sigma_x)^2}{(n-2)\Delta(\Sigma_y)^2} \tag{6.3.17}$$

According to the recommendation, the value of ω_A should not exceed about 10% and the values of ω_C and m about 20%. These conditions prevent situations in which the size range is insufficient compared with the scatter of test results. Such a situation is illustrated in Fig. 6.3.3a in which a unique slope can still be obtained, but is highly uncertain. Fig. 6.3.3b illustrates the case where large scatter of test results necessitates the use of a very broad size range, while Fig. 6.3.3c illustrates the case where a small scatter of test results permits the use of a narrow size range. Obviously, the necessary breadth of the size range is proportional to the scatter band width and can be reduced by carefully controlled testing which yields low scatter.

From the coefficients of variation of the parameters of the regression line, the coefficients of variation of the fracture parameters can be estimated. The basic equation is that, if $\zeta = f(\xi_j)$ is a function of N *uncorrelated* random variables ξ_j, $(j = 1, 2, \cdots, N)$, then the coefficient of variation of ζ is

$$\omega_\zeta^2 = \frac{1}{\zeta^2}\sum_{j=1}^{N}\left[\frac{\partial f(\xi_i)}{\partial \xi_j}\right]^2 \xi_j^2\omega_{\xi_j}^2 \tag{6.3.18}$$

Thus, for K_{Ic} and G_f which, according to (6.3.6), depend respectively on A, and on A and E (which can be viewed as uncorrelated because they are obtained in different tests), the coefficients of variation are simply given by

$$\omega_{K_{Ic}} = \frac{1}{2}\omega_A\,, \qquad \omega_G = \sqrt{\omega_A^2 + \omega_E^2} \tag{6.3.19}$$

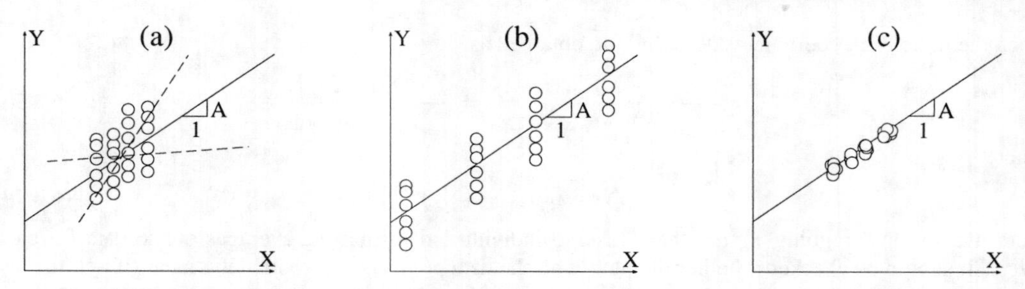

Figure 6.3.3 Unacceptable (a) and acceptable (b–c) size ranges relative to the scatter of the results (adapted from Bažant and Pfeiffer 1987).

For c_f, the matter is a little more complicated, because C and A are not uncorrelated. It is well known that if we write the regression line in the form $Y = A(X - \bar{X}) + \bar{C}$ where $\bar{X} = \Sigma_x/\Sigma$ is the abscissa of the centroid of the data points, then A and $\bar{C}$ are uncorrelated. Obviously, $\bar{C}$ is the Y intercept at the centroid, and it is easy to show that $\bar{C}$ does indeed coincide with the Y coordinate of the centroid of the data points:

$$\bar{C} = \frac{\Sigma_y}{\Sigma} \tag{6.3.20}$$

Furthermore, the coefficient of variation of $\bar{C}$ is given by

$$\omega_{\bar{C}} = \frac{\chi^2 \Sigma}{(n-2)(\Sigma_y)^2} \tag{6.3.21}$$

Now, since $C = A\bar{X} + \bar{C}$, it turns out that $C/A = \bar{X} + \bar{C}/A$, and so, from (6.3.6) and (6.3.18), we get

$$\omega_{c_f}^2 = \frac{1}{(C/A)^2}\left[\left(\frac{\bar{C}}{A}\right)^2 \omega_A^2 + \left(\frac{\bar{C}}{A}\right)^2 \omega_{\bar{C}}^2\right] = \frac{\bar{C}^2}{C^2}(\omega_A^2 + \omega_{\bar{C}}^2) \tag{6.3.22}$$

Example 6.3.2 Consider again Bažant and Pfeiffer's results analyzed in example 6.3.1, and apply to them the foregoing analysis. We first construct the sums in (6.3.12): $\Sigma = 12, \Sigma_x = 1.7145, \Sigma_{xx} = 0.37015, \Sigma_y = 1.1335, \Sigma_{yy} = 0.13099, \Sigma_{xy} = 0.21544$, and then compute $\Delta = 1.5023$ from (6.3.14). Next the values of A and C follow from (6.3.13): $A = 0.427, C = 0.0334$; they coincide, as they should, with the values obtained with the commercial program in example 6.3.1. Now, according to the data in Table 1.5.1, the span-to-depth ratio was $S/D = 2.5$ for series B1. Thus, from (6.3.15) $p_{2.5}(\alpha) = 1.6p_4(\alpha) - 0.6p_\infty(\alpha)$, and using the expressions (3.1.2) and (3.1.3) for p_4 and p_∞, we get

$$p_{2.5}(\alpha) = 1.847 - \alpha[-0.1424 + 0.6960(1-\alpha) - 0.4308(1-\alpha)^2 + 1.221(1-\alpha)^3] \tag{6.3.23}$$

Since the relative notch depth was $\alpha_0 = 1/6$ (see Table 1.5.1), the value of k_0 is readily obtained from (6.3.15): $k_0 = 0.687$. To obtain k_0' we use the numerical approximation $k_0' \approx [k(1/6 + 0.005) - k(1/6 - 0.005)]/0.01$ and get $k_0' = 2.10$. Next, (6.3.6) yields $K_{Ic} = k_0/\sqrt{A} = 0.687/\sqrt{0.427} = 1.05$ MPam$^{1/2}$, $c_f = (k_0/2/k_0')C/A = (0.687/2/2.1) \times 0.0334/0.427 = 0.0128$ m, or $c_f = 12.8$ mm. To determine G_f, we need E, which was not directly determined in Bažant and Pfeiffer's work. We can approximate E by ACI formula $E = 4734\sqrt{f_c'}$, which, using the value $f_c' = 34.1$ MPa from Table 1.5.2, provides $E = 27.6$ GPa, from which $G_f \approx 39.9$ N/m. For the statistics, we first compute χ^2 from the last of (6.3.16), which gives $\chi^2 = 1.060 \times 10^{-3}$; then, from (6.3.17), we get $\omega_A = 6.8\%$, $\omega_C = 15.4\%$ and $m = 14.7\%$, and from (6.3.21) $\omega_{\bar{C}} = 3.2\%$, with $\bar{C} = 0.09446$ from (6.3.21). According to this and Eq. (6.3.19), the coefficient of variation of K_{Ic} is $\omega_{K_{Ic}} = 3.4\%$; finally, from (6.3.22) we get $\omega_{c_f} = 22\%$. The coefficient of variation of G_f is *at least* 6.8%; the coefficient of variation ω_E of E cannot be ascertained from the original Bažant and Pfeiffer's data; so neither with certainty can the coefficient of variation ω_{G_f} of G_f. □

Table 6.3.1 Best fits of G_f and c_f for the results of Bažant and Pfeiffer (1987).

Series	Material	Specimen[a] type	E^b (MPa)	$k_0{}^c$	K_{Ic} (MPa$\sqrt{\text{m}}$)	G_f (N/m)	ω_A (%)	$\omega_{K_{Ic}}$ (%)
B1	concrete	SEN-TPB	27.7	0.673	1.03	38.4	6.8	3.4
B2	concrete	DEN-EC	29.0	1.30	1.14	44.5	7.0	3.5
B3	concrete	DEN-T	25.5	0.832	0.96	35.9	14.8	7.4
C1	mortar	SEN-TPB	32.9	0.673	0.84	21.4	4.1	2.1
C2	mortar	DEN-EC	32.8	1.30	0.86	22.6	4.9	2.5
C3	mortar	DEN-T	32.4	0.832	0.86	22.9	10.0	5.0

[a]See specimen types in Fig. 1.5.1

[b]Calculated from ACI formula $E = 4735\sqrt{f'_c}$

[c]Finite element calculation by Bažant and Pfeiffer (1987)

6.3.5 Performance of the Size Effect Method

The fact that determination of G_f and c_f from size effect tests based on Eqs. (6.2.3) and (6.2.2) yields approximately the same results for specimens of very different geometries has been verified by Bažant and Pfeiffer (1987). In their tests, three-point-bend specimens, double edge-notched tension specimens and eccentric compression specimens of size ratios 1:2:4:8 were tested. The essential information about these tests is given in Chapter 1; they correspond to Series B1–B3 for concrete, and C1–C3 for mortar. The optimal fits of the test results are shown in Fig. 1.5.3. The values of the fracture toughness and fracture energy obtained by the regressions of the test results are given in Table 6.3.1. The critical effective crack extension was not computed in Bažant and Pfeiffer's original work.

It is remarkable that, despite the very different specimen geometries used, the coefficients of variation of K_{Ic} values obtained for specimens of various geometries were, on average, less than 5% (which is better than the scatter of strength test results). It may also be noted that studies of Karihaloo and Nallathambi (1991), and Swartz and Refai (1989), further indicated that this method yields systematic results free of size and shape effect, and that it also yields for the fracture energy similar results as Jenq and Shah's two-parameter method (see Section 5.5).

Karihaloo and Nallathambi (1991) listed most of the fracture test results known at that time, in which they compared their method (see Section 5.4.3) to Bažant's size effect method. For this purpose, they converted G_f to K_{Ic} according to the LEFM relationship $K_{Ic} = \sqrt{G_f E}$. For the beam tests they reported, the K_{Ic} values were 0.847—0.892 kNm$^{-3/2}$ for the size effect method and $K_{Ic} \approx 0.867$ kNm$^{-3/2}$ for their method. A similar agreement was found for the data of Bažant and Pfeiffer (1987). The size effect method was applied by Brühwiler (1988), Saouma, Broz et al. (1991), and He et al. (1992) to measure the fracture energy of dam concrete (using wedge-splitting specimens of size up to 1.8×1.8 m).

On the other hand, the fracture energy value obtained by these methods is generally quite different from that obtained from the RILEM work of fracture method (RILEM 1985; Hillerborg 1985a); see Chapter 7 for a discussion on the sources of the discrepancy.

6.3.6 Improved Regression Relations

As already pointed out, the regressions in Eqs. (6.3.1), (6.3.4), and (6.3.7) are not completely equivalent and do not yield exactly the same results because these regressions imply different weighting of the data points. If consistent weighting is used, then the difference between the three regressions is only marginal, as we show next.

The basic idea in the following is that, in the frame of the maximum likelihood theory, the classical least-square fitting in which the function $\chi^2 = \sum_k (Y_k - \hat{Y}_k)^2$ is minimized, assumes that the (standard) errors of the various measurements are identical ($\hat{Y}$ is the theoretical expression for Y as a function of X and the regression parameters). However, in practice this condition may very well not hold, because, for example, the errors are proportional to the measured quantity (constant coefficient of variation), or different load cells have been used to perform the measurements for different X, or else Y is not the measured value but a nonlinear function of it, etc. In such an event, *weighted* least-square fitting must be performed. In such weighted regression, the function to be minimized is

$$\chi^2 = \sum_{k=1}^{n} w_k (Y_k - \hat{Y}_k)^2 \tag{6.3.24}$$

where the weights w_k must be inversely proportional to the square of the error (or variance) of the corresponding measurement:

$$w_k \propto \frac{1}{s_k^2} \approx \frac{1}{\omega_k^2 Y_k^2} \tag{6.3.25}$$

in which s_k is the standard deviation for the kth measurement and ω_k its coefficient of variation (the last expression is only approximate because we use the approximate value Y_k instead of the unknown true value of the variable).

The determination of s_k or ω_k for a particular experimental setup is not a trivial matter, and is very rarely carried out in detail. The usual assumption is that the directly measured quantity (in our case, the nominal strength σ_{Nu}) has a constant relative error, i.e., a constant coefficient of variation $\omega_{\sigma_{Nu}}$. Taking this assumption for granted, the weights for the various regressions become fixed. Indeed, using the formula (6.3.18) for the propagation of errors, we get the following relationships for the coefficients of variation of the regression variables $y = \ln \sigma_{Nu}$, $Y = 1/\sigma_{Nu}^2$, and $Y' = 1/(D\sigma_{Nu}^2)$:

$$\omega_y = \frac{\omega_{\sigma_{Nu}}}{y}\ , \qquad \omega_Y = 2\omega_{\sigma_{Nu}}\ , \qquad \omega_{Y'} = 2\omega_{\sigma_{Nu}} \tag{6.3.26}$$

Thus, the weights to be used for the various regressions are

$$w_k \propto \frac{y_k^2}{\omega_{\sigma_{Nu}} y_k^2} \propto 1 \quad \text{for logarithmic nonlinear regression} \tag{6.3.27}$$

$$w_k \propto \frac{1}{2\omega_{\sigma_{Nu}} Y_k^2} \propto \frac{1}{Y_k^2} \quad \text{for linear regression I} \tag{6.3.28}$$

$$w_k \propto \frac{1}{2\omega_{\sigma_{Nu}} Y'^2_k} \propto \frac{1}{Y'^2_k} \quad \text{for linear regression II} \tag{6.3.29}$$

The foregoing analysis shows that the equal-weight assumption is valid only for the logarithmic nonlinear regression; for the other two regressions, the weights must be inversely proportional to the square of the dependent variable Y or Y'. More generally, it shows how the weights in the three regressions are inter-related, and that given any one set of weights, the other two sets are fixed.

For the reasons just explained, it is better to use either a logarithmic nonlinear regression or a weighted linear regression to obtain G_f and c_f by the size effect method. If a linear regression of type I is used, then we introduce for these points the weights $w_k = 1/Y_k^2$ $(k = 1, \ldots n)$.

The weighted regression formulas are entirely similar to those for equal weights. The only change is that the sums Σ, Σ_x, etc., defined in (6.3.12) are now redefined as weighted sums as follows:

$$\Sigma = \sum_{k=1}^{n} w_k\ , \quad \Sigma_x = \sum_{k=1}^{n} w_k X_k\ , \quad \Sigma_y = \sum_{k=1}^{n} w_k Y_k$$

$$\Sigma_{xx} = \sum_{k=1}^{n} w_k (X_k)^2\ , \quad \Sigma_{xy} = \sum_{k=1}^{n} w_k Y_k X_k\ , \quad \Sigma_{yy} = \sum_{k=1}^{n} w_k (Y_k)^2 \tag{6.3.30}$$

This is the only modification. The expressions for Δ, A, C, $\bar{C}$, χ^2, and the coefficients of variation are identical to the formulas in Section 6.3.4, except that the Σs appearing in those formulas are now given in (6.3.30).

Example 6.3.3 Consider again Bažant and Pfeiffer's results analyzed in example 6.3.1, and apply to them the weighted linear regression I. We first construct the sums in (6.3.30): $\Sigma = 2620.5, \Sigma_x =$

189.25, $\Sigma_{xx} = 23.448$, $\Sigma_y = 160.75$, $\Sigma_{yy} = 12$, $\Sigma_{xy} = 16.097$, and then compute $\Delta = 25630$ from (6.3.14). Next the values of A and C follow from (6.3.13): $A = 0.459$, $C = 0.0282$; they are sensibly different from the values obtained with the equal-weight regression in examples 6.3.1 and 6.3.2. Now, using the values for k_0 and k_0' found in example 6.3.2, we get $K_{Ic} = k_0/\sqrt{A} = 0.687/\sqrt{0.459} = 1.01$ MPam$^{1/2}$, $c_f = (k_0/2/k_0')C/A = (0.687/2/2.1) \times 0.0282/0.459 = 0.01005$ m, or $c_f = 10.0$ mm. Using for E the ACI estimate $E = 27.6$ GPa, computed in example 6.3.2, we get $G_f = K_{Ic}^2/E \approx 37.3$ N/m. For the statistics, we first compute χ^2 from the last of (6.3.16), which gives $\chi^2 = 7.9857 \times 10^{-2}$; then, from (6.3.17), we get $\omega_A = 6.2\%$ and $\omega_C = 9.6\%$, and from (6.3.21) $\omega_{\bar{C}} = 2.8\%$, with $\bar{C} = 0.061343$ from (6.3.21). According to this and Eq. (6.3.19), the coefficient of variation of K_{Ic} is $\omega_{K_{Ic}} = 3.1\%$; finally, from (6.3.22) we get $\omega_{c_f} = 14.9\%$.

Note how the coefficients of variation have decreased with respect to the equal-weight solution: this is because the weighted regression minimizes the relative deviations from the theoretical curve rather than the absolute deviations. □

Example 6.3.4 It is instructive to compare the solutions for the various regressions applicable to Bažant and Pfeiffer series B1. Without detailing the computations (which follow the steps illustrated in the preceding examples), the results for K_{Ic} and c_f and their coefficients of variation are included in the following table:

	Regression				
Parameter	Nonlinear	Linear I	Linear II	Weighted I	Weighted II
K_{Ic} (MPa m$^{1/2}$)	1.00	1.05	0.98	1.01	1.01
c_f (mm)	9.8	12.8	9.00	10.0	10.0
ωK_{Ic} (%)	3.4	3.4	3.4	3.1	3.1
ω_{c_f} (%)	16	22	14	15	15

We see that the two weighted linear regressions and the nonlinear regression deliver nearly the same values and coefficients of variation. Indeed, it is easy to prove that the two weighted linear regressions are, in fact, identical, with $C' = A$ and $A' = C$. The proof is left as an exercise. □

Exercises

6.3 Compute k_0 and k_0' for the test series run on SEN-TPB specimens of those included in Section 1.5. Use the size effect parameters shown in Table 1.5.2 to deduce the fracture properties K_{Ic}, G_f, and c_f for the various materials. Sort the materials in increasing order of (a) strength, (b) toughness, and (c) brittleness.

6.4 The experimental results of Bažant, Gettu and Kazemi for the limestone specimens described in Section 1.5 were as shown in the following table:

		Specimen No.											
Var.	Units	#1	#2	#3	#4	#5	#6	#7	#8	#9	#10	#11	#12
D	mm	13	13	13	25	25	25	51	51	51	102	102	102
P_u	N	82	85	78	134	140	140	238	243	243	418	405	394

(a) Determine Bf_t' and D_0 from the linear regression I. (b) Same for weighted regression I. (c) Same for weighted regression II. (d) Compare with the results of the nonlinear logarithmic regression shown in Table 1.5.2. (e) Calculate K_{Ic}, G_f, and c_f.

6.5 Prove that the weighted versions of the linear regressions I and II are identical, with $C' = A$ and $A' = C$. (Hint: write the weighted sum of squares to be minimized for regressions I and II; then set the variables of the second in terms of the variables of the first.)

6.6 Consider the nonlinear logarithmic regression written as $y = -\frac{1}{2}\ln[A\exp(x) + C]$ with $y = \ln \sigma_{Nu}$ and $x = \ln D$. Show that this approaches the weighted linear regression I (or II) if $Y_k - AX_k - C \ll Y_k$. (Hint: write the weighted sum of squares to be minimized for the nonlinear regression; then set the variable (y, x) in terms of (Y, X) and write that $\ln(1 + \epsilon) \approx \epsilon$ for $\epsilon \ll 1$.)

6.4 Determination of R-Curve from Size Effect

As discussed in the previous chapter, the R-curves, which serve as a basis of the equivalent linear elastic fracture analysis, can be determined in various ways which give similar results but are not completely equivalent (Section 5.6.5). The most recent method is the determination from the size effect data on the maximum loads of geometrically similar specimens of different sizes.

One advantage of the size effect method is that, by definition, the R-curve obtained is size independent, yet it seems to have a very broad applicability, and to be particularly good for prediction of the maximum loads because it is determined on the basis of the maximum loads. All R-curves strongly depend on the structural geometry, although the same R-curve can be used only as an approximation for different but very close geometries.

There is a more profound advantage of the size effect method. If data on the load deflection or other similar curves are used to determine the R-curves, such data are very sensitive to specimen geometry and size, and so there is no way to go from one geometry to another except by repeating measurements. The size effect law, on the other hand, has not only, by definition, size-independent parameters, but has also the same shape for very different geometries. This fact has been extensively proven by experiments as well as numerical calculations (within the size range of 1:20, which suffices for most practical purposes). Thus, using the size effect law, one can get at least the shapes of the R-curves for all geometries, in fact, without any further testing. Only those parameters of the size effect law which are translated into the parameters of the R-curve need to be calibrated for different geometries. Let us now derive the equations for the R-curve from the size effect curve.

6.4.1 Determination of R-Curve from Size Effect

Consider that the maximum load P_u has been measured for a set of geometrically similar specimens of different size D. For each size, and each effective crack length a, the energy release rate at peak load $\mathcal{G}_u$ can be calculated from Irwin's relationship K_I^2/E' as

$$\mathcal{G}_u(a) = \frac{1}{E'}\sigma_{Nu}^2 D k^2(\alpha) \tag{6.4.1}$$

where $\alpha = a/D$ and $k(\alpha)$ is the shape factor for K_I, same for all sizes D (Section 2.3.2). A different curve $\mathcal{G}_u(a)$ is obtained for each specimen size, as indicated in Fig. 6.4.1a. On each curve $\mathcal{G}_u(a)$, there is normally one, and only one, point that represents the failure point (critical state). At that point, the $\mathcal{G}(a)$ must, for every size, be tangent to the R-curve as indicated in Fig. 6.4.1a (see Section 5.6.3). In general, the curves $\mathcal{G}_u(a)$ for different sizes will have distinct tangent points. Consequently, if we draw the curves $\mathcal{G}_u(a)$ for all the sizes, we will obtain the R-curve as the envelope of these curves as shown in Fig. 6.4.1b for a particular case (see a subsequent example for details).

To describe the envelope mathematically, let us assume that the size effect law is known. This means that the ultimate nominal stress σ_{Nu} is a known function of D, thus

$$\sigma_{Nu} = \sigma_{Nu}(D) \tag{6.4.2}$$

At peak load we, of course, have the condition of equilibrium fracture propagation $\mathcal{G} = R$ which, with the aid of (6.4.1) and (6.4.2), can be written as

$$\mathcal{G}_u(\Delta a, D) = R(\Delta a) \quad \text{with} \quad \mathcal{G}_u(\Delta a, D) = \frac{D}{E'}\sigma_{Nu}^2(D)\, k^2\!\left(\alpha_0 + \frac{\Delta a}{D}\right) \tag{6.4.3}$$

Now, this equation holds for every D, so that differentiation with respect to D must give

$$\frac{\partial \mathcal{G}_u(\Delta a, D)}{\partial D} = 0 \tag{6.4.4}$$

where $\partial R(\Delta a)/\partial D$ vanishes because, by definition, the resistance curve depends only on Δa, not on D. Also, note that α_0 must be taken as a constant during differentiation because, by hypothesis, we consider only geometrically similar bodies.

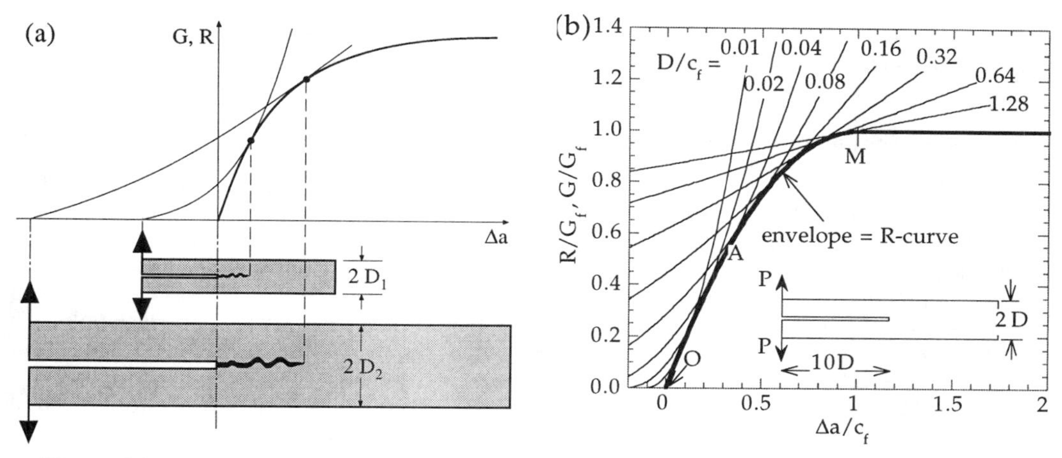

Figure 6.4.1 R-curve as the envelope of the energy release rate curves at peak load for various sizes.

Mathematically, the system of two equations (6.4.3) and (6.4.4) represents the parametric equations of the envelope of the family of curves $\mathcal{G}_u(\Delta a, D)$. Elimination of D from the two sets of equations yields $R(\Delta a)$.

Thus, it has been proven that the knowledge of the size effect law for a given geometry permits the determination of a unique R-curve (for that geometry). We now apply this property to Bažant's size effect law, first in an example, then in general form.

Example 6.4.1 Consider a double-cantilever beam specimen as sketched in Fig. 6.4.1. Note that $a_0 = 10D$ so that $\alpha_0 = 10$ and all the specimens are geometrically similar. The expression for $\mathcal{G}$ found for this specimen in Chapter 2 —Eq. (2.1.29)— can be rewritten as

$$\mathcal{G} = 12\sigma_N^2 D(a/D)^2/E' \tag{6.4.5}$$

where D replaces h in the original equation and the nominal stress is defined as $\sigma_N = P/bD$. Comparing this to (6.4.1), we see that $k^2(\alpha) = 12\alpha^2$. If we further assume that Bažant's size effect law holds, i.e., $\sigma_{Nu} = Bf_t'(1 + D/D_0)^{-1/2}$, the energy release curve at peak load is given by:

$$\mathcal{G}_u(a) = \frac{(Bf_t')^2 D_0}{E'} \frac{D/D_0}{1 + D/D_0} 12 \left(\frac{a}{D}\right)^2 \tag{6.4.6}$$

We now substitute Bf_t' in terms of G_f using (6.2.3), and D_0 in terms of c_f using (6.2.2), for which we compute $2k_0'/k_0 = 2/\alpha_0 = 1/5$. The final expression for $\mathcal{G}_u(a)$ thus becomes

$$\mathcal{G}_u(a) = G_f \frac{5D/c_f}{1 + 5D/c_f} \left(\frac{a}{10D}\right)^2 = G_f \frac{5D/c_f}{1 + 5D/c_f} \left(1 + \frac{\Delta a}{10D}\right)^2 \tag{6.4.7}$$

in which we set $a = a_0 + \Delta a$ and $a_0 = 10D$. Fig. 6.4.1 shows various $\mathcal{G}_u(a)$ curves for a number of sizes D plotted in a $\mathcal{G}/G_f$ vs. $\Delta a/c_f$ plot. It is obvious that these curves define an envelope: the R-curve. To obtain the equation of the R-curve, we set the partial derivative of (6.4.8) with respect to D equal to zero and solve for D. The result is

$$D = 2.5 \frac{\Delta a}{c_f - \Delta a} \qquad (0 \le \Delta a < c_f) \tag{6.4.8}$$

where the inequalities in parenthesis simply state that neither Δa nor D can be negative. Substitution of the foregoing equation into expression (6.4.7) for $\mathcal{G}_u$ gives the equation for the R-curve:

$$R(\Delta a) = G_f \, 2\frac{\Delta a}{c_f} \left(1 - \frac{\Delta a}{2c_f}\right) \qquad (0 \le \Delta a < c_f) \tag{6.4.9}$$

It represents a parabolic arc OAM, shown in Fig. 6.4.1. The R-curve is continued by a horizontal line $R(\Delta a) = G_f$ for $\Delta a \geq c_f$, a segment of which has also been drawn in the figure. □

6.4.2 Determination of R-Curve from Bažant's Size Effect Law

The previous example deals with a very simple case in which the analytic determination of the R-curve is feasible. This is generally not the case, because of the complexity of the expressions for $k(\alpha)$ and its derivative. In the initial work of Bažant, Kim and Pfeiffer (1986), the R-curve was computed numerically. The numerical values of $\mathcal{R}$ so obtained were then fitted with an analytical expression. It was found that the expression $R(\Delta a) = G_f[1 - (1 - k\Delta a)^n]$ for $\Delta a < 1/k$ and $R(c) = G_f$ for $\Delta a \geq 1/k$, with optimized constants n and k, gave very good results. Subsequently, however, an analytical determination of the R-curve from the size effect law has been discovered (Bažant and Kazemi 1990a). The method, which provides the R-curve in an explicit parametric form, is as follows.

According to Bažant's size effect law and the relation (6.2.3) between G_f and Bf'_t, Eq. (6.4.2) can be explicitly written as

$$\sigma_{Nu}^2(D) = \frac{G_f E'}{(D_0 + D)k_0^2} \tag{6.4.10}$$

Substitution into the second of (6.4.3) gives the expression for $\mathcal{G}_u(\Delta a, D)$

$$\mathcal{G}_u(\Delta a, D) = G_f \frac{D}{(D_0 + D)k_0^2} k^2\!\left(\alpha_0 + \frac{\Delta a}{D}\right) \tag{6.4.11}$$

Before proceeding any further, it is convenient to define the parameter α' such that

$$\Delta a = (\alpha' - \alpha_0)D \tag{6.4.12}$$

The meaning of α' in this context is the value of $\alpha = a/D$ for which a specimen of size D reaches the peak load; thus, the straight solution of the problem would require solving for α' from the peak-load condition to get the failure point for a given size. However, since we, in general, cannot solve for α', we circumvent the problem by writing every equation in terms of α' itself, i.e., using a parametric representation of all the equations in the problem.

The first equation to write is the tangency condition (6.4.4); thus, we differentiate (6.4.11) with respect to D, substitute Δa from (6.4.12), and solve for D; the result is

$$D = \frac{D_0 k(\alpha')}{(\alpha' - \alpha_0)2k'(\alpha')} - D_0 \tag{6.4.13}$$

which explicitly gives the size of the specimen for which the peak load is reached when $\alpha = \alpha'$. We substitute this value back into (6.4.12) and get Δa as a function of α'

$$\Delta a = \left[\frac{k(\alpha')}{2k'(\alpha')} - (\alpha' - \alpha_0)\right] D_0 \tag{6.4.14}$$

from which, inserting D_0 from expression (6.2.2), $D_0 = 2k'_0 c_f / k_0$, we get the final expression for Δa

$$\frac{\Delta a}{c_f} = f_1(\alpha') = \left[\frac{k(\alpha')}{2k'(\alpha')} - (\alpha' - \alpha_0)\right] \frac{2k'_0}{k_0} \tag{6.4.15}$$

This equation gives the abscissa of the R-curve as a function of parameter α'. To get the ordinate, we solve for $\mathcal{R}$ by setting, in accordance with (6.4.3), $\mathcal{R} = \mathcal{G}_u$ with $\mathcal{G}_u$ given by (6.4.11). After replacing D by its expression (6.4.13) and rearranging, the following expression for $\mathcal{R}$ ensues:

$$\frac{\mathcal{R}}{G_f} = f_2(\alpha') = \frac{k(\alpha')k'(\alpha')}{k_0 k'_0} f_1(\alpha') \tag{6.4.16}$$

where $f_1(\alpha')$ is the function defined in (6.4.15). Eqs. (6.4.15) and (6.4.16) define the R-curve in explicit parametric form with parameter α' (Bažant, Kazemi and Gettu 1989; Bažant and Kazemi 1990a).

We note that the right-hand sides of these equations depend only on geometrical properties $k(\alpha')$ and its derivative. The only relevant material properties are G_f and c_f. They enter the equations by scaling, respectively, $\mathcal{R}$ and Δa. In fact, elimination of α' from the foregoing equations leads to an R-curve of the type

$$\frac{\mathcal{R}}{G_f} = f\left(\frac{\Delta a}{c_f}\right) \tag{6.4.17}$$

where function f is completely defined by the shape of the specimen (including the initial relative notch depth).

In general, eliminating α' is not feasible analytically. So it is better to use the parametric form directly. To draw the R curve (scaled by G_f), we select a set of α' values. For each of them, we evaluate $f_1(\alpha')$ from Eq. (6.4.15) and then calculate $f_2(\alpha')$ from Eq. (6.4.16). To obtain the R-curve for a particular material, the size effect method explained in the previous section is used to determine G_f and c_f. Then, the R-curve points $(\Delta a, \mathcal{R})$ for the various α' follow by setting $\Delta a = c_f f_1(\alpha')$ and $\mathcal{R} = G_f f_2(\alpha')$. Let us illustrate this by an example.

Example 6.4.2 Let us consider the three-point bend specimens used by Bažant and Pfeiffer (1987) defined as series B1 and C1 in Table 1.5.1 and Fig. 1.5.1, and let us determine the R-curve. We use a graphics program to work out the curve. We first generate an array with various values of α', starting with $\alpha' = \alpha_0 = 1/6$; a selection of the evaluation points is shown in the fist row of the table below (in reality, we use points spaced $\delta\alpha' = 0.005$, but there is no point in giving all the values here; the calculations for the 10 values shown can be obtained with a hand-held programmable calculator within reasonable time). The first step is to program function $k(\alpha')$ using the expressions (6.3.15) and (6.3.16). The function is run for the set of values α', stored in a second array (the second row of the table). The third row contains the values of $k'(\alpha')$ which are obtained by calculating $k'(\alpha') \approx [k(\alpha'+0.005) - k(\alpha'-0.005)]/0.01$. Then, it is an easy matter to compute $\Delta a/c_f = f_1(\alpha')$ from (6.4.15) (fourth row of the table) and $\mathcal{R}/G_f = f_2(\alpha')$ from (6.4.16) (fifth row of the table).

α'	1/6	0.2	0.22	0.24	0.26	0.28	0.30	0.32	0.34	0.35
$k(\alpha')$	0.687	0.756	0.798	0.842	0.887	0.933	0.982	1.03	1.09	1.12
$k'(\alpha')$	2.10	2.10	2.14	2.20	2.29	2.39	2.52	2.67	2.85	2.95
$\Delta a/c_f$	1.00	0.897	0.815	0.720	0.614	0.498	0.375	0.244	0.109	0.039
$\mathcal{R}/G_f$	1.00	0.989	0.967	0.927	0.865	0.774	0.645	0.470	0.235	0.090

The dimensionless R-curve can be plotted by taking the fourth row as abscissas and the fifth row as ordinates. Such a plot is shown in Fig. 6.4.2 (the symbols indicate the points included in the table). The curve so obtained has parabolic form (as for the case in example 6.4.1) and is continued for $\Delta a/c_f > 1$ with a plateau $\mathcal{R} = G_f$. □

As is obvious from the foregoing examples, the R-curve obtained from the size effect law starts from zero, which means that the process zone forms right at the beginning of loading and that there is never any singularity at the mathematically sharp crack tip. (This type of R-curve is obtained by calculations from cohesive crack models with no nonlinearity in the surrounding material, or from crack-bridging models that have no other toughening mechanism; Horii, Zihai and Gong 1989; Elices and Planas 1992, 1993; Planas, Elices and Ruiz 1993). Some models for composite materials consider the R-curve to start from a certain initial non-zero value, interpreted as a certain small-scale value of the fracture energy. However, this kind of R-curve implies that the crack tip would be able to sustain, up to some value of the stress intensity factor, a singular stress field without showing any damage, which does not seem reasonable.

The R-curve obtained in this manner, as well as the load-deflection diagrams calculated from such R-curves, have been found to be in good agreement with numerous data on concretes and rocks (Bažant and Kazemi 1990a; Bažant, Gettu and Kazemi 1991) as well as tough aluminum alloys (Bažant, Lee and Pfeiffer 1987).

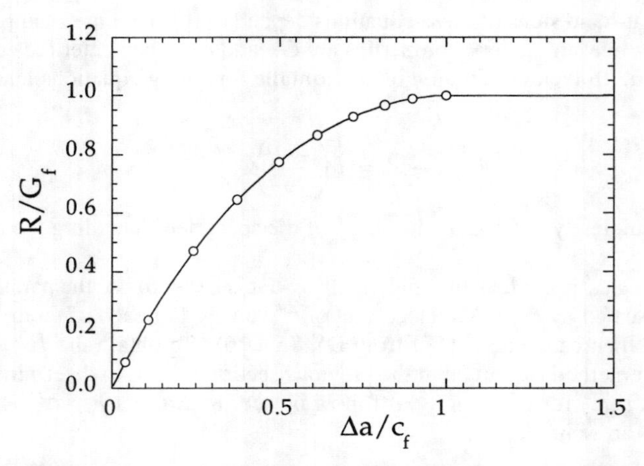

Figure 6.4.2 *R*-curve obtained from size effect for the three-point-bend notched beams tested by Bažant and Pfeiffer (1987). See Tables 1.5.1 and 1.5.2, and Fig. 1.5.2 for details.

Determination of the R-curve from size effect data does not work in all circumstances. It obviously fails when $k'(\alpha_0) = 0$, and does not work when $k'(\alpha_0) \leq 0$, because $k'(\alpha_0) > 0$ ($D_0 > 0$) was implied in the derivation of Eq. (6.4.16). It also fails when $k'(\alpha_0)$ is too small, because of the scatter of test results. So this method must be limited to the positive specimen geometries, for which $k'(\alpha_0) > 0$. This nevertheless comprises most practical situations.

6.4.3 Determination of the Structural Response from the R-Curve

Once the R-curve has been determined, as just explained, the determination of the load-displacement or load-CMOD curve for a particular specimen size is easy. The first step is to determine the load corresponding to a point on the R-curve. Let us assume that we have obtained a point $(\Delta a, \mathcal{R})$ as illustrated in the previous paragraph. The load corresponding to this point for a static crack growth for a specimen of size D is obtained by setting $\mathcal{G} = \mathcal{R}$ or

$$\mathcal{R} = \frac{1}{E'}\sigma_N^2 D k^2(\alpha_0 + \Delta a/D) \tag{6.4.18}$$

which can immediately be solved for σ_N as the load parameter:

$$\sigma_N = \frac{\sqrt{E'\mathcal{R}}}{\sqrt{D}\,k(\alpha_0 + \Delta a/D)} \tag{6.4.19}$$

Now, the expression for the displacement u follows from (3.5.6) as

$$u = \frac{\sigma_N}{E'} D\,v(\alpha_0 + \Delta a/D) \tag{6.4.20}$$

where $v(\alpha)$ can either be found in closed form in a handbook, or computed using (3.5.7), as derived in Section 3.5.1. Similar expressions can be used for the CMOD (see Section 3.5.3).

Remark: *Question Concerning Possible Modification of the Postpeak Region of R-Curves.* When the R-curve given by Eqs. (6.4.14) and (6.4.16) is used in the calculation of the load-deflection curve, it must give, of course, maximum loads that exactly agree with the size effect law. This has been confirmed numerically. It has been also checked that such calculations give the correct shapes of the load-deflection diagrams up to the peak point (Bažant and Kazemi 1990a,b)). However, significant deviations from the observed load-deflection curves have been encountered in the postpeak regions for small specimens.

To explain these observations, the following hypothesis has been suggested by Bažant, Gettu and Kazemi (1991): the size effect law yields only a master R-curve whose entire length is followed only for an infinitely large specimen. In actual specimens, the master R-curve would be followed only up to the maximum load

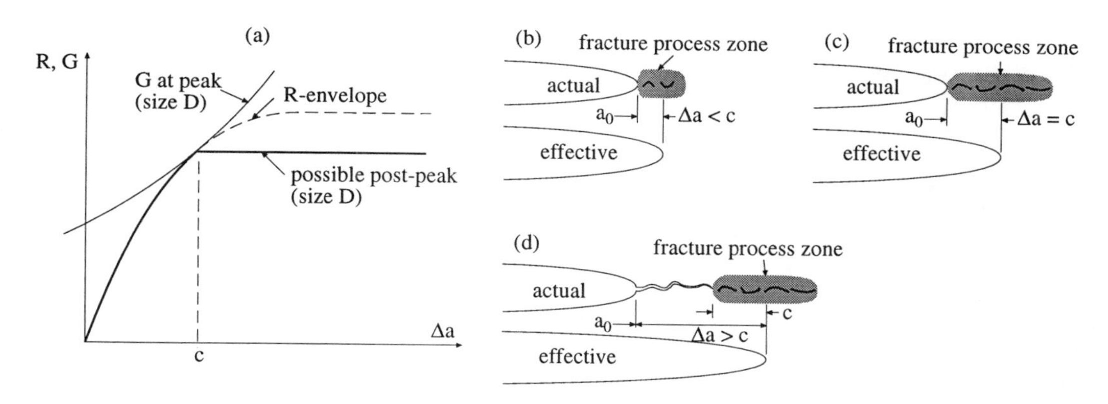

Figure 6.4.3 $\mathcal{R}$ is kept constant after peak load (a). Process zone at different stages: (b) before peak load; (c) at peak load; (d) after peak load (adapted from Bažant and Kazemi 1990b).

point, after which one would assume, in order to get correct predictions of the postpeak deflection, that the value of R remains constant and equal to the value reached at the peak load, i.e., the R-curve after the peak load is a horizontal line as depicted in Fig. 6.4.3a (Bažant, Gettu and Kazemi 1991; Bažant and Kazemi 1990b). The reason for this behavior has not been firmly established, but, apparently, it might consist in the fact that, in the postpeak regime, the fracture process zone, due to decreasing load, cannot keep growing; instead, it detaches itself from the tip of the notch or initial crack and travels forward without growing in size (see Fig. 6.4.3b–d). However, further scrutiny of this hypothetical explanation is needed. △

Exercises

6.7 Prove that for a double-cantilever beam, the R-curve deduced from Bažant's size effect law is independent of the initial crack length.

6.8 Prove that for a crack of length $2a_0$ in an infinite panel subjected to uniform remote tensile stress, the envelope of the $\mathcal{G}_u(\Delta a)$ curves degenerates into a point of coordinates (c_f, G_f). Show that the R-curve can be considered to be defined by the segments of the limiting $\mathcal{G}_u(a)$ curves for $D = 0$ and $D = \infty$ which define the bilinear R-curve $R = G_f \Delta a / c_f$ for $0 \leq \Delta a \leq c_f$ and $R = G_f$ for $\Delta a \geq c_f$.

6.9 Find the R-curve based on Bažant's size effect law (given G_f and c_f) for a double-cantilever beam whose arms are subjected to a uniformly distributed load p per unit length. Take $\sigma_N = p/b$, b = thickness, D = arm depth, $a_0 = \alpha_0 D$.

6.10 Determine the load-displacement curve of the double-cantilever beam analyzed in example 6.4.1 for the case $D = c_f/4$.

6.11 Determine the load-displacement curve of the double-cantilever beam analyzed in example 6.4.1 for the case $D = c_f$.

6.12 Determine the load-displacement curve of the double-cantilever beam analyzed in example 6.4.1 for the case $D = 4c_f$.

6.13 Show that the load-displacement curve expressed as a σ_N–u when the behavior is governed by an R-curve deduced from Bažant's size effect law always takes the form

$$\sigma_N = \sigma_1 \phi\left(\frac{u}{u_1}, \frac{c_f}{D}\right) , \quad \sigma_1 = \sqrt{\frac{E' G_f}{c_f}} , \quad u_1 = c_f \frac{\sigma_1}{E'} \tag{6.4.21}$$

where ϕ is a nondimensional function, which depends implicitly on the shape only.

7

Cohesive Crack Models

In the previous chapters, the fracture process was always considered to occur at the tip of a sharp crack, either real (in LEFM) or equivalent. However, such models are strictly applicable only when the fracture process zone is small compared to the relevant dimensions of the specimen. Furthermore, they neglect a detailed description of what is happening in the fracture process zone because they lump it all into the crack tip.

This chapter introduces the simplest models that describe the fracture process in full, albeit with many simplifications. Such full description is essential when the crack length or other dimensions of the specimen are small relative to a fully developed fracture process zone, which is a common situation for structures of concrete and tough ceramics. Even if the fracture process zone is small, describing what is going on inside it may be convenient for the purpose of understanding the fracture mechanisms and designing appropriate modifications of the material, such as toughening by reinforcement.

The fracture process zone can be described by two simplified approaches: (1) The entire fracture process zone is lumped into the crack line and is characterized in the form of a stress-displacement law which exhibits softening; or (2) the inelastic deformations in the fracture process zone are smeared over a band of a certain width, imagined to exist in front of the main crack. In this chapter we focus on the first approach, which may be found in the literature under a variety of names, e.g., cohesive crack model, fictitious crack model, Dugdale-Barenblatt model, and crack with bridging stresses. The second approach will be studied in the next chapter. Both approaches can resolve the detail of fracture process in one dimension —in the direction of the crack. None of them can resolve the fracture process detail across the width of the fracture process zone. The exposition of the more sophisticated nonlocal models, which can give some information on the distribution of fracture deformation across the width of the process zone, will be deferred until Chapter 13.

In the simplest, most usual formulation of the cohesive crack model, all the body volume remains elastic, and linear elasticity may be used to solve the response of the parts of structure on each side of the crack. The nonlinearity is included in the boundary conditions along the crack line. Most of the chapter deals with such a simplified model, although some more sophisticated issues, still open to debate, are briefly discussed. Applications of the cohesive crack models to particular structures are postponed until later chapters, where applications of fracture mechanics are discussed in a wider context.

The first section presents the features common to all cohesive crack models. The second section gives an overview of the application of cohesive cracks to concrete, although most of the concepts can also be applied to other materials; in the presentation of that section, most details are skipped to grasp only the essentials, since more details on many topics are given in the remaining sections and in the chapters that follow. The third section deals with the details of the experimental determination of cohesive crack properties, the main focus being on concrete. The two last sections give an extended account of numerical methods specially developed to carry out fast computations for cohesive crack growth in mode I.

7.1 Basic Concepts in Cohesive Crack Model

As already pointed out, the cohesive crack is the simplest model that describes in full the progressive fracture process. In its origin, a cohesive crack is nothing else but a fictitious crack able to transfer stress from one face to the other. This model was introduced in the early sixties by Barenblatt (1962) and Dugdale (1960) to represent very different nonlinear processes located at the front of a pre-existent crack. Cohesive cracks have since been used by many researchers to describe the near-tip nonlinear zone for cracks in most materials: metals, polymers, ceramics, and geomaterials.

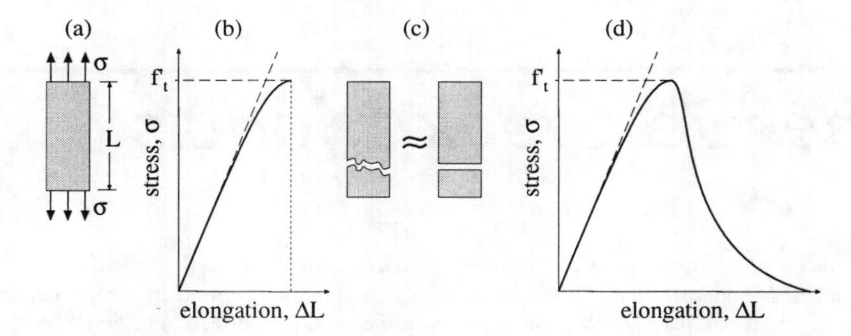

Figure 7.1.1 Uniaxial test on a concrete bar (a). Typical unstable stress-elongation curve with single crack failure (b). A rough crack is idealized as a flat crack (c). Stress-elongation curve for a stable tension test (d).

In the late seventies, Hillerborg, Modéer and Petersson (1976) extended the concept of cohesive crack for concrete by proposing that the cohesive crack may be assumed to develop anywhere, even if no pre-existing macrocrack is actually present. They called this extension of the cohesive crack the *fictitious crack model* (Hillerborg, Modéer and Petersson 1976; Modéer 1979; Petersson 1981; Gustafsson 1985; Hillerborg 1985a,b). It is easily shown that when the fictitious crack model is used to describe the behavior of a pre-existing crack, the mathematical formalism is identical to that for the classical cohesive cracks.

The reverse is not true in general: a near-tip cohesive approximation cannot be extended to uncracked situations for all materials. This is quite obvious for metals, in which the Dugdale approach may give a rough yet reasonable estimate of the plastic processes close to a crack tip, but cannot approximate *at all* the behavior of an uncracked specimen subjected, say, to pure tension: the specimen would experiment plastic flow and necking well before a cohesive crack develops.

In this chapter, and in general in this book, we use the cohesive crack concept in the sense of Hillerborg because we have in mind any quasibrittle material, such as concrete, that breaks in a pure tensile test by cracking before any substantial plastic flow and necking take place. That is why we start this section by deriving the cohesive crack model from the analysis of an ideal tension test, as done by Hillerborg (Petersson 1981, Hillerborg 1985a), even though the development of cohesive cracks followed chronologically a different path. Later we will give an overview of the Barenblatt and Dugdale models, as well as other cohesive zone models.

7.1.1 Hillerborg's Approach: The Cohesive Crack as a Constitutive Relation

Consider a uniaxial tension test on a concrete bar (Fig. 7.1.1a). If no special care is exerted, a nearly linear stress-elongation curve will be recorded up to the peak stress. At this point, catastrophic (dynamic) failure will occur, usually through a single crack (Fig. 7.1.1b) . This crack is very rough and is perpendicular to the bar only on the average, because of the high heterogeneity of concrete. However, one can ignore this fact and approximate the final failure by a smooth crack perpendicular to longitudinal axis of the bar (Fig. 7.1.1c). Before going any further, one can easily imagine that the permanent strain of each of the two pieces into which the specimen broke is likely to be very small, at most of the order of the inelastic strain measured just before peak. Thus, roughly speaking, one can say that, after the load peak, all the deformation localizes into the crack that, ultimately, cuts the specimen in two. This is the first important ingredient in Hillerborg's approach. The second ingredient is the description of the evolution of this crack from no crack at all to a fully broken bar.

The second ingredient was justified by the experimental evidence set forth by various researchers in the late sixties (e.g., Hughes and Chapman 1966; Evans and Marathe 1968; Heilmann, Hilsdorf and Finsterwalder 1969) which showed that if the tensile specimen is short enough and a very stiff testing machine is used, the crack can evolve in a stable manner, making it possible to obtain the complete load-elongation curve, including the post-peak region, as shown in Fig. 7.1.1d. This unambiguously showed that the transition from full-strength to zero-strength was gradual rather than instantaneous, as was often assumed before. The experimental results of Heilmann, Hilsdorf and Finsterwalder (1969),

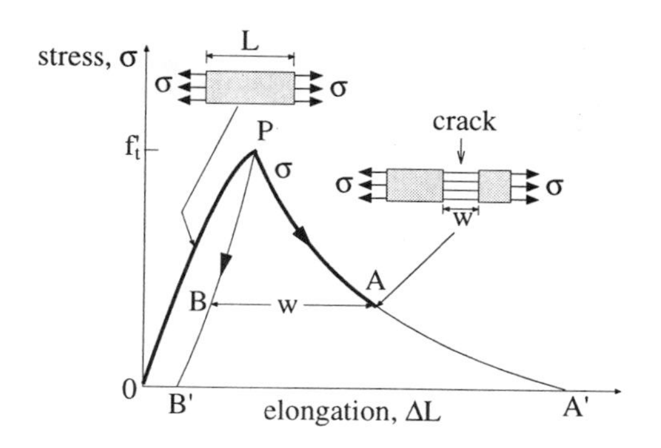

Figure 7.1.2 An idealized tensile test.

further showed that after the peak load, the strain localizes into a very narrow region — which will later become the visible crack — while the rest of the specimen unloads. They detected this effect by gluing a set of overlapping strain gages along the axis of concrete prisms subjected to tension. The records of the strain measured by the gages showed that at or near the peak, the strain was rapidly increasing in one or two gages while it decreased in the remaining gauges; towards the end of the test, it was evident that the gages experimenting the strain increase were cut by the final crack.

Thus, Hillerborg described the tensile test in the following idealized manner (Fig. 7.1.2). Up to the peak load, the load increases while the bar strain remains distributed uniformly along the specimen (arc OP). At the peak load, a cohesive crack normal to the axis of the bar appears somewhere in the specimen (at the weakest cross section). After the peak, the crack develops a finite opening w while still transferring stress, and at the same time the remainder of the specimen unloads and its strain decreases uniformly along the arc PB. Thus, the total elongation at point A is the addition of a uniform strain corresponding to point B and the crack opening w:

$$\Delta L = L\varepsilon_B + w \tag{7.1.1}$$

where L is the length of the specimen and ε_B is the strain of the material elements not containing a crack which unload from peak load.

After setting the foregoing picture of the evolution of fracture in an ideal tensile test, Hillerborg assumed that the stress transferred through the cohesive crack was a function of the crack opening, and wrote

$$\sigma = f(w) \tag{7.1.2}$$

where $f(w)$ is a function characteristic of the material that must be determined from experiments. This function, called the softening curve, can be extracted from Fig. 7.1.2 and plotted separately as shown in Fig. 7.1.3a. By its very definition, the softening curve verifies the property that the limit stress for a zero crack opening is the tensile strength, as indicated in the figure.

To simplify the computations while retaining the essentials of the model, Hillerborg further assumed that the inelastic strain in the loading-unloading path was negligible, i.e., the behavior of the bulk material was linear elastic, so that, given the softening curve in Fig. 7.1.3a, the load-elongation curve is constructed as Fig. 7.1.3b shows. Thus, the post-peak elongation can be computed from (7.1.1) as

$$\Delta L = L\frac{\sigma}{E} + w = L\frac{f(w)}{E} + w \tag{7.1.3}$$

where E is the elastic modulus. It is evident that, if E and $f(w)$ are given, the load-elongation curve in tensile tests under monotonic extension is completely determined by E and the softening function $f(w)$.

It may be pointed out that these ideas are analogous to those introduced simultaneously by Bažant (1976) in his analysis of the failure of a tensioned concrete bar. A figure analogous to Fig. 7.1.2 and an equation analogous to Eq. (7.1.3) were presented by Bažant (1976) with the only difference that a cohesive crack was replaced by a localized crack band in which the material is loading and softening, while the rest of the specimen is unloading. The crack band, which will be described in detail in the

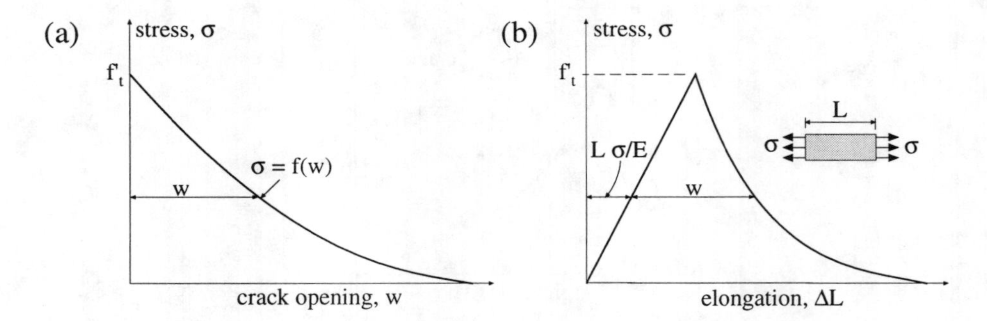

Figure 7.1.3 Softening curve (a), and resulting stress-elongation curve when the bulk material behavior is assumed to be linear elastic (b).

next chapter, is more realistic at the beginning of softening, while a cohesive crack is more realistic at the end of softening (as the cracking is getting more localized). But mathematically both approaches are approximately equivalent.

Summarizing the essential ideas in the foregoing analysis, we see that to describe the quasi-static failure of a bar in tension we introduced three main ingredients (Elices and Planas 1989): **(1)** the stress-strain behavior of the bulk material including loading-unloading behavior; **(2)** the cracking criterion indicating the condition for the appearance of the cohesive crack and its orientation; and **(3)** the evolution law for the cohesive crack.

Of course, the preceding discussion referred only to pure tension, and an extension to other situations must be done to obtain a general model. As pointed out by Elices and Planas (1989) and Planas, Elices and Guinea (1995), the extensions of the cohesive crack model can be very general, including nonlinear behavior of the bulk, effect of triaxiality on the cracking criterion and crack evolution law, and effect of shear displacement of the crack lips. However, most of the computations and analysis existing in the literature have been performed for the simplest cohesive model which assumes the following hypotheses (and restrictions):

1. The bulk material can be approximated by an isotropic linear elastic material, with elastic modulus E and Poisson's ratio ν.
2. A crack is supposed to form at a point when the maximum principal stress at that point reaches the tensile strength f'_t; the crack forms perpendicular to the maximum stress direction.
3. When the analysis is restricted to pure opening (no relative slip between the crack faces) and to monotonic opening, then the stress transferred between the faces of the crack is given by (7.1.2).

With this very simple generalization, the model is ready for solving many problems, although not all the problems can be handled with the restrictions imposed in point **3** above. Before proceeding with the analysis of the cohesive crack models, we briefly review other similar approaches.

7.1.2 Other Approaches to Cohesive Cracks

As already pointed out, cohesive cracks were born in the early sixties to take into account the basic aspects of the nonlinear behavior of the material ahead of the tip of a pre-existent crack. Barenblatt (1962) introduced a cohesive zone to account for the nonlinear behavior of the atomic bonds breaking during crack propagation (Fig. 7.1.4a). His analysis showed that the cohesive zone made it possible to relieve the crack tip singularity (an unrealistic feature of LEFM) while for large cracks, the LEFM equations were preserved. The fracture energy G_f was related to the interatomic potential (characterizing the atomic binding forces, i.e, the softening curve). His analysis was limited to cracks very large compared to the cohesive zone itself, and he analyzed only the critical state (i.e., the onset of steady crack growth).

Simultaneously to Barenblatt's, Dugdale (1960) proposed a relatively simple model to deal with plasticity at a crack tip. He assumed that the stress on the crack line ahead of the crack tip was limited by

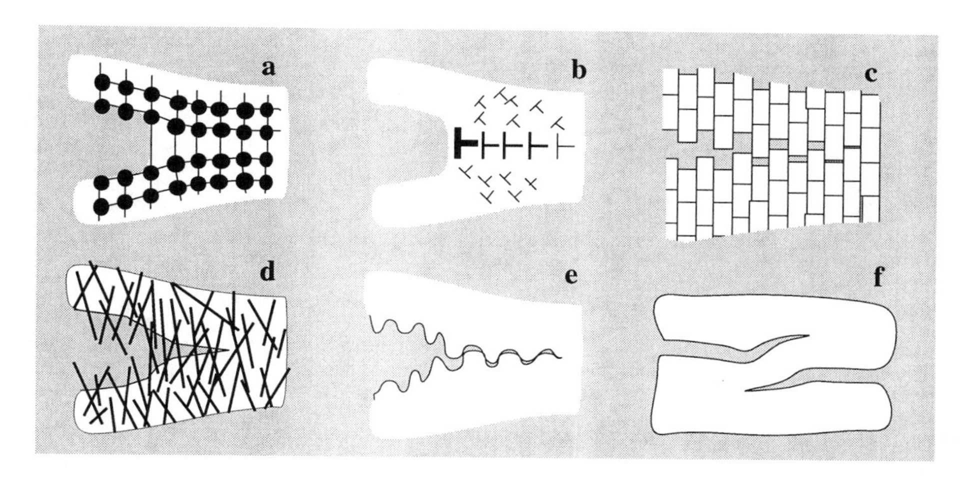

Figure 7.1.4 Various physical sources of cohesive cracks: (a) atomic bonds, (b) yield (dislocation) strip, (c) grain bridging, (d) fiber bridging, (e) aggregate frictional interlock, and (f) crack overlap (from Planas, Elices and Guinea 1995).

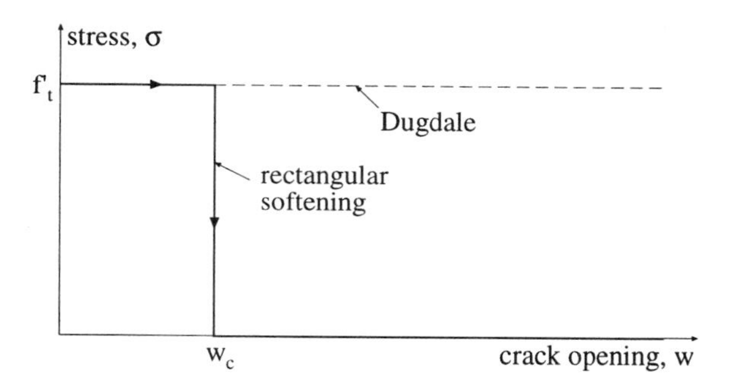

Figure 7.1.5 Cohesive stress vs. crack opening for the original Dugdale model (dashed line) and for the rectangular softening. In the Dugdale model, the stress limit is usually regarded as a plastic flow stress.

the yield strength, and that the plastic deformation was concentrated along the crack line, thus generating a displacement discontinuity similar to a crack (although material continuity was preserved). This approximation can be justified based on flow of dislocations that combine and concentrate on the crack line (Fig. 7.1.4b). Originally, Dugdale's model did not include any softening or fracture criterion: it was a purely 'plastic' model; in fact, it is also known as the *strip yield model* (see, e.g., Tada, Paris and Irwin 1985).

When a cut-off is added so that the material breaks (completely) as the crack opening reaches a critical value w_c, Dugdale's model is formally equivalent to a cohesive crack model —in the sense of Hillerborg— with a rectangular softening function, i.e., a material in which $\sigma = f(w) = \sigma_y =$ const. for $0 < w < w_c$, and $\sigma = f(w) = 0$ for $w > w_c$. Fig. 7.1.5 shows the original Dugdale model together with the rectangular softening function. The rectangular softening is usually inadequate for accurate results. However, it has the important advantage that it accepts analytical treatment in various interesting cases and that it correctly captures the main trends of the fracture behavior. That is why we are going to use it repeatedly in this chapter.

Since the pioneering studies of Barenblatt and Dugdale, cohesive zones have often been used to describe the nonlinear behavior near the crack tip in metals, polymers, and ceramics. It is remarkable that essentially the same models hold for many distinct micromechanisms, as sketched in Fig. 7.1.4, with scales ranging from nanometers (Fig. 7.1.4a-b) to centimeters (Fig. 7.1.4e-f, for concrete with large aggregates).

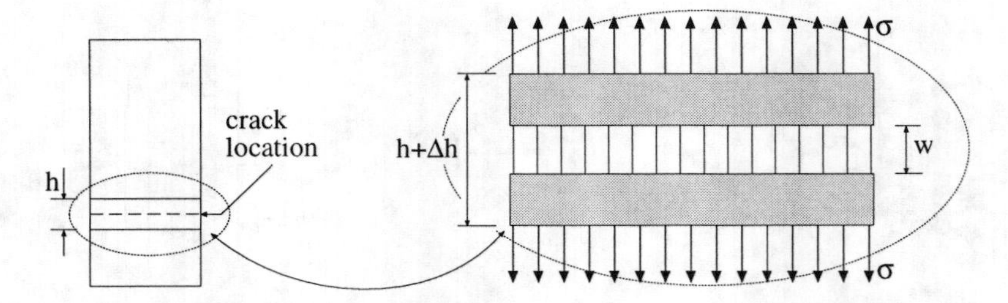

Figure 7.1.6 Thin strip containing the cohesive crack.

7.1.3 Softening Curve, Fracture Energy, and Other Properties

The softening curve $\sigma = f(w)$ is the basic ingredient of the cohesive crack model. Two properties of the softening curve are most important: the tensile strength f'_t and the cohesive fracture energy G_F. As already stated, the tensile strength is the stress at which the crack is created and starts to open; we thus have

$$f(0) = f'_t \tag{7.1.4}$$

The cohesive fracture energy G_F is the external energy supply required to create and fully break a unit surface area of cohesive crack. To calculate it, consider the tensile specimen in Fig. 7.1.2 and a thin element of initial length h centered at the cohesive crack location. The stress that the rest of the specimen transmit upon the faces of this element is σ (Fig. 7.1.6). During the test, the elongation of such an element is Δh and the incremental external work is

$$d\mathcal{W} = \sigma S d(\Delta h) \tag{7.1.5}$$

where S is the area of the cross-section of the specimen. According to (7.1.1), the total elongation of the element is

$$d(\Delta h) = h d\varepsilon + w \tag{7.1.6}$$

where $d\varepsilon$ is the variation of strain of the bulk. Thus, the total work on an element of length h that includes the crack is

$$\mathcal{W} = hS \int_{OPB'} \sigma d\varepsilon + S \int_{OPA'} \sigma dw \tag{7.1.7}$$

In this expression, the first term represents the energy required to deform inelastically the bulk material and the second term the energy required to open the cohesive crack. Obviously, if we consider a vanishingly thin element, so that $h \to 0$, the bulk energy becomes vanishingly small and only the second term, representing the cohesive energy, remains; dividing this energy by the crack surface (the specimen cross section S), we get

$$G_F = \int_0^\infty \sigma \, dw = \int_0^\infty f(w) \, dw \tag{7.1.8}$$

where we have used (7.1.2) to express the stress in terms of w. Note that we have extended the limit of integration to infinity even though we will usually consider softening curves that give complete failure (i.e., $f(w) = 0$) for w larger than some finite opening w_c called the *critical crack opening.* However, there is nothing fundamental against considering softening curves with infinitely long tails, as long as they are integrable, as shown by the following example.

Example 7.1.1 The exponential softening is one of the analytically simpler softenings. Its general expression is $\sigma = c_1 \exp(-c_2 w)$ where c_1 and c_2 are material constants. These constants can be easily related to f'_t and G_F. From (7.1.4), we immediately get $c_1 = f'_t$ and from (7.1.8), we get

$$G_F = f'_t \int_0^\infty \exp(-c_2 w)\, dw = \frac{f'_t}{c_2} \tag{7.1.9}$$

and so, $c_2 = f'_t / G_F$. Thus, we can rewrite the exponential softening in the form

$$\sigma = f'_t \exp\left(-\frac{f'_t}{G_F} w\right) \tag{7.1.10}$$

which shows that the exponential softening is completely determined by f'_t and G_F. ☐

Geometrically, G_F coincides with the area under the softening curve in Fig. 7.1.3a (more precisely, the area defined by the curve and the axes); it also coincides with the area under the curve σ–ΔL in Fig. 7.1.3b (area $OPAA'$). This indicates that when the bulk of the material behaves elastically, as usually assumed, the energy supplied to totally break the specimen is G_F times the area of the broken surface. However, in the more general case of Fig. 7.1.2 in which energetic dissipation takes place in the bulk of the specimen, the energy supply to completely break the specimen is larger than the cohesive energy, and G_F in Fig. 7.1.2 amounts only to the area $B'BPAA'$, not to the total area.

The tensile strength and the fracture energy can be used to make the softening curve nondimensional. The tensile strength f'_t is used to reduce the stress to a nondimensional form. The magnitude that reduces w to a nondimensional form is the ratio G_F/f'_t, which has the dimension of length; for our future use, we set

$$w_{ch} = \frac{G_F}{f'_t} \tag{7.1.11}$$

and then define the reduced stress $\hat{\sigma}$ and reduced crack opening $\hat{w}$ as

$$\hat{\sigma} = \frac{\sigma}{f'_t}, \quad \hat{w} = \frac{w}{w_{ch}} \tag{7.1.12}$$

Then, the softening curve can be written in the form

$$\hat{\sigma} = \hat{f}(\hat{w}) \tag{7.1.13}$$

where $\hat{f}(\hat{w})$ is now a dimensionless function related to the usual softening function by the obvious relations

$$f(w) = f'_t \hat{f}\left(\frac{w}{w_{ch}}\right), \quad \hat{f}(\hat{w}) = \frac{1}{f'_t} f(w_{ch}\hat{w}) \tag{7.1.14}$$

For example, the dimensionless version of the exponential softening described by Eq. (7.1.10) in the previous example reduces to

$$\hat{\sigma} = e^{-\hat{w}} \tag{7.1.15}$$

Note that the dimensionless function has, in this case, no free parameters; in general, the dimensionless function will have two free parameters less than the complete softening function. Moreover, from (7.1.4) and (7.1.8), the dimensionless softening function is easily shown to verify

$$\hat{f}(0) = 1, \quad \int_0^\infty \hat{f}(\hat{w})\, d\hat{w} = 1 \tag{7.1.16}$$

This nondimensionalization is only a normalization procedure that allows a systematic way of comparing softening curves. It is no more than a change of scale on both axes, so that by choosing appropriate drawing scales the actual σ–w curve and the reduced one look exactly the same, as shown in Fig. 7.1.7. The second of (7.1.16) implies that the area between the dimensionless softening curve and the axes is 1, as indicated in the figure.

Recently it became apparent (Planas and Elices 1992b; Bažant and Li 1995b; Planas, Guinea and Elices 1995, 1997) that one essential parameter controlling interesting structural properties is the initial slope

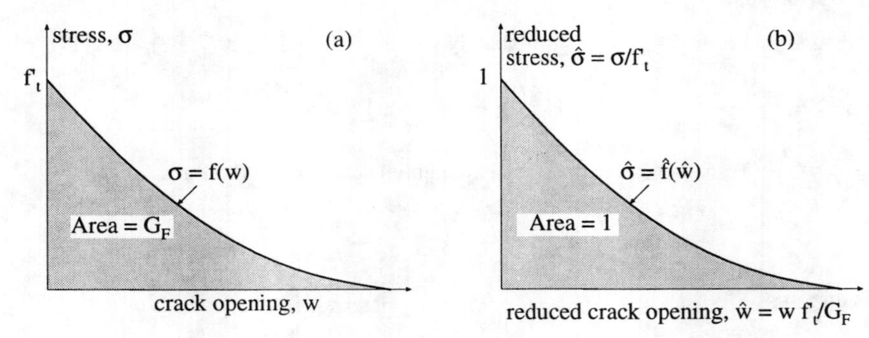

Figure 7.1.7 The actual softening curve (a) and its dimensionless version (b) look exactly the same for appropriate drawing scales.

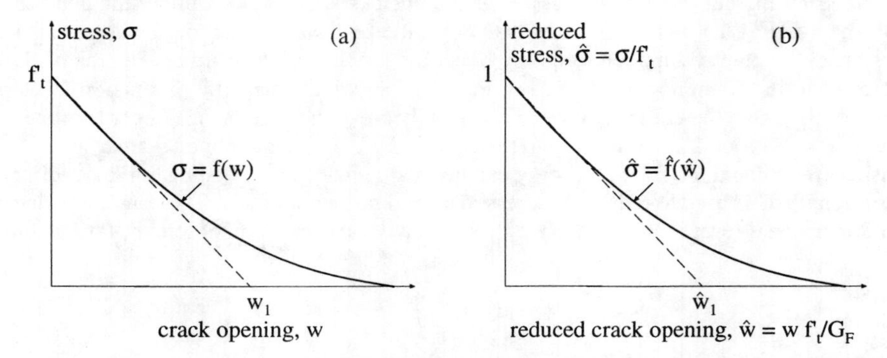

Figure 7.1.8 Softening curve and its initial linear approximation: (a) in natural scale; (b) in dimensionless representation.

the softening curvecurve. This is so because in situations where no point in the structure undergoes large softening, a linear approximation to the softening curve, shown in Fig. 7.1.8, is very accurate.

This might seem a fact of only rather academic interest, but it turns out that for notched specimens (or structures) of not too large sizes, and for unnotched structures of any size, the peak load occurs before any large softening occurs. This was observed in calculations of the modulus of rupture for bilinear softening by Alvaredo and Torrent (1987) and confirmed numerically by Planas and Elices (1992b) and Li and Bažant (1994a) for other cases; a definitive proof was given in two recent papers by Planas, Guinea and Elices (1995, 1997) and will be discussed later in Section 7.2.2 and Chapter 6.

The initial linear approximation to the softening is completely defined by its vertical and horizontal intercepts, f_t' and w_1. From these two values a parameter similar in nature to ℓ_{ch} (Eq. 5.2.11) was defined by Planas and Elices (1992b) as

$$\ell_1 = \frac{Ew_1}{2f_t'} \tag{7.1.17}$$

As a rule of the thumb, for ordinary concrete $w_1 \approx G_F/f_t$ (although it can be as low as $0.6G_F/f_t$ for a particular concrete; see exercise 7.5 at the end of the next section); accordingly, $\ell_1 \approx 0.5\ell_{ch}$.

7.1.4 Extensions of the Cohesive Crack Model

The foregoing formulation of the cohesive crack model is its simplest version, but it is by no means the only one suitable. There are at least five possible extensions that are worth considering: **(1)** extend the formulation so that a singularity at the cohesive crack tip be admissible; **(2)** accept a behavior other than isotropic linear elastic in the material around the crack; **(3)** introduce dependence on stress triaxiality; **(4)** introduce a fully consistent mixed-mode formulation; and **(5)** introduce time-dependence in the cracking behavior and in the bulk material.

The first extension has already been in use in many models dealing with brittle matrix composites, particularly with fiber reinforced ceramics and cements. This extension went under the name of *bridged cracks*, in which the long range cohesive stresses are called the *bridging stresses*. It will be analyzed with some detail in the remainder of this section together with the second aforementioned extension, which may be essential in some materials displaying large inelastic strains before the peak in tension, such as some ductile fiber reinforced brittle matrix composites.

To the authors' knowledge, the third extension has never been considered at a theoretical level, and is one of the most frequently heard criticisms to the cohesive crack model. However, there is no basic objection to introducing a dependence of the softening curve on the triaxial stress and strain state. This may require further hypotheses regarding the loading-unloading behavior, and may lead to a model that may be difficult to verify experimentally, because the only data so far available in biaxial tests concern the effect on the strength, not on the post-peak (stable crack opening) behavior. Interesting research in this direction has been done by Tschegg, Kreuzer and Zelezny (1992) on wedge-split cubes subjected to compression parallel to the crack plane. However, their results have not yet led to a theoretical formulation, and further work is needed.

The fourth aforementioned extension is essential since a complete vectorial formulation is required before general purpose programs may be confidently used. Even though there is at present a general agreement that the fracture initiates in pure opening mode (mode I, no shear), it is obvious that crack sliding displacement may take place after the crack opens and before the fracture is complete. The testing procedures in this field are tremendously difficult, but they go ahead of the theoretical formulation (van Mier, Nooru-Mohamed and Schlangen 1991; Hassanzadeh 1992). To the authors' knowledge a complete cohesive crack model relating the traction vector on the crack faces to the history of the crack displacement vector is just starting to emerge, coming from the continuum field, specifically from the microplane and multiplane theories that we will analyze in Chapter 14. Indeed, Červenka and Saouma (1995) and Weihe and Kröplin (1995) have proposed general cohesive models relating the traction vector $\mathbf{t}$ on the crack surface to the evolution of the displacement jump vector $\mathbf{w}$ based on the developments of Carol and Prat (1991) and Carol, Bažant and Prat (1992).

The last extension, i.e., the time dependent cohesive crack models, is dealt with in Chapter 11.

7.1.5 Cohesive Cracks with Tip Singularity

At the first glance it may seem that a cohesive crack with tip singularity is a model radically different from Hillerborg's, in which no singularity is present. However, Elices and Planas (1989) pointed out that such a singularity may very well be just a mathematical approximation of a softening curve consisting of an initial spike with large strength, followed by a long tail with a much lower cohesive stress as shown in Fig. 7.1.9a. This kind of softening is typical for fiber reinforced composites, where the spike corresponds to the failure of the matrix, and the tail to the fiber-bridging stress. Mathematically speaking, the singularity arises when the initial spike is approximated by a Dirac δ-function, as depicted in Fig. 7.1.9b. Smith (1995) showed that there is indeed a smooth transition from the nonsingular model to the singular one. Therefore, this extension is sound and does not require further developments, except for quantifying how narrow the spike must be for the singular approximation to work within a certain degree of accuracy.

In the cohesive crack model with stress singularity, the total fracture energy G_F is split into two parts, one associated with the singularity G_{cc} (the area of the spike in Fig. 7.1.9) and the other associated with the remainder of the softening curve, G_{Fs}. The fracture energy of the singularity is related to a critical stress intensity factor K_{Icc} by Irwin's relationship, and so we can write

$$G_F = G_{Fs} + G_{cc} = G_{Fs} + \frac{K_{Icc}^2}{E'} \tag{7.1.18}$$

The fracture energy G_{Fs} is the area enclosed by the regular part of the softening curve, which we still write as $f(w)$.

7.1.6 Cohesive Cracks with Bulk Energy Dissipation

As is clear from Fig. 7.1.2, deviation from elastic behavior of the bulk material was considered from the very beginning by Hillerborg as a realistic possibility, although the slight increase in accuracy did

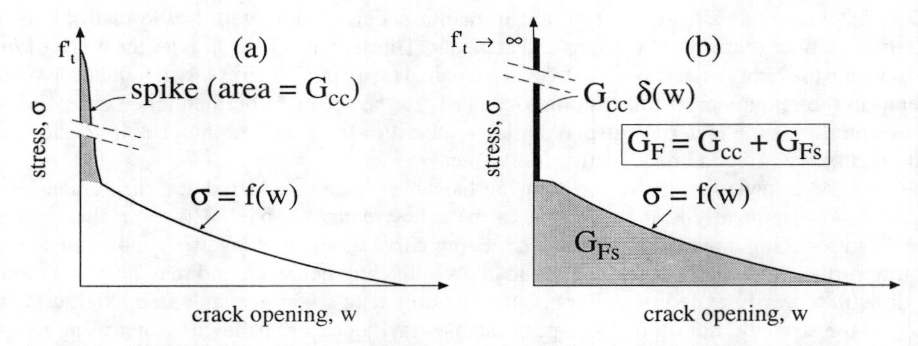

Figure 7.1.9 Cohesive crack with crack tip singularity: (a) actual softening curve with a strong initial spike; (b) mathematical approximation with a Dirac δ-function.

not generally compensate for the increase in computational complexities, at least for materials such as concrete. However, some highly ductile fiber reinforced materials may require considering the prepeak bulk dissipation, even if the matrix is brittle.

A cohesive crack model with bulk dissipation can be readily formulated by retaining the hypothesis regarding the cohesive crack model, as previously described, while relaxing the hypothesis of linear elastic behavior of the bulk material. Unfortunately, in doing so, one of the most appealing features of the cohesive crack model, namely, the coincidence of G_F with mean energy dissipation per unit area of complete fracture is lost, because in the total energy dissipation there are contributions from both the crack surface and the bulk of the material surrounding it.

The general aspects of the energy balance in such kind of materials were analyzed by Elices and Planas (1989) and are summarized next. Consider first the ideal tensile test in Fig. 7.1.2, in which we can calculate the total work required to fracture the specimen from the area under load-elongation curve. This area may be split in two parts, the area $OPBB'O$ which corresponds to the energy uniformly distributed through the bulk of the material, and the area $B'BPAA'B'$ which corresponds to the cohesive crack itself. Since only one crack is formed, the second part is just the cross section area times the fracture energy of the cohesive crack G_F. Since the first part is uniformly distributed in the volume, we can write:

$$\mathcal{W}_F = LA\,\gamma_U + AG_F \tag{7.1.19}$$

where L and A are the length and cross sectional area of the specimen, respectively, and γ_U is the energy supplied to a unit volume of the material when it is loaded up to peak load and then unloaded. Obviously γ_U is a material property, independent of G_F, with dimensions of energy per unit volume. For convenience, we can rewrite it in the form

$$\gamma_U = \alpha_U \frac{G_F}{\ell_{ch}} = \alpha_U \frac{f_t'^2}{E} \tag{7.1.20}$$

where α_U is dimensionless and also a material property; $\alpha_U = 0$ for the traditional cohesive crack models without bulk dissipation. For concrete, where the inelastic strain at peak load is typically less than the corresponding elastic strain, $\alpha_U < 1$, although in many cases it can be less than 0.3.

If we substitute the foregoing expression for γ_U into the equation for the fracture work, and then calculate the mean fracture energy G_{Fm}, we get

$$G_{Fm} = G_F\left(1 + \alpha_U \frac{L}{\ell_{ch}}\right) \tag{7.1.21}$$

in which we see that the mean fracture energy for direct tension is the true surface energy G_F plus a second term that depends on the size of the specimen. For other geometries, the split is similar and, for geometrically similar specimens, we have (Elices and Planas 1989)

$$G_{Fm} = G_F\left[1 + \alpha_U\,\eta\!\left(\frac{D}{\ell_{ch}}\right)\right] \tag{7.1.22}$$

where the size effect function $\eta(D/\ell_{ch})$ depends on geometry as well as on material properties and must be calculated for every particular case. Only two limiting cases allow a general treatment : the asymptotic limits for very small and very large sizes. For very small sizes, since the volume vanishes, $\eta(D/\ell_{ch})$ should also vanish. For very large size in notched specimens, the LEFM limit must be fulfilled. In such a limit, the volume of material undergoing unloading when the crack propagates selfsimilarly is finite and equal (for a given material) to a constant *independent of geometry.* This is so because, if the size is large enough, the fields surrounding the cohesive zone and the inelastic unloading zone are the asymptotic ones, which depend only on K_I, not on the details of the geometry. Therefore, for similar notched specimens of infinite size, we must have (Elices and Planas 1989):

$$J_c = \frac{K_{Ic}^2}{E'} = G_F\,(1 + \alpha_U \eta_\infty) \tag{7.1.23}$$

where η_∞ is a geometry-independent constant, which can be calculated for any particular combination of the softening curve and the prepeak stress-strain relation.

Such a model was formulated using an elastoplastic hardening model for the material with a Rankine criterion and an associative flow rule (Guinea 1990; Guinea, Planas and Elices 1990; Planas, Elices and Guinea 1992). The computations using an approximate perturbation method showed that, for a typical concrete, the influence of the inelastic strain around the cohesive crack was small. When expressed as the dissipated energy, the influence was below 5% of the total energy for notched specimens of all sizes (i.e., $\alpha_U \eta_\infty < 0.05$).

For ductile brittle matrix composites, the use of perturbation approximation is beyond question and a full nonlinear analysis is required. Even if the deviation from linearity is small, elastic isotropy might not be adequate for some unidirectional fiber reinforced composites. However, these situations do not rule out the cohesive crack concept. They only call for an extension of the equations and of the methods of analysis.

Exercises

7.1 Give ℓ_1 as a function of ℓ_{ch} for the case of exponential softening, (a) using as the linear approximation the initial tangent to the curve, and (b) using a secant approximation defined by the point $(f'_t, 0)$ and a point with the stress $\sigma = 0.75 f'_t$. [Answer: (a) $0.500\ell_{ch}$; (b) $0.575\ell_{ch}$.]

7.2 A softening curve has the form $\sigma = c_1(w_c - w)^m$ for $w \le w_c$, where c_1, w_c, and m are constants. Rewrite the equation in terms of f'_t, G_F, and w_c. Rewrite the equation in dimensionless form $\hat{\sigma} = \hat{f}(\hat{w})$.

7.3 For the softening curve in the previous exercise, determine ℓ_1 by both the tangent approach and the secant approach as in the first exercise, in terms of $\hat{w}_c = w_c f'_t / G_F$. Give numerical values for $\hat{w}_c = 11$.

7.4 The softening curve $\sigma = f'_t[1 - (w/w_c)^m]$ for $w < w_c$ was used for concrete by Reinhardt (1984). Determine the fracture energy. Rewrite the equation in a dimensionless form. What is the initial linear approximation for this particular softening?

7.2 Cohesive Crack Models Applied to Concrete

7.2.1 Softening Curves for Concrete

The main ingredient of the cohesive crack model is the softening curve of the material, which, in a sense, replaces the stress-strain curve in theories such as plasticity. Strictly speaking, every concrete mix and, in general, every material has its own softening curve which should be determined from experiments on this particular material. Therefore, a softening curve valid for all concrete does not exist.

However, a number of experiments show that the softening curve is similar in shape for different mixes of ordinary concrete. This similarity was first noted by Petersson (1981) who was also the first to carry out stable tensile tests to determine the softening curve of concrete. Fig. 7.2.1a shows the softening curves found by Petersson for four different concrete mixes, which, as expected, differ one from another. When these results are plotted nondimensionally (by dividing σ by f'_t and w by w_{ch}), the curves in

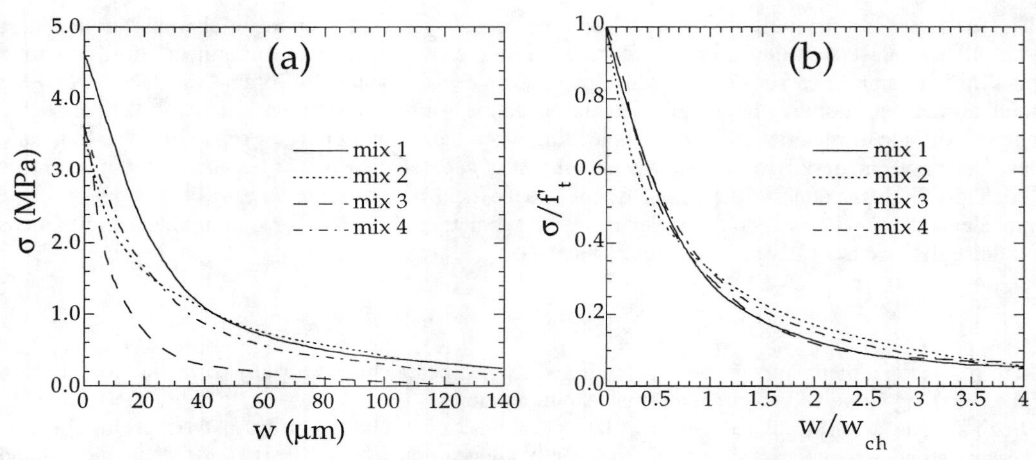

Figure 7.2.1 Experimental results for four concrete mixes in tension: (a) softening curves, and (b) dimensionless plots of the experimental softening curves (adapted from Petersson 1981).

Fig. 7.2.1b are obtained. They show that the shapes of the various curves are very close to each other. An analogous conclusion was later reached by the research group at Delft University (Cornelissen, Hordijk and Reinhardt 1986a).

Most of the experimentally determined softening curves look similar to those in Fig. 7.2.1b (they are rather smooth and nonlinear). However, to perform the analysis and to infer general trends for particular structures, it is convenient to introduce simplified analytical expressions that approximate the true softening curve. Many such curves have been used in theoretical, analytical, and numerical analysis ranging from very simple to rather sophisticated equations. We review the most important of them in the following, having in mind that any one of them, even the roughest, gives much better predictions of the observed structural behavior than either plasticity or LEFM.

The simplest softening available is the rectangular softening introduced in Fig. 7.1.5 and reproduced, just for comparison, in Fig. 7.2.2a. The rectangular softening captures the main trends of the fracture processes and the size effect, but it usually overestimates the strength of normal-sized specimens and structures (depths of the order of decimeters). Next in simplicity is the linear softening depicted in Fig. 7.2.2a. This kind of softening is the obvious choice when no further data regarding the actual material behavior are available. It was first used by Hillerborg, Modéer and Petersson (1976) in the pioneering paper on cohesive cracks applied to concrete. Later it turned out that, for concrete, this softening was not very realistic either, because it causes the predicted structure strength to be too high. Thus, Petersson (1981) proposed the bilinear softening curve shown in Fig 7.2.2a. This curve has its kink point fixed at $(0.8, 1/3)$ (in the dimensionless representation) and becomes zero for $\hat{w} = \hat{w}_c = 3.6$ (i.e., for $w = 3.6G_F/f_t$).

Since 1981, the bilinear curves have been accepted as reasonable approximations of the softening curve for concrete, although there is no agreement about the precise location of the kink point (which is not unusual because different concrete mixes will have different softening curves). Today there are methods, some quite simple, to identify any bilinear softening to fit particular experimental data (see Section 7.3). Fig. 7.2.2b shows some of the bilinear curves proposed in the literature, including those proposed in the CEB-FIP Model Code (CEB 1991). The curves proposed by Wittmann, Roelfstra et al. (1988) and Rokugo et al. (1989) are nearly coincident and display a tail longer than that of Petersson's curve. The curves proposed by the CEB-FIP Model Code (CEB 1991) depend on the maximum aggregate size and have even longer tails (the labels 8, 16, and 32 in the curves stand for the maximum aggregate size of the concrete mix, in mm).

Although a bilinear curve may suffice for most practical purposes, other curves, some smooth, have also been proposed. Various mathematical forms have been proposed, among which the exponential (Gopalaratnam and Shah 1985; Cornelissen, Hordijk and Reinhardt 1986a,b ; Planas and Elices 1986a, 1990a), the power-law (Reinhardt 1984), and the trilinear (short horizontal plateau followed by bilinear softening; Cho et al. 1984; Liaw et al. 1990) are the most common.

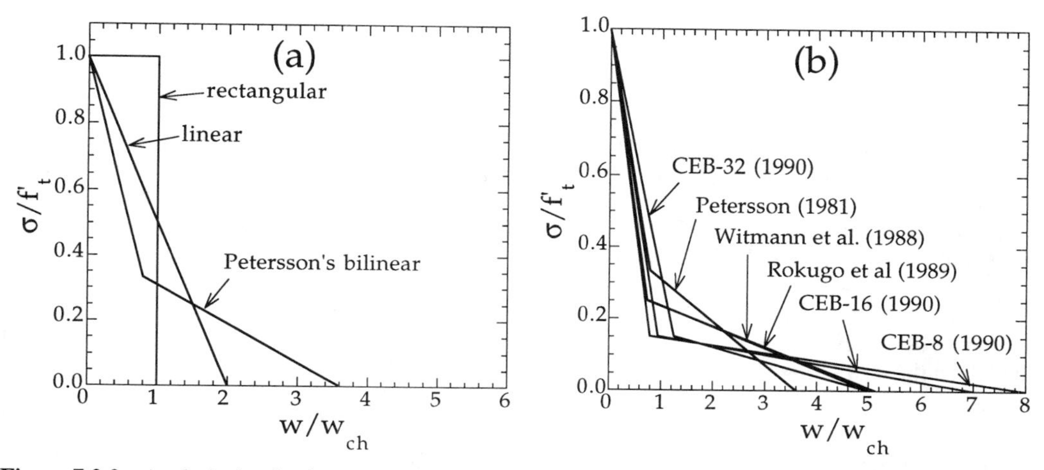

Figure 7.2.2 Analytical softening curves. (a) Rectangular, linear, and Petersson's bilinear curves. (b) Various bilinear curves proposed in the literature; the curves labeled CEB-8, 16, 32 correspond to different maximum aggregate sizes, namely 8,16, and 32 mm.

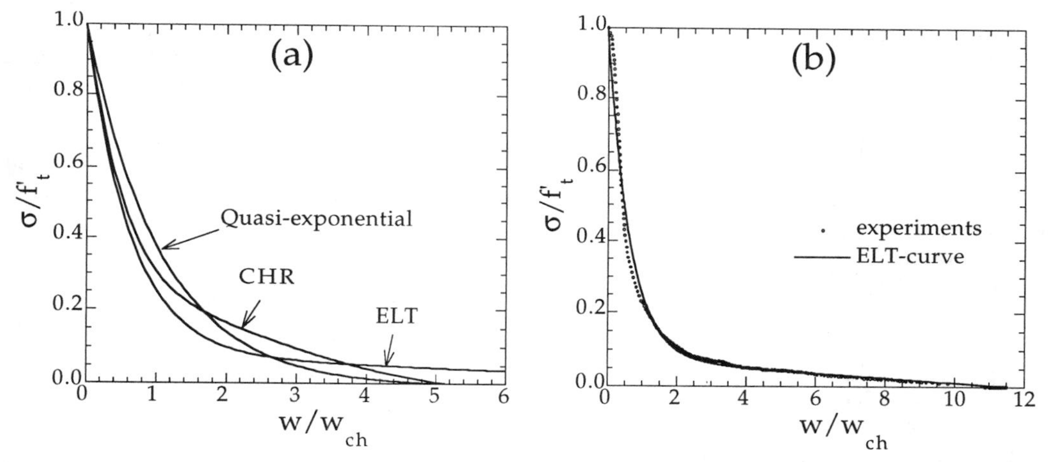

Figure 7.2.3 (a) Quasi-exponential, Cornelissen-Hordijk-Reinhardt (CHR) and extra-long tail (ELT) curve. (b) Comparison of the experimental results of Rokugo et al. (1989) with the analytical ELT-curve.

Cornelissen, Hordijk and Reinhardt (1986a,b) proposed the following curve of the exponential type (that we call CHR-curve, Fig. 7.2.3a):

$$\hat{\sigma} = \left(1 + 0.199\hat{w}^3\right) e^{-1.35\hat{w}} - 0.00533\hat{w}\,, \qquad \text{for} \quad 0 < \hat{w} \leq 5.14 \tag{7.2.1}$$

for which it is understood that $\hat{\sigma} = 0$ for $\hat{w} > \hat{w}_c = 5.14$ (i.e., $w > 5.14 G_F/f'_t$). The area under this dimensionless curve is very approximately 1 (the small difference is due to roundoff error in writing the constants appearing in the equation). Hordijk (1991) analyzed the experimental results from 12 different sources and concluded that this equation can yield a reasonable approximation for all of them.

The quasi-exponential softening curve proposed by Hillerborg and used systematically by Planas and co-workers to investigate the influence of the length of the tail of the softening curve on various kinds of structural response is given by

$$\hat{\sigma} = (1 + c_1)e^{-c_2\hat{w}} - c_1 \qquad \text{for} \quad \hat{w} < \hat{w}_c \tag{7.2.2}$$

for which it is understood that $\hat{\sigma} = 0$ for $\hat{w} \geq \hat{w}_c$; $\hat{w}_c$ is a material constant defining the length of the softening curve, and c_1 and c_2 are constants related to each other and to $\hat{w}_c$ by the conditions

$$c_2 = 1 - c_1 \ln \frac{1 + c_1}{c_1}, \quad \hat{w}_c = \frac{1 - c_2}{c_1 c_2} \tag{7.2.3}$$

These conditions just state that the area enclosed by the curve and the axes is 1 and that the stress becomes 0 for $w = w_c$. With this, the foregoing quasi-exponential softening becomes a uniparametric family of curves with parameter $\hat{w}_c$ which has the interesting property that when $\hat{w}_c$ spans the interval $[1, \infty)$, the curves smoothly change from the rectangular softening ($\hat{w}_c = 1$) to linear softening ($\hat{w}_c = 2$) and to full exponential ($\hat{w}_c = \infty$). Fig. 7.2.2b shows the shape of the curve for $\hat{w}_c = 5$.

Some experimental data indicate that the tail of the softening curve can be extremely long, with w_c as large as $12G_F/f'_t$. For such cases, a bilinear curve seems awkward in the central part. Then a smooth curve consisting of the sum of a straight line and an exponential seems preferable. The equation for such a curve— called extra long tail curve, ELT (Planas and Elices 1992b)— is

$$\hat{\sigma} = 0.0750 - 0.00652\hat{w} + 0.9250e^{-1.614\hat{w}} \qquad \text{for } \hat{w} \leq 11.5 \tag{7.2.4}$$

for which it is understood that $\sigma = 0$ for $\hat{w} > 11.5$. Fig. 7.2.3b shows a comparison of the experimental results by Rokugo et al. (1989) with the foregoing equation.

7.2.2 Experimental Aspects

The foregoing curves are dictated, in general, by analytical convenience and by limited amounts of experimental data, and must be used with care. There is no general agreement about the "best" softening curve and, in principle, such a curve should be determined for each particular concrete (or other quasi-brittle material).

In principle, a direct test analogous to the *ideal* tensile test previously used to introduce the cohesive crack model (Figs. 7.1.1 and 7.1.2) should provide the means of getting the softening curve experimentally. However, such an approach is more complicated in practice. In the first place, it is difficult to test prismatic or cylindrical specimens because the location of the cohesive crack is not known *a priori*; moreover, concrete heterogeneity may promote multiple cracking as sketched in Fig. 7.2.4a; this may happen even in tapered specimens, as shown in Fig. 7.2.4b (after Planas and Elices 1985). Thus, although a few results have been obtained for unnotched dog bone specimens (Planas and Elices 1985; Guo and Zhang 1987; Phillips and Binsheng 1993), most of the published results are for notched specimens, either with shallow sharp notches, as shown in Fig. 7.2.4c (Cornelissen, Hordijk and Reinhardt 1986a,b ; Wecharatana 1986; Phillips and Binsheng 1993) or with deeper smooth notches as shown in Fig. 7.2.4d (Petersson 1981, Carpinteri and Ferro 1994).

However, this does not completely solve the problem. The reason is twofold:

1. Even if a single central crack is formed, the specimens tend to shift to an asymmetric mode of fracture, as sketched in Fig. 7.2.4e–f, due either to rotations of the supports or to internal elastic rotations in the specimen itself. The existence of such rotations have been demonstrated both experimentally and theoretically (Rots 1988; Hordijk 1991; van Mier and Vervuurt 1995; van Mier, Schlangen and Vervuurt 1996). A simple analytical solution of the asymmetric fracture evolution was given by Bažant and Cedolin (1993).

2. Even if the rotations are avoided using very short specimens and a very stiff machine —or a special servo-controlled system as used by Carpinteri and Ferro (1994)— the two cracks growing from both sides of the (Fig. 7.2.4g) tend to avoid each other when they approach and overlap, and so one never has a true single crack (van Mier and Vervuurt 1995).

Since *true* tensile tests are so difficult to carry out, alternative, indirect methods have been put forth by many researchers. In such methods, a single crack is made to run in a notched specimen (typically three-point bending or compact specimens) while measuring load and displacement (load-point displacement or CMOD), and from the analysis of the results a softening curve is inferred.

Among these methods, that proposed by Li, Chan and Leung (1987) is based on a J-integral technique; however, although theoretically elegant, it can hardly be applied to concrete because it is based on subtracting the load-displacement curves for specimens with only slightly different notches. This procedure multiplies the experimental errors so much that the final result becomes meaningless.

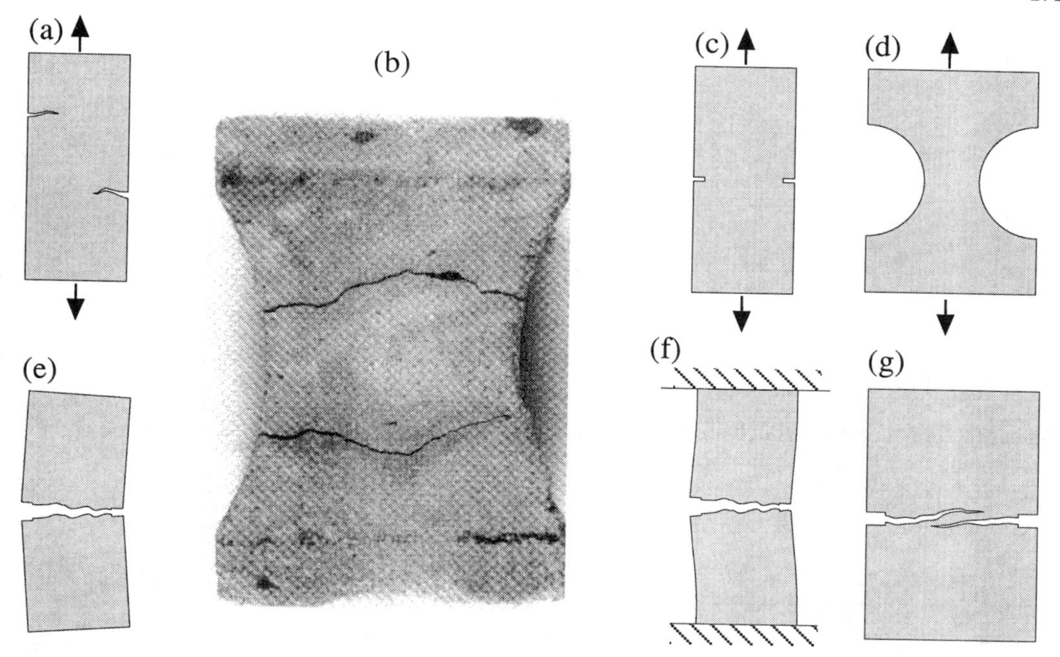

Figure 7.2.4 (a) Sketch of multiple cracking in prismatic specimens. (b) Example of double crack in slightly tapered specimens (after Planas and Elices 1985). Notched specimens: (c) shallow sharp notches; (d) deep smooth notches. Rotation of crack faces: (e) with rotation of specimen ends; (f) accommodated by internal deformation. (g) Cracks growing from notches to the center and their overlap when they meet.

The remaining indirect methods are in essence inverse procedures that identify the softening curve by comparing the load-displacement curve obtained from the test to the load-displacement curves theoretically predicted; the softening curve is modified, by trial and error or by some kind of optimization technique, until the fit is good enough. The difference between the various proposed methods is operational rather than conceptual. Wittmann and Mihashi postulate a 4-parameter bilinear softening curve and determine its parameters by optimum fitting of load-deflection curves measured in stable tests of notched specimens (Roelfstra and Wittmann 1986; Wittmann, Rokugo et al. 1987; Mihashi 1992). A similar procedure has been used by Ulfkjær and Brincker (1993). Another procedure, based again on a cohesive crack model, was recently proposed by Uchida et al. (1995); in this case, a multi-linear softening function is obtained from the load-displacement data from a three-point bending test of a notched beam by means of an optimization algorithm.

A simplified inverse procedure has recently been devised by Guinea, Planas and Elices (1994b). This procedure is based on a bilinear softening whose four parameters are deduced from two tests —cylinder splitting and three-point bending of notched beams. The application of this method has been brought to a completely analytical form and involves quasi-independent determination of f_t, G_F, w_1 and the center of gravity of the softening curve from which w_c can be immediately computed. This method is explained in detail in Section 7.3.

Although a complete characterization of the cohesive crack model requires the full softening curve, partial information about it sometimes suffices for making predictions of engineering value. This is the case for peak load determination for very large sizes and for relatively small sizes. In those situations, simplified experiments can provide valuable information.

When the size of the structure is much larger than the cohesive zone size, the limit case of LEFM is a reasonable first order approximation to the cohesive crack model. In this limit (in which both the crack length and the uncracked ligament must be very large), the peak load is fully controlled by the fracture energy G_F, and thus an experimental set-up to determine only this parameter is of practical interest.

The fracture energy can be measured using the procedure described in the RILEM (1985) draft recommendation which consists in measuring the external work supply required to break the specimen of a

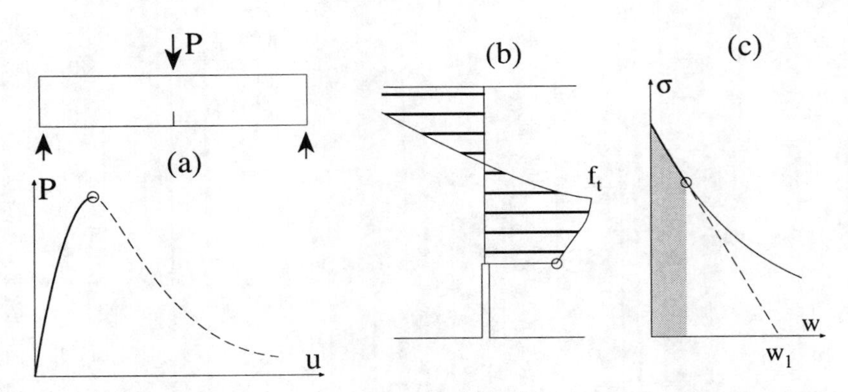

Figure 7.2.5 Notched beam in three-point bending: (a) peak load on the load-displacement curve; (b) stress profile along the ligament at peak load; (c) softening experienced at peak load by the material at the notch tip (from Elices and Planas 1996).

stable test on a notched beam. The work supplied divided by the initial ligament area equals G_F. Details about the experimental procedures are given in Section 7.3.

However, it appears that a direct application of this procedure delivers values of G_F that are size-dependent. Elices, Guinea and Planas (1992) have showed that a number of spurious energy dissipation sources may be present in a typical test: hysteresis of the testing arrangement, dissipation in the bulk of the specimen, dissipation by crushing at the supports, and neglect of the final portion of the load-displacement curve. When all these factors are properly taken into account, the size effect of the experimental values appears to be strongly reduced (Guinea, Planas and Elices 1992; Planas, Elices and Guinea 1992; Elices, Guinea and Planas 1992).

The size-dependence of G_F observed for some concretes deserves further research. Three aspects deserve study. First the cohesive crack model may have to be extended by including parameters not yet considered, such as multiaxiality; second, the possibility of an additional size effect due to the random nature of the material (statistical size effect) should be examined; and third, the cohesive crack model might be intrinsically inadequate and a completely new approach might be needed. At this moment, it seems that the experimental evidence is not sufficient to decide these questions. More extensive and accurate experimentation is needed.

In many practical cases, the peak load is completely determined by the tensile strength and the slope of the initial portion of the softening curve. This is so for notched specimens of laboratory sizes (Guinea, Planas and Elices 1994a) and for unnotched specimens of all sizes (Planas and Elices 1992b; Planas, Guinea and Elices 1995). Fig. 7.2.5 sketches the reason for this property: for peak loads of relatively small specimens, the stress profile along the ligament is as shown in Fig. 7.2.5a, which means that no point on the specimen has softened further than the circled point in Fig. 7.2.5b. Therefore, at the peak load, the results for the actual softening curve (full line) and for the linear approximation (dashed line) are indistinguishable.

This property can be used to determine the initial straight part of the softening curve (which depends on only two parameters) from the measurement of the peak load of two specimen types. Up to now, the authors have used the cylinder splitting strength (which appears to give a good estimate of the tensile strength) and the peak load for a notched three-point bend beam, which yields the horizontal intercept w_1 (see Section 7.3; Guinea, Planas and Elices 1994b).

7.2.3 Computational Procedures for Cohesive Crack Analysis

The growth of a cohesive crack is a highly nonlinear process which usually requires numerical procedures to find the solution. Two approaches can, in principle, be used: **(1)** special purpose programs devised to deal with mode I problems, and **(2)** general purpose methods devised to deal with arbitrary crack growth.

The mode I problem, sketched in Fig. 7.2.6, is often encountered in experimental research. Although

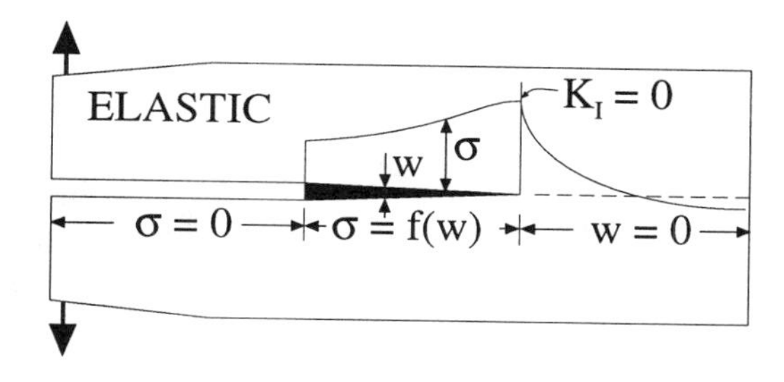

Figure 7.2.6 Cohesive crack model in mode I with the boundary conditions along the central line: zero stress on the notch, softening equation over the cohesive zone (black triangle), zero crack opening along the uncracked ligament, and zero stress intensity factor K_I at cohesive zone front.

general purpose approaches can be used, inverse methods usually call for fast computation, so that a large number of hypotheses could be tested in a short time. The fastest computational procedures use boundary integral (BI) formulations which can be reduced to systems of equations with a small number of degrees of freedom compared to volume methods (finite elements, for example).

Some of the methods of this family, that we call pseudo-boundary-integral methods (PBI), do not use explicit integral formulation with posterior discretization of the integrals, but use an *a priori* discretization of the kernels relying on elastic computations usually coming from finite element calculations. Due to relative simplicity and ease of use, these methods are briefly described in Section 7.4, which starts with the method first used by Petersson (1981) and develops further refinements by Planas and Elices (1991a) and the practical implementation of the *smeared-tip superposition* method which, based on an initial analytical approach by Planas and Elices (1986b, 1992a), produces a very fast computational tool. This method has been incorporated into a program called *Splitting-Lab*, which deals with the most usual laboratory testing geometries and softening curves.

True boundary integral methods, with analytically defined kernels, have been used extensively to analyze cohesive cracks in many materials (ceramics, composites, polymers) in very special cases, basically semi-infinite cracks and center cracks in an infinite body. Recently, new developments have been made that make use of general properties of the BI formulation to provide results of practical as well as theoretical interest, with applications to concrete (Planas and Elices 1986b, 1991a, 1992a, 1993a; Li and Bažant 1994a, 1996; Bažant and Li 1994a, 1995a, 1995b). These methods will be reviewed in Section 7.5.

Among the general methods, the boundary element method (BEM) is very effective to deal with the cohesive crack since the bulk material remains elastic. However, only few attempts are recorded in the literature using such method (Ohtsu and Chahrour 1995; Saleh and Aliabadi 1995), and all of them may be considered somewhat preliminary. In contrast, a huge amount of computational effort has been devoted to deal with cohesive cracks using the finite element method (FEM). Three basic approaches may be found in the FEM field: **(1)** discrete interelement crack approach, **(2)** smeared crack approach, and **(3)** discrete intraelement crack approach.

In the so-called discrete crack approach, the crack extends between elements as shown in Fig. 7.2.7a and the cohesive forces are simulated either by nonlinear mixed boundary conditions —which requires special purpose routines not available in most conventional codes— or using interface elements connecting the nodes on both sides of the crack. Many commercial FEM codes have nonlinear spring elements with user-defined stress strain curves that can be easily used to simulate crack propagation when the crack path is known *a priori*, which is the case for mode I fracture specimens. These elements can even be used for mixed mode crack growth if a reasonable guess of the crack path can be made, as illustrated in Fig. 7.2.7b for nonsymmetric three-point bending of a notched beam.

As pointed out, this simple procedure requires a good guess of the crack path. If the crack path is not known in advance, remeshing techniques are required for this kind of approach. Unfortunately, to the authors' knowledge a fully general FEM code implementing discrete interelement cohesive cracks is not yet available, although the basic technology seems to exist since automatic remeshing and crack growth have been implemented in a few FEM codes.

Ingraffea and co-workers at Cornell University developed various programs to analyze crack growth in LEFM with automatic remeshing (FEFAP, in the eighties, and later FRANC, which is free software;

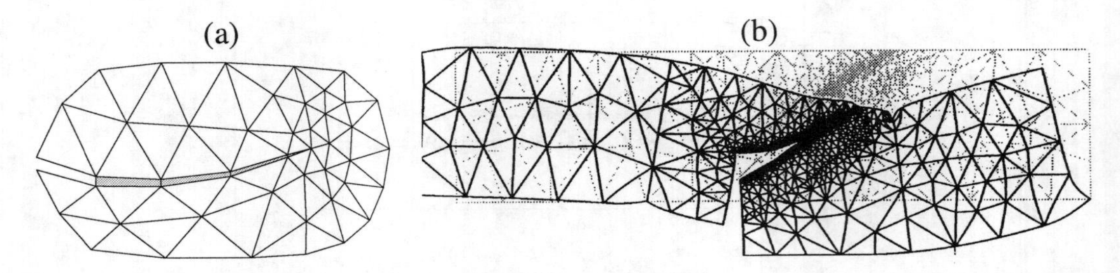

Figure 7.2.7 (a) Interelement discrete crack model. (b) Simulation of cohesive crack growth in mixed mode using a commercial FEM code (ABAQUS) with nonlinear spring elements (after Gálvez, personal communication, 1996).

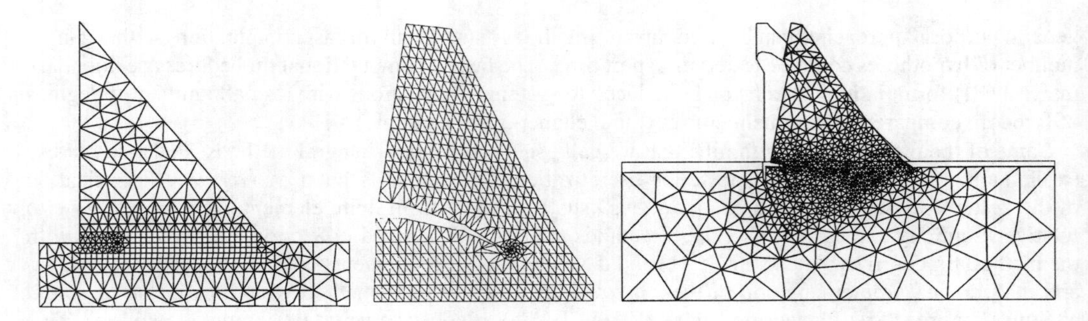

Figure 7.2.8 Typical finite element meshes at intermediate computing steps for analysis of crack growth in a concrete dam: Left from Gálvez, Llorca and Elices 1996, using FRANC 2D; middle from Valente 1995; right from Červenka and Saouma 1995, using MERLIN.

Ingraffea and Gerstle 1985; Ingraffea et al. 1984; Ingraffea and Saouma 1984; Wawrzynek and Ingraffea 1987). Although a special algorithm to simulate arbitrary cohesive crack propagation in FRANC was proposed by Bittencourt, Ingraffea and Llorca (1992), the procedure is not yet fully operative.

MERLIN, a program developed for EPRI by Saouma at Boulder also deals with propagation of LEFM cracks and cohesive cracks using remeshing and interface elements (Reich, Červenka and Saouma 1994; Červenka 1994; Červenka and Saouma 1995). A special purpose FEM program was developed by Carpinteri and co-workers at the Politecnico di Torino (Carpinteri and Valente 1989; Bocca, Carpinteri and Valente 1990, 1991, 1992; Valente 1995). Fig 7.2.8 shows typical finite element meshes at intermediate computing steps for various cases analyzed with the foregoing programs. Other research groups have also developed interface elements to analyze discrete crack growth (e.g., Larsson and Runesson 1995).

A different approach to cohesive crack analysis is the smeared crack concept. In this method, conventional finite element formulations are used with element-dependent stress-strain relations obtained by smearing the crack opening displacement w in the element intersected by the crack, as illustrated in Fig. 7.2.9a. This procedure has the advantage, at least in principle, that the mesh topology is unchanged during crack growth, as shown in the example in Fig. 7.2.9b. The formulation resulting from the smeared crack concept is very close to the numerical implementation of the so-called crack-band models, which have a wider scope and have been implemented in several commercial finite element codes; their analysis, together with a comparison with the cohesive crack model, will be deferred to Chapter 8.

Recently, a discrete crack approach in which the crack runs through the element as sketched in Fig. 7.2.10a has been developed. This approach emerged as a means to deal with general strain-localization phenomena (such as shear bands in metals) of which the cohesive crack is but a particular case (Belytschko, Fish and Englemann 1988; Simo and Rifai 1990; Dvorkin, Cuitiño and Gioia 1990; Klisinski, Runesson and Sture 1991; Simo, Oliver and Armero 1993; Simo and Oliver 1994; Lofti and Shing 1994; Oliver 1995; Larsson, Runesson and Åkesson 1995).

The theoretical foundations of the various approaches to intraelement cracks can be different, owing

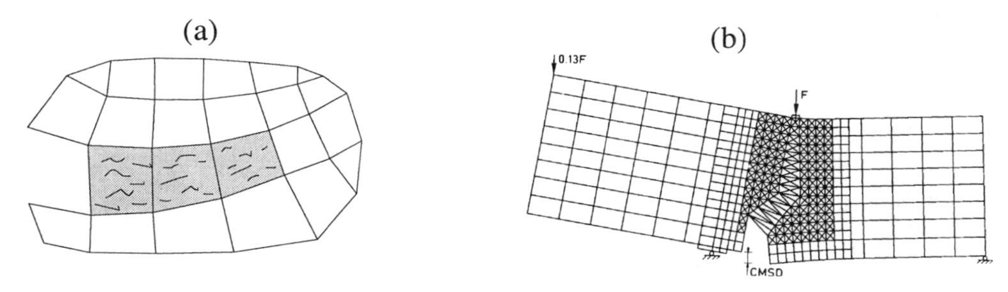

Figure 7.2.9 (a) Smeared crack approach. (b) Simulation of cohesive crack growth in mixed mode using smeared crack concepts (from Rots 1989).

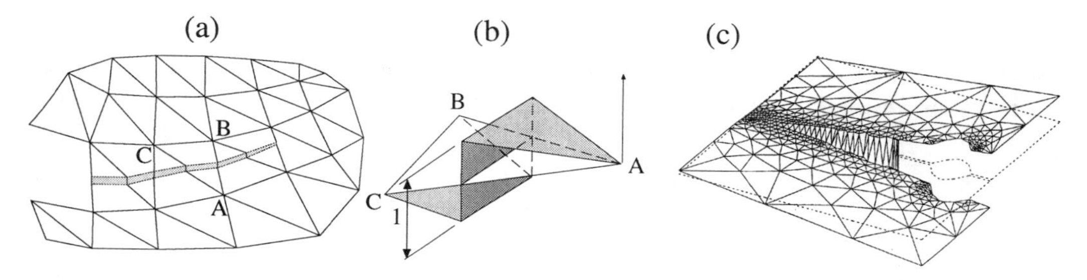

Figure 7.2.10 (a) Intraelement crack. (b) Shape function including a jump discontinuity (after Oliver 1995). (c) Three-dimensional representation of the displacements in a compact specimen (displacements plotted along the third direction; from Oliver 1995).

to the differences in the basic underlying continuum model (plasticity, damage, coupled damage and plasticity, and so on). But one feature is common: after localization, the stress-strain constitutive relation is replaced in one way or another by a softening relation between stress and relative displacement, as in the cohesive crack model. This is a basic ingredient in continuum problems with strong discontinuities (displacement jumps).

The numerical implementation of the intraelement cracks have minor differences, but they essentially consist in introducing discontinuous shape functions into an element. For example, Oliver (1995) uses triangular elements in which, apart from the classical continuous shape functions, a shape function with a jump is introduced as shown in Fig. 7.2.10b. This approach allows a precise representation of the crack opening as shown in Fig. 7.2.10c.

7.2.4 Size Effect Predictions

One of the main aspects of fracture which can be analyzed through cohesive crack models is the size effect. In the few last decades, many simulations and theoretical analyses have been carried out and today the understanding of the size effect and its description seems quite complete. Since this is an important topic in engineering, the size effect caused by cohesive crack behavior will be analyzed in a broader framework in Chapter 9. Here, we summarize only the most important results.

The first general result is that notched and unnotched structures behave very differently, as illustrated in Fig. 7.2.11 for three-point bend beams. For notched structures with positive geometries, the overall trend is similar to Bažant's size effect law: in a log σ_{Nu}-log D plot, the size effect curve has a horizontal asymptote for small sizes and an inclined asymptote of slope $-1/2$ for large sizes, corresponding respectively, to the plastic and LEFM limits. However, the details of the equation are different.

For large sizes, a notched specimen behaves close to LEFM; Planas and Elices (Planas and Elices 1992a, 1993a) proved that the limiting size effect behavior is given by an expression identical to that resulting from an equivalent elastic crack model. More generally, these authors were able to prove that, for large enough sizes, the values of stresses and displacements at points far from the cohesive zone are identical

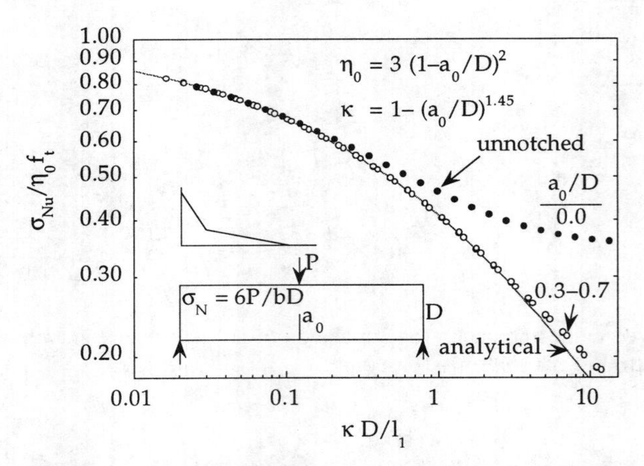

Figure 7.2.11 Size effect curves for notched and unnotched three-point bend beams. The symbols correspond to computed values using a bilinear softening curve and the full line to the analytical expression (7.2.11) (adapted from Planas, Guinea and Elices 1997).

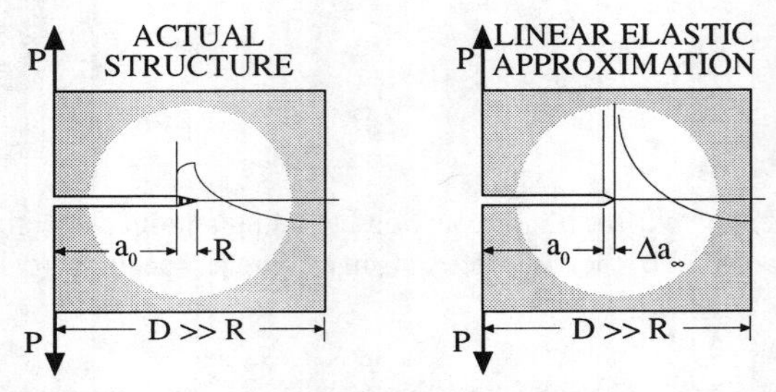

Figure 7.2.12 Effective crack extension in the infinite size limit; the remote fields (shaded areas) coincide up to order R/D (adapted from Planas and Elices 1991a).

to those given by a suitably chosen effective crack model. More specifically, in a cracked specimen (as sketched in Fig. 7.2.12) having an initial crack a_0 and loaded in mode I so that a cohesive zone of size R extends ahead of the initial tip, every far-field (stress, displacement) may be approximated, up to order R/D, by the far-field corresponding to an equivalent elastic crack of length $a_0 + \Delta a_\infty$.

The symbol Δa_∞ stands for the effective crack extension for large sizes. Given the softening curve of the material, Δa_∞ can be computed at any loading step using the formulas deduced by Planas and Elices (1992a, 1993a), which are given in Section 7.5.7. Particularly important is the critical effective crack extension $\Delta a_{\infty c}$ (the value of Δa_∞ at peak load). It turns out that its value is strongly dependent of the softening curve, particularly on the length of its tail. Planas and Elices (1993a) found the following values:

$$\text{rectangular softening:} \quad \Delta a_{\infty c} = \frac{\pi}{24}\ell_{ch} \tag{7.2.5}$$

$$\text{linear softening:} \quad \Delta a_{\infty c} = 0.419\ell_{ch} \tag{7.2.6}$$

$$\text{quasi-exponential softening } (\hat{w}_c = 5)\text{:} \quad \Delta a_{\infty c} = 2.48\ell_{ch} \tag{7.2.7}$$

Another theoretical result of Planas and Elices is that a closed form lower bound exists for the effective crack extension in general, and for $\Delta a_{\infty c}$ in particular. Surprisingly, the actual value of $\Delta a_{\infty c}$ is very close to the lower bound for any practical softening function. We have

$$\Delta a_{\infty c} \approx \frac{\pi}{32} w_c^2 \frac{E'}{G_F} \tag{7.2.8}$$

where the right-hand side is the lower bound, which reflects the fact that the critical effective crack extension varies quadratically with the critical crack opening w_c (the value of the crack opening at which the cohesive stress becomes zero).

Since, for large sizes, the cohesive crack tends to an effective crack model, as discussed in Chapters 5 (Section 5.3.2), and 6 (Section 6.1.1), the asymptotic size effect curve coincides with Bažant's size effect law, with c_f replaced by $\Delta a_{\infty c}$, which can be recast in the form given by Planas and Elices:

$$\frac{E'G_F}{K_{INu}^2} = 1 + \frac{\Delta a_{\infty c}}{\overline{D}} \tag{7.2.9}$$

where $\overline{D}$ is the intrinsic size defined in Eq. (5.3.12).

For small sizes, the size effect is known to disappear, and so a horizontal asymptote is approached in the $\log \sigma_{Nu}$-$\log D$ plot. What was not too clear until very recently is the rate at which the actual curve deviates from the asymptote: while Bažant's law implies that the deviation is linear, i.e., that $\sigma_{Nu} = \sigma_0 - c_1 D +$ (higher order terms), recent analyses showed that this deviation is much faster (see Section 9.2; also in Planas, Guinea and Elices 1997):

$$\sigma_{Nu} = \sigma_0 - c_1\sqrt{D} + o(\sqrt{D}) \tag{7.2.10}$$

where σ_0 is the strength for very small size, which can be computed from plastic limit analysis, c_1 is a constant and $o(x)$ is a function vanishing faster than its argument.

With the foregoing asymptotic trends, Planas, Guinea and Elices (1997) were able to find a single function that fits the size effect curve for notched specimens of small and medium sizes. For three-point bending specimens with relatively depth notches ($a_0/D \geq 0.3$), a single equation represents very well all the results for sizes $D \leq 4\ell_1$, if a Petersson's bilinear softening is used; see Fig. 7.2.11. This equation is

$$\sigma_{Nu} = \eta_0 f_t' \left[\frac{1 + 4.23\sqrt{\kappa D/\ell_1}}{1 + 0.622\sqrt{\kappa D/\ell_1}} + 2.7\frac{\kappa D}{\ell_1}\right]^{-\frac{1}{2}}, \quad \eta_0 = 3(1-\alpha_0)^2, \quad \kappa = 1 - \alpha_0^{1.45} \tag{7.2.11}$$

where $\alpha_0 = a_0/D$ and ℓ_1 was defined in (7.1.17). Further numerical and experimental support for this equation is given in Section 9.2.

The foregoing considerations apply to notched specimens of positive geometry. Unnotched specimens following the cohesive crack model behave in a different way as shown by the corresponding curve in Fig. 7.2.11. The size effect displays two horizontal asymptotes, one for small sizes, corresponding to plastic analysis, and one for large sizes that corresponds to an elastic-brittle model (failure occurs as soon as the maximum principal stress at any point reaches f_t'). A detailed analysis corresponding to unnotched beams in bending (i.e., modulus of rupture) is given in Chapter 9.

7.2.5 Cohesive Crack Models in Relation to Effective Elastic Crack Models

The size of the most usual structural concrete elements is not large enough to admit the simplified LEFM approach. In general, one must resort to involved numerical techniques for solving the nonlinear problems posed by the cohesive crack model. The equivalent elastic crack approach (EEC) emerges as an intermediate technique to avoid cumbersome computing while still using the simple LEFM procedures with an associate R-curve. This approach is not always legitimate and its justification and limitations are briefly summarized.

The equivalent elastic crack greatly simplifies the computations by shifting from a cohesive material (and, hence, a nonlinear structural analysis) to a linear elastic material (which requires only linear analysis). The unified concept of equivalent elastic crack, as described here, was triggered by the asymptotic results described in the previous subsection, particularly the far field equivalence sketched in Fig. 7.2.12, and was further developed to include most of the useful equivalences (Planas and Elices 1990c, 1991b; Elices, Planas and Guinea 1993; Planas, Elices and Ruiz 1993).

The concept of elastic equivalence is best illustrated on a particular example. Let us choose the force-crack mouth opening displacement equivalence or P-CMOD equivalence. As shown in Fig. 7.2.13, two geometrically identical cracked samples are loaded under CMOD control. One specimen is made from a cohesive material and the other (the equivalent one) from a linear elastic material. The measured loads in

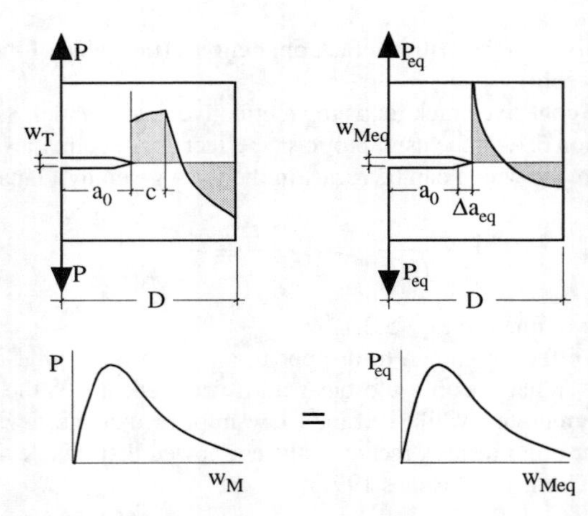

Figure 7.2.13 Definition of the P-CMOD equivalence, in which the P–w_M response for the actual cohesive material (left) and an equivalent linear elastic specimen (right) are forced to be identical (from Elices, Planas and Guinea 1993).

the two samples —P and P_{eq}— for the same crack mouth opening displacement w_M will be, in general, different, but we can force the loads to match each other ($P = P_{eq}$), at each value of w_M, by choosing a suitable equivalent crack length a_{eq} and a suitable crack growth resistance $\mathcal{R}_{eq}$.

In this way we force both samples to exhibit the same P-CMOD behavior, but, generally, the equivalence ends here; the stress field, for example, need not be the same, as seen in Fig. 7.2.13. The price paid for the simplification is that the linear elastic material does not have a constant resistance to fracture growth (G_F is constant for the cohesive material and for the LEFM approach), but a variable resistance described by the R-curve (an $\mathcal{R}$-Δa curve). This curve, in general, is not a material property but depends on the geometry and specimen size.

As an example of the application of the load-CMOD equivalence, imagine we want to predict the load-CMOD curve of a cracked panel in tension made from a cohesive material, using the test results of a three-point notched beam made of the same material. The procedure to compute the load-CMOD curve for the cracked panel is as follows: the P-CMOD is first measured from the beam test, from which the corresponding $\mathcal{R}$-Δa curve is derived. Then, assuming that the R-curve is a material property, a P-CMOD curve for the notched panel is predicted under the simplified assumption of linear elastic behavior. The prediction is shown in Fig. 7.2.14a for a panel of the same depth as the beam used to determine the R-curve. The agreement of the prediction and the experimental result is excellent (in this example, the experimental results are numerically computed using the cohesive crack model, as the experimental values are assumed to be indistinguishable from the numerical computations based on cohesive cracks). However, if the relative sizes and notch depths of both structures are quite different, the agreement can be rather poor, as shown in Fig. 7.2.14b. More details about this example and the validity of the assumptions implicit to the R-curve approach can be found in Elices and Planas (1993) and Planas, Elices and Ruiz (1993).

Apart from the P-CMOD equivalence, other P-Y equivalences may be sought (where Y stands for other possible magnitudes). For example, P-u, P-CTOD, or P-J equivalences, where u, CTOD and J are, respectively, the load-point displacement, the crack tip opening displacement and the J-integral.

The concept of P-Y equivalence can be generalized to any controlling variable, not necessarily the load P, and the same procedure extended to any pair of variables X-Y. This generalization permits one to consider apparently unrelated procedures, such as Bažant's $\mathcal{R}$-Δa approach (Bažant, Kim and Pfeiffer 1986) or the J-$CTOD$ approach (Planas, Elices and Toribio 1989), within the same framework.

7.2.6 Correlation of Cohesive Crack with Bažant's and Jenq and Shah's Models

The size effect results provide a sound basis to establish a correlation between the parameters of the various models; the method requires just finding which set of parameters gives approximately the same peak load for notched specimens of a typical laboratory size when the various models are used. The

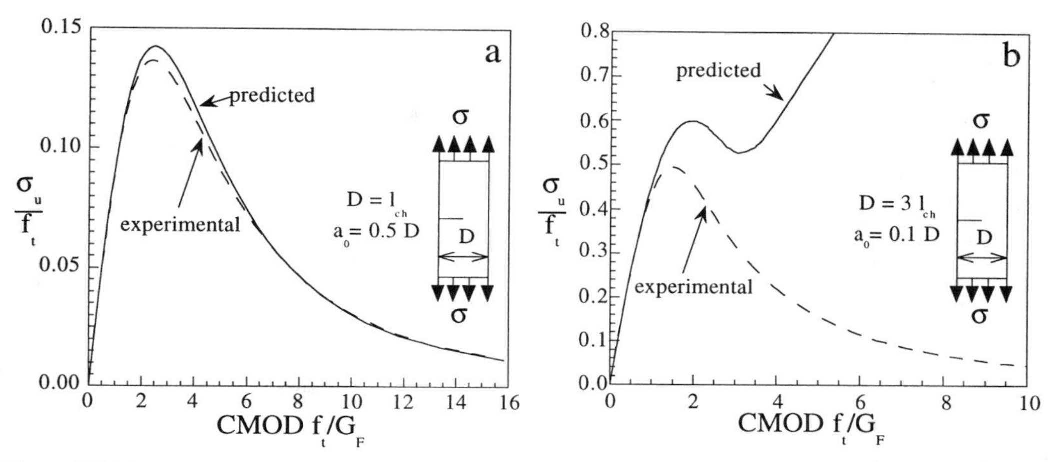

Figure 7.2.14 Comparison of the experimental response (simulated by cohesive crack model) of a SEN panel in tension to its response predicted using the $\mathcal{R}$-Δa curve determined from a P-CMOD equivalence for a three-point bend beam (TPB) of a depth $D = \ell_{ch}$ and a notch-to-depth ratio of 0.5: (a) SENT specimen with dimensions similar to those of the TPB specimen; (b) SENT specimen with dimensions differing greatly. (From Elices and Planas 1996.)

correlation is, of course, not exact because the range of sizes is relatively vague; however, for ordinary concrete, the variability is narrow enough for these correlations to be useful as first approximations.

Planas and Elices (1990a) established the correlation between the various parameters using G_F and ℓ_{ch} as the basic parameters for the cohesive model. It turned out then, that the correlation depended on the shape of the softening curve. Subsequently it appeared that, for the usual experimental range of sizes, the peak load for the cohesive model is completely defined by the initial linear softening.

Therefore, if one seeks to correlate the results with the tensile strength f'_t and the horizontal intercept w_1 (Fig. 7.1.8), the resulting correlation needs to be nearly universal. It turns out that the results in Planas and Elices (1990a) can be readily transformed by using the result (derived in Planas and Elices 1992b) that, for quasi-exponential softening, the fracture energy is related to f'_t and w_1 by

$$G_F = 0.86 f'_t w_1 \tag{7.2.12}$$

from which the characteristic length follows as

$$\ell_{ch} = 0.86 \frac{E' w_1}{f'_t} = 1.72 \ell_1 \tag{7.2.13}$$

Inserting the preceding relations in the results in Planas and Elices (1990a), the following approximate relations are found for the parameters G_f and c_f of Bažant's size effect model (Chapter 6), and the parameters K_{Ic} and w_{Tc} of Jenq-Shah two parameter model (Chapter 5):

$$G_f \approx 0.45 f'_t w_1 \;, \qquad c_f \approx 0.19 \ell_1 \qquad \text{for Bažant's model} \tag{7.2.14}$$

$$K_{Ic} \approx 0.91 f'_t \sqrt{\ell_1} \;, \qquad w_{Tc} \approx 0.52 w_1 \qquad \text{for Jenq-Shah model} \tag{7.2.15}$$

These correlations hold within a few percent for size ranges of the order of 12 to 40 cm for ordinary concrete and notched specimens. If they have to be used for other materials (high strength concrete or mortar), the appropriate range is expressed in terms of the intrinsic size as

$$\overline{D} \approx 0.1\text{–}0.4\ \ell_1 \tag{7.2.16}$$

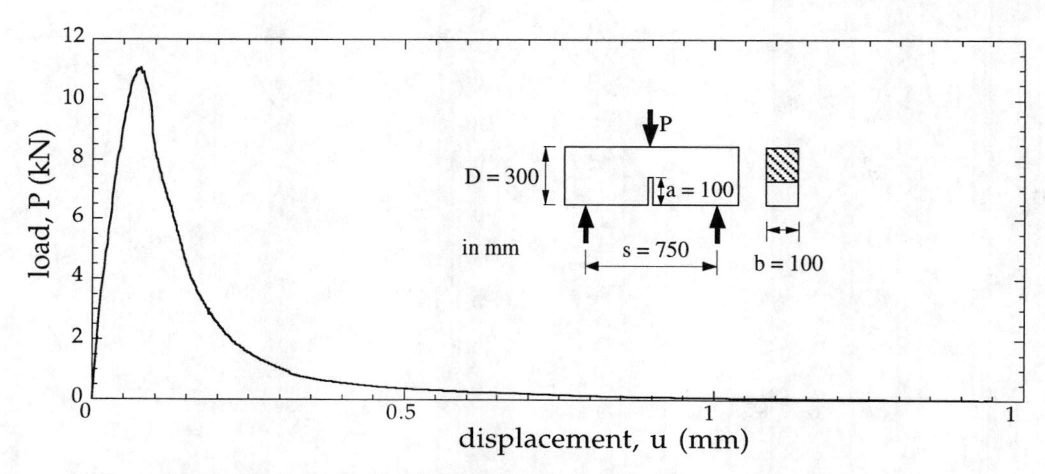

Figure 7.2.15 Experimental load-displacement curve for a three-point bend notched beam.

Exercises

7.5 Give an estimate of the horizontal intercept of the initial tangent approximations for the various mixes tested by Petersson (Fig. 7.2.1). Give both the absolute values and the values in terms of $w_{ch} = G_F/f_t'$.

7.6 Compute the lower bound for $\Delta a_{\infty c}$ for the rectangular, linear, and quasi-exponential softening ($\hat{w}_c = 5$) and compare with the more exact values given in Eqs. (7.2.5)–(7.2.7).

7.7 Rewrite the expression (7.2.8) for $\Delta a_{\infty c}$ as a function of ℓ_{ch}.

7.8 Fig. 7.2.15 shows the load-displacement record for a three-point bend notched beam whose dimensions, in mm, are also indicated on the figure. Complementary tests were performed to determine the tensile strength and the elastic modulus, with the results $f_t' = 3.0$ MPa and $E = 27$ GPa. (a) Estimate the peak load according to a plastic limit analysis of the using an unidimensional rigid-perfectly plastic model with a plastic yield strength in tension equal to f_t' and infinite strength in compression, and compare the result with the peak load that has been actually recorded. (b) Improve the estimate by considering a plastic limit in compression equal to $10 f_t'$. Is plastic limit accurate at all for this problem? [Partial answer: (a) 32 kN; (b) 29 kN.]

7.9 Analyze the results of Fig. 7.2.15 using LEFM. (a) If LEFM applies, the total work supply (the area under the load-displacement curve) must be equal to the fracture energy G_c times the initial area of the ligament. Use a rough triangulation (two or three triangles) to estimate the work supply and G_c. (b) Use Irwin's relationship to deduce the fracture toughness of the material K_{Ic}. (c) Use the foregoing result and the expressions given in Chapter 3 for K_I to determine the peak load predicted by LEFM, and compare to the actually recorded peak load. Is plastic limit accurate at all for this problem? [Partial answer: (a) $\approx$ 80 J/m^2; (c) 20 kN.]

7.10 Analyze the results of Fig. 7.2.15 using cohesive models, based on the value of the fracture energy determined in the previous exercise, which now must coincide with G_F. (a) Estimate the peak load in the assumption that the softening is linear; (b) same for Petersson's bilinear softening; (c) same for the bilinear softening of Wittmann et al.; (d) Same for the ELT softening (here you will have to make a reasonable estimate of w_1 for this softening); (e) which of the foregoing softening curves gives a better agreement with the actual recorded peak load?

7.3 Experimental Determination of Cohesive Crack Properties

In this section we examine the simplest test methods to determine cohesive crack properties. Although, in principle, these methods can be applied to any material whose fracture behavior can be approximated by a cohesive crack model, the focus is essentially on concrete. So, a certain care must be exerted when extending these methods to other materials; in particular, the physical size of the specimens and the size relative to the characteristic material size ℓ_1 do play a role in making and interpreting the experiments.

The essential information for a cohesive crack model is the complete softening curve. Ideally, this seems best determined from direct tension tests, but, as discussed in Section 7.2.2, tension tests are extremely difficult to carry out, and are out of the question in routine testing. The alternative is to obtain the softening curve by indirect methods together with complementary hypotheses. Here we do not discuss the general inverse procedures commented on in Section 7.2.2; we give only an account of the simplest methods that estimate the softening curve from relatively simple test data and require few straightforward calculations.

The first routine test proposed for determining fracture properties of concrete was based on the cohesive crack model and described the determination of the fracture energy by the work-of-fracture method (Hillerborg 1985a). It was described in a RILEM recommendation (RILEM 1985) and has been the subject of debate since then, because the experimental values of G_F so obtained were seen to be size dependent. In this debate, it is sometimes difficult to understand whether the objections are directed against the experimental method or against the model itself. Certainly if the experimental method is sound *and* the model is perfect for concrete, then the experimental values of G_F must be constant. If they are not, either the experimental method has flaws *or* the model is valid only approximately *or both*.

A careful analysis of the test procedure was performed by Planas and co-workers (Planas and Elices 1989b; Guinea, Planas and Elices 1992; Planas, Elices and Guinea 1992; Elices, Guinea and Planas 1992, 1997). They showed that the experimental method could be enhanced to reduce the apparent size effect on G_F. If a size effect still remains, it is because the cohesive crack model, as usually formulated, is only an approximation, and because further improvements of the model are required to get a better match between experiment and theory. Particularly, the effect of triaxiality could explain the size dependence. Of course, this would improve the performance of the model at the expense of an increased complexity.

Therefore, the method whose description follows must be understood as the means of finding approximate values of parameters of an approximate model. This is similar to the determination of the elastic modulus of concrete: every civil engineer knows that concrete is *not* a linear elastic material and has no true elastic modulus. However, a standard E is defined operatively. Everybody knows it cannot be blindly used to predict concrete strain, yet it is an excellent indicator of the deformability of the concrete. G_F as determined using the work-of-fracture method is certainly not a physical constant, but does provide an excellent indication of the toughness of concrete and, if used consistently (i.e., within the frame of the cohesive crack models), provides a good basis for meaningful predictions.

Besides the foregoing considerations, it is important to realize that G_F alone does not provide sufficient information to analyze structures except in very special cases, namely, very large cracks in very large bodies. To become useful, more information about the softening curve is required. For example, the tensile strength f'_t is another essential parameter. Even if both G_F and f'_t are known, no useful predictions can be made without information about the shape of the softening curve. This is a key point in understanding the approximation to the cohesive crack model of Hillerborg and co-workers in the late seventies and early eighties: f'_t and G_F were to be measured and the shape of the softening curve was postulated (i.e., guessed on the basis of previous extensive testing on other concretes that were assumed to be "similar" to any other concrete). Later, however, it turned out that the similitude of the softening curves is not so sharp. Most importantly, it became apparent that for some important structural properties, such as the strength of unnotched structures, the total fracture energy is not too important: only the initial part of the softening curve is essential, while the tail —and G_F— is largely irrelevant.

In the following we analyze the simplest tests for tensile strength f'_t, for the initial part of the softening curve, for G_F, and, finally, for a bilinear softening.

7.3.1 Determination of the Tensile Strength

Although the tensile strength is a key parameter in a cohesive model, relatively little attention has been devoted to its experimental determination. The most common method in international standards is the splitting cylinder test (Brazilian test). However, doubts were recently cast on its applicability because several researchers showed the results of such a test to be size-dependent (see Section 9.4).

A recent in-depth study of the Brazilian test has nevertheless indicated that although the size effect does exist, it can be explicated by the cohesive crack model itself (Fig. 7.3.1, after Rocco 1996). Moreover, the cylinder splitting strength appears to be very close to the 'true' tensile strength if the following two conditions are met in the test: **(1)** the specimen diameter is large enough, and **(2)** the loading strips are narrow enough.

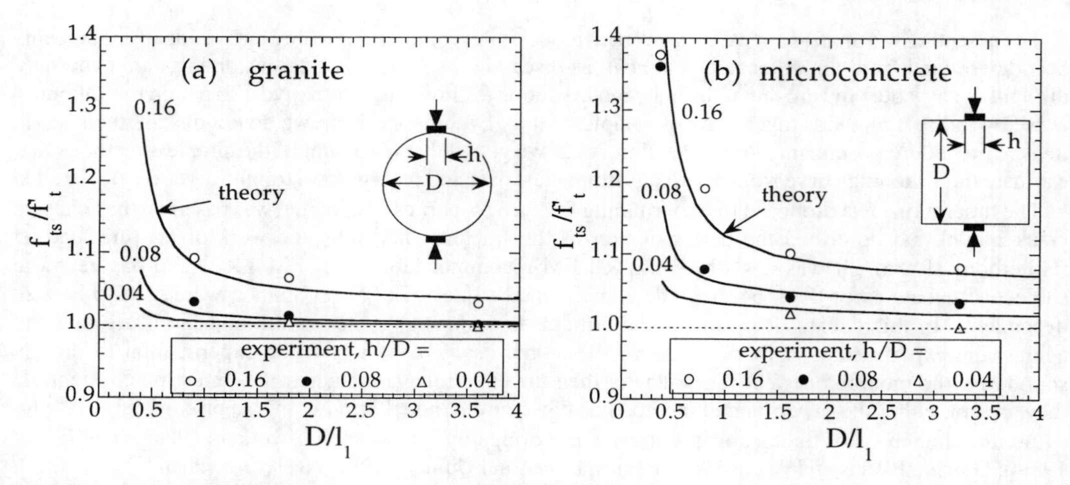

Figure 7.3.1 Ratio between the splitting strength and the 'true' tensile strength as a function of size and width of loading strip, (a) for cylindrical specimens, and (b) for cubic specimens. The symbols correspond to experimental results for (a) granite, and (b) a microconcrete. The lines are predictions of the cohesive crack model. (After Rocco 1996.)

Indeed, Fig. 7.3.1a shows that if the cylinder diameter D is larger than $0.6\ell_1$ and the loading strip width h is less than $0.08D$, the deviation from the true tensile strength is less than 4%, which is adequate for most practical purposes. For cube splitting (a specimen accepted by some standards, such as the British), the trend is very similar, but the cube splitting strength is about 4% higher than the cylinder splitting strength when the cube side is equal to the cylinder diameter size (Fig. 7.3.1b).

Therefore, the simplest method is the Brazilian cylinder splitting test with a loading strip width of about 1/12 of the diameter. Note that the ASTM standard (ASTM C496) recommends strips twice as wide, which leads to splitting strengths significantly larger for similar sizes (Fig. 7.3.1a).

The results of Rocco (1996), combined with the technique described next to determine the initial part of the softening curve, provide a means of quantifying the accuracy of the tensile strength produced by the cylinder splitting test: the relationship D/ℓ_1 can be computed *a posteriori* and the curves in Fig. 7.3.1 can be used to verify the closeness of f_{ts} to f'_t. The process of determination can even be iterated to obtain a better estimate.

7.3.2 Determination of the Initial Part of the Softening Curve

As pointed out in Section 7.2.4, the peak load of not too large specimens is completely controlled by the initial portion of the softening curve which, for concrete, can usually be approximated by a straight line (Fig. 7.2.5). This was exploited by Guinea, Planas and Elices (1994b) for determining the initial slope of the softening curve from the determination of the peak load, as follows.

Assume that we select a particular testing geometry, for example, a three-point bent beam with a span-to-depth ratio of 4 and an initial notch-to-depth ratio $\alpha_0 = 0.5$, and that we make a bending test and measure only the peak load P_u (with appropriate corrections to take into account the specimen self-weight), from which we compute the ultimate nominal strength $\sigma_{Nu} = 3P_u s/(2bD^2)$. Assume further that we perform cylinder splitting tests on the same material so that we have a good estimate f_{ts} of the tensile strength f'_t. According to the theory, the ultimate strength can be written in the form

$$\frac{\sigma_{Nu}}{f'_t} = \phi\left(\frac{D}{\ell_1}\right) \tag{7.3.1}$$

where D is the specimen size (say beam depth) and $\phi(D/\ell_1)$ is a function that can be computed once and for all for the particular testing geometry using an appropriate numerical method. Then this function

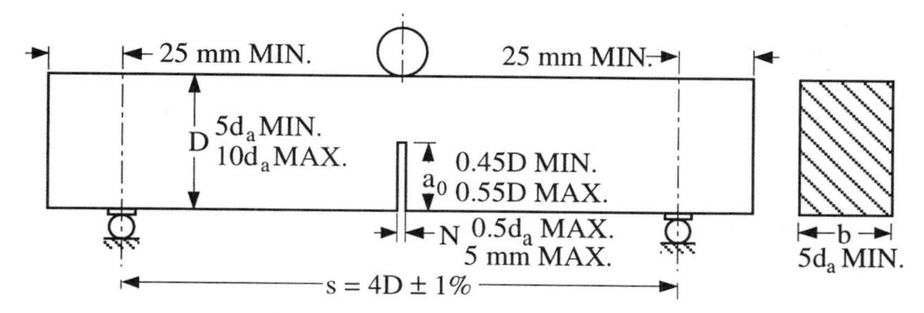

Figure 7.3.2 Standard specimen dimensions proposed by Planas, Guinea and Elices (1994a,b); d_a is the maximum aggregate size.

can be numerically (or graphically) inverted to yield

$$\frac{D}{\ell_1} = \chi\left(\frac{\sigma_{Nu}}{f'_t}\right) \tag{7.3.2}$$

From this, ℓ_1 is determined as

$$\ell_1 = D\frac{1}{\chi(\sigma_{Nu}/f'_t)} \tag{7.3.3}$$

Since D is basic experimental data, and σ_{Nu} and $f'_t \approx f_{ts}$ have been measured, ℓ_1 is determined from the foregoing equation. Once ℓ_1 is known, w_1 can be determined from (7.1.17), assuming that the elastic modulus has been identified experimentally using appropriate methods.

Planas, Guinea and Elices (1994a,b) proposed a draft test method to determine w_1 based on three-point-bend tests on notched specimens with the geometry shown in Fig. 7.3.2. They provide closed form expressions to deduce the elastic modulus from the initial slope of the load–CMOD curve, as well as closed form expressions for w_1, which is calculated as follows:

1. Compute the intermediate variable Y, with the meaning of a reduced flexural strength, as

$$Y = \frac{\sigma_{Nu}}{f'_t}(1 + 3.9934\Delta\alpha + 12.220\Delta\alpha^2 + 32.409\Delta\alpha^3) \tag{7.3.4}$$

where σ_{Nu} and f'_t are the mean values determined from the bending and flexural tests, respectively, and $\Delta\alpha = a_0/D - 0.5$ is the deviation of the relative notch depth from the standard value 0.5.

2. Check that $0.25 \le Y \le 0.54$, which is the range of validity of the closed-form equations.

3. Determine the intermediate variable X, with the meaning of a reduced intrinsic size, as

$$X = \frac{0.053107}{Y^2} - \frac{0.0081523}{Y} - 0.55999 + 1.0820Y - 0.60473Y^2 \tag{7.3.5}$$

4. Determine ℓ_1 and w_1 as

$$\ell_1 = \frac{D}{X}(0.15755 - 0.25677\Delta\alpha - 0.22136\Delta\alpha^2)\,, \qquad w_1 = \frac{2f'_t\ell_1}{E} \tag{7.3.6}$$

where D is the beam depth and E the elastic modulus; the remaining symbols have already been defined.

The foregoing formulae were developed by curve fitting, based on the intrinsic size concept, before the more general and accurate size effect equation (7.2.11) was discovered; using it will probably simplify the foregoing equations and expand the range of applicability. (At any rate, the foregoing equations are simple to program in a hand-held calculator.)

Note that although the foregoing applies to notched beams in bending, similar expressions can be easily developed for other specimens such as the compact or the wedge splitting specimens.

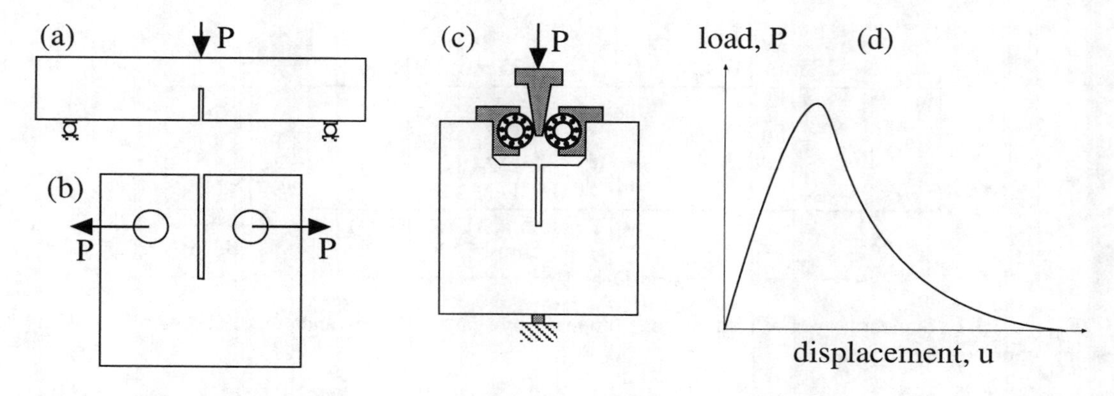

Figure 7.3.3 Typical laboratory specimens: (a) three-point bend, (b) compact tension, and (c) wedge splitting. Test output: (d) complete load-displacement curve.

7.3.3 Determination of Fracture Energy G_F

Conceptually, the simplest means of determining the fracture energy is the work-of-fracture method, proposed for concrete by Hillerborg (1985a) and the RILEM Committee 50 (RILEM Draft Recommendation 1985).

According to the cohesive crack model, the fracture energy is the energy required to produce a unit area of crack (fully broken). Therefore, if we statically break a specimen, such as any of those depicted in Fig. 7.3.3a-c, and measure the work W_F supplied to fracture it, then the approximate fracture energy is calculated as

$$G_F = \frac{W_F}{b(D - a_0)} \tag{7.3.7}$$

where b is the specimen thickness and $D - a_0$ the initial ligament length. Note that the area considered is the mean or projected area, i.e., the surface roughness is neglected, in accordance with the macroscopic nature of the model which must ignore microstructural details (note that the roughness is of the order of the aggregate size, the largest microstructural feature).

As simple as it seems, the application of the foregoing concept is not easy. First we have to measure the work supply statically: this requires a stable test in which static equilibrium is maintained all the time, so that the complete load-displacement curve could be measured, as sketched in Fig. 7.3.3d. This may require servocontrolled testing, which is normally carried out under CMOD control; if servocontrol is not available, the test can be stable if a stiff loading system is used. In this respect, the wedge-splitting test (Fig. 7.3.3c), initially proposed by Linsbauer and Tschegg (1986) and later modified by Brühwiler (1988) and Brühwiler and Wittmann (1990), is very stable if low angle wedges are used, and does not require servocontrol.

Second, and most important, it must be ensured that all the work done by the applied (measured) force is used in producing the cohesive crack, *and* that this is the only external work. If this is not so, we must correct the results by subtracting the spurious work dissipation and adding the work done by forces other than the measured one.

The sources of error in the measurement of the work supply have been extensively analyzed by the group at Madrid Technical University (Planas and Elices 1989b; Guinea 1990; Guinea, Planas and Elices 1992; Planas, Elices and Guinea 1992; Elices, Guinea and Planas 1992). Their main conclusions are as follows:

1. The hysteretic behavior of the loading and measuring system can be kept low, but must be checked. This is done by testing elastic specimens (steel or aluminum specimens well in the elastic range) subjected to slow cyclic loading up to the maximum load to be used in the actual tests.
2. The friction at the supports is a source of important errors. Steel-on-steel polished and rolling surfaces must be used in the supports. Free-rolling must be ensured by cleanliness. The friction

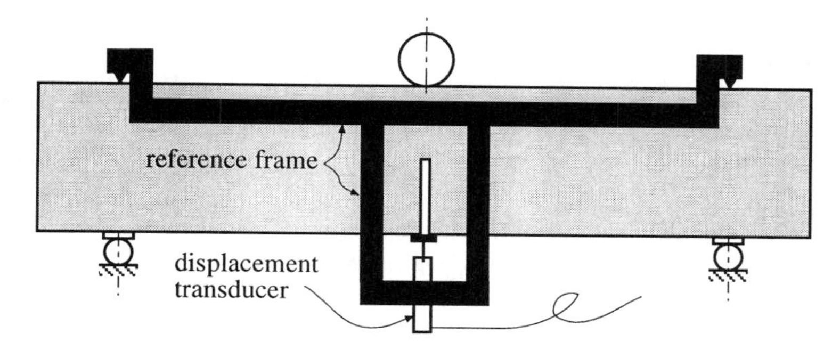

Figure 7.3.4 Measuring displacement relative to opposite points over the supports.

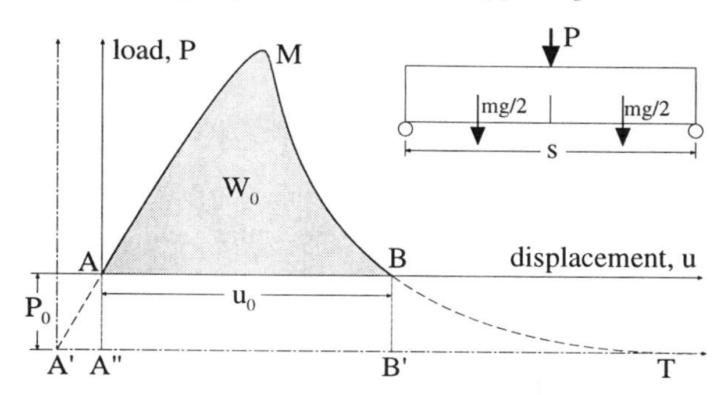

Figure 7.3.5 Load-displacement curves for uncompensated three-point bending test.

effect is magnified by wedge loading in the wedge-splitting test (see, e.g., report by ACI Committee 446 1992), and so special care must be exerted when using this type of test.

3. Crushing below the supports is an energy sink that can lead to overestimation of the fracture energy. A relatively easy way of excluding the crushing energy is to measure the displacements relative to points directly above and below the supports on the opposite faces, as depicted in Fig. 7.3.4.

4. Energy dissipation in the bulk of the material around the main crack, due to inelastic strains at stresses below f_t', is likely to be small in concrete, less than 2% for ordinary sizes and less than 5% for very large sizes.

5. The self weight of the specimen and the energy not measured in the far end of the test must be properly taken into account. This aspect deserves further comments because it may lead to large errors, and also because it is at the root of the method to determine the bilinear softening curve.

In testing notched beams in three-point bending as well as in other testing configurations, the weight of the specimen contributes to the overall loading of the system. It is not accounted for by the measuring devices unless special methods, which we call *weight compensation,* are used. The RILEM Draft Recommendation (1985) does not recommend weight compensation. Instead, it uses an analytical correction described next.

For the simple test configuration shown in Fig. 7.3.5, the load cell is zeroed when the self weight is already acting on the specimen. Thus, at zero applied load, a bending moment already exists at the notch cross section which is equal to $mgs/8$ where mg is the weight of the specimen and s the loading span (m stands for the mass and g for the acceleration of gravity). This moment is equivalent to a constant central load $P_0 = mg/2$, so that in performing the test the total load-displacement curve would be represented by the dashed line in Fig. 7.3.5. However, the *recorded* curve is only the full line portion. When point B is reached, the beam breaks unstably under its own weight.

Now, the work-of-fracture to be substituted in (7.3.7) is the area under the *total* load-displacement curve, i.e., area $A'AMBTB'A"A'$. However, the area actually recorded is area $AMBA$ (shaded in the figure) which can be considerably less; we call it W_0. To obtain the total area, we must add to W_0 the area $A'ABTB'A"A'$. Now, except for the small triangle $A'AA"$, the area $A'ABB'A"A'$ is equal to

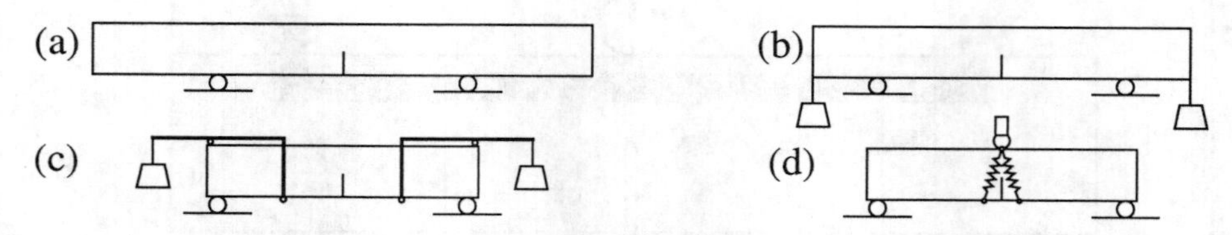

Figure 7.3.6 Weight compensation devices: (a) longer specimens, (b) dead weights, (c) lever with dead weights, (d) spring attachment.

$P_0 u_0$, where u_0 is the recorded load-point displacement when the recorded load becomes zero and the specimen suddenly fails. Hillerborg and Petersson evaluated the remaining area $B'BTB'$ following an approximate procedure to be described later and found that it was also equal to $P_0 u_0$ (Petersson 1981). Therefore, the total work of fracture is

$$W_F = W_0 + 2P_0 u_0 \tag{7.3.8}$$

We have seen that, for the simple case sketched in Fig. 7.3.5, $P_0 = mg/2$. In a more general case, P_0 must be obtained as the central load that produces the same central bending moment as the system of dead loads:

$$P_0 = \frac{4M_0}{s} \tag{7.3.9}$$

where M_0 is the central bending moment generated by the dead-loads (the self-weight of the specimen plus fixtures resting on the specimen) and s is the loading span.

The correction term found by Petersson (1981) is based on assuming that at point B the specimen is very close to complete failure. This is not the case for large specimens, in which P_0 can be an appreciable fraction of the peak load. Therefore, it is much better to obtain the full softening curve instead of calculating (as done by Petersson) the last portion. The way of doing so is to use *weight compensation*. This can be done in a variety of ways as sketched in Fig. 7.3.6: by using specimens twice as long as the loading span (Fig. 7.3.6a), by using dead-weights(Fig. 7.3.6b-c), or by attaching the central part of the specimen to the loading head by means of springs (Fig. 7.3.6d). When using these springs, it is essential to ensure that the tension force in the springs be larger than P_0 as determined before, and that the springs be placed so that their length does not change during specimen deformation (if the spring length changes and no correction is done, the work required to stretch the springs will incorrectly get manifested as fracture work).

Although it is theoretically possible to compensate for the weight *exactly*, in practice it is better to work with a slight overcompensation so that the dead forces produce a negative bending moment; this means that at the end of the test the force P is equal to a certain positive constant load P_0', and the load displacement curve looks as that depicted in Fig. 7.3.7, in which the relative value of P_0' is grossly exaggerated. It is now essential to realize that P_0' is not well known in advance, but can be measured after the test on the load displacement record itself as shown in Fig. 7.3.7. The work of fracture would be the shaded area, determined by the curve and the line $A'T$, drawn parallel to the displacement axis through the last point of the record T, where it is assumed that the specimen is fully broken. Note that if the area between the curve and the original OT' axis is measured, an important error is produced. This is so mainly when P_0' is small and the computations are made by computer, making it possible to overlook the error.

The foregoing picture seems to close the case, but this is not so, because in a bending test complete failure is approached asymptotically. This means that point T is infinitely far, and one is forced to stop the test before most energy is dissipated. Therefore, the actual picture is like that in Fig. 7.3.8, in which point B is the last recorded point of the test, which lies a small distance P_B' above the true complete failure asymptote. Now, we can draw line AB and compute the area above it (dashed area in the figure). We can also compute the area $A'ABTB'A$ if P_B' is known using the method of Hillerborg and Petersson (because in this case P_B' *is* small). A formula identical to (7.3.8) holds with W_0 replaced by W_m, P_0 by P_B', and u_0 by u_B':

$$W_F = W_m + 2P_B' u_B' \,, \qquad P_B' = P_B - P_0' \tag{7.3.10}$$

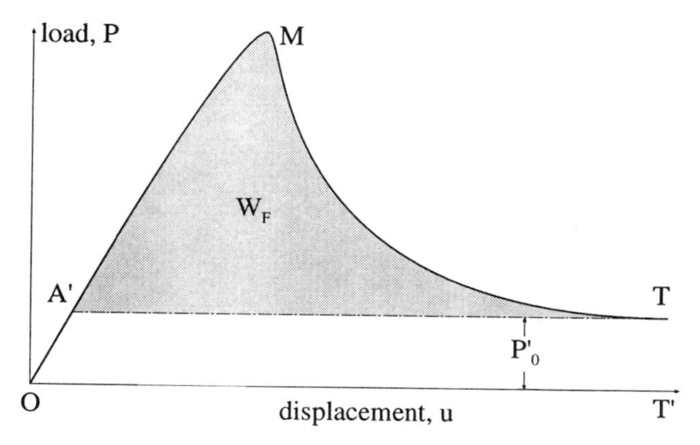

Figure 7.3.7 Load-displacement curves for overcompensated three-point bending test in an ideal situation (at T, the specimen is fully broken).

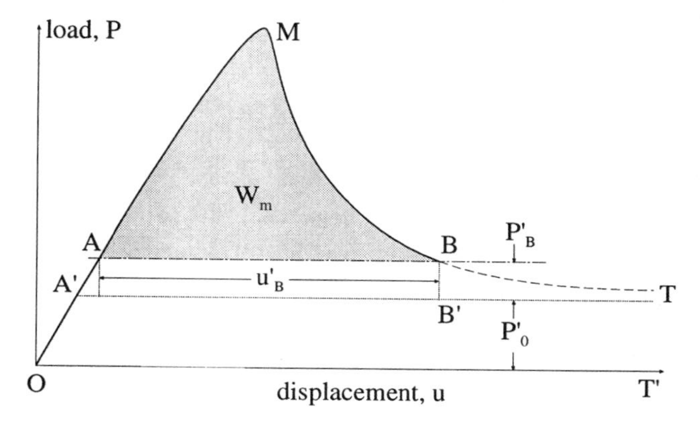

Figure 7.3.8 Load-displacement curves for overcompensated three-point bending test in a real situation. The test is stopped at B, before the specimen is fully broken.

where $u'_B = u_B - u_{A'} \approx u_B - u_A$ (note that the points A and A' lie on the elastic line, and so $u_A - u_{A'} \ll u_B$). Unfortunately, we do not know P'_B. But we can estimate it as described next.

One of the results of Petersson (1981), which will be proven with extended generality in the next subsection, is that the load-displacement curve behaves asymptotically as u^{-2}, i.e.,

$$P - P'_0 = \frac{A}{(u - u_A)^2} \qquad \text{for large } u \tag{7.3.11}$$

where A is a constant (we shall see later how this constant is related to the softening curve; for the time being, it suffices to see how to calculate it). In this equation P and u are the measurements, and u_A is easily determined by interpolation. Then, if the test has proceeded long enough, we can expect this equation to hold for the last part of the curve. So, P'_0 and A can be calculated by least-square fitting to the data.

Since, in reality, we do not need P'_0, we can force the curve to be exactly satisfied by the last point B, so that $P_B - P'_0 = P'_B = A/(u_B - u_A)^2$. Eliminating P'_0, we get

$$P - P_B = A\left[\frac{1}{(u - u_A)^2} - \frac{1}{(u_B - u_A)^2}\right] \tag{7.3.12}$$

Therefore, if we plot $P - P_B$ vs. $(u - u_A)^{-2} - (u_B - u_A)^{-2}$, we can determine A by least square fitting of a straight line through the origin. (Notice, in doing so, that A has dimension $[FL^2]$.) Fig. 7.3.9 displays this plot for a particular case revealing the trend of the experimental data and the fitted line. Nonlinear

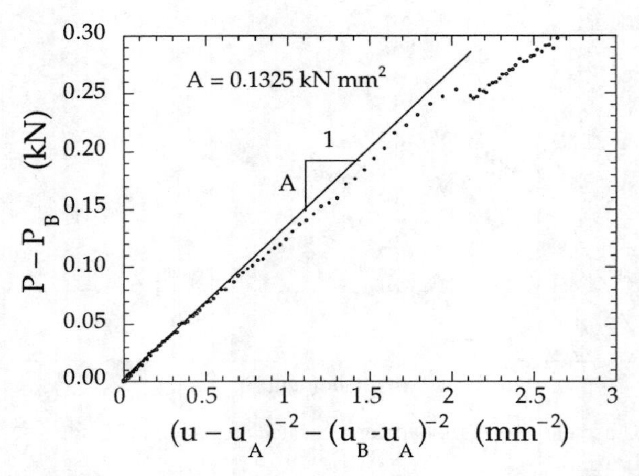

Figure 7.3.9 Example of the determination of constant A in Eq. (7.3.13). The plot corresponds to the test results in Fig. (7.2.6).

regression can be used as an alternative, but the graphic representation in Fig. 7.3.9 is recommended to identify the fitting interval and to check that the test has been well behaved. Once A has been determined, (7.3.10) can be rewritten as

$$W_F = W_m + 2\frac{A}{u_B - u_A} \tag{7.3.13}$$

Elices, Guinea and Planas (1992) found that the correction term in the foregoing equation, which is the part of the work not measured, may be as large as 20% of the total, for small specimens. So, due care must be exerted in calculating it.

7.3.4 Determination of a Bilinear Softening Curve

Using the data from the foregoing stable fracture test and from a cylinder splitting test, it is possible to determine a bilinear curve that approximates the softening behavior of concrete (Guinea, Planas and Elices 1994b). A good estimate of f'_t is obtained from the Brazilian test. Next, from the peak load of the stable test, the value of w_1 is determined. From the stable test, two more values are obtained as indicated in the previous subsection: that of G_F, which coincides with the area under the softening curve, and that of a certain constant A that can be related to the abscissa of the center of gravity of the area defined by the softening curve $\overline{w}$. With these four values the complete bilinear curve is completely determined.

The relationship between the constant A that defines the tail of the bending test and the center of gravity of the softening curve is established following Petersson's rigid-body approximation of the kinematics of the beam at the late stages of loading depicted in Fig. 7.3.10. It is assumed that the two halves of the beam are rigid and that they are bridged by a cohesive zone, so that the crack opening is given by

$$w = 2\theta x \tag{7.3.14}$$

where x is measured as shown in Fig. 7.3.10 and θ is the rotation of each half of the specimen (θ is assumed to be small). We now write the condition of equilibrium of moments with respect to the load-point:

$$\frac{Ps}{4} = \int_0^D \sigma\, x\, b\, dx \tag{7.3.15}$$

Since the stress σ is the cohesive stress, we have $\sigma = f(w)$, and substituting x as a function of w from (7.3.14), we finally get

$$P = \frac{b}{s\theta^2} \int_0^{w_T} f(w)\, w\, dw \tag{7.3.16}$$

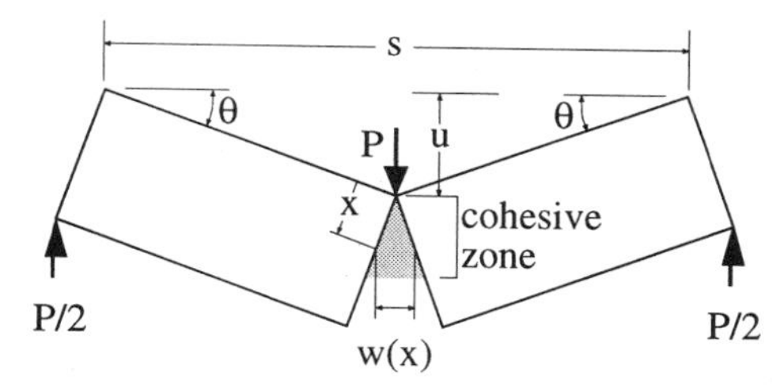

Figure 7.3.10 Rigid body kinematics towards the end of the test (adapted from Elices, Guinea and Planas 1992).

where w_T is the opening at the initial notch tip. If the rotation is large enough, then $w_T > w_c$ and the integral is the first order moment of the softening curve, which is equal, by definition, to the abscissa $\overline{w}$ of the center of gravity of the area defined by the curve and the axes, times this area, which is G_F also by definition. Therefore, if we also write that $\theta = u/(s/2)$, we finally get the expression (Elices, Guinea and Planas 1992)

$$P = \frac{bs}{4u^2} G_F \overline{w} \tag{7.3.17}$$

Now, comparing this to (7.3.11) with $P_0' = 0$ and $u_A = 0$ (since we assumed a perfect weight compensation), we see that the value of $\overline{w}$ in terms of A is:

$$\overline{w} = \frac{4A}{bsG_F} \tag{7.3.18}$$

Once f_t', w_1, G_F, and $\overline{w}$ have been determined, it is a simple geometrical problem to completely define the bilinear softening curve, as depicted in Fig. 7.3.11a. When nondimensional values, labeled by hats, are used as defined in (7.1.12), $\hat{w}_c$ is obtained from the quadratic equation

$$\hat{w}_c^2 - 2\hat{w}_c \frac{3\hat{\overline{w}} - \hat{w}_1}{2 - \hat{w}_1} + 2\hat{w}_1 \frac{3\hat{\overline{w}} - 4}{2 - \hat{w}_1} = 0 \tag{7.3.19}$$

and the coordinates of the kink point are given by

$$\hat{w}_k = \hat{w}_1 \frac{\hat{w}_c - 2}{\hat{w}_c - \hat{w}_1}, \qquad \hat{\sigma} = \frac{2 - \hat{w}_1}{\hat{w}_c - \hat{w}_1} \tag{7.3.20}$$

Planas and co-workers have found a good agreement of experiments and prediction corresponding to the bilinear curve so determined (Guinea, Planas and Elices 1992; Planas, Elices and Guinea 1995). As an example, Fig. 7.3.11b shows the comparison of the theoretical and experimental results for *independent* tests of *unnotched* beams in bending; as can be seen, the agreement is satisfactory.

Exercises

7.11 To analyze the conditions for crack instability in an unnotched structural element, specimens of concrete from a precast concrete factory were taken. Brazilian tests on cylinders 150 mm in diameter were carried out with a loading strip 12 mm width. The result was a splitting tensile strength $f_{ts} = 3.23$ MPa. Moreover, three-point bending tests were carried out on beams 100 mm in depth and 100 mm thick, with notches 48 mm deep (following the tolerances in Fig. 7.3.2). The mean failure load was 2.53 kN. Determine the characteristic length ℓ_1 and the horizontal intercept of the initial part of the softening curve w_1 if the measured elastic modulus was 36 GPa.

7.12 To refine the estimates made in the previous exercise, use the ℓ_1 value found to evaluate D/ℓ_1 for the Brazilian test, and use the curves in Fig. 7.3.1 to get a better estimate of f_t'. Recompute the values of ℓ_1 and w_1 and decide whether further iterations are required.

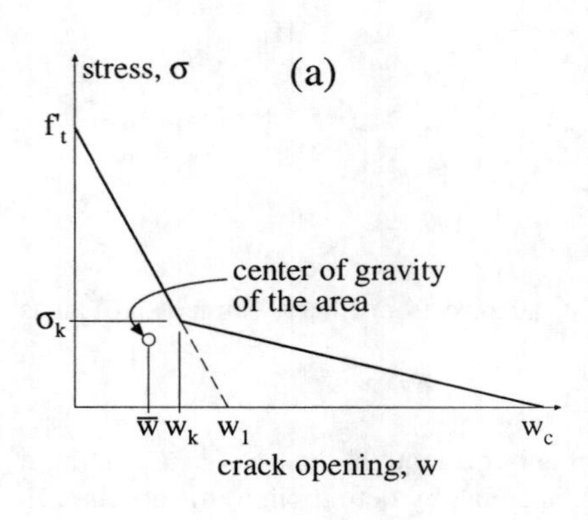

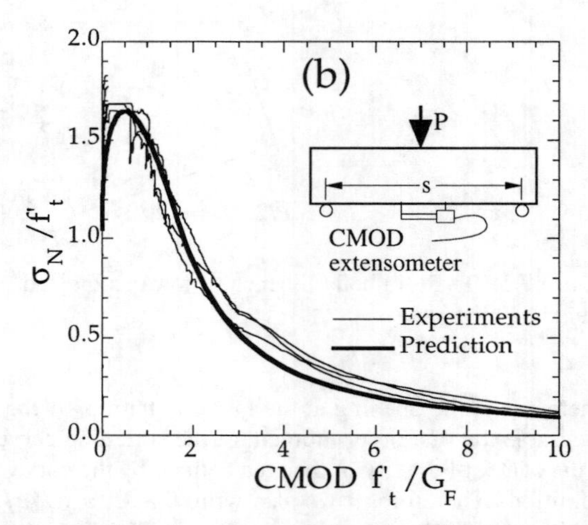

Figure 7.3.11 (a) Geometry of the bilinear softening curve; (b) predicted and measured load-CMOD curves of unnotched bent beams; the softening curve was independently determined from cylinder splitting and notched beam tests. (Adapted from Planas, Elices and Guinea 1995.)

7.13 Repeat the preceding analysis under the assumption that the Brazilian tests were carried out using loading strips of 25 mm width.

7.14 In the test record of Fig 7.2.15, the origin has been set at point A as defined in Fig. 7.3.8, so that the last point of the record has apparent load equal to zero, and displacement $u'_B = 1.477$ mm. The plot of Eq. (7.3.12) corresponding to this test is shown in Fig. 7.3.9. (a) Determine the position of the actual zero load (i.e., the position of the asymptote in Fig. 7.3.8, defined by P'_B). How much does this affect the value of the peak load in percent? (b) Determine the correction to be done for the fracture work. How much does this affect the value of the fracture energy in percent?

7.15 For the concrete of the previous exercise, whose characteristics were defined in exercise 7.8, give a bilinear approximation of the softening curve. (Note: the work enclosed by the curve in Fig. 7.2.15, which corresponds to area W_m in Fig. 7.3.8, has been precisely evaluated as 1.64 J.)

7.4 Pseudo-Boundary-Integral Methods for Mode I Crack Growth

This section presents a collection of methods with common features that reduce the cohesive crack problem to a system of equations defined on the cracked cross-section. They are similar to boundary integral methods, but the kernel of the integral equation is discretized *a priori* and determined by other computational means, typically the finite element method (FEM). We call them the pseudo-boundary-integral methods (PBI).

7.4.1 The Underlying Problem

The PBI methods that we are going to discuss now are based on a discretization of the problem as sketched in Fig. 7.4.1: it is assumed that the crack section (a symmetry plane) is discretized by dividing it into M' elements, each of size $h = D/M'$ where D is the depth of cross section, and the governing equations are written for the nodal crack openings and stresses. We assume, in general, that the structure has an initial notch spanning the nodes $1, \cdots, N$, and that the cohesive zone has extended from the node at the initial crack tip $T = N + 1$ to node C; then, the nodes from $U = C + 1$ on pertain to the uncracked ligament.

The equations that must be satisfied along the central cross-section are indicated in Fig. 7.4.1, and consists simply on the three following sets: (1) $\sigma = 0$ on the notch faces, (2) $\sigma = f(w)$ on the cohesive

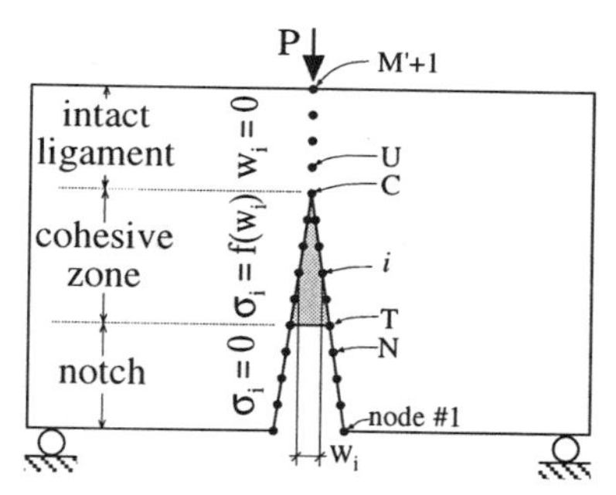

Figure 7.4.1 Node layout for the pseudo-boundary-integral (PBI) methods.

zone ($f(w)$ is the softening function), and (3) $w = 0$ on the intact ligament. This is written in term of the nodal values as

$$\sigma_i = 0 \qquad \text{for } i = 1, \cdots, N \tag{7.4.1}$$

$$\sigma_i = f(w_i) \qquad \text{for } i = T, \cdots, C \tag{7.4.2}$$

$$w_i = 0 \qquad \text{for } i \geq C \tag{7.4.3}$$

where σ_i are the nodal stresses and w_i the nodal stresses. Note that both the second and third conditions are to be satisfied at the cohesive crack tip ($i = C$).

Note also that the stresses at the initial crack tip are usually discontinuous : zero on the notch side and nonzero on the cohesive zone side; therefore, σ_i for $i = T$ refers to the stress on the element on only one of its sides.The relation between stresses and nodal forces p_i is thus approximated as

$$\sigma_i = \begin{cases} p_i/bh & \text{for} \quad i \neq T \\ 2p_i/bh & \text{for} \quad i = T \end{cases} \tag{7.4.4}$$

where b is the thickness. The formula indicates that for all but the initial tip the nodal force is distributed over one half element on each side, while, at the tip, the nodal force is distributed only over the half element on the cohesive side.

In the foregoing system, σ_i and w_i are explicit unknowns, and the load P is an implicit unknown. The PBI methods use elastic relations to link σ_i, w_i, and P.

7.4.2 Petersson's Influence Method

In Petersson's (1981) method, the solution is sought by superposing the elastic cases shown in Fig. 7.4.2: a crack is let into the structure with its tip at node $M + 1$; the resulting geometry is subjected to the external load P (Fig. 7.4.2a) and to closing nodal forces p_j (Fig. 7.4.2b). P and the nodal forces must be computed to satisfy the governing equations (7.4.1)–(7.4.3). To do so, we write the crack opening at node i resulting from the action of P and p_j:

$$w_i = C_i P - \sum_{j=1}^{M} K_{ij} p_j \tag{7.4.5}$$

where C_i is the crack opening at node i produced by a unit external force, and K_{ij} is the crack opening produced at node i by a unit *opening* nodal force applied at node j.

C_i can be obtained from a single elastic analysis using a standard finite element program; the determination of K_{ij} requires M such analyses. In total $M + 1$ elastic analyses must be performed. However, as we will see in the following, these $M + 1$ elastic analyses provide information for geometrically similar specimens of any size, which is one powerful feature of the method (Section 7.4.5).

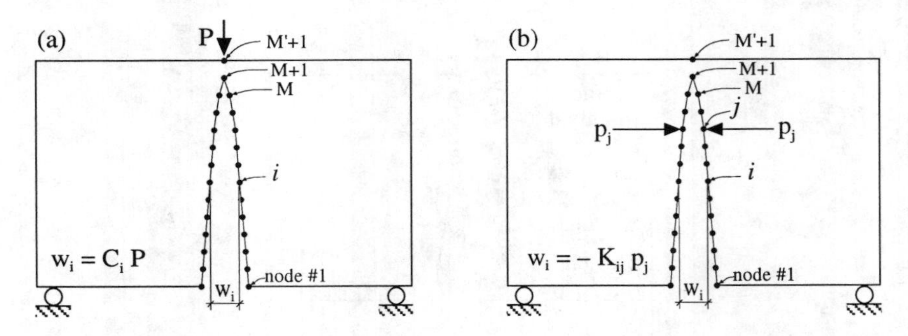

Figure 7.4.2 Distribution of crack openings in a deeply cracked structure due to: (a) external load; (b) nodal forces. (Note that the nodal force is taken positive when closing because this corresponds to tensions on the crack faces; although these forces produce crack closure, the crack is shown open for clarity; the sign is incorporated into the expression $w_i = -K_{ij}p_j$.)

The foregoing equations, which involve nodal forces, are immediately written as a function of stresses using (7.4.4), with the result

$$w_i = C_i P - \sum_{j=1}^{M} K'_{ij}\sigma_j \tag{7.4.6}$$

where

$$K'_{ij} = \begin{cases} K_{ij}bh & \text{for} \quad j \neq T \\ K_{ij}bh/2 & \text{for} \quad j = T \end{cases} \tag{7.4.7}$$

Note that while K_{ij} is symmetric by virtue of the reciprocity theorem, K'_{ij} is not due to the presence of the notch (the T-th column is divided by 2 as indicated in (7.4.7), while the T-th row is not).

The system of the M equations (7.4.6) and the $M + 1$ equations (7.4.1)–(7.4.3) defines a solution for the $2M + 1$ unknowns w_i, σ_i and P if the extension of the cohesive crack is given (Petersson 1981). To solve this problem, Petersson used a polygonal approximation for the softening curve, so that the system of equations would be piecewise linear. For nonlinear softening, an iterative method is required which, for a large number of equations (say over 150) may slow down the computation on a personal computer significantly. The number of equations can be reduced by the analytical approach described next.

7.4.3 Improved Solution Algorithm of Planas and Elices

Planas and Elices (1991a) found that the solution of the foregoing equations can be considerably accelerated if a partial solution of the system is carried out before the step-by-step crack growth analysis starts. The procedure is based on the observation that, since Eqs. (7.4.1) directly give the stresses on the notch faces, we can partition the unknown column matrices $\boldsymbol{\sigma}$ and $\mathbf{w}$ as follows:

$$\boldsymbol{\sigma} = [\underbrace{[0, \cdots, 0]}_{\text{notch}}, \underbrace{[\sigma_T, \cdots, \sigma_M]}_{\text{initial ligament}}]^T = [\boldsymbol{\sigma}_N(\equiv \mathbf{0}), \boldsymbol{\sigma}_L]^T \tag{7.4.8}$$

$$\mathbf{w} = [\underbrace{[w_1, \cdots, w_N]}_{\text{notch}}, \underbrace{[w_T, \cdots, w_M]}_{\text{initial ligament}}]^T = [\mathbf{w}_N, \mathbf{w}_L]^T \tag{7.4.9}$$

Then, the column matrix C_i and the square matrix K'_{ij} are accordingly partitioned, so that Eq. (7.4.6) is written as

$$\begin{bmatrix} \mathbf{K}'_{NN} & \mathbf{K}'_{NL} \\ \mathbf{K}'_{LN} & \mathbf{K}'_{LL} \end{bmatrix} \begin{Bmatrix} \mathbf{0} \\ \boldsymbol{\sigma}_L \end{Bmatrix} + \begin{Bmatrix} \mathbf{w}_N \\ \mathbf{w}_L \end{Bmatrix} = P \begin{Bmatrix} \mathbf{C}_N \\ \mathbf{C}_L \end{Bmatrix} \tag{7.4.10}$$

Now, given the notch depth, this system can be solved for $\boldsymbol{\sigma}_L$ before starting the analysis of cohesive crack so that we get

$$\boldsymbol{\sigma}_L = P\mathbf{L}_L - \mathbf{M}_{LL}\mathbf{w}_L \tag{7.4.11}$$

where

$$\mathbf{M}_{LL} = \left(\mathbf{K}'_{LL}\right)^{-1}, \qquad \mathbf{L}_L = \mathbf{M}_{LL}\mathbf{C}_L \tag{7.4.12}$$

Once this is done, at each crack propagation step $\boldsymbol{\sigma}_L$ and $\mathbf{w}_L$ are partitioned again into the parts corresponding to the cohesive zone and to the uncracked ligament:

$$\boldsymbol{\sigma}_L = [\underbrace{[\sigma_T, \cdots, \sigma_C]}_{\text{cohesive zone}}, \underbrace{[\sigma_U, \cdots, \sigma_M]}_{\text{uncracked ligament}}]^T = [\boldsymbol{\sigma}_C, \boldsymbol{\sigma}_U]^T \tag{7.4.13}$$

$$\mathbf{w}_L = [[w_T, \cdots, w_C], [0, \cdots, 0]]^T = [\mathbf{w}_C, \mathbf{w}_U (\equiv \mathbf{0})]^T \tag{7.4.14}$$

and the same is done for the vector $\mathbf{L}_L$ and the matrix $\mathbf{M}_{LL}$ in (7.4.11), so that the equation becomes

$$\begin{bmatrix} \boldsymbol{\sigma}_C \\ \boldsymbol{\sigma}_U \end{bmatrix} = P \begin{bmatrix} \mathbf{L}_C \\ \mathbf{L}_U \end{bmatrix} - \begin{bmatrix} \mathbf{M}_{CC} & \mathbf{M}_{CU} \\ \mathbf{M}_{UC} & \mathbf{M}_{UU} \end{bmatrix} \begin{bmatrix} \mathbf{w}_C \\ \mathbf{0} \end{bmatrix} \tag{7.4.15}$$

Developing the equation for the upper submatrices, we find that the stresses on the cohesive zone depend only on the crack openings over this zone. Then, imposing the conditions (7.4.2), and the additional condition that $w_C = 0$, we get the system of equations:

$$f(w_i) = P\ \{\mathbf{L}_L\}_i - \sum_{j=T}^{C-1} \{\mathbf{M}_{LL}\}_{ij}\, w_j \qquad \text{for } i = T, \cdots, C-1 \tag{7.4.16}$$

$$f(0) = P\ \{\mathbf{L}_L\}_C - \sum_{j=T}^{C-1} \{\mathbf{M}_{LL}\}_{Cj}\, w_j \tag{7.4.17}$$

which can be solved for P from the last equation, substituted back into Eqs. (7.4.16) and then iteratively solved for w_i on the cohesive zone. Once $\mathbf{w}_C$ has been determined, $\boldsymbol{\sigma}_L$ follows from (7.4.11) and $\mathbf{w}_N$ from (7.4.10).

Overall, this algorithm is about 3 times faster than Petersson's in the worst case (no notch), but it is 12 times faster when the notch spans half the total number of nodes, and can be 100 times faster for very deep notches.

7.4.4 Smeared-Tip Method

The solution algorithm can be further improved using a method first proposed by Planas and Elices (1986b, 1992a) in a semianalytical formulation of the crack growth analysis and later adopted, with some modifications, by Bažant (1990d) and Bažant and Beissel (1994), who called it the smeared-tip method. The virtue of the method is that it leads to a triangular system of equations that can be directly solved by forward substitution. Here we give a version based on the discretization of the ligament described at the beginning of this section (Planas 1993).

In essence, the method uses a decomposition of the problem into LEFM problems such as that shown in Fig. 7.4.3: the solution is sought as a superposition of $M < M'$ elastic cases, each with a stress-free crack with its tip at node j ($j = 1, \cdots, M$), and an external load equal to ΔP_j. Then, the distributions of nodal forces p_i and crack openings w_i, and the total external load P are given by

$$p_i = \sum_{j=1}^{M} R_{ij}\, \Delta P_j \tag{7.4.18}$$

$$w_i = \sum_{j=1}^{M} D_{ij}\, \Delta P_j \tag{7.4.19}$$

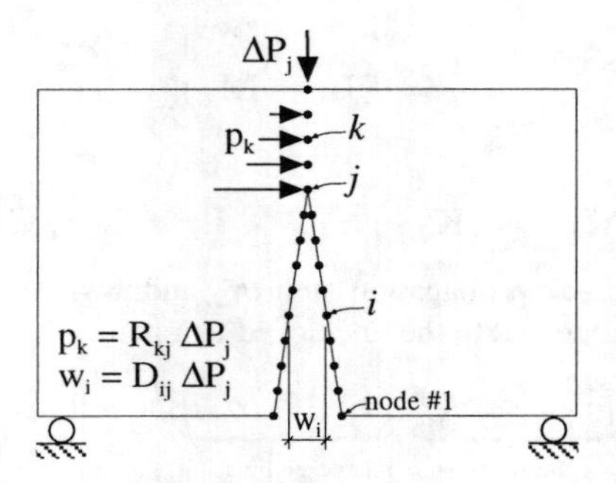

Figure 7.4.3 Basic linear elastic case for the smeared-tip superposition.

$$P = \sum_{j=1}^{M} \Delta P_j \tag{7.4.20}$$

in which R_{ij} is the reaction (i.e., nodal force) at node i produced by a unit external load when the crack tip is at node j, and D_{ij} is the crack opening at node i due to a unit applied force when the crack tip is located at node j. The *loading vector* ΔP_j includes the set of unknowns in this formulation. They vanish identically except on the cohesive zone.

Since the crack faces are stress-free, we have $R_{ij} = 0$ for $i < j$, and so $\mathbf{R}$ is a lower-triangular matrix:

$$\mathbf{R} = \begin{bmatrix} R_{1\,1} & 0 & 0 & \cdots & 0 \\ R_{2\,1} & R_{2\,2} & 0 & \cdots & 0 \\ \vdots & \vdots & \vdots & \vdots & \vdots \\ R_{M-1\,1} & R_{M-1\,2} & \cdots & R_{M-1\,M-1} & 0 \\ R_{M\,1} & R_{M\,2} & \cdots & R_{M\,M-1} & R_{M\,M} \end{bmatrix} \tag{7.4.21}$$

In a similar way, since the crack opening is zero for $i \geq j$, $\mathbf{D}$ is an upper-triangular matrix with zero main diagonal:

$$\mathbf{D} = \begin{bmatrix} 0 & D_{1\,2} & D_{2\,2} & \cdots & D_{1\,M} \\ 0 & 0 & D_{2\,3} & \cdots & D_{2\,M} \\ \vdots & \vdots & \vdots & \vdots & \vdots \\ 0 & 0 & \cdots & 0 & D_{M-1\,M} \\ 0 & 0 & \cdots & 0 & 0 \end{bmatrix} \tag{7.4.22}$$

Note that $\mathbf{R}$ and $\mathbf{D}$ can be stored in the same full square matrix, to minimize memory requirements. To obtain $\mathbf{R}$ and $\mathbf{D}$, M elastic computations for a unit load must be performed, with crack tips at nodes $1, 2, \cdots, M$. Alternatively, $\mathbf{R}$ and $\mathbf{D}$ can be calculated from Petersson's influence matrices $\mathbf{K}$ and $\mathbf{C}$ (see Section 7.4.5).

Assuming that $\mathbf{M}$ and $\mathbf{D}$ have been computed using finite elements or any other method, we can solve the cohesive crack problem by substituting (7.4.18) first into (7.4.4), and then the result in (7.4.1) and (7.4.2); next, we substitute (7.3.19) into (7.4.2) and (7.4.3). The resulting set of equations is:

$$\sum_{j=1}^{i} R'_{ij}\,\Delta P_j = 0 \qquad \text{for} \qquad i = 1, \cdots, N \tag{7.4.23}$$

$$\sum_{j=1}^{i} R'_{ij}\,\Delta P_j = f\left(\sum_{j=i+1}^{M} D_{ij}\,\Delta P_j\right) \qquad \text{for} \qquad i = T, \cdots, C \tag{7.4.24}$$

$$\sum_{j=i+1}^{M} D_{ij}\,\Delta P_j = 0 \qquad \text{for} \qquad i = C, \cdots, M \tag{7.4.25}$$

where

$$R'_{ij} = \begin{cases} R_{ij}/bh & \text{for} \quad j \neq T \\ 2R_{ij}/bh & \text{for} \quad j = T \end{cases} \tag{7.4.26}$$

and the limits of the sums have been set in accordance with the triangular properties of $\mathbf{R}'$ and $\mathbf{D}$. Now, the first and last subset of equations are readily solvable, with the result:

$$\Delta P_j = 0 \qquad \text{for } i \notin (T, \cdots, C) \tag{7.4.27}$$

and so the set of equations is reduced to

$$\sum_{j=T}^{i} R'_{ij}\,\Delta P_j = f\left(\sum_{j=i+1}^{C} D_{ij}\,\Delta P_j\right) \qquad \text{for } i = T, \cdots, C \tag{7.4.28}$$

This is a nonlinear system of $C - T + 1$ equations, which can be solved iteratively. Given an estimate of the loading vector ΔP_j^{α}, we substitute in the right-hand side and solve for a better estimate $\Delta P_j^{\alpha+1}$ from the left-hand side:

$$\sum_{j=T}^{i} R'_{ij}\,\Delta P_j^{\alpha+1} = f\left(\sum_{j=i+1}^{C} D_{ij}\,\Delta P_j^{\alpha}\right) \qquad \text{for } i = T, \cdots, C \tag{7.4.29}$$

Since the system of equations is triangular, it is readily solved by forward substitution. Once convergence is achieved, the stresses, crack openings, and applied load are computed from (7.4.18)–(7.4.19).

This algorithm has been implemented in a program called *Splitting-Lab* by J. Planas and his colleagues at the University of Madrid (a demonstration version of the program running on an Apple Macintosh is available from the author). They have solved hundreds of cases with success; although incremental propagation of the cohesive tip decreases the number of iterations at each step, large steps can be used without any convergence problems. Iteration can even be initiated with $\Delta P_j = 0$.

7.4.5 Scaling of the Influence Matrices

In the foregoing, we assumed the elastic influence matrices and vectors K_{ij}, C_i, R_{ij}, and D_{ij} to be determined computationally. Were they to be calculated for every particular case, the procedure would be very expensive (exactly as much as using statically condensed superelements in advanced FEM codes). The advantage of the PBI methods is that the influence matrices need only to be computed once for each family of geometrically similar specimens. A similarity of notches is not required. The reason is that elastic similitude holds, according to which nodal forces are proportional to the load, but independent of the elastic constants and size, while the displacements are proportional to the load, and inversely proportional to thickness and elastic modulus (for plane stress), and again independent of the size. Therefore, it suffices to make a computation for $P = 1, b = 1, D = 1$, and $E = 1$ (in appropriate units and plane stress). If the result of such calculations are K^1_{ij} and C^1_i, and (or) R^1_{ij} and D^1_{ij}, the corresponding values for any size, thickness, and material are

$$K_{ij} = K^1_{ij}\frac{1}{bE}\,, \qquad C_i = C^1_i\frac{1}{bE}\,, \qquad R_{ij} = R^1_{ij}\,, \qquad D_{ij} = D^1_{ij}\frac{1}{bE} \tag{7.4.30}$$

Since the pairs $\mathbf{K}$-$\mathbf{C}$ and $\mathbf{R}$-$\mathbf{D}$ can be used to solve the same problem, they must be related. The relationship is easily found by applying (7.4.5) to the problem of a unit load ($P = 1$) applied to a structure with a crack having its tip at node k (similar to Fig. 7.4.3, but with a unit load). Then, by definition, $p_j = R_{jk}$ and $w_i = D_{ik}$ and, assuming that C_i and K_{ij} are given, we must have

$$D_{ik} = C_i - \sum_{j=1}^{M} K_{ij} R_{jk} \tag{7.4.31}$$

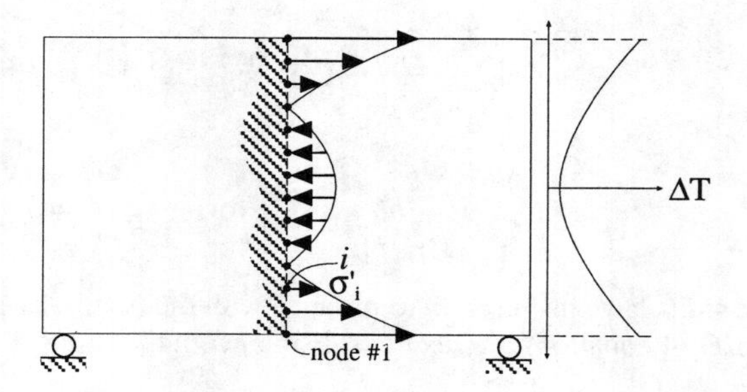

Figure 7.4.4 Internal stress field.

From these equations, we can completely determine $\mathbf{D}$ and $\mathbf{R}$. Indeed, owing to the triangularity of D_{ik} and R_{jk}, we can split the foregoing system in two, one for $i < k$ and the other for $i \geq k$. The resulting two systems of equations are

$$D_{ik} = C_i - \sum_{j=k}^{M} K_{ij} R_{jk} \qquad \text{for } i < k \tag{7.4.32}$$

$$0 = C_i - \sum_{j=k}^{M} K_{ij} R_{jk} \qquad \text{for } i \geq k \tag{7.4.33}$$

These systems can be solved sequentially. First the second system is solved for R_{jk} (once for each k). Next, D_{ik} is obtained from the first system by direct substitution. Therefore, from a mathematical point of view, the smeared-tip method is just a transformation of the classical influence method which leads to an *a priori* triangularization of the matrices. This has the obvious advantage that the triangularization is made once and for all, and so the time savings are tremendous.

7.4.6 Inclusion of Shrinkage or Thermal Stresses

To include internal stresses due to internal strain fields created by shrinkage or thermal strains (or even steel-concrete interaction), one only needs to know the elastic stress distribution that the imposed strain fields would produce on the *uncracked* structure (Fig. 7.4.4). This field is directly superposed to that in Fig. 7.4.3. Since the internal strain loading contributes neither to the crack openings (there is no crack) nor to the external load, Eqs. (7.4.19) and (7.4.20) do not change. The only changes are in the stresses, and so, after inserting (7.4.4) and using the triangularity of $\mathbf{R}'$, Eq. (7.4.18) is replaced by

$$\sigma_i = \sum_{j=i}^{M} R'_{ij} \Delta P_j + \sigma'_i \tag{7.4.34}$$

where σ'_i is the internal stress at node i due to the internal strain field. These stresses are assumed as data here, and must be determined by solving the corresponding thermal or shrinkage elastic problem.

Now, proceeding as we did to obtain Eqs. (7.4.23)–(7.4.25), we get the system

$$\sum_{j=1}^{i} R'_{ij}\, \Delta P_j + \sigma'_i = 0 \qquad \text{for} \qquad i = 1, \cdots, N \tag{7.4.35}$$

$$\sum_{j=1}^{i} R'_{ij}\, \Delta P_j + \sigma'_i = f\left(\sum_{j=i+1}^{M} D_{ij}\, \Delta P_j \right) \qquad \text{for} \qquad i = T, \cdots, C \tag{7.4.36}$$

$$\sum_{j=i+1}^{M} D_{ij}\,\Delta P_j = 0 \qquad \text{for} \qquad i = C, \cdots, M \tag{7.4.37}$$

These equations are a little bit more complicated than before, but still can be rearranged to solve a system similar to that for the simple case without internal stresses. First, from the last system of equations, we find as before that

$$\Delta P_j = 0 \qquad \text{for } j > C \tag{7.4.38}$$

Next we solve (7.4.35) numerically by forward substitution for ΔP_j $(j = 1, \cdots, N)$ and substitute in (7.4.36). Note that this operation is independent of the loading step and needs to be carried out only once at the beginning of the calculation if the internal stress field does not change during the loading progress. After some algebra, the resulting equation is obtained:

$$\sum_{j=T}^{i} R'_{ij}\,\Delta P_j = f\left(\sum_{j=i+1}^{M} D_{ij}\,\Delta P_j\right) - \overline{\sigma}_i \qquad \text{for} \qquad i = T, \cdots, C \tag{7.4.39}$$

where the vector $\overline{\sigma}_i$ is given by

$$\overline{\sigma}_i = \sum_{j=1}^{N} R'_{ij}\,\Delta P_j + \sigma'_i \tag{7.4.40}$$

in which everything is known from previous calculations.

Now we can summarize the steps required to perform a calculation, assuming that the internal stress distribution σ'_j is known:

1. Solve for ΔP_j $(j = 1, \cdots, N)$ by forward substitution from (7.4.35).
2. Compute $\overline{\sigma}_i$ $(i = T, \cdots, M)$ from (7.4.40).
3. Solve iteratively (7.4.39) for ΔP_j $(j = T, \cdots, C)$.
4. Compute the stress distribution σ_i from (7.4.34)
5. Compute the stress distribution w_i from (7.4.19)
6. Compute the external load P from (7.4.20)

In the last three steps, Eq. (7.4.38) must be used. This is automatically done by extending the sums up to $j = C$ only (instead of up to $j = M$).

7.4.7 Inclusion of a Crack-Tip Singularity

Including a crack tip singularity in the smeared-tip method is easy (although limited by the accuracy with which ordinary finite elements can represent the singular stress field). Assume that we have a cohesive crack model with singularity, and let K_{Icc} be the critical stress intensity factor at the cohesive zone tip. Let the cohesive zone tip be at node C as in Fig 7.4.1. Then we can seek a solution as a superposition of the case shown in Fig. 7.4.5a, in which the external load P_C is such that it produces a stress intensity factor exactly equal to K_{Icc}, and the case in Fig. 7.4.5b, which has no singularity, and can thus be treated as in the previous analyses (except that P is replaced by $P - P_C$).

Now we notice that the first case in Fig. 7.4.5a is the same as that in Fig. 7.4.3 with $j = C$ and $\Delta P_j = P_C$. Therefore, the expressions for the nodal stresses and displacements can be written in general as

$$\sigma_i = \begin{cases} \sum_{j=1}^{i} R'_{ij}\,\Delta P_j + \sigma'_i + R'_{iC} P_C & \text{for } i \neq C \\ \sum_{j=1}^{i} R'_{ij}\,\Delta P_j + \sigma'_i & \text{for } i = C^- \\ \infty & \text{for } i = C^+ \end{cases} \tag{7.4.41}$$

$$w_i = \sum_{j=i+1}^{M} D_{ij}\,\Delta P_j + D_{iC} P_C \tag{7.4.42}$$

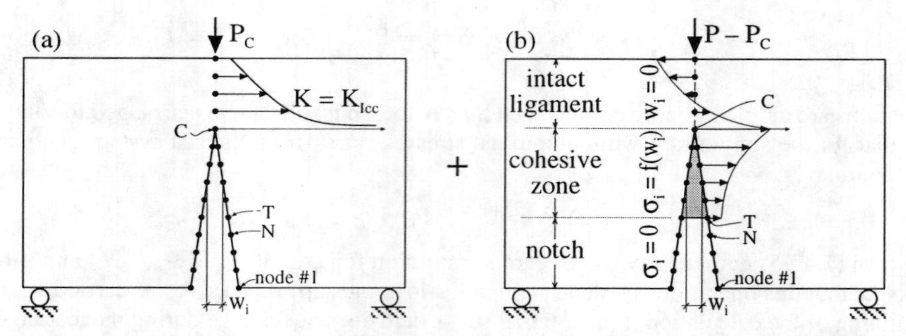

Figure 7.4.5 Cohesive crack with singularity: elastic decomposition in a case with singular stress field (a) and a regular stress distribution (b).

$$P = \sum_{j=1}^{M} \Delta P_j + P_C \tag{7.4.43}$$

where we have also included the internal stresses, for the sake of generality. Note that the expressions for the stress are now discontinuous at the cohesive crack tip ($i = C$), and so the stress is finite at the cohesive side ($i = C^-$) and infinite at the intact ligament side ($i = C^+$).

The unknowns in these equations are again ΔP_j ($j = 1, \cdots, M$) because P_C is the datum. Indeed, if the expression for the stress intensity factor is known in the form $K_I = P/(bD^{1/2})\hat{k}(\alpha)$, then

$$P_C = \frac{1}{\hat{k}(\alpha_C)} K_{Icc} b\sqrt{D}\,, \qquad \alpha_C = \frac{C}{M'} \tag{7.4.44}$$

where M' is the total number of equal elements along the central cross section.

Since the condition that the stress intensity factor must be equal to K_{Icc} is verified automatically, the only equations that remain to be fulfilled are again (7.4.1)–(7.4.3) where now function $f(w)$ has the meaning of bridging stresses and (7.4.2) extends from T to C^-. Then substitution of expressions (7.4.41)–(7.4.43) leads to a system of equations for ΔP_j. Since $R_{iC} = 0$ for $i < C$ and $D_{iC} = 0$ for $i \geq C$, it turns out that the resulting system is identical to the system (7.4.35)–(7.4.37) in the previous analysis, except for (7.4.36) which changes to

$$\sum_{j=1}^{i} R'_{ij}\, \Delta P_j + \sigma'_i = f\left(\sum_{j=i+1}^{M} D_{ij}\, \Delta P_j + D_{iC} P_C\right) \qquad \text{for } i = T, \cdots, C^- \tag{7.4.45}$$

Therefore, (7.4.38), (7.4.35), and (7.4.40) still hold, and (7.4.39) is substituted by

$$\sum_{j=T}^{i} R'_{ij}\, \Delta P_j = f\left(\sum_{j=i+1}^{M} D_{ij}\, \Delta P_j + D_{iC}\, P_C\right) - \overline{\sigma}_i\,, \qquad \text{for } i = T, \cdots, C^- \tag{7.4.46}$$

7.4.8 Computation of Other Variables

In the foregoing, we have focused on the determination of the load, stresses, and crack openings as the cohesive crack grows. Obviously, any other variable can be computed if the influence factors for the basic elastic cases are computed (they usually are; it is just a matter of recording them). For example, if we refer to the load-point displacement u, we shall have, for Petersson's method,

$$u = C\,P + \sum_{j=1}^{M} Q_j p_j \tag{7.4.47}$$

where, referring to the geometry in Fig. 7.4.2, C is the displacement for $P = 1$, and Q_i is the displacement produced by a unit opening nodal force applied at node j. The reciprocity theorem assures that $Q_j = C_j$.

For the smeared-tip method, we have

$$u = \sum_{j=1}^{M} S_j \, \Delta P_j \tag{7.4.48}$$

where S_j is the load-point displacement produced by a unit external load when the structure has a crack up to node j (Fig. 7.4.3).

7.4.9 Limitations of the Pseudo-Boundary Integral (PBI) Methods

The PBI methods are not intended to solve a single particular case, because the time savings in computing the crack growth cannot compensate for the computation of the influence matrices. They are very powerful for the analysis of repetitive geometries (laboratory specimens, typically) of various sizes and different material properties. These methods are particularly suited for determining the size effect (peak load) for a wide range of sizes.

However, since the number of elements is fixed, the size of an element is proportional to the size of the structure, and becomes very large for very large sizes. Since the size of the cohesive zone at peak load is bounded, the number of elements over the cohesive zone decreases with size. Therefore, the method loses accuracy as soon as too few elements are left over the cohesive zone. From the experience of many computations, it appears that the method is dependable as long as $h < 0.08\ell_1$, approximately.

Exercises

7.16 Write down the matrix equation for (7.4.25) and verify that it implies that $\Delta P_j = 0$ for $j > C$.

7.17 Consider a center-cracked infinite panel with an initial crack of length $2a_0$ subjected to remote stress σ_∞. Take σ_∞ as a generalized force and consider the right half panel. Place equally spaced nodes along the crack line so that the half-crack is divided into N equal elements. Let the node at the center of the crack be the node number 1, so that the right crack tip coincides with node $T = N + 1$. Make a simple estimate of R'_{ij} and D_{ij} based on the known elastic solution for this case. (Hint: for D_{ij} and R'_{ij}, with $i \neq j$, use the value of the crack opening and the stress at the node as given by the elastic solution; for R_{ii}, integrate over the right half-element.) [Partial answer: $R'_{ii} = \sqrt{i - 3/4}$ for $i \neq T$, $R'_{TT} = 2\sqrt{T - 3/4}$.]

7.5 Boundary-Integral Methods for Mode I Crack Growth

7.5.1 A Basic Boundary Integral Formulation

We consider a cohesive crack growing in mode I as shown in Fig. 7.5.1a. We assume proportional loading characterized by a generalized force P. Assuming that the crack grows monotonically from its initial length a_0 to length a, we can take the crack length a as the independent variable and seek the value of P and of all the remaining variables (displacements, stresses, and so on) for any given value of a.

Since the bulk of the body is linear elastic, we can express the overall nonlinear solution as a superposition of known elastic solutions. This can be done by the decomposition shown in Fig. 7.5.1a-c in which the overall solution (a) is obtained as the superposition of the case (b) in which the external forces act and the cohesive stresses have been set to zero, and the cases (c) in which the cohesive stresses alone act on the cohesive zone. With this, the crack opening over the cohesive zone can be written as

$$w(x) = C_x(a)P - \int_{a_0}^{a} C_{xx'}(a) b\sigma(x') dx' \tag{7.5.1}$$

where $C_x(a)$ is the cross-compliance function giving the crack opening profile for a unit external load and no cohesive stresses; $C_{xx'}(a)$ is the cross-compliance relating the crack opening at x produced by unit

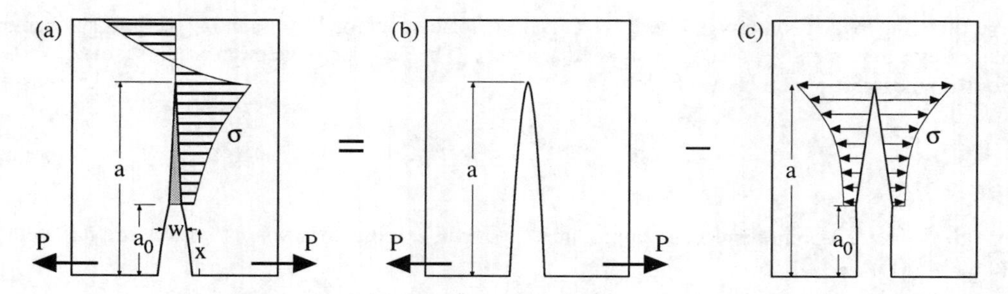

Figure 7.5.1 Decomposition of the cohesive crack problem.

opening forces at x'. The compliance functions C_x and $C_{xx'}$ can be obtained by classical elastic analysis, in particular by the method explained in Section 3.5.5 (we will apply this method after completing the system of equations).

Stipulating now that, in the cohesive zone, the stress and crack opening are related by the softening equation $\sigma = f(w)$, we get

$$\sigma(x) = f\left[C_x(a)P - \int_{a_0}^{a} C_{xx'}(a)b\sigma(x')dx'\right] \qquad \text{for} \quad a_0 < x < a \tag{7.5.2}$$

This is a nonlinear integral equation from which σ_x can be solved for any given *a and* P. This last point is very important and has to do with the fact that the foregoing equation gives stresses only inside the cohesive zone, not at the points ahead of the (cohesive) crack tip. A further condition is required. This condition is obtained by requiring that the stress ahead of the crack tip must be finite, i.e., that the stress intensity factor caused by the applied load and the cohesive stresses must vanish. This can be written as

$$\frac{P}{b\sqrt{D}}\hat{k}(\alpha) - \frac{1}{\sqrt{D}}\int_{a_0}^{a} k_G(\alpha, x/D)\sigma(x)dx = 0 \tag{7.5.3}$$

in which the first term is the stress intensity factor created by the applied load, written in the standard form (2.3.11), and the second term is the stress intensity factor caused by the cohesive stresses and written in terms of Green's function—see (3.1.10) and example 3.1.2, particularly Eq. (3.1.11). Note that this formulation can be automatically extended to cohesive cracks with singularity by simply setting the right-hand member of the last equation equal to K_{Icc}.

Now the equations are complete. We can solve for P from (7.5.3) and substitute it into (7.5.2) to get

$$\sigma(x) = f\left[C_x(a)\frac{b}{\hat{k}(\alpha)}\int_{a_0}^{a} k_G(\alpha, x'/D)\sigma(x')dx' - \int_{a_0}^{a} C_{xx'}(a)b\,\sigma(x')dx'\right] \quad \text{for } a_0 < x < a \tag{7.5.4}$$

or, combining the two integrals,

$$\sigma(x) = f\left[\int_{a_0}^{a}\left[\frac{C_x(a)}{\hat{k}(\alpha)}k_G(\alpha, x'/D) - C_{xx'}(a)\right]b\sigma(x')dx'\right] \qquad \text{for} \quad a_0 < x < a \tag{7.5.5}$$

This is a nonlinear integral equation which can be solved for stresses σ_x in the cohesive zone. Then the applied load follows from (7.5.3); other variables of interest (such as the displacement) can be obtained from the superposition in Fig. 7.5.1 by computing the appropriate cross-compliances of the basic cases.

Although this formulation is not always suitable for practical applications (because the required functions, particularly k_G, are not known in a closed form), it is interesting from a theoretical point of view because it helps disclose the dependence of the equations on various parameters. To show this dependence, we are going to reduce this equation to a nondimensional form by introducing the relative coordinates

$$\xi = \frac{x}{D}, \qquad \xi' = \frac{x'}{D} \tag{7.5.6}$$

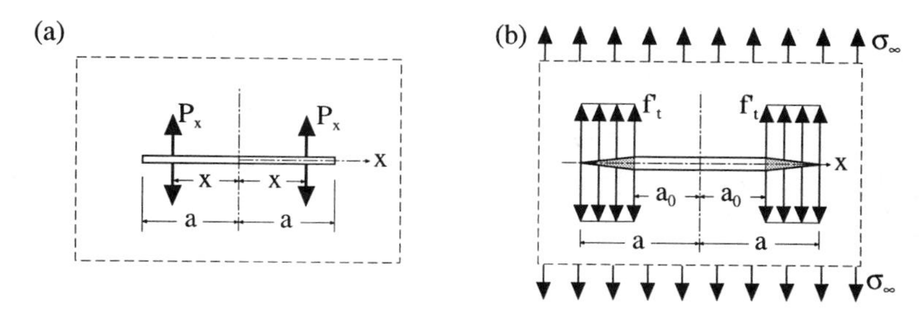

Figure 7.5.2 (a) Center crack in an infinite plate with two symmetric pairs of concentrated forces. (b) Center crack in an infinite plate with Dugdale cohesive zones.

With this, the expression for $C_x(a)$ found in (3.5.41) can be rewritten as

$$C_x(a) = \frac{1}{bE'}\hat{v}_x(\xi,\alpha)\,, \qquad \hat{v}_x(\xi,\alpha) = 2\int_{\xi}^{\alpha} \hat{k}(\alpha')k_G(\alpha',\xi)d\alpha' \tag{7.5.7}$$

The expression for $C_{xx'}(a)$ can be computed in a form analogous to the foregoing case. We replace the shape factor for the applied load by the shape factor for the concentrated load on the faces of the crack (taking special care with regard to the integration interval). The result is

$$C_{xx'}(a) = \frac{1}{bE'}\hat{v}_{xx'}(\xi,\xi',\alpha)\,, \qquad \hat{v}_{xx'}(\xi,\xi',\alpha) = 2\int_{\xi_m}^{\alpha} k_G(\alpha',\xi')k_G(\alpha',\xi)d\alpha' \tag{7.5.8}$$

where the lower integration limit $\xi_m = \max(\xi,\xi')$ follows from condition (3.5.38).

With the foregoing equations, we can rewrite the integral equation as

$$\sigma(\xi) = f\left[\frac{D}{E'}\int_{\alpha_0}^{\alpha} N(\xi,\xi',\alpha)\sigma(\xi')d\xi'\right] \qquad \text{for} \quad \alpha_0 < \xi < \alpha \tag{7.5.9}$$

where the expression within the square brackets is the crack opening $w(x)$ and

$$N(\xi,\xi',\alpha) = \frac{\hat{v}_x(\xi,\alpha)}{\hat{k}(\alpha)}k_G(\alpha,\xi') - \hat{v}_{xx'}(\xi,\xi',\alpha) \tag{7.5.10}$$

In many practical situations we are required to express the loading in terms of the nominal stress σ_N rather than in terms of the load P. Then the stress intensity factor is written as $\sigma_N\sqrt{D}\,k(\alpha)$. As we saw in Section 2.3.2, $k(\alpha)$ and $\hat{k}(\alpha)$ are proportional, so we can easily rewrite all the foregoing equations in terms of $k(\alpha)$ instead of $\hat{k}(\alpha)$. In particular, $N(\xi,\xi',\alpha)$ is rewritten as

$$N(\xi,\xi',\alpha) = \frac{v_x(\xi,\alpha)}{k(\alpha)}k_G(\alpha,\xi') - \hat{v}_{xx'}(\xi,\xi',\alpha) \tag{7.5.11}$$

where now $v_x(\xi,\alpha)$ is given by

$$v_x(\xi,\alpha) = 2\int_{\xi}^{\alpha} k(\alpha')k_G(\alpha',\xi)d\alpha' \tag{7.5.12}$$

Example 7.5.1 The determination of the kernel can be done analytically in only a few cases, particularly those corresponding to the infinite elastic space with straight cracks. To illustrate how the work proceeds, consider an infinite panel with a center crack of initial length $2a_0$ subjected to a remote stress σ_∞ normal to the crack plane. Since the only relevant dimension is the initial crack size, we take $a_0 \equiv D$; we also take $\sigma_\infty \equiv \sigma_N$. Then, since we know that $K_I = \sigma_\infty\sqrt{\pi a}$ for arbitrary crack length, we immediately get $k(\alpha) = \sqrt{\pi\alpha}$. Now, for the point loads on the faces, we enforce symmetry and consider two equal pairs of forces P_x located at x and $-x$ as shown in Fig. 7.5.2a. The stress intensity factor at the right-hand tip due to these forces, for arbitrary crack length $2a$, and the corresponding function $k_G(\alpha,\xi)$ are given by (e.g., Tada, Paris and Irwin 1985):

$$K_I = \frac{2P_x\sqrt{a}}{\sqrt{\pi}}\frac{1}{\sqrt{a^2-x^2}} \quad\Rightarrow\quad k_G(\alpha,\xi) = \frac{2\sqrt{\alpha}}{\sqrt{\pi}}\frac{1}{\sqrt{\alpha^2-\xi^2}} \tag{7.5.13}$$

Then, from (7.5.12), we get

$$v_x(\xi,\alpha) = 4\int_\xi^\alpha \frac{\alpha'\,d\alpha'}{\sqrt{\alpha'^2-\xi^2}} = 4\sqrt{\alpha^2-\xi^2} \tag{7.5.14}$$

which, except for the factor σ_∞/E, gives the crack opening profile, Eq. (2.2.7). Next, we compute $\hat{v}_{xx'}$ from (7.5.8):

$$\hat{v}_{xx'}(\xi,\xi',\alpha) = \frac{8}{\pi}\int_{\xi_m}^\alpha \frac{\alpha'\,d\alpha'}{\sqrt{\alpha'^2-\xi^2}\sqrt{\alpha'^2-\xi'^2}} = \frac{8}{\pi}\ln\frac{\sqrt{\alpha^2-\xi^2}+\sqrt{\alpha^2-\xi'^2}}{\sqrt{\left|\xi^2-\xi'^2\right|}} \tag{7.5.15}$$

where the integration is easily performed if we assume, for example, $\xi > \xi'$, make the substitution $t=\sqrt{\alpha'^2-\xi^2}$ from which a standard integral is obtained, and then make the function symmetric with respect to ξ and ξ'. With the foregoing results, the kernel $N(\xi,\xi',\alpha)$ is obtained from (7.5.10):

$$N(\xi,\xi',\alpha) = \frac{8}{\pi}\left(\frac{\sqrt{\alpha^2-\xi'^2}}{\sqrt{\alpha^2-\xi'^2}} + \ln\frac{\sqrt{\alpha^2-\xi^2}+\sqrt{\alpha^2-\xi'^2}}{\sqrt{\left|\xi^2-\xi'^2\right|}}\right) \tag{7.5.16}$$

Note that this kernel has a logarithmic singularity in its second term. This is a general feature of these kernels based on the elastic solution for concentrated loads on the crack faces. ▯

7.5.2 Size-Dependence of the Equations

Note that, in the foregoing equations, the kernel $N(\xi,\xi',\alpha)$ and the integration limits are independent of the size of the specimen, and are identical as long as the geometry and loading are similar (i.e., if all the length ratios are the same). The size of the specimen enters the argument of function f only as a factor playing the same role as the λ parameter in a linear Fredholm equation of the second kind (see, e.g., Press et al. 1992):

$$g(\xi) = \lambda\int_a^b K(\xi,\xi')g(\xi')d\xi' + h(\xi) \tag{7.5.17}$$

where $g(\xi)$ is the unknown function, a, b are constants and $h(\xi)$ is a given function.

Example 7.5.2 For the case of linear softening, the integral equation (7.5.9) becomes linear, as long as the crack opening at the initial tip does not exceed the critical crack opening. Setting $f(w) = f'_t[1 - wf'_t/(2G_F)]$ in (7.5.9), we get

$$\sigma(\xi) = f'_t - \frac{D}{2\ell_{ch}}\int_{\alpha_0}^\alpha N(\xi,\xi',\alpha)\sigma(\xi')d\xi' \qquad \text{for} \quad \alpha_0 < \xi < \alpha \tag{7.5.18}$$

where $\ell_{ch} = E'G_F/f_t'^2$ is the characteristic size. We see that this integral equation coincides with (7.5.17) with $\lambda = -D/2\ell_{ch}$ and $h(\xi) = f'_t$. ▯

In the foregoing example, we see that the integral equation depends on D through the nondimensional ratio D/ℓ_{ch}. This is general property, which can be shown by using the dimensionless stress and softening curve defined in (7.1.12)–(7.1.14) so that (7.5.9) is transformed to the form

$$\hat{\sigma}(\xi) = \hat{f}\left[\frac{D}{\ell_{ch}}\int_{\alpha_0}^{\alpha} N(\xi,\xi',\alpha)\hat{\sigma}(\xi')d\xi'\right] \qquad \text{for} \quad \alpha_0 < \xi < \alpha \tag{7.5.19}$$

Solving this integral equation for any given cohesive zone (i.e., for any given α) delivers the dimensionless stress profile over this zone. Once the profile is known, the load is obtained from (7.5.3), which can be rewritten as

$$P = f_t' b D \frac{1}{\hat{k}(\alpha)}\int_{\alpha_0}^{\alpha} k_G(\alpha,\xi)\hat{\sigma}(\xi)d\xi \tag{7.5.20}$$

or else, in terms of the nominal strength

$$\sigma_N = f_t' \frac{1}{k(\alpha)}\int_{\alpha_0}^{\alpha} k_G(\alpha,\xi)\hat{\sigma}(\xi)d\xi \tag{7.5.21}$$

where $k(\alpha)$ is the shape factor when the stress intensity factor is written in terms of σ_N.

If required, the crack opening profile can be computed from (7.5.1) or from its transformed expression appearing in the square brackets in (7.5.9). The corresponding dimensionless expressions are:

$$\hat{w}(\xi) = \frac{D}{\ell_{ch}} v_x(\xi,\alpha)\hat{\sigma}_N - \frac{D}{\ell_{ch}}\int_{\alpha_0}^{\alpha} N(\xi,\xi',\alpha)\hat{\sigma}(\xi')d\xi' \tag{7.5.22}$$

$$\hat{w}(\xi) = \frac{D}{\ell_{ch}}\int_{\alpha_0}^{\alpha} N(\xi,\xi',\alpha)\hat{\sigma}(\xi')d\xi' \tag{7.5.23}$$

7.5.3 The Dugdale and Rectangular Softening Cases

The only case known in which the integral equation (7.5.3) is solvable analytically is that corresponding to the Dugdale model (or to the rectangular softening). In this case, $\hat{f}(\hat{w}) = 1$ and, trivially, (7.5.3) yields the solution $\hat{\sigma} = 1$. Then (7.5.20) or (7.5.21) give the relationship between the load and the cohesive crack extension, and (7.5.23) the crack opening distribution.

Of particular interest is to note that for the Dugdale model there is no softening, and so the load increases monotonically as the crack grows. This is not so, however, for the rectangular softening depicted in Fig. 7.1.5; for positive geometries, the peak load is reached as soon as the opening at the initial crack tip w_T reaches the critical crack opening w_c. Now, this condition is found by setting $\xi = \alpha_0$ in (7.5.23), to get $\hat{w}_T$, and then setting resulting expression equal to $\hat{w}_c = 1$. The result is an equation with a single unknown, namely a_u, the cohesive crack length at which the peak occurs. Solving for this value and substituting the results into (7.5.21), we obtain the nominal strength. Let us fix these ideas with an example.

Example 7.5.3 Consider an infinite panel with a center crack of length $2a_0$, subjected to a remote stress σ_∞ normal to the crack plane, as shown in Fig. 7.5.2b. Substituting the expressions for $k(\alpha)$ and $k_G(\alpha,\xi)$ given in that example into (7.5.21) in which $\alpha_0 = a_0/a_0 = 1$ and $\hat{\sigma}(\xi) = 1$, we find σ_N/f_t' for any given cohesive crack size:

$$\hat{\sigma}_N = \frac{1}{\sqrt{\pi\alpha}}\int_1^{\alpha}\frac{2\sqrt{\alpha}}{\sqrt{\pi}}\frac{1}{\sqrt{\alpha^2-\xi^2}}d\xi = \frac{2}{\pi}\cos^{-1}\left(\frac{1}{\alpha}\right) = \frac{2}{\pi}\cos^{-1}\left(\frac{a_0}{a}\right) \tag{7.5.24}$$

This can be inverted to obtain the position of the cohesive crack tip as a function of the applied stress:

$$a = a_0 \sec\left(\frac{\pi\sigma_\infty}{2f_t'}\right) \tag{7.5.25}$$

The expression for the crack opening profile is much more complicated and will not be given here; it can be found, for example, in Planas and Elices (1992a). However, a simple expression is found for w_T, the opening at the initial crack tip $w_T = w(\alpha_0)$. It is obtained by setting $\xi = \alpha_0 = 1$ in (7.5.23) with the kernel given by (7.5.16). The result is:

$$\hat{w}_T = \frac{8a_0}{\pi \ell_{ch}} \ln \alpha \tag{7.5.26}$$

The relative crack length α_u at which the peak load occurs is found by setting $\hat{w}_T = \hat{w}_c = 1$ in the preceding equation:

$$\alpha_u = \exp\left(\frac{\pi \ell_{ch}}{8a_0}\right) \tag{7.5.27}$$

Then the nominal strength follows by substitution in (7.5.24):

$$\hat{\sigma}_{Nu} = \frac{2}{\pi} \cos^{-1}\left[\exp\left(-\frac{\pi \ell_{ch}}{8a_0}\right)\right] \tag{7.5.28}$$

This is the size effect law for this particular geometry and model. It is interesting that this conforms to the general form of the size effect law for cohesive models in which σ_{Nu}/f'_t depends explicitly only on the ratio D/ℓ_{ch} (remember that, in this example, $D \equiv a_0$). □

7.5.4 Eigenvalue Analysis of the Size Effect

Li and Bažant put forth an effective method to determine the size effect law without the need to compute the full load-crack length curve (Li and Bažant 1994a, 1996; Bažant and Li 1994a, 1995a, 1995b). Their derivation was based on an energetic variational formulation. Here, we use a less fundamental, but simpler derivation based on the previous development. The idea is to impose directly the condition of maximum and, instead of seeking the depth of the crack for which the peak occurs for a given size, seek the size at which the peak occurs for a given relative crack depth. In this way, the equation to be satisfied becomes an eigenvalue problem that can be solved with great computational economy.

The basic result upon which the Bažant-Li formulation rests is that the variation of Eq. (7.5.1) preserves its form, i.e., that

$$\delta w(x) = C_x(a)\,\delta P - \int_{a_0}^{a} C_{xx'}(a) b\,\delta\sigma(x')dx' \tag{7.5.29}$$

where $\delta w(x)$, δP, and $\delta\sigma(x')$ are the variations of the corresponding variables when the crack undergoes an infinitesimal increase δa. Note that δa does not enter the foregoing equation, which is not obvious. Indeed, in taking variations in Eq. (7.5.1), an extra term appears which is due to the variations with a of the kernels and of the interval of integration. The expression of this term, say δm, is easily found by differentiation of (7.5.1) as

$$\delta m = \left[-C_{xa}(a) b\sigma(a) + \frac{\partial C_x(a)}{\partial a} P - \int_{a_0}^{a} \frac{\partial C_{xx'}(a)}{\partial a} b\,\sigma(x')dx'\right] \delta a \tag{7.5.30}$$

where the first term arises from the differentiation of the integral with respect to its upper limit, and the other two from the differentiation of the compliance functions. We can now show that $\delta m = 0$. Since $C_{xx'}(a)$ is the crack opening at x produced by a unit force pair at x', and it is symmetric by virtue of the reciprocity theorem, i.e., $C_{xx'}(a) = C_{x'x}(a)$, it turns out that $C_{xa}(a)$ is the crack opening at the crack tip, which is zero. So we have

$$C_{xa}(a) = 0 \tag{7.5.31}$$

Next, we can compute the derivative of $C_x(a)$ from (7.5.7) taking into account that $\partial/\partial a = \partial/(D\partial\alpha)$:

$$\frac{\partial C_x(a)}{\partial a} = \frac{2}{bDE'} \hat{k}(\alpha) k_G(\alpha, \xi) d\alpha' \tag{7.5.32}$$

Similarly, the derivative of $C_{xx'}(a)$ is determined from (7.5.8) as

$$\frac{\partial C_{xx'}(a)}{\partial a} = \frac{2}{bDE'} k_G(\alpha, \xi) k_G(\alpha, \xi') \tag{7.5.33}$$

Substituting the foregoing results into the expression for δm, we get

$$\delta m = \frac{2}{bDE'} k_G(\alpha, \xi) \left[k(\alpha) P - \int_{a_0}^{a} k_G(\alpha, \xi') \sigma(x') dx' \right] \delta a \tag{7.5.34}$$

But the bracketed factor vanishes by virtue of (7.5.3), expressing the condition that $K_I = 0$ (which also shows that this condition is essential and cannot be omitted). Thus, $\delta m = 0$ and the variation of $w(x)$ reduces to (7.5.29). Q.E.D.

Then, the incremental form of (7.5.2) is

$$\delta\sigma(x) = f'[w(x)] \left[C_x(a)\, \delta P - \int_{a_0}^{a} C_{xx'}(a) b\, \delta\sigma(x') dx' \right] \tag{7.5.35}$$

where $f'(w) = df(w)/dw$. Since at peak load $\delta P = 0$ by definition, it turns out that at the maximum load, the following condition must be satisfied:

$$\delta\sigma(x) = -f'[w(x)] b \int_{a_0}^{a} C_{xx'}(a)\, \delta\sigma(x') dx' \tag{7.5.36}$$

This is a homogeneous integral equation for the stress variation $\delta\sigma(x)$ whose nontrivial solutions are the eigenvectors (or eigenfunctions) of the equation. It is convenient to cast this equation in a dimensionless form using (7.5.6) and (7.5.8) together with the dimensionless variables $\hat{\sigma}$ and $\hat{w}$:

$$\delta\hat{\sigma}(\xi) = -\hat{f}'[\hat{w}(\xi)] \frac{D}{\ell_{ch}} \int_{\alpha_0}^{\alpha} \hat{v}_{xx'}(\xi, \xi', \alpha)\, \delta\sigma(\xi') d\xi' \tag{7.5.37}$$

which can be seen to take the form of an eigenvalue problem, in which D/ℓ_{ch} is the eigenvalue. The peak load condition is obtained for the smallest eigenvalue, i.e., for the smallest size D (Li and Bažant 1994a, 1996; Bažant and Li 1994a, 1995a, 1995b). Note that in the theory of integral equations, the eigenvalue is usually taken, as we also do, as the factor multiplying the integral term, which is the inverse of what is customary in the matrix eigenvalue problem.

The solution is particularly easy for the case of linear softening in which the eigenvalue problem is decoupled from the static problem. Indeed, for linear softening, $\hat{f}'(\hat{w}) = -1/2$ and the foregoing equation becomes a standard eigenvalue problem for a symmetric integral equation. Setting $\tau(\xi) = \delta\hat{\sigma}(\xi)/\delta\alpha$ (or any other proportional function), the resulting eigenequation is

$$\tau(\xi) = \frac{D}{2\ell_{ch}} \int_{\alpha_0}^{\alpha} \hat{v}_{xx'}(\xi, \xi', \alpha)\, \tau(\xi') d\xi' \tag{7.5.38}$$

This equation can be solved for the smallest eigenvalue $D/(2\ell_{ch})$ and its corresponding eigenvector $\tau(\xi)$.

Next, the cohesive crack equation can be written as $\hat{\sigma}(\xi) = 1 - \hat{w}/2$. Substituting expression (7.5.22) for $\hat{w}$, multiplying the resulting equation by the eigenvector and integrating on the cohesive zone, we get

$$\int_{\alpha_0}^{\alpha} \tau(\xi)\hat{\sigma}(\xi)\, d\xi = \int_{\alpha_0}^{\alpha} \tau(\xi)\, d\xi - \frac{D}{2\ell_{ch}} \hat{\sigma}_{Nu} \int_{\alpha_0}^{\alpha} \tau(\xi) v_x(\xi, \alpha)\, d\xi + \\ + \frac{D}{2\ell_{ch}} \int_{\alpha_0}^{\alpha} \tau(\xi) \int_{\alpha_0}^{\alpha} \hat{v}_{xx'}(\xi, \xi', \alpha) \hat{\sigma}(\xi')\, d\xi'\, d\xi \tag{7.5.39}$$

Inverting now the order of integration in the double integral and using the eigenvalue equation (7.5.38), it turns out that the term containing the double integral cancels out with the term in the left-hand member, and so the unknown function $\sigma(\xi)$ is eliminated. Then we can solve for the peak load and get

$$\hat{\sigma}_{Nu} = \frac{2\ell_{ch}}{D} \frac{\int_{\alpha_0}^{\alpha} \tau(\xi)\, d\xi}{\int_{\alpha_0}^{\alpha} \tau(\xi) v_x(\xi, \alpha)\, d\xi} \tag{7.5.40}$$

Note that the advantage of this method is that for each relative crack depth α, the solution involves obtaining only the smallest eigenvalue of (7.5.38) and its corresponding eigenfunction. Then the maximum load follows through simple quadratures.

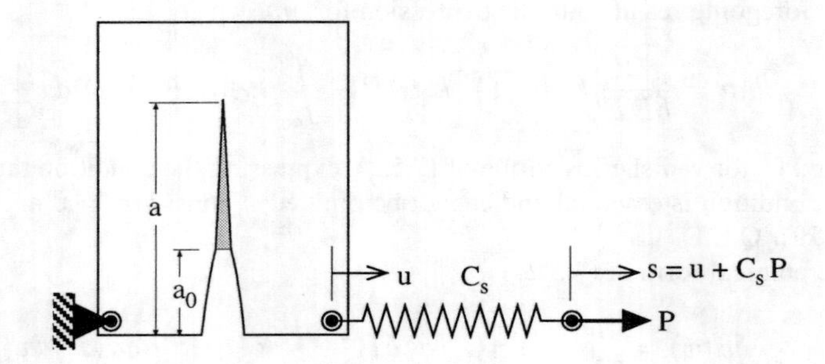

Figure 7.5.3 Specimen loaded through a spring.

Details of a numerical implementation to exploit the foregoing method is given in Li and Bažant (1994a), where the procedure to handle nonlinear softening is also described. In that case, it is no longer possible to solve the peak load without solving the stresses and displacements and it is necessary to solve the static problem defined by Eq. (7.5.19) and the eigenvalue problem defined by Eq. (7.5.38) simultaneously. This is done iteratively by Li and Bažant (1994a).

7.5.5 Eigenvalue Analysis of Stability Limit and Ductility of Structure

The foregoing analysis was developed by Bažant and Li (1995a, 1995b) in a wider energy-based framework dealing with stability analysis. In particular, they considered the case of the cracked structure loaded through a spring, as depicted in Fig. 7.5.3. In this case, we call s the total displacement, and u the displacement of the point-load of the specimen. Then the total displacement s and its variation δs are given by

$$s = u + C_s\, P\,, \qquad \delta s = \delta u + C_s\, \delta P \tag{7.5.41}$$

where C_s is the compliance of the spring. The stability at constant s is ensured if the only solution for $\delta s = 0$ is the null solution; if a nonzero solution can be found for the variation of the other variables, then the critical state of instability occurs: that is where eigenvalue analysis comes into play.

To take the spring into account, we need to add to the equations already presented, the equation for the displacement of the load-point. This equation is obtained from the decomposition in Fig. 7.5.1 and is written as

$$u = C(a)\, P - \int_{a_0}^{a} C_{x'}(a) b \sigma(x') dx' \tag{7.5.42}$$

where $C(a)$ is the displacement (in the sense of u) produced by a unit applied load, and $C_{x'}(a)$ is the displacement due to a unit pair of forces applied at the crack faces at location x'. Note that the reciprocity theorem ensures that this function is the same as that appearing in (7.5.1). $C(a)$ was determined in Chapter 3, Eq. (3.5.4), and is given by

$$C(a) = \frac{1}{bE'} \hat{v}(\alpha)\,, \qquad \hat{v}(\alpha) = \hat{v}_0 + 2\int_0^{\alpha} \hat{k}^2(\alpha')\, d\alpha' \tag{7.5.43}$$

It is a simple exercise to show that δu keeps the form (7.5.42) with P and $\sigma(x')$ replaced, respectively, by δP and $\delta\sigma(x')$. Substituting this expression of δu into (7.5.41), solving for δP, and inserting the result in (7.5.29), we get a formally identical expression:

$$\delta w(x) = \tilde{C}_x(a)\, \frac{\delta s}{C_s} - \int_{a_0}^{a} \tilde{C}_{xx'}(a) b\, \delta\sigma(x') dx' \tag{7.5.44}$$

in which

$$\tilde{C}_x(a) = \frac{C_x(a) C_s}{C(a) + C_s}\,, \qquad \tilde{C}_{xx'}(a) = C_{xx'}(a) - \frac{C_x(a) C_{x'}(a)}{C(a) + C_s} \tag{7.5.45}$$

The condition of stability limit (critical state) is obtained by setting $\delta s = 0$ in the preceding equation, and performing the analysis exactly as in the case for the peak load. After making all the equations dimensionless and performing the associated algebra, the resulting eigenvalue equation for the linear softening case is found as

$$\tau(\xi) = \frac{D}{2\ell_{ch}} \int_{\alpha_0}^{\alpha} \tilde{v}_{xx'}(\xi, \xi', \alpha)\, \tau(\xi') d\xi' \tag{7.5.46}$$

where

$$\tilde{v}_{xx'}(\xi, \xi', \alpha) = \hat{v}_{xx'}(\xi, \xi', \alpha) - \frac{\hat{v}_x(\xi, \alpha)\hat{v}_{x'}(\xi', \alpha)}{\hat{v}(\alpha) + C_s bE'} \tag{7.5.47}$$

Once the smallest eigenvalue and its associated eigenfunction have been determined, operations similar to those leading to (7.5.40) furnish the total displacement s_{cr} at which instability the stability limit is reached:

$$s_{cr} = \frac{2G_F}{f'_t}[\hat{v}(a) + bE'C_s] \frac{\int_{\alpha_0}^{\alpha} \tau(\xi)\, d\xi}{\int_{\alpha_0}^{\alpha} \tau(\xi)\hat{v}_x(\xi, \alpha)\, d\xi} \tag{7.5.48}$$

The ratio $\delta_s = s_{cr}/s_e$ where s_e is the displacement of an identical elastic structure with $a = 0$ under the same load, can be regarded as a measure of the ductility of the structure.

7.5.6 Smeared-Tip Superposition Method

We formulate now the analytical counterpart of the smeared-tip superposition developed in the previous section for PBI methods. The smeared-tip superposition consists in writing the solution of the problem as a sum, in fact, an integral, of all the cases corresponding to cracks of different lengths s, ranging from a_0 to a, each case been subjected to an infinitesimal load $dP(s)$ (Fig. 7.5.4). Then, the stress and crack opening displacements can then be written as (Planas and Elices 1986b, 1992a, 1993a; Bažant 1990d; Bažant and Beissel 1994):

$$\sigma(x) = \int_{a_0}^{a} S(x, s)\, dP(s) \tag{7.5.49}$$

$$w(x) = \int_{a_0}^{a} W(x, s)\, dP(s) \tag{7.5.50}$$

where $S(x, s)$ and $W(x, s)$ are, respectively, the stress and crack opening at point x for a crack of length s in a body subjected to a unit load $P = 1$. It is obvious that, for the elastic crack of length s, the stress is zero over the crack itself and the opening is zero over the ligament. So we always have

$$S(x, s) = 0 \quad \text{for} \quad x < s\,, \qquad \text{and} \qquad W(x, s) = 0 \quad \text{for} \quad x \geq s \tag{7.5.51}$$

Moreover, it is known from LEFM that $S(x, s)$ is singular for $x - s \rightarrow 0^+$, where it behaves as $(x - s)^{-1/2}$, and that $W(x, s)$ behaves as $(s - x)^{1/2}$ for $s - x \rightarrow 0^+$.

It is useful to interpret the foregoing integrals in the sense of the theory of distributions, and write $dP(s)$ as $p(s)ds$ where $p(s)$ is the density of load for cracks in the neighborhood of s. Then the integrals are rewritten as

$$\sigma(x) = \int_{a_0}^{a} S(x, s)\, p(s)\, ds \tag{7.5.52}$$

$$w(x) = \int_{a_0}^{a} W(x, s)\, p(s)\, ds \tag{7.5.53}$$

In this way, almost any kind of distribution of stresses and displacements on the cohesive zone can be built, even those containing singularities. For example, an effective elastic crack with its tip at a_1 (such that $a_0 \leq a_1 \leq a$) is simply represented by a load density function $p(s) = \delta(s - a_1)$, where $\delta(s - a_1)$ is Dirac's δ-function located at $s = a_1$.

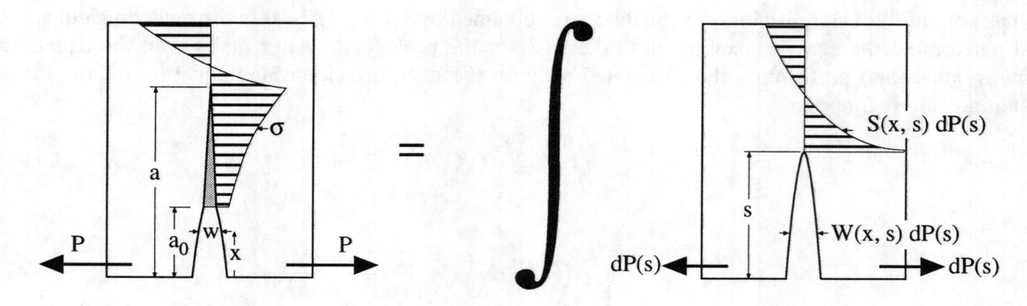

Figure 7.5.4 Smeared-tip superposition.

This having been said, the application to a classical cohesive crack problem consists in stipulating that, in the cohesive zone, $\sigma(x) = f[w(x)]$. Substituting the foregoing expressions for $\sigma(x)$ and $w(x)$, we get the following governing integral equation:

$$\int_{a_0}^{x} S(x,s)\,p(s)\,ds = f\left[\int_{x}^{a} W(x,s)\,p(s)\,ds\right] \tag{7.5.54}$$

where the limits of integration have now been set in accordance with the properties (7.5.51). Notice that the kernels of this integral equation are of Volterra type. This means that if the discretization is done properly, triangular matrices are obtained, a lower-triangular matrix for the stresses, and an upper-triangular matrix for the crack openings, as it happened for the PBI smeared-tip method.

Thus, assuming that the kernels $S(x,s)$ and $W(x,s)$ are known, the problem consists in solving the foregoing integral equation for $p(s)$ given the cohesive crack length a. Then the stresses and crack openings at any point follow from (7.5.52) and (7.5.53), and the load from the integral obviously implied in Fig 7.5.4:

$$P = \int_{a_0}^{a} p(s)\,ds \tag{7.5.55}$$

Example 7.5.4 Before proceeding, let us see an example of the kernels. Consider the center cracked panel of Fig. 7.5.2, and let us find the kernels and write the corresponding integral equation. First, it is convenient to replace P by σ_∞ and $p(s)$ by $\zeta(s)$, so that (7.5.55) is written as

$$\sigma_\infty = \int_{a_0}^{a} \zeta(s)\,ds \tag{7.5.56}$$

Next, the kernels $S(x,s)$ and $W(x,s)$ are taken directly from (2.2.2) and (2.2.7) by replacing a with s and setting $\sigma_\infty = 1$. Then the integral equation reduces to

$$\int_{a_0}^{x} \frac{x\,\zeta(s)\,ds}{\sqrt{x^2-s^2}} = f\left[\frac{4}{E'}\int_{x}^{a} \sqrt{s^2-x^2}\,\zeta(s)\,ds\right] \tag{7.5.57}$$

In general, this equation must be solved numerically. One exception is the rectangular softening case in which the right-hand member is constant (equal to f'_t). Then the equation reduces to

$$\int_{a_0}^{x} \frac{x\,\zeta(s)\,ds}{\sqrt{x^2-s^2}} = f'_t \tag{7.5.58}$$

This integral equation can be solved analytically (Planas and Elices 1991a), with the result

$$\zeta(s) = (2a_0 f'_t/\pi)[s^2(s^2-a_0^2)]^{-1/2} \tag{7.5.59}$$

Substituting this solution into (7.5.56) and integrating, the expression (7.5.24) is recovered, as it should. The crack tip opening displacement is calculated by computing the expression in the square brackets in (7.5.57), with $x = a_0$. The result again coincides with (7.5.26). □

The foregoing example shows that the solution for the load density function (be it $p(s)$ or $\zeta(s)$) is singular near the initial crack tip. It turns out (Planas and Elices 1991a) that it always tends to $(s - a_0)^{-1/2}$, which must be taken into account in the numerical implementations if the fields are to be reproduced accurately. Planas and Elices (1986b, 1992a, 1993a) and Bažant and Beissel (1994) use the expedient of changing the unknown function $p(s)$ to another $g(s)$ that is not singular by setting

$$p(s) = \frac{g(s)}{\sqrt{s - a_0}} \tag{7.5.60}$$

Bažant and Beissel (1994) used this approximation within a wider scope, for time-dependent cohesive cracks and aging viscoelastic materials. They used a central finite difference formulation in time and space, with analytical integration of the kernels when the singular point was at one of the ends of the integration interval (this corresponds to the diagonal terms of the triangular matrices).

For finite size specimens, Planas and co-workers use the PBI method described in Section 7.4.4, in which the integration of the kernels is replaced by the finite element calculation of the influence matrices. For large size, an asymptotic procedure was developed as described next.

7.5.7 Asymptotic Analysis

Planas and Elices (1986b, 1992a, 1993a) used the smeared-tip method to analyze the large size asymptotic behavior. The full development is too lengthy to be presented here, and only the zeroth-order asymptotics will be described. Suffice to say that the basic idea is to rewrite the foregoing equations for a very large crack in a very large body using, as the origin of measurements of coordinates, the initial crack tip, i.e., replacing $x - a_0$ by x, $s - x$ by x and using the crack extension $R = a - a_0$ as the measure of the crack advance (Fig 7.5.5). Then the kernels, the integrals, and the integral equation itself are expanded in series of powers of R/D (which tends to zero for large sizes) and the series is truncated according to the required order of analysis. Although the analysis of the peak load has been performed using two terms (which leads to the asymptotic size effect summarized in Section 7.2.4), the full analysis has been performed only to order zero, which means that only the first term is retained in the series expansion. This is equivalent to considering a semi-infinite crack in an infinite body, in which case the kernels are simple well-known functions.

Referring to Fig. 7.5.5, in which we assume that $R \ll D$ where D is the size of the body, we first note that we can approximate the stress and crack opening displacements of a crack of length $a_0 + s$ by its near-tip approximations, which, in turn, are proportional to the stress intensity factor. Using the expressions (2.2.4) and (2.2.9), we find the kernels as

$$S(x,s) = \frac{K_I(P, a_0+s)}{P}\,(x-s)^{-1/2}\,, \qquad W(x,s) = \frac{K_I(P, a_0+s)}{P}\,\frac{8}{\sqrt{2\pi}\,E'}(s-x)^{1/2} \tag{7.5.61}$$

where $K_I(P, a_0 + s)$ indicates the stress intensity factor for load P and crack length $a_0 + s$ and it is understood that the equations define the kernels wherever their values are real, while the values elsewhere are taken as zero. The stress intensity factor can be written as

$$\frac{K_I(P, a_0+s)}{P} = \frac{1}{b\sqrt{D}}\,\hat{k}(\alpha_0 + s/D) \approx \frac{1}{b\sqrt{D}}\,\hat{k}(\alpha_0) = \frac{K_{IN}}{P} \tag{7.5.62}$$

In this approximation, we have noted that since in the cohesive zone $s < R$ and since $R \ll D$, then also $s \ll D$, and the stress intensity factor can thus be approximated by the nominal stress intensity factor, as indicated in the equation. Substituting now the foregoing expressions into the smeared-tip formulas, we

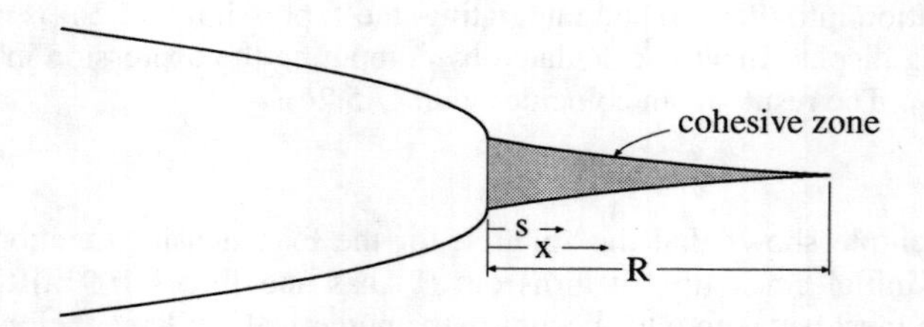

Figure 7.5.5 Cohesive zone ahead of the tip of a very large crack.

get the asymptotic governing equations:

$$\int_0^x \frac{q_1(s)\, ds}{\sqrt{x-s}} = f\left[\frac{8}{E'}\int_x^R q_1(s)\sqrt{s-x}\, ds\right] \tag{7.5.63}$$

$$K_{IN} = \sqrt{2\pi}\int_0^R q_1(s)\, ds \tag{7.5.64}$$

where the new unknown function $q_1(s)$ has been defined as

$$q_1(s) = \frac{\hat{k}(\alpha_0)}{b\sqrt{2\pi D}}\, p(s) \tag{7.5.65}$$

To solve this equation numerically, two methods are possible: (1) In the classical method, a set of elements is defined ahead of the initial crack tip. The cohesive zone grows element by element, as it did in the PBI methods of the previous section, and so the number of equations increases at each step. (2) In an alternative method, we may profit by the fact that the foregoing expressions are defined analytically to reduce the equation to a fixed interval, so that the number of equations is kept constant. This second method was selected by Planas and Elices (1986b, 1991a, 1993a) to solve the equation. To do so, we first introduce the following relative coordinates:

$$\xi = \frac{x}{R}, \qquad \eta = \frac{s}{R} \tag{7.5.66}$$

The governing equations (7.5.63) and (7.5.64) are readily transformed into

$$\int_0^\xi \frac{q(\eta)\, d\eta}{\sqrt{\xi-\eta}} = f\left[\frac{8R}{E'}\int_\xi^1 q(\eta)\sqrt{\eta-\xi}\, d\eta\right] \tag{7.5.67}$$

$$K_{IN} = \sqrt{2\pi R}\int_0^1 q(s)\, ds \tag{7.5.68}$$

in which $q(\eta) = q_1(\eta R)\sqrt{R}$. We see that the cohesive zone, and thus the integration interval, has been projected onto the segment $(0, 1)$.

To solve this integral equation, we must take into account, as said before, that the solution for $q(\eta)$ is singular. That this must be so was shown by Planas and Elices (1991a) based on the following argument: the right-hand side of (7.5.68) is the stress over the cohesive zone, and so we can write

$$\sigma(\xi) = \int_0^\xi \frac{q(\eta)\, d\eta}{\sqrt{\xi-\eta}} \tag{7.5.69}$$

This is a Volterra integral equation whose solution is (Planas and Elices 1991a)

$$q(\eta) = \frac{\sigma(0^+)}{\pi\sqrt{\eta}} + \frac{1}{\pi}\int_0^\eta \frac{d\sigma(\xi)}{\sqrt{\eta-\xi}} \tag{7.5.70}$$

in which the first singular term vanishes only if the cohesive stress vanishes at the initial crack tip (strictly speaking, on its right, $x = \xi = 0^+$).

Therefore, Planas and Elices (1986b, 1991a, 1992a, 1993a) sought the solution of the equation expressed by means of an intermediate unknown function $h(\eta)$ defined by $q(\eta) = h(\eta)\eta^{-1/2}$. They introduced a mesh of equal size elements in the cohesive zone, with linear interpolation for $h(\eta)$, and used collocation at the nodes with analytical integration to set up the system of equations. The resulting system of equations takes the form

$$\sum_{j=1}^{N} L_{ij} h_j = f\left[\frac{8R}{E'}\sum_{j=1}^{N} U_{ij} h_j\right] \tag{7.5.71}$$

where N is the number of nodes, h_j are the nodal values of $h(\eta)$, and L_{ij} and U_{ij}, respectively, lower- and upper-triangular matrices. Both of these matrices are constant, and are computed at the beginning of the calculation. They change only if the number of elements is changed, so they can be stored for use in various computations, for example, with various softening curves. The solution algorithm is very simple: based on a previous solution for h_j (which may be zero) the right-hand side is evaluated and a new estimate is done by solving for h_j from the left-hand member of the resulting triangular equation (which requires only forward substitution).

In calculating the stiffness matrix, there is one special feature that deserves special comment: for the first node, located at $\xi = 0$, the collocation for the integral in the first member must be understood in the sense of a limit, i.e.:

$$L_{11} h_1 = \lim_{\epsilon \to 0}\left[\int_0^{\epsilon} \frac{h(\eta)\, d\eta}{\sqrt{\eta(\epsilon - \eta)}}\right] = \pi\, h(0) = \pi\, h_1 \tag{7.5.72}$$

which means that $L_{11} = \pi$ whatever the mesh.

This method is useful not only because of its computational efficiency but also, and mainly, because it naturally leads to a closed-form expression for the effective crack extension Δa_∞ introduced in Section 7.2.4. The expression in terms of the solution for the function $q(\eta)$ is

$$\Delta a_\infty = R\frac{\int_0^1 \eta\, q(\eta)\, d\eta}{\int_0^1 q(\eta)\, d\eta} \tag{7.5.73}$$

which shows the remarkable property that the position of the elastic crack tip coincides with the center of gravity of the density function $q(\eta)$. The solution (7.5.70) can be used to express the effective crack extension in terms of the cohesive stress distribution (Smith 1995).

Exercises

7.18 Show that the logarithmic term in (7.5.16) can be also written as

$$\frac{1}{2}\ln\frac{\sqrt{\alpha^2 - \xi^2} + \sqrt{\alpha^2 - \xi'^2}}{\left|\sqrt{\alpha^2 - \xi^2} - \sqrt{\alpha^2 - \xi'^2}\right|}$$

(This is the form furnished by a symbolic mathematics package.)

7.19 Use (7.5.28) to show that when the crack is very large compared to the characteristic length, the fracture load of the panel is approximately given by LEFM.

7.20 Use (7.5.28) to show that when the crack is large compared to the characteristic length, the fracture load of the panel can be approximated by the effective elastic crack theory, Eq. (5.3.9). [Hint: The power series expansion of $cos^{-1}(1 - x)$ is $\sqrt{2x}(1 + x/12 + \cdots)$.]

7.21 Prove that the variation of (7.5.42) is $\delta u = C(a)\, \delta P - \int_{a_0}^{a} C_{x'}(a)\, b\, \sigma(x')\, dx'$.

7.22 Use the results in example 7.5.3 to find the following asymptotic relations ($a_0 \to \infty$) for the Dugdale model: (a) between the nominal stress intensity factor K_{IN} and the cohesive crack extension R, (b) between the crack tip opening displacement w_T and R, and (c) between K_{IN} and w_T. [Hint: Set $\alpha = a/a_0 = 1 + R/a_0$

and $\sigma_\infty = K_{IN}/\sqrt{\pi a_0}$ in (7.5.25) and (7.5.26) and take the limits for $a_0 \to \infty$. The first terms of Maclaurin's series expansion of $\sec x$ and $ln(1+x)$ are $1 + x^2/2 + \cdots$, and $x + \cdots$, respectively.]

7.23 Consider a Dugdale model for the asymptotic limit of large crack in large body. (a) Use (7.5.70) to determine the function $q(\eta)$; (b) calculate K_{IN} as a function of R; (c) calculate the crack tip opening displacement w_T as a function of R; and (d) compare the results of parts (b) and (c) with those of parts (a) and (b) in the previous exercise.

7.24 For a Dugdale model, show that the crack opening profile in the cohesive zone for a semi-infinite crack in an infinite body is given by

$$w(\xi) = \frac{8Rf'_t}{\pi E'} \int_\xi^1 \eta^{-1/2}(\eta - \xi)^{1/2} d\eta = \frac{8Rf'_t}{\pi E'} \left[\sqrt{1-\xi} - \xi \ln \frac{1+\sqrt{1-\xi}}{\sqrt{\xi}} \right] =$$

$$= \frac{8Rf'_t}{\pi E'} \left[\sqrt{1-\xi} - \frac{\xi}{2} \ln \frac{1+\sqrt{1-\xi}}{1-\sqrt{1-\xi}} \right], \quad \text{with} \quad \xi = \frac{x}{R} \tag{7.5.74}$$

7.25 Show that for the Dugdale model, $\Delta a_\infty = R/3$.

7.26 For a rectangular softening, determine the asymptotic values R_c and $\Delta a_{\infty c}$ at peak load as a function of ℓ_{ch}.

8

Crack Band Models and Smeared Cracking

Modeling of fracture by discrete line cracks, which has been discussed in the preceding chapters, is not the only viable approach. Another approach, which has gained wide popularity in finite element analysis of concrete structures (Meyer and Okamura, Eds., 1986) and is used almost exclusively in design practice, is to represent fracture in a smeared manner. In this approach, introduced by Rashid (1968), infinitely many parallel cracks of infinitely small opening are imagined to be continuously distributed (smeared) over the finite element. This can be conveniently modeled by reducing the material stiffness and strength in the direction normal to the cracks after the peak strength of the material has been reached. Such changes of the stiffness matrix are relatively easy to implement in a finite element code, and, hence, the appeal of smeared cracking. The evolution of the cracking process down to full fracture implies *strain softening*, a term which describes the postpeak gradual decline of stress at increasing strain.

The term evolved from the terminology of plasticity where work hardening describes the gradual increase of yield stress resulting in a rising stress-strain diagram of a slope that is positive but smaller than the elastic slope. After it was realized that the hardening is not merely a function of the plastic work, a scalar, but depends on all the components of the strain tensor, the term strain hardening has been adopted. From the viewpoint of plasticity, the postpeak decline of stress may be regarded as a gradual decrease of the yield limit, i.e., softening. This phenomenon again is not just a function of work (in which case we could speak of work softening) but of all the strain components; hence, strain softening.

The smeared cracking (with strain softening), however, leads to certain theoretical difficulties which were initially unknown or unappreciated. They consist of the so-called localization instabilities and spurious mesh sensitivity of finite element calculations. After years of controversies and polemics, it has now been generally accepted that these difficulties can be adequately tackled by supplementing the material model with some mathematical condition that prevents localization of smeared cracking into arbitrarily small regions. The simplest way to attain this goal is the crack band model, which is the object of this chapter.

Since it is essential to understand why fracture cannot be consistently and objectively described just by postulating a stress-strain curve with softening and nothing else, we first analyze in this chapter the strain localization in systems displaying softening. We start with the series coupling of discrete elements (Section 8.1), which serves as the starting point for the analysis of the localization of strain in a softening bar (Section 8.2). From this, it follows that some kind of localization limiter must be associated with the softening stress-strain curve to get meaningful results. Next, we analyze the basic issues in the crack band model, in the simplest uniaxial approximation (Section 8.3). Then we deal with the underlying stress-strain relations with softening, first in the simple uniaxial version (Section 8.4) and then in full three-dimensional analysis (Section 8.5). After this, we discuss the triaxial features of the crack band models and smeared cracking, with emphasis placed on the numerical issues (Section 8.6). A comparison of the crack band and cohesive crack approaches closes the chapter (Section 8.7).

8.1 Strain Localization in the Series Coupling Model

Whenever a structure contains elements that may soften, localization of the strain can take place. This section analyzes this phenomenon for the simple, yet important, quasi-static uniaxial case. The case of two nominally identical elements coupled in series is first presented and studied from the point of view of the imperfection approach to bifurcation (no two elements can be exactly identical), and then from the

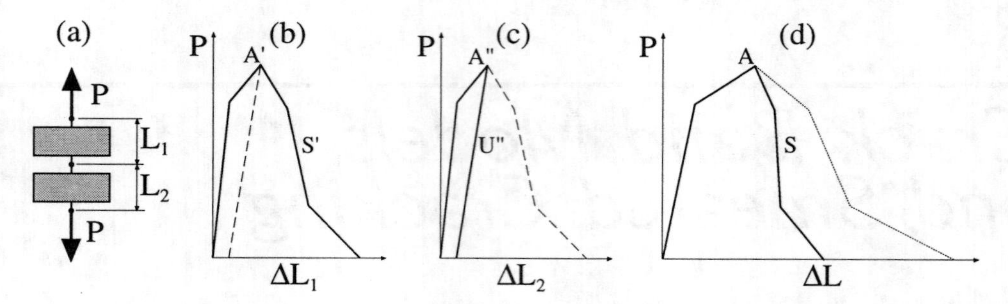

Figure 8.1.1 (a) Series coupling of two softening elements. (b) Load-displacement curve of one element. (c) Resulting load-displacement curve (full line).

point of view of the more general thermodynamic analysis of bifurcations. Next, as a simple extension, a chain of many softening elements is analyzed to show that, after reaching the peak load, only one element will be stretched further, while all the remaining elements unload. This is the starting point for the analysis of a softening continuous homogeneous bar, considered in the next section.

8.1.1 Series Coupling of Two Equal Strain Softening Elements: Imperfection Approach

Consider two nominally identical elements **1** and **2** coupled in series as shown in Fig. 8.1.1a. Assume that each element has a load-elongation (P-ΔL) curve displaying softening as sketched in Fig. 8.1.1b for element **1**. In this plot, the full line is for monotonic extension, and the dashed line corresponds to unloading (shortening) right at the peak. The question is: What is the load-elongation response of the series coupling of the two elements?

A quick answer, extrapolated from the more usual cases of hardening structures would be: Just multiply the elongation by a factor of two. Wrong! Softening breaks down the usual rules. To clarify this, we take first the imperfection approach to bifurcation. In this approach, one realizes that no two elements can be really identical. One of them must have a strength (peak load) slightly smaller than the other one. Assume that such is the case for element **1**. So the element **2**, whose curve is depicted in Fig. 8.1.1c, has a strength only slightly larger than element **1**. The difference is so slight that it cannot be discerned at the scale of the drawing.

As the series coupling is extended, both elements **1** and **2** load up until the peak A' of element **1** is reached. Upon further extension, element **1** must begin to soften, following path A'-S' with decreasing load. Since the load on both elements is identical, the load on element **2** must decrease, too. But since element **2** has not yet reached the peak, it is not going to soften. It is going to *unload* following the path A''-U''.

Therefore, as soon as one element reaches the peak, further straining leads to softening of this element and to unloading of the other. We say that strain *localizes* into one element due to softening. Fig. 8.1.1d shows the resulting P-ΔL curve as a full line. The dotted line represents the (wrong) result obtained by assuming that both elements go into the softening regime (we call it homogeneous deformation, same extension in each element). Note that the rising portion of the curve (the hardening part) displays a displacement that is twice the displacement for a single element, the classical result. The difference lies only in the softening portion of the curve.

The foregoing result (see also Bažant and Cedolin 1991, Sec. 13.2) is based on the idea that the strength of the two elements cannot be identical. Note that the amount by which they differ is immaterial. The same will happen if the difference were only one part in 10^{12}, which is much less than what can be experimentally detected.

We have assumed that element **1** was the weaker element. In practice, we cannot know *a priori* which of the two elements is going to break. We can only state that, if the loading system is perfectly symmetric, the probabilities of failure through one or other element must be equal, so that 50% of tests will show failure of element **1** and 50% failure of element **2**.

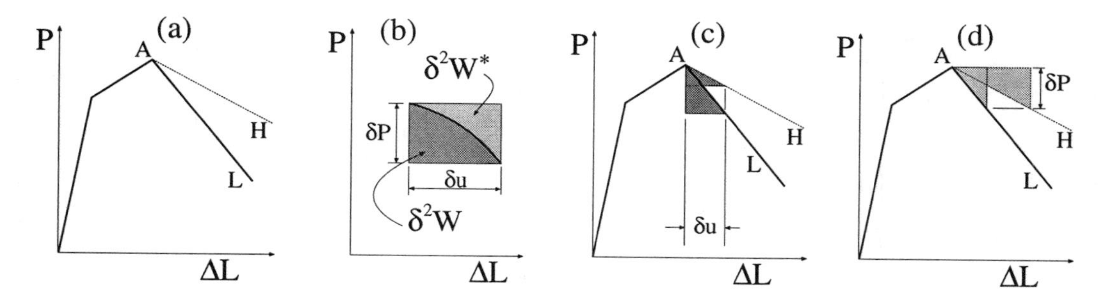

Figure 8.1.2 Series coupling of two strictly identical softening elements: (a) the two possible paths, (b) graphical representation of second-order work and second-order complementary work, (c) postpeak, second-order work, (d) postpeak, second-order complementary work.

8.1.2 Series Coupling of Two Equal Strain Softening Elements: Thermodynamic Approach

The foregoing discussion makes use of inhomogeneities or imperfections to get a general conclusion. However, this result may be also obtained on the basis of thermodynamics. To do so, we consider a series coupling of two identical elements, and consider the possibility of bifurcation at the peak load. The two possible resulting paths are depicted in Fig. 8.1.2a. Path A-H (dotted line) corresponds to a homogeneous deformation, while path A-L (full line) corresponds to softening that localizes into one of the elements, while the other unloads. Which is the preferred path? Following Bažant and Cedolin (1991, Sec. 10.2), for the correct path, the second-order work $\delta^2\mathcal{W} = \frac{1}{2}\delta P\,\delta u$ for imposed displacement increment δu must be minimum, or, alternatively, the second-order complementary work $\delta^2\mathcal{W}^* = \delta P\,\delta u - \delta^2\mathcal{W}$ for imposed load increment δP must be maximum.

Fig. 8.1.2b shows the graphical representation of the second-order work and second-order complementary work for a softening incremental process. Note that the values of the second-order areas are negative because $\delta P < 0$. Therefore, the foregoing principles may be restated by expressing that the second-order area below the P-u curve must be maximum at fixed δu, and that the second-order area over the P-u curve must be minimum at fixed δP.

Figs. 8.1.2c-d show the application of the foregoing principles to our case. It is obvious that the correct path is that for which the localization occurs (see also Bažant and Cedolin 1991, Sec 13.2).

8.1.3 Mean Stress and Mean Strain

Whatever the nature of the foregoing elements, we can define the mean uniaxial stress as the load per unit representative area of the cross section, and the mean strain of each element and the mean strain of the whole coupling as the elongation per unit initial length, i.e., we set, in general,

$$\sigma = \frac{P}{A}, \qquad \varepsilon = \frac{\Delta L}{L} \tag{8.1.1}$$

where A is the representative area.

The advantage of this representation is that the hardening portions of the (mean) stress-strain curves are identical for each of the elements and for the series coupling. This is not so, however, for the softening part of the curves. Let ε_{h1} be the mean strain on the hardening branch of the curve for any one of the two elements. Further, let ε_{u1} be the strain at the same stress level on the unloading branch emanating from the peak, and let ε_{s1} be the strain at the same stress level on the softening part of the curve, as indicated in Fig. 8.1.3c–d. The curves in this figure are the same as those in Fig. 8.1.1, with a change of scale. The resulting (mean) stress-strain curve is shown in Fig. 8.1.3d, in which the mean strain at the given stress level is given by

$$\varepsilon = \begin{cases} (L_1\varepsilon_{h1} + L_2\varepsilon_{h2})/(L_1 + L_2) = \varepsilon_{h1} & \text{for hardening} \\ (L_1\varepsilon_{s1} + L_2\varepsilon_{u2})/(L_1 + L_2) = \varepsilon_{u1} + \frac{1}{2}(\varepsilon_{s1} - \varepsilon_{u1}) & \text{for softening} \end{cases} \tag{8.1.2}$$

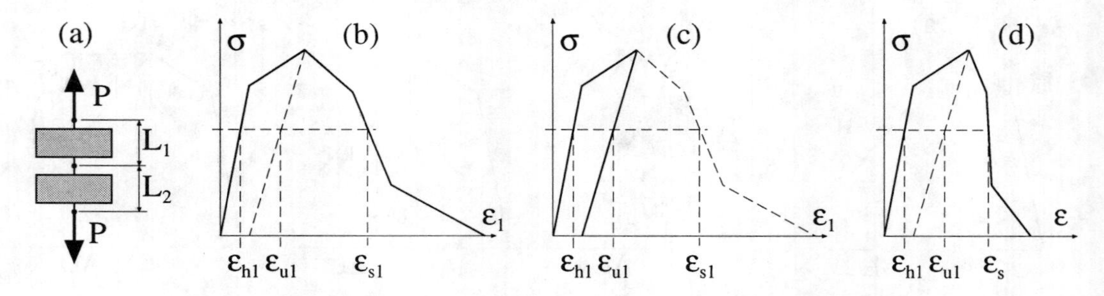

Figure 8.1.3 Series coupling of two softening elements: load-average strain curves.

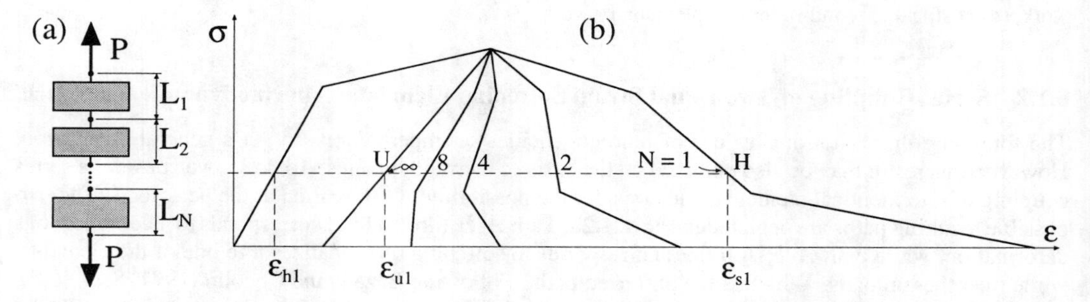

Figure 8.1.4 (a) Series coupling of N equal softening elements. (b) Stress vs. average-strain curves for various values of N.

8.1.4 Series Coupling of N Equal Strain Softening Elements

Consider now a chain of N nominally identical softening elements (Fig. 8.1.4a). Following the same reasoning as in the previous analysis, it is immediately obvious that after reaching the peak, only one of the elements, say element **1**, will soften, while the remaining $N-1$ will unload. Keeping this in mind, we consider how the mean strain will evolve as a function of N.

The mean strain of the whole chain is

$$\varepsilon = \frac{\sum_{i=1}^{N} L_i \varepsilon_i}{\sum_{i=1}^{N} L_i} = \frac{1}{N} \sum_{i=1}^{N} \varepsilon_i \tag{8.1.3}$$

where i indicates the element number, and ε_i the strain of that element. Expressing the fact that on the hardening branch all the strains are identical and equal to ε_{h1} and that on the softening branch the strain of the first element is ε_{s1} while the strain of the remaining $N-1$ elements is ε_{u1}, we get the following result for the mean stress-strain curve:

$$\varepsilon = \begin{cases} \varepsilon_{h1} & \text{for hardening} \\ \varepsilon_{u1} + \frac{1}{N}(\varepsilon_{s1} - \varepsilon_{u1}) & \text{for softening} \end{cases} \tag{8.1.4}$$

Fig. 8.1.4b plots the foregoing analytical results for N=1, 2, 4, 8, and ∞ based on the curve of Fig. 8.1.3b (note that the horizontal scale has been expanded). The construction of the softening branch is very easy to perform graphically: at each stress level, take the segment UH where U and H are the points, respectively, on the unloading and softening branches for a single element. Then, take a segment N times smaller with origin at U. The other end of the segment determines the point of the softening branch of the series coupling of N elements.

One essential result of this analysis is that, while the peak load does not change with the number of elements, the brittleness does so in the sense that the larger the number of elements, the steeper the softening branch gets. In the limit of an infinite number of elements, the behavior is perfectly brittle.

Exercises

8.1 Analyze the response of a series coupling of two equal elements whose load-displacement curve shows a perfect plateau at peak load. For simplicity, assume that the load-elongation curve has the shape of a trapezium, rising linearly from $(0, 0)$ to (P_u, u_0), then extending horizontally to (P_u, u_1), and finally descending linearly to $(0, u_2)$, where $u_0 < u_1 < u2$.

8.2 Consider the series coupling of elements that have a triangular load-displacement curve and are identical except for small imperfections. Assume that for one element the peak occurs at 1.2 kN for an elongation of 5 μm, and that a zero load is reached for an elongation of 200 μm. Determine and make a sketch of the load-displacement curve for (a) 2 elements, (b) 10 elements, (c) 100 elements, and (d) determine the number of elements for which the load drops vertically just after the peak.

8.3 Consider the series coupling of elements with exponential softening. The load-displacement curve for one single element is given by the equations

$$\Delta L = \begin{cases} C_0 P & \text{for} \quad \Delta L \le C_0 P_u \\ C_0 P + u_0 \ln(P_u/P) & \text{for} \quad \Delta L \ge C_0 P_u \end{cases} \tag{8.1.5}$$

in which $C_0 = 1.1$ μm/kN, $P_u = 3.1$ kN, $u_0 = 68.2$ μm. Determine: (a) the energy required to break one element, (b) the load-elongation curve for a coupling of 10 elements, (c) same for 100 elements (draw the curve). (d) Determine also the lowest number of elements for which the resulting softening branch displays a vertical tangent.

8.2 Localization of Strain in a Softening Bar

In the preceding section we obtained some general results concerning a series coupling of discrete elements, for which the reasoning is somewhat easier than for a continuous bar. Now let us discuss the behavior of a uniaxially stressed bar of a homogeneous material.

It may appear that using a classical stress-strain formulation including softening is a natural way to introduce fracture (loss of strength down to zero). However, this is not straightforward. If no other precaution is taken, the resulting model is both physically incorrect and numerically ill-posed. Let us see why.

8.2.1 Localization and Mesh Objectivity

Consider a homogeneous bar of initial length L (Fig. 8.2.1a) made of a material whose stress-strain curve (uniaxial) is assumed to exhibit softening, as sketched in Fig. 8.2.1b. Because of the hypothesis of homogeneity, we can imagine the bar to be subdivided in N identical shorter bars (N being arbitrary) which then act as N equal elements coupled in series, as sketched in Fig. 8.2.1c. We have seen in the previous section that when N elements are coupled in series, the strain localizes after the peak in only one of them, so that the resulting σ-ε curves look similar to that in Fig. 8.1.4b.

Therefore, the postpeak softening of the bar depends totally on the assumed subdivision, as indicated in Fig. 8.2.1d. This has two direct consequences: on purely mechanical grounds, the result is absurd because the physical result cannot depend on the *imagined* subdivision; on numerical grounds, it implies that the result one would obtain by using finite elements would completely depend on the number of elements or element size. This is a subjective choice of the analyst, and, thus, is not an objective property, as pointed out by Bažant (1976). This last property is referred to as lack of *mesh objectivity,* or as *spurious mesh sensitivity.*

Keeping the numerical point of view, we must realize that the response of the foregoing model is reached upon infinite mesh refinement, i.e., for $N \to \infty$. This means that strain localization is predicted to occur only within one infinitely thin element, i.e., an element of infinitely small length and volume. Now, is that consistent with the principles of thermodynamics? Yes, it is. We stated in the previous section that among many possible equilibrium paths, the actual one is that in which the second-order complementary work $\delta^2\mathcal{W}^*$ at constant δP is maximum. Since, in the softening branch, δP is negative, the maximum $\delta^2\mathcal{W}^*$ occurs for the path with the largest (positive) inverse slope. As shown in the figure, this slope does indeed correspond to considering infinitely many elements (Fig. 8.2.1d).

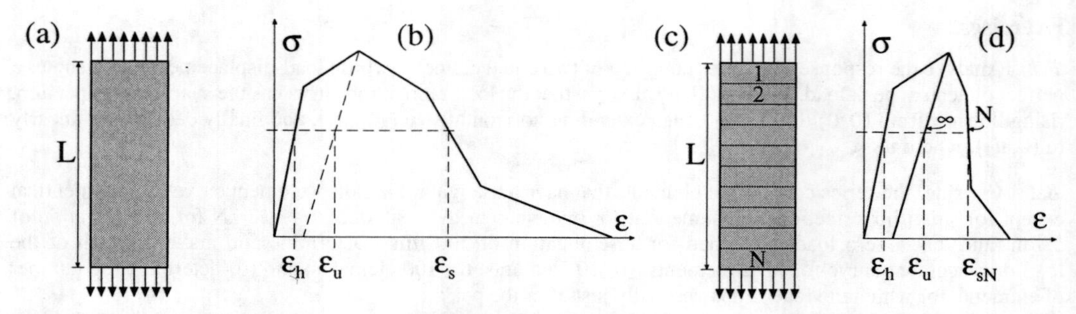

Figure 8.2.1 (a) Homogeneous bar. (b) Stress-strain curve of the material. (c) Subdivision of the bar into N equal elements. (d) Resulting stress-strain curve.

A further implication is that any variable that is ultimately bounded by the length or the volume will vanish. (We say that a variable ϕ is ultimately bounded by volume V if $|\phi| < MV$, for some finite $M > 0$.) This is so for the inelastic displacement and the energy dissipation after the peak. We say that the corresponding physical quantities have *measure zero.* Let us take a closer look at this problem for one special, yet important case.

8.2.2 Localization in an Elastic-Softening Bar

Consider a homogeneous bar of length L (Fig. 8.2.2a) that has a stress-strain curve of the elastic-softening type, as depicted in Fig. 8.2.2b, and is characterized by a linear elastic behavior up to the peak, followed by strain-softening. We can then write the strain on the softening branch as

$$\varepsilon_s = \frac{\sigma}{E} + \varepsilon^f \tag{8.2.1}$$

where E is the elastic modulus and ε^f is the inelastic *fracturing* strain, graphically defined as shown in Fig. 8.2.2b. Unloading from the peak is assumed to be fully elastic. We further assume that the softening branch is unique, i.e., that a unique relationship exists between σ and ε^f as long as ε^f increases monotonically:

$$\sigma = \phi(\varepsilon^f) \tag{8.2.2}$$

This function can be extracted from the σ-ε curve and plotted independently as shown Fig. 8.2.2c. We can also compute the work γ_F required to fully break a unit volume of material (the fracture energy density): it is the area under the σ-ε curve, and so the area under the σ-ε^f curve:

$$\gamma_F = \int_0^\infty \sigma \; d\varepsilon^f = \int_0^\infty \phi(\varepsilon^f) \; d\varepsilon^f \tag{8.2.3}$$

Note that Figs. 8.2.2 and 7.1.3 and the foregoing integral are similar to the definition of G_F in (7.1.8). It might seem that the correspondence is immediate and logical. It is not.

Consider a quasi-static process in which the bar is monotonically stretched. Up to the peak, the strain is uniform, equal to the elastic strain. At peak, just as seen before, a bifurcation can occur so that a portion of the bar, of length $h \le L$, continues stretching, while the rest of the bar unloads elastically (Fig. 8.2.2d). The total elongation of the bar in the softening branch is thus:

$$\Delta L = \frac{\sigma}{E}(L-h) + \left[\frac{\sigma}{E} + \varepsilon^f\right] h = \frac{\sigma}{E} L + \varepsilon^f h \tag{8.2.4}$$

where we see that the first term in the last inequality is the elastic elongation. Therefore, we can define the fracturing elongation as

$$\Delta L^f = \varepsilon^f \, h \tag{8.2.5}$$

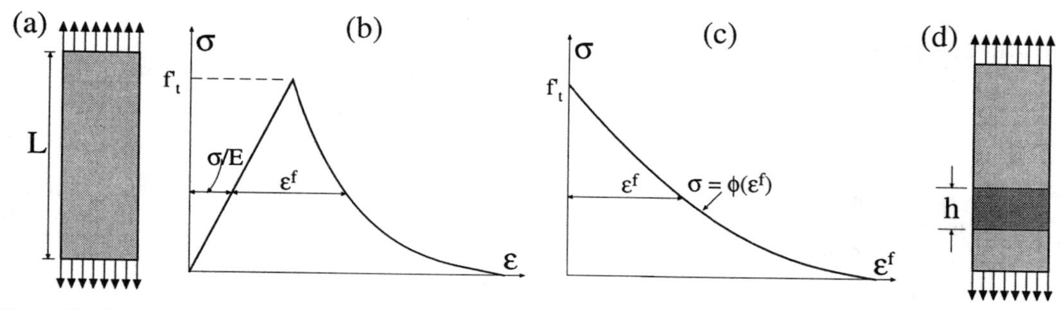

Figure 8.2.2 (a) Homogeneous bar. (b) Elastic-softening stress strain curve. (c) Stress-fracturing strain curve. (d) Bar with a softening band of length h.

On the other hand, the total work supply required to break the whole specimen is just the work required to break the softening portion (the remainder is always elastic) so that

$$\mathcal{W}_F = A \int_0^\infty \sigma \, d(h\varepsilon^f) = Ah \int_0^\infty \sigma \, d\varepsilon^f = Ah\gamma_F \tag{8.2.6}$$

where A is the area of the cross section of the bar.

Up to now h has been arbitrary, but what is its preferred value? To find it, we apply again the maximum second-order work condition ($\delta^2 W^* = \max$) and find immediately that the thermomechanical solution is $h = 0$, in complete concordance with the previous result $N = \infty$ for equally sized elements. It follows from this essential result and from (8.2.5) and (8.2.6) that, according to this model, both the inelastic displacement and the fracture work are zero. This is physically unacceptable and contrary to experiment.

8.2.3 Summary: Necessity of Localization Limiters

The foregoing simple analysis corresponds to static loading and shows that the simple stress-strain model with strain softening leads to unacceptable behavior both physically and computationally: (**1**) the softening zone has a zero width and volume; (**2**) the inelastic strain and fracture work are zero; and (**3**) the computational results are mesh-unobjective.

Further analyses indicate that similar conclusions apply to dynamic situations. For example, Bažant and Belytschko (1985) analyzed the problem of two converging elastic waves propagating from the ends of a bar towards its center, where they add up to exceed the tensile strength. The results show that failure is instantaneous and occurs again over a zone of zero width, and with zero energy dissipation. Belytschko, Bažant et al. (1986) reached similar conclusions for converging elastic waves in a sphere or a cylinder; although the fracture pattern was chaotic, with fracture occurring at many locations, the results still had zero measure and were mesh unobjective (see also Bažant and Cedolin 1991, Sec. 13.1).

The conclusion is that these models are not suitable at all because they allow localization in a region of zero volume. Therefore, if a continuum formulation based on stress-strain curves with strain softening is to be used, it is necessary to complement it with some conditions that prevent the strain from localizing into a region of measure zero. Such conditions are generically called *localization limiters* (Bažant and Belytschko 1985).

The model with the simplest localization limiter is the crack band model that we introduce next. Models with more general limiters are the nonlocal models that are presented in Chapter 13.

Exercises

8.4 Consider a bar with a triangular stress-strain curve defined as $E\varepsilon = \sigma$ for $E\varepsilon \leq f'_t$, and $E\varepsilon = (1+m)f'_t - m\sigma <$ for $f'_t \leq E\varepsilon \leq (1+m)f'_t$, where $E = 30$ GPa, $f'_t = 3$ MPa, $E = 30$ GPa, and $m = 21$; the stress is zero for $E\varepsilon > (1+m)f'_t$. Determine the load-elongation curve and the energy supplied to break the bar if its length is 0.5 m and the softening localizes in a zone of width (a) $h = 25$ cm, (b) $h = 10$ cm, (c) $h = 3$ cm, (d) $h = 1$ cm.

8.5 In the previous exercise, determine the width of the softening zone for which the stress drops vertically to zero right after the peak.

8.6 Consider a bar that has an exponential stress-strain curve defined as

$$E\varepsilon = \begin{cases} \sigma & \text{for} \quad E\varepsilon \leq f'_t \\ \sigma + m f'_t \ln(f'_t/\sigma) & \text{for} \quad E\varepsilon \leq f'_t \end{cases} \tag{8.2.7}$$

in which $E = 27$ GPa, $f'_t = 3.1$ MPa, $m = 12$. Determine the load-elongation curve and the energy supplied to break the bar if its length is 0.5 m and the softening localizes in a zone of width (a) $h = 25$ cm, (b) $h = 10$ cm, (c) $h = 3$ cm, (d) $h = 1$ cm.

8.7 In the previous exercise, determine the width of the softening zone for which the tangent to the stress-elongation curve right after the peak becomes vertical.

8.3 Basic Concepts in Crack Band Models

From the preceding analysis it is clear that, in order to make strain softening an acceptable constitutive relation, localization of strain softening into arbitrarily small regions must be prevented. This is, in general, achieved by some mathematical concept, called the localization limiter. There are various such concepts of varying degrees of generality and complexity. The most general concept is the nonlocal continuum concept, which will be discussed in Chapter 13. Now we describe a rather simple albeit less general concept, known as the crack band model, which was proposed in general terms in Bažant (1976), and was developed in full detail for sudden cracking in Bažant and Cedolin (1979, 1980, 1983) and Cedolin and Bažant (1980), and for gradual strain softening in Bažant (1982) and Bažant and Oh (1983a). The basic attribute of the crack band model is that the given constitutive relation with strain softening must be associated with a certain width h_c of the crack band, which represents a reference width and is treated as a material property. Here we discuss the most basic aspects only, with the help of simple uniaxial models. The three-dimensional analysis is deferred until Section 8.6.

8.3.1 Elastic-Softening Crack Band Models

As for cohesive cracks, the prepeak stress-strain relation can be nonlinear, but for many purposes it is enough to assume a linear behavior up to the peak followed by softening (Section 8.2.2). Then, the stress-elongation curve is given by (8.2.4), for arbitrary h. In Bažant's approach, the width of the band cannot be less than a certain characteristic value h_c. Thus, substituting $h = h_c$ in (8.2.4), we get an expression that is formally identical to the corresponding expression for the cohesive crack if we identify $h_c\varepsilon^f$ with the cohesive crack opening displacement w:

$$h_c\varepsilon^f = w \tag{8.3.1}$$

Thus, the stress-elongation curve for the band model and for the cohesive model will coincide if we relate the softening curve of stress vs. fracturing strain $\phi(\varepsilon^f)$ to the softening curve of stress vs. crack opening of the cohesive crack, i.e.,

$$\phi(\varepsilon^f) = f(w) = f(h_c\varepsilon^f) \qquad \text{or} \qquad f(w) = \phi(w/h_c) \tag{8.3.2}$$

where $f(w)$ is the equation of the softening curve for the cohesive crack model. Therefore, there is a unique relationship between the crack band model and the cohesive crack model, at least for the simple elastic-softening case that we are analyzing. The correspondence is illustrated in Fig. 8.3.1 which shows the softening curve for the cohesive crack (Fig. 8.3.1a) and the corresponding stress-strain curve for the crack band (Fig. 8.3.1b). Also shown is the correspondence for the initial linear approximation to the curve, the horizontal intercept of which satisfies $\varepsilon_1 = w_1/h_c$. It follows that a linear approximation for the softening of crack bands will be a good approximation in the same circumstances as it was for the cohesive crack model, principally for peak loads of not too large specimens, if notched, but any size specimens, if unnotched). This explains why the use of linear softening was very successful in the work of Bažant and Oh (1983a).

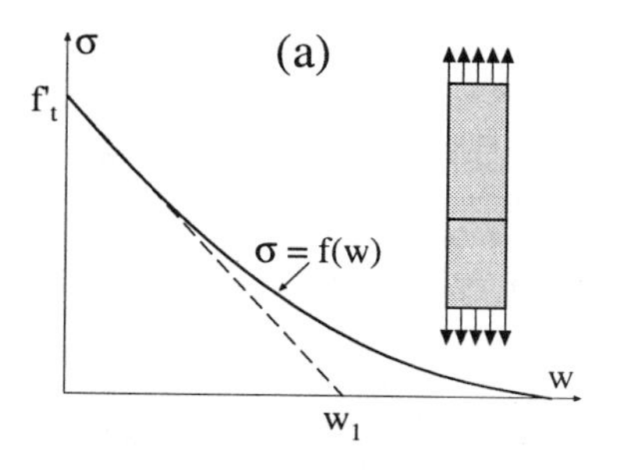

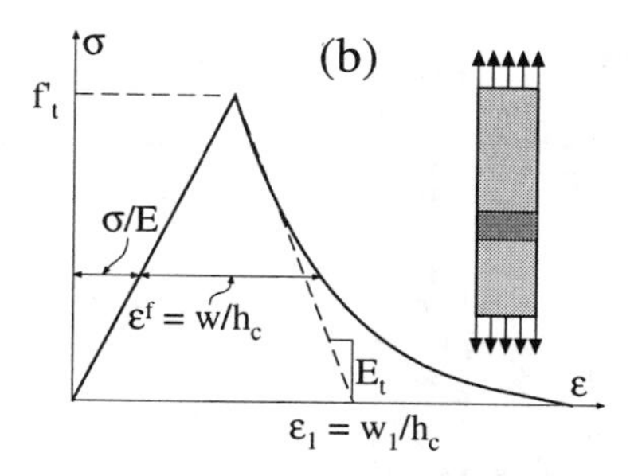

Figure 8.3.1 Correspondence between the softening curve of the cohesive crack model (a), and the stress-strain curve of the crack band model (b).

The correspondence is obviously maintained for the specific fracture energy G_F. Indeed, from (8.2.6) it follows that the energy required to form a complete crack (or a fully softened band) is

$$G_F = \frac{\mathcal{W}_F}{A} = h_c \gamma_F \tag{8.3.3}$$

From this, the characteristic size ℓ_{ch} can be easily obtained in terms of the properties of the crack band model as

$$\ell_{ch} = \frac{E G_F}{f_t'^2} = h_c \frac{E \gamma_F}{f_t'^2} \tag{8.3.4}$$

The characteristic size, ℓ_1, based on the initial linear softening is then

$$\ell_1 = \frac{E w_1}{2 f_t'} = h_c \frac{E \varepsilon_1}{2 f_t'} \tag{8.3.5}$$

where ε_1 is the horizontal intercept of the initial tangent (Fig. 8.3.1b).

A parameter of interest in numerical calculations using the crack band model is the softening modulus E_t for the linear approximation (Fig. 8.3.1b). It is a simple matter to show that

$$2\ell_1 = h_c \left(1 - \frac{E}{E_t}\right) \tag{8.3.6}$$

The correspondence between the two models can further be systematized by defining a characteristic strain ε_{ch} and a reduced fracturing strain $\hat{\varepsilon}^f$ as

$$\varepsilon_{ch} = \frac{h_c f_t'}{G_F} = \frac{f_t'}{\gamma_F} \qquad \text{and} \qquad \hat{\varepsilon}^f = \frac{\varepsilon^f}{\varepsilon_{ch}} \tag{8.3.7}$$

With this, the nondimensional expression for the softening function is identical to that for the cohesive model, with the obvious change $\hat{\varepsilon}^f \leftrightarrow \hat{w}$.,i.e.:

$$\hat{\sigma} = \hat{f}(\hat{\varepsilon}^f) \tag{8.3.8}$$

Therefore, all the softening curves discussed in the previous chapter can be directly implemented in the crack band model. The only difference between the results for one and other model is in the strain and displacement distribution. Figs. 8.3.2a and b show the comparison of the axial displacement distribution in a bar for a cohesive crack and a crack band model. Figs. 8.3.2c and d show the corresponding strain distributions. Obviously, the difference is nil for engineering purposes if $h_c \ll L$. This is almost invariably true in practical situations because, as we will see later in more detail, h_c is of the order of a few maximum aggregate sizes (Bažant and Oh 1983a).

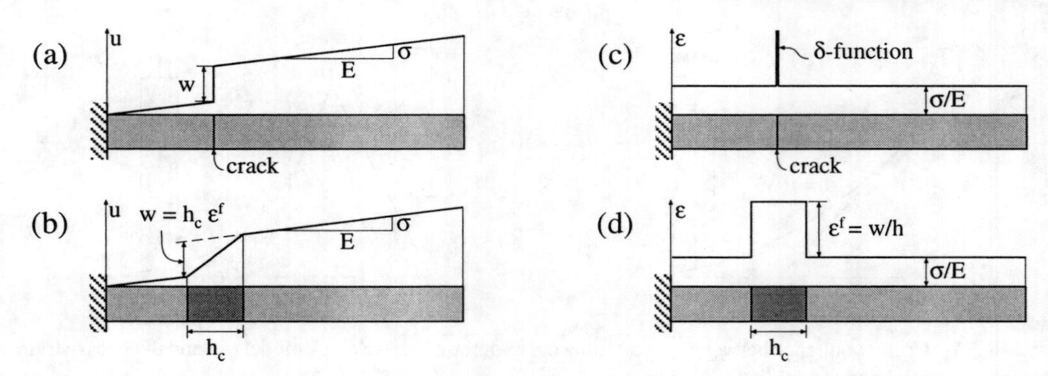

Figure 8.3.2 Comparison of the distributions of axial displacement and of strain in a bar for the cohesive crack model (a, c) and the crack band model (b, d).

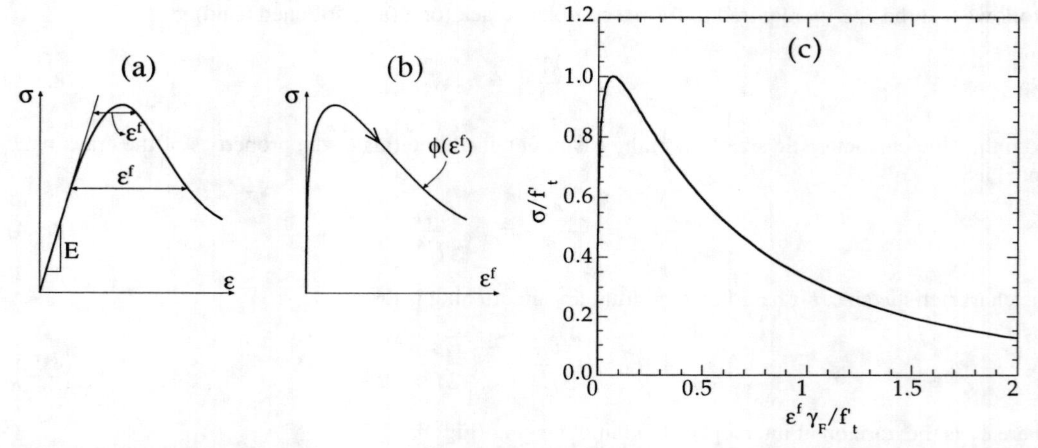

Figure 8.3.3 (a) Curvilinear stress-strain curve. (b) Curvilinear stress-fracturing strain curve. (c) Plot of the curve defined in Eq. (8.3.9).

8.3.2 Band Models with Bulk Dissipation

The cracking in reality does not begin upon reaching the tensile strength, but earlier, and so the diagram of stress vs. strain should have the form shown in Fig. 8.3.3a. As discussed in Section 7.1.6 for cohesive crack models, such behavior can be incorporated into the computational models with relative conceptual simplicity, but it considerably complicates the numerical treatment and experimental interpretation. Moreover, at least for concrete, neglecting the prepeak nonlinearity is generally acceptable for practical engineering use, and so the elastic-softening models we previously discussed are those most used. This seems to be clearly established (Planas, Elices and Guinea 1992) when there is one main crack, i.e., sharp localization occurs, because then the large postpeak strains dominate over the prepeak deformation. However, for situations where the localization is not sharp, the prepeak nonlinearity may play a dominant role, and its inclusion might be necessary. This may be the case in the prelocalization stages when there is reinforcement or when the stress field has a high gradient (as in the case for shrinkage stresses).

In our discussion of cohesive crack models, the inclusion of bulk dissipation (prepeak inelasticity) required defining an inelastic constitutive equation for the bulk in addition to the softening curve for the cohesive crack. One of the appealing features of the crack band model is that such a dichotomy is not necessary. Indeed, it is enough to define a single curvilinear stress-strain curve such as that shown in Fig. 8.3.3a. Then we can split the strain into the elastic and inelastic or fracturing part and use the curve of stress vs. inelastic strain as shown in Fig. 8.3.3b. For example, Bažant and Chern (1985a) proposed

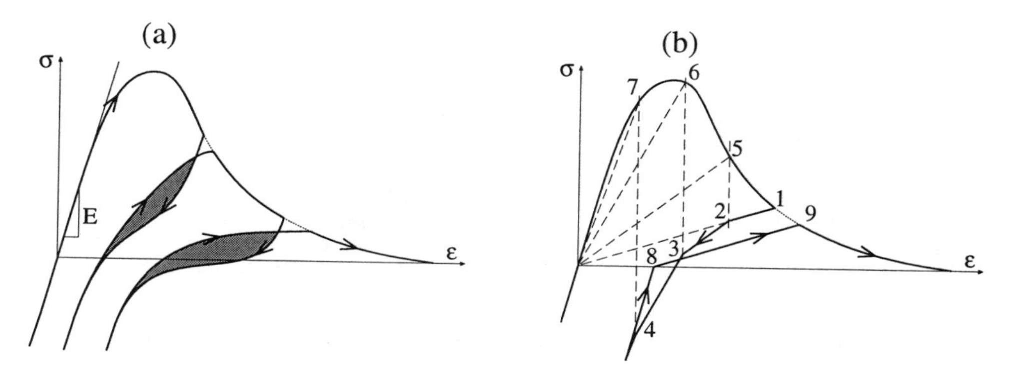

Figure 8.3.4 (a) General trend for the unloading-reloading branches found in experiments. (b) Construction of the unloading branch according to the secant-tangent rule (Bažant and Chern 1985a,b).

the following power-exponential curve:

$$\sigma = \phi(\varepsilon^f) = C_1 \varepsilon^{f\,p} e^{-b\varepsilon^{f\,q}} \tag{8.3.9}$$

where C_1, p, b, and q are constants. Fig. 8.3.3c shows the appearance of this curve for $p = 1/3, q = 0.55$ as derived by Bažant and Chern (1985a) by fitting of various experimental data. Note that the curve has been nondimensionalized so that its peak and area are equal to one.

Alternatively, the complete stress-strain curve can be given in the form $\sigma = \psi(\varepsilon)$. This is equivalent to giving the $\sigma(\varepsilon^f)$ curve in parametric form as:

$$\begin{aligned} \sigma &= \psi(\varepsilon) \\ \varepsilon^f &= \varepsilon - \tfrac{1}{E}\psi(\varepsilon) \end{aligned} \tag{8.3.10}$$

Among the formulas of this kind we have the power-exponential form (Bažant 1985a):

$$\sigma = E\varepsilon\, e^{-b\varepsilon^q} \tag{8.3.11}$$

where E is the elastic modulus and b, and q are constants. Another expression is

$$\sigma = \frac{E\varepsilon}{1 + a\varepsilon + b\varepsilon^q} \tag{8.3.12}$$

which was introduced by Saenz (1964) for compression strain softening, and in which a, b and q are constants.

8.3.3 Unloading and Reloading

For general applications in finite element programs, one must also specify what happens when, after partial or full cracking, the material is unloaded or reloaded. Experimentally observed behavior at unloading and reloading is rather complicated and looks approximately as sketched in Fig. 8.3.4a which is characterized by hysteretic loops of considerable area (Reinhardt and Cornelissen 1984; Hordijk 1991). In most finite element programs, however, it is assumed that unloading and reloading are linear. In the next section, devoted to the uniaxial softening models, we show how these linear unloading-reloading curves are generated within theoretical frameworks that can be easily generalized to the general three-dimensional models.

If the detailed uniaxial unloading-reloading curves need to be reproduced, the expressions developed to generate realistic unloading-reloading curves in cohesive crack models (Section 11.7.4) are easily incorporated into the crack band model through the basic relationship (8.3.1).

A simpler rule, called the secant-tangent rule, was proposed in the frame of crack band models by Bažant and Chern (1985a,b), as illustrated in Fig. 8.3.4b. Given the stress-strain diagram for monotonic

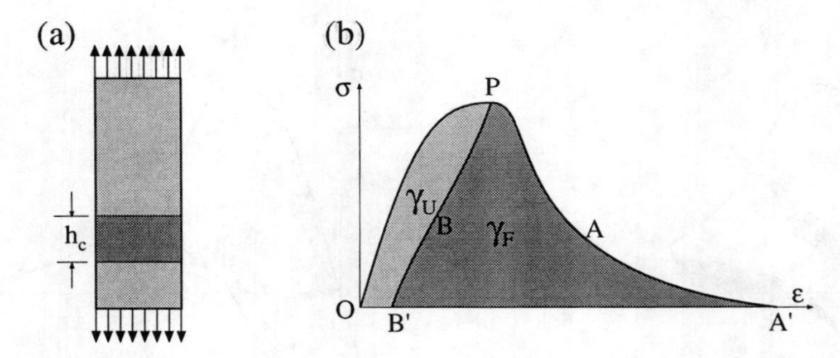

Figure 8.3.5 Energy dissipation in a crack band model with prepeak inelasticity.

stretching, $\sigma = \psi(\varepsilon)$, the secant-tangent rule assumes that the unloading has always the same slope as the secant-modulus for virgin loading at the same strain value, i.e.,

$$d\sigma = \frac{\psi(\varepsilon)}{\varepsilon} d\varepsilon \qquad \text{if } d\varepsilon < 0 \tag{8.3.13}$$

Graphically, this means (Fig. 8.3.4b) that segment $\overline{23}$ is parallel to the secant $\overline{05}$, segment $\overline{34}$ is parallel to the secant $\overline{06}$, etc., where points 5, 6, 7 are obtained from points 2, 3, 4 by vertical projections onto the virgin stress-strain curve.

For reloading one may assume either the same path as for unloading, or, better, a straight line reloading up to point 8 on the strain axis and then either a straight line back to point 1 where unloading started or a straight line $\overline{89}$ parallel to the secant $\overline{01}$. The tangent-secant rule underestimates the area of the hysteresis loops, but it has the advantage that it yields, without any additional material parameters, an approximately correct location of point 4 at which the initial elastic slope is resumed. Furthermore, using the rule shown by curve $\overline{489}$, point 9 at which the virgin curve is reached again is approximately correct. The tangent-secant rule was applied in a finite element program for combined smeared cracking, creep, and shrinkage of concrete (Bažant and Chern 1985a,b).

8.3.4 Fracture Energy for Crack Bands With Prepeak Energy Dissipation

For stress-strain curves with prepeak inelasticity, the energy dissipation consists of two terms. One corresponds to the energy dissipated in the prepeak range, which is proportional to the volume, and the second corresponds to the energy dissipated after peak which, in the cases of localization in a single band, is proportional to the volume of the band $h_c A$, where A is the area of the main surface of the crack band. Therefore, the problem is identical to that for the cohesive crack with bulk dissipation (Section 7.1.6).

An analysis analogous to that for the cohesive cracks was performed by Elices and Planas (1989): for a uniform bar in tension (Fig. 8.3.5a), with the stress-strain curve shown in Fig. 8.3.5b, the material follows initially the path OP up to the peak. Then the bifurcation occurs and the material outside the crack band follows the unloading path PBB' while the material in the crack band follows the path PAA'. Therefore, the total work of fracture is

$$\begin{aligned} \mathcal{W}_F = A(L-h_c)\ \text{area}(OPBB'O) + Ah_c\ \text{area}(OPAA'O) = \\ AL\ \text{area}(OPBB'O) + Ah_c\ \text{area}(BB'AA'B) \end{aligned} \tag{8.3.14}$$

The area $OPBB'O$ (lightly shaded in the figure) is the energy supplied to a unit volume when it is loaded up to the peak and then unloaded; we represent it by γ_U. The area $BB'AA'B$ (darker shading in the figure) represents the *extra* energy supply required to break a unit volume of material in the crack band. Therefore, we may write the foregoing equation as

$$\mathcal{W}_F = AL\gamma_U + Ah_c\gamma_F \tag{8.3.15}$$

Now, identifying the second term as the surface energy dissipation, we can apply (8.3.3) and get an

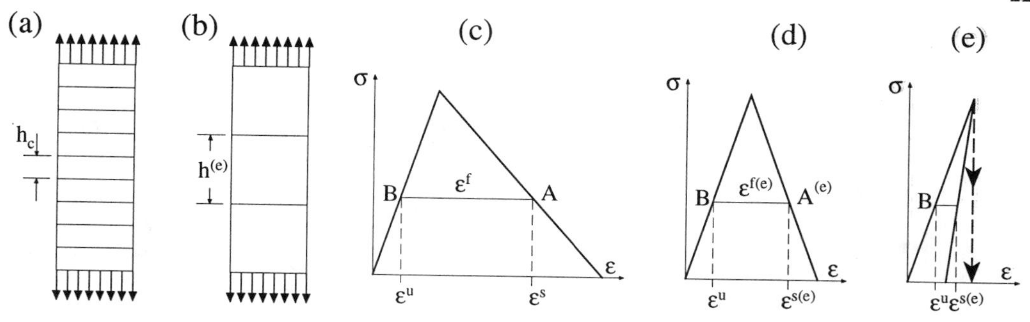

Figure 8.3.6 (a) Bar discretized in finite elements of size h_c. (b) Bar discretized in finite elements of arbitrary size $h^{(e)}$. (c) Triangular stress-strain curve for a physical band of width h_c. (d) Corresponding stress-strain curve for an element of size $h^{(e)}$, in which $\overline{BA^{(e)}} = \overline{BA}\, h^{(e)}/h_c$. (e) Same for too large an element, leading to snapback; the arrows indicate sudden failure at peak load.

expression identical to (7.1.19). From this point on, the analysis is identical to that for the cohesive cracks with bulk dissipation, which leads to a dependence of the mean fracture energy given by (7.1.22). This can obviously be recast in terms of the properties of the stress-strain curve using (8.3.3):

$$G_{Fm} = h_c \gamma_F \left[1 + \alpha_U \eta \left(\frac{D}{\ell_{ch}} \right) \right], \qquad \alpha_U = \frac{\gamma_U E}{f_t'^2}, \qquad \ell_{ch} = \frac{E h_c \gamma_F}{f_t'^2} \tag{8.3.16}$$

8.3.5 Simple Numerical Issues

Strict application of the crack band model, as formulated by Bažant, with h_c equal to a material constant, would require a finite element mesh in which the cracking band has exactly a width h_c. Thus, if the crack-band location is not known in advance, all the finite elements would have to be of width h_c as depicted in Fig. 8.3.6a for the uniaxial case. This is unpractical and, fortunately, unnecessary. The fundamental reason is that h_c does not enter explicitly the essential macroscopic parameters, of which the most important is $G_F = h_c \gamma_F$. Therefore, if finite elements larger than h_c need to be used, it is possible to keep the essential response if we preserve the fracture energy. To do so, an adequate approximation is to distribute uniformly the fracturing strain over the element and rescale the softening part of the stress-strain curve to keep G_F constant. The resulting stress-strain curve will depend on the element size, and must be scaled so that

$$h^{(e)} \gamma_F^{(e)} = h_c \gamma_F \qquad \Rightarrow \qquad \gamma_F^{(e)} = \frac{h_c}{h^{(e)}} \gamma_F \tag{8.3.17}$$

where $h^{(e)}$ is the size of the element and $\gamma_F^{(e)}$ the density of fracture energy to be used for this element.

For models of the elastic-softening type with stress-strain curves such as the one shown in Fig. 8.3.1b, the scaling is easy: just multiply ε^f by the factor in the preceding formula, i.e.,

$$\varepsilon^{f\,(e)} = \frac{h_c}{h^{(e)}} \varepsilon^f \tag{8.3.18}$$

Note that only the fracturing part of the strain must be scaled. The result of the scaling is shown in Figs. 8.3.6c-e for the simple linear softening. Note also that if the size of the element is too large, as deliberately shown in Fig. 8.3.6e, the resulting softening branch for the stress-strain curve of the element will show a snapback. In these cases, the finite element will become unstable at the snapback point, and the stress will drop suddenly to zero. Then the energy dissipated cannot be made equal to G_F, since all the elastic energy in the element is released. Therefore, either the element must be kept small so that no snapback would occur, or the curve must be modified to preserve the energy dissipation. This problem will be addressed in Section 8.6 in a wider three-dimensional framework.

The matters are a bit more complicated if the stress-strain curve has a prepeak nonlinearity as in Fig. 8.3.7a. In that case, the strain in the hardening and unloading branches must not be scaled; only the

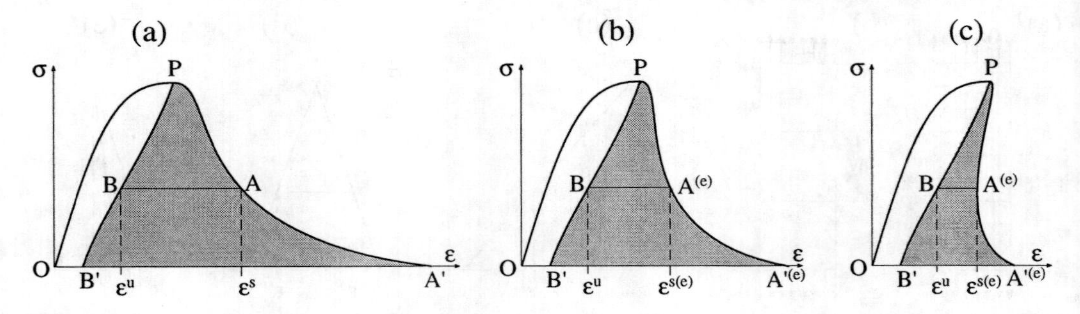

Figure 8.3.7 (a) Definition of the unloading and softening branches, for a physical band, of width h_c; (b) unloading and softening branches to be used in a finite element of a larger width $h^{(e)}$, for which $\overline{BA^{(e)}} = \overline{BA}\, h^{(e)}/h_c$; (c) same, but for an element so large that it leads to snapback.

strain contribution to the surface component of energy must. Therefore, the strain on the softening branch has to be scaled so that

$$\varepsilon^{s(e)} - \varepsilon^u = \frac{h_c}{h^{(e)}}(\varepsilon^s - \varepsilon^u) \tag{8.3.19}$$

where ε^s and ε^u are the strains on the softening branch and on the unloading branch, respectively, for the same level of stress (Fig. 8.3.5a–c). This can be rewritten as

$$\varepsilon^{s(e)} = \frac{h_c}{h^{(e)}}\varepsilon^s - \left(\frac{h_c}{h^{(e)}} - 1\right)\varepsilon^u \tag{8.3.20}$$

Thus, as shown in Fig. 8.3.7b–c, only the part of the softening curve on the right of the unloading branch is to be scaled. This may substantially complicate the use of otherwise simple stress-strain curves (i.e., with simple, beautiful expressions). Note, again, that snapback may occur in this case, too, if the element is too large, as shown in Fig. 8.3.7c.

In the foregoing it is implicitly assumed that the fracture will localize in a single element, since we showed that this is *the* solution in the preceding two sections. However, if a homogeneous case, such as the tensioned bar (or, in three dimensions, a pure bend beam), is numerically analyzed, and if the elements are given exactly the same properties, a normal finite element code will not catch the bifurcation. The reason is that the program will search for a solution by extrapolating from the previous step, and thus all elements will go through the peak into the softening branch simultaneously. And they will stay there! To avoid sophisticated bifurcation analysis (which is more elegant and more robust, but much more complicated), a simple expedient may be used: put imperfections into the material. Then either one element selected at random is taken to be a few percent weaker than the rest, or the strength of each element is assigned at random using a narrow strength distribution function. This is necessary only for structures with a nominally homogeneous distribution of elastic stresses and strains (laboratory specimens, typically). In most structures the elastic fields have stress concentrations which trigger localization without the need for introducing imperfections. (However, in some situations, the danger remains that the loading step is too large for the imperfections assumed to trigger localization. Theoretically, without bifurcation analysis, one is not sure, in general, that a localization has not been missed.)

We have addressed here two basic aspects of the numerical computation: the stress-strain curves to use, and the way to trigger the localization. This, of course, does not exhaust the discussion on the numerical models, but the other important aspects are fundamentally three-dimensional and are discussed later in Section 8.6, after presenting the three-dimensional softening models.

8.3.6 Crack Band Width

From the foregoing analysis it transpires that, in a finite element formulation with a free element size, the strain-softening curve must be adjusted according to the element size so that the calculations would yield macroscopically consistent results whatever the element size. This is close to saying that the crack band width h_c is arbitrary since it is replaced by $h^{(e)}$ without a noticeable effect (as long as the element size is kept small). This means that h_c cannot be determined from fracture tests in which a single crack (or crack band) is formed.

The value of h_c, however, does have an effect in those situations where cracking does not localize but remains distributed over large zones. This may happen as a consequence of a dense reinforcement grid or in problems such as shrinkage, where the mass of concrete in front of the drying zone restrains the cracking zone and may (but need not) force the cracking to remain distributed. Thus, the value of h_c can be identified only by comparing the results of fracture tests with the results of tests in which the cracking is forced to be distributed. The problem is the same as that in determining the characteristic length for nonlocal models, and we will discuss it in more detail in Chapter 13.

In a crude manner, the value of h_c can be approximately identified from fracture tests for specimens of various geometries, in which the cracking is localized to a different extent. This has been done in Bažant and Oh (1983a), with the conclusion that the crack band width $h_c = 3d_a$ where d_a = maximum aggregate size, is approximately optimal. However, the optimum was weak, and crack band width anywhere between $2d_a$ and $5d_a$ would have given almost equally good results.

A better test for determining h_c was conceived by Bažant and Pijaudier-Cabot (1989). Localization was prevented by gluing parallel thin rods on the surface of a uniaxially tensioned prism. However, a uniform field of strain-softening was still not achieved. For details, see Section 13.2.4.

Exercises

8.8 Give a detailed proof of Eq. (8.3.6).

8.9 Determine the uniaxial stress-strain curves for a concrete which, according to experimental measurements, has an elastic modulus of 25 GPa, a tensile strength of 2.8 MPa, and a fracture energy of 95 N/m, and is assumed to display an elastic-softening behavior with triangular softening and a crack band width of 50 mm (approximately equal to $3d_a$ with $d_a = 16$ mm).

8.10 Determine the uniaxial stress-strain curves to be used for the same material as defined in the previous exercise if the numerical analysis is to be performed using finite elements 20 cm in size.

8.11 Determine the maximum size of the finite elements to be used in a numerical analysis with the same material in order for the stress-strain curve to be stable.

8.12 Determine the uniaxial stress-strain curves for a concrete which, according to experimental measurements, has an elastic modulus of 25 GPa, a tensile strength of 2.8 MPa, and a fracture energy of 95 N/m, and is assumed to display an elastic-softening behavior with exponential softening and a crack band width of 50 mm (approximately equal to $3d_a$ with $d_a = 16$ mm).

8.13 Determine the uniaxial stress-strain curves to be used for the material defined in the previous exercise if the numerical analysis is to be performed using finite elements 20 cm in size.

8.14 Determine the maximum size of the finite elements to be used in a numerical analysis with the foregoing exponential material in order for the stress-strain curve to be stable (not to exhibit snapback).

8.15 For the material defined by Eq. (8.3.9), determine the fracturing strain ε_u^f and stress f_t' at which the peak occurs as a function of the constants C_1, p, b, and q. Show that the equation can be rearranged to read $\sigma = f_t' \xi^p \exp[-p(\xi^q - 1)/q]$, in which $\xi = \varepsilon^f/\varepsilon_u^f$.

8.16 For the material defined by Eq. (8.3.9), show that the fracture energy density γ_F can be written as

$$\gamma_F = \frac{C_1}{q b^{(p+1)/q}} \Gamma\left(\frac{p+1}{q}\right)$$

where $\Gamma(n)$ is the Eulerian Gamma function defined as $\Gamma(n) = \int_0^\infty x^{n-1} e^{-x} dx$.

8.17 Consider a material with a stress-strain curve given by $\sigma = E\varepsilon e^{-b\varepsilon}$. Show that E is indeed the elastic modulus. Determine b in terms of E and f_t'. Determine, as a function of E and f_t', (a) the total energy density absorbed by a material element that follows the softening branch; (b) the energy density absorbed by a material element loaded up to the peak and then unloaded, if it is assumed that the unloading is linear with the same elastic modulus as for the initial loading; (c) the density of fracture energy γ_F; and (d) show that for this model the ratio h_c/ℓ_{ch} is constant and determine its value. [Answers: (a) $7.39 f_t'^2/E$, (b) $1.45 f_t'^2/E$, (c) $5.94 f_t'^2/E$, (d) $h_c/\ell_{ch} \approx 0.17$.]

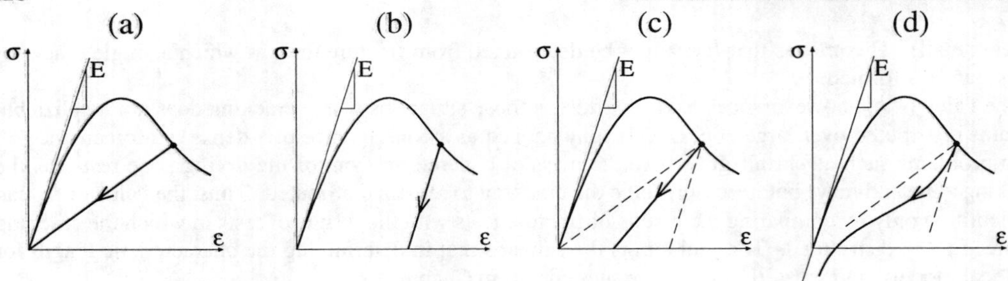

Figure 8.4.1 Types of stress-strain curves: (a) stiffness degradation; (b) yield limit degradation; (c) mixed behavior; (d) more realistic behavior with nonlinear unloading.

8.4 Uniaxial Softening Models

After explaining the basic concepts in crack band models, we can now discuss in detail the uniaxial version of various simple constitutive models for strain softening. Depending on the behavior at unloading, one may distinguish three basic types of models :

1. Continuum damage mechanics, in which strain-softening is due solely to degradation of elastic moduli and no other inelastic behavior takes place. The basic characteristic of such theory is that the material unloads along a straight line pointed toward the origin (Fig. 8.4.1a).

2. Plasticity with yield limit degradation, in which the constitutive relation is the same as in plasticity except that the yield limit is decreasing, rather than increasing. The elastic moduli remain constant (Fig. 8.4.1b). The basic characteristic is that the unloading slope is constant, equal to the elastic modulus E.

3. A mixed theory, in which both the elastic moduli and the yield limit suffer degradation. This behavior, for which the unloading slope is intermediate (as shown Fig. 8.4.1c), is normally a better description of experimental reality.

The foregoing classification neglects the fact that the unloading-reloading response is actually nonlinear, as discussed in Section 8.3.3 and depicted in Fig. 8.4.1d. Models including such behavior can be generated, but they are considerably more complex, and are a subject for specialized studies that will not be treated here. As an exception, the microplane model, which implements this kind of behavior naturally, will be discussed at length in Chapter 14.

8.4.1 Elastic-Softening Model with Stiffness Degradation

As a simple continuum model of a material fracturing in tension, we can adopt the elastic-softening model whose behavior for monotonic stretching was described in Fig. 8.2.2. To give a physical content to the model, we can assume that this behavior corresponds to an elastic matrix with an array of densely distributed cracks normal to the load direction (Fig. 8.4.2a). Thus, we assume that the total strain is the sum of the elastic strain of the elastic matrix, ε^{el}, and the strain contributed by the crack opening, ε^f:

$$\varepsilon = \varepsilon^{el} + \varepsilon^f = \frac{\sigma}{E} + \varepsilon^f \tag{8.4.1}$$

where E is the elastic modulus of the matrix (i.e., of the virgin material between the cracks). For monotonic straining the assumed behavior is that shown in Fig. 8.2.2. Consider now the unloading behavior after the specimen has been loaded until a certain maximum inelastic strain $\tilde{\varepsilon}^f$ (Figs. 8.4.2b–d). Let us further assume that unloading is straight to the origin. This means that during unloading, the cracks close so that they are completely closed at zero stress. As depicted in (Fig. 8.4.2b), $\tilde{\varepsilon}^f$ represents the maximum cracking strain reached before unloading, and ε^f the actual cracking strain. Obviously, for such unloading

$$\varepsilon^f = \frac{\tilde{\varepsilon}^f}{\phi(\tilde{\varepsilon}^f)}\sigma \tag{8.4.2}$$

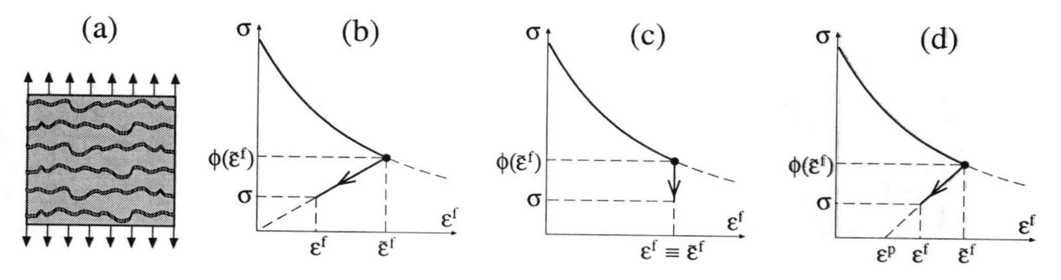

Figure 8.4.2 (a) Densely cracked elastic material. (b) Model with cracks completely closed after full unloading. (c) Model with totally prevented crack closure. (d) Mixed type model with partial crack closure upon unloading.

But this equation also holds if the loading is monotonic (i.e., if fracture is taking place) because then $\varepsilon^f = \tilde{\varepsilon}^f$ and $\sigma = \phi(\tilde{\varepsilon}^f)$. However, we must impose the condition that the line that corresponds to monotonic loading can never be crossed. This can be expressed in various ways, but two are most useful: one in terms of σ and $\tilde{\varepsilon}^f$, and the other in terms of ε^f and $\tilde{\varepsilon}^f$:

$$\sigma - \phi(\tilde{\varepsilon}^f) \leq 0 \qquad \text{or} \qquad \varepsilon^f - \tilde{\varepsilon}^f \leq 0 \tag{8.4.3}$$

Note that while ε^f can decrease, $\tilde{\varepsilon}^f$ is a nondecreasing variable.

Now, the foregoing results can be reformulated so as to look as a genuine continuum damage model. To this end, one may *define* a derived variable ω, the *damage*, as

$$\frac{\omega}{1-\omega} = E\frac{\tilde{\varepsilon}^f}{\phi(\tilde{\varepsilon}^f)} \qquad \text{or} \qquad \omega = \frac{E\tilde{\varepsilon}^f}{E\tilde{\varepsilon}^f + \phi(\tilde{\varepsilon}^f)} \tag{8.4.4}$$

Then we just insert this definition into (8.4.2) and the result into (8.4.1), which yields the classical form of continuum damage mechanics for an elastic matrix:

$$\varepsilon = \frac{\sigma}{E(1-\omega)} \tag{8.4.5}$$

We will introduce this expression in a more standard way after we analyze the strength degradation model and the mixed model for this elastic softening model.

8.4.2 Elastic-Softening Model with Strength Degradation

Consider now the same basic parallel crack model of Fig. 8.4.2a, but assume that upon unloading the cracks cannot close, as depicted in Fig. 8.4.2c. This is, of course, a tremendous simplification, but frictional grain interlock, as well as debris and surface mismatch, can prevent cracks to a large extent from closing in materials such as concrete. Obviously, the resulting model is of a plastic type with softening such that $\varepsilon^f \equiv \varepsilon^p$ where ε^p is the plastic strain.

Now we need only to specify that, since we do not consider compression, $\varepsilon^f = \tilde{\varepsilon}^f$ at all times and that the monotonic curve cannot be exceeded. As before, we thus have

$$\sigma - \phi(\tilde{\varepsilon}^f) \leq 0 \tag{8.4.6}$$

as the plastic criterion. Note that in this case we cannot use the criterion in terms of ε^f. Note also that it may seem that making a distinction between ε^f and $\tilde{\varepsilon}^f$ is superfluous. However, for the three-dimensional case the distinction will be essential because ε^f is a second-order tensor, while $\tilde{\varepsilon}^f$ will remain a scalar (known as the equivalent uniaxial inelastic strain).

8.4.3 Elastic-Softening Model with Stiffness and Strength Degradation

We consider the same model as before, but consider now that, upon unloading, some permanent strain ε^p remains at zero stress, as depicted in Fig. 8.4.2d. We can simply write that, on the unloading-reloading

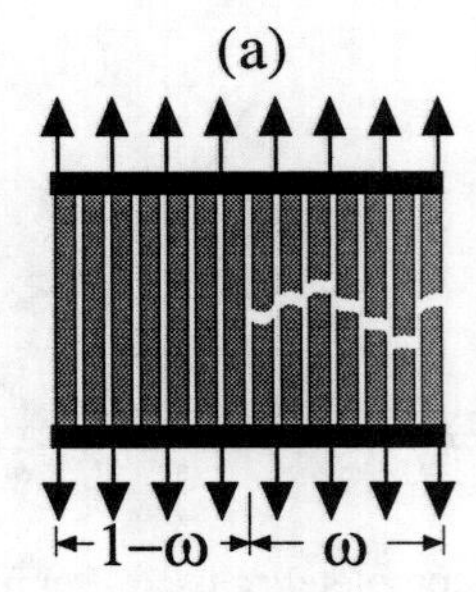

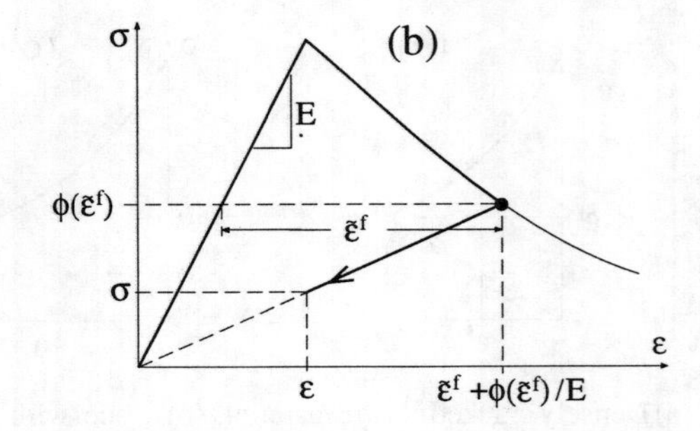

Figure 8.4.3 (a) Parallel micro-rod coupling (Dougill 1976). (b) Determination of the damage parameter in terms of the maximum inelastic strain.

branch,

$$\varepsilon^f = \varepsilon^p + \frac{\tilde{\varepsilon}^f - \varepsilon^p}{\phi(\tilde{\varepsilon}^f)}\sigma \tag{8.4.7}$$

The only extra information required is the evolution of ε^p. The simplest assumption is that it is a unique function of $\tilde{\varepsilon}^f$. For example, Ortiz (1985) assumed (in a much more complex framework) that the permanent strain is a fixed fraction of the (maximum) inelastic strain, i.e.,

$$\varepsilon^p = \alpha\tilde{\varepsilon}^f \tag{8.4.8}$$

where α is a constant between 0 (for pure damage) and 1 (for pure plasticity).

8.4.4 A Simple Continuum Damage Model

Let us now briefly review a very simple damage model. We base it on Dougill's approach in which a material element is considered to be formed, ideally, by many infinitesimal rods connected in parallel (Fig. 8.4.3a). We assume that the rods are identical in all but strength, which is randomly distributed. Upon stretching, the weaker rods fail first. At a given strain level ε, a fraction ω of the rods have failed. Then, the resulting stress is given by

$$\sigma = E(1 - \omega)\varepsilon \tag{8.4.9}$$

Note that this is the average stress. The stress in the surviving rods is that for the virgin, undamaged material. Therefore, the relationship between the macrostress σ and the microstress τ (also called the true-stress) is

$$\tau = \frac{\sigma}{1 - \omega} \tag{8.4.10}$$

This is the basic relation in continuum damage mechanics, initiated by Kachanov (see, e.g., Lemaitre and Chaboche 1985). This relation applies not only to brittle materials in which the relationship between the true stress and the strain is linear, as we have here, but to any other (true) stress-strain relationship.

The model must specify the evolution of damage. This is done on the basis of the uniaxial stress-strain curve as shown in Fig. 8.4.3b. Assuming that the data are in the form of the $\sigma(\tilde{\varepsilon}^f)$ curve, we get the damage at each point by writing

$$1 - \omega = \frac{\sigma}{E\varepsilon} = \frac{\phi(\tilde{\varepsilon}^f)}{\phi(\tilde{\varepsilon}^f) + E\tilde{\varepsilon}^f} \tag{8.4.11}$$

from which (8.4.11) follows. So, the two formulations are fully equivalent, even though the underlying micromodels seem to be completely different.

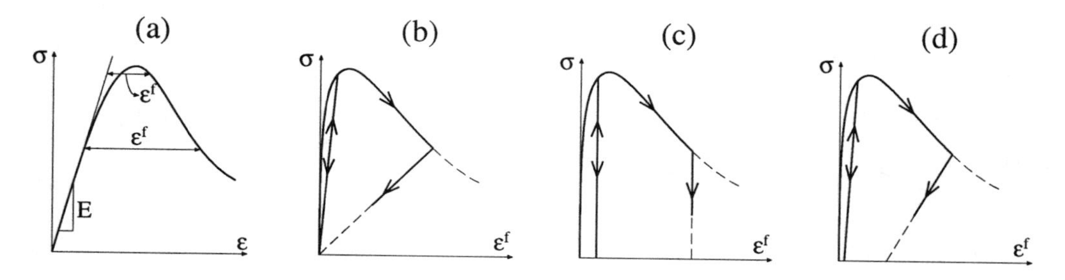

Figure 8.4.4 Models with prepeak inelasticity: (a) stress-strain curve for monotonic loading; (b) stiffness degradation; (c) strength degradation; (d) mixed behavior.

8.4.5 Introducing Inelasticity Prior to the Peak

Although, for concrete in tension, the inelastic strain prior to the peak is relatively small, for some reinforced materials the prepeak nonlinearity can be important and must be taken into account. This can be done exactly as before, with the only assumption that the cracking strain (or damage) starts before the peak is reached. This is illustrated in Fig. 8.4.4a, which shows the full $\sigma(\varepsilon)$ curve, and Figs. 8.4.4b-d, which show the three possibilities of unloading behavior.

Therefore, to get a model incorporating the prepeak nonlinearity, it suffices to use the adequate expression for the function $\phi(\tilde{\varepsilon}^f)$. Some candidates for such a model were given in Section 8.3.2.

We recall here that, when used in finite element formulations in which the element width h is greater than the characteristic crack band width h_c, the softening part of the curve must be scaled as indicated in Section 8.3.5.

8.4.6 Crack Closure in Reverse Loading and Compression

For concrete as well as for other quasibrittle materials, the basic inelastic deformation mechanism in tension is cracking. If the material is subjected to tensile stress producing a crack, then is unloaded and the stress reversed into compression, the crack closes and the stiffness in compression is recovered to a large extent. As already pointed out, a nonlinear unloading behavior such as that sketched in Fig. 8.4.1d is observed (except at strong lateral confinement). However, the simpler models based on damage mechanics may be more convenient for computational purposes, and then some mechanism must be devised to ensure that the compliance reduction due to damage, as shown in Fig. 8.4.5a, would not appear on the compression side. Then, the split form (8.4.1) together with (8.4.2) is most efficient in handling the problem. It suffices to write that, for $\sigma < 0$, the crack opening must be zero; this may be compactly written as

$$\varepsilon = \frac{\sigma}{E} + \frac{\tilde{\varepsilon}^f}{\phi(\tilde{\varepsilon}^f)} \langle\sigma\rangle^+ \tag{8.4.12}$$

where $\langle\sigma\rangle^+$ is the positive part of σ, defined as σ for positive values and zero for negative values, or, in algebraic terms:

$$\langle\sigma\rangle^+ = \frac{\sigma + |\sigma|}{2} \tag{8.4.13}$$

The behavior becomes elastic, characterized by the initial elastic modulus, as soon as the stress becomes negative (Fig. 8.4.5b).

For the case of pure strength degradation, no special precaution needs to be taken since, by definition, the crack opening is fully irrecoverable. For the case of mixed unloading behavior (Fig. 8.4.4d), the positive part must include the entire expression (8.4.7), and so the total strain may be written as:

$$\varepsilon = \frac{\sigma}{E} + \left\langle \varepsilon^p + \frac{\tilde{\varepsilon}^f - \varepsilon^p}{\phi(\tilde{\varepsilon}^f)} \sigma \right\rangle^+ \tag{8.4.14}$$

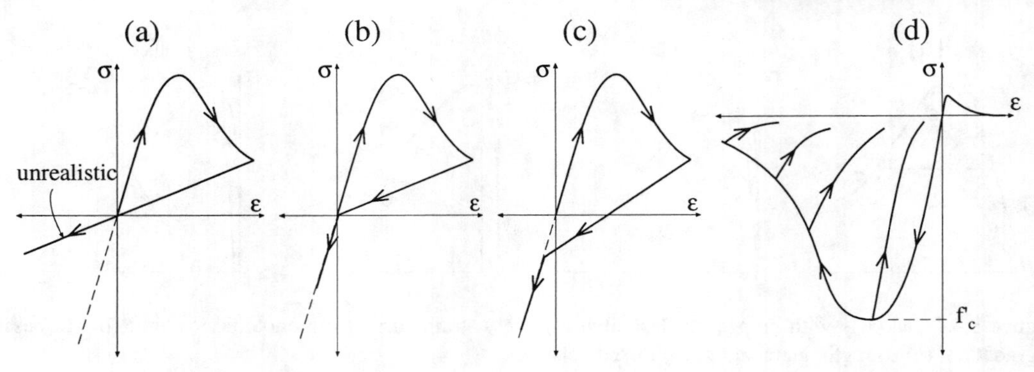

Figure 8.4.5 Reversing the stress sign: (a) invalid result with stiffness degradation also in compression; (b) model with crack closure; (c) stress-strain curve showing softening in compression as well as in tension.

In this way, the material recovers the undamaged behavior in compression as soon as the unloading branch reaches the initial elastic line, as shown in Fig. 8.4.5c.

Certainly, inelastic strain and cracking occur in compression, too. In practical analysis of concrete structures, especially in the analysis of beams and plates based on a uniaxial or biaxial stress-strain diagram, it is normally assumed that the stress-strain diagram of concrete in uniaxial compression also exhibits a peak followed by strain softening (Fig. 8.4.5d). As a consequence of this hypothesis, all localization phenomena described for tension occur for compression as well.

This means that one needs to also use fracture mechanics for compression behavior. Similar to tension, one needs to introduce either a softening band in compression (Bažant 1976) or one might postulate a compressive fictitious crack, as suggested by Hillerborg (1989). If, however, triaxial stress-strain relations are considered, such assumptions do not reflect realistically the actual mechanism of compression failure. Compression strain softening is not due to large strain in the direction of compression, unlike tensile strain softening, but is due to volume expansion of the material which causes large strains in the directions *transverse* to the direction of compression. So, compression softening is a strictly triaxial phenomenon, while tensile strain softening can, to a large extent, be treated as a uniaxial phenomenon. If volume expansion (transfer of strains) is prevented, e.g., by strong enough confining reinforcement or encasement of concrete in a strong enough pipe, then there is no compression softening and the stress-strain relation has no peak in compression.

A realistic triaxial stress-strain relation for compression strain softening must reflect these features. But many existing triaxial constitutive models do not, and the biaxial ones, in fact, cannot because they do not involve volume expansion as a variable. In any case, whether compression softening is modeled directly as a function of the compression strain or as a triaxial phenomenon associated with volume expansion, the fracture mechanics aspects associated with localization of compression softening need to be taken into account. Much research remains to be done in this direction.

8.4.7 Introducing Other Inelastic Effects

In the foregoing we have adopted the simplest approach and assumed that the inelastic behavior is completely due to cracking. This allows building models with a minimum of information. Indeed, for the pure damage or pure strength-degradation models all that is needed is the function $\phi(\varepsilon^f)$ deduced from a tensile test. For the mixed model, a further function relating ε^p to ε^f is required.

However, this simple approach neglects other sources of inelastic behavior that may take place in the bulk material between the cracks, such as plastic-type strains, creep (viscoelasticity or viscoplasticity), or shrinkage. A simple, yet effective way of modeling more complex behaviors is to relax the assumption of elastic material between cracks implicit in (8.4.1) and allow the bulk to suffer inelastic strains, too. This can be conveniently sketched as a series coupling of the cracking strain, that we represent by a fracturing element in Fig. 8.4.6, with a bulk element that can be purely elastic as in Fig. 8.4.6a, or include inelastic strains as in 8.4.6b, which are represented by a black-box where we can introduce the desired inelastic

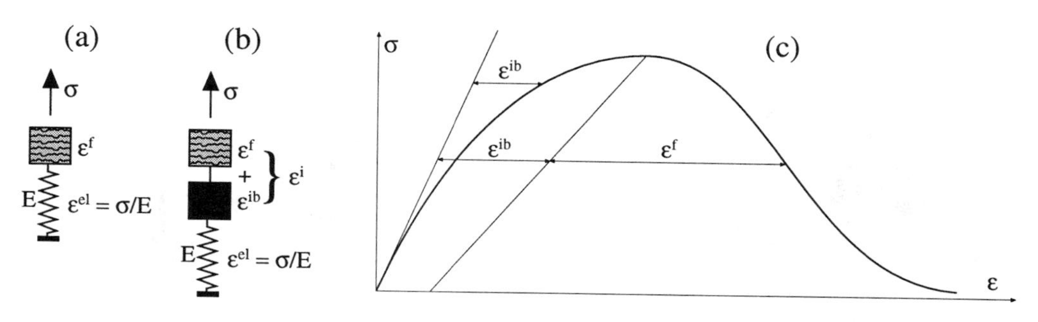

Figure 8.4.6 (a) Elastic-fracturing series coupling. (b) Elastic-bulk-inelastic-fracturing model. (c) Stress-strain curve split in which all the prepeak inelasticity is confined to the bulk and all the postpeak softening is confined into the fracturing element.

behavior. For example, Bažant and Chern (1985a) used a fracturing element coupled with a viscoelastic element and a shrinkage element. This means that the inelastic strain is now split into an inelastic strain associated with bulk behavior ε^{ib} and an inelastic strain associated to fracturing ε^f, that is,

$$\varepsilon = \frac{\sigma}{E} + \varepsilon^{ib} + \varepsilon^f \tag{8.4.15}$$

Focusing on time-independent models, the bulk inelasticity and the fracturing strain can each be modeled as done in the foregoing analysis in which a single inelasticity mechanism was assumed. Of course, more information is required to model the behavior. In particular, one function $\phi_b(\tilde{\varepsilon}^{ib})$ is required to describe the growth of the inelastic bulk strain, in addition to the function $\phi_f(\tilde{\varepsilon}^f)$ that describes the evolution of the fracturing strain.

Obviously, the experimental determination of the two functions is very difficult. One particular simplifying hypothesis may help to get an easy-to-handle model. It consists in assuming that all the prepeak inelasticity in tension is due to the bulk inelasticity. Then the stress-strain curve in the softening branch may be split as shown in Fig. 8.4.6c, so that the fracturing part would be the only part that has to be scaled according to the size of the element. So, Eq. (8.3.20) is reduced to the simple form

$$\varepsilon^{f(e)} = \frac{h_c}{h^{(e)}}\varepsilon^f \tag{8.4.16}$$

However, this is a split for mathematical convenience only, since most material scientists will agree that cracking (fracturing) starts before the peak. Nevertheless, since the amount of cracking prior to the peak is only a small fraction of the total, the *ad hoc* split can be justified on practical grounds.

Due to the enormous variety of combinations that arise as soon as one combines the two inelastic strain mechanisms, the following analysis will be restricted only to the fracturing mechanism. The other mechanism can be added as convenient based on classical inelasticity models *without* softening.

Exercises

8.18 Consider the uniaxial constitutive equation $\sigma = E\varepsilon e^{-b\varepsilon}$ and assume that it unloads to the origin. Determine the evolution law for the damage parameter ω. [Answer: $\omega = 1 - e^{-b\tilde{\varepsilon}}$.]

8.19 Consider the uniaxial elastic-softening model defined by an exponential softening curve $\sigma = f'_t e^{-(\varepsilon^f/\varepsilon_0)}$ for monotonic straining, and assume that the unloading is to the origin. (a) Show that the full stress-strain curve can be written as $\varepsilon = (1/E + C^f)\sigma$ in which C^f is a function of $\tilde{\varepsilon}^f = \max(\varepsilon^f)$, and determine this function. (b) Determine the fracturing work supply per unit volume of material γ^f, defined as the external work supply density when the fracturing strain increases up to $\tilde{\varepsilon}^f$ and then the stress is fully released. [Answer: $\gamma^f = f'_t\varepsilon_0[1 - (1 + \tilde{\varepsilon}^f/2\varepsilon_0)\exp(-\varepsilon/\varepsilon_0)$.] (c) Determine γ_F, the fracture work per unit volume for complete rupture.

8.20 For the model in the previous exercise, (a) determine the rate of the fracturing work density; (b) show that $\dot{\gamma}^f = \sigma^2\dot{C}^f/2$; (c) generalize the result to *any* model that unloads to the origin.

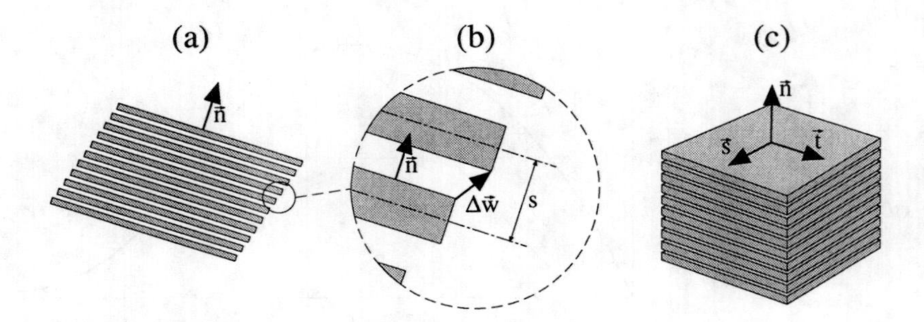

Figure 8.5.1 (a) Idealized crack band. (b) Detail of crack displacements. (c) Base vectors.

8.21 Write the rate of fracturing work (as defined in the previous two exercises) for a model with stiffness degradation as a function of the damage parameter ω.

8.5 Simple Triaxial Strain-Softening Models for Smeared Cracking

The smeared cracking models have the advantage that they can capture the influences of all the triaxial stress and strain components on the fracture process provided that the triaxial stress-strain relation is known. Unfortunately, formulation of this relation, which is needed for finite element programs, is a difficult problem of constitutive modeling. There is in the literature a huge number of models trying to adequately model fracture of concrete and similar materials in general triaxial situations. A whole book would be required to describe all of them and their modifications and possible extensions. In this section, we just touch the simplest models for smeared cracking, focusing on tensile stress states. Although they do not suffice to describe the complete behavior of concrete at complex triaxial stress states and histories, such as those with high compression stresses parallel to the crack planes, they are adequate for many practical instances of tensile cracking.

8.5.1 Cracking of Single Fixed Orientation: Basic Concepts

For many purposes one can assume that the cracks in concrete are parallel and have a fixed direction, which does not change during the loading process. This is a reasonable approximation for those loading processes in which the axes of principal stress and strain directions do not change drastically since first cracking. We can then consider, as in Section 8.4.1, that a cracked zone has formed consisting of an elastic material intersected by an array of densely distributed parallel cracks. Our task is to describe how the equations must be arranged to incorporate the response of this array of cracks to triaxial loading.

We consider that in an initially isotropic elastic material the maximum principal stress reached the tensile strength, and so a zone of cracked material formed. After that, the cracking process can be idealized as shown in Fig. 8.5.1a, which shows a crack band, formed by layers of elastic material separated by parallel cracks whose normal is defined by the unit vector $\vec{n}$. This normal vector is fixed and coincides with the maximum principal stress direction at the onset of cracking. Let s be the mean spacing of the cracks and $\Delta\vec{w}$ the mean displacement between the faces of the cracks (Fig. 8.5.1b). We can define the mean cracking strain *vector* $\vec{\varepsilon}^f$ as the crack opening per unit crack band width, i.e.,

$$\vec{\varepsilon}^f = \frac{\Delta\vec{w}}{s} \tag{8.5.1}$$

Now, take point O in the cracked zone as a reference (Fig. 8.5.1a). Let $\mathbf{x}_0$ be its position vector, and let A be any other point within the cracked zone, with position vector $\mathbf{x}$. The distance between O and A measured normal to the cracks is (Fig. 8.5.1a) $OA' = (\mathbf{x} - \mathbf{x}_0) \cdot \vec{n}$, where the dot indicates the scalar (internal) product. Hence, the number of cracks between the two points is $(\mathbf{x} - \mathbf{x}_0) \cdot \vec{n}/s$ and, correspondingly, the displacement generated by the cracks is

$$\mathbf{u}^f = \frac{(\mathbf{x}-\mathbf{x}_0)\cdot\vec{n}}{s}\Delta\vec{w} = [(\mathbf{x}-\mathbf{x}_0)\cdot\vec{n}]\Delta\vec{w} \tag{8.5.2}$$

Taking the gradient of the displacement function, and then its symmetric part, we get the macroscopic small strain tensor that corresponds to the cracking:

$$\boldsymbol{\varepsilon}^f = \frac{1}{2}\left(\vec{n}\otimes\vec{\varepsilon}^f + \vec{\varepsilon}^f\otimes\vec{n}\right) = \left(\vec{\varepsilon}^f\otimes\vec{n}\right)^S \tag{8.5.3}$$

where from now on we use $\mathbf{T}^S$ to indicate the symmetric part of an arbitrary second-order tensor $\mathbf{T}$. The foregoing result indicates that the cracking strain is not a general symmetric tensor, since it has three of the six possible components identically zero. Indeed, taking an orthonormal base $\{\vec{n},\vec{s},\vec{t}\}$ so that the unit vectors $\vec{s}$ and $\vec{t}$ are parallel to the cracks (Fig. 8.5.1c), we easily find that

$$\varepsilon^f_{tt} = \varepsilon^f_{ss} = \varepsilon^f_{ts} = 0 \tag{8.5.4}$$

The fracturing strain tensor thus has only three degrees of freedom, corresponding to the three components of the vector $\vec{\varepsilon}^f$.

The foregoing equations define the kinematics of the problem. Before getting any further, we must emphasize that a consistent set of rules must be used to properly define the foregoing vectors. It is implied in our sketch in Fig. 8.5.1b that one of the two faces of the crack is taken as reference; then $\vec{n}$ is taken as the unit normal to that face external to the uncracked material, and $\delta\vec{w}$ is the displacement of the other face of the crack relative to the first. The reader can easily verify that, upon changing the reference to the other face of the crack, $\vec{n}$ and $\vec{\varepsilon}^f$ change sign but $\boldsymbol{\varepsilon}^f$ remains unchanged.

The total strain tensor is obtained by adding up the elastic strain to the fracturing strain:

$$\boldsymbol{\varepsilon} = \frac{1+\nu}{E}\boldsymbol{\sigma} - \frac{\nu}{E}\text{tr}\,\boldsymbol{\sigma}\,\mathbf{1} + \left(\vec{\varepsilon}^f\otimes\vec{n}\right)^S \tag{8.5.5}$$

where E and ν are, respectively, the elastic modulus and Poisson's ratio of the bulk (uncracked) material. This is one of the basic equations of the fixed crack models. Note that it provides six equations, while we need nine equations. Given the strain tensor, we need to compute the stress tensor and the fracturing strain *vector*, nine components in all. The remaining three equations must relate the crack opening to the stress.

Since the basic internal variable is the vector $\vec{\varepsilon}^f$, rather than the fracturing strain tensor $\boldsymbol{\varepsilon}^f$, it is natural to look for a relationship between $\vec{\varepsilon}^f$ and the traction vector $\vec{\sigma}$ on the crack faces, rather than trying to directly connect $\boldsymbol{\varepsilon}^f$ to the stress tensor $\boldsymbol{\sigma}$. Therefore, we assume that the cracking behavior of the material is defined by a vectorial relationship of the form:

$$\vec{\sigma} = \boldsymbol{\sigma}\vec{n} = \vec{\Phi}(\vec{\varepsilon}^f,\vec{n},\cdots) \tag{8.5.6}$$

where $\vec{\Phi}$ must be understood as a functional which, given crack orientation, evolution of $\vec{\varepsilon}^f$ and, possibly, some other variables acting as parameters, yields the traction vector on the crack faces. Various definitions of this functional have been proposed and more could be invented. We discuss next some of the possibilities.

8.5.2 Secant Approach to Cracking of Fixed Orientation

For application to problems with monotonic crack opening close to mode I, Bažant and Oh (1983a) proposed a crack band model in which the stress-strain relations have a secant form, with varying secant compliances. Here we give an enhanced version that explicitly considers the crack sliding (crack shearing) and that naturally leads to a damage formulation of the cracking problem.

Consider first proportional paths in which the microcrack opening and shearing increase monotonically. For these very particular paths we can assume the traction vector to be a function of the fracturing strain vector, i.e.,

$$\vec{\sigma} = \vec{F}(\vec{\varepsilon}^f,\vec{n}) \tag{8.5.7}$$

where $\vec{F}(\cdot)$ is a vector-valued function of two vector arguments. Note that $\vec{\sigma}$ is made to depend on the orientation of the crack; this is essential to disclose the structure of $\vec{F}(\cdot)$. The material containing the crack band is assumed isotropic. Therefore, we require that, if we rotate simultaneously the crack and the crack strain vector, we must obtain a traction vector rotated by the same angle. This means that the function $\vec{F}(\vec{\varepsilon}^f, \vec{n})$ must be isotropic, i.e., that

$$\vec{F}(\mathbf{Q}\vec{\varepsilon}^f, \mathbf{Q}\vec{n}) = \mathbf{Q}\vec{F}(\vec{\varepsilon}^f, \vec{n}) \tag{8.5.8}$$

for any orthogonal second-order tensor $\mathbf{Q}$. A classical representation theorem (Spencer 1971) then requires the traction vector to have the form

$$\vec{\sigma} = S_N(\varepsilon_N^f, \varepsilon_T^f)\varepsilon_N^f\vec{n} + S_T(\varepsilon_N^f, \varepsilon_T^f)\vec{\varepsilon}_T^f \tag{8.5.9}$$

where $S_N(\varepsilon_N^f, \varepsilon_T^f)$ and $S_T(\varepsilon_N^f, \varepsilon_T^f)$ are scalar functions that have the meaning of normal and tangent secant stiffnesses; ε_N^f is the normal component of $\vec{\varepsilon}^f$, $\vec{\varepsilon}_T^f$ is the vectorial component of $\vec{\varepsilon}^f$ in the plane of the cracks, and ε_T^f is the magnitude of that component. Algebraically,

$$\varepsilon_N^f = \vec{\varepsilon}^f \cdot \vec{n}\,, \qquad \vec{\varepsilon}_T^f = \vec{\varepsilon}^f - \varepsilon_N^f\vec{n}\,, \qquad \varepsilon_T^f = |\vec{\varepsilon}_T^f| = \sqrt{\vec{\varepsilon}_T^f \cdot \vec{\varepsilon}_T^f} \tag{8.5.10}$$

If we similarly define the normal and shear components of the traction vector, i.e.,

$$\sigma_N = \vec{\sigma} \cdot \vec{n}\,, \qquad \vec{\sigma}_T = \vec{\sigma} - \sigma_N\vec{n}\,, \qquad \sigma_T = |\vec{\sigma}_T| = \sqrt{\vec{\sigma}_T \cdot \vec{\sigma}_T} \tag{8.5.11}$$

the foregoing equations reduce to

$$\sigma_N = S_N(\varepsilon_N^f, \varepsilon_T^f)\varepsilon_N^f \qquad \text{and} \qquad \vec{\sigma}_T = S_T(\varepsilon_N^f, \varepsilon_T^f)\vec{\varepsilon}_T^f \tag{8.5.12}$$

which has a beautiful uncoupled form, with $\vec{\sigma}_T$ parallel to $\vec{\varepsilon}_T^f$. Certainly we could have assumed this from the onset. But we now have proved that this is the *most general* possibility consistent with the initial assumption (8.5.7) and the condition of isotropy. This means that we need to specify two functions of two variables to determine the material behavior for proportional monotonic loading.

No doubt, many simplifications will be required to characterize limited experimental evidence. However, before attempting such simplifications, let us find the general structure of the stress-strain relations. First we solve for the components $\vec{\varepsilon}^f$ from (8.5.12) and substitute them into (8.5.3) to get the expression of the fracturing strain tensor:

$$\boldsymbol{\varepsilon}^f = C_N\sigma_N\,\vec{n}\otimes\vec{n} + C_T(\vec{\sigma}_T\otimes\vec{n})^S \tag{8.5.13}$$

C_N and C_T are the normal and shear compliances, defined as

$$C_N = \frac{1}{S_N(\varepsilon_N^f, \varepsilon_T^f)}\,, \qquad C_T = \frac{1}{S_T(\varepsilon_N^f, \varepsilon_T^f)} \tag{8.5.14}$$

where a dependence on ε_N^f and ε_T^f is implied (although hidden from now on). This expression is now substituted into (8.5.5) to get the total strain tensor as

$$\boldsymbol{\varepsilon} = \frac{1+\nu}{E}\boldsymbol{\sigma} - \frac{\nu}{E}\text{tr}\,\boldsymbol{\sigma}\,\mathbf{1} + C_N\sigma_N\,\vec{n}\otimes\vec{n} + C_T(\vec{\sigma}_T\otimes\vec{n})^S \tag{8.5.15}$$

For computational purposes this relation is best expressed in a component form relative to the base $\{\vec{n}, \vec{s}, \vec{t}\}$ in Fig. 8.5.1c, arranging the stresses in the six-dimensional column matrix $(\sigma_{nn}, \sigma_{ss}, \sigma_{tt}, \sigma_{ns}, \sigma_{st}, \sigma_{tn})^T$ and the strains in the corresponding column matrix $(\varepsilon_{nn}, \varepsilon_{ss}, \varepsilon_{tt}, \gamma_{ns}, \gamma_{st}, \gamma_{tn})^T$ where $\gamma_{\mu\nu} = 2\varepsilon_{\mu\nu}$. It turns out that the equations for the normal and shear components are mutually uncoupled and can be written as

$$\left\{\begin{array}{c}\varepsilon_{nn}\\ \varepsilon_{ss}\\ \varepsilon_{tt}\end{array}\right\} = \frac{1}{E}\begin{bmatrix}1+EC_N & -\nu & -\nu\\ -\nu & 1 & -\nu\\ -\nu & -\nu & 1\end{bmatrix}\left\{\begin{array}{c}\sigma_{nn}\\ \sigma_{ss}\\ \sigma_{tt}\end{array}\right\} \tag{8.5.16}$$

and

$$\begin{Bmatrix} \gamma_{ns} \\ \gamma_{st} \\ \gamma_{tn} \end{Bmatrix} = \frac{1}{G} \begin{bmatrix} 1 & 0 & 0 \\ 0 & 1 + GC_T & 0 \\ 0 & 0 & 1 + GC_T \end{bmatrix} \begin{Bmatrix} \sigma_{ns} \\ \sigma_{st} \\ \sigma_{tn} \end{Bmatrix} \tag{8.5.17}$$

where $G = E/2(1 + \nu)$ is the shear modulus.

The foregoing secant equations are particularly simple: they depend only on the two functions C_N and C_T that appear in only three diagonal elements. Now we must specify how these functions evolve. In the early times of the smeared crack applications (see Rots 1988, for the basic references), before fracture mechanics concepts became widely accepted, the structural finite element codes used to set both the normal and tangential stiffnesses to zero just after cracking starts. This is equivalent to setting $C_N = C_T = \infty$ in the foregoing equations. This turned out to lead to numerical problems because of the sudden energy release implied by such approximation, and a certain amount of shear stiffness was retained, such that $\gamma_{sn} = \sigma_{sn}/(\beta_s G)$, where β_s was called the shear retention factor (Suidan and Schnobrich 1973; Yuzugullu and Schnobrich 1973), whose value was of a few tenths, typically 0.2. This is equivalent to setting the shear compliance C_T equal to a constant of value

$$C_T = \frac{1 - \beta_s}{\beta_s G} \tag{8.5.18}$$

The introduction of the shear retention factor, however, is not satisfactory for four reasons: (1) it has no physical interpretation, (2) it is difficult to measure experimentally (if possible at all), (3) it leads to a behavior in which the material always has a stiffness in shear even if the crack is widely open, which is completely unrealistic, and (4) it seems a variable made to play with in numerical simulations. At any rate, for cases in which the cracking occurs close to mode I, the importance of the shear retention factor is not great. However, the results by Rots (1988) indicate that very low values of β_s —even zero— give better results in most cases.

For the normal compliance, Bažant and Oh (1983a) introduced a progressively degrading compliance as dictated by the uniaxial tension data. This is the equivalent to postulating that shearing the crack does not contribute to degradation in the normal direction. This is certainly a simplification, but can be realistic if the magnitude of shear is limited. This hypothesis allows a complete determination of the normal compliance C_N from the uniaxial stress-strain curve. Then, for monotonic straining normal to the crack, we have

$$C_N = \frac{\varepsilon_N^f}{\phi(\varepsilon_N^f)} \tag{8.5.19}$$

where $\phi(\varepsilon_N^f)$ is the function $\sigma = \phi(\varepsilon^f)$ which is deduced from uniaxial tests and has been repeatedly analyzed in the two previous sections (only the name of the argument changes, since for uniaxial loading $\varepsilon^f \equiv \varepsilon_N^f$).

The foregoing hypothesis implies that the cracking process is controlled by ε_N^f. For very small ε_N^f very little damage exists and the shear compliance should be small; for very large ε_N^f the damage is large and the shear compliance should be correspondingly large. Thus it seems logical, as done by Rots (1988) in a slightly different formulation to be defined later, to take C_T as an increasing function of C_N. The simplest of all is to assume C_T as proportional to C_N, i.e.,

$$C_T = c_T C_N(\varepsilon_N^f) \tag{8.5.20}$$

where c_T is a constant that should be determined by experiment.

This is a very simplified model, devised for monotonic crack opening, that can be easily brought to a more general formulation involving unloading to the origin, i.e., to a damage model as described next.

8.5.3 Scalar Damage Model for Cracking of Fixed Orientation

To convert the foregoing model into a damage model with unloading to the origin, it suffices to postulate that C_N and C_T are functions, not of the instantaneous value of ε_N^f, but of its maximum past value, defined as $\tilde{\varepsilon}_N^f$. Thus, C_N and C_T get "frozen" as soon as ε_N^f starts to decrease.

Although not strictly necessary, we may define a normal damage function similar to (8.4.4):

$$\omega_N = \frac{EC_N}{1 + EC_N} \quad \text{and} \quad \omega_T = \frac{GC_T}{1 + GC_T} \tag{8.5.21}$$

from which (8.5.16) and (8.5.17) reduce to

$$\begin{Bmatrix} \varepsilon_{nn} \\ \varepsilon_{ss} \\ \varepsilon_{tt} \end{Bmatrix} = \frac{1}{E} \begin{bmatrix} 1/(1-\omega_N) & -\nu & -\nu \\ -\nu & 1 & -\nu \\ -\nu & -\nu & 1 \end{bmatrix} \begin{Bmatrix} \sigma_{nn} \\ \sigma_{ss} \\ \sigma_{tt} \end{Bmatrix} \tag{8.5.22}$$

and

$$\begin{Bmatrix} \gamma_{ns} \\ \gamma_{st} \\ \gamma_{tn} \end{Bmatrix} = \frac{1}{G} \begin{bmatrix} 1 & 0 & 0 \\ 0 & 1/(1-\omega_T) & 0 \\ 0 & 0 & 1/(1-\omega_T) \end{bmatrix} \begin{Bmatrix} \sigma_{ns} \\ \sigma_{st} \\ \sigma_{tn} \end{Bmatrix} \tag{8.5.23}$$

If we adopt the simple form (8.5.20) and select $C_T = 2(1+\nu)C_N$, the model further simplifies, because then

$$\omega_N = \omega_T = \omega = \frac{E\tilde{\varepsilon}_N^f}{E\tilde{\varepsilon}_N^f + \phi(\tilde{\varepsilon}_N^f)} \tag{8.5.24}$$

where we noted that the expression is identical to that for the uniaxial case (8.4.4). This is the simplest possible model with unloading to the origin that is based only on the information from the uniaxial test.

8.5.4 Incremental Approach to Cracking of Fixed Orientation

The foregoing formulation can obviously be rewritten in incremental form, which is obtained by differentiating the equations. However, it is possible to directly formulate an incremental formulation that, in general, is not equivalent to a secant formulation because it does not satisfy the integrability conditions. This happens with the incremental approach proposed by Rots (1988) which is briefly outlined now.

The incremental form is obtained by establishing relationships between the rates of the variables. Eqs. (8.5.5) and (8.5.6) are replaced by

$$\dot{\boldsymbol{\varepsilon}} = \frac{1+\nu}{E}\dot{\boldsymbol{\sigma}} - \frac{\nu}{E}\mathrm{tr}\,\dot{\boldsymbol{\sigma}}\,\mathbf{1} + \left(\dot{\vec{\varepsilon}}^f \otimes \vec{n}\right)^S \tag{8.5.25}$$

$$\dot{\vec{\sigma}} = \dot{\boldsymbol{\sigma}}\vec{n} = \mathbf{S}^t \dot{\vec{\varepsilon}}^f \tag{8.5.26}$$

where the superimposed dot indicates the time rate, and $\mathbf{S}^t$ is a second-order tensor defining the tangent stiffnesses for the cracks. The structure of this tensor depends on the details of the model. The tensor need not be symmetric. It may have up to nine independent components, which are reduced to five if one assumes that the tensor depends only on the crack orientation $\vec{n}$ and the instantaneous cracking strain vector $\vec{\varepsilon}^f$. At any rate, much more information is required than is currently available from the experimental knowledge, and strong simplifications are introduced. Rots (1988) assumed that the normal and tangent components of stress and strain were mutually proportional, with no mixed stiffness terms. With this assumption, the equations are similar to those for the secant formulation. In particular, (8.5.12) is replaced by

$$\dot{\sigma}_N = S_N^t(\varepsilon_N^f)\dot{\varepsilon}_N^f \quad \text{and} \quad \dot{\vec{\sigma}}_T = S_T^t(\varepsilon_N^f)\dot{\vec{\varepsilon}}_T^f \tag{8.5.27}$$

where S_N^t and S_T^t are incremental (tangent) stiffnesses. Rots (1988) further assumed that both S_N^t and S_T^t depended only on the normal cracking strain ε_N^f, but introduced independent functions to describe them. S_N^t is derived from the uniaxial tensile test and is related to the secant stiffness and the softening function $\phi(\varepsilon_N^f)$ by

$$S_N^t = \frac{\partial(S_N\,\varepsilon_N^f)}{\partial\varepsilon_N^f} = \frac{\partial\phi(\varepsilon_N^f)}{\partial\varepsilon_N^f} \tag{8.5.28}$$

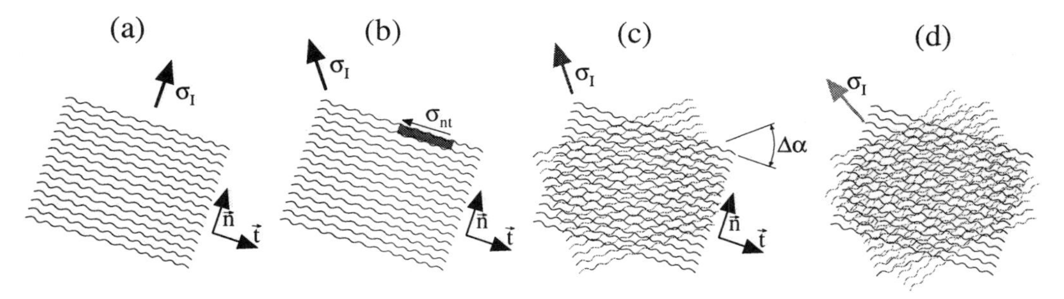

Figure 8.5.2 Multiple cracking: (a) primary cracks; (b) shear stress built up due to principal stress rotation; (c) secondary cracking formed; (d) tertiary cracking.

For the incremental shear stiffness S_T^t, Rots introduced a decreasing function that was infinite for zero crack opening and decreased progressively to vanish for the normal cracking strain at which the normal stress drops to zero. He showed that this is equivalent to using an incremental shear retention factor varying from 1 just after crack creation, down to 0 for a fully broken material.

At first glance, it may appear that this formulation is equivalent to the secant formulation. It is not; indeed, if we differentiate (8.5.12) to get the rate equations for the secant model (with the assumption that the stiffnesses depend only on the normal cracking strain), we find that the equation for the normal component is equivalent, but the equation for the shear component is

$$\dot{\vec{\sigma}}_T = S_T(\varepsilon_N^f)\dot{\vec{\varepsilon}}_T^f + \frac{\partial S_T(\varepsilon_N^f)}{\partial \varepsilon_N^f}\vec{\varepsilon}_T^f\,\dot{\varepsilon}_N^f \tag{8.5.29}$$

This is certainly not equivalent to (8.5.27) except for proportional straining (see exercises), and gives a lower tangential stiffness than (8.5.27) because S_T is decreasing. No comparative analysis of the two approaches has been performed to date. The difference is analogous to that between the incremental and the total-strain theories of plasticity. The latter is known to give better (softer) stiffness predictions for the first deviation from a proportional loading (predicting the so-called vertex effect), and the same probably applies here.

8.5.5 Multi-Directional Fixed Cracking

A difficult question with the foregoing formulation is the orientation of cracking. The practice which has been and is still typical of most large finite element codes is to set the crack direction to be normal to the maximum principal stress at the moment the tensile strength (or the tensile yield surface) is reached (Fig. 8.5.2a). During the subsequent loading process, the direction of the maximum principal stress can rotate. At the moment the cracks begin to form, there is, by definition, no shear stress on the cracking planes. However, due to keeping the cracking orientation fixed and assuming shear interlocking of the opposite crack faces, shear stresses can arise later if the principal stress direction rotates (Fig. 8.5.2b). It was for this reason that the diagonal compliances or stiffnesses for shear had to be included in Eqs. (8.5.17) and (8.5.23).

When the principal stress direction rotates, it is possible that the tensile strength f_t' is reached again in another direction that is inclined with regard to the normal of the originally formed cracks. In that case, it is assumed that a second system of smeared cracks forms in the direction normal to the current principal stress (Fig. 8.5.2c). This system is inclined at some general angle $\Delta\alpha$ with regard to the orientation of the primary cracks. The cracking strain due to the formation of secondary cracks is then superposed on the original cracking strain, which means that another fracturing strain tensor $(\vec{\varepsilon}_2^f \otimes \vec{n})^S$ is added to the right-hand side of Eq. (8.5.5). The orientation of the secondary cracks is also kept fixed even when the principal stress directions subsequently rotate during the loading process. Thus, it may happen that the tensile strength is again reached in a third direction (Fig. 8.5.2d), in which case tertiary smeared cracks begin to form and the corresponding cracking strain needs to be again superimposed in Eq. (8.5.5).

The laws governing the secondary and tertiary cracking strains may be assumed to be the same as

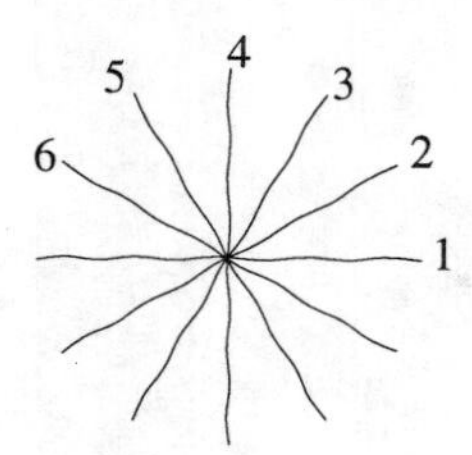

Figure 8.5.3 Multi-crack system with fixed angular separation.

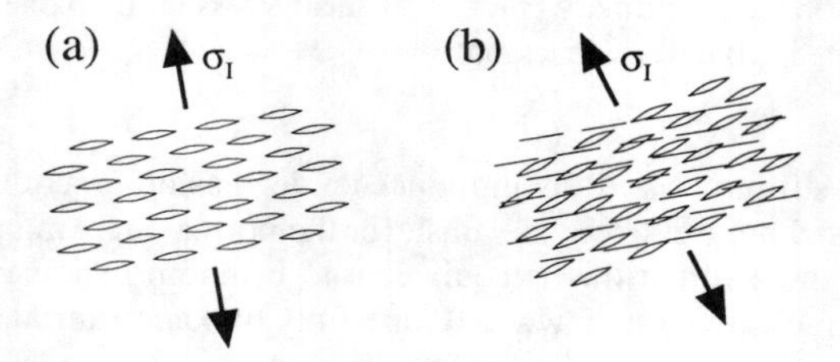

Figure 8.5.4 Rotating crack process: (a) Primary cracks form; (b) secondary cracks form and become dominant.

for the primary cracking, although some formulations allow for interaction between the various crack systems. The formulation of multiple cracking with fixed directions, which has been worked out in the greatest generality perhaps by de Borst (1986), can obviously get quite complicated when some crack systems open and cause others to close. Special computational strategies must then be devised to follow the possible bifurcation paths.

In the method just described, the secondary (and tertiary) cracks can have arbitrary orientations with regard to the primary ones. In this manner, cracks of many directions can form. In that case, it may be more convenient to assume that the cracks may form only in certain specified spatial orientations which are uniformly distributed among all spatial directions (Fig. 8.5.3). Such an assumption is also more realistic because it prevents the angle between intersecting cracks from being too small (say 10°, which is unlikely to occur). This approach, in which, again, the cracking strains from all the assumed discrete crack orientations are superimposed, makes it possible to describe the fact that a principal stress of a certain direction may cause microcracking with various intensities at various orientations (this is captured more systematically by the context of the microplane approach to be discussed in Chapter 14; see Carol and Prat 1990, and Carol and Bažant 1997).

8.5.6 Rotating Crack Model

When the smeared cracks of the primary direction start forming, which is the start of strain softening, there is actually only a system of discontinuous microcracks. If the maximum principal stress direction rotates, these microcracks partially close and microcracks of a new orientation begin to form (Fig. 8.5.4a). Eventually the secondary microcracks may become the major ones and the primary ones may get nearly closed. Although a precise description would be rather difficult (and would perhaps be best done in terms of the microplane model, Chapter 14), the fact that the previously formed microcracks may to a large extent close and microcracks of a new orientation may become dominant can be better described by assuming that the direction of smeared cracking rotates (Fig. 8.5.4b) and remains always normal to the maximum principal stress. In reality, of course, a crack, once formed, cannot actually rotate.

The notion of a rotating crack (also called swinging crack), originally proposed by Cope et al. (1980) and reformulated by Gupta and Akhbar (1984), and Crisfield and Wills (1987), is just a computational convenience. The reality is more complicated than the preceding discussion suggests. Even for a constant principal stress direction, the microcracks during the process of crack formation do not have the same orientation; due to heterogeneity of the microstructure, microcracks arise in all directions, and one can only say that the microcracks that are normal to the maximum principal stress direction are the statistically dominant ones.

The rotating crack model can be formulated in a way very similar to the fixed crack model, although the resulting equations are simpler. Consider first a single crack system. By the very definition of the model, the normal to the cracks $\vec{n}$ is now coincident with $\mathbf{p}_\text{I}$, the unit vector in the direction of the maximum principal strain, which coincides with the principal stress direction. Then, the crack displacements are in pure opening and we can write, referring to Fig. 8.5.1b

$$\Delta\vec{w} = \Delta w\,\mathbf{p}_\text{I} \qquad \Rightarrow \qquad \vec{\varepsilon}^f = \varepsilon^f\,\mathbf{p}_\text{I} \tag{8.5.30}$$

Proceeding again as in Eqs. (8.5.3)–(8.5.5) with $\vec{n}$ replaced by $\mathbf{p}_\text{I}$, we get the total strain tensor as

$$\boldsymbol{\varepsilon} = \frac{1+\nu}{E}\boldsymbol{\sigma} - \frac{\nu}{E}\text{tr}\,\boldsymbol{\sigma}\,\mathbf{1} + \varepsilon^f\,\mathbf{p}_\text{I}\otimes\mathbf{p}_\text{I} \tag{8.5.31}$$

in which the fracturing strain tensor now depends on only one scalar variable, ε^f.

As for the equation governing microcracking, Eq. (8.5.6) now becomes a scalar equation since, by definition $\boldsymbol{\sigma}\mathbf{p}_\text{I} = \sigma_\text{I}\mathbf{p}_\text{I}$, where σ_I is the principal stress with principal direction $\mathbf{p}_\text{I}$. Thus, we need only a relationship between ε^f and σ_I. This coincides with the uniaxial stress-fracturing strain relation (for the monotonic loading case), i.e.,

$$\sigma_\text{I} = \phi(\varepsilon^f) = S_N(\varepsilon^f)\varepsilon^f \tag{8.5.32}$$

in which we keep the nomenclature in the previous sections to keep the meaning clear. From the foregoing equation, we can solve for ε^f (at constant secant stiffness) and get the secant formulation:

$$\boldsymbol{\varepsilon} = \frac{1+\nu}{E}\boldsymbol{\sigma} - \frac{\nu}{E}\text{tr}\,\boldsymbol{\sigma}\,\mathbf{1} + C_N\sigma_\text{I}\,\mathbf{p}_\text{I}\otimes\mathbf{p}_\text{I} \tag{8.5.33}$$

The component expression for this equation is particularly simple if the axes are taken along the principal stress and strain directions:

$$\begin{Bmatrix} \varepsilon_\text{I} \\ \varepsilon_\text{II} \\ \varepsilon_\text{III} \end{Bmatrix} = \frac{1}{E}\begin{bmatrix} 1+EC_N & -\nu & -\nu \\ -\nu & 1 & -\nu \\ -\nu & -\nu & 1 \end{bmatrix}\begin{Bmatrix} \sigma_\text{I} \\ \sigma_\text{II} \\ \sigma_\text{III} \end{Bmatrix} \tag{8.5.34}$$

in which the similarity with (8.5.16) is blatant.

In the foregoing, Eq. (8.5.32) is valid for virgin loading. Unloading and reloading require further rules. The simplest is a damage model in which C_N is taken to be a function of $\tilde{\varepsilon}^f$, the maximum fracturing strain ever reached. Then, if $\varepsilon^f < \tilde{\varepsilon}^f$, the unloading-reloading proceeds at constant C_N and, for virgin loading, $\varepsilon^f = \tilde{\varepsilon}^f$ and C_N increases.

The model can further be extended by considering three mutually orthogonal jointly rotating systems of cracks, normal to the three principal stress and strain directions. Each system is allowed to follow an independent cracking process, same as described before, characterized by fracturing strains ε^f_ν, with $\nu = \text{I}$, II, or III. The resulting equation, which incorporates the three fracturing strains, can be written as

$$\boldsymbol{\varepsilon} = \frac{1+\nu}{E}\boldsymbol{\sigma} - \frac{\nu}{E}\text{tr}\,\boldsymbol{\sigma}\,\mathbf{1} + \sum_{\nu=\text{I}}^{\text{III}} C_N(\tilde{\varepsilon}^f_\nu)\,\sigma_\nu\,\mathbf{p}_\text{I}\otimes\mathbf{p}_\nu \tag{8.5.35}$$

whose component form is

$$\begin{Bmatrix} \varepsilon_\text{I} \\ \varepsilon_\text{II} \\ \varepsilon_\text{III} \end{Bmatrix} = \frac{1}{E}\begin{bmatrix} 1+EC_N(\tilde{\varepsilon}^f_\text{I}) & -\nu & -\nu \\ -\nu & 1+EC_N(\tilde{\varepsilon}^f_\text{II}) & -\nu \\ -\nu & -\nu & 1+EC_N(\tilde{\varepsilon}^f_\text{III}) \end{bmatrix}\begin{Bmatrix} \sigma_\text{I} \\ \sigma_\text{II} \\ \sigma_\text{III} \end{Bmatrix} \tag{8.5.36}$$

Note that although there are three different damage components, the behavior of the material is described with only one material function which can, in principle, be determined from the uniaxial test.

8.5.7 Generalized Constitutive Equations with Softening

It is possible to put the constitutive equations with softening into a very general framework that embraces most known models. The most general thermomechanical approach is outside the scope of this book, and the reader may refer to the book by Lemaitre and Chaboche (1985).

A general constitutive equation may be based on three fundamental concepts:

1. A set of independent internal variables, p_k, which together with the infinitesimal strain tensor ε (or the stress tensor $\boldsymbol{\sigma}$) are assumed to characterize uniquely the instantaneous state of the body at a given point. The internal variables may represent a physical magnitude or be abstract in nature. They can be related to kinematic events or to structural features. For example, the vector $\vec{\varepsilon}^f$ in the smeared crack models is intended to represent the internal kinematics of cracks and $\vec{n}$, the crack orientation, is a structural internal variable. It must be noted that, when a set of internal variables is chosen, any other set, uniquely related to the first, is strictly equivalent and, consequently, can be used instead of the first. This makes the physical interpretation of a given set of internal variables somewhat ambiguous.

2. A system of equations relating the stress to the strain and to the internal variables:

$$\varepsilon = \mathbf{E}(\boldsymbol{\sigma}, p_k) \tag{8.5.37}$$

where $\mathbf{E}(\cdot)$ is a symmetric tensor-valued function. In modern thermodynamic formulations $\mathbf{E}(\cdot)$ is derived from a free energy function, which is a scalar function to be specified instead of $\mathbf{E}(\cdot)$. Usually, $\mathbf{E}(\cdot)$ is assumed to be linear in the infinitesimal strain tensor, that is,

$$\varepsilon = \boldsymbol{C}(p_k)\boldsymbol{\sigma} + \varepsilon^p(p_k) \tag{8.5.38}$$

where $\boldsymbol{C}(p_k)$ is the secant fourth-order compliance tensor, depending only on the internal variables; $\varepsilon^p(p_k)$ is the irrecoverable or plastic strain tensor, which again depends only on the internal variables. When $\varepsilon^p(p_k) \equiv \mathbf{0}$ and $\boldsymbol{C}(p_k) \equiv \boldsymbol{C}^{el}$ = constant, the elastic behavior is obtained. When $\varepsilon^p(p_k)$ varies, and $\boldsymbol{C}(p_k) \equiv \boldsymbol{C}^{el}$ = constant, a model displaying strength degradation is obtained. When $\varepsilon^p(p_k) \equiv \mathbf{0}$ and $\boldsymbol{C}(p_k)$ is variable, one obtains a model displaying stiffness degradation, which always unloads to the origin ($\boldsymbol{\sigma} = \mathbf{0}$ for $\varepsilon = \mathbf{0}$, and vice versa). When both $\boldsymbol{C}(p_k)$ and $\varepsilon^p(p_k)$ are variable, a general damage model with mixed behavior is obtained.

3. A set of *flow rules,* which specify the way in which the internal variables increase during loading. This is a delicate yet essential point, since assigning different flow rules to models having the same set of internal variables and the same structure for the stress-strain relation will lead to different behaviors. Moreover, the flow rules must be consistent with the irreversibility condition posed by the second law of thermodynamics. A detailed discussion of this important aspect is outside the scope of this book, so only general aspects will be mentioned. (By analogy with plasticity, the term "flow rule" is used even though "cracking rule" would be more logical in models in which cracking dominates.)

The flow rules may be specified at many different levels of generality. One relatively simple way is to use one or more loading functions obtained by direct generalization of the theory of classical plasticity. For the simplest case of a single yield surface, a loading function $F(p_k, \boldsymbol{\sigma}, \mu)$ is specified, in which μ is one further internal variable that governs hardening and softening (it can be singled out from the beginning). The loading function defines the region in which the behavior is elastic (i.e., in which $d\mu = 0$ and $dp_k = 0$ for any k) which can be written as

$$F(p_k, \boldsymbol{\sigma}, \mu) \leq 0 \tag{8.5.39}$$

The associated flow rules are:

$$\dot{p}_k = H_k(p_k, \boldsymbol{\sigma}, \mu)\dot{\mu} \qquad \text{with} \qquad \dot{\mu} \geq 0 \tag{8.5.40}$$

where μ is the hardening-softening variable which takes the place of the plastic multiplier and $H_k(p_k, \boldsymbol{\sigma}, \mu)$ are the hardening-softening functions. The flow rule for μ itself is deduced from the consistency condition requiring that $F(p_k, \boldsymbol{\sigma}, \mu)$ remain equal to 0 if $\dot{\mu} > 0$.

Although this is a rather general formulation, it is not the only one possible. Formulations of the endochronic type and multi-yield surface type are also possible.

In the literature, one can find many models meant to describe more or less general softening behaviors, from the simplest, with a single scalar internal variable (plus the hardening-softening variable), to very sophisticated models in which the internal variables are tensors of second or fourth order (even eighth-order tensors have been proposed as internal variables). For example, a very interesting model was proposed by Ortiz (1985) in which the full fourth-order compliance tensor is one internal variable (equivalent to a set of 21 internal variables, the independent components of the compliance tensor), and the second-order tensor of plastic strain a further internal variable (equivalent to a set of six scalar variables). However, this model is too complex to be described here in detail. Only two very simple models will be briefly discussed: Mazars' scalar (isotropic) damage model and Rankine's associated plastic model with strain softening.

8.5.8 Mazars' Scalar Damage Model

Mazars (1981, 1984, 1986) developed a series of damage models, which aim at damage in tension and compression. When specialized for tension, the only primary internal variable is the scalar damage ω, varying form $\omega = 0$ (for no damage), to $\omega = 1$ (for complete failure). The hardening-softening variable is $\mu \equiv \tilde{\varepsilon}$ where $\tilde{\varepsilon}$ has the meaning of an equivalent uniaxial strain. The equations for this model are

$$\boldsymbol{\varepsilon} = \frac{1}{1-\omega}\left(\frac{1+\nu}{E}\boldsymbol{\sigma} - \frac{\nu}{E}\mathrm{tr}\,\boldsymbol{\sigma}\,\mathbf{1}\right) \tag{8.5.41}$$

$$\omega = Q(\tilde{\varepsilon}) \tag{8.5.42}$$

$$\tilde{\varepsilon} = \max\left(\varepsilon^{+}\right) \qquad \text{with} \qquad \varepsilon^{+} = \sqrt{\langle\boldsymbol{\varepsilon}\rangle^{+}\cdot\langle\boldsymbol{\varepsilon}\rangle^{+}} \tag{8.5.43}$$

$Q(\tilde{\varepsilon})$ is a scalar function characterizing the material and $\langle\boldsymbol{\varepsilon}\rangle^{+}$ is the positive (or tensile) part of the strain tensor, defined as the tensor possessing the same principal directions as $\boldsymbol{\varepsilon}$ and having principal values that coincide with those of $\boldsymbol{\varepsilon}$ when positive and are set to zero when negative. This model is restricted to tensile damage since, by its very definition, no damage is introduced if the principal strains are negative. Note also that, due to its simplicity, brought about by the scalar nature of the internal variable, the flow rule takes an integrated form.

The function $Q(\tilde{\varepsilon})$ is uniquely determined from the uniaxial stress-strain curve. Indeed, taking the axis x_1 to lie along the specimen axis and the axes x_2 and x_3 to be normal to it, it turns out that in uniaxial tension $\langle\boldsymbol{\varepsilon}\rangle^{+}{}_{ij} = \varepsilon_{11}\delta_{1i}\delta_{1j}$, so that for monotonic straining $\tilde{\varepsilon} = \varepsilon^{+} = \varepsilon_{11} = \varepsilon$ (where ε denotes the axial strain). Then, if the stress strain curve is given by $\sigma = \psi(\varepsilon)$, we substitute this into (8.5.41), solve for ω, and equate the result to (8.5.42) to get

$$\omega = Q(\tilde{\varepsilon}) = 1 - \frac{\psi(\tilde{\varepsilon})}{E\tilde{\varepsilon}} \tag{8.5.44}$$

The main problem with this model is that the prediction for the transverse strain is unrealistic. Indeed, it is easily verified that, for uniaxial tension in the direction of x_1, we have $\varepsilon_{22} = \varepsilon_{33} = -\nu\varepsilon_{11}$ at all times. This means that, for full fracture, when $\varepsilon_{11} \to \infty$, we get $\varepsilon_{22} = \varepsilon_{33} \to -\infty$, which is unrealistic. Therefore, a directional scalar damage model such as that given by (8.5.33) in which C_N is the scalar damage variable and $\tilde{\varepsilon}^{f}$ the hardening-softening variable, may often be more suitable to describe the fracture behavior.

8.5.9 Rankine Plastic Model with Softening

This is a very simple model with strength degradation, which exhibits a certain analogy with the rotating crack model. In the rotating crack model the inelastic strain has the same principal directions as the stress tensor, while in Rankine plasticity this holds for the inelastic strain *increments*.

The formulation is classical. The compliance tensor is fixed, and so we write

$$\boldsymbol{\varepsilon} = \boldsymbol{\varepsilon}^{el} + \boldsymbol{\varepsilon}^{p} \tag{8.5.45}$$

where the only primary internal variable is taken to be ε^p. The loading function is taken to be the Rankine yield criterion,

$$\sigma_I - \phi(\tilde{\varepsilon}^p) \leq 0 \tag{8.5.46}$$

where σ_I is the largest principal strain, $\tilde{\varepsilon}^p$ the equivalent plastic uniaxial strain, and ϕ is a function defining the evolution of the strength. This means that the inelastic strain occurs when the major principal stress attains the instantaneous strength. The flow rule simply states that the inelastic strain takes place in the direction of the maximum principal strain, i.e.:

$$\dot{\boldsymbol{\varepsilon}}^p = \mathbf{p}_I \otimes \mathbf{p}_I \, \dot{\tilde{\varepsilon}}^p \tag{8.5.47}$$

where $\mathbf{p}_I$ is a unit eigenvector corresponding to the principal stress σ_I.

The reader can easily check that for uniaxial tension under monotonic straining, $\tilde{\varepsilon}^p = \varepsilon^p$ and $\sigma_I = \sigma$. So, the function $\phi(\tilde{\varepsilon}^p)$ is nothing more than the curve of stress vs. inelastic strain for uniaxial tension.

8.5.10 A Simple Model with Stiffness and Strength Degradation

It is relatively simple to build a model with mixed properties, by combining the simple rotating crack model given by Eq. (8.5.33) and the foregoing plastic Rankine model. We assume the total strain to be split into the elastic and fracturing parts,

$$\boldsymbol{\varepsilon} = \boldsymbol{\varepsilon}^{el} + \boldsymbol{\varepsilon}^f \tag{8.5.48}$$

and assume further that the fracturing strain $\boldsymbol{\varepsilon}^f$ can be split into a term linear in the stress and a permanent (irreversible) strain tensor $\boldsymbol{\varepsilon}^p$:

$$\boldsymbol{\varepsilon}^f = C^f \, \sigma_I \, \mathbf{p}_I \otimes \mathbf{p}_I + \boldsymbol{\varepsilon}^p \tag{8.5.49}$$

in which C^f is the inelastic unloading-reloading compliance (which replaces the secant compliance C_N). The evolution of C^f is given in integrated form as a function of the hardening variable $\tilde{\varepsilon}^f$, while the evolution of $\boldsymbol{\varepsilon}^p$ is given in incremental form as before:

$$C^f = \eta(\tilde{\varepsilon}^f) \tag{8.5.50}$$

$$\dot{\boldsymbol{\varepsilon}}^p = \mathbf{p}_I \otimes \mathbf{p}_I \, \frac{dH(\tilde{\varepsilon}^f)}{d\tilde{\varepsilon}^f} \, \dot{\tilde{\varepsilon}}^f \tag{8.5.51}$$

where $\eta(\tilde{\varepsilon}^f)$ and $H(\tilde{\varepsilon}^f)$ are material functions which will be related in the sequel to the uniaxial stress-strain curve. The evolution of the hardening-softening parameter $\tilde{\varepsilon}^f$ is deduced from the loading function and the consistency condition which is taken according to the Rankine criterion:

$$\sigma_I - \phi(\tilde{\varepsilon}^f) \leq 0 \tag{8.5.52}$$

Considering the uniaxial tensile test with monotonic stretching, let us call σ, ε^f and ε^p the axial components of the stress, and fracturing and permanent strain tensors, respectively, we obviously have $\sigma_I = \sigma$ and, by definition, $\varepsilon^f = \tilde{\varepsilon}^f$, and so $\phi(\tilde{\varepsilon}^f)$ is again the softening function. On the other hand, $\mathbf{p}_I$ is a vector coinciding with the axis of the uniaxial tension. And according to (8.5.51), the permanent strain is readily integrated; it has only a nonzero component, namely the axial one,

$$\varepsilon^p = H(\tilde{\varepsilon}^f) \tag{8.5.53}$$

Substituting this into (8.5.48) and identifying the axial components (the remaining ones are all zero), we get

$$\tilde{\varepsilon}^f = \eta(\tilde{\varepsilon}^f)\phi(\tilde{\varepsilon}^f) + H(\tilde{\varepsilon}^f) \tag{8.5.54}$$

from which we can solve for $\eta(\tilde{\varepsilon}^f)$:

$$\eta(\tilde{\varepsilon}^f) = \frac{\tilde{\varepsilon}^f - H(\tilde{\varepsilon}^f)}{\phi(\tilde{\varepsilon}^f)} \tag{8.5.55}$$

Therefore, given $\phi(\tilde{\varepsilon}^f)$ and $H(\tilde{\varepsilon}^f)$, the properties of the material are completely determined. Note that, according to (8.5.53) $H(\tilde{\varepsilon}^f)$ is nothing else than the permanent strain which is obtained when the specimen in a uniaxial test is stretched up to $\tilde{\varepsilon}^f$ and then unloaded. If no further information is available, it may be assumed that this is a fixed proportion α of the maximum inelastic strain, i.e., $H(\tilde{\varepsilon}^f) = \alpha\,\tilde{\varepsilon}^f$ (in which $\alpha \leq 1$). With this, the flow rules can be rewritten as

$$C^f = (1-\alpha)\,\frac{\tilde{\varepsilon}^f}{\phi(\tilde{\varepsilon}^f)} \tag{8.5.56}$$

$$\dot{\varepsilon}^p = \alpha\,\mathbf{p}_\text{I} \otimes \mathbf{p}_\text{I}\,\dot{\tilde{\varepsilon}}^f \tag{8.5.57}$$

This constitutes the simplest triaxial generalization of the uniaxial model described in Section 8.4.3, and may also be viewed as a strongly simplified version of Ortiz's model (1985). However, one useful feature of Ortiz's model is that it describes softening in compression as well as tension, which is obviously not the case with this simplified version.

Exercises

8.22 Show that ε^f in (8.5.13) can be written in a general tensorial form as

$$\varepsilon^f = (C_N \boldsymbol{A} + C_T \boldsymbol{B})\boldsymbol{\sigma} \tag{8.5.58}$$

where $\boldsymbol{A}$ and $\boldsymbol{B}$ are (for constant $\vec{n}$) constant fourth-order tensors which are given in cartesian components by

$$A_{ijkl} = n_i n_j n_k n_l\,, \qquad B_{ijkl} = \frac{1}{2}\delta_{ik} n_l n_j + \frac{1}{2}\delta_{jk} n_l n_i - n_i n_j n_k n_l \tag{8.5.59}$$

8.23 Consider a fixed-direction crack model with elastic-softening behavior defined by exponential softening in uniaxial tension, $\sigma = f'_t e^{-\varepsilon^f/\varepsilon_0}$. Determine the evolution of the axial and transverse stress components in uniaxial extension, in which $\varepsilon_{11} = \varepsilon$ increases monotonically and all the remaining components of the strain tensor are zero.

8.24 Show that the response for the uniaxial extension in the preceding exercise is identical for fixed and rotating crack models as long as the strain-softening curve is the same.

8.25 Consider a fixed-direction crack model with elastic-softening behavior defined by exponential softening in uniaxial tension, $\sigma = f'_t e^{-\varepsilon^f/\varepsilon_0}$. Determine the evolution of the axial and transverse strains in a plane stress tension test, in which $\varepsilon_{11} = \varepsilon$ increases monotonically and $\varepsilon_{33} = \sigma_{22} = 0$, the shear components being zero.

8.26 Consider a fixed-direction crack model that exhibits elastic-softening behavior defined by exponential softening in uniaxial tension, $\sigma = f'_t e^{-\varepsilon^f/\varepsilon_0}$, and is amenable to the scalar damage model described in Section 8.5.3. Referring to cartesian axes $\{x_1, x_2, x_3\}$, consider a process in which the stress components $\sigma_{22} = \sigma_{33} = \sigma_{23} = \sigma_{13} = 0$, while the fracturing strain tensor evolves such that $\varepsilon^f_{11} = \varepsilon_0\lambda$, $\varepsilon^f_{12} = \varepsilon_0\lambda^2$ in which λ increases monotonically starting at $\lambda = 0$. All the remaining components are zero. Assuming that $\nu = 0.2$, determine: (a) the evolution of the stress components; (b) the evolution of the maximum principal stress; and (c) whether a secondary crack forms. [Hint: Note that for $\lambda \to 0$, the shear component is negligible, and thus the crack band forms normal to x_1.]

8.27 Same as the preceding exercise except that $\varepsilon^f_{12} = \varepsilon_0\lambda\mu$ in which λ increases monotonically starting at $\lambda = 0$ and μ varies in a way to be determined. Determine the upper bound for μ as a function of λ so that secondary cracking would not occur. Determine the evolution of the stress for the limiting case. [Answer: $|\mu| < 2(1-\nu)e^{2\lambda}\sqrt{1-e^{-\lambda}}$.]

8.28 Generalize the foregoing result to any softening function $\phi(\tilde{\varepsilon}^f)$.

8.29 Show that the tangent approach with S_N and S_T depending only on ε^f_N cannot be distinguished from Rots' incremental approach for proportional microcracking, i.e., for loadings such that $\vec{\varepsilon}^f_T = \varepsilon^f_N \vec{m}$, where $\vec{m}$ is an arbitrary vector in the crack plane. Find S^t_T in terms of S_T for this particular case.

8.30 Consider a material conforming to Mazars' isotropic damage model with elastic-softening behavior defined by exponential softening in uniaxial tension, $\sigma = f'_t e^{-\varepsilon^f/\varepsilon_0}$. Determine the evolution of the axial and transverse stress components in uniaxial extension, in which $\varepsilon_{11} = \varepsilon$ increases monotonically and all the remaining components of the strain tensor are zero.

8.31 Consider a thin layer of a material conforming to Mazars' isotropic damage model with elastic-softening behavior defined by exponential softening in uniaxial tension, $\sigma = f_t' e^{-\varepsilon^f/\varepsilon_0}$. This layer is sandwiched between two thick plates of an elastic material with the same elastic moduli as the adjacent material, and the sandwich is subjected to uniaxial tension normal to the layer. Neglecting end effects, determine the evolution of the stress tensor in the layer as a function of the strain of the layer in the normal direction (assume that the transverse strain in the layer is dictated by the transverse strain in the elastic plates, which is, in turn, dictated by the elastic Poisson effect).

8.32 Carry out the algebra leading to (8.5.56) and (8.5.57).

8.33 Determine the strain evolution in a material element following the general model in Section 8.5.10 if the response is elastic-softening with an exponential softening curve $\sigma = f_t' e^{-\varepsilon^f/\varepsilon_0}$. Consider that the element is subjected to proportional loading with $\sigma_{11} = 2\sigma_{22}$ and all the remaining stress components are equal to zero, while $\varepsilon_{11} \equiv \varepsilon$ increases monotonically up to $1.4\varepsilon_0$ and then the element is unloaded. Take $\varepsilon_0 = 10 f_t'/E$ and $\nu = 0.2$.

8.6 Crack Band Models and Smeared Cracking

In Section 8.3.5, we discussed a simple way to determine the stress-strain curve for a finite element of any size based on the stress-strain curve for the band. However, this applied only for the uniaxial case, which is extremely simple. In this section, we will address the complexities raised by the triaxial nature of most practical problems, although we consider principally two-dimensional problems. We start by seeking the triaxial strain-softening equations for a finite element of any size.

8.6.1 Stress-Strain Relations for Elements of Arbitrary Size

To be precise, we limit the analysis to elastic-softening materials and consider a fracturing model with one definite cracking orientation (fixed or rotating). Consider first a case in which the crack band evolves with the cracks oriented parallel to one of the directions of the finite element mesh, as depicted in Fig. 8.6.1a–c. It is intuitively clear that the stress-strain relations for the direction normal to the band must be very close to the uniaxial formulation deduced in Section 8.3.5. However, in that uniaxial formulation, the transversal strains were ignored, while in the actual three- or two-dimensional elements, a mismatch of strains parallel to the cracks can occur between the crack band that softens and the remainder of the element that unloads. In view of the other simplifications involved, this might not cause a serious error, but it is not difficult to enforce the proper interface continuity requirements at this interface. In fact, the formulation in Section 8.4.2 ensures compatibility automatically.

Consider the simple case in Fig. 8.6.1d. After the stress peak, the material inside the crack band softens, while the rest of the element unloads (for the elastic-softening case considered here, unloading means elastic behavior). We want to enforce that the strain components in the plane of the cracks be the same for the unloading and softening regions, which, with reference to the base vectors in Fig. 8.5.1c, is written as

$$\varepsilon^u_{tt} = \varepsilon^s_{tt}\,, \qquad \varepsilon^u_{ss} = \varepsilon^s_{ss} \qquad \text{and} \qquad \varepsilon^u_{ts} = \varepsilon^s_{ts} \tag{8.6.1}$$

where superscripts u and s refer to the unloading region and to the softening band. Now, according to the hypothesis of elastic-softening behavior, the strain tensor in the unloading region is related to the stress tensor by the elastic relations, while the strain tensor in the softening band is given by (8.5.5), and so we have:

$$\boldsymbol{\varepsilon}^u = \frac{1+\nu}{E}\boldsymbol{\sigma}^u - \frac{\nu}{E}\text{tr}\,\boldsymbol{\sigma}^u\,\mathbf{1} \tag{8.6.2}$$

$$\boldsymbol{\varepsilon}^s = \frac{1+\nu}{E}\boldsymbol{\sigma}^s - \frac{\nu}{E}\text{tr}\,\boldsymbol{\sigma}^s\,\mathbf{1} + \left(\vec{\varepsilon}^f \otimes \vec{n}\right)^S \tag{8.6.3}$$

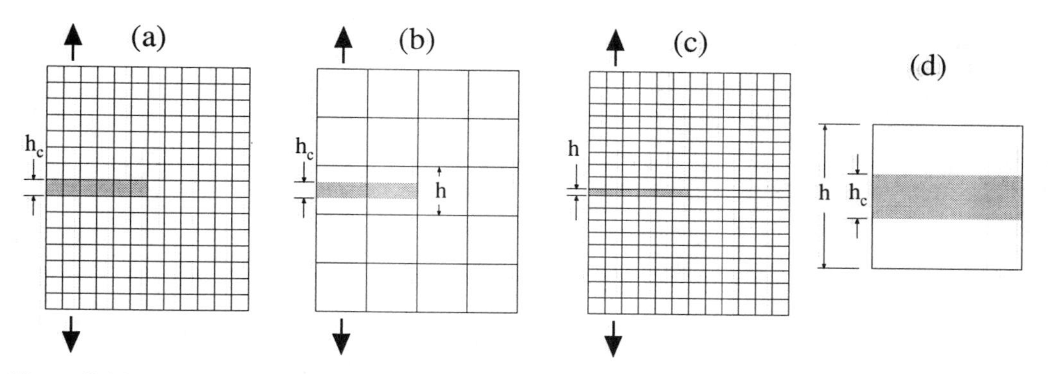

Figure 8.6.1 Rectangular panel with various mesh sizes, identical (a), larger (b), or smaller (c) than the crack band width h_c. (d) Detail of an element with an embedded crack band.

Therefore, writing the components appearing in (8.6.1), and taking into account (8.5.4), we can reduce the strain conditions to identical conditions in stresses, i.e.,

$$\sigma_{tt}^u = \sigma_{tt}^s, \qquad \sigma_{ss}^u = \sigma_{ss}^s \qquad \text{and} \qquad \sigma_{ts}^u = \sigma_{ts}^s \tag{8.6.4}$$

If we now take into account that the traction vectors on the interface of the softening and unloading parts must be equal, i.e.,

$$\boldsymbol{\sigma}^u \vec{n} = \boldsymbol{\sigma}^s \vec{n} \tag{8.6.5}$$

it turns out that the remaining three components of the stress tensors in the two regions must also be mutually equal. Therefore, the compatibility and continuity equations are satisfied if the stress tensors in the softening band and in the unloading region are identical, so that we can write

$$\boldsymbol{\sigma}^u = \boldsymbol{\sigma}^s = \boldsymbol{\sigma} \tag{8.6.6}$$

Since the stress tensors are the same, the two regions are fully coupled in series. The average strain in the element can be obtained by stipulating that the virtual work of the mean fields is equal to the sum of the virtual works in the unloading and softening portions, that is

$$\boldsymbol{\sigma} \cdot \delta\boldsymbol{\varepsilon}^{(e)} V^{(e)} = \boldsymbol{\sigma}^u \cdot \delta\boldsymbol{\varepsilon}^u V^u + \boldsymbol{\sigma}^s \cdot \delta\boldsymbol{\varepsilon}^s V^s \tag{8.6.7}$$

where $V^{(e)}, V^u$, and V^s are, respectively, the volumes of the element, the unloading region and the softening region. By virtue of (8.6.6), this condition is identically satisfied for all the stress states and virtual strain increments if

$$\boldsymbol{\varepsilon}^{(e)} = (1-f)\boldsymbol{\varepsilon}^u + f\boldsymbol{\varepsilon}^s \tag{8.6.8}$$

where f is the volume fraction of the crack band. Substituting (8.6.6) into (8.6.2) and (8.6.3) and the results in (8.6.8) we get the final expression

$$\boldsymbol{\varepsilon}^{(e)} = \frac{1+\nu}{E}\boldsymbol{\sigma} - \frac{\nu}{E}\text{tr}\,\boldsymbol{\sigma}\,\mathbf{1} + f\left(\vec{\varepsilon}^f \otimes \vec{n}\right)^S \tag{8.6.9}$$

which shows that the equation for the element has a structure identical to the original stress-strain model, except that the fracturing strain is affected by the factor f. This factor is trivially equal to h_c/h for the simple cases shown in Fig. 8.6.1 in which the elements are rectangular and the crack band is perpendicular to one pair of sides.

This case occurs frequently in the analysis of test results for mode I crack growth, and its use was pioneered by Bažant and Oh (1983a), who analyzed with success a tremendous amount of experimental data. Bažant and Oh used a crack band model with a finite strain slope in a finite element analysis with square meshes. In computations, small increments of the load-point displacement were prescribed, and the reaction, representing the load P, was calculated in each loading step. The same stress-strain relation was assumed to hold for all the finite elements, although only some of them entered nonlinear behavior.

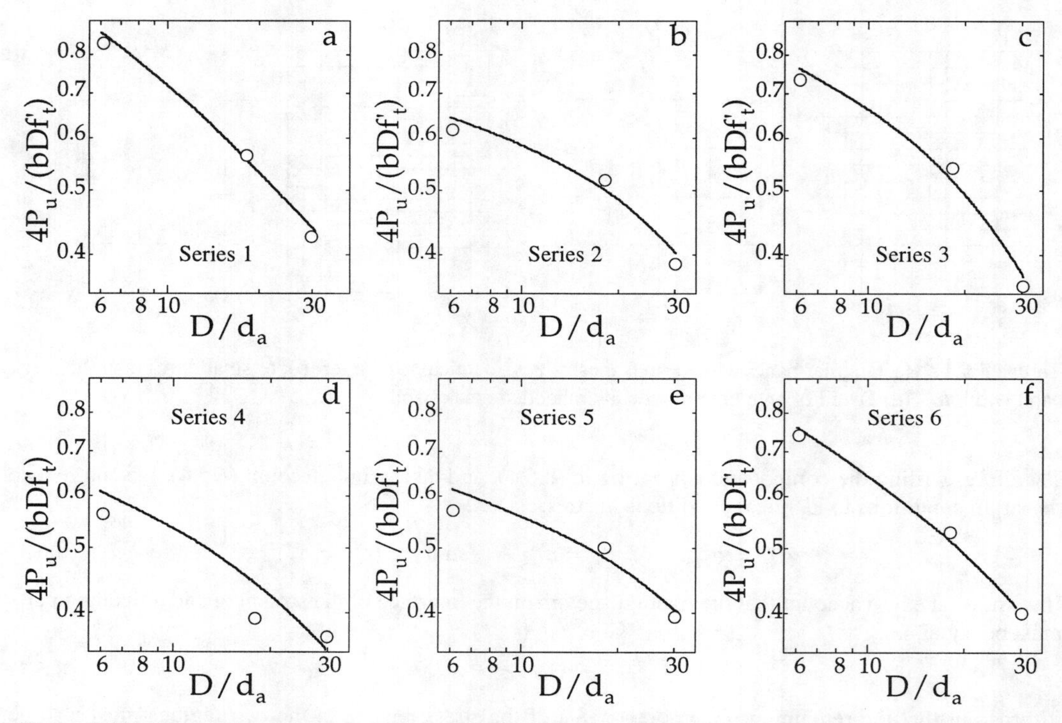

Figure 8.6.2 Comparison of the peak load predictions of the crack band model of Bažant and Oh (1983a) with the experimental data of Walsh (1972).

A plane stress state was assumed for all the calculations. Although the width of the crack band (size of the square elements) was found to have very little effect, its optimum was approximately $w_c = 3d_a$ (d_a = maximum aggregate size), and this value was used throughout the computations.

The crack band theory was able to reproduce with accuracy the experimental results of Naus (1971), Walsh (1972), Kaplan (1961), Mindess, Lawrence and Kesler (1977), Huang (1981), Carpinteri (1980), Shah and McGarry (1971), Gjørv, Sorensen and Arnesen (1971), Hillerborg, Modéer and Petersson (1976), Sok, Baron and François (1979), Wecharatana and Shah (1980), Brown (1972), and Entov and Yagust (1975). As an example, we plot in Fig. 8.6.2a–f the results for the six Walsh's series (1972) described in Section 1.5 (see Tables 1.5.1 and 1.5.1, Series A1-A6, and Figs. 1.5.1 and 1.5.2).

8.6.2 Skew Meshes: Effective Width

The foregoing considerations are, of course, applicable even when a crack band of width h_c is embedded in a finite element of size h and is inclined with respect to the size of the element (Fig. 8.6.3a), an issue that has been used repeatedly in various fields. The idea of embedding a band of strain softening in a finite element was first developed for plastic shear bands (Pietruszczak and Mróz 1981), and the subsequent development of a finite element with an embedded crack band (Willam, Bićanić and Sture 1986; Willam, Pramono and Sture 1989) was mathematically analogous. Recently, a general and fully consistent three-dimensional formulation for an embedded strain-softening band in general finite elements was presented by Dvorkin, Cuitiño and Gioia (1990).

In the present formulation, the only modification that is necessary is to substitute a proper value for the volume fraction of the crack band within the element f. For square meshes, Bažant and Oh (1983a) proposed to use an effective bandwidth for the element h_b such that

$$f = \frac{h_c}{h_b} \qquad \text{with} \qquad h_b = \frac{h}{\cos\theta} \tag{8.6.10}$$

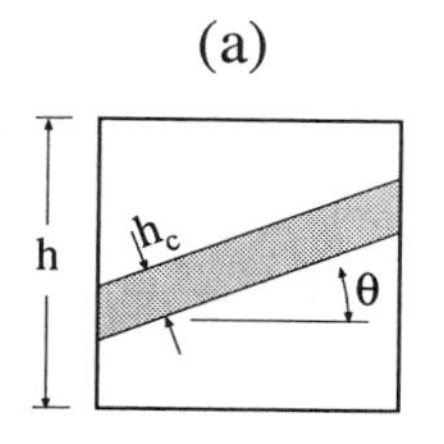

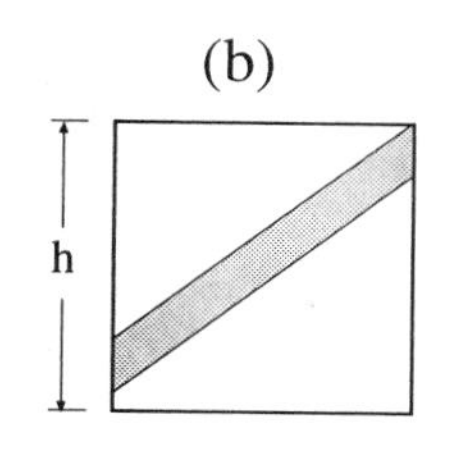

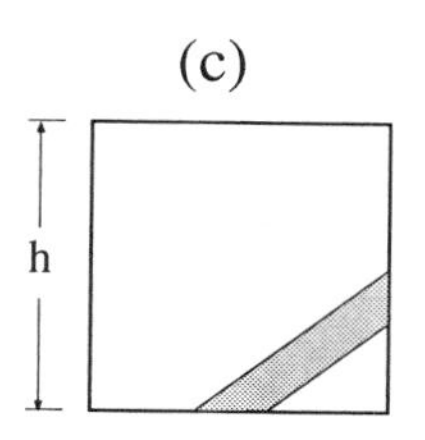

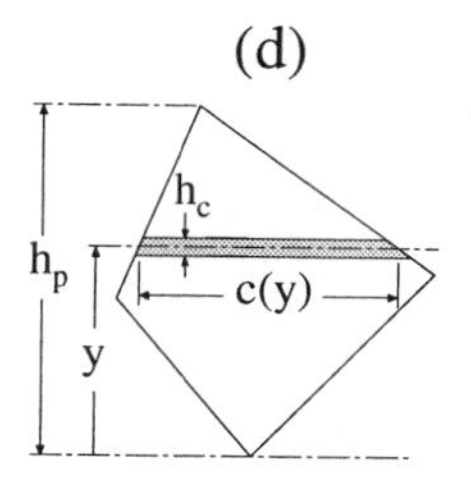

Figure 8.6.3 (a) Element skew to the crack band. (b) Centered band, with large volume fraction. (c) Lateral band, with small volume fraction. (d) Sketch to define the average volume fraction.

where θ is the angle between the band and the base of the element. An approximate generalization of this rule to irregular elements was proposed in Bažant (1985a). In general, however, such extrapolations to irregular elements can hardly be satisfactory, and a more general approach is needed.

The problem, however, is not trivial. The reason is that the volume fraction, when the crack band is inclined with respect to the element side, or the element is irregular, is not well defined: it depends on the precise position of the crack band with respect to the element. This is illustrated in Figs. 8.6.3b and c, for which the volume fractions are in the proportion 2:1, approximately. Therefore, either information on the position of the band within the element must be given —which is not possible if ordinary elements are used— or else, the bandwidth must be defined in an average sense. The average can be obtained in the following manner. Consider a bidimensional element of thickness b, arbitrary size and shape, and arbitrarily oriented with respect to the band, which is drawn horizontal (Fig. 8.6.3d). Let y be the axis normal to the band, with its origin at the lowest point of the element. If the band is located at distance y, as shown in the figure, the volume of the band is approximately given by $V(y) = h_c b c(y)$ where $c(y)$ is the length of the intercept of the center of the band with the element. Let $\varphi(y)dy$ be the probability that the band lies at a distance between y and $y + dy$. Then the average volume of the band is given by

$$V_b = h_c b \int_0^{h_p} \varphi(y) c(y)\, dy \tag{8.6.11}$$

where h_p is the maximum ordinate of the element, which coincides with the *projected element size*. If the probability density is uniform, then $\varphi = 1/h_p$ and we get

$$V_b = \frac{h_c}{h_p} b \int_0^{h_p} c(y)\, dy \tag{8.6.12}$$

but $b \int_0^{h_p} c(y)\, dy$ is the volume of the element, and therefore,

$$f = \frac{h_c}{h_p} \tag{8.6.13}$$

This indicates that for equally probable distributions, the element bandwidth coincides with the projected size of the element (projected on the normal of the crack band). This coincides with the formula proposed by Bažant (1985a) for rectangular meshes, and has been implemented in commercial finite element codes (e.g., SBETA; Červenka and Pukl 1994). However, it must be clearly understood that it is an average value, which may differ appreciably from the actual value for a particular element in a particular mesh.

The foregoing calculation is based on the assumption that the strain is uniform within the element, which is generally not the case because quadratic or higher order elements are used with various possible integration schemes (i.e., distribution of integration points). In such cases, the analysis would have to be redone starting from the virtual work equation (8.6.7), a task that is not straightforward. Rots (1988) used a trial-and-error method to determine the effective bandwidths of the elements for a particular problem, and Oliver (1989) proposed an objective formulation of an integral to define h_b. Červenka et al. (1995) empirically found that using the projected element size gave a still larger dissipation for inclined bands, and proposed a correction factor γ so that

$$h_b = \gamma h_p \tag{8.6.14}$$

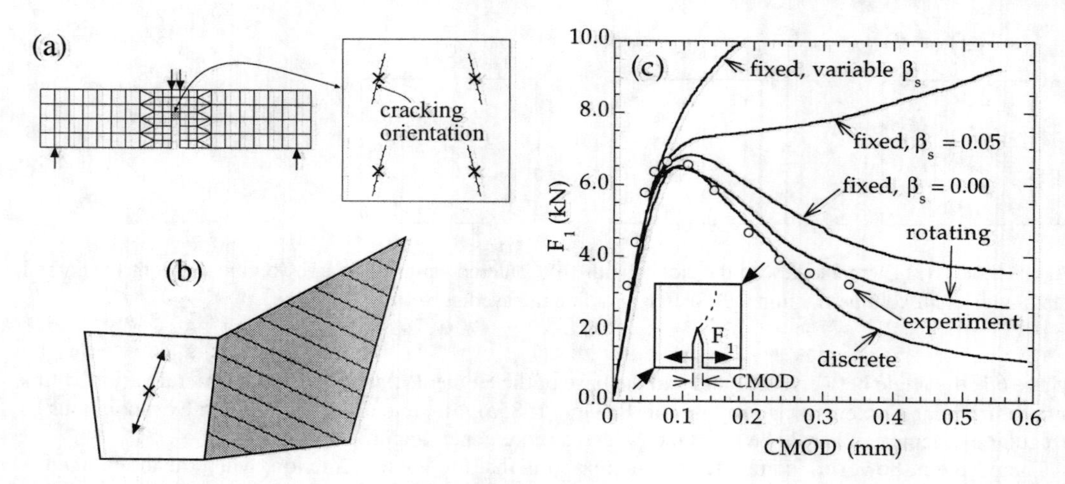

Figure 8.6.4 (a) Skew cracking orientation at the integration points. (b) Inclined cracking in the shaded element induces stress in the uncracked element (from Rots 1989). (c) Load-CMOD curves for a compact specimen under mixed mode loading (from Rots 1989).

Here γ varies linearly from 1 for element sides perpendicular to the band to $\gamma = 1.5$ for sides at 45°; in the case of irregular elements, an average side angle needs to be used. With this correction, they obtained results approximately independent of the orientation of the mesh, for mode I cases (bending and tension).

It seems that using conventional finite elements with plain smearing, as used in most finite element codes, implies a variable degree of uncertainty in the definition of the element bandwidth for meshes skew to the band. There are two alternatives to circumvent this problem: (1) use remeshing techniques to achieve a mesh in which the band runs parallel and perpendicular to the sides of the elements it crosses (see Fig. 8.7.3b), or (2) use enriched elements with embedded strain discontinuities similar to those described in Section 7.2.3. In fact, the displacement discontinuity in Oliver's element depicted in Fig. 7.2.10 is numerically implemented as a thin band with large, but finite, constant strain, very similar to an embedded crack band.

8.6.3 Stress Lock-In

As pointed out before, the crack band analysis for mode I and elements aligned with the crack path gives, in general, good results. Various details of numerical modeling, however, deserve attention. Leibengood, Darwin and Dodds (1986) modeled the crack band by square elements with four integration points straddling the line of symmetry. They showed that the results for the crack band and the discrete crack closely match each other if the cracking directions at the integration points within the finite elements are forced to be parallel to each other and to the symmetry line of the crack band. But if this parallelism is not enforced, the cracks form at different orientations at each integration point of the same element, as shown in Fig. 8.6.4a. Then the response predicted by the crack band model is somewhat stiffer than that predicted by the discrete crack model, even if the element sides are parallel to the true crack.

The reason for this behavior is that the integration points lying out of the symmetry plane sense the shear components, and so the cracks form at an angle. This results in spurious fracturing strain components parallel to the crack path, which cause large strains in the neighboring elements parallel to the main crack path, which results in overall stiffening. Although the crack growth is actually in mode I, the problem is aggravated because, at the integration points, the loading is interpreted as mixed mode. This is manifested by a spurious sensitivity of the solution to the shear retention factor β_s, which, for mode I, should be nil. A solution to this problem is to determine the crack normal and the cracking strain on the average or at a single integration point at the center of the element. This is actually the only way consistent with the hypothesis that the cracking strain is uniformly distributed over the element.

The foregoing problem is related to the phenomenon known as *stress lock-in*, that appears in mixed mode problems, when the crack grows skew to the mesh (Rots 1988, 1989). The stress lock-in consists

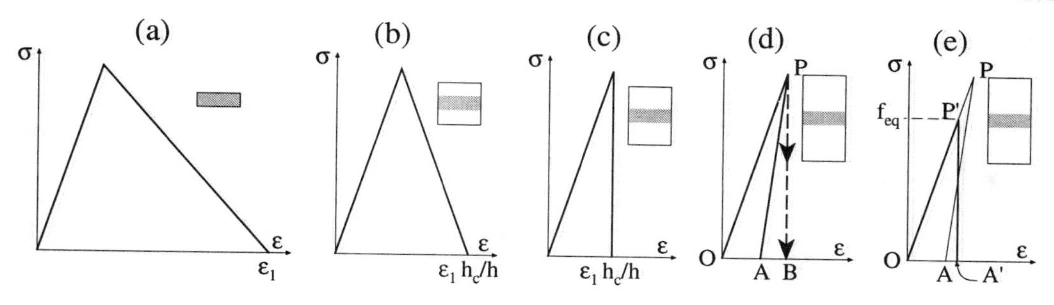

Figure 8.6.5 Linear softening and tensile stress modification.

in the effect that the elements near the crack band remain stressed after a nearly complete failure of the elements in the crack band, because the inelastic strain in the band induces stresses in the neighbors, as sketched in Fig. 8.6.4b. Rots' (1988, 1989) results indicate that the degree of stress lock-in is very sensitive to the shear retention factor, as illustrated by the results in Fig. 8.6.4c, corresponding to a double cantilever beam tested by Kobayashi et al. (1985). The fixed crack must be used with a zero shear retention factor to get a better approximation. In the example shown, the smeared crack model with the best behavior is the rotating crack model, but in other cases (single-notch sheared beams, for example), the fixed crack with a zero shear retention factor may be better (Rots 1988, 1989).

The only effective solution to this problem (keeping classical finite elements) is to first run the calculations with a standard mesh to get the approximate crack path, and then remesh to get mesh lines aligned with the crack path and run the calculation again (see also Section 8.7.4).

8.6.4 Use of Elements of Large Size

In all the foregoing analyses, it is assumed that the finite element is small enough for the resulting stress-strain curve to be stable. If the element is too large, then the resulting stress-strain curve has a snapback, as shown in Fig. 8.3.6d. If this occurs, the finite element analysis will be very difficult to stabilize and will dissipate more energy than it actually should. It may be argued that the problem should be solved by using smaller elements, but this may be computationally too expensive and it may be worth using larger elements if the accuracy is not greatly sacrificed.

To simplify the problem, let us consider the simple linear softening depicted in Fig. 8.6.5a for the actual crack band thickness h_c. If the element size is $h > h_c$, the stress-strain curve for the element is as shown in Fig. 8.6.5b. The softening branch becomes vertical when $\varepsilon_1 h_c/h = f'_t/E$ (see Fig. 8.6.5c), i.e., for $h = h_c \varepsilon_c E/f'_t$. Since in this linear case $G_F = h_c f'_t \varepsilon_1/2$, eliminating h_c leads to the simple condition

$$h = 2\ell_{ch} \tag{8.6.15}$$

Thus, for $h > 2\ell_{ch}$, a snapback occurs as shown in Fig. 8.6.5d. Because in the finite element computation the nodal displacements are controlled, the stress will drop to zero as soon as the peak is reached and the dissipated energy will appear to be the area OPB instead of the area OPA which is the correct value. This means that using larger elements will make the material appear tougher than it actually is.

A solution to this problem (Bažant 1985b,c) is to replace the actual stress-strain curve with snapback by a stress-strain diagram of the same area having a vertical stress drop (Fig. 8.6.5e). To keep the same area, one must reduce the tensile strength from f'_t to f_{eq} so that

$$h\frac{{f_{eq}}^2}{2E} = G_F \qquad \Rightarrow \qquad f_{eq} = f'_t\sqrt{\frac{2\ell_{ch}}{h}} \tag{8.6.16}$$

Thus, the strength must be reduced in inverse proportion to the square root of the element size. For the case of vertical stress drop, the fracture process zone has the smallest length permitted by the finite element subdivision. Therefore, this represents the closest possible approximation to LEFM. Since the element size is normally taken proportional to the structure size, this means that the crack band model with a vertical drop yields an approximate equivalent LEFM behavior for structures of large sizes. However,

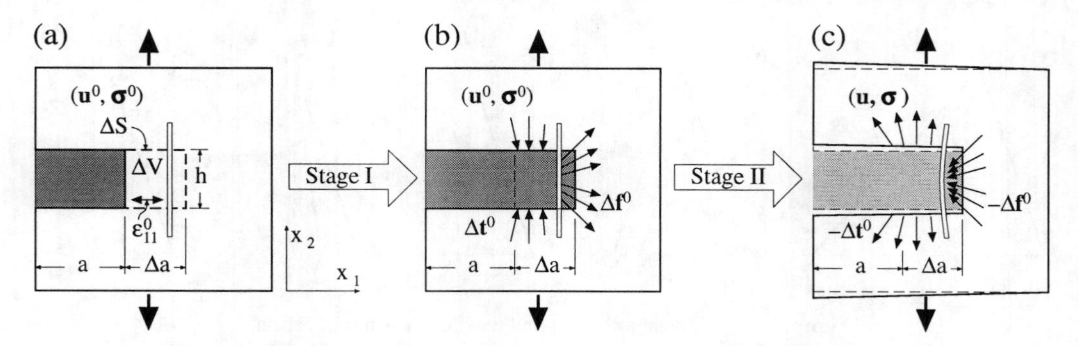

Figure 8.6.6 Sketch of the computational procedure of Bažant and Cedolin (1979, 1980): (a) initial state; (b) intermediate state; (c) final state.

this is not the only way to handle the problem of brittle behavior with large elements. In the following, an energy-based analysis is presented as a possible alternative.

8.6.5 Energy Criterion for Crack Bands with Sudden Cracking

As just described, if the cracks are assumed to form suddenly, i.e., the stress to drop suddenly to zero, a spurious mesh sensitivity and lack of objectivity appears because of the dependence of the apparent energy dissipation on the element size. This effect is eliminated by the previous equivalent strength method, but can also be eliminated by directly applying an energy criterion analogous to linear elastic fracture mechanics. The proper form of the energy criterion, which was obtained by Bažant and Cedolin (1979, 1980) by generalization of Rice's (1968b) energy analysis of the extension of a notch, can be formulated as follows. The crack band extension by length Δa into volume ΔV (of the next finite element, Fig. 8.6.6) may be decomposed for calculation purposes into two stages.

Stage I. Smeared cracks are created in concrete inside volume ΔV of the element ahead of the crack in the direction of tensile principal stress (Fig. 8.6.6b), while at the same time, the deformations and stresses in the rest of the body are imagined to remain fixed (frozen). This means that one must introduce surface tractions $\Delta \mathbf{t}^0$ applied on the boundary ΔS of volume ΔV, and distributed forces $\Delta \mathbf{f}^0$ applied at the concrete-steel interface, such that they replace the action of concrete that has cracked upon the remaining volume $V - \Delta V$ and upon the reinforcement within ΔV.

Stage II. Next, forces $\Delta \mathbf{t}^0$ and $\Delta \mathbf{f}^0$ (Fig. 8.6.6c) are released (unfrozen) by gradually applying the opposite forces $-\Delta \mathbf{t}^0$ and $-\Delta \mathbf{f}^0$, reaching in this way the final state.

Let $\mathbf{u}^0$ and $\boldsymbol{\varepsilon}^0$ be the displacement vector and strain tensor before the crack band advance, and let $\mathbf{u}$ and $\boldsymbol{\varepsilon}$ be the same quantities after the crack band advance. For the purpose of analysis, the reinforcement may be imagined to be smeared in a separate layer coupled in parallel and undergoing the same strains as concrete in the crack band. The interface forces between steel and concrete, $\Delta \mathbf{f}^0$, then appear as volume forces applied on the concrete layer.

Upon passing from the initial to the intermediate state (Stage I), the strains are kept unchanged while cracking goes on. Thus, the corresponding stress changes within concrete in ΔV are given by $\Delta\sigma_{11} = \sigma_{11}^0 - E'\varepsilon_{11}^0 = (\varepsilon_{11}^0 + \nu'\varepsilon_{22}^0)E'/(1-\nu'^2) - E'\varepsilon_{11}^0$; $\Delta\sigma_{22} = \sigma_{22}^0$; $\Delta\sigma_{12}^c = 0$ (cracks are assumed to propagate in the principal stress direction). Here, σ_{ij}^0 denote the components of stress carried before cracking by the concrete alone (they are defined as the forces in concrete per unit area of the steel-concrete composite); and E and ν are the Young's modulus and Poisson's ratio of concrete. The values $E' = E$ and $\nu' = \nu$ apply to plane stress and $E' = E/(1-\nu^2)$ and $\nu' = \nu/(1-\nu)$ to plane strain. The change in strain energy of the system during Stage I in Fig. 8.6.6b is given by the elastic energy initially stored in ΔV and released by cracking, i.e.,

$$\Delta U_1 = -\int_{\Delta V} \frac{1}{2}(\boldsymbol{\sigma}^0 \cdot \boldsymbol{\varepsilon}^0 - E'{\varepsilon_{11}^0}^2)dV. \tag{8.6.17}$$

The change in strain energy during Stage II is given by the work done by the forces $\Delta \mathbf{t}^0$ and $\Delta \mathbf{f}^0$ while they are being released, i.e.,

$$\Delta U_2 = \int_{\Delta S} \frac{1}{2}\Delta \mathbf{t}^0 \cdot (\mathbf{u} - \mathbf{u}^0) dS + \int^{\Delta V} \frac{1}{2}\Delta \mathbf{f}^0 \cdot (\mathbf{u} - \mathbf{u}^0) dV. \qquad (8.6.18)$$

Coefficients $1/2$ must be used because forces $\mathbf{t}$ and $\mathbf{f}$ at the end of Stage II are reduced to zero.

If the concrete is reinforced, part of the energy is consumed by the bond slip of reinforcing bars during cracking within volume ΔV. This part may be expressed as $\Delta W_b = \int_s F_b \delta_b ds$, in which δ_b represents the relative slip between the bars and the concrete which is required to accommodate fracture advance; F_b is the average bond force during displacement δ_b per unit length of the reinforcing bar (force during the slip); and s is the length of the bar segment within the actual fracture process zone of width h_c (and not within volume ΔV, since the energy consumed by bond slip would then depend on the element size).

The energy criterion for the crack band extension may now be expressed as follows:

$$\Delta U = U'\Delta a = G_f \Delta a - \Delta U_1 - \Delta U_2 - \Delta W_b \begin{cases} > 0 & \text{stable, no propagation} \\ = 0 & \text{equilibrium propagation} \\ < 0 & \text{unstable} \end{cases} \qquad (8.6.19)$$

Here ΔU is the energy that must be externally supplied to the structure to extend the crack band of width h by length Δa. (ΔU = total energy in the case of rapid, or adiabatic fracture, and Helmholtz free energy in the case of slow, isothermal fracture.) If $\Delta U > 0$, then no crack extension can occur without supplying energy to the structure, and so the crack band is stable and cannot propagate. If $\Delta U < 0$, crack band extension causes a spontaneous energy release by the structure, which is an unstable situation, and so the crack extension must happen dynamically, the excess energy $-\Delta U$ being transformed into kinetic energy. If $\Delta U = 0$, no energy needs to be supplied and none is released, and so the crack band can extend in a static manner.

For practical calculations, the volume integral in Eq. (8.6.17) needs to be expressed in terms of nodal displacements using the distribution functions of the finite element. The boundary integral in Eq. (8.6.18) is evaluated from the change of nodal forces acting on volume ΔV from the outside. The energy ΔU_2 released from the surrounding body into ΔV may also be alternatively calculated as the difference between the total strain energy contained in all the finite elements of the structure before and after the crack band advance. According to the principle of virtual work, the result is exactly the same as that from Eq. (8.6.18). This calculation, however, is possible only if the structure is perfectly elastic whereas Eq. (8.6.18) is correct even for inelastic behavior outside the process zone, providing Δa is so small that $\mathbf{t}$ and $\mathbf{f}$ vary almost linearly during Stage I.

It may also be noted that Pan, Marchertas and Kennedy (1983) calculated ΔU_2 in their crack band finite element program by means of the J-integral, keeping the integration contour the same for various crack lengths. This calculation must yield the same ΔU_2 if the integration contour passes through only the elastic part of the structure, except for crossing the crack band behind the fracture process zone where the stresses are already reduced to zero.

Under general loading, the crack band may propagate through a mesh of finite elements in an inclined direction, in which case the band has a zig-zag shape. This means that the crack length increment during the breakage of the next element is not well defined, and an effective crack extension Δa_e must then be used for the element. This crack extension is easily determined based on the effective bandwidth of the element, by writing that (for two dimensions) the area of the element ΔA_e must be identical to the effective bandwidth h_e times the effective crack extension Δa_e, and thus

$$\Delta a_e = \frac{\Delta A_e}{h_e} \qquad (8.6.20)$$

where, if no further analysis is available, it may be assumed that the effective bandwidth is the projected element size h_p. For rectangular meshes this reduces to the formula proposed by Bažant (1985a).

The ability of the energy balance and equivalent strength methods to describe the fracture processes in large structures was demonstrated in a series of papers by Bažant and Cedolin (1979, 1980, 1983) and by Bažant and Oh (1983a). As an example of their results, we consider here the problem of a plain concrete panel with a center crack, as depicted in Fig. 8.6.7a. Bažant and Cedolin (1980) analyzed the results for

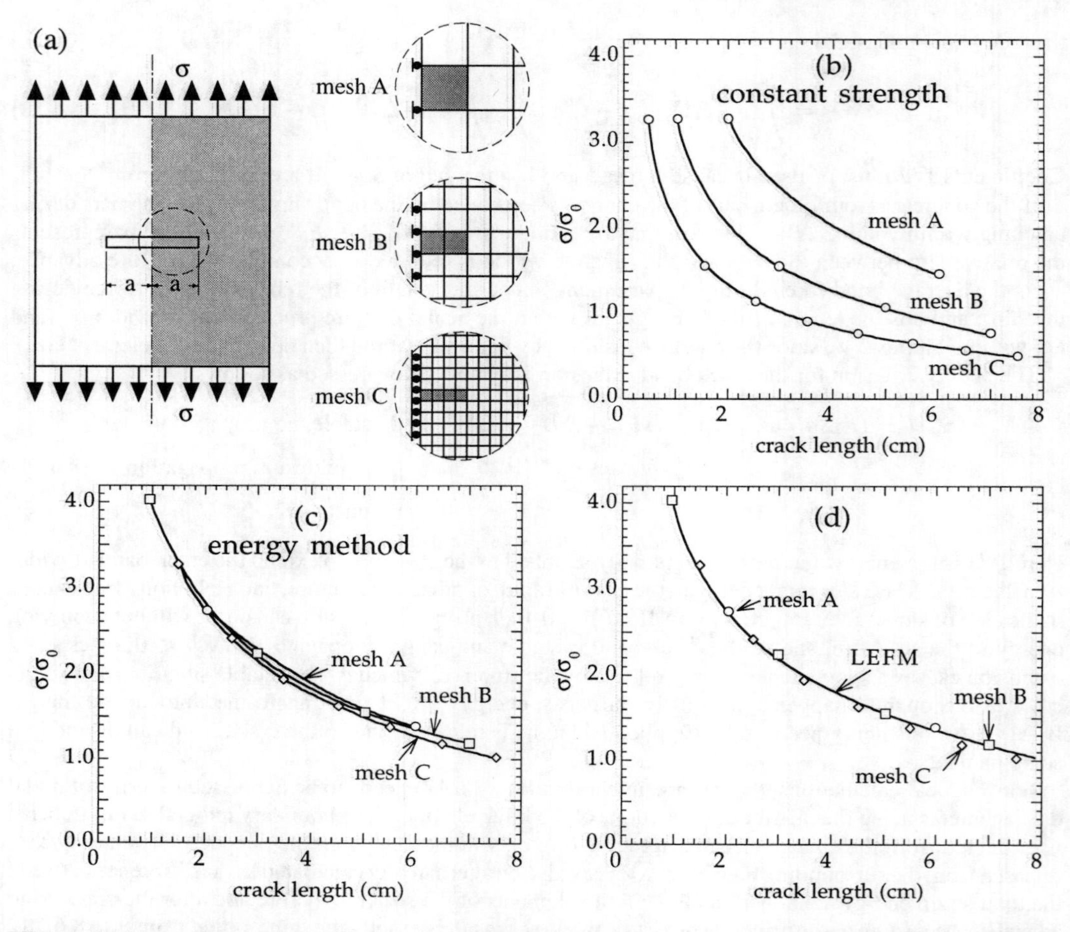

Figure 8.6.7 (a) Center-cracked panel analyzed by Bažant and Cedolin (1980). (b) Mesh-dependent results derived from constant-strength formulations. (c) Mesh independent results based on the energy formulation. (d) Comparison of the finite element results with the exact LEFM predictions.

three finite element meshes with element sizes in the relation 1:2:4, as sketched in Fig. 8.6.7a. If the classical tensile strength criterion (i.e., constant tensile strength) with sudden drop is used, the results shown in Fig. 8.6.7b are found, where it appears that the effect of the element size is tremendous. The strength for each crack length is seen to be smaller, the smaller the element size. This result may be expected because the results must converge to LEFM with an apparent fracture energy equal to the elastic energy density at fracture $f_t'/2E$ times the element width h. This means that $G_{F\text{apparent}} \propto h \to 0$ for $h \to 0$, and the strength tends to zero for infinite mesh refinement, which is obviously wrong.

On the contrary, Fig. 8.6.7c shows the result obtained following the energy balance method, which shows very little influence of the mesh size. The curves for the equivalent strength method closely follow the results of the energy method (Bažant and Cedolin 1980). To check that the results of the crack band analysis are not only mesh independent, but also accurate, it suffices to compare them with the prediction deduced from LEFM analysis, which can be obtained in closed form for this case (using the solutions for the center cracked panel and $K_{Ic} = \sqrt{EG_F}$). The comparison in Fig. 8.6.7d shows that the correspondence is excellent.

From the foregoing, we can conclude that if the mesh refinement is feasible so that $h < 2\ell_{ch}$, and if each element displays progressive softening, the classical finite element analysis suffices to get consistent results. For larger elements, either the equivalent strength approximation in Section 8.6.4 or the energy balance method just described, will give mesh-independent and accurate results.

Exercises

8.34 Consider a rectangular uniform mesh, with elements of dimensions h_x and h_y in the horizontal and vertical directions, respectively, and a crack band extending at an angle θ from the horizontal. Show that the mean or effective width for these elements is $h_b = h_x \sin\theta + h_y \cos\theta$. Show that for square meshes of size h, this reduces to $h\sqrt{2}\cos(45° - \theta)$ (Bažant 1985a).

8.35 Consider a rectangular uniform mesh, with elements of dimensions h_x and h_y in the horizontal and vertical directions, respectively, and a crack band extending at an angle θ from the horizontal. Show that the effective crack extension Δa_{eff} when a crack band extends by one element is given by $\Delta a_{\text{eff}} = h_x h_y/(h_x \sin\theta + h_y \cos\theta)$. Show that for square meshes of size h, this reduces to $h/[\sqrt{2}\cos(45° - \theta)]$ (Bažant 1985a).

8.36 To make a simple and fast analysis of a concrete gravity dam, a bidimensional model is generated, having approximately square elements with 3 m sides. The elastic modulus, strength, and fracture energy are estimated to be $E = 19$ GPa, $f_t' = 21$ MPa and $G_F = 92$ N/m. Determine the stress-strain curve with sudden strength drop that should be used.

8.7 Comparison of Crack Band and Cohesive Crack Approaches

During the 1980s, there was a lively debate on the relative merits and deficiencies of the crack band and cohesive crack representations of concrete fracture. There are four aspects of comparison to consider: **(1)** the ability to describe fracture that has localized in a single crack; **(2)** the ability to describe distributed fracture; **(3)** the ability to describe the micromechanical level; and **(4)** the possibility to predict fracture of arbitrary direction.

8.7.1 Localized fracture: Moot Point Computationally

First, we should recall that the cohesive crack model (i.e., the fictitious crack model of Hillerborg) and the crack band model yield about the same results (with differences of only about 1%, for $h = h_c$) if the stress-displacement relation in the fictitious crack model and the stress-strain relation in the crack band model are calibrated through Eq. (8.3.1), that is, if the crack opening displacement w is taken as the fracturing strain ε^f that is accumulated over the width h_c of the crack band. This equivalence, for example, follows from the fact that in the crack band model the results are almost insensitive to the choice of h_c, as well as h, and in the limit for $h \to 0$ the crack band model becomes identical to the fictitious crack model (provided that the fracture energy equivalence is preserved, of course).

Thus, the question "Discrete crack or crack band?" is moot from the viewpoint of computational modeling. The only point worthy of debate is computational effectiveness and convenience. In the cases where boundary integral methods can be applied, the use of the cohesive crack model can be computationally more efficient. When the general finite element method is used, these two models appear to be about equal when the fracture propagates along the mesh lines. However, the programming of the crack band model is generally easier, and that is why it has been preferred in the industry. For other fracture paths, there are various differences but special methods must be used for both models.

8.7.2 Nonlocalized Fracture: Third Parameter

As we recall from Section 7.1.3, if the shape of the tensile softening curve is fixed, then the cohesive crack (fictitious crack) model is defined by two material parameters, G_F and f_t'. The crack band model, on the other hand, is defined by three material parameters, G_F, f_t', and h_c. For the fictitious crack model, too, a third material parameter with dimensions of length, namely ℓ_{ch}, has been defined (see Section 7.1.3); however, this is a derived parameter, not an independent one, while h_c is an independent material parameter. Why does the crack band model, in its simplest form, have one more material parameter?

In answer to this question, we must first recall from Section 8.3.6 that, for localized fracture, the effect of the value of h_c on the results is almost negligible, provided, of course, that the softening part of the stress-strain diagram is adjusted so as to always yield the same fracture energy G_F for any value of

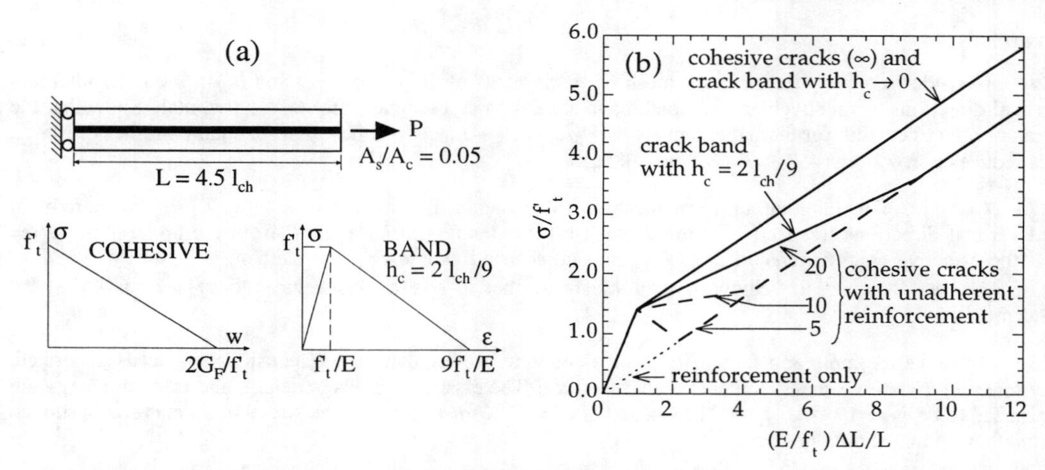

Figure 8.7.1 (a) Reinforced concrete bar and definition of linear cohesive crack model and the corresponding band model. (b) Resulting stress–mean strain curves for several possibilities (full lines are for adherent bar, dashed lines for unbonded bar).

h_c. Therefore, in the case of isolated fracture, the crack band model has, in effect, only two material parameters, G_F and f'_t, just like the fictitious crack model.

The value of crack band width, h_c, however, does make a difference in the case of nonlocalized fracture, that is, when densely spaced parallel cracks can form. Such situations, in which the strain softening state is stable against localization (in the macroscopic sense), can arise in various situations; for example, when there is sufficient reinforcement that can stabilize distributed cracking against localization (this occurs when the reinforcement is so strong that the tangential stiffness matrix of the composite of steel and cracked concrete is positive definite even though that of cracked concrete alone is not). Another possibility is the parallel cracks caused by drying shrinkage, which may be stabilized (against localization into isolated fractures) by the intact concrete in front of the cracks, due to shear stiffness of the material. The same situation can arise in bending, if the beam is sufficiently reinforced.

From these examples it transpires that the physical significance of h_c is not really the width of the actual cracking zone at the fracture front but the minimum possible spacing of parallel cohesive cracks, each of which is equivalent to one crack band. Since adjacent crack bands cannot overlap (the material cannot be cracked twice), the distance between the symmetry lines of the adjacent crack bands is at least h_c.

Now, is it necessary that the minimum possible spacing of parallel cracks be a material fracture parameter? In the early analysis of the problem it seemed, based on some examples, that it was so. For example, Bažant (1985b, 1986a) discussed the problem of a reinforced concrete bar loaded in centric tension, see Fig. 8.7.1, where the reinforcing bar represents five percent of the cross section area. In that case, smeared cracking is stable against localization. Bažant's (1985b, 1986a) interpretation was that the cohesive cracks could form at any spacing, s, and as far as the fictitious crack model is concerned, these cracks could be arbitrarily close. He concluded that the number of cracks per unit length can approach infinity while each crack can have a finite opening width w. But this would mean that, according to the fictitious crack model, the energy dissipated by the cracking of concrete in the bar could approach infinity – a paradoxical result. On the other hand, if there is such a condition as the minimum spacing s, then, of course, the energy dissipated by the cracking in the bar is bounded, even according to the computations based on the fictitious crack model.

However, Bažant's theoretical example can be reinterpreted in different terms, as done by Planas and Elices (1993b). For these authors, the cracks can be infinitely close while having a vanishingly small crack opening. This implies that no appreciable softening takes place, and thus the concrete deforms at $\sigma = f'_t$. Therefore, the heavily reinforced concrete behaves in an elastic-perfectly plastic fashion. Moreover, this solution, with infinitely many cracks, is consistent with Bažant's simplified analysis, which assumed that the cross-sections of the bar remained plane during the stretching process and that full bond existed between the bar and the concrete. However, it is easily shown that if two cracks form at any finite spacing

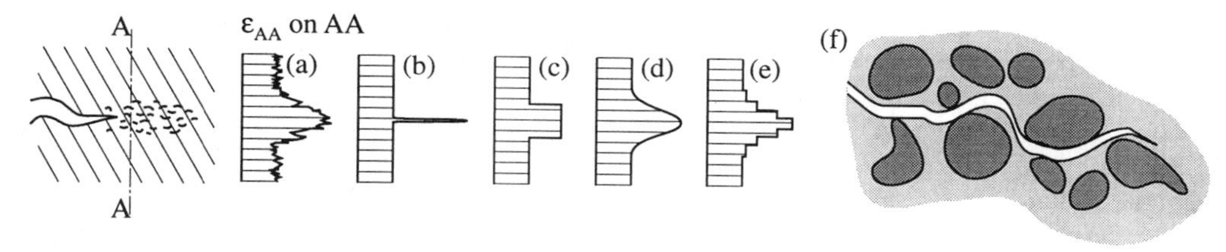

Figure 8.7.2 Strain distribution across the fracture process zone: (a) actual distribution; (b) cohesive crack approximation; (c) crack band model approximation; (d) nonlocal approximation; (e) finite element approximation (adapted from report of ACI Committee 446 1992). (f) Tortuous crack path.

s, the subsequent infinitesimal loading step causes the tensile strength to be exceeded at the middle point between them. Therefore, a third crack must form and we have cracks spaced at $s/2$. Repeating the reasoning *ad infinitum* it turns out that if the bar is bonded *and* if we assume plane cross sections to remain plane, then the only solution consistent with the cohesive crack model is that cohesive cracks form infinitely close to each other. However, we know by experiment that, at the end, a collection of discrete cracks appear, even for strongly reinforced bars. The key point in the explanation of this effect is that, upon localization, the sections cease to be plane, an effect that cannot be caught by the simple classical analysis.

It is worth to note that the solution based on the assumption of Planas and Elices does converge to the crack band solution for $h_c \to 0$, as shown in Fig. 8.7.1 by the full lines. The dashed lines correspond to localized crack solutions valid only if bond is neglected, in which case the reinforcement is not interacting with concrete except at the ends of the bar, and then the bifurcation analysis given in the first section indicates that both the cohesive crack model and the crack band model predict that a single crack will occur.

Therefore, the simplified analysis of this problem seems to show that, in accordance with Planas and Elices (1993b) hypothesis, it is still possible to use the cohesive crack model for fully distributed cracking in conjunction with an associated elastic-perfectly plastic Rankine model. The problem still remains, however, of determining when the fracture will localize. In their work on shrinkage, Planas and Elices made some special assumptions for the localization point, but pointed out that the actual localization must be determined by bifurcation analysis, which could be based on the principle of minimum second-order work (as done in the simple case of the series coupling model in Section 8.1.2 and justified thermodynamically in Bažant and Cedolin 1991, Sec. 10.1).

8.7.3 Relation to Micromechanics of Fracture

The normal microstrains across the fracture process zone may be distributed roughly as shown in Fig. 8.7.2a. The discrete fictitious crack model simplifies this random strain distribution as a Dirac delta function, Fig. 8.7.2b. The crack band model simplifies it as a rectangular strain distribution, Fig. 8.7.2c. The nonlocal continuum model, which we will discuss in Chapter 13, describes this strain distribution as a smooth bell-shaped profile across the crack band, as shown in Fig. 8.7.2d (cf. Bažant and Pijaudier-Cabot 1988). The finite element approximation to the nonlocal continuum simplifies the strain distribution in the form of several steps as in Fig. 8.7.2e. Now, which representation is more correct?

Among the simple distributions, i.e., those for the fictitious cohesive crack and the crack band (Figs. 8.7.2b-c) neither one is better or worse, as an approximation to the true distribution in Fig. 8.7.2a. Efforts have been made to physically observe the microcracks and strains throughout the fracture process zone. In optical microscopic observations, distributed cracking has not been seen in concrete (although it has been clearly observed in ceramics). The explanation might be that it is difficult to distinguish new very fine microcracks from the pre-existing ones, or that the microcracks on the fringes of the fracture process zone have extremely small openings while being extremely numerous and thus still contributing significantly to the overall relative displacement across the width of the fracture process zone.

With regard to the optical observations, it must be noted that fracture in concrete is normally highly tortuous, meandering to each side of the fracture axis by a distance approximately equal to the maximum aggregate size (Fig. 8.7.2f). Now, even if all the microcracking is concentrated on a line, but this line is highly tortuous, the fracture is represented by a straight line crack no better than by a crack band of width of about two aggregate sizes. So, even if cohesive cracks are a reality for concrete, one still cannot claim that a *straight* fictitious crack is a more realistic model than a crack band.

It is also pertinent to mention that measurements of the localizations of the acoustic emission during the fracture process in concrete (Labuz, Shah and Dowding 1985; Maji and Shah 1988) indicate, despite considerable scatter, that the locations of the emission sources are spread over a relatively wide band in the frontal region of fracture. This tends to support the crack band model. On the other hand, various measurements of strains on the surface, for example by interferometry (Cedolin, DeiPoli and Iori 1983, 1987) or by laser holography (Miller, Shah and Bjelkhagen 1988), indicate that very high strains are concentrated within a very narrow zone at the front of fracture. This might be better modeled by a cohesive crack than a crack band. It should be noted, though, that the fracture strains might be localized at the surface of concrete specimens to a greater extent than in the invisible interior, due to the wall effect as well as other effects.

In view of the foregoing three arguments, there seems to be no compelling reason for rejecting either the crack band model or the cohesive (fictitious) crack model. The choice seems to be a matter of convenience of analysis. When the fracture shape is known in advance, both formulations appear to be about equally convenient. However, if the shape of the fracture is unknown in advance, the crack band model might be more convenient.

8.7.4 Fracture of Arbitrary Direction

Finite element modeling of fracture is easy and accurate only if the fracture runs along the mesh lines. If the fracture path is known in advance, either from experience or some preliminary calculations, then it is possible to design the mesh to accommodate the fracture path as a smooth path along the mesh lines. In general practical problems, however, the fracture path is not known. It is one of the unknowns to be found by analysis. In such general situations, which need to be tackled in general purpose finite element programs, there are basically two possibilities to proceed: either to modify the finite element mesh each time the fracture advances, or to keep a fixed mesh and allow the fracture to have a rugged boundary and zig-zag shape (Fig. 8.7.3c). The second possibility is not possible with the fictitious crack model, since it would cause serious problems with interlocking in the case of shear. On the other hand, the first possibility, that is, remeshing, exists both for the discrete fictitious crack and for the crack band, although in practice it has so far been used apparently only for the cohesive crack approach. The automatic remeshing (Fig. 8.7.3a,b) at crack advance is not simple to program; however, various research groups have nevertheless succeeded in developing finite element programs which do just that (see Section 7.2.3). So far, however, the remeshing approach has not gained a wide popularity, due to the complexity of remeshing.

Although remeshing has not yet been used in conjunction with the crack band modeling of fracture, one must realize that this is a possibility which would be no more complex than remeshing for the cohesive crack. The algorithm for remeshing as developed by Ingraffea and co-workers (Section 7.2.3) could, no doubt, be easily adapted for remeshing in the case of crack bands (Fig. 8.7.3b).

As normally perceived, however, one of the advantages of the crack band approach is that fracture of arbitrary direction can be represented without any remeshing. The next element that undergoes cracking is decided on the basis of either the strength criterion (for the tensile sudden stress drop) or the stress-strain relation with strain softening. The zig-zag band is normally found to propagate roughly in the direction of previous cracking, however, it is possible under certain situations (for example a strip of concentrated shear stress) to obtain propagation of the band of cracked elements in a direction that is inclined to the direction of cracking in the elements. This represents shear fracture or mixed mode fracture (e.g., Bažant and Pfeiffer 1986).

However, as discussed in Section 8.6.2 and Section 8.6.3, there are certain errors associated with a zig-zag crack band. Due to the inevitable development of shear stresses on the planes parallel to the overall direction of the zig-zag band, there is some degree of interlocking, i.e., an increased resistance to shear, larger than that obtained with a smooth crack (cohesive crack) or a smooth band with remeshing (Fig. 8.7.3a,b). Although, to a large extent, the errors are tolerable compared to other errors involved in the analysis of fracture, remedies are needed to obtain accurate results.

The problem can be partially alleviated by using a square mesh in which each square is subdivided

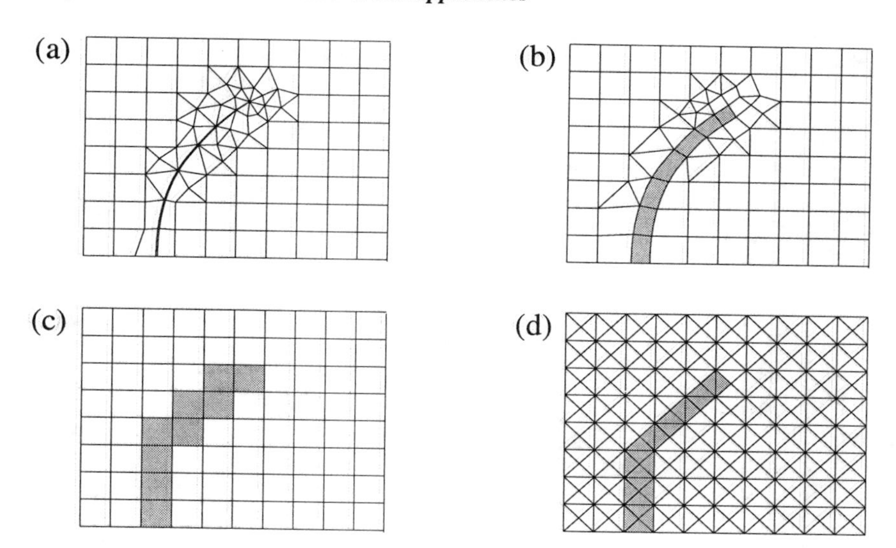

Figure 8.7.3 Description of fracture path inclined with respect to initial mesh lines: (a) cohesive crack with remeshing; (b) crack band with remeshing; (c) zig-zag crack band in a square mesh; (d) mesh allowing better representation of inclined fracture.

into four triangular elements. In this case, there are not only horizontal and vertical mesh lines, but also mesh lines at 45° inclinations (Fig. 8.7.3d). This kind of mesh, which allows approximating an arbitrary fracture propagation direction more closely, should always be used with the crack band model when the propagation direction is unknown.

A better remedy is to employ a nonlocal approach, to be discussed in the next chapter. This, however, brings the penalty that there must be several finite elements across the width of the crack band, which in turn necessitates either a more refined mesh in the fracture zone or an artificial increase of the width of the crack band (with the corresponding modification of the postpeak stress-strain relation). Probably the simplest solution is to use a standard mesh (of the type shown in Fig. 8.7.3d) to get the approximate crack path, and then remesh to fit the mesh lines to the crack path, as indicated in Fig. 8.7.3b, and recalculate.

9

Advanced Size Effect Analysis

As shown in Chapter 1, for many concrete specimens and structures, the size effect law of Bažant describes the influence of the structure size on its strength quite well. This simple law is, in general, applicable over size ranges 1 to 20 for notched plain concrete beams and brittle reinforced concrete elements. However, in some special cases, the size effect law can become too crude or inapplicable and a special analysis is required to get the right size effect. In this chapter we deal with such special cases.

First, we present recent generalizations of Bažant's size effect law that give good descriptions of the size effect over much wider size ranges, and include the failure at crack initiation from a smooth surface (Section 9.1). Next, we present the latest summary of the size effect predicted by the cohesive models for notched specimens (Section 9.2). Then, we analyze the applicability of the various models encountered in this book for two particular but interesting cases: the modulus of rupture (unnotched beams in bending; Section 9.3) and the cylinder splitting test (Brazilian test; Section 9.4). Next, we discuss compression failure due to propagation of splitting crack band (Section 9.5). We close the chapter with an introduction to the size effect in fracture of large floating plates of sea ice (Section 9.6).

9.1 Size Effect Law Refinements

The theory of size effect was presented in the initial chapters of this book in various simplified forms. Here, we present the latest reformulation that generalizes the size effect expression and extends it to situations other than those initially envisaged, particularly to structures without initial cracks or notches which fail at nonproportional crack extension.

The essential results were derived by Bažant in a series of increasingly refined works (Bažant 1983, 1984a, 1985b, 1987a, 1993a, 1995a,b, 1997a) based on an energetic approximation coupled with dimensional analysis. Here, we derive the same results following a more abstract analysis.

9.1.1 The Generalized Energy Balance Equation

In Section 2.1 we analyzed fracture of elastic materials from an energetic perspective. The basic idea of fracture mechanics is that, for fracture to happen, the available energy must be equal to the energy required to crack the material (Section 2.1.1). Of course, this basic idea still holds even when the material behavior at the crack tip is inelastic. Therefore, the quasi-static fracture condition (2.1.5) can be conceptually generalized to any other situation incorporating inelastic behavior as long as the concept of crack length a makes sense. Therefore, we understand a as an effective crack length (in a material that is generally not elastic).

It is logical to generalize the balance equation to

$$\tilde{\mathcal{G}} = \tilde{\mathcal{R}} \tag{9.1.1}$$

where the tildes denote the generalized energy release rate and the generalized crack growth resistance. Let us now see how the generalization can be implemented, having as a model the LEFM case.

In the LEFM limit, the crack growth resistance must be constant and equal to G_f:

$$\tilde{\mathcal{R}} = \mathcal{R} = G_f \tag{9.1.2}$$

For the energy release rate, the LEFM expression (2.3.12) indicates that:

$$\tilde{\mathcal{G}} = \mathcal{G} = \frac{\sigma_N^2}{E'} D g(\alpha) \tag{9.1.3}$$

where $g(\alpha)$ depends on the shape of the body and the load distribution, but not on the size.

The simplest way to take explicitly the nonlinear behavior of the material into account is to assume that the foregoing expressions depend on both the structural size D and the crack extension Δa. However, dimensional analysis requires that the dependence be on dimensionless ratios. Therefore, we assume that $\tilde{\mathcal{R}}$ and $\tilde{\mathcal{G}}$ depend on the ratios

$$\eta = \frac{\Delta a}{\ell}, \quad \theta = \frac{\ell}{D} \tag{9.1.4}$$

where ℓ is some material length. Different expressions can be adopted for this length, depending on the underlying fracture model, but it can always be reduced to an expression proportional to Irwin's approximation for the plastic zone size (Section 5.2) or to Hillerborg's characteristic size $E'G_f/f_t'^2$. For certain applications, it may be convenient to select ℓ to coincide with c_f, the critical effective crack extension for infinite size, which is known to be proportional to Irwin's and Hillerborg's lengths.

Thus, we write the generalized energy release and crack resistance as

$$\tilde{\mathcal{R}} = G_f \, \tilde{r}(\alpha_0, \eta, \theta) \tag{9.1.5}$$

$$\tilde{\mathcal{G}} = \frac{\sigma_N^2}{E'} D \tilde{g}(\alpha_0, \eta, \theta) \tag{9.1.6}$$

Note that since $\alpha = \alpha_0 + \eta\theta$, the dependence on α can be (and has been) replaced by the dependence on α_0. This explicitly shows the influence of the initial geometry.

The foregoing formulation is a generalization of the R-curve concept (Section 5.6), and the peak load condition is obtained by requiring that for fixed D (or θ) and fixed initial notch depth (fixed α_0) the $\tilde{\mathcal{G}}$-η and the $\tilde{\mathcal{R}}$-η curves be tangent, i.e., $\left[\partial\tilde{\mathcal{G}}/\partial\eta\right]_{\sigma_N} = \partial\tilde{\mathcal{R}}/\partial\eta$. Dividing this equation by $\tilde{\mathcal{G}} = \tilde{\mathcal{R}}$, we have

$$\frac{1}{\tilde{\mathcal{G}}}\left[\frac{\partial\tilde{\mathcal{G}}}{\partial\eta}\right]_{\sigma_N} = \frac{1}{\tilde{\mathcal{R}}}\frac{\partial\tilde{\mathcal{R}}}{\partial\eta} \tag{9.1.7}$$

Substituting (9.1.5) and (9.1.6) we see that the tangency condition can be written as

$$\frac{1}{\tilde{g}(\alpha_0, \eta, \theta)} \frac{\partial\tilde{g}(\alpha_0, \eta, \theta)}{\partial\eta} = \frac{1}{\tilde{r}(\alpha_0, \eta, \theta)} \frac{\partial\tilde{r}(\alpha_0, \eta, \theta)}{\partial\eta} = \tag{9.1.8}$$

which can be solved, in principle, for η to get its critical value

$$\eta = \eta_c(\alpha_0, \theta) \tag{9.1.9}$$

Substituting this value back into the condition $\tilde{\mathcal{G}} = \tilde{\mathcal{R}}$ and solving for the nominal strength, we get

$$\sigma_{Nu} = \frac{\sqrt{E'G_f}}{\sqrt{D\, h(\alpha_0, \theta)}} \tag{9.1.10}$$

where

$$h(\alpha_0, \theta) = \frac{\tilde{g}[\alpha_0, \eta_c(\alpha_0, \theta), \theta]}{\tilde{r}[\alpha_0, \eta_c(\alpha_0, \theta), \theta]} \tag{9.1.11}$$

Equation (9.1.10), derived by Bažant (1995a, 1997a) appears to be the most general expression for the size effect. Note, however, that the function $h(\alpha_0, \theta)$ depends on hidden parameters: the geometrical ratios (shape) and, maybe, the ratios of material constants (e.g., shape of the softening curve).

9.1.2 Asymptotic Analysis for Large Sizes

For very large sizes $\theta \rightarrow 0$. So, assuming that $h(\alpha_0, \theta)$ can be expanded in a power series of θ, we can use a linear approximation of $h(\alpha_0, \theta)$:

$$h(\alpha_0, \theta) = h_0 + h_1\theta + o(\theta) \qquad \text{for } \theta \rightarrow 0 \tag{9.1.12}$$

where

$$h_0 \equiv h(\alpha_0, 0)\ , \qquad h_1 \equiv \left. \frac{\partial h(\alpha_0, \theta)}{\partial \theta} \right|_{\theta=0} \tag{9.1.13}$$

and $o(\theta)$ is a function vanishing faster than its argument, i.e., satisfying

$$\lim_{\theta \rightarrow 0} \frac{o(\theta)}{\theta} = 0 \tag{9.1.14}$$

Inserting the asymptotic linear expression for $h(\alpha_0, \theta)$ into the general size effect equation, we get

$$\sigma_{Nu} = \frac{\sqrt{E'G_f}}{\sqrt{h_0 D + h_1 \ell}}\ [1 + o(\theta)] \qquad \text{for } \theta \rightarrow 0 \tag{9.1.15}$$

Neglecting the term $o(\theta)$ for large sizes, we can bring this equation to the form of Bažant's size effect law

$$\sigma_{Nu} = \frac{B f'_t}{\sqrt{1 + D/D_0}} \tag{9.1.16}$$

with the notation

$$B f'_t = \sqrt{\frac{E'G_f}{h_1 \ell}}\ , \qquad D_0 = \frac{h_1}{h_0}\ell \tag{9.1.17}$$

Note that B and D_0 depend both on geometrical properties (shape of the structure and loading arrangement) and on material properties (G_f, ℓ and other hidden dimensionless properties such as the shape of the softening curve). Next we try to relate the parameters B and D_0, hence, also h_0 and h_1, to the classical fracture properties.

9.1.3 Matching to the Effective Crack Model

For infinite size, in which the relative size of the fracture process zone vanishes, the foregoing asymptotic size effect must match the LEFM size effect deduced from $\mathcal{G} = G_f$. However, such limit is reached asymptotically, and for large, but not strictly infinite sizes, the behavior tends to be that of the simplest model —Irwin's effective crack model. This has been rigorously proved for cohesive crack models and for R-curve models (Planas and Elices 1992a, 1990b; Elices and Planas 1992). In this model, the energy release rate can be approximated by the elastic energy release rate (9.1.3), and the ultimate load occurs when the crack has extended by a fixed length c_f that is a material property. Thus, the peak load is given by

$$\sigma_{Nu} = \frac{\sqrt{E'G_f}}{\sqrt{D g(\alpha_0 + c_f/D)}} \tag{9.1.18}$$

It can be shown that this simple asymptotic behavior is accurate up to and including first order in c_f/D for cohesive crack models and R-curve models (Planas and Elices 1991a, 1992a, 1993a; Elices and Planas 1992). Postulating that this is generally so, we take two terms of the Taylor series expansion of $g(\alpha_0 + c_f/D)$ and get

$$\sigma_{Nu} = \frac{\sqrt{E'G_f}}{\sqrt{D[g_0 + g'_0 c_f/D]}}\ [1 + o(c_f/D)] \qquad \text{for } c_f/D \rightarrow 0 \tag{9.1.19}$$

where g_0 and g_0' stand for the values of $g(\alpha)$ and its first derivative $g'(\alpha)$ for the initial crack or notch length $\alpha = \alpha_0$. Now, identification with (9.1.15) is straightforward and we get

$$h_0 = g_0 \; , \qquad \ell = g_0' c_f / h_1 = \kappa c_f \tag{9.1.20}$$

Where $\kappa = g_0'/h_1$ must, at this time be calibrated empirically. From this and (9.1.17), the relationships between Bažant's size effect parameters and the fracture parameters G_f and c_f are easily and rigorously established. As expected, the results coincide with those derived in Section 6.1.1, Eq. (6.1.4).

9.1.4 Asymptotic Formula for Small Sizes and Its Asymptotic Matching with Large Sizes

For small sizes, we have $\theta \longrightarrow \infty$ and the asymptotic series expansion requires a different treatment. We now set

$$\xi = \frac{1}{\theta} = \frac{D}{\ell} \tag{9.1.21}$$

and so (9.1.10) is rewritten as

$$\sigma_{Nu} = \frac{\sqrt{E'G_f}}{\sqrt{\ell \, p(\alpha_0, \xi)}} \tag{9.1.22}$$

where

$$p(\alpha_0, \xi) = \xi \; h(\alpha_0, 1/\xi) \tag{9.1.23}$$

Now the point is to analyze the behavior of $p(\alpha_0, \xi)$ near $\xi = 0$. Bažant (1995b, 1997a) assumes that this function is smooth enough to use a linear (Taylor) approximation at $\xi = 0$. Then the asymptotic size effect takes the form

$$\sigma_{Nu} = \frac{\sqrt{E'G_f}}{\sqrt{\ell \; [p_0 + p_1 \xi]}} \; [1 + o(\xi)] \qquad \text{for } \xi \to 0 \tag{9.1.24}$$

where

$$p_0 = p(\alpha_0, 0) = \lim_{\xi \to 0} [\xi \; h(\alpha_0, 1/\xi)] \tag{9.1.25}$$

$$p_1 = \left. \frac{\partial p(\alpha_0, \xi)}{\partial \xi} \right|_{\xi=0} = \lim_{\xi \to 0} \left[h(\alpha_0, 1/\xi) - \frac{1}{\xi} \left. \frac{\partial h(\alpha_0, \theta)}{\partial \theta} \right|_{\theta = 1/\xi} \right] \tag{9.1.26}$$

The foregoing asymptotic expression can, for small values of ξ, be written again in Bažant's form

$$\sigma_{Nu} = \frac{B_1 f_t'}{\sqrt{1 + D/D_1}} \tag{9.1.27}$$

where

$$B_1 f_t' = \sqrt{\frac{E'G_f}{p_0 \ell}} \; , \qquad D_1 = \frac{p_0}{p_1} \ell \tag{9.1.28}$$

Note that $B_1 f_t'$ is the strength for very small size, which is expected to follow from plasticity analysis.

Bažant (1995b, 1997a) assumes that $B_1 = B_0$ and $D_1 = D_0$. This is, of course, a simplification, since, in general, (9.1.17) and (9.1.28) need not coincide. Based on this simplification, Bažant further concluded that since the small and large size asymptotics were amenable to the same expression, such expression also provided a formula for intermediate cases. This kind of approximation is akin to the theory of intermediate asymptotics or matched asymptotics, which has been enormously successful in many branches of physics, for example, the boundary layer theory of fluid mechanics (Hinch 1991; Bender and Orszag 1978; Barenblatt 1979). Thus, the size effect law of Bažant may be approximately regarded as an asymptotic matching formula, which explains why this law has such a broad applicability.

Despite the reasonable assumptions made in the foregoing derivation of the small size asymptotics, recent results by Planas, Guinea and Elices (1997) show that for certain types of softening behavior of the cohesive crack model, the small size asymptotics differ from that in (9.1.23). Indeed, for cohesive cracks with initially linear softening, the asymptotic behavior for $p(\alpha_0, \xi)$ includes half-integer powers of ξ (see Section 9.2.3). Although this feature is not directly apparent in Bažant's (1995b, 1997a) work, it is indirectly included in his extended size effect law that we introduce next.

9.1.5 Asymptotic Aspects of Bažant's Extended Size Effect Law

In view of the inevitable random scatter, Bažant's size effect law fits test data as well as might be desired for a range up to about 1:20. Computer results, however, do not have random scatter, and so small deviations become noticeable. Bažant's size effect law can fit reasonably well the size effect predictions of the cohesive crack model with initially linear softening for a range 1:4 (Bažant 1985d). However, when the range is extended, the fit (of data calculated by finite elements) loses accuracy. To cope with this fact, Bažant (1985b) extended his size effect to include noninteger powers of D.

In his latest formulation, Bažant (1995b, 1997a) followed a development parallel to that in the preceding sections, but he made the following assumptions:

1. For the large size range, he assumed that function $h(\alpha_0, \theta)$ can be written as

$$h(\alpha_0, \theta) = \left[\hat{h}(\alpha_0, \zeta)\right]^{1/r}, \quad \zeta = \theta^r = \left(\frac{\ell}{D}\right)^r \tag{9.1.29}$$

where r is some constant.

2. He further assumed that function $\hat{h}(\alpha_0, \zeta)$ accepts the MacLaurin series expansion:

$$\hat{h}(\alpha_0, \zeta) = \hat{h}_0 + \hat{h}_2\zeta + o(\zeta) \tag{9.1.30}$$

3. For the small size asymptotics, he assumed that the function $p(\alpha_0, \xi)$ can be written as

$$p(\alpha_0, \xi) = \left[\hat{p}(\alpha_0, \chi)\right]^{1/r}, \quad \chi = \xi^r = \left(\frac{D}{\ell}\right)^r \tag{9.1.31}$$

where r is the *same* constant exponent as that which appears in the large size asymptotics.

4. The function $\hat{p}(\alpha_0, \chi)$ accepts a MacLaurin series expansion:

$$\hat{p}(\alpha_0, \chi) = \hat{p}_0 + \hat{p}_1\chi + o(\chi) \tag{9.1.32}$$

5. The coefficients of the MacLaurin expansions of $\hat{h}$ and $\hat{p}$ satisfy

$$\hat{h}_1 = \hat{p}_0, \qquad \hat{h}_0 = \hat{p}_1 \tag{9.1.33}$$

With these hypotheses, it is easy to show that the size effect law can be written in the form first proposed by Bažant (1985b):

$$\sigma_{Nu} = \sigma_P \left[1 + \left(\frac{D}{D_0}\right)^r\right]^{-1/2r} \tag{9.1.34}$$

where σ_P and D_0 are constants characterizing the material and the geometrical shape of the structure (with the geometry of the load). Fig. 9.1.1 shows the influence of parameter r on the shape of the size effect law: the larger the value of r, the faster the approach to the asymptote.

It is interesting to note that for the usual experimental range and scatter, the fits are good with the classical value $r = 1$. However, to fit finite element computations for notched beams using the cohesive crack model over a wide range of sizes, values of r close to 0.5 are required (Bažant 1985b). The aforementioned results of Planas, Guinea and Elices (1997) regarding the asymptotic behavior for small sizes explain why this is so; see Section 9.2.3.

When the extended size effect is used, the fracture energy is easy to obtain by identification with the effective crack model, as done for the classical effective crack extension. The result is identical to that obtained for the classical size effect with $r = 1$. However, the value of c_f is shown to be infinite; this implies that the R-curve associated with this kind of size effect approaches G_f asymptotically.

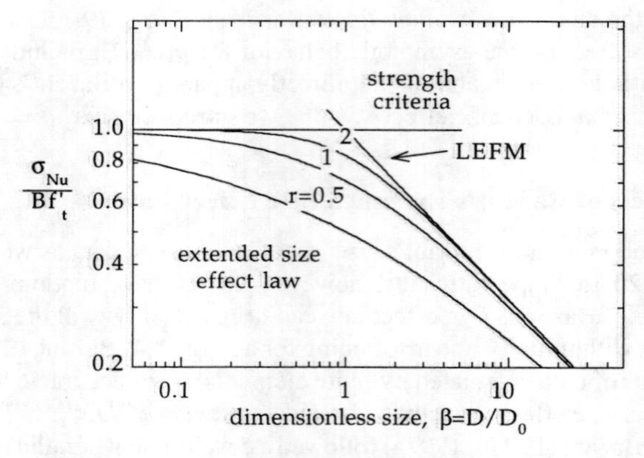

Figure 9.1.1 Extended Bažant's size effect laws.

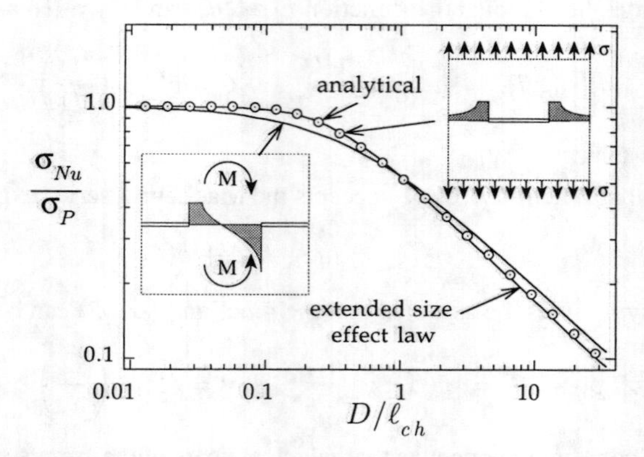

Figure 9.1.2 Analytical solutions and their approximation by the extended size effect law: for the center-cracked panel (right) the best fit is for $r = 1.99$, for the doubly notched panel subjected to bending (left) the fit is exact, with $r = 1$.

On the other hand, if the softening curve in the cohesive crack model starts with a long yield plateau, or if the geometry is clearly different from bending, it appears that widely different values of r can be found, each giving an excellent fit to numerical results. This is illustrated in Fig. 9.1.2 which shows the size effect curves determined analytically for two geometries when the softening is rectangular (Dugdale model with a cutoff). The first geometry is the center-cracked panel subjected to remote uniaxial stress, whose size effect curve was calculated in Chapter 7 as Eq. (7.5.28). This analytical curve can be accurately approximated by (9.1.34) (with $D \equiv a_0$) for $\sigma_P = f'_t, r = 1.99$, and $D_0 = 0.293\ell_{ch}$. The second geometry is the infinite panel with two symmetric semi-infinite cracks (see the inset in Fig. 9.1.2). This problem was analytically solved by Planas and Elices (1991c), who found that the size effect is *exactly* given by (9.1.34) with $\sigma_P = 3f'_t, r = 1$, and $D_0 = (\pi/8)\ell_{ch}$ when D is identified with the initial ligament and $\sigma_N = 6M/BD^2$.

9.1.6 Size Effect for Failures at Crack Initiation from Smooth Surface

Consider now the failure of quasibrittle structures that have no notches and reach the maximum load when a cohesive crack initiates from a smooth surface. This happens, for example, in the bending tests of plain concrete beams, which will be analyzed in detail in Section 9.3. The essential fact that comes both from experiment and from theoretical considerations is that in such a case the large size asymptotics is different from that obtained in the previous analysis.

The basic result is that, for very large sizes, the strength approaches a horizontal asymptote dictated by the classical strength of materials theory for elastic-brittle materials (materials that fail completely when the stress at any point reaches the tensile strength). This kind of behavior was analyzed by Bažant and Li (1995c) based on the idea that a densely microcracked (quasi-plastic) zone developed in a boundary layer of the specimen and that the peak load was reached when the thickness of such layer reached a critical constant value ℓ_f. Their basic result was that the asymptotic size effect is given by

$$\sigma_{Nu} = \sigma_B \left(1 + 2\frac{\ell_f}{D}\right) + o(\ell_f/D) \tag{9.1.35}$$

where σ_B is the strength dictated by the elastic-brittle analysis, and ℓ_f is the critical thickness of the microcracked boundary layer.

In a more recent work (Bažant 1995b), the foregoing asymptotic law was derived following the line of reasoning of the preceding sections. Thus, starting from the general equation (9.1.10), it is noted that $h(0,0)$ must vanish, because the energy release rate of a crack of zero length vanishes and $\tilde{\mathcal{G}} = 0$. Hence, for $\alpha_0 = 0$, the first term of the MacLaurin expansion of h in powers of θ identically vanishes. Then, to keep a two-term approximation, the quadratic term must be included, and so

$$h(0,\theta) = h_{01}\theta + h_{02}\theta^2 + o(\theta^2) \tag{9.1.36}$$

with the notation

$$h_{01} = \left.\frac{\partial h(0,\theta)}{\partial \theta}\right|_{\theta=0}, \qquad h_{02} = \left.\frac{\partial^2 h(0,\theta)}{2\partial \theta^2}\right|_{\theta=0} \tag{9.1.37}$$

From this, the size effect equation turns out to be

$$\sigma_{Nu} = \frac{\sqrt{E'G_f}}{\sqrt{h_{01}\ell + h_{02}\ell\theta}} \tag{9.1.38}$$

which can be rewritten as

$$\sigma_{Nu} = \frac{\sigma_B}{\sqrt{1 - 4\ell_f/D}} \tag{9.1.39}$$

where

$$\sigma_B = \sqrt{\frac{E'G_f}{h_{01}\ell}}, \qquad \ell_f = \frac{-h_{02}}{4h_{01}}\ell \tag{9.1.40}$$

where it is assumed that h_{01} and $-h_{02}$ are both positive. Noting that, for $D \to \infty$, $(1 - 4\ell_f/D)^{-1/2} \approx 1 + 2\ell_f/D$, one concludes that the asymptotic trend of this equation coincides with the Bažant-Li equation (9.1.35). As a simplified approximation, Bažant (1995b) assumes that $h_{01} \propto g'(0)$ and $h_{02} \propto g''(0)$ and so one can use the following simplified expression for ℓ_f, instead of the second of (9.1.40):

$$\ell_f \approx \frac{-g''(0)}{4g'(0)}\ell \tag{9.1.41}$$

where, for a crack starting from a smooth surface, $g'(0) = 1.12^2\pi$ regardless of the structure shape if σ_N is taken to be the elastically computed stress parallel to the surface at the point of cracking.

Both of the foregoing asymptotic expressions have limitations: for vanishing size D, the Bažant-Li equation becomes infinite, and (9.1.39) becomes imaginary. Even though a size D smaller than about 3 times the size of material inhomogeneity (aggregate size) makes no physical sense, it may be preferable to avoid these abnormal behaviors for $D \to 0$ while keeping the right asymptotic trend for $D \to \infty$ and nearly the same for not too small D. Bažant (1995b, 1997a) proposes to use the following rational approximation so that the size effect function become finite for $D \to 0$:

$$\sigma_{Nu} = \sigma_B \left(1 + \frac{1}{\gamma + D/(2\ell_f)}\right) \tag{9.1.42}$$

Here γ is a small positive constant that must be calibrated experimentally.

At this point it must be noted that an asymptotic equation having exactly the same structure as the Bažant-Li formula (9.1.35) was proposed by Gustafsson (1985) in the frame of the cohesive crack model, and that formulas identical to (9.1.42) were proposed by Gustafsson (1985), Uchida, Rokugo and Koyanagi (1992) and, more recently, by Planas, Guinea and Elices (1995). We will discuss such approaches and their generalization in Section 9.3 dealing with the size effect on the modulus of rupture.

9.1.7 Universal Size Effect Law for Cracked and Uncracked Structures

Bažant (1995b, 1997a) derived a general asymptotic matching formula that satisfies all the asymptotic properties analyzed in the foregoing, i.e., the large size and small size asymptotic properties for both a large notch and no notch. Its expression is as follows:

$$\sigma_{Nu} = \frac{Bf_t'}{\sqrt{1 + D/D_0}}\left[1 + \frac{2\ell_f D_0}{(2\gamma\ell_f + D)(D_0 + D)}\right] \tag{9.1.43}$$

in which the first term is the classical Bažant law for notched specimens, while the second term contains the formula for unnotched specimens. The values of Bf_t' and D_0 are identical to the classical values:

$$Bf_t' = \sqrt{\frac{E'G_f}{c_f g_0'}}, \qquad D_0 = \frac{c_f g_0'}{g_0} \tag{9.1.44}$$

Parameter ℓ_f, however, must now be generalized, by contrast to (9.1.41), as a variable:

$$\ell_f = \frac{\langle -g''(\alpha_0)\rangle}{4g'(0)}\ell = \frac{\langle -g''(\alpha_0)\rangle}{4g'(0)}\kappa c_f \tag{9.1.45}$$

where the angle (Macauley) brackets denote the positive part, and the second equality follows from (9.1.20).

To prove the asymptotic properties, first note that $g_0 = g(\alpha_0) = 0$ for $\alpha_0 = 0$. Therefore, according to the second equation in (9.1.44), the parameter D_0 tends to infinity for unnotched specimens, and so, for this case, the result becomes identical to (9.1.42) if we express the elastic-brittle limit in the form

$$\sigma_B = Bf_t' \tag{9.1.46}$$

The constant γ determines the plastic limit σ_P, i.e., the limit for $D \to 0$:

$$\sigma_P = Bf_t'(1 + 1/\gamma) \tag{9.1.47}$$

Eq. (9.1.43) can be generalized, while keeping the asymptotic behavior untouched, by introducing exponents r and s as follows (Bažant 1995b, 1997a):

$$\sigma_{Nu} = Bf_t'\left[1 + \left(\frac{D}{D_0}\right)^r\right]^{-1/2r}\left[1 + s\frac{2\ell_f D_0}{(2\gamma\ell_f + D)(D_0 + D)}\right]^{1/s} \tag{9.1.48}$$

This formula was called *universal size effect law* by Bažant (1995b, 1997a). Although the parameters r and s are not determined by the theory and any value is, in principle, possible, Bažant suggest that the values $r = s = 1$ seem to be the most appropriate, which corresponds to the simplest formula (9.1.43). Fig. 9.1.3 shows the foregoing expression (for $r = s = 1$) plotted as a surface in which the logarithm of the size is drawn along the X axis, the logarithm of the initial relative notch depth (which is an effect of structure geometry) along the Y axis, and the logarithm of the nominal strength along the Z axis. The surface clearly shows how the inclined asymptote for large sizes and notched structures becomes horizontal when the structure has no notch nor traction-free crack.

Remark: The sudden change of slope seen in Fig. 9.1.3 is caused by the use of the Macauley brackets in (9.1.45). The surface could be made smooth α_0 by replacing $\langle -g''(\alpha_0)\rangle$ in (9.1.45) with a suitable smooth function that nearly coincides with $\langle -g''(\alpha_0)\rangle$ for large $|g''(\alpha_0)|$. △

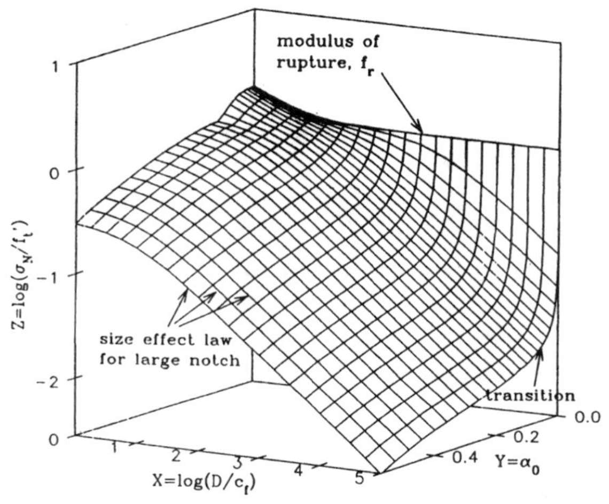

Figure 9.1.3 Surface of universal size effect law amalgamating the size effects for structures with large cracks (or notches) and structures failing at a crack initiation from a smooth surface (modulus of rupture for bent beams) (from Bažant 1995b).

9.1.8 Asymptotic Scaling Law for Many Loads

The foregoing asymptotic theory may be easily extended to the case of many loads P_i, characterized by nominal stresses $\sigma_i = P_i/bD$. Here we use the simplest approach based on the effective crack model, as done in Section 6.1.1. We simply write that at peak load the crack has extended by c_f and that the corresponding stress intensity factor is equal to K_{Ic} or to $\sqrt{E'G_f}$. The stress intensity factors of the individual loads, K_{Ii}, are additive. Therefore, by superposition, $K_I = \sum_i \sigma_{Ni}\sqrt{Dg_i(\alpha)}$, where $g_i(\alpha) = k_i^2(\alpha)$ are the shape factors for the energy release rate of the ith load, see Eq. (2.3.12). Setting then $\alpha = \alpha_0 + c_f/D$ and $K_I = K_{Ic}$ we get the peak load condition for the many loads case:

$$\sum_i \sigma_{Ni}\sqrt{Dg_i\left(\alpha_0 + \frac{c_f}{D}\right)} = \sqrt{E'G_f} \tag{9.1.49}$$

Expanding functions $g_i(\alpha_0 + c_f/D)$ into a Taylor series in terms of c_f/D about point $\theta = 0$, we obtain the relation $g_i(\alpha_0 + c_f/D) = g_i(\alpha_0) + g'(\alpha_i)c_f/D + \frac{1}{2}g_i''(\alpha_0)(c_f/D)^2 + ...$ For the case of a large crack, this series may be truncated after the second (linear) term. Furthermore, we may set $\sigma_{Ni} = \mu\sigma_{Di}$ where σ_{Di} are the given (fixed) design loads and μ = safety factor. After rearrangements:

$$\mu = \sqrt{EG_f}\,(\rho_1\sigma_{D1} + \rho_2\sigma_{D2} + ... + \rho_n\sigma_{Dn})^{-1/2} \tag{9.1.50}$$

$$\rho_i = \sqrt{g_i(\alpha_0)D + g_i'(\alpha_0)c_f} \tag{9.1.51}$$

This equation gives the size effect as well as the geometry effect for the case of a large crack. At the same time, it may be regarded as the interaction diagram (failure envelope) for many loads.

For the case of macroscopic crack initiation from a smooth surface, we have $g_i(0) = 0$. Therefore, same as for one load, the series cannot be truncated after the linear term. We may truncate them after the quadratic terms. A similar procedure then leads for μ to the same expression as (9.1.50), but with

$$\rho_i = \sqrt{g_i'(0)c_f + \frac{1}{2}g_i''(0)\frac{c_f^2}{D}} \tag{9.1.52}$$

Equations (9.1.50) and (9.1.52) represent the large-size asymptotic approximations of size effect. Small-size asymptotic approximations can be derived similarly, replacing the variable c_f/D with D/c_f. Similar to the case of one load, it is possible to find a universal size effect law for many loads that has the correct asymptotic properties for large and small sizes and large cracks or crack initiation. It is analogous to (9.1.48) and may be written in the form of (9.1.50) but with

$$\rho_i = r_i \left[1 + \left(\frac{D_{0i}}{D}\right)^r\right]^{1/2r} \left\{1 + s\left[\left(\bar{\eta} + \frac{D}{D_{bi}}\right)\left(1 + \frac{D}{D_{0i}}\right)\right]^{-1}\right\}^{-1/s} \tag{9.1.53}$$

$$r_i = [c_f g_i'(\alpha_0)]^{-1/2}, \qquad D_{0i} = c_f g_i'(\alpha_0)/g_i(\alpha_0), \qquad D_{bi} = c_f\langle -g_i''(\alpha_0)\rangle/4g_i'(\alpha_0) \tag{9.1.54}$$

Here r, s and $\bar{\eta}$ are empirical constants whose values may probably be taken as 1 for most practical purposes.

9.1.9 Asymptotic Scaling Law for a Crack with Residual Bridging Stress

In the case of compression fracture due to lateral propagation of a band of axial splitting cracks (Section 9.5), a residual stress given by the critical stress for internal buckling in the band remains. Lumping the fracturing strains in the band into a line, one may approximately treat such a fracture as a line crack in which interpenetration of the opposite faces is allowed and the softening compressive stress-displacement law terminates with a plateau of residual constant stress σ_Y. Likewise, a constant residual stress σ_Y may be assumed for characterizing the tensile stress-displacement law for a crack in fiber-reinforced composite (e.g., concrete).

The preceding asymptotic formulae for the case of many loads can be applied to this case because the uniform pressure σ_Y along the crack can be regarded as one of two loads on the structure. We write the stress intensity factors due to the applied load P and the uniform crack pressure σ_Y as $K_I^2 = \sigma_N^2 Dg(\alpha)$, and $K_I^2 = \sigma_Y^2\gamma(\alpha)$, respectively, where g and γ are dimensionless functions taking the role of g_1 and g_2 in the preceding formulae. In this manner, (9.1.50) and (9.1.51) yield, after rearrangements, the following formula for the size effect (and shape effect) in the case of a large crack:

$$\sigma_N = \frac{\sqrt{EG_f} + \sigma_Y\sqrt{\gamma'(\alpha_0)c_f + \gamma(\alpha_0)D}}{\sqrt{g'(\alpha_0)c_f + g(\alpha_0)D}} \tag{9.1.55}$$

For geometrically similar structures and size-independent α_0, this formula yields a size effect curve that terminates, in the doubly logarithmic scale, with a horizontal asymptote.

In the case of initiation of a crack with uniform residual stress σ_Y, equations (9.1.50) and (9.1.52) can be reduced to the following size (and shape) effect formula:

$$\sigma_N = \frac{\sqrt{EG_f} + \sigma_Y\sqrt{\gamma'(0)c_f + \frac{1}{2}\gamma''(0)\frac{c_f^2}{D}}}{\sqrt{g'(0c_f + \frac{1}{2}g''(0)\frac{c_f^2}{D}}} \tag{9.1.56}$$

whose double logarithmic plot also terminates with a horizontal asymptote.

The last two formulae can also be applied to the compression kink bands in wood or in composites reinforced by parallel fibers. This problem has so far been treated by elasto-plasticity, and solutions of failure loads which give good agreement with the existing test data have been presented (Rosen 1965, Argon 1972, Budianski 1983, Budianski et al. 1997, Budianski and Fleck 1994, Kyriakides et al. 1995, Christensen and DeTeresa 1997). Measurements of the size effect over a broad size range, however, appear to be unavailable, and there is a reason to suspect that a size effect exists. The reason is that shear slip and fracture along the fibers in the kink band probably exhibits softening, i.e., a gradual reduction of the shear stress to some final asymptotic value, and that the kink band does not form simultaneously throughout the cross section but has a front that propagates, in the manner of the band of parallel compression splitting cracks (Section 9.5).

Exercises

9.1 Show that for $\tilde{\mathcal{G}}$ to reach the LEFM limit for $D \to \infty$ ($\theta = 0$), the following condition must hold for $\tilde{g}$:

$$\tilde{g}(\alpha_0, \eta, 0) = g(\alpha_0) \quad \text{for } \eta < \infty \tag{9.1.57}$$

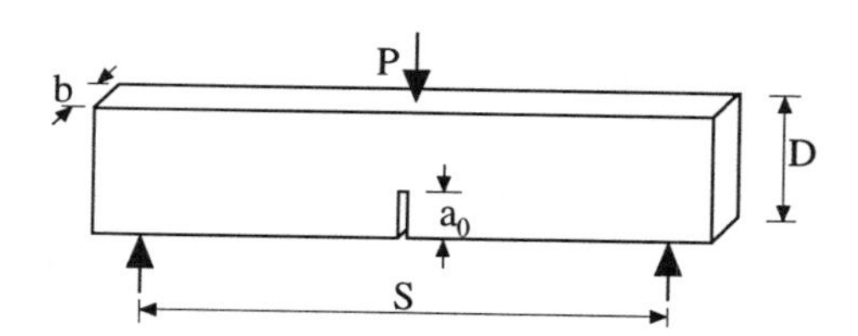

Figure 9.2.1 Precracked structure in mode I.

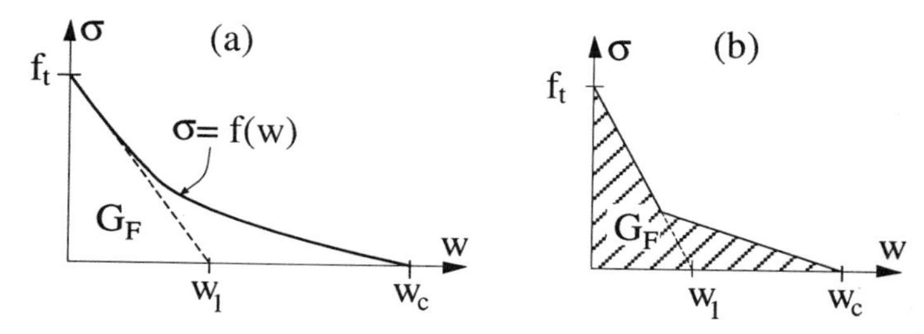

Figure 9.2.2 (a) Nonlinear softening curve and initial linear approximation. (b) Bilinear approximations of the softening curve for concrete.

9.2 Show that, in order for $\tilde{\mathcal{R}}$ to approach, for very large sizes, an R-curve independent of size and geometry such that the value G_f for $\Delta a = c_f$ be also reached, $\tilde{r}$ must satisfy the following conditions:

$$\tilde{r}(\alpha_0, \eta, 0) = r(\eta) \ , \quad r(\eta) < 1 \text{ for } \eta < 1, \quad \text{and} \quad r(\eta) = 1 \text{ for } \eta \geq 1 \tag{9.1.58}$$

9.3 Show that, when $r = 0.5$, the extended Bažant size effect law (9.1.34) for very small sizes tends to behave as

$$\sigma_{Nu} = \frac{\sigma_P}{1 + \sqrt{D/D_0}} \tag{9.1.59}$$

9.2 Size Effect in Notched Structures Based on Cohesive Crack Models

In this section, we focus attention on size effect implied by the cohesive crack model for notched (or precracked) structures, subjected to mode I loading which is exemplified by the three-point bend beam in Fig. 9.2.1.

The analysis summarizes the most recent results obtained with concrete in mind, although they are applicable to all materials that have a softening curve whose initial part can be approximated by a descending straight line, as depicted in Fig. 9.2.2a. Although the softening curve is recognized to be usually smooth, it is generally accepted (see Section 7.2.1) that for concrete, a bilinear approximation, as shown in Fig. 9.2.2b, can describe the results within engineering tolerances (Petersson 1981; Gustafsson 1985; Roelfstra and Wittmann 1986; Wittmann, Rokugo et al. 1987; Rokugo et al. 1989; Mihashi 1992; Guinea, Planas and Elices 1994b; Uchida et al. 1995). Usually the first segment of the bilinear softening extends down to stresses as low as $1/3$ to $1/5$ of f_t'; see Section 7.2.1.

9.2.1 The General Size Effect Equation

In the work of Hillerborg and collaborators (Petersson 1981; Gustafsson 1985), it was already clear that the peak load for geometrically similar specimens was expressible as a function of the specimen size relative to the material characteristic size:

$$\sigma_{Nu} = f_t' \, \phi\left(\frac{D}{\ell_{ch}}\right) \tag{9.2.1}$$

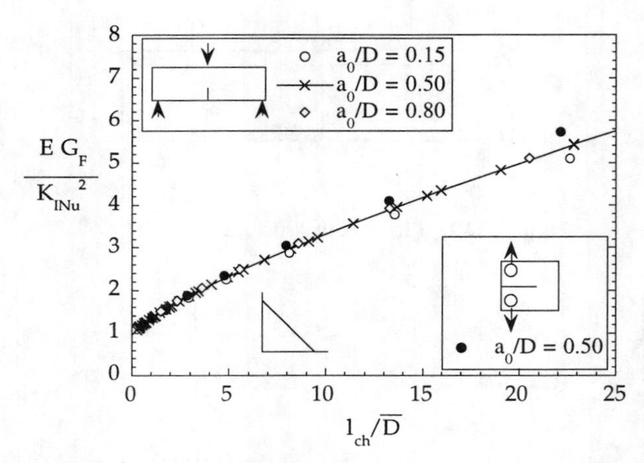

Figure 9.2.3 Size effect curves for various geometries expressed as a function of intrinsic size (from Planas, Guinea and Elices 1997).

where ϕ is a dimensionless function depending implicitly on the geometrical shape of the structure and on the shape of the softening function. The general proof of this property is a direct corollary of the result, obtained in Section 7.5.2, that the governing integral equation (7.5.19) depends only on D/ℓ_{ch}.

Planas and Elices (1989a, 1990a, 1991a) and Llorca, Planas and Elices (1989) sought the general dependence of this equation on geometry and found that for the usual testing geometries (notched beams in bending, compact tension specimens, and single-edge-notch specimens in tension), the size effect equation could be expressed as

$$\sigma_{Nu} = f'_t \, \psi\left(\frac{\overline{D}}{\ell_{ch}}\right) \tag{9.2.2}$$

where $\overline{D}$ is the intrinsic size defined in (5.3.11), and ψ is a dimensionless function *approximately* independent of geometry for not too small sizes (including most laboratory sizes). This is illustrated in Fig 9.2.3 which shows the size effect for various notch-to-depth ratios in a notched three-point bend beam and for a compact tension specimen as a plot of the nominal stress intensity factor vs. the intrinsic size. As can be seen, the results are very close, so it appears that most of the geometry dependence is covered by the definition of the intrinsic size. However, the resulting size effect was seen to be strongly dependent on the shape of the softening curve.

Further work showed that, for not too large specimens, the size effect does not depend on the entire softening curve, but only on its initial portion (Planas and Elices 1992b; Guinea, Planas and Elices 1994a,b; Planas, Guinea and Elices 1995). The reason for this is shown in Fig 9.2.4 where the maximum softening at peak load is plotted vs. the specimen size for bending of notched specimens. It is seen that for not too large sizes (up to beam depths of about 2.7 times ℓ_1, roughly 35 cm for concrete) the softening is less than 70% of the tensile strength, a range over which most softening curves used in practice can be approximated by a linear segment.

Based on this observation, Guinea, Planas and Elices (1994a) determined an approximate size effect expression that can describe the size effect for different geometries and softening curves in a unified way. But the resulting equation is rather complicated and cannot fit the size effect for small sizes. This is due to the fact that most of the work, and especially the definition of intrinsic size (5.3.11) relies on the analysis of the asymptotic trends for large sizes and loses applicability to small sizes. In a further work, Planas, Guinea and Elices (1997) found a formulation that can fit both the large and small size limits, and also gives a very good description of the size effect for notched specimens in the practical size range. This formulation, which we consider next, is based on the independent analysis of the large and small size asymptotes followed by an empirical interpolation.

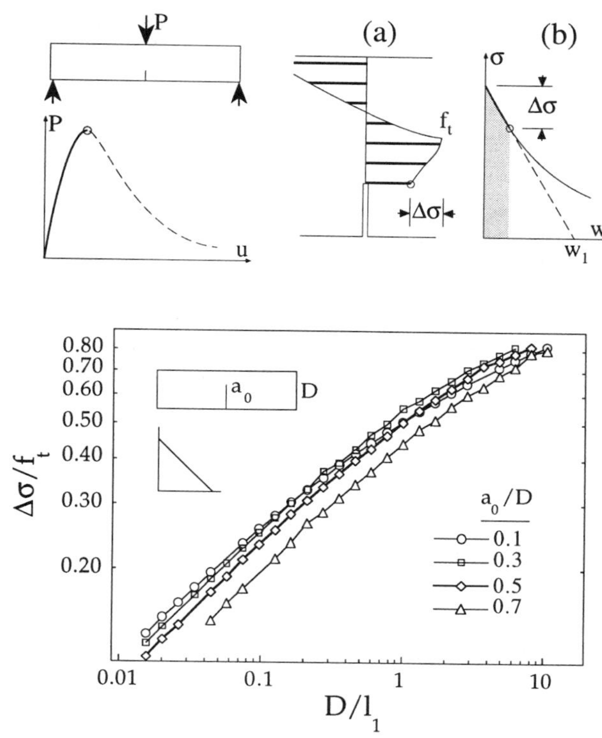

Figure 9.2.4 Maximum stress drops at peak load as a function of specimen size; (a) shows the stress profile along the specimen ligament at peak load and (b) the segment of the softening curve that has been spanned upon reaching the peak load (from Planas, Guinea and Elices 1997).

9.2.2 Asymptotic Analysis for Large Sizes

The asymptotic analysis for large size was initially based on the general expression (9.2.1) rewritten in terms of the nominal stress intensity factor as

$$\frac{EG_F}{K_{INu}^2} = \chi\left(\frac{\ell_{ch}}{D}\right) \tag{9.2.3}$$

Planas and Elices (1986a) assumed that function χ can be expanded in a power series of its argument:

$$\frac{EG_F}{K_{INu}^2} = c_0 + c_1\frac{\ell_{ch}}{D} + c_2\left(\frac{\ell_{ch}}{D}\right)^2 + \cdots \tag{9.2.4}$$

and postulated that, for $D \to \infty$ ($\ell_{ch}/D \to 0$), the equation converged to the LEFM solution $k_{INu} = \sqrt{E'G_F}$, thus implying $c_0 = 1$; the asymptotic size effect was then obtained by truncation of the series after the first term:

$$\frac{EG_F}{K_{INu}^2} = 1 + c_1\frac{\ell_{ch}}{D} \quad \text{for } \frac{\ell_{ch}}{D} \to 0 \tag{9.2.5}$$

Following an asymptotic analysis based on the smeared tip superposition technique (see Section 7.5.7), Planas and Elices (1992a) proved that the second coefficient acquires the simple form

$$c_1 = \frac{2k_0'}{k_0}\,\frac{\Delta a_{c\infty}}{\ell_{ch}} \tag{9.2.6}$$

Here $\Delta a_{c\infty}$ is the critical effective crack extension for infinite size defined in Section 7.2.4, which can be computed for any softening curve with great accuracy following the numerical method by Planas and

Elices (1991a, 1993a), described in Section 7.5.7. When this last expression is substituted back into (9.2.5), and the definition (5.3.11) of the intrinsic size is introduced, the asymptotic form (7.2.9) emerges; we rewrite it here in a slightly different form:

$$\frac{EG_F}{K_{INu}^2} = 1 + \frac{\Delta a_{c\infty}}{\ell_{ch}} \frac{\ell_{ch}}{\overline{D}} \quad \text{for } \frac{\ell_{ch}}{D} \to 0 \tag{9.2.7}$$

where the ratio $\Delta a_{c\infty}/\ell_{ch}$ is a constant for a given shape of the softening curve (see Section 7.2.4 for the values for some particular cases). This result, which shows that for large sizes the dependence on geometry is completely taken into account by using the intrinsic size, is at the root of the previously mentioned approximation (9.2.2) of Planas et al. As said before, this property applies exactly only for very large sizes but is reasonably fulfilled to practical sizes, but not too small sizes (note in Fig. 9.2.3 that for $\ell_{ch} \approx 20\overline{D}$, the points for different geometries spread about 10% from the reference value).

As pointed out in Section 7.2.4, the foregoing large size asymptotics can be cast in the form of Bažant's size effect law

$$\sigma_{Nu} = \frac{\eta_\infty f_t'}{\sqrt{1 + \theta_\infty D/\ell_{ch}}}, \quad \text{for } D \to \infty \tag{9.2.8}$$

where η_∞ and θ_∞ are constants given by

$$\eta_\infty = \sqrt{\frac{\ell_{ch}}{2k_0 k_0' \Delta a_{c\infty}}}, \quad \theta_\infty = \frac{k_0 \ell_{ch}}{2k_0' \Delta a_{c\infty}} \tag{9.2.9}$$

Here it is evident that η_∞ and θ_∞ depend on the geometry through the LEFM shape function $k(\alpha)$, and on the softening function through the ratio $\Delta a_{c\infty}/\ell_{ch}$.

9.2.3 Asymptotic Analysis for Small Sizes

For small sizes, the analysis requires a different approach, because there is no suitable analytic treatment to disclose the asymptotic behavior (there are analytic approximations for some particular geometries and rectangular softening, but not for initially linear softening).

Since it is generally accepted that for very small sizes σ_{Nu} approaches a constant value, one could write as a starting point an integer power series expansion in D:

$$\frac{\sigma_{Nu}}{f_t'} = \eta_0 + \eta_1 \frac{D}{\ell_{ch}} + \eta_2 \left(\frac{D}{\ell_{ch}}\right)^2 + \cdots \tag{9.2.10}$$

where η_i are coefficients depending on the geometrical shape (particularly on the initial notch depth). For $D \to 0$ the asymptotic behavior of the foregoing is identical (up to linear terms) to that of Bažant's size effect law, and so we can write:

$$\sigma_{Nu} = \frac{\eta_0 f_t'}{\sqrt{1 + \theta_0 D/\ell_{ch}}}, \quad \text{for } D \to 0 \tag{9.2.11}$$

where $\eta_0 f_t'$ is the strength for a very small size, which is expected to follow from plasticity analysis, and $\theta_0 = -2\eta_1/\eta_0$.

Now, in the classical Bažant size effect law it was implicitly assumed that $\eta_\infty = \eta_0$ and $\theta_\infty = \theta_0$. However, according to Planas, Guinea and Elices (1997), this is a strong assumption without physical support since (9.2.8) and (9.2.11) refer to completely different situations. The plastic-like behavior at small sizes is expected to be only mildly correlated to the brittle behavior at large sizes. Therefore, even if the small size asymptotic behavior is dictated by (9.2.11), a transition to the large size asymptotics must be provided.

Moreover, for bending, the foregoing small size asymptotics can be shown to be valid for rectangular softening functions (Dugdale model with cutoff). Indeed, this follows from the analytical results of of Planas and Elices (1991c) for the infinite double edge notch panel subjected to bending previously illustrated in Fig. 9.1.2. Numerical results show that this is also the case for more realistic geometries, as shown by the open circles in Fig. 9.2.5 for a three-point bend notched beam. Note that other geometries

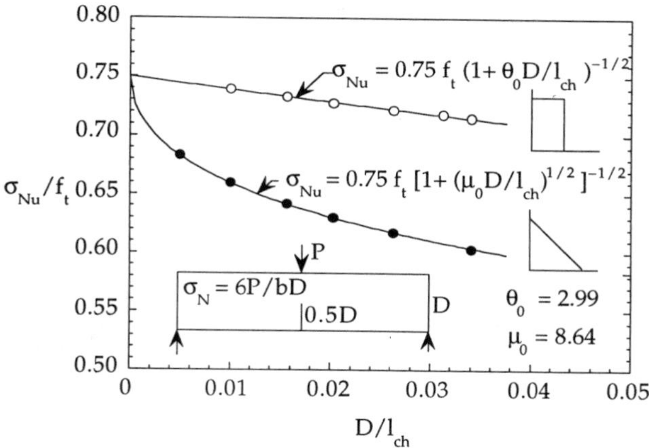

Figure 9.2.5 Small size asymptotic behavior for the strength of a three-point bend notched beam for rectangular softening (open symbols) and initially linear softening (full symbols) (from Planas, Guinea and Elices 1997).

(for example, the center-cracked panel) may display a completely different small size asymptotic behavior; see exercise 9.5.

For an initially linear softening, better suited for concrete than the rectangular softening, the asymptotic trend is completely different. The asymptotic curve does converge to the plasticity solution, but the deviation from it increases with the square root of the size, $D^{1/2}$, rather than linearly. This is shown by the full circles in Fig. 9.2.5 for a three-point bend notched beam. This means that, for small sizes, the asymptotic behavior for linear softening can be approximated by an equation of the type

$$\sigma_{Nu} = \frac{\eta_0 f_t'}{\sqrt{1 + \sqrt{\mu_0 D/\ell_{ch}}}} \quad \text{for } D \to 0 \tag{9.2.12}$$

where $\eta_0 f_t'$ still corresponds to the plastic limit strength and μ_0 is a dimensionless constant. Fig 9.2.5 shows that this equation describes very well the small size results for the material with initially linear softening.

Remark: The limit analysis for a notched beam of a material with elastic-perfectly plastic behavior in tension (with yield strength f_t') and purely elastic in compression, is known to give an ultimate moment $M_u = f_t'(D-a)^2/2$, from which $\sigma_{Nu} = 3f_t'(1-\alpha_0)^2$. This means that $\eta_0 = 3(1-\alpha_0)^2$, and so $\eta_0 = 0.75$ for $\alpha_0 = 0.5$, as shown in Fig 9.2.5. $\triangle$

Obviously a transition is required to shift from (9.2.8) for small sizes to (9.2.12) for large sizes. This transition is addressed next.

9.2.4 Interpolation Formula

Fig. 9.2.6 sketches the two asymptotic curves (9.2.8) and (9.2.12) in a log–log plot. The solid line depicts a transition curve that satisfies both asymptotic trends and provides an interpolation between them. Of course, infinitely many curves exist that satisfy this condition, but one of them, found by Planas, Guinea and Elices (1997), is both simple and realistic enough for most practical purposes. The equation of such a curve is given by

$$\sigma_{Nu} = n_0 f_t' \left[\frac{m_1 + (m_1+1)\sqrt{m_0 D/\ell_{ch}}}{m_1 + \sqrt{m_0 D/\ell_{ch}}} + m_2 \frac{D}{\ell_{ch}} \right]^{-\frac{1}{2}} \tag{9.2.13}$$

where n_0, m_0, m_1, and m_2 are dimensionless parameters directly related to the constants $\eta_0, \mu_0, \eta_\infty$, and θ_∞. Indeed, matching the asymptotic behavior of the foregoing curve to the asymptotic formulas (9.2.8)

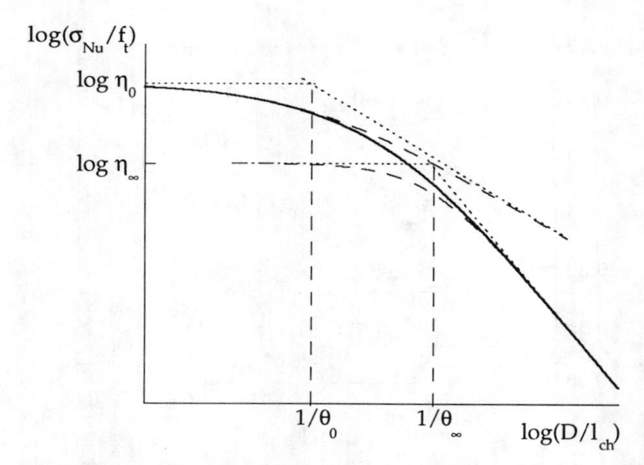

Figure 9.2.6 Asymptotic curves (dashed line) and intermediate interpolating formula (full line); dotted lines show the asymptotes (from Planas, Guinea and Elices 1997).

and (9.2.12), the following relations are found:

$$\eta_0 = n_0 \; , \qquad \mu_0 = m_0 \, , \qquad \eta_\infty = \frac{n_0}{\sqrt{1+m_1}} \; , \qquad \theta_\infty = \frac{m_2}{1+m_1} \tag{9.2.14}$$

The parameters n_0 and m_i have to be determined from comparison of the foregoing equation to numerical results through some kind of least square fitting. This means, first, that the foregoing relationships will hold only approximately (except in the unlikely event that the foregoing equation were exact); and second, that they have to be determined for each geometry and softening. The fitting task may appear to be difficult to systematize, but this is not so. Planas, Guinea and Elices (1997) found that the dependence of the foregoing parameters on the material properties and geometry is relatively simple.

For example, as it turned out, if n_0 is determined from curve fitting to numerical results for sizes over three orders of magnitude, and η_0 is evaluated analytically from plastic analysis, then the relationship $n_0 = \eta_0$ is satisfied within a few tenths of a percent. Thus, n_0 can be taken to coincide with its asymptotic value η_0 deduced from plastic analysis. For example, as described in the Remark at the end of Section 9.2.3, for a bent specimen we have

$$n_0 = \eta_0 = 3(1-\alpha_0)^2 \tag{9.2.15}$$

and similar formulas can be easily derived for other usual geometries.

Planas, Guinea and Elices (1997) also found that for softening functions with a linear initial portion, the parameter m_0 can be rewritten as

$$m_0 = p_0 \kappa \, \frac{\ell_{ch}}{\ell_1} \tag{9.2.16}$$

where p_0 is a constant and κ is a function depending only on the geometry and satisfying the condition $\kappa = 1$ at $\alpha_0 = 0$. For three-point bending, we will show that fitting of numerical results requires

$$p_0 = 13 \; , \qquad \kappa = 1 - \alpha_0^{1.45} \tag{9.2.17}$$

Similar (although not so neat) properties apply to the other two parameters. Furthermore, Planas, Guinea and Elices (1997) found that for a given geometry and relatively deep notches, one can give for the size effect a closed-form expression that has *no free parameters* and is *independent of the softening function* for usual laboratory sizes. This was done for beams in bending as described next, but the same kind of analysis can be used for other typical geometries.

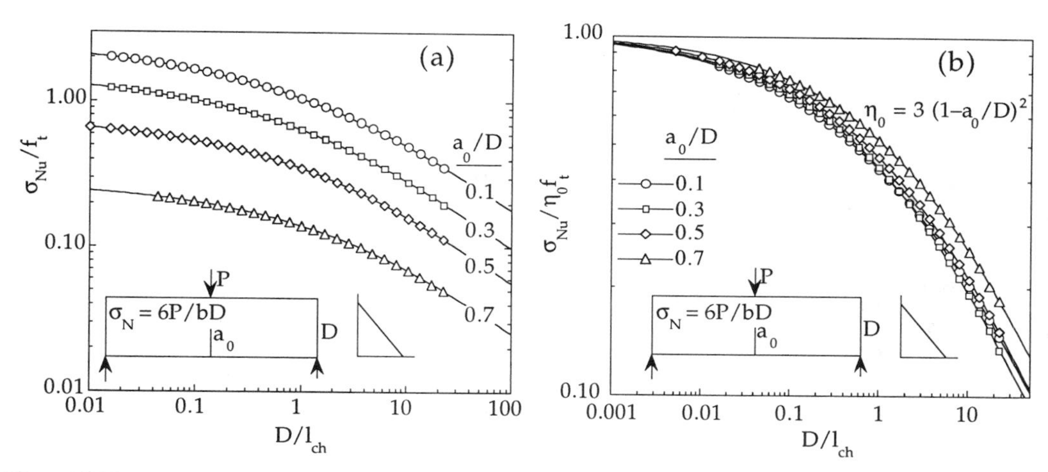

Figure 9.2.7 (a) Size effect computed for bending beams with various notch-to-depth ratios; linear softening and fit by Eq. (9.2.13). (b) Same results transformed by normalizing the stress to the plastic limit. (From Planas, Guinea and Elices 1997.)

9.2.5 Application to Notched Beams with Linear Softening

The linear softening is of basic interest not in itself, but because for not too large sizes (particularly for laboratory sizes) most softenings can be approximated by a linear equation (Fig. 9.2.2). Therefore, the size effect curve for linear softening must closely describe the observed size effect for not too large specimens for any kind of softening (as long as it begins linearly).

Fig. 9.2.7a shows a bilogarithmic plot of the computed size effect curves for various notch-to-depth ratios, which shows that the ability of the interpolation formula to fit the numerical results is equally good for all notch depths, and discloses the horizontal asymptote towards the left side.

Now, plotting $\sigma_{Nu}/\eta_0 f_t'$ vs. the size, the horizontal asymptotes of all the curves merge, as shown in Fig. 9.2.7b. This figure also shows that, in the log-log plot, the left part of any curve can be brought on top of any other curve by a horizontal translation, which is equivalent to a change of scale in the abscissas (i.e., in the size). Planas, Guinea and Elices (1997) determined this scale factor graphically and rewrote Eq. (9.2.13) using (9.2.15) and (9.2.16) as follows:

$$\sigma_{Nu} = \eta_0 f_t' \left[\frac{m_1 + (m_1 + 1)\sqrt{p_0 \kappa D/\ell_1}}{m_1 + \sqrt{p_0 \kappa D/\ell_1}} + p_2 \frac{\kappa D}{\ell_1} \right]^{-\frac{1}{2}}, \qquad \kappa = 1 - \alpha_0^{1.45} \tag{9.2.18}$$

where $p_2 = m_2/\kappa$.

Fig. 9.2.8 shows the plot of $\sigma_{Nu}/\eta_0 f_t'$ vs. $\kappa D/\ell_1$. It may be concluded that the size effect curves merge in a single curve for deep notches ($a_0/D \geq 0.3$). Note that although the curves for shallow cracks deviate from this *deep notch master curve,* they do so slowly, and even the unnotched case is very close to this curve for small sizes (roughly, less than 5% deviation for $D <\approx 0.5\ell_1$).

9.2.6 Application to Notched Beams with Bilinear Softening

To verify the applicability of the foregoing interpolation formula to other kinds of softenings, Planas, Guinea and Elices (1997) used Petersson's (1981) bilinear softening depicted in Fig 7.2.2 and reproduced with more detail in Fig. 9.2.9a. It must be realized that this is only an example of a bilinear softening curve and that other bilinear curves can better represent the behavior of a particular concrete. However, Petersson's curve is the worst among all the bilinear curves used in the literature, because (as Fig 7.2.2 clearly shows) its initial linear portion is the shortest: if the linear approximation holds for Petersson's curve, it will certainly hold for all other proposed bilinear softenings.

Fig. 9.2.9b shows the size effect curves computed for this kind of softening compared to the results

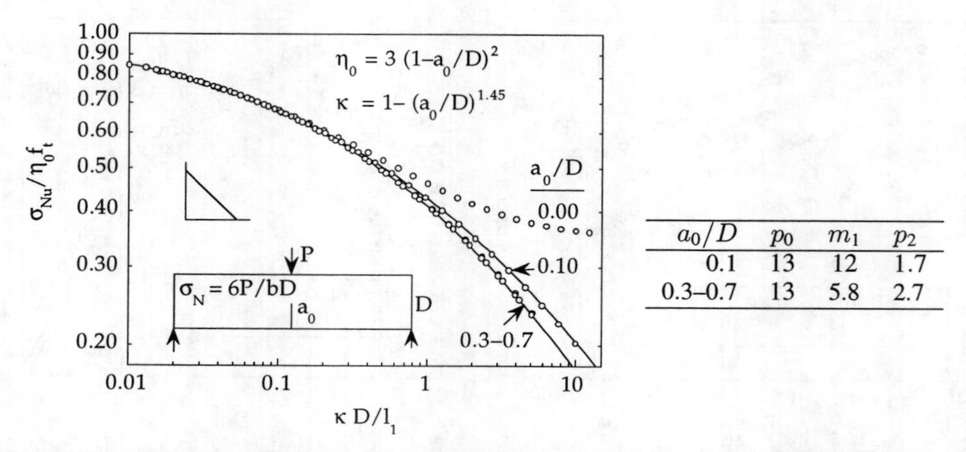

a_0/D	p_0	m_1	p_2
0.1	13	12	1.7
0.3–0.7	13	5.8	2.7

Figure 9.2.8 Size effect computed for bending beams with various notch-to-depth ratios; linear softening and fit by Eq. (9.2.18) (from Planas, Guinea and Elices 1997).

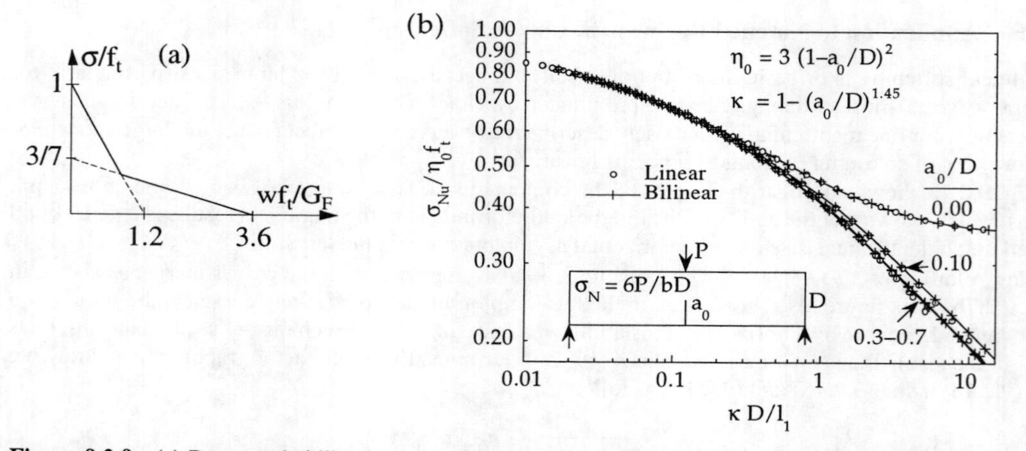

Figure 9.2.9 (a) Petersson's bilinear curve. (b) Size effect curves for linear and bilinear softening. (From Planas, Guinea and Elices 1997.)

previously obtained for the linear softening. The curves are clearly coincident for not too large values of the size. The figure also contains the curves for unnotched beams. They obviously do *not* display the same trend as those for notched beams, and, remarkably enough, they coincide *exactly* for any size. This important fact will be analyzed in detail in Section 9.3.

It is thus clear that the analytical expression found for deep notches and linear softening can be used for a bilinear softening and ordinary laboratory sizes. The final size effect expression then reads

$$\sigma_{Nu} = \eta_0 f_t' \left[\frac{1 + 4.23\sqrt{\kappa D/\ell_1}}{1 + 0.622\sqrt{\kappa D/\ell_1}} + 2.7\frac{\kappa D}{\ell_1} \right]^{-\frac{1}{2}} , \quad \kappa = 1 - \alpha_0^{1.45} , \quad \eta_0 = 3(1-\alpha_0)^2 \qquad (9.2.19)$$

which approximates the numerical results very well for deep notches ($\alpha_0 \geq 0.3$) and $\kappa D/\ell_1 < 2$ as shown more clearly in Fig. 9.2.10. More detailed analysis shows that the error is less than 1% for this range.

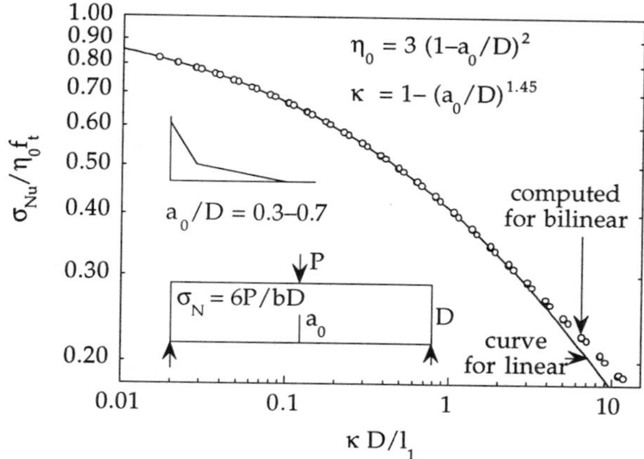

Figure 9.2.10 Approximation of the size effect curve for bilinear softening by the best fit for linear softening, Eq. (9.2.19) (from Planas, Guinea and Elices 1997).

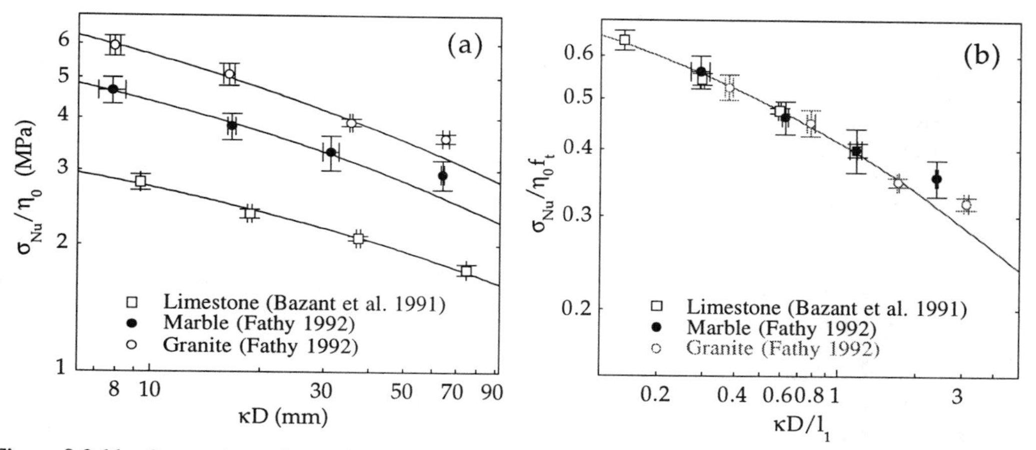

Figure 9.2.11 Comparison of experimental results with the approximate formula (9.2.19) (from Planas, Guinea and Elices 1997).

9.2.7 Experimental Evidence

Due to the inherent experimental scatter, along with the practical limitation of the range of sizes that can be accurately tested, it is very difficult to provide strong support for any given model based on size effect tests alone (Planas and Elices 1989a, 1990a). Therefore, as stressed by Planas, Guinea and Elices (1997), the following results are not intended to prove that the foregoing formula is *the* formula, but just to show that it gives realistic results.

Among the results available in the literature, Planas, Guinea and Elices (1997) selected three series that (a) use three-point bend tests, (b) use deep notches, (c) provide an independent estimation of the tensile strength, and (d) include at least three sizes that span a minimum range of 1:4 and are small enough ($\kappa D <\approx 1.5\ell_1$) to be fitted by (9.2.19).

Three series of tests on rocks were found that satisfy the foregoing conditions: marble and granite tested by Fathy (1992) and Indiana limestone tested by Bažant, Gettu and Kazemi (1991). The essential characteristics of these tests series are summarized in Section 1.5.1, in Tables 1.5.1 and 1.5.2, and Fig. 1.5.1 (Series E1, E2, and F1, respectively).

Fig. 9.2.11a shows the plot of σ_N/η_0 vs. κD for these tests, whose size range is 1:8. Shown in this plot are also the best fits of Eq. (9.2.19); the last point in the series for granite and marble have not been included in the least square fitting because for these $\kappa D/\ell_1$ exceeds 1.5. Fig. 9.2.11b shows the results in a nondimensional plot in which the theoretical curve has the same representation for all cases. The values of the best estimates of ℓ_1 are 63, 27, and 21 mm for limestone, marble, and granite, respectively. The corresponding estimates for f'_t are 4.4, 8.3, and 11.3 MPa, which compares very well with the values of the Brazilian splitting tensile strength determined by the authors of the tests: 3.45, 7.7, and 12.3 MPa, respectively.

Exercises

9.4 Use the result (7.5.28) to analyze the asymptotic behavior of the size effect of a center-cracked panel with rectangular softening in the limit of large sizes. Show that the intrinsic size is a_0 and find $\Delta a_{\infty c}$.

9.5 Use the result (7.5.28) to analyze the asymptotic behavior of the size effect of a center-cracked panel with rectangular softening in the limit of small sizes ($a_0 \to 0$). Note that for small x, $\cos^{-1}(x) \to \pi/2 - x$. Make a sketch of the resulting function in a plot similar to that in Fig. 9.2.5. Show that the limiting size effect curve deviates from the plastic limit $\sigma_{Nu} = f'_t$ and follows a curve with a horizontal tangent and zero curvature.

9.3 Size Effect on the Modulus of Rupture of Concrete

The modulus of rupture, measured for beams in either three- or four-point bending, provides a measure of the strength of a brittle or quasi-brittle material which is experimentally convenient because the tests are relatively easy to carry out. This has been recognized and various standard procedures for such tests are available. Unfortunately, the modulus of rupture does not, in general, coincide with the tensile strength. For example, the ACI code assumes that, on the average, the rupture modulus f_r is 25% higher than the tensile strength f'_t. However, the problem cannot be reduced to such a simple relationship. It is widely accepted that the modulus of rupture is size dependent.

The size dependence of the modulus of rupture has been experimentally demonstrated many times (Reagel and Willis 1931; Wright 1952; Nielsen 1954; Lindner and Sprague 1953; Walker and Bloen 1957; Petersson 1981; Alexander 1987; Elices, Guinea and Planas 1995) and has been incorporated, for example, into the CEB-FIP Model Code 1990 (CEB 1991). To justify and evaluate such a size effect, various theories have been used. The earliest theories were based on the statistical approach of Weibull (1939) developed for brittle materials whose strength is totally dependent on the characteristics of the worst flaw (weakest link theory). However, it turned out later that this theory is not directly applicable to concrete because in concrete many initial flaws grow and coalesce before the peak load is reached (see Chapter 12). This is not to say that concrete is immune to the statistical size effect; of course, it is not, but the dependence is milder than for brittle materials failing due to a single flaw, and in most cases, the fracture mechanics size effect dominates (see, e.g., Bažant, Xi and Reid 1991, with further details in Chapter 12).

This section, based on a recent analysis by Planas, Guinea and Elices (1995), reviews and compares various theories developed to explain the size effect on the modulus of rupture, all based on nonlinear fracture models simplified to various degrees.

9.3.1 Notation and Definition of the Rupture Modulus

For bending, either three- or four-point (Fig. 9.3.1), the nominal stress is conveniently defined as

$$\sigma_N = \frac{6M}{bD^2} \tag{9.3.1}$$

where M is the bending moment in the central cross-section, b the specimen thickness, and D the beam depth. The nominal strength σ_{Nu} is, by definition, the modulus of rupture f_r:

$$f_r = \sigma_{Nu} = \frac{6M}{bD^2} \tag{9.3.2}$$

where M_u is the ultimate bending moment.

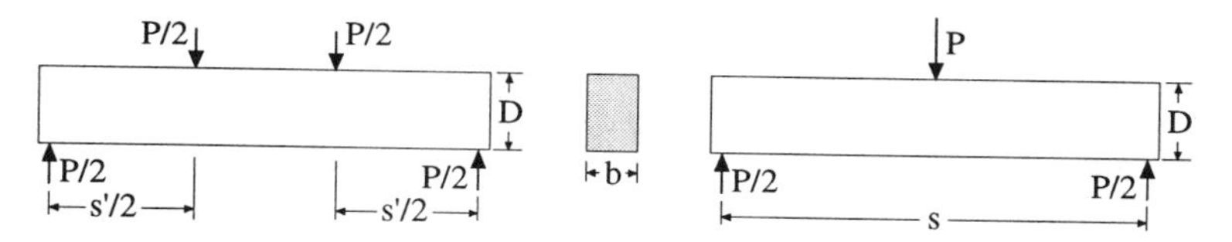

Figure 9.3.1 Four- and three-point bending beams.

The modulus of rupture was originally assumed to be a material property coinciding with the tensile strength. This is basically true for an elastic-brittle material, defined as a material that remains elastic until, at some point, the tensile stress reaches the tensile strength f_t, at which moment catastrophic failure occurs. Then, for pure bending the rupture modulus of an elastic-brittle material must coincide with the tensile strength. However, for three-point bending, the rupture modulus must be expected to be larger by about 5% even for an elastic-brittle material. This is so because, in three-point bending, the elastic stress distribution along the central cross-section is slightly different from that in pure bending. In particular, the maximum elastic tensile stress $\sigma^{el}_{\max}$ is not identical to σ_N (as for pure bending), but is slightly lower. According to Timoshenko and Goodier (1951), the peak stress is given by

$$\sigma^{el}_{\max} = \sigma_N \left(1\text{–}0.1773\frac{D}{s}\right) \tag{9.3.3}$$

where s is the loading span. Thus, for an elastic-brittle material, the modulus of rupture would be

$$f_r = \beta f'_t \qquad \text{with} \qquad \beta = \frac{1}{1\text{–}0.1773 D/s} \tag{9.3.4}$$

According to this formula, $\beta = 1.046$ for $s/D = 4$. Gustafsson (1985) numerically found $\beta = 1.053$ for such a case; Planas, Guinea and Elices (1995), using finite elements with 100 equal elements over the central cross-section, found a value coincident with (9.3.4), within the expected error, which we retain here. Note that Gustafsson redefined the modulus of rupture to include the factor β, and so the newly defined rupture modulus would be equal to the tensile strength not only for pure bending but also for three-point bending. Here, we stick to the classical definition (9.3.2), as used in most standards.

9.3.2 Modulus of Rupture Predicted by Cohesive Cracks

The bending test was one of the early applications of the cohesive or fictitious crack model (Hillerborg, Modéer and Petersson 1976). After this initial work, which used linear softening and a relatively coarse finite element mesh, further, more precise computations were performed by Hillerborg and co-workers to reveal the influence of the shape of the softening curve (Modéer 1979) and to examine the influence of shrinkage (Petersson 1981; Gustafsson 1985). The results showed that the rupture modulus was size-dependent and that it also depended, quite sensitively, on the shape of the assumed softening curve. Subsequently, many other researchers have used the cohesive crack to describe the fracture in bending. In no case, however, is the numerical solution completely consistent with the cohesive crack hypotheses. Indeed, as pointed out qualitatively by Hillerborg and co-workers and shown quantitatively by Olsen (1994), the tensile strength is exceeded in finite element computations over relatively large regions outside the assumed main crack. This means that secondary cracking must occur. This is neglected in the solutions produced so far for the modulus of rupture. Planas and Elices (1993b) included the secondary cracking in a relatively simple way for the cases where strong shrinkage is present. Such a way can be extended to the no-shrinkage case with additional hypotheses which, however, require relatively sophisticated computations.

All the solutions for cohesive crack models discussed in the following use the classical computational approach in which the secondary cracking is neglected. These classical computations are relatively easy to carry out using the methods described in Section 7.4. To summarize the results delivered by the numerical computations, various formulas have been proposed, that we analyze next.

Gustafsson (1985) analyzed the rupture modulus in detail, and, although he did not produce a general formula, he derived two asymptotic expressions for pure bending ($\beta = 1$) and large sizes:

$$\frac{f_r}{f'_t} = 1 + \frac{1}{b_1 D/\ell_{ch}} \quad \text{first order approximation} \tag{9.3.5}$$

$$\frac{f_r}{f'_t} = 1 + \frac{1}{1 + b_1 D/\ell_{ch}} \quad \text{second order approximation} \tag{9.3.6}$$

where b_1 is a dimensionless constant depending on the shape of the softening curve, and ℓ_{ch} is the classical Hillerborg's characteristic size (see Section 7.1.3). Gustafsson (1985) found $b_1 = 3.7$ for Petersson's bilinear softening depicted in Fig. 9.2.9a (and also in Fig. 7.2.2).

Recently, Eo, Hawkins and Kono (1994) analyzed again the evolution of the modulus of rupture using bilinear softening curves of various shapes. Their results confirmed that the modulus of rupture depends on the shape of these curves. They proposed to fit the results by a modification of Bažant's (1984a) size effect law, by adding to it a constant term as previously suggested by Bažant (1987a) in a different context:

$$\frac{f_r}{f'_t} = \frac{B}{\sqrt{1 + D/D_0}} + C \tag{9.3.7}$$

where B, D_0, and C are constants to be obtained by curve fitting procedures for each softening. The fits to numerical results obtained by Eo, Hawkins and Kono were excellent over the size range from 0.1 to 1.5 m. However, this curve has three parameters and we will see later that, for cohesive cracks, only one parameter should be allowed to vary freely with the softening shape. Moreover, the values obtained for C were around 0.15, which means that, for this curve, the asymptotic value for $D \to \infty$ is $f_r = 0.15 f'_t$, while it is known that the correct asymptotic value for a cohesive crack model is $f_r = f'_t$ (for pure bending).

Uchida, Rokugo and Koyanagi (1992) proposed the following formula, which has the correct asymptotic behavior for large sizes:

$$\frac{f_r}{f'_t} = 1 + \frac{1}{b_2 + b_1 D/\ell_{ch}} \tag{9.3.8}$$

where b_1 and b_2 are constants depending on the softening curve. For the bilinear softening curve of Rokugo et al. (1989) (Fig. 7.2.2b), the numerical results are approximated quite well by (9.3.8) for $D \geq 0.1\ell_{ch}$, if $b_1 = 4.5$ and $b_2 = 0.85$ (Uchida, Rokugo and Koyanagi 1992).

The fact that all the constants in the foregoing fitting equations depend on the shape of the softening curve is a drawback for the generalization of these equations. To achieve a formulation independent (in a sense to be made more precise later) of the shape of the softening curve, one needs a different choice of the shape parameters of the softening curve. The way towards such a formulation was opened by Alvaredo and Torrent (1987), who noticed that for bilinear curves the rupture modulus depends only on the initial softening segment, rather than on the entire softening curve. As shown by Planas and Elices (1992b, 1993b) this property extends to other nonlinear softening curves and to other situations (presence of shrinkage stresses). The basic result of their analysis is that the softening curve influences the modulus of rupture only through its initial part, which can, in most cases, be approximated by a straight line, as discussed in Chapter 7 and depicted in Fig. 7.1.8. This means that only two parameters of the softening curve are relevant: the tensile strength f'_t and the horizontal intercept of the initial tangent w_1 (Fig. 7.1.8). Thus, from the basic equations governing the cohesive crack growth, it turns out that the relationship between the rupture modulus and the size must take the form

$$\frac{f_r}{f'_t} = H\left(D/\ell_1\right) \quad \text{with} \quad \ell_1 = \frac{E w_1}{2 f'_t} \tag{9.3.9}$$

where H is a dimensionless function that depends implicitly on the kind of loading, e.g., the three- or four-point bending, and on span-to-depth ratio. The foregoing property is corroborated by Fig. 9.3.2a, which shows that the curves of the modulus of rupture vs. D/ℓ_1 for the three different softening curves depicted in Fig. 9.3.2b (solid curves) approximately coincide. We will discuss in more detail in Section 9.3.3 why and for which kind of softening this property holds.

A closed-form expression for function $H(D/\ell_1)$ was found by Planas, Guinea and Elices (1995). This expression is approximately valid for the entire range of sizes, satisfying the condition that $f_r \to 3f'_t$ for

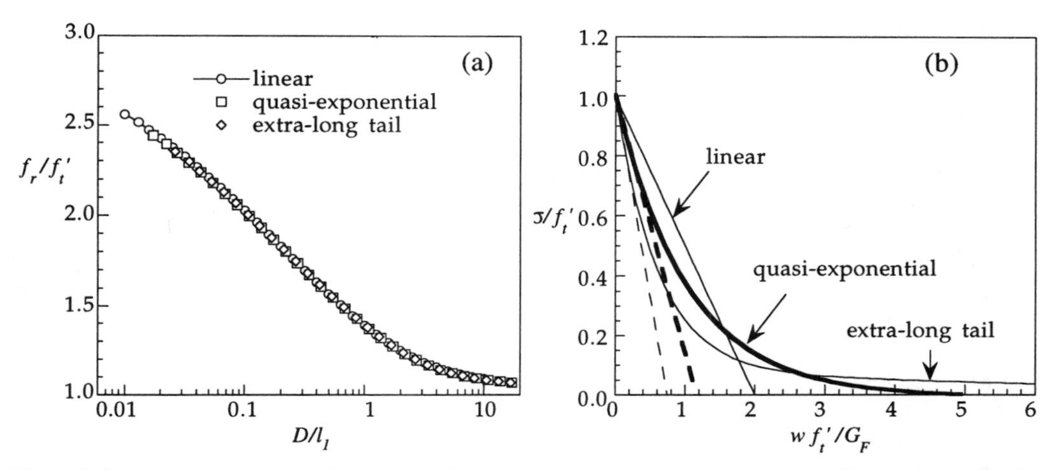

Figure 9.3.2 Size effect curves for the modulus of rupture (a) computed for various softening shapes (b) (from Planas, Guinea and Elices 1995).

$D \to 0$ (plastic limit solution for cohesive cracks) and $f_r = \beta f'_t$ for $D \to \infty$. The expression proposed is a generalization of Gustafsson's expression (9.3.6) and reads:

$$\frac{f_r}{f'_t} = \beta + \frac{3 - \beta + 99D^*}{(1 + 2.44D^*)(1 + 87D^*)}, \qquad D* = \frac{D}{\ell_1} \tag{9.3.10}$$

Expressions such as (9.3.5), (9.3.6), and (9.3.8) that use the characteristic size, can be transformed to the form (9.3.9), independent of the softening curve, by making use of the relationship between ℓ_1 and ℓ_{ch}:

$$\ell_1 = \frac{w_1 f'_t}{2G_F} \ell_{ch} = c_0 \ell_{ch} \tag{9.3.11}$$

where, obviously, the factor c_0 depends on the shape of the softening curve. For concrete, c_0 is usually in the range 0.4–0.6 (the steeper the initial softening, the smaller c_0). In this way, (9.3.5), (9.3.6), and (9.3.8) can be recast in the general form

$$\frac{f_r}{f'_t} = \beta \left(1 + \frac{1}{c_2 + c_1 D/\ell_1}\right) \tag{9.3.12}$$

where, for successively better approximation levels, the constants take the values

$$\begin{aligned} c_2 &= 0.00, \quad c_1 = 2.3 \quad \text{(level I)} \\ c_2 &= 1.00, \quad c_1 = 2.3 \quad \text{(level II)} \\ c_2 &= 0.85, \quad c_1 = 2.3 \quad \text{(level III)} \end{aligned} \tag{9.3.13}$$

Fig. 9.3.3a shows the comparison of the numerical predictions of the cohesive model with the expressions (9.3.10) and (9.3.12)–(9.3.13) for three-point bending with a span-to-depth ratio of 4 ($\beta = 1.046$). All give a good approximation for large sizes ($D > 3\ell_1$). The third order approximation gives a very good approximation for $D > 0.1\ell_1$, a range sufficient for most practical purposes (for a typical concrete $\ell_{ch} \approx 300$ mm, so $\ell_1 \approx 150$ mm and $0.1\ell_1 \approx 15$ mm, a very small size for most applications).

At this point, it is important to note that, at the time Planas, Guinea and Elices (1995) carried out the foregoing analyses, they had not yet discovered the small size asymptotic behavior described in Section 9.2.3, which has been found to hold for the unnotched beams, too. Therefore, although the foregoing formulas remain useful (they are accurate enough for any practical purpose), they satisfy only the large size asymptotic behavior. For small size they should have the form

$$\frac{f_r}{f'_t} \approx 3 - m_1\sqrt{D/\ell_1} + \cdots \approx \frac{3}{1 + m_1\sqrt{D/\ell_1}} \tag{9.3.14}$$

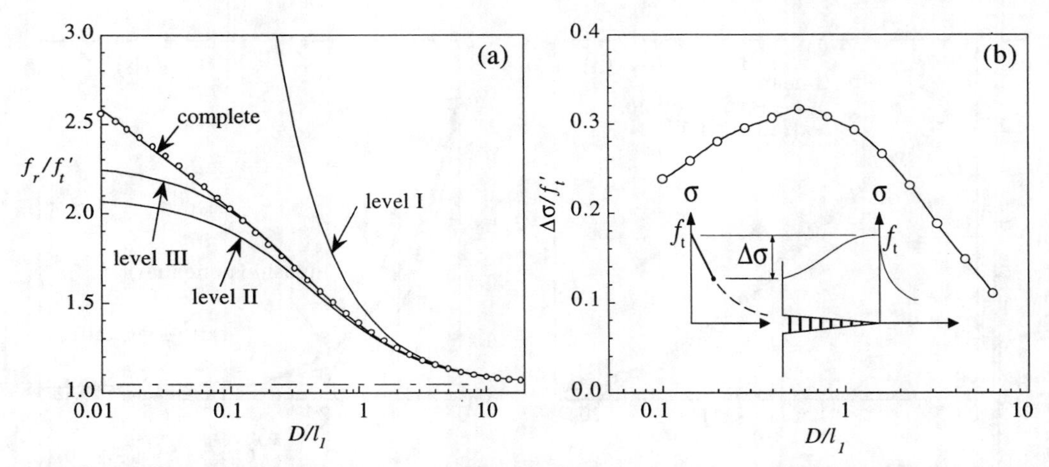

Figure 9.3.3 (a) Computed size effect curve for the modulus of rupture in three-point bending with s/D ratio of 4, and analytical approximations (9.3.10)–(9.3.12). (b) Stress-softening at the cohesive crack mouth vs. specimen size. (From Planas, Guinea and Elices 1995.)

which is not true of the foregoing equations. Further work is needed to get a simple formula satisfying both asymptotic trends and good accuracy in the intermediate range.

Before closing this subject, it should be stressed that the foregoing conclusions are predicated on the hypothesis that the cohesive crack model describes crack initiation from a smooth surface well. Because initially the cracking damage is spread over a relatively large zone and gradually localizes into a single crack, the nonlocal damage model would probably be more realistic.

9.3.3 Further Analysis of the Influence of the Initial Softening

The idea that the modulus of rupture depends only on the initial softening slope was initially based by Planas and Elices on the examination of a number of cases for which this happens to be the case. However, there is a logical reason why this should be so: for the bending tests on unnotched specimens, the peak load occurs before any point in the cohesive zone softens very much. This can be verified by recording the amount of softening experienced at peak load by the material element situated at the cohesive crack mouth, which is the one experiencing most softening.

Fig. 9.3.3b shows the results for a range of specimen sizes; they demonstrate that the softening experienced at peak load never exceeds 32% of the tensile strength. Therefore, any softening that can be reasonably approximated by a straight segment in the range of cohesive stresses from the maximum of f_t' down to $0.78 f_t'$ may be expected to give size effect curves well described by the equations in the foregoing section.

9.3.4 Modulus of Rupture According to Bažant and Li's Model, Bažant's Universal Size Effect Law, and Zero-Brittleness Method

The approximation of Bažant and Li (1995c) is based on the assumption that prior to the peak load, the cracking in concrete is distributed rather than localized (Fig. 9.3.4a), although the reality is doubtless between uniform distribution of cracking and its full localization into one discrete crack. Bažant and Li assume the peak load to occur when the greatest depth of the microcracked zone reaches a certain critical value ℓ_f. The problem is further simplified by assuming that, up to peak, the beam can be analyzed by the elementary beam theory (cross-sections remain plane, shear is neglected), and that the uniaxial stress-strain behavior of concrete is elastic with linear softening, as depicted in Fig. 9.3.4b. Writing the axial and moment equilibrium conditions for the stress distribution in Fig. 9.3.4, and setting $M = bD^2 f_r/6$,

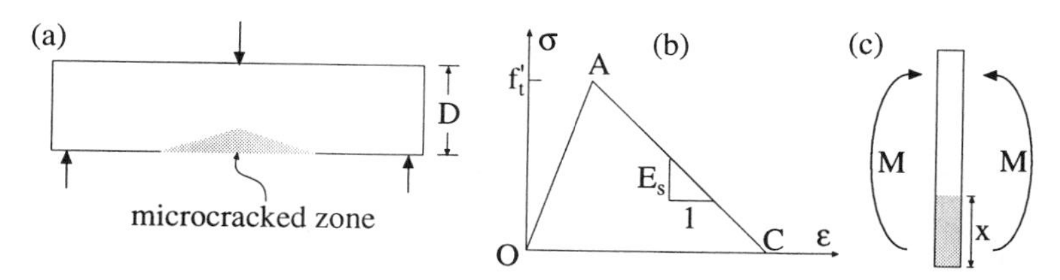

Figure 9.3.4 (a) Microcracked zone in Bažant-Li model. (b) Softening curve. (c) Microcracked zone at the central cross-section. (From Planas, Guinea and Elices 1995.)

one obtains (Planas, Guinea and Elices 1995):

$$\frac{f_r}{f'_t} = 1 + 2\frac{\ell_f}{D} - \frac{4k}{k + (D/\ell_f - 1)^2} \tag{9.3.15}$$

where

$$k = E_s/E \tag{9.3.16}$$

and E_s is the softening modulus (Fig. 9.3.4b). The foregoing is the exact solution of the problem as defined by Bažant and Li (1995c), who worked out only a truncated series expansion (which is quicker to solve), and suggested that for relatively large sizes ($\ell_f/D << 1$) only two terms were sufficient. Since the third term in (9.3.15) is quadratic in ℓ_f/D, the size effect proposed by Bažant and Li (1995c) reads

$$\frac{f_r}{f'_t} = 1 + 2\frac{\ell_f}{D} \tag{9.3.17}$$

Fig. 9.3.5a shows the evolution of the modulus of rupture with size as predicted by Bažant and Li's (1995c) model. Note that the first-order approximation is achieved over most of the range if the value of k is very small (i.e., if the softening is very gradual). For larger values of k, the model predicts a descending branch for small sizes, which is unrealistic. As initially formulated, Bažant and Li's model did not take into account the effect of the concentrated load. This can be included by multiplying the right-hand side of equation (9.3.17) by β. It then turns out that Bažant and Li's asymptotic equation coincides with Gustafsson's first order approximation (9.3.5) and its reformulation (9.3.12)-(9.3.13). Thus, ℓ_f and ℓ_1 can be related so that the asymptotic behavior would be coincident; setting the parenthesis in (9.3.12) —with $c_2 = 0$ and $c_1 = 2.3$— equal to (9.3.17) we get

$$\ell_f = \frac{\ell_1}{2c_1} = 0.217\ell_1 \qquad \text{or} \qquad \ell_1 = 4.6\ell_f \tag{9.3.18}$$

Fig. 9.3.5b compares the predictions of Bažant and Li's and the cohesive crack models. Note that, selecting an appropriate value for k (≈ 0.35), one achieves a very good coincidence for $D >\approx 0.8\ell_1$. If the first order approximation is used (equivalent to $k = 0$) the range over which the two models coincide within 10% is reduced to $D >\approx \ell_1$. However, no good correspondence is found for the small sizes typical of laboratory specimens.

Bažant and Li's model is based on a very simple hypothesis. It has been, in theory, superseded by Bažant's asymptotic analysis presented in the first section of this chapter, which led to Bažant's universal size effect law (9.1.48). For unnotched specimens, $D_0 \to \infty$; this reduces to (9.1.42) which has obviously the form (9.3.12) and can thus be made to coincide with the level III approximation of the cohesive crack model if we take

$$\gamma = 0.85 \qquad \text{and} \qquad \ell_f = 0.217\ell_1 \tag{9.3.19}$$

The corresponding curve has been plotted in Fig. 9.3.5a and b. As expected, the fit to the numerical results for the cohesive crack model is excellent.

For quality control in the field, it would be convenient to use specimens of only one size (cast in one and the same mould), achieving a sufficient range of brittleness number by varying only the notch length.

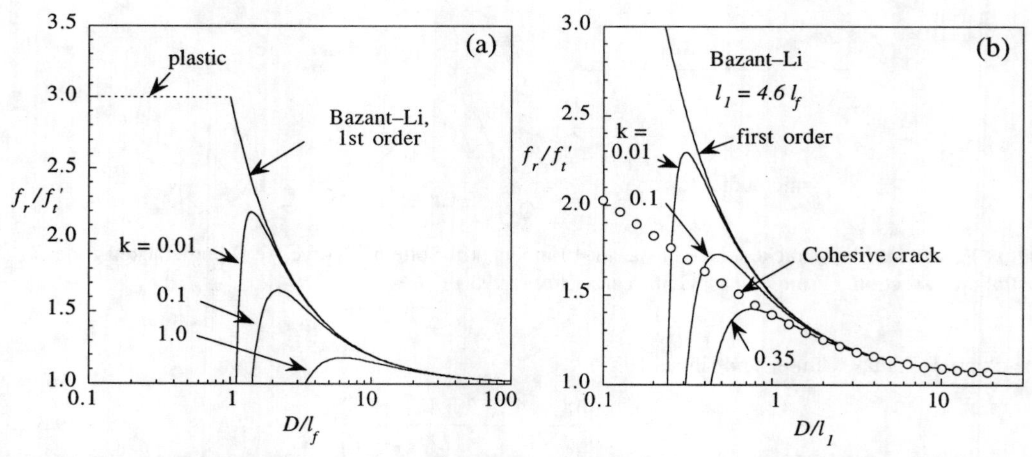

Figure 9.3.5 Evolution of size effect curve according to Bažant and Li's (1995c) model (from Planas, Guinea and Elices 1995.)

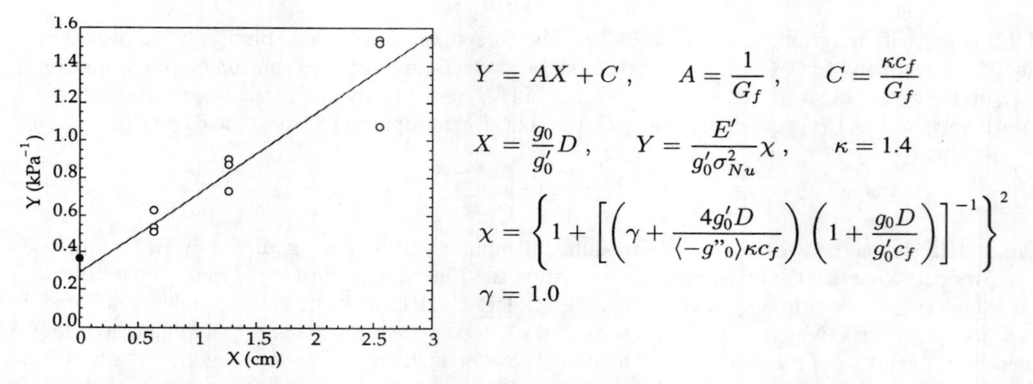

Figure 9.3.6 Linear regression plot by universal size effect law of the data of Bažant and Gettu (1992), including zero brittleness point (adapted from Bažant and Li 1996).

However, as shown by Bažant and Li (1996), varying notch length cannot achieve a brittleness range broader than 1:3, which is not enough in view of the typical experimental scatter of concrete.

A sufficient range can nevertheless be achieved if, in addition to notched specimens, unnotched specimens of the same size and shape are also included among the maximum load tests, in the so-called *zero-brittleness method.* Fitting of the universal size effect law, Eq. (9.1.48), can then provide the fracture parameters G_f and c_f, by means of a linear regression (Bažant and Li 1996). This is illustrated in Fig. 9.3.6 which includes the basic equations (Eqs. 11 and 12 in Bažant and Li 1996) which follow from Eq. (9.1.48), with $r = s = 1$). However, there is one question that needs to be studied deeper. Parameter γ in the universal size effect law Eq. (9.1.48), defining the ratio between the effective size c_f of the fracture process zone around notch tip and the thickness l_f of the boundary layer of cracking in unnotched specimen, is empirical and needs to be known in advance. The value of γ has been determined by Bažant and Li for one type of concrete, and it was shown that for different specimen shapes the point of zero brittleness (Fig. 9.3.6), based on unnotched specimen test, lies then always near the size effect regression line of the data points (empty points in Fig. 9.3.6) for geometrically similar specimens. Thus, if only the black point and the empty points for the largest size D in this figure were known, one would get about the same results for G_f and c_f. But it is not clear whether the same γ value could be applied to other concretes. Further study is needed.

Apart from the universal size effect law, there might be another way to include the case of zero brittleness among the data—determine in advance the value of Bf_t' in the size effect law. This value represents the nonbrittle (plastic) limit of σ_N for $D/D_0 \to 0$, which suggests one could calculate it by plastic limit analysis assuming the material to follow the Mohr-Coulomb failure criterion. This can be done by simple formulae (Eqs. 6 and 7 in Bažant and Li 1996), provided the value of the tensile yield limit f_t^* of the material is determined in advance. It turns out that f_t^*, corresponding to extrapolation to specimen sizes smaller than any realistic size, is not equal to f_t'. Moreover, the f_t^*-values for specimens of different shapes cannot even be considered the same, as indicated by the shape dependence of parameters of the size effect law of Bažant and Kazemi (see Eq. 8 in Bažant and Li 1996, giving the shape dependence of f_t^*). A suitable value of f_t^* for one type of concrete and certain specimen types was determined by Bažant and Li (1996) but it is not clear whether this could be used generally.

9.3.5 Modulus of Rupture Predicted by Jenq-Shah Model

The Jenq-Shah model (Section 5.5) assumes that, starting from a preexisting crack which may be taken to be vanishingly small, a macrocrack grows until the peak load is reached. At that moment both the stress intensity factor K_I and the crack tip opening displacement w_T (CTOD) reach their critical values K_{Ic} and w_{Tc}; the corresponding crack length at the peak is denoted as a_c. For a vanishingly small initial crack, the stress intensity factor and CTOD can be written as

$$K_I = \sigma_N \sqrt{D}\, k(a/D)\,, \qquad w_T = \frac{\sigma_N}{E'} D\, v_M(a/D) \tag{9.3.20}$$

where $k(a/D)$ and $v_M(a/D)$ are shape functions that have been approximated by closed form expressions for both three-point and four-point bending (see e.g., Tada, Paris and Irwin 1985 for general formulas valid for $s/D = 4$; see Pastor et al. 1995, for enhanced expressions valid for any $s/D \geq 2.5$). Writing (9.3.20) for the peak load condition ($K_I = K_{Ic}, w_T = w_{Tc}, a = a_c$, and $\sigma_N = f_r$) and assuming that the two material parameters K_{Ic} and w_{Tc} have been determined by other experiments, we get two equations with two unknowns, a_c and f_r, which can be solved for f_r for any given size. When we want to obtain the size effect curve, i.e., let D vary, it is better to use $\alpha_c = a_c/D$, D, and f_r as variables, so that solving for f_r and D as a function of α_c, we get:

$$\frac{D}{\ell_0} = \frac{k^2(\alpha_c)}{v_M^2(\alpha_c)} \qquad \text{where} \qquad \ell_0 = \frac{E'^2 w_{Tc}^2}{K_{Ic}^2} \tag{9.3.21}$$

$$\frac{f_r}{f_0} = \frac{2v_M(\alpha_c)}{3k^2(\alpha_c)} \qquad \text{where} \qquad f_0 = 1.5\frac{K_{Ic}^2}{E' w_{Tc}} \tag{9.3.22}$$

These two equations are the parametric representation of the size effect curve, with parameter α_c. Plotting the pairs (D, f_r) for all α_c, the size effect plot is obtained. This has been done for three-point bending, using the following expressions of $k(\alpha)$ and $M(\alpha)$:

$$k(\alpha) = \sqrt{\alpha}\frac{1.9 - \alpha\left[-0.089 + 0.60(1-\alpha) - 0.44(1-\alpha)^2 + 1.22(1-\alpha)^3\right]}{(1+2\alpha)(1-\alpha)^{3/2}} \tag{9.3.23}$$

$$v_M(\alpha) = 4\alpha\left[(0.76 - 2.28\alpha + 3.87\alpha^2 2 - 2.04\alpha^3 + \frac{0.66}{(1-\alpha)^2}\right] \tag{9.3.24}$$

The expression for $v_M(\alpha)$ has been taken from Tada, Paris and Irwin (1985), and the expression for $k(\alpha)$ from Pastor et al. (1995). This latter expression is preferred to the more usual Srawley expression because it gives the correct limit for K_I for short cracks —i.e., 4.4% less than the Srawley limit $1.12\sigma_N\sqrt{\pi a}$, as required by (9.3.3). Fig. 9.3.7a shows the resulting size effect curve. Note that the size effect curve for the Jenq-Shah model exhibits a descending branch for small sizes, which is unrealistic.

To compare the rupture moduli predicted by the Jenq-Shah model with the cohesive model, we again force the asymptotic behaviors to coincide. Taking two terms of the power expansion of functions $k(\alpha)$ and $v_M(\alpha)$, one easily finds the asymptotic Jenq-Shah prediction is

$$\frac{f_r}{f_0} \to 1.049\left(1 + \frac{1}{5.3D/\ell_0}\right) \qquad \text{for} \qquad D \to \infty \tag{9.3.25}$$

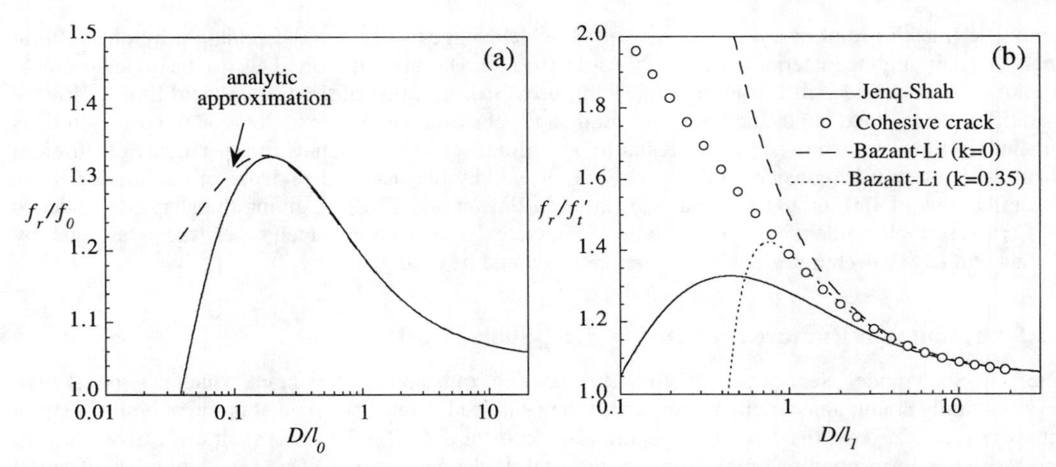

Figure 9.3.7 Size effect curve for the modulus of rupture according to Jenq–Shah two-parameter model (from Planas, Guinea and Elices 1995.)

which can be made to coincide with the first order approximation of the cohesive crack in (9.3.12)–(9.3.13), with $\beta = 1.046$, if we set

$$f_0 = 0.997 f_t' \approx f_t' \qquad \ell_0 = 2.3\ell_1 \tag{9.3.26}$$

Fig. 9.3.7b compares the predictions of the cohesive model, the Bažant-Li model and the Jenq-Shah model with the foregoing values of the parameters. An analytical expression with the correct asymptotic limit was fitted by Planas, Guinea and Elices (1995) to describe the prediction for sizes $D >\approx 0.15\ell_0$. The expression, drawn as a dashed line in Fig. 9.3.7a, is as follows:

$$\frac{f_r}{f_0} = 1.049\left(1 + \frac{6.1D/\ell_0}{(1+6.1D/\ell_0)(1+5.3D/\ell_0)}\right) \tag{9.3.27}$$

This expression was used by Planas, Guinea and Elices to determine the parameters ℓ_0 and f_0 by curve fitting to experimental data.

9.3.6 Carpinteri's Multifractal Scaling Law

Recently, a scaling law for strength supposedly based on the consideration of the fractal nature of the fracture process has been proposed by Carpinteri, Chiaia and Ferro (1994). Strictly speaking, this so-called multifractal scaling law (MFSL) does not have a deterministic basis. This is so, first because, in its present form, it is not the outcome of any mechanical model of fracture, i.e., the parameters of the law have to be fitted for every particular material and geometry, and if they are known for a material and geometry, there is no way to compute them for the same material but a different geometry. Second, the model is based on considering the microscopic features of the material, in particular, its random and fractal nature.

This is directly connected to the statistical theory of fracture that is developed in Chapter 12, where the discussion of the so-called multifractal scaling law best finds its place. Indeed, we will see there that the classical Weibull theory, based on the idea of a failure originating at a single, worst flaw, predicts the strength to vary proportional to a power of the size $D^{3/m}$ (for full three-dimensional similarity) where m is the Weibull modulus. In pure fractal theories a similar law is obtained, in which the strength is proportional to D^{1-d_f} where d_f is the fractal dimension. If $d_f = 1$, then no size effect is obtained. The multifractal scaling law assumes a continuous transition from the macroscopic scale in the large-size range (fractal dimension equal to 1, constant strength) to the microscopic, fully disordered limit, in which the theoretical fractal dimension is 1/2 (strength decreasing as $D^{-1/2}$). Note that these asymptotic trends are assumed to hold whatever the shape of the structure, whether or not it has a notch.

Since all the available evidence and the LEFM asymptotics indicate that the size effect in notched

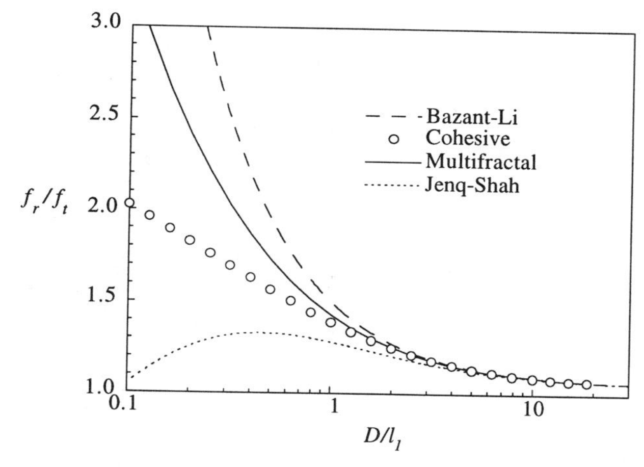

Figure 9.3.8 Comparison of the multifractal scaling law with the other theories for identical asymptotic behavior (from Planas, Guinea and Elices 1995).

structures must tend to $D^{-1/2}$, not to D^0, it is certain that this law cannot hold for notched structures. For unnotched structures, the issue of applicability of this law requires a more detailed analysis for which a more appropriate framework is provided in Chapter 12.

Here we limit ourselves to present the comparison with other formulas carried out by Planas, Guinea and Elices (1995). Following these authors, the multifractal scaling law can be written in the following way:

$$\frac{f_r}{f_t'} = \beta\sqrt{1 + \frac{\ell_M}{D}} \tag{9.3.28}$$

where f_t' is the tensile strength in the macroscopic limit (large size) and ℓ_M is a constant length characteristic of the material and of the geometry. The factor β was introduced by Planas, Guinea and Elices (1995) to provide consistency with the other theories. To push the comparison further, the asymptotic expressions of this model and the cohesive crack model are again made to coincide. The result is now

$$\ell_1 = 1.15\ell_M \qquad \text{or} \qquad \ell_M = 0.87\ell_1 \tag{9.3.29}$$

Fig. 9.3.8 shows that the so-called multifractal law lies between the size effect curves deduced from the cohesive crack model and Bažant and Li's model.

9.3.7 Comparison With Experiments and Final Remarks

Fig. 9.3.9 shows the experimental results from 9 experimental series and the theoretical curves found by nonlinear optimization of fit assuming equally weighted data (Planas, Guinea and Elices 1995). Results a–g correspond to concrete, h to mortar, and i to microconcrete. It appears that the Jenq-Shah model is the one experiencing most difficulty in describing the experimental results, since only for series b, c, d, and e are the slopes of the experimental curve and of the model curve in relatively good agreement. The fits with the Bažant-Li and the multifractal scaling law are essentially coincident for all experimental series. The cohesive, Bažant-Li, and multifractal models agree equally well with the experimental series b, c, d, and e; the Bažant-Li and multifractal models give better fits than the cohesive crack model for the series a, f, and g; the cohesive crack model gives a better agreement for series i.

When all circumstances are taken into account, it is difficult to conclude that any of the models is clearly superior to the others. Even if some of the cohesive crack fits seem to be slightly less accurate than the Bažant-Li and multifractal equations, the cohesive model has the advantage of being a general fracture model that can be calibrated by other tests. Moreover, the cohesive crack model can be extended to include statistical effects. This feature, as shown by Gustafsson (1985), can make it fit very well the

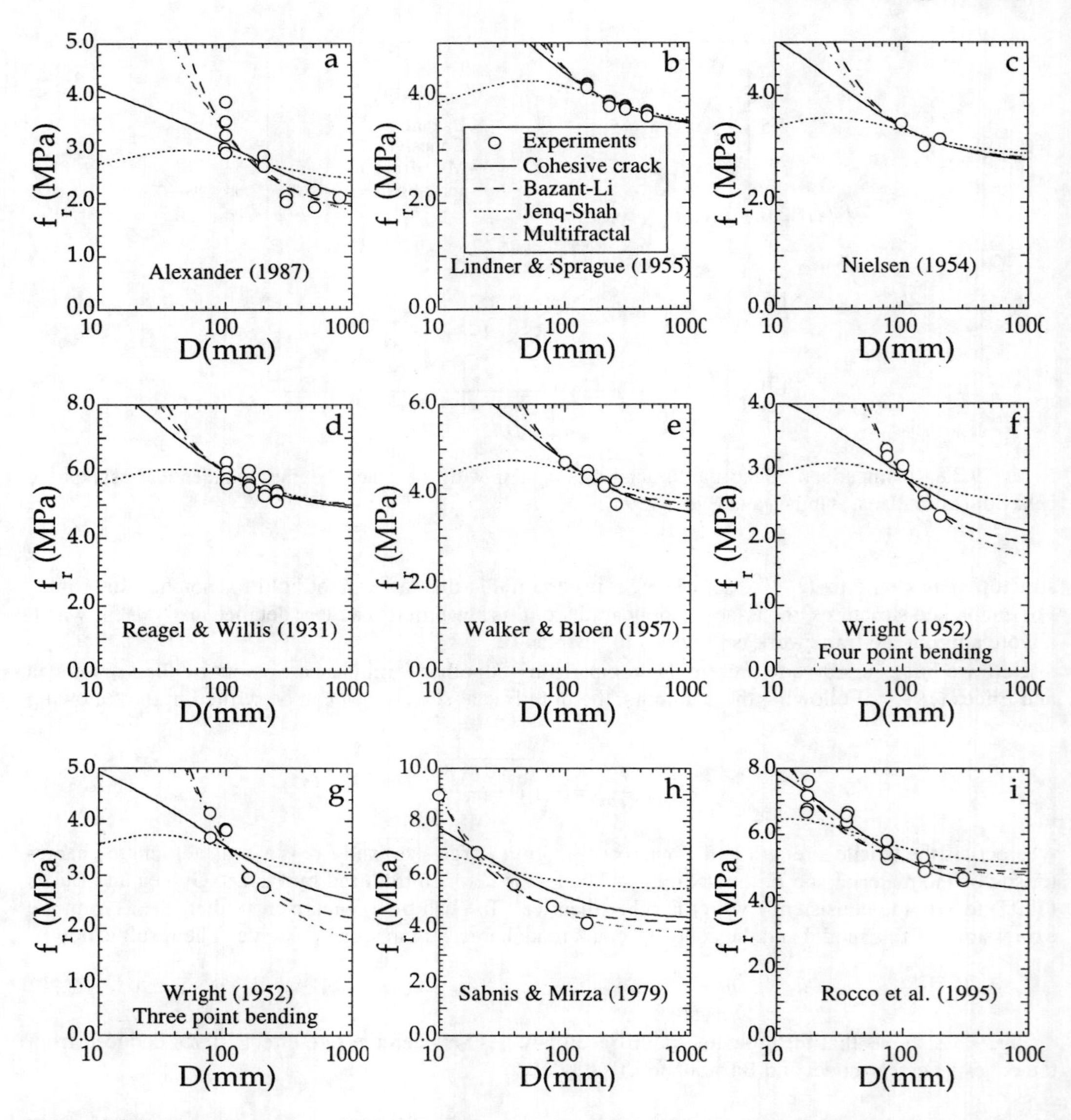

Figure 9.3.9 Experimental results from various sources and best fits for the various models (from Planas, Guinea and Elices 1995).

results showing a stronger size effect, such as those of Wright (1952) and Sabnis and Mirza (1979), series f, g, and h in Fig. 9.3.9.

In conclusion, it seems that the cohesive crack model provides a consistent and general framework for the analysis of the modulus of rupture. The results depend on only two material constants for an initially linear softening and can be conveniently summarized by Eq. (9.3.10). However, further work is required to incorporate the right asymptotic behavior for small sizes and to take into account secondary microcracking (outside the cohesive crack) in the cohesive crack numerical algorithm. A deeper understanding of the statistical effects would also be of great interest. The results of the cohesive crack model are consistent with Bažant's universal size effect law in which, however, the analysis of the small size asymptotics needs to be extended to include a square-root initial deviation and to connect the parameters ℓ_f and γ quantitatively to the fracture characteristics. Based on the analysis of Planas, Guinea and Elices, the values defined in (9.3.19) could be a convenient starting point to do that.

9.4 Compression Splitting Tests of Tensile Strength

Under compression, concrete as well as other quasibrittle materials fail either by inclined shear cracks or axial splitting. A prototype of the axial splitting failure is the Brazilian split-cylinder test. Introduction of this test by Carneiro and Barcellos (1953) in Brazil and others was motivated by the fact that the elastic solution for diametrically opposite concentrated forces acting on a circular cylinder yields a nearly uniform distribution of transverse normal stress along the line of the load, except for concentrated transverse compressive stresses in small regions under the load. For this reason, the test has been used as a measure of tensile strength (even though the stress-state is not uniaxial) and has been included in most standards as a routine method of estimating the tensile strength (ASTM C496, BS 1881-117, ISO 4108). The test method was extended by some standards to include square prisms or even general prisms, as depicted in Fig. 9.4.1. The splitting tensile strength f_{st} is determined from the peak load P_u as

$$f_{st} = \frac{2P_u}{\pi b D} \tag{9.4.1}$$

where b is the length of the cylinder (or prism) and D the diameter of the cylinder (or the depth of the prism).

As in all the failures due to cracking of concrete, a size effect must be expected, and indeed, most experimental series display a size effect. However, a unique trend cannot be deduced from the available data. For example, among the six experimental series shown in Fig. 9.4.2, we find the following types of behavior: (a) the tests of Sabnis and Mirza (1979) and Kim et al. (1989) show a continuous decrease of strength over the range of sizes of testing (so do the tests by Ross, Thompson and Tedesco 1989, not shown in the figure); (b) the tests by Chen and Yuan (1980), by contrast, show an increase, similar to the tests reported by Hondros (1959) (not shown); (c) the tests by Hasegawa, Shioya and Okada (1985), covering a broad size range, from 100 to 3000 mm in diameter, reveal a decrease of strength followed by a plateau where the size effect disappears or even a slight increase is discerned (a similar trend was observed in the tests of granite by Lundborg 1967); (d) the tests on square prisms reported by Rocco et al. (1995) show a strength constant within experimental scatter; (e) finally, the tests by Bažant, He et al. (1991) show first a marked decrease of strength, followed by a slight increase (although a plateau could also be assumed, in view of the overall scatter).

The complexity and disparity of the observed behavior may be due mainly to three factors. (**1**) Except for the tests of Rocco et al. (1995), there is no independent determination of the fracture properties, and, hence, there is no material scale reference; the results could probably be made consistent with each other by appropriately rescaling the axes. (**2**) The splitting strength is sensitive to the boundary conditions, particularly to the relative width h/D of the loading strip, which should be kept constant to assure full similarity; since some of the reported tests might have not used loading strips, or used strips of constant width, the similarity is not guaranteed. (**3**) As with all phenomena in concrete, the splitting strength is rate dependent; if the loading rates are not properly scaled, appreciable deviation can be found (see Section 9.4.4 below).

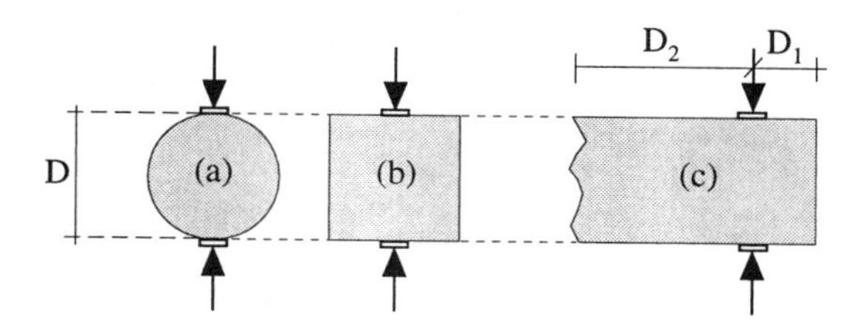

Figure 9.4.1 Various types of standard specimens: (a) cylinder, (b) square prism, (c) prism or beam.

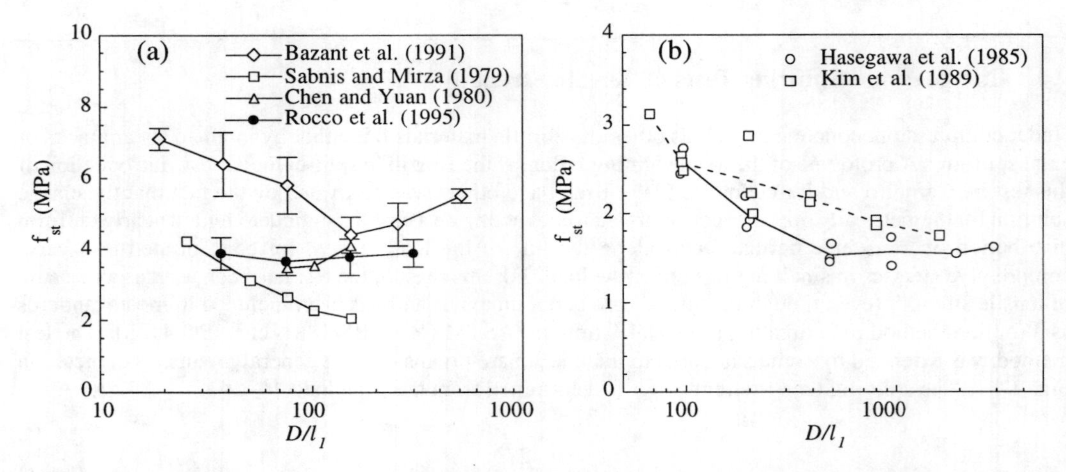

Figure 9.4.2 Examples of experimental results for size effect of Brazilian test. (a) Series with D up to 500 mm. (after Rocco et al. 1995). (b) Series with huge specimens (after Kim and Eo 1990).

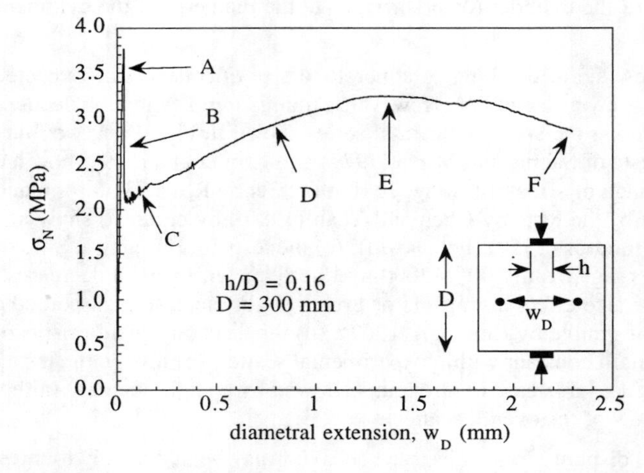

Figure 9.4.3 Load vs. transverse extension for a test on a square prism of microconcrete (after Rocco 1996).

9.4.1 Cracking Process in Stable Splitting Tests

In this section we review the approaches to the description of size effect in the splitting tests, but, before doing so, we present recent experimental evidence concerning the fracture mechanisms involved in the process based on stable tests by Rocco (1996).

Rocco (1996) performed stable tests on granite cylinders and microconcrete square prisms with careful control of the boundary conditions. The diametral extension was measured during the test by an extensometer over a gauge length $0.75D$ across the expected main crack, and this was used as a feedback signal for the servocontrol. Since the test was completely stable, the evolution of the cracks could be recorded in detail by means of a video camera. A typical diagram of load vs. diametral extension is shown in Fig. 9.4.3 corresponding to a relative strip width $h/D = 0.16$ as recommended by ASTM C496. Note that the sharp peak at very small diametral extension is followed by a relatively smooth curve with a secondary peak.

Points $A, B, \cdots, F$, marked on the curve can be related to the photographs of Fig. 9.4.4. In this figure, one can discern the crack growth sequence. Right after the peak (point A), a small crack can be seen just behind the extensometer (along with the trace of the transmitted moisture on a film of fluorescent powder; see Rocco 1996 for details). Immediately after the peak (point B), the crack extends rapidly until the second hardening zone is reached at point C at which the main fracture stops (the crack crosses the

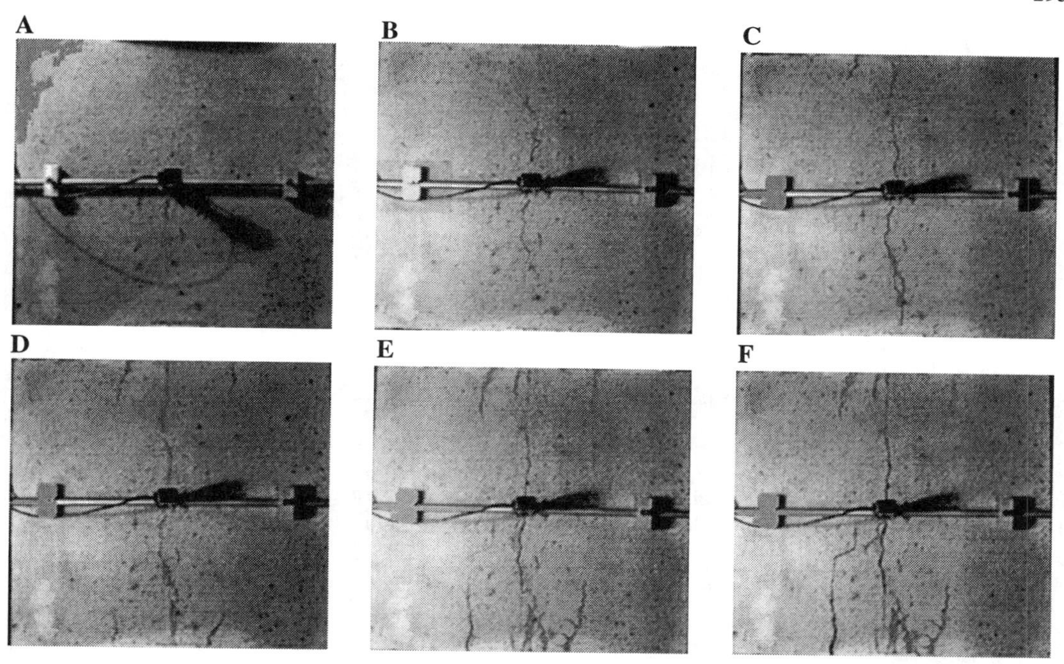

Figure 9.4.4 Cracking sequence of a square prism of microconcrete. The photograms are in correspondence with equally labeled points on the load-diametral extension curve in Fig. 9.4.3 (from Rocco 1996).

specimen) and a secondary crack is already visible on the top to the left of the loading strip (a symmetric shorter crack can be guessed at the bottom-right). Upon reaching point D, four secondary cracks become clearly visible at either side of the loading strips. At point E, the bottom-left secondary crack is opening more than the others and diffuse cracking appears below the lower loading strip. At point F, the bottom-left secondary crack has grown up to the center of the specimen and a failure mechanism is formed.

This sequence is typical; small differences in the symmetry pattern may appear, but all follow the sequence: main crack $\longrightarrow$ main peak $\longrightarrow$ approximately symmetric secondary cracks $\longrightarrow$ secondary peak $\longrightarrow$ loss of symmetry and failure. Slip-wedges (considered in plastic analysis of the problem) were never seen. The secondary cracks appear roughly at the points of maximum tensile strength as determined by an elastic finite element calculation assuming a stress-free (noncohesive) crack across the central diameter.

One of the important results of Rocco (1996) is that, on purely experimental grounds, the relative height of the secondary peak depends on the ratio h/D and on the size of the specimen itself (which is expressed relative to the characteristic length ℓ_1 in Rocco's work). This is clearly shown in Fig. 9.4.5 where the curve of load vs. diametral extension is plotted for fixed $h/D = 0.16$ and various sizes. The secondary peak clearly rises when the size is decreased and is very close to the primary peak for $D \approx 0.9\ell_1$. For smaller sizes the overall maximum corresponds to the secondary peak and, therefore, the maximum load recorded in the test does *not* correspond to the propagation of the central crack, which is the crack most directly related to tensile fracture. Rocco's results also indicate that the wider the loading strips, the higher the secondary peak.

A practical consequence is that the loading strip width recommended by ASTM C496 is probably too wide and an h/D ratio of about 0.08 (or 1/12) would be better to ensure that the overall maximum in the test correspond to the first peak. In the future, one should set a lower limit for the size, to guarantee the validity of the test.

These results may be useful in the ongoing discussion of the size effect theories of various authors. As in previous sections, we consider next the modifications of Bažant's size effect law, the prediction of Jenq-Shah model, and the predictions of the cohesive crack model.

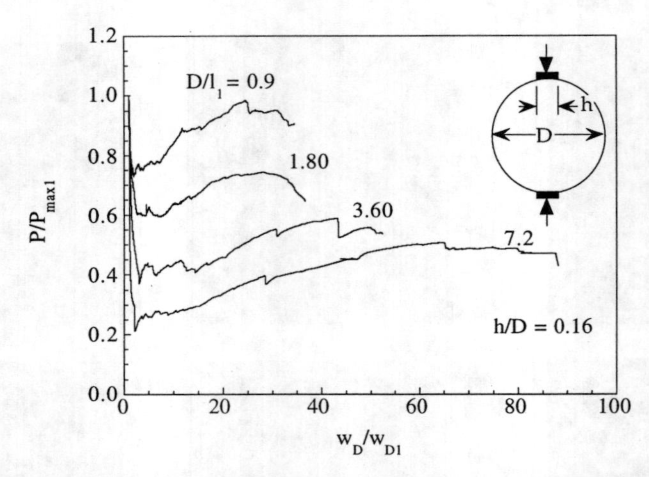

Figure 9.4.5 Evolution of the load-diametral extension curve with the specimen diameter for a loading strip 16% of the diameter for granite cylinders (from Rocco 1996); $P_{\max_1}$ is the load of the first peak and w_{D1} the corresponding diametral extension.

9.4.2 Modified Bažant's Size Effect Law

Bažant, He et al. (1991) observed that the original size effect law (6.1.5) could not hold for split-cylinder tests over a broad range of sizes, and tried to modify it. They suggested two possible mechanisms modifying the basic size effect: (**1**) the length of the crack along the diameter at maximum load ceases to be proportional to the diameter and becomes constant; and (**2**) the failure mechanism may change at large sizes to one in which the maximum load is reached by frictional plastic slip in the small highly confined wedge-shaped zones under the loading platens, which are in a plastic state due to high confinement and thus are better described by plasticity which exhibits no size effect.

The second mechanism had been already addressed by Bažant (1987a) who suggested two equations involving a plateau of size effect curve for large sizes:

$$\sigma_{Nu} = \frac{Bf'_t}{\sqrt{1 + D/D_0}} + Cf'_t \tag{9.4.2}$$

$$\sigma_{Nu} = \max\left[\frac{Bf'_t}{\sqrt{1 + D/D_0}}, Cf'_t\right] \tag{9.4.3}$$

where B and D_0 are the usual constants in Bažant's size effect law and C is a further constant giving the large size nominal strength, attributable to a plastic-like mechanism operating in parallel with the crack (this analogy is strictly valid for the first formula).

For the first mechanism —a nonsimilar crack growth— Bažant, He et al. (1991) derived a formula that we can deduce here by starting with the first equality (6.1.3) which (in a little expanded form to show the arguments) reads as follows:

$$\sigma_{Nu} = \frac{K_{Ic}}{\sqrt{k^2(\alpha_0) + 2k(\alpha_0)k'(\alpha_0)c_f}} \tag{9.4.4}$$

where $\alpha_0 = a_0/D$; a_0 is the initial crack for notched specimens with positive geometry, or the crack length at peak for specimens in which slow crack growth occurs prior to the peak. For the classical Bažant size effect law to hold, the ratio a_0/D must be constant. Bažant, He et al. (1991) assumed that in the Brazilian test, a_0 is indeed proportional to the size for sizes smaller than D_t (transition size), after which

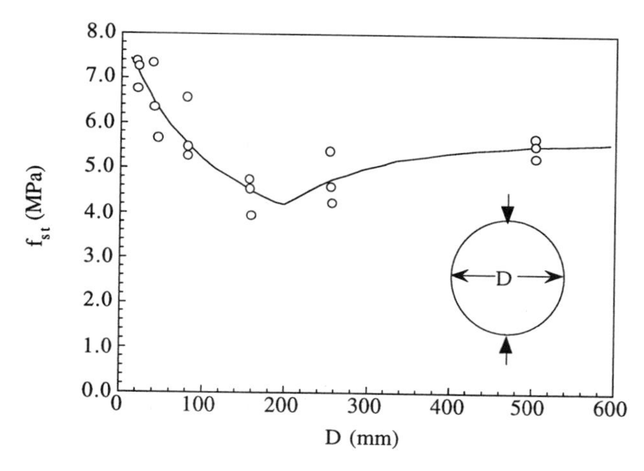

Figure 9.4.6 Experimental data and best fit by Eq. (9.4.5), after Bažant, He et al. (1991)

a_0 remains constant and equal to a_t. Calling $a_t/D_t = \alpha_0$, the foregoing equation is now split as follows:

$$\sigma_{Nu} = \begin{cases} \dfrac{K_{Ic}}{\sqrt{k^2(\alpha_0) + 2k(\alpha_0)k'(\alpha_0)c_f}} & \text{for} \quad D < D_t \\ \dfrac{K_{Ic}}{\sqrt{k^2(\alpha_0 D_t/D) + 2k(\alpha_0 D_t/D)k'(\alpha_0 D_t/D)c_f}} & \text{for} \quad D > D_t \end{cases} \tag{9.4.5}$$

Bažant, He et al. (1991) determined a closed-form expression for $k(\alpha)$ and $k'(\alpha)$ and analyzed the performance of the foregoing formulas (9.4.2), (9.4.3), and (9.4.5). The second and last were definitely superior to the first in all cases. The last two gave an equally good fit of the results of Hasegawa, Shioya and Okada (1985), but only (9.4.5) was able to fit the change in slope of the results by Bažant, He et al. (1991), as depicted in Fig. 9.4.6.

Although the fits were close, these formulas must be regarded as tentative because the sharp rise in strength for large sizes has not been confirmed by further experimental results. Also, the values of K_{Ic} and c_f that were obtained by fitting of the data would have to be confirmed by independent tests. Finally, Bažant's universal size effect law (9.1.48) will probably supersede these previous results and explain satisfactorily the experimental observations without recourse to further hypotheses.

9.4.3 Size Effect Predicted by Jenq-Shah Model

Tang, Shah and Ouyang (1992) analyzed the split cylinder test using the two-parameter model of Jenq and Shah. They assumed symmetric crack growth starting from a central notch of various sizes and considered various widths of the loading strip. With this model they produced the size effect curves shown in Fig. 9.4.7a, for a particular concrete. Note that each curve corresponds to a loading strip of constant width (h = const.) rather than constant relative width (h/D = const.).

A size effect analysis for full similarity can be performed in a way similar to that carried out for the modulus of rupture in Section 9.3.5. The size effect curve is given by the same parametric equations (9.3.21) and (9.3.22), with the appropriate shape functions $k(\alpha)$ and $v_M(\alpha)$. Expressions for the shape functions for cylinders are given in Tang, Shah and Ouyang (1992). Rocco (1996) produced closed-form expressions for split square prisms and performed the corresponding size effect analysis. The resulting curves are shown in Fig. 9.4.7b. Also shown in this figure are the experimental data of Rocco which do not conform to the reverse size effect predictions. Note that no curve fitting was allowed in this representation, because the values of the two parameters K_{Ic} and w_{Tc} were determined by Rocco from independent tests following the RILEM Draft Recommendation (RILEM 1990a).

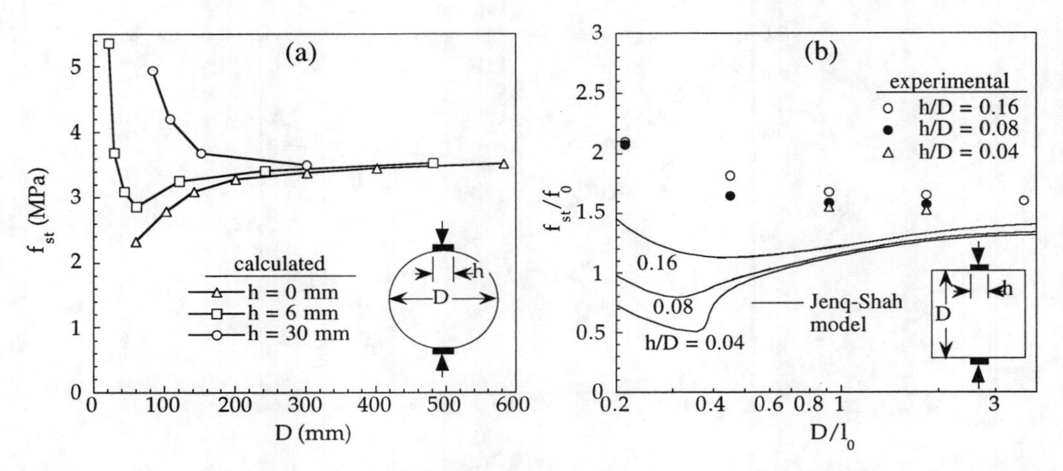

Figure 9.4.7 (a) Size effect predicted by Jenq-Shah two-parameter model for cylinder splitting strength with constant width of loading strips (after Tang, Shah and Ouyang 1992). (b) Prediction of size effect on the splitting strength of square prisms predicted by Jenq-Shah two-parameter model, and comparison with experimental results (after Rocco 1996).

9.4.4 Size Effect Predicted by Cohesive Crack Models

The splitting of square prisms can also be analyzed using cohesive crack models. This was done by Modéer (1979) using a linear softening function, for a loading strip of width $h = 0.1D$. He assumed a single crack along the loading plane and considered the material to be elastic. He found a strong size effect for small dimensions ($D < 0.5\ell_{ch}$) and a mild size effect for larger dimensions. He expressed the reservation that his analysis should be viewed merely as indicative of the possibility of strong size effect because during the computation, the tensile strength was exceeded at points outside the main crack path. Although he carried out splitting tests on paste and mortar, he did not try to correlate quantitatively the experimental results to the theoretical predictions.

Recently Rocco (1996) undertook calculations using basically the same hypotheses as Modéer, but with a finer mesh and an improved numerical method (smeared crack tip method, Section 7.4.4). The analysis of the stable splitting tests summarized in Section 9.4.1 indicates that, for not too small specimens, and not too large widths of the loading strips, the failure occurs with a sharp peak by growth of a single central crack. Thus for this kind of specimen, the hypotheses of the computations are reasonable and their results must be expected to be good. This, of course, refers only to the determination of the primary peak in the curves in Figs. 9.4.3 and 9.4.5, not to the relatively flat secondary peak, accompanied by multiple cracking and an appreciable amount of crushing.

Rocco verified that the peak was reached before much softening took place, similar to what happens for the modulus of rupture. So, whatever the softening curve, its initial linear approximation suffices to predict the peak load, and so the results are universal when expressed as f_{ts}/f'_t vs. D/ℓ_1. The essential results of Rocco (1996) were already presented in Fig. 7.3.1. It should be emphasized that, as with the previous results for Jenq-Shah model, no parameter fitting was done to achieve better match. The values of f'_t and ℓ_1 were determined in independent tests as explained in Section 7.3.2. In this light, the correspondence between experiment and prediction is excellent.

Rocco (1996) also experimentally compared the results of stable splitting tests to those of the more usual unstable tests. He found that the results of unstable tests and stable tests were similar if the loading rate was kept low. However, a substantial increase of strength can be obtained if the loading rate is increased, as shown in Fig. 9.4.8. Although these results do not allow a direct quantitative generalization, one further conclusion of Rocco's results is that the loading rate must be kept low, which should be taken into account in formulating test standards.

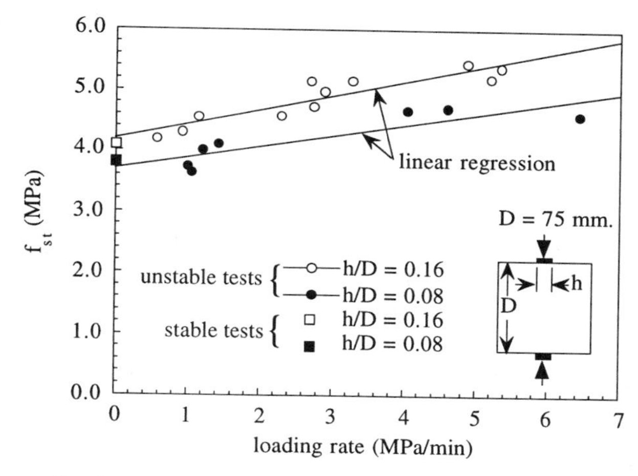

Figure 9.4.8 Influence of loading rate of unstable tests upon the measured splitting strength for square prisms of microconcrete (from Rocco 1996).

9.5 Compression Failure Due to Propagation of Splitting Crack Band

Similar to the tensile failure, many types of compression failure of quasibrittle materials such as concrete, rock, ice, ceramics, and composites exhibit a size effect (e.g., van Mier 1986; Gonnermann 1925; Blanks and McNamara 1935; Marti 1989; Jishan and Xixi 1990). However, the compression failure, and especially its size effect, is more complex and less understood. Yet it often is the more important and dangerous mode of failure, which is highly brittle, lacking ductility. The reason is that compression failure is not controlled by a material strength criterion, as assumed in nearly all practical applications up to now. Rather, as suggested or implied by some researchers (e.g., Ingraffea 1977; Bažant, Lin and Lippmann 1993; Bieniawski 1974; Hoek and Bieniawski 1965; Cotterell 1972; Paul 1968) and described mathematically in this section, the compression failure in quasibrittle materials is caused predominantly by the release of stored energy from the structure, similar to tensile fracture.

Whenever the consequence of material failure is the postpeak softening or the lack of ductility, which occurs in concrete under compression, a size effect must be expected. The size effect is the most important practical consequence of fracture phenomena, and so the size effect in compression failure ought to be taken into account in engineering practice.

We will present the analysis of the effect of structure size on the nominal strength of quasibrittle structures failing in compression, following the work of Bažant and Xiang (1997), which expanded a 1993 conference presentation (Bažant 1994a). The objective is to give a simple, intuitively clear explanation of the phenomenon of size effect, to formulate a simplified model for the global mechanism of propagation of compression fracture, and, finally, to use this model to determine the size effect and compare the results to test data.

9.5.1 Concepts and Mechanisms of Compression Fracture

In ductile metals, compression failure (as well as tensile failure) is caused by plastic slip on inclined shear bands. This type of failure is ductile, without any significant postpeak decrease of the applied load. It does not cause any size effect.

In quasibrittle materials, however, such ductile compression failure is possible only under extremely high lateral confining pressures, which is a case beyond the scope of this book. Such pressures lacking, no shear slip can develop in quasibrittle materials such as concrete. The interlock of rough surfaces of cracks inclined to the principal compressive stress direction prevents any slip, unless the cracks are already widely opened and the material near the crack is heavily damaged. Macroscopically, of course, shear failures are often observed, but their microscopic physical mechanism is not the slip. It normally consists of tensile microcracking in an inclined direction.

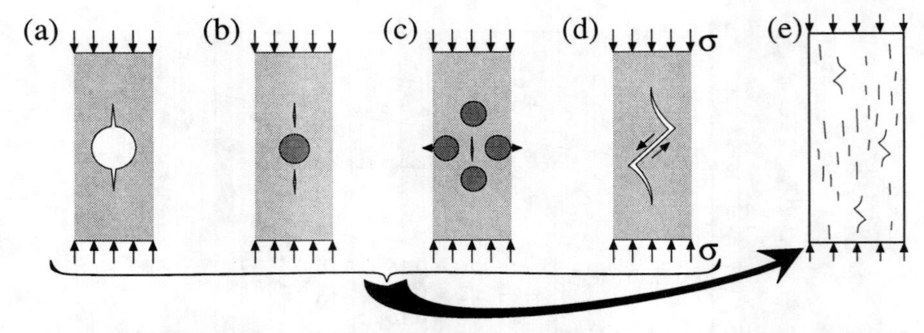

Figure 9.5.1 (a)–(d)Microscopic mechanisms of compression fracture; (e) test specimen with distributed damage caused by these mechanisms, which must be macroscopically smeared (from Bažant and Xiang 1997).

On the microscale, one can discern three different mechanisms triggering compression fracture:

1. *Pores with microcracks.* It has long been known that porosity is the main controlling factor for compression strength of various materials. The linear elastic fracture mechanics (LEFM) was used to show that pores cause axial tensile splitting microcracks to grow from the pore under a compression load of increasing magnitude (Fig. 9.5.1a); see, e.g., Cotterell (1972); Sammis and Ashby (1986); Ashby and Hallam (1986); Kemeny and Cook (1987, 1991); Steif (1984); Ingraffea (1977); Zaitsev and Wittmann (1981); Wittmann and Zaitsev (1981); Zaitsev (1985); Fairhurst and Cornet (1981); Ingraffea and Heuzé (1980); Kemeny and Cook (1987, 1991); Shetty, Rosenfield and Duckworth (1986); Nesetova and Lajtai (1973); Carter, Lajtai and Yuan (1992); Carter (1992); Yuan, Lajtai and Ayari (1993). An important point to note is that these axial cracks can grow for only a certain finite distance from the pore, which is of the same order of magnitude as the pore diameter. Therefore, this mechanism cannot explain the global fracture. A similar conclusion applies to various configurations of several pores which enhance the local transverse stresses or produce shear stresses on axial planes.
2. *Inclusions with microcracks.* A stiff inclusion, for example, a rigid piece of stone aggregate in a softer mortar matrix, causes tensile stresses at a certain distance above and below the inclusion, which can produce short tensile splitting microcracks (Fig. 9.5.1b). More effective generators of transverse tensile stress in a macroscopically uniform uniaxial compression field are various groups of inclusions, such as a group of two inclusions pressed between two others. Such a failure mechanism (proposed for concrete long ago by Brandtzaeg and by Baker) can be shown to produce short tensile splitting microcracks between the inclusions (Fig. 9.5.1c). Again, an important point is that the cracks remain short, of the same order of magnitude as the inclusion, and so the global fracture cannot be explained.
3. *Wing-tip microcracks.* In a material without pores and without inclusions, cracks in a macroscopically uniform uniaxial compression field can be produced by weak inclined interfaces between crystals. Slip on an inclined crack causes the growth of curved cracks gradually turning into the direction of compression, called wing-tip cracks (Fig. 9.5.1d). The wing-tip crack models have been extensively analyzed by fracture mechanicists, both numerically and analytically (Hawkes and Mellor 1970; Ingraffea 1977; Ashby and Hallam 1986; Nemat-Nasser and Obata 1988; Horii and Nemat-Nasser 1982, 1986; Kachanov 1982; Lehner and Kachanov 1996; Sanderson 1988; Schulson 1990; Batto and Schulson 1993; Costin 1991; and Schulson and Nickolayev 1995), and critically examined by Nixon (1996). Curved crack growth under compression has been clarified (Cotterell and Rice 1980). A fully realistic analysis of wing-tip cracks would have to be three-dimensional, which has apparently not yet been accomplished. Important to note, the length of the wing-tip cracks is again of the same order of magnitude as the length of the inclined slipping crack, and so the global fracture cannot be modeled.

In regard to the global failure mechanisms on the macroscale, one may distinguish those that cause a global energy release with size effect from those that do not. A mechanism that does not cause global energy release is represented by the propagation of a continuous macroscopic splitting crack (Fig. 9.5.2c),

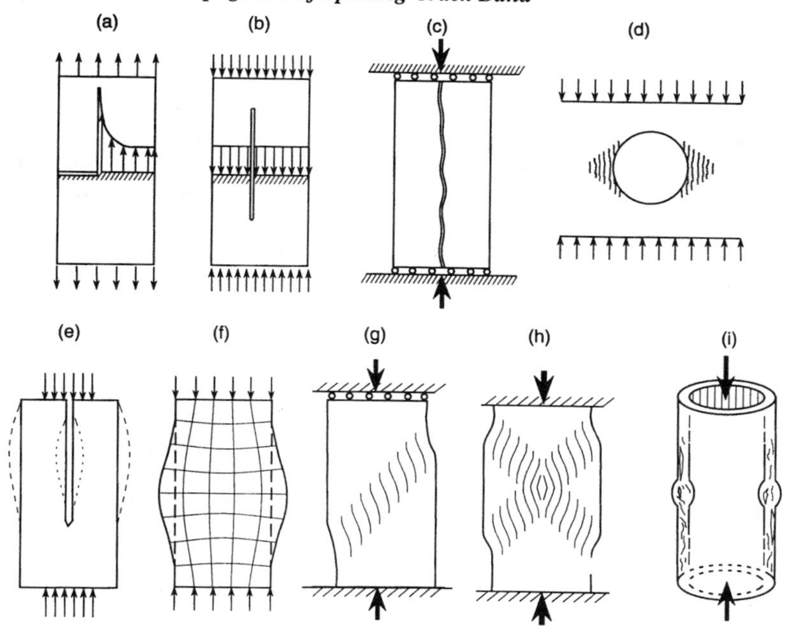

Figure 9.5.2 Global mechanisms and hypotheses on compression fracture (from Bažant and Xiang 1997).

which is known to occur in small laboratory test specimens when the ends are sliding with negligible friction. While a transverse tensile crack causes a change in the macroscopic stress field (Fig. 9.5.2a), a splitting macrocrack does not change the macroscopic stress field (Fig. 9.5.1b), and so it causes no global release of energy. The energy to form the crack and propagate it must come from a local mechanism, such as the release of stored energy from the fracture process zone (and must be calculated from its triaxial constitutive relation; Bažant and Ožbolt 1992). Because of the absence of a global energy release, this type of compression failure cannot cause any size effect, as confirmed by the numerical results of Bažant and Ožbolt (1992) and Droz and Bažant (1989). Since a size effect has been observed and is of main interest here, the splitting macrocrack will not be analyzed.

If the load required to drive the local mechanism of axial splitting crack propagation is higher than that required to drive a global mechanism of failure due to energy release, the global mechanism will occur. As we will see, the global mechanism is accompanied by size effect, and so it will prevail for sufficiently large sizes.

The mechanism of global energy release must obviously involve some sort of transverse propagation of a cracking band (or damage band). Such a band may logically be supposed to consist of densely distributed axial splitting microcracks. (That a band of discontinuous splitting microcracks can propagate sideways has been observed microscopically; e.g., Davies 1992, 1995.) The weakening of the material by microcracks may be expected to cause internal buckling.

Although important contributions have been made by mathematical modeling of the aforementioned mechanisms in which cracks are engendered by pores, inclusions, and slips on inclined interfaces, it must be recognized that they explain only the microscopic initiation of compression fracture. They do not describe the global, macroscopic compression failure. The microcracks can grow in the compression direction only for a limited distance under increasing load, but the maximum load is not reached according to these mechanisms. In the axial cross-sections through a specimen under a uniform uniaxial compression stress field, each of these three mechanisms produces a profile of self-equilibrated, alternatively tensile and compressive, microstresses which averages to a zero transverse stress on the macroscale (Fig. 9.5.2d).

It cannot be denied that the compression splitting fracture begins microscopically as a series of straight, or wing-tip, or other microcracks shown in Fig. 9.5.1a–d, but how these microcracks connect and propagate macroscopically is not explained by the aforementioned microscopic mechanisms. This needs to be described by a global mechanical model. A simple form of such a model, simple enough to allow a straightforward analytical solution, was proposed by Bažant (1994a) and refined by Bažant and Xiang

(1997), and is described here. The model is based on the hypothesis that the axial straight or wing-tip microcracks can become stacked in a lateral direction to produce a transverse (inclined or orthogonal) compression-shear band (in rock mechanics also called 'en-echelon' cracks).

Relatively little work has been done on the global mechanism of compression failure. Biot (1965), in relation to his previous model of internal instabilities such as strata folding in geology, proposed that compression failure involves internal buckling of a three-dimensional continuum, and pioneered elastic continuum solutions of such instabilities (Fig. 9.5.2g). Biot's studies, however, were limited to elastic materials without damage, and consequently the predicted critical stresses for such instabilities were much too high. Bažant (1967) applied finite strain analysis to the bulging and other internal instabilities of thick compressed solids made orthotropic by microcracking damage (Fig. 9.5.2i) and showed that such instabilities can explain the failure of an axially compressed fiber-reinforced composite tube, describing realistically the dependence of the failure stress on the ratio of the wall thickness to the diameter.

The role of buckling was further clarified in an important contribution by Kendall (1978), who studied the axial splitting fracture of a prism compressed on only a part of its end surfaces. He managed to obtain rather simple formulas. Simple formulas were also derived for axially compressed fiber-reinforced laminates in which internal buckling is engendered by the waviness of fibers in the layers of fabric (Bažant 1968; Bažant and Cedolin 1991, Sec. 11.9). In Kendall's model, however, the buckling of the specimen halves was caused by load eccentricity. His model could not explain the axial splitting fracture of a compressed specimen uniformly loaded over the entire end surface, for which the critical buckling stress obtained from his model is much too high. Nevertheless, the notion that instability of a specimen weakened by axial cracks is part of a global compression failure mechanism has been clearly established.

Another phenomenon that drives the compression failure is a release of the stored energy, the same as in tensile fracture. This concept was introduced in the analysis of stopes in very deep mines in Transwaal in the 1960s, and an empirical failure criterion based on the energy release from the rock mass as a function of the length of the stope was established (and simulated by an electric analog model at the Chamber of Mines in Pretoria); Hoek and Bieniawski (1965), Bieniawski (1974).

The global energy release aspect was brought into the modeling of compression failure in a study of the compression breakout of boreholes in rock (Bažant, Lin and Lippmann 1993). A band of parallel splitting cracks was considered to propagate from the sides of the borehole, driven by the release of strain energy from the surrounding rock mass. It was shown that such a model predicts a size effect, which basically agrees with the recent test results of Haimson and Herrick (1989), Carter (1992), and Carter, Lajtai and Yuan (1992). This solution contrasts with previous plasticity solutions of borehole breakout which predict no size effect. The stored energy release due to propagation of a band of axial splitting cracks, coupled with buckling of the slabs of material between the cracks, have been two principal aspects of a model proposed by Bažant, Lin and Lippmann (1993) which was the basis of Bažant's (1994a) and Bažant and Xiang's (1997) analysis.

A nonlocal constitutive damage model capable of capturing the energy release was used by Droz and Bažant (1989) and Bažant and Ožbolt (1992) for finite element modeling of compressed rectangular specimens. These studies predicted no significant size effect for such specimens. The absence of size effect does not contradict the available test data for normal size laboratory specimens. These specimens are too small to exhibit size effect. For the size effect due to energy release to get manifested, the compressed structure must be much larger than the compression cracking zone, which is not the case for normal test specimens.

The delamination fracture, in which models involving buckling and fracture propagation are well established, is a rather special and broad topic which lies beyond the scope of the following analysis. So does a possible fractal aspect of the problem, for example, fractal comminution and fragmentation of crushed sea ice (Palmer and Sanderson 1991).

9.5.2 Energy Analysis of Compression Failure of Column

The plastic limit state analysis of compression failure is based on the hypothesis that a system of yielding surfaces creating a single-degree-of-freedom failure mechanism develops at maximum load. Except in the case of strongly confined concrete under very high pressure, such a hypothesis, however, appears to be unrealistic for the following reasons. **(1)** The load-deflection diagram for plastic failure of a stocky column would have to end with a horizontal yield plateau, but in reality there is postpeak softening. **(2)** Material

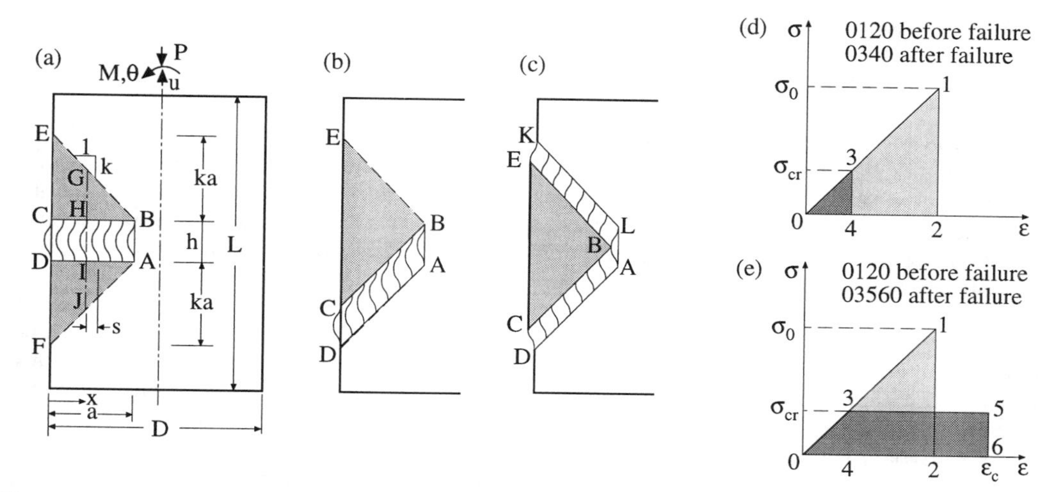

Figure 9.5.3 (a-c) Splitting cracks, buckling of microslabs, and stress relief zone, (d-e) stress-strain diagrams with and without buckling, and areas representing strain energy changes (from Bažant and Xiang 1997).

tests indicate that concrete is incapable of plastic deformation except under high confining pressure. (**3**) In experiments, the compression failure is actually seen to be caused by fractures. (**4**) According to the recent reduced-scale tests (Bažant and Kwon 1994), there is a strong size effect, which is of the type associated with the propagation of crack bands.

In brittle or quasibrittle materials, compression failure begins by the formation of axial splitting cracks. However, the axial splitting cracks do not change the macroscopic continuum stress state due to uniaxial compression. Consequently, they cause no energy release. So they cannot, by themselves, be the mechanism of compression failure, and cannot control the failure load. They can only be the mechanism that triggers the macroscopic compression failure. As already suggested in Bažant (1994a), it is proposed that the principal mechanism of compression failure of a concrete column is sideways propagation of a band of parallel axial splitting cracks, in a direction either orthogonal or inclined with respect to the direction of compression, as shown in Fig. 9.5.3a. This figure shows several alternative geometries of the crack band, which lead, according to the approximate analysis to be presented, to equivalent results.

Following Bažant (1994a) and Bažant and Xiang (1997), we will now analyze compressed columns of different sizes D, geometrically similar in two dimensions (Fig. 9.5.3a). The column length is L, the thickness is $b = 1$, and the width is taken as the size, or characteristic dimension, D. The bottom of column is fixed. The top of column is loaded by axial compressive force P of eccentricity e. To compare the load capacity of columns of different sizes, we define the nominal strength of column, $\sigma_{Nu} = P_u/bD$, where P_u is the maximum value of load P. We are interested in failure under load-control conditions (dead load), and so the failure occurs at maximum load. Therefore, P remains constant during the small deformation increment representing failure.

The initial normal stress in the cross sections before any fracturing is

$$\sigma_0(x) = -\frac{E}{L}\left[u + \theta\left(\frac{D}{2} - x\right)\right] \tag{9.5.1}$$

where E = Young's elastic modulus, u = load-point axial displacement, θ = rotation of top end, and x = transverse coordinate measured from the compressed face (Fig. 9.5.3a). We now assume that, at a certain moment of loading, axial cracks of spacing s and length h, forming a band as shown in Fig. 9.5.3a,b,c, suddenly appear and the microslabs of the material between the axial cracks, behaving as beams of depth s, lose stability and buckle. This can happen in any one of the three mechanisms shown in Fig. 9.5.3a,b,c, and for all of them, the mathematics turns out to be identical. If the length of the cracks in the two inclined bands in Fig. 9.5.3c is denoted as $h/2$, the critical stress for the microslab buckling for all the cases shown in Fig. 9.5.3a,b,c is, according to Euler's formula for columns,

$$\sigma_{cr} = -\frac{\pi^2 E s^2}{3h^2} \tag{9.5.2}$$

The key idea is the calculation of the change in the stored strain energy caused by buckling (Bažant 1994a). On both sides of the crack band, there is obviously a zone in which the initial stress σ_0 is reduced. For the sake of simplified analysis, we assume that the stress in the shaded triangular zones of Fig. 9.5.3a,b,c is relieved to the value of σ_{cr}. Further, we assume that outside these zones the initial stress does not change (this is, of course, a simplification of reality, because stress concentration arises ahead of the crack band).

The triangular stress relief zones are limited by the so-called "stress diffusion lines" of slope k, whose magnitude is of the order of 1. The value of k can be determined by experiment or by solution of the elastic stress field. Approximately, the value of k may be taken as the value that gives the exact stress intensity factor of the edge-cracked half-space according to linear elastic fracture mechanics, which is $k = \pi \{[f_1 D + f_2 6(e+w)]/[D + 6(e+w)]\}^2$, in which $f_1(a/D)$ and $f_2(a/D)$ can be found in handbooks (Tada, Paris and Irwin 1985; Murakami 1987). However, the value of k need not be known for our purposes. The only important point is that k is a constant having the same value for geometrically similar columns of different sizes.

The strain energy density in the shaded triangular stress relief zones before and after fracture is equal to the areas of triangles 0120 and 0340 in Fig. 9.5.3d. So, the loss of strain energy density at points on a vertical line of coordinate x is

$$\Delta\overline{\mathcal{U}}_r = \frac{\sigma_0^2(x)}{2E} - \frac{\sigma_{cr}^2}{2E} \tag{9.5.3}$$

The situation is more complicated in the crack band. The microslabs buckle, and the energy associated with the postbuckling deflections must be taken into account, which is an important point for the solution (Bažant 1994a). The strain energy density before buckling of the microslabs is given by the area 0120 in Fig. 9.5.3e. The well-known solution of postbuckling behavior of columns (Bažant and Cedolin 1991, Sec. 1.9 and 5.9) indicates that the stress in the axis of the microslab follows, after the attainment of the critical load of the microslab, the straight line 35 which has a very small positive slope (precisely equal to $\sigma_{cr}/2$). This slope is far smaller than the slope E before buckling and can, therefore, be neglected. So the postbuckling behavior in Fig. 9.5.3e is approximately a horizontal plateau 35 (however, this is not the same as plastic behavior because unloading proceeds along the path 530). Because the microslabs remain elastic during buckling, the stress-strain diagram 035 is fully reversible and the energy represented by the area under this diagram is the stored elastic strain energy. The triangular area 0340 in Fig. 9.5.3e represents the axial strain energy density of the microslabs and the rectangular area 35643 represents the bending energy density. The change in strain energy density in the microslabs is the difference of areas 0120 and 03560 in Fig. 9.5.3e, that is,

$$\Delta\overline{\mathcal{U}}_c = \frac{\sigma_0^2(x)}{2E} - \left[\sigma_{cr}\epsilon_c(x) - \frac{\sigma_{cr}^2}{2E}\right] \tag{9.5.4}$$

where ϵ_c is the axial strain of the microslabs in the crack band after buckling (it is important to realize that this strain is generally not equal to 04 or 02 in Fig. 9.5.3e).

Integration of (9.5.3) and (9.5.4) yields the total loss of strain energy at constant u and σ:

$$\Delta\mathcal{U} = -\int_0^a \left(\frac{\sigma_0^2(x)}{2E} - \frac{\sigma_{cr}^2}{2E}\right) 2k(a-x)\,dx - \int_0^a \left\{\frac{\sigma_0^2(x)}{2E} - \left[\sigma_{cr}\epsilon_c(x) - \frac{\sigma_{cr}^2}{2E}\right]\right\} h\,dx \tag{9.5.5}$$

where a = horizontal length of the crack band (Fig. 9.5.3a,b,c). The rate of this energy loss must be equal to the rate at which the energy is consumed by formation of the axial splitting cracks. Thus, the energy balance criterion of fracture mechanics may be written as:

$$-\left[\frac{\partial \Delta\mathcal{U}}{\partial a}\right]_{\theta,u} = \frac{\partial}{\partial a}\left(G_f h \frac{a}{s}\right) = G_f \frac{h}{s} \tag{9.5.6}$$

where G_f is the fracture energy of the axial splitting cracks, assumed to be a material property, and $G_f^* = G_f h/s$ represents the fracture energy of the crack band.

The axial strain in the crack band can be determined from the compatibility condition. The blank zone outside the shaded triangular stress relief zone (Fig. 9.5.3a,b,c) behaves during buckling as a rigid body because the load, and thus also the stress in this zone, are constant during the deformation increment representing failure. So the line segment GJ in Fig. 9.5.3a at any x does not change length during buckling. Expressing the change of length of this segment on the basis of σ_{cr}, ϵ_c, and σ_0 and setting this change equal to zero, one acquires the following compatibility condition:

$$\epsilon_c(x) = \frac{\sigma_0(x)}{Eh}\left[h + 2k(a-x)\right] - \frac{2k}{h}(a-x)\frac{\sigma_{cr}}{E} \tag{9.5.7}$$

The length h of the axial cracks, representing the width of the crack band in Fig. 9.5.3a,b or double the crack band width in Fig. 9.5.3c, is an important parameter that must be determined. The critical stress according to (9.5.2) would decrease with increasing h, and so the largest energy release would be obtained for $h \to \infty$. Since the largest energy release is what must happen (because of thermodynamic considerations; Bažant and Cedolin 1991, chapters 10 and 12), the prediction would be $\sigma_{cr} = 0$, which, however, is unreasonable.

In a recent study of the role of axial splitting cracks in borehole breakout (Bažant, Lin and Lippmann 1993; see a subsequent section), the microslab buckling was assumed to be opposed by shear stresses on the microcracks taken as proportional to the slip on the microcracks. That assumption leads to a more complicated formula for σ_{cr} than (9.5.2), but it is noteworthy that the σ_{cr} attains a minimum for a certain finite value of h. A similar approach would be possible for the present problem. However, to keep the solution simple, we will not specifically consider shear stress transmission across the cracks and will simply assume that h is a constant to be determined empirically.

In tied reinforced concrete columns, there is also another feature affecting h—the ties, whose spacing probably poses an upper limit on the crack length.

Now we can substitute (9.5.1)–(9.5.4) into (9.5.5) and integrate. The result is

$$\Delta\mathcal{U} = \frac{D^2}{6E}\left[2k\sigma_2^2\alpha^4 + 4(\bar{h}\sigma_2^2 - k\sigma_2\sigma_3)\alpha^3 + 3(k\sigma_3^2 - 2\bar{h}\sigma_2\sigma_3)\alpha^2 + 3\bar{h}\sigma_3^2\alpha)\right] \tag{9.5.8}$$

in which,

$$\sigma_1 = E\frac{u}{L} + \sigma_{cr}, \qquad \sigma_2 = E\frac{\theta D}{2L}, \qquad \sigma_3 = \sigma_1 + \sigma_2$$
$$\alpha = \frac{a}{D}, \qquad \bar{h} = \frac{h}{D} \tag{9.5.9}$$

Substituting this into the energy criterion (9.5.6) of crack band propagation and noting that $\partial/\partial a = (1/D)\partial/\partial\alpha$, we get

$$8k\sigma_2^2\alpha^3 + 12(\bar{h}\sigma_2^2 - k\sigma_2\sigma_3)\alpha^2 + 6(k\sigma_3^2 - 2\bar{h}\sigma_2\sigma_3)\alpha + 3\bar{h}(\sigma_3^2 - 2EG_f/s) = 0 \tag{9.5.10}$$

This equation represents the failure condition, that is, the condition of crack band propagation at maximum load. It is a condition expressed in terms of displacements u and θ because σ_1, σ_2, and σ_3 are functions of u and θ. However, we are interested in the nominal strength σ_{Nu}, which is a parameter of the maximum load P_u. To obtain the failure condition in terms of the load, we substitute the expressions:

$$\sigma_1 = \sigma_{cr} - \sigma_N, \qquad \sigma_2 = -\sigma_M, \qquad \sigma_3 = \sigma_{cr} - \sigma_N - \sigma_M \tag{9.5.11}$$

in which

$$\sigma_N = \frac{P}{bD}, \qquad \sigma_M = \frac{MD}{2I} = \frac{6M}{bD^2}, \qquad M = P(e+w) \tag{9.5.12}$$

Here, P is positive when compressive, M is the bending moment at the location of the crack band, and w is the column deflection (Fig. 9.5.4c) at that location (maximum deflection within the column length). For a stocky column, one may use $w \approx 0$. For a slender column, one may approximate w on the basis of

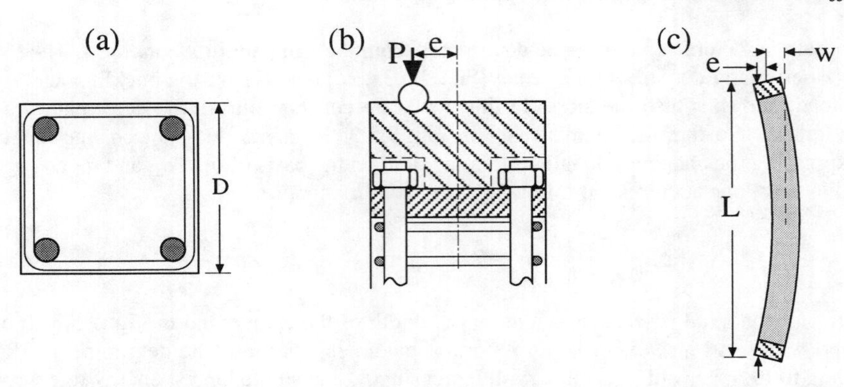

Figure 9.5.4 Reduced scale columns with reduced-size aggregate tested by Bažant and Kwon (1994): (a) typical cross section; (b) detail of loading system; c) sketch of overall loading and deflection.

the amplification factor, i.e.,

$$e + w = e\left(1 - \frac{P}{P_{cr}}\right)^{-1} \tag{9.5.13}$$

where P_{cr} = first critical load of the column according to the theory of elasticity; e.g., in the case of a simply supported (hinged) column, $P_{cr} = EI\pi^2/L^2$ where $I = bD^3/12$ = centroidal moment of inertia of the cross section.

Substituting (9.5.11) into (9.5.10), one gets the crack band propagation criterion in terms of P, which has the general form:

$$F(k, \alpha, h, s, G_f; P) = 0. \tag{9.5.14}$$

To solve for P, one needs to know the five parameters listed above. Of these, only two are free, namely α and s, while G_f and h are material properties which must be given, and k can be determined from elasticity as already mentioned. The constancy of h is the central hypothesis of the present analysis.

The relative crack band length α at maximum load could, in principle, be determined from the criterion of stability loss of the crack band (in the R-curve model, this criterion represents the condition that the energy release curve at constant P be tangent to the R-curve). Experience with other problems shows that application of this criterion usually gives α-values that are nearly independent of D. For the sake of simplicity, we prefer not to complicate the solution with the condition of crack band stability, and simply assume that the values of the relative crack band length $\alpha = a/D$ for specimens of different sizes are the same. They can be determined by experiments.

Similar to the solution of borehole breakout (Bažant, Lin and Lippmann 1993), we may assume that s will be such that the column fails at the first opportunity it has. This means that the value of s can be determined so that P be minimized. The necessary condition of minimum is

$$\frac{\partial F}{\partial s} = 0 \tag{9.5.15}$$

Substituting function F according to (9.5.10) and differentiating with respect to s, we obtain

$$-2k\sigma_2\alpha^2 + 2(k\sigma_3 - \bar{h}\sigma_2)\alpha + \bar{h}\sigma_3 + \frac{3h^3G_f}{2\pi^2 s^3 D} = 0 \tag{9.5.16}$$

Equations (9.5.10) and (9.5.16) define the solution of the size effect plot of σ_{Nu} vs. D. After substituting (9.5.11), (9.5.12), and (9.5.2), equation (9.5.10) can be rearranged to an equation that is quadratic in σ_{Nu} and of fifth degree in s, and (9.5.16) can be rearranged to an equation that is linear in σ_{Nu} and of fifth degree in s. To obtain a point of the plot, we assume a value of s. Then, σ_{Nu} can be solved from (9.5.16) if parameters G_f, a/D, and h are known. With σ_{Nu} known, s can be reevaluated from (9.5.10). This calculation cycle is then iterated, yielding a pair of σ_{Nu} and P_u values.

Alternatively, one may solve (9.5.10) and (9.5.16) with the help of the Levenberg–Marquardt nonlinear optimization algorithm. If experimental data on the size effect curve of σ_{Nu} vs. D are given, one can use this algorithm to find the values of G_f, a/D, and h that minimize the sum of squares of the differences between the experimental data and the solution of (9.5.10) and (9.5.16).

9.5.3 Asymptotic Effect for Large Size

Let us now examine the asymptotic size effect for columns of very large sizes, $D \to \infty$, under the assumption that h, α, and e/D for columns of different sizes are constant. G_f, as a material property, must, of course, be constant, too. Analysis of (9.5.10) and (9.5.16) with (9.5.11), (9.5.12), and (9.5.2) shows that, for $D \to \infty$, $P \propto bsD^{1/2}$ and

$$s = c_m D^{-1/5} \qquad (9.5.17)$$

where c_m is a certain constant and $\propto$ is the proportionality sign. Because $\sigma_{Nu} = P_u/bD$, the asymptotic size effect on the nominal strength is

$$\sigma_{Nu} \propto s D^{-1/2} \qquad (9.5.18)$$

or

$$\sigma_{Nu} \propto D^{-2/5} \qquad (9.5.19)$$

It is also found that $\sigma_1 \propto \sigma_2 \propto \sigma_3 \propto D^{-2/5}$. For the buckling of microslabs, the large-size asymptotic behavior is

$$\sigma_{cr} = \pi^2 E s^2 / 3h^2 \propto D^{-2/5}. \qquad (9.5.20)$$

For the nominal bending strength σ^0_{Nu} of short columns (for which $w \ll e$) with similar load eccentricities (same e/D), we obtain $\sigma_M = MD/2I = 6PeD/bD^3 \propto sD^{1/2}DD/bD^3$ or

$$\sigma^0_M \propto s D^{-1/2} \propto D^{-2/5} \qquad (9.5.21)$$

To verify (9.5.19), note that the first and second terms in (9.5.16) are, for large D, of the order of $D^{-2/5}$ (note that $\bar{h}\sigma_2$ in the second term is of the order of $D^{-9/5}$, which is a higher order small term than $D^{-2/5}$ and can, therefore, be neglected in comparison). The last term in (9.5.16), proportional to $1/(D^{-3/5}D)$, is also of the order of $D^{-2/5}$. The same type of analysis can be applied to (9.5.10).

It is important to note that the asymptotic size effect in compression failure, as indicated in (9.5.19), is weaker than in linear elastic fracture mechanics (LEFM), for which $\sigma_{Nu} \propto D^{-1/2}$. This difference in the asymptotic size effect, which is the same as previously found for the compression breakout of boreholes in rock (Bažant, Lin and Lippmann 1993), is caused by the fact that the spacing s of the axial splitting microcracks is not constant but, according to (9.5.17), decreases with size D asymptotically as $D^{-1/5}$. This variation of s is the consequence of our minimizing the failure load P with respect to the microcrack spacing. Such minimization is physically correct only if the material inhomogeneities are sufficiently fine. We will comment on that more.

9.5.4 Size Effect Law for Axial Compression of Stocky Column

The size effect according to (9.5.10) and (9.5.16) is given implicitly. However, for centric axial compression of a column of negligible slenderness, a simple explicit formula for the size effect can be obtained. For $M = 0$, we have $\sigma_2 = 0$ and $\sigma_1 = \sigma_3$. From (9.5.10), we get

$$(2k\alpha + \bar{h})\sigma_3^2 - \frac{2\bar{h}EG_f}{s} = 0 \qquad (9.5.22)$$

Then, from (9.5.16)

$$(2k\alpha + \bar{h})\sigma_3 + \frac{3h^3 G_f}{2\pi^2 s^3 D} = 0 \qquad (9.5.23)$$

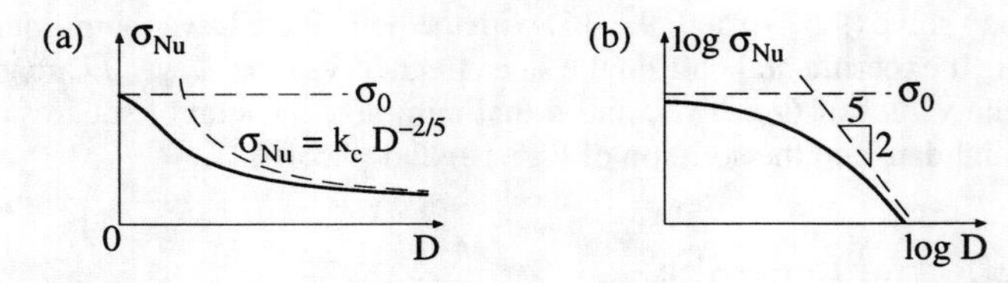

Figure 9.5.5 Size effect deduced for compression failures (a) in linear scale, and (b) in logarithmic scale (adapted from Bažant and Xiang 1997).

From the last two equations, we obtain

$$\sigma_3 = -2.21 \left[\frac{E^3 G_f^2}{(2ka + h)^2} \right]^{1/5} \tag{9.5.24}$$

$$s = 0.41h \left[\frac{G_f}{E(2ka + h)} \right]^{1/5} \tag{9.5.25}$$

and consequently,

$$\sigma_{Nu} = \sigma_{cr} - \sigma_3 = \frac{\pi^2 E s^2}{3h^2} + 2.21 \left[\frac{E^3 G_f^2}{(2ka + h)^2} \right]^{1/5} \tag{9.5.26}$$

Finally, upon rearrangement,

$$\sigma_{Nu} = 2.76 \left[\frac{E^3 G_f^2}{(2ka + h)^2} \right]^{1/5} \tag{9.5.27}$$

Now consider the limit cases of size effect. For $D \to 0$ (which implies $a \to 0$), we have $\sigma_{Nu} = 2.76(E^3 G_f^2/h^2)^{1/5} = \sigma_0$, representing the material strength limit. For $D \to \infty$ (which implies $a \to \infty$), we have $\sigma_{Nu} \to 0$. Aside from that, we already know that the large size asymptotic size effect is $\sigma_{Nu} \propto D^{-2/5}$. Based on these results, the size effect plot has the shape shown in Fig. 9.5.5.

An explicit formula for σ_{Nu} as a function of D cannot be obtained. It turns out, however, that an explicit formula can be constructed for the inverse relation (as already indicated in Bažant 1994a). Indeed, denoting $\bar{\sigma} = \sqrt{2EG_f h/s}$, we can rearrange (9.5.22) to the following formula for the inverse size effect law:

$$a = \frac{h}{2k}\left(\frac{\bar{\sigma}^2}{\sigma_3^2} - 1\right) = \frac{h}{2k}\frac{(\bar{\sigma} - \sigma_3)(\bar{\sigma} + \sigma_3)}{\sigma_3^2}$$
$$= \frac{h}{2k}\frac{(\bar{\sigma} - \sigma_{cr} + \sigma_{Nu})(\bar{\sigma} + \sigma_{cr} - \sigma_{Nu})}{(\sigma_{Nu} - \sigma_{cr})^2} \tag{9.5.28}$$

It may be noted that this formula is quite similar to the following formula obtained by Bažant (1994a) after making deeper simplifications:

$$D = D_0 \frac{(\sigma_0 - \sigma_P)(\sigma_P + \sigma_0 - 2\sigma_r)}{(\sigma_0 - \sigma_r)^2} \tag{9.5.29}$$

Remark: If the compressed prism is quite short and the ends are not sliding freely, as is typical of standard compression tests, then the tensile splitting band cannot develop and cannot release any significant amount of energy from the surrounding undamaged concrete. It is for this reason that no clear deterministic size effect is observed in standard compression tests, and none is indicated by finite element simulations (Droz and Bažant1989; Bažant and Ožbolt 1992). △

9.5.5 Effect of Buckling Due to Slenderness

A more slender column deflects more under the same axial load, and so it stores more strain energy. Consequently, it can also release more energy to drive the crack band, which means one should expect a size effect closer to LEFM, i.e., stronger. Indeed, the reduced scale laboratory experiments of Bažant and Kwon (1994) showed that the size effect in columns becomes more pronounced with increasing slenderness, D/L. The question now is how the influence of slenderness on the size effect should be incorporated into (9.5.10) and (9.5.16). There are two ways to do that.

a) Simpler Approach Based on Magnification Factor. The size effect implied in (9.5.10) with (9.5.16) can be described as $\sigma_{Nu} = f(D)$ where f is the function implicitly defined by these equations. The consequence of slenderness is to magnify the lateral deflection. The magnification can be approximately calculated as μe in which $\mu = (1 - P/P_{cr})^{-1}$ = magnification factor (e.g., Bažant and Cedolin 1991, Ch. 1), P_{cr} = first critical load of the column, whose value decreases with increasing slenderness ℓ/D. Writing now the same definition of the nominal stress as for small slenderness, and imposing the condition that the stress given by the size effect law be the maximum stress in the deflected slender column, we have

$$\sigma_N = \frac{P}{D}\left(1 + \frac{6e}{D}\right), \qquad \frac{P}{D}\left[1 + \frac{6e}{D}\left(1 - \frac{P}{P_{cr}}\right)^{-1}\right] = f(D) \tag{9.5.30}$$

The size effect plot of σ_{Nu} vs. D is the solution of this system of two equations, in which P figures as a parameter to be eliminated. The solutions indicate that, indeed, the size effect is more pronounced for higher slenderness.

b) Fundamental Approach Based on Additional Energy Release. From the fracture mechanics viewpoint, the effect of slenderness should be calculated on the basis of the additional energy release engendered by slenderness. This approach is not as simple as the previous one, but is not excessively complicated either. If the column is slender, the release of strain energy from the column must be taken into account. As will now be described, this can be done in the manner outlined in Bažant (1994a). We begin by identifying with the column a short segment of length L (Fig. 9.5.3) confined between the cross sections at the end of the triangular stress relief zones, having relative displacement u and relative rotation θ.

To make the analysis simple, we may assume that, during the advance of crack band length, δa, the values of u and θ remain constant. This means that the applied load, P, the load-point displacement u at column end, and the mid-height deflection w all change. In that case, the change of stresses and deformations due to column buckling does not interfere with the triangular energy release zones we considered earlier (Fig. 9.5.3). Of course, one could calculate the energy release at fixed load-point displacement or at fixed load. But, in that case, the stresses and strains in the unshaded area of column in Fig. 9.5.3 would not remain constant during the advance of the band of splitting cracks, but would change. This would make it impossible to build on our preceding solution without slenderness effect, and would thus complicate the solution.

The fact that u and θ, rather than the column ends, are considered to be fixed is not objectionable. It is well known in fracture mechanics that the energy release of a fracture specimen can be calculated for different types of load control, e.g., for constant load, or constant deflection, or constant ratio of load and deflection, and always with the same result (e.g., Bažant and Cedolin, Sec. 12.1).

Considering the ends of the column to be supported on hinges, we may approximate the deflection curve as $z = w\sin(\pi y/\ell)$, where w = midheight deflection and y = longitudinal coordinate. The change in the axial force at midheight can be calculated from the change of the stress distribution due to the extension of the band of splitting cracks by δa:

$$\delta P = [\sigma_{cr} - \sigma_0(a)]\delta a, \qquad \delta M = [\sigma_{cr} - \sigma_0(a)]\left(\frac{P}{2} - a\right)\delta a \tag{9.5.31}$$

where σ_{cr} is the critical stress of the microslabs. The axial load, P, is assumed to have constant eccentricity e at both ends of the column, and so $M = P(e + w)$ or $w = (M/P) - e$. Differentiating, we have

$$\delta w = [\delta M - (e + w)\delta P]/P \tag{9.5.32}$$

The axial shortening due to deflection w is $u = \int_0^\ell (z')^2/2\,dy = \pi^2 w^2/4\ell$, and so the work of P during

δa is

$$\delta W = P\delta u = \frac{\pi^2 P w}{2\ell}\delta w \tag{9.5.33}$$

The change of stored bending energy during δa is $\delta U = \delta \int_0^\ell EI(z'')^2/2\,dy = \delta(\pi^4 EI w^2/4\ell^3)$, that is,

$$\delta U = \pi^2 EI w\,\delta w/2\ell^3 \tag{9.5.34}$$

where $I = bD^3/12$ = centroidal moment of inertia of the cross section of column.

The change of strain energy due to axial strains is $\delta \mathcal{U}_a = -\delta(P^2\ell/2EA)$ where $A = bD$ = cross section area of the column. Now the change of strain energy due to column deformation during crack band advance da is given by:

$$\delta(\Delta\mathcal{U}_a) = \mathcal{G}_2\delta a = -\frac{\pi^2}{2\ell}(P_{cr} - P)w\,\delta w - \frac{\ell}{EA}P\,\delta P \tag{9.5.35}$$

where $P_{cr} = \pi^2 EI/\ell^2$ = first critical load of hinged column. Integrating (9.5.35), we obtain the following expression for the additional energy release that needs to be added to that calculated before in (9.5.8):

$$\Delta\mathcal{U}_a = \left[\frac{\pi^2}{2}e\left(e + w - a + \frac{D}{2}\right) + \frac{P\ell^2}{EA}\right](\alpha\sigma_2 - \sigma_3) \tag{9.5.36}$$

It may now be noted that, if the column is axially very stiff and $P = P_{cr}$, there is no energy release due to column deformation, as expected. When $P < P_{cr}$, there is a positive energy release because $P\delta w$ and $P\delta P$ are negative during crack band extension. The additional energy release must obviously promote fracture, and thus it must intensify the size effect.

The subsequent calculation is the same as that which led to (9.5.10) and (9.5.16). One finds that the following terms need to be added to the left-hand sides of (9.5.10) and (9.5.16), respectively:

$$A_1 = -6A_2(\alpha\sigma_2 - \sigma_3) \tag{9.5.37}$$

$$A_2 = -\left[\frac{E\pi^2}{2\ell^2}e(e + w + \frac{D}{2} - a) + \frac{P}{A}\right] \tag{9.5.38}$$

The size effect curves for a slender column calculated on the basis of the additional energy release and on the basis of the magnification factor will be further discussed when they are compared with the test data in the next section. The size effect curve obtained when the slenderness is neglected is also shown.

9.5.6 Comparison with Experimental Data

The foregoing theory was compared to the test data of Bažant and Kwon (1994). The investigators tested, at Northwestern University, reduced-scale reinforced concrete columns of square cross section (Fig. 9.5.4). Three different sizes D for each of three different slendernesses ℓ/D were tested. The size ratio was 1:2:4; the cross section sides were $D = 0.5$, 1, 2 in. (12.7, 25.4, 50.8 mm); and the slendernesses were $\ell/D = 19.2$, 35.8, 52.5. The reinforcement, which was scaled in proportion to D, consisted of four longitudinal bars in the corners of the square cross section (diameters 1/16, 1/8, 1/4 in., or 1.59, 3.18, 6.35 mm), and of ties (diameters 1/32, 1/16, 1/8 in., or 0.79, 1.59, 3.18 mm) spaced at 0.3, 0.6, 1.2 in. (7.62, 15.2, 30.5 mm). Portland cement microconcrete of maximum aggregate size 0.132 in. (3.35 mm) was used. The concrete cover of the bars was also scaled. For further details, see Bažant and Kwon (1994).

The test data were analyzed by Bažant and Xiang (1997) under the assumption that, for columns of different sizes as well as different slendernesses, the crack band width h, rather than the relative width h/D, is constant. A reasonable assumption is that the crack band width is roughly proportional to the material length, ℓ (the same assumption is reasonable for tensile crack bands). The material length is related to the maximum inhomogeneity size, i.e., the maximum aggregate size in the case of concrete, or to the combination of fracture energy and tensile strength f'_t of the material having a dimension of length, namely, Irwin's length $\ell_{ch} = EG_f/f'^2_t$. Bažant and Xiang (1997) checked that the assumption of constant h agreed with the test results much better than the assumption of constant h/D.

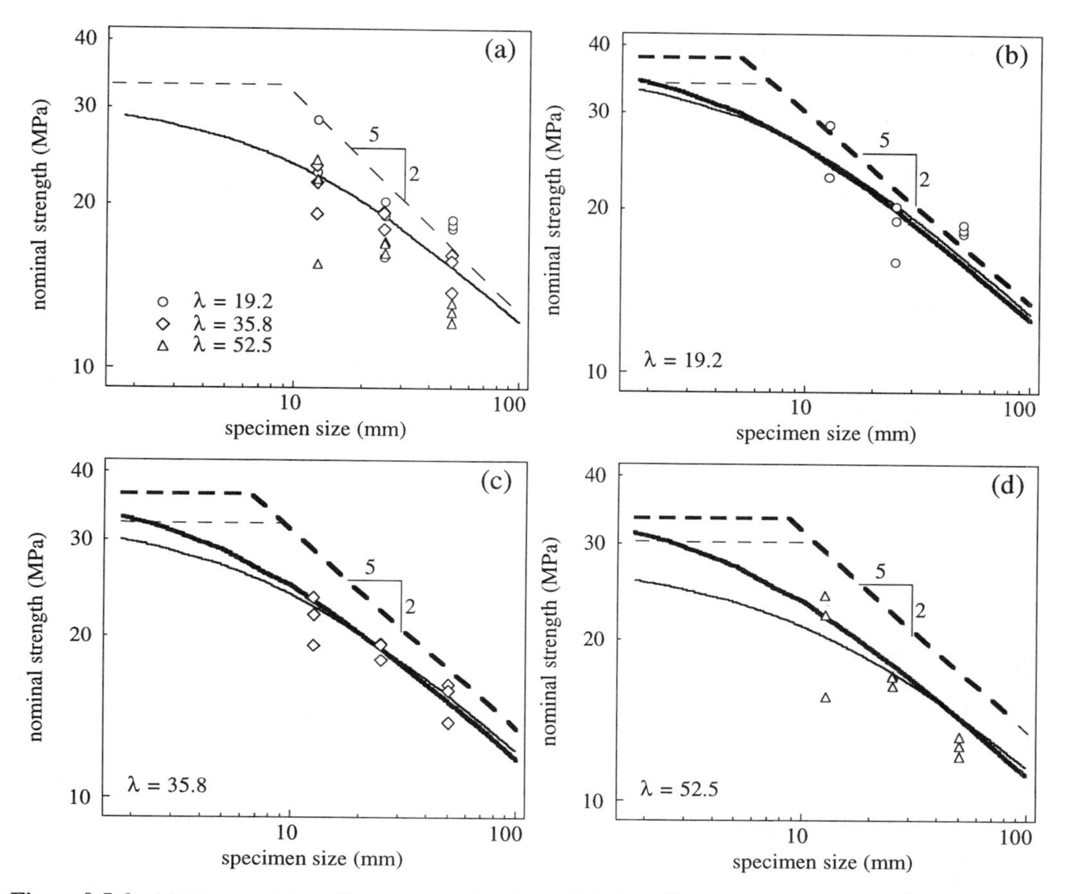

Figure 9.5.6 (a) Measured size effect compared to theoretical size effect curves when slenderness is neglected. (b)–(d) Comparison of measured size effect compared to theoretical size effect curves when slenderness is taken into account by simple analysis based on magnification factor (thin line) and by analysis based on the additional energy release (thick line). Dashed lines are the asymptotes. (After Bažant and Xiang 1997.)

The values a/D were also assumed constant for columns of different sizes, including those of different slendernesses. In this regard, it may be noted that differences in slenderness cannot represent important deviations from geometrical similarity of failure mode because, according to Saint-Venant's principle, addition of a mass to the column end to make the column longer cannot appreciably affect the stress field in the fracturing zone in the middle of the column.

The comparison of the theory with these data is presented in Fig. 9.5.6. The lines show the theoretical results, and the data points show the test results. Fig. 9.5.6a gives the results when the slenderness is neglected. Fig. 9.5.6b–d show the comparisons for the simpler analysis of slenderness based on magnification factor (thin lines) and the analysis of slenderness based on the additional energy release (thick line). The optimized parameters of the three different methods of analysis giving the best fit of the test data are listed in Table 9.5.1.

As seen from Figs. 9.5.6b–d, both the simplified and the additional energy release approaches to slenderness effect can represent the test results for the reduced scale columns quite well. It may also be noted that the size effect, which is neglected by the present code specifications, is quite significant.

Size effect tests of columns of normal sizes and with normal size aggregate are, of course, needed. A series of such tests has just been completed by B. I. G. Barr and S. Şener at the University of Wales in Cardiff (private communication to Z. P. Bažant, August 1996). The results are similar and agree with the foregoing theory.

Table 9.5.1 Parameters for optimization of test data (after Bažant and Xiang 1997)

Approaches	Slenderness neglected	Magnification factor	Additional energy release
a/D	0.221	0.276	0.255
h (mm)	33.6	28.6	23.1
G_f (N/m)	15.5	19.7	16.7
ω	0.137	0.127	0.122

9.5.7 The Question of Variation of Microcrack Spacing with Size D

By minimizing load capacity P, we found in (9.5.17) that the spacing of the splitting microcracks $s = c_m D^{-1/5}$ where c_m = constant. Strictly speaking, however, a continuous variation of crack spacing with the structure size is possible only for homogeneous materials. For highly heterogeneous materials such as concrete, the dominant spacing of the main splitting microcracks must be an integer multiple of the dominant spacing of the largest aggregate pieces, s_0. Therefore, spacing s can vary only by jumps from one integer multiple of s_0 to the next. Its value may be taken as the positive integer multiple as close to $c_m D^{-1/5}$ as possible, i.e.,

$$s = s_0 \operatorname{Int}[(0.5 + (c_m D^{-1/5}/s_0)] \qquad \text{but} \qquad s \geq s_0 \tag{9.5.39}$$

This means that for finite large intervals of the size range, the dominant microcrack spacing s cannot change in the case of a highly heterogeneous material such as concrete. It is easy to adjust the preceding analysis for the case of constant spacing s. A constant rather than variable s needs to be substituted into (9.5.18), (9.5.20), and (9.5.21). The main difference is that, instead of (9.5.19), the asymptotic size effect then becomes

$$s \propto D^{-1/2} \tag{9.5.40}$$

which is the same as for all the types of tensile failure studied before (e.g., Bažant 1984a).

The materials that are not sufficiently heterogeneous for being characterized by (9.5.39) and (9.5.40) probably include fine-grained rocks such as limestone, mortar, ceramics, and pure ice, and probably exclude concrete.

In summary: under the assumption of small enough material inhomogeneities, the spacing s of the splitting cracks is calculated by minimizing the failure load and is found to decrease with structure size D as $D^{-1/5}$. The size effect on the nominal strength of geometrically similar columns is found to disappear asymptotically for small sizes D, and to asymptotically approach the power law $D^{-2/5}$ for large sizes D (where D = cross section dimension). However, when the material inhomogeneities are so large that they preclude the decrease of s with increasing D, the asymptotic size effect changes to $D^{-1/2}$.

9.5.8 Special Case of Compression with Transverse Tension

A simple special modification (Bažant 1996b) of the mechanism just explained is necessary to cope with the failure under shear loading or, which is equivalent, under compression loading accompanied by large transverse tensile strain. In that case, a system of parallel cracks in the direction of compression may develop before the maximum compression load (Fig. 9.5.7c) because these cracks are not axial splitting cracks but are produced by transverse tension. If a sufficient restraint is provided, the opening of these cracks does not localize into a single crack. An example is the formation of diagonal cracks under shear loading of a concrete beam in which restraint is provided by the longitudinal reinforcement and stirrups (Fig. 9.5.7d), which is analyzed in detail in Section 10.3, following Bažant (1996b). The transverse tensile stress and the diagonal cracks caused by it have a large effect on the compression behavior in the direction of the cracks (Hsu 1988, 1993).

When a transverse tensile crack forms in an isotropic specimen, the stress is relieved approximately from the shaded triangular areas shown in Fig. 9.5.7a whose height is about the same as the width. But when a transverse tensile crack forms in a highly orthotropic specimen under tension, for example, a unidirectional fiber composite, the triangular stress relief zones from which the strain energy is released

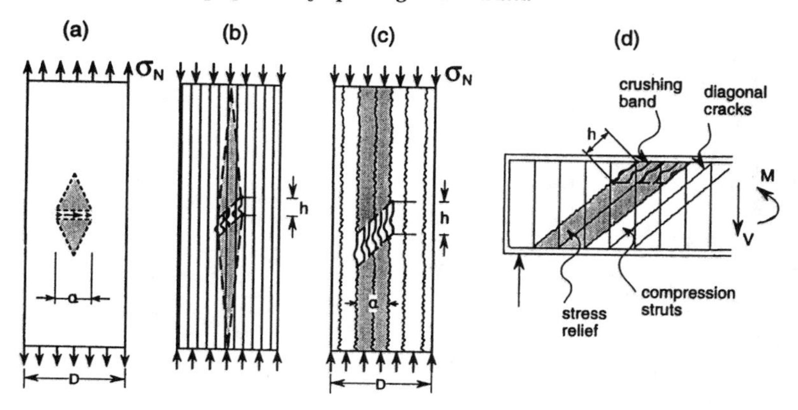

Figure 9.5.7 Energy release zones of (a) a tensile crack in an isotropic material; (b) a tensile crack in a highly orthotropic elastic material; (c) a band of compression splitting cracks in a concrete with a system of parallel macrocracks due to transverse tension loading; and (d) application to size effect in failure of compression struts in reinforced concrete beam with stirrups under shear loading (after Bažant 1996b; see also Section 10.3).

are extremely elongated, with the sides almost parallel to the direction of fibers (Fig. 9.5.7b). The same is true for compression loading when a transverse slit is cut out. From this analogy it is clear that a system of continuous parallel tensile cracks can also cause the material such as concrete to become highly orthotropic on the macroscale. The stress relief zone of a transversely propagating band of axial splitting microcracks can then become nearly a strip, which is limited by parallel tensile cracks and is shown as the shaded strip in Fig. 9.5.7c (Bažant 1996b).

The energy released from the shaded strip in Fig. 9.5.7c, from which the stress is relieved by the band, is expressed easily as:

$$\Delta\mathcal{U} = \frac{\sigma_N^2}{2E} baL \tag{9.5.41}$$

in which σ_N is the applied, initially uniform, axial compressive stress and L is the specimen length. For the sake of simplicity, the residual stress in the crack band is neglected. The energy consumed (dissipated) by the band of axial splitting microcracks is $W_f = G_f bha/s$. The energy balance during the propagation of the band requires that $-\partial\Delta\mathcal{U}/\partial a = \partial W_f/\partial a$. This yields the relation $(\sigma_N^2/2E)(L/D)D = G_f h/s$, where L/D is constant. Hence, (Bažant 1996b)

$$\sigma_N = C_P D^{-1/2}, \qquad C_P = \sqrt{2EG_f h/s} \tag{9.5.42}$$

in which C_P is a constant. So we see that this special mechanism of energy release, which is characteristic of the shear failure of concrete, also yields a size effect.

The size effect obtained, being of LEFM type, is very strong, however. This is due to the fact that the width h of the band of splitting microcracks has been considered constant. In reality, h may be expected to become constant only after a certain initial growth, as approximately described by the equation $h = h_0 a/(c_0 + a)$ where h_0 and c_0 are material constants. In that case, the size effect obtained for the shear failure of reinforced concrete beams with or without stirrups is obtained the same as Bažant's (1984a) original size effect law. See Section 10.3 in which the truss model (or strut-and-tie model) is modified to calculate the energy release due to localized crushing of compression struts and simple formulas are derived for the size effect in beams with or without stirrups.

9.5.9 Distinction Between Axial Splitting and Failure Appearing as Shear

Compression failures of two types have traditionally been distinguished: (**1**) axial splitting failure; and (**2**) a failure that may appear to be caused by shear, but, in reality, is caused by transverse propagation of a band of axial splitting cracks (as already discussed), shear slip being only the terminal postpeak mechanism of final break.

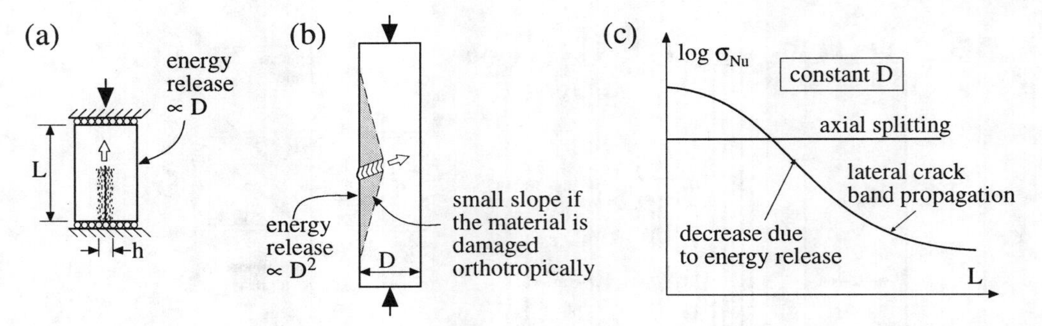

Figure 9.5.8 Axial splitting vs. shear banding: (a) axial splitting band; (b) compression microbuckling band (appearing as a shear band); (c) size effect curves for variable specimen length and constant specimen width.

Which type of failure will actually occur? This depends on the geometry, size, and boundary conditions of the structure, as well as the type of material. But what actually decides? To answer this question, Bažant (1996d) proposed the following energy concept of the splitting failure.

If the axial crack were a line, without any damage band surrounding it, then the formation of an axial splitting crack in a uniform uniaxial compressive stress field would cause no change in the stress field, and would thus cause no release of strain energy. Yet, energy is needed to create a crack. It follows that there must exist a damage band of finite width in which the axial stress gets reduced (Fig. 9.5.8a).

Formation of the axial crack band releases energy only from the band itself, but not from the adjacent material. Consequently, there can be no size effect. The laterally propagating band, however, also releases energy from the zone of material adjacent to the band (Fig. 9.5.8b. If geometric similarity is preserved, this zone grows quadratically with the specimen size, and so does the energy release from this zone, which is the source of the size effect, as already explained. If the width D is fixed and length L increases (Fig. 9.5.8), the development of the energy release zone will magnify the energy release. Therefore, the plots of σ_N vs. L for these two types of failure are a horizontal line and a descending curve (Fig. 9.5.8c). The fact that the axial splitting failure can be observed means that these two lines must cross at some realistic size.

For small enough specimens in which the fracture process zone occupies nearly the entire specimen, the size effect vanishes for both the laterally and axially propagating crack bands. The axial splitting normally occurs for small specimens if the boundary conditions do not resist the splitting (i.e., if lubricated end platens or brushes are used). Thus it follows that the fracture energy of the axially propagating band, G_f^A, must be smaller than the fracture energy of the laterally propagating band, G_f^L.

9.6 Scaling of Fracture of Sea Ice

Different types of size effect are exhibited by sea ice failures. The scaling of failure of floating sea ice plates in the Arctic presents some intricate difficulties. One practical need is to understand and predict the formation of very long fractures (of the order of 10 km to 100 km) which cause the formation of open water leads or serve as precursors initiating the build-up of pressure ridges.

Large fractures can be produced in sea ice as a result of the thermal bending moment caused by cooling of the surface of the ice plate (Fig. 9.6.1). The floating plate behaves exactly as a plate on elastic Winkler foundation, with the foundation modulus equal to the unit weight of sea water. Assuming the plate to be infinite and elastic, of constant thickness h, the temperature profiles across the thickness h to be similar for various thicknesses h, and the thermal fracture to be semi-infinite and propagate statically (i.e., with insignificant inertia forces), Bažant (1992a) found that the critical temperature difference

$$\Delta T_{cr} \propto h^{-3/8} \tag{9.6.1}$$

Since the thermal stresses are proportional to the temperature differences, this means that the critical

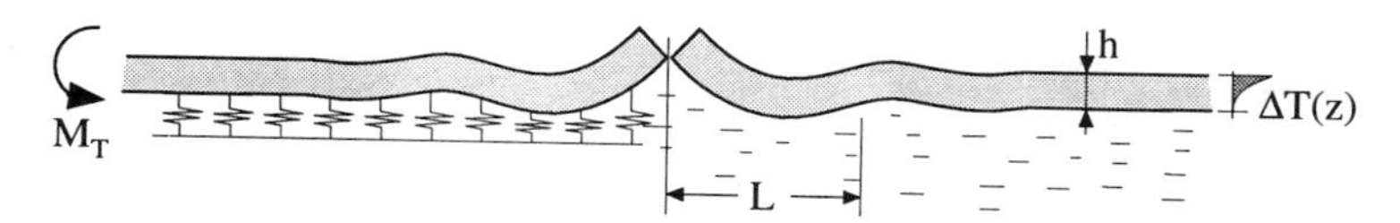

Figure 9.6.1 Cross section of an ice plate normal to a thermal crack (adapted from Bažant 1992a).

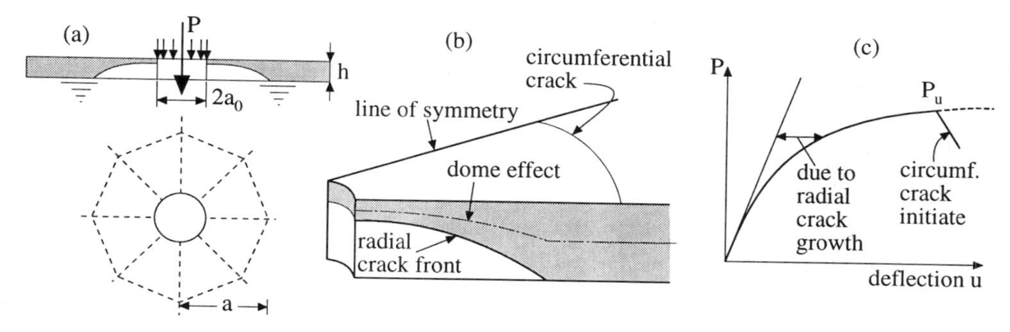

Figure 9.6.2 Crack-system in an ice plate deflected by a load P: (a) loading and crack system (for 8 radial cracks); (b) three-dimensional sketch showing dome effect; (c) load-deflection curve (adapted from Bažant, Kim and Li 1995).

nominal thermal stress $\sigma_N \propto h^{-3/8}$, too. In this approach, LEFM type analysis was used. This is justified despite the existence of a large fracture process zone. The reason is that, because a steady-state propagation must develop in the thermal bending fracture of an infinite plate, the fracture process zone does not change as it travels with the fracture front, and thus dissipates energy at a constant rate.

It has been shown that the scaling law in (9.6.1) must apply more generally to failures caused by any type bending cracks, provided that they are full-through cracks propagating along the plate (created by any type of loading, e.g., by vertical load; Slepyan 1990; Bažant and Li 1994b; Li and Bažant 1994b).

It may be surprising that the exponent of this large size asymptotic scaling law is not $-1/2$. However, this apparent contradiction may be resolved if one realizes that the plate thickness is merely a parameter but not actually a dimension in the plane of the boundary value problem, that is, the horizontal plane. In that plane, there is only one characteristic length in this problem—namely, the well-known flexural wavelength of a plate on elastic foundation, L. It turns out that L is not proportional to h but to $h^{3/4}$. Thus, it follows that the exponent of L in the scaling law is $(-3/8)(4/3) = -1/2$. So the scaling of thermal bending fracture obeys the law

$$\Delta T_{cr} \propto L^{-1/2}, \tag{9.6.2}$$

which agrees with what we have shown previously.

Simplified calculations (Bažant 1992a) have shown that, in order to propagate such a long thermal bending fracture through a plate 1 m thick, the temperature difference across the plate must be about 25°C, while for a plate 6 m thick, the temperature difference needs to be only 12°C. This is a large size effect, which may explain why very long fractures in the Arctic Ocean are often seen to run through the thickest floes rather than through the thinly refrozen water leads between and around the floes (as observed by Assur in 1963).

An important practical problem is the scaling of failure caused by vertical (downward or upward) penetration through the floating ice plate (Fig. 9.6.2). In that case, the fractures are known to form a star pattern of radial cracks (Fig. 9.6.2a–b) which propagate outward from the load, and the failure occurs when the circumferential cracks begin to form, as indicated by the load-deflection diagram in Fig. 9.6.2c.

This problem was initially analyzed under the assumption of full-through bending cracks, in which case, the asymptotic scaling law for large cracks again appears to be of the type $h^{-3/8}$ (Slepyan 1990; Bažant 1992a). However, experiments as well as finite element analyses show that the radial cracks before

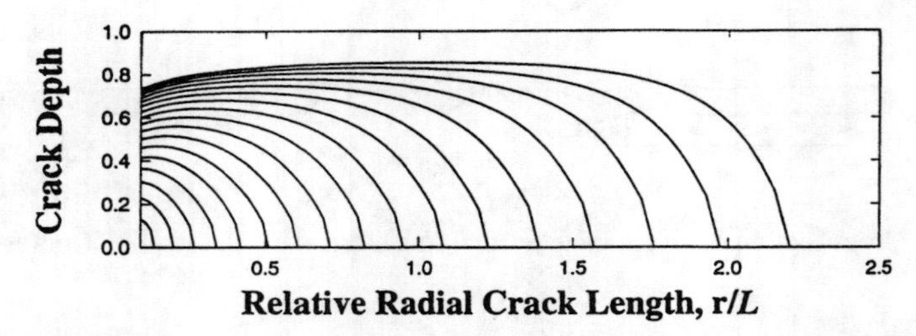

Figure 9.6.3 Radial crack profiles. The last profile corresponds to peak load (from Bažant and Kim 1996a).

failure do not reach through the full thickness of the ice plate, as shown in Fig. 9.6.2b. The normal forces transmitted across the radial cross section with the crack are significant and cause a dome effect which helps to carry the vertical load. This enormously complicates the analysis.

To solve this problem, Bažant and Kim (1996a) characterized the elasticity of one half of the sector of the floating plate between two cracks by a compliance matrix obtained numerically. The radial cracked cross section is subdivided into narrow vertical strips. In each strip, the crack is assumed to initiate through a plastic stage (representing an approximation of the cohesive zone). This is done according to a strength criterion (in the sense of Dugdale model), with a constant in-plane normal stress assumed through the portion of cross section where the strain corresponding to the strength limit is exceeded. For the subsequent fracture stage, the relationship of the bending moment M and normal force N in each cracked strip to the additional rotation and in-plane displacement caused by the crack is assumed to follow the nonlinear line spring model of Rice and Levy (1972). The transition from the plastic stage to the fracture stage is assumed to occur as soon as the fracture values of M and N become less than their plastic values (to do this consistently, the plastic flow rule is assumed such that the ratio M/N would be the same as for fracture). This analysis provided the profiles of crack depth shown in Fig. 9.6.3, where the last profile corresponds to the maximum load (the plate depth is greatly exaggerated in the figure).

An important question in this problem is the number of radial cracks that form. The solution (Bažant and Li 1995d) shows that the number of cracks depends on the thickness of the plate and has a significant effect on the scaling law.

Numerical solution of the integral equation along the radial cracked section, expressing the compatibility of the rotations and displacements due to crack with the elastic deformation of the plate wedge between two cracks, provided the size effect plot shown in Fig. 9.6.4. The numerical results shown by data points can be relatively well described by the generalized size effect law of Bažant, Eq. (9.1.34), shown in the figure as a full line. The stepped curve in the figure indicates the number of radial cracks for each range of ice thickness. The deviation of the numerical results from the smooth curve, seen in the middle of the range in the figure, is probably caused by insufficient density of nodal points near the fracture front.

As confirmed by Fig. 9.6.4, the asymptotic size effect does not have the slope $-3/8$, but the slope $-1/2$. Obviously, the reason is that, at the moment of failure, the cracks are not full-through bending cracks but grow vertically through the plate thickness.

9.6.1 Derivation of Size Effect for Thermal Bending Fracture of Ice Plate

Consider an infinitely extending ice plate of thickness h floating on water of specific weight ρ (Fig. 9.6.5a). Formation of a long (semi-infinite) crack along axis x releases the initially existing thermal bending moment M_T (per unit length) and causes the fracture edge to rotate through angle θ (Fig. 9.6.5b). Deflection w in the direction of vertical axis z causes the supporting pressure of sea water to decrease by $\Delta p = -\rho w$ (Archimedes' law). This means that water acts exactly as the Winkler elastic foundation of foundation modulus ρ (Fig. 9.6.5b, left). As is well known (e.g., Hetényi 1946), the edge rotation is:

$$\theta = \left[\frac{dw}{dx}\right]_{y=0} = \frac{4M_T}{\rho L^3} \tag{9.6.3}$$

where $L = (4D/\rho)^{1/4}$, with $D = Eh^3/12(1-\nu^2)$ = cylindrical stiffness of the plate, E = Young's modulus, and ν = Poisson ratio.

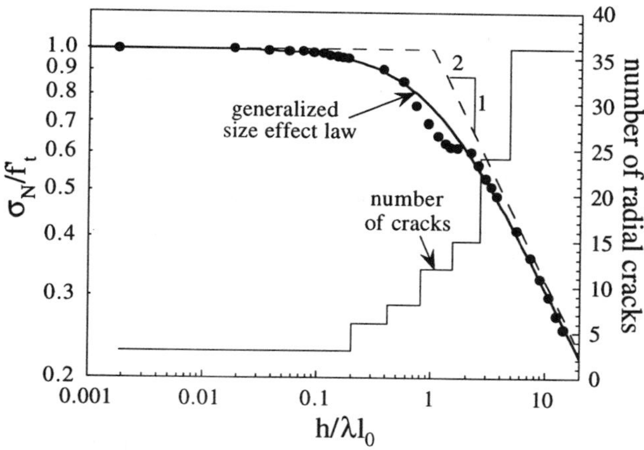

Figure 9.6.4 Size effect in thermal bending of ice plates. The heavy dots are the finite element results, and the smooth curve is the best fit by Bažant's extended size effect law ($r = 1.20$). The stepped curve indicates the number of active radial cracks. (Adapted from Bažant and Kim 1996a.)

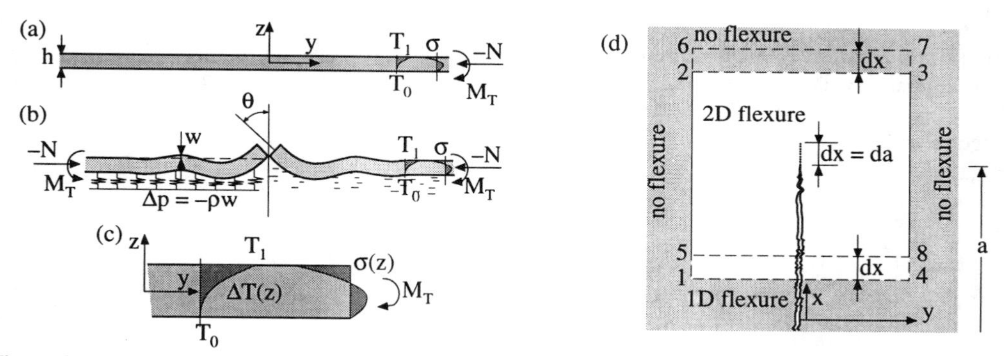

Figure 9.6.5 Thermal cracking of a floating ice plate: (a) floating ice plate; (b) plate deformation after fracture showing the modeling as a Winkler elastic foundation (left part); (c) detail of the temperature and stress profiles; (d) view of propagating fracture from top, with rectangular contours used to compute the energy release rate (adapted from Bažant 1992a).

Attention is restricted to steady-state propagation of a large crack in a large plate (formally, a semi-infinite crack in an infinite plate; Fig. 9.6.5 d). Behind the crack tip, the bending fracture cuts through the whole plate thickness h, and is opened so widely there is no crack bridging. Around the crack tip we can imagine a sufficiently large rectangle 1234, such that behind this rectangle the plate bending is essentially one-dimensional, in the direction y normal to the crack, and that ahead of this rectangle and on its sides, there is essentially no bending at all. (The width of the rectangle can be taken equal to about $5L$, which is a distance at which the bending disturbance from the crack decays by factor $e^{-y/L} \approx 0.01$.)

Although the deformation of the plate in rectangle 1234 is very complex, we do not need to worry about it if we consider steady state propagation. In that case, the state of the rectangle 1234 remains constant as it moves ahead with the fracture front, and so the strain energy released in advancing the crack by da is equal to the strain energy in the rectangle 2673 minus the strain energy in the rectangle 1584. But this is the energy released from the floating beam and the elastic foundation when a floating beam of width da is fully cracked, i.e.,

$$- d\mathcal{U} = \frac{M_T \, da \, \theta}{2} = \frac{2M^2}{\rho L^3} \tag{9.6.4}$$

where the second equation follows from (9.6.3). Now, the energy balance requires the energy released to be equal to $G_f h \, da$, the energy necessary for fracture to happen. Thus, setting $-d\mathcal{U} = G_f h da$, we find that the critical thermal bending moment at which the fracture propagates is (Bažant 1992a):

$$M_f = \sqrt{\frac{\rho h L^3 G_f}{2}} \tag{9.6.5}$$

The difference of ice temperature from the water temperature may be written as $\Delta T(z) = \Delta T_1 f(\zeta)$ where $\zeta = z/h$ = relative vertical coordinate, ΔT_1 = difference in temperature between the top and bottom of the plate, and f = function defining the temperature profile (Fig. 9.6.5c), which must be calculated in advance. Taking into account that the normal strains in the x and y directions as well as the vertical normal stresses are zero, we find that the thermal bending moment locked in the plate before fracture is

$$M_T = \int_{-h/2}^{h/2} \hat{E}\alpha\Delta T(z) z dz \,, \qquad \hat{E} = \frac{E}{1-\nu} \tag{9.6.6}$$

where α = coefficient of linear thermal expansion of ice, and the values of E and α are taken as the averages over the plate thickness (although they vary as a function of temperature). Substituting for $\Delta T(z)$, we obtain:

$$M_T = \hat{E}\alpha\Delta T_1 h^2 I_T \,, \qquad I_T = \int_{-1/2}^{1/2} f(\zeta)\zeta d\zeta \tag{9.6.7}$$

where I_T is a constant characterizing the temperature profile. (To be able to judge the size effect, we must, of course, consider similar temperature profiles, even though, for the same profile, the thicker the plate, the longer the duration of cooling to establish that profile.)

Fracture will propagate if $M_T = M_f$. From this, the critical temperature difference required for crack propagation is

$$\Delta T_{cr} = \Delta T_1 = \frac{1}{\hat{E}\alpha I_T h^2}\sqrt{\frac{\rho h L^3 G_f}{2}} \tag{9.6.8}$$

Substituting now the foregoing expressions for L and D, we obtain the result (Bažant 1992a, Eq. 7):

$$\Delta T_{cr} = C_1 h^{-3/8} \,, \qquad \text{with} \qquad C_1 = \frac{(1-\nu)^{5/8}\rho^{1/8}\sqrt{G_f}}{\sqrt{2}[3(1-\nu^2)^{3/8}E^{5/8}\alpha I_T} \tag{9.6.9}$$

This solution has further been extended to take into account in-plane compressive force N_y in the ice plate, which reduces the bending stiffness of the plate. Instead of (9.6.9), one obtains:

$$\Delta T_{cr} = C_1 h^{-3/8}(1-2\gamma)^{1/2}(1-\gamma)^{-1/4} \,, \qquad \text{with} \qquad \gamma = -\frac{N_y}{h\sqrt{h}}\sqrt{\frac{3(1-\nu^2)}{E\rho}} \tag{9.6.10}$$

Practical application to sea ice further necessitates taking into account the creep of ice during the period of cooling. This has been done approximately on the basis of the effective modulus for creep. Various numerical calculations, some of which we already mentioned, have been made (Bažant 1992a).

9.6.2 General Proof of 3/8-Power Scaling Law

Before fracture, the floating plate is undeflected ($w = 0$). Thermal fracture relaxes the thermal bending moments in the vicinity of the fracture and causes release of the strain energy due to the thermal bending moments. The governing differential equation for the two-dimensional deflection surface $w(x, y)$ is $D\nabla^4 w + \varrho w = 0$ (e.g., Hetényi 1946). The plate is infinite, with the boundary condition $w = 0$ at infinity. At the crack faces Γ_0, which can in general be curved, the boundary conditions are $Dw_{,nn} = M_T$ and $w_{,nnn} + (1-\nu)w_{,ntt} = 0$, where subscripts following a comma denote partial derivatives, and n and t are the coordinate axes normal and tangential to the crack face.

Introduce now dimensionless variables $\xi = x/L$, $\eta = y/L$, and $\zeta = w/L$, and note the transformation of derivatives: $\partial/\partial x = L^{-1}\partial/\partial\xi$, etc. Transforming the boundary value problem to these variables, we obtain the governing partial differential equation:

$$\bar{\nabla}^4 \zeta + 4\zeta = 0 \tag{9.6.11}$$

with boundary conditions $\zeta = 0$ at infinity; here $\bar{\nabla}$ is the gradient operator with respect to reduced variables ξ, ζ. At crack faces Γ_0, the boundary conditions in dimensionless coordinates are:

$$\text{On } \Gamma_0: \quad \zeta_{,\nu\nu} = \frac{L}{D} M_T \quad \text{and} \quad \zeta_{,\nu\nu\nu} + (1-\nu)\zeta_{,\nu\tau\tau} = 0 \tag{9.6.12}$$

where ν and τ are the coordinate axes normal and tangential to the crack faces Γ_0 in the dimensionless space (ξ, η).

Due to linearity of equations (9.6.11) and (9.6.12), the solution ζ is proportional to M_T. Therefore, it is convenient to define: $\zeta = F(\xi, \eta; \overline{a})$ = solution of differential equation (9.6.11) for the relative crack length $\overline{a} = a/L$ and for the aforementioned boundary conditions except that the first boundary condition in (9.6.12) is replaced by $\zeta_{,\nu\nu} = 1$. In such a boundary value problem, there are no physical constants, and so the solution F is independent of the size and material properties, and depends only on $\overline{a}$. Then the solution for the actual boundary conditions (9.6.12) is

$$\zeta = \frac{L}{D} M_T F(\xi, \eta, \overline{a}) \tag{9.6.13}$$

Now, by transformations of coordinates, we get for the crack faces Γ_0: $\zeta_{,\nu} = F_{,\nu} M_T L/D, \zeta_{,n} = \zeta_{,\nu}/L$, and $\vartheta = w_{,n} = L\zeta_{,n} = F_{,\nu} M_T L/D$ = rotation at the crack face about the tangential axis τ. The total strain energy release due to fracture, $-\Delta\mathcal{U}$, is equal to the work of the released thermal bending moment, as it is reduced to zero, on rotation ϑ, i.e., $-\Delta\mathcal{U} = \int_a \frac{1}{2} M_T \vartheta da$. From this we get the energy release per unit length of fracture: $-\partial\mathcal{U}/\partial a = \frac{1}{2} M_T \vartheta = G_f h$. Substituting now the foregoing expression for ϑ, solving the resulting equation for M_T, and expressing M_T in terms of ΔT_{cr}, and L and D in terms of h, we obtain the following general result (Bažant 1992a; Bažant and Li 1994b):

$$\Delta T_{cr} = \frac{2^{1/4}(1-\nu)^{3/8}}{\sqrt{F_{,n}(\xi, \eta; \overline{a})}} C_1 \, h^{-3/8} \tag{9.6.14}$$

in which C_1 was defined in (9.6.9).

This proves, in general, that thermal bending fracture of floating ice plate exhibits a $(-3/8)$-power size effect, provided that either F is independent of the crack length (which occurs for a semi-infinite crack in an infinite plate) or the crack length a is proportional to the flexural wave length L (rather than to thickness h) (Bažant 1992a,c).

In terms of the flexural wave length L, the size effect is of the type $\Delta T_{cr} \propto L^{-1/2}$, similar as before. This simple conclusion is not surprising since, due to Kirchhoff's assumption, the plate problem is two-dimensional and, in the plane (x, y) of the boundary value problem, length L is the only characteristic length present (h enters only indirectly, through D).

It is interesting to note that if we carried out the dimensional analysis according to Buckingham's Π theorem in the usual manner, taking the only geometric dimension present as the characteristic dimension of the structure, we would have found for σ_N a result different from $h^{-3/8}$. However, we could as well have taken the flexural wave length L as the characteristic dimension, and then we would have obtained the correct size effect $h^{-3/8}$, so the result would have been inconclusive. It is necessary to know something about the mechanics of the problem in order to realize that L rather h must be taken as the characteristic dimension, and only then the correct size effect $h^{-3/8}$ is obtained from Buckingham's Π theorem.

10

Brittleness and Size Effect in Structural Design

Except for a terse review in the introduction to the book, we have so far focused on fracture specimens and have not given adequate attention to real structures, reinforced as well as unreinforced. We will focus attention on them in this chapter, however, with a deeper focus on some structures than others. After dealing in the first section with the general aspects of brittleness and size effect and the general procedures to introduce them in practical formulations, we will devote three sections to a relatively thorough analysis of two important types of structural failure: the diagonal shear of longitudinally reinforced beams (Section 10.2 and Section 10.3), and the reinforced beams in bending —with particular emphasis on lightly reinforced beams (Section 10.4). The last section of the chapter will concisely review some of the main issues for other structural elements, from torsion of beams to reinforced columns.

10.1 General Aspects of Size Effect and Brittleness in Concrete Structures

In the preceding chapter, it became obvious that the strength of geometrically similar specimens of a quasibrittle material —particularly concrete— can be written in the general form

$$\sigma_{Nu} = f_t' \phi \left(\frac{D}{\ell}, \text{geometry} \right) \tag{10.1.1}$$

where f_t' is the tensile strength, D a characteristic structural dimension, and ℓ a characteristic material size; we explicitly indicate that the function depends on geometry, which is equivalent to say that the dependence on D/ℓ is different for different structural types and loading.

The material characteristic size ℓ (as well as the function ϕ itself) is different for the various existing models. However, as shown in the previous chapter, all the models can be set to give very similar size effect predictions over the practical experimental range; thus, there is a strong correlation between the fracture parameters of the various models for a given material.

In principle, the foregoing equation can be computed for every geometry and material model. In practice, computations can be very complex except for some simple cases. Thus, simplified expressions are convenient to extrapolate the experimental results. The simplest of these expressions is Bažant's size effect law expounded in Chapters 1 and 6 —Eqs. (1.4.10) or (6.1.5). As discussed in the previous chapter, this law can be generalized to give more precise descriptions over a broader range of sizes and a broader range of geometries. However, the extended size effect laws, including those derived from cohesive models, require information that is usually lacking for the classical tests on which the formulas for the codes were based. Therefore, in this chapter, we mostly use the simplest (Bažant's) law in comparing the experimental trends and the theoretical size effect. The correlations in the previous chapter can then be used to shift to other models.

In this section, we first discuss the conditions under which Bažant's size effect law is expected to give a good description of the size effect; we then analyze the existing proposals to characterize the structural brittleness through a brittleness number. We conclude the section by examining the general methodology proposed by Bažant to generate size effect corrections to ultimate loads in codes, including the effect of reinforcement.

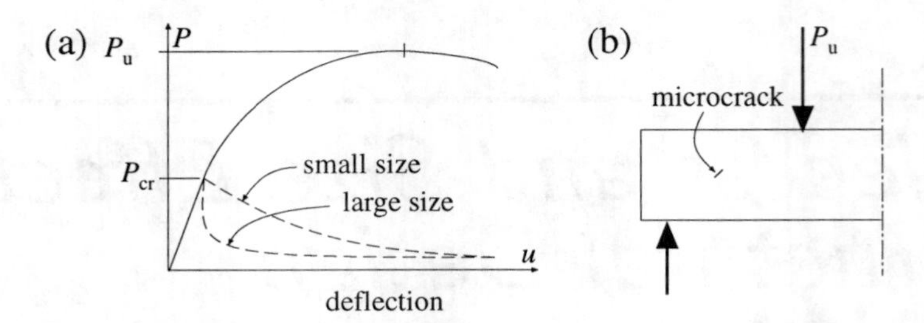

Figure 10.1.1 (a) Load-deflection curves for a relatively ductile structure (full line), and for a brittle structure that fails at first cracking (dashed line). (b) A brittle structure failing at crack initiation, the crack at maximum load still being microscopic.

10.1.1 Conditions for Extending Bažant's Size Effect Law to Structures

As briefly mentioned in Section 1.4.3, extension of the size effect law to real structures that have no notches is valid only if the following two additional hypotheses are fulfilled:

1. The structure must not fail at macrocrack initiation.

2. The shapes and lengths of the main fracture at the maximum loads of similar structures of different sizes must also be geometrically similar.

According to the available experimental evidence as well as finite element simulations, the foregoing assumptions appear to be satisfied for many types of failure of reinforced concrete structures within the size range that has been investigated so far. Let us examine the reasons for this, and the exceptions, more closely.

In a structure failing at crack initiation, the maximum load P_u is equal to the initial cracking load P_{cr} as indicated by the dashed load-deflection curve in Fig. 10.1.1a; in such a failure, the crack at maximum load is still microscopic, as shown in Fig. 10.1.1b. The case $P_u \approx P_{cr}$ can occur for metallic structures with initial flaws. But since the main purpose of reinforcing concrete is to prevent failure at crack initiation, good practice requires designing concrete structures in such a manner that $P_u \gg P_{cr}$ as illustrated by the solid load-deflection curve in Fig. 10.1.1a. For some types of failure this is explicitly required by the design codes (for example the ACI code requires that, for the bending failure, the maximum load, after applying the capacity reduction factor, be at least $1.25P_{cr}$, and for a good design it is normally much larger); furthermore, this is indirectly enforced by many other design code provisions on reinforcement layout. Then, the major cracks at P_u necessarily intersect a major portion of the cross section (say 30% – 90%).

There are, of course, cases in which the first condition is not met. This is the case for unnotched unreinforced structures such as the beams for rupture modulus tests (see Section 9.3), and some cases of footings, retaining walls, and pavement slabs. Except for these, and some cases of more theoretical than practical interest involving large under-reinforced structures, there is hardly any case of a structure failing at crack initiation, and so it is of little interest to develop the size effect formulation for failures at small cohesive cracks for other structure types.

The second hypothesis is illustrated in Fig. 10.1.2. This hypothesis means that the main fracture at the maximum load has the shape AB and $A'B'$. Point B' is located at same relative distance to the boundaries as point B. If the fracture front at the maximum load of the larger structure were at point C' rather than B', the size effect law could not apply. Likewise, it could not apply if the main fracture at maximum load were $A'B'$ or $A'E'$ for the larger structure in Fig. 10.1.2.

It appears that a deviation from this similarity of the main fractures at the maximum load is the main reason for the deviations from the size effect law which are observed in the Brazilian split-compression failure of cylinders of very large sizes (see Section 9.4).

The large major crack in a typical concrete structures at maximum load has the same effect as the notches in fracture specimens. In effect, well-designed structures develop, in a stable manner, large cracks which behave the same as notches. However, there is a small difference. In fracture specimens, the notches are cut precisely. In real structures, the growth of large major cracks is influenced by the

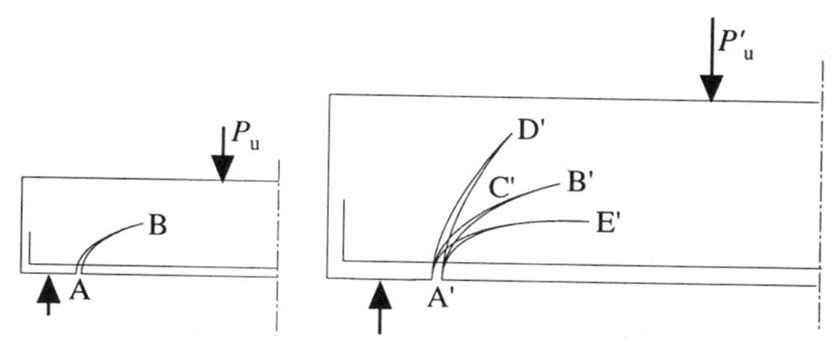

Figure 10.1.2 Illustration of the condition of crack similarity at peak load. Crack $A'B'$ in the large structure is similar to crack AB in the small structure. Cracks $A'C'$, $A'D'$, and $A'E'$ are not.

randomness of material properties, originating from material heterogeneity. Thus, the major fractures in similar structures of various sizes can be geometrically similar only on the average, in the statistical sense. In individual cases, there are deviations. For example, point B' can have a slightly smaller relative distance to the top boundary than point B. The consequence of this randomness is that in real structures in which there are no notches the measured maximum load values are more scattered than in fracture specimens. Further randomness is, of course, caused by environmental effects and their random fluctuations, by inferior quality control, etc. For this reason, the question of the precise shape of the size effect curve [for example, the question whether exponent r in Eq. (9.1.34) should be different from 1] is not practically very important.

It may be useful to also recall some other previously introduced assumptions. If the size effect should not be mixed with other influences, we must consider structures made of the same material, which means the same mix proportions and the same aggregate size distribution. If the maximum aggregate size d_a were increased in proportion to the structure size, the material in structures of different sizes would be different. This is not only because of the increased d_a, but also because a change in d_a requires a change in the mix proportions, particularly in the specific cement content.

10.1.2 Brittleness Number

The concept of brittleness of structural failure, which is the opposite of ductility, is an old one, but for a long time, the definition of brittleness has been fuzzy and has not stabilized. One of the fundamental reasons is that the apparent brittleness depends simultaneously on the material, the geometry of the structure and loading, and the size of the structure. Therefore, it is not easy to find a single figure properly incorporating all these influences.

The first idea in quantifying the brittleness is to look for a quantity that is vanishingly small for the perfect plasticity limit and infinitely large for the elastic-brittle limit. A number with these properties is the ratio D/ℓ appearing in the general size effect law (10.1.1). Therefore, any variable proportional to it is a good candidate to be a brittleness number:

$$\beta_n \propto \frac{D}{\ell} \tag{10.1.2}$$

Over the years, there have been various proposals for brittleness numbers of these forms. Well known in the metals community is the brittleness characterization based on Irwin's estimate of the nonlinear zone (see Section 5.2.2) which is at the basis of the ASTM E 399 condition for validity of the fracture toughness test. This brittleness number, say β_K, can be written as

$$\beta_K = \frac{D f_t'^2}{K_{Ic}^2} \tag{10.1.3}$$

With this definition, the condition for valid fracture toughness measurements reduces to $\beta_K \geq 2.5$.

In the field of concrete, probably the first ratio used as a brittleness number was put forward by Hillerborg and co-workers. It was defined as

$$\beta_H = \frac{D}{\ell_{ch}}, \qquad \ell_{ch} = \frac{EG_F}{f_t'^2} \tag{10.1.4}$$

in which G_F refers to the fracture energy of the underlying cohesive crack model. Note that the foregoing two equations are essentially identical, because of Irwin's relationship $K_{Ic} = \sqrt{EG_F}$.

The foregoing brittleness numbers are useful to compare various materials and sizes for a given structural shape and loading, but they cannot be used directly to compare the brittleness of different structural geometries, because the dependence of brittleness on geometry is not included in their definition.

Pertaining to this category, but with a slightly different definition, is Carpinteri's brittleness number s_C (Carpinteri 1982):

$$s_C = \frac{G_F}{Df_t'} \tag{10.1.5}$$

We notice that it is the inverse of a brittleness (the more brittle, the smaller s_C), and should better be called a ductility number. Note also that it is related to the Hillerborg brittleness by

$$s_C = \frac{f_t'}{E}\frac{1}{\beta_H} \tag{10.1.6}$$

This means that Carpinteri's brittleness number can be used to compare brittleness of various materials only as long as they have the same f_t'/E ratio. It has the same limitations as β_H in not giving comparable results for different geometries.

To get a brittleness number that embodies the influences of material, geometry, and shape, we may recall the concept of intrinsic size defined in Section 5.3.3 and use the brittleness number defined as

$$\beta_n \propto \frac{\overline{D}}{\ell} \tag{10.1.7}$$

in which $\overline{D}$ is given by (5.3.11) and (5.3.12). The first brittleness number of this category was introduced by Bažant (1987a; also Bažant and Pfeiffer 1987), although the concept of intrinsic size was still to come. Bažant's brittleness number was defined as

$$\beta = \frac{\overline{D}}{c_f} = \frac{D}{D_0} \tag{10.1.8}$$

where the second expression is the original definition, which is equivalent to the first because of (6.2.2).

Brittleness numbers similar in concept, but based on the cohesive crack model have also been extensively used. Planas and Elices (1989a, 1991a) introduced the obvious extension of Hillerborg's brittleness number as

$$\beta_P = \frac{\overline{D}}{\ell_{ch}} \tag{10.1.9}$$

which gives a unified representation of the size effect and brittleness properties in the medium and large range of sizes ($\beta_P > 0.04$) for most laboratory geometries (Llorca, Planas and Elices 1989; Guinea, Planas and Elices 1994a).

The foregoing definition of β_P is, however, sensitive to the shape of the softening curve. Although (as discussed in Chapters 7 and 9) there is not much variability of shapes for ordinary concrete, it turned out to be better to use a brittleness number that refers the intrinsic size to the properties of the initial portion of the softening curve, characterized by the tensile strength and the horizontal intercept w_1 (Fig. 7.1.8); its definition is

$$\beta_1 = \frac{\overline{D}}{\ell_1}, \qquad \ell_1 = \frac{Ew_1}{2f_t'} \tag{10.1.10}$$

This brittleness number adequately captures the fracture properties for small and medium sizes, including most practical situations ($0.1 < \beta_1 < 1$). Moreover, from the Planas-Elices correlation (7.2.14), we get

$$\beta \approx 5.3\beta_1 \tag{10.1.11}$$

and it turns out that the two brittlenesses can be interchangeably used since the factor 5.3 is independent of the shape and size of the structure as well as of the material, as long as the material softening curve can be approximated by a straight line in its first part. This is usually the case for concrete. For granite, there is also evidence of this fact (Rocco et al. 1995). In the remainder of this chapter we will mainly use Bažant's brittleness number β.

10.1.3 Brittleness of High Strength Concrete

High strength concrete (HSC) is known to be more brittle than normal strength concrete (NC). This is so because c_f is smaller for HSC and then, for a given geometry $D_0^{HSC} < D_0^{NC}$. Therefore, considerable care must be exerted in extrapolating the results obtained for NC to HSC.

There are few data to establish general correlations. Gettu, Bažant and Karr (1990), based on size effect tests on notched beams, proposed the following approximate formula for HSC:

$$\frac{c_f^{HSC}}{c_f^{NC}} = \frac{D_0^{HSC}}{D_0^{NC}} \approx \left(\frac{f_c'^{NC}}{f_c'^{HSC}} \right)^{1/3} \tag{10.1.12}$$

where it is understood that the aggregate size is identical for both NC and HSC. However, this equation is a rough approximation. The aggregate shape, strength, and stiffness, or whether a crushed aggregate or river aggregate is used, etc. may have significant influence, too. Further studies are needed, but it is clear that the transitional size is substantially less for HSC and so the behavior is more brittle for HSC than for NC.

10.1.4 Size Effect Correction to Ultimate Load Formulas in Codes

In principle, plastic limit analysis is a wrong theory for the majority of the design code provisions which deal with brittle failures, such as diagonal shear, torsion, punching, pullout, etc. So, in fact, is LEFM. The theoretically best approach would be to base the design on nonlinear fracture mechanics taking into account the large size of the fracture process zone. However, as pointed out in the previous section, this would be quite complicated for the basic design problems covered by the code, and not really necessary because a highly accurate fracture analysis is not necessary for most situations. A simple way to obtain the load capacities corresponding to nonlinear fracture mechanics is to exploit the size effect law (1.4.10). Two kinds of formulas are possible:

1. One can start from the formula based on plastic limit analysis which now exists in the code, and introduce in it a correction due to the size effect law.

2. Alternatively, one can set up the ultimate load formula based on LEFM, and again introduce into it a correction according to the size effect law.

The first kind is no doubt preferable to the concrete engineering societies, because it makes it possible to retain the formulas that now exist in the codes, and introduce in them only a relatively minor correction (of course, the formula needs to be slightly scaled up because, for normal sizes, it must give about the same load capacity as before, even after the reduction for the typical structure size according to the size effect law has been introduced). Obviously, the accuracy of this type of correction would decrease with increasing size, as the behavior is getting more brittle and more remote from the size to which plastic limit analysis approximately applies. Some structures of normal sizes exhibit failures that are closer to limit analysis than to LEFM. For such structures, the accuracy by the first type of correction is adequate.

However, for very large structures or for certain types of failure (anchor pullout, diagonal shear), the failure is known to be very brittle, actually closer to LEFM than to plastic limit analysis. In that case, the second kind of formula, based on LEFM, must be expected to give a more realistic result. The error of this correction increases with a decreasing structure size and is the smallest for large sizes close to the LEFM asymptote.

In the remainder of this section, we discuss how to introduce the size effect correction into the formulas existing in the codes. We consider first the ideal case of plain concrete structures (or structures for which the steel does not contribute appreciably to strength, such as anchor pullout). Then we analyze how these formulas must be modified to include the contribution of the reinforcement.

10.1.5 Size Effect Correction to Strength-Based Formulas

For reinforced structures in which the steel does not contribute appreciably to the overall strength, one can expect a structural size effect approximately given by (1.4.10). This equation contains two parameters, Bf_t' and D_0, which would need to be specified for the new design formulas in the codes. Now, for very small sizes we must have $\sigma_{Nu} = \sigma_{Nu}^p$ in which σ_{Nu}^p is the plastic limit (i.e., the strength computed from plasticity or limit analysis). Therefore, in (1.4.10) we must have $Bf_t' = \sigma_{Nu}^p =$ and we can rewrite that equation as

$$\sigma_{Nu} = \frac{\sigma_{Nu}^p}{\sqrt{1+\beta}}, \qquad \beta = \frac{D}{D_0} \tag{10.1.13}$$

Since the design formulae in the code have been based both on limit analysis and experiments, one can assume that the code formula provides a prediction of the ultimate strength σ_{Nu}^C which coincides with the foregoing formula for the size used in the experiments that served to validate the code formula. We thus must have

$$\sigma_{Nu}^C = \frac{\sigma_{Nu}^p}{\sqrt{1+\beta_r}}, \qquad \beta = \frac{D_r}{D_0} \tag{10.1.14}$$

where D_r is the size of specimens used in the calibration tests (on average).

Solving for σ_{Nu}^p from (10.1.14) and substituting in (10.1.13) we get the size effect correction to the formulas in the code as

$$\sigma_{Nu} = \sigma_{Nu}^C \sqrt{\frac{1+\beta_r}{1+\beta}}, \qquad \beta = \frac{D}{D_0}. \tag{10.1.15}$$

where σ_{Nu}^C is the value obtained from the current formula in the code.

Assuming that D_r can be estimated from the data on the test series in the literature, the only parameter that needs to be estimated is D_0. Its theoretical calculation is more difficult because it depends both on the geometry of the structure and on the fracture properties of the material. Indeed, from (6.1.4) we have

$$D_0 = \frac{2k_0'}{k_0} c_f \tag{10.1.16}$$

which shows that D_0 consists of two factors. The first factor, $2k_0'/k_0$, is purely geometrical and can be easily determined by elastic calculations for notched specimens of positive geometry in which the relative crack length is well determined. For unnotched specimens, the problem is not well posed because the substitution of $\alpha_0 = 0$ leads to $D_0 = \infty$. Thus, slow crack growth must take place, as described in the previous section, and then α_0 is an unknown. Therefore, for these geometries the geometric factor must be determined either experimentally or by numerical simulation using a nonlinear fracture model.

The second factor is a material property which should, in principle, be experimentally determined for each concrete, but this would be impractical. The optimum approach would be to get a sound correlation between c_f and the basic characteristics of concrete, particularly f_c' and the aggregate size d_a. Unfortunately, such correlation is still unavailable.

Certain approximations, however, exist for some particular cases. For example, although the theoretical and experimental support is limited, Bažant and co-workers suggested that approximately $c_f \propto d_a$, where d_a is the maximum aggregate size. Thus, according to (10.1.16), for a fixed structural geometry also $D_0 \propto d_a$. The proportionality factor can be obtained by analysis of the existing experimental data. For example, from the tests of diagonal shear failure of beams, one can recommend the value $D_0 = 25d_a$. These values are only estimates based on seven classical data series studied by Bažant and Kim (1984) and Bažant and Sun (1987); see Section 10.2.2. However, the size of the beam was not the only parameter varied and the size range was not broad enough; because of other influences, such as differences in the shear span and the overall span to length ratio, as well as the use of different concretes, the scatter of these data was very large. Nevertheless, the size effect trend is clearly evident and makes it possible to obtain the aforementioned value of D_0, valid, of course, only for diagonal shear (see Section 10.2).

10.1.6 Effect of Reinforcement

To counteract brittle failure, one may use densely distributed reinforcement such as shear (or torsional) stirrups in beams. Together with the longitudinal reinforcement of beams, the shear reinforcement alone, at its yield limit, can resist a certain load, characterized by a nominal stress σ^s_{Nu}. When the structure is sufficiently large, there may be enough strain energy stored in the structure to drive a crack through the entire cross section at a load that only slightly exceeds the load carried by plastic reinforcement. However, there can be no size effect if that load is not exceeded. So, the size effect law must be applied only to the portion of the load capacity or nominal stress that is in excess of σ^s_{Nu}, that is

$$\sigma_{Nu} = \sigma^s_{Nu} + \sigma^c_{Nu}, \qquad \sigma^c_{Nu} = \frac{\sigma^{cp}_{Nu}}{\sqrt{1+\beta(\rho)}}, \qquad \beta(\rho) = \frac{D}{D_0(\rho)} \tag{10.1.17}$$

where σ^{cp}_{Nu} is a possible contribution of concrete to the overall strength at the plastic limit, $D_0(\rho)$ is a function with the dimension of length, and ρ is the steel reinforcement ratio. Note the explicit dependence of the brittleness number on the steel ratio.

The procedure to determine σ^{cp}_{Nu} is analogous to that sketched in the previous section, and requires no more than using the classical formulas of the code in (10.1.15). The determination of $D_0(\rho)$ is more difficult. For plain concrete, Eq. (10.1.16) shows that D_0 is a constant that is determined by the material fracture property c_f and the structure geometry. Now, the structure geometry is altered by the presence of the reinforcement. Therefore, the geometrical factor must depend on the steel ratio as well as on other dimensionless ratios defining the layout of the reinforcement.

Let us now sketch a possible unified framework for the influence of the steel ratio on the value of D_0. We write β in the form (10.1.8) and substitute the general form (5.3.11) for $\overline{D}$:

$$\beta = \frac{\overline{D}}{c_f} = \frac{K_I}{c_f 2\partial_a K_I} \tag{10.1.18}$$

where ∂_a denotes partial differentiation with respect to a. Next, we use the superposition principle and write the condition that the stress intensity factor is the sum of the stress intensity without reinforcement, minus the stress intensity factor caused by the steel-concrete interaction. The negative sign is due to the fact that the steel forces tend to close the crack. The resulting equation for K_I can always be written in the form

$$K_I = \sigma_N \sqrt{D} k(\alpha) - \rho \sigma_s \sqrt{D} k_s(\alpha) \tag{10.1.19}$$

where σ_s is the stress in the steel bar and $k_s(\alpha)$ a shape factor taking into account the steel distribution.

The simplest behavior one can encounter is that in which the effect of the reinforcement is exactly equivalent to decreasing the externally applied load (a pure parallel coupling). For such cases, $k_s(\alpha) \propto k(\alpha)$ and

$$\beta(\rho) = \beta(0) = \beta\,, \qquad D_0(\rho) = D_0(0) = D_0 \tag{10.1.20}$$

Pure cases of this kind are difficult to find in practice, but the results of Bažant and Kim (1984) seem to indicate that this is approximately valid for longitudinal reinforcement in diagonal shear.

For a densely distributed reinforcement, the behavior is quite different and one cannot assume that $k_s \propto k$. The result for $D_0(\rho)$ is then obtained as

$$D_0(\rho) = \frac{1 - m_1\rho}{1 - m_0\rho} D_0\,, \qquad m_1 = \frac{\sigma_s k_s(\alpha_0)}{\sigma_N k(\alpha_0)}\,, \qquad m_0 = \frac{\sigma_s k'_s(\alpha_0)}{\sigma_N k'(\alpha_0)} \tag{10.1.21}$$

For small values of ρ, we can take a MacLaurin expansion and write

$$D_0(\rho) \approx (1 + m\,\rho)\, D_0\,, \qquad m = m_0 - m_1 \tag{10.1.22}$$

This is the form postulated by Bažant and Sun (1987) for the influence of the stirrups on the size effect in diagonal shear. As it transpires from the foregoing derivation, m is a geometrical factor that can be, in principle, determined either from experiment or from numerical simulation. As an example, for

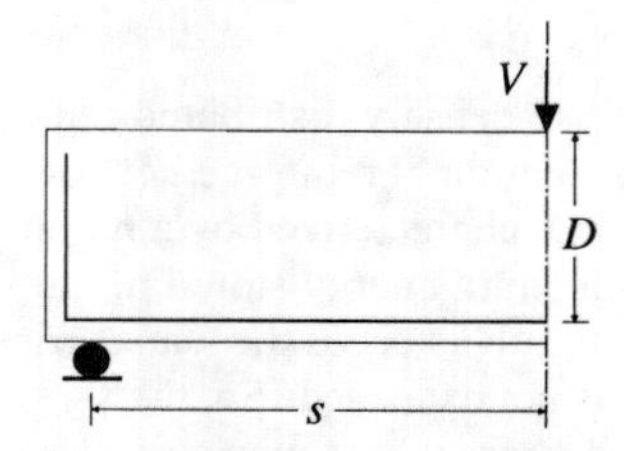

Figure 10.1.3 Longitudinally reinforced beam subjected to constant shear.

the diagonal shear of beams, Bažant and Sun (1987) proposed the following formula, determined by optimization of data fits and the condition that $m \to 0$ for short spans and $m \to$ constant for large spans:

$$m = 400\left[1 + \tanh\left(2\frac{s}{D} - 5.6\right)\right] \tag{10.1.23}$$

where s is the shear span and D the effective depth of the reinforcement (see Fig. 10.1.3).

10.2 Diagonal Shear Failure of Beams

10.2.1 Introduction

In the current ACI Standard 318 (Sec. 11.3), the nominal shear strength is not based on the ultimate load data but on data on the load that causes the formation of the first large cracks. The current ACI formula can be written:

$$\sigma_{Nu} \equiv v_c = \sigma_1\left(1.9\sqrt{\frac{f'_c}{\sigma_1}} + 2500\rho_w\frac{D}{s'}\right) \le 3.5\sigma_1\sqrt{\frac{f'_c}{\sigma_1}}\,, \qquad \sigma_1 = 1 \text{ psi} = 6.895 \text{ kPa} \tag{10.2.1}$$

where $v_c = V_c/(b_w D)$ nominal shear strength provided by concrete, D = effective depth of the longitudinal steel, b_w = width of the beam web, f'_c = compressive strength of concrete, ρ_w = longitudinal steel reinforcement ratio, and

$$s' = \frac{M_u}{V_u} \tag{10.2.2}$$

in which V_u = factored shear force at ultimate, M_u = factored moment at ultimate. For the case of constant shear of Fig. 10.1.3 $s' = s$.

If the first diagonal shear crack were considered to be very small compared to beam depth D, no size effect would occur, as implied by Eq. (10.2.1). However, it seems that most data refer to the formation of relatively large cracks, in which the size effect ought to occur even though it is ignored in Eq. (10.2.1).

The fact that the strength-based failure criterion used in contemporary design codes is not very realistic is, for example, confirmed by the extremely large scatter of the vast amount of test data available in the literature (Park and Paulay 1975; Bažant and Kim 1984; Bažant and Sun 1987). Moreover, in the commentary to the ACI Code (Sec. R11.3.2.1) it is acknowledged that the diagonal shear failure experiments of Kani (1966, 1967) reveal a decrease of the shear strength with the depths of the beam. These results are not considered in the code ACI 318-89, which is justified by assuming the code to be based on the load at initiation of very small cracks rather than formation of first large cracks or the ultimate load. For deep beams such that $L/D < 5$ (L = clear span of the beam), the nominal shear strength is obtained by multiplying Eq. (10.2.1) with the factor $(3.5 - 2.5s'/D)$, which is intended to introduce the increase of the shear strength from the first cracking load to the ultimate load in deep beams. (This is explained by assuming that the mode of shear resistance changes from flexure to arch action or the action of diagonal compression struts.)

Some revisions to the code that partially addressed some concerns stemming from fracture mechanics were proposed by ACI–ASCE Committee 426 (1973, 1974, 1977) and by MacGregor and Gergely (1977),

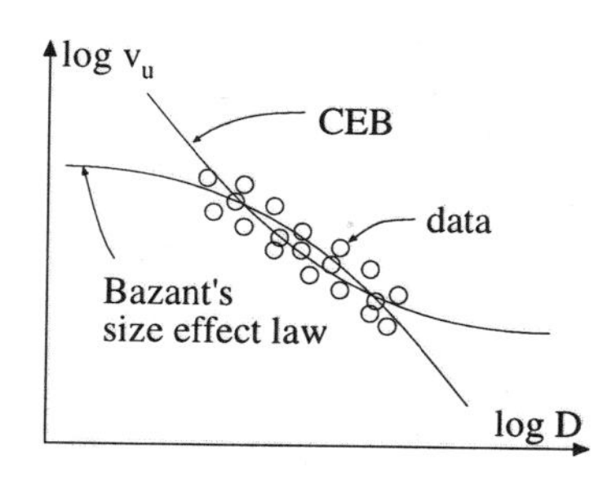

Figure 10.2.1 Experimental data are available only for a range over which both the CEB formula and Bažant's equation describe the results adequately, given the large experimental scatter.

but have not been incorporated into the ACI Code (they were proposed on the basis of an exhaustive study of experimental data obtained prior to 1974). Reinhardt (1981a) analyzed some suitably chosen test data and in 1981 found that there was a size effect and that it agreed quite well with LEFM. Later, however, the LEFM size effect was found to be too strong by Bažant and Kim (1984).

These authors, and also Bažant and Sun (1987), concluded from a statistical analysis of over 400 test series that the code approach to design, which is not based on the maximum load, does not provide a uniform margin of safety against failure of beam of various sizes because it ignores size effect. They noted that introduction of the size effect law leads to a better agreement with the ultimate load test data compared to the current ACI Code formulas which lack the size effect (as well as an LEFM-type formula proposed by Reinhardt, in which the nominal strength decreases inversely to the $\sqrt{D}$, which is too strong).

An empirical formula for the size effect in diagonal shear has been introduced in the CEB Model Code design recommendations (CEB 1991). It has the form $v_u = v_0(1 + \sqrt{D_0/D})$, where v_0 and D_0 are constants. This formula, however, has the opposite asymptotic behavior than the size effect law. For large sizes, it approaches a horizontal asymptote, and for small sizes it approaches an inclined asymptote of slope -1/2, which cannot be logically justified. The reason that this formula compared acceptably with the test data is that the data used pertained only to the middle of the size range. Due to scatter, distinguishing various laws without any theory is impossible for such limited data, as illustrated in Fig. 10.2.1.

The diagonal shear strength was also investigated using the cohesive crack model by Gustafsson (1985) and Gustafsson and Hillerborg (1988), but not with the aim to produce code formulas. Rather, their objective was to show that a size effect was theoretically predicted and to investigate how the shear strength is influenced by the fracture properties, particularly the fracture energy.

Other models have also been used to analyze the diagonal shear of beams. Jenq and Shah (1989) extended their two parameter model to describe crack growth in mixed mode and applied it to diagonal shear. A nonlocal microplane has also been used to analyze the size effect in diagonal shear of beams (Bažant, Ožbolt and Eligehausen 1994; Ožbolt and Eligehausen 1995)

In this section, some of the most important results of the aforementioned works are summarized. In the next section, a recent modification of the classical truss model (or strut-and-tie model) is described which approximately captures the effect of energy release and explains the physical mechanism of size effect in a simple, easily understandable way.

10.2.2 Bažant-Kim-Sun Formulas

Bažant and Kim (1984) and Bažant and Sun (1987) developed a set of phenomenological equations to describe the dependence of the diagonal shear strength on the size, shape, and steel ratios of beams failing in diagonal shear. The work of Bažant and Kim has three essential ingredients. The first one is the general structure of the formula which is based on the approach described in Section 10.1.5, and thus takes the form (10.1.13). The second is the development of a rather general expression for σ^p_{Nu} derived by analyzing the arch action and the composite beam action and summing their contributions. The combination of these

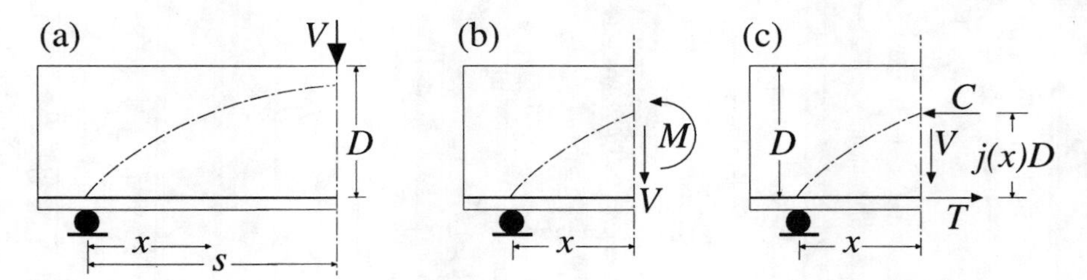

Figure 10.2.2 Diagonal shear strength analysis of Bažant and Kim (1984): (a) geometry; (b) actions at intermediate cross section; (c) decomposition of normal forces.

two ingredients lead the authors to the general formula for the shear strength effect:

$$\sigma_{Nu} \equiv v_u = \frac{v_u^p}{\sqrt{1 + D/D_0}} \tag{10.2.3}$$

$$v_u^p = \sigma_1 k_1 \rho^p \left[\left(\frac{f'_c}{\sigma_1} \right)^q + k_2 \sqrt{\rho} \left(\frac{D}{s} \right)^r \right], \qquad \sigma_1 = 1 \text{ psi} = 6.895 \text{ kPa} \tag{10.2.4}$$

in which we recognize the numerator v_u^p as equivalent to σ_{Nu}^p, and where p, q, r, k_1, and k_2 are dimensionless constants.

The expression for v_u^p is similar to that used in the ACI Code (10.2.1), but it is to a greater extent based on mechanics analysis and contains more empirical parameters, namely k_1, k_2, p, q, and r. As pointed out in Section 10.1.5, the parameter D_0, characterizing the size effect, must also be empirically determined.

Bažant and Kim's derivation of the general expression for v_u^p follows the classical trends in simplified structural analysis. Consider the end portion of the beam as depicted in Fig. 10.2.2a. The overall actions V and M at a cross-section at x are as shown in Fig. 10.2.2b. The normal forces at the cross-section can be decomposed into the steel force $T(x)$ and the compressive resultant on the concrete, $C(x)$, located at distance $j(x)D$ above the reinforcement (Fig. 10.2.2c). Then, from the horizontal equilibrium condition, we have $C(x) = T(x)$, and from the condition of equilibrium of moments,

$$M(x) = T(x) j(x) D \tag{10.2.5}$$

The overall equilibrium equation for the beam requires that $V = \partial M(x)/\partial x$ and thus

$$V = V_1 + V_2 , \qquad V_1 = \frac{dT(x)}{dx} j(x) D , \qquad V_2 = \frac{dj(x)}{dx} T(x) D \tag{10.2.6}$$

where V_1 and V_2 are known as the composite beam action and arch action, respectively. Bažant and Kim empirically approximated $j(x)$ by a power law function:

$$j(x) = k\rho^{-m} \left(\frac{x}{s} \right)^r \frac{dj(x)}{dx} T(x) D \tag{10.2.7}$$

The value of dT/dx is obtained from the equilibrium condition along the reinforcement which requires $dT/dx = n_b \pi D_b \tau_b$, where n_b is the number of bars, πD_b their perimeter, and τ_b the shear bond stress. Since the perimeter of the bars is proportional to the square root of their area —hence, proportional to $\sqrt{\rho}$— and the ultimate bond strength is roughly proportional to $f_c'^q$ with $q = 1/2$, Bažant and Kim were able to write V_1 at the critical section $x = s$ as

$$V_1 = k_0 \rho^{1/2 - m} f_c'^q b D \tag{10.2.8}$$

where k_0 is a constant. Next, using (10.2.7) and assuming that the critical cross section for arch action is at $x = D$, and that, at failure, the steel stress is a constant, they found

$$V_2 = c_0 \rho^{1-m} \left(\frac{x}{s} \right)^r b D \tag{10.2.9}$$

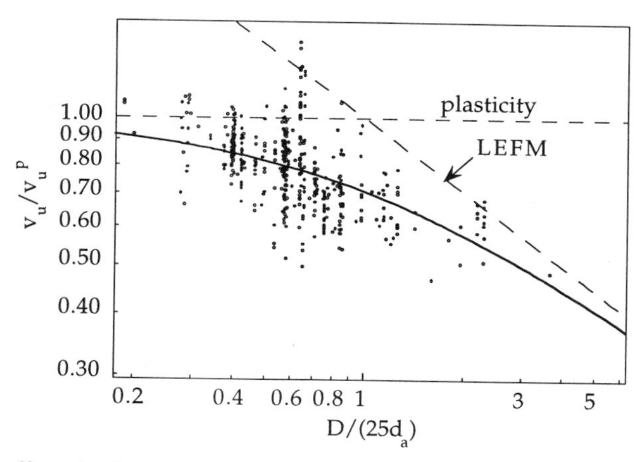

Figure 10.2.3 Size-effect plot for Bažant-Kim-Sun formula, compared to 461 available data points for beams without stirrups (data from Bažant and Sun 1987).

Substituting the last two expressions into the first of (10.2.6) and rearranging leads to Eq. (10.2.4) for v_u^p (where the dummy stress σ_1 has been introduced for dimensional compatibility).

Bažant and Kim compared this formula to a large number of tests from the literature in order to get average values for the foregoing empirical parameters. It was shown that Bažant's size effect law was able to describe the size dependence of the classical data by Kani (1966, 1967) and Walraven (1978). As shown in Section 1.5, this finding was further supported by the tests by Bažant and Kazemi (1991); Fig. 1.5.7, series K1 and K2.

Comparison of equation (10.2.3) to the results from seven classical data series was used by Bažant and Kim to optimize the parameters in that equation. The values of the parameters so determined were as follows :

$$p = \frac{1}{3}\,, \quad q = \frac{1}{2}\,, \quad r = \frac{5}{2}\,, \quad k_1 = 10, \quad k_2 = 3000, \quad D_0 = 25d_a \tag{10.2.10}$$

With this formula, Bažant and Kim were able to fit 296 experimental data points with a coefficient of variation of 30%, much better than the ACI formula.

Later, Bažant and Sun (1987) further improved Eq. (10.2.4) by introducing the effect of maximum aggregate size d_a. This led to the replacement of the value 10 for the factor k_1 in (10.2.10) with the expression

$$k_1 = 6.5\left(1 + \sqrt{c_0/d_a}\right)\,, \qquad c_0 = 0.2 \text{ in} = 5.1 \text{ mm} \tag{10.2.11}$$

Bažant and Sun also collected and tabulated a still larger set of data than Bažant and Kim (1984), involving 461 test data, and showed that the improved formula gives still better results, reducing the coefficient of variation to 25%. Fig. 10.2.3 shows the size effect plot for the 461 data points.

Bažant and Sun further introduced in the formula the influence of the stirrups that the ACI code neglects. Although in the original work the approach was completely empirical, a theoretical background is now provided by the analysis in Section 10.1.6. The equations (10.1.22) and (10.1.23) introduce the modification of the size effect due to the stirrups. Thus, the final formula taking all factors into account is

$$v_u = v_u^s + v_u^{pc}\left[1 + \frac{D}{D_0}\right]^{-1/2} \tag{10.2.12}$$

in which v_u^s, v_u^{pc} and D_0 are given by

$$v_u^s = \rho_v f_{yv}(\sin\alpha + \cos\alpha) \tag{10.2.13}$$

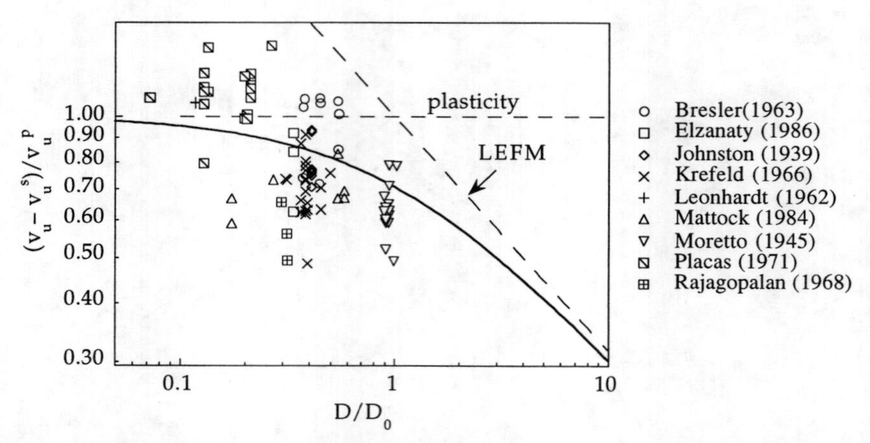

Figure 10.2.4 Size-effect plot for Bažant-Kim-Sun formula, compared to 87 available data points for beams of rectangular cross section with stirrups (data from Bažant and Sun 1987).

$$v_u^{pc} = \sigma_1 c_1 \sqrt[3]{\rho} \left[\sqrt{\frac{f_c'}{\sigma_1}} + 3000 \sqrt{\frac{\rho}{(s/D)^5}} \right], \qquad \sigma_1 = 1 \text{ psi} = 6.895 \text{ kPa} \tag{10.2.14}$$

$$D_0 = 25 d_a (1 + m\rho_v) \tag{10.2.15}$$

in which ρ_v is the steel ratio of stirrups ,f_{yv} the yield strength of the stirrups, $c_1 = 6.5$, and m is given by (10.1.23). The foregoing value of c_1 is adequate to obtain the best fit on average. For design, c_1 is reduced to $c_1 = 4.5$.

Fig. 10.2.4 shows the size effect plot for Bažant-Kim-Sun formula and 87 available data points for beams of a rectangular cross section with stirrups from the test data listed by Bažant and Sun (1987). Although the scatter is large, the experimental results lie relatively close to the Bažant-Kim-Sun formula, closer than to other expressions including the ACI formulas.

10.2.3 Gustafsson-Hillerborg Analysis

Gustafsson (1985) and Gustafsson and Hillerborg (1988) performed approximate analysis of diagonal shear failure of reinforced beams of various depths. In their analysis, they assumed that a single polygonal cohesive crack with linear softening was formed as depicted in Fig. 10.2.5a, while the bulk of the concrete remained linear elastic. The interaction between steel and concrete was represented by the curve of shear stress vs. bond slip displacement, which was assumed to be of the elastic-perfectly plastic type (Fig. 10.2.5b). The behavior of the reinforcing steel was assumed to be linear elastic all the time. To determine the strength of the beam, 5 possible crack paths, as shown in Fig. 10.2.5c, were analyzed using finite elements. A typical, albeit idealized, load-displacement curve is shown in Fig. 10.2.5d. There is a maximum M caused by the failure of the concrete in tension followed, eventually, by a snapback. However, since in the computation the material surrounding the crack is assumed to be linear elastic, the load starts to increase again due to progressive stressing of the reinforcement. If the material behavior were really elastic, the load would increase forever along the dashed curve, approaching an asymptote (dash-dot line) corresponding to a fully cracked concrete sewed up by the reinforcement.

Of course, this is not actually possible, and failure does occur either by yielding of reinforcement or by crushing of concrete in the compressed ligament. In the analysis of Gustafsson and Hillerborg, only the crushing of concrete is considered. To this end, at each crack growth step Gustafsson and Hillerborg make a check of the integrity of the ligament based on the criterion described below. The computation ends at a certain point C in Fig. 10.2.5d when the crushing criterion is satisfied. Gustafsson and Hillerborg found that point C lies above point M for the cracks closer to the loading cross-section (path 1) and goes down as the path deviates more and more from the vertical. Fig. 10.2.5e sketches the values of the load

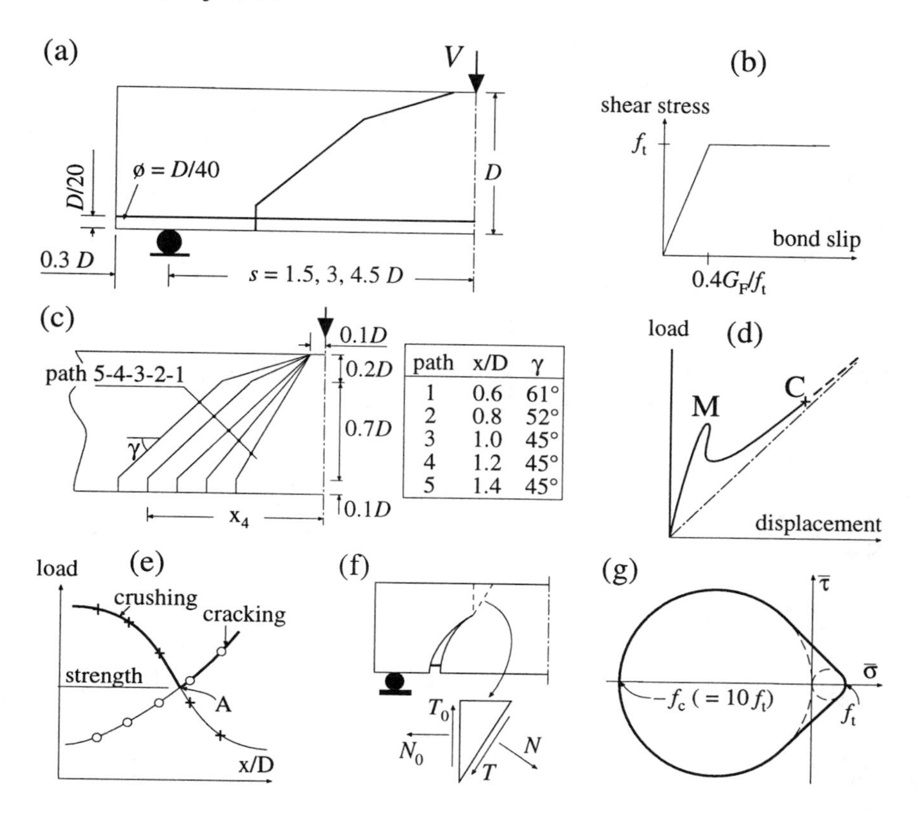

Figure 10.2.5 Gustafsson-Hillerborg model. (a) Geometry of the problem with polygonal cohesive crack path. (b) Bond stress-slip relationship. (c) Crack paths considered in the calculation. (d) Idealized load-displacement curve. (e) Cracking and crushing load vs. crack path mouth position and definition of beam strength. (f) Normal and shear forces across the ligament. (g) Crushing criterion. (Adapted from Gustafsson 1985.)

corresponding to points M (circles) and C (crosses) vs. the relative position of the crack mouth x/D (see Fig. 10.2.5a). Since the strength for a given path is given by the upper branch (heavy line), Gustafsson and Hillerborg assumed that the actual strength of the beam corresponds to the path with less strength, given by point A in the figure. For this path, the loads for point M and C are identical (identical cracking and crushing strength).

A few words regarding the crushing failure criterion. At each step in the calculation, the resultant normal and shear forces(N_0, T_0) across the uncracked ligament were computed (Fig. 10.2.5f). From them and the equilibrium condition, the normal and shear forces (N, T) across any plane at an arbitrary angle could be computed, as well as the corresponding average stresses $(\overline{\sigma}, \overline{\tau})$. Gustafsson and Petersson postulated that crushing failure occurred as soon as, for some orientation, a criterion defined by a condition $F(\overline{\sigma}, \overline{\tau}) = 0$ was met, where the criterion was graphically defined as depicted in Fig. 10.2.5g.

Using the foregoing approach, Gustafsson and Hillerborg analyzed the influence of the size (beam depth), the steel ratio ρ, and the shear span ratio s/D. Fig. 10.2.6 summarizes the results of their computations. Although Gustafsson and Hillerborg proposed a size effect in which $v_u \propto D^{-1/4}$ for $0.4 \leq D/\ell_{ch} \leq 5$, as suggested by other researchers, it appears that an exponent of -0.3 instead of -0.25 fits the results better (dashed lines in Fig. 10.2.6).

10.2.4 LEFM Analyses of Jenq and Shah and of Karihaloo

Jenq and Shah (1989) analyzed diagonal shear fracture using LEFM. They considered the idealized diagonal crack shown in Fig. 10.2.7a and approximated the solution as the superposition of the cases

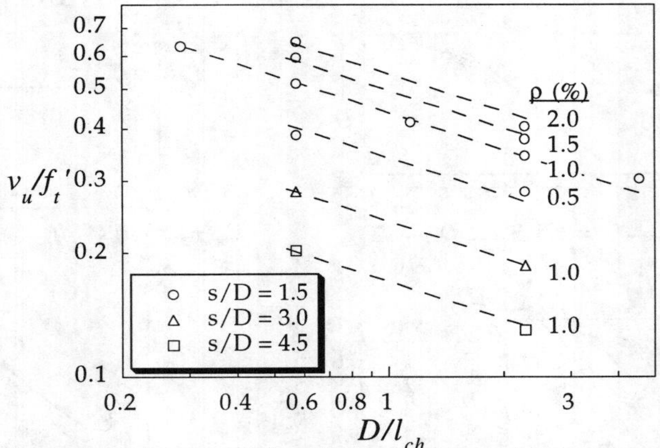

Figure 10.2.6 Nondimensional shear strength vs. beam depth for various span-to-depth ratios and steel ratios according to Gustafsson and Hillerborg model (data from Gustafsson 1985). The dashed lines correspond to power law expressions of the form $v_u \propto D^{-0.3}$.

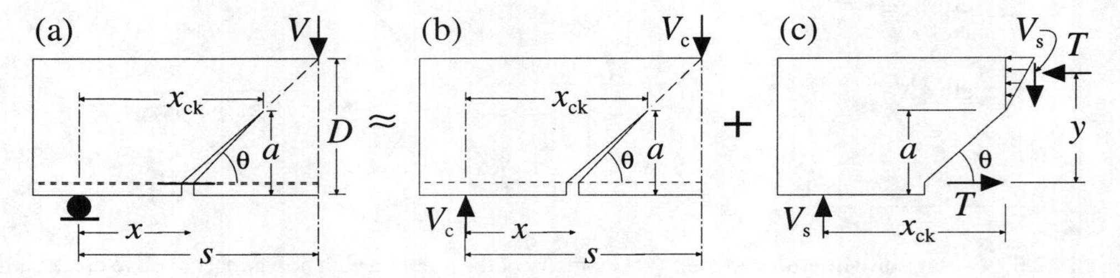

Figure 10.2.7 Jenq and Shah's (1989) analysis of diagonal shear.

shown in Figs. 10.2.7b-c. The first case corresponds to the concrete taking a load V_c such that the stress intensity factor at the crack tip is K_{Ic}. The second case corresponds to concrete plus reinforcement taking a load V_s computed from classical no-tension strength of material analysis (no crack singularity, neutral axis at the crack tip, linear stress distribution along the ligament). Note that in this model only the situation at ultimate load is considered, and the equations that follow cannot be used to analyze the crack growth. This means, in particular, that a_c is the critical crack length (actually its vertical projection), understood as the crack length at peak load.

Because there is no closed-form expression for the precise geometry in Fig. 10.2.7b, Jenq and Shah (1989) assumed that the stress intensity factor can be approximated by the stress intensity factor of a pure bend beam with a symmetric edge notch of depth a subjected to the bending moment corresponding to the cross section at the mouth of the crack (i.e.: $M = V_c x$; note that it is not clear why the moment should not be taken at the cross section at the tip of the crack, $M^* = V_c x_{ck}$). According to this, the crack growth condition is

$$K_{Ic} = \frac{6V_c x}{bD^2}\sqrt{D}k(\alpha)\,, \qquad \alpha_c = \frac{a_c}{D} \tag{10.2.16}$$

where in the first fraction we recognize the expression for the nominal stress in bending ($6M/bD^2$) and $k(\alpha)$ is given, for example, by Srawley's expression (5.4.8). From this we get

$$V_c = \frac{K_{Ic} b D^{3/2}}{6xk(\alpha_c)}\,, \qquad \alpha_c = \frac{a_c}{D} \tag{10.2.17}$$

On the other hand, the equilibrium of moments for the case in Fig. 10.2.7c requires

$$V_s = T(x)\frac{y}{x_{ck}}\,, \qquad y = D - c - \frac{1}{3}(D - a_c) = \frac{2D}{3} + \frac{a}{3} - c \tag{10.2.18}$$

Given x and a_c (together with the initial geometry of the beam), the foregoing equations determine V if the distribution of the steel force $T(x)$ is known. Jenq and Shah (1989) assume as a simplification that this distribution can be approximated by a power law:

$$T(x) = T_{\max}\left(\frac{x}{s}\right)^N \tag{10.2.19}$$

where $T_{\max}$ is the value of the steel force below the concentrated load. Based on test data by Ferguson and Thompson (1962, 1965), Jenq and Shah (1989) proposed a formula for $T_{\max}$. They made an intensive numerical analysis of the influence of the exponent N, from which they recommended the value $N = 2.5$. The recommended expression for $T_{\max}$ is:

$$T_{\max} = 2.509 s f_t' \sqrt{\frac{\rho}{D}} \tag{10.2.20}$$

where the result is in kN if s and D are in mm and f_t' in MPa. This result is strictly valid only for beam thicknesses of 10 in (254 mm) and for a single steel bar. To obtain a dimensionally correct equation, it is better to express the force carried out by the steel at the central section as the length of the steel bars s, times their perimeter $n_b \pi D_b$, times the average bond shear stress (Karihaloo 1992) $\overline{\tau_b}$:

$$T_{\max} = s n_b \pi D_b \overline{\tau_b} \tag{10.2.21}$$

Setting now $n_b \pi D_b^2/4 = \rho b D$ and $\overline{\tau_b} \propto f_t'/D$, as deduced by Jenq and Shah from the Ferguson and Thompson data (1962, 1965), we get the result

$$T_{\max} = L_b s \sqrt{\frac{n_b \rho b}{D}}\, f_t'\,, \qquad L_b = 2.509 \text{ m} \tag{10.2.22}$$

The value of L_b is determined so that this formula coincides with (10.2.20) for $n_b = 1$ and $b = 254$ mm.

Given x (or θ) and a_c, the shear strength is determined from the foregoing equations setting $v_u = (V_c + V_s)/bD$; the result is

$$v_u = \frac{K_{Ic}\sqrt{D}}{6xk(\alpha_c)} + L_b \frac{s}{D}\sqrt{\frac{n_b \rho}{bD}}\, f_t' \left(\frac{x}{s}\right)^N \frac{y}{x_{ck}} \tag{10.2.23}$$

Karihaloo analyzed this model and improved it in a series of papers (Karihaloo 1992, 1995; So and Karihaloo 1993). First he modified the way the model is applied and used it as a forensic engineering tool by using the values of x and a_c as measured (optically) in a test. Using this method on two beam tests, he concluded that the Jenq-Shah model predicted shear strengths that were too low (see exercise 10.3). So and Karihaloo (1993) extended the range of applicability of (10.2.20) to include other parameters. They reevaluated the results of Ferguson and Thompson (1962, 1965) and proposed a new formula that takes into account the bar diameter and the number of bars; for an embedment length L_e, the formula reads

$$T_{\max} = F_1 L_e \sqrt{4 n_b \pi \rho b D}\; \overline{\tau_b} \tag{10.2.24}$$

where F_1 is a reduction factor for $n_b = 2$ (the formula is strictly applicable only for one or two bars):

$$F_1 = \frac{93 + 135A_2 - 7A_2^2}{93 + 135A_1 - 7A_1^2}\,, \qquad A_1 = \frac{b}{D_b} - 1\,, \qquad A_2 = \frac{b}{2D_b} - 1 \tag{10.2.25}$$

The average ultimate bond shear strength $\overline{\tau_b}$ is given by

$$\overline{\tau_b} = F_2\left[0.4684\sqrt{f_c'}\left(\frac{\rho L_e}{D_b}\right)^n + \langle 0.0271(c - 1.5D_b)\rangle\right], \qquad n = -0.8205 D_b^{-0.2933} \tag{10.2.26}$$

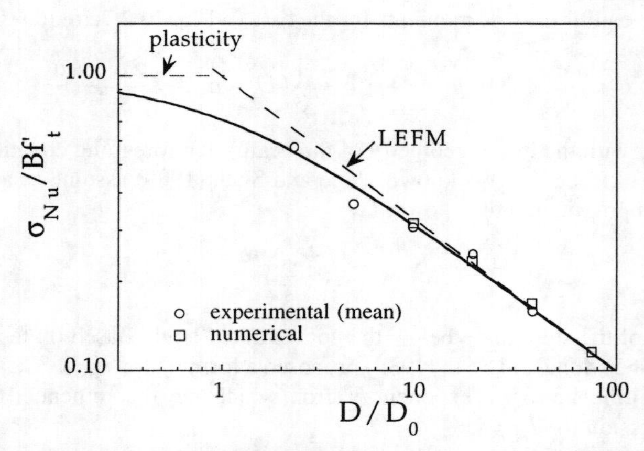

Figure 10.2.8 Comparison of experimental results for diagonal shear with computed values obtained using a nonlocal microplane model (data from Bažant, Ožbolt and Eligehausen 1994). The fit by Bažant's size effect law is also shown.

where all the dimensions must be in millimeters and f_c' in MPa. The factor F_2 is given by

$$F_2 = \begin{cases} 0.3889 + 0.1184\dfrac{s}{D} - 0.0068\left(\dfrac{s}{D}\right)^2, & \frac{s}{D} \leq 17.5 \\ 0.9\,, & \frac{s}{D} > 17.5 \end{cases} \tag{10.2.27}$$

Although this equation substantially improves the Jenq-Shah expression for T_{max}, the problem of the shape of the distribution (particularly the value of exponent N in (10.2.19)), still remains. Karihaloo (1995) further improved the treatment of the LEFM crack growth condition by explicitly considering the mixed mode condition at the crack tip. But even with this enhancement, the strength predictions of the model were too low, even for exponents N as low as 1.25.

10.2.5 Finite Element Solutions with Nonlocal Microplane Model

Additional insight and even partial validation of the design formulas for brittle failures of concrete structures, including diagonal shear, can be gained from careful finite element analysis based on a realistic material model verified over a broad range of experimentally observed behavior. One such model appears to be the nonlocal finite element model combined, on the material level, with the microplane model for the stress-strain relation (Bažant, Ožbolt and Eligehausen 1994). These models, which will be discussed in detail in Chapters 13 and 14, provided, for the diagonal shear tests of Bažant and Kazemi, the results shown by the triangular data points in Fig. 10.2.8. The figure also shows the test data and the best fit by the size effect law. It is seen that the agreement is quite close.

10.2.6 Influence of Prestressing on Diagonal Shear Strength

The effect of longitudinal prestressing on the diagonal shear of longitudinally reinforce beams was addressed by Bažant and Cao (1986) and also by Gustafsson (1985). Similar to the procedure of Bažant, Kim and Sun, Bažant and Cao first used simple equilibrium considerations to get an approximate expression for v_u^p and then applied the size effect correction (10.1.13). Their result is:

$$v_u = v_u^p\left(1 + \frac{D}{D_0}\right)^{-1/2}, \quad v_u^p = \frac{D}{s}\left(c_1\sigma_1\sqrt{\frac{f_c'}{\sigma_1}} + c_2\sigma_{cc}\right) \tag{10.2.28}$$

where D_0, c_1, and c_2 are empirical constants, s is the shear span, and σ_{cc} the uniaxial stress due to prestressing; $\sigma_1 = 1$ psi $= 6.895$ MPa is introduced for dimensional compatibility. From the analysis of

235 test results from the literature, Bažant and Cao proposed the following values of the constants:

$$D_0 = 25d_a \,, \qquad c_1 = 4.9 \,, \qquad c_2 = 0.54 \tag{10.2.29}$$

These values yield the average shear strength; for design, they proposed the values $c_1 = 4$, $c_2 = 0.4$.

It must be realized that although the analysis of data showed a clear size effect, the scatter was so large that the values of the parameters are merely roughly indicative. Nevertheless, the study of Bažant and Cao also shows that the proposed formula provides a better agreement with test results than other formulas found in the literature such as the ACI formula or the formula proposed by Sozen, Zwoyer and Siess (1958).

In contrast to nonprestressed beams, the arch action was not considered in the derivation of the foregoing formula. However, in prestressed beams it is difficult to distinguish the arch action from the effect of prestress. To some extent, the separate consideration of shear force $c_2\sigma_{cc}D/s$ associated with prestress substitutes for the consideration of arch action. However, improvements might be in order.

The foregoing formula did not anchor the size effect into a complete plasticity solution, which should be applicable for an infinitely small size. However, according to the size effect data, the plasticity solution would be applicable, in theory, for beam depths smaller than the aggregate size, and thus it is not clear whether the application of plasticity is permitted. It calls for further research to determine whether this might be so and, if it would, then one could draw on various elegant plasticity solutions for diagonal shear (for example, the recent developments in truss analogy, see Section 10.3).

Exercises

10.1 Karihaloo (1992) reported tests on two reinforced concrete beams tested in three-point bending. The dimensions of the beams were as follows (refer to Fig 10.2.7 for notation): s = 800 mm, D = 150 mm, b = 100 mm, c = 25 mm. Beam number 1 was reinforced with one ribbed bar 12 mm in diameter and beam number 2 with two ribbed bars of the same characteristics, giving steel ratios of 0.0075 and 0.015, respectively. The steel had a yield strength f_y = 463 MPa. The concrete mix had the following characteristics: d_a = 20 mm, f'_c = 38 MPa, E = 30 GPa, f_t = 3.4 MPa. The fracture toughness was estimated (from tests on similar mixes) as K_{Ic} = 1.27 MPa$\sqrt{\text{m}}$. Beam number 1 failed in bending with a main crack close to the central cross-section, while beam number 2 failed in diagonal shear. The failure loads were approximately equal to 24 kN and 33 kN, respectively (note that this is the total load $P = 2V$). (a) Determine the expected strength of the beams according to the Bažant, Kim and Sun's formula whenever applicable. (b) Determine the design strength according to the Bažant, Kim and Sun's model, and determine the actual safety factor for the beam that failed in diagonal shear.

10.2 Consider the beams in the previous example. (a) Make an estimate of the values of G_F and ℓ_{ch} to be used in the Gustafsson-Hillerborg model. (b) Plot the experimental results on a copy of Fig. 10.2.6. (c) Give at least two reasons for each beam (not necessarily the same) why the results of Hillerborg and Petersson cannot be applied directly.

10.3 Consider again the beams in the previous examples. (a) Determine the shear strength of beam number 2 using the Jenq-Shah model with $N = 2.5$, with the following assumptions (based on the observed failure, taken from Karihaloo 1992): $x \approx 250$ mm, $a_c \approx 125$ mm, $\theta \approx 45°$. Compare the result to the experimental value. (b) Determine the value of exponent N that should be used to make the model deliver the observed strength.

10.3 Fracturing Truss Model for Shear Failure of Beams

The truss model of Ritter (1899) and Mörsch (1903), also called the strut-and-tie model, has been widely used in successively refined versions to analyze the failure of beams in diagonal shear (Nielsen and Braestrup 1975; Thürlimann 1976; Collins 1978; Collins and Mitchell 1980; Marti 1980, 1985; Schlaich, Schafer and Jannewein 1987; Hsu 1988, 1993; Collins, Mitchell et al. 1996). A fracture model retaining the basic hypotheses of the truss model has recently been proposed. It explains the size effect observed in this type of structures based on the concepts previously analyzed in Section 9.5, particularly the generation and growth of a band of axial splitting cracks parallel to the compressive principal stress (Bažant 1996b), or alternatively a shear compression crack propagating across the strut. In this section we present the

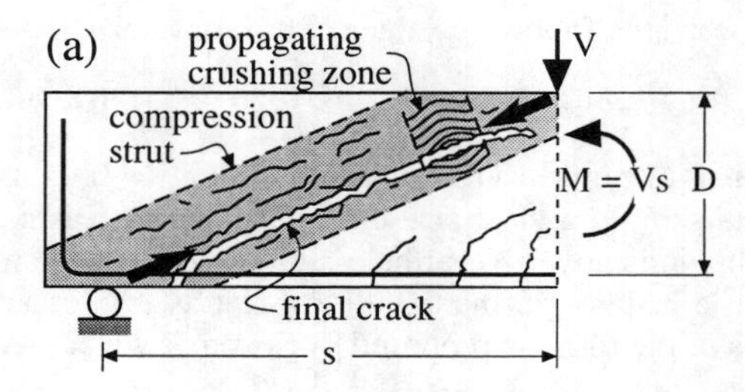

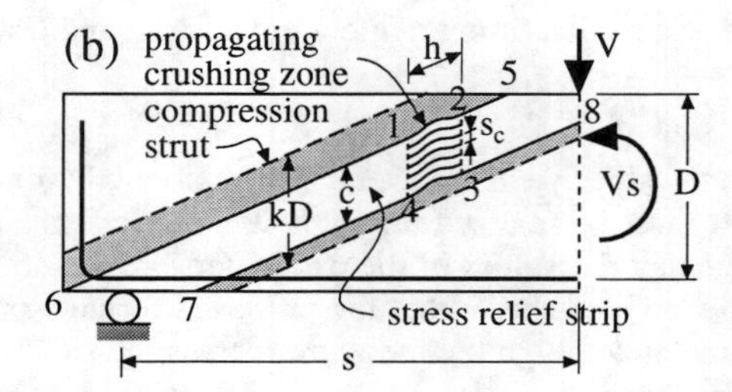

Figure 10.3.1 (a) Compression strut in a beam without stirrups and crushing zone propagating across the compression strut during failure, (b) stress relief zones caused by crushing band propagating across compression strut in beams of different sizes.

basic hypotheses underlying the model and two alternative theoretical analyses, one based on the stress relief zone and strain energy release, and the other based on the stress redistribution and complementary energy. The section closes by discussing the size effect on the cracking load, which is sometimes claimed to be free from size effect.

10.3.1 Basic Hypotheses of Fracturing Truss Model

Consider the sheared beam in Fig. 10.1.3 (which shows only the left-end portion of the beam). For the sake of simplicity, the beam is considered to have a rectangular cross section (a generalization to flanged cross sections, however, would not be difficult). The analysis of the size effect performed by Bažant (1996b) rests on the two following hypotheses:

Hypothesis I: *The failure modes at maximum load of beams of different sizes are geometrically similar.*
This means that, for example, the shear span s and the length c of the material failure zone at maximum load are geometrically similar (Fig. 10.3.1). In other words, the ratios s/D and c/D are assumed to be constant. The hypothesis is applicable only within a certain range of sizes. However, experience from testing as well as finite element analysis indicates that this range covers the size range of practical interest.

Hypothesis II: *The maximum load is determined by the compression failure in the inclined compression struts.*
The compression failure must be interpreted as a temporary incremental strain-softening in compression (or progressive crushing) of concrete in the strut, characterized by a negative slope of the stress-strain diagram. Hypothesis II means that the concrete in the compression strut is suffering splitting cracks in the direction of compression only during a certain, possibly short, portion of the loading history during which the applied load is reaching its maximum. It does not mean that the concrete will get crushed completely once the load will be reduced to zero (such complete crushing is seen only in T-beams; Leonhardt 1977). During the postpeak softening, the splitting cracks may interconnect to produce what looks like compression shear cracks in the horizontal or vertical direction (Fig. 10.3.2) (however, if the failure were assumed to be caused by propagation of a horizontal or vertical shear crack across the strut, the calculation results would be the same). Thus, after the failure is completed, the failure might not look as crushed concrete but as a diagonal crack and a shear crack. The lack of complete crushing may be caused by the failure process taking place under a decreasing load, after the maximum load. The concrete in the strut may have been partially damaged by compression splitting but need not have disintegrated.

Denying the existence of progressive failure of the compression strut at maximum load would be tantamount to denying the validity of the truss model (strut-and-tie model). If this model is valid, then (**1**) diagonal tensile cracks must form before the maximum load, (**2**) the tensile and shear stresses (crack-bridging or cohesive stresses) transmitted across these cracks must be negligible compared to compression stresses in the struts, and (**3**) the compression struts between these cracks must be aligned in the direction of the compressive principal stress in concrete. Only under these conditions, the concrete, stirrups, and

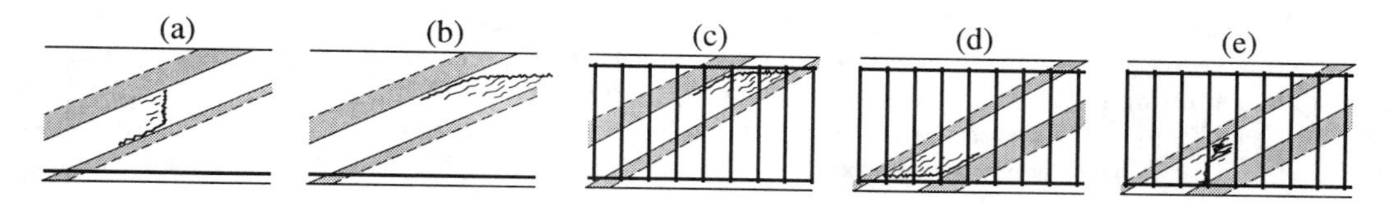

Figure 10.3.2 Splitting crack interconnection to form horizontal or vertical compression-shear cracks: (a,b) beams without stirrups; (c–e) beams with stirrups.

longitudinal bars may be treated as a truss. Assuming that the stirrups and longitudinal bars are designed strong enough, the truss can fail only in concrete. Because the concrete is in uniaxial compression, the failure must be compression failure.

The stresses transmitted across the diagonal cracks are, of course, nonzero, because the cracks are not open widely enough at maximum load. But the important point, which justifies the truss model, is that these stresses are much smaller in magnitude than the compression stresses in the struts.

The energy release due to fracture propagation can be calculated in two ways: (**1**) from the change of the strain energy of the structure-load system at constant displacement, or (**2**) from the change of the complementary energy of the structure at constant load (see Chapter 3). We will examine both approaches in a simplified manner and show that they give approximately the same results.

10.3.2 Analysis Based on Stress Relief Zone and Strain Energy for Longitudinally Reinforced Concrete Beams Without Stirrups

The typical pattern of cracks forming during the failure of a simply supported beam is seen in Fig. 10.3.1a. Although after the failure only one final crack emerges, cracks of various orientations form during the loading process. The first cracks caused by shear loading are tensile cracks of inclination approximately 45°. On approach to the maximum load, these cracks interconnect and form a larger crack running approximately along the line connecting the application point of the load V to the support in Fig. 10.3.1a. This major crack is free of shear stresses and has approximately the direction of the maximum principal compression stress, σ_{II}.

According to the truss model (or strut-and-tie model), we may imagine that most of the load is transferred through the shaded zone called the compression strut (in the case of distributed load, it would be a compressed arch). The normal stress in the direction orthogonal to the strut is essentially zero and the material can expand freely in that direction.

The failure behavior is approximately idealized as shown in Fig. 10.3.1b for two geometrically similar beams of different size. Although for calculation purposes the compression strut is assumed to represent a one-dimensional bar connecting the point of application of V and the support, it has a finite effective width, denoted as kD (Fig. 10.3.1b) where D is the depth to the reinforcement and k is approximately a constant, independent of the beam size.

According to experimental evidence, supported by finite element results, a beam (with a positive bending moment) fails at maximum load due to compression failure of the concrete, usually near the upper end of the compression strut, provided that the longitudinal bar is anchored sufficiently so that it cannot slip against concrete near the support. Aside from the fact that the compression failure (axial splitting or compression shear crack) occurs only within a portion of the length of the strut, the basic premise of the present analysis is that the width h of the cracking zone in the direction of the strut is for a given concrete approximately a constant (which is probably approximately proportional to the maximum aggregate size and also depends on Irwin's characteristic length and on other material characteristics).

The fact that h, in contrast to the length and width of the stress-relieved strip in the strut (the white strip 56785 in Fig. 10.3.1b), is not proportional to the beam size is the cause of the size effect. If the width h of the crushing zone were proportional to the beam size, there would be no size effect. For calculation purposes, we will assume that the compression failure of the material consists of a band 12341 of splitting cracks (Fig. 10.3.1b) growing vertically across the strut upward or downward, or both (which of these is immaterial for the present analysis). These cracks may interconnect after the peak load to produce what looks as a shear crack.

Microscopically, the compression failure may be regarded as internal buckling of an orthotropically damaged material (Bažant and Xiang 1994, 1997; see Section 9.5). The failure begins by formation of dense axial splitting microcracks in the direction of maximum compression, which greatly reduces the transverse stiffness of the material, thus causing the microslabs of the material between the microcracks to buckle laterally. The details of the process are, however, not needed for the present analysis. Neither is it important that the crushing band is pictured to propagate vertically. If it propagated across the strut in an inclined or horizontal direction, the calculation results would be about the same.

The growth of the splitting crack band, which causes the load-deflection curve to reach a maximum load and subsequently decline, relieves the compression stress from the strip 56785 shown in Fig. 10.3.1b. The reason that the boundaries of the stress relief zone, that is, the lines 16, 25, 38, and 47, are parallel to the direction of the strut is that the material is heavily weakened by cracks parallel to the strut. Otherwise, a more realistic assumption would be a triangular shape of the stress relief zone, as considered in the case of tensile failures (Section 1.4; see also the remark at the end of Section 9.5).

Now, how to make the size effect intuitively clear with minimum calculations? To this end, note that the area of the stress relief zone 56785 in Fig. 10.3.1b is proportional to ca, where c is the length of the crack band at failure. Since $ca = (c/D)(a/D)D^2$, and c/D and a/D are constants, independent of D, the area of the stress relief zone is proportional to D^2. Because the average strain energy density in the strut is proportional to the nominal shear stress at ultimate load, v_u^2, the total energy release from the stress-relieved strip 56785 of the strut is proportional to $v_u^2 D^2$. However, assuming the energy dissipation per unit volume of the crack band to be constant, the energy dissipation in the entire cracking band is proportional to D, because the area of the crushing band is proportional to $ch = (c/D)hD$. Therefore, varying the beam size D, $v_u^2 D^2$ must be proportional to D, which means that v_u must be proportional to $1/\sqrt{D}$. This represents a very strong size effect corresponding to LEFM.

In summary, the cause of the size effect is simply the fact that the energy release from the structure is approximately proportional to $v_u^2 D^2$ whereas the energy consumed by fracture is approximately proportional to D.

Let us now do the calculations in detail, following the stress relief zone approximation illustrated in Section 3.2.2. The condition that the entire shear force V must be transmitted by the compression strut yields, for the axial compression stress in the strut, the following expression:

$$\sigma_c = \frac{1}{bkD} \frac{V}{\sin\theta\cos\theta} = \frac{v_u}{k}\left(\frac{s}{D} + \frac{D}{s}\right) \tag{10.3.1}$$

in which θ is the inclination angle of the compression strut from the horizontal (note that $\tan\theta = D/s$). The strain energy density in the strut is $\sigma_c^2/2E_c$, where E_c is the elastic modulus of concrete. The volume of the strut is sbc (where b = beam width). Therefore, the loss of strain energy from the beam caused by stress relief during the formation of the crack band at constant load-point displacement is, approximately:

$$\Delta\mathcal{U}_c = -\frac{\sigma_c^2}{2E_c} sbc = -\frac{v_u^2}{2E_c k^2}\left(\frac{s}{D} + \frac{D}{s}\right)^2 sbc \tag{10.3.2}$$

The minus sign expresses the fact that this is an energy loss rather than gain. The energy release rate due to the growth of the cracking band is obtained from (2.1.15) as

$$\mathcal{G} = -\frac{1}{b}\left[\frac{\partial\mathcal{U}_c}{\partial c}\right]_u = -\frac{1}{b}\left[\frac{\partial\Delta\mathcal{U}_c}{\partial c}\right]_u = -\frac{v_u^2 s}{2E_c k^2}\left(\frac{s}{D} + \frac{D}{s}\right)^2 \tag{10.3.3}$$

The energy dissipated by the cracking zone may be expressed on the basis of the fracture energy G_f characterizing the axial splitting microcracks in the crack band. The length of these cracks is h (width of the band), and their average spacing is denoted as s_c. The number of axial splitting cracks in the band is c/s_c. Thus, the total energy dissipated by the crack band is $W_f = (c/s_c)bhG_f$. Differentiating with respect to c, we find that the energy dissipation in the crack band per unit length of the band and unit width of the beam (which we call $\mathcal{R}$ because it has the meaning of a crack growth resistance) is:

$$\mathcal{R} = \frac{h}{s_c} G_f \tag{10.3.4}$$

In this equation, however, it would be too simplistic to consider h to be a constant through the entire evolution of the crack band. Naturally, the crack band must initiate from a small zone of axial splitting cracks. The length of these cracks first extends in the direction of the strut until they reach a certain characteristic length h_0. After that the crack band grows across the strut at roughly constant width $h = h_0$ (see the intuitive picture of the subsequent contours of the crack zone in Fig. 10.3.1a). Such behavior may be simply described by the equation

$$h = \frac{c}{w_0 + c} h_0 \tag{10.3.5}$$

in which h_0, w_0 = positive constants; h_0 represents the final width of the crack band. Thus, strictly speaking, our hypothesis of a constant width of the cracking zone means that the final width rather than h is a constant.

The increase of $\mathcal{R}$ with c, as described by (10.3.4) with (10.3.5), represents an R-curve behavior (because $\mathcal{R}$ represents the resistance to fracture). The R-curve behavior in tensile fracture is also caused by the growth of the fracture process zone size. Here, however, this growth is expressed indirectly in terms of the length of the axial splitting cracks in the cracking band.

It is also conceivable that, instead of a band of parallel splitting cracks, a shear crack would propagate in a direction inclined to the compression strut (Fig. 10.3.2). In that case

$$\mathcal{R} = G_{fs} \frac{c}{h_0 + c} \tag{10.3.6}$$

where G_{fs} = fracture energy of the shear crack and h_0 now characterizes the R-curve behavior of the shear crack. This is mathematically identical to (10.3.4) if one sets $G_f = G_{fs} s_c / h_0$, and so we will not pursue it further.

The balance of energy during equilibrium propagation of the crushing band requires that $\mathcal{G} = \mathcal{R}$. Substituting here the expressions in (10.3.3)–(10.3.5), one obtains the result:

$$v_u = v_p \left(1 + \frac{D}{D_0}\right)^{-1/2} \tag{10.3.7}$$

in which the following notations have been made

$$D_0 = w_0 \frac{D}{c} \tag{10.3.8}$$

$$v_p = c_p K_c \left(\frac{s}{D} + \frac{D}{s}\right)^{-1} \tag{10.3.9}$$

$$K_c = \sqrt{E_c G_f}, \qquad c_p = k \sqrt{\frac{2h_0}{w_0 s_c} \frac{c/D}{s/D}} \tag{10.3.10}$$

Here the expression for K_c is that for the fracture toughness (the critical stress intensity factor) of the axial splitting microcracks. An important point is that, because of our assumptions (constant c/D and s/D), the values of D_0, v_p, and c_p are constant, independent of size D. The value v_p is the limiting (asymptotic) value of the nominal shear strength for a very small size D.

Eq. (10.3.7) represents the size effect law discussed in Chapters 1 and 6. This law was introduced into the analysis of diagonal shear failure by Bažant and Kim (1984), however, on the basis of a more general and less transparent argument (see Section 10.2.2).

By the same calculation procedure, it can also be easily shown that if and only if, contrary to our hypothesis, the width h of the crushing band were proportional to D instead of obeying (10.3.5), there would be no size effect. If constant w_0 were taken as 0, one would have $v_u \propto D^{-1/2}$, which is the size effect of linear elastic fracture mechanics (LEFM), representing the strongest size effect possible. However, the experimental data exhibit a weaker size effect, which implies that the constant w_0 should be considered finite.

As seen in Chapters 1 and 6, the size effect curve given by (10.3.7) represents a smooth transition from a horizontal asymptote corresponding to the strength theory or plastic limit analysis to an inclined asymptote of slope $-1/2$, corresponding to LEFM. The approach to the horizontal asymptote means that the plasticity approach, that is, the truss model (or strut-and-tie model), can be used only for sufficiently small beam sizes D.

For very small beam sizes D, we may substitute in (10.3.1) $\sigma_c = f_c^b$ = compression strength of the strut, and replace v_u by plastic nominal strength v_p. From this we can solve:

$$v_p = k f_c^b \left(\frac{s}{D} + \frac{D}{s} \right)^{-1} \tag{10.3.11}$$

which is an alternative to (10.3.9). Thus, the size effect law in (10.3.7) can be alternatively written as

$$v_u = k f_c^b \left(\frac{s}{D} + \frac{D}{s} \right)^{-1} \left(1 + \frac{D}{D_0} \right)^{-1/2} \tag{10.3.12}$$

which also shows the effect of the relative shear span s/D on the nominal shear strength. Note that f_c^b cannot be expected to represent the uniaxial compression strength f_c' of concrete because the progressively fracturing concrete in the strut is under high transverse tensile strain in the other diagonal direction and has been orthotropically damaged by cracking parallel to the strut due to previous high transverse tensile stress (Hsu 1988, 1993). So f_c^b is a certain biaxial strength of concrete, depending both on the uniaxial compression strength f_c' and the direct tensile strength f_t'. This dependence needs to be calibrated by shear tests of beams.

It is interesting to determine the ratio to the nominal strength for bending failure, σ_{Nu}^b. The ultimate bending moment in the cross section under the load V is $M_u = V_u s = \sigma_N^b bsD$. From the moment equilibrium condition of the cross section under the load V, we also have $M_u = (f_y \rho b D) k_b D$, in which f_y is the yield strength of the longitudinal reinforcing bars, ρ is the reinforcement ratio (which means that $\rho b D$ is the cross section area of the longitudinal reinforcing bars), and $k_b D$ represents the arm of the internal force couple at the ultimate load. As is well known, k_b is approximately constant. Equating the expressions for M_u, we obtain $\sigma_N^b = \rho f_y k_b D/s$. Considering now (10.3.12), we conclude that:

$$\frac{\sigma_N^b}{v_u} = \frac{\rho f_y k_b}{k f_c^b} \left(\frac{s}{D} + \frac{D}{s} \right) \sqrt{1 + \frac{D}{D_0}} \tag{10.3.13}$$

This equation shows that the ratio of the nominal bending strength to the nominal shear strength of the beam decreases when the relative shear span s/D increases, which confirms a well-known fact. It means that slender beams, for which s/D is large, fail by bending, while deep beams, for which s/D is small, fail by shear. However, as is clear from (10.3.13), the relative shear span $(s/D)_{tr}$ at the transition between the shear and bending failures is not constant but is larger for a larger beam size D. To express it precisely, one sets $\sigma_N^b = v_u$ in (10.3.13), and needs to solve (10.3.13) for s/D, which is a cubic equation. The transitional shear span obviously exhibits a size effect.

The foregoing analysis assumes the reduction of the compressive stress σ_c all the way to zero. Similar to the analysis of compression fracture in Section 9.5, it could be that the compression stress σ_c is reduced to some small but finite residual strength σ_r. However, the residual stress is anyway likely to be smaller than for uniaxial compression, due to the existence of large tensile strain. A finite σ_r seems more realistic when we consider beams with stirrups, which provide some degree of confinement. If σ_r were nonzero for the present case, it would have the effect of adding a constant term to the right hand side of (10.3.7).

The tensile strength of concrete, f_t', has played no direct role in the foregoing analysis. The tensile strength is not a material parameter in LEFM, nor in the R-curve model of nonlinear fracture. It does appear in the cohesive (fictitious) crack model or the crack band model. However, those models are too complicated for achieving a simple analytical solution. The tensile strength, of course, controls the initiation of the inclined shear cracks, however, their growth is governed by fracture energy. In the present analysis we take the view that the inclined cracks due to shear loading have already formed before the maximum load.

Does shear stress transmission across cracks due to friction and aggregate interlock play any role? It could, although according to the present analysis, it cannot be significant. As shown in Fig. 10.3.1a, only cracks rather curved within the area of the compression strut can be subjected to shear and normal loading. Their capability of shear stress transmission decreases with the crack width, and the crack width

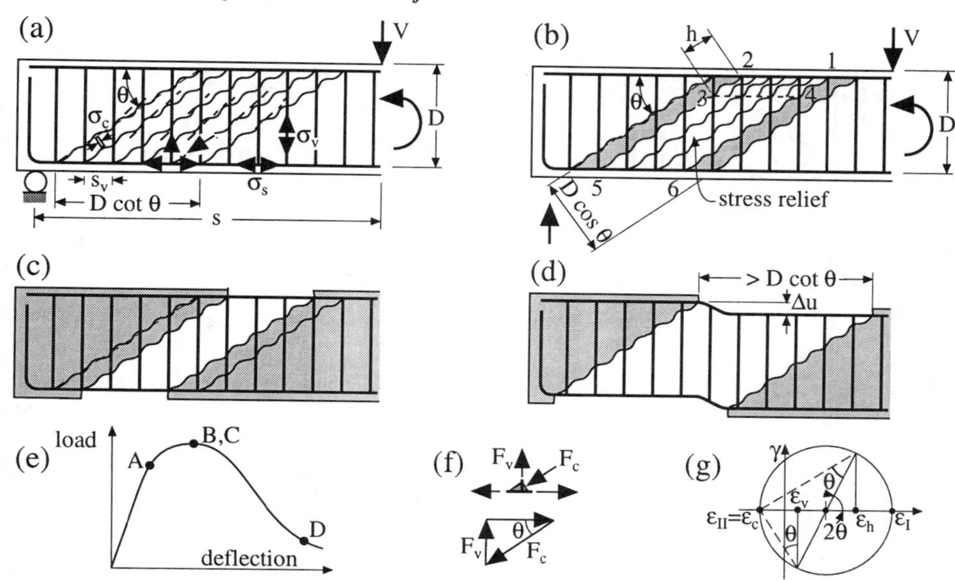

Figure 10.3.3 Evolution of diagonal cracks in beam with stirrups under shear loading: (a) diagonal crack formation before maximum load, (b) growth of crushing band across compression strut during failure at maximum load, (c) beam at maximum load with the crushed and stress-relieved parts of the compression strut removed, (d) state of beam without crushed and stress-relieved concrete after collapse (i.e., when the load has been reduced to a small value), (e) location of the states represented in Figs. a–d on the load deflection curve, (f) equilibrium of forces in stirrups and struts, and (g) Mohr circle of strains.

may be assumed to increase with an increasing beam size, which obviously would also introduce a size effect (this idea was proposed by Reineck 1991). The cracks are most inclined to the compression strut direction and are opened the most widely at the bottom of the beam. However, the maximum load appears to be controlled by progressive compression crushing near the major crack at the top of the beam. For this reason, the effect of crack opening on the shear stress transmission across cracks can hardly play a major role in the size effect on the maximum load.

10.3.3 Analysis Based on Stress Relief Zone and Strain Energy for Longitudinally Reinforced Concrete Beams With Stirrups

Consider now a beam with stirrups (Fig. 10.3.3a). The stirrups cause the diagonal cracks due to shear to be more densely distributed. The first hairline cracks, shown by the dashed lines in Fig. 10.3.3a, form near the neutral axis, with inclination about 45° before the maximum load. These cracks later interconnect and form continuous major cracks at inclination angle θ with the horizontal (Fig. 10.3.3a). These cracks run in the direction of the maximum principal compressive stress σ_{II}, transmitting no shear stresses. They are, of course, cohesive cracks transmitting tensile bridging stresses. These stresses will probably be less than one half of the tensile strength, f_t', while the compression stresses in the truss will be equal or nearly equal to the compressive strength of concrete, f_c'. So, it is safe to assume that the tensile principal stress is negligible ($sigma_{\text{I}} \approx 0$) compared to the compressive principal stress, which justifies treating the beam approximately as a truss. This makes the truss statically determinate. It is this circumstance that makes the well-known simple analysis of the truss model (or strut-and-tie model) possible. If σ_I were not negligible, the truss model would be invalid.

The failure at maximum load is assumed to be caused by the progressive crushing of concrete in the compression struts between the major inclined cracks. Similar to beams without stirrups, a crack band which consists of dense axial splitting microcracks first widens to its full width h and then propagates sideways as shown in Fig. 10.3.3b. For the case of a positive bending moment, this crack band probably forms near the top of the beam and may be assumed to propagate horizontally, to the left or to the right,

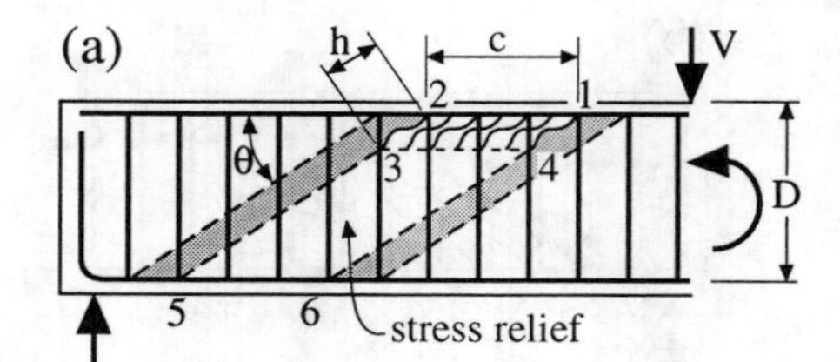

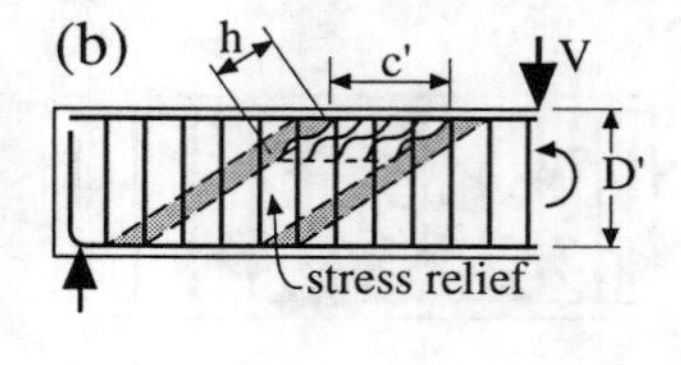

Figure 10.3.4 Stress relief zones caused by a crack band propagating across the compression strut in beams of different sizes with stirrups. The beam in (b) is similar to the beam in (a) except that the crack band width h is the same.

or both. The location and direction of the propagation of the crack band is actually not important for the present analysis, and the same results would be obtained if the band propagated at other inclinations to the compression strut. An important point, however, is that the final length h_0 of the axial splitting cracks, that is, the final width h_0 of the band, is a material property, independent of the size of the beam. If the width h_0 of the band were proportional to beam depth D, there would be no size effect. Since it is less than proportional to D, there must be size effect.

Thus, the cause of the size effect is the localization of the compression failure of the strut into a crack band of a fixed width, and the growth of this band across the strut.

An important point is that the stirrups as well as the longitudinal steel bars are not necessarily yielding during the failure at maximum load. They might not have yielded before the crushing of the strut began, or they may have yielded and unloaded. There is no reason why the yielding of steel should occur simultaneously with the progressive compression crushing.

The formation of the crack band 12341 (Fig. 10.3.3b) may again be assumed to relieve the compression stress from the entire length of the compression struts in the region 12561 (Fig. 10.3.3b). This causes a release of strain energy from the compression struts, which is then available to drive the propagation of the crack band. This represents the mechanism of failure at maximum load.

With the stress relieved from the compression struts, the beam acts essentially as shown in Fig. 10.3.3c, as if there were a gap in concrete (provided the residual strength of crushed concrete is neglected). However, since the steel is not in general yielding, this does not represent a failure mechanism. A failure mechanism can be created only when a sufficient number of compression struts fail as shown in Fig. 10.3.3d, in which case even nonyielding bars permit free movement because the bending resistance of the bars is negligible. However, this type of collapse mechanism corresponds to a postpeak state at which the load is already reduced to a very small value (such as state D in Fig. 10.3.3e). Thus, the stress relief at maximum load does not imply the structure has become a mechanism.

First, let us explain the size effect mechanism in the simplest possible terms. The area of the compression struts from which the compression stress is relieved, that is, area 12561 in Fig. 10.3.4, is proportional to cD, which is equal to $(c/D)D^2$. But since the failure is assumed to be geometrically similar for beams of different sizes (shown in Fig. 10.3.4), c/D is a constant, and so the area of the stress relief zone is proportional D^2. The strain energy density before the stress relief is proportional to $v_u^2/2E_c$, and so the total energy release is proportional to $v_u^2 D^2$. The area of the crack band is proportional to $ch = (c/D)hD$. Since both h and c/D are constant for beams of different sizes, the area of the crack band is proportional to D, and so is the energy dissipated in the crack band. So, considering the failures of geometrically similar beams of different sizes, $v_u^2 D^2$ must be proportional to D, which means that v_u must be proportional to $1/\sqrt{D}$. Again, same as for the beam without stirrups, we thus obtain a size effect, and it is the strong size effect of LEFM. In practice, the size effect for smaller beam sizes is weaker because of the R-curve behavior of the crack band 12341.

We assume the stirrups to be uniformly distributed (smeared). Equilibrium on a vertical cross section of the beam (Figs. 10.3.4 and 10.3.3f) requires that

$$\sigma_c = -\frac{F_c}{bD\cos\theta} = -\frac{v_u bD}{\sin\theta}\frac{1}{bD\cos\theta} = -\frac{2v_u}{\sin 2\theta} \tag{10.3.14}$$

in which θ is the inclination of the compression struts, F_c = compression force in the strut per length D, and σ_c is the compression stress transmitted by the strut (which, in general, is not equal to the standard compression strength f_c' of concrete and depends on the size of the beam in a manner to be determined). Equilibrium on an inclined cross section of the beam parallel to the compression struts further requires that

$$\sigma_v = (V s_v / A_v D) \tan\theta = v_u s_v b \tan\theta / A_v \tag{10.3.15}$$

in which A_v = cross section of the stirrups, s_v = spacing of the stirrups, and σ_v = tensile stress in the stirrups, which, in general, is not equal to the yield stress. The stress in the longitudinal bars is obtained from the moment equilibrium condition in a cross section and is $\sigma_s = M/A_s k_b D$, in which M = bending moment, A_s = cross section area of the longitudinal bars, and $k_b D$ = arm of the internal force couple in the cross section.

We do not attempt to determine the angle θ of the diagonal cracks and the struts by fracture analysis. The diagonal cracks delineating the struts start to form before the maximum load, and not during failure. For the sake of simplicity, we assume the orientation of the major diagonal cracks not to rotate and adopt the method introduced into the truss model by Mitchell and Collins (1974) in their compression field theory, in which they used the compatibility condition for the average strains in the truss in a similar way as Wagner (1929) used the compatibility condition for approximate analysis of the shear buckling of the webs of steel beams. The average strains of the truss are defined as the strains of a homogeneously deforming continuum that is attached to the joints of the truss at the nodes (tops and bottoms of the stirrups). According to the Mohr circle shown in Fig. 10.3.3g (in which ε denotes the strains, and ε_h is the strain in the longitudinal bars), the overall compatibility of the average strains of the struts, the stirrups, and the longitudinal bars requires that

$$\tan^2\theta = \frac{\varepsilon_v - \varepsilon_c}{\varepsilon_h - \varepsilon_c} = \frac{(\sigma_v/E_s) - f(\sigma_c)}{(\sigma_s/E_s) - f(\sigma_c)} \tag{10.3.16}$$

Here the strains have been expressed in terms of the stresses assuming the steel not to be yielding and denoting by $f(\sigma_c)$ the stress-strain diagram of concrete. (For the precise method in which the strains entering (10.3.16) are calculated, see Mitchell and Collins 1974.) The foregoing calculation, of course, requires that the diagonal cracks and the struts be aligned with the direction of the compressive principal strain, which coincides with the direction of the compressive principal stress.

Fracture analysis begins by expressing the strain energy change (Fig. 10.3.4) caused by the formation of the crack band of length c at constant load-point displacement:

$$\Delta\mathcal{U}_c = -\frac{(\sigma_c - \sigma_r)^2}{2E_c} cD \tag{10.3.17}$$

The minus sign reflects the fact that this is an energy loss rather than gain.

The stress σ_r in the foregoing equation represents the residual compression strength of the crack band of concrete. In this study, the residual compression strength σ_r is considered to be an empirical property. However, it can be mathematically expressed on the basis of the concept of internal buckling of a material heavily damaged by axial splitting microcracks, as proposed in Bažant (1994a) and Bažant and Xiang (1997); see Section 9.5.

The energy release rate may be calculated as:

$$\mathcal{G} = -\frac{1}{b}\left[\frac{\partial \Delta\mathcal{U}_c}{\partial c}\right]_u = \frac{(\sigma_c - \sigma_r)^2}{2E_c} D \tag{10.3.18}$$

The energy dissipation rate (fracture resistance) of the crack band is again given by (10.3.4), i.e., $\mathcal{R} = G_f h / s_c$, in which the width of the crack band may be assumed to evolve again according to (10.3.6), i.e., $h = h_0 c/(w_0 + c)$.

Substituting now (10.3.14) and (10.3.15) into (10.3.18), and using the fracture propagation criterion $\mathcal{G} = \mathcal{R}$, we obtain an equation which can be easily solved for v_u. This provides the result:

$$v_u = v_p \left(1 + \frac{D}{D_0}\right)^{-1/2} + v_r \tag{10.3.19}$$

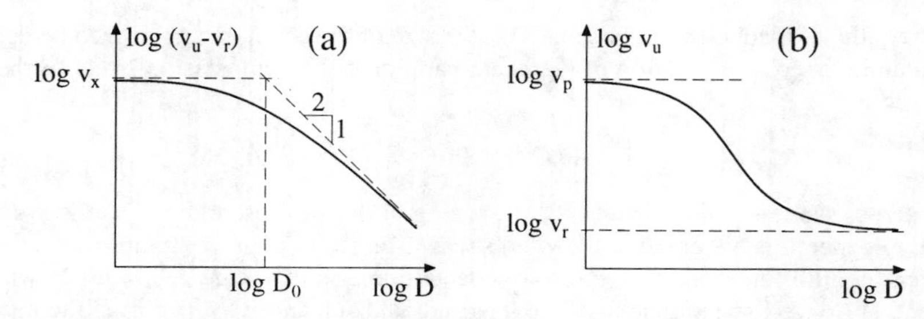

Figure 10.3.5 Size effect in shear failure of concrete beam in terms of the logarithm of either v_u or $v_u - v_r$.

in which we introduced the notations:

$$D_0 = w_0 \frac{D}{c} \tag{10.3.20}$$

$$v_r = \frac{\sin 2\theta}{2} \sigma_r, \qquad v_p = K_c \sqrt{\frac{h_0}{2 s_c w_0}} \sqrt{\frac{c}{D}} \sin 2\theta \tag{10.3.21}$$

The size effect described by (10.3.19) is plotted in Fig. 10.3.5 in two ways, in terms of $\log(v_u - v_r)$ and in terms of $\log v_u$. By virtue of the residual compression strength, the nominal shear strength of the beam tends at infinite size to a finite value. An equation of the form of (10.3.19) was proposed on the basis of general considerations in Bažant (1987a).

The question whether the confinement of concrete by stirrups suffices to cause the residual compression strength σ_r, and thus the residual nominal strength v_r, to be nonzero needs to be studied further. It is on the safe side to take $v_r = 0$, in which case, the effect of stirrups on the residual nominal strength provided by concrete is neglected.

10.3.4 Analysis Based on Stress Redistribution and Complementary Energy

The truss model also allows an easy alternative calculation of the energy release on the basis of complementary energy $\mathcal{U}_c^*$. For the sake of simplicity, we now consider the residual strength $v_r = 0$, although a generalization to finite v_r would be feasible.

In the truss model, we isolate the representative cell limited by the shaded zone in Fig. 10.3.4. This cell must alone be capable to resist the applied shear force V. The compression failure of concrete in the band 12341 (Fig. 10.3.4) is considered to completely relieve the stress from the inclined strip 12561. If the applied shear force V is kept constant, the stress in the cell must redistribute such that all of the compression force in the inclined strut is carried by the remaining strips, shaded in Fig. 10.3.4. After that, all of the complementary energy in concrete in the cell is contained in the shaded strips. According to Bažant (1996b), the energy density is given by the shaded area in Fig. 10.3.6, and so the complementary energy density may be expressed as $\mathcal{U}_c^* = (\bar{\sigma}_c^2/2E_c)\,\mathcal{V}$ in which $\mathcal{V} = b(D\cos\theta - c\sin\theta)D/\sin\theta =$ volume of the shaded strips (Fig. 10.3.4), $\bar{\sigma}_c = F_c/b(D\cos\theta - c\sin\theta) =$ average normal stress in the direction of the strut, and $F_c = V/\sin\theta = v_u bD/\sin\theta =$ compression force transmitted by the strut. This yields for the complementary energy, after the stress redistribution at constant shear force V, the expression:

$$\mathcal{U}_c^* = \left(\frac{v_u D}{\sin\theta}\right)^2 \frac{bD}{2E_c(D\cos\theta - c\sin\theta)\sin\theta} \tag{10.3.22}$$

According to (2.1.21), the energy release rate is obtained by differentiation of the complementary energy at constant load (or constant shear force V):

$$\mathcal{G} = \frac{1}{b}\left[\frac{\partial \mathcal{U}_c^*}{\partial c}\right]_V = \frac{v_u^2 bD^3}{2E_c \sin^2\theta(D\cos\theta - c\sin\theta)^2} \tag{10.3.23}$$

This must be equal to the energy dissipation rate, which is given by the following equations, same as

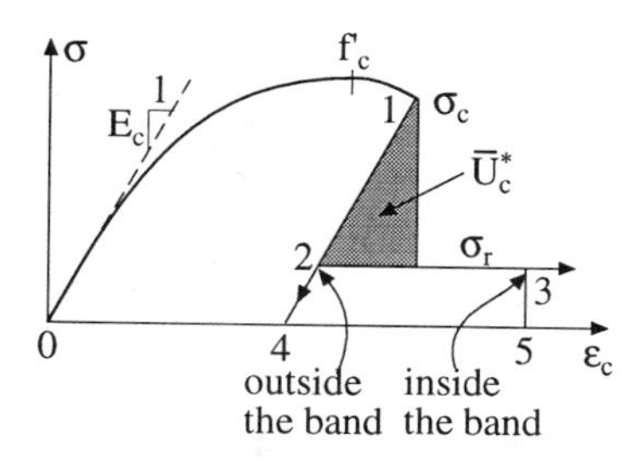

Figure 10.3.6 Compression stress-strain diagram of concrete with unloading after peak stress and area representing the strain energy release.

before:

$$\mathcal{R} = \frac{h}{s_c} G_f, \quad h = \frac{h_0 c}{w_0 + c} \tag{10.3.24}$$

There is now one difference from the previous approach. In (10.3.18), the energy release rate was constant, while in (10.3.23) it increases with c. This difference should not surprise since both solutions are approximate. In the case of variable $\mathcal{G}$, which is a typical case in fracture mechanics, the crack length at maximum load, that is, at a loss of stability, need not be considered as empirical, as done in our previous calculation based on the strain energy change, but can be calculated from the stability criterion. It is well known that, at the limit of stability, the curve of energy release rate at constant load must be tangent to the R-curve (see Section 5.6.3):

$$\frac{\partial \mathcal{G}}{\partial c} = \frac{d\mathcal{R}}{dc} \tag{10.3.25}$$

(This stability criterion could not be applied to the previous case with (10.3.18), because in that case, due to the approximations made, we had $\partial \mathcal{G}/\partial c = 0$ and thus c was indeterminate.) Because $\mathcal{G} = \mathcal{R}$, an equivalent condition is

$$\frac{1}{\mathcal{G}} \frac{\partial \mathcal{G}}{\partial c} = \frac{1}{\mathcal{R}} \frac{d\mathcal{R}}{dc} \tag{10.3.26}$$

which is more convenient. We may now substitute here the expressions in (10.3.23) and (10.3.24), and carry out the differentiations. This leads to a quadratic equation for c/D, whose only real solution is

$$\frac{c}{d} = \frac{3w_0}{4d} \left(-1 + \sqrt{1 + \frac{8d}{9w_0} \cot\theta} \right) \tag{10.3.27}$$

This represents a theoretical expression for the length of the crack band at maximum load (i.e., at stability loss).

It may now be observed that c/D tends to zero as the size $D \to \infty$. In that limiting case, the stress relief region would become an infinitely narrow strip, which would not be a realistic model. Therefore, (10.3.27) is meaningful only for sufficiently small sizes. For this reason, and for the sake of simplicity, we consider the second term under the square root in (10.3.27) to be small compared to 1. Because $\sqrt{1+2x} \approx 1 + x$ when $x \ll 1$, (10.3.27) for small D yields the approximation:

$$\frac{c}{D} = \frac{\cot\theta}{3} \tag{10.3.28}$$

Substituting this into the fracture propagation criterion $\mathcal{G} = \mathcal{R}$, along with (10.3.23) and (10.3.24), we obtain an equation whose solution furnishes the simple result:

$$v_u = v_p \left(1 + \frac{D}{D_0} \right)^{-1/2} \tag{10.3.29}$$

in which we have introduced the notations:

$$D_0 = 3w_0 \tan\theta \tag{10.3.30}$$

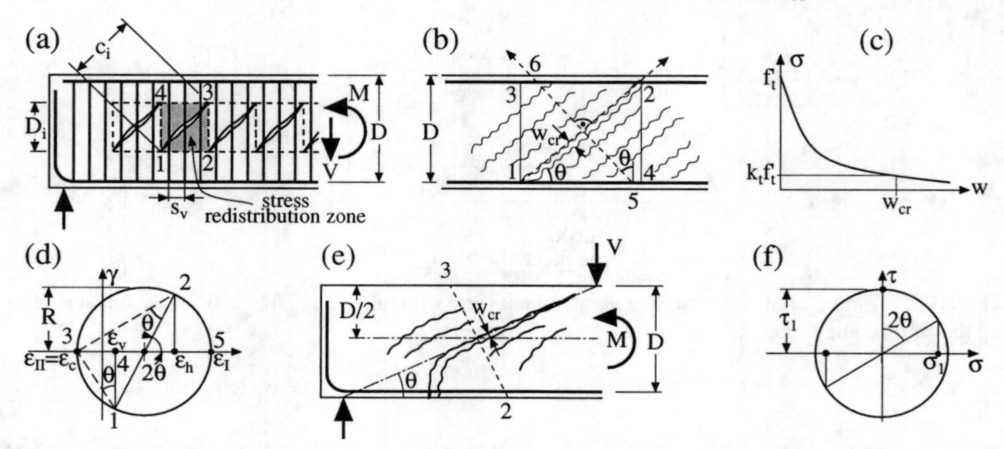

Figure 10.3.7 (a) Stress redistribution zones for initial diagonal shear cracks. (b) Localization of the openings of diagonal cracks into one major diagonal crack in a beam with stirrups. (c) Tensile stress-displacement diagram for the opening of a cohesive crack. (d) Mohr circle of strains. (e) Localization of openings of diagonal cracks into one major crack in a beam without stirrups. (f) Mohr circle of stresses.

$$v_p = K_c \sqrt{\frac{2h_0}{27 s_c w_0}} \sin 2\theta \sqrt{\cot\theta} \tag{10.3.31}$$

The result we have obtained has the same form as (10.3.19), although the expressions for the size effect constants D_0 and v_p are partly different. The differences reveal the degrees of uncertainty caused by the simplifications of analysis we made. The comparison of (10.3.19) and (10.3.29) indicates that the general form of the size effect we obtained ought to be realistic although the coefficients D_0 and v_p cannot be fully predicted by the present theory but must be corroborated on the basis of experiments.

10.3.5 Size Effect on Nominal Stress at Cracking Load

It has been suggested that the size effect might not be of concern because the current ACI Code (ACI Committee 318, 1992) and other codes are intended to provide safety against the cracking load at which large diagonal cracks form, rather than against the collapse load, which is considerably higher. However, the nominal stress corresponding to the cracking load also exhibits size effect. There are two possibilities to define the cracking load.

Load Causing Cracks of Given Relative Depth

One possibility is to define the cracking load as the load that produces initial diagonal shear cracks of a depth D_i representing a given percentage of beam depth D, i.e., such that the ratio D_i/D is a given constant (Fig. 10.3.7a), say 0.5. We imagine an array of the initial cracks, as shown in Fig. 10.3.7a. The formation of each initial crack causes stress redistribution in triangular zones 1321 and 1341, shaded in Fig. 10.3.7a. (In contrast to Fig. 10.3.4, the stress relief zones are not strips, nor elongated triangles, because the material is not orthotropically damaged before the initial cracks form.) For the sake of simplicity, these zones may be assumed to consist of triangles with angles roughly $\theta = 45°$, each two triangles making a square. The shape of these zones and the length of the initial cracks obviously determines their spacing.

Before the initial diagonal cracks form, the vertical stress in the beam is 0, and so the stirrups have no stress, while shear force V is resisted by shear stresses in concrete taken approximately as $v = V/bd$. The complementary energy initially contained in the shaded square cell in Fig. 10.3.7a is $\mathcal{U}_0^* = (v^2/2G_c)b(c_i\cos\theta)(c_i\sin\theta) = v^2(1+\nu)bc_i^2\sin\theta\cos\theta/E_c$, where $G_c = E_c/2(1+\nu)$ = elastic shear modulus of concrete, ν = Poisson ratio ($\nu \approx 0.18$), and c_i is defined in Fig. 10.3.7a. After the initial cracks form, the diagonal tensile stress in the shaded square zone is reduced to 0 and the applied shear stress v is then carried by truss action in the cell, i.e., by tensile stress σ_v in the vertical stirrups, given by (10.3.15), and by diagonal compressive stress σ_c, given by (10.3.14). So the complementary energy contained in the cell after the initial cracks form is approximately calculated as $\mathcal{U}_1^* = (\sigma_c^2/2E_c)b(c_i \sin\theta)(c_i \cos\theta) + (\sigma_v^2/2E_s)A_v(c_i^2/s)\sin\theta\cos\theta$, where $\sigma_v = vbs\tan\theta/A_v, \sigma_c = -v/\sin\theta\cos\theta$. For the sake of simplicity, we assume $\theta = 45°$. The complementary energy change per crack at constant V is $\Delta\mathcal{U}^* = \mathcal{U}_1^* - \mathcal{U}_0^*$, which yields

$$\Delta\mathcal{U}^* = \frac{bc_i^2v^2}{E_c}\left(\frac{1-\nu}{2} + \frac{bs}{4nA_v}\right), \qquad n = \frac{E_s}{E_c} \tag{10.3.32}$$

Consider now the final infinitesimal crack length increment δc_i by which the crack size c_i is reached (the shaded square zone in Fig. 10.3.7a grows with c_i, and at the end of this increment, it touches the square zone corresponding to the adjacent crack). During this increment, the change of complementary energy is $[\partial(\Delta\mathcal{U}^*)/\partial c_i]\delta c_i$. This must be equal to the energy consumed and dissipated by the crack, which is $b\mathcal{R}\delta c_i$; $\mathcal{R}$ is the crack resistance, which represents the critical energy release rate required for crack growth. In general, $\mathcal{R}$ depends on c_i, representing an R-curve behavior. This dependence may be approximately described as

$$\mathcal{R} = G_f\frac{c_i}{c_0 + c_i} \tag{10.3.33}$$

where c_0 is a positive constant. For large enough c_i, $\mathcal{R} = G_f$ = fracture energy of the material. The balance of energy during the crack length increment requires that

$$\frac{\partial(\Delta\mathcal{U}^*)}{\partial c_i}\delta c_i = b\mathcal{R}\delta c_i \tag{10.3.34}$$

Substituting (10.3.32) here, we obtain an equation whose solution yields, for the size effect on the applied nominal shear stress v_{cr} at initial cracking, the following equation:

$$v_{cr} = v_{cr0}\left(1 + \frac{D}{D_{cr0}}\right)^{-1/2} \tag{10.3.35}$$

in which the following constants have been introduced:

$$D_{cr0} = c_0\frac{D}{c_i}, \qquad v_{cr0} = \left[\frac{E_cG_f}{c_0}\left(2 + \frac{bs}{2nA_v} - \frac{1+\nu}{2}\right)^{-1}\right]^{1/2} \tag{10.3.36}$$

Note that the ratio D/c_i is assumed to be a given constant by which the cracking load is defined. Equation (10.3.35) shows that the applied nominal shear stress at cracking follows again Bažant's size effect law. As a special case, this equation applies to a beam without stirrups ($A_v = 0$).

Load Causing Cracks of Given Opening Width

Another possibility is to define the cracking load as the load that produces cracks of a given critical width w_{cr}. Consider first the beams with stirrups. Under a certain load, a number of parallel diagonal cracks may initiate. The cracks are cohesive. This means that crack-bridging stresses are transmitted across the cracks (due to aggregate pullout and other phenomena). Reduction of the crack-bridging stress to zero requires a considerable opening displacement of the crack, as is clear from the typical stress-displacement diagram used in the cohesive (fictitious) crack model; see Chapter 7. Furthermore, it is known that when many parallel cracks form, only one of them may open widely, while the others unload and close. In fact, such a localization of crack openings into one among many parallel cracks is a necessity unless there is enough reinforcement to ensure a stiffening rather than softening behavior (see Chapter 8; also Chapter 12 in Bažant and Cedolin 1991). Thus, unless the stirrups are extremely strong, the situation as shown in Fig. 10.3.7b must be expected.

Since the reduction of the crack-bridging stress to zero requires a very large opening, we consider that the stress is reduced only to a certain small but finite fraction k_t of the tensile strength f_t' of concrete.

Consider now the relative displacement between points 5 and 6 at the bottom and top of the beam, lying on a line normal to the cracks after one large crack forms. This displacement may be approximately expressed as $\Delta u_I = (D/\cos\theta)(k_t f_t'/E_c) + w_{cr}$, in which $D/cos\theta$ is the length of the line segment 56, and w_{cr} is a critical crack opening displacement at which the crack bridging stress is reduced from f_t' to $k_t f_t'$ (Fig. 10.3.7c). Dividing this by the length of segment 56, we obtain the average normal strain in the direction orthogonal to the diagonal cracks:

$$\bar{\varepsilon}_{I_{cr}} = \frac{\Delta u_I}{D/\cos\theta} = \frac{k_t f_t'}{E_c} + \frac{w_{cr}\cos\theta}{D} \tag{10.3.37}$$

Displacement Δu_1 or strain $\bar{\varepsilon}_{Icr}$ must be compatible with the overall deformation of the truss. Imagining the nodes of the truss to be attached to a homogeneously deforming continuum, this condition means that strain $\bar{\varepsilon}_{Icr}$ must be tensorially compatible with the normal strains ε_c in the inclined struts and ε_v in the vertical stirrups, as well as with the principal direction angle θ. This strain compatibility condition may be easily deduced from the Mohr circle in Fig. 10.3.7d. Noting that $14 = (\varepsilon_v - \varepsilon_c)\cot\theta$, $R = 05 = 01 = 14/\sin 2\theta = (\varepsilon_v - \varepsilon_c)\cot\theta/\sin 2\theta$, $\bar{\varepsilon}_I = \varepsilon_c + 2R$, we obtain the following expression for the average strain in the direction orthogonal to the diagonal cracks:

$$\bar{\varepsilon}_I = \varepsilon_c + \frac{\varepsilon_v - \varepsilon_c}{\sin^2\theta} = \frac{\varepsilon_v}{\sin^2\theta} - \varepsilon_c\cot^2\theta \tag{10.3.38}$$

In terms of the stresses, $\varepsilon_v = \sigma_v/E_s$, $\varepsilon_c = \sigma_c/E_c$, in which E_s = elastic modulus of steel and E_c = secant modulus for the compression strut at the moment the diagonal cracks form, which is less than the initial elastic modulus but larger than the secant modulus for the peak stress point of the compression stress-strain diagram. Here the stresses may be expressed from the equilibrium conditions of the truss: $\sigma_v = v_{cr}s_v b\tan\theta/A_v$, $\sigma_c = -2v_{cr}/\sin 2\theta$, where A_v = cross section area of one stirrup, and $v_{cr} = V_{cr}/bD$ = nominal stress corresponding to the shear force at the moment of formation of large diagonal cracks. Substituting these expressions into (10.3.38), we obtain:

$$\bar{\varepsilon}_I = \frac{2}{\sin 2\theta}\left(\frac{s_v b}{A_v E_s} + \frac{\cot^2\theta}{E_c^{\text{sec}}}\right) v_{cr} \tag{10.3.39}$$

Setting this expression equal to (10.3.37), we obtain an equation for v_{cr}, the solution of which furnishes the result:

$$v_{cr} = v_\infty + v_0\frac{w_{cr}}{D} \tag{10.3.40}$$

Here we introduced the notations:

$$v_0 = \sin\theta\cos^2\theta\left(\frac{s_v b}{A_v E_s} + \frac{\cot^2\theta}{E_c^{\text{sec}}}\right)^{-1}, \qquad v_\infty = \frac{k_t f_t' v_{cr}^0}{E_c\cos\theta} \tag{10.3.41}$$

Equation (10.3.40) describes a size effect which is an alternative to (10.3.35). The asymptotic constant value v_∞ exists because we assume that the critical crack opening w_{cr} corresponds to nonzero crack bridging stress $k_t f_t'$; if this stress were neglected, we would obtain $v_\infty = 0$.

Consider now a beam without stirrups. This problem is more complicated because there is no truss model that could give the value of the average strain along the line 23 in Fig. 10.3.7c. Other simplifications are, therefore, needed to obtain a simple result. We will assume that the normal strains along the line segment 23 in Fig. 10.3.7e may be approximated according to the beam theory. The shear stress in the vertical plane is distributed parabolically, and so, at point 1 at mid depth of the beam (neutral axis), it has the value $\tau_1 = 1.5v_{cr}$. From the Mohr circle in Fig. 10.3.7f, we then obtain the normal stress σ_1 in the direction 23 at point 1 and the corresponding strain: $\varepsilon_1 = 1.5v_{cr}\sin 2\theta/E_c$. The normal strain in the direction 23 may also be assumed distributed parabolically, in which case the average normal strain along this line is $\bar{\varepsilon}_1 = v_{cr}\sin 2\theta/E_c$. Multiplying this by the length of segment 23, we obtain the relative displacement between points 2 and 3 in the direction 23:

$$\Delta u_{23} = \bar{\varepsilon}_1\frac{D}{\cos\theta} = \frac{v_{cr}}{E_c}\sin 2\theta\frac{D}{\cos\theta} \tag{10.3.42}$$

At the same time, in analogy to (10.3.37):

$$\Delta u_{23} = \frac{D}{\cos\theta}\frac{k_t f_t'}{E_c} + w_{cr} \tag{10.3.43}$$

Equating the last two expressions, we obtain the same equation as (10.3.41), that is, $v_{cr} = v_\infty + v_0(w_c/D)$, in which we now make the notations:

$$v_\infty = \frac{2k_t f_t'}{3\sin 2\theta}, \quad v_0 = \frac{E_c w_{cr}}{3\sin\theta} \tag{10.3.44}$$

10.3.6 Conclusions

1. The fracture modification (Bažant 1996b) of the classical widely used truss model (or the strut-and-tie model) for the shear failure of reinforced concrete beams describes the energy release and localization of damage into a band of compression splitting cracks (or a compression-shear crack) within a portion of the compressed concrete strut.

2. If the analysis of the maximum load based on the truss model is valid (and if the stirrups are designed sufficiently strong), the concrete strut must undergo compression softening (with progressive fracture) during the portion of loading history in which the maximum load is reached.

3. Analysis of the energy release into the crack band shows that a size effect on the nominal strength at shear failure of a reinforced concrete beam must occur and that it should approximately follow Bažant's size effect law. Conversely, the fracture behavior of the truss model (strut-and-tie model), particularly the damage localization with energy release, provides an explanation of the size effect widely observed in many tests as shown in the previous section.

4. The applied nominal shear stress that causes the initial large diagonal cracks also exhibits a size effect. The law of this size effect depends on how the large diagonal cracks are defined.

5. The foregoing size effect formulae have not yet been calibrated and verified by the available test results for beams. The expressions for the coefficients in these formulas need to be studied further in order to develop a design procedure incorporating the size effect.

10.4 Reinforced Beams in Flexure and Minimum Reinforcement

In this section, we examine with some detail the existing approaches to the failure of beams in bending. In general, it is accepted that strongly reinforced beams that fail by steel yielding are mostly fracture-insensitive. So, structures of this type have not been much investigated from the viewpoint of fracture mechanics. However, there are situations in which fracture plays a role; two extreme cases have been investigated by various authors: (1) failure of concrete in compression for normally reinforced concrete, and (2) failure of lightly reinforced beams.

The first type of failure, which was investigated by Hillerborg (1990), will be discussed in Section 10.5.12, mainly to show that even if normally reinforced beams are not sensitive to fracture in tension, they do show size effect due to fracture in compression. Investigations on the second case tremendously expanded in recent years. They will be discussed now.

10.4.1 Lightly Reinforced Beams: Overview

The question of minimum reinforcement calls for answering two problems: (**1**) stability of a system of interacting cracks, which ensures that the cracks will remain densely spaced, and (**2**) avoidance of snapback in bending at a cross section with only one crack. The first problem controls the spacing of bending cracks in beams which in turn controls their width. A certain minimum reinforcement is required to prevent a large crack spacing, causing a large crack opening. This problem is important for serviceability under normal loads and small overloads. The second problem, which is important for preventing sudden catastrophic failure without warning, will be discussed now.

In recent years, the analysis of lightly reinforced beams and of minimum reinforcement received considerable attention. This is probably due to the widespread feeling that this is a problem that can be handled with relative ease using fracture mechanics. In particular, lightly reinforced beams in three-point bending fail by a single crack across the central cross section, as opposed to normally reinforced beams in which multiple or distributed cracking occurs prior to collapse.

Before reviewing the various theoretical approaches to this problem, let us describe the main empirical facts. Figs. 10.4.1a–c show load-displacement curves measured for various reinforcement ratios. Fig. 10.4.1a shows the results by Bosco, Carpinteri and Debernardi (1990b), for concrete reinforced with standard ribbed steel. Fig. 10.4.1b shows some results reported by Hededal and Kroon (1991) on a similar material; note that in these tests, the beams had a short notch —5% of beam depth— on the tension side of the beam, which explains the sharper peak. Fig. 10.4.1c shows very recent results of Planas, Ruiz and Elices (1995) for lightly reinforced beams made of microconcrete. Although the materials and the test arrangements were quite different, the results are clearly similar.

From their tests and the theoretical analysis to be described later (Section 10.4.4), Ruiz, Planas and Elices (1993, 1996) suggested that for steel with low strain hardening, the load-deflection curve can generally be sketched as shown in Fig. 10.4.1d. A linear portion OL is followed by a nonlinear zone up to the peak LM after which a U-shaped portion MNP follows, ending at point P at which the reinforcement yields (if the reinforcement is elastic-perfectly plastic). This is followed by a relatively long tail PT with mild softening which theoretically has no end (for ideal steel) but in practice ends by steel necking and fracture. Since the steel never follows exactly an ideal plastic behavior with a sharp transition from elastic to plastic, the actual curve may look closer to the dashed curve $NP'T$ which rounds the corner at P due to strain-hardening.

In an ideal situation (no internal stresses due to shrinkage, no thermal gradients, nor chemical reactions) the linear limit depends only on the tensile strength of concrete, with $\sigma_N^L \approx f'_t$ (we do not care here about the 5% difference due to the concentrated load that was discussed in Section 9.3 with reference to the rupture modulus). After that limit, a fracture zone starts to grow towards the reinforcement across the cover, and the load-displacement curve for similar unreinforced beams is approximately followed (dashed line). When the fracture zone approaches or reaches the steel, two phenomena occur simultaneously: **(1)** the fracture zone is *sewed up* by the reinforcement, which is still elastic thus requiring an extra load to continue cracking; and **(2)** steel pullout and slip takes place. Therefore, the peak and near postpeak in the load-displacement curve and its neighborhood is controlled by three factors: **(a)** The steel ratio; **(b)** the bond-slip properties; and **(c)** the steel cover.

The influence of the steel ratio on the peak load was already illustrated in Figs. 10.4.1a–c. The influence of the bond is illustrated in Fig. 10.4.1e which shows the results of Planas, Ruiz and Elices (1995) for a fixed steel ratio and for two different types of reinforcement: ribbed bars with strong bond, and smooth bars with weak bond. It is clear that the bond strength modifies substantially the response. Finally, the influence of the cover is not so evident and little experimental support is available. However, that the cover must play a role can be inferred by the following reasoning. If the cover is large enough, the specimen load will exhibit a peak before the fracture zone reaches the reinforcement; then, after some load decrease, the growing fracture zone will reach the reinforcement and will be arrested, thus engendering hardening followed by a second peak and further softening. Therefore, the cover must influence the response. Indeed, Ruiz and Planas (1995) have detected, both experimentally and theoretically, the existence of the double peak, as shown in Fig. 10.4.1f. Certainly there also must be an indirect effect of the cover thickness and bar spacing because these variables are known to modify the bond strength, but this is a secondary influence in the usual analysis.

Several models have been proposed to describe the foregoing results; they can be classified as pertaining to three wide groups: **(1)** models that make use of LEFM as the basic tool; **(2)** models that use a simplified smeared cohesive cracks; and **(3)** models that use cohesive cracks. In the following, we describe the mean features of these three groups of models.

10.4.2 Models Based on LEFM

All the LEFM models are rooted in the model first proposed by Carpinteri (1981, 1984; also 1986, Sec. 6.2). Figs. 10.4.2a–c show the basic superposition in Carpinteri's approach: the reinforced beam with a crack of length a subjected to bending (Fig. 10.4.2a) is approximated by a beam subjected to the bending

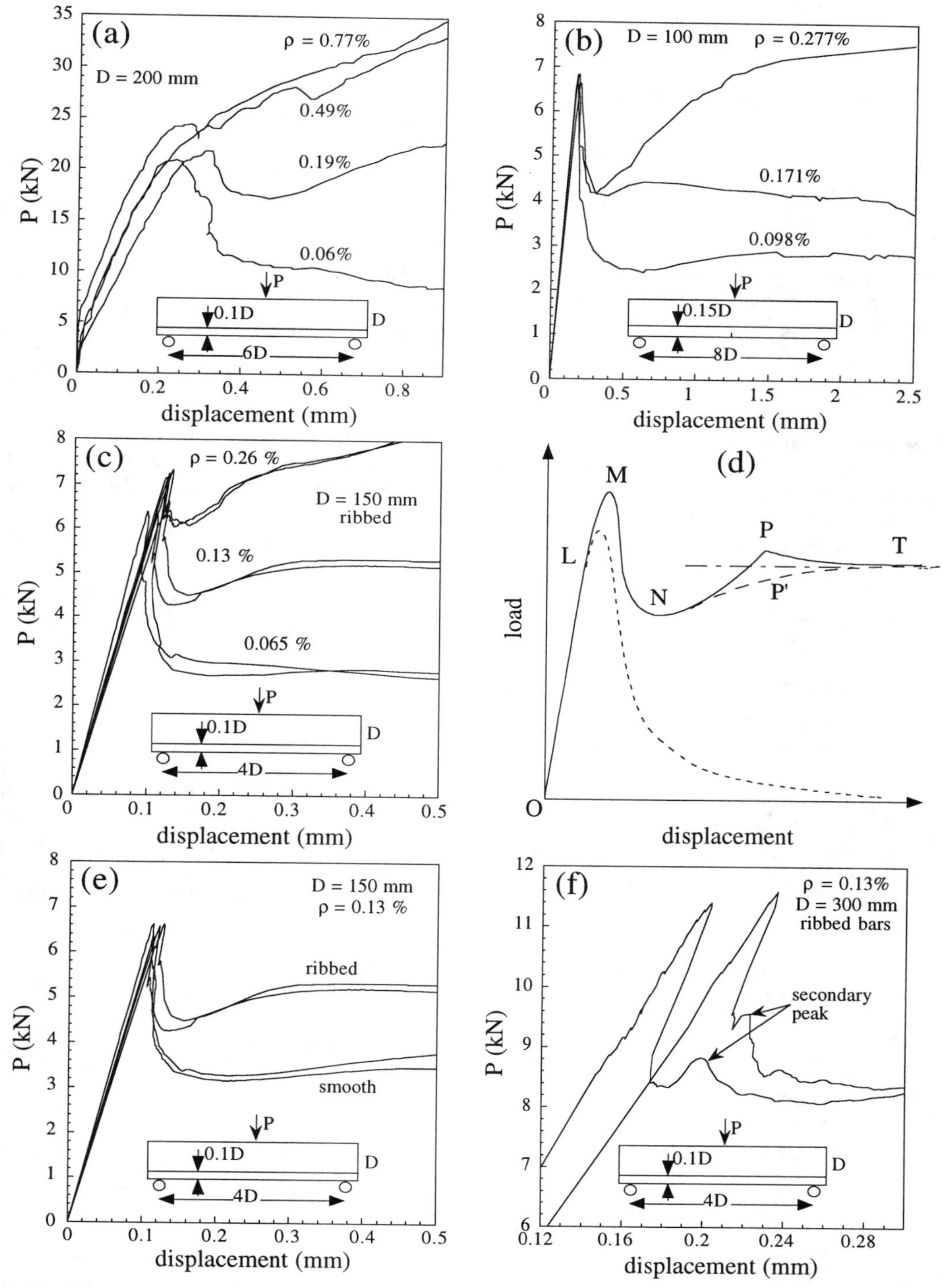

Figure 10.4.1 Influence of steel ratio on the load-displacement curves: (a) data from Bosco, Carpinteri and Debernardi (1990b); (b) data from Hededal and Kroon (1991); (c) data from Planas, Ruiz and Elices (1995) for microconcrete. (d) General trend of the load-displacement curve according to Ruiz, Planas and Elices (see the text for details). (e) Influence of bond on the response (after Planas, Ruiz and Elices 1995). (d) Experimental double peaks for a relatively thick cover (data from Ruiz and Planas 1995).

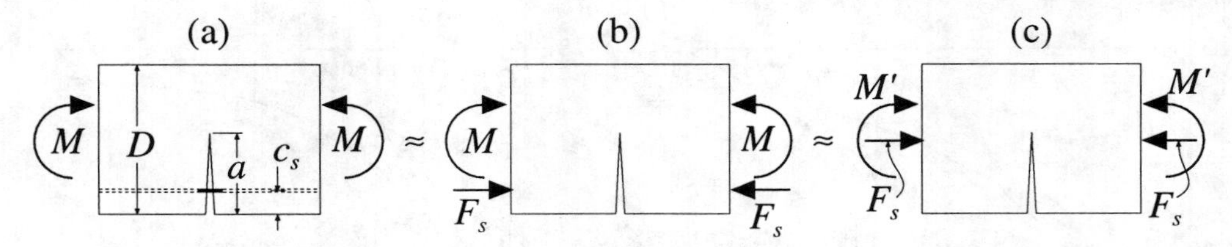

Figure 10.4.2 Carpinteri's LEFM approximation.

moment and to the steel force applied *remotely* from the crack plane (Fig. 10.4.2b). Next, the steel action is decomposed in a standard way into a bending moment and a centric force (Fig. 10.4.2c).

With this decomposition, we can easily write the stress intensity factor as:

$$K_I = \frac{6M'}{bD^2}\sqrt{D}k_M(\alpha) - \frac{F_s}{bD}\sqrt{D}k_\sigma(\alpha)\,, \qquad \alpha = \frac{a}{D}\,, \qquad M' = M - F_s\left(\frac{D}{2} - c_s\right) \tag{10.4.1}$$

where $k_M(\alpha)$ and $k_\sigma(\alpha)$ are the shape function for pure bending and for a uniform remote tension, respectively, and c_s is the steel cover (see Fig. 10.4.2). An approximate, but accurate, expression for $k_M(\alpha)$ is given, for example, by (3.1.1) for $S/D = \infty$; an expression for $k_\sigma(\alpha)$ can be found in most stress intensity factor manuals (e.g., Tada, Paris and Irwin 1985).

Carpinteri also calculated the additional rotation caused by to the crack θ_c according to the method described in Section 5.5.2 with $P_1 \equiv M$, $\theta_c \equiv u_1$, and $P_2 \equiv F$, with the result

$$\theta_c = \frac{6M'}{E'bD^2}v_{MM}(\alpha) - \frac{F_s}{E'bD}v_{M\sigma}(\alpha)\,, \qquad M' = M - F_sD\left(\frac{1}{2} - \gamma\right) \tag{10.4.2}$$

where $\gamma = c_s/D$ is the relative cover thickness and

$$v_{MM}(\alpha) = 12\int_0^\alpha k_M^2(\alpha')d\alpha'\,, \qquad v_{M\sigma}(\alpha) = 12\int_0^\alpha k_M(\alpha')k_\sigma(\alpha')d\alpha' \tag{10.4.3}$$

Carpinteri assumed that the steel behavior was elastic-perfectly plastic, and that the crack was closed ($\theta_c = 0$) while the steel remained elastic. Therefore, the crack growth takes place only when the steel yields and, simultaneously, $K_I = K_{Ic}$. With these conditions, it is easy to obtain the parametric equations of the moment-rotation curves (with parameter α). Indeed, setting $F_s = \rho bDf_y$ (where ρ is the steel ratio and f_y the steel yield stress) and $K_I = K_{Ic}$ in the foregoing equations, the solutions can be written as

$$\frac{\sigma_N\sqrt{D}}{K_{Ic}} = \frac{1}{k_M} + \left(3 - 6\gamma + \frac{k_\sigma}{k_M}\right)N_p \tag{10.4.4}$$

$$\frac{\theta_c E'\sqrt{D}}{K_{Ic}} = \frac{v_{MM}}{k_M} + \frac{k_\sigma v_{MM} - k_M v_{M\sigma}}{k_M}N_p \tag{10.4.5}$$

where, as always, $\sigma_N = 6M/bD^2$, and N_p is Carpinteri's brittleness number for reinforced beams in bending defined as

$$N_p = \rho\sqrt{\frac{D}{\ell_p}}\,, \qquad \ell_p = \left(\frac{K_{Ic}}{f_y}\right)^2 \tag{10.4.6}$$

Here we have introduced the length ℓ_p to emphasize the similitude of this brittleness number to those based on Irwin's or Hillerborg's characteristic size: the only change is to replace the tensile strength of concrete by the tensile strength of the reinforcement f_y.

One of the limitations of this model is that, due to the simplifications involved in its derivation, the crack cannot grow while the steel remains elastic and does not slip (in reality, it must slip). This limitation was removed, using very different methods, by Baluch, Azad and Ashmawi (1992) and by Bosco and Carpinteri (1992).

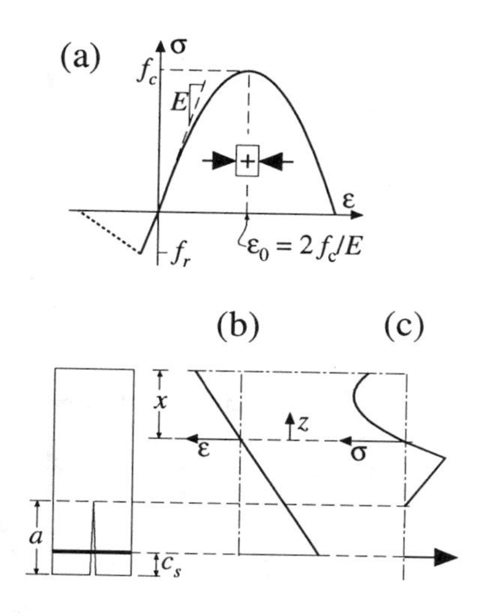

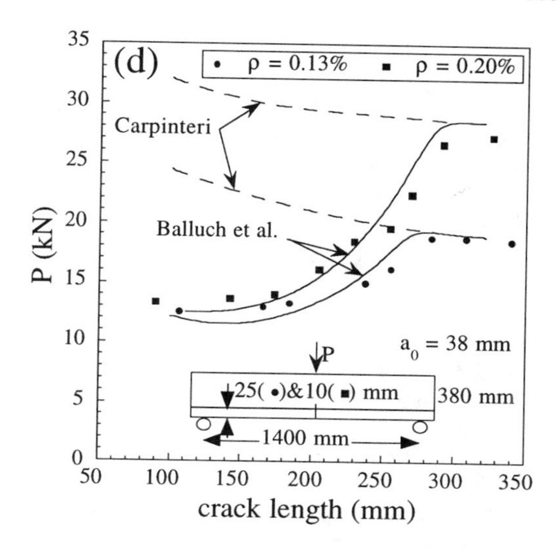

Figure 10.4.3 Model of Baluch, Azad and Ashmawi (1992): (a) stress-strain curve for concrete; (b) strain distribution; (c) stress distribution; (d) comparison of experimental and theoretical curves of load vs. crack length for lightly reinforced beams with a notch.

The model of Baluch, Azad and Ashmawi keeps Carpinteri's solution after steel yielding, but relaxes the assumption that the crack remains closed while the steel is elastic. In the elastic regime for steel, the model retains the stress intensity equation (10.4.1) and the condition $K_I = K_{Ic}$, which can be rewritten as

$$K_{Ic} = \frac{6M}{bD^2}\sqrt{D}\, k_M(\alpha) - \frac{F_s}{bD}\sqrt{D}\left[\left(3 - 6\frac{c}{D}\right) k_M(\alpha) + k_\sigma(\alpha)\right] \tag{10.4.7}$$

which provides one equation with two unknowns, namely, M and F_s (α is given in this context). To determine F_s, Baluch et al. introduced a classical analysis based on a stress-strain formulation with the following assumptions: **(1)** the stress-strain curve is as depicted in Fig. 10.4.3a, parabolic in compression and linear in tension down to the failure stress f_r which is taken to coincide with the modulus of rupture rather than with the tensile strength; **(2)** the softening in tension is linear, as depicted by the dotted line in Fig. 10.4.3a, but the softening slope depends on the geometry as indicated later; **(3)** the strain distribution is linear (Fig. 10.4.3b); **(4)** the fracture process zone is represented by a linear distribution of stress which is zero at the crack tip as shown in Fig. 10.4.3c. Note that the essential difference with respect to other formulations is that here the softening curve for concrete in tension is *not* related to the strain in a predefined way; rather, the form of the spatial distribution of stress is postulated. With these hypotheses, and given the stress-strain curve of the steel, it is possible to determine a relationship between F_s and M. For elastic behavior of the steel, the strain distribution is obtained from Fig. 10.4.3b:

$$\varepsilon = \frac{F_s}{A_s E_s}\,\frac{z}{D - c_s - x} \tag{10.4.8}$$

where A_s is the steel cross section and E_s its elastic modulus; x is the depth of the neutral axis. From this, the stress distribution can be determined as sketched in Fig. 10.4.3c. Then, the equilibrium of forces provides an equation with the two unknowns F_s and x, and the equilibrium of moments a further equation with the three unknowns F_s, x, and M. Complementing these two equations with (10.4.7), we get a system that determines the three aforementioned unknowns. This system must be solved numerically. Baluch, Azad and Ashmawi (1992) use two iteration loops: given a, they assume a value for F_s and solve iteratively for x from the condition of equilibrium of forces (inner iteration loop); then they compute M from the equilibrium of moments and from (10.4.7); if the two values coincide, this is the solution for the given crack depth; if not, they start over with a new value of F_s (outer iteration loop).

Baluch, Azad and Ashmawi (1992) checked their model by comparing the load-crack length curves for two lightly reinforced notched beams tested in three-point bending (the determination of load-displacement

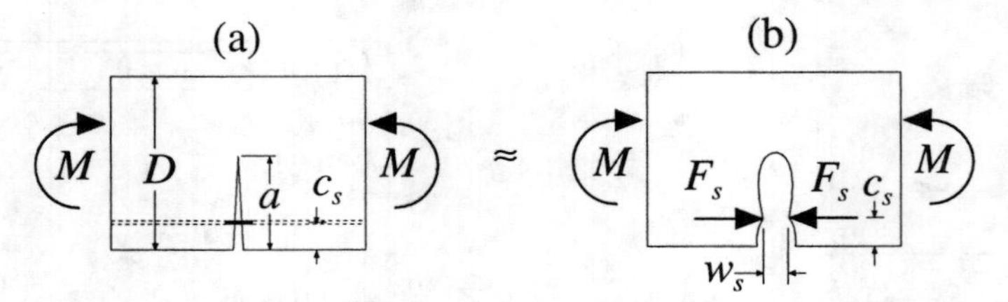

Figure 10.4.4 LEFM approximation of Bosco and Carpinteri (1992).

or moment-rotation curves were not included as part of the formulation). Fig. 10.4.3d shows the experimental results and the theoretical predictions (the dashed curves correspond to Carpinteri's model; note that the prediction of both models coincide after the yielding of steel). Note also that, much like what we indicated for the load-displacement curves in Fig 10.4.1c, the theory predicts a sharp change of slope upon steel yielding, while the experiments show a rounded transition.

Bosco and Carpinteri (1992) adopted an approach radically different from the one just discussed. They modified the initial Carpinteri's model by letting the force of the reinforcement act on the crack faces rater than remotely from the crack plane, as shown in Figs. 10.4.4a–b. However, the slip of reinforcement which must occur near the crack faces was neglected (even though elasticity indicates infinite stress at the point of intersection of the steel bar with the crack face). With this, the expression for the stress intensity factor reads

$$K_I = \frac{6M}{bD^2}\sqrt{D}\, k_M(\alpha) - \frac{F_s}{bD}\sqrt{D}\, k_F(\alpha,\gamma)\,, \qquad \alpha = \frac{a}{D}\,, \qquad \gamma = \frac{c_s}{D} \tag{10.4.9}$$

where $k_M(\alpha)$ is the same as in Eq. (10.4.1), and $k_F(\alpha,\gamma)$ is the shape factor for a pair of forces acting on the faces of the crack; a closed form expression for this shape factor can be found in Tada, Paris and Irwin (1985).

Now it is no longer necessary to assume that the crack is closed everywhere while the steel is elastic; it is enough to assume that the crack is closed at the point where the reinforcement crosses it. Allowance for bond slip could also be made; however, Bosco and Carpinteri did not consider this possibility. The method in Section 5.5.2 is used to determine expressions for the rotation and the crack opening at the reinforcement level with $P_1 \equiv M, u_1 \equiv \theta_c$ and $P_2 \equiv -F_s, u_2 \equiv w_s$, where w_s is the crack opening at the reinforcement level. The resulting expressions are as follows:

$$\theta_c = \frac{6M}{E'bD^2}v_{MM}(\alpha) - \frac{F_s}{E'bD}v_{MF}(\alpha,\gamma) \tag{10.4.10}$$

$$w_s = \frac{M}{E'bD}v_{MF}(\alpha,\gamma) - \frac{F_s}{E'b}v_{FF}(\alpha,\gamma) \tag{10.4.11}$$

where $v_{MM}(\alpha)$ is given by the first of (10.4.3), while $v_{MF}(\alpha,\gamma)$ and $v_{FF}(\alpha,\gamma)$ are given by

$$v_{MF}(\alpha,\gamma) = 12\int_\gamma^\alpha k_M(\alpha',\gamma)k_F(\alpha',\gamma)\, d\alpha'\,, \qquad v_{FF}(\alpha,\gamma) = 2\int_\gamma^\alpha k_F^2(\alpha',\gamma)\, d\alpha' \tag{10.4.12}$$

The integration is carried out over the crack portion in excess of cover thickness. This is so because for shorter cracks the stress intensity factor caused by the point loads is zero, i.e., $k_F(\alpha,\gamma) = 0$ for $\alpha < \gamma$.

Assuming that the functions v_{MM}, v_{MF} and v_{FF} have been determined, Eqs. (10.4.9)–(10.4.11) completely solve the problem of crack propagation. Two cases can arise:

Case 1: The steel is still in elastic state. In this situation, we set $w_s = 0$ in (10.4.11), solve for F_s from the resulting equation and substitute it in (10.4.9), simultaneously setting $K_I = K_{Ic}$; then we solve

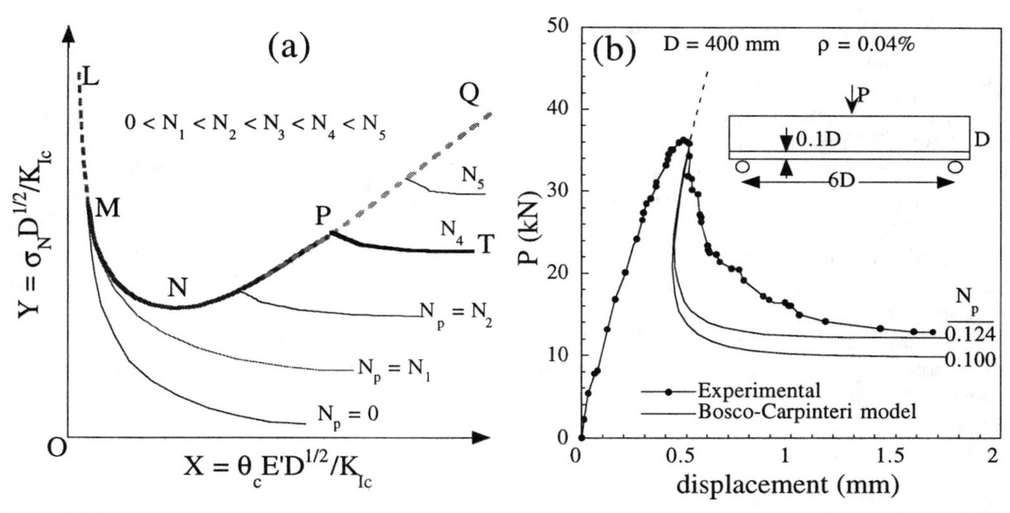

Figure 10.4.5 Bosco-Carpinteri model: (a) dimensionless load-curvature plot; (b) Massabò's (1994) fit to experimental curves by Bosco, Carpinteri and Debernardi (1990b)

the resulting equation for M. Finally, the rotation follows from (10.4.9). The final results are:

$$\left.\begin{aligned} \frac{\sigma_N\sqrt{D}}{K_{Ic}} &= \frac{v_{FF}}{k_M v_{FF} - 6k_F v_{MF}} \\ \frac{\rho\sigma_s\sqrt{D}}{K_{Ic}} &= \frac{v_{MF}}{k_M v_{FF} - 6k_F v_{MF}} \\ \frac{\theta_c E'\sqrt{D}}{K_{Ic}} &= \frac{1}{6}\,\frac{6v_{FF}v_{MM} - v_{MF}^2}{k_M v_{FF} - 6k_F v_{MF}} \end{aligned}\right\} \quad \text{for} \quad \sigma_s < f_y \qquad (10.4.13)$$

where the arguments of the shape functions have been dropped for brevity, and σ_s is the stress in the steel. Note that the right hand sides of these equations are independent of the brittleness number N_p defined in (10.4.6). If the value of σ_s resulting from the foregoing equations exceeds the steel yield stress f_y, then we move to the next case.

Case 2: The steel yields. In this case, we set $F_s = bDf_y$ in (10.4.9)–(10.4.11) and solve for M (or σ_N) and θ_c (it might be useful to also check that $w_s > 0$; otherwise, we are in case 1). The resulting equations then are

$$\left.\begin{aligned} \frac{\sigma_N\sqrt{D}}{K_{Ic}} &= \frac{1}{k_M} + \frac{k_F}{k_M}N_p \\ \frac{\theta_c E'\sqrt{D}}{K_{Ic}} &= \frac{v_{MM}}{k_M} - \frac{k_F v_{MM} + v_{MF}k_M}{k_M}N_p \\ \frac{6w_s E'}{K_{Ic}\sqrt{D}} &= \frac{1}{k_M} - \frac{k_F v_{MF} + 6v_{FF}k_M}{k_M}N_p \end{aligned}\right\} \quad \text{for} \quad w_s > 0 \qquad (10.4.14)$$

The plot of the $\sigma_N(\theta_c)$ curves in terms of the nondimensional variables $X = \theta_c E'\sqrt{D}/K_{Ic}$ and $Y = \sigma_N\sqrt{D}/K_{Ic}$ consists of two parts as sketched in Fig. 10.4.5a. The arc MNP is a part of the fixed curve $LMNPQ$ which is given by (10.4.13); this curve depends only on the relative steel cover γ, but is independent of the beam size, of the amount and quality of steel, and of the properties of concrete; it is a pure geometrical property. The arc PT corresponds to the solution for yielded steel and its shape is concave with a horizontal asymptote which corresponds to fully broken concrete. This branch PT depends only on the brittleness number N_p as sketched in the figure (and also on the relative cover thickness which is constant for geometrically similar beams).

In the foregoing equations, the rotation includes only the additional rotation caused by the crack at remote cross sections. When dealing with the load-displacement curves, one can either subtract the elastic displacement from the total displacement to isolate the rotation due to the crack, or conversely, one can add the elastic displacement (analytically computed) to the concentrated rotation determined by the foregoing theory. This last approach was used in a recent work by Massabò (1994) to analyze the experimental results of Bosco, Carpinteri and Debernardi (1990a,b).

Massabò (1994) determined the values of v_{MM}, v_{MF}, and v_{FF} for each α and γ by numerically performing the integrations in (10.4.3) and (10.4.12). The integration for v_{FF} deserves further comments because as written in (10.4.12), its value is infinite. This is so because, for simplicity, the load was assumed to be applied at a single point, which always gives a logarithmic singularity at the load-point. In reality, the action of the reinforcement is distributed over a certain area which is of the order of the diameter of the bars. This problem can be easily handled by assuming, for example, that the force is uniformly distributed and using the general formulas to determine the crack opening profile as given in Section 5.2. But this requires a double integration which greatly complicates the solution of the problem. Therefore, Massabò (1994) proposed to take this effect into account by performing the integral over an interval that does not include the load-point, as follows:

$$v_{FF}(\alpha,\gamma) = 2\int_{\gamma+\epsilon}^{\alpha} k_F^2(\alpha',\gamma)\, d\alpha' \qquad (10.4.15)$$

Here, for a single layer of steel bars, ϵ is a small value proportional (but not identical) to the ratio D_b/D, with D_b = diameter of the bars.

Fig. 10.4.5b shows the kind of agreement with the experiments attained with the model of Bosco and Carpinteri. Note that the postpeak behavior is reasonably well predicted, but the model predicts a very large initial strength (for short cracks). This is a general limitation for the LEFM-based models because the stress intensity factor is always zero for uncracked specimens, implying that, in strict LEFM, a crack can never initiate in an unnotched specimen. Therefore, all these models must be interpreted as approximately describing the evolution of the fracture after the crack has formed; the crack initiation itself can be described only by recourse to nonlinear fracture mechanics, as in the models to be described next.

10.4.3 Simplified Cohesive Crack Models

In the literature, there are two slightly different approaches that use the cohesive crack model in a simplified form that avoids finite element computations. One model was put forward by the research group at Aalborg University (Ulfkjær, Brincker and Krenk 1990; Ulfkjær et al. 1994) and the other by the group at the University of New Mexico (Gerstle et al. 1992).

The basic idea is to describe the fictitious crack as a smeared crack of width h as shown in Fig. 10.4.6a. It is further assumed, based on the hypothesis of plane cross sections remaining plane, that the strain distribution along the beam depth is linear (Fig. 10.4.6b). Then, assuming a linear softening curve, the stress distribution can be computed as sketched in Fig. 10.4.6c. The stress-strain curve in tension is completely determined by the tensile strength f_t' and the critical crack opening w_c, as shown in Fig. 10.4.6d. The stress-strain curve in compression is assumed to be linear all the way to complete fracture. The essential difference between the two approaches is that Ulfkjær et al. consider a smeared band width proportional to the size, $h = \eta D$, while Gerstle et al. consider a width twice the instantaneous cohesive crack length ($h = 2y$, see Fig. 10.4.6).

Although the analytical approaches vary, the essential steps are the same: (**1**)write the strain distribution as a function of the curvature κ and the position of the neutral axis x; (**2**) use the stress-strain curves for concrete and steel to express the stress distribution and the steel force as a function of κ and x (this includes the determination of y); (**3**) write the equation of equilibrium of forces and solve for x as a function of κ; (**4**) write the equation of equilibrium of moments; (**5**) finally, from the three equations deduced in (2)-(4), solve for y, x, and M for any given κ. Obviously, the system of equations depends on the load level because of the discontinuity in the derivatives of the stress-strain curves for concrete and steel.

As pointed out before, the smearing band width assumed in these models is different. Ulfkjær et al. assume $h = \eta D$, with $\eta = 0.5$. This value of η was based on comparisons of the load-displacement (moment-curvature) diagrams for unreinforced beams predicted by the approximate model and by accurate finite element computations. Gerstle et al. assume *a priori* that $h = 2y$ and verify that, for an unreinforced

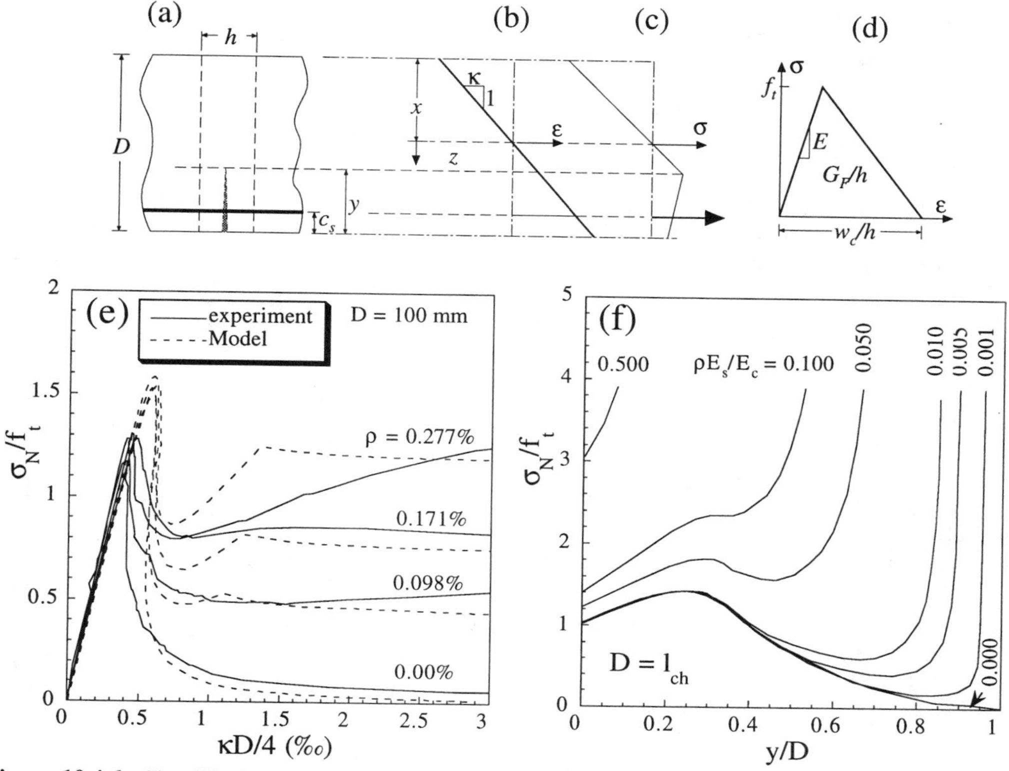

Figure 10.4.6 Simplified cohesive crack models. (a) Smearing band. (b) Strain distribution. (c) Stress distribution. (d) Approximate stress-strain curve. (e) Experimental and theoretical nominal stress-rotation curves computed by Ulfkjær et al. (1994). (f) Theoretical nominal stress-cohesive crack length curves computed according to the model by Gerstle et al. (1992).

beam, the moment vs. cohesive crack length curve predicted by their model is reasonably close to the curve obtained by finite elements. However, this verification has been done only for the ascending part of the curve and only for one size. Further validation is necessary.

Secondary differences between these models are that, in the formulation of Ulfkjær et al., the steel cover is arbitrary and the steel is allowed to yield in a perfectly plastic manner, while Gerstle et al. consider only a vanishing cover thickness to obtain simpler expressions and assume the steel to be always elastic, as is manifested by the increasing load value at the tail of the curves in Fig 10.4.6f.

10.4.4 Models Based on Cohesive Cracks

The cohesive crack model has been used by several investigators to analyze lightly reinforced beams in three-point bending. All the analyses up to now simplify the problem by assuming that a single cohesive crack forms at the central cross section while the concrete in the bulk behaves elastically and the steel is elastic-perfectly plastic. The various analyses differ in the computational method and in the way they incorporate the effect of the reinforcement.

Hawkins and Hjorsetet (1992) use a commercial finite element code to simulate the experiment of Bosco, Carpinteri and Debernardi (1990b). They use the method described in Section 7.2.3 in which the cohesive zone is modeled by an array of elastic-softening springs. Although they do not explicitly consider bond slip, they made two kinds of analysis: one standard (called P-MAX) in which perfect adherence was assumed, and another (called P-MIN) in which the cross sections were forced to remain plane. In the first case, a large strain is generated in the reinforcement as soon as the crack tip reaches it, which causes the steel to yield. In the second case, the strain is smeared over the element width, which is similar (although

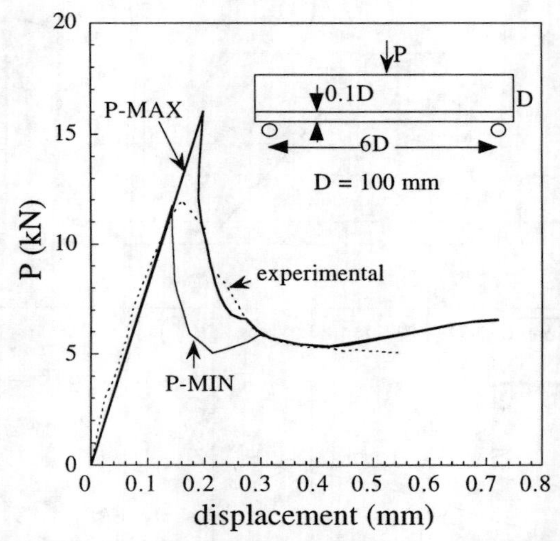

Figure 10.4.7 Numerical load-displacement curves computed by Hawkins and Hjorsetet (1992) using a cohesive crack model; P-MAX curve corresponds to perfect bond; P-MIN to plane cross sections.

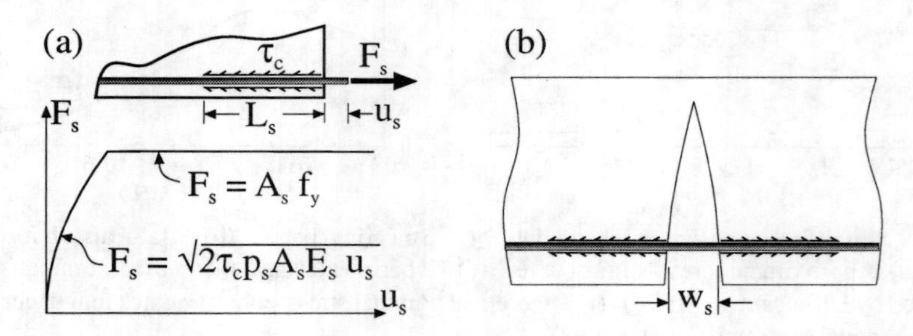

Figure 10.4.8 Simple pullout model (a) and application to the reinforced beam (b) (from Ruiz 1996).

not identical) to having a slip length equal to the element width. Fig. 10.4.7 shows the results of the computations together with the experimental results of Bosco, Carpinteri and Debernardi (1990b). Note that the P-MAX and P-MIN predictions differ appreciably, especially as far as the peak load is concerned; this indicates that the bond must play an important role in defining the minimum reinforcement.

Hededal and Kroon (1991) and Ruiz et al. (Ruiz, Planas and Elices 1993, 1996; Ruiz and Planas 1994, 1995; Planas, Ruiz and Elices 1995; Ruiz 1996) use very similar computational procedures, traceable to Petersson's influence matrix method (see Section 7.4). The two groups consider bond slip in a very similar fashion, but use a different way to implement it numerically. They both consider the same classical load-displacement curve which is obtained for pullout from a rigid half-space, depicted in Fig. 10.4.8a. They also assume that, in the actual test, the steel displacement u_s is given by half the crack opening at the reinforcement, w_s (Fig. 10.4.8b); therefore, the force-crack opening displacement is given by

$$F_s = \begin{cases} \sqrt{\tau_c p_s A_s E_s w_s} & \text{for } w_s < w_y = \dfrac{A_s f_y^2}{p_s E_s \tau_c} \\ A_s f_y & \text{for } w_s < w_y \end{cases} \tag{10.4.16}$$

where F_s is the resultant tensile force in the steel at the central cross section, τ_c is the bond shear strength (rigid-plastic behavior assumed), p_s and A_s are the perimeter and the area of the reinforcement, and E_s, and f_y are the elastic modulus and the yield limit of steel (elastic-perfectly plastic behavior is assumed).

Hededal and Kroon (1991) introduce the action of the steel on the concrete as the force F_s concentrated at the surface of the cohesive crack and treat it as a cohesive force with a load-crack opening curve as deduced from (10.4.16). Their theoretical predictions compare quite realistically with their experimental

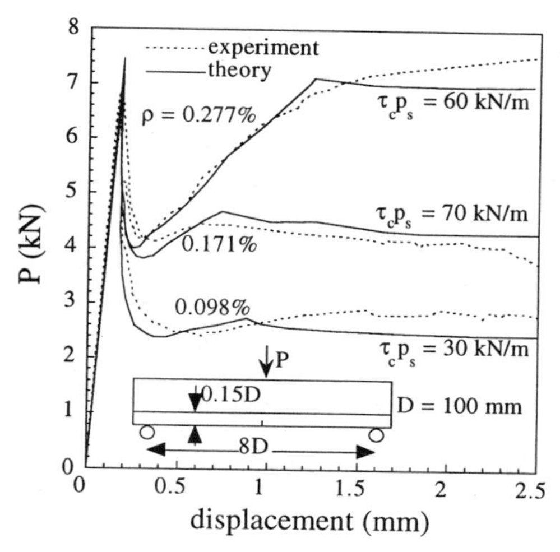

Figure 10.4.9 Comparison of the numerical and experimental results of Hededal and Kroon (1991).

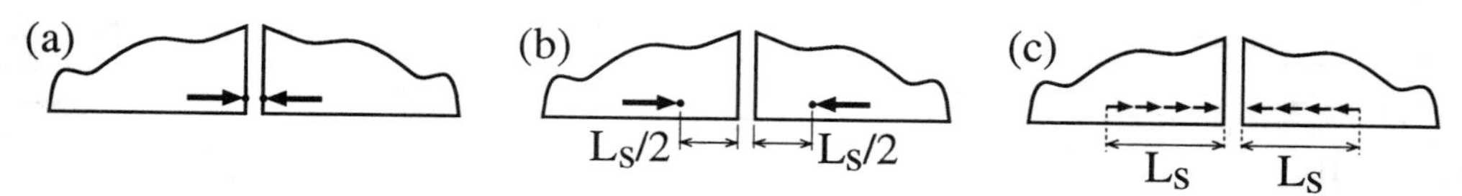

Figure 10.4.10 Approximations analyzed by Planas, Ruiz and Elices (1995): (a) concentrated forces on the crack faces; (b) concentrated forces at the center of gravity of the bond stresses; (c) distributed bond stresses. (From Planas, Ruiz and Elices 1995.)

results as shown in Fig. 10.4.9. In making the predictions, Hededal and Kroon use material parameters determined from independent experiments, except for the bond strength which they select in each case to give a good fit of the postpeak values. The softening curve for concrete is assumed to be bilinear and is determined from tests on notched plain concrete specimens. The steel bars are threaded bars rather than conventional reinforcing steel bars. The ultimate load and the apparent elastic modulus are determined from tensile tests. Note that Hededal and Kroon use the product $\tau_c p_s$ instead of τ_c to characterize the bond strength; $\tau_c p_s$ is the shear force per unit length of reinforcement.

Ruiz and Planas (1994) and Ruiz, Planas and Elices (1993) use a different numerical approach which incorporates the effect of the reinforcement by means of internal stresses. This allows considering the steel-concrete interaction to be located within the concrete rather than at the surface. They analyze the three options depicted in Fig. 10.4.10a–c. In their first approach, they analyze the case of perfect bond with the steel-concrete interaction represented by two forces acting on the crack faces (Fig. 10.4.10a; Ruiz, Planas and Elices 1993; Ruiz 1996).

This analysis reveals that the cohesive crack growth process follows the stages shown in Fig. 10.4.11a-e: in stage (a), the cohesive zone extends through the cover and may go through the first peak if the cover is thick; then the cohesive crack is pinned by the steel and hardening occurs until, as shown in (b), the tensile strength is reached at points ahead of the reinforcement; from then on, two separate cracks exist at both sides of the reinforcement until the yield strength is reached in the steel as shown in (c); then a softening phase begins, with an open crack extending across the reinforcement as shown in (d).

The analysis confirms Hawkins and Hjorsetet's (1992) conclusion that perfect adherence implies a very sharp and high peak. However, it also turns out that this peak depends strongly on the width (diameter) of the reinforcement or, if the steel force is concentrated at a node, on the width of the elements used in the computations. The reason is that, in this approach, the steel force is modeled as a nodal force, which causes that the computational procedure smears this force roughly over an element width, and thus one never deals with a concentrated force but with a distributed force; if the force were really concentrated at a

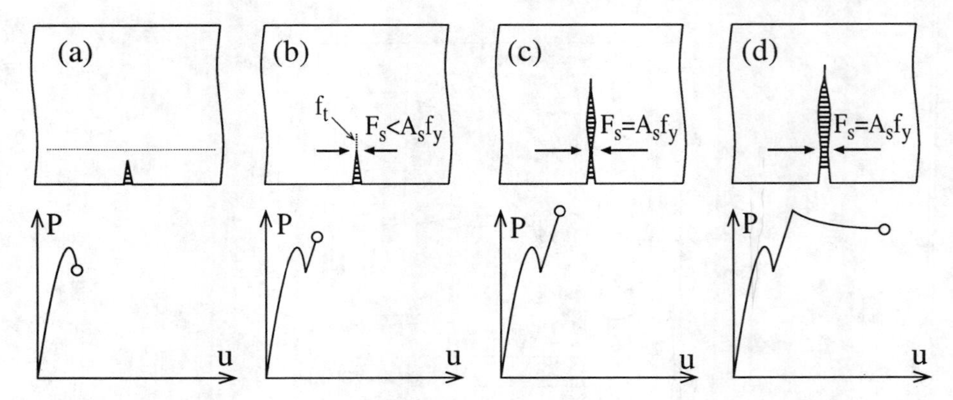

Figure 10.4.11 Crack growth process for full bond of reinforcement: (a) The cover cracks; (b) new crack forms ahead of the reinforcement; (c) steel yields; (d) full crack formed. (Adapted from Ruiz 1996.)

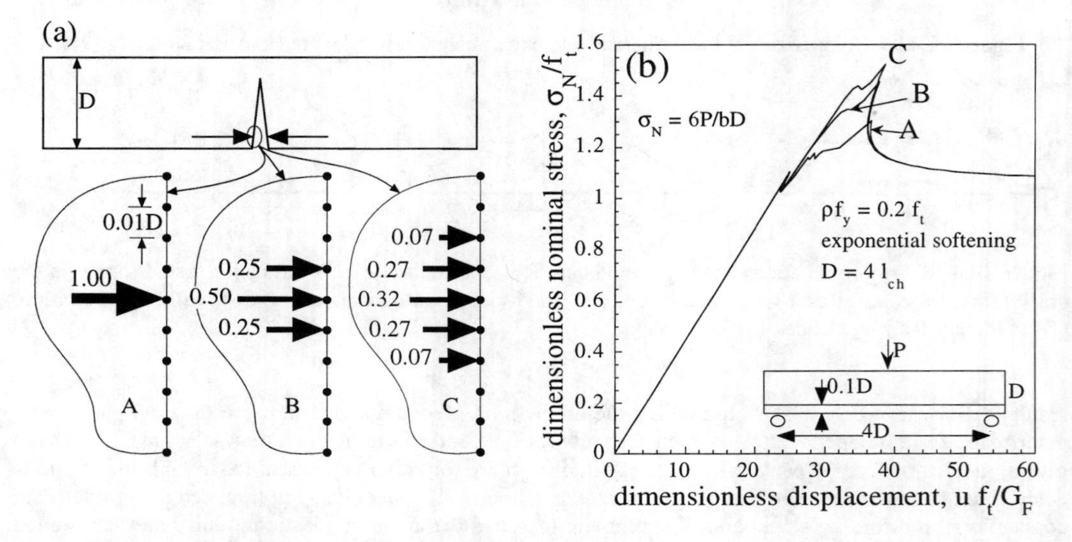

Figure 10.4.12 Influence of distributing the steel force on the crack faces: (a) three possible distributions; (b) corresponding load-displacement curves. (After Ruiz 1996.)

point, the compliance would be infinite and the peak would decrease. This effect is shown in Fig. 10.4.12, where the effect of smearing the force over 1, 3, or 5 nodes is shown for equal elements 1/100 of the beam depth in size (Ruiz 1996). It is clear that the wider the reinforcement, the stiffer the response.

The foregoing problem —the effect of the element size or reinforcement width— appears whenever the steel force is concentrated at the crack faces, even if the bond slip is taken into account. To avoid introducing the width of the reinforcement as a further variable, Planas, Ruiz and Elices (1995) and Ruiz (1996) let the steel-concrete interaction occur inside the concrete as shown in Fig. 10.4.10b–c. The simplest approach uses a concentrated force acting at the center of gravity of the shear stress distribution, much like in the approach by Bažant and Cedolin (1980), called the effective slip-length model. The location of the concentrated forces varies with the crack opening at the level of steel as follows:

$$L_e = \frac{L_s}{2} = \sqrt{\frac{A_s E_s}{4\tau_c p_s} w_s} \tag{10.4.17}$$

where L_s is the slip length, which is readily obtained from the simple pullout model.

To solve the numerical problem in a computationally inexpensive way, Ruiz et al. first write the actual

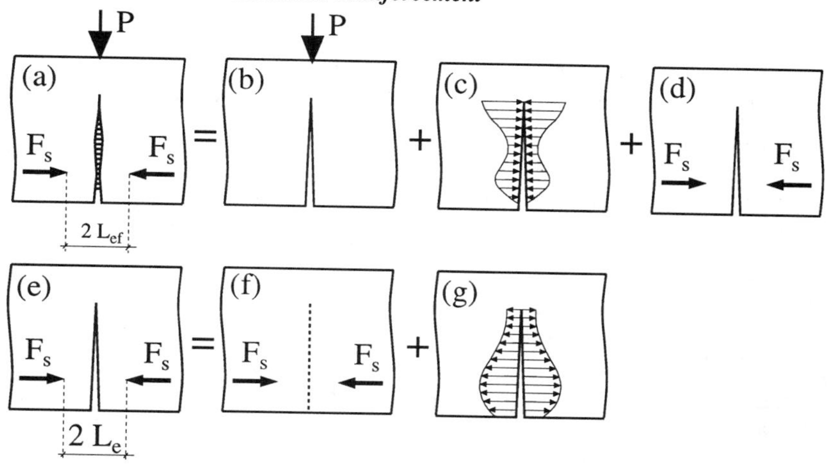

Figure 10.4.13 Successive decomposition of the problem as a sum of elastic problems (from Ruiz 1996).

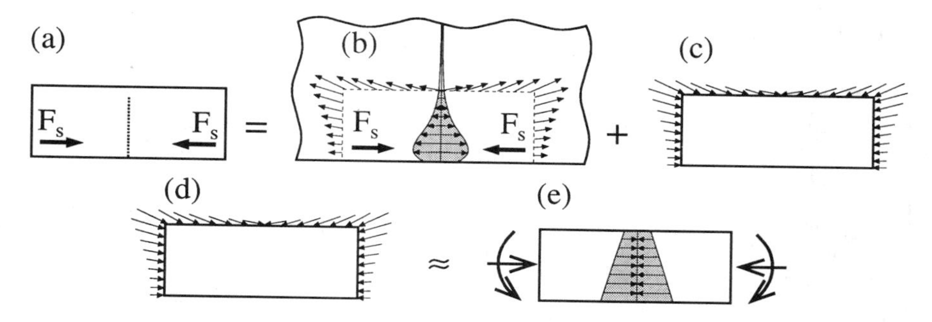

Figure 10.4.14 Approximate closed form solution for the internal pair of forces (from Ruiz 1996).

problem (Fig. 10.4.13a) as the superposition of the three elastic cases shown in Figs. 10.4.13b-d. The first two cases are the classical cases appearing in plain concrete. The third case —introducing crack openings, but no stresses— is handled as shown in Figs. 10.4.13e-g which involve the determination of the stresses engendered on the central cross section in an uncracked beam: this is an internal stress field which is then handled in a way similar to thermal or shrinkage stresses (Petersson 1981; Planas and Elices 1992b, 1993b). The only problem is the determination of the stresses in the auxiliary problem in Fig 10.4.13f. This is approximately solved in closed form as follows (Ruiz and Planas 1994; Ruiz 1996).

The actual problem —Fig. 10.4.14a— is considered as the elastic solution for two concentrated loads parallel to the surface of an elastic half-space (Fig. 10.4.14b) plus the elastic solution for the beam subjected to surface tractions canceling those in the previous solution (Fig. 10.4.14c). The last problem is approximately solved by replacing it with a mechanically equivalent linear stress distribution at the cross-section, as sketched in Fig. 10.4.14d-e. The complete stress distribution can thus be obtained in a closed form from Melan's (1932) elastic solution for a point load parallel to the surface of the half-space. The integration of the surface tractions and their moments, required to find the solution in Fig. 10.4.14e, can be performed analytically (a symbolic mathematical package was used by Ruiz to get the closed form quadratures; see Ruiz 1996 for details).

The model was further refined by Ruiz (1996) to allow distributed bond stresses to be directly used as shown in Fig. 10.4.13a. The stress distribution caused by the reinforcement can be obtained in a closed form by integrating the solution for the concentrated load. This is cumbersome but feasible if one of the modern symbolic mathematical packages is used. However, detailed comparisons showed that the differences with respect to the effective slip-length approach are negligible for most practical cases (Ruiz 1996).

This model was successfully used to describe the tests on microconcrete performed by Ruiz et al. as

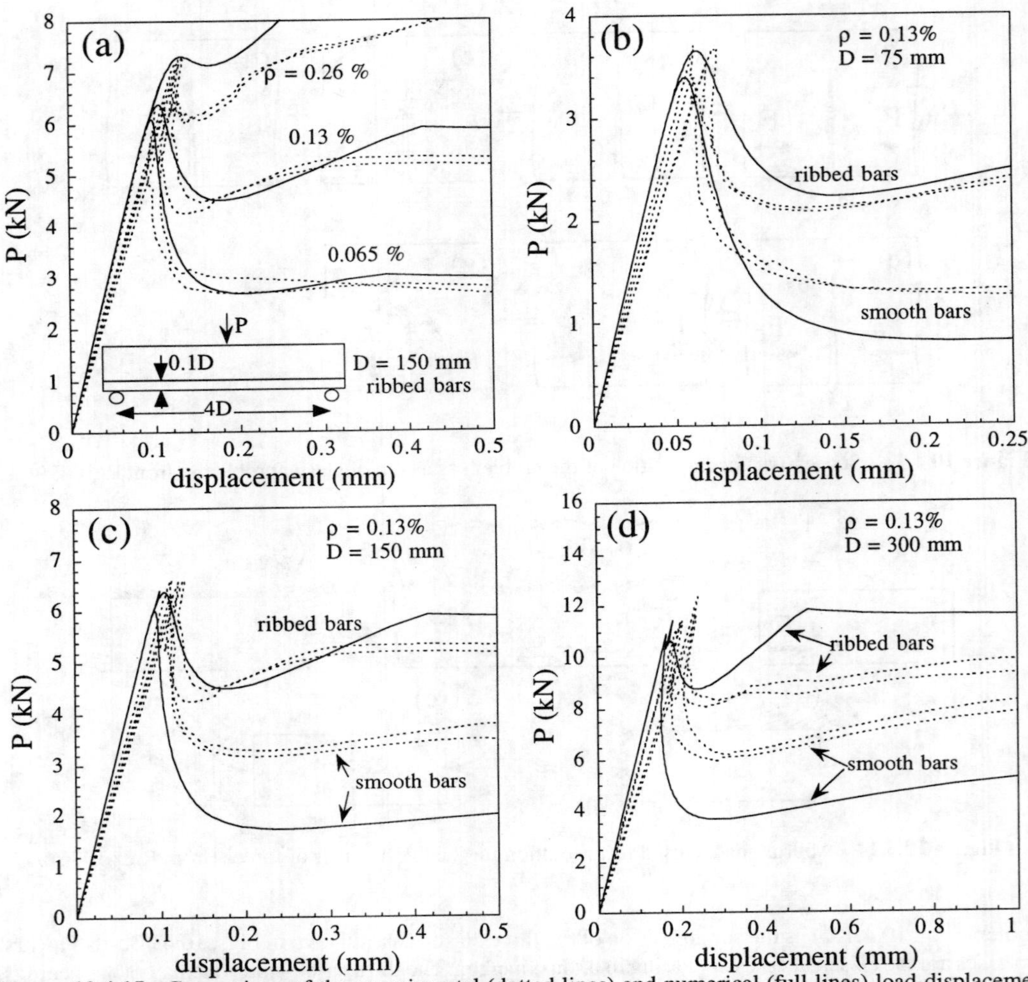

Figure 10.4.15 Comparison of the experimental (dotted lines) and numerical (full lines) load-displacement curves (from Planas, Ruiz and Elices 1995): (a) influence of the reinforcement ratio; (b–d) influence of the bond strength on beams with the same reinforcing ratio but different depth. Note that the bond was determined from independent pullout tests: no parameter fitting has been done.

illustrated in Fig 10.4.15a (Ruiz and Planas 1995; Planas, Ruiz and Elices 1995; Ruiz 1996). The important point in this comparison is that all the parameters required to make the predictions were determined by independent tests. In particular, the bond strength τ_c was determined form pullout tests; much better fits can be achieved if the value of τ_c is adequately selected for each test. Moreover, the model, conceptually simple as it is, shows that the problem is governed by four dimensionless parameters. These parameters are the following:

$$D^* = \frac{D}{\ell_1}, \qquad \rho = \frac{A_s}{A_c}, \qquad f_y^* = \frac{f_y}{f_t'}, \qquad \mu = \sqrt{\frac{n\tau_c p_s \ell_1}{f_t' A_s}} \tag{10.4.18}$$

where ℓ_1 is the characteristic size based on the initial linear softening defined in (10.1.10), ρ is the steel ratio, f_y^* is the relative yield strength of steel, μ is a dimensionless parameter which characterizes the bond, and $n = E_s/E_c$ is the ratio of the elastic moduli of steel and concrete.

The foregoing model was used to investigate the influence of various parameters on the behavior of lightly reinforced beams. The most important result is that a closed-form expression has been found for the first peak of the load-displacement curve. Since this first peak occurs at the initial stages of cracking, before much softening take place, the peak load is controlled by the characteristics of the initial straight portion of the softening curve. Consequently, as discussed in Section 7.2.4, the size effect is controlled by ℓ_1, rather than by $\ell_c h$; see (7.1.17). Moreover, since the steel remains elastic at this stage, the yield strength of steel cannot influence the value of this first peak. This means that the nominal stress at the first peak σ_{Nc} can be written as

$$\sigma_{Nc} = f'_t \phi \left(\frac{D}{\ell_1}, \frac{c_s}{\ell_1}, \rho, \mu \right) \tag{10.4.19}$$

where ϕ is a dimensionless function and c_s is the steel cover. Numerical simulations showed that this function may be, in a crude approximation, expressed as

$$\sigma_{Nc} = f_r + \rho f'_t \mu \, 6 \left(1 - \frac{c_s}{D}\right) \psi \tag{10.4.20}$$

where f_r is the rupture modulus and ψ a factor depending on the beam depth and cover thickness, approximately given by

$$\psi = \left(\frac{D}{\ell_1} \right)^{1/4} - 3.61 \frac{c_s}{\ell_1} \geq 0 \tag{10.4.21}$$

where the last inequality defines the range of application of the formula.

Note that the modulus of rupture in the foregoing formulas is itself size-dependent and can be approximated by the formula (9.3.12) due to Planas, Guinea and Elices (1995).

10.4.5 Formulas for Minimum Reinforcement Based on Fracture Mechanics

In most cases, the minimum reinforcement must ensure that the ultimate (collapse) load (point T in Fig. 10.4.1d) be equal to the first peak load (point M in Fig. 10.4.1d). Based on purely experimental grounds, Bosco and Carpinteri (1992) proposed a formula which correlates the brittleness number $N_{P\min}$ at which the minimum reinforcement condition is met, to the compressive strength of the concrete, f_c:

$$N_{P\min} = 0.1 + 0.23 \frac{f_c}{\sigma_1}, \qquad \sigma_1 = 100 \text{ MPa} \tag{10.4.22}$$

From this and the definition (10.4.6), we can solve for $\rho_{\min}$ and get

$$\rho_{\min} = \frac{K_{Ic}}{f_y \sqrt{D}} \left(0.1 + 0.23 \frac{f_c}{\sigma_1} \right), \qquad \sigma_1 = 100 \text{ MPa} \tag{10.4.23}$$

Note that this formula is purely empirical since the first peak of the load cannot be adequately predicted with the model of Bosco and Carpinteri (1992).

Baluch, Azad and Ashmawi (1992) proposed the following formula for minimum reinforcement based on the model described in Section 10.4.2:

$$\rho_{\min} = \frac{1.9134 K_{Ic}^{0.82}}{f_y^{0.9922} (1.7 - 2.6 c_s / D)} \tag{10.4.24}$$

where K_{Ic} and f_y must be expressed, respectively, in MPa$\sqrt{\text{m}}$ and MPa. Note that the lack of dimensional consistency indicates that there is a certain degree of empiricism in this equation. (In the original paper, there is a misprint in the formula using units of N and mm —the factor in the numerator should appear in the denominator; here the standard IS units are used, and the formula has been checked against the tabulated values in that paper.)

Gerstle et al. (1992) pushed the definition of $\rho_{\min}$ further by requiring that the load increase monotonically all the time during the test (e.g., curve for $\rho E_s / E_c = 0.10$ in Fig. 10.4.6f). The formula they proposed is

$$\rho_{\min} = \frac{E_c}{E_s} \left(\sqrt{0.0081 + 0.0148 \frac{f'_t D}{E_c w_c}} - 0.0900 \right)^{1/2} \tag{10.4.25}$$

Note that this formula, due to the particular definition of the authors, does not depend on the strength of the reinforcement, but only on its elastic properties. We will see that this formula gives values far larger than the other models and larger than the currently accepted values in the codes.

Hawkins and Hjorsetet (1992) also proposed a formula for minimum reinforcement based on the cohesive crack model as well as some concepts derived from Carpinteri's approach. Their final formula reads

$$\rho_{\min} = 0.175 \frac{f_r D}{f_y (D - c_s)} \tag{10.4.26}$$

where f_r is the rupture modulus for an unreinforced beam of the same dimensions as the actual reinforced beam. According to these authors the modulus of rupture can be computed using a cohesive crack model with the appropriate softening curve. Therefore, it is possible to use the equation (9.3.12) to get a closed-form expression as

$$\rho_{\min} = 0.175\zeta \left(1 + \frac{1}{0.85 + 2.3D/\ell_1}\right) \frac{f_t' D}{f_y (D - c_s)}, \qquad \ell_1 = \frac{E_c w_1}{2 f_t'} \tag{10.4.27}$$

where $\zeta = 1.046$ for three-point bending and 1 for four-point bending.

None of the preceding formulas take into account the bond strength. Ruiz, Planas and Elices (1996) and Ruiz (1996) have proposed a formula taking this effect into account. The formula is based on the cohesive crack model, more specifically on the expression (10.4.20) giving the first peak load. The final plastic collapse load is obtained from elementary considerations of equilibrium of moments as

$$\sigma_{Np} = \rho f_y \, 6 \left(1 - \frac{c_s}{D}\right) \tag{10.4.28}$$

Then, setting $\sigma_{Nc} = \sigma_{Np}$, solving for $\rho = \rho_{\min}$, and inserting (9.3.12), one gets

$$\rho_{\min} = \frac{\zeta}{6(1 - c_s/D)} \frac{1 + (0.85 + 2.3D/\ell_1)^{-1}}{f_y/f_t' - \mu \left[(D/\ell_1)^{1/4} - 3.61 c_s/\ell_1 \right]} \tag{10.4.29}$$

Comparing the aforementioned models is not straightforward because they are based on different assumptions and depend on different parameters. This means that the predictions can be similar for certain conditions and differ for other conditions; no exhaustive comparison has been done to date. Fig. 10.4.16 shows the dependence on size of the minimum reinforcement for the various models in a particular case defined as follows:

1. The concrete is assumed to be characterized by $f_t' = 4$ MPa, $f_c = 40$ MPa, $E_c = 30$ GPa, $G_F = 160$ N/m (total fracture energy, as determined from the work-of-fracture test). For Carpinteri's and Baluch's model, it is assumed that Irwin's relation holds: $K_{Ic} = \sqrt{E_c G_F} = 2.19$ MPa$\sqrt{\text{m}}$. For Gerstle's formula, linear softening is assumed with $w_c = w_1 = 2G_F/f_t' = 80$ μm. For Hawkins' formula, Petersson's bilinear softening is assumed with $w_1 = G_F/f_t' = 48$ μm. For Ruiz's models, bilinear softening is assumed with $w_1 = G_F/f_t' = 40$ μm.
2. The steel is assumed to have $E_s = 210$ GPa, $f_y = 480$ MPa.
3. The cover of concrete reinforcement is assumed to be of constant thickness $c_s = 24$ mm.
4. For Ruiz's model, the bond parameter μ is taken to be constant. This is achieved by using bars of the same diameter for all beam sizes. Two values are considered: $\mu = 10$ (weak bond,16 mm diameter bars with $\tau_c \approx 0.4 f_t'$) and $\mu = 40$ (strong bond, 8 mm diameter bars with $\tau_c \approx 3 f_t$).

Fig. 10.4.16a shows the entire set of curves with a vertical logarithmic scale making it clear that the formulas of Baluch, Azad and Ashmawi (1992) and of Gerstle et al. (1992) give too large values. Fig. 10.4.16b shows an enlarged plot (with a linear scale) in which some code specifications have been included for comparison. From the plots it is evident that, for small beam depths, the models of Bosco and Carpinteri, Hawkins and Hjorsetet, and Ruiz, Planas and Elices give very similar results, slightly higher than those specified by ACI 318(92), and higher that the specifications of the Model Code and Eurocode 2. For medium and large sizes, the model of Bosco and Carpinteri gives values sharply below those of ACI and the other models, while the formula of Hawkins and Hjorsetet gives values very close to those given

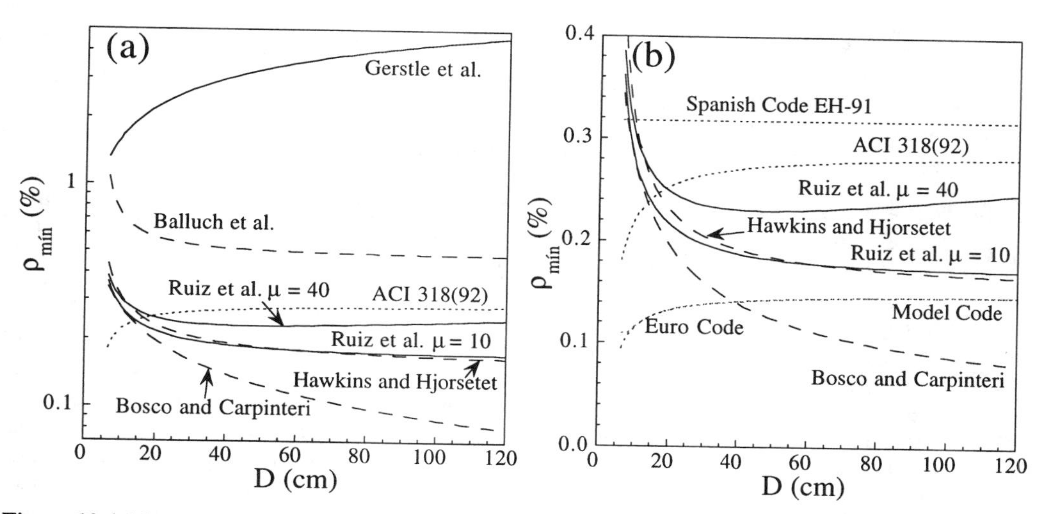

Figure 10.4.16 Comparison of minimum reinforcement formulas for various models (data from Ruiz 1996).

by the formula of Ruiz et al. in the case of weak bond, both giving values between those in the Model Code and the ACI Code. For a strong bond, the model of Ruiz et al. predicts a minimum reinforcement that first decreases with the size and then increases, with values between those recommended by the ACI code and the Spanish Code.

Exercises

10.4 Show that in the plot of Fig. 10.4.5a the position of the asymptotes on the right is given by $Y = 6(1-\gamma)N_p$, where $1-\gamma = 1 - c_s/D$ is the relative depth of the reinforcement.

10.5 Derive Eq. (10.4.28) in detail.

10.5 Other Structures

In this section, we give a brief account of some of the existing results for structures or structural elements not included in the preceding analysis: torsion of beams, punching of slabs, anchor pullout, bond-slip of reinforcing bars, beam and ring failure of pipes, concrete dams, footings, pavements, keyed joints, failure of joints, break-out of boreholes, and compression failure of concrete beams.

10.5.1 Torsional Failure of Beams

Torsion leads to another type of brittle failure of reinforced concrete beams. The classical test data existing in the literature, which were analyzed by Bažant and Şener (1987), and particularly the data by Humphrey (1957), Hsu (1968), and McMullen and Daniel (1975), reveal that a size effect exists, but cannot indicate which equation should describe it because the data were too scattered, the size range was too narrow, and geometrical similarity was not maintained.

Geometrically similar tests of size range 1:4 were conducted on microconcrete beams with reduced maximum aggregate size by Bažant, Şener and Prat (1988). The tests were made both on unreinforced beams and beams reinforced longitudinally. These tests clearly revealed a strong size effect and were shown to agree well with the size effect law. The results were briefly described in Section 1.5, Fig. 1.5.7 (series J1 and J2). From these tests it appears that the size effect in torsion is very strong, and the behavior is quite close to the LEFM asymptote. However, the scatter of the limited experimental data is quite large, and more extensive tests are needed. The scatter is larger for longitudinally reinforced beams, which may be attributed to the fact that bond failure must accompany a torsional crack, and bond failure is a phenomenon of high random scatter.

The code formulas for torsion of beams with rectangular cross section are based on the plastic limit analysis solution, which indicates that the nominal shear strength in torsion (Park and Paulay 1975) is $v_u = T/(\alpha_p b^2 D)$, $\alpha_p = [1 - (b/3D)]/2$, where T = torque, b = length of the shorter side of the rectangular cross section, D = length of the longer side (depth). Since the small size limit of the size effect law should coincide with the plastic solution, Bažant, Şener and Prat (1988) proposed the correction indicated, in general, by Eq. (10.1.13) and showed that it agrees well with the data. Calculations with the microplane model by Bažant, Ožbolt and Eligehausen (1994) also agree quite well with the experimental points and the size effect correction.

No test data seem to exist on the size effect in torsional failure of reinforced concrete beams with stirrups. However, it may be expected that the stirrup effect would be similar to that discussed for diagonal shear, and that the size effect would disappear beyond a certain critical reinforcement ratio of the stirrups.

Torsion in beams is normally combined with bending, and so the interaction diagram between the maximum torque and the maximum bending moment is of considerable interest for design. Hawkins (1985) examined the test results of Wiss (1971) on diagonal tension cracking combined with torsion and bending. Using an energy based fracture criterion for failure under combined loading, he calculated the interaction diagram and showed it to be circular (when the maximum shear force and the maximum torque are normalized with respect to their values for pure torsion or pure shear). He suggested this was an argument for applicability of fracture mechanics, pointing out that the strength-based criteria yield a straight-line interaction diagram. However, this is not sufficient proof of fracture mechanics applicability because the lower-bound plastic limit analysis also gives a circular interaction diagram (e.g., Hodge 1959).

10.5.2 Punching Shear Failure of Slabs

Quasibrittle behavior accompanied by a transitional size effect is also characteristic of the punching shear failure of reinforced concrete slabs. For the nominal shear strength in to punching shear, ACI currently uses the formula

$$v_u^p = k_1\sigma_1[1 + (k_2 D/b)]\sqrt{\frac{f_c'}{\sigma_1}}\,, \qquad \sigma_1 = 1 \text{ psi } = 6.895 \text{ kPa} \tag{10.5.1}$$

in which k_1, k_2 = empirical constants, D = thickness of the slabs, and b = punch diameter (ACI Committee 318, 1989). This equation was derived by strength analysis based on a modified Coulomb yield criterion, which exhibits no size effect. Based on a series of displacement-controlled punching shear tests on geometrically similar two-way reinforced circular slabs of three different sizes (1:2:4), made of concrete of reduced aggregate size, Bažant and Cao (1987) proposed a size-dependent generalization of this formula based on (10.1.13):

$$\sigma_{Nu} = \sigma_{Nu}^p\left(1 + \frac{D}{D_0}\right)^{-1/2} \quad \text{with } \sigma_{Nu}^p = c_1 f_t'\left(1 + c_2\frac{D}{b}\right) \tag{10.5.2}$$

where $c_1 f_t'$, c_2, and D_0 are empirical constants. The test results by which this formula was calibrated are shown in Section 1.5, Fig. 1.5.7, along with the optimum fit by the size effect law (series L1). The size effect was considerably milder than in the diagonal shear tests, which might be due to the fact that the largest slab was not sufficiently large. Fig. 1.5.8 shows the load-deflection diagrams measured on the small, medium, and thick plates. This figure illustrates how the postpeak softening is getting steeper with an increasing size and thus confirms a transition from relatively ductile behavior (the small slab with mild postpeak slope) to very brittle behavior (the largest slab, with a very steep postpeak drop).

The fact that the size effect should be considered in calculating the punching shear strength of slabs was also confirmed by the study of Broms (1990), which was focused on punching shear under high biaxial (radial) compressive stresses and suggested a formula of the type $v_c = v_c^0(k/D)^{1/3}$. The exponent $1/3$, according to the present theory, cannot be right for extrapolations to very large sizes; however, in the middle of the size effect transition, it works well. From the test data alone, it is not possible to say what should be the exact form of the size effect formula. Nevertheless, the presence of the size effect, and thus inapplicability of plastic limit analysis, is clearly verified by the test results.

Cryptodome failure of nuclear reactor vessel slab. The failure of thick prestressed concrete nuclear reactor vessels (primary vessels for gas-cooled reactors which were intensely researched between 1960–1980) is known to occur through a conical surface similar to the punch failure (called cryptodome), rather than by bending. The design of these nuclear reactor vessels has been done according to strength criteria, however, it now appears that, because of the similarity to the punching shear failure, a fracture behavior exhibiting a size effect should be expected (Bažant 1989b). If nuclear power is revived, this question should be researched further.

10.5.3 Anchor Pullout

The current ACI Code provisions for the pullout failures of bars and anchors are based on plastic limit analysis (ACI Committee 408, 1979; ACI Committee 349, 1989). However, the size effect is very strong in this kind of failure, and considerable work has been done in the last decade to increase the understanding of anchor pullout in terms of fracture mechanics, a topic of considerable interest because it is at the base of the design of anchors and of the recently introduced nondestructive test method for concrete strength based on pullout of a headed stud (the Swedish "lok" test).

The ACI Code provision (Sec. 15.8.3) requires that the "anchor bolts and mechanical connections shall be designed to reach their design strength prior to anchorage failure or failure of surrounding concrete". This means the anchor bar must yield before fracture occurs, but this can be ensured only if the load causing the fracture is correctly predicted. ACI Committee 349 (1989) recommends that "the design pullout strength of concrete, P_d, for any anchorage shall be based on a uniform tensile stress of $4\sqrt{f_c'}$ acting on an effective stress area which is defined by the projected area of stress cones radiating toward the attachment from the bearing edge of the anchors". This gives the pullout force P_u and nominal strength

$$P_u = k_1\sigma_1\sqrt{\frac{f_c'}{\sigma_1}}\pi D^2 , \qquad \sigma_{Nu} \equiv \frac{P_u}{\pi D^2} = k_1\sigma_1\sqrt{\frac{f_c'}{\sigma_1}} , \qquad \sigma_1 = 1 \text{ psi} = 6.895\text{kPa} \tag{10.5.3}$$

in which k_1 = empirical constant, f_c' = standard compression strength of concrete, and D = the embedment depth of the anchor bolt. This expression obviously corresponds to plastic limit analysis, the size effect being ignored.

A clear confirmation of a strong size effect in the pullout failure of reinforced concrete bars without anchors was provided by the tests of Bažant and Şener (1988); see Fig. 1.5.6 (series I1). They tested microconcrete cubes of size ratio 1:2:4. The bar diameter and the embedment length were scaled so as to maintain geometric similarity. As is seen from Fig. 1.5.6, in the logarithmic size effect plot, the test results lie very close to the LEFM asymptote. This reveals an extremely high brittleness number for this type of failure. Eligehausen and Ožbolt (1990) and Bažant, Ožbolt and Eligehausen (1994) further showed that these test results agreed closely with nonlocal finite element solutions using a realistic material model for concrete (the microplane model).

Eligehausen and Sawade (1989) proposed a LEFM-based formula for pullout strength. This formula was written as

$$P_u = 2.1\sqrt{EG_F}D^{3/2} \quad \text{or } \sigma_{Nu} = 0.67 f_t'\sqrt{\frac{\ell_{ch}}{D}} \tag{10.5.4}$$

in which the fracture parameters appear explicitly. To avoid the explicit use of fracture parameters — which have not been measured in most of the available test series in the literature, Eligehausen et al. (1991) proposed the following formula based only on the cube compression strength:

$$P_u = a_1\sqrt{f_{cc}}\, D^{3/2} \tag{10.5.5}$$

They evaluated the results of 209 pullout tests of headed anchors carried out at different laboratories. In all tests, the failure occurred by a conical crack surface. The tests were done on concretes of various strengths, and, therefore, the measured maximum loads were normalized to the cube compression strength f_{cc} = 25 MPa, by multiplying them with the factor $\sqrt{25/f_{cc}}$. The normalized failure loads are plotted in Fig. 10.5.1 as a function of the effective embedment depth D, together with the fit of Eq. (10.5.5). From this fit it turns out that $a_1 \approx 15.5$ for P_u in N, f_{cc} in MPa, and D in mm. The formula is seen to closely describe the experimental results, which means that the nominal strength almost follows LEFM. This was confirmed numerically by Eligehausen and Ožbolt (1990) using a microplane nonlocal model: the LEFM formula and Bažant's size effect law differed less than 6% up to embedment length of 400 mm.

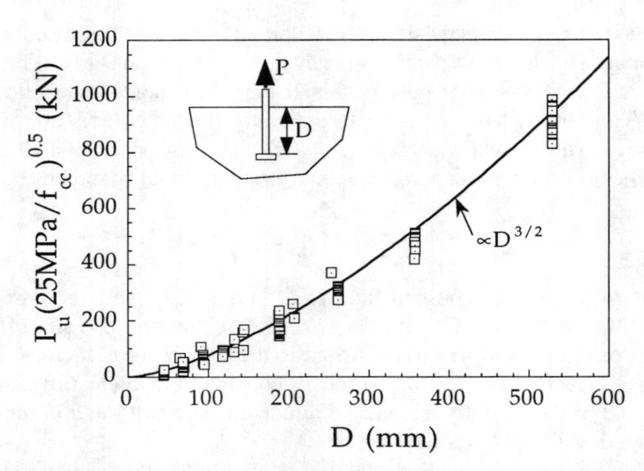

Figure 10.5.1 Size effect in the pullout of headed studs (data from Eligehausen et al. 1991).

Apart from the nonlocal microplane model just mentioned, many different approaches have been used in the last decade to analyze this interesting problem. A two-dimensional LEFM analysis with a mixed-mode crack was used by Ballarini, Shah and Keer (1985) to study the pullout of rigid anchor bolts. They used the Green's function for a concentrated force in an elastic half space, represented the crack opening by means of dislocations and thus reduced the problem to a system of singular integral equations, whose numerical solution yielded the mixed-mode stress intensity factor. The calculations provided the crack profiles and crack growth. Stability checks were made and the results were compared with anchor pullout experiments. An interesting point was that if the support reactions are sufficiently removed from the axis of the anchor, crack propagation becomes unstable and the load capacity is reduced (this is obviously due to the higher stored energy when the support reactions act farther away). On the other hand, for support reactions close to the anchor axis, as well as for sufficiently deep embedments, the crack propagation was found to be stable.

The pullout of circular disc-shaped anchors was studied by Elfgren, Ohlsson, and Gylltoft (1989). They used the finite element discrete crack approach, in which the tensile and shear softening were taken into account according to the formulation of Gylltoft (1984). They studied straight cracks inclined by 45° and 67° from the pullout axis as well as a crack starting at angles 73° and curving according to the principal tensile stress direction. They found the lowest pullout strength to occur for the 45° straight crack. They did not study the size effect, nor the effect of geometry.

The plane stress and axisymmetric problems of anchor pullout were analyzed by numerous researchers in a recent round-robin contest (Elfgren 1990) using various numerical procedures based on various fracture mechanics models, from LEFM to lattice models. Elfgren and Swartz (1992) published summaries of the contributions and a state-of-the-art report is in preparation by Elfgren, Eligehausen, and Rots.

Some of the results can be found in the proceedings of a special seminar on anchorage engineering held at Vienna Technical University in 1992 (Rossmanith, ed., 1993).

10.5.4 Bond and Slip of Reinforcing Bars

The bond and slip of reinforcing bars embedded in concrete is an important and certainly difficult phenomenon. In the previous sections, the bond strength was considered a secondary "internal" problem which was treated by means of very simple models of perfectly plastic shear-slip behavior. The ACI Code, too, gives simple provisions for the so-called development length of reinforcing bars, representing the length of embedment in concrete required for ensuring that the bar can develop its full yield capacity. These formulas are also based on plasticity. The development length is obtained by balancing the steel force at yield $F_s = f_y A_s$, against the bond strength $\tau_c p_s L_s$, where p_s is the perimeter of the reinforcement and L_s the slip (development) length. Taking then $\tau_c \propto f_t' \propto \sqrt{f_c'}$ leads to ACI empirical formulas in which the development length is $C_1 A_s f_y / \sqrt{f_c'}$ or $C_2 f_y / \sqrt{f_c'}$, where C_1 and C_2 are empirical constants

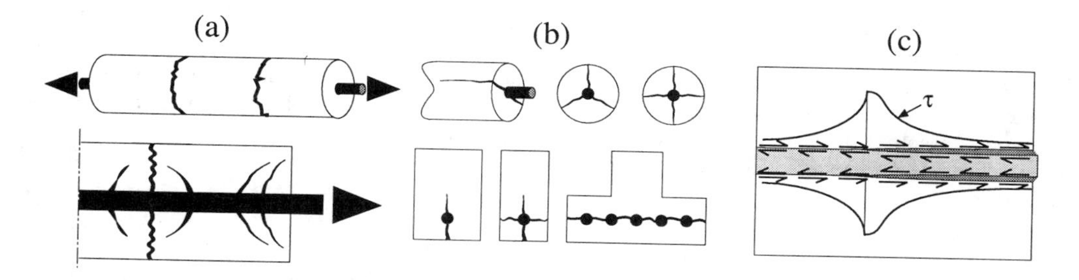

Figure 10.5.2 Cracking in bar pullout: (a) Transverse cracks; (b) longitudinal (splitting cracks); (c) shear-slip interfacial crack.

defined in the code. The code also gives various modifications of these formulas to take into consideration the clear spacing of parallel bars, thickness of concrete cover, and corrections for the case of lightweight aggregate concretes and for epoxy coated reinforcement. The formulas, however, do not consider any size effect associated with the brittleness of concrete.

In reality, the problem of bond and slip is a fracture problem; even more, it is a multiple fracture problem, and a three-dimensional one. Fig. 10.5.2 sketches the three kinds of cracks or crack systems involved in the pullout of a bar from a concrete cylinder or prism. The first system of cracks is transverse to the prism, as shown in Fig. 10.5.2a; the cracks are conical in shape (secondary cracks) or plane (principal cracks). The second system of cracks is longitudinal to the prism; the cracks are generated by hoop stresses and vary in number depending of the morphology of the cross-section (Fig. 10.5.2b). The third system of cracks consists of the shear cracks generated at the interface of steel and concrete; the jump in displacements at the interface is the crack sliding (Fig. 10.5.2c).

The role of transverse fracture of concrete in the bond between concrete and deformed reinforcing bars was first studied by Gylltoft (1984) and by Ingraffea et al. (1984). Gylltoft examined the role of fracture in axisymmetric pullout of bars from concrete blocks. He considered both monotonic and cyclic loading, carried out experiments, and was able to successfully predict the load-slip diagrams observed in the pullout tests. Special crack elements that involve linear strain softening of concrete in tension and a linear strain hardening in shear were used to model the interface. Ingraffea et al. (1984) used a discrete mixed-mode nonlinear fracture model in axisymmetric finite element analysis. They applied the cohesive crack model to characterize the tensile softening at each bond crack, and adopted the aggregate interlock model of Fenwick and Paulay (1968) to characterize the shear softening. The study of Ingraffea et al. (1984) indicated that secondary cracking around the primary cracks contributed to bond slip. Placing special interface elements at the primary crack locations, and comparing numerical results to test results for a center-cracked reinforced concrete plate under uniform tension, Ingraffea et al. (1984) calculated the degradation of stiffness and the crack opening profiles. Ingraffea et al. used in their finite element program (FRANC) a sophisticated technique for remeshing around the crack tip as the crack tip advances. Rots (1988, 1992) analyzed the problem of transverse cracking concomitant with longitudinal cracking. He used a smeared crack approach for the secondary cracks, and a discrete crack approach modeled by interface elements for the primary cracks.

A problem in the foregoing analysis is to correctly handle longitudinal cracking that occurs simultaneously with transverse cracking. In reality, a three-dimensional formulation ought to be used to analyze the problem in detail; however, this would require an enormous computational effort. Axisymmetric formulations have been used, with the expedient of using a circumferential stress-strain relationship that is a smeared version of a cohesive crack. This is done as in Chapter 8 for the uniaxial case, except that now the stress is the circumferential stress σ_θ and the cracking strain is ε_θ^c. These are related to each other and to the crack opening by

$$\varepsilon_\theta^c = n^c \frac{w(r)}{2\pi r} \Rightarrow \sigma_\theta = f[w(r)] = f\left(\frac{2\pi r}{n^c}\varepsilon_\theta^c\right) \tag{10.5.6}$$

where $w(r)$ is the opening of each crack at a distance r from the rotational axis, n^c the number of cracks, and $f(w)$ the softening function (for a single crack). Note that the number of cracks n^c must be assumed before the calculation and cannot be inferred from the analysis. It is usually selected between 2 and 4, based on experience.

An entire family of simplified analyses of longitudinal splitting cracks taking into account the fracture behavior of concrete was developed based on modifications of the initial approach by Tepfers (1973, 1979). Tepfers assumes that the rise of interfacial shear stresses τ is accompanied by a rise of a contact normal stress σ. He further postulates that at splitting failure, σ and τ are related by a Coulomb-type law that he writes as

$$\sigma = \tau \ \tan\phi \tag{10.5.7}$$

where ϕ is a constant complementary friction angle. Tepfers then reduces the analysis to an axisymmetric problem of a thick-walled concrete tube subjected to inner pressure σ. Tepfers considers only elastic-brittle behavior and elastic-perfectly plastic behavior. Keeping this approach, several researchers extended Tepfers' analysis to include softening and fracture. All these analyses use further simplifications, such as neglecting Poisson's effect, and use the circumferential smeared cracking as given by (10.5.6). The main difference is in the kind of softening curve used by the various authors: van der Veen (1991) uses a power-law softening (Reinhardt 1984); Reinhardt and Van der Veen (1992) and Reinhardt (1992) use the CHR softening curve (Cornelissen, Hordijk and Reinhardt 1986b; see Section 7.2.1); Rosati and Schumm (1992) use a hyperbolic law; and Noghabai (1995a,b) uses a linear softening. As a further difference, Rosati and Schumm (1992) consider, instead of the Coulomb criterion (10.5.7), a Mohr-Coulomb condition given by $\sigma = (\tau - \tau_0) \ \tan\phi$, in which τ_0 is a constant "cohesion".

The problem of discrete longitudinal splitting cracks was directly addressed by Choi, Darwin and McCabe (1990). They used a three-dimensional finite element method to analyze test results and design code provisions on the bond failure of epoxy-coated or uncoated steel bars as a function of the bar size, variations in interface characteristics, and specimen geometry. Splitting fractures observed in the tests were well reproduced by the computational model —based on a cohesive crack model— and the results provided support to some empirical code provisions. The computational results described well the increase of pullout strength with the cover thickness.

Recently, Noghabai (1995a,b) considered the numerical analysis of longitudinal splitting cracks based on the boundary conditions in Tepfers' approach (concrete thick-walled tube with internal stress σ) and analyzed localized cracking using three numerical approaches with the same underlying material model (a cohesive crack with linear or nonlinear softening). The first numerical procedure —the so-called discrete crack approach— was carried out by placing 28 radial layers of interface elements incorporating the stress-crack opening relationship. The strength of the layers was randomly assigned to promote localization into a small number of cracks. The second numerical procedure was the classical smeared crack approach, with the crack opening distributed within the elements. The third procedure used enriched shape functions to describe the displacement jump within each element —the so-called inner softening band finite elements (Klisinski, Runesson and Sture 1991; Klisinski, Olofsson and Tano 1995). The three procedures gave similar results for the curves of pressure vs. radial deformation, although none were able to continue into the structural softening branch. The inner softening band method seems very promising for capturing the cracking pattern. Still, the weakest link in the model is the relationship between the normal and shear stresses. A realistic relationship between the normal and shear stresses at the interface must somehow be related to the slip between steel and concrete.

The third type of crack involved in the pullout process is the shear-slip (mode II) crack occurring at the steel-concrete interface (see Fig. 10.5.2c, where the separation between the bar and the concrete is grossly exaggerated). A straightforward approach is to treat this shear-slip crack as a cohesive crack, i.e., to postulate that a certain relationship exists between the transferred shear stress τ and the relative slip s:

$$\tau = t(s) \tag{10.5.8}$$

where $t(s)$ is the softening function for shear-slip. Introduced by Bažant and Desmorat (1994), this is a very simplified model which does not take into account friction and dilatancy occurring at the interface. More sophisticated models involving the crack opening due to dilatancy and the influence of the normal stress may be formulated, but will not be further described (for an overview of models and a thorough discussion of the coupling between normal and shear stresses and displacements, see Cox 1994). The simple model defined by (10.5.8) can, however, suffice to get a rough picture of the influence of the bond degradation on the overall response in bar pullout.

A simple and crude mathematical model which, nevertheless, realistically captures some aspects of fracture and the size effect was used by Bažant and Desmorat (1994), who considered a uniaxially stressed bar (or fiber) embedded in a concrete bar also behaving in an uniaxial manner, each with the cross section remaining plane and orthogonal to the bar axis. The interface between the bar and the concrete tube is characterized by the $\tau - s$ relation (10.5.8). For the sake of simplicity, this relation was assumed linear (triangular stress-displacement diagram). The solution can then be obtained analytically, by integration of the differential equation of equilibrium in the axial coordinate x. The solution yields simple formulas. It is found that, during failure, zones of slip initiate at the beginning of embedment of the bar or at the bar end, or both, and spread along the bar as the end displacement of the bar is increased, as sketched (for a more general case) in Fig. 10.5.2c. For geometrically similar situations, a strong size effect is observed. The size effect is caused by the fact that the ratio of the length of the slipped zone to the bar diameter decreases with increasing diameter, i.e., the slip zone localizes. For a sufficiently small size, the slip zone at the maximum load extends over the entire length of the bar embedment, and for a size approaching infinity, the relative length of the slipped zone tends to zero. The calculated size effect curve turned out to be very close to the generalized size effect law proposed by Bažant [Eq. (9.1.34)], with the exponent $r = 1.25$ for a concave nonlinear softening law. The one-dimensional solution may, of course, be expected to be good only when the slip zone is very long or very short compared to the bar diameter and the concrete cover around the bar is not too thick. In general, three-dimensional fracture analysis is, of course, required. Nevertheless, despite the one-dimensional simplification, it seems that the generalized size effect law indicated by this analysis may be applied as a simple approach to practical problems.

Further tests of bar pullout from normal and high strength concrete cubes were conducted by Bažant, Li and Thoma (1995). In these tests, it was tried to separate the effect of radial fractures emanating from the bar and the bond crack along the bar. The tests were designed so that no radial fractures would form and the bar would fail only due to bond fracture and slip. This was achieved by using a relatively short embedment of the bar in the concrete cube. The results again revealed a strong size effect. Because of the absence of radial cracks, it was admissible to compare the results to the aforementioned one-dimensional solution of Bažant and Desmorat (1994). The comparison was satisfactory although large scatter prevents considering this as a validation of the Bažant and Desmorat's equation.

A similar degree of brittleness as in pullout occur in the failure of splices of reinforcing bars in which the lapped bars are not connected and the tensile force in the bars is transmitted through the concrete in which the bars are embedded. The codes provide empirical provisions for the length of overlap and for the so-called development length over which the yield force of the bar can be transmitted from concrete to the bar. These formulas are of the strength theory type which exhibit no size effect. Şener (1992) reports experiments which confirm that splices indeed exhibit a strong size effect which may be well described by Bažant's size effect law and is rather close to LEFM (in more detail, Şener, Bažant and Becq-Giraudon 1997). The aforementioned type of correction of the existing formula —Eq. (10.1.13)— is also needed in this case.

10.5.5 Beam and Ring Failures of Pipes

In the failure of pipes, it has for a long time been recognized that the apparent strength is different for the transverse bending that leads to ring-type failure (Fig. 10.5.3a) and for longitudinal bending of the whole pipe that leads to beam-type failure (Fig. 10.5.3b). Gustafsson and Hillerborg (1985) analyzed such failures using the fictitious (cohesive) crack model. A plot of the size effect that they obtained is shown in Fig. 10.5.3c. In this plot, σ_N is defined as the maximum elastic stress according to mechanics of materials theory. Thus, for the ring failure, we have

$$\sigma_{Nu} = \frac{3F_u D}{\pi t^2}, \qquad D = \frac{D_i + D_o}{2} \tag{10.5.9}$$

where F_u is the maximum force per unit length of pipe, D_i and D_o the inner and outer diameters, respectively, and t the pipe thickness. The nominal stress for the beam failure is

$$\sigma_{Nu} = \frac{M_u D_0}{2I}, \qquad I = \frac{\pi}{64}(D_0^4 - D_i^4) \tag{10.5.10}$$

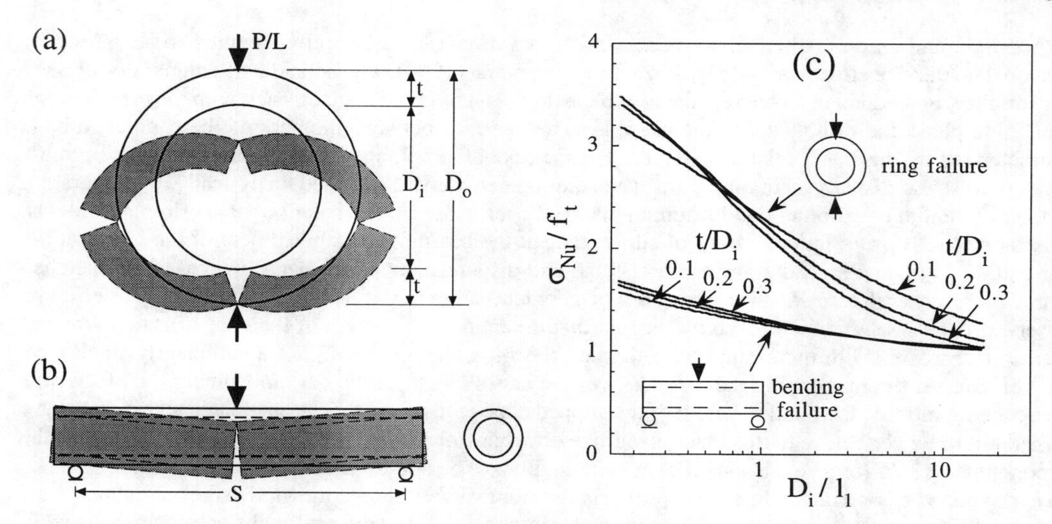

Figure 10.5.3 Size effect in unreinforced concrete pipes according to the computations of Gustafsson and Hillerborg (1985): (a) scheme of ring (crushing) failure; (b) scheme of bending (beam) failure; (c) strength vs. size for the two types of failure. (σ_{Nu} = is the maximum elastically-computed stress at peak load, see the text.)

in which M_u is the ultimate bending moment (at the failure cross section), and I is the centroidal moment of inertia of the ring cross section. It is apparent that the size effect displays the same general trends as the size effect for the modulus of rupture (Section 9.3), and, thus, is expected to have similar properties. Indeed, the plot in Fig. 10.5.3c is a modification of Gustafsson and Hillerborg's results which uses the property that, for unnotched specimens, the size effect curve is independent of the softening when plotted as a function of D/ℓ_1 where ℓ_1 is the characteristic size associated to the initial linear softening. Gustafsson and Hillerborg performed the computations using Petersson's bilinear softening curve (Section 7.2.1), and produced plots of σ_{Nu}/f_t vs. D/ℓ_{ch}; the plot in Fig. 10.5.3c has been rescaled by taking into account that for such softening $\ell_1 = 0.6\ell_{ch}$ and in the given form can be applied to any softening curve with initially linear softening.

From the foregoing results, it follows that smaller pipes are seen to be stronger and more ductile in their postpeak response than larger pipes (D_i in the figure is the inner pipe diameter). It follows from this analysis that a size independent "modulus of rupture" currently used in design (ACI Committee 318, 1989) is unconservative for large pipes. Gustafsson and Hillerborg also observed that the ring-type failure is more size sensitive than the beam-type failure (Fig. 10.5.3c).

The failure of pipes was also studied by Bažant and Cao (1986), who considered the test results from Gustafsson (1985) and Brennan (1978). They compared the available test results to Bažant's size effect law and concluded that the size effect is strong and that Bažant's size effect law could be used. However, it must also be cautioned that the size effect law should not be fully applicable in this case, because the pipes reach their maximum load after only a small crack growth (that is, a large crack does not develop before failure). Thus, the size effect is primarily due to the formation of the fracture process zone, as characterized, for example, by the cohesive crack model or crack band model. Therefore, Bažant's size effect law might not work well if the range of sizes is increased or the scatter of measurements reduced.

10.5.6 Concrete Dams

Concrete dams typically fail by fracture. However, even though they are unreinforced, they do not fail at crack initiation. Rather, very large cracks, typically longer than one-half of the cross section, grow in a stable manner before the maximum load is reached. Therefore, if geometrically similar dams of different sizes with geometrically similar cracks are considered, a strong size effect, essentially following Bažant's size effect law [Eq. (1.4.10)], must be expected.

Even though the large aggregate size used in dams (up to 250 mm in older dams and about 75 mm in recent dams) forces the fracture process zone to be considerably larger than in normal structural concretes (with aggregates up to 30 mm in size), most dams are so large that their global failure may be, in most situations, analyzed by LEFM (Ingraffea, Linsbauer and Rossmanith 1989; Linsbauer et al. 1988a,b; Saouma, Ayari and Boggs 1989). Large cracks are often produced in dams as a result of thermal and shrinkage stresses or differential movements in the foundations and abutments, and, in an earthquake, as a result of large inertial forces and dynamic reactions from the reservoir. Cracking is often promoted by weak construction joints. Currently, the design, its computer evaluation, and analysis of seismic response, are being done on the basis of the strength theory; however, fracture mechanics should, in principle, be introduced. This is particularly needed for evaluating the performance of dams that have already developed large cracks, which is known to occur frequently. Evaluation of the effectiveness of repair methods also calls for fracture mechanics.

LEFM analysis with mixed-mode cracks was applied by Linsbauer et al. (1988a,b) to determine the profile and growth of a crack from the upstream and downstream faces of a doubly curved arched dam. On the basis of their anisotropic mixed-mode fracture analysis, Saouma, Ayari and Boggs (1989) found that the classical method of analysis is normally much more conservative than fracture analysis. This conclusion suggests that fracture analysis might not be needed to obtain safe designs, but there is an opportunity to optimize the design. The U.S. Army Corps of Engineers (1991) have issued guidelines that require applying fracture mechanics for the safety and serviceability analysis of existing cracked dams (Saouma, Broz et al. 1990). The existing computational studies considered only two dimensional cracks (Ingraffea, Linsbauer and Rossmanith 1989; Linsbauer et al. 1988a,b; Saouma, Ayari and Boggs 1989). Three-dimensional cracks (Martha et al. 1991) still need to be studied, and so does the propagation of cracks along interfaces between concrete and rock or along construction joints.

For analyzing dam fracture, the proper value of fracture energy (or fracture toughness), and of the effective length of the fracture process zone c_f, needs to be known for concretes with very large aggregates. This question was experimentally studied by Brühwiler and Wittmann (1990), Saouma, Broz et al. (1991), Bažant, He et al. (1991), and He et al. (1992). The last mentioned study utilized geometrically similar wedge-splitting fracture specimens with maximum cross section dimension 6 ft., and exploited the size effect method for determining the fracture energy of the material. The effect of moisture content and water pressure in the crack on the fracture energy was found by Saouma, Broz et al. (1991) to be important. Zhang and Karihaloo (1992) studied the stability of a large vertical crack extending from the upstream phase of a buttress-type dam. They treated concrete as a viscoelastic material, took into account tensile strain softening, and demonstrated feasibility of the fracture analysis.

Since large fractures often grow in dams slowly, over a period of many years, the effects of loading rate and duration need to be understood. These effects were studied by Bažant, He et al. (1991). Testing dam concrete as well as normal concretes, Bažant (1991a) and Bažant and Gettu (1990) observed that the slower the loading, the more brittle the response (in the sense that in the logarithmic size effect plot, the response points move to the right, i.e., closer to the LEFM asymptote, as the load duration is increased or the loading rate is decreased; see Chapter 11 for details and mathematical modeling.

One interesting question, which was provoked by Bažant (1990b) is the question of safety or the so-called "no-tension" design. It has been a widespread opinion that fracture analysis of dams can be avoided by using the so-called "no-tension" design, which is based on an elasto-plastic analysis with a yield criterion in which the tensile yield limit is zero or nearly zero (Rankine criterion or a special case of Mohr-Coulomb criterion).

However, it was demonstrated (Bažant 1991a) that such a design is not guaranteed to be safe. The stress intensity factor at the tip of a large crack that satisfies the no-tension criterion according to the elasto-plastic analysis can be, and often is, non-zero and positive. For the latter, Bažant's size effect law ought to apply, and thus it follows that, for a given crack and dam geometry and a fixed nominal stress characterizing the loading, there always exists a certain critical dam size such that for larger sizes the critical value of the stress intensity (fracture toughness) is exceeded. Examples of this have been given by Gioia, Bažant and Pohl (1992) and Bažant (1996a). The detailed study by Bažant (1996a) led to the following conclusions:

1. For a brittle (or quasi-brittle) elastic structure, the elastic-perfectly plastic analysis with a zero value of the tensile yield strength of the material is not guaranteed to be safe because it can happen that: (a) the calculated length of cracks or cracking zones corresponds to an unstable crack propagation, (b) the uncracked ligament of the cross section, available for resisting horizontal sliding due to shear loads, is predicted much too large, compared to the fracture mechanics prediction, (c) the

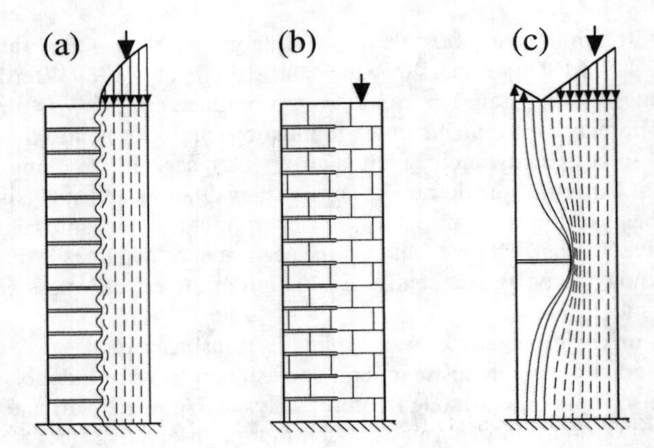

Figure 10.5.4 Crack patterns and lines of principal stress. (a) Closely spaced cracks and trajectories of minimum principal stress for no-toughness design; (b) closely spaced cracks for dry masonry; (c) approximate trajectories of minimum principal stress for $K_{Ic} > 0$. (Adapted from Bažant 1996a.)

calculated load-deflection diagram lies lower than that predicted by fracture mechanics, or (d) the load capacity for a combination of crack face pressure and loads remote from the crack front is predicted much too large, compared to the fracture mechanics prediction.

2. Due to the size effect, the preceding conclusions are true, not only for zero fracture toughness (no-toughness design), but also for finite fracture toughness, provided the structure is large enough.
3. The no-tension limit design cannot always guarantee the safety factor of the structure to have the specified minimum value. Fracture mechanics is required for that.
4. Increasing the tensile strength of the material can cause the load capacity of a brittle (or quasi-brittle) structure to decrease or even drop to zero.
5. The no-tension limit design would be correct if the tensile strength of the material were actually zero throughout the whole structure. This is true for dry masonry with sufficiently densely distributed joints, but not for concrete (or for jointed rock masses).

One simple explanation of the foregoing conclusions is that the finiteness of the tensile strength of the material at points farther away from the cracks or rock joints (or construction joints) of negligible tensile strength causes the structure to store more strain energy. Thus, energy can be released at a higher rate during crack propagation.

The reason that an increase of strength of the material from zero to a finite value causes a crack to propagate is illustrated in Fig. 10.5.4. For zero tensile strength (which is the case of dense cracking, Fig. 10.5.4a, or dry masonry, Fig. 10.5.4b), there are many cracks and the tensile principal stress trajectories are essentially straight. But for finite strength, these trajectories get compressed at the crack tip as shown in Fig. 10.5.4c, which causes stress concentration and crack propagation.

The results of the finite element study by Gioia, Bažant and Pohl (1992) are summarized in Fig. 10.5.5. The geometry of the cross section of the Koyna dam, which was stricken by an earthquake in 1967, was considered. Fig. 10.5.5a shows the finite element mesh and the shape of the critical crack for the loading considered. Finite element solutions were compared according to no-tension plasticity and according to fracture mechanics. The yield surface of no-tension plasticity was a particular case of Otossen's (1977) yield surface (described also in Chen 1982, Sec. 5.7.1) for the tensile strength approaching zero. Because the origin of the stress space must lie inside the yield surface, the calculations have actually been run for a very small but nonzero value of the tensile yield strength of concrete, approximately 10 times smaller than the realistic value. Similarly, the no-toughness design was approximated in the finite element calculations by taking the K_{Ic}-value to be approximately 10 times smaller than the realistic value. The crack length obtained by fracture mechanics is very insensitive to the K_{Ic} value because K_I represents a small difference of two large values: K_I due to water pressure minus K_I due to gravity load.

In the calculations, some of whose results are plotted in Fig. 10.5.5b-c, the height of the water overflow

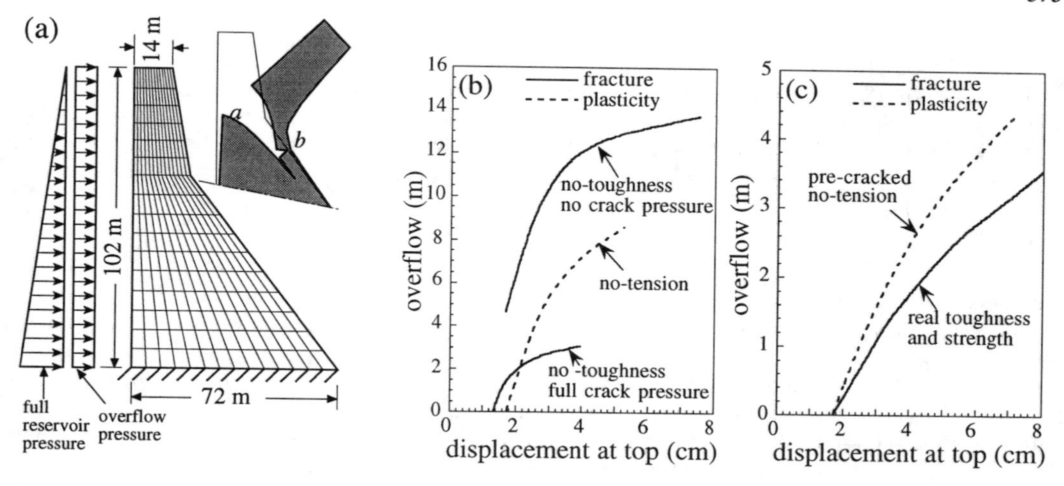

Figure 10.5.5 Koyna dam analyzed by Gioia, Bažant and Pohl (1992): (a) finite element mesh and failure mode; (b) comparison of curves of overflow height vs. deflection for no-tension plasticity analysis and no-toughness fracture analysis; (c) curves for no-tension limit analysis and fracture analysis with realistic values of strength and toughness. (Adapted from Bažant 1996a; the peak and postpeak branches obtained by Jirásek for the branched crack b are not shown.)

above the crest of the dam was considered as the load parameter. A downward curving crack, which was indicated by calculations to be the most dangerous crack, was considered (Fig. 10.5.5a).

The differences between the no-tension limit design and fracture mechanics have been found to be the most pronounced for the case when water penetrates into the crack and applies pressure on the crack faces, as shown in Fig. 10.5.5b. Because plastic analysis cannot describe crack growth, the dam has been assumed to be precracked and loaded by water pressure along the entire crack length.

From the results in Fig. 10.5.5c, it is seen that the diagram of the load parameter vs. the horizontal displacement at the top of the dam lies lower for fracture mechanics than it does for no-tension plasticity. In other words, the resistance offered by the dam to the loading by water is lower according to the fracture mechanics solution, with a realistic value of K_{Ic}, than it is according to no-tension plasticity. It should be added that, for these finite element calculations, the maximum of the load-deflection diagram could not be reached for realistic heights of overtopping of the dam. The reason has been found by Jirásek and Zimmermann (1997). A descent of the load is caused by crack branching due to the formation of a secondary crack (crack b in Fig. 10.5.5), the possibility of which was not checked by Gioia, Bažant and Pohl (1992). If this is considered, a maximum load point occurs on curves in Figs. 10.5.5b–c, and another curve descends from that point.

10.5.7 Footings

One well-documented case of fracture in a footing comes from the collapse of the New York State Thruway Bridge over the Schoharie Creek in 1987, which caused the death of 10 people. A flood produced scouring under the foundation plinth supporting a pier, which caused fracture of the plinth which was very weakly reinforced. Although crack stability and propagation were not analyzed (Wiss, Janney, Eltsner Associates, Inc., and Mueser Rutledge Consulting Engineers 1987), finite element analysis based on nonlinear fracture mechanics of discrete cracks was used by Swenson and Ingraffea (1991) who concluded that although the bridge failed primarily because of scouring beneath the foundation plinths (pier footing), a necessary complementary cause was unstable fracture in the pier. It was recommended that foundation plinth bridges of this type should be designed with consideration of crack propagation stability and crack arrest. In the disaster, the crack must have become unstable before it ceased to be small compared to the cross section size, and, therefore, the behavior described by the size effect law due to Bažant probably did not play a significant role. Rather, a boundary layer type size effect due to formation on the fracture process zone of a cohesive crack, same as in the case of the modulus of rupture, must have played a significant role.

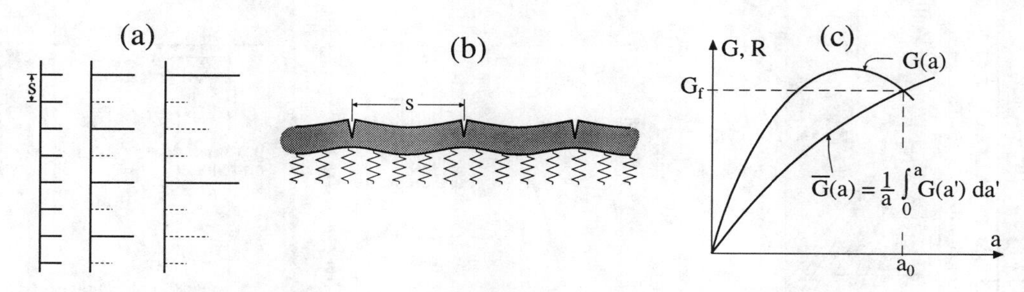

Figure 10.5.6 (a) Crack distribution at the surface of a half space. (b) Crack distribution in a pavement. (c) Determination of the initial crack jump by intersection of instantaneous and average energy release curves.

10.5.8 Crack Spacing and Width, with Application to Highway Pavements

The prediction of the spacing and opening of cracks in asphalt or concrete pavements of roads and runways is important for their durability assessment. This problem is similar to the ice plate—a plate resting on Winkler elastic foundation—but the foundation is much stiffer. Similar to the problem of sea ice penetration (Section 9.6), the crack spacing also is important for the size effect.

One basic problem is the spacing s of parallel planar cracks initiating from a half space surface (Fig. 10.5.6a), which was solved approximately by Bažant and Ohtsubo (1977) and Bažant, Ohtsubo and Aoh (1979), and rigorously by Li, Hong and Bažant (1995) (see also Bažant and Cedolin 1991, Ch. 13). The crack opening at the crack mouth is approximately $w = -s\varepsilon^0$, where ε^0 is the free shrinkage strain or thermal (cooling) strain ($\varepsilon^0 < 0$).

The problem of crack spacing in pavements has been solved according to the theory of plate (beam) on Winkler elastic foundation (Fig. 10.5.6b) by Hong, Li and Bažant (1997). The calculated values of crack spacing were in relatively good agreement with the previously reported observations on asphalt concrete pavements.

The theory of initiation of parallel equidistant cracks from a smooth surface, developed in Li and Bažant (1994b) as an extension of the approximate crack spacing criterion proposed by Bažant and Ohtsubo (1977) and Bažant, Ohtsubo and Aoh (1979) (see also Bažant and Cedolin 1991, Ch. 13), was applied in the aforementioned study. Although the strength concept must be applied for the crack initiation stage, the cracks are considered simply as LEFM cracks afterward. The theory, which was studied rigorously in Li, Hong and Bažant (1995), rests on the following three conditions:

1. Just before crack initiation from a smooth surface, the stress at the surface reaches the material strength limit, f_t'.
2. After initial cracks of a certain initial length a_0 form (by a dynamic jump), the energy release rate must be equal to the fracture energy of the material or the R-curve value.
3. The total energy release due to the initial crack jump must be equal to the energy needed to produce the initial cracks, according to the fracture energy G_f or the R-curve (an equivalent statement is that the average of energy release rate during the initial crack formation must be equal to the value of the fracture energy G_f or the average value of the R-curve, as illustrated in Fig. 10.5.6c).

The problem can be solved if the stress intensity factor (or energy release rate) as a function of the crack length, the crack spacing and the load parameter (e.g., the penetration depth of the cooling or drying front) is known. For the elastic halfspace, the stress intensity factor has been solved from a Cauchy integral equation (Li, Hong and Bažant 1995). The solution of conditions 2 and 3 graphically represents the intersection of the curves giving the energy release rate and the average energy release rate (the intersection always exists if the fracture geometry is, or becomes, positive); see Fig. 10.5.6c. All three conditions together allow solving three unknowns: the initial crack spacings, the initial crack length, and the load level (load parameter) at which the cracks initiate. Generalization to the full cohesive crack model is possible.

A different basic problem is how a system of parallel cracks evolves after it has initiated. Often it happens that every other crack stops growing and closes when a certain critical length a_{cr} is reached

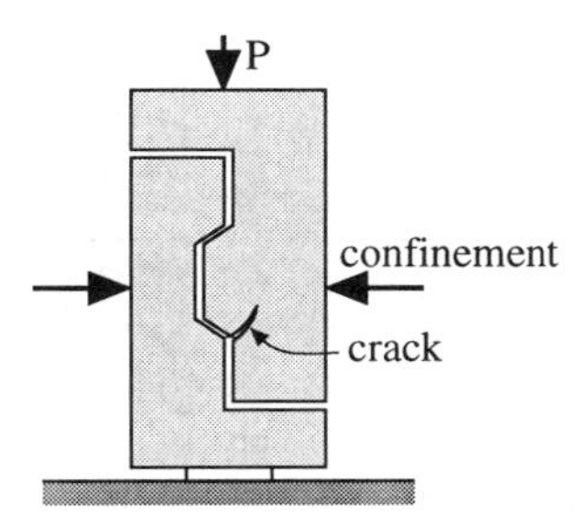

Figure 10.5.7 Specimen used in studying cracking in keyed joints in segmental box girder sections (after Buyukozturk, Bakhoum and Beattie 1990).

(Fig. 10.5.6a). The value of a_{cr} is decided by stability and bifurcation analysis of the interacting crack system (Bažant and Ohtsubo 1977; Bažant, Ohtsubo and Aoh 1979; Bažant and Cedolin 1991, Ch. 13; Bažant and Wahab 1979, 1980). The increase of spacing of the opened crack causes their opening width (due to shrinkage or strain) to increase. Although this problem has been analyzed only two-dimensionally so far, the crack pattern viewed orthogonally to the surface of halfspace is often hexagonal or random, calling for three-dimensional analysis.

10.5.9 Keyed Joints

Rectangular sheared keys are used to improve the resistance against shear slip of joints between the segments of prestressed box-girder bridges. Buyukozturk and Lee (1992a) showed that this is a very brittle type of failure, exhibiting a strong size effect close to LEFM. They used LEFM mixed-mode fracture analysis (Swartz and Taha 1990, 1991) to study the failure of typical shear keys used in bridge construction (Fig. 10.5.7). In contrast to the diagonal shear cracks in beams, which can be counteracted by shear reinforcement (stirrups), a diagonal crack which is initiated at the shear key may also propagate parallel to the joint. Such a path is not crossed by any shear reinforcement (Buyukozturk, Bakhoum and Beattie 1990).

The design provisions of the Post-Tensioning Institute (1988) for the segments of prestressed box-girder bridges are at present empirical and follow the strength theory, exhibiting no size effect. They are based on the shear capacities determined by tests of prestressed beams failing by flexure-shear cracks or web-shear cracks. An enhanced formula was proposed by Buyukozturk, Bakhoum and Beattie (1990), but this was still free from size effect.

Based on their mixed-mode LEFM analysis, Buyukozturk and Lee calculated design charts corresponding to the failure criterion

$$C_k^2 K_I^2 + K_{II}^2 = K_{Ic}^2 \tag{10.5.11}$$

where K_I, K_{II} = stress intensity factors in Mode I and Mode II, K_{Ic} = Mode I fracture toughness of concrete, and C_k = empirical constant (which obviously represents the ratio of Mode II fracture toughness to K_{Ic}). The high brittleness of failure is further compounded by the use of high strength concrete in these bridges. Another aggravating factor for brittleness is the presence of large uniaxial compressive stresses normal to the joint, which are beneficial by increasing friction but detrimental increasing the brittleness. Thus, even though this relatively small size of the shear keys would indicate the use of nonlinear fracture mechanics with a transitional size effect (following the size effect law), it appears that LEFM is applicable.

10.5.10 Fracture in Joints

Cracks in joints differ from cracks in bulk material in three respects. **(1)** Cracks in concrete usually (albeit probably not always) propagate in the direction normal to the maximum principal stress as in Mode I, but a crack in a joint is subjected to normal as well as shear loading and is of a mixed-mode. **(2)** The roughness of a crack in a smooth joint can be much smaller than the roughness of a crack in the bulk of concrete. Thus, aggregate interlock plays a lesser role and friction dominates as a means of transferring

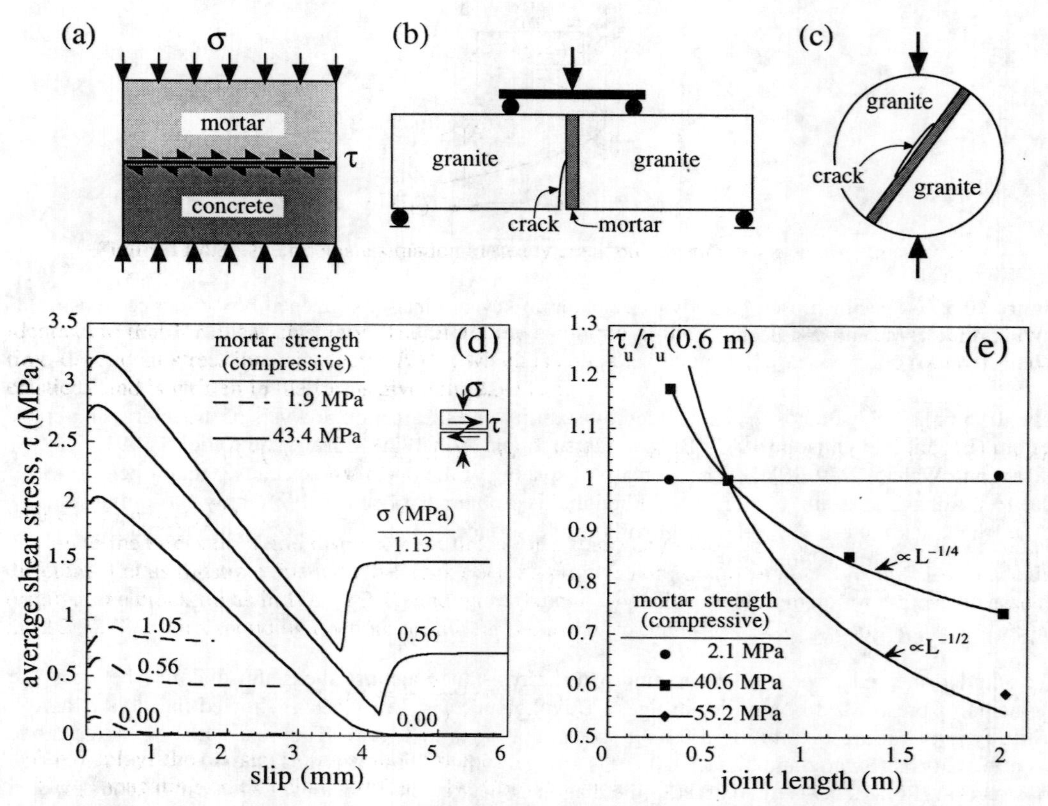

Figure 10.5.8 Fracture of joints: (a) sketch of tests by Reinhardt (1982); (b–c) tests by Buyukozturk and Lee (1992b); (d) average shear stress-slip curves for Reinhardt's tests; (e) size effect in Reinhardt's tests.

shear stresses across a joint. (**3**) A crack in concrete exhibits considerable dilatancy associated with shear slip, but in a smooth joint, the dilatancy may be quite small.

The behavior of joints of dissimilar materials was investigated experimentally by Reinhardt (1982) and by Buyukozturk and Lee (1992b). Reinhardt tested joints of strong concrete and mortar of variable strength subjected to various compressive normal stresses (Fig. 10.5.8a). Buyukozturk and Lee tested sandwich specimens of granite and mortar with an interfacial crack (Fig. 10.5.8a) as a means of characterizing the aggregate-mortar interface, although their results could be useful for macroscopic structures as well.

Fig. 10.5.8d shows shear stress-slip curves for the same joint length, two different mortar strengths and, in each case, three different compressive normal stresses across the joint. It is seen that the stress first rises abruptly with very little slip up to a certain maximum and then, for the joints made in high strength mortar, a steep drop of stress follows, while for the joints made in low strength mortar only a mild drop is seen. After the development of a full crack, a frictional plateau gets established with the residual shear capacity determined by the compressive stress across the joint. This capacity does not depend on the slip magnitude, nor on the mortar strength. (It might be noted, though, that the response shown could have been influenced by the stiffness of the loading frame as well as the response frequency of the servo-controller.)

A plot of the normalized shear strength of the three different joints vs. the length of the joint is shown in Fig. 10.5.8e. The joint made with low strength mortar was found to exhibit no dependence of the shear strength on the joint length, i.e., no size effect. On the other hand, the joint made with a mortar of high strength exhibited a strong size effect close to LEFM (in this case, the joint strength decreased as $L^{-1/2}$, with l = joint length). The joint made with a mortar of medium strength was found to exhibit an intermediate size effect.

The interfacial crack propagation was interpreted by Reinhardt (1982) on the basis of the LEFM

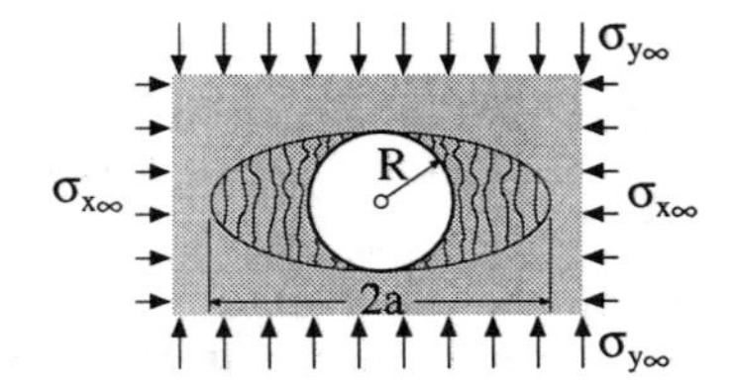

Figure 10.5.9 Break-out of boreholes; $|\sigma_{y\infty}| \gg |\sigma_{x\infty}|$.

solutions of Sih (1973), Rice and Sih (1965), and Erdogan (1963) for particular geometries of cracks at the interface of two dissimilar halfspaces. Buyukozturk and Lee (1992b) also interpreted their tests in terms of LEFM, based on solutions by Suo and Hutchinson (1989). The LEFM treatment of interfacial crack theory is outside the scope of this book; it is a conceptually involved topic, because the power series expansion that is relatively simple to handle for the single material problem (Section 4.3) has complex exponents for the bimaterial case. The dominant solution of the displacement field still decays as $r^{1/2}$ near the tip, but displays an oscillating behavior near the origin. For example, it can be shown that the dominant term for the crack opening at a distance r from the crack tip takes the form

$$w(r) \propto \sqrt{r}\ \cos(\phi + \omega \ln r) \tag{10.5.12}$$

where ϕ and ω depend on the loading, geometry, and elastic properties of the two materials; $\omega = 0$ if the two bodies have identical elastic constants. Note that when r approaches zero (the crack tip), the factor $\cos(\phi + \omega \ln r)$ oscillates between -1 and 1 with frequency tending to infinity (because $\lim_{r\to 0}(\ln r) = -\infty$). This means that the solution always includes negative crack openings, which are not physically admissible (there would be interpenetration of the opposite faces of the crack). This is but one of the difficulties involved in the interfacial crack problem. The interested reader is referred to the original papers and to recent papers by Rice (1988), Hutchinson (1990), and He and Hutchinson (1989).

10.5.11 Break-Out of Boreholes

When the mass of rock (or concrete) in which a borehole has been drilled is subjected to large compressive stresses, it may suddenly collapse in a brittle manner. This type of failure is called the break-out. The classical approach to the break-out has been by plasticity. However, because the failure occurs by cracking, fracture mechanics appears to be more appropriate. Its use, of course, inevitably leads to size effect, which is known to occur in the break-out of boreholes in rock, as experimentally demonstrated by Nesetova and Lajtai (1973), Carter (1992), Carter, Lajtai and Yuan (1992), Yuan, Lajtai and Ayari (1993), and Haimson and Herrick (1989).

An approximate energy-based analytical solution of the break-out has been obtained (Bažant, Lin and Lippmann 1993) under the simplifying assumption that the splitting cracks occupy a growing elliptical zone as sketched in Fig. 10.5.9 (although in reality this zone is narrower and closer to a triangle). The assumption of an elliptical boundary permits the energy release from the surrounding infinite solid to be easily calculated (Bažant, Lin and Lippmann 1993) according to Eshelby's (1956) theorem for uniform eigenstrains in ellipsoidal inclusions in infinite medium. According to the theorem (see, e.g., Mura 1987), the following approximate expression for the energy release from the infinite rock mass has been derived:

$$\Delta\mathcal{U} = -\frac{\pi(1-\nu^2)}{2E}[(a+2R)R\sigma_{x\infty}^2 + (2a+R)a\sigma_{y\infty}^2 - 2aR\sigma_{x\infty}\sigma_{y\infty} - 2a^2\sigma_{cr}^2] \tag{10.5.13}$$

in which R = borehole radius, a = principal axis of the ellipse (Fig. 10.5.9), $\sigma_{x\infty}$ and $\sigma_{y\infty}$ = remote principal stresses, E = Young's modulus of the rock, and ν = Poisson ratio. A similar analysis as that for the propagating band of axial splitting cracks, already explained in Section 9.5, has provided a break-out stress formula of the type

$$\sigma_{Nu} \approx C_0 D^{-2/5} + C_1 \tag{10.5.14}$$

where C_0 and C_1 are constants.

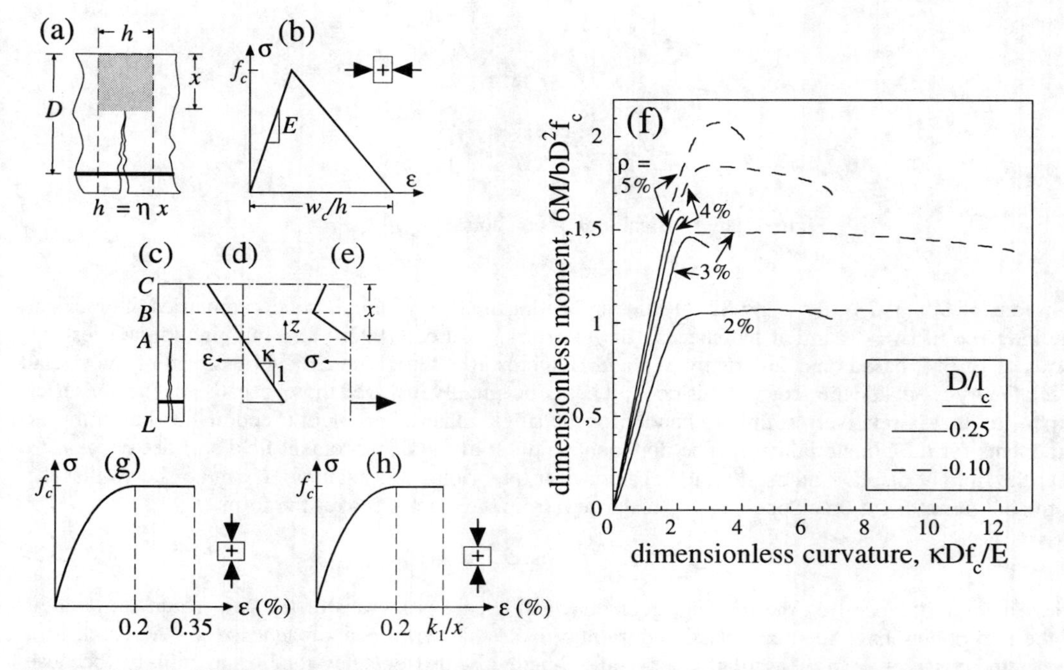

Figure 10.5.10 Hillerborg's (1990) analysis of bending of reinforced concrete beam: (a) sketch of the failure cross section; (b) stress-strain curve in compression; (c) no-tension zone (L–A), elastic zone (A–B), and softening zone (B–C); (d) strain profile; (e) stress profile; (f) moment-curvature diagrams for various sizes; (g) stress-strain curve in compression according to the CEB-FIP Model Code; (h) stress-strain curve with size-dependent cut-off proposed by Hillerborg (1990).

10.5.12 Hillerborg's Model for Compressive Failure in Concrete Beams

Hillerborg's (1990) model for compressive failure in concrete beams follows, in the formal aspects, the classical bending theory for concrete: a uniaxial stress-strain relation with plane cross sections remaining plane and no-tension for concrete. The essential difference is that he introduces softening and strain localization in compression to explain the size effect on ductility.

Fig. 10.5.10a shows the central section of the beam where the inelastic behavior is represented by a no-tension crack and a compressed zone (shaded in the figure) of width h into which the strain will localize. Hillerborg assumes that

$$h = \eta x \tag{10.5.15}$$

as indicated in the figure, where x is the depth of the compressed zone, and η is a constant (approximately equal to 0.8). Hillerborg further assumes linear softening expressed by a stress-displacement $\sigma(w)$ curve, where w here has the meaning of an inelastic displacement in compression, equivalent to the crack opening in tension. Fig. 10.5.10b shows the corresponding stress-strain curve. Note that the slope depends on the depth of the compression zone x, hence also on the size of the beam. w_c is assumed to be a material property, and so is the compressive strength f_c and the elastic modulus E.

According to these hypotheses, the beam depth (Fig 10.5.10c) is divided into three parts: over part LA no stress is transferred (except across the reinforcement), over part AB the concrete is compressed and elastic, and over BC the concrete undergoes crushing and strain localization. The strain is assumed to vary linearly as shown in Fig. 10.5.10d, so that

$$\varepsilon = \kappa z \tag{10.5.16}$$

where κ is the curvature. From this, together with the stress-strain curves of the steel and of the concrete already defined, the stress profile can be computed as a function of the position of the neutral line x and the curvature κ, as sketched in Fig. 10.5.10e. Then x is computed from any given κ from the equilibrium of forces, and next the bending moment is computed from the equilibrium of moments. In this way, the full moment-curvature diagram is obtained.

The essential feature is that, since the stress-strain curve of concrete is made to depend on the depth of the compression zone, the resulting moment-curvature diagrams are size-dependent. More specifically, they depend on the dimensionless size D^* defined as

$$D^* = \frac{D}{\ell_c}, \qquad \ell_c = \frac{w_c E}{2 f_c} \tag{10.5.17}$$

ℓ_c represents the characteristic material length for fracture in compression. Fig. 10.5.10f illustrates Hillerborg's results, which clearly display an increase of brittleness with the size and with an increasing steel ratio.

Certainly this model is crude, but offers one simple way of taking into account the softening behavior in compression to predict a size-dependent response of concrete in bending. This is not actually included in the codes, which take size-independent stress-strain curves for concrete in compression, such as the parabola-rectangle diagram of the CEB-FIP Model Code shown in Fig. 10.5.10g. Hillerborg suggests a simple way of using this kind of diagrams to include the size effect: the strain cut-off ε_u is made to depend on the depth of the compression zone as

$$\varepsilon_u = \frac{k_1}{x} \tag{10.5.18}$$

where k_1 is a parameter, with dimensions of length, which includes the fracture properties in compression —roughly proportional to ℓ_c in (10.5.17)— and which might, eventually, depend on the geometrical details of the beam. However, further research is required, both on the experimental and theoretical sides, to settle on the best model that should go into the code provisions.

11

Effect of Time, Environment, and Fatigue

In the previous chapters, fracture has been treated as if it were independent of time and of environmental factors. This is, of course, only an approximation. In reality, the fracture behavior depends on the rate of loading, as well as on temperature and, for concrete, on moisture content. Moreover, since fracture is inherently nonlinear, the loading history affects the observed behavior, and the response to cyclic loading is different from the response under monotonic loading, which has been the subject of the previous chapters.

The time, temperature, and moisture dependence may be included in various ways. The simplest way is to include them as parameters in the time-independent formulation. Thus, the classical time-independent elastic and fracture parameters (E, K_{IC}, G_F, softening curve, etc.) are simply considered to depend on the rate, temperature, moisture content, or number of cycles. This may be a useful expedient for some kind of loading histories in which the rates and temperatures are kept within a narrow range (close to constant rate and isothermal conditions). However, extrapolation to widely variable loading rates and nonisothermal conditions is very difficult. More detailed models are required for these kinds of situations. An overview of such models, with emphasis placed on concrete, is the object of this chapter.

Most of the chapter is devoted to time-dependent fracture behavior. Section 11.1 discusses the main sources of time-dependence and explores the available evidence of the influence of time on the fracture of concrete.

Section 11.2 deals with the activation energy theory and the rate processes, which describes time-dependent fracture by means of a relationship between the crack growth rate and the stress intensity factor. This is the simplest approach to time-dependent fracture and assumes that the fracture zone is very small and that the bulk of the material behaves linearly and elastically.

Section 11.3 discusses the applications of the rate theory to the fracture of concrete, including the effect of temperature and moisture content. It also presents an extension of the R-curve concept to time-dependent processes. The essential features of the viscoelastic solutions are considered, both for vanishingly small fracture process zone and for certain simple solutions of time-independent cohesive cracks. The foundations for time-dependent R-curves and cohesive cracks are presented and discussed. Finally, generalizations of the cohesive crack model and distributed (smeared) cracking models are also explained.

Section 11.4 briefly introduces linear viscoelastic fracture mechanics. The essential concepts in viscoelasticity are first introduced, with special attention to the creep functions for concrete. Then the stress and displacement fields for a running crack are analyzed and the crack growth resistance is examined, first from a phenomenological point of view and then based on a time-independent cohesive zone model with rectangular softening. The general approach to crack growth analysis in a viscoelastic plate closes this section.

Section 11.5 describes the rate-dependent R-curve model with creep developed for concrete by Bažant and Jirásek (1993), which explains many of the experimental aspects of the fracture of concrete. However, the explanation of all the main aspects of time-dependent fracture requires using models that take into account the finite size of the fracture process zone. Rate-dependent cohesive crack and crack band models have been developed to analyze in detail the evolution of time-dependent fracture. These relatively sophisticated models are discussed in Section 11.6, which closes the presentation of the available tools for time-dependent fracture analysis.

Section 11.7, the last of the chapter is devoted to fatigue. Fatigue fracture has been studied to a great depth in the field of metals and composites, since it plays an essential role in the safety and lifetime of machines, aircraft, ships, etc. For concrete, however, fatigue has been little studied. In concrete, though, fatigue phenomena are of lesser importance. This is explained by the lack of plasticity, especially the

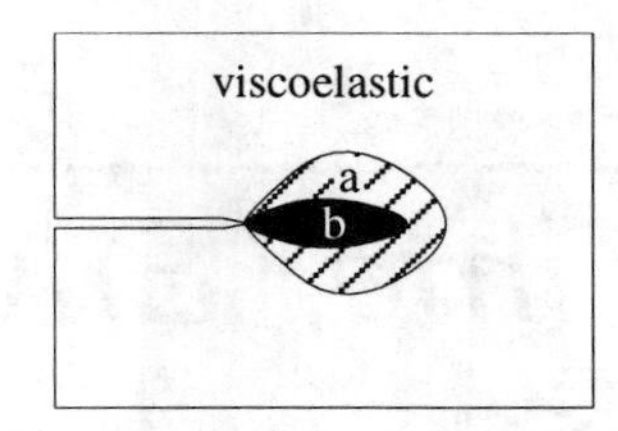

Figure 11.1.1 Time dependent behavior of the zones surrounding a growing crack tip: (a) nonlinear hardening creep; (b) rate-dependent softening.

lack of cyclic plastic hysteretic loops that cause buildup of self equilibrating residual microstresses near fracture tips, but recently it has been recognized that, in some cases, fatigue growth of fracture in concrete may be very important, too.

11.1 Phenomenology of Time-Dependent Fracture

In the previous chapters, we have specifically assumed that the mechanical behavior of the materials was time-independent. This is, of course, not true; rather, all kinds of mechanical responses are time dependent. This section first discusses the types of time-dependent behavior involved in crack growth, and then presents a number of experimental results illustrating the rate effect on the fracture process in concrete.

11.1.1 Types of Time-Dependent Fracture

The picture given in Chapter 5 of the various zones surrounding the crack tip can be generalized to bring time-dependence into play. The generalization of the three zones depicted in Fig. 5.2.1 is shown in Fig. 11.1.1. The outermost zone is assumed to be linear, but time dependent: it is linear viscoelastic. The intermediate zone is time-dependent, displaying nonlinear hardening. The innermost zone is the fracture process zone in which the material is getting disrupted and softens; in general, the fracture process will also be time-dependent and thus we face time-dependent fracture growth.

Despite the fact that all three kinds of mechanical rate-dependent zones are present, their relative importance can be quite variable. For example, in most metals, the contributions of the viscoelastic and fracturing zones to the time-dependent response are negligible compared to the contribution of the nonlinear hardening creep zone (which in metals is viscoplastic). In such a case, the outermost zone can be assumed to be linear elastic, and the fracture zone very small and completely controlled by the surrounding viscoplastic zone. This kind of situation is the subject of the theory of *creep fracture*, which is outside of the scope of this book. The interested reader may find the basic analysis of the topic in Chapter 7 of Kanninen and Popelar (1985) and in a review by Riedel (1989).

For quasi-brittle materials, defined as those in which the plastic (or viscoplastic) zone is very small, three possibilities must be envisaged, apart from that in which time-dependence can be neglected altogether:

1. The outer zone is approximately linear elastic (time-independent) and the fracture process is time-dependent.
2. The outer zone is linear viscoelastic and the fracture process is time-independent.
3. Both the outer zone and the fracture zone are time-dependent.

This division is made for methodological convenience, because the theories used to describe the first two cases follow rather different approaches. When coupled together, these theories provide the tools for the description of the third, general case.

Apart from the division implied by the possible combinations of the various kinds of behavior, there is also the dichotomy introduced by the size of the fracture process zone. For a very small time-dependent fracture process zone in a linear material, the rate-theory can be applied to relate the crack growth rate $\dot{a}$

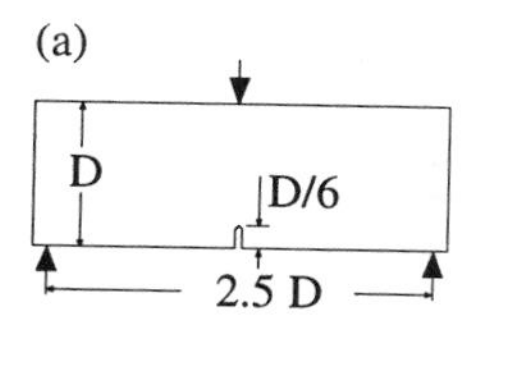

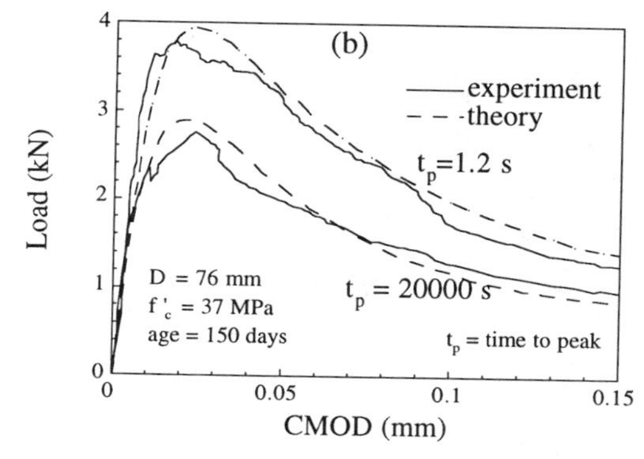

Figure 11.1.2 (a) Specimen geometry used by Bažant and Gettu (1992). (b) Comparison of load-CMOD curves for two CMOD rates differing by 4 orders of magnitude (after Bažant and Gettu 1992). Dashed lines correspond to theoretical predictions of Wu and Bažant (1993).

to the crack driving force $\mathcal{G}$, or to K_I; this approach consistently incorporates the effect of temperature and moisture content and will be developed in Section 11.2 in its general aspects and in Section 11.3 for the specific application to concrete.

When the fracture process zone is very small and is surrounded by a viscoelastic material, the crack growth turns out to be also rate-dependent, even if the fracture process is fully rate-independent. Section 11.4 is devoted to the basic theory of viscoelastic fracture, and Section 11.5 to its application to concrete.

When the fracture process zone is large, the problem must be handled numerically. This requires a specific time-dependent model for the fracture process zone, which may be coupled to viscoelastic behavior of the material outside the fracture process zone. Some models, including cohesive cracks in the time-dependent framework, will be discussed in Section 11.6.

Before undertaking the description of the models, a short account of the available experimental evidence for time-dependent behavior of concrete is given in the following paragraphs.

11.1.2 Influence of Loading Rate on Peak Load and on Size Effect

The presence of rate effect in concrete fracture is clearly demonstrated by the experimental results reported in Bažant and Gettu (1992). These experiments dealt with simultaneous rate and size effects for three-point-bend concrete fracture specimens (Fig. 11.1.2a). Each experiment was performed under a constant CMOD rate. Fig. 11.1.2b compares two load-CMOD curves for two very different rates of loading, giving a time to peak $t_p = 1.2$ s for the fastest test, and $t_p = 20000$ s (5.6 hr) for the slowest. The rate effect is self-evident.

Specimens of three different sizes ($D = 38$mm, 76mm, 152mm) and CMOD rates ranging from 4×10^{-11} m/s to 10^{-5} m/s were used, with the corresponding times to peak ranging from 3 days to 1 second (for a table of the measured peak loads, see Bažant and Gettu 1992). Fig. 11.1.3 shows the effect of the loading rate on the peak load for each size. The lines correspond to the theoretical model of Bažant and Jirásek (1993) to be discussed later. Fig. 11.1.4 shows the log-log size effect plot, in which the full line represents Bažant's size effect equation analyzed in Chapters 1 and 6. As is clear from Fig. 11.1.3, the measured values of the peak load suffer from considerable scatter, which could be explained by the fact that the specimens were cast from several batches of concrete. Nevertheless, some general trends can still be observed: (1) the peak loads increase with increasing rate of loading; (2) the rate dependence of the peak loads is stronger for large specimens than for small ones; (3) the nominal strength decreases with increasing size, approximately following the size effect law proposed by Bažant (1984a); (4) the size effect on the peak loads is stronger for slow loading rates than for fast ones; and (5) a decrease of loading rate causes a shift towards more brittle behavior, as is clear from Fig. 11.1.4, where the points corresponding to slower rates lie closer to the LEFM asymptote.

The latter point —which implies a rate dependence of D_0 in Bažant's size effect law— is not easy to model. A model based on a rate equation for the crack growth, which includes the R-curve effect and

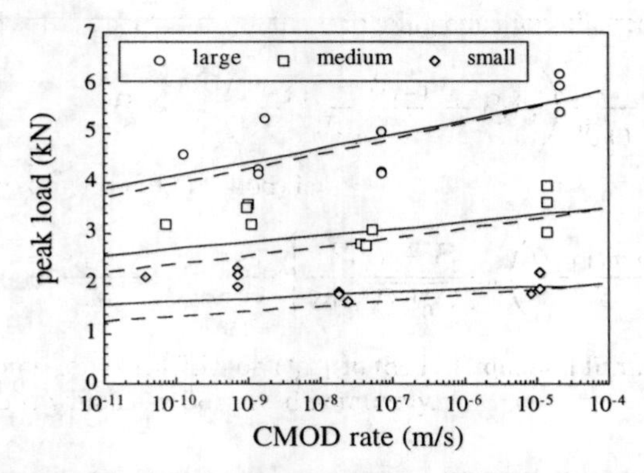

Figure 11.1.3 Effect of loading rate and specimen size on the peak load (after Bažant and Gettu 1992). Full and dashed lines correspond to theoretical predictions by Bažant and Jirásek (1993).

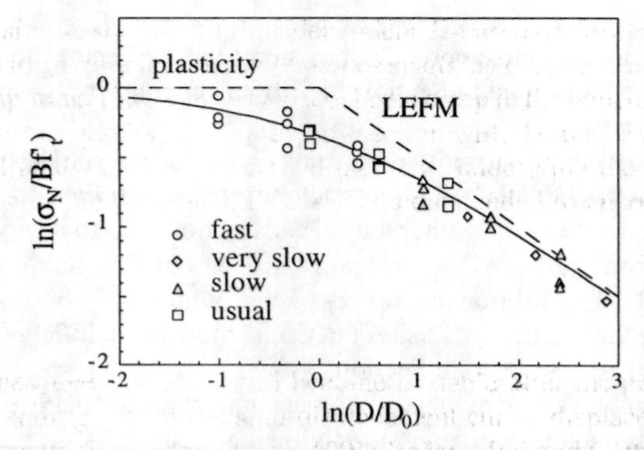

Figure 11.1.4 Log–log size effect plot showing the rate-dependent shift of brittleness (after Bažant and Gettu 1992).

creep of concrete, was used by Bažant and Jirásek to analyze the foregoing results. This model, which will be introduced in Section 11.3.3 and further developed in Section 11.5, predicts the rate-dependence shown as the dashed lines in Fig. 11.1.3. As can be observed, the model predicts too strong a rate dependence for small sizes in concrete; this is contrary to what is observed for limestone, for which the model gives very good predictions (see Section 11.3.3).

The model was modified to include an explicit rate-dependence in the R-curve by making the size of the critical crack extension depend on the crack growth rate (Section 11.5). With this, the predictions are enhanced, but they still give too strong a rate effect for small sizes (full line in Fig. 11.1.3).

The difficulty in modeling the observed behavior by means of rate-dependent effective crack models is probably due to the fracture process zone size being too large for this kind of model to be applicable to specimens of small sizes. This means that the behavior of the fracture process zone must be modeled. A model of this kind, which includes a time dependent cohesive zone and takes into account the viscoelastic creep of concrete, has been put forward by Wu and Bažant (1993), Bažant and Li (1997), and Li and Bažant (1997), and will be discussed in Section 11.6.

11.1.3 Load Relaxation

Another type of experiment illuminating the rate effect are the relaxation tests. The CMOD rate of loading of three-point bend specimens was first held constant and, at a certain moment, was suddenly decreased to zero (Bažant and Gettu 1992). The specimens were as in the previous tests (Fig. 11.1.2a) with a depth of 76 mm.

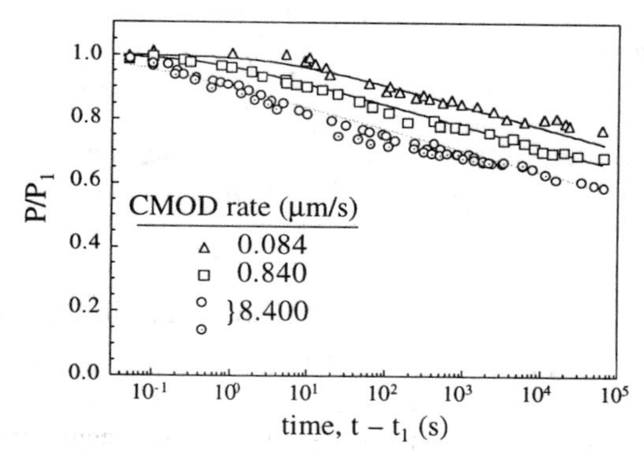

Figure 11.1.5 Load relaxation for various initial loading rates (after Bažant and Gettu 1992).

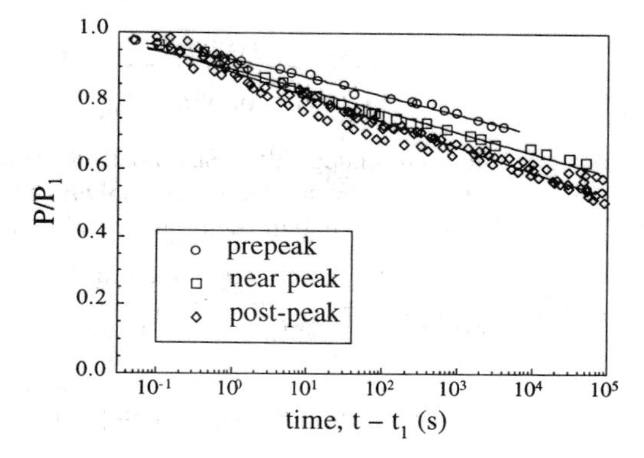

Figure 11.1.6 Load relaxation for initial loads in the prepeak, near to the peak and in the postpeak (after Bažant and Gettu 1992).

In the first series of experiments, the initial rates were different and relaxation started in the postpeak range at about 85 percent of the peak load. Denoting the time at which relaxation started by t_1 and the corresponding load by P_1, one can plot the relaxation curves $P(t)/P_1$ vs. $t - t_1$. The relaxation curves are shown in Fig. 11.1.5. The curves corresponding to different initial rates have the same final slope in a logarithmic plot and are shifted with respect to each other. The curves shown correspond to the fits by the semiempirical equation (Bažant and Gettu 1992)

$$\frac{P(t)}{P_1} = 1 - A \ \ln \left[1 + B(t - t_1)^N\right] \tag{11.1.1}$$

The second series of experiments was conducted with the same initial rate ($8.5\ 10^{-6}$ m/s) but relaxation started at different stages — in the prepeak range, at peak, and at different load levels in the postpeak range. The results showed that the relaxation is less for initial loading in the prepeak, while for the postpeak, the relaxation is only mildly dependent on the load level. Fig. 11.1.6 shows the results for initial loading in the prepeak, close to the peak and in the postpeak. Also shown is the purely viscoelastic relaxation function (determined on unnotched specimens). It appears that the purely viscoelastic relaxation is a substantial part of the total relaxation, which means that both time-dependent fracture and viscoelastic creep must be included in modeling the rate-dependent behavior of concrete.

We shall see in Section 11.5 that the Bažant and Jirásek's model can describe qualitatively this behavior, but that to get an accurate prediction, a detailed modeling of the fracture zone is required, in the manner of Wu and Bažant (1993); Section 11.6.

Relaxation tests in tension have been performed at the University of Lund (Zhou and Hillerborg 1992; Zhou 1992). The tests were performed on cylinder specimens 64 mm in diameter and 60 mm in length,

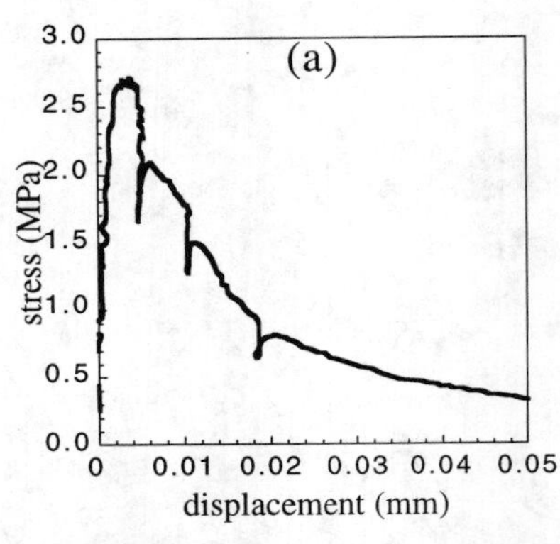

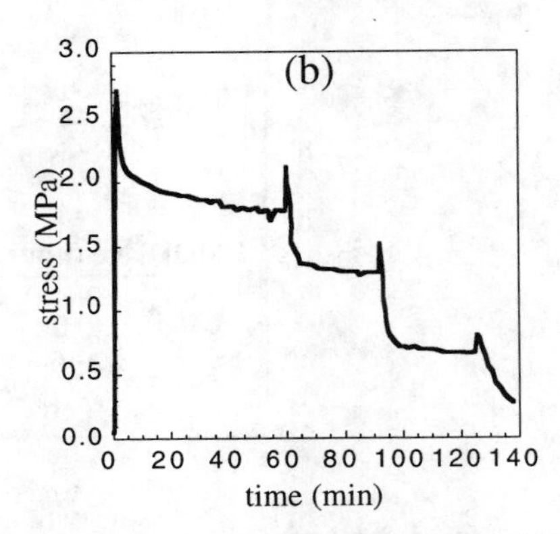

Figure 11.1.7 Tensile relaxation tests on notched cylinders of Zhou and Hillerborg (1992). (a) Stress vs. displacement (notch mouth opening). (b) Stress vs. time.

with 12 mm depth circumferential notch. In these tests, the displacement (notch mouth opening) was controlled. The displacement was increased at constant rate up to a point just after the peak and then the displacement was held constant for 60 min while the load relaxed (Fig. 11.1.7). After this hold time, the displacement was increased again up to a second hold point at which the displacement was kept constant for 30 min. This sequence was repeated a third time, as shown in Fig. 11.1.7. Note the fast rise in stress upon resuming stretching at the end of each holding period. The model of Zhou and Hillerborg (1992) based on these results will be discussed in Section 11.6.

11.1.4 Creep Fracture Tests

If a notched beam is subjected to a long-time constant load, a crack may slowly grow and the material creeps. Zhou and Hillerborg (1992) and Zhou (1992,1993) performed three-point bending creep tests on notched concrete beams $100 \times 100 \times 800$ mm (thickness $\times$ depth $\times$ loading span), with a notch-to-depth ratio of 0.5. Fig. 11.1.8a shows a typical creep curve in which the CMOD rate first decreases (primary creep), goes through a constant-rate zone (secondary creep), and then starts to increase at an accelerated rate up to failure (tertiary creep). The experimental results are shown by the data points. The dashed line is the theoretical prediction of the model of Zhou and Hillerborg (1992), which will be described in Section 11.6.

The data points in Fig. 11.1.8b represent the experimental results of Zhou (1992,1993) for the stress-lifetime curve. The solid line is the power-law regression line, and the dashed line shows the theoretical prediction of the model of Zhou and Hillerborg (1992).

11.1.5 Sudden Change of Loading Rate

Still another time-dependent effect is seen after a sudden change of loading rate. In the tests of Bažant, Gu and Faber (1995), the initial CMOD rate was held constant up to a certain point in the postpeak range. After the load decreased from its peak value P_p to some lower value P_1, the CMOD rate was suddenly increased or decreased by several orders of magnitude and the test continued with the new value of a constant CMOD rate. This resulted into a sudden change of slope of the load-CMOD diagram. For a sufficiently large increase of the loading rate, the load started increasing again and a second peak P_2 could be observed (Fig. 11.1.9, curve A). Furthermore, as expected, a decrease of the loading rate was followed by a fast drop in the slope of the load-CMOD curve (Fig. 11.1.9, curve B.).

Again, the Bažant-Jirásek model to be described in Sections 11.3.3 and 11.5 can qualitatively describe this behavior, but for an accurate prediction the model must include the analysis of the fracture process zone and the viscoelastic behavior of concrete (Section 11.6).

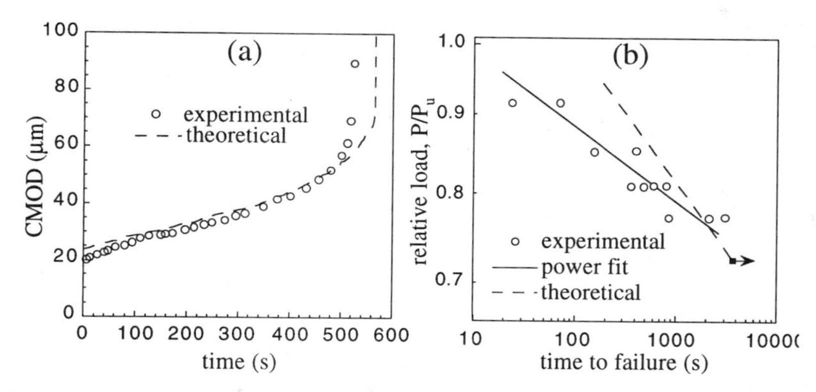

Figure 11.1.8 Creep tests on notched beams: (a) CMOD-creep results of Zhou and Hillerborg (1992); (b) stress-lifetime results (after Zhou 1992). Dashed lines are theoretical predictions of the model of Zhou and Hillerborg (Section 11.6).

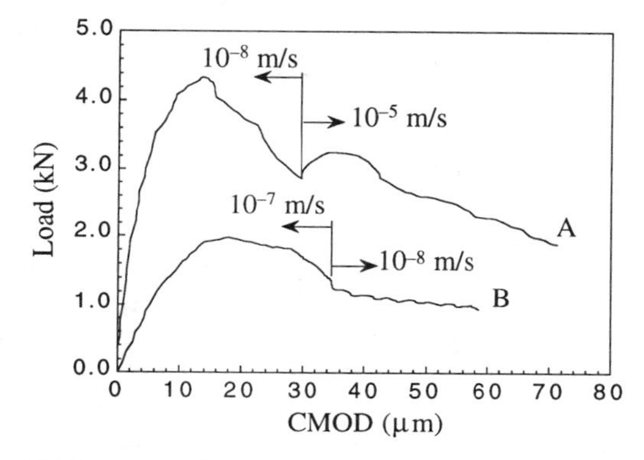

Figure 11.1.9 Load-CMOD curves with a sudden change of rate (after Bažant, Gu and Faber 1995. Curves A and B correspond to specimens of depths 76 and 38 mm, respectively.

11.1.6 Dynamic Fracture

Dynamic fracture of materials, concrete included, is a complex subject whose detailed treatment is beyond the scope of this book. For comprehensive information on the current state of knowledge in dynamic fracture, albeit focused mainly on metals and with no particular attention to concrete, the reader may consult the excellent book by L.B. Freund (1990). Valuable state-of-art reviews also exist (e.g., ACI Committee 446, subcommittee IV (Chaired by J. Isenberg), Dynamic Fracture, State-of-Art Report in preparation; Mihashi and Wittmann 1980; Mindess and Shah 1986). In this section, we will content ourselves with a brief discussion of the essentials.

Dynamic fracture is the process of fracture in which the inertia effects with the associated wave propagation phenomena are important. In elastic materials, the maximum possible crack propagation velocity is equal to the velocity of Rayleigh surface waves (which is about 0.6 of the pressure wave velocity). This classical result was established by Stroh (1957). In the dynamic tests of concrete, however, the observed crack velocities have typically been of the order of 100 m/s, which is about 10% of the Rayleigh wave velocity. For such crack propagation rates, the inertia effects become unimportant. Wave propagation analysis is nevertheless important if the loading is very fast, as in impact or blast, and in that case, the wave diffraction at the crack tip needs to be carefully analyzed.

Aside from the inertia effects, two other physical phenomena, which have already been discussed, play an essential role in dynamic fracture: (1) dependence of the fracture model on either the velocity of the crack tip or the rate of crack opening, which is explained principally by the activation energy theory for the rate process of rupture of interatomic or intermolecular bonds, and (2) viscoelastic or viscoplastic effects

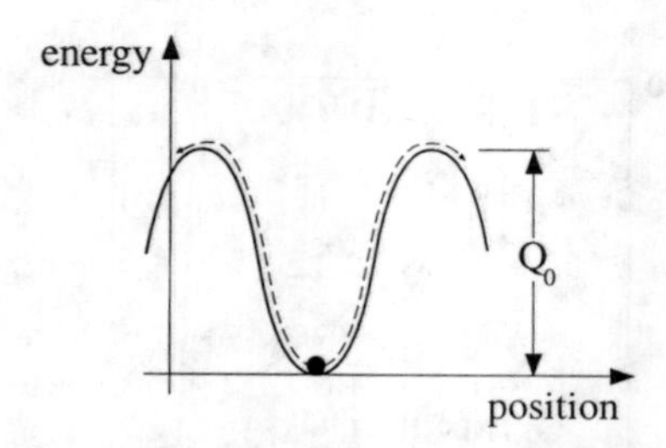

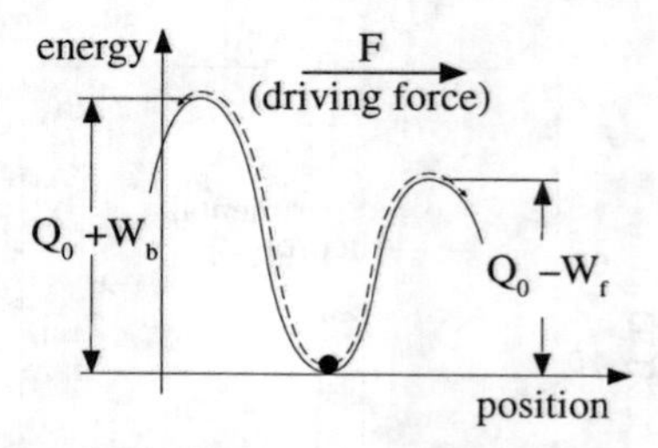

Figure 11.2.1 Energy barriers in absence of driving forces (left) and their modification when a driving force F is applied (right).

in the bulk of the structure and in the fracture process zone surrounding the crack tip (creep). These creep effects cause stress relaxation in the fracture process zone. The relaxation can explain (as already pointed out) why extremely slow static fracture tests show that the size of the fracture process zone decreases with a decreasing rate of loading, which means that the behavior is getting more brittle, closer to LEFM.

It is interesting that an opposite phenomenon, namely, a decrease of the size of the fracture process zone with an increasing rate of loading, is observed in very fast dynamic fracture (John and Shah 1986; Du, Kobayashi and Hawkins 1989). A possible explanation is that, while the stress relaxation effect becomes negligible in fast dynamic fracture, the activation-energy-controlled character of microcrack growth rate does not allow enough time for the fracture process zone to grow to its full size. So it appears that concrete fracture at both very high and very low loading rates is more brittle than it is in normal tests (1 to 10 minutes to maximum load).

Related to the activation energy aspect and the wave propagation phenomena is another interesting phenomenon. While at low strain rates the cracks tend to meander and pass along the surfaces of weakness, predominantly along the mortar-aggregate interfaces, at very fast fracture, the cracks tend to propagate more straight and cut through the pieces of aggregate. This again means that the fracture process zone is smaller at very fast rates (this may provide an additional explanation of the decrease of the process zone size at very high rates).

The principal devices for testing the dynamic fracture of materials, including concrete, have been the Hopkinson split bar (in which fracture is produced by a wave propagating along a bar, produced by impact loading or sudden unloading), the Charpy impact tester (a mass mounted on a pendulum) and the drop weight machine.

In analytical modeling, some basic elastodynamic wave propagation solutions have been worked out (Freund 1990). The near-tip stress and displacement fields of a propagating crack have been solved, both for isotropic and anisotropic materials, by asymptotic analysis (e.g., Bažant, Glazik and Achenbach 1976, 1978; Achenbach, Bažant and Khetan 1976a, 1976b). Many experimental results on the rate effect in the response of structures to rapid or dynamic loading have been interpreted in terms of the rate dependence of a stress-strain relation, rather than by fracture models. Such results are useful, but of limited validity, since they do not address the fracture problem itself (see a review in Bažant and Oh 1982).

11.2 Activation Energy Theory and Rate Processes

At a molecular or atomic level, fracture, as many other processes, implies relative movement of two particles of solids (atoms or molecules). This is generally a thermally activated process. The particles are in permanent random vibratory motion about their equilibrium positions representing the minimum potential of the binding forces. The potential energy surface of the force, sketched in Fig. 11.2.1, exhibits maxima representing energy barriers whose height is called the activation energy, Q. If the energy of the atom or molecule exceeds Q, the atom or molecule can jump over the activation energy barrier. In the case of fracture, this is the rupture of the bond. The frequency of the jumps of atoms or molecules over their activation energy barrier Q controls the rate of the rupture process. It may be interesting to note that it also controls many other processes, such as chemical reactions, diffusion, adsorption, creep, etc. The derivation of the rate theory from first principles is out of the scope of this book. Approaches to the rate theory specifically intended for mechanical processes are given in Krausz and Krausz (1988) and Krausz and Eyring (1975).

11.2.1 Elementary Rate Constants

The basic equation of the rate processes is the expression of the so-called *elementary rate constant* $\mathcal{K}$, equal to the number of jumps over the potential barrier per unit time in a given direction (usually denoted as forward or backward). The elementary rate constant is given by (Krausz and Krausz 1988)

$$\mathcal{K} = \nu_T \exp\left(-\frac{q_0}{kT}\right) \tag{11.2.1}$$

where T is the absolute temperature, $k = 1.3805\ 10^{-23}$ J K^{-1} is the Boltzmann constant, q_0 is the activation energy of the barrier (energy per particle), and ν_T the frequency of the thermal oscillations given by

$$\nu_T = \frac{kT}{h} \tag{11.2.2}$$

in which $h = 6.6256 \times 10^{-34}$ J s is Planck's constant.

In the foregoing equation, the energies are expressed per particle. The equation is often presented to express the energies per mole, in which case

$$\mathcal{K} = \nu_T \exp\left(-\frac{Q_0}{RT}\right) \tag{11.2.3}$$

where $R = 8.314$ J mole^{-1} K^{-1} and Q_0 is the activation energy in J mole^{-1}. Q_0 and q_0 are related by $Q_0 = N_A q_0$, where $N_A = 6.0225 \times 10^{23}$ mole^{-1} is Avogadro's number.

In general, the particles may jump the barrier in Fig. 11.2.1a from left to right (say, forward direction) or from right to left (backward direction). The forward direction may be imagined to correspond to bond breakage, and the backward direction to bond restoration. In equilibrium situations, the forward and backward rates are identical and no macroscopic effect is noticed. However, if a driving force is applied, then the barrier is modified as shown in Fig. 11.2.1b and the energy barrier for the forward jump is decreased by a certain amount W_f, equal to the work of the external driving force in going up to the activated state. Similarly, the energy barrier for the backward jump is increased by a certain amount W_b. Therefore, the forward and backward elementary rate constants may be written as

$$\mathcal{K}_f = \nu_T \exp\left(-\frac{Q_0 - W_f}{RT}\right) \tag{11.2.4}$$

$$\mathcal{K}_b = \nu_T \exp\left(-\frac{Q_0 + W_b}{RT}\right) \tag{11.2.5}$$

The net elementary forward rate is then

$$\mathcal{K}_{fn} = \nu_T \left[\exp\left(-\frac{Q_0 - W_f}{RT}\right) - \exp\left(-\frac{Q_0 + W_b}{RT}\right)\right] \tag{11.2.6}$$

which gives the resulting net rate in the forward direction.

Now, since these equations are general, they do not refer to any particular physical process. K_{fn} have the dimensions of s^{-1} and give the fraction of particles making a jump per unit time. Before applying these quantities, we need to relate them to the macroscopic observable variables.

11.2.2 Physical Rate Constants

From the view point of its manifestation, the process may usually be linked to some observable physical magnitude: the degree of reaction in a chemical reaction, the plastic shear deformation in a metal, or the crack growth in a brittle material. If the underlying elementary processes are unique (e.g., a single

Figure 11.2.2 Representation of thermally activated mechanisms: (a) single mechanism; (b) kinematic hardening; (c) series coupling; (d) consecutive coupling.

chemical reaction $A + B \rightarrow C$; slipping of a single family of dislocations in plastic shear flow; or breaking of a single class of bonds in fracture), then the rate of the observable magnitude is proportional to the elementary rate. In general, if s is the macroscopic variable of interest (graphically represented as a the sliding distance in Fig. 11.2.2), and F the associated macroscopic driving force associated (correctly giving the infinitesimal work $\delta\mathcal{W} = F\delta s$), the equation will read

$$\dot{s} = N s_{00} \mathcal{K}_{fn} \tag{11.2.7}$$

where N is the number of active elementary processes and s_{00} is a proportionality coefficient, with the same dimensions as s, representing the contribution of each elementary process to the macroscopic variable. Of course s_{00} is a property of the system being studied.

The terms, W_f and W_b, represent the work done by the force driving the process. In our representative sketch of Fig. 11.2.2, it is the force acting on the slider. The simplest assumption is to take W as a linear function of the driving force and write:

$$W_f = L_{0f} F \ ; \quad W_b = L_{0b} F \tag{11.2.8}$$

where L_{0f} and L_{0b} may be called the activation lengths, over which the force F does work in reaching the activated state.

The basic equation (11.2.7), coupled to (11.2.6) and (11.2.8), may lead to a various final rate equation, depending on the structure of the barriers and their evolution. Some of the possibilities are reviewed next, mainly to show that relatively complex macroscopic behaviors may be modeled with simple underlying mechanisms.

(a) Symmetric forward and backward barriers. If the potential energy barriers are symmetric and repeat periodically, the behavior is fully symmetric and the rate takes the form

$$\dot{s} = N s_{00} \mathcal{K}_{fn} = 2N s_{00} \exp\left(-\frac{Q_0}{RT}\right) \sinh\frac{L_0 F}{RT} = \dot{s}_{0T} \sinh(c_T F) \tag{11.2.9}$$

where the last expression follows by setting

$$\dot{s}_{0T} = 2N s_{00} \nu_T \exp\left(-\frac{Q_0}{RT}\right) \ , \quad c_T = L_0/RT \tag{11.2.10}$$

The expression (11.2.9) shows that the resulting isothermal creep is steady state creep ($\dot{s}$ = const. for F = const. and T = const.). Note also that the factor $\dot{s}_{0T}$ is strongly dependent on the temperature because it contains an exponential factor, $\exp(-Q_0/RT)$, representing the classical empirical Arrhenius equation for chemical reactions.

(b) No backward flow. If the behavior is fully irreversible, no backward flow is possible ($Q_{0b} \rightarrow \infty$) and then $\mathcal{K}_b = 0$. Dropping, for brevity, the subscript f (forward), we get the equation

$$\dot{s} = N s_{00} \mathcal{K} = \frac{1}{2} \dot{s}_{0T} e^{c_T F} \tag{11.2.11}$$

This equation has the unpleasant and illogical feature that the forward flow rate is nonzero even if the driving force is zero. However, this initial rate is typically very small, and equation (11.2.11) tends very quickly to (11.2.9). For example, for a driving force such that the rate is 10 times the initial rate (at a given temperature), the difference between the two equations is only 1%. That is why, in many theoretical works, the backward rate is neglected. It simplifies the resulting expressions.

(c) Kinematic hardening. Microstructural features may cause the local force on the slider not to coincide with the external applied force, because of the build up of internal stresses due to the sliding itself. This has been represented in Fig. 11.2.2b as an internal spring. The net force acting on the slider is thus $F - Es$, where E is the spring constant. Nonlinear springs may also be considered for certain mechanisms (e.g., pile-up of dislocations against an obstacle). If all the elements are identical, then the equation for this structure reduces to

$$\dot{s} = N s_{00} \mathcal{K}_{fn} = \dot{s}_{0T} \sinh[c_T(F - Es)] \tag{11.2.12}$$

In this model, isothermal creep is not a steady state. The creep rate decreases and vanishes asymptotically as the slip distance s approaches F/E.

(d) Isotropic hardening by barrier modification. The local fields of the individual mechanisms may be modified due to subsequent internal rearrangement. This modifies the activation energy, and, eventually, the activation length. To take into account the fact that the microstructural modification does not depend on the sense of movement, we may use a "cumulative sliding" (similar to the cumulative plastic strain) defined as $\dot{\overline{s}} = |\dot{s}|$, and assume that the activation energy is raised by $Q_h(\overline{s})$. The resulting equation is

$$\dot{s} = N s_{00} \mathcal{K}_{fn} = \dot{s}_{0T} \exp\left(-\frac{Q_0 + Q_h(\overline{s})}{RT}\right) \sinh(c_T F) \tag{11.2.13}$$

In this case, the hardening also induces an unsteady isothermal creep.

(e) Series coupling of various mechanisms. In some cases, it is possible to find various elementary processes of different kind participating in a single macroscopic process. Depending on the interaction of the various mechanisms, the resulting rate equation may look very different. The simplest interactions are the series coupling and the consecutive barriers. If two different mechanisms are coupled in series as shown in Fig. 11.2.2c, their individual contributions add up so the resulting rate is

$$\dot{s} = \dot{s}_1 + \dot{s}_2 \tag{11.2.14}$$

If, for simplicity, we consider that the backward flows for these mechanisms can be neglected, we get from (11.2.11)

$$\dot{s} = \frac{1}{2}\dot{s}_{0T1}\, e^{c_{T1}F} + \frac{1}{2}\dot{s}_{0T2}\, e^{c_{T2}F} \tag{11.2.15}$$

where subscript 1 and 2 refer to the two mechanisms. If $c_{T1} > c_{T2}$, we can recast the equation as

$$\dot{s} = \frac{1}{2}\dot{s}_{0T1}\, e^{c_{T1}F}\left[1 + \frac{\dot{s}_{0T2}}{\dot{s}_{0T1}}\, e^{-(c_{T1}-c_{T2})F}\right] \tag{11.2.16}$$

and it turns out that for large driving forces, the exponential term in square brackets vanishes and the rate is completely dominated by the first mechanism (the fastest one). This is also valid for the symmetric barrier because, as already pointed out, the hyperbolic sine tends rapidly to an exponential.

(f) Consecutive mechanisms. In some cases, for the deformation to proceed, each particle must jump two consecutive barriers of different kinds. This may be valid at an atomic scale, but may also happen at a larger scale (micro- or meso-scales). This is particularly true for composite materials in which, for the process to proceed, the dislocation, crack tip, or whatever other deformation mechanism (such as the slider of Fig. 11.2.2d) must cross materials of different characteristics. Let the path of the slider be composed, on average, of a fraction α of material 1 and a fraction $1 - \alpha$ of material 2. The time required to slide a total length L is, on average,

$$t_L = \frac{\alpha L}{\dot{s}_1} + \frac{(1-\alpha)L}{\dot{s}_2} \tag{11.2.17}$$

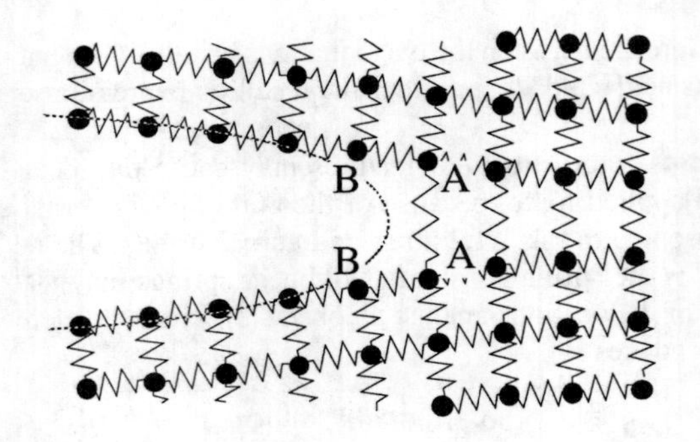

Figure 11.2.3 Sketch of bond stretching near the crack tip.

and so the mean sliding rate $s = L/t_L$ verifies the rule

$$\frac{1}{\dot{s}} = \frac{\alpha}{\dot{s}_1} + \frac{1-\alpha}{\dot{s}_2} \tag{11.2.18}$$

This is immediately generalized to any number N of processes by writing

$$\frac{1}{\dot{s}} = \sum_{i=1}^{N} \frac{\alpha_i}{\dot{s}_i} \quad \text{with} \quad \sum_{i=1}^{N} \alpha_i = 1 \tag{11.2.19}$$

11.2.3 Fracture as a Rate Process

The rate of fracture is proportional to the number of bond ruptures per unit time, which is equal to the number of particle-pairs at or near the crack tip whose energy exceeds the binding potential energy. In the analysis of such a problem, the main issue is the definition of the driving force for bond breaking. It seems obvious that, as far as the nonlinear zone remains very small, the driving force must be related in some way to the energy release ratio $\mathcal{G}$ or to the stress intensity factor K_I. In other cases, it may be better related to other parameters such as the J-integral. The problem of selecting the right driving force is considered by Krausz and Krausz (1988) as an integral part of the problem of identifying the rate mechanisms in a particular case.

In brittle fracture, the usual parameter used as the driving force is the stress intensity factor K_I. This may be justified, in a rough manner, by relating the force on the pair of atoms located at the crack tip with the elastic stress distribution as follows: consider the simple atomic lattice shown in Fig 11.2.3; let the equivalent crack tip lie midway between the pairs of atoms BB (whose bond has been already broken) and AA, the latter being the most stressed pair. If we assume that the behavior is essentially linear elastic, the smoothed stress distribution is given by the elastic solution $\sigma = K_I/\sqrt{2\pi x}$, where x is the distance to the crack tip. Then, the force on the pair AA is the resultant of this stress distribution over a square of side b, where b is the interatomic distance:

$$F = b \int_0^b \frac{K_I}{\sqrt{2\pi x}} dx = 2b^2 \frac{K_I}{\sqrt{2\pi b}} \tag{11.2.20}$$

Thus, for brittle fracture, it is reasonable to assume the driving force for bond breaking to be proportional to K_I. In such a case, the expressions for the crack growth rate may be found, as in the preceding sections, just by substituting $\dot{a}$ in place of $\dot{s}$. For example, for a symmetric barrier, Eqs. (11.2.9) and (11.2.10) are transformed into

$$\dot{a} = \dot{a}_{0T} \sinh(c_T K_I) \tag{11.2.21}$$

$$\dot{a}_{0T} = 2 l_0 \nu_T \exp\left(-\frac{Q_0}{RT}\right) \quad , \quad c_T = \kappa_0/RT \tag{11.2.22}$$

where l_0 is a constant with dimensions of length and κ_0 a constant with dimensions of length$^{3/2}$ mole^{-1}.

Note that Eq. (11.2.21) is contingent upon the assumption that there is a single, physically distinct process governed by activation energy. In reality, several such processes may be proceeding simultaneously, in which case several terms of the type of Eq. (11.2.21) get superimposed as was shown in the foregoing sections. This explains why, in reality, more complicated laws can be observed in nature. For the stress dependence, a power-law has often been used. For example, by analogy with the Arrhenius equation, Evans and Fu (1984) and Thouless, Hsueh and Evans (1983) proposed and experimentally verified an empirical power-law formula for ceramics:

$$\dot{a} = C_0 K_I^n e^{-Q/RT} \tag{11.2.23}$$

where C_0, n and Q are constants. Within the limited range of experiments, the optimum fit of test results by the power function of K_I in this formula is almost indistinguishable from the fit by the sinh-function. There is a difference, however; Eq. (11.2.23) has a K_I-dependent term that is independent of temperature T, in contrast to Eq. (11.2.21). The power function is particularly convenient for the evaluation of test data, since it allows identifying exponent n easily by linear regression in a logarithmic plot. Another appealing feature of the power function is that (11.2.23) bears similarity to Paris' law for fatigue crack growth (Section 11.7) as well as other phenomena such as creep in which a power dependence on stress is used. But the hyperbolic sine function is theoretically better justified.

On the other hand, if we admit that there might be several distinct rate processes controlled by different activation energies, then the sinh-function in Eq. (11.2.21) cannot describe the dependence on K_I accurately, and might not be any better than a power function. This is particularly true for environmentally assisted cracking, where diffusional, chemical, and mechanical processes are interwoven; see Krausz and Krausz (1988) for the techniques of analysis of these kinds of processes.

The foregoing analysis deals with situations in which the fracture process zone length is very limited and the bulk of the material remains linearly elastic. For concrete, this may be a first approximation for the analysis of the dependence of crack growth rate on temperature and other factors. A few steps toward this end are described in the next section. However, full understanding of the rate-dependence of concrete fracture will require inclusion of the viscoelastic creep of the material surrounding the fracture process zone and description of the rate behavior of a sizable fracture zone. This will be addressed in later sections.

11.2.4 General Aspects of Isothermal Crack Growth Analysis

Let us analyze the general methodology for isothermal crack growth analysis when the crack growth is governed by a rate equation of the type (11.2.21) or (11.2.23). We can write the rate equation in the general form

$$\dot{a} = \dot{a}_{0T}\, f\!\left(\frac{K_I}{K_{0T}}\right) \tag{11.2.24}$$

where $\dot{a}_{0T}$ depends on temperature and has dimensions of velocity, and K_{0T} can also depend on temperature and has dimensions of stress intensity factor; f is a dimensionless function.

The foregoing equation relates the crack growth rate to the stress intensity factor. However, in most tests the directly measured variables are the load P and some displacement (e.g., the load-point displacement, or CMOD). The equation relating the load to the stress intensity factor may always be written in the form (2.3.11)

$$K_I = \frac{P}{b\sqrt{D}}\hat{k}(\alpha) \tag{11.2.25}$$

where b is the specimen thickness, D an in-plane dimension of the specimen (called the size, for short), $\alpha = a/D$, and $\hat{k}(\alpha)$ is a nondimensional function depending on the geometry of the specimen.

Let us call the measured displacement u_E. In LEFM, the displacement can be written in the form (3.5.24)

$$u_E = \hat{v}_E(\alpha)\frac{P}{bE'} \tag{11.2.26}$$

where $\hat{v}_E(\alpha)$ is a dimensionless function.

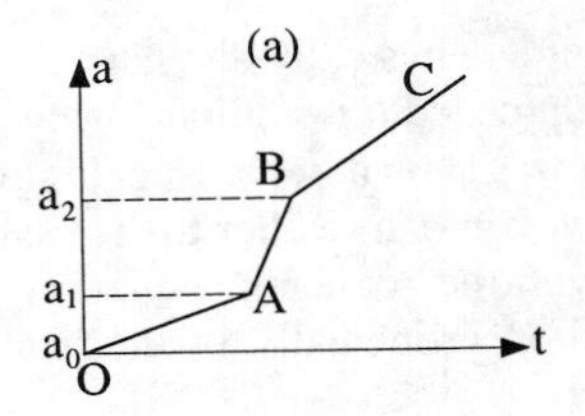

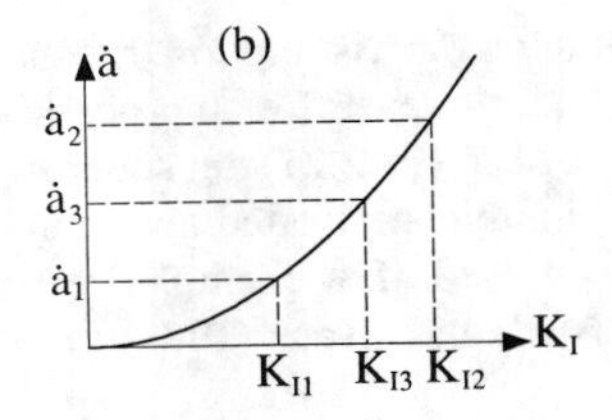

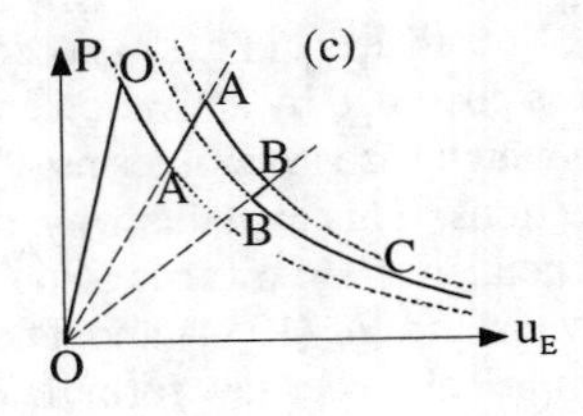

Figure 11.2.4 Fracture controlled by crack-rate. (a) Crack evolution. (b) Rate equation. (c) Load-displacement curve.

Equations (11.2.24)–(11.2.26) completely determine the evolution of $a(t)$, $K_I(t)$, $P(t)$, and $u_E(t)$ with the appropriate initial conditions, e.g., $a(0) = a_0$, and the boundary conditions. In normal testing, the boundary conditions are usually formulated by imposing the load history —$P(t)$ given— or the displacement history —$u_E(t)$ given. These cases will be examined in forthcoming sections for the particular case of the power-law rate equation. Here we examine the ideal case of a test in which the crack length is controlled and the crack growth rate can be selected arbitrarily.

If $\dot{a}$ is specified and constant over a certain time interval, K_I can be determined from (11.2.24) and is also constant over that time interval. Thus, in a plot of P vs. u_E, the process follows an iso-K curve which is obtained eliminating α from equations (11.2.25) and (11.2.26). Consider, for example, the process depicted in Fig. 11.2.4a in which the crack is made to run initially at rate $\dot{a}_1$ until the crack reaches the length a_1. Then the rate is increased to $\dot{a}_2$ up to length $a = a_2$; after that, the crack rate is decreased to $\dot{a}_3$ and kept constant. If the rate equation (11.2.24) is plotted as in Fig. 11.2.4b, the stress intensity factor for each interval of constant rate is determined as shown in that figure. The resulting $P(u_E)$ curve is sketched in Fig. 11.2.4c. Note that this ideal test involves discontinuities in the loading, i.e., instantaneous changes of load and displacement at the points where the rate changes. However, even in a crude way, it illustrates the results found in experiments with sudden changes of the rate of loading as detected in the tests by Bažant, Gu and Faber (1995).

11.2.5 Load-Controlled Processes for Power-Law Rate Equation

Consider a test starting at $t = 0$, in which the load history is imposed, i.e., $P(t)$ is known. Assume that the rate equation for the material is a power-law equation written as

$$\dot{\alpha} = \frac{\dot{a}_{0T}}{D} \left(\frac{K_I}{K_{0T}} \right)^n \tag{11.2.27}$$

where n is a positive constant. Substituting (11.2.25) for K_I, separating variables and integrating, we find the equation relating the crack length to time:

$$\int_{\alpha_0}^{\alpha} [\hat{k}(\alpha')]^{-n} \, d\alpha' = \frac{\dot{a}_{0T}}{D(K_{0T} b\sqrt{D})^n} \int_0^t [P(t')]^n \, dt' \tag{11.2.28}$$

Once the integrals are solved for a particular case, the evolution of the displacement with time can be computed from (11.2.26). The following example applies the methodology to a very simple particular case.

Example 11.2.1 Consider a very large panel with a center crack of initial length $2a_0$. We want to determine time t_1 required for the crack to reach length a_1 under a constant remote stress σ, under the assumption of the isothermal rate equation of the power-law type (11.2.27). To solve the problem, we follow the foregoing procedure writing that, for the panel, $K_I = \sigma\sqrt{\pi a}$. We substitute this into (11.2.27) and separate variables to get the equation

$$\int_{a_0}^{a_1} a^{-n/2} \, da = \dot{a}_{0T} \left(\frac{\sqrt{\pi}}{K_{0T}} \right)^n \int_0^{t_1} \sigma^n \, dt' \tag{11.2.29}$$

In this simple case the integrals are readily evaluated and, after some manipulation,

$$t_1 = \frac{(n-2)a_0}{2\dot{a}_{0T}} \left(\frac{K_{0T}}{\sigma\sqrt{\pi a_0}} \right)^n \left(1 - \alpha_1^{-\frac{n}{2}+1} \right) \tag{11.2.30}$$

where $\alpha_1 = a_1/a_0$. ▯

11.2.6 Displacement-Controlled Processes for Power-Law Rate Equation

Consider a test starting at $t = 0$ in which the displacement history is imposed, i.e., $u_E(t)$ is known. Assume that the rate equation for the material is the power-law equation (11.2.27). Eliminating P from (11.2.25) and (11.2.26) and solving for K_I, we get

$$K_I = E' D^{-1/2} \frac{\hat{k}(\alpha)}{\hat{v}_E(\alpha)} u_E \tag{11.2.31}$$

Now we substitute this into (11.2.27), separate variables, and integrate to get

$$\int_{\alpha_0}^{\alpha} \left[\frac{\hat{k}(\alpha')}{\hat{v}_E(\alpha')} \right]^{-n} d\alpha' = \frac{\dot{a}_{0T}}{D} \left(\frac{E'}{K_{0T}\sqrt{D}} \right)^n \int_0^t [u_E(t')]^n \, dt' \tag{11.2.32}$$

Once the integrals in the preceding equations are solved for a particular case, the evolution of the displacement with time can be computed from (11.2.26). The following example gives an illustration.

Example 11.2.2 Consider a double cantilever beam (DCB) specimen of thickness b, depth $2h$, and initial crack length a_0, subjected to a test that is run at a constant mouth opening displacement rate $\dot{u}$ (note that in this case the displacement and the CMOD are coincident). We adopt the bending theory approximation —see example 2.1.2, Eqs. (2.1.29)— in which the expressions for u and the stress intensity factor are

$$u = \frac{8P\alpha^3}{E'b} \tag{11.2.33}$$

$$K_I = 2\sqrt{3} \frac{P\alpha}{b\sqrt{h}} \tag{11.2.34}$$

where $\alpha = a/h$ and h plays the role of D. It thus appears that $\hat{k}(\alpha) = 2\sqrt{3}\alpha$ and $\hat{v}_E(\alpha) = 8\alpha^3$. Substituting this directly into (11.2.32), we get

$$\frac{4^n}{3^{(n/2)}} \int_{\alpha_0}^{\alpha} \alpha'^{2n} \, d\alpha' = \frac{\dot{a}_{0T}}{h} \left(\frac{E'}{K_{0T}\sqrt{h}} \right)^n \int_0^u u'^n \frac{du'}{\dot{u}} \tag{11.2.35}$$

in which we set $dt' = du'/\dot{u}$. The integrals are readily computed to give

$$\alpha = \left[\alpha_0^{2n+1} + \beta \left(\frac{u}{h} \right)^{n+1} \right]^{\frac{1}{2n+1}} \tag{11.2.36}$$

in which

$$\beta = \frac{\dot{a}_{0T}}{\dot{u}} \left(\frac{E'\sqrt{3h}}{4K_{0T}} \right)^n \tag{11.2.37}$$

We now substitute this into (11.2.33) and solve for P to get

$$P = \frac{1}{8} E' b u \left[\alpha_0^{2n+1} + \beta \left(\frac{u}{h} \right)^{n+1} \right]^{-\frac{3}{2n+1}} \tag{11.2.38}$$

which is the analytical expression of the $P(u)$ curve. ▯

Exercises

11.1 The elementary rate constant of a thermally activated process is 1000 s^{-1} at 20°C. Determine the frequency ν_T of the thermal vibrations, the activation energy per particle q_0, and the activation energy per mole Q_0. What is the elementary rate constant at 50°C?

11.2 Crack growth tests on a ceramic material performed at 20°C for various stress intensity factors yield the results shown in the table below. Fit these results to (a) a power-law; (b) a sinh-law; (c) an exponential law. Given that the coefficient of variation of the results is about 30%, can you distinguish the "right" model from these experimental results? [The answer is obviously NO. Note that to be able to distinguish, the independent variable K_I should be varied over at least one order of magnitude, but this is practically not feasible. It is very difficult to measure crack rates below 10^{-11} m/s (about 6 microns per week), and it is very difficult to control the test for rates above 10^{-3} m/s (1 mm/s), because the specimen breaks within a few seconds.]

K_I (MPa m$^{1/2}$)	0.4	0.43	0.46	0.49
$\dot{a}$ (m s^{-1})	1.4 10^{-9}	2.5 10^{-8}	3.8 10^{-7}	4.7 10^{-6}

11.3 A large panel of a material with a rate equation given by $\dot{a} = 10^{13} K_I^{44}$, with $\dot{a}$ in m/s and K_I in MPa m$^{1/2}$, is subjected to a constant tensile stress of 1 MPa. If the panel is to be replaced when a crack reaches a total length of 100 mm, find the expected time between replacements if the panel contains through cracks of (a) 1 mm; (b) 0.1 mm; (c) 10 microns.

11.4 A large panel with a center crack of initial length $2a_0$ is subjected to a uniform constant stress σ. If the time to reach a crack length $a_1 = \alpha a_0$ ($\alpha > 1$) is t_1 and the time to reach a length $a_2 = \alpha^2 a_0$ is t_2, and the rate equation is assumed to be of the power type, find the exponent and the coefficient of the equation in terms of α, t_1, t_2, a_0, and σ. (Hint: Compare the time required to grow from a_0 to a_1 with that required to grow from a_1 to a_2.)

11.5 For a double-cantilever beam of a material with a power-law rate equation of exponent n, subjected to a constant CMOD rate, find the relative crack extension at which the peak load is reached.

11.6 For a double-cantilever beam of a material with a power-law rate equation of exponent n, subjected to a constant CMOD rate $\dot{u}$, determine the peak nominal stress intensity factor K_{INu}, defined as the value of K_I computed for the initial crack length and the peak load. Prove that if two specimens of sizes (arm depths) h_1 and h_2 are tested, respectively, at speeds $\dot{u}_1$ and $\dot{u}_2$, the relationship between the peak nominal stress intensity factors is $K_{INu1}/K_{INu2} = (u_1\sqrt{h_1}/u_2\sqrt{h_2})^{1/(1+n)}$.

11.3 Some Applications of the Rate Process Theory to Concrete Fracture

In this section we present some basic applications of the rate process theory to concrete. We first review the influence of temperature and humidity on the fracture energy as described in terms of the rate theory by Bažant and Prat (1988a), complemented by some experimental results at low temperatures by Maturana, Planas and Elices (1990). We next introduce the basic idea of extending the R-curve theory to rate-dependent crack growth, following the development of Bažant and Jirásek (1993), which, in turn, rests on previous experimental results by Bažant and Gettu (1989, 1990, 1992), Bažant, Bai and Gettu (1993) and Bažant, Gu and Faber (1995). The extension of the Bažant-Jirásek theory to take into account creep effects is, however, deferred until Section 11.5.

11.3.1 Effect of Temperature on Fracture Energy of Concrete

For the purpose of theoretical modeling, Bažant and Prat (1988a), upon noting that $K_I^2 = \mathcal{G}E'$, rewrote Eq. (11.2.23) in the form

$$\dot{a} = \nu_c \left(\frac{\mathcal{G}}{G_f^0}\right)^{n/2} \exp\left[-\frac{Q}{R}\left(\frac{1}{T} - \frac{1}{T_0}\right)\right] \tag{11.3.1}$$

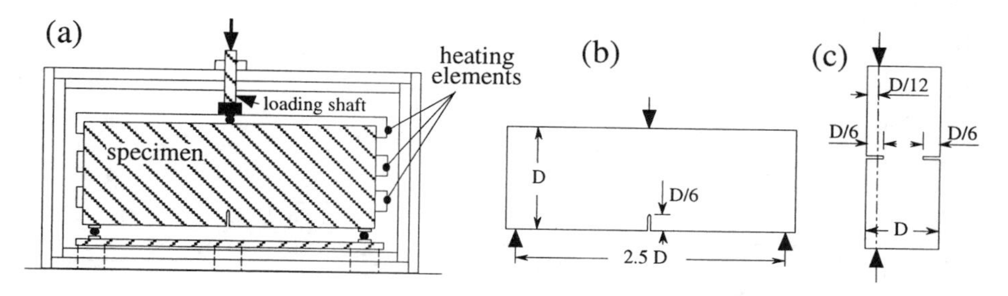

Figure 11.3.1 Bažant and Prat (1988a) tests on predried specimens: (a) bending test arrangement in an oven; (b) three-point bending specimen; (c) eccentric compression specimen.

in which T_0 is a chosen reference temperature, ν_c is a constant, and G_f^0 is the fracture energy at the reference temperature T_0. In this form, the exponential term becomes 1 at reference temperature.

The choice of the reference temperature T_0 in Eq. (11.3.1) is, of course, arbitrary. There is no reason why one could not take as the reference temperature any other temperature T used in Eq. (11.3.1). In that case, Eq. (11.3.1) reduces to the form

$$\dot{a} = \nu_c (G/G_f)^{n/2} \tag{11.3.2}$$

where G_f is the fracture energy value at temperature T.

The crack growth rate expressions based on T and on T_0 must be equivalent. Setting Eqs. (11.3.1) and (11.3.2) equal, one obtains Bažant's (1987a) expression:

$$G_f = G_f^0 \exp\left(\frac{\gamma}{T} - \frac{\gamma}{T_0}\right), \quad \gamma = \frac{2Q}{nR} \tag{11.3.3}$$

in which γ is a constant. The last equation has been used by Bažant and Prat (1988a) to describe the results of their tests of predried concrete fracture specimens tested in a heated oven as shown in Fig. 11.3.1. Two types of specimens were used as shown in Fig. 11.3.1b–c: three-point bent notched beams (TPB), and eccentric compression doubly notched specimens (ECS). The size effect method was used at each temperature to obtain the fracture energy; the size effect law was followed reasonably well at each temperature (for details of the tests, see Bažant and Prat 1988a).

The values of the fracture energy obtained for each temperature are plotted in a semi-logarithmic diagram in Fig. 11.3.2, in which the inverse of the temperature is plotted along the horizontal axis. In such a plot, the Bažant-Prat equation Eq. (11.3.3) appears as a straight line. The optimum fit was obtained by the linear regression shown in Fig. 11.3.2, in which Eq. (11.3.3) becomes a straight line. It is seen that Eq. (11.3.3) describes these data quite well for a broad range of temperatures.

11.3.2 Effect of Humidity on the Fracture Energy of Concrete

The humidity or the water content of concrete affects all the mechanical properties, including fracture. A strong effect of pore relative humidity h on the fracture properties must be expected on the basis of what is known about the hardened cement paste microstructure. The tricalcium silicate hydrate particles and other particles in the hardened cement paste are strongly hydrophilic, adsorbing water molecules to the pore surfaces. There are many pores in cement gel of the order of 10^{-9} m, which is only about 4 water molecules thick. In such layers, there is a phenomenon of the so-called hindered adsorption. It is manifested by the fact that a layer of water molecules confined between two gel particles (which hinder development of the full thickness of adsorption layer) develops a large transverse pressure, called the disjoining pressure. This pressure tries to separate the adjacent particles, i.e., to open up the gel pores (for a mathematical formulation, see Bažant 1972a,b). The disjoining pressures increase with increasing relative humidity in the capillary pores, or with increasing pore water content. Since the disjoining pressures must be resisted by the solid framework of the hardened cement paste and concrete, large tensile microstresses, which are in addition to those from the applied load, are introduced into the solid framework by an increase in the pore water content.

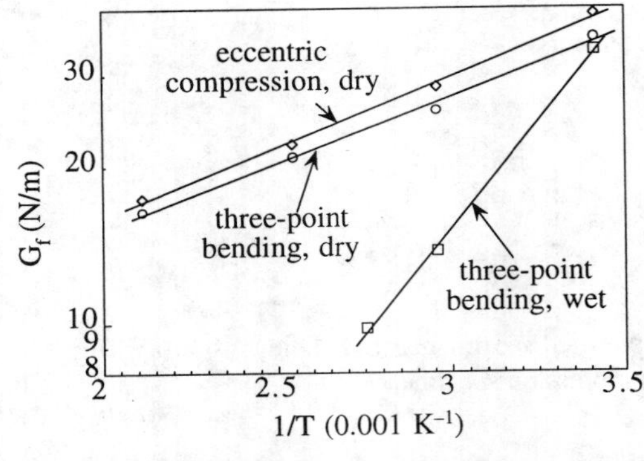

test		γ (K)	G_f^0 (N/m)
dry	TPB	581	34.5
	ECS	623	38.4
wet	TPB	1875	33.0

Figure 11.3.2 Experimental results and theoretical fits for the effect of temperature on fracture energy, for both dry and wet (saturated) specimens. (After Bažant and Prat 1988a.)

From the foregoing consideration, it becomes clear that the effect of changing the pore water content must be equivalent to changing the value of the activation energy Q. A change in the value of Q as a function of pore relative humidity h is then manifested in all the foregoing derivation, and particularly in Eq. (11.3.1) for the crack propagation rate and in Eq. (11.3.3) for the effective fracture energy G_f. Thus, the effect is introduced by replacing constant γ by function $\gamma(h)$. This function has previously been assumed to be linear (Bažant and Prat 1988a);

$$G_f = G_f^0 \exp\left(\frac{\gamma(h)}{T} - \frac{\gamma(h)}{T_0}\right) \qquad \gamma = \gamma_0 + (\gamma_1 - \gamma_0)\frac{w}{w_1} \tag{11.3.4}$$

in which w is the specific evaporable water content of concrete (per 1 m^3); $\gamma_0, \gamma_1, w_1, G_f^0$ and T_0 are constants; and w_1 is the specific evaporable water content of concrete at saturation.

The variation of the effective activation energy corresponding to the linearity assumption in Eq. (11.3.4) is

$$Q = Q_0 + (Q_1 - Q_0)\frac{w}{w_1} \tag{11.3.5}$$

in which $Q_0 = nR\gamma_0/2$ and $Q_1 = nR\gamma_1/2$ are constants.

The fitting of Bažant and Prat's (1988a) fracture tests of predried specimens ($w = 0$) yields the value $\gamma = \gamma_0$. Further fracture tests, using the size effect method, have been conducted on specimens submerged in a heated water bath. The test results are included in Fig. 11.3.2. The theoretical formula for the effect of temperature again fits well. However, the range is limited because wet specimens cannot be tested at temperatures exceeding 100°C, as the pore water rapidly evaporates (100°C could be exceeded only by conducting the fracture tests at very high atmospheric pressures, which, however, seems difficult).

From Fig. 11.3.2 one should note the extremely strong effect of humidity. As we see, wet concrete at higher temperatures becomes much weaker for fracture than dry concrete (this is true, of course, for the states of concrete at which thermodynamic equilibrium of water has been reached and no large residual stresses have been produced by drying or wetting; otherwise, the effect might be more complex). If we could dry (well-hydrated) concrete structures without damaging them, we would make them much stronger.

Fig. 11.3.3 shows a plot of the values of G_f relative to the fracture energy measured at room temperature. This plot includes the results of Maturana, Planas and Elices (1990) for low temperature tests on saturated concrete (full symbols). These tests used the work-of-fracture method, which gives values roughly double those delivered by the size effect method. However, the relative variations are very similar for both kinds of measurement. The plot shows again that the temperature effect is much more marked for the saturated specimen. For this situation, it appears that the Bažant-Prat formula (11.3.4) fits the results well down to about –10°C. For lower temperatures, the measured fracture energy is much less than that predicted by the equation.

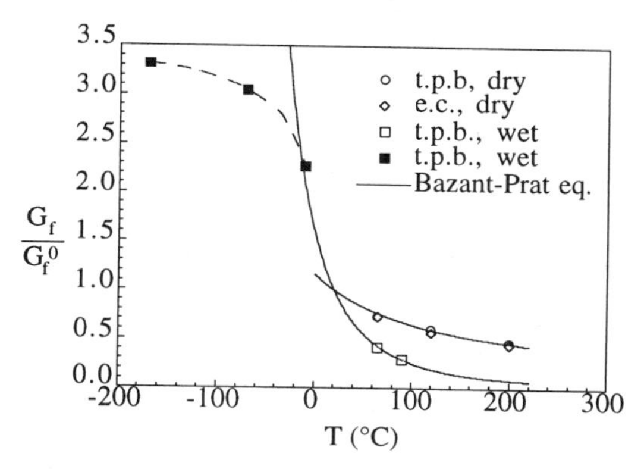

Figure 11.3.3 Fracture energy relative to that at 20°C. Open symbols from Bažant and Prat (1988a); full symbols from Maturana, Planas and Elices (1990).

This deviation can be explained by the progressive change of phase (freezing of water) that is known to occur in saturated concrete in the interval between, approximately, −10 and −70°C, as manifested in dilatometric tests (Planas, Corres et al. 1984; Elices, Planas and Corres 1986; Corres, Elices and Planas 1986). During this process, two competitive mechanisms take place: microcrack healing due to solid phase formation, and microcrack formation (damage) due to the water expansion during freezing. The intermediate stages and the final product (concrete with completely frozen pore water) may be expected to display a thermal behavior completely different from the ordinary concrete.

To sum up, the effective fracture energy of concrete significantly depends on temperature. It decreases monotonically and smoothly as the temperature increases. For wet, saturated concrete this decrease is much stronger than for predried concrete. The dependence of the fracture energy on temperature agrees with a formula derived from the concept of activation energy over the temperature range in which the capillary pore water remains liquid. For lower temperatures, the dependence on temperature is much weaker. Comparisons with tests further confirm that the size effect law discussed before is applicable at various temperatures and water contents.

11.3.3 Time-Dependent Generalization of R-Curve Model

Equations (11.2.21) and (11.2.23) are valid for very small fracture process zones, in which the fracture process is completely governed by the value of K_I. When the fracture process zone becomes larger, further modifications must be done. As discussed in Chapter 5, the equivalent LEFM models, particularly the R-curve approximations, are one of the ways to extend the use of LEFM to larger fracture zones. The basic idea in doing so is to modify the crack growth rate equation, which for large sizes depends only on K_I and T, and let it depend on the crack extension $\Delta a = a - a_0$:

$$\dot{a} = f(K_I, \Delta a, T). \tag{11.3.6}$$

Bažant and Jirásek (1993) introduce this dependence through the time-independent R-curve. Their starting point is the time-independent R-curve (resistance curve) $K_R(\Delta a)$, in which the crack growth is governed by an equation of the type

$$K_I = K_R(\Delta a) \tag{11.3.7}$$

To incorporate an R-curve behavior into the rate equation, it is enough to let the crack growth rate depend not only on K_I and T, but also on Δa. Bažant and Jirásek (1993) propose to introduce the dependence on Δa through K_R in the form

$$\dot{a} = f\left[K_I, K_R(\Delta a), T\right]. \tag{11.3.8}$$

where it is understood that K_R is a known function of Δa. It is clear that $\dot{a}$ should increase with increasing K_I and with decreasing K_R. But what should be the actual form of the crack growth rate function $f(K, K_R, T)$? Experimental evidence indicates that changing the loading rate by several orders of magnitude causes the peak loads to change by only a factor less than 2 (Bažant and Gettu 1992; Bažant, He et al. 1991). Therefore, the crack growth rate function should allow for a very large variation of $\dot{a}$ with only moderate changes of its arguments. This can be achieved by one of the following formulas, which are the direct generalization of (11.2.23) or (11.2.21):

$$\dot{a} = \kappa_0 \left(\frac{K_I}{K_R}\right)^n \exp\left[\frac{Q}{R}\left(\frac{1}{T_0} - \frac{1}{T}\right)\right], \tag{11.3.9}$$

or

$$\dot{a} = \kappa_0 \sinh\left(\frac{K_I T_0}{K_R T}\right) \exp\left[\frac{Q}{R}\left(\frac{1}{T_0} - \frac{1}{T}\right)\right] \tag{11.3.10}$$

where κ_0, Q, and n are constants; T_0 is the reference temperature. In their work on time dependent R-curves, Bažant and Jirásek (1993) used the isothermal version of the empirical equation (11.3.9) which reads

$$\dot{a} = \dot{a}_{0T} \left(\frac{K_I}{K_R}\right)^n \tag{11.3.11}$$

where $\dot{a}_{0T}$ is constant for isothermal conditions. Experiment indicates that the exponent $n \gg 1$; typically $n = 30$ to 60.

The analysis of crack growth for this kind of model can be performed in a way analogous to that developed in Sections 11.2.4–11.2.6 for the case where the rate equation is independent of a. The only modification is that Eq. (11.3.11) replaces Eq. (11.2.24). This introduces some computational complexities that are better explained through examples.

Example 11.3.1 Consider a very large panel with a center crack of initial length $2a_0$. We want to determine the elapsed time t_1 until the crack reaches a half-length a_1 under a constant remote stress σ, under the assumptions that the isothermal rate equation of the power-law type (11.3.11) and that the R-curve $\mathcal{R}(a - a_0)$ is known. To solve the problem we just substitute $K_R = \sqrt{E'\mathcal{R}(a - a_0)}$ and $K_I = \sigma\sqrt{\pi a}$ into (11.3.11), separate variables, and integrate to get the following equation:

$$\int_{a_0}^{a_1} \left[\frac{\mathcal{R}(a - a_0)}{a}\right]^{n/2} da = \dot{a}_{0T} \left(\frac{\pi\sigma^2}{E'}\right)^{n/2} \int_0^{t_1} dt \tag{11.3.12}$$

From this, the time required for the crack to reach a length a_1 is obtained as:

$$t_1 = \frac{1}{\dot{a}_{0T}} \left(\frac{\pi\sigma^2}{E'}\right)^{-n/2} \int_{a_0}^{a_1} \left[\frac{\mathcal{R}(a - a_0)}{a}\right]^{n/2} da \tag{11.3.13}$$

Note that this solution has the same structure as that in example 11.2.1, except that the integral, in general, will have to be evaluated numerically. ▯

The foregoing example can be generalized to any kind of geometry by writing the stress intensity factor in the form (11.2.25). Proceeding as in the example, the general result for a power-law isothermal rate function under constant load is

$$t_1 = \frac{D}{\dot{a}_{0T}} \left(\frac{P^2}{b^2 E' D}\right)^{-n/2} \int_{\alpha_0}^{\alpha_1} \left[\frac{\mathcal{R}[D(\alpha - \alpha_0)]}{k^2(\alpha)}\right]^{n/2} d\alpha \tag{11.3.14}$$

For other loading conditions, the equations can be manipulated in a similar way to get a differential equation in the appropriate variables. In particular, the case where the boundary conditions are imposed in terms of displacements, rather than load, is interesting for testing, and can be handled as illustrated in the following example.

Example 11.3.2 Consider, as in example 11.2.2, a double-cantilever beam (DCB) of thickness b, depth $2h$, and initial crack length a_0, subjected to a test that is run at a constant CMOD rate $\dot{u}$. Working as in example 11.2.2, we get K_I as a function of u, Eq. (11.2.35). We substitute this into the power-law rate equation (11.3.11) together with $K_R = \sqrt{E'\mathcal{R}(a - a_0)}$, and after separating variables, we get

$$\int_{\alpha_0}^{\alpha} \left[\alpha'^4 \mathcal{R}[h(\alpha' - \alpha_0)]\right]^{n/2} d\alpha' = \frac{\dot{a}_{0T}}{h}\left(\frac{3E'}{16h}\right)^{n/2} \int_0^t [u(t')]^n dt' \tag{11.3.15}$$

Setting $t' = u'/\dot{u}$ in the right-hand side integral, and evaluating the resulting integral, we finally get the relationship between u and α:

$$u^{n+1} = \frac{(n+1)h\dot{u}}{\dot{a}_{0T}}\left(\frac{3E'}{16h}\right)^{-n/2} \int_{\alpha_0}^{\alpha} \left[\alpha'^4 \mathcal{R}[D(\alpha' - \alpha_0)]\right]^{n/2} d\alpha' \tag{11.3.16}$$

It thus appears that, for a given crack length, the crack mouth opening displacement can be found from (11.3.16). The corresponding load can then be solved from (11.2.33), and thus the $P(u)$ curve can finally be plotted for the given CMOD rate. ▯

The foregoing example can, again, be generalized to any other geometry over which a displacement u_E (not necessarily the load-point displacement) is increased at constant rate. The resulting equation is

$$u_E^{n+1} = \frac{(n+1)D\dot{u}_E}{\dot{a}_{0T}}\left(\frac{E'}{D}\right)^{-n/2} \int_{\alpha_0}^{\alpha} \left[\frac{\hat{v}_E^2(\alpha')\mathcal{R}[D(\alpha' - \alpha_0)]}{k^2(\alpha')}\right]^{n/2} d\alpha' \tag{11.3.17}$$

In the foregoing, the equations relating the load to the displacements have been based on the assumption of linear elasticity. This approximation can be acceptable for materials exhibiting low creep or for relatively narrow intervals of loading rates. However, for concrete and for loading rates spanning over several orders of magnitude, creep effects can play an important role. The theory required to deal with such viscoelastic effects will be introduced in the next section. In this section we present the basic results of the application of the foregoing theory to a low-creep rock.

11.3.4 Application of the Time-Dependent R-Curve Model to Limestone

Bažant and Jirásek (1993) applied the foregoing model to the experimental results of Bažant, Bai and Gettu (1993). These tests were conducted on notched beams of various sizes subjected to three-point bending at controlled CMOD rate. The toughness curve K_R was obtained from the size effect using the procedures defined in Chapter 6. The R-curve takes the form

$$\mathcal{R} = \frac{K_f^2}{E'}\rho\left(\frac{D(\alpha - \alpha_0)}{c_f}\right) \tag{11.3.18}$$

where K_f is the fracture toughness obtained by the size effect method, E' the elastic modulus, and $\rho[D(\alpha - \alpha_0)/c_f]$ a dimensionless function; c_f is the critical effective crack extension which is obtained, together with ρ, from the size effect analysis (Chapter 6).

With K_f, E', and c_f determined from size effect tests at the usual testing rates, the analysis in the foregoing section for other loading rates requires only the knowledge of the constants $\dot{a}_{0T}$ and the exponent n.

Fig. 11.3.4 shows the experimental results of Bažant, Bai and Gettu (1993) for the dependence of peak load on the CMOD rate for limestone. Also shown are the predictions of the time-dependent R-curve model made by Bažant and Jirásek (1993), which fit very well the experimental results. A very large value of the exponent ($n = 55$) is required to achieve this fit.

Exercises

11.7 Consider a material with an R-dependent power-law rate of exponent n, with an R-curve given by $\mathcal{R} = G_f/2$ for $\Delta a < c_f$, and $\mathcal{R} = G_f$ for $\Delta a > c_f$. For a center cracked panel under constant load:

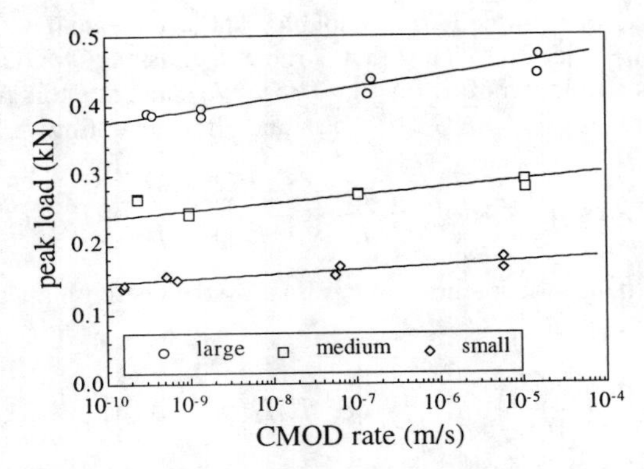

Figure 11.3.4 Effect of the rate on the peak load for notched beams of Indiana limestone of various sizes, tested in three-point bending. Experimental results of Bažant, Bai and Gettu (1993) and theoretical fits of Bažant and Jirásek (1993).

(1) determine the time t_1 required for the crack to grow from a_0 to $a_0 + c_f$; (2) determine the time required for the crack to grow from a_0 to $a_0 + mc_f$, with $m > 1$.

11.8 Give a detailed derivation of Eq. (11.3.14).

11.9 Consider a material with an R-dependent power-law rate of exponent n, with an R-curve given by $\mathcal{R} = mG_f \; \Delta a/c_f$ for $\Delta a \ll c_f$, where $m > 1$ is a constant related to the initial slope of the R-curve. For a double-cantilever beam driven at constant displacement rate $\dot{u}$, give an estimate of the evolution of the crack length, displacement, and load during the first stages of loading (use a single-term approximation of the expressions for the shape functions, i.e., use $a \approx a_0$ in all functions not vanishing for $a = a_0$).

11.10 Generalize the preceding result to any geometry tested at constant displacement rate.

11.11 Give a detailed derivation of Eq. (11.3.17).

11.4 Linear Viscoelastic Fracture Mechanics

Fracture of viscoelastic materials has been developed mainly to deal with the fracture of brittle polymers (Knauss 1970, 1973, 1974, 1976, 1989; Mueller and Knauss 1971; Schapery 1975a–c; Kanninen and Popelar 1985). Only recently the fracture models for concrete have incorporated the analysis of the viscoelastic behavior of the material (Bažant and Jirásek 1993; Wu and Bažant 1993; Bažant and Li 1994a, 1997). Even though the behavior of polymers and concrete is very different, the theory required to analyze the crack growth is similar.

The purpose of this section is to introduce the basic tools required to perform such an analysis. The presentation is far from exhaustive. Many important topics in linear viscoelasticity and in polymer fracture are left out because they are not used in the applications to concrete, which are of the main interest in the remainder of the chapter.

11.4.1 Uniaxial Linear Viscoelasticity

A time-dependent material is said to be linear viscoelastic if the superposition principle holds, i.e., if the response to the superposition of any two loadings is the sum of the responses for the individual loadings. For these materials, the uniaxial strain ε caused by any uniaxial stress history $\sigma(t)$, may be written in terms of the response in a creep test, in which a unit constant stress is applied at time t' (Fig. 11.4.1). Such response is called the uniaxial compliance function $J(t, t')$. For an aging material such as concrete, the origin of time is usually fixed at concrete casting time, and the response (the compliance function) depends on the age of the concrete at loading and on the elapsed time. For nonaging materials, there is

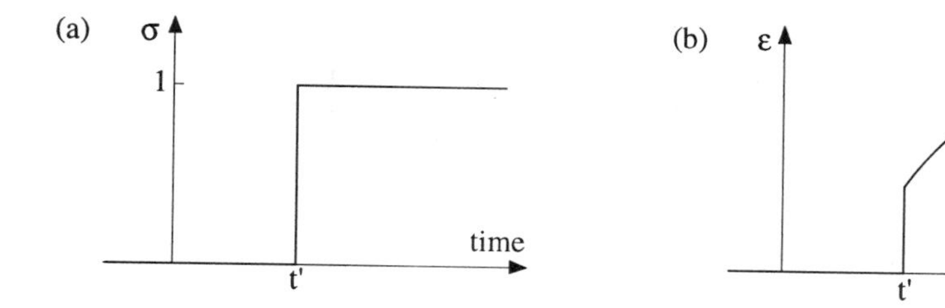

Figure 11.4.1 Definition of the compliance function: (a) unit step loading; (b) strain response.

no particular origin of time to be chosen, and the response can depend only on the time elapsed from the instant of loading, i.e., on $t - t'$. Given $J(t, t')$, and any arbitrary stress history $\sigma(t')$, defined for any time $t' \leq t$, the resulting strain at time t may be written as

$$\varepsilon(t) = \int_{-\infty}^{t} J(t, t') \, d\sigma(t') \tag{11.4.1}$$

where $d\sigma(t') = \dot{\sigma}(t') \, dt'$ and $\dot{\sigma}(t')$ is the classical time-derivative of the stress history if the stress is continuous and piecewise differentiable. If the stress history includes discontinuities (jumps), the derivative must be interpreted in the sense of distributions and include Dirac's δ-functions to represent the jumps. For concrete, where the origin of time is taken to coincide with the casting time, it is usually understood that $\sigma(t') = 0$ for $t' \leq 0$, so that the lower limit in the integral may be set to zero.

For a nonaging material, the compliance function takes the form of a function of one variable, i.e., $J(t - t')$. If a uniaxial constant stress σ is suddenly applied at time t', the instantaneous response is elastic and is given by $\varepsilon(t') = J(0)\sigma$. Therefore, we can identify $J(0)$ with the inverse of the (constant) instantaneous modulus E_0. Likewise, for long-term response, we have $\varepsilon(\infty) = J(\infty)\sigma$ and we can identify $J(\infty)$ with the inverse of the long-term elastic modulus E_∞:

$$J(0) = \frac{1}{E_0} \; , \qquad J(\infty) = \frac{1}{E_\infty} \tag{11.4.2}$$

The nonaging uniaxial viscoelastic behavior is often represented by mechanistic models composed by linear springs and linear dashpots (Figs. 11.4.2a–b), as illustrated by the Kelvin element (Fig. 11.4.2c) and the Maxwell element (Fig. 11.4.2d). Figs. 11.4.2e–f show two equivalent representations of the so-called standard solid obtained by using, respectively, Kelvin elements and Maxwell elements. Figs. 11.4.2g–h show the corresponding generalizations, the Kelvin chain and the Maxwell chain. It may be shown that for any given Kelvin chain there is an equivalent Maxwell chain. The forces and elongations of the springs represent, respectively, stresses and strains, and so the stiffnesses of the springs represent the elastic moduli.

Example 11.4.1 The compliance function for a Kelvin unit (Fig. 11.4.2c) is easily obtained by writing that the total stress acting on the element is the sum of the stress acting on the spring plus the stress acting on the dashpot:

$$\sigma = E_1 \varepsilon + \eta \dot{\varepsilon} \tag{11.4.3}$$

where E_1 is the modulus of the spring and η the viscosity coefficient of the dashpot. Now, $J(t)$ is the strain produced by a unit stress applied at time $t' = 0$, and thus it is the solution of the foregoing differential equation for $\sigma = 1$. That equation is easily integrated, and for the initial condition $\varepsilon = 0$ at $t = 0$:

$$J(t) = \frac{1}{E_1} \left(1 - e^{-E_1 t / \eta} \right) \tag{11.4.4}$$

This result shows that the Kelvin unit displays an infinite instantaneous stiffness ($J(0) = 1/E_0 = 0$) and that the modulus of the spring is the long-term modulus ($J(\infty) = 1/E_1$). It also shows that the viscosity

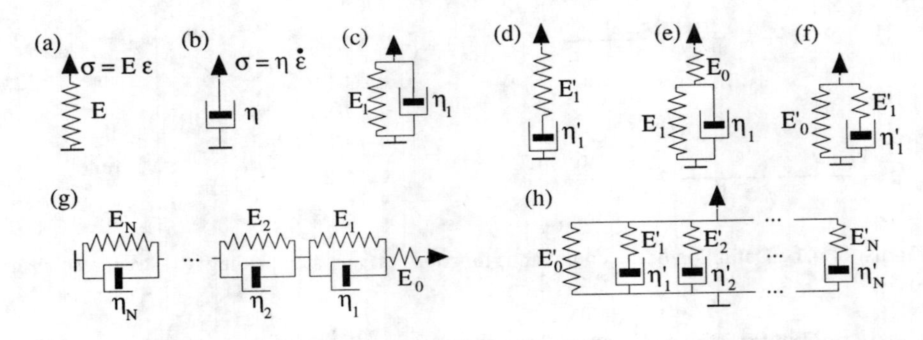

Figure 11.4.2 Mechanistic uniaxial viscoelastic models: (a) elastic element (spring), (b) viscous element (dashpot), (c) Kelvin element, (d) Maxwell element, (e) standard solid using Kelvin elements, (f) standard solid using Maxwell elements, (g) Kelvin chain, and (h) Maxwell chain.

η and the modulus E_1 appear in the exponent combined in the form $\eta/E_1 = \tau_1$, which has the dimension of time; τ_1 can be specified directly, instead of η, so that one writes

$$J(t) = \frac{1}{E_1}\left(1 - e^{-t/\tau_1}\right) \tag{11.4.5}$$

τ_1 is called the *relaxation time* of the Kelvin unit. □

Example 11.4.2 To remove the infinite instantaneous stiffness of the Kelvin unit, we may couple a Kelvin unit in series with a linear spring: this yields the standard solid (Fig. 11.4.2e). In this case the strain is just the sum of the strain of the spring and the strain of the Kelvin unit. Thus:

$$J(t) = \frac{1}{E_0} + \frac{1}{E_1}\left(1 - e^{-t/\tau_1}\right) \tag{11.4.6}$$

Now the instantaneous modulus is E_0 and the long-term modulus corresponds to the series coupling of the springs E_0 and E_1:

$$\frac{1}{E_\infty} = \frac{1}{E_0} + \frac{1}{E_1} \tag{11.4.7}$$

The foregoing model may be extended to any number of Kelvin units in series, which leads to a Kelvin chain (Fig. 11.4.2g). The compliance function is just the sum of the compliance functions:

$$J(t) = \sum_{\mu=0}^{N} \frac{1}{E_\mu}\left(1 - e^{-t/\tau_\mu}\right) \tag{11.4.8}$$

where E_μ and τ_μ are the elastic moduli and relaxation times of the units, and N is their number. □

For an aging material, the instantaneous response to a sudden loading at time t' is given by $\varepsilon(t') = J(t', t')\sigma$. So, $1/J(t', t')$ is equal to the instantaneous modulus $E_0(t')$ at age t'. For long time response to a constant stress σ applied at time t', we have $\varepsilon(\infty) = J(\infty, t')\sigma$ and we can identify $J(\infty, t')$ with the inverse of the long-term elastic modulus for loading at age t', $E_\infty(t')$:

$$J(t', t') = \frac{1}{E_0(t')}, \quad J(\infty, t') = \frac{1}{E_\infty(t')} \tag{11.4.9}$$

11.4.2 Compliance Functions for Concrete

Compliance functions for concrete have been extensively analyzed by Bažant and co-workers. The simplest is the double power-law formulated by Bažant (1975, p. 15) and Bažant and Osman (1976) and calibrated by Bažant and Panula (1978) which is specially suited for short-term loading. For basic creep (sealed specimens, the drying of which is prevented) the BP equation is:

$$J(t,t') = \frac{1}{E_0}\left[1 + \varphi_1\left(t'^{-m} + \alpha\right)\left(t - t'\right)^n\right] \tag{11.4.10}$$

where E_0, φ_1, α, m, and n are constants depending on the concrete characteristics (see Bažant and Panula 1978, for the correlations between these constants and other concrete characteristics such as strength, water-cement ratio, etc.).

A new compliance function based on solidification theory was developed recently (Bažant and Prasannan 1989; Carol and Bažant 1993). In this theory it is recognized that the aging must be physically caused by the volume growth of hydrated cement, and that the creep properties of the basic constituent of the material, that is, the cement gel, must be age-independent. From the fact that a newly solidified material must be stress-free at the time of solidification, a certain particular form of the compliance function for creep results. This formulation has allowed a better description of experimental data than the previous empirical models. It also allowed certain thermodynamic restrictions to be satisfied.

In solidification theory, the compliance rate, that is $\dot{J}(t,t') = \partial J(t,t')/\partial t$, is expressed in the form

$$\dot{J}(t,t') = \frac{\dot{C}(t-t')}{v(t)} + \frac{q_4}{t} \tag{11.4.11}$$

in which $\dot{C}(\xi)$ is the rate of the compliance function of the nonaging material constituent; q_4 is an empirical constant representing a viscous flow rate; $\eta = t/q_4$ can be regarded as a viscosity growing with age t; and $v(t)$ is the volume fraction of the solidified constituent, representing both the actual volume of hydrated cement as well as the effect of the so-called polymerization, causing formation of new bonds to increase the load-bearing volume fraction of the hydrated cement. The following empirical functions have been shown to lead to good agreement with test data:

$$\frac{1}{v(t)} = \left(\frac{\lambda_0}{t}\right)^n + \alpha, \qquad C(t-t') = q_2 \ln\left[1 + \left(\frac{t-t'}{\lambda_0}\right)^n\right] \tag{11.4.12}$$

in which q_2, λ_0, and n are empirical constants (from tests, it was determined that $\lambda_0 = 1$ day). The foregoing expression for the creep compliance function $C(t-t')$ of solidified constituent represents the log-power-law. Denoting $q_3 = \alpha q_2$, we can write

$$\dot{J}(t,t') = \frac{n(q_2 t^{-m} + q_3)}{t - t' + (t-t')^{1-n}} + \frac{q_4}{t} \qquad (m = 0.5, n = 0.1) \tag{11.4.13}$$

Three empirical constants, q_2, q_3, and q_4, must be calibrated by tests for each particular concrete, and the remaining empirical constants m and n can be fixed for all concretes once for all, as indicated. The term multiplying q_2 yields a binomial integral that is not integrable in a closed form, and for this reason a more complicated approximate (but very accurate) expression for $J(t,t')$ has been derived (Bažant and Prasannan 1989). However, for numerical step-by-step analysis of structures, the values of $J(t,t')$ are not needed. The algorithm given in Bažant and Prasannan (1989) (or in Bažant, Ed., 1988a, Chapter 2) involves only the increments ΔJ which can be calculated as $\dot{J}\Delta t$.

An optimal empirical model based on analysis of extensive data in a computerized data bank, called the B3 model, was recently developed by Bažant and Baweja (1995a,b). It was accepted as a RILEM Recommendation (and also received an unanimous positive vote in ACI committee 209, Creep and Shrinkage, for standard recommendation). A simplified but less accurate version of this model was also developed (Bažant and Baweja 1995c), as an update of a previous formulation by Bažant, Xi and Baweja (1993). In this approximation, the compliance function for basic creep is given by

$$J(t,t') = q_1 + q_2\, Q(t,t') + q_3\, \ln\left[1 + \left(\frac{t-t'}{\tau_1}\right)^n\right] + q_4\, \ln\left(\frac{t}{t'}\right) \tag{11.4.14}$$

where, in absence of data for a particular concrete, the following approximations may be used:

$$q_1 = \frac{0.60}{E_{28}}, \quad E_{28} = 4721\sqrt{\sigma_0 f'_c}, \quad \sigma_0 = 1 \text{ MPa} = 145 \text{ psi} \tag{11.4.15}$$

$$q_2 = \frac{1.853 \times 10^{-3}}{\sigma_0}\left(\frac{c}{c_0}\right)^{0.5}\left(\frac{f'_c}{\sigma_0}\right)^{-0.9}, \quad c_0 = 100 \text{ kg/m}^3 = 6.238 \text{ lb. ft.}^{-3} \tag{11.4.16}$$

$$q_3 = 0.29\left(\frac{w}{c}\right)^4 q_2, \quad n = 0.1, \quad \tau_1 = 1 \text{ day}, \quad q_4 = \frac{2.03 \times 10^{-5}}{\sigma_0}\left(\frac{a}{c}\right)^{-0.7} \tag{11.4.17}$$

where f'_c is the average standard 28-day cylinder strength, c the specific cement content (cement mass per unit volume), w/c the water–cement ratio by weight, and a/c is the aggregate–cement ratio by weight. The constants τ_1, σ_0, and c_0 have been introduced for dimensional compatibility. The function $Q(t, t')$ in (11.4.14) is a binomial integral which cannot be expressed analytically. It is given by Bažant and Baweja (1995c) in tabular form and, approximately, by the following expression, valid with less than 1% error for $m = 0.5$ and $n = 0.1$:

$$Q(t, t') = Q_f(t')\left[1 + \left(\frac{Q_f(t')}{Z(t, t')}\right)^{r(t')}\right]^{1/r(t')} \quad \text{with} \quad r(t') = 8 + 1.7\left(\frac{t'}{\tau_1}\right)^{0.12} \tag{11.4.18}$$

and

$$Z(t, t') = \left(\frac{t'}{\tau_1}\right)^{-m} \ln\left[1 + \left(\frac{t - t'}{\tau_1}\right)^n\right], \quad Q_f(t') = \frac{11.63}{\left(\frac{t'}{\tau_1}\right)^{2/9} + 14.07\left(\frac{t'}{\tau_1}\right)^{4/9}} \tag{11.4.19}$$

The foregoing equations suffice for sealed specimens or for short-term loading for which drying is negligible. However, if drying is important, the time-dependent strain due to drying shrinkage, to stress-induced shrinkage, and to the contributions to drying creep caused by microcracking must be added (Bažant and Chern 1985b; Bažant, Ed., 1988a, Chapter 2; Xi and Bažant 1993; Bažant and Baweja 1995a–c).

11.4.3 General Linear Viscoelastic Constitutive Equations

For an isotropic linear viscoelastic material, the relationship between the infinitesimal strain tensor and the stress tensor history can be written in the form:

$$\varepsilon_{ij} = \int_0^t J(t, t')\left\{[1 + \nu(t, t')]\, d\sigma_{ij}(t') - \nu(t, t')\, d\sigma_{kk}(t')\delta_{ij}\right\} \tag{11.4.20}$$

where $J(t, t')$ is the compliance function for uniaxial stress discussed in the previous sections, and $\nu(t, t')$ is the Poisson ratio function. They give the evolution of the strain tensor when the stress is applied at a time t' and held constant thereafter. In general, $\nu(t, t')$ is not constant. However, for concrete, as well as for many other materials, it can be taken to be approximately constant; such an approximation considerably simplifies viscoelastic analysis.

For plane cases, either plane stress or generalized plane strain, a generalized compliance function $J'(t, t')$ can be defined similar to the generalized Young modulus E':

$$\begin{aligned} J'(t, t') &= J(t, t') && \text{for plane stress} \\ J'(t, t') &= J(t, t')\left\{1 - [\nu(t, t')]^2\right\} && \text{for plane strain} \end{aligned} \tag{11.4.21}$$

11.4.4 The Correspondence Principle (Elastic-Viscoelastic Analogy)

Under some quite general conditions, the time-dependent linear viscoelastic solution can be sought as a time integral of classical linear elastic solutions, as shown for nonaging materials by Alfrey (1944) and Biot (1955), and for concrete by McHenry (1943) (see also Bažant, Ed., 1988a). For nonaging materials, this correspondence (also called the elastic-viscoelastic analogy) is analyzed for standard boundary conditions by Christensen (1971), and for varying boundary regions (as applicable to fracture mechanics) by Graham (1968) and Kanninen and Popelar (1985). Here we develop a simple analysis for aging materials, plane cases, and constant Poisson ratio.

Let us assume that we have a plane body subjected to a time-dependent generalized force $P(t)$, with a mode I crack whose length a can vary arbitrarily (independently). Assume further that we can obtain the elastic solution corresponding to an elastic material of generalized elastic modulus E' and Poisson's ratio ν. The instantaneous elastic stress fields must have the form:

$$\sigma_{ij}(x_k, t) = \sigma_{ij}[x_k, a(t), P(t)] \tag{11.4.22}$$

where $\sigma_{ij}(x_k, a, P)$ is the stress at point x_k when the (elastic) body has a crack length a and is subjected to a load P. Note that there is a double time-dependence: one through the load P and another through the crack length a. This stress distribution satisfies the equilibrium equation and the traction-defined boundary conditions at time t. The elastic displacement field has the form

$$u_i^{el}(x_k, t) = u_i^{el}[x_k, a(t), P(t)] \tag{11.4.23}$$

where $u_i^{el}(x_k, a, P)$ is the stress at point x_k when the (elastic) body has a crack length a and is subjected to a load P. These displacements satisfy the displacement-imposed boundary conditions. According to Hooke's law,

$$\varepsilon_{ij} = \frac{1}{2}\left(u_{i,j}^{el} + u_{i,j}^{el}\right) = \frac{1+\nu}{E}\overline{\sigma}_{ij} - \frac{\nu}{E}\overline{\sigma}_{kk}\delta_{ij} \tag{11.4.24}$$

Let us investigate under which conditions the linear viscoelastic solution may be sought in terms of the foregoing elastic solution so that the stress field is identical to the elastic stress field (11.4.22) and the displacement field is given by

$$u_i(x_k, t) = \int_{-\infty}^{t} J(t, t')E\frac{d}{dt'}\left\{u_i^{el}[x_k, a(t'), P(t')]\right\}\, dt' \tag{11.4.25}$$

or, equivalently (because ν is constant and identical for the elastic and viscoelastic solutions),

$$u_i(x_k, t) = \int_{-\infty}^{t} J'(t, t')E'\dot{u}_i^{el}(x_k, t')\, dt' \tag{11.4.26}$$

If one determines the strain tensor field from this equation, and uses the condition (11.4.24), it turns out that it satisfies the linear viscoelastic constitutive equation (11.4.20), with constant ν. But what it is not so obvious is whether the displacement boundary conditions are also automatically satisfied. We need to examine two kinds of boundary conditions: the boundary condition at the supports and the boundary conditions along the crack plane.

The boundary conditions at the supports are automatically satisfied if they consist, as usual, in prescribed displacements over a fixed area: $u_i = 0$ over $\mathcal{S}_u$, where $\mathcal{S}_u$ is the boundary region over which the displacements are imposed. This is easily seen by realizing that in such a case the elastic solution verifies $u_i^{el} = 0$ over $\mathcal{S}_u$, and thus so does the viscoelastic solution (11.4.26).

For the crack plane, we have a displacement continuity condition along the uncracked ligament. This may be rewritten as a zero jump condition along the uncracked ligament. The only difference with the zero displacement condition at the supports is that the ligament area increases as the crack propagates. However, it is easily seen that if a point on the uncracked ligament at time t has also been on the uncracked ligament at all previous time, then the zero jump condition is also satisfied at time t. This means that (11.4.26) satisfies automatically the boundary condition along the crack ligament for growing cracks, but does not for the unlikely event of healing cracks, or for the (intrinsically nonlinear) case in which, after some opening, the crack faces again come in contact (crack closure). With these restrictions in mind, we are now ready to analyze the behavior of cracks in linear viscoelastic plates.

11.4.5 Near-Tip Stress and Displacement Fields for a Crack in a Viscoelastic Structure

Subject to the restrictions described in the previous sections, namely, (1) constant Poisson's ratio, (2) fixed displacement condition at the supports, and (3) open, nonhealing crack, the near-tip stress field is

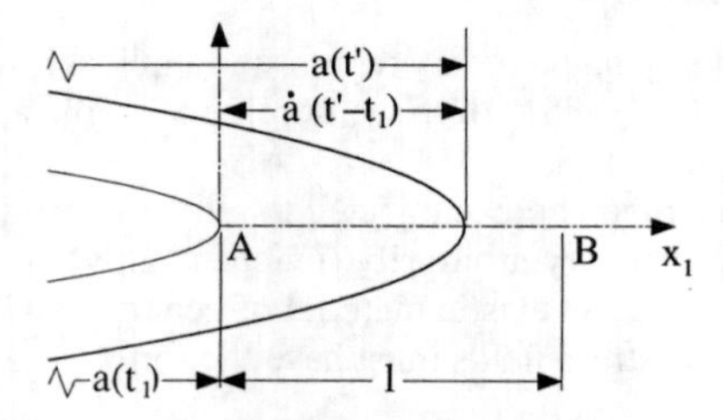

Figure 11.4.3 Intermediate situation in steady crack propagation from point A to point B.

identical to that for linear elasticity. Therefore, the evolution of the crack tip stress field is defined by a time-dependent stress intensity factor $K_I(t)$ which is computed in a standard way in the frame of linear elasticity and is written in the forms given in (2.3.11), i.e.,

$$K_I(t) = \frac{P(t)}{b\sqrt{D}}\,\hat{k}[\alpha(t)] \quad \text{or} \quad K_I(t) = \sigma_N(t)\sqrt{D}\,k[\alpha(t)] \tag{11.4.27}$$

where $\alpha(t) = a(t)/D$.

Unlike the stress field, the displacement fields for a stationary crack and for a propagating crack are different. Let us illustrate this with the crack opening profile. For a stationary crack, let the axes have the origin at the crack tip as in Fig. (2.2.2), and write the crack opening displacement for the elastic case as in (2.2.9). The corresponding opening profile for the linear viscoelastic case is, according to (11.4.26):

$$w(r) = \frac{8}{\sqrt{2\pi}}\sqrt{r}\int_{-\infty}^{t} J'(t,t')\dot{K}_I(t')\,dt' \tag{11.4.28}$$

which displays the classical $\sqrt{r}$ parabolic shape.

The propagating crack behaves differently. Consider the crack shown in Fig. 11.4.3 and assume that its tip is at point A at time t_1, and that the crack travels under constant K_I at constant speed $\dot{a}$ to point B, located at a short distance $l \ll D$ from A, where D is the size of the specimen. Over such a small distance, the elastic stress and displacement fields are very accurately given by the first term of the near-tip expansion. Let us find the crack opening profile over the segment AB for the viscoelastic case. Taking fixed axes at A, the elastic opening at an arbitrary point at distance x_1 from A at time $t' > t_1$ follows from (2.2.9) as

$$w^{el}(x_1,t') = \begin{cases} 0 & \text{for } t' < t_1 + x_1/\dot{a} \\ \dfrac{8K_I}{\sqrt{2\pi}E'}\sqrt{\dot{a}(t'-t_1)-x_1} & \text{for } t' > t_1 + x_1/\dot{a} \end{cases} \tag{11.4.29}$$

From this we may write the crack profile as the crack reaches B at time $t = t_1 + l/\dot{a}$;

$$w(x_1) = \frac{8K_I}{\sqrt{2\pi}}\int_{t_1+x_1/\dot{a}}^{t_1+l/\dot{a}} \frac{J(t_1+l/\dot{a},t')\dot{a}}{2\sqrt{\dot{a}(t'-t_1)-x_1}}\,dt' \tag{11.4.30}$$

Setting $u = \dot{a}(t'-t_1) - x_1$ and $l - x_1 = r$, we get

$$w(r) = \frac{8K_I}{\sqrt{2\pi}}\int_0^r \frac{J'[t_1+l/\dot{a},t_1+(u+l-r)/\dot{a})]}{2\sqrt{u}}\,du \tag{11.4.31}$$

If the crack growth rate is not very slow, the aging during such a small crack length increment can be neglected, and then we may set $J'(t,t') = J'(t-t')$. So the resulting expression for the profile takes the form

$$w(r) = \frac{8K_I}{\sqrt{2\pi}}\int_0^r J'\left(\frac{r-u}{\dot{a}}\right)\frac{du}{2\sqrt{u}} \tag{11.4.32}$$

which is the expression for the near-tip opening profile for a crack running at constant speed under constant stress intensity factor. Examples for particular compliance functions are given next.

Example 11.4.3 Consider a particular nonaging material with power-law creep given by:

$$J'(t,t') = \frac{1}{E'_0}\left[1 + \left(\frac{t-t'}{\tau_0}\right)^n\right] \tag{11.4.33}$$

where τ_0 is a constant with the dimension of time. Such an equation may be a good approximation for concrete under short-term loading, as discussed in the preceding section. The near-tip opening profile for a sustained K_I suddenly applied at time $t' = 0$ is, according to (11.4.28):

$$w(r) = \frac{8K_I}{\sqrt{2\pi}E'_0}\sqrt{r}\left[1 + (t/\tau_0)^n\right] \tag{11.4.34}$$

which shows, as stated before, that the $\sqrt{r}$ parabolic shape of the near-tip opening profile is preserved, and that it is identical to a LEFM crack with a time-dependent effective elastic modulus given by:

$$E'(t) = \frac{E'_0}{1 + (t/\tau_0)^n} = \frac{1}{J'(t,0)} \tag{11.4.35}$$

On the other hand, if the crack is propagating at constant rate under constant stress intensity factor, the crack opening profile is found by inserting (11.4.33) into (11.4.32)

$$w(r) = \frac{8K_I}{\sqrt{2\pi}E'_0}\int_0^r \left[1 + \left(\frac{r-u}{\tau_0 \dot{a}}\right)^n\right]\frac{du}{2\sqrt{u}} \tag{11.4.36}$$

This integrates to give

$$w(r) = \frac{8K_I}{\sqrt{2\pi}E'_0}\sqrt{r}\left[1 + c_n\left(\frac{r}{\tau_0 \dot{a}}\right)^n\right] \tag{11.4.37}$$

where c_n, which is a positive constant depending on n, can be expressed in terms of the Eulerian Γ- and β-functions:

$$c_n = \frac{1}{2}\int_0^1 (1-u)^n u^{-1/2} du = \frac{1}{2}\beta(n+1, 1/2) = \frac{\sqrt{\pi}\,\Gamma(m+1)}{(2m+1)\,\Gamma(m+1/2)} \tag{11.4.38}$$

The foregoing equation shows that, as $r \to 0$, the first term in the square brackets of (11.4.37) dominates and the $\sqrt{r}$ near-tip shape is preserved. Moreover, the dominant term is identical to that in LEFM with the effective $E' = E'_0$. For this first term to be dominant, however, r must be small. Specifically, one must have $r \ll \tau_0 \dot{a}$. Thus, we see that the viscoelastic behavior introduces into the problem of a propagating crack a new parameter $\ell_\tau = \tau_0 \dot{a}$ with the dimension of length, which quantifies the deviation from LEFM. For large values of ℓ_τ, the effects of time dependence are noticeable only at points far from the crack tip. Appreciable deviations from LEFM come closer to the crack tip as ℓ_τ decreases. ▯

Example 11.4.4 Consider now the standard solid defined in Fig. 11.4.2e, whose compliance function is given by (11.4.2). Proceeding as in the preceding example, we find similar results. For the stationary crack under sustained K_I, we get a crack opening profile identical to the linear elastic one except that the a time dependent effective modulus $E'(t) = 1/J'(t,0)$ must be used. For the crack running at constant speed $\dot{a}$ under constant K_I, substitution of (11.4.2) into (11.4.32) followed by the change $u = s/r$ leads to the expression

$$w(r) = \frac{8K_I}{\sqrt{2\pi}E'_0}\sqrt{r}\left[1 + \frac{E'_0}{2E'_1}\varphi\left(\frac{r}{\tau_1 \dot{a}}\right)\right] \tag{11.4.39}$$

where

$$\varphi(\xi) = 2 - \int_0^1 (1-s)^{-1/2} e^{-\xi s}\, ds \tag{11.4.40}$$

Although this expression cannot be integrated in closed form, we can get an approximation for small values of r (or ξ) by taking two terms of the power series expansion of the exponential, with the result:

$$\varphi(\xi) = \frac{4}{3}\xi \qquad (11.4.41)$$

Thus, the deviation from LEFM is again negligible for $r \ll \ell_\tau = \tau_1 \dot{a}$. For the reverse situation, $\ell_\tau \ll r$ (or $\xi \to \infty$), the second term in (11.4.40) vanishes, and the crack opening profile far from the crack tip becomes

$$w(r) = \frac{8K_I}{\sqrt{2\pi}}\sqrt{r}\left(\frac{1}{E'_0} + \frac{1}{E'_1}\right) = \frac{8K_I}{\sqrt{2\pi}\,E'_\infty}\sqrt{r} \qquad (11.4.42)$$

where the last expression follows from (11.4.7). This indicates that, at points far from the crack tip (as compared to ℓ_τ), the crack profile coincides with that for a linear elastic crack with an effective modulus equal to E'_∞. Note that the distances at which this occurs do not need to be large in physical terms. It suffices that ℓ_τ be small, i.e., that either the relaxation time or the crack velocity be small enough. This is consistent with the intuitive idea that, for very small velocities, the material must behave as an elastic material with an elastic modulus equal to E_∞. However, the foregoing analysis shows that this is true only at a certain distance from the crack tip, and that a small region around the crack tip always sees the loading as instantaneous. □

The important point to note from the foregoing examples is that the near tip crack opening profile for a crack advancing at constant speed $\dot{a}$ is always given by the instantaneous elastic profile, and that the deviation from this profile is measured by a function $\varphi(r/\tau\dot{a})$ that vanishes with its argument (linearly for the standard solid and as a power for a power-law creep material). This means that the effect of viscoelasticity is measured by a creep length ℓ_τ defined by

$$\ell_\tau = \tau\dot{a} \qquad (11.4.43)$$

where τ is the relaxation time for the viscoelastic model in use.

11.4.6 Crack Growth Resistance in a Viscoelastic Medium

The LEFM fracture criteria based on the energy release ratio or the stress intensity factor were seen in Chapter 5 to be applicable whenever the fracture process zone size R is much smaller than any characteristic in-plane dimension of the structure D. In viscoelastic fracture mechanics, the situation is different. In this section we show on phenomenological grounds that, in the case $R \ll D$, one may expect a rate-dependent crack growth, such that the crack growth rate depends on the stress intensity factor. In the next section, we introduce a simple model to show that this simplified analysis is justified by more precise approaches.

The important point in the present approach is that, as we have just seen, the time dependence introduces a new length ℓ_τ which, in a sense, measures the influence of the time-dependent part of the strain on the response to the crack advance. Therefore, the case $R \ll D$ may be split into three different cases:

1. $R \ll \ell_\tau \ll D$, in which case the fracture process zone effect is fully negligible.
2. $\ell_\tau \approx R \ll D$, in which case the evolution of the fracture process zone is completely coupled to the viscous behavior.
3. $\ell_\tau \ll R \ll D$, in which case the evolution of the fracture process zone is controlled mainly by the time-dependent fields.

Moreover, since the value of ℓ_τ varies continuously with the crack growth rate, the crack growth model must provide a smooth transition among the three foregoing cases. In the first case, the fracture process zone, being vanishingly small, is completely embedded in a region whose stress and strain correspond to instantaneous loading, as we saw in the preceding section. Therefore, the fracture process zone sees the surrounding field as an instantaneous loading and behaves according to LEFM. So we can define a critical stress intensity factor K_{Ic} and an instantaneous fracture energy G_f related by Irwin's equation:

$$K_{Ic}^2 = E'_0 G_f \qquad (11.4.44)$$

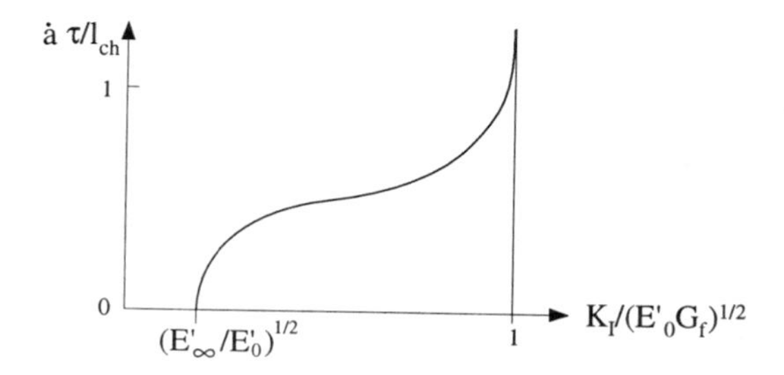

Figure 11.4.4 General trend of the crack growth rate vs. K_I curves.

where E'_0 is the instantaneous elastic modulus. We express the crack growth condition in this limiting case as $K_I = K_{Ic}$, and rewrite the condition $R \ll \ell_\tau$ so that the fracture properties and the crack velocity appear explicitly. Because in this case the behavior is instantaneous elastic, the fracture process zone size R must be proportional to $\ell_{ch} = E'_0 G_f / f_t^2$, where f_t is the instantaneous tensile strength or yield stress. Thus, the condition $R \ll \ell_\tau$ may be rewritten as

$$\dot{a} \gg \ell_{ch}/\tau \tag{11.4.45}$$

Therefore, this first case may be summarized by expressing that for large crack growth rates the stress intensity factor must be equal to the critical stress intensity factor:

$$K_I \to \sqrt{E'_0 G_f} \qquad \text{for} \quad \dot{a} \gg \ell_{ch}/\tau \tag{11.4.46}$$

For the third case, which corresponds to very low values of the crack speed, one may expect that the fracture process zone feels a nearly complete relaxation, so that, if we assume (without any solid base at this point) that the energy required to break the material is rate independent, we can write

$$K_I \to \sqrt{E'_\infty G_f} \qquad \text{for} \quad \dot{a} \ll \ell_{ch}/\tau \tag{11.4.47}$$

For the intermediate range, over which neither full relaxation or instantaneous loading are good approximations, the fracture process zone should be taken into account explicitly. However, for the time being, it will suffice to point out that a solution satisfying the foregoing limiting cases is to assume that, in general, the crack growth velocity is uniquely related to the applied stress intensity factor by a relationship of the form

$$\frac{\dot{a}\tau}{\ell_{ch}} = \varphi\left(\frac{K_I}{\sqrt{E'_0 G_f}}\right) \tag{11.4.48}$$

where $\varphi(K^*)$ is a dimensionless function that satisfies the conditions:

$$\lim_{K^* \to 1} \varphi(K^*) = \infty\,, \quad \varphi\left(\sqrt{\frac{E'_\infty}{E'_0}}\right) = 0\,, \quad \varphi(K^*) = 0 \quad \text{for} \quad K^* < \sqrt{\frac{E'_\infty}{E'_0}} \tag{11.4.49}$$

Fig. 11.4.4 illustrates this kind of behavior. The exact definition of the function $\varphi(K^*)$ is by now arbitrary. To determine it, the exact behavior of the fracture process zone must be defined. For a very simple model, this is done next.

11.4.7 Steady Growth of a Cohesive Crack with Rectangular Softening in an Infinite Viscoelastic Plate

Let us now deduce the relationship between the applied stress intensity factor and the crack growth rate, considering a cohesive crack with rectangular softening in the limit of very small relative size of the cohesive zone.

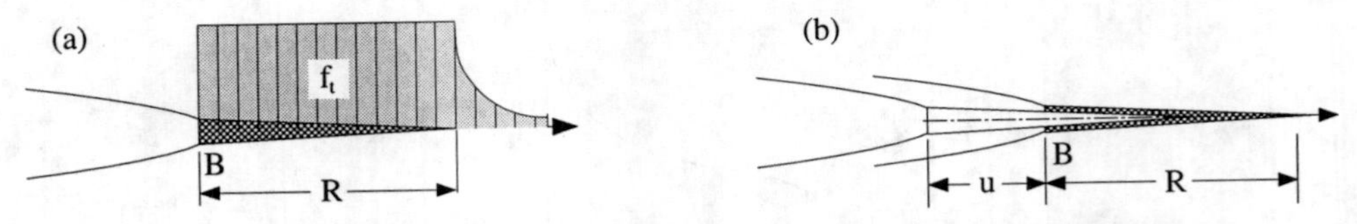

Figure 11.4.5 Steady growth of a cohesive crack with rectangular softening.

Consider a semi-infinite crack in an infinite plate running at constant speed $\dot{a}$ under a constant remote stress intensity factor, and assume that a cohesive zone of size R exists ahead of the crack tip. Let the softening behavior of this cohesive zone be time-independent, so that the definitions introduced in Chapter 7 for the cohesive cracks hold. Finally, assume the softening to be rectangular, with tensile strength f'_t and fracture energy G_F. Then the critical crack opening w_c at which the stress drops to zero is, thus $w_c = G_F/f'_t$.

To solve the viscoelastic problem, we first note that according to the previous results, the stress distribution may be taken to coincide with that which occurs in an elastic medium. In particular, the stress distribution along the crack line would be as shown in Fig. 11.4.5a, with R such that the stress intensity factor at the cohesive zone tip is zero. This condition was imposed in Chapter 7, with the result:

$$R = \frac{\pi}{8}\left(\frac{K_I}{f'_t}\right)^2 \tag{11.4.50}$$

The relationship of K_I to the crack velocity is established by requiring that, at the trailing edge of the cohesive zone (point B), the crack opening (w_B) must coincide with the critical crack opening w_c. To compute w_B, we use the general form (11.4.26), for which we need to compute the time derivative of the elastic crack opening at point B. To do so, we note that if the crack tip location B in Fig. 11.4.5a corresponds to time t, the crack was a distance u behind it at time $t' = t - u/\dot{a}$, as shown in Fig. 11.4.5b. So, the elastic crack opening at point B was:

$$w_B^{el} = w^{el}(x)\big|_{x=u}\,, \quad \text{with} \quad u = \dot{a}\,(t - t') \tag{11.4.51}$$

where $w^{el}(x)$ is the elastic crack opening at a distance x from the tip of the stress-free crack (Fig 7.5.5). Thus, the time derivative of the elastic crack opening is

$$\frac{\partial w_B^{el}}{\partial t'} = -\dot{a}\;\frac{dw^{el}(x)}{dx}\bigg|_{x=u} \tag{11.4.52}$$

(which is analogous to the material time derivative). Taking the expression of the elastic crack opening profile from (7.5.74), differentiating and substituting the result into the foregoing equation, we finally get

$$\frac{\partial w_B^{el}}{\partial t'} = \frac{4f'_t}{\pi E'}\dot{a}\ln\frac{\sqrt{R}+\sqrt{R-(t-t')\dot{a}}}{\sqrt{R}-\sqrt{R-(t-t')\dot{a}}} \quad \text{for} \;\; t' > t - \frac{R}{\dot{a}} \;\; \text{and 0 otherwise} \tag{11.4.53}$$

Inserting this expression into (11.4.26), we get

$$w_B = \frac{4f'_t}{\pi}\dot{a}\int_{t-R/\dot{a}}^{t} J'(t,t')\ln\frac{\sqrt{R}+\sqrt{R-(t-t')\dot{a}}}{\sqrt{R}-\sqrt{R-(t-t')\dot{a}}}\,dt' \tag{11.4.54}$$

Since B is at the trailing end of the cohesive zone, $w_B = w_c = G_F/f_t$. For short-term loading, the aging may be neglected, and so $J'(t,t') = J'(t-t')$, and making in the integral the change $t' = t + (R/\dot{a})s$, we get:

$$1 = \frac{4f_t'^2}{\pi G_F}R\int_0^1 J'\left(\frac{R}{\dot{a}}s\right)\ln\frac{1+\sqrt{1-s}}{1-\sqrt{1-s}}\,ds \tag{11.4.55}$$

In this expression, we may substitute (11.4.50) for R, and rewrite $J'(t-t')$ in the form

$$J'(t-t') = \frac{1}{E_0'}\left[1+\phi\left(\frac{t-t'}{\tau}\right)\right] \tag{11.4.56}$$

where τ is the relaxation time of the material and $\phi(\xi)$ is a dimensionless function that vanishes for $\xi = 0$ and approaches the value $1 - E_0/E_\infty$ for $\xi \to \infty$. With this, the following expression is obtained:

$$1 - K^{*2} = K^{*2}\, Q\!\left(\frac{\pi K^{*2}}{8\dot{a}^*}\right) \tag{11.4.57}$$

where we substituted the identity

$$\frac{1}{2}\int_0^1 \ln\frac{1+\sqrt{1-s}}{1-\sqrt{1-s}}\, ds = 1 \tag{11.4.58}$$

(proven by integration by parts), and used the symbols

$$K^* = \frac{K_I}{\sqrt{E_0 G_F}}, \quad \dot{a}^* = \frac{\tau\dot{a}}{\ell_{ch}}, \quad Q(u) = \frac{1}{2}\int_0^1 \phi(us)\ln\frac{1+\sqrt{1-s}}{1-\sqrt{1-s}}\, ds \tag{11.4.59}$$

Since $\phi(0) = 0$, we have $Q(0) = 0$, and so, for $\dot{a} \to \infty$, $K^* = 1$, and the condition for large crack velocity set forth in the preceding section is fulfilled. At the other end, since $\phi(\infty) = 1 - E_0'/E_\infty'$, it follows from (11.4.58) that $Q(\infty) = 1 - E_0'/E_\infty'$. From (11.4.57) we get $K_I = \sqrt{E_\infty' G_F}$ for the limit of very small crack velocities, as assumed in the previous section. Indeed, assuming $Q(u)$ to be invertible over the interval $0 \le Q \le 1 - E_0'/E_\infty'$, we may rewrite (11.4.57) in the form

$$\dot{a}^* = \frac{\pi K^{*2}}{8Q^{-1}[(1-K^{*2})/K^{*2}]} \tag{11.4.60}$$

where $Q^{-1}(\cdot)$ stands for the inverse of function $Q(u)$, i.e., $Q^{-1}[Q(u)] = u$.

Example 11.4.5 Consider the power-law compliance function defined by (11.4.33). The expressions for Q and Q^{-1} are easily found to be

$$Q = c_n u^n, \quad u = Q^{-1}(Q) = (Q/c_n)^{1/n} \tag{11.4.61}$$

where

$$c_n = \frac{1}{2}\int_0^1 s^n \ln\frac{1+\sqrt{1-s}}{1-\sqrt{1-s}}\, ds = \frac{1}{2(n+1)}\int_0^1 s^n(1-s)^{-1/2}\, ds = \\ = \frac{\sqrt{\pi}\,\Gamma(n+1)}{2(n+1)\Gamma(n+3/2)} \tag{11.4.62}$$

where the second expression follows integrating by parts, and $\Gamma(n)$ is the Eulerian Γ-function. Substitution into (11.4.60) delivers the crack growth rate as a function of the stress intensity factor:

$$\dot{a}^* = c_n^{1/n}\frac{\pi K^{*2+2/n}}{8(1-K^{*2})^{1/n}} \tag{11.4.63}$$

Fig. 11.4.6 shows a plot of this equation for various values of exponent n. Note that, for K_I values less than about $0.2\sqrt{E_0' G_F}$, the rate equation can be approximated by a power-law. ▯

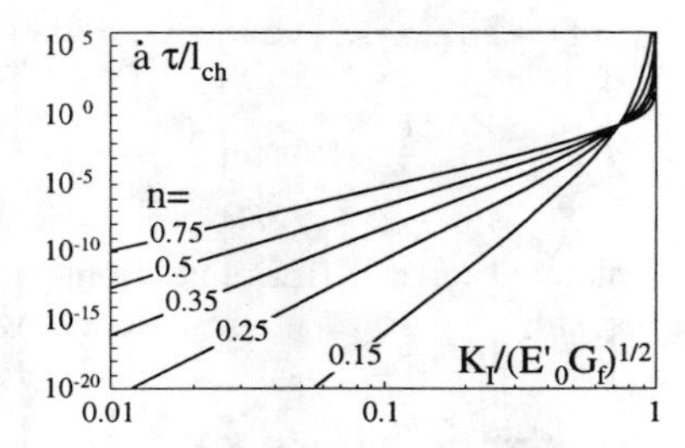

Figure 11.4.6 Crack growth rate vs. applied K_I for a cohesive crack with rectangular softening in a power-law viscoelastic solid.

11.4.8 Analysis of Crack Growth in a Viscoelastic Plate

According to the analysis in the previous two paragraphs, it appears that, for small fracture process zones, the steady crack growth is governed by an equation of the rate type (11.4.48). For an arbitrary compliance function and a time-independent cohesive zone with rectangular softening, the rate equation is given by (11.4.60). The rate equation for a power-law creep is given by (11.4.63). Conceptually similar results are obtained for other types of time-independent softening (Schapery 1975c).

These rate equations are strictly valid for only a steady state. However, if the state changes are slow, the steady-state equation can approximately describe the actual crack growth. Knauss (1976) analyzed the error involved in such approximation, and he concluded that the condition for applicability was that

$$\frac{\dot{K}_I}{K_I} \ll \frac{\dot{a}}{2R} \tag{11.4.64}$$

In view of (11.4.50), this can be reduced to the simple expression

$$\dot{R} \ll \dot{a} \tag{11.4.65}$$

Assuming the foregoing conditions are met, we are ready to analyze the crack growth given by the rate equation (11.4.48) and the initial and boundary conditions. The problem is very similar to that analyzed in Section 11.2, except that now the displacement is governed by a viscoelastic law.

To be specific, let us consider a viscoelastic plate with an initial crack of length a_0, growing in mode IN under load P. Introducing expression (11.2.25) for K_I in (11.4.48), and setting $a = \alpha D$, we get the differential equation

$$\dot{\alpha} = \frac{\ell_{ch}}{\tau D}\varphi\left[\frac{P\hat{k}(\alpha)}{b\sqrt{DE_0'G_F}}\right] \tag{11.4.66}$$

which relates P and α. Given the initial condition $\alpha = \alpha_0$ and the load evolution $P(t)$, it is a standard problem to find the solution of this differential equation. Once $\alpha(t)$ is known, the displacement u_E can be computed by writing first the elastic displacement u_E^{el} as in (11.2.26). Then (11.4.26) which yields the expression:

$$u_E = \frac{1}{b}\int_{-\infty}^{t} J'(t,t')\frac{d}{dt'}\left\{\hat{v}_E[\alpha(t')]P(t')\right\}dt' \tag{11.4.67}$$

In general, the solution of the differential equation (11.4.66) requires numerical procedures. However, in the basic case of constant load, the problem reduces to a quadrature. Indeed, if P is applied at time $t = 0$ and kept constant, the time for the crack to grow to a relative length α can be obtained by separation of variables:

$$t = \frac{\tau D}{\ell_{ch}}\int_{\alpha_0}^{\alpha}\frac{d\alpha'}{\varphi\left[P\hat{k}(\alpha')/(b\sqrt{DE_0G_F})\right]} \tag{11.4.68}$$

A particularly interesting application of this equation is the determination of the time to failure t_f. We define it as the time in which the crack growth rate becomes infinite. According to the first condition for function φ in (11.4.49), this happens when $K_I = \sqrt{E'G_F}$ or, for constant load P, when the crack length reaches failure value α_f, which is obtained from the equation

$$\frac{P}{b\sqrt{D}}\hat{k}(\alpha_f) = \sqrt{E_0'G_F} \tag{11.4.69}$$

After solving this equation, the time to failure is obtained by setting in (11.4.68) the upper integration limit equal to α_f. Upon using (11.4.69), the result is:

$$t_f = \frac{D\tau}{\ell_{ch}}\int_{\alpha_0}^{\alpha_f} \frac{d\alpha'}{\varphi\left[\hat{k}(\alpha')/\hat{k}(\alpha_f)\right]} \tag{11.4.70}$$

Example 11.4.6 Let us find the time to failure of a large plate with a center crack of initial length $2a_0$ subjected to a constant stress σ applied at time $t = 0$. In this case $K_I = \sigma\sqrt{\pi a}$ and the critical crack size is $a_f = E_0'G_F/(\pi\sigma^2)$. Taking bD to be the cross section of the plate, $P = \sigma bD$ and $\hat{k}(\alpha) = \sqrt{\pi\alpha}$. Substitution into (11.4.70) leads to

$$t_f = \frac{\tau}{\ell_{ch}}\int_{a_0}^{f} \frac{da'}{\varphi\left(\sqrt{a/a_f}\right)} \tag{11.4.71}$$

and setting $a/a_f = s$ provides

$$t_f = \frac{\tau a_f}{\ell_{ch}}\Phi\left(\frac{a_f}{a_0}\right), \quad \text{with} \quad \Phi(\zeta) = \int_{1/\zeta}^{1} \frac{ds}{\varphi(\sqrt{s})} \tag{11.4.72}$$

For the case of the rectangular softening analyzed in the previous example, the rate equation is given by (11.4.63); from this, after setting $u = 1/s - 1$ in the integral, the function $\Phi(\zeta)$ is reduced to

$$\Phi(\zeta) = \frac{8}{\pi c_n^{1/n}}\int_0^{\zeta-1} \frac{u^{1/n}}{1+u}du \tag{11.4.73}$$

where c_n is given by (11.4.62). ▯

11.4.9 Crack Growth Analysis at Controlled Displacement

In some practical situations, and in many experiments, the structure is loaded controlling the growth of one displacement, say u_E. In such situations, Eqs. (11.4.67) and (11.4.66) still govern the process, but they are coupled, because the boundary condition is given as $u_E = u_E(t)$. Thus, a set of two simultaneous integrodifferential equations needs to be solved. In general, this must be done numerically.

A general treatment of this problem, to which many approaches are possible, is out of the scope of this book. Here we limit ourselves to give a simple numerical algorithm used by Bažant and Jirásek (1993) to solve a similar problem (although more complicated, see Section 11.5.2).

Take the initiation of the loading to coincide with $t = t_0$, and divide the time in discrete intervals defined by instants $t_i = t_0 + i\Delta t$ $(i = 0, 1, 2, \dots N)$. For brevity, set $u_E = u$, and assume that the values of the displacements $u_i = u(t_i)$ are known. Suppose that the approximate values $\alpha_i = \alpha(t_i)$, $P_i = P(t_i)$ for $i = 0, 1, 2, \dots j$ have already been calculated and the new values α_{j+1}, P_{j+1} are to be found. The discrete approximation of Eq. (11.4.67) in the interval (t_j, t_{j+1}) is:

$$bu_{j+1} = \sum_{i=0}^{j} J(t_{j+1}, \frac{t_{i+1}+t_i}{2})[P_{i+1}\hat{v}_E(\alpha_{i+1}) - P_i\hat{v}_E(\alpha_i)], \tag{11.4.74}$$

For convenience, let us denote $J_{j,i} = J[t_{j+1}, (t_{i+1}+t_i)/2]$, $S_{j-1} = \sum_{i=0}^{j-1} J_{j,i}(P_{i+1}\hat{v}_{i+1} - P_i\hat{v}_i)$

and $\hat{v}_i = \hat{v}_E(\alpha_i)$, and solve for P_{j+1} from the foregoing equation; the result is

$$P_{j+1} = \left(\frac{bw_{j+1} - S_{j-1}}{J_{j,j}} + P_j \hat{v}_j \right) \frac{1}{\hat{v}_{j+1}} \tag{11.4.75}$$

Now, the discrete approximation of (11.4.66) is

$$\frac{\alpha_{j+1} - \alpha_j}{\Delta t} = \frac{\ell_{ch}}{\tau D} \phi \left[\frac{P_{j+1} + P_j}{2b\sqrt{DE'_0 G_F}} k \left(\frac{\alpha_{j+1} + \alpha_j}{2} \right) \right] \tag{11.4.76}$$

Substitution of the expression (11.4.75) for P_{j+1} in the last equation furnishes a nonlinear equation with only one unknown α_{j+1}, which can be solved iteratively.

11.5 Rate-Dependent R-Curve Model with Creep

Bažant and Jirásek's simple solution of time dependent R-curve performed in Section 11.3.3 was limited to linear elastic behavior. However, for loading rates spanning over several orders of magnitude, the creep effects can play an important role. Linearly viscoelastic creep in the bulk of the specimen can be taken into account by combining the results of the previous section with the rate-dependent R-curve. This section describes the general analytical and numerical methods, underlying the analysis of Bažant and Jirásek (1993). The main results are presented, discussed, and extended to an R-curve explicitly dependent on the crack growth rate.

11.5.1 Basic Equations

Experiments performed under load control become unstable after the peak load has been reached. To achieve a stable descending part of the load-displacement curve, displacement control must be adopted. The available experiments (Bažant and Gettu 1992; Bažant, Bai and Gettu 1993; Bažant, He et al. 1991) have been performed under a constant CMOD rate. In such a case, the time history of CMOD is described by a linear function $w_M(t) = \dot{w}_M(t - t_0)$, where t_0 = time at the beginning of the experiment, and $\dot{w}_M$ = constant (representing the prescribed CMOD rate). The analysis of the rate-dependent R-curve model for a linear elastic material was performed in Section 11.3.3. A similar analysis was carried out in the previous section for a viscoelastic material with a constant crack growth resistance. The model of Bažant and Jirásek (1993) is a generalization of the latter, except that the general displacement in Eq. (11.4.67) is replaced by the crack mouth opening displacement w_M (with corresponding dimensionless compliance v_M), and Eq. (11.4.66) is replaced by (11.3.11).

Bažant and Jirásek (1993) write the toughness curve K_R required by (11.3.11) in the form (11.3.18) in which E' must now be interpreted as a conventional elastic modulus, and K_R is obtained from Irwin's relation $K_R = \sqrt{E'\mathcal{R}}$. With this, the problem is similar to that discussed in Section 11.4.8, in which the governing equations (11.4.67) and (11.4.66) are replaced by

$$w_M = \frac{1}{b} \int_{-\infty}^{t} J'(t, t') \frac{d}{dt'} \{ \overline{C}_M[\alpha(t')] P(t') \} \, dt' \tag{11.5.1}$$

$$\dot{\alpha} = \frac{\dot{a}_{0T}}{D} \left[\frac{P\hat{k}(\alpha)}{b\sqrt{D} K_f \sqrt{\rho[D(\alpha - \alpha_0)/c_f]}} \right]^n \tag{11.5.2}$$

The unknown functions $P(t)$ and $\alpha(t)$, describing the variation of the applied load and evolution of the crack length, can be determined by solving this system for specified CMOD history $w_M(t)$. The initial conditions for loading starting at time t_0 are $\alpha(t_0) = \alpha_0$, $P(t_0) = 0$, and $w_M(t_0) = 0$. As in previous sections, the input can be specified as the load history $P(t)$ and then Eq. (11.5.2) is first solved for $\alpha(t)$ and, second, $w_M(t)$ is evaluated from Eq. (11.5.1).

The numerical treatment of this problem by Bažant and Jirásek (1993) is that explained in Section 11.4.9. However, special treatment is necessary in the first few time steps when the process zone is very small and K_R is, therefore, close to zero. In fact, at time t_0 the ratio K_I/K_R is not defined because $K_R = 0, K_I = 0$. Even though one does not need to evaluate this ratio at t_0, α is very close to α_0 at the beginning of crack propagation and numerical problems arise due to strong sensitivity of the high power $(K_I/K_R)^n$ to even very small changes of α. To overcome these problems, Bažant and Jirásek (1993) make use of an approximate analytical solution for $\alpha - \alpha_0 \ll 1$. This approximation is explained next.

11.5.2 Approximate Solution for Small Crack Extensions

For constant CMOD rate, $w_M = \dot{w}_M(t - t_0)$, Bažant and Jirásek (1993) use for the initial state an approximate analytical solution, which can be derived for $\alpha - \alpha_0 \ll 1$ when P is approximated by a linear function of time; then $P(t) = \dot{P}(t - t_0)$, where $\dot{P}$ is the mean loading rate over the interval (t_0, t). For small $\alpha - \alpha_0$, $\hat{v}_M(\alpha)$ can be replaced by $\hat{v}_M(\alpha_0)$, so that (11.5.1) is transformed into $b\dot{w}_M(t - t_0) = \dot{P}\hat{v}_M(\alpha_0)\int_{t_0}^{t} J'(t,t')dt'$, from which

$$\dot{P} = \frac{b\dot{w}_M(t - t_0)}{\hat{v}_M(\alpha_0)\int_{t_0}^{t} J'(t,t')dt'} \qquad (\text{for} \quad \alpha - \alpha_0 \ll 1) \tag{11.5.3}$$

The dependence of the right-hand side of this equation on time indicates that $\dot{P}$ is not constant for finite intervals of time. So, as previously pointed out, $\dot{P}$ represents the mean loading rate over (t_0, t). It tends, however, to a constant as $t \to t_0$ because, in that case, $J'(t,t') \approx J'(t_0, t_0) = 1/E'(t_0)$. So,

$$\dot{P} = \frac{b\dot{w}_M(t - t_0)}{\hat{v}_M(\alpha_0)\int_{t_0}^{t} J'(t_0,t_0)dt'} = \frac{b\dot{w}_M E'(t_0)}{\hat{v}_M(\alpha_0)} \qquad \text{for} \quad \alpha - \alpha_0 \ll 1 \tag{11.5.4}$$

which is indeed a constant.

Once P is known for the initial stages of loading, the rate equation (11.5.2) is simplified by writing that, for small $\alpha - \alpha_0$,

$$\rho[D(\alpha - \alpha_0)/c_f] \approx \rho'(0)D(\alpha - \alpha_0)/c_f \tag{11.5.5}$$

where $\rho'(0)$ is the initial derivative of function ρ. With this, Eq. (11.5.2) can be transformed to

$$\dot{\alpha} = \beta_0\left(\frac{t - t_0}{\sqrt{\alpha - \alpha_0}}\right)^n, \qquad \beta_0 = \frac{\dot{a}_{0T}}{D}\left(\frac{\dot{P}\hat{k}(\alpha_0)\sqrt{c_f}}{bDK_f\sqrt{\rho'(0)}}\right)^n \tag{11.5.6}$$

Solving the approximate crack propagation equation (11.5.6) by separation of variables, we get

$$\alpha = \alpha_0 + \beta_1(t - t_0)^q \qquad \beta_1 = \left(\frac{\beta_0}{q}\right)^p \qquad p = \frac{2}{n+2}, \qquad q = p(n+1) \tag{11.5.7}$$

Note that if n is large, $\alpha - \alpha_0 \propto (t - t_0)^2$.

Bažant and Jirásek use this analytical approximation for the first time-steps and then shift to a numerical algorithm similar to that described in Section 11.4.9.

11.5.3 Comparison with Tests

Bažant and Jirásek (1993) compared the performance of the foregoing model with the experimental results reported in Bažant, He et al. (1991), Bažant and Gettu (1992), and Bažant, Bai and Gettu (1993) which have been described in Section 11.1.2. The creep compliance function $J(t,t')$ was approximated by the well-known double-power-law (used in the BP model): $J(t,t') = E_0^{-1}\left[1 + \phi_1(t'^{-\bar{m}} + \alpha)(t - t')^{\bar{n}}\right]$; its parameters were: $E_0 = 48.4$ GPa, $\phi_1 = 3.93$, $\bar{m} = 0.306$, $\bar{n} = 0.133$, $\alpha = 0.00325$.

For the fitting of data, it is helpful to make the following observations. (1) Parameters $\dot{a}_{0T}$ and K_f are mutually dependent, and so only one of them can be regarded as a free parameter. By increasing K_f or decreasing $\dot{a}_{0T}$, the peak loads are increased for all the rates and sizes in the same ratio. (2) Parameter

n affects mainly the rate sensitivity (and does so for all the sizes in the same manner). By increasing n, one can decrease the slope of the rate effect curve, which is indicated by experiments to be roughly linear when the CMOD rate is plotted in a logarithmic scale. (3) Parameter c_f affects brittleness, and does so for all the rates in roughly the same manner. Increasing c_f causes a shift towards the left in the size effect plot of Fig. 11.1.4, i.e., to a more ductile behavior.

To avoid excessive rate sensitivity, a very large exponent n is needed. For example, $n = 38$ for the concrete studied by Bažant and Gettu (1992, see Fig. 11.1.3a) and $n = 55$ for the limestone studied by Bažant, Bai and Gettu (1993, see Fig. 11.3.4).

The rate-dependent R-curve model has also been used to model wedge-splitting tests of very large concrete specimens reported in Bažant, He et al. (1991) and in more detail by He et al. (1992). Similar trends have been observed.

The predictions of the model are compared with the experiments in Fig. 11.1.3 (dashed curves). As pointed out in Section 11.1.2, the foregoing rate-dependent generalization of the R-curve model yields no shift of brittleness. Thus, it was possible to get a close agreement between theory and experiments at different sizes for limestone (Fig. 11.3.4), but not for concrete (Fig. 11.1.3). Therefore, a modification of the model to include a rate-dependent brittleness was required for concrete.

11.5.4 Rate-Dependence of Process Zone Length

The rate-dependent R-curve model just expounded suffers by a serious drawback: it is incapable of modeling the rate-dependent shift of brittleness observed in experiments (Bažant and Gettu 1992). To model it, we may replace the constant value of c_f (process zone length at peak load for an infinitely large specimen) by a rate-dependent function $c_f(\dot{\alpha})$. The rate-dependence of c_f is not illogical. Stress relaxation in the fracture process zone may be expected to cause the stress profile along the crack extension line to develop a steeper drop to zero, spanning a shorter length. This means that the effective fracture process zone length should be smaller at slower crack propagation.

As explained in Chapter 5, c_f is the basic parameter affecting brittleness. Because brittleness is seen to decrease with increasing rate, c_f should be an increasing function of $\dot{\alpha}$. The brittleness number, and thus also c_f, should change by only one order of magnitude as the loading rate changes over five orders of magnitude. It is therefore reasonable to use a power function with a low exponent:

$$c_f = k_a \dot{a}^{1/m} \tag{11.5.8}$$

where $\dot{a} = da/dt$ = rate of crack length growth and k_a, m = constants, with $m \gg 1$.

With c_f dependent on $\dot{a}$, Eq. (11.5.2) now becomes an implicit law for $\dot{a}$. If the model is to be physically reasonable, there must exist a unique nonnegative solution $\dot{a}$. This condition imposes a restriction on the value of m. A simple analysis of this restriction has been performed by Bažant and Jirásek (1993) by approximating $\rho(\Delta_a/c_f)$ with a piecewise linear function. It transpired that the parameter m in (11.5.8) must be larger than $n/2$, n being the exponent in (11.5.2). But to obtain a large shift of brittleness, one needs a small m value. The limitation $m > n/2$ unfortunately prevents obtaining as large a shift of brittleness as that observed in the tests of Bažant and Gettu (1992), as shown by the solid lines in Fig. 11.1.3.

11.5.5 Sudden Change of Loading Rate and Load Relaxation

Another time-dependent effect is seen after a sudden change of loading rate as in the tests of Bažant, Gu and Faber (1995) discussed in Section 11.1.2 (Fig. 11.1.5). The rate-dependent R-curve model described here exhibits qualitatively the same behavior, as Fig. 11.5.1 shows. The tests further suggest that, after a rate change, the curve for the new rate asymptotically approaches the curve for a constant rate test with a rate equal to the new rate. The theory agrees with this behavior, too.

Quantitative agreement between theory and experiments can be verified by plotting the measured ratio P_2/P_p vs. P_1/P_p, where P_p is the first peak load, P_1 the load at which the change of rate is produced, and P_2 is the secondary peak load. The points marked by different symbols in Fig. 11.5.2 correspond to tests on specimens with three sizes (d = 38mm, 76mm, 152mm) in which the rate increased by three orders of magnitude (on the average from 10^{-8} m/s to 10^{-5} m/s). The test results seem to be independent of size. Calculations based on the present theory agree with this behavior.

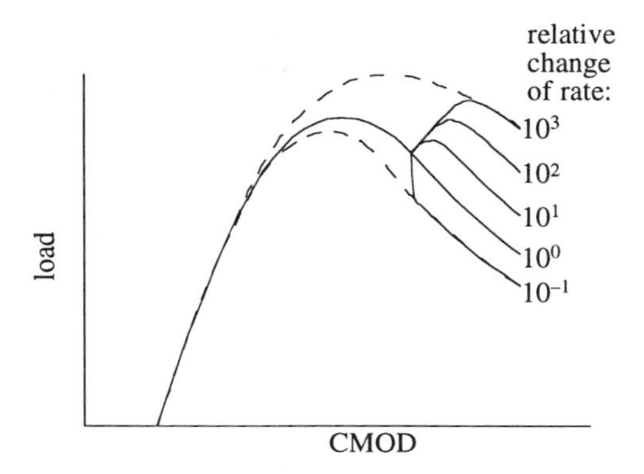

Figure 11.5.1 Theoretical prediction of the effect of a sudden change in CMOD rate (after Bažant and Jirásek 1993).

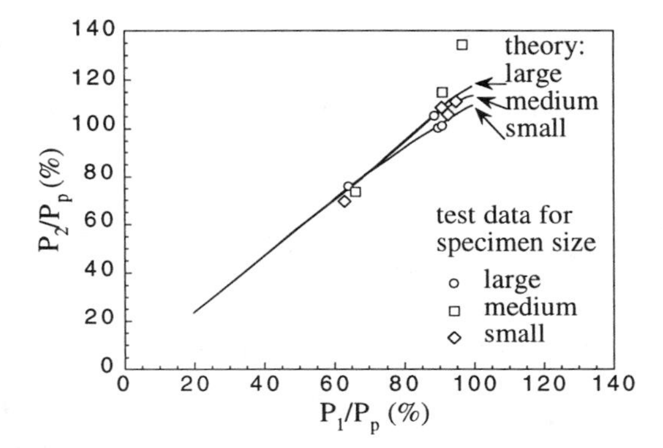

Figure 11.5.2 Relative values of second peak (P_2) for a thousandfold increase in the CMOD rate at load P_1.

Bažant and Jirásek applied their model to interpret the relaxation tests reported by Bažant and Gettu (Section 11.1.2). Fig. 11.5.3 gives the predicted relaxation curves for various initial CMOD rates, to be compared with the experimental curves in Fig. 11.1.5. A qualitative agreement is found — the curves corresponding to different initial rates have the same final slope in a logarithmic plot and are shifted with respect to each other. However, the theoretical curves are steeper than the experimental ones.

Fig. 11.5.4 shows the theoretical relaxation curves for loading that has been stopped before the peak,

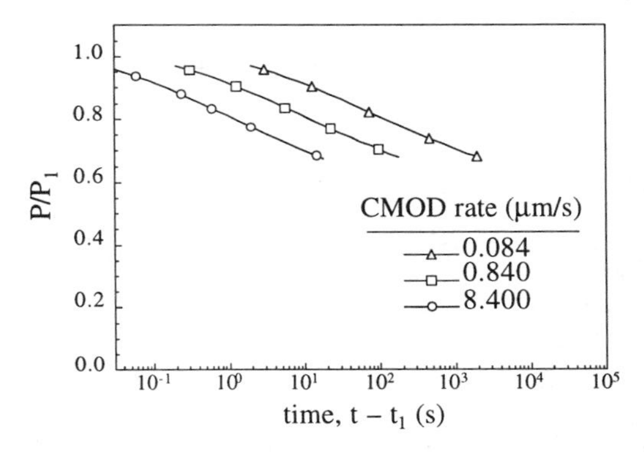

Figure 11.5.3 Relaxation curves predicted by the Bažant and Jirásek's model for various initial CMOD rates.

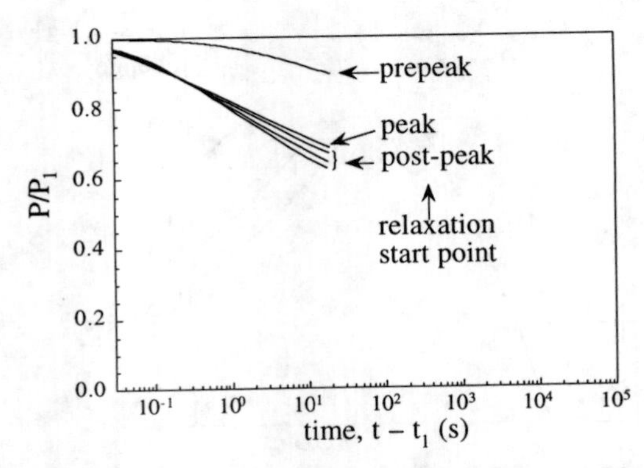

Figure 11.5.4 Relaxation curves predicted by the Bažant and Jirásek's model for an initial CMOD rate of 8.5 μm/s at various load levels.

close to the peak, and after the peak. It reveals that there is again only a qualitative agreement with the corresponding experimental results of Fig. 11.1.6 — the relaxation curves starting in the postpeak range lie below the curve starting approximately at peak, which in turn lies below the curve starting in the prepeak range. The theoretical curves are again steeper than the experimental ones.

11.5.6 Summary

The equivalent linear elastic fracture model based on an R-curve can be generalized to the rate effect if the crack propagation velocity is assumed to depend on the stress intensity factor and on its critical value based on the R-curve. This dependence may be assumed in the form of an increasing power function with a large exponent.

The creep in the bulk of a concrete specimen must also be taken into account. This can be done by replacing the elastic constants in the LEFM formulas with a linear viscoelastic operator in time. For rocks, which do not creep, this is not necessary.

The experimental observation that the brittleness of concrete increases with a decreasing loading rate (i.e., the response shifts in the size effect plot closer to linear elastic fracture mechanics) can be at least approximately modeled by assuming the effective fracture process zone length in the R-curve expression to decrease with a decreasing rate. This dependence may again be described by a power function. Good agreement with the previous test results for concrete and limestone, recently measured at very different loading rates (with times to peak ranging from 1 second to 250,000 seconds) has been achieved.

The model can also predict several other phenomena observed in the laboratory. (1) When the loading rate is suddenly increased, the slope of the load-displacement diagram suddenly increases. For a sudden rate increase, the slope becomes positive even in the postpeak range, and, later in the test, a second peak, lower or higher than the first peak, is observed. (2) When the rate suddenly decreases, the slope suddenly decreases and the response approaches the load-displacement curve for the lower rate. (3) When the displacement increase is arrested, relaxation causes a drop of load, approximately following a logarithmic time curve.

11.6 Time-Dependent Cohesive Crack and Crack Band Models

The time dependence of fracture has in the past been studied mainly for materials in which the fracture process zone is very small relative to the cross section dimension. Such a simplification, as we now know, is inadequate for quasibrittle materials with coarse heterogeneous microstructure, such as concretes, rocks, ice, and toughened ceramics.

The simplest way to take the effects of a large fracture process zone into account is the R-curve model, whose generalization for time dependence was presented in the preceding section. However, the R-curve model, which is essentially an equivalent LEFM approximation, cannot be accurate enough in all situations and is insufficient for situations where the fracture process zone is so large that it interferes or interacts with the specimen boundaries.

A more general fracture model that can handle such situations is the cohesive (or fictitious) crack model analyzed in Chapter 7, as well as its distributed counterpart, the crack band model expounded in Chapter 8. In the section devoted to viscoelastic fracture mechanics we have already used a cohesive crack in a viscoelastic body, although only for the simplest rectangular softening and for the large size limit. We showed that even if the cohesive stresses are time-independent, the resulting crack growth is time-dependent because of the viscoelastic behavior of the bulk material. However, the cohesive crack itself can be made time dependent by making the cohesive stress depend not only on the crack opening, but also on the rate. Thus, three categories of the time-dependent cohesive crack models may be distinguished: (1) time-independent cohesive stresses in a viscoelastic (time-dependent) body; (2) time-dependent cohesive stresses in an elastic (time-independent) body; and (3) time-dependent cohesive stresses in a viscoelastic body.

11.6.1 Time-Independent Softening in a Viscoelastic Body

Section 11.4.7 shows that a cohesive crack with time-independent softening displays a time-dependent growth when growing in a viscoelastic body. This kind of models in which the time-dependent behavior is entirely due to the viscous creep of the bulk of the body, has been extensively used to describe time-dependent fracture of polymers (Knauss 1970, 1973, 1974, 1976, 1989; Mueller and Knauss 1971; Schapery 1975a–c; Kanninen and Popelar 1985). These models, however, consider only small scale fracturing (i.e., cohesive zones very small compared to the size of the specimen). The formulation of this kind of model, which was sketched in Sections 11.4.6–11.4.9, is hardly applicable to concrete, except perhaps to the largest structures such as cracked dams. In particular, it can never be used to analyze macroscopic crack formation in an initially uncracked structure.

A systematic analysis of a time-independent cohesive stress zone in a material with viscoelastic creep seems to be unavailable in the literature. Such an analysis is needed to clarify the relative importance of creep of the bulk in the observed time-dependent behavior of cracked structures. There exists, however, a simplified analysis by Hillerborg (1991) and by Bažant (1992a), which helps explain the general implications of viscoelastic creep. In particular, it qualitatively explains the rate-dependent shift of brittleness found by Bažant and Gettu (1992) and illustrated in Fig. 11.1.4.

Hillerborg's (1991) approach consists in making a pseudo-elastic analysis with a time-dependent effective elastic modulus. The starting point is the result (Chapter 7) that the nominal strength σ_{Nu} of geometrically similar bodies can be written in the form

$$\sigma_{Nu} = f_t' \, Q\left(\frac{D}{\ell_{ch}}\right) \tag{11.6.1}$$

where f_t' is the tensile strength, $Q(D^*)$ a dimensionless function, and $\ell_{ch} = E'G_F/f_t'^2$ the characteristic size. Hillerborg, ignoring the dependence on crack velocity, introduces the time dependence by replacing E' with the effective modulus defined as

$$E'(t) = \frac{1}{J(t, t_0)} \tag{11.6.2}$$

where t_0 is the age at loading and $J(t, t_0)$ the compliance function. With this, the time-independent strength equation (11.6.1) is transformed into a strength that depends on the loading duration:

$$\sigma_{Nu} = f_t' \, Q\left(\frac{D f_t'^2 J(t, t_0)}{G_F}\right) \tag{11.6.3}$$

in which σ_{Nu} is now the maximum load that can be sustained by the structure until time t if loaded at time t_0.

Bažant's (1992a) paper (based on a 1991 report), dealt with sea ice rather than concrete. It also used the effective modulus for assumed linear viscoelastic creep in the bulk of the material and it also led to

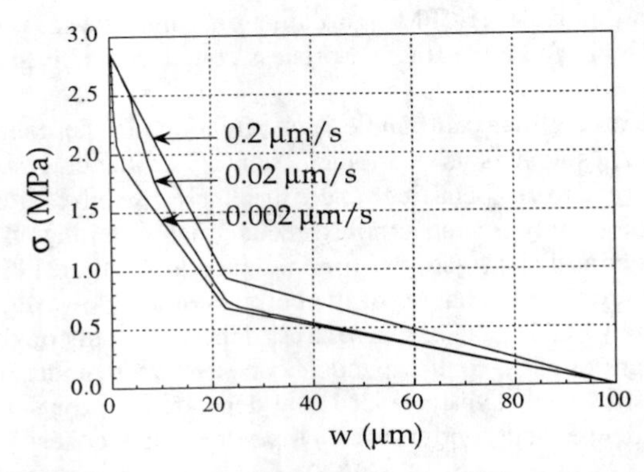

Figure 11.6.1 Cohesive stress vs. crack opening for various crack opening rates in the approximation of Zhou and Hillerborg (1992) (from Zhou 1993).

the conclusion that the creep caused a shift of brittleness. Furthermore, the effect of the rate of crack propagation was approximately considered in this work. This was done by decreasing the value of G_f with a decreasing velocity.

11.6.2 Time-Dependent Softening in an Elastic Body

A more realistic approach is to consider a time-dependent cohesive zone in an elastic medium. Zhou and Hillerborg (1992) and Zhou (1992, 1993) have developed a model of this type based on the stress-relaxation tests in tension described in Section 11.1. The model has some features of plasticity and some of viscoelasticity. The plastic-like feature is that there is a stress limit $f(w)$ defined as a function of the crack opening. It coincides with the classical softening curve for an infinite rate $\dot{w}$. For stress levels below the softening curve, two stress-rate components are considered: an instantaneous one, which is taken to be linear through the origin, and a relaxation term, which is taken to be proportional to an over-stress $\sigma - \sigma_0(w)$ where $\sigma_0(w)$ is a threshold stress depending on the instantaneous crack opening. The governing equations are as follows:

$$\dot{\sigma} = -\frac{\langle \sigma - \sigma_0(w) \rangle}{\tau} + \frac{\sigma}{w}\dot{w} \quad \text{if} \quad \sigma \leq f(w) \text{ or } [\sigma = f(w) \text{ and } \dot{w} > h(w)] \tag{11.6.4}$$

$$\dot{\sigma} = f'(w)\dot{w} \quad \text{if} \quad \sigma = f(w) \text{ and } \dot{w} \leq h(w) \tag{11.6.5}$$

where τ is a constant relaxation time, $\sigma_0(w)$ is the threshold stress function, and $\langle \cdot \rangle$ denote the Macauley brackets, defined so that $\langle u \rangle = u$ for $u > 0$ and $\langle u \rangle = 0$ for $u \leq 0$; $f(w)$ is the softening curve for (infinitely) fast loading; and

$$h(w) = \frac{[f(w) - \sigma_0(w)]w}{[f(w) - wf'(w)]\tau} \tag{11.6.6}$$

is the critical crack opening rate that makes the response follow the softening curve. This rate is obtained by setting $\dot{\sigma} = f'(w)\dot{w}$ and $\sigma = f(w)$ in (11.6.4), and solving for $\dot{w}$. These two equations, with the initial cracking condition $\sigma = f(0) = f'_t$, can be integrated to give the evolution of the cohesive stress and crack opening for any given loading.

Zhou and Hillerborg (1992) use a threshold value $\sigma_0(w)$ proportional to the softening function. In particular, they assume $\sigma_0(w) = \alpha f(w)$, in which α is a constant. For the function $f(w)$ they use a bilinear softening. The predictions shown in Figs. 11.1.8 were obtained for $f'_t = 2.8$ MPa, $G_F = 82$ N/m, $\tau = 25$ s, and $\alpha = 0.7$. Fig. 11.6.1 shows the σ–w curves for three different rates of loading. Note that each curve initiates from the tensile strength value (so the tensile strength is time-independent in this model).

The predictions of this model correctly capture the order of magnitude and the qualitative trend of the experimental results (Fig. 11.6.1). However, the lifetime predictions show too strong a power-law dependence, and, moreover, the correlation with experiments has been investigated only for a time-span of about one order of magnitude. It is very likely that a relaxation term proportional to $\sigma - \sigma_0(w)$ cannot hold over many decades.

However, the model can be easily modified to include a power-law or exponential relaxation, and then equations (11.6.4) and (11.6.6) are generalized to

$$\dot{\sigma} = -\phi(\sigma, w) + \frac{\sigma}{w}\dot{w} \quad \text{if} \quad \sigma \leq f(w) \quad \text{or } [\sigma = f(w) \quad \text{and} \quad \dot{w} > h(w)] \tag{11.6.7}$$

$$h(w) = \frac{\phi(\sigma, w)}{f(w) - wf'(w)} \tag{11.6.8}$$

where function $\phi(\sigma, w)$, the relaxation rate, can take any suitable form, for example, a power-law:

$$\phi(\sigma, w) = \frac{1}{\tau}\left(\frac{\sigma}{f(w)}\right)^n \tag{11.6.9}$$

where τ and n are constants. This is only speculative, and more research is needed to find out the most suitable form for $\phi(\sigma, w)$.

11.6.3 Time-Dependent Cohesive Crack Model

A rate-dependent model based on the activation energy concept was proposed by Bažant (1993b) to describe the evolution of the cohesive stresses. In this model, the rate of crack opening $\dot{w}$ is assumed to correspond to a single thermally activated process with symmetric barriers, which gives rise to a dependence on the driving force of the sinh-type as in Eq. (11.2.12). Obviously, the driving force must be related to the applied stress. However, as the crack opening increases, the number of active bonds decreases. The driving force is the stress acting on the surviving highly stressed bonds σ_b, rather than the average stress (the cohesive stress σ). Thus, we can rewrite Eq. (11.3.10) in the form

$$\dot{w} = \dot{w}_{0T} \sinh(c_T \sigma_b), \qquad c_T = \frac{V_b}{RT}, \qquad \dot{w}_{0T} = \dot{w}_1 \nu_T e^{-Q_0/RT} \tag{11.6.10}$$

where V_b is the activation volume, and C, $\dot{w}_1$, and Q_0 are constants.

The stress σ_b in the bonds that undergo fracturing needs to be related to the overall crack-bridging (cohesive) stress σ. To do this precisely, we would need to take into account the number of the resisting bonds, which decreases with increasing crack opening, as well as the local variation of the stress distribution along the crack surfaces, which depends on the local deformation of the crack surfaces, which, in turn, depends on the local stiffness of the microstructure at various parts of the crack surface. This aspect of modeling would obviously be complicated, and a simplification is inevitable.

Therefore, Bažant introduced for σ_b a semi-empirical function $\sigma_b \propto \sigma - \phi(w)$, where $\phi(w)$ is a function that describes the stress-displacement relation $\sigma = \phi(w)$ of the cohesive crack model for an infinitely slow rate $\dot{w} \longrightarrow 0$ of the crack opening. This expression for σ_b is the simplest way to satisfy two conditions: **(1)** for $\sigma = \phi(w)$, we must have $\dot{w} = 0$; and **(2)** at constant σ, an increase of w must cause a decrease in the number of bonds per unit area, and this, in turn, must cause an increase of force per bond.

But the need to achieve a reasonable behavior at the tail of the stress-displacement curve suggests a more complicated semi-empirical expression, as proposed by Bažant (1993b) and used by Wu and Bažant (1993), Bažant and Li (1997), and Li and Bažant (1997):

$$c_T \sigma_b = (T_0/T)\psi(\sigma, w)\ , \qquad \psi(\sigma, w) = \frac{\langle \sigma - \phi(w) \rangle}{k_0 f'_{t0} + k_1 \phi(w)} \tag{11.6.11}$$

where k_1, k_0, and f'_{t0} are constants; f'_{t0} is the tensile strength at zero opening rate, introduced for the sake of dimensionality; $\langle \cdot\cdot \rangle$ denote the Macauley brackets, $\langle u \rangle = \max(u, 0)$. This implies that the behavior for $\sigma < \phi(w)$ is time-independent.

The form of the denominator of Eq. (11.6.11) is suggested, according to Wu and Bažant (1993), by the need to prevent the denominator from approaching 0 ($k_1 > 0$), and also by the need to model the long

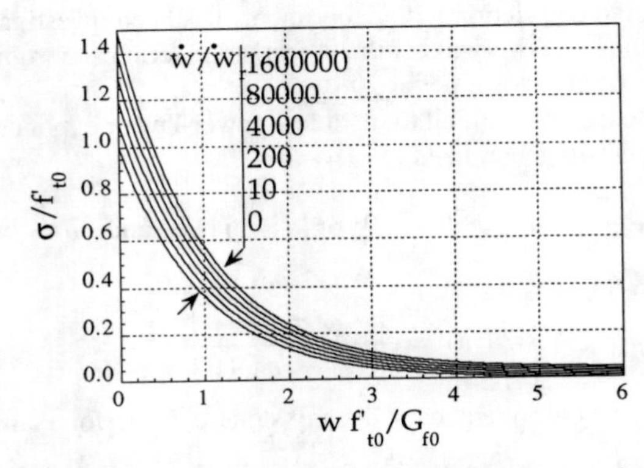

Figure 11.6.2 Stress-displacement relations for various displacement rates in the model of Wu and Bažant (1993).

plateau at the tail of the stress-displacement diagram, as seen in tests. In a subsequent study, focused mainly on the size effect on σ_{Nu}, Bažant and Li (1997) used $k_1 = 0$ with good results.

Eq. (11.6.11) leads to the relation (proposed in Bažant 1993b):

$$\dot{w} = \dot{w}_r \sinh\left[\frac{T_0}{T}\psi(\sigma, w)\right] \exp\left(\frac{Q_0}{RT_0} - \frac{Q_0}{RT}\right) \tag{11.6.12}$$

where $\dot{w}_r$ is an empirical constant. For the reference temperature T_0, this simplifies as

$$\dot{w} = f(\sigma, w) = \dot{w}_r \sinh[\psi(\sigma, w)] \qquad (\text{ at } T = T_0) \tag{11.6.13}$$

in which f is a function of σ and w. For the purpose of numerical calculations, it is convenient to express the crack-bridging stress σ explicitly from (11.6.11) and (11.6.13). The result for constant temperature is

$$\sigma = F(w, \dot{w}) = \phi(w) + [k_0 f'_{t0} + k_1\phi(w)] \sinh^{-1}(\dot{w}/\dot{w}_r) \qquad (\text{ at } T = T_0) \tag{11.6.14}$$

The range of possible values of coefficient k_1 in the model was estimated by Wu and Bažant (1993) to be 0.01–0.05, based on the experimental fact that the peak stress is increased by approximately 25% when the loading rate is increased by 4 to 8 orders of magnitude.

Function $\phi(w)$ for an infinitely small rate of crack opening cannot be measured directly. It is convenient to relate it somehow to the stress-displacement relation $\sigma = f(w)$ for fracture in normal static tests in the laboratory. It seems that a simple but realistic assumption might be $\phi(w) = \kappa_1 f(w)$, in which $\sigma = f(w)$ is the stress-displacement relation of the cohesive crack model for the normal loading rate in a static test, that is, the loading rate that leads to the maximum load within about 3 to 10 minutes, and κ_1 is a constant that is close to 1 but less than 1. Roughly, $\kappa_1 = 0.8$ because the long-time strength of concrete structures is known to be about 80% of the short-time strength (e.g., Rüsch 1960).

The constants involved in Eq. (11.6.14) were calibrated according to the general experimental trends by Wu and Bažant (1993), with the results $k_0 = 0.003$, $k_1 = 0.03$, $\dot{w}_r = 25.4$ μm/day. The corresponding plots of the stress-displacement relations for various displacement rates are shown in Fig. 11.6.2. The areas under these curves represent the apparent fracture energies of the material for various opening rates. The area for $\dot{w} \to 0$ represents G_f, the basic material property.

This model was applied by Wu and Bažant (1993) to analyze the crack growth in concrete taking into account the aging creep behavior. A summary of this application is given next.

11.6.4 Analysis of Viscoelastic Structure with Rate-Dependent Cohesive Crack by Finite Elements

According to Bažant (1993b) and Wu and Bažant (1993), the time dependence of fracture is significant over many orders of magnitude of the loading rates and crack growth rates, with the times to the peak load in a constant loading rate test ranging between 0.001 s to at least 10 days (at least nine orders of

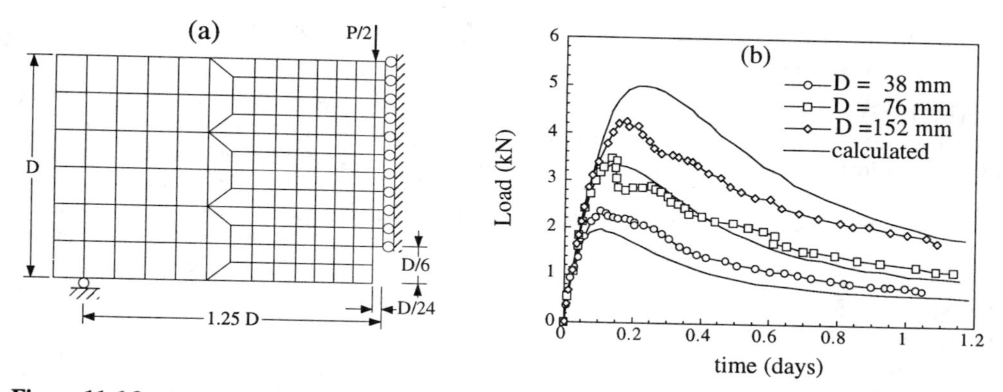

Figure 11.6.3 (a) Finite element mesh used by Wu and Bažant (1993) in the simulations of three-point bending tests on notched concrete specimens. (b) Comparison of test results for three sizes from Bažant and Gettu (1992) and theoretical predictions of Wu and Bažant (1993).

magnitude), and probably even over 50 years. Furthermore, a realistic model applicable throughout such a broad range requires distinguishing properly between the rate effects in the fracture process zone, which are highly nonlinear and are described by the model just presented, and the rate effects in the rest of the structure, which are approximately linear and hereditary, describable by a viscoelastic stress-strain relation in which the aging due to the progress of hydration in concrete must be included for longer load durations.

Thus, the complete model of Bažant (1993b) and Wu and Bažant (1993), discussed in more detail by Bažant (1995c), is built by using the foregoing time-dependent cohesive crack model in a viscoelastic material with a creep function defined by the solidification theory (Section 11.4.2). The numerical implementation of the viscoelastic equations is made in two steps. The first step is semianalytical, and consists in using an algorithm devised by Bažant and Prasannan (1989) to approximate the log-power-law for $C(t-t')$ in the constitutive equation (11.4.12) by a Kelvin chain with N elements. As a consequence of this approximation, the viscoelastic constitutive equation can be reduced to a set of linear differential equations of the first order in the variables $\boldsymbol{\sigma}$, $\boldsymbol{\varepsilon}$, and $\boldsymbol{\varepsilon}_\mu$, where $\boldsymbol{\sigma}$ is the stress tensor, $\boldsymbol{\varepsilon}$ is the strain tensor, and $\boldsymbol{\varepsilon}_\mu$ ($\mu = 1, \cdots, N$) are internal variables representing the strains of the viscous Kelvin elements. The second step consists in discretizing these differential equations to get a system of linear algebraic incremental equations.

The cohesive crack model is implemented in the finite element code by using the classical smeared crack formulation in which the crack opening is related to the mean fracturing strain across an element by setting $w = h\varepsilon_f$, where ε_f is the fracturing strain in the stress-strain relation with strain-softening, and h is the size of the element. For the details of the tensorial formulation, see Chapter 8 or Wu and Bažant (1993).

At the structural level, the problem leads to a system of nonlinear algebraic equations which are solved by a Newton-Raphson iterative method within each time-step increment (for details, see Wu and Bažant 1993).

Several examples of finite element analysis using the model just described have been presented by Wu and Bažant (1993). Of primary interest here is the simulation of the three-point bending tests of Bažant and Gettu (1992) reported in Section 11.1. Fig. 11.6.3a shows the finite element mesh used in the computations. The performance of the model is compared with the test results of Bažant and Gettu (1992) in Fig. 11.1.2 ($k_1 = 0.03, \dot{w}_r = 10^{-3}$ in./day, $G_f = 0.2$ lb./in.; the tensile strength f_t' and Young's modulus are calculated from the ACI formula). Fig. 11.6.3b shows the calculated load-CMOD curves for the specimens of three different sizes at approximately the same CMOD rate. As shown by these comparisons, the model of Wu and Bažant seems promising, and it can be probably further refined to improve the predictions, describe the size effect at various rates, model the effect of changes of loading rate, etc.

11.6.5 Analysis of Viscoelastic Structure with Rate-Dependent Cohesive Crack by Compliance Functions

Bažant and Li (1997) recently developed a boundary-integral model which greatly increases computational efficiency. The integral formulation is then discretized following a method which leads to a pseudo-boundary-integral formulation similar to that described in Section 7.4 for time-independent cohesive cracks. Here we merely extend to the viscoelastic case the results obtained for elastic bodies in Section 7.4. The extension is based on the correspondence principle (elastic-viscoelastic analogy) expounded in Section 11.4.4. According to this principle, if the history of tractions acting on the boundary of the structure (which includes the cohesive crack faces) is given, then **(1)** the instantaneous stress field coincides with the elastic field caused by the instantaneous tractions, and **(2)** the displacement field is given by the convolution integrals (11.4.25) or (11.4.25).

The second property implies that the crack opening displacement at a nodal point (see Figs. 7.4.1 and 7.4.2) can be written as

$$w_i(t) = \int_{-\infty}^{t} J(t,t')E\frac{d}{dt'}\{w_i^{el}[P(t')]\}\,dt' \qquad (i = 1, \cdots, M) \tag{11.6.15}$$

Note that here and in what follows, the derivative must be interpreted in the sense of the theory of distributions, i.e., if w_i^{el} has a jump Δw_i at $t = t_0$, its derivative must include a Dirac's δ-function: $\Delta w_i\,\delta(t - t_0)$.

Now, we can express the elastic crack opening at the nodes in terms of the compliances using (7.4.6). The derivative of w_i^{el} follows at once as

$$\frac{d}{dt'}\{w_i^{el}[x, a(t'), P(t')]\} = C_i\dot{P}(t') - \sum_{j=1}^{M} K'_{ij}\dot{\sigma}_j(t') \tag{11.6.16}$$

Inserting this expression into (11.6.15) we get the following expression for the nodal crack openings (Li and Bažant 1997):

$$w_i(t) = C_i P^{\text{eff}}(t) - \sum_{j=1}^{M} K'_{ij}\sigma_j^{\text{eff}}(t) \tag{11.6.17}$$

in which

$$P^{\text{eff}}(t) = \int_{-\infty}^{t} J(t,t')E\,\dot{P}(t')\,dt'\,, \qquad \sigma_j^{\text{eff}}(t) = \int_{-\infty}^{t} J(t,t')E\,\dot{\sigma}_j(t')\,dt' \tag{11.6.18}$$

Thus, the formula for the crack opening based on the influence matrix is similar to that for the elastic case, except that the load and the nodal stresses are replaced by their effective values given in (11.6.18).

The conditions to be satisfied when the cohesive crack tip reaches node C are the same as (7.4.1)–(7.4.3), except that (7.4.2) must be replaced by its time-dependent version

$$\sigma_i = F(w_i, \dot{w}_i) \tag{11.6.19}$$

where $F(w, \dot{w})$ is given in (11.6.14).

To solve the problem, the time-dependent boundary condition must be given. For example, Li and Bažant (1997) consider a test driven at constant CMOD rate. The corresponding condition can be written as

$$w_1 = \dot{w}_0\,t \tag{11.6.20}$$

where $\dot{w}_0$ is a constant, and the loading is assumed to start at time $t = 0$. Li and Bažant (1997) adopt a method of solution in which, instead of assuming that the time step is given (as is usual in time-dependent models), the crack growth step is fixed and equal to the finite element size. In this way, they force the cohesive crack to grow *exactly* from node to node. This considerably improves the accuracy of crack representation and eliminates roughness in the response curves.. The time t_α at which the crack reaches node α is obtained as part of the solution.

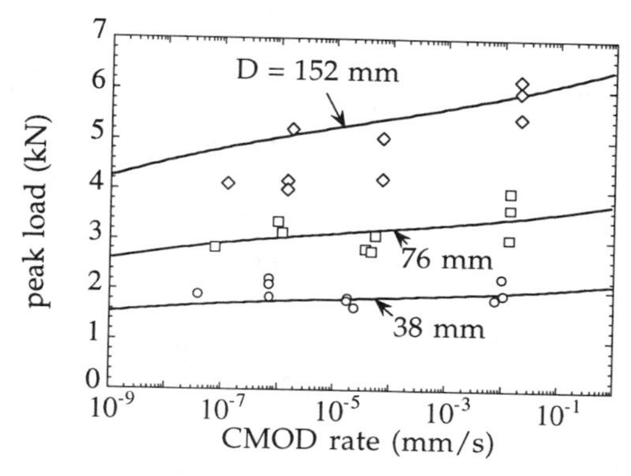

Figure 11.6.4 Influence of specimen size and loading rate on the peak load of notched concrete specimens: comparison of the experimental results of Bažant and Gettu (1992) with the predictions of the model of Li and Bažant (1997).

To see how the solution may proceed in each loading step, consider that we know the solution for the moment at which the crack tip has reached node $T-1$ and want to compute the solution at the instant the crack will reach node T. Then system of simultaneous equations (11.6.17)–(11.6.20), (7.4.1), and (7.4.3), provide in all $3M+3$ equations with $3M+3$ unknowns, namely σ_i, σ_i^{eff}, w_i, P, P^{eff}, and t (the derivatives $\dot{w}_i$ are directly expressible in terms of w_i and t). Li and Bažant (1997) use a semianalytical technique similar to that used by Planas and Elices (1991a) for the time-independent formulation (Section 7.4.3) to reduce the number of nonlinear simultaneous equations to be solved at each step from $3M+3$ to $(T-N+1)+2$, where $(T-N+1)$ is the number of nodes in the cohesive zone at the corresponding stress. This enormously enhances the computational efficiency. For further details, see the original paper by Li and Bažant (1997).

Li and Bažant applied this method to solve a number of theoretical cases using a time-dependent softening function (11.6.14) in which $\phi(w)$ is linear and $k_1 = 0$. They further applied the method to the analysis of the experimental results of Bažant and Gettu (1992). The results, shown in Fig. 11.6.4, indicate that the model can describe reasonably well the effect of size and rate on the maximum load. The model also clarifies the differences in the effects of viscoelasticity and rate-dependent opening considered separately, as seen in Fig. 11.6.5. The effect of viscoelasticity alone is shown in Fig. 11.6.5a, and the effect of time-dependent opening alone in Fig. 11.6.5b. Fig. 11.6.5c further shows the simulation of the response to a sudden change of loading rate, for which the rate dependence of crack opening is essential if the slope reversal from negative to positive should be captured. Viscoelasticity alone cannot model the slope reversal. Comparison with further experimental results can be found in Tandon et al. (1995).

11.7 Introduction to Fatigue Fracture and Its Size Dependence

Crack growth due to load repetition, or cyclic loading, is the cause of fatigue. After a certain number of load cycles, a crack can reach a critical length at which the structure becomes unstable and fails. From a macroscopic (continuum) point of view, the growth of cracks is caused by irreversible inelastic behavior in the fracture process zone and the surrounding nonlinear hardening zone. This irreversible behavior generates a cumulative damage which causes the material to soften at loading levels below those required for softening under monotonic (static) loading.

From a microscopic point of view, there are many possible micromechanisms of fatigue, but they are radically different for metals (in which irreversible dislocation motion is the main source of cumulative damage) and for brittle materials such as ceramics; see, e.g., Suresh (1991).

In this section we limit the presentation to a concise macroscopic description of crack growth in metals, brittle materials, and especially, concrete. For the cases where the fracture zone is small compared to the specimen dimensions (*all* dimensions, crack length included), the crack growth rate can be correlated to

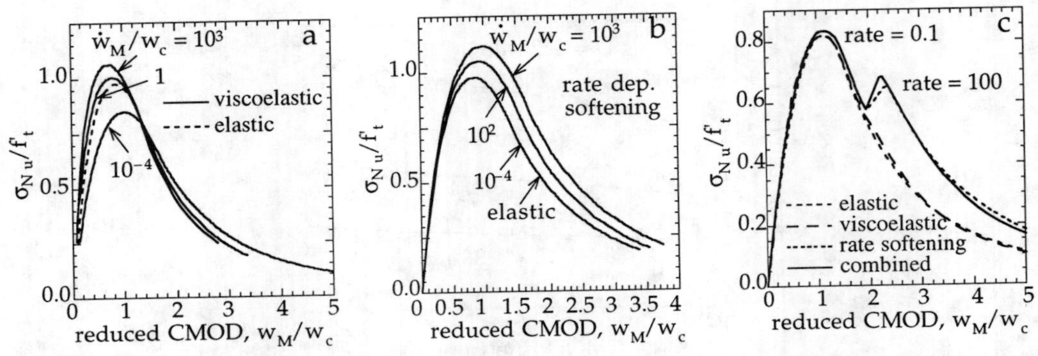

Figure 11.6.5 Load vs. CMOD curves predicted by Bažant and Li's model: (a) model including viscoelasticity only; (b) curves including time-dependent crack opening only; (c) curves for a sudden increase of loading rate predicted by models with viscoelasticity only, time-dependent softening only, and with their combined effect. The elastic curve (with time-independent softening) is also included for comparison. (Adapted from Li and Bažant 1997.)

the amplitude of the cyclic stress intensity factor. However, for concrete, in which the fracture process zone is large, the results of Bažant and Xu (1991) and Bažant and Schell (1993) show that this correlation also depends on the size of the specimen. This dependence can be taken into account by using a size effect law mathematically similar to the classical Bažant's size effect law. Useful though the phenomenological correlations may be, a deeper knowledge of what is going on in the fatigue process requires modeling of the fracture process zone under general loading histories. A few approaches in this direction have been proposed recently, and will be briefly discussed in closing the section.

11.7.1 Fatigue Crack Growth in Metals

Fatigue fracture has been extensively studied for metals, for which it is of paramount importance. A basic result has been obtained by Paris, Erdogan, and co-workers (Paris, Gomez and Anderson 1961; Paris and Erdogan 1963). They observed that the crack growth rate, characterized by $\Delta a/\Delta N$ where a = crack length and N = number of cycles, depends mainly on the amplitude of the stress intensity factor ΔK_I calculated by LEFM, but, in the first approximation, is independent of the maximum and minimum values of K_I attained in the cycles. Furthermore, they showed that the dependence of the crack growth rate on ΔK_I is approximately a power-law, usually called Paris law (or Paris-Erdogan law):

$$\frac{\Delta a}{\Delta N} = A(\Delta K_I)^m, \quad \Delta K_I = K_{max} - K_{min} \tag{11.7.1}$$

in which A and m are empirical constants. The exponent m for metals is usually in the range 3 to 4, with theoretical models of cumulative damage predicting a value of 4 (Suresh 1991). The proportionality factor A is strongly dependent on the microstructure of the metal, and is influenced by temperature and by the load ratio R, defined as

$$R = \frac{K_{min}}{K_{max}} \tag{11.7.2}$$

In pure fatigue (independent of loading duration and rate), the fatigue crack growth is independent of frequency and wave form. In practice, time-dependent crack growth is superimposed, and these factors also affect the parameters in Paris law. The influence of environment can be particularly strong (corrosion-fatigue).

The Paris law is a good approximation for what is called the intermediate range of the fatigue crack growth (Fig. 11.7.1a). Tests with a very broad range of ΔK_I indicate that, for large ΔK_I, the values of $\Delta a/\Delta N$ deviate from the straight line of slope n upward, and for small ΔK_I downward, producing an S-shaped curve (Fig. 11.7.1a). This curve shows that stable crack growth can take place only for values of ΔK_I larger than a certain value called the fatigue threshold ΔK_{th}, and that (because the crack rate tends

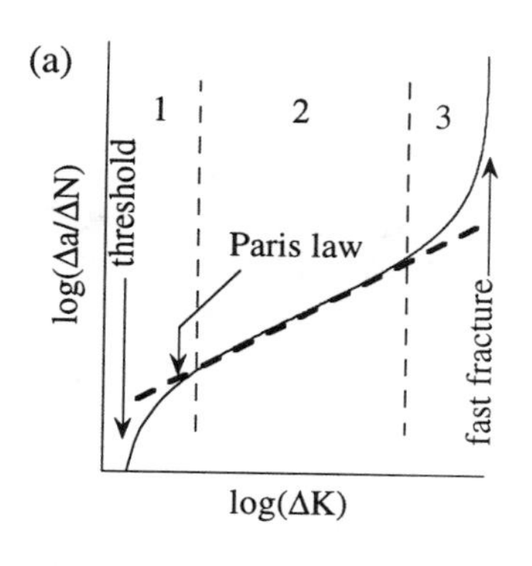

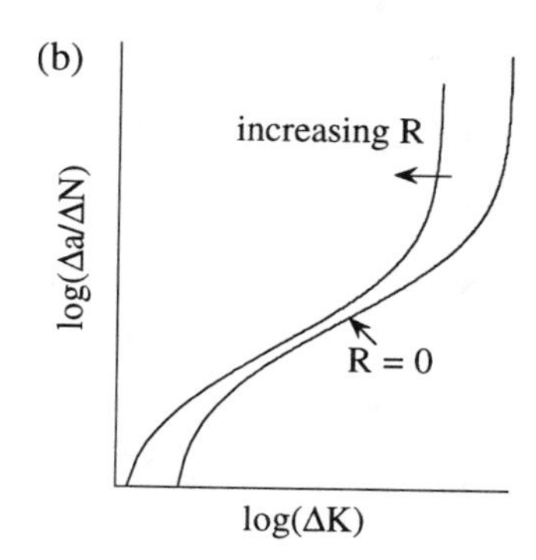

Figure 11.7.1 (a) General trend of the fatigue curve for a metal. (b) Sketch of the dependence of the fatigue curve on the load ratio R.

to infinity) the crack becomes unstable when the maximum stress intensity factor K_{max} approaches the material fracture toughness K_{Ic}.

Many equations have been proposed to describe the complete fatigue crack growth curve (see, e.g., Kanninen and Popelar 1985). That proposed by Priddle (1976) captures the S-shaped form of the curve:

$$\frac{\Delta a}{\Delta N} = A_1 \left(\frac{\Delta K_I - \Delta K_{th}}{(1-R)K_{Ic} - \Delta K_I} \right)^{m_1} \tag{11.7.3}$$

in which ΔK_{th} is not a material constant, but can depend upon R. Empirical values for the threshold are given, for example, in Rolfe and Barsom (1987). A power-law function of the type $\Delta K_{th} = B(1-R)^\gamma$ can be a good approximation for steels, with γ within 0.5 and 1.0 depending on the type of steel (Schijve 1979).

In summary, the fatigue crack growth curve depends on the load ratio R in a way that depends strongly on the material, and, in general, the precise dependence must be determined empirically. The general dependence of the fatigue curves on this ratio is sketched in Fig. 11.7.1b. It appears that the central part of the curve (the part described by Paris law) is only mildly dependent on R, but the upper and lower parts (the threshold) strongly depend on it.

11.7.2 Fatigue Crack Growth in Brittle Materials

Brittle materials, particularly ceramics, were thought to be insensitive to fatigue loading. In recent years this view has changed and it is now accepted that fatigue crack growth can take place in ceramics and other brittle materials, as well as concrete (Suresh 1991; Mai 1991). Fig. 11.7.2 shows fatigue curves for some materials. The most remarkable aspect of these plots is that the Paris exponent is very large for these materials (10 to 45). This explains why the fatigue measurements appeared so difficult: the range of K_{max} for which the crack growth per cycle is noticeable is very narrow and close to K_{Ic}.

The micromechanisms of fatigue crack growth can be quite varied (Suresh 1991). However, from the macroscopic point of view, the process seems to be describable by means of a fracture process zone with cumulative damage. A cyclically degrading fracture zone has been analyzed in the macroscopic sense for ceramics by Mai (1991) who used a cohesive model with degradation (called the bridging zone in the ceramists' terminology). In Mai's model, the cohesive (or bridging) stress is assumed to depend both on the crack opening and on the number of cycles in the form

$$\sigma = \sigma_R \left(1 - \frac{w}{w_c(N)} \right)^p \tag{11.7.4}$$

where σ_R and p are constants and $w_c(N)$ is a function of the number of cycles. The dependence of the softening function on the number of cycles is somewhat arbitrary and does not give any insight into the unloading-reloading behavior. On the other hand, it is not clear how to take into account the fact that different points on the cohesive zone have experienced a different number of cycles.

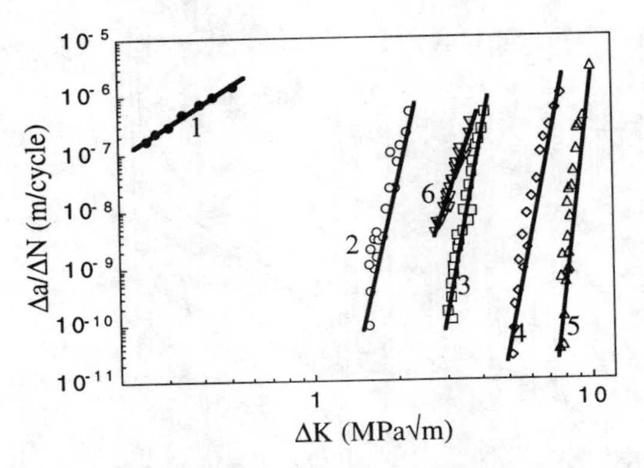

Figure 11.7.2 Fatigue crack growth curves for various brittle and quasi-brittle materials at room temperature. [(a) $R = 0.3$, after Baluch, Qureshy and Azad (1989). (b) $R = 0.1$, after Dauskardt, Marshall and Ritchie (1990). (c) $R = -1$, after Reece, Guiu and Sammur (1989).]

For concrete, a different approach, based also on the cohesive crack concept, has been proposed by Hordijk and Reinhardt (1991, 1992) and, with strong simplifications, by Horii, Shin and Pallewatta (1990) and Horii (1991). These models describe the essential of the unloading-reloading behavior and do not depend explicitly on the number of cycles.

We will discuss these models in more detail later. Before that, however, we should note that the representation of crack growth rate in terms of ΔK_I is completely meaningful only if the LEFM limit is valid, i.e., if the fracture process zone is very small compared to all the specimen dimensions, including the crack length (this is one of the reasons why the fatigue curves are different for short cracks and for long cracks). If that condition is not met, the fatigue curves must be expected to depend on the size of the specimen. This is particularly true for concrete, in which the fracture process zone is usually far from negligible. This aspect is the object of the following paragraph.

11.7.3 Size Effect in Fatigue Crack Growth in Concrete

The applicability of Paris law to fatigue crack growth for concrete was verified by Swartz, Hu and Jones (1978); Swartz and Go (1984); Baluch, Qureshy and Azad (1989), and Perdikaris and Calomino (1989). The results of Bažant and Xu (1991) and especially Bažant and Schell (1993) also confirm a power-law for the cyclic crack growth rate, as Fig. 11.7.3 illustrates.

The tests of Bažant and Xu (1991) and Bažant and Schell (1993) were conducted on concrete with a reduced maximum aggregate size (0.5 in.). The specimens were similar three-point-bend beams of depths $D = 1.5, 3$, and 6 in. and spans $L = 2.5D$, with notch depths $a_0 = D/6$ (Fig. 11.1.2a). Because it is next to impossible to determine the effective crack length by optical observations, the crack length was determined from the unloading stiffness in each load-deflection cycle. For further details of the tests, see Bažant and Xu (1991) and Bažant and Schell (1993).

As shown in Fig. 11.7.3, these tests displayed a strong dependence of the result on the specimen size. It was found (see Bažant and Xu 1991 for normal concrete, and Bažant and Schell 1993 for high strength concrete) that a correction for the size effect can be introduced into the Paris law by rewriting it as

$$\frac{\Delta a}{\Delta N} = \kappa \left(\frac{\Delta K_I}{K_{Ic}} \right)^m \tag{11.7.5}$$

where κ and m are size-independent constants, but K_{Ic}, unlike the original Paris equation, is size dependent. The dependence of K_{Ic} on the size turns out to follow an equation similar to the Bažant size effect law for the apparent fracture toughness in monotonic loading, that is,

$$K_{Ic} = K_{If} \sqrt{\frac{\beta'}{1+\beta'}}, \qquad \beta' = \frac{D}{D_0'} \tag{11.7.6}$$

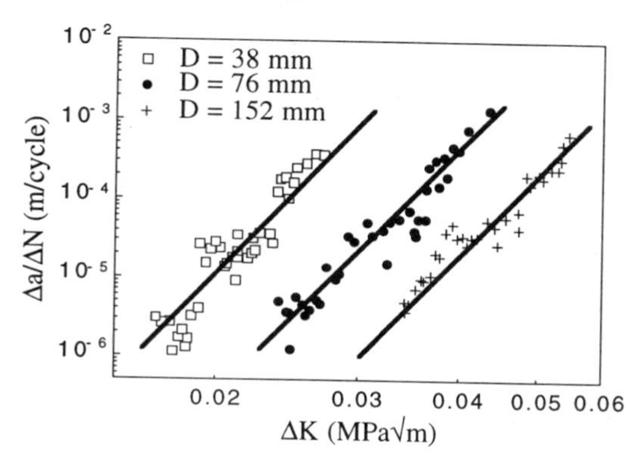

Figure 11.7.3 Logarithmic plots of crack length increment per cycle vs. stress increment amplitudes, with regression lines for specimens of three sizes (after Bažant and Xu 1991)

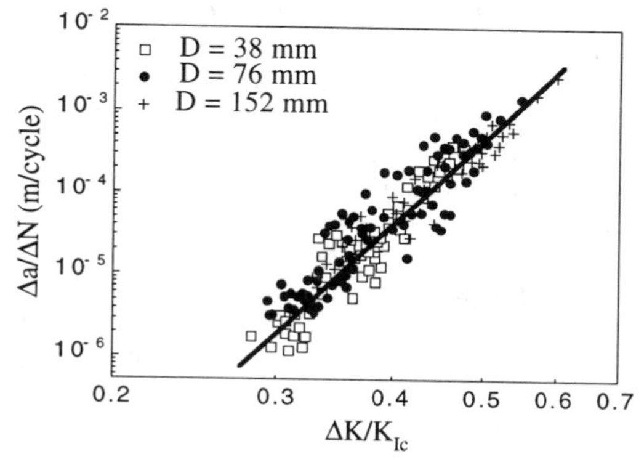

Figure 11.7.4 Logarithmic plot of crack length increment per cycle vs. the amplitude of size-adjusted stress intensity factor (after Bažant and Xu 1991).

in which K_{If} and D_0' are empirical constants, and β' is the brittleness number for cyclic loading. The primes are attached to β and D_0 because, as experiments indicated, the value of the transitional size D_0' is larger than the value of D_0 for monotonic loading.

It has been found that the optimum fits of test data, for normal strength concrete, are obtained for $D_0' = 10D_0$, and those for high strength concrete for $D_0' = 1.5D_0$. For large sizes, i.e., $\beta' \to \infty$, Eq. (11.7.6) approaches the classical Paris law with $K_{Ic} \to K_{If}$. For high strength concrete, the classical Paris law is approached at smaller sizes than for normal strength concrete. This is explained by a smaller size of the fracture process zone in high strength concrete, which, in turn, is caused partly by a smaller size of the aggregate and partly by greater homogeneity (due to a smaller difference between the elastic moduli of the aggregate and the cement mortar matrix). Note also that, because of the difference between D_0' and D_0, the brittleness number of the same structure is higher for monotonic loading than for cyclic loading, and is higher for high strength concrete than for normal strength concrete.

Eq. (11.7.5) with K_{Ic} given by Eq. (11.7.6) is compared to test data in Fig. 11.7.4 for normal concrete. It is seen that this extension of the Paris law correctly describes the shift of the straight lines as a function of the size. Similar results are obtained for high strength concrete (Bažant and Schell 1993).

To sum up, Paris law is applicable to concrete but only as long as a very narrow range of sizes is considered. For a broad size range, it needs to be adjusted by combining it with the size effect law previously developed for monotonic loading. The transitional size, separating predominantly brittle failures from predominantly ductile (plastic or strength-governed) failures appears to be larger than for monotonic loading, and more so for normal concrete than for high strength concrete. Accordingly, the brittleness number for a structure under cyclic loading is smaller than it is for monotonic loading.

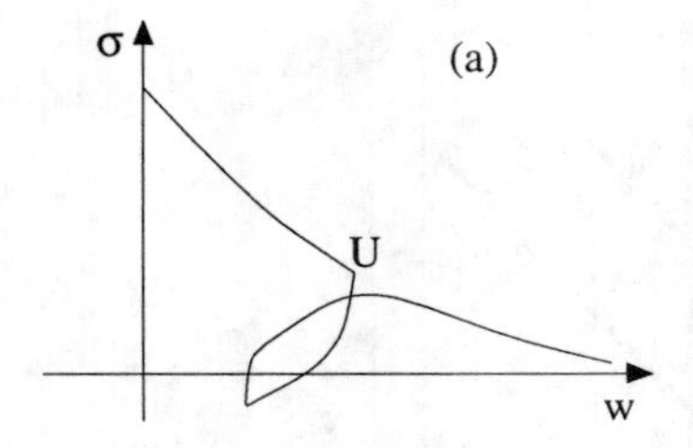

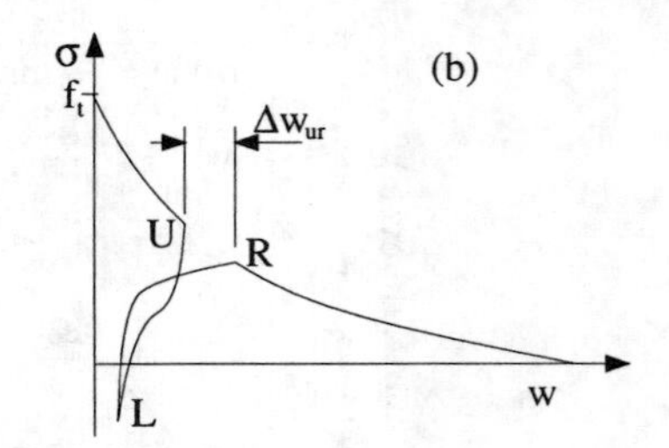

Figure 11.7.5 (a) Sketch of an unloading-reloading cycle. (b) Sketch of the cyclic behavior in the Hordijk-Reinhardt model.

11.7.4 Fatigue Description by History-Dependent Cohesive Models

Fatigue crack growth in quasibrittle materials can be described by a cohesive crack model with cumulative damage in unloading-reloading cycles (Gylltoft 1983; Hordijk 1991; Hordijk and Reinhardt 1991, 1992; Horii, Shin and Pallewatta 1990; Horii 1991). The basic characteristic of these models is that, as sketched in Fig. 11.7.5a, the end-point of an unloading-reloading loop does not coincide with the initial unloading point U. The differences between the models of the various authors concerns the way in which the unloading-reloading behavior is defined. Gylltoft (1983) used polygonals to define the curves, and assumed a behavior that deviates from the experimental results found later at Delft University (Reinhardt and Cornelissen 1984; Yankelevsky and Reinhardt 1989; Hordijk 1991). These results showed that the envelope of the cyclic stress-crack opening (σ–w) curve was essentially coincident with the monotonic σ–w curve. Both the models of Horii and of Hordijk and Reinhardt make use of this fact, but Horii's model drastically simplifies the description, as described later.

The Hordijk-Reinhardt model describes the cyclic behavior as sketched in Fig. 11.7.5b. The envelope curve coincides with the monotonic curve and is given by the equation $\sigma = f(w)$. The unloading curve depends on the unloading point U, in particular of the opening w_U. It has an equation of the form $\sigma = f_u(w, w_U)$. The reloading branch has an equation which depends on w_U and on the minimum point L reached during the unloading. Its equation has the form $\sigma = f_r(w, w_U, w_L)$. The complete expressions for these functions are quite complex and will not be reproduced here. They can be found in Hordijk and Reinhardt (1991) and Hordijk (1991). It is, however, interesting to note that the unloading-reloading loop ends at a point R which is situated to the right of the initial unloading point. The crack opening increase due to the unloading-reloading cycle, Δw_{UR} (Fig. 11.7.5), depends, in the Hordijk-Reinhardt approach, on the initial unloading point, and on the stress amplitude of the loop:

$$\Delta w_{UR} = 0.1 w_U \ \ln\left(1 + 3\frac{\sigma_U - \sigma_L}{f_t'}\right) \tag{11.7.7}$$

This equation correctly captures the general trend and magnitude observed in experiments, which, on the other hand, display quite a large scatter.

The Hordijk-Reinhardt cyclic model was called by the authors the continuous function model, in contrast to the focal point model of Yankelevsky and Reinhardt (1989). It has been implemented computationally in two ways. In the first, carried out by Janssen (1990), notched three-point bend beams were analyzed using a finite element code with special interface elements incorporating the σ—w equations. This procedure turned out to require a very large amount of computer time. In the second (Hordijk 1991; Hordijk and Reinhardt 1992), a simplified method was used similar to that described in Section 10.4.3 in which the cohesive crack is smeared over a central band of width h (see Figs. 10.4.6a–c). In this approach, the band width h was taken to be proportional to the beam depth. For the range of geometries and sizes investigated by Hordijk, both ways gave similar results for $h = 0.5D$.

One of the essential implications of this model is that, for a given geometry, the static curve gives an upper bound envelope of the cyclic curve. This is sketched in Fig. 11.7.6, based on the numerical predictions of Janssen (1990). The full lines in the figure show a number of loading-reloading loops produced under cyclic loading of constant amplitude. The dashed curve represents the quasi-static monotonic load-displacement curve. The cyclic loading becomes unstable when the reloading branch of the last loop reaches the downward branch of the curve of load vs. displacement (load-point deflection or CMOD).

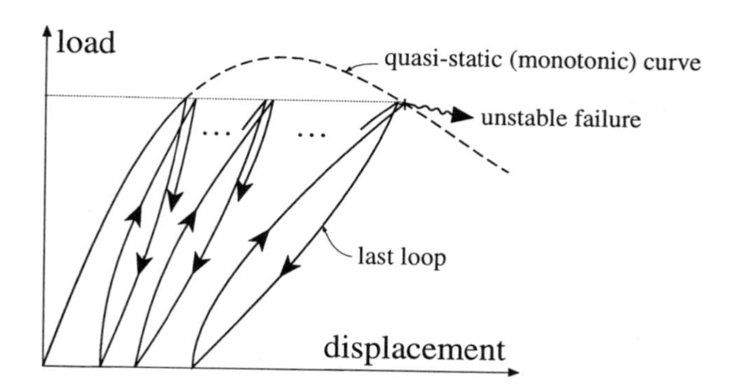

Figure 11.7.6 Sketch of load-displacement evolution under cyclic loading of constant amplitude (full line) predicted by the Hordijk-Reinhardt model. The dashed line represents the quasi-static (monotonic) load-displacement curve. Instability is reached when the cyclic curve meets the quasi-static curve.

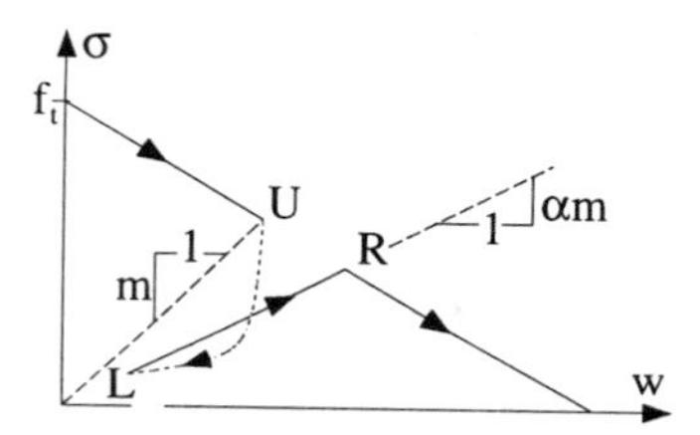

Figure 11.7.7 Monotonic softening curve and reloading curve in Horii's model. The unloading portion is not required in this model.

Horii's model keeps the general philosophy of the Hordijk-Reinhardt model. However, only the reloading portion of the curve is defined. Fig. 11.7.7 sketches the basic assumptions: (1) the softening curve for monotonic loading is linear, and (2) the reloading curve is a straight line through the origin defined by

$$\sigma = \alpha \frac{\sigma_U}{w_U} w \tag{11.7.8}$$

where α is a constant of the order of 0.75. This simplification implies that the reloading curve is independent of the lowest stress reached, σ_L, which, apparently, is not supported by the experimental results. Although the simplifications involved are strong, the model captures the essentials of the effect of fatigue.

The fundamental point in Horii's model is that the crack situation near the maximum load of the cycle can be computed without knowledge of the unloading part because it is assumed that the reloading curve is independent of the unloading level. The limitation to linear softening could be removed with little extra computational effort. The fact that the loading curve is not a straight line through the origin could also be implemented without much trouble. However, the simplifying hypothesis that the reloading curve is independent of the lowest stress reached is essential for computational efficiency. If this hypothesis is removed, then the model requires computation of the entire loading history and becomes conceptually identical to the Hordijk-Reinhardt model.

12

Statistical Theory of Size Effect and Fracture Process

So far we have explained the size effect on the nominal strength of structures in purely deterministic terms. However, the classical explanation, completed by Weibull (1939, 1951), has been statistical.

The question of scaling has occupied a central position in many problems of physics and engineering. It acquired a particularly prominent role in fluid mechanics more than a hundred years ago, providing the impetus for the development of the boundary layer theory, initiated by Prandtl (1904).

In solid mechanics, the scaling problem of main interest is the effect of structure size on its nominal strength. This is a very old problem, older than the mechanics of materials and structures. The question of size effect was discussed already by Leonardo da Vinci (1500s), who stated that "Among cords of equal thickness the longest is the least strong" . He also wrote that a chord "is so much stronger ... as it is shorter". This rule implies inverse proportionality of the nominal strength to the length of a cord, which is of course a strong exaggeration of the actual size effect.

More than a century later, the exaggerated rule of Leonardo was rejected by Galileo (1638) in his famous book in which he founded mechanics of materials. He argued that cutting a long cord at various points should not make the remaining part stronger. He pointed out, however, that a size effect is manifested in the dissimilar shapes of animal bones when small and large animals are compared.

Half a century later, a major advance was made by Mariotte (1686). He experimented with ropes, paper and tin and made the observation, from today's viewpoint revolutionary, that "a long rope and a short one always support the same weight unless that in a long rope there may happen to be some faulty place in which it will break sooner than in a shorter". He proposed that this results from the principle of "the Inequality of the Matter whose absolute Resistance is less in one Place than another". In qualitative terms, he thus initiated the statistical theory of size effect, two and half centuries before Weibull. The probability theory, however, was at its birth at that time and not yet ready to handle the problem.

Mariotte's conclusions were later rejected by Thomas Young (1807). He took a strictly deterministic viewpoint and stated that "a wire 2 inches in diameter is exactly 4 times as strong as a wire 1 inch in diameter", and that "the length has no effect either in increasing or diminishing the cohesive strength". This was a setback, but he obviously did not have in mind the random scatter of material strength. Later more extensive experiments clearly demonstrated the presence of size effect for many materials.

The next major advance was the famous paper of Griffith (1921). In that paper, he not only founded fracture mechanics but also introduced fracture mechanics into the study of size effect. He concluded that "the weakness of isotropic solids...is due to the presence of discontinuities or flaws... The effective strength of technical materials could be increased 10 or 20 times at least if these flaws could be eliminated". He demonstrated this conclusion by his experiments showing that the nominal strength of glass fibers was raised from 42,300 psi for the diameter of 0.11 mm to 491,000 for the diameter of 0.0033 mm. In Griffith's view, however, the flaws or cracks deciding failure were only microscopic, which is not true for quasibrittle materials. Their random distribution determined the local macroscopic strength of the material. Thus, Griffith's work represented a physical basis of Mariotte's statistical concept, rather than a discovery of a new type of size effect.

With the exception of Griffith, theoreticians in mechanics of materials paid hardly any attention to the question of scaling and size effect—an attitude that persisted into the 1980s. The reason doubtless was that all the theories that existed prior to the mechanics of distributed damage and quasibrittle (nonlinear) fracture use a failure criterion expressed in terms of stresses and strains (including the elasticity with allowable stress, plasticity, fracture mechanics with only microscopic cracks or flaws) exhibit no size effect (Bažant 1984a). Therefore, it was universally assumed (until about 1980) that the size effect, if observed, was inevitably statistical. Its study was supposed to belong to the statisticians and experimentalists, not mechanicians. For example, the subject was not even mentioned by Timoshenko in 1953 in his comprehensive treatise, *History of Strength of Materials.*

Progress was nevertheless achieved in probabilistic and experimental investigations. Peirce (1926) formulated the weakest-link model for a chain and introduced the extreme value statistics originated by Tippett (1925), which was later refined by Fréchet (1927), Fischer and Tippett (1928), von Mises (1936) and others (see also Freudenthal 1968). This progress culminated with the work of Weibull (1939) in Sweden (see also Weibull 1949, 1956).

Weibull (1939) reached a crucial conclusion: The tail distribution of extremely small strength values with extremely small probabilities cannot be adequately described by any of the known distributions. He proposed for the extreme value distribution of strength a power law with a threshold. Others (see, e.g., Freudenthal 1968, 1981) then justified this distribution theoretically, by probabilistic modeling of the distribution of microscopic flaws in the material. This distribution came to be known in statistics as the Weibull distribution.

With Weibull's work, the basic framework of the statistical theory of size effect became complete. Most subsequent studies until the 1980s dealt basically with refinements, justifications and applications of Weibull's theory (e.g., Zaitsev and Wittmann 1974; Mihashi and Izumi 1977; Zech and Wittmann 1977; Mihashi and Wittmann 1980; Mihashi and Zaitsev 1981; Mihashi 1983; see also Carpinteri 1986, 1989; Kittl and Diaz 1988, 1989, 1990). It was generally assumed that, if a size effect was observed, it had to be of Weibull type. Today we know this is not the case, as has been made clear in the preceding chapters.

Weibull statistical theory of size effect applies to structures that (1) fail (or must be assumed to fail) right at the initiation of the macroscopic fracture and (2) have at failure only a small fracture process zone causing negligible stress redistribution. This is the case especially for metal structures embrittled by fatigue. But this is not the case for quasibrittle materials, as we have already seen.

In this chapter we start by presenting the basics of Weibull's theory and the size effect that results from it. We follow by a thorough discussion of the limitations of the classical Weibull theory when applied to quasibrittle materials (Sections 12.1–12.3). Our analysis of the limitations will show that the Weibull theory cannot be applied to concrete, mainly because the structures made of it do not fail at first crack growth; rather, stable microcracking and, possibly, large cracks develop over large zones before the peak load is reached and the stress redistributions so generated mitigate (smear out) the effect of the weak spots. It will be seen that Weibull's approach fails when a brittle-elastic solid contains a sharp macrocrack: it predicts a zero strength for most practical situations, a prediction that, of course, is not realized in practice. Indeed, extensive testing as well as theoretical considerations indicate that the size effect in concrete structures is predominantly deterministic and the statistical influence appears to be minor in most situations. To match the basic Weibull's ideas to the observed size effect behavior, various expedients have been proposed: Sections 12.4 and 12.5 present the modifications introduced by Bažant and Xi (1991) and Planas that eliminate part of the limitations by including (1) a special treatment for the near tip statistics (core statistics), (2) nonlocal models (averaging), and (3) interpolating formulas for the size effect. The essential outcome is that for large sizes the dominant size effect is deterministic (LEFM).

The next section (12.6) is devoted to a recent different view of the random fracture process for structures with a preexisting macrocrack, which is modeled as an effective elastic crack with random crack growth resistance. This model again shows that the size effect in this type of statistical behavior is also primarily deterministic.

The last section (12.7) briefly discusses the recent studies of fracture and size effect based on the fractal approach to the geometry of cracks and microcracks.

The scope of the chapter is limited to time-independent fracture, and to uncoupled statistics, i.e., to models in which the stress distribution is independent of the random nature of the material on the microscale —in other words, models in which the randomness is smeared out for the purpose of stress analysis, but is singled out for the purpose of failure analysis. The stochastic models that explicitly incorporate the randomness of the material properties in the computation of the structural response —such as numerical concrete, stochastic finite elements, lattice models, particulate models, and random field analysis— will be briefly described in Chapter14. The recent developments in which the randomness of crack path is simulated by Markovian random process (Xi and Bažant 1992; Bažant and Xi 1994) are beyond the scope of this book.

12.1 Review of Classical Weibull Theory

Let us begin by reviewing the basic principles of the Weibull theory. We first introduce the weakest link model for discrete systems such as chains and then, for continuous, homogeneously stressed, systems. The Weibull probability density of strength and its analysis follow. Then the theory is extended to systems with nonhomogeneous uniaxial stress. Finally, the theory is generalized to systems with nonhomogeneous multiaxial stress. In the presentation the statistical tools are kept to a minimum; for further details, see Freudenthal (1968), and Kittl and Diaz (1988, 1989, 1990) for recent works and reviews.

12.1.1 The Weakest-Link Discrete Model

We consider a one-dimensional structure consisting of many elements coupled in series, for example, a chain (Fig. 12.1.1a). All of the elements (links of the chain) have the same distribution of strength σ, characterized by the cumulative probability distribution $P_1(\sigma)$, which represents the probability of failure of one element, i.e., the probability that the strength in the element is less than the applied stress σ. The survival probability of one element for this stress level is $1 - P_1(\sigma)$. If the whole chain should survive, all of its elements must survive. This means that the probability of survival of the chain is the joint probability of survival of all of the elements. According to the joint probability theorem, the survival probability of a chain of N elements is

$$1 - P_f = \underbrace{(1 - P_1)(1 - P_1)\ldots(1 - P_1)}_{N \text{ times}} = (1 - P_1)^N \tag{12.1.1}$$

where P_f is the failure probability of the chain as a whole. Now, taking natural logarithms of these expressions, we have

$$\ln(1 - P_f) = N \ln(1 - P_1) \tag{12.1.2}$$

Since P_1 is extremely small in practical situations, we may set $\ln(1 - P_1) \approx -P_1$ as a very good approximation. Therefore, after solving for P_f, the foregoing equation reduces to

$$P_f(\sigma) = 1 - \mathrm{e}^{-NP_1(\sigma)} \tag{12.1.3}$$

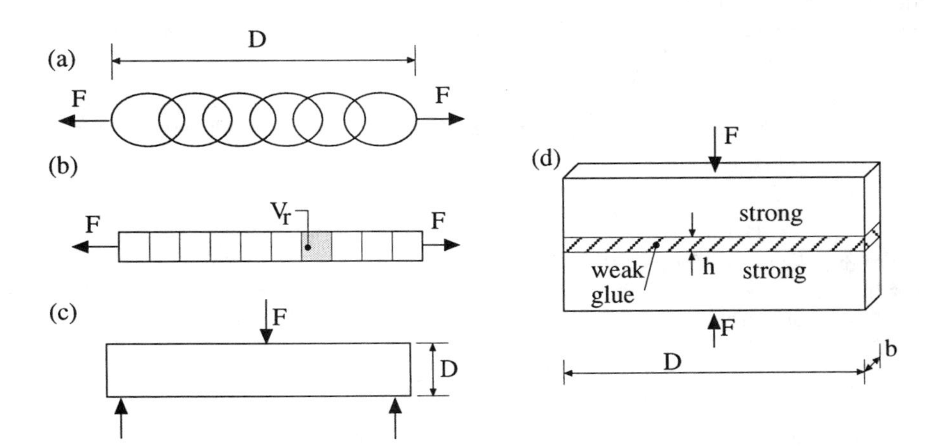

Figure 12.1.1 Various cases for Weibull distribution: (a) chain; (b) unidimensional bar; (c) unnotched beam; (d) layer of weak glue (after Bažant, Xi and Reid 1991).

12.1.2 The Weakest-Link Model for Continuous Structures under Uniaxial Stress

The last formula may be extended to a continuous, homogeneously stressed body (e.g., a long fiber, Fig. 12.1.1b) by setting $N = V/V_r$, where V is the volume of the body and V_r is a representative volume of the material. Substitution in the formula (12.1.3) immediately leads to

$$P_f(\sigma) = 1 - \exp\left[-\frac{V}{V_r}P_1(\sigma)\right] \tag{12.1.4}$$

where now $P_1(\sigma)$ is the failure probability of a certain reference volume V_r for the given stress level σ.

In the greatest generality, V_r should be interpreted as the smallest volume for which the material can be treated as a continuum (and for which the concept of stress on the macroscopic scale makes sense). However, in the classical theory of statistical strength, applicable to brittle materials with some kind of randomly distributed flaws or "defects", there is no intrinsic size scale and, for practical purposes, V_r may be interpreted as the volume of a small test specimen. Only when the theory is extended to materials with an intrinsic length scale, which do not follow Weibull's classical formulation, there is a real need to introduce the representative volume in all its significance. This is the case for concrete (as well as rocks, sea ice, and toughened ceramics), and will be dealt with beginning with Section 12.4.

The aforementioned way of deducing the strength probability distribution is simple but introduces some unnecessary conditions in the process. One, used in the simplification leading to the final equation of the discrete model, is that $P_1(\sigma) \ll 1$. The other, as already mentioned, is the representative volume itself because it does not enter the theory of random strength of brittle materials in some experimentally measurable way (even though its existence is appealing on physical grounds). If V_r cannot be determined, neither can the strength distribution P_1 of elements of volume V_r.

To avoid this conceptual difficulty, we can obtain the global strength distribution (12.1.4) in a strictly mathematical way. To do so, we first note that the essential implication of Eq. (12.1.4) is that the probability of failure for a given stress level depends on the volume

$$P_f = P_f(V) \tag{12.1.5}$$

where $P_f(V)$ is some function whose dependence on σ is known while its dependence on V must be determined. Next, we consider a uniformly stressed body of arbitrary total volume V_T and decompose it (in our imagination) into two parts, one of volume V and another of volume $V_T - V$. The probabilities of failure of the three bodies are, respectively, $P_f(V_T)$, $P_f(V)$, and $P_f(V_T - V)$. Under the hypotheses of series coupling and independent behavior, the survival probability of the whole body is the product of the survival probabilities of the two sub-bodies:

$$1 - P_f(V_T) = [1 - P_f(V)][1 - P_f(V_T - V)] \tag{12.1.6}$$

This equation must be satisfied for any V_T and any V, which means that P_f cannot be arbitrary. To obtain its form, we differentiate with respect to V and find the following differential equation:

$$0 = -P_f'(V)[1 - P_f(V_T - V)] + [1 - P_f(V)]P_f'(V_T - V) \tag{12.1.7}$$

where P_f' is the derivative of P_f with respect to its argument. This equation may be rearranged as

$$\frac{P_f'(V)}{1 - P_f(V)} = \frac{P_f'(V_T - V)}{1 - P_f(V_T - V)} \tag{12.1.8}$$

This means that the value of a function is the same for any two values (V and $V_T - V$) of its argument. The only function to satisfy this condition is, obviously, a constant function, so that the above equation is equivalent to the differential equation

$$\frac{P_f'(V)}{1 - P_f(V)} = c \tag{12.1.9}$$

in which c is a constant with respect to the volume (but it depends on the stress). This differential equation is easily integrable:

$$P_f(V) = 1 - Ae^{-cV} \tag{12.1.10}$$

where A is the integration constant to be determined. To do so, we substitute the general solution (12.1.10) back into the initial functional equation (12.1.6) and immediately obtain that $A = 1$. Since c is constant with respect to the volume but depends on the stress level, we finally have the result:

$$P_f(\sigma, V) = 1 - \mathrm{e}^{-c(\sigma)V} \tag{12.1.11}$$

Obviously, this has exactly the same form as Eq. (12.1.4) if one defines

$$c(\sigma) = P_1(\sigma)/V_r \tag{12.1.12}$$

Now, what is the physical meaning of $c(\sigma)$? Note that P_1 is nondimensional, and so $c(\sigma)$ has the dimension of the inverse of volume. Thus, in view of Eq. (12.1.12), $c(\sigma)$ represents the *spatial* density —or concentration— of material failure probability (distinguish this from what is called the probability density, which is $P_1'(\sigma)$, the derivative of the probability distribution). In the classical treatment of brittle materials such as steel, $c(\sigma)$ has been interpreted as the concentration of flaws or defects (their number per unit volume) having a strength lower than σ. In this context, the "strength of a defect" is short for "the macro-stress value that makes a microcrack grow to macroscopic dimensions and thus cause fracture". In the following, we call $c(\sigma)$ the *concentration function*, for short. In this classical sense, $P_1(\sigma)$ then means the number of defects of a strength lower than σ found in the volume V_r.

For a material such as concrete, however, the concept of a "flaw" or "defect" might be questioned, because the material has a totally disordered microstructure and no continuum (nor crystalline) matrix. For such a material, the "flaw" or "defect" should be interpreted as a microcrack on the scale up to the the maximum aggregate size, or a spot that is overstressed due to localized chemical or hygral volume changes in the microstructure.

We adopt the form (12.1.11) for $P_f(\sigma, V)$, but in order to simplify the notation we drop the argument V and keep in mind that P_f depends parametrically on V as well as on σ. Then, from (12.1.11), the strength density function $\phi(\sigma)$, mean strength $\overline{\sigma}_u$, and variance s^2 can be obtained using the well-known expressions:

$$\phi(\sigma) = \frac{dP_f(\sigma)}{d\sigma} \tag{12.1.13}$$

$$\overline{\sigma}_u = \int_{-\infty}^{\infty} \sigma\phi(\sigma)d\sigma = \int_{-\infty}^{\infty} \sigma dP_f(\sigma) \tag{12.1.14}$$

$$s^2 = \int_{-\infty}^{\infty} (\sigma - \overline{\sigma}_u)^2\, \phi(\sigma)d\sigma = \int_{-\infty}^{\infty} \sigma^2 dP_f(\sigma) \;-\; \overline{\sigma}_u^2 \tag{12.1.15}$$

where it is understood that all these variables and functions depend also on V. We will compute the mean and variance after introducing Weibull's specific function for $c(\sigma)$.

12.1.3 The Weibull Statistical Probability Distribution

Weibull (1939, 1951) introduced the following empirical formula for the concentration function:

$$c(\sigma) = \frac{1}{V_0}\left\langle \frac{\sigma - \sigma_1}{\sigma_0} \right\rangle^m \tag{12.1.16}$$

where $\langle\,\rangle$ denotes the positive part of the argument, i.e., $\langle\sigma\rangle = \sigma$ if $\sigma > 0$, and $\langle\sigma\rangle = 0$ if $\sigma \leq 0$; σ_0, σ_1, and m are three empirical material parameters; m is called the shape parameter or Weibull's modulus, σ_1 is the strength threshold, and σ_0 is the scale parameter. V_0 is a reference volume introduced for dimensional reasons, which is to be understood as the volume of the specimens used to experimentally determine the parameters of the equation (especially σ_0), as shown below.

For calculations it is convenient to assume that $\sigma_1 = 0$, and then the results of direct tensile tests of concrete indicate approximately $m = 12$ (Zech and Wittmann 1977). In reality, the threshold σ_1 is of course nonzero, but it is hard anyway to determine σ_1 unambiguously, due to scatter of test results. Unless the strength range of data is very broad and the random scatter is very small, very different σ_1-values

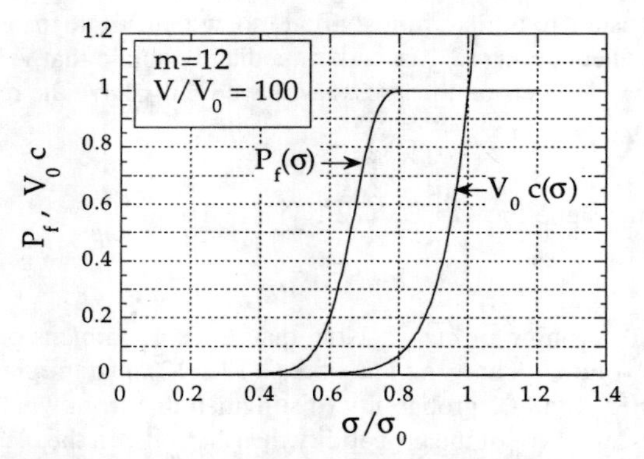

Figure 12.1.2 Weibull's statistical functions for a Weibull modulus $m = 12$, $V/V_0 = 100$, and $\sigma_1 = 0$ (note that P_f is experimentally indistinguishable from 0 up to $\sigma \approx 0.4\,\sigma_0$).

(with very different corresponding values of σ_0 and m) allow almost equally good fits of the data. For $\sigma_1 = 0$, Weibull's concentration function then takes its simplest form

$$c(\sigma) = \frac{1}{V_0}\left\langle \frac{\sigma}{\sigma_0} \right\rangle^m \tag{12.1.17}$$

Substitution of Weibull's simplified distribution (12.1.17) into (12.1.11) leads to the following statistical probability distribution of strength called the Weibull distribution:

$$P_f(\sigma) = 1 - \exp\left[-\frac{V}{V_0}\left\langle \frac{\sigma}{\sigma_0} \right\rangle^m \right]. \tag{12.1.18}$$

The shapes of this function and the function $c(\sigma)$ in (12.1.17) are shown in Fig. 12.1.2 for Weibull modulus $m = 12$, which is appropriate for concrete.

Weibull's distribution (12.1.18) contains all the information required to predict experimental (or observable) results, such as the mean strength, $\overline{\sigma}_u$, the standard deviation, or the median σ_u^{50} or any other desired probability cutoff, e.g., σ_u^{95} (defined as the strength that is not exceeded in 95 percent of realizations). Calculations are more direct and simple for the specified probability cutoff, because one does not need to integrate over P_f. For example, for the 50 percent cutoff, the median, one directly solves for σ_u^{50} from Eq. (12.1.18) with $P_f = 0.5$:

$$\sigma_u^{50} = (\ln 2)^{\frac{1}{m}} \left(\frac{V_0}{V} \right)^{\frac{1}{m}} \sigma_0 \tag{12.1.19}$$

However, it must be kept in mind that a good experimental determination of the median requires many more data than a good experimental determination of the mean.

The analytical determination of the mean requires integration over all P_f, which is more complicated as it leads to a Γ function. Indeed, the mean strength $\overline{\sigma}_u$ of a specimen of volume V is obtained by calculating the integral in Eq. (12.1.14) with the distribution $P_f(\sigma)$ in (12.1.18) :

$$\overline{\sigma}_u = \int_{-\infty}^{\infty} \sigma dP_f(\sigma) = \frac{Vm}{V_0\sigma_0^m}\int_0^{\infty} \sigma^m \exp\left[-\frac{V\sigma^m}{V_0\sigma_0^m} \right] d\sigma = \sigma_0\Gamma\left(1 + \frac{1}{m}\right)\left(\frac{V_0}{V}\right)^{\frac{1}{m}} \tag{12.1.20}$$

where the integration has been performed using the substitution $u = (V/V_0)(\sigma/\sigma_0)^m$, and $\Gamma(p)$ is the Eulerian gamma function:

$$\Gamma(p) = \int_0^{\infty} u^{p-1} e^{-u}\, du \tag{12.1.21}$$

Values of the Γ function may be found in many mathematical handbooks, but Planas found that for $5 \leq m \leq 50$, which is usually the practical range, the following power law gives an accuracy better than

0.5 percent:

$$\Gamma\left(1+\frac{1}{m}\right) \approx 0.6366^{1/m} \qquad (5 \le m \le 50) \tag{12.1.22}$$

Equation 12.1.20 clearly shows that the mean strength of a uniformly stressed specimen does depend on volume. It also shows that, in this particular case, one may determine the average strength for any specimen volume V from tests of specimens of volume V_0, which give mean strength $\overline{\sigma}_0$. Then, (12.1.20) may be rewritten as

$$\overline{\sigma}_u = \left(\frac{V_0}{V}\right)^{\frac{1}{m}} \overline{\sigma}_0 \tag{12.1.23}$$

and, obviously, the scale factor σ_0 is related to the measured mean strength of specimens of volume V_0 by

$$\sigma_0 = \frac{\overline{\sigma}_0}{\Gamma(1+\frac{1}{m})} \approx 0.6366^{-1/m}\overline{\sigma}_0 \tag{12.1.24}$$

where the error is within 0.5 percent for $5 \le m \le 50$, as already stated.

The variance is obtained from Eqs. (12.1.15) and (12.1.18):

$$s^2 = \frac{V}{V_0\sigma_0^m} m \int_0^\infty \sigma^{m+1} \exp\left[-\frac{V}{V_0\sigma_0^m}\sigma^m\right] d\sigma - \overline{\sigma}_u^2 = \Gamma\left(1+\frac{2}{m}\right)\left(\frac{V_0}{V}\right)^{\frac{2}{m}} \sigma_0^2 - \overline{\sigma}_u^2 \tag{12.1.25}$$

which, after using (12.1.23), becomes

$$s^2 = \overline{\sigma}_u^2 \left[\frac{\Gamma(1+\frac{2}{m})}{\Gamma^2(1+\frac{1}{m})} - 1\right] \tag{12.1.26}$$

Since, as already mentioned, the mean strength depends on the volume, so does the variance. However, the coefficient of variation $\omega = s/\overline{\sigma}_u$ is independent of V; it depends only on the Weibull modulus, since from the previous equation, one immediately finds

$$\omega^2 = \frac{\Gamma(1+\frac{2}{m})}{\Gamma^2(1+\frac{1}{m})} - 1 \tag{12.1.27}$$

This is a well-known remarkable property which is widely exploited to determine the Weibull modulus from a set of experimental results on specimens of a single size. Figure 12.1.3 shows the graphic representation of the relationship between the Weibull modulus and the coefficient of variation, which indicates that the Weibull modulus is an inverse measure of the random scatter of material strength; the larger is m, the smaller is ω. The approximation (12.1.24) is not accurate enough for determining the dependence of ω on m. As Planas found, this dependence may be approximated within 0.25 percent by the following hyperbolic law in the range $5 \le m \le 50$:

$$\omega \approx (0.462 + 0.783m)^{-1} \qquad (5 \le m \le 50) \tag{12.1.28}$$

12.1.4 Structures with Nonhomogeneous Uniaxial Stress

To generalize Eq. (12.1.4) for a structure with nonuniform stress (e.g., a beam, Fig. 12.1.1c), one may imagine the structure to consist of many parts of small volumes $\Delta V_{(j)}$, each with uniform stress $\sigma_{(j)} (j = 1, ..., N)$. Obviously, the stress $\sigma_{(j)}$ applied to the part j must depend on the load applied to the structure in some way. To be specific, assume that the structure is loaded proportionally, and that the load level is characterized, as usual, by means of a nominal stress σ_N (see Section 1.4.1). We may then perform the structural analysis to find the stress distribution as a function of σ_N, i.e., $\sigma_{(j)} = \sigma_{(j)}(\sigma_N)$.

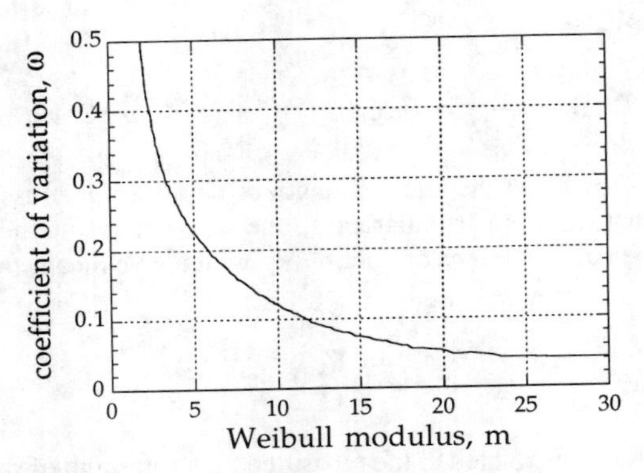

Figure 12.1.3 Coefficient of variation versus Weibull's modulus.

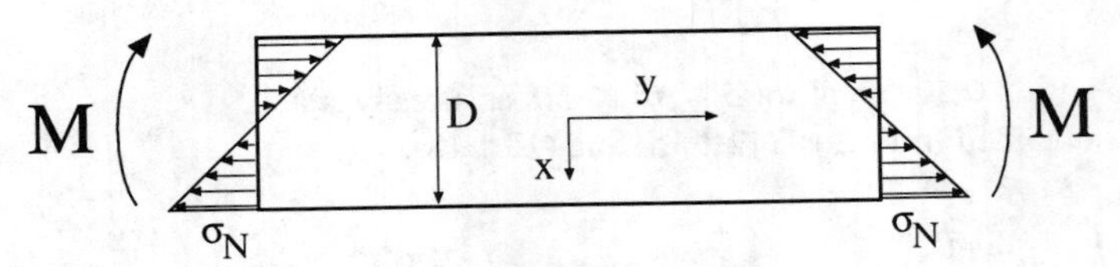

Figure 12.1.4 Beam in pure bending.

The probability of survival of the structure when a nominal stress σ_N is applied to it is the joint probability of survival of all its parts, and so

$$1 - P_f(\sigma_N) = \prod_{j=1}^{N} \exp\left[-c[\sigma_{(j)}(\sigma_N)]\Delta V_{(j)}\right] = \exp\left[-\sum_{j=1}^{N} c[\sigma_{(j)}(\sigma_N)\Delta V_{(j)}]\right] \tag{12.1.29}$$

Now, if the volume of each part tends to zero and the number of the parts tends to infinity, one obtains a structure with continuously variable stress $\sigma(\mathbf{x}, \sigma_N)$, where $\mathbf{x}$ is the coordinate vector. The failure probability for the specified applied nominal stress thus becomes

$$P_f(\sigma_N) = 1 - \exp\left[-\int_V c[\sigma(\mathbf{x}, \sigma_N)]\, dV(\mathbf{x})\right] \tag{12.1.30}$$

Example 12.1.1 Consider an elastic-brittle beam of depth D, length L, and thickness b subjected to pure bending as shown in Fig. 12.1.4. Let M be the bending moment and let the nominal stress be defined as the maximum stress according to elastic bending theory;

$$\sigma_N = \frac{6M}{bD^2} \tag{12.1.31}$$

Using the coordinates shown in the figure, the elastic stress distribution depends only on $\mathbf{x}$ and is given by

$$\sigma(\mathbf{x}, \sigma_N) = \frac{2x}{D}\sigma_N \tag{12.1.32}$$

Using the simplified Weibull distribution function, the integral in the exponent of Eq. (12.1.30) is found to be

$$\int_V c[\sigma(\mathbf{x}, \sigma_N)]\, dV(\mathbf{x}) = \frac{Lb}{V_0}\left(\frac{2\sigma_N}{D\sigma_0}\right)^m \int_0^{D/2} x^m\, dx = \frac{V}{2(1+m)V_0}\left(\frac{\sigma_N}{\sigma_0}\right)^m \tag{12.1.33}$$

where the volume of the specimen has been substituted in place of LbD. Comparing this with the exponent for the uniform tension case, Eq. (12.1.18), we see that, from the statistical point of view, a beam in pure bending is equivalent to a specimen in pure tension (direct tension) with a volume $2(1+m)$ times smaller. Therefore, the mean strength derived from pure bending (flexure), $\overline{\sigma}_u^B = \sigma_{Nu}$, and for pure tension, $\overline{\sigma}_u$ on specimens of the same volume are related by

$$\frac{\overline{\sigma}_u^B}{\overline{\sigma}_u} = [2(1+m)]^{1/m} \quad (\approx 1.31 \text{ for } m = 12) \tag{12.1.34}$$

(According to ACI, this ratio is $7.5\sqrt{f_c'}/6\sqrt{f_c'} = 1.25$, which would correspond to $m = 16$ if Weibull theory were the correct explanation. However, experiments show that this ratio is actually dependent on the specimen size, an effect that can be taken into account by neither Weibull's nor ACI formulas; see Chapter 9.) ▯

12.1.5 Generalization to Triaxial Stress States

Eq. 12.1.30 needs to be further generalized to triaxial stress states. Since we already know how to shift from homogeneous states to inhomogeneous ones, let us analyze a structure consisting of a volume V subjected to homogeneous triaxial stress $\boldsymbol{\sigma}$. A reasoning analogous to that which led to Eq. (12.1.11) shows that the failure probability under the stress tensor $\boldsymbol{\sigma}$ must take the form

$$P_f(\sigma, V) = 1 - \mathrm{e}^{-c(\boldsymbol{\sigma})V} \tag{12.1.35}$$

where, now, $c(\boldsymbol{\sigma})$ is a scalar-valued function of a second order tensor. This function again has the meaning of the concentration of failure probability for a given stress tensor, and it must be specified to be able to perform the statistical analysis.

The mathematical structure of this function is similar to that for plasticity or failure criteria. In particular, for isotropic solids (where it is understood that isotropy also applies to the micro-flaw distribution) this function must depend only on the principal invariants of the stress tensor or, equivalently, on the principal stresses $\sigma^I, \sigma^{II}, \sigma^{III}$:

$$c(\boldsymbol{\sigma}) = c\left(\sigma^I, \sigma^{II}, \sigma^{III}\right) \tag{12.1.36}$$

The detailed structure of $c(\boldsymbol{\sigma})$ must be derived either from phenomenological analysis or from micromechanical considerations. As the simplest example we can consider the case of materials with brittle failure only in tension. This means that $c(\boldsymbol{\sigma})$ is nonzero only if at least one of the principal stresses is positive. We may want, also, that for uniaxial tension ($\sigma^I = \sigma, \sigma^{II} = \sigma^{III} = 0$), $c(\boldsymbol{\sigma})$ reduce to simplest Weibull's form (12.1.17). One function satisfying these general conditions is (Freudenthal 1968)

$$c(\boldsymbol{\sigma}) = \frac{1}{V_0\,\sigma_0^m}\left(\langle\sigma^I\rangle^m + \langle\sigma^{II}\rangle^m + \langle\sigma^{III}\rangle^m\right) = \frac{1}{V_0\,\sigma_0^m}\sum_{P=I}^{III}\langle\sigma^P\rangle^m \tag{12.1.37}$$

This simple concentration function adequately takes into account the obvious fact that a given volume of material subjected to biaxial or triaxial tension is more prone to failure than if it is subjected to uniaxial tension, as shown in the following example.

Example 12.1.2 To illustrate Eq. (12.1.37), let us examine the predictions for biaxial tension $\sigma_I = \sigma^{II} = \sigma$, $\sigma^{III} = 0$ as compared to uniaxial tension. The c-function for biaxial tension $c_2(\sigma)$, turns out to be, according to (12.1.37), just the double of that for uniaxial tension. Therefore, the strength distribution function exponent is also doubled, which is equivalent to saying that the strength distribution function for biaxial tension on volume V is the same as the strength distribution function for uniaxial tension for a volume $2V$. Accordingly, the mean biaxial strength $\overline{\sigma}_{u2}$ is related to the mean uniaxial strength $\overline{\sigma}_u$ by

$$\overline{\sigma}_{u2}(V) = \overline{\sigma}_u(2V) = 2^{-\frac{1}{m}}\overline{\sigma}_u(V) \tag{12.1.38}$$

For $m = 12$ (concrete), this gives $\overline{\sigma}_{u2} = 0.94\overline{\sigma}_u$. ▯

The physical basis for a larger failure probability in biaxial or triaxial tension is that the uniaxial stress essentially triggers flaws normal to that stress, while the triaxial state triggers flaws at any orientation. This intuitive reasoning can be further justified based on the additivity of the concentration function, which is described next.

12.1.6 Independent Failure Mechanisms: Additivity of the Concentration Function

Consider a volume of material V subjected to a uniform stress tensor $\boldsymbol{\sigma}$, and assume that the material contains N families of flaws that can independently cause fracture. We label the failure through a particular family of flaws by superscript α ($\alpha = 1, 2, \cdots, N$). Let us assume that these flaws do not interact, and that the law for the probability of the volume failing through flaw family α is analogous to (12.1.35):

$$P_f^\alpha = 1 - \mathrm{e}^{-c_\alpha(\boldsymbol{\sigma})V} \tag{12.1.39}$$

where $c_\alpha(\boldsymbol{\sigma})$ is the concentration function corresponding to flaw family α.

Now, if all the possible flaw families ($\alpha = 1, 2, \cdots, N$) are present, the survival of the whole structure is possible only if the failure of none of the flaw families is activated; so, invoking again the theorem of joint probabilities, we have

$$1 - P_f = \prod_{\alpha=1}^{N} (1 - P_f^\alpha) \tag{12.1.40}$$

Then, inserting (12.1.39) for the individual failure probabilities, and solving for P_f, we get

$$P_f = 1 - \exp\left[-\sum_\alpha c_\alpha(\boldsymbol{\sigma})\, V\right] \tag{12.1.41}$$

It follows that, when various *independent* failure mechanisms are present, the failure probability distribution has the same structure as for a single mechanism, except that the concentration function is the sum of the corresponding functions for the various mechanisms:

$$c(\boldsymbol{\sigma}) = \sum_\alpha c_\alpha(\boldsymbol{\sigma}) \tag{12.1.42}$$

The foregoing result can be exploited to find realistic triaxial concentration functions based on the knowledge of the microscopic flaw systems. One example is given next, based on Petrovic's (1987) work.

Example 12.1.3 Consider a brittle material that can fail only by cracking through planes with a discrete distribution of orientations. Let N be the number of possible orientations and $\mathbf{n}^\alpha$ ($\alpha = 1, 2, \cdots, N$) the unit normal to the planes of the various orientations. Assume further that, in deterministic terms, the failure through a plane occurs when the stress normal to that plane reaches a critical value. The statistical counterpart states that the concentration function depends only on the stress normal to that plane, and so, assuming the same density of potential failure planes for all possible orientations, we have:

$$c_\alpha(\boldsymbol{\sigma}) = c_1(\boldsymbol{\sigma}\mathbf{n}^\alpha \cdot \mathbf{n}^\alpha) = c_1(\sigma_{ij} n_i^\alpha n_j^\alpha) \tag{12.1.43}$$

where $c_1(\sigma)$ is a scalar function of a scalar variable. Taking this function to coincide with Weibull's distribution without threshold (Eq. 12.1.17)— we have

$$c_\alpha(\boldsymbol{\sigma}) = \frac{1}{V_0 \sigma_0^m} \langle \boldsymbol{\sigma}\mathbf{n}^\alpha \cdot \mathbf{n}^\alpha \rangle^m = \frac{1}{V_0 \sigma_0^m} \langle \sigma_{ij} n_i^\alpha n_j^\alpha \rangle^m \tag{12.1.44}$$

and then the total concentration function is

$$c(\boldsymbol{\sigma}) = \frac{1}{V_0 \sigma_0^m} \sum_{\alpha=1}^{N} \langle \boldsymbol{\sigma}\mathbf{n}^\alpha \cdot \mathbf{n}^\alpha \rangle^m = \frac{1}{V_0 \sigma_0^m} \sum_{\alpha=1}^{N} \langle \sigma_{ij} n_i^\alpha n_j^\alpha \rangle^m \tag{12.1.45}$$

which follows from (12.1.42). ◻

The simplest application of the foregoing example is to assume only three families of failure planes, each perpendicular to one principal stress direction. Then (12.1.42) leads to the simple expression (12.1.37).

Note that, although we have assumed in this example a discrete distribution of failure planes, a continuous distribution is also possible. In such a case, the summation becomes an integral over all directions in space. The resulting equations are out of the intended scope of this chapter. We shall keep the simpler formulation in (12.1.37) as the first approximation.

12.1.7 Effective Uniaxial Stress

The foregoing results for triaxial stresses may be summarized and somewhat simplified by introducing a scalar measure of the stress tensor called the effective uniaxial stress, $\tilde{\sigma}$. To do so note that we may define a function homogeneous of the first degree in stress by taking the m-th root of the sum in the formulas (12.1.37) and (12.1.42). Thus, based on (12.1.37) we define $\tilde{\sigma}$ as:

$$\tilde{\sigma} = \left[\sum_{P=I}^{III} \langle \sigma_P \rangle^m \right]^{\frac{1}{m}} \tag{12.1.46}$$

and similarly for (12.1.42). Then the concentration function always takes the same form, namely

$$c(\boldsymbol{\sigma}) = \frac{1}{V_0} \left(\frac{\tilde{\sigma}}{\sigma_0} \right)^m \tag{12.1.47}$$

This can be easily generalized as long as the underlying failure criterion is homogeneous of the first degree in the stress tensor. In such cases, we can define $\tilde{\sigma}$ as a function homogeneous of the first degree in the stress tensor, i.e.,

$$\tilde{\sigma} = \tilde{\sigma}(\boldsymbol{\sigma}, m) \qquad \text{with} \qquad \tilde{\sigma}(\lambda\boldsymbol{\sigma}, m) = \lambda\tilde{\sigma}(\boldsymbol{\sigma}, m) \quad \text{for} \quad \lambda \geq 0 \tag{12.1.48}$$

which explicitly shows that, in general, the effective uniaxial stress $\tilde{\sigma}$ depends on m, as is clear from the the particular case (12.1.46).

12.1.8 Summary: Nonhomogeneous States of Stress

For structures subjected to arbitrary stress fields, computation of the failure probability follows exactly the same steps as in the uniaxial case. In general, the triaxial counterpart of (12.1.30) is obtained by simply writing that $c()$ is now a function of the stress tensor, as discussed in Section 12.1.5:

$$P_f = 1 - \exp\left[- \int_V c[\boldsymbol{\sigma}(\mathbf{x})]\, dV(\mathbf{x}) \right] \tag{12.1.49}$$

Note that in all the foregoing equations, the stress $\boldsymbol{\sigma}(\mathbf{x})$ is understood to be expressed in terms of the load P or of the nominal stress σ_N as described in Section 12.1.4.

If the structure of the concentration function $c(\boldsymbol{\sigma})$ is such that the concept of effective uniaxial stress is applicable, as usually assumed, then the computation can be split into two parts. First, one computes the stress distribution and from it the effective uniaxial stress $\tilde{\sigma}(\mathbf{x})$. Then one computes the failure probability distribution as

$$P_f = 1 - \exp\left\{ - \int_V \left[\frac{\tilde{\sigma}(\mathbf{x})}{\sigma_0} \right]^m \frac{dV(\mathbf{x})}{V_0} \right\} \tag{12.1.50}$$

Example 12.1.4 Consider a spherical cavity of radius R in a brittle infinite solid subjected to internal pressure p. The distribution of radial elastic stresses is known to be $\sigma_r = -pR^3/r^3$, where r is the distance to the center of the cavity; this stress component is a compressive principal stress ($\sigma^{III} = \sigma_r < 0$). The circumferential components are tensile and equal; they are principal stresses and their distribution is also

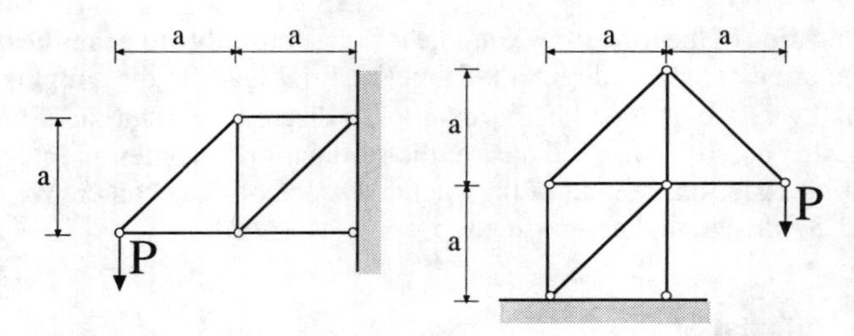

Figure 12.1.5 Trusses of exercise 12.2.

known from elastic analysis: $\sigma^I = \sigma^{II} = \sigma = pR^3/(2r^3)$. Substituting this into (12.1.46), we get

$$\tilde{\sigma} = 2^{1/m}\, p\frac{R^3}{2r^3} \tag{12.1.51}$$

Inserting this into (12.1.50) and integrating over concentric spherical membranes, one gets the following value for the exponent:

$$\int_V \left(\frac{\tilde{\sigma}}{\sigma_0}\right)^m \frac{dV}{V_0} = 2^{1-m} 4\pi R^{3m} \frac{p^m}{V_0\sigma_0^m} \int_R^\infty r^{2-3m} dr = \frac{p^m}{V_0\sigma_0^m} 2^{1-m} \frac{4\pi R^3}{3(m-1)} \tag{12.1.52}$$

Since $4\pi R^3/3$ is the volume of the cavity V_c (not of the material!), the problem has the same solution as the case of uniaxial stress for $\sigma = p$ and volume $V = 2^{1-m}V_c/(m-1)$. Therefore, according to (12.1.23) the mean failure pressure $\overline{p}_u$ is given by

$$\overline{p}_u = 2\overline{\sigma}_0 \left(\frac{(m-1)V_0}{2V_c}\right)^{\frac{1}{m}} \tag{12.1.53}$$

where, as we may recall, $\overline{\sigma}_0$ is the mean uniaxial strength for tensile specimens of volume V_0. ▯

Note that all the nonhomogeneous field examples lead to statistical distributions that are always equivalent to an uniaxial problem with a certain equivalent volume proportional to some physical volume (not necessarily that of the material, as we just showed). This fact will be systematically exploited in the next section to analyze size effect. The coefficient of variation is then independent of the volume and of the particular geometry analyzed. This result may be shown to be exactly valid only if the threshold stress σ_1 in the general Weibull equation is zero, but it must be close to reality even for nonzero σ_1.

Exercises

12.1 Tensile tests have been performed following ASTM A 370 on brittle steel specimens. The specimens were cylindrical, of 12.7 mm diameter and 76.2 mm effective length. The mean strength was 293 MPa and the standard deviation 7.3 MPa. (a) Make an estimate of Weibull modulus assuming zero threshold strength. (b) Find the characteristic strength σ_0 for the reference volume $V_0 = 10^{-3}$ m^3. (c) Find the mean strength and the 99.9 percent probability strength of a tensioned bar of 30 mm diameter and 6 m length.

12.2 Determine the probability of failure of the trusses in Fig. 12.1.5 with $a = 3$ m, P= 10 kN. All the bars have identical cross section area, and the design was done for an allowable stress in tension of 20 MPa. Assume that brittle failure occurs only in tension and that a Weibull distribution applies with $V_0 = 0.001$m^3, $\sigma_0 = 21$ MPa, $\sigma_1 = 0$, and $m = 25$.

12.3 Determine the mean strength for a three-point-bend beam with a span-to-depth ratio of 16, assuming an elastic-brittle material for which Weibull's analysis applies. Give it as a function of the beam volume, the mean strength of a tensile specimen of same volume, and the Weibull modulus. Particularize the results for (a) $m = 12$ and (b) $m = 25$. Use beam bending theory (neglecting shear), and take the nominal strength defined in (12.1.31), with M equal to the maximum bending moment.

12.4 Determine the mean strength for a cantilever beam with a length-to-depth ratio of 10, subjected to a uniformly distributed load. Assume an elastic-brittle material for which Weibull's analysis applies. Give the mean strength as a function of the beam volume, the mean strength of a tensile specimen of same volume, and the Weibull modulus. Consider (a) $m = 12$ and (b) $m = 25$. Use beam bending theory (neglecting shear), and take the nominal strength defined in (12.1.31), with M equal to the maximum bending moment.

12.5 Consider a material with the triaxial concentration function of Eq. (12.1.37). Determine the mean biaxial failure locus in plane stress for specimens of a given volume (i.e., the combinations of σ^I and σ^{II}) that cause failure on average, σ^{III} being kept zero). Assume that the tests are performed using proportional loading.

12.6 Consider the concentration function

$$c(\boldsymbol{\sigma}) = \frac{1}{V_0\,\sigma_0^m}\left\langle \sigma^I + \sigma^{II} + \sigma^{III} \right\rangle^m = \frac{1}{V_0\,\sigma_0^m}\left\langle \sigma_{kk} \right\rangle^m \tag{12.1.54}$$

Show that it leads to the classical Weibull function for uniaxial stress and that failure can occur only if at least one of the principal stresses is tensile. Find the mean biaxial failure locus for plane stress. Is this failure locus realistic to describe failure of brittle materials?

12.7 Determine the mean torsional strength of a cylinder of radius R and length L. Define the nominal stress σ_N as the maximum tensile stress in the specimen, and relate the strength to the mean strength of a tensile specimen of the same volume. Assume elastic-brittle behavior and applicability of Weibull's analysis based on the effective uniaxial stress (12.1.42) with (a) $m = 12$ and (b) $m = 25$.

12.8 Determine the mean torsional strength of a hollow cylinder of radius R, length L, and radius of the hole $R_1 = 3R/4$. Define the nominal stress σ_N as the maximum tensile stress in the specimen, and relate the strength to the mean strength of a tensile specimen of the same volume. Assume elastic-brittle behavior and applicability of Weibull's analysis based on the effective uniaxial stress (12.1.42) with (a) $m = 12$ and (b) $m = 25$.

12.2 Statistical Size Effect due to Random Strength

The size effect is defined by comparing geometrically similar structures of different characteristic dimensions (sizes) D. In a deterministic approach (Section 1.4 and Chapter 6), one introduces the nominal stress σ_N and analyzes the influence of the size on the strength σ_{Nu} of the structure, defined as the peak (maximum) nominal stress before failure:

$$\sigma_{Nu} = c_N \frac{P_u}{bD} \tag{12.2.1}$$

where P_u is the failure load (ultimate load), b = thickness of the structure, which may be either constant (two-dimensional similarity) or proportional to D (three-dimensional similarity); c_N is a constant chosen for convenience in each kind of loading. It is usually selected in such a way that the nominal stress coincides with the maximum tensile stress throughout the structure. Now, as already shown in Chapter 2, the basic property of all structures is that, according to elastic analysis with allowable stress limit as well as plastic analysis or any analysis based on some failure criterion in terms of stress or strain or both, σ_{Nu} is independent of the structure size D, i.e., there is no size effect.

This is not true, however, when the material properties are random. Several examples have already been presented in the previous section and it appeared that the mean strength depended on the volume of the structure. In this section we give a systematic analysis leading to a general size effect for Weibull-type of statistical behavior.

12.2.1 General Strength Probability Distribution and Equivalent Uniaxial Volume

In this section we prove that for the case of proportional loading, the strength probability distribution coincides with that of a uniformly stressed specimen in uniaxial tension of a certain volume, called the equivalent uniaxial volume V_N. To do so, we first note that, whatever the constitutive equation relating stresses and strains, the stress at a point of position vector $\mathbf{x}$ of coordinates (x_1, x_2, x_3) may be written in the form

$$\boldsymbol{\sigma}(\mathbf{x}, \sigma_N) = \sigma_N \, \mathbf{s}\left(\frac{\mathbf{x}}{D}\right) = \sigma_N \, \mathbf{s}\left(\frac{x_1}{D}, \frac{x_2}{D}, \frac{x_3}{D}\right) \tag{12.2.2}$$

where $\mathbf{s}$ is a function with dimensionless tensorial values which describes the form of the stress distribution and D is a characteristic dimension of the body. The intensity of the stress distribution is contained in the load factor represented by the nominal stress, σ_N. Because of the homogeneity condition, the equivalent uniaxial stress $\tilde{\sigma}$ is (12.1.48)

$$\tilde{\sigma} = \tilde{\sigma}\left[\sigma_N \, \mathbf{s}\left(\frac{\mathbf{x}}{D}\right), m\right] = \sigma_N \, \tilde{\sigma}\left[\mathbf{s}\left(\frac{\mathbf{x}}{D}\right), m\right] = \sigma_N \, \tilde{s}\left(\frac{\mathbf{x}}{D}, m\right) \tag{12.2.3}$$

where $\tilde{s}(\mathbf{x}/D, m)$ is a function that depends only on geometry and Weibull modulus, but not on the load level, and σ_N has been defined to be always positive.

The strength probability distribution is obtained by substituting the foregoing expression into (12.1.50):

$$P_f(\sigma_N) = 1 - \exp\left\{-\frac{1}{V_0}\left(\frac{\sigma_N}{\sigma_0}\right)^m \int_V \left[\tilde{s}\left(\frac{\mathbf{x}}{D}, m\right)\right]^m dV(\mathbf{x})\right\} \tag{12.2.4}$$

The integral has the dimension of volume, and is independent of the load level. Thus, it is a geometrical property that we call the equivalent uniaxial volume, V_N:

$$V_N = \int_V \left[\tilde{s}\left(\frac{\mathbf{x}}{D}, m\right)\right]^m dV(\mathbf{x}) \tag{12.2.5}$$

where V_N depends, in general, on the Weibull modulus m.

With this definition, the strength probability distribution becomes

$$P_f(\sigma_N) = 1 - \exp\left[-\frac{V_N}{V_0}\left(\frac{\sigma_N}{\sigma_0}\right)^m\right] \tag{12.2.6}$$

Since σ_N is chosen to be positive, this distribution coincides with the distribution (12.1.18) for an uniaxial homogeneous situation if σ and V are replaced by σ_N and V_N. Therefore, all the results derived for the uniaxial homogeneous case in Section 12.1.3 are applicable to the general case, with the aforementioned changes. In particular, the mean strength $\overline{\sigma}_{Nu}$ is found to be

$$\overline{\sigma}_{Nu} = \left(\frac{V_0}{V_N}\right)^{\frac{1}{m}} \overline{\sigma}_0 \tag{12.2.7}$$

and the coefficient of variation is still given by (12.1.27), because it is independent of the volume.

Example 12.2.1 Consider a pure bent beam of rectangular cross section, length L, depth D, and thickness b. Assume that the uniaxial stress-strain relationship is symmetric and of the power type

$$\sigma = \sigma_c |\varepsilon|^{n-1} \varepsilon \tag{12.2.8}$$

where σ_c and n are constants.

The classical hypothesis of cross sections remaining planes implies a linear variation of strains $\varepsilon = Cx$. So, with the coordinate axes shown in Fig. 12.1.4,

$$\sigma = \sigma_c C^n \, |x|^{n-1} x \tag{12.2.9}$$

The equilibrium equation may be written as

$$M = \int_{-D/2}^{D/2} \sigma b x \, dx = \sigma_c C^n \, 2b \int_0^{D/2} x^{n+1} \, dx = \sigma_c C^n \frac{2b}{n+2} \left(\frac{D}{2}\right)^{n+2} \tag{12.2.10}$$

Solving for C from the last equation and substituting in (12.2.9), one obtains the stress distribution as a function of the applied loading:

$$\sigma = \frac{6M}{bD^2}\frac{n+2}{3}\left|\frac{2x}{D}\right|^{n-1}\frac{2x}{D} = \sigma_N\frac{n+2}{3}\left|\frac{2x}{D}\right|^{n-1}\frac{2x}{D} \tag{12.2.11}$$

where the classical definition (12.1.31) for σ_N in bending has been used. Obviously, this stress distribution has the form (12.2.2). In this simple case the stress state is uniaxial, and so the uniaxial equivalent stress coincides with the positive part of the stress and the function $\tilde{s}$ of Eq. (12.2.3) is

$$\tilde{s} = \frac{n+2}{3}\left\langle\frac{2x}{D}\right\rangle^n \tag{12.2.12}$$

According to Eq. (12.2.5), the equivalent volume is (with $x_1 = x, x_2 = y, x_3 = z$)

$$V_N = Lb\int_0^{D/2}\frac{n+2}{3}\left(\frac{2x}{D}\right)^{nm}dx = \frac{n+2}{6(nm+1)}LbD = \frac{n+2}{6(nm+1)}V \tag{12.2.13}$$

Note that this is proportional to the actual beam volume, and the proportionality coefficient depends on both Weibull modulus m and hardening exponent n. Of course, for linear elasticity $n = 1$ and one recovers the results given in the example 12.1.1 in Section 12.1.4. ▯

12.2.2 Statistical Size Effect Laws

The size effect laws for the mean strength of the structure are easily obtained from Eq. (12.2.7). This equation can be rewritten in the following way: Let D_0 be a reference size and $\overline{\sigma}^0_{Nu}$ the corresponding measured mean strength. Then, the mean strength for any other size D is

$$\overline{\sigma}_{Nu} = \overline{\sigma}^0_{Nu}\left(\frac{V_N(D_0)}{V_N(D)}\right)^{1/m} \tag{12.2.14}$$

This shows that the dependence of the strength on the size is obtained indirectly through the dependence of the effective volume on D. Now, one may guess that for geometrically similar structures the effective volume is proportional to the physical volume, with a fixed proportionality coefficient. To prove it formally, consider first a three-dimensional similitude. The body occupies a certain region which can be defined by an equation of the type $v(\mathbf{x}/D) < 0$, where $v(\mathbf{x}/D)$ is a dimensionless function and D a characteristic dimension of the body. Then (12.2.5) can be written as

$$V_N = \int_{v(\mathbf{x}/D)<0}\left[\tilde{s}\left(\frac{\mathbf{x}}{D}, m\right)\right]^m dx_1dx_2dx_3 \tag{12.2.15}$$

and then transformed by making the change of variable $\mathbf{x} = D\boldsymbol{\xi}$. The effective volume then becomes

$$V_N = D^3\int_{v(\xi)<0}[\tilde{s}(\boldsymbol{\xi}, m)]^m\, d\xi_1\, d\xi_2\, d\xi_3 = \beta D^3 = \beta' V \tag{12.2.16}$$

where β denotes the value of the integral, which is seen to be independent of D. Therefore, in three-dimensional similitude, the statistical size effect law is

$$\overline{\sigma}_{Nu} = \overline{\sigma}^0_{Nu}\left(\frac{D_0}{D}\right)^{3/m} \tag{12.2.17}$$

Second, consider plane states and bodies with two-dimensional similitude in which the thickness of the body is kept constant. This implies that all the fields are constant with respect to the coordinate normal to the structure plane (say x_3). In particular, this holds true for the effective uniaxial stress field in Eq. (12.2.5), and so the integral in that equation can be performed by integrating first through the thickness and then over the plane. The result, after making the substitutions $x_1 = D\xi_1, x_2 = D\xi_2$, is

$$V_N = bD^2\int_{a(\xi_1,\xi_2)<0}[\tilde{s}(\xi_1, \xi_2, m)]^m\, d\xi_1\, d\xi_2 = \beta b D^2 = \beta' V \tag{12.2.18}$$

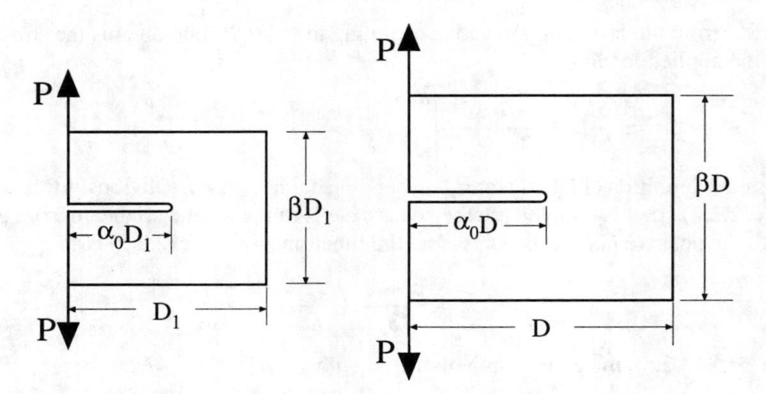

Figure 12.2.1 Geometrically similar structures with a macroscopic crack.

where $a(\xi_1, \xi_2) < 0$ defines the plane region (area) of the structure and all the other symbols are as before. So, the size effect law for two-dimensional similitude reads

$$\overline{\sigma}_{Nu} = \overline{\sigma}_{Nu}^0 \left(\frac{D_0}{D}\right)^{2/m} \tag{12.2.19}$$

The foregoing statistical size effect equations may be summarized by writing that for a n_d-dimensional similitude, the dependence of the mean strength on size is

$$\overline{\sigma}_{Nu} = \overline{\sigma}_{Nu}^0 \left(\frac{D_0}{D}\right)^{n_d/m} \tag{12.2.20}$$

The last equation is also valid for one-dimensional similitude, for which $n_d = 1$. This case is obtained, for a long chain, cable, or bar under uniaxial stress, in which the whole cross section fails simultaneously. Another example is the body in Fig. (12.1.1d) which consists of two rigid blocks joined by a thin layer of weak deformable glue. The blocks cannot fail and failure is assumed to occur in the glue layer as soon as one elementary volume of the glue fails. In that case, $n_d = 1$ provided that the block thickness b and the glue layer thickness h are not varied.

12.2.3 Divergence of Weibull Failure Probability for Sharply Cracked Bodies

The foregoing size effect laws apply to bodies with regular stress distributions. However, when linear elastic-brittle behavior is assumed, they do not apply for specimens having a sharp macrocrack such as the one in Fig. 12.2.1. This is so because (for normal value of m) the singularity at the crack tip causes the integral in (13.2.5) to diverge , hence predicting an infinite effective volume and zero strength.

To prove it, we consider the plane specimen of Fig. 12.2.1 and recall from Chapter 2 that the two-dimensional stress distribution in polar coordinates (r, θ) must have (for proportional loading) the form

$$\sigma_{kl} = \sigma_N \left(\frac{r}{D}\right)^{-1/2} S_{kl}\left(\frac{r}{D}, \theta\right) \tag{12.2.21}$$

where functions $S_{kl}\,(r/D, \theta)$ are bounded and smooth (except at the points of concentrated loads and at reentrant boundary corners).

From the foregoing stress distribution, and according to the homogeneity property, the effective uniaxial stress takes the form

$$\tilde{\sigma} = \sigma_N \; \tilde{s}(\rho, \theta, m) \; \rho^{-1/2} \tag{12.2.22}$$

in which $\rho = r/D =$ relative polar coordinate, and $\tilde{s}(\rho, \theta, m)$ is a dimensionless function. Now, repeating the step that led from Eq. (12.2.3) to Eq. (12.2.5) we obtain the expression for the effective volume V_N:

$$V_N = bD^2 \int_{-\pi}^{\pi} \int_0^{\hat{\rho}(\theta)} \rho^{-m/2} \left[\tilde{s}(\rho, \theta, m)\right]^m \rho \, d\rho \, d\theta \tag{12.2.23}$$

where $\hat{\rho}(\theta)$ is the relative radial distance from the crack tip to the boundary points and defines the geometry of the structure.

The integral in the foregoing expression diverges because of the $\rho^{-m/2}$ factor. Indeed, since the effective stress, and thus also $\tilde{s}(\rho, \theta, m)$, are always positive, one can write the condition that the integral extending over the whole structure be larger than the integral extending over a small enough circle centered at the crack tip and having relative radius $\rho_1 = r_1/D \ll 1$:

$$V_N \geq bD^2 \int_{-\pi}^{\pi} \int_0^{\rho_1} \rho^{-m/2} \left[\tilde{s}(\rho, \theta, m)\right]^m \rho \, d\rho \, d\theta \tag{12.2.24}$$

which can be integrated with respect to θ to give

$$V_N \geq bD^2 \int_0^{\rho_1} \rho^{-m/2} s(\rho, m) \rho \, d\rho \tag{12.2.25}$$

where $s(\rho, m) = \int_{-\pi}^{\pi} \left[\tilde{s}(\rho, \theta, m)\right]^m \, d\theta$ is a regular function of ρ which may be assumed to admit a MacLaurin series expansion for $\rho \leq \rho_1$. For small enough ρ_1, the expansion may be truncated after the first term and one gets

$$V_N \geq bD^2 s(0, m) \int_0^{\rho_1} \rho^{1-m/2} \, d\rho \tag{12.2.26}$$

This integral diverges for any $m \geq 4$. Since the measured Weibull modulus is greater than 4 for any practical case, we have proved that Weibull's analysis leads to a divergent effective uniaxial volume when a sharp macrocrack is present in a brittle-elastic body.

This divergence may be avoided by using a nonlinear stress-strain relationship. For incompressible elastic material with a power law function $\sigma \propto \varepsilon^n$, with $n \leq 1$, Hutchinson (1968) and Rice and Rosengren (1968) found that the stress field was still singular, but the degree of the singularity in the Eq. (12.2.21) was no longer $-1/2$, but $-n/(n+1)$:

$$\sigma_{kl} = \sigma_N \left(\frac{r}{D}\right)^{-n/(n+1)} S_{kl}\left(\frac{r}{D}, \theta, n\right) \tag{12.2.27}$$

This is known as the HRR near-tip field (for Hutchinson-Rice-Rosengren). Repeating all the foregoing calculations, one obtains, instead of Eq. (12.2.26),

$$V_N \geq bD^2 s(0, m) \int_0^{\rho_1} \rho^{1-mn/(n+1)} \, d\rho \tag{12.2.28}$$

The integral diverges whenever $m \geq 2(n+1)/n$. This brings about convergence (finite V_N) for a broader, more realistic, range of Weibull's modulus. For example, for $n = 0.1$ there is convergence as long as $m \leq 21$.

Now, if one assumes that the divergence may be avoided using nonlinear models to compute the stress distribution, the obvious result of the analysis is that the structures free from singularities exhibit exactly the same size effect as sharply precracked structures. This contradicts recent experimental observations which clearly show that for large sizes the two-dimensional size effect law shows a $D^{-1/2}$ trend as predicted by deterministic linear elastic fracture mechanics, rather than a $D^{-2/m}$ trend as obtained by Weibull analysis. This is particularly noticeable in brittle metallic alloys for which the value of m obtained from tensile tests is such that the exponent $2/m$ is much less than $1/2$.

A consistent way of handling a cracked body problem will be presented beginning in Section 12.4 after summarizing the basic criticisms of the classical Weibull-type approach.

12.2.4 The Effect of Surface Flaws

In the classical approach to statistical fracture, based on a perfect material with dilute small defects, one may also easily treat the case of defects concentrated in a narrow layer at the structure surface. This is the case of flaws introduced by machining wear, drying shrinkage stresses (before shrinkage penetrates too deep) or corrosive processes that affect only the surface, not the bulk. In general, it is possible that in a particular structure both bulk and surface flaws are present. The essential point is that their strength and statistical distribution will differ, and so their effect on the structural scale dependence will be different.

Let us consider first how to take into account the effect of surface flaws alone (i.e., no flaws are present in the bulk). Following the spirit of Weibull's theory, it is assumed that failure initiates at the surface of the specimen, and that complete failure occurs right at initiation of growth of a flaw. With these hypotheses, the Weibull analysis for surface flaws is completely analogous to that for bulk flaws performed in the previous section for bulk defects, except that the dependence on the volume V is substituted by the dependence on surface S. Then, the equation describing the failure probability is analogous to (12.1.50):

$$P_{fS} = 1 - \exp\left\{ -\int_S \left[\frac{\tilde{\sigma}_S(\mathbf{x})}{\sigma_{0S}}\right]^{m'} \frac{dS(\mathbf{x})}{S_0} \right\} \tag{12.2.29}$$

where P_{fS} stands for failure probability at a surface defect, S_0 is a reference surface, σ_{0S} and m' are constants analogous to σ_0 and m for the bulk; $\tilde{\sigma}_S$ is the effective stress for the surface flaws, which is identical in concept to the effective stress for bulk flaws.

Following a reasoning parallel to that in Section 12.2.1, the failure probability can be cast in terms of the nominal stress σ_N as

$$P_{fS}(\sigma_N) = 1 - \exp\left[-\frac{S_N}{S_0}\left(\frac{\sigma_N}{\sigma_{0S}}\right)^{m'} \right] \tag{12.2.30}$$

in which the equivalent surface S_N is given by an expression similar to (12.2.5)

$$S_N = \int_S \left[\tilde{s}_S\left(\frac{\mathbf{x}}{D}, m'\right)\right]^m dS \tag{12.2.31}$$

where $\tilde{s}_S = \tilde{\sigma}_S/\sigma_N$ is a purely geometrical function. Since the foregoing expression for the failure probability is formally identical to that for bulk statistics, we can proceed by analogy and end up with the equivalent of (12.2.14):

$$\overline{\sigma}_{Nu} = \overline{\sigma}^0_{Nu}\left(\frac{S_N(D_0)}{S_N(D)}\right)^{1/m'} \tag{12.2.32}$$

Then, the size effect due to surface flaws is simple for full three-dimensional similitude, for which $S_N \propto D^2$ and

$$\overline{\sigma}_{Nu} = \overline{\sigma}^0_{Nu}\left(\frac{D_0}{D}\right)^{2/m'} \tag{12.2.33}$$

For two-dimensional similitude the expression is more complicated because of different similitude laws for differently oriented surfaces. For planar specimens, one must split the integral defining S_N into two parts: one S_{N1} defined on in-plane surfaces, and another S_{N2} defined on out-of-plane surfaces (the edges of the structure). The first part is proportional to D^2 and the second part is proportional to bD, where b is the thickness of the specimen; thus one can write

$$S_N \propto D^2\left(1 + \gamma_1\frac{b}{D}\right) \tag{12.2.34}$$

where $\gamma_1 = S_{N2}D/(S_{N1}b)$ is constant for a given structural shape. Then, if we take the reference size D_0 to be such that $D_0 = \gamma_1 b$, the size effect implied in this similitude is

$$\overline{\sigma}_{Nu} = \overline{\sigma}^0_{Nu}\left(\frac{D_0}{D}\right)^{2/m'}\left(\frac{2}{1 + D_0/D}\right)^{1/m'} \tag{12.2.35}$$

This is a curve that behaves as $D^{-1/m'}$ for small values of D ($D \ll D_0$) and as $D^{-2/m'}$ for large values of D ($D \gg D_0$). In a $\log \sigma_{Nu}$–$\log D$ plot, it is a curve concave downward with negative slope smoothly increasing from $1/m'$ on the far left toward $2/m'$ on the far right.

Now, the foregoing analysis refers to cases in which no flaws are present in the bulk. Let us now analyze the coupling of both surface and bulk flaws. Because the structure can fail through either a surface flaw or a bulk flaw, the survival probability is the product of the survival probabilities for these two independent cases:

$$(1 - P_f) = (1 - P_{fB})(1 - P_{fS}) \tag{12.2.36}$$

where P_{fB} is the probability of failure through a bulk flaw. Substitution of (12.2.6) for P_{fB} and (12.2.30) for P_{fS} leads to the joint failure probability

$$P_f(\sigma_N) = 1 - \exp\left[-\frac{V_N}{V_0}\left(\frac{\sigma_N}{\sigma_0}\right)^m - \frac{S_N}{S_0}\left(\frac{\sigma_N}{\sigma_{0S}}\right)^{m'}\right] \tag{12.2.37}$$

A closed-form expression for the mean strength is possible only in the case $m = m'$. For such a case the mathematical structure of the probability distribution is identical to that found previously, and one can easily conclude that the size effect can be written as

$$\overline{\sigma}_{Nu} = \overline{\sigma}^0_{Nu}\left(\frac{V_N(D_0) + \lambda_1 S_N(D_0)}{V_N(D) + \lambda_1 S_N(D)}\right)^{1/m'} \tag{12.2.38}$$

in which $\lambda_1 = \sigma_0^m V_0/(\sigma_{0S}^m S_0)$.

For three-dimensional similarity, $V_N \propto D^3$ and $S_N \propto D^2$ so the foregoing law can be manipulated to read

$$\overline{\sigma}_{Nu} = \overline{\sigma}^0_{Nu}\left(\frac{D_0}{D}\right)^{3/m'}\left(\frac{1 + \gamma_2\lambda_1/D_0}{1 + \gamma_2\lambda_1/D}\right)^{1/m'} \tag{12.2.39}$$

where $\gamma_2 = S_N(D)D/V_N(D)$ is a constant. To simplify this expression we can select the reference size as $D_0 = \gamma_2\lambda_1$ so that the expression reduces to

$$\overline{\sigma}_{Nu} = \overline{\sigma}^0_{Nu}\left(\frac{D_0}{D}\right)^{3/m'}\left(\frac{2}{1 + D_0/D}\right)^{1/m'} \tag{12.2.40}$$

where it must be remembered that D_0 is no longer arbitrary.

Now, if two-dimensional similitude is considered, $V_N \propto bD^2$. Therefore, for an appropriate choice of D_0, the size effect law can be written again as in (12.2.35), but with D_0 given by

$$D_0 = \frac{b\lambda_1\gamma_1}{\gamma_3 b + \lambda_1} \tag{12.2.41}$$

where $\gamma_3 = V_N/(bS_{N1})$ is a constant (note, furthermore, that $\gamma_3 = 1$ if $\tilde{\sigma} = \tilde{\sigma}_S$).

Comparing (12.2.40) to (12.2.35), it turns out that the statistical size effect curve for bulk and surface flaws distributed with the same Weibull modulus can be unified and written as

$$\overline{\sigma}_{Nu} = \overline{\sigma}^0_{Nu}\left(\frac{D_0}{D}\right)^{n_d/m'}\left(\frac{2}{1 + D_0/D}\right)^{1/m'} \tag{12.2.42}$$

where n_d is the dimensionality of the similitude and D_0 a *fixed* reference size whose value depends on the parameters of the statistical distributions and on the shape of the structure.

Exercises

12.9 Determine the equivalent uniaxial volume of the following three structures and compare it to the volume of the material: (a) a thin-walled pipe of length L, inner radius R, and thickness t, subjected to internal pressure p (assume free-ends); (b) a thin-walled sphere of inner radius R and thickness t, subjected to internal pressure p; and (c) a thin-walled tubular vessel (with spherical end caps) of length L, inner radius R, and thickness t

subjected to internal pressure p (neglect end effects and assume $L \gg R$ so that the probability of failure through the caps is negligible). Define the nominal strength σ_N as the maximum principal stress in the structure. Assume elastic-brittle behavior and applicability of Weibull's analysis based on the effective uniaxial stress (12.1.42).

12.10 Determine the equivalent uniaxial volume of a thick-walled cylinder of length L, outer radius R, and inner radius R_1 subjected to torsion. Define the nominal strength σ_N as the maximum principal stress in the cylinder. Assume elastic-brittle behavior and applicability of Weibull's analysis based on the effective uniaxial stress (12.1.42).

12.11 Determine the equivalent uniaxial volume of a solid cylinder of length L and radius R subjected to an uniaxial tensile force F and to a torque $4RF$. Define the nominal strength σ_N as the maximum principal stress in the cylinder. Assume elastic-brittle behavior and applicability of Weibull's analysis based on the effective uniaxial stress (12.1.42).

12.12 Determine the equivalent volume for a three-point bent beam of length L, depth D, thickness b, and a span-to-depth ratio of 16. The beam is made of a material with a stress-strain curve given by $\sigma = \sigma_c|\varepsilon|^{n-1}\varepsilon$. Analyze the effect of the hardening exponent n on the resulting equivalent volume. Assume applicability of Weibull's analysis based on the effective uniaxial stress (12.1.42).

12.13 Determine the equivalent uniaxial volume of a solid cylinder of length L and radius R subjected to pure torsion. Define the nominal strength σ_N as the maximum principal stress in the cylinder. Assume applicability of Weibull's analysis based on the effective uniaxial stress (12.1.42), but take the material to be nonlinear elastic with a shear stress-shear strain relationship given by $\tau = \tau_c|\gamma|^{n-1}\gamma$, in which τ is the shear stress, τ_c a constant, and γ the engineering shear strain.

12.14 Determine the equivalent uniaxial volume of a solid cylinder of length L and radius R subjected to an uniaxial tensile force F and to a torque $2RF$. Define the nominal strength σ_N as the maximum principal stress in the cylinder. Assume applicability of Weibull's analysis based on the effective uniaxial stress (12.1.42). Consider the material to be incompressible nonlinear elastic with a stress-strain relation given by

$$\varepsilon_{kk} = 0\,, \qquad \varepsilon_{ij} = \sigma_c^{-n}\hat{\sigma}^{n-1}\sigma'_{ij}$$

where σ_c is a constant, $\sigma'_{ij} = \sigma_{ij} - (1/3)\sigma_{kk}\delta_{ij}$ is the deviatoric stress tensor, and $\hat{\sigma}$ is defined as

$$\hat{\sigma} = \sqrt{\frac{3}{2}\sigma'_{ij}\sigma'_{ij}}$$

12.15 Analyze the size effect for direct tension on solid cylinders of diameter D and length L in the following cases: (a) D and L vary proportionally; (b) D varies while L is kept constant; (c) D is kept constant while L varies. Assume classical Weibull analysis.

12.16 Analyze the size effect for direct tension on solid cylinders of diameter D and length L if they contain both bulk flaws and surface flaws. Assume that the Weibull moduli are the same for bulk and surface flaws. Consider three cases: (a) D and L vary proportionally; (b) D varies while L is kept constant; and (c) D is kept constant while L varies.

12.3 Basic Criticisms of Classical Weibull-Type Approach

12.3.1 Stress Redistribution

The key to the calculation of failure probability of a structure is the function $\mathbf{s}(\mathbf{x}/D)$, characterizing the stress at point $\mathbf{x}$. In this regard, one must distinguish two types of structures: (1) those failing at the initiation of the macroscopic crack growth (i.e., the structure just before failure contains only microscopic cracks or other flaws, as is typical of structures made of brittle materials); and (2) those failing only after a large stable macroscopic crack growth. The latter is the case of brittle failures of reinforced concrete structures or compressed plain concrete structures such as dams, of rock failures (tunnels, mine openings, boreholes, deep excavations, mountain slides), as well as other structures made of all quasi-brittle materials.

For the first type of structure, the key point is that function $\mathbf{s}(\mathbf{x}/D)$ just before failure is known, since microscopic flaws have negligible influence on the overall stress distribution within the structure. In such

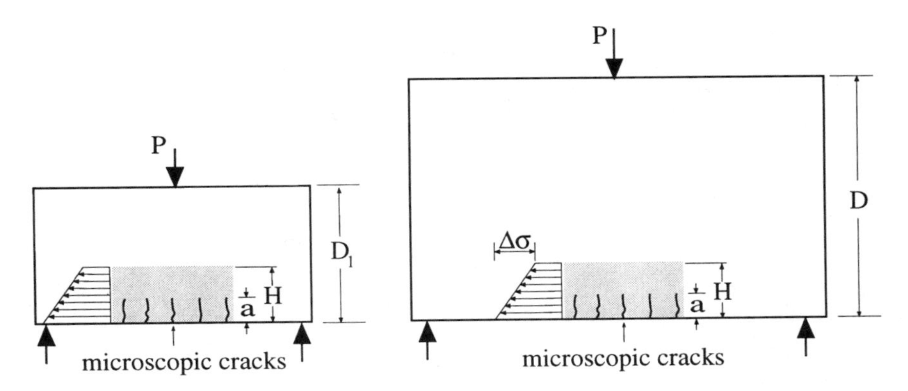

Figure 12.3.1 Geometrically similar structures with microscopic cracks (after Bažant, Xi and Reid 1991).

structures there exists a region of size H (Fig. 12.3.1) such that

$$a \ll H \ll D \tag{12.3.1}$$

where D = structure dimension and a = crack or flaw size. The condition $H \ll D$ means that the stress distribution within region H would be nearly uniform if the flaw did not exist ($|\Delta\sigma| \ll |\sigma|$ in Fig. 12.3.1). If the size of the flaw or initial crack is very small, $a \ll D$, it is a characteristic of the state of the material. It is related to the inhomogeneity size and is independent of the structure dimension D. Since $H \ll D$, the presence of the flaw of size a affects the stress distribution only locally, and the situation is nearly the same as that of a crack in an infinite space with a uniform-stress state at infinity equal to the stress in region H. Thus, the only effect of the flaw of size a is a local reduction of the effective macroscopic strength of the material. This permits the random variation of the initial flaw sizes to be related to the random variation of the material strength, as described by Weibull distribution.

For the second type of structure, for example, reinforced concrete structures, the behavior is completely different. Due to reinforcement as well as the existence of strain softening in a large zone of microcracking and crack bridging near the front of a continuous fracture, reinforced concrete structures do not fail at crack initiation. Fig. 12.3.2 shows experimentally observed macroscopic crack patterns for various structure sizes, which form before the maximum load is attained. In fact, design codes require the failure load to be significantly higher than the crack-initiation load (for bending, at least 1.25 times higher, according to ACI Standard 318, but, in practice, this ratio is usually much higher). Consequently, a reinforced concrete structure normally undergoes pronounced inelastic deformation with large macroscopic stable crack growth prior to reaching the failure load (maximum load). This inevitably engenders stress redistributions, such that the stress distribution $\boldsymbol{\sigma}(\mathbf{x})$ at incipient failure is very different from the elastic stress distribution, which has commonly been assumed in the previous studies of the statistical size effect. The same is true for a dam, tunnel, borehole, mine opening, rock slide, or rock wall. Even in bending of a plain concrete beam or slab, a zone of microcracking not negligible compared to the beam depth develops before the maximum load unless the beam is extremely deep. The stress redistribution, moreover, is of the localization type, causing the stresses outside the localization domain to become insignificant, while the size of the localization domain becomes, for large D, almost independent of D.

To sum up, Weibull theory cannot be applied to reinforced concrete structures as well as many plain concrete structures (dams, not to deep beams) and rock structures unless the effect of stable macroscopic crack growth on the stress distribution function $\mathbf{s}(\mathbf{x}/D)$ is taken into account (Bažant 1988b).

12.3.2 Equivalence to Uniaxially Stressed Bar

Another limitation of the existing Weibull-type formulations is revealed by realizing that, if the stress distribution function $\mathbf{s}(\mathbf{x}/D)$ is known *a priori*, every structure is equivalent to an uniaxially stressed bar. Bažant (1988b)and Bažant, Xi and Reid (1991) proved that in the Weibull statistical formulation any structure is equivalent to an uniaxially stressed bar of variable cross section. In the previous sections, we

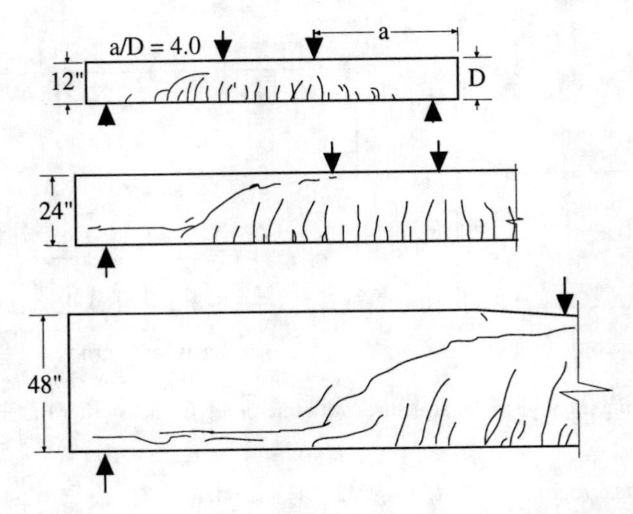

Figure 12.3.2 Example of geometrically similar reinforced concrete structures with stable macroscopic fracture growth before failure (from Bažant, Xi and Reid 1991, adapted from Kani 1967).

showed that the equivalence may be advanced one step further, showing that there is always an equivalent uniaxially stressed bar of *uniform* cross section to which the actual problem reduces.

Obviously, in this approach, all information about the mechanics of failure is lost, and the structural geometry becomes irrelevant. Of course, this cannot be true for any kind of structure and material. So the Weibull-type approach cannot be regarded as realistic, even if the stress distribution function would realistically describe the stress field at imminent failure.

12.3.3 Differences between Two- and Three-Dimensional Geometric Similarities

Another questionable aspect of the classical Weibull-type approach is the effect of the number of dimensions, n_d, implied by Eq. (12.2.20) (Bažant 1988b). Consider, for example, that the beam dimensions are increased in the ratio D_2/D_1 according to either two-dimensional similarity, in which case the beam thickness b is kept constant, or according to three-dimensional similarity, in which case the beam thickness is also increased in proportion to D. According to Eq. (12.2.20), the nominal strength σ_{Nu} should change in the ratio $(D_2/D_1)^{2/m}$ or $(D_2/D_1)^{3/m}$, respectively. Now, although systematic data on the effect of thickness are unavailable, it appears from experience that there is no significant difference between these two cases.

In particular, no significant difference is manifested by comparing the slopes of the plots of $\log(\sigma_{Nu})$ vs. $\log(D)$ for tests with two-dimensional similarity and three-dimensional similarity. The effect of the number of dimensions, n_d, can be checked by using Bažant and Kazemi's (1991) tests of diagonal shear failure of concrete beams (reinforced by longitudinal bars with hooks at the ends), which were similar in two dimensions (same thickness), and the pullout tests of bars by Bažant and Şener (1988), which were similar in three dimensions; see the data points in Fig. 12.3.3. Taking the diagonal shear test as reference, the fact that the slope of the mean trend of the data is approximately $-1/2$ implies that $m = 4$ (we pretend we do not know any uniaxial test data that indicate a much larger m). Then, for three-dimensional similarity, the slope of the line should be $-n/m = -3/4$. But the pullout test data made with the same concrete indicate the slope to be also $-1/2$, which does clearly disagree with the classical Weibull-type analysis (but agrees with the modified statistical theory presented in the forthcoming sections).

The classical Weibull-type theories are further put in question when one tries to compare the results of tests on bars of different sizes failing in uniaxial tension with the diagonal shear tests. The former tests indicate that, approximately, $m = 12$ (Zech and Wittmann 1977). But if $m = 12$, then the slope of the line in Fig. 12.3.3 on the left would have to be $-1/6$ rather than $-1/2$. This is a serious discrepancy indeed (it is remedied in the forthcoming sections).

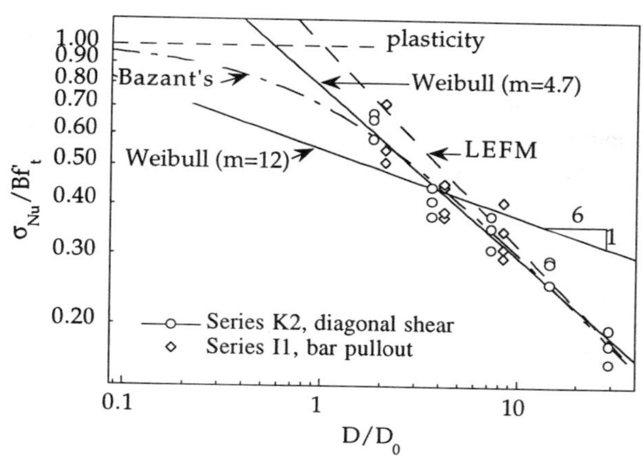

Figure 12.3.3 Comparison of test results with size effect lines obtained from various theories, including classical Weibull-type theory. ○ — two-dimensional similitude, after Bažant and Kazemi (1991); ◇ — three-dimensional similitude, after Bažant and Şener (1988). See Tables 1.5.1–1.5.2 and Fig. 1.5.1 for details on these tests.

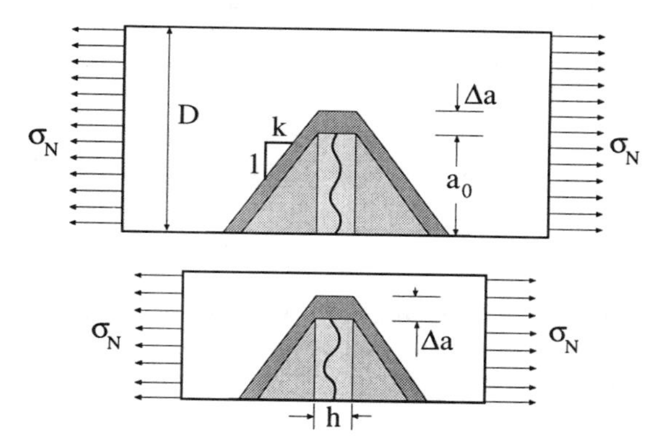

Figure 12.3.4 Stress relief zone in geometrically similar panels with large similar fractures (adapted from Bažant, Xi and Reid 1991).

12.3.4 Energy Release Due to Large Stable Crack Growth

From the mechanics viewpoint, the basic problem with the classical Weibull-type approaches to reinforced concrete structures is that they generally ignore energy release from the structure which is inevitably caused by macroscopic fracture growth. Experiments confirm that in concrete structures the fracture length at maximum load is usually proportional to the structure dimension, while the width of the fracture process zone, h, is almost the same for any size and is a material property, as shown by the example of a rectangular panel in Fig. 12.3.4. Fractures of length a_0 release energy before failure of this panel, which may be imagined to come from the sparsely shaded triangular areas in the figure. When, during failure, the fracture extends by Δa, the stress is further relieved from the densely cross-hatched narrow strips whose area gets larger, the larger the structure size. This means that the release of the stored energy of the structure into the fracture extension Δa, which comes from the strip, is larger for a larger structure if the nominal stress is the same. However, fracture extension requires roughly the same energy per unit length of extension, regardless of the structure size. Therefore, the nominal stress at failure must get smaller if the structure gets larger, so that the strain energy density in the densely cross-hatched strip would be smaller in a larger structure, thus making it possible to obtain the same energy release per unit length of the fracture. The foregoing argument has been used in Chapter 1 to derive the deterministic size effect law proposed by Bažant (1983, 1984a) that is different from that in Eq. (12.2.20) and agrees quite well with a broad range of test results. That law has also been supported by certain other, more general, arguments.

12.3.5 Spatial Correlation

Another questionable aspect of classical Weibull-type theories based on the weakest-link hypothesis is the neglect of spatial correlation. This might be justified for the links in a chain, but not for continuous bodies in general, and not for monolithic (cast at one time) concrete structures in particular. If the strength value realized in one small material element is on the low side of the average strength, the strength value realized in the adjacent material element is more likely to be also on the low side than on the high side of the average strength. The standard way to deal with spatial correlation would be to introduce a spatial autocorrelation function for material behavior, such as strength, but that approach would be rather complicated for the present purpose. There are, nevertheless, other simpler ways to introduce spatial correlation. A well justified way is the nonlocal concept, advanced in the subsequent sections. An even simpler, but less realistic, way is to introduce some phenomenological rule for load sharing after a local break, analogous to stress redistribution in a cable after one of its wires breaks.

12.3.6 Summary of the Limitations

The classical applications of Weibull theory to reinforced concrete structures suffer from several serious shortcomings:

1. The stress distribution function used in the applications is assumed to be the elastic stress distribution, although it properly must take into account the stress redistributions caused by large macroscopic stable crack growth prior to reaching maximum load. That growth causes a strong deterministic (or systematic) size effect, which prevails over the statistical size effect due to random strength.

2. According to the classical applications of Weibull theory, every structure is mathematically equivalent to an uniaxially stressed bar, which means that no information on the failure mechanism is taken into account.

3. According to the classical theories, the differences in the size effect between two-dimensional similarity and three-dimensional similarity appear to be too strong, contradicting experience.

4. Tests of geometrically similar concrete structures, e.g., diagonal shear tests, show a much stronger size effect than that predicted by classical Weibull theory (provided that the Weibull modulus value is taken the same as that obtained from direct tension test).

5. The classical Weibull-type theories neglect spatial statistical correlation of material failure probabilities at various points.

Modifications of the Weibull theory required to eliminate the aforementioned shortcomings will occupy us next.

12.4 Handling of Stress Singularity in Weibull-Type Approach

As argued in the preceding section, the stress distribution function to be used in the integral for failure probability of the structure must be the stress distribution at incipient failure, rather than some stress distribution that exists long before failure. This distribution must reflect localization of strains and stresses that occurs prior to reaching the maximum load. In the extreme case of a complete localization of cracking, a sharp crack develops upon reaching the maximum load, as illustrated in Fig. 12.4.1. But according to the analysis in Section 12.2.3, when a sharp crack is present, the stress singularity causes the Weibull approach to predict unrealistic statistical strength distributions.

The problem arises because the deterministic failure criterion underlying the Weibull approach is strength-based and *local.* In this context, "local" means that the failure of an *infinitesimal* element will occur when the stress in it reaches a threshold value — the material strength at the element location (a point). This kind of criterion cannot be applied when stress singularities occur, because in this case the stress at the singularity always exceeds the material strength and, thus, failure would occur at vanishingly small loads.

To overcome this problem, the classical simple approaches in strength-based (deterministic) fracture

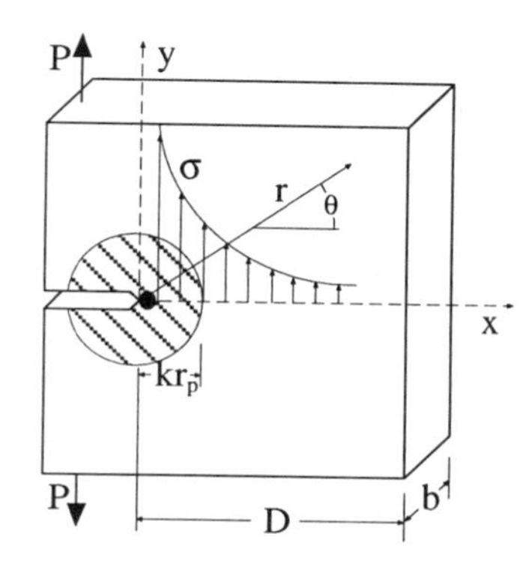

Figure 12.4.1 Stress distributions and crack process zone in precracked specimens (after Bažant and Xi 1991).

mechanics, use an *a posteriori* nonlocal formulation: the stress field is computed in a classical, local, framework, and then the global fracture is assumed to occur when the tensile strength has been overcome at all the points on a finite region of the body. The size of the region over which the tensile strength has to be exceeded, say λ, is assumed to be a material parameter and is one of the basic properties of the material. In this approach, nonlocality affects only the statement of the global fracture condition and is related to the microstructure of the body (Knott 1973, Section 8.9). The statistical treatment of this model has been done by Beremin (1983). However, his approach has been essentially devoted to model cleavage fracture of low carbon steel and its applicability to concrete is dubious.

In a different approach, Bažant and Xi (1991) used a nonlocal formulation to overcome the problem of the singularity. In their model, the fracture criterion is made nonlocal by assuming that the driving force for fracture of an infinitesimal element is not the stress over this element, but rather an average stress computed on a finite neighborhood of the element. The size (radius) of this neighborhood is, again, a basic material property.

The development of Bažant and Xi (1991) is very technical and relies on somewhat special hypotheses regarding the average stress distribution. In this section, we use a recent approach (Planas and Bažant 1997) which uses very mild assumptions regarding the material behavior to derive the essential conclusions of Bažant and Xi's theory. This approach is simpler because the asymptotic case of large size is analyzed right away instead of obtaining it as the large-size limit of the general nonlocal formulation for all sizes.

12.4.1 A Simplified Approach to Crack Tip Statistics

To be specific, consider the planar, mode I specimen shown in Fig. 12.4.1. Let the specimen thickness be b and its in-plane size be D. We specifically consider the case where the specimen is very large compared to the zone of nonlinear behavior. Let r_p be the critical size of the nonlinear zone for a certain underlying deterministic approach. As seen in Chapter 5, this can always be written as proportional to $(K_{Ic}/f'_t)^2$, where K_{Ic} is the critical stress intensity factor and f'_t the tensile strength.

If D is large enough, we may find a circle centered at the crack tip with a radius $r_1 = kr_p$ which is: (1) so much greater than r_p that the effect of the nonlinear zone is vanishingly small at that distance (in view of Saint Venant's principle); and (2) so much smaller than the in-plane size D of the structure that the stress field in the neighborhood of the circumference $r = r_1$ is accurately given by the singular term of the near-tip field expansion of the elastic stress solution. This requires that

$$r_p \ll kr_p \ll D \quad \text{or} \quad 1 \ll k \ll \frac{D}{r_p} \tag{12.4.1}$$

Now, we consider separately the core part of the structure defined by the circle of radius $r < kr_p$ and the remaining part. Guided by the experimental fact that failure in a deeply cracked large specimen rarely occurs by failure far from the crack tip, we postulate that the probability of failure taking place outside the core is negligible. Therefore, the statistics of the whole specimen must be very close to the statistics of the core.

According to our previous hypotheses, the boundary conditions for the core are controlled by the stress intensity factor. This means that, whatever the mechanical behavior of the material, the stress distribution must be given by an expression of the form:

$$\sigma(\mathbf{x}) = \frac{K_I}{\sqrt{kr_p}}\mathbf{s}\left(\frac{\mathbf{x}}{kr_p}, \frac{1}{k}\right) \tag{12.4.2}$$

in which the only explicit dependence on the material properties is through r_p which, as previously indicated, is a material property dependent on the fracture toughness and the tensile strength. The explicit dependence of the stress distribution on $1/k$ comes from the fact that this ratio is the relationship between the size of the inelastic zone and the size of the core.

The foregoing equation is the equivalent of Eq. (12.2.2) in which σ_N is replaced by $K_I/\sqrt{kr_p}$ and D by kr_p. Thus, all the results of Section 12.2 are directly usable if we further assume the singularity to be relieved so that the integral in (12.2.5) would converge for the core. The result for the failure probability distribution is

$$P_f(K_I) = 1 - \exp\left[-\frac{V_C}{V_0}\left(\frac{K_I}{\sigma_0\sqrt{kr_p}}\right)^m\right] \tag{12.4.3}$$

where V_C is the effective uniaxial volume for the core. Recalling that the core properties depend only on r_p and k, and noting that we have a planar situation,

$$V_C = br_p^2 c(k, m) \tag{12.4.4}$$

where $c(k, m)$ is a dimensionless function. Since k is arbitrary as long as it satisfies (12.4.1), it is now reasonable to impose the condition that the failure probability of the core must be insensitive to small variations of k. According to (12.4.3), this is exactly achieved if we assume

$$c(k, m) = c_0 k^{m/2} \tag{12.4.5}$$

where c_0 is a constant. If we now substitute this equation into Eq. (12.4.4) and the result back into (12.4.3) we get

$$P_f(K_I) = 1 - \exp\left[-\frac{b}{b_0}\left(\frac{K_I}{K_0}\right)^m\right] \tag{12.4.6}$$

where b_0 is some reference thickness and K_0 is a material constant with the dimension of stress intensity factor.

The foregoing equation shows that, under the aforementioned reasonable hypotheses, the statistics for the critical (failure) stress intensity factor is analogous to that of the classical strength, except that the specimen thickness replaces the volume. Therefore, the mean critical stress intensity factor is given by

$$\overline{K_{Ic}} = \Gamma\left(1 + \frac{1}{m}\right)\left(\frac{b_0}{b}\right)^{\frac{1}{m}} K_0. \tag{12.4.7}$$

The relationship between the mean critical stress intensity factors for specimens of thicknesses b and b_0 is

$$\overline{K_{Ic}} = \left(\frac{b_0}{b}\right)^{\frac{1}{m}} \overline{K_{Ic0}}. \tag{12.4.8}$$

12.4.2 Generalization of the Thickness Dependence of the Crack Tip Statistics

As noted by Bažant and Xi (1991), the dependence of the crack tip (or core) statistics on the specimen thickness must be interpreted as the possibility of crack extension taking place over arbitrarily small segments along the crack front line. This is not very realistic because the crack front remains macroscopically smooth. This may be achieved by generalizing the above results to accept that, for small thickness, the crack must grow simultaneously through the whole thickness, but may grow over only a part of the thickness for very thick specimens.

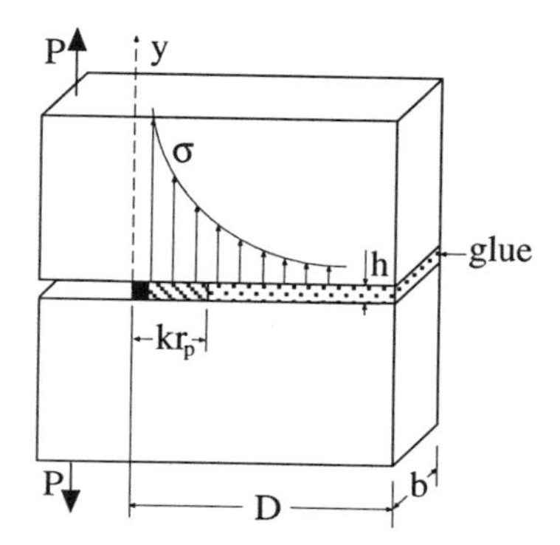

Figure 12.4.2 Stress distributions and crack process zone in a glue layer (after Bažant and Xi 1991).

Qualitatively, this may be taken into account by modifying Eq. (12.4.6) as

$$P_f(K_I) = 1 - \exp\left[-\left(\frac{b}{b_0}\right)^p \left(\frac{K_I}{K_0}\right)^m\right] \tag{12.4.9}$$

where $p = 0$ in the limit of propagation with straight crack front and $p = 1$ for crack propagation in which the front can become arbitrarily curved (Bažant and Xi 1991).

12.4.3 Asymptotic Size Effect

The foregoing analysis is valid for very large sizes ($D \gg r_p$) and for planar (two-dimensional) similarity. A similar analysis may be performed in the large size limit for axisymmetric similarity, and also for the one-dimensional similarity exemplified by the failure of a thin glue layer between two blocks (Fig. 12.4.2). In the case of an axisymmetric crack, the core is a torus around the crack front, and in the case of the glue layer, the core is a strip in front of the crack tip (Fig. 12.4.2).

The obvious result is that all the equations remain the same, except that the specimen thickness in the axisymmetric case must be replaced by the crack front length. Therefore, we reinterpret b as the crack front length and keep all the foregoing equations.

Consider now $p = 0$, which means that the crack is assumed to propagate simultaneously everywhere along its front. Then, the distribution equations become independent of b and the mean critical stress intensity factor is independent of the specimen thickness. Since

$$K_I \propto \sigma_N \sqrt{D} \tag{12.4.10}$$

it thus appears that the asymptotic size effect for any kind of similarity (three-, two- or one-dimensional) is the same and coincides with that predicted by LEFM:

$$\overline{\sigma}_{Nu} \propto D^{-1/2} \quad \text{for large } D \tag{12.4.11}$$

12.4.4 Extending the Range: Bulk Plus Core Statistics

In the previous approach, we neglected the probability of failure outside the core. This may be introduced, at least qualitatively, by stating that the failure may occur either by crack growth (to which the previous results apply) or by rupture in the bulk of the specimen outside the core (to which the classical Weibull analysis applies). Assuming that the two failure processes are statistically independent, we start from the condition that the joint survival probability is the product of the survival probabilities for both processes. After similar calculations as in Section 12.1.6 one gets the result

$$P_f = 1 - \exp\left[-\left(\frac{b}{b_0}\right)^p \left(\frac{K_I}{K_0}\right)^m + \frac{V_B}{V_0}\left(\frac{\sigma_N}{\sigma_0}\right)^m\right] \tag{12.4.12}$$

where V_B is the effective uniaxial volume corresponding to the failure in the bulk (outside the core).

Now, in view of (12.4.10) and the fact that $V_B \propto bD^2$ (where b keeps the meaning of the length of the crack front edge) and $K_I \propto \sigma_N \sqrt{D}$, the foregoing equation may be rewritten as

$$P_f = 1 - \exp\left\{-\left[c_1 \left(\frac{b}{b_0}\right)^p \left(\frac{D}{D_0}\right)^{m/2} + c_2 \frac{b}{b_0} \left(\frac{D}{D_0}\right)^2\right] \left(\frac{\sigma_N}{\sigma_0}\right)^m\right\} \tag{12.4.13}$$

where c_1 and c_2 are dimensionless constants and D_0 is the reference size introduced for convenience. Note that, according to this equation, the term corresponding to the core statistics dominates for large D if $m > 4$, as previously postulated.

12.4.5 More Fundamental Approach Based on Nonlocal Concept

The last equation (12.4.13) is remarkably close to that derived by Bažant and Xi (1991) based on a nonlocal model, the only difference being that in Bažant and Xi, the factor c_2 depends on the ratio r_p/D, while in our asymptotic approach, it is constant; in fact, it is the limit of the Bažant-Xi equation for $r_p/D \to 0$.

Even though the nonlocal approach to fracture has not yet been presented (this will be done in Chapter 13), it is helpful to at least mention this basic idea in the present context. The basic assumption in the nonlocal approach (in the frame of the Weibull theory) is that the failure probability at a point $\mathbf{x}$ of the material does not depend on stress $\boldsymbol{\sigma}$ at that point, but on the average stress (or average strain) over a certain characteristic neighborhood of that point. It is intuitively clear that this should remove the aforementioned problems of the classical theory. The failure probability in this theory is still given by Eq. (12.1.50), but the expression (12.1.46) for the effective uniaxial stress $\tilde{\sigma}$ is replaced by

$$\tilde{\sigma} = \left[\sum_{P=I}^{III} \langle \overline{\sigma}_P \rangle^m \right]^{\frac{1}{m}} \tag{12.4.14}$$

where $\overline{\sigma}_P$ $(P = I, II, III)$ is an average of the principal stress σ_P over the characteristic neighborhood of the point:

$$\overline{\sigma}_P(\mathbf{x}) = \int_V \sigma_P(\mathbf{s}) \alpha(\mathbf{s}, \mathbf{x})\, dV(\mathbf{s}) \tag{12.4.15}$$

Here $\alpha(\mathbf{s}, \mathbf{x})$ is an empirical weight function centered at point $\mathbf{x}$ for which the average is computed. This function usually has a bell-shape, and the width of the bell is proportional to the characteristic length of the material. Bažant and Xi used a planar formulation in which the volume is $V = bA$ and $dV = bdA$, where A is the area of the plane body. They used the following weight function:

$$\alpha(\mathbf{s}, \mathbf{x}) = \frac{1}{\overline{\alpha}(\mathbf{x})} \left\langle 1 - \frac{|\mathbf{s} - \mathbf{x}|^2}{\lambda^2} \right\rangle^2 \tag{12.4.16}$$

where $\mathbf{s}$ and $\mathbf{x}$ are now in-plane vectors (i.e., with only two components different from zero), λ is a constant with dimensions of length giving the radius of the neighborhood over which the averaging is made, and $\overline{\alpha}(\mathbf{x})$ is a normalizing factor given by

$$\overline{\alpha}(\mathbf{x}) = \int_A \left\langle 1 - \frac{|\mathbf{s} - \mathbf{x}|^2}{\lambda^2} \right\rangle^2 b\, dA(\mathbf{s}) \tag{12.4.17}$$

This factor assures that

$$\int_A \alpha(\mathbf{s}, \mathbf{x}) b\, dA(\mathbf{s}) = 1 \tag{12.4.18}$$

for any location of the averaging point.

Using this nonlocal approximation, it is not necessary to subdivide *a priori* the specimen volume into the crack-tip core and the rest. Such a subdivision is obtained in principle as a consequence, although the complexity of handling the nonlocal formulation based on Eqs. (12.4.14)–(12.4.16) necessitated some approximations in order to obtain Eq. (12.4.13).

Exercises

12.17 On which of the following aspects can the constant c_0 in (12.4.5) depend: (a) geometry; (b) size; (c) material?

12.18 Show that the constant K_0 appearing in (12.4.6) can be written in terms of the parameters appearing in Eqs. (12.4.3)–(12.4.5) as

$$K_0 = \sigma_0 \sqrt{r_p} \left(\frac{V_0}{c_0 b_0 r_p^2} \right)^{1/m}$$

12.5 Approximate Equations for Statistical Size Effect

12.5.1 Bažant-Xi Empirical Interpolation Between Asymptotic Size Effects

The asymptotic size effect law for very large sizes in Eq (12.4.11) is very simple. More importantly, it is also independent of the geometry of the structure. For very small sizes, we can write another equation in which the nonlinear zone extends over the whole specimen and the stress singularity completely disappears. This case leads to classical Weibull size effect, and thus Eq. (12.2.20) holds, again independently of the geometry.

As shown in Section 1.5, it appears that, as a good approximation relative to the scatter of the test result, the form of the size effect law can be considered as shape-independent through a range of sizes up to about 1:20. In this range, Bažant's size effect law (1.4.10) gives a good description of the size effect. We remind that this law can be written as

$$\sigma_{Nu} = \frac{B f_t'}{\sqrt{1 + \frac{D}{D_0}}}, \tag{12.5.1}$$

where $B f_t'$ and D_0 are two empirical constants (f_t' is the direct tensile strength, in ACI notation, introduced to make B nondimensional).

In view of the foregoing observations, it is not inappropriate to obtain the complete statistical size-effect law for bodies with large fractures by a simple empirical interpolation formula that agrees with the asymptotic cases —(12.4.11) for large sizes and (12.2.20) for small sizes— and, at the same time, reduces to (12.5.1) for $m \to \infty$, the deterministic case. A simple formula satisfying these three asymptotic conditions is (Bažant and Xi 1991)

$$\sigma_{Nu} = \frac{B' f_t'}{\sqrt{\beta^{2n_d/m} + \beta}}, \qquad \beta = \frac{D}{D_0'} \tag{12.5.2}$$

where $B' f_t'$ and D_0' are again empirical constants. The prime is introduced just to distinguish these parameters from those corresponding to deterministic analysis. Note that in this approximation the influence of the thickness is neglected, i.e., the value of p for the thickness dependence in (12.4.9) is assumed to be 0.

12.5.2 Determination of Material Parameters

Before attempting to determine the parameters $B' f_t'$ and D_0', the Weibull modulus m will normally be obtained from other tests. Such tests may be based on the relationship (12.1.27) between m and the coefficient of variation of specimen strength, ω.

Once m is known, it is possible to determine the parameters $B' f_t'$ and D_0' from σ_{Nu}-data for geometrically similar specimens of various sizes, in a way similar as that for Eq. (12.5.1). Although logarithmic nonlinear regression of data is recommended whenever possible (see Sections 6.3.2 and 6.3.6), it is also feasible to reduce Eq. (12.5.2) to a linear form and get the estimates simply by linear regression.

To this end, we may algebraically rearrange Eq. (12.5.2) to the linear plot $Y = AX + C$ in which

$$X = D^{1-\frac{2n_d}{m}}, \quad Y = \frac{1}{\sigma_{Nu}^2 D^{\frac{2n_d}{m}}}, \quad A = \frac{1}{(B'f_t')^2 D_0'}, \quad C = \frac{1}{(B'f_t')^2 D_0'^{\frac{2n_d}{m}}} \tag{12.5.3}$$

Thus, if m is known, A and C, along with the coefficients of variation of A and of the deviations of the data points from the regression line, can be determined from the plot of Y vs. X by linear regression. $B'f_t'$ and D_0' follow from Eq. (12.5.3). The linear regression previously introduced for the deterministic size effect law, Eq. (12.5.1), is the special case of Eq. (12.5.2) for $m \to \infty$.

Similar to the previous use of Eq. (12.5.1) in Chapter 6, the regression results for $B'f_t'$ can be used to determine K_{Ic} and G_f, defined as the critical stress intensity factor and critical energy release rate in an infinitely large fracture specimen. According to LEFM, one has $G_f = K_{Ic}^2/E' = \sigma_{Nu}^2 Dk^2(\alpha_0)/E'$. So, from Eq. (12.5.2), the asymptotic behavior for large D is $\sigma_{Nu} = B'f_t'(D_0'/D)^{1/2}$. Setting these two expressions equal, one gets

$$K_{Ic} = k_0 B' f_t' \sqrt{D_0'}, \qquad G_f = k_0^2 \frac{(B'f_t')^2 D_0'}{E'} \tag{12.5.4}$$

which happens to coincide with the expression (6.2.3) obtained previously from Eq. (12.5.1). This means that the statistical strength effects on G_f are nil, which is not surprising.

In analogy to the size effect regression plot introduced by Planas and Elices (1989a, 1990a), Eq. (12.5.2) can also be algebraically rearranged to the plot $Y' = A'X' + C'$ in which

$$X' = \frac{1}{D^{1-\frac{2n_d}{m}}}, \quad Y' = \frac{1}{\sigma_{Nu}^2 D}, \quad A' = \frac{1}{D_0'^{\frac{2n_d}{m}} (B'f_t')^2}, \quad C' = \frac{1}{(B'f_t')^2 D_0'} \tag{12.5.5}$$

After determining A' and C' by linear regression, the constants D_0 and B follow from the last two equations. This plot is advantageous for extrapolation to large sizes, because $D \to \infty$ corresponds to $X' = 0$. On the other hand, the plot according to Eq. (12.5.3) shows better results for small sizes, which dominate in the testing of concrete.

12.5.3 The Question of Weibull Modulus m for the Fracture-Process Zone

In our derivation, Weibull's distribution of material strength has been used in a somewhat different sense than in the classical problem of failure of a long fiber or chain. Is our value of Weibull modulus m the same as for direct tension tests of long fibers or bars? Probably not, and probably it is larger.

The reason is that Weibull modulus m increases with increasing uniformity of the distribution of the flaws in the material (Freudenthal 1968), and the distribution of flaws that must be considered is that just before failure. The uniformity of flaws in the fracture process zone is probably much higher than in the initial state of the material. Therefore, the value of m may be larger than the value obtained from direct tension tests of bars of different lengths, which is about 12. However, due to the absence of test results that would suffice for determining m, we will consider, in the numerical examples that follow, the value $m = 12$ because it is conservative. The larger the m-value, the weaker is the size effect. In future research, however, an estimate of the effective m-value for the fracture process zone should be obtained.

12.5.4 Comparison with Test Results

To be able to check the asymptotic trend defined by (12.4.11) and (12.2.20) and the empirical interpolation formula in (12.5.2), test data of a rather broad size range are needed. Furthermore, to check statistics, a large number of identical tests is desirable. No data are known which would be perfect in this regard. However, the recent data in Fig. 12.5.1 are interesting. They show the optimum fits by Eq. (12.5.1) (deterministic) and Eq. (12.5.2) (statistical) for the test results of Bažant and Kazemi (1991) on diagonal shear failure of longitudinally reinforced concrete beams without stirrups. The size range of these data, 1:16, was quite broad. We see that the statistical Eq. (12.5.2) fits these data somewhat better than the deterministic Eq. (12.5.1). However, the difference between these two size effect laws is discernible only

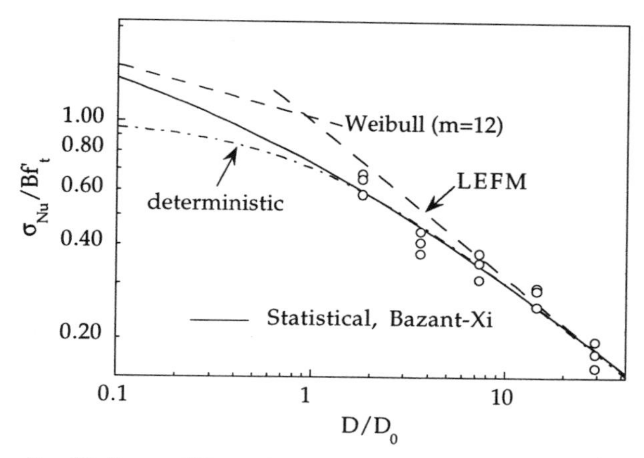

Figure 12.5.1 Test results of Bažant and Kazemi (1991) on diagonal shear failure of reinforced concrete beams without stirrups; $D_0 = 11$ mm and $Bf'_t = 4.6$ MPa, which correspond to the deterministic best fit to the five size case (after Bažant and Xi 1991; see Series K2 in Tables 1.5.1–1.5.2 and Fig. 1.5.1 for details on the tests).

for very small specimen sizes, and the small-size asymptotic slope $-2n_d/m$ seems to be acceptable. The difference becomes insignificant only for extrapolation to sizes smaller than the aggregate size.

Further support is provided by numerical solutions of the nonlocal statistical model of Bažant and Xi (1991, see Section 12.4.5) for the center-cracked panel shown in Fig. 12.5.2a. The results of these numerical calculations for different sizes are shown by the open circles in Fig. 12.5.2b, which also shows a comparison with the generalized size effect-law in Eq. (12.5.2). We see that the numerical results are very close to the analytical curve, which provides further supporting evidence for this simple law, although the fit provided by the deterministic law is slightly better than for the statistical law (the regression coefficients are 0.9999 for the deterministic and 0.9990 for the statistical law).

An interesting point is to see whether the predictions for K_{Ic} of the two fits are sensibly different; they are not. Indeed, the best fit values of the parameters of the two equations (obtained by a logarithmic nonlinear regression) are related as follows:

$$B'f'_t = 0.92Bf'_t\,, \qquad D'_0 = 1.3D_0 \tag{12.5.6}$$

so from (12.5.4), it follows that

$$K_{Ic}(\text{statistical fit}) = 1.05K_{Ic}(\text{deterministic fit}) \tag{12.5.7}$$

Therefore, the difference is only 5%, which is less than the accuracy of the estimated results.

12.5.5 Planas' Empirical Interpolation Between Asymptotic Size Effects

The Bažant-Xi formula (12.5.2) is not the only possible equation displaying the asymptotic trends given by (12.4.11)—for large sizes— and (12.2.20)—for small sizes. A more general equation satisfying these conditions is

$$\sigma_{Nu} = \frac{B'f'_t}{\left(\beta^{\frac{2n_d r}{m}} + \beta^r\right)^{\frac{1}{2r}}}\,, \qquad \beta = \frac{D}{D'_0} \tag{12.5.8}$$

where r is some constant, analogous to that considered in Bažant (1985b) and Bažant and Pfeiffer (1987) for the size effect —see Section 9.1, Eq. (9.1.34). There is, however, an important difference: in the Bažant and Pfeiffer's approach, r is an empirical constant, to be determined from experiment, while in Planas' approach, r is inferred from theoretical considerations.

These considerations are as follows: in Chapter 5 we saw that, when the effective crack model was applicable (which was the case for relatively large specimens), the failure condition was given by (5.3.6).

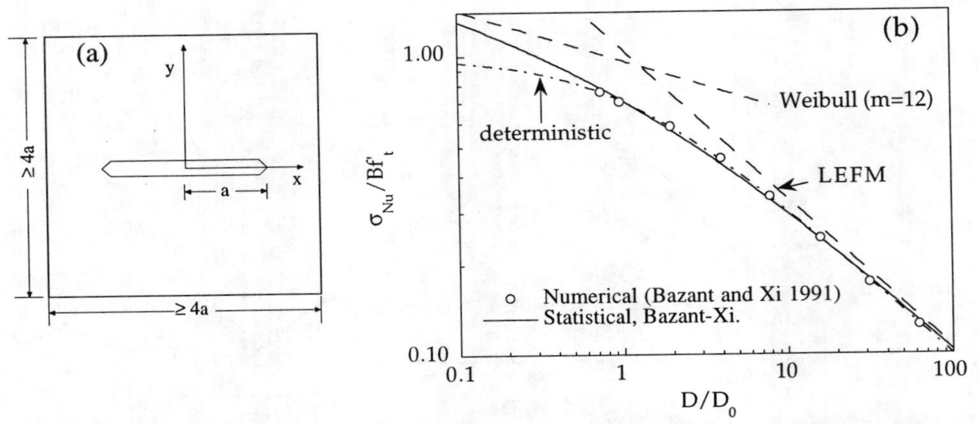

Figure 12.5.2 Numerical solution of Bažant-Xi nonlocal statistical model: (a) definition of geometry, and (b) size effect results (after Bažant and Xi 1991).

This equation can be solved for σ_{Nu} to give

$$\sigma_{Nu} = \frac{K_{Ic}}{\sqrt{D}\, k(\alpha_{ec})} \tag{12.5.9}$$

where $\alpha_{ec} = \alpha_0 + c_f/D$. Now, for $D \gg c_f$, we can use a linear approximation for k as a function of c_f/D, and write

$$\sigma_{Nu} = \frac{K_{Ic}}{\sqrt{D}\,(k_0 + k_0' c_f/D)} \quad \text{for} \quad D \gg c_f \tag{12.5.10}$$

where k_0 and k_0' stand, as always, for the value of $k(\alpha)$ and its first derivative at the initial notch depth α_0.

Now, we look for the values of $B'f_t$, D_0', and r that make (12.5.8) fit the form (12.5.10). To this end, we first rewrite (12.5.8) as

$$\sigma_{Nu} = \frac{B' f_t' \sqrt{D_0'}}{\sqrt{D}\left[1 + (D_0'/D)^{r - \frac{2n_d r}{m}}\right]^{\frac{1}{2r}}} \tag{12.5.11}$$

and take the first two terms of the binomial expansion of the denominator:

$$\sigma_{Nu} = \frac{B' f_t' \sqrt{D_0'}}{\sqrt{D}\left[1 + \frac{1}{2r}(D_0'/D)^{r - \frac{2n_d r}{m}}\right]} \tag{12.5.12}$$

Identification with (12.5.10) requires the expression in the square brackets to be linear in $1/D$, i.e., $r - 2n_d r/m = 1$; this condition furnishes the value

$$r = \frac{m}{m - 2n_d} \tag{12.5.13}$$

Moreover, completing the identification gives the relationship of the fracture parameters K_{Ic} and c_f to the size effect parameters $B'f_t'$, D_0', and m:

$$K_{Ic} = k_0 B' f_t' \sqrt{D_0'}\,, \qquad c_f = \frac{k_0}{2r k_0'} D_0' \tag{12.5.14}$$

Note that only the expression for c_f is affected by the value of r (and m). Furthermore, for $m \to \infty$, we get $r \to 1$ and we recover the deterministic case, as expected.

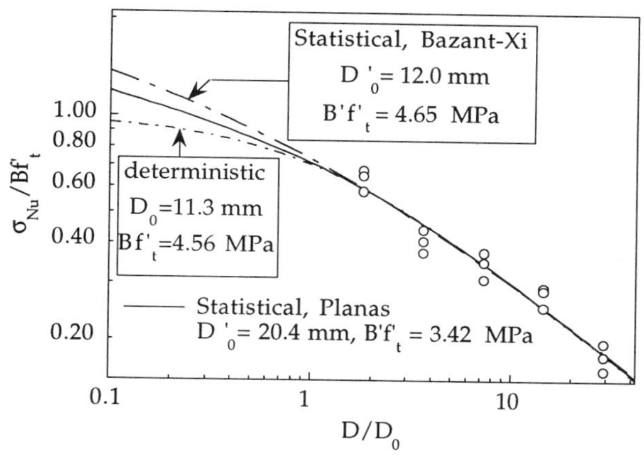

Figure 12.5.3 Planas' optimum fit of the experimental results of diagonal shear of Bažant and Kazemi (1991). See also Fig. 12.5.1.

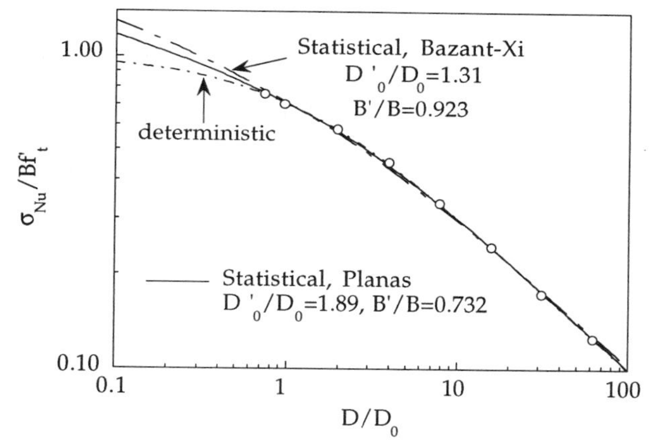

Figure 12.5.4 Planas' optimum fit of the numerical nonlocal results of Bažant and Xi (1991). See also Fig. 12.5.2.

Example 12.5.1 Consider the case of two-dimensional similarity ($n_d = 2$) and $m = 12$ (typical for concrete). Then, $r = 1.5$ and the size effect law reads

$$\sigma_{Nu} = \frac{B' f'_t}{\left(\beta^{1/2} + \beta^{3/2}\right)^{\frac{1}{3}}} \tag{12.5.15}$$

where $\beta = D/D'_0$. ▯

Fig. 12.5.3 shows the optimum fit of the foregoing equation to the results of Bažant and Kazemi previously shown in Fig. 12.5.1. The deterministic and Bažant-Xi fits are also shown for comparison. A similar comparison is shown in Fig. 12.5.4 for the nonlocal statistical case of Fig. 12.5.2.

Both figures show that Planas' interpolating formula lies between the other two, and that for D larger than about $0.6D_0$ (the deterministic value), the three equations differ less than 3% from the mean. Moreover, the fracture parameters K_{Ic} and c_f obtained from Planas' fitting are very close to those obtained from the deterministic formula: within 1% for K_{Ic} and larger by 20–26% for c_f.

12.5.6 Limitations of Generalized Weibull Theory

The survival probability of the structure is the joint probability of survival of all its elementary parts. The basic hypothesis of the classical Weibull-type theory is that the probability of survival of each part depends only on the stress in that part and is independent of the stresses as well as of the loading history of all structure parts. None of this, of course, can be expected to be exactly true.

The survival probabilities of various elementary parts are not, in reality, independent. For example, the survival probability of one elementary part may depend also on the stress in the adjacent parts. This dependence is approximately described, in the foregoing theory, by the failure probability of the core, which considers the zone surrounding the crack tip as a unit, thus enforcing a kind of nonlocal behavior. In Bažant and Xi (1991), this dependence was somewhat better reflected through nonlocal averaging.

Moreover, the failure is not a sudden event. Rather, it is a process, and a fully realistic theory would have to consider the probabilistic nature of the steps in this process, for example, the question how the survival probability of one elementary part is influenced by the previous failure of another elementary part. In particular, this would mean following the incremental jumps of the fracture process in a probabilistic manner, and formulating a stochastic process model for the progression of failure.

In the present theory, the failure probability is calculated from an instantaneous picture of the strain field at one critical moment of the loading process. Probability, however, does not enter the calculation of the process that leads to this critical moment. This is certainly a simplification.

For a fully realistic method of probabilistic analysis, one would have to consider probabilities in each loading step. When the major crack extends to a certain point, one needs to consider the probabilities of various propagation directions, and the probabilities of the lengths of the jump of the crack tip for a given load increment. Weibull-type probabilistic considerations have been made for the crack jump process, e.g., by Bruckner and Munz (1984) and Chudnovsky and Kunin (1987). Bogdanoff and Kozin (1985) —for metals— and Xi and Bažant (1992) and Bažant and Xi (1994)) —for concrete— described the probabilistic nature of the directions and lengths of the jumps of the fracture process more realistically by a Markov process model. Numerical simulations have been made as well. But it seems hardly possible to obtain in this manner general trends such as Eq. (12.4.11).

Exercises

12.19 Follow a procedure parallel to that in Section 12.5.5 to show that the value of c_f for the Bažant-Xi statistical size effect formula is infinite. Hint: show first from (12.5.8) that

$$c_f = \frac{K_{Ic}}{k_0'} \lim_{D\to\infty} \left[\frac{\partial}{\partial(1/D)} \left(\sigma_{Nu}\sqrt{D}\right)^{-1} \right]$$

where the dependence of σ_{Nu} needs to be made explicit before differentiation.

12.20 Use the values of the size effect parameters included in Fig. 12.5.3 to determine the values of the fracture parameters K_{Ic} and c_f obtained for the statistical fits relative to those for the deterministic fit.

12.21 Use the relative values of the size effect parameters included in Fig. 12.5.4 to determine the values of the fracture parameters K_{Ic} and c_f obtained for the statistical fits relative to those for the deterministic fit.

12.6 Another View: Crack Growth in an Elastic Random Medium

In this section we address the problem of a crack growing in a random medium as recently proposed by Planas (1995) and by Planas and Bažant (1997). First, the simplest problem of a crack growing in a medium with random crack growth resistance is introduced, and it is shown that the random nature of the material generates R-curve behavior.

Next, the strength distribution and the size effect is analyzed for a particularly simple distribution of crack growth resistances, and a deterministic R-curve is coupled to the statistical description. The resulting size effect is analyzed for a simple case to show that this kind of random properties do not modify appreciably the deterministic size effect, except for very large sizes.

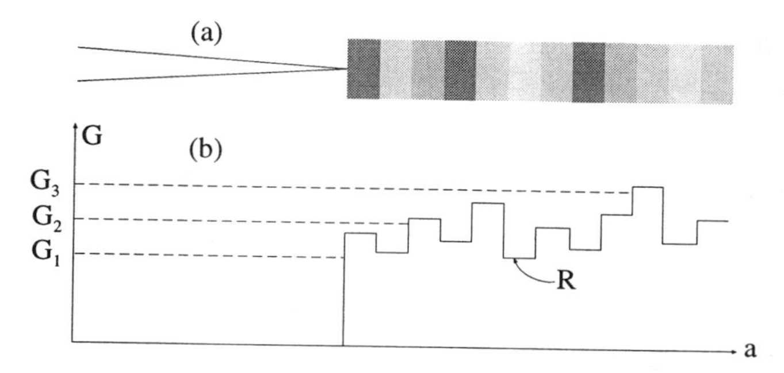

Figure 12.6.1 Crack growth in an idealized discrete random medium: (a) initial crack must grow across the discrete cells represented by shaded rectangles; (b) the cell's crack growth resistance $\mathcal{R}$ is random, which implies a randomly stepped R-curve.

12.6.1 The Strongest Random Barrier Model

Consider first a very long crack in a very large body, as depicted in Fig. 12.6.1a, and assume, for simplicity, that the elastic modulus is homogeneous and that the crack grows in pure mode I under a fixed crack driving force $\mathcal{G}$. Assume further that the material ahead of the crack tip is divided into small cells with crack growth resistances $\mathcal{R}$ uniform within each cell, but varying randomly from cell to cell. The crack growth problem can be illustrated in a $\mathcal{G}$–a plot as shown in Fig. 12.6.1b, in which the crack growth resistance of the various cells is drawn as a full line, and the applied $\mathcal{G}$ level as a dashed line.

The figure shows that for $\mathcal{G} = \mathcal{G}_1$, the crack cannot grow. For a larger value $\mathcal{G} = \mathcal{G}_2$, the crack breaks 2 cells and stops. For a still larger value, the crack runs through 9 elements before being stopped by a stronger cell. This behavior is logical, because the larger the crack jump, the larger is the probability of finding a stronger cell.

It appears, then, that this problem is the converse (or dual) of the Weibull weakest link problem. Instead of studying the distribution of defects (weak links) and the probability of failure, we are bound to analyze the distribution of strong cells (that we call *barriers*, for short) and the probability of the crack being stopped.

The foregoing analysis, which assumes discrete, equally sized cells, is unnecessarily schematic. The general problem can be addressed similarly to the weakest link concept, just by assuming that the resistance $\mathcal{R}$ along the crack path varies randomly and independently (i.e., no correlation exists between the resistances of neighboring cells).

Let us call $P_s(L, \mathcal{G})$ the probability of the crack being stopped within distance L from the initial tip location when a constant crack driving force $\mathcal{G}$ is applied. To find the dependence on L, we use the same method as in Section 12.1.2. We imagine a segment of length L_T divided into two segments of lengths $L < L_T$ and $L_T - L$. Since we assumed that there is no spatial correlation, we just state that the probability of the crack *not* being stopped on L_T is the product of the probabilities of it not being stopped on L, nor on $L_T - L$. We thus have

$$1 - P_s(L_T) = [1 - P_s(L)][1 - P_s(L_T - L)] \tag{12.6.1}$$

which turns out to be formally identical to (12.1.6). Following an analysis identical to that in equations (12.1.7)–(12.1.11), we arrive at the following structure for $P_s(L, \mathcal{G})$:

$$P_s(L, \mathcal{G}) = 1 - e^{-b(\mathcal{G})\, L} \tag{12.6.2}$$

where $b(\mathcal{G})$ has the meaning of the concentration of barriers (elements) with a crack growth resistance *larger* than $\mathcal{G}$.

Despite the resemblance of the probability functions for the weakest link theory and for the strongest random barrier model, their use is completely different, as shown in the following paragraphs. We first show that the model embodies the concept of a statistical R-curve.

12.6.2 The Statistical R-Curve

The experimental interpretation of the foregoing equation is as follows. Suppose we test many specimens by applying a constant crack driving force $\mathcal{G}$, and measure the crack jump Δa experienced by each specimen. Then $P_s(L, \mathcal{G})$ gives the probability that $\Delta a \leq L$. Therefore, (12.6.2) is the (cumulative) probability distribution of crack growth for a given $\mathcal{G}$. The corresponding probability density function $\phi_{\Delta a}(L, \mathcal{G})$ is given by

$$\phi_{\Delta a}(L, \mathcal{G}) = \frac{dP_s}{dL} = e^{-b(\mathcal{G})\,L}\, b(\mathcal{G}) \tag{12.6.3}$$

from which we can compute the mean crack jump $\overline{\Delta a}$:

$$\overline{\Delta a} = \int_0^\infty L\phi_{\Delta a}(L, \mathcal{G})dL = \int_0^\infty Le^{-b(\mathcal{G})\,L}\, b(\mathcal{G})dL \tag{12.6.4}$$

This is easily integrated by substituting $u = b(\mathcal{G})L$. The result is

$$\overline{\Delta a}(\mathcal{G}) = \frac{1}{b(\mathcal{G})} \tag{12.6.5}$$

A simple calculation shows that, for this model, the coefficient of variation of the crack extension is 1:

$$\omega_{\Delta a} = \frac{s_{\Delta a}}{\overline{\Delta a}} = 1 \tag{12.6.6}$$

In principle, there is no restriction on the barrier concentration function, except that, for a given L, P_s must decrease with $\mathcal{G}$ from 1 for $\mathcal{G} = 0$ to 0 for $\mathcal{G} \to \infty$. This requires

$$b(0) = \infty\,, \qquad b(\infty) = 0 \tag{12.6.7}$$

This is satisfied by an inverse power law of the type

$$b(\mathcal{G}) = \frac{1}{L_0}\left\langle \frac{\mathcal{G} - R_1}{R_0} \right\rangle^{-q} \tag{12.6.8}$$

where R_1 and R_0 are constants with dimensions of fracture energy, and L_0 is an arbitrary length introduced for dimensional consistency; $q > 0$ will be called the Fréchet exponent, because of the similarity of the resulting failure probability function $P_f = 1 - P_s$ to the Fréchet distribution (see, e.g., Castillo 1988)

$$F(x) = e^{-\langle x \rangle^{-q}} \tag{12.6.9}$$

where x is a random variable.

The similitude between $b(\mathcal{G})$ and the Weibull expression (12.1.16) for $c(\sigma)$ is apparent, but the change in the sign of the exponent completely changes the trends of the two functions. Here, as in the Weibull case, we will develop all the solutions for the simplest case, namely, $R_1 = 0$. This leads to an R-curve with a zero threshold, as shown in Fig. 12.6.2 for various values of the Fréchet exponent. Note that, for $q \to \infty$, we obtain the deterministic case with $\mathcal{R} = R_0 =$ constant.

12.6.3 Finite Bodies

Consider now a crack in a finite body, such as that depicted in Fig. 12.6.3, with an initial crack of length a_0. Let us find the probability $P_s(L)$ of the crack being stopped within a distance L of the initial crack tip when subjected to a given loading, and assume that the body is subjected to constant load, although this assumption is not essential for the derivation that follows.

Because $\mathcal{G}$ is not constant as the crack grows, we must extend the previous formulation to varying $\mathcal{G}$. To do so, we divide length L in N short segments of length $\Delta L = L/N$, short enough that $\mathcal{G}$ could be taken as constant in each of the segments. Let $\mathcal{G}_i$ be the value of $\mathcal{G}$ in segment i $(i = 1, \cdots, N)$. Then

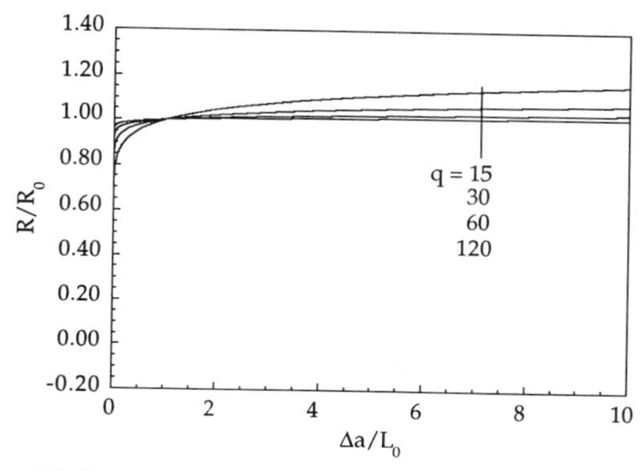

Figure 12.6.2 Statistical R-curves for various Fréchet exponents.

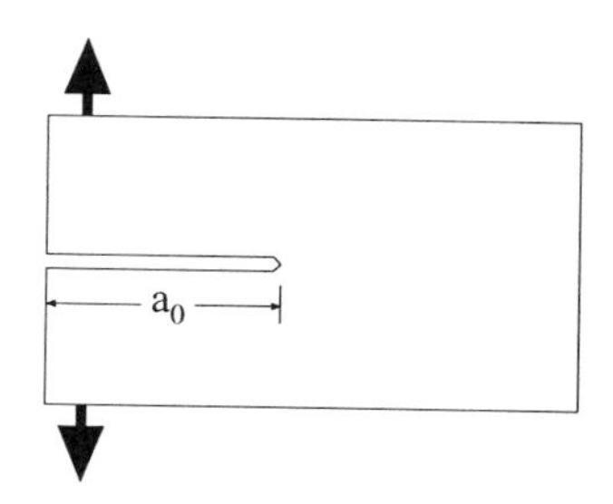

Figure 12.6.3 Generic finite size geometry.

the probability of the crack not being stopped within distance L is the probability of it not being stopped on any of the segments, i.e.,

$$1 - P_s(L) = \prod_{i=1}^{N} [1 - P_s(\Delta L, \mathcal{G}_i)] \tag{12.6.10}$$

Substituting now expressions (12.6.2) for the probabilities on the right-hand side, we get

$$P_s(L) = 1 - \exp\left[-\sum_{i=1}^{N} b(\mathcal{G}_i)\Delta L \right] \tag{12.6.11}$$

If we now take the limit $N \to \infty$, the foregoing sum becomes an integral, and stating explicitly that $\mathcal{G}$ depends on a, we have

$$P_s(L) = 1 - \exp\left[-\int_{a_0}^{a_0+L} b[\mathcal{G}(a)]da \right] \tag{12.6.12}$$

Example 12.6.1 Consider a double-cantilever beam (see Fig 2.1.3b), with an initial crack of length a_0. Assume that $b(\mathcal{G})$ is given by the inverse power law (12.6.8) with $R_1 = 0$. For constant load P, the expression for $\mathcal{G}(a)$ is given by (2.1.29), and so the probability of the crack being stopped within distance L is

$$P_s(L) = 1 - \exp\left[-\left(\frac{12P^2}{R_0 E b^2 h^3} \right)^{-q} \int_{a_0}^{a_0+L} a^{-2q} \frac{da}{L_0} \right] \tag{12.6.13}$$

Integrating, we obtain, after some algebraic rearrangement,

$$P_s(L) = 1 - \exp\left[-\left(\frac{\mathcal{G}_N}{R_0}\right)^{-q} \frac{a_0}{(2q-1)L_0}\left(1 - \frac{1}{(1+L/a_0)^{2q-1}}\right)\right] \tag{12.6.14}$$

where $\mathcal{G}_N$ is the nominal energy release (computed for the initial crack length) and it is assumed that $q > 1$ (in fact, we will show that for practical cases $q \gg 1$). The failure probability is the probability of the crack never being stopped, i.e.,

$$P_f = 1 - P_s(\infty) = \exp\left[-\left(\frac{\mathcal{G}_N}{R_0}\right)^{-q} \frac{a_0}{(2q-1)L_0}\right] \tag{12.6.15}$$

which (except for a change of scale) is a Fréchet distribution for $\mathcal{G}_N$. □

In general, the failure probability, according to (12.6.12), is

$$P_f = 1 - P_s(\infty) = \exp\left[-\int_{a_0}^{\infty} b[\mathcal{G}(a)]\, da\right] \tag{12.6.16}$$

where it is understood that $b(\mathcal{G}) = 0$ becomes zero in the event that the crack grows out of the body.

A special expression can be obtained if a particular form is assumed for $b(\mathcal{G})$. In the following sections we analyze the case of the inverse power law.

12.6.4 Fréchet's Failure Probability Distribution

For an arbitrary geometry we can write the energy release ratio in the forms defined in Eq. (2.3.12). They can be rearranged to read

$$\mathcal{G} = \mathcal{G}_N \frac{k^2(\alpha)}{k_0^2} \tag{12.6.17}$$

where, as already indicated, $\mathcal{G}_N$ is the nominal energy release ratio that can be written as

$$\mathcal{G}_N = \frac{K_{IN}^2}{E'} = \frac{\sigma_{IN}^2 D}{E'} k_0^2 \tag{12.6.18}$$

Then, if one assumes that the inverse power law (12.6.8) holds for $b(\mathcal{G})$, the failure probability distribution (12.6.16) can be easily written as

$$P_f = \exp\left[-\frac{D}{L_0}\left(\frac{\mathcal{G}_N}{\kappa^2 R_0}\right)^{-q}\right] \tag{12.6.19}$$

where

$$\kappa = \kappa(\alpha_0, q) = k_0 \left[\int_{\alpha_0}^{\alpha_l} k^{-2q}(\alpha)\, d\alpha\right]^{\frac{1}{2q}} \tag{12.6.20}$$

is a dimensionless and size-independent expression depending on α_0 and q (and, implicitly, on geometry); and α_l is the value of $\alpha = a/D$ for which the crack reaches the boundary of the body (complete failure). α_l is constant for geometrically similar bodies. For some structures, such as the center-cracked infinite panel, α_l can be infinite.

Eq. (12.6.19) gives the failure probability in terms of $\mathcal{G}_N$. Similar expressions can be written for K_{IN} and for σ_N. Setting $K_0 = \sqrt{E' R_0}$ we get

$$P_f(K_{IN}) = \exp\left[-\frac{D}{L_0}\left(\frac{K_{IN}}{\kappa K_0}\right)^{-2q}\right] \tag{12.6.21}$$

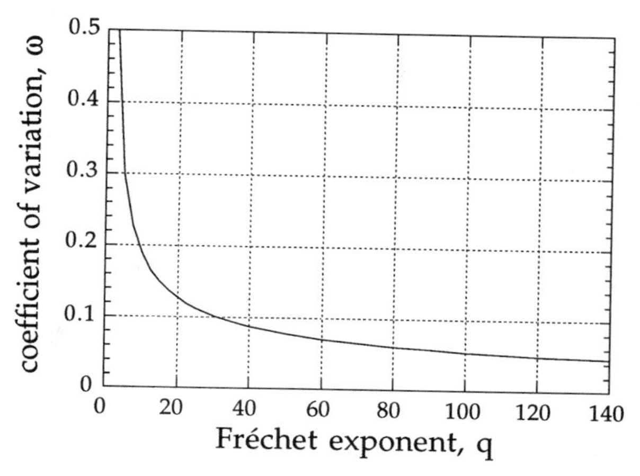

Figure 12.6.4 Coefficient of variation vs. Fréchet exponent.

$$P_f(\sigma_N) = \exp\left[-\frac{D}{L_0}\left(\frac{\sigma_N\sqrt{D}k_0}{\kappa K_0}\right)^{-2q}\right] \tag{12.6.22}$$

From the last equation we see that the strength distribution is again a Fréchet distribution, but its exponent is $2q$ rather than q. An immediate consequence of this equation is that the size effect predicted by this model is proportional to $D^{-\frac{1}{2}+\frac{1}{2q}}$. Indeed, the median of the ultimate stresses σ_{Nu}^{50} is easily found by setting $P_f = 0.5$:

$$\sigma_{Nu}^{50} = \left(\frac{D}{L_0 \ln 2}\right)^{1/2q} \frac{\kappa K_0}{k_0\sqrt{D}} \qquad \Rightarrow \qquad \sigma_{Nu}^{50} \propto D^{-\frac{1}{2}+\frac{1}{2q}} \tag{12.6.23}$$

A similar relationship is found for the mean strength; indeed, following a procedure identical to that in Section 12.1.3 for Weibull's distribution, we get the following results for the mean strength:

$$\overline{\sigma_{Nu}} = \left(\frac{D}{L_0}\right)^{\frac{1}{2q}} \Gamma\left(1 - \frac{1}{2q}\right) \frac{\kappa K_0}{k_0\sqrt{D}} \tag{12.6.24}$$

from which it follows that

$$\overline{\sigma_{Nu}}(D) = \overline{\sigma_{Nu}}(D_1)\left(\frac{D_1}{D}\right)^{\frac{1}{2}-\frac{1}{2q}} \tag{12.6.25}$$

Now the point is, how large can one expect q to be. This can be estimated based on the coefficient of variation ω. This is again calculated similar to Weibull's distribution with a similar result —just m is replaced by $-2q$:

$$\omega^2 = \frac{\Gamma(1 - \frac{1}{q})}{\Gamma^2(1 - \frac{1}{2q})} - 1 \tag{12.6.26}$$

Fig. 12.6.4 shows the resulting curve of ω vs. q. It follows from it, that, in order to achieve an experimental coefficient of variation below 10% (a reasonable limiting value), the value of q must be larger than about 30. This leads to a very small deviation of the size effect from the LEFM limit. Indeed, the negative slope in a log-log plot would shift from 0.5 to more than $0.5 - 1/60 \approx 0.48$.

Fig. 12.6.5 shows the comparison of both the deterministic and the random barrier statistical size effect curves (for $q = 30$) with the experimental results of Bažant and Pfeiffer (1987) for notched specimens of concrete and mortar. The random barrier model can fit very well the large size limit, approached closely only for mortar. For small sizes, however, this statistical model cannot fit the experimental results for realistic values of q. For series B2, for instance, the best fit of a power law requires $q = 2.7$, which,

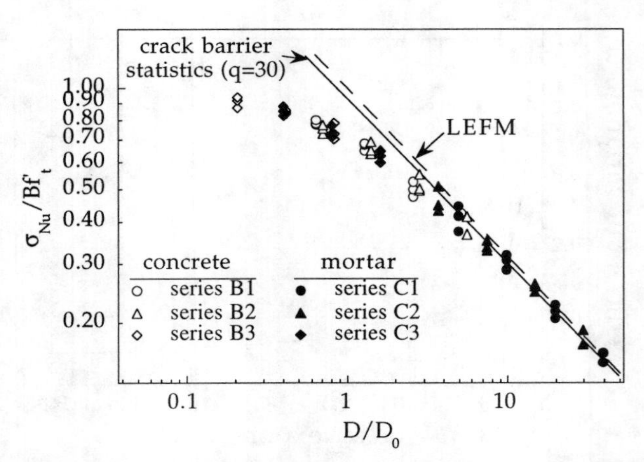

Figure 12.6.5 Comparison of experimental results with the asymptotic prediction of the random barrier model.

according to Fig. 12.6.4, would lead to a coefficient of variation close to 50%, which is far larger than what is found in practice.

Therefore, this statistical model must be modified to account for the observed size effect. The simplest and obvious modification is to introduce a deterministic R-curve behavior coupled to the barrier statistics. This leads to a random R-curve. We next analyze a simple way to introduce this modification into the Fréchet distribution.

12.6.5 Random R-curve

To introduce an intrinsic R-curve effect (i.e., an effect present even for negligible randomness), we just let the barrier concentration function $b(\mathcal{G})$ depend explicitly on the crack extension $\Delta a = a - a_0$, that is,

$$b(\mathcal{G}) \equiv b(\mathcal{G}, \Delta a) \tag{12.6.27}$$

This simple change does not modify the basic analysis leading from (12.6.10) to (12.6.16), and so the failure probability is given by the obvious extension of (12.6.16):

$$P_f = \exp\left\{-\int_{a_0}^{\infty} b[\mathcal{G}(a), a - a_0]\, da\right\} \tag{12.6.28}$$

A simple way to generate a barrier concentration function leading to a given R-curve for the deterministic case is to use the inverse power law function (12.6.8) with the constants R_0 and R_1 replaced by functions of the crack extension; the simplest is the function

$$b(\mathcal{G}, \Delta a) = \frac{1}{L_0}\left[\frac{\mathcal{G}}{R(\Delta a)}\right]^{-q} \tag{12.6.29}$$

To compactly write the probability distribution for strength and the statistical parameters, it is convenient to refer the applied stress σ_N to the *deterministic* failure load, σ_{N0}. We recall from Section 5.6.3 that, in the deterministic case, the peak load occurs when the $\mathcal{G}$–curve becomes tangent to the R-curve (Fig 12.6.6), and that, furthermore, $\mathcal{G} \propto \sigma_N^2$. Thus, we can rewrite the barrier density in the form

$$b(\mathcal{G}, \Delta a) = \frac{1}{L_0}\left[\frac{\sigma_N^2\, \mathcal{G}_0(a)}{\sigma_{N0}^2\, R(\Delta a)}\right]^{-q} \tag{12.6.30}$$

Note that σ_{N0} is size-dependent. Substitution of this expression into (12.6.28) leads to the following

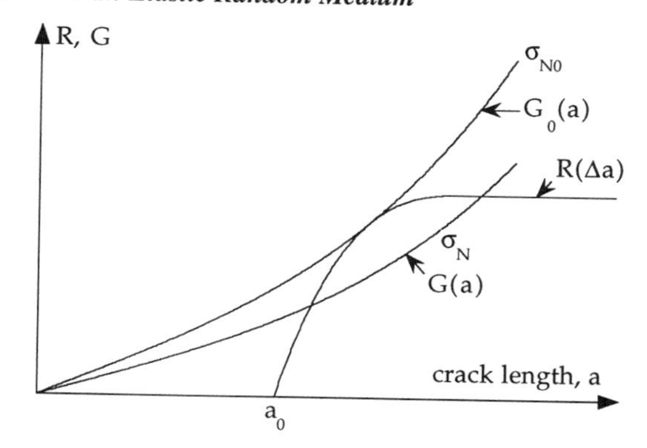

Figure 12.6.6 R-curve and $\mathcal{G}$ curves (deterministic).

equation for the failure probability:

$$P_f(\sigma_N) = \exp\left[-\left(\frac{\sigma_N}{\eta\,\sigma_{N0}}\right)^{-2q}\right] \tag{12.6.31}$$

where η is a *size-dependent* function defined as

$$\eta^{2q} = \int_{a_0}^{\infty}\left[\frac{R(a-a_0)}{\mathcal{G}_0(a)}\right]^q \frac{da}{L_0} \tag{12.6.32}$$

Thus, it turns out that, for a fixed size, the statistical distribution has the same form as that for the case of no R-curve. In particular, the coefficient of variation depends only on q as shown in Fig. 12.6.4. The mean strength is given by

$$\overline{\sigma_{Nu}} = \Gamma\left(1-\frac{1}{2q}\right)\eta\,\sigma_{N0} \tag{12.6.33}$$

No universal expression for η as a function of size seems to exist, but the following example shows that we can expect the randomness to modify the deterministic size effect only slightly, at least for ordinary sizes and realistic values of q, i.e., realistic values of the coefficient of variation.

Example 12.6.2 Consider an infinite panel with a center crack of length $2a_0$ subjected to a remote stress σ normal to the crack faces. Assume that the R-curve is bilinear, described by the equations

$$\mathcal{R} = R_0\,\frac{\Delta a}{c_f} \qquad \text{for} \quad \Delta a \le c_f \tag{12.6.34}$$

$$= R_0 \qquad \text{for} \quad \Delta a \ge c_f \tag{12.6.35}$$

where R_0 and c_f are constants (c_f is the critical effective crack extension).

For this geometry we have $\mathcal{G} = \sigma^2\pi a/E'$. The peak load condition is satisfied when $\mathcal{G} = R_0$ and $a = a_0 + c_f$. So for the deterministic strength σ_0 we have:

$$\sigma_0 = \sqrt{\frac{E'R_0}{\pi(a_0+c_f)}} \tag{12.6.36}$$

which has the classical Bažant's form. The equation for $\mathcal{G}_0$ thus is

$$\mathcal{G}_0 = \frac{\sigma_0^2\pi a}{E'} = R_0\frac{a}{a_0+c_f} \tag{12.6.37}$$

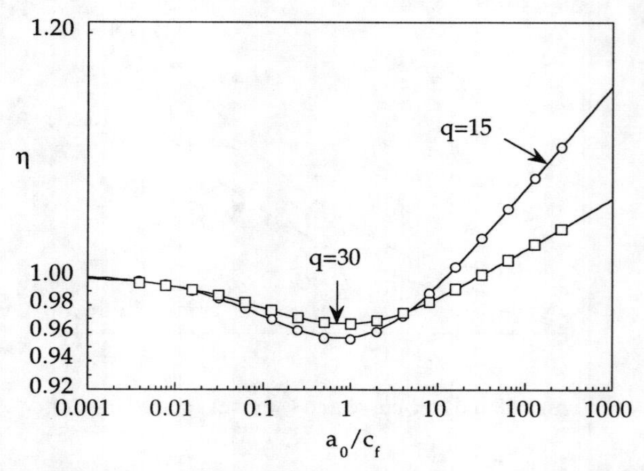

Figure 12.6.7 Variation of factor η with size and Fréchet exponent.

To compute η, we first make the length L_0, so far arbitrary, to coincide with c_f, which is always possible by an appropriate rescaling of R_0. Then we must split the integral in (12.6.32) into two, one extending from a_0 to $a_0 + c_f$ and the second from $a_0 + c_f$ to ∞:

$$\eta^{2q} = \int_{a_0}^{a_0+c_f} \left[\frac{(a-a_0)(a_0+c_f)}{c_f a} \right]^q \frac{da}{c_f} + \int_{a_0+c_f}^{\infty} \left[\frac{(a_0+c_f)}{a} \right]^q \frac{da}{c_f} \tag{12.6.38}$$

The second integral is straightforward, and the first can be transformed by setting $u = (a - a_0)/c_f$. Introducing the notation $\beta = a_0/c_f$ (Bažant's brittleness number), we get

$$\eta^{2q} = (1+\beta)^q \int_0^1 \left[\frac{u}{u+\beta} \right]^q du + \frac{1+\beta}{q-1} \quad , \qquad \beta = \frac{a_0}{c_f} \tag{12.6.39}$$

The above integral cannot be expressed in a closed form. It was performed numerically using a mathematical software package. The results for η are shown in Fig. 12.6.7. Note that, except for very large values of β, the result is very close to 1. From (12.6.39) the asymptotic behaviors for large and small β are easily determined:

$$\eta^{2q} \to \frac{2q}{q^2-1} + \frac{\beta}{q-1} \qquad \text{for} \qquad \beta \to \infty \tag{12.6.40}$$

$$\eta^{2q} \to \frac{q}{q-1} \qquad \text{for} \qquad \beta \to 0 \tag{12.6.41}$$

A curve satisfying these asymptotic limits is

$$\eta^{2q} = \frac{q}{(q+1)\left[1+(q-1)\beta^r\right]} + \frac{2q}{q^2-1} + \frac{\beta}{q-1} \tag{12.6.42}$$

where r is the only adjustable parameter, which depends slightly on q: $r = 0.793$ for $q = 15$ and $r = 0.856$ for $q = 30$.

The full lines in Fig. 12.6.7 show the fits of the numerical results by the foregoing equation. They look excellent. Using the equation for η, and the expression for the deterministic size effect, the size effect curves in Fig. 12.6.8 are obtained. The main result is that the size effect on the *mean* strength is not substantially modified by the barrier statistics, except, perhaps, for very large sizes, for which the asymptotic size effect is proportional to $D^{-(1/2)+(1/2q)}$. What is very much affected by the Fréchet exponent is the coefficient of variation, which is about 10% for $q = 30$ and 15% for $q = 15$. □

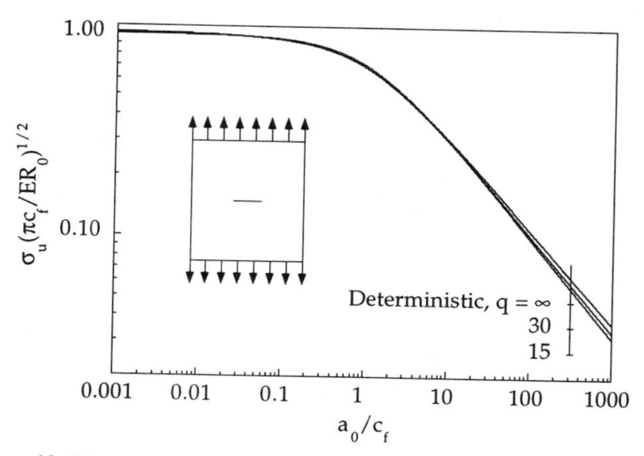

Figure 12.6.8 Size effect curves for various values of the Fréchet exponent.

12.6.6 Limitations of the Random Barrier Model

By its very conception, the random barrier model as formulated in the foregoing can be valid only for notched or precracked specimens. It also requires this failure model to be combined with an R-curve. These limitations do not seem to be too stringent for quasibrittle materials.

A further limitation in the development of the model is the assumption of a barrier concentration that is an inverse power law. A different law might work better, but the results cannot be much different when one requires the values of the coefficient of variation to be realistic.

The strongest limitation of the model is that there is no spatial correlation. This, of course, may be a crude approximation. To avoid it, the model could be extended to include a stochastic R-curve based on the random field theory. Such an approach has apparently not been undertaken but it is doubtful that closed form solutions could be found, except by fitting the results of Monte Carlo simulations.

The model nevertheless corroborates the conclusion that, for propagation of a single crack in a quasibrittle material, the modification of size effect (on the mean strength) induced by the randomness of the crack growth resistance is very small.

Exercises

12.22 Show that, for positive geometries, the function $\kappa(\alpha_0, q)$ in (12.6.20) satisfies the two conditions: (a) $\kappa \leq 1$, and (b) $\lim_{q\to\infty} \kappa = 1$.

12.23 Show that, for an infinite panel with a center crack of length $2a_0$ subjected to remote stress σ, $\kappa = 1/(2q-1)^{1/2q}$ if one takes $D \equiv a_0$. Write the failure probability function for $\mathcal{G}_N$, K_{IN}, and σ.

12.24 Give a detailed proof of Eqs. (12.6.40) and (12.6.41).

12.7 Fractal Approach to Fracture and Size Effect

An intriguing idea was injected into the study of size effect by Carpinteri (1994a,b, 1996), Carpinteri and Chiaia (1995, 1996), Carpinteri and Ferro (1993, 1994), and Carpinteri, Chiaia and Ferro (1994, 1995a,b). It was motivated by numerous recent studies of the fractal characteristics of crack surface in various materials (e.g., Mandelbrot, Passoja and Paullay 1984 for metals; Saouma, Barton and Gamaleldin 1990 for concrete; see Bažant 1997d for further references). Carpinteri et al. proposed that the difference in fractal characteristics of the fracture surfaces or microcrack distributions at different scales of observation is the principal source of size effect in concrete.

The arguments they offered, however, were not based on mechanical analysis and energy considerations.

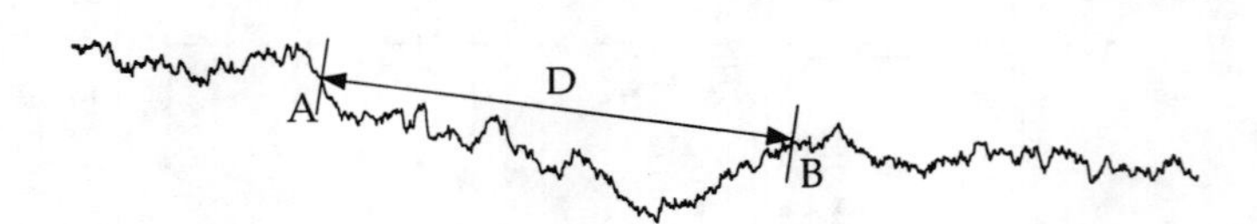

Figure 12.7.1 Self-affine fractal line and secant (apparent) size of its portion of it (the fractal line was generated by the midpoint displacement method with $\phi = 0.5$ using the program *Fractals* included in the book by Russ 1994).

Rather they were strictly geometrical and partly intuitive. Recently, Bažant (1994c, 1995e,f, 1996e, 1997d) presented a mechanical analysis of the problem which casts doubt on Carpinteri's conclusions. However, the fact that the surface roughness of cracks in many materials can be described by fractal concepts, at least over a certain limited range, is not in doubt. In this section we briefly introduce the fractal approach and discuss the main implications and limitations.

12.7.1 Basic Concepts on Fractals

Many geometrical features in nature are endowed with the property of self-similarity. This property implies that the geometrical features are similar whatever the scale of observation. This is a well-known property of mountain outlines, cloud contours, coastal profiles, river patterns, streaks of lightning, and many other natural lines. Fracture surfaces also display the property of self-similarity over various orders of magnitude, and so do the linear profiles obtained from fracture surfaces. In all these examples, the measured length of the line depends on the resolution of the measure, i.e., of the length of the 'measuring stick' (ruler). For the case of surfaces in which the area is measured by triangularization, the measured area depends on the length of the side of the triangles used in the triangularization. If self-similarity is complete, the dependence of the measured length or area is proportional to the length of the measuring stick u raised to a *noninteger* power ϕ. We say then that the geometry is fractal and write

$$Y_u = Y_0 \left(\frac{u_0}{u}\right)^{\phi} \tag{12.7.1}$$

where ϕ is a dimensionless constant, u_0 is a constant with the dimension of length, and Y_0 is a constant with the same dimensions as Y, either length, surface, or volume. For classical smooth sets (euclidean sets) $\phi = 0$. If $\phi > 0$, the fractal is called *invasive* (or densifying). If $\phi < 0$, the fractal is called *lacunar* (or rarefying). Note that Y_0 and u_0 are not independent: the equation is characterized by the product $Y_0 u_0^{\phi}$, and so u_0 can be understood as an arbitrary reference length. However, in the following we assume that u_0 has been set once and for all as a *conventional* measuring length, ideally of the order of the interatomic distances. In this way, Y_0 is a conventional euclidean measure of the fractal set.

The foregoing equation can be written in a number of equivalent ways. Beginning with a fractal line, such as that sketched in Fig. 12.7.1, the number of times N_u that the unit of measure u can be fitted in a given portion of the fractal line is obtained by dividing the length L_u (here $Y \equiv L$) by u, thus obtaining

$$N_u = \frac{L_u}{u} = \frac{L_0}{u_0}\left(\frac{u_0}{u}\right)^{1+\phi} = N_0 \left(\frac{u_0}{u}\right)^{d_f} \tag{12.7.2}$$

where $d_f = 1 + \phi$ is the *fractal dimension*. Consider now a portion of the fractal line, say AB in Fig. 12.7.1, of arbitrary apparent (secant) length D, and call $L_u(D)$ its measured length for resolution u. Then, by definition, we have $L_u = D$ for $u = D$. Inserting this condition in the foregoing equation, we can solve for $L_0(D)$ as

$$L_0(D) = u_0 \left(\frac{D}{u_0}\right)^{d_f} \quad \text{or} \quad \frac{L_0(D)}{D} = \left(\frac{D}{u_0}\right)^{\phi} \tag{12.7.3}$$

and we find that the measure of length in this kind of curves is size dependent unless $\phi = 0$, i.e., unless the line is an ordinary, euclidean curve.

Similar considerations apply to any geometrical dimension (line, surface, or volume). The general result is that if, within a fractal structure of dimension d_Y ($d_Y = 1, 2, 3$), we take similar regions defined

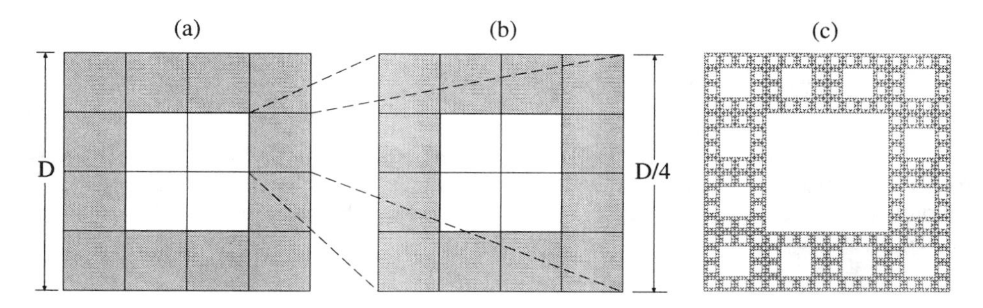

Figure 12.7.2 Fractal cross section: (a) basic pattern, (b) similar pattern, but 4 times smaller, (c) appearance of the fractal cross section after 4 iterations.

by a linear size D, the conventional measure of these regions as a function of D is given by

$$Y_0 = u_0^{d_Y}\left(\frac{D}{u_0}\right)^{d_Y+\phi} = u_0^{d_Y}\left(\frac{D}{u_0}\right)^{d_f} \quad \text{or} \quad \frac{Y_0}{D^{d_Y}} = \left(\frac{D}{u_0}\right)^{\phi} \tag{12.7.4}$$

where $d_f = d_Y + \phi$ is again the fractal dimension.

There are several methods to produce fractal lines and surfaces. For lines, the simplest method is the so called midpoint displacement procedure in which one starts with a segment which is bisected and the midpoint displaced vertically (up or down) randomly in proportion to the length of the segment. After this, two segments are left, which are in turn bisected and so on, down to the required resolution (or down to the computer's precision). The vertical (or parallel) displacements produce so-called self-affine fractals. If the midpoints are displaced in direction normal to the segment, one obtains the so called self-similar fractals. Although the latter have been pictured as models for fractal cracks, they do not permit kinematic separation of the surfaces because recessive and spiral segments can occur. Only self-affine fractals (Fig. 12.7.1) are admissible as cracks.

A similar method exists for surfaces. Starting from a plane triangle, the midpoints of the sides of the triangle are displaced randomly in proportion to the length of each side. This defines 4 smaller triangles for which the midpoint of each side is in turn displaced randomly, and so on. Other methods to obtain fractal lines and surfaces exist, as well as methods to determine the fractal dimension of lines or surfaces determined experimentally. A presentation of such methods is outside the scope of this book, and the interested reader is referred to the specialized literature (e.g., Russ 1994).

Rough surfaces and profiles are examples of invasive fractals in which $\phi > 0$. Lines or surfaces in which gaps or holes are punched selfsimilarly are examples of lacunar fractals. Of course, a line with gaps (exemplified by the Cantor set, see Carpinteri 1994a or Russ 1994) becomes a discontinuous set. However, a plane surface with holes can still be continuous and can be illustrated by a plane cross section of a spongy solid. In the following example we examine a very simple set which will later be viewed as the cross section of a fractal wire.

Example 12.7.1 Consider the square region of side D in Fig. 12.7.2a which is composed of 16 squares of side $D/4$. A square hole of side $D/2$ is punched in the center of the square. Each of the remaining 12 squares (gray in the figure) are similar to the first as depicted in Fig. 12.7.2b, and each contains 12 further gray squares similar to the first (but 16 times smaller), and so on as many times as allowed by the best resolution available. Fig. 12.7.2c shows the resulting cross section after 4 iterations. To find the size effect for the surface area, we can use a recurrent strategy: calling $A_0(D)$ the net (conventional) surface area of a (gray) square of side D, we write the condition that the net area for a square of size D must be equal to 12 times the net area of a square of side $D/4$, that is:

$$A_0(D) = 12A_0\left(\frac{D}{4}\right) \tag{12.7.5}$$

Guided by the form of (12.7.5), we try a solution of the form $A_0(D) = cD^{d_f}$, where c is a constant, and get:

$$cD^{d_f} = 12c\frac{D^{d_f}}{4^{d_f}} \qquad \Rightarrow \qquad d_f = \frac{\ln 12}{\ln 4} = 1.792 \tag{12.7.6}$$

from which $\phi = d_f - 2 = -0.2075$. The set so defined is lacunar, as expected. This result can be generalized to a subdivision in $N \times N$ squares from which M squares are punched off. A similar calculation leads to the result $d_f = \ln(N^2 - M)/\ln N$ (note that the exact situation of the punched squares is irrelevant). ▯

12.7.2 Invasive Fractal and Multifractal Size Effect for G_F

As a first approximation, Carpinteri and Ferro (1993, 1994) and Carpinteri (1994a,b) considered that the fracture energy was best described by a fractal-type of structure. The basic idea behind their analysis is that, if the surface of fracture has a fractal structure and the work of fracture is proportional to its conventional area $A_0(D)$, then

$$\mathcal{W}_F(D) \propto A_0(D) \propto D^{d_f} \tag{12.7.7}$$

Now, for full three-dimensional similarity, the projected cross section of the specimen is proportional to D^2, and so $\mathcal{W}_F(D) \propto G_F(D)D^2$ in which $G_F(D)$ is the apparent (mean) fracture energy. From this chain of proportionalities, we find that the apparent fracture energy can be written as

$$G_F(D) = G_{F0}\left(\frac{D}{u_0}\right)^{\phi_G} = \breve{G}_F\, D^{\phi_G}\,, \qquad \phi_G = d_f - 2 \tag{12.7.8}$$

where G_{F0} is the conventional fracture energy based on the conventional measuring length u_0, and $\breve{G}_F$ is a fractal fracture energy that has the dimensions F L$^{-(1+\phi_G)}$ instead of F L^{-1}. Obviously, $\breve{G}_F$ is size independent. Here, subscript G has been given to ϕ to distinguish it from other fractal exponents. Indeed, it must be realized that, in the foregoing approach, all the surface area must be accounted for, not only that due to roughness, but also the areas inaccessible to a profilometer, such as branching of the main crack and separated microcracks. This means that ϕ_G might be sharply larger than the value of ϕ obtained from profilometry.

This simple model was applied by Carpinteri and co-workers to several sets of data in the literature with partial success. They were able to fit the results well over a size range 1:4, but the experimental results deviated from the power law for larger ranges. This was attributed to the fact that the fractal description may break down for large sizes, where a macroscopic (euclidean) scale dominates. They thus designed a transition formula from a fractal to a euclidean regime, which they called *multifractal* (Carpinteri and Chiaia 1995, 1996; Carpinteri 1996). They proposed the monofractal regime, valid for small sizes, to be characterized by $\phi_G = 0.5$, without sound justification (some models based on uncorrelated random-walk generation give $\phi = 0.5$; however, other techniques can produce fractal lines and surfaces with any ϕ between 0 and 1). Their equation, called the multifractal scaling law for G_F, reads

$$G_F(D) = \frac{G_F^\infty}{\sqrt{1 + \ell_G/D}} \tag{12.7.9}$$

where G_F^∞ is the asymptotic fracture energy for infinite size and ℓ_G a size that characterizes the transition from the fractal to the euclidean regime. This formula fits reasonably well various experimental data sets in the literature, although most of them are limited to size ranges in the proportion 1:4 (for details, see Carpinteri and Chiaia 1995). ℓ_G and G_F^∞ have to be determined by optimum fit to experimental data and, in principle, are specimen dependent. This limits very much the applicability of the model, because it cannot be used predictively.

12.7.3 Lacunar Fractal and Multifractal Size Effect for σ_{Nu}

Simultaneous with the development of the fractal theory for G_F, Carpinteri (1994a,b) and Carpinteri and Ferro (1993, 1994) applied similar concepts to the nominal strength of structures, although in this case the fractality turned out to be of the lacunar type. The lacunar character may be explained by assuming a material of the plastic or elastic-brittle type (i.e., which fails when the net stress reaches a certain limiting value). If the net area of a cross section has fractal nature, then the apparent strength is also fractal. Consider, for instance, a fractal wire whose cross section has the fractal structure defined in example 12.7.1. If the wire, with a cross section of dimensions $D \times D$, is subjected to an axial force F, the apparent (nominal) stress is $\sigma_N = F/D^2$. However, the net (conventional) area of the cross section is given by $A_0 = cD^{2+\phi_\sigma}$, where c is a constant. Then, if the conventional strength (net stress at failure) is σ_{R0} (a material constant) the maximum force is given by $F_u = \sigma_{R0} c D^{2+\phi_\sigma}$. From this it follows that the apparent (nominal) strength is given by

$$\sigma_{Nu} = \sigma_{R0} c D^{\phi_\sigma} = \breve{\sigma}_R D^{\phi_\sigma} \tag{12.7.10}$$

where $\breve{\sigma}_R$ is a fractal strength (a material property) with dimensions of FL$^{-2+\phi_\sigma}$, and ϕ_σ is the fractal exponent for the stress. Since ϕ_σ is negative for a lacunar cross section such as that assumed, the nominal strength decreases with size following a power law formally similar to Weibull's law. According to (12.2.20), the two size effects are equivalent for three-dimensional similarity for a Weibull's modulus $m = -3/\phi_\sigma$.

The foregoing power law size effect was applied by Carpinteri and Ferro to describe the size effect on the nominal strength. Same as for G_F, they were able to fit the results well over a size range 1:4, but the experimental results deviated from the power law for larger ranges. An empirical formula called multifractal was then devised by Carpinteri and co-workers to describe the transition from a fractal regime to a euclidean regime (Carpinteri, Chiaia and Ferro 1994, 1995a,b; Carpinteri and Chiaia 1996; Carpinteri 1996). They proposed the monofractal regime for small sizes to be characterized by $\phi_\sigma = -0.5$, again without sound justification. The transition equation for σ_{Nu} then reads

$$\sigma_{Nu} = \sigma_R^\infty \sqrt{1 + \frac{\ell_\sigma}{D}} \tag{12.7.11}$$

where σ_R^∞ is the asymptotic strength for infinite size and ℓ_σ a size that characterizes the transition from the fractal to the euclidean regime. This formula fits reasonably well various experimental data sets in the literature; for example, those for the rupture modulus shown in Fig. 9.3.9. However, the constants ℓ_σ and σ_R^∞ have to be determined by optimum fit to experimental data and, in principle, are specimen dependent. Again, this limits very much the applicability of the model, because it cannot be used predictively.

12.7.4 Fracture Analysis of Fractal Crack Propagation

The approach of Carpinteri et al. does not analyze the process of crack propagation as the mechanism that ultimately leads to fracture, because no mechanical formulation is included in it. Bažant (1994c, 1995e,f, 1996e, 1997d) presented a mechanical analysis of the size effect for fractal cracks that includes the crack propagation. His starting point is a formula for the work of fracture due to Borodich (1992), which followed a physical approach to fractal fracture by Mosolov and Borodich (1992). In this formula, which is similar, but not identical, to that underlying (12.7.8), the work of fracture for a body with constant fracture energy ($\mathcal{R} = G_f$ = constant) has fractal nature and, for two-dimensional similarity, can be written as

$$\frac{\mathcal{W}_F}{b} = \breve{G}_f a^{1+\phi} \tag{12.7.12}$$

where b is the specimen thickness, a the macroscopic projected crack length (Fig. 12.7.3a), ϕ the fractal exponent, and $\breve{G}_f$ a fractal fracture energy with dimensions FL$^{1+\phi}$. The foregoing equation corresponds to the total dissipated work. The crack growth resistance is the derivative of this function with respect to a, and so:

$$\mathcal{R} = \frac{\partial \mathcal{W}_F}{b \partial a} = \breve{G}_f (1 + \phi) a^\phi \tag{12.7.13}$$

Bažant assumes that for large cracks the macroscopic (euclidean) regime is reached. He assumes that this occurs for some crack length a_f for which $\mathcal{R}$ reaches the macroscopic value G_f, i.e., $G_f = \breve{G}_f (1+\phi) a_f^\phi$,

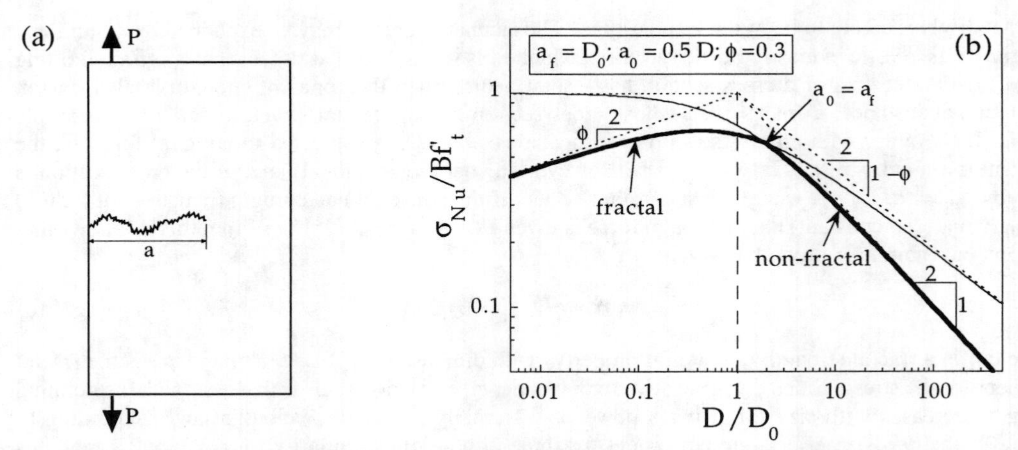

Figure 12.7.3 (a) Fractal crack and its projected length. (b) Fractal and nonfractal size effect curves according to Bažant's analysis (1997d) for notched structures or structures that fail at similar crack lengths.

and thus one can rewrite the expression for $\mathcal{R}$ as

$$\mathcal{R} = \begin{cases} G_f \left(\dfrac{a}{a_f}\right)^{\phi} & \text{for} \quad a \leq a_f \\ G_f & \text{for} \quad a > a_f \end{cases} \tag{12.7.14}$$

This formulation can further be generalized by allowing the material to exhibit a general R-curve behavior similar to that introduced in Section 9.1. Thus, the generalized crack resistance adopts the form of Eq. (9.1.5) for the macroscopic regime and the form of that equation multiplied by $(a/a_f)^{\phi}$ for the fractal regime, i.e.,

$$\tilde{\mathcal{R}} = \begin{cases} G_f\, \tilde{r}(\alpha_0, \eta, \theta) \left(\dfrac{a}{a_f}\right)^{\phi} & \text{for} \quad a \leq a_f \\ G_f\, \tilde{r}(\alpha_0, \eta, \theta) & \text{for} \quad a > a_f \end{cases} \tag{12.7.15}$$

where α_0, η, and θ are the reduced variables defined in Section 9.1.

The second fundamental hypothesis in Bažant's approach to fractal crack propagation is that the energy release rate, or its generalized version $\tilde{\mathcal{G}}$, keeps its macroscopic expression (9.1.6). Therefore, for the macroscopic regime, an analysis identical to that in Section 9.1 leads to the classical Bažant's size effect law. For the fractal regime, however, the equation is initially slightly different since the condition $\tilde{\mathcal{G}} = \tilde{\mathcal{R}}$, after substituting (9.1.6) and the first of (12.7.15), requires that

$$\frac{\sigma_N^2}{E'} D\tilde{g}(\alpha_0, \eta, \theta) = G_f\, \tilde{r}(\alpha_0, \eta, \theta) \left(\frac{a}{a_f}\right)^{\phi} \tag{12.7.16}$$

A complete asymptotic treatment together with asymptotic matching is given in Bažant (1997a). Here we make a simplified derivation based on the similitude to the analysis in Section 9.1. The foregoing equation is easily transformed to a form similar to that in the macroscopic regime by substituting into the equation the equality $a = (\alpha_0 + \eta\theta)D$ and rearranging:

$$\frac{\sigma_N^2}{E' D^{\phi}} D\tilde{g}(\alpha_0, \eta, \theta) = \frac{G_f \alpha_0^{\phi}}{a_f^{\phi}}\, \check{\tilde{r}}(\alpha_0, \eta, \theta) \tag{12.7.17}$$

where

$$\check{\tilde{r}}(\alpha_0, \eta, \theta) = \tilde{r}(\alpha_0, \eta, \theta) \left(1 + \frac{\eta\theta}{\alpha_0}\right)^{\phi} \tag{12.7.18}$$

Here we note that the second factor is very close to 1 for large sizes and notched specimens ($a_0 \neq 0$).

Now, it turns out that the mathematical form of this equation is identical to that for the macroscopic regime except that σ_N is replaced by $\sigma_N/D^{\phi/2}$, G_f is replaced by $G_f\alpha_0^\phi/a_f^\phi$, and the function $\tilde{r}$ replaced by the function $\check{\tilde{r}}$. Thus, if the asymptotic analyses for large and small sizes and its asymptotic matching is performed as in Section 9.1, the solution is found as follows:

$$\sigma_{Nu} = \left(\frac{\alpha_0 D}{a_f}\right)^{\phi/2} m_\phi \frac{Bf'_t}{\sqrt{1 + n_\phi D/D_0}} \tag{12.7.19}$$

where Bf'_t and D_0 are identical to the macroscopic values, and m_ϕ and n_ϕ are constants depending on geometrical ratios and on the fractal exponent ϕ. The dependence of these constants on ϕ is engendered by the factor $(1 + \eta\theta/\alpha_0)^\phi$ in 12.7.18. Obviously, $m_\phi = n_\phi = 1$ for $\phi = 0$ (the macroscopic case) and the deviation from 1 may be expected to be very small if the failure occurs for very small values of the effective crack extension, as assumed by Bažant (1997b). In such a case we can set $m_\phi = n_\phi = 1$. With this, the size effect in the fractal regime reads

$$\sigma_{Nu} = \left(\frac{\alpha_0 D}{a_f}\right)^{\phi/2} \frac{Bf'_t}{\sqrt{1 + D/D_0}} \tag{12.7.20}$$

In this equation, the macroscopic regime is reached when the first factor reaches unity, i.e., when $\alpha_0 D = a_0 = a_f$, which seems logical: when the initial notch or crack length at peak is larger than the transition size, the behavior is macroscopic. Fig. 12.7.3b shows the size effect curves for the fractal and nonfractal regimes, for a particular case ($\phi = 0.3$, $a_f = D_0$, and $a_0 = 0.5D$), but other cases are similar. For small sizes the regime is fractal and the strength is initially *increasing* with size. For a certain size, the curves for the fractal and nonfractal regimes meet (at $a_0 = a_f$) and for larger sizes the regime is macroscopic. The complete size effect curve is indicated by a thick line.

One salient feature of the foregoing result is that nominal strength is an increasing function of D when D is not too large. This is a strange feature, not supported by the available test results for quasibrittle materials. So it appears that, at very small scale, the fractal nature of the crack surfaces cannot be the major cause of size effect, except perhaps if the size c_f of the fracture process zone vanishes, which is an unrealistic assumption for quasibrittle materials (Bažant 1997d).

A second salient feature of the foregoing result is that, if no transition to macroscopic scale is assumed, the fractal size effect curve in the log-log plot approaches an asymptote of a slope much less that $-1/2$ (about -0.25 to -0.4, according to the ϕ–values reported for concrete; Fig. 12.7.3b). But there exist many test results for concrete and rock which clearly exhibit a close approach to an asymptote of slope $-1/2$ as demonstrated by the various examples in Section 1.5.

From these two features, Bažant (1997d) concluded that the size effect in these test data cannot be caused by crack surface fractality.

12.7.5 Bažant's Analysis of Fractal Crack Initiation

As already explained in Section 9.1.6, to treat the crack initiation from a smooth surface, as in the modulus of rupture (flexural strength) test, it is necessary to include the second (quadratic) term of the large-size asymptotic expansion because the first term is zero. Following a procedure similar to that in Section 9.1.6, but with a fractal fracture energy as in the preceding section, the following law of the size effect for crack initiation may then be deduced (Bažant 1997d):

$$\sigma_{Nu} = \sigma_B \left(\frac{D}{D_f}\right)^{(d_f-1)/2} \left(1 + 2\frac{\ell_f}{D}\right) = \sigma_B \left(\frac{D}{D_f}\right)^{\phi/2} \left(1 + 2\frac{\ell_f}{D}\right) \tag{12.7.21}$$

in which σ_B is the strength for infinite size, ℓ_f is the characteristic size given by (9.1.41); D_f is the size for which the behavior ceases to be fractal (proportional to a_f).

Equation (12.7.21) is plotted in the logarithmic scales of D and σ_{Nu} in Fig. 12.7.4, for both the fractal and nonfractal cases. For the fractal case, the size effect predicted from the fractal hypothesis represents a transition from a declining to a rising asymptote. Such behavior is contradicted by the experimentally observed trend. According to Bažant (1997d), this is another reason to conclude that the hypothesis of a fractal source of size effect is not defensible.

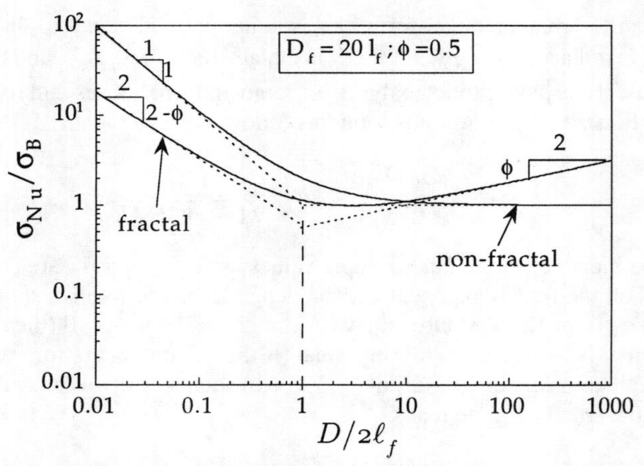

Figure 12.7.4 Fractal and nonfractal size effect curves according to Bažant's analysis (1997d) for structures that fail at crack initiation.

12.7.6 Is Fractality the Explanation of Size Effect?

In the recent polemics concerning the influence of fractality on the size effect for concrete, there are only two points of agreement: **(1)** the fracture surfaces have fractal structure over various orders of magnitude, and **(2)** there must exist a transition to macroscopic (nonfractal) behavior for large enough structural sizes.

An important fact in regard to the second point is that the material is the same, i.e., the aggregate size distribution does not change when the structure size increases or decreases. If the aggregate size increased in proportion to the structure size, then the material would change together with its basic fracture properties.

Given the foregoing agreement, the first essential point is to clarify whether the transition occurs for ordinary laboratory sizes or for much larger sizes. According to Bažant's (1997d) analysis, the transition is complete for many laboratory sizes because the size effect curve approaches the asymptote of slope $-1/2$, and thus fractality is not dominant. According to Carpinteri et al.'s so-called multifractal model, the large size asymptote should in reality be horizontal, and be reached for sizes much larger that those in usual laboratory testing. One might think that the question can be decided based on experiments, but it cannot.

The problem of the experimental verification is illustrated in Fig. 12.7.5 for the test data for notched three-point-bend beams of mortar of Bažant and Pfeiffer (1987) (Series C1 in Section 1.5). As seen in the figure, both Bažant's size effect law and the multifractal scaling law approximate the experimental data within the experimental scatter, as do most tests in the literature (for an extensive comparison to experimental data, see Carpinteri, Chiaia and Ferro 1995c). However, the results of the multifractal model indicate that the macroscopic range is reached only for sizes approaching 10 m ($\ell_\sigma \approx 1.3$ m from optimum fitting of data; this requires $D \approx 13$ m to approach the asymptote within 5%). It is difficult to believe that so large a size should be needed to reach the macroscopic regime in a mortar with a maximum aggregate size of only a few millimeters.

Furthermore, it appears that the multifractal scaling law does not display the right asymptote for large sizes in *notched* specimens. This is so because, for notched specimens in the macroscopic regime, LEFM must hold for large enough sizes (the near-notch-tip fields being K-dominated). Since it is agreed that for large sizes the macroscopic regime holds, then LEFM must hold for large sizes and the asymptote of slope $-1/2$ must be approached.

There also exist size effect test data that do not approach an asymptote of slope $-1/2$ and exhibit a positive overall curvature in the logarithmic plot, as shown by the examples in Chapter 9 for the modulus of rupture and the Brazilian test. These data suggest that, beyond a certain size range, the descending size effect curve of $\log \sigma_N$ vs. $\log D$ might exhibit a transition to a horizontal line, i.e., the size effect might disappear for sufficiently large sizes. This phenomenon cannot occur if the size effect is due to a monofractal structure, but might be described by the multifractal scaling law (12.7.11), as shown in Section 9.3. However, the evidence that multifractality is the only explanation is only circumstantial

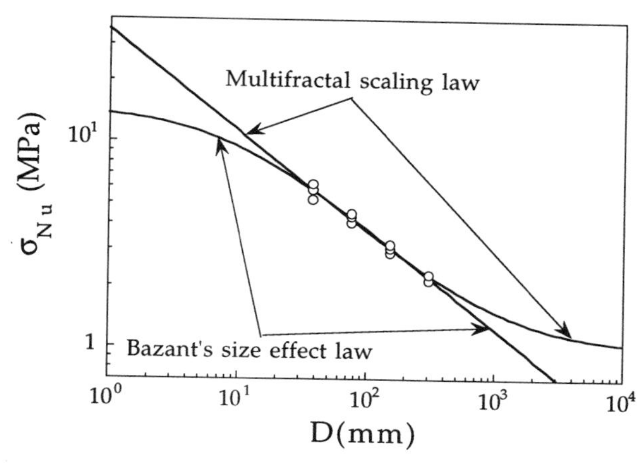

Figure 12.7.5 Test data by Bažant and Pfeiffer (1987) for notched beams in three-point-bend tests, and their optimum fits by Bažant's size effect law and the multifractal scaling law (for details on the tests, see Series C1 in Tables 1.5.1 and 1.5.2).

and disputable, since we have seen in Chapter 9 and Section 9.3 that other approaches can also describe the existence of a horizontal asymptote for large sizes: Bažant's and Kazemi's size effect law (if a_0 ceases to increase in proportion to D), the size effect law for failure at crack initiation (incorporated into Bažant's universal size effect described in Section 9.1), the cohesive crack model, the Jenq and Shah's two-parameter model, and, certainly, all the more sophisticated continuum nonlocal models.

Therefore, with the scarce evidence now available, it appears that the behavior of concrete structures of laboratory sizes can be adequately described without recourse to fractal models. This does not mean that fractals play no role at all. They probably do, but only for the smallest sizes, and only for the statistics of the fracture. Indeed, we have already seen that the monofractal lacunar model for size effect is equivalent to a Weibull formulation. It may also be shown that the random R-curve model discussed in Section 12.6 is connected to a fractal structure: its integral is a fractal line when plotted vs. a.

In view of the foregoing, the second essential point is how to connect fractality and mechanics, i.e., how to build a fracture mechanics of fractal media. A first attempt has been done by Bažant as described in the previous section, but this is not the only way, and other possibilities exist. For example, it appears that, for notched specimens, the fractal dependence of the fracture work might be on the crack extension Δa rather than on the full crack length a. More importantly, it might be that the energy release rate, which is assumed by Bažant as being macroscopic, is also fractal. Note indeed that in the hypothetical wire of fractal cross section analyzed in the example 12.7.1, the apparent elastic modulus is also fractal. Therefore, all the mechanical properties and the balance equations themselves might need to be formulated in a consistent fractal framework. This would be a task for the years to come, but an essential one if fractality should ever lead to rigorous *predictive* models.

13

Nonlocal Continuum Modeling of Damage Localization

In concrete and other quasibrittle materials, fracture develops as a result of localization of distributed damage due to microcracking. In discrete fracture models which have been discussed in previous chapters, the damage due to distributed cracking is lumped into a line, but this is not sufficiently realistic for all applications. The width and microcracking density distribution at the fracture front may vary depending on structure size, shape, and type of loading.

Such behavior can be captured only by continuum damage models. However, such models cannot be implemented in the sense of the classical, local continuum, i.e., a continuum in which the stress at a point depends only on the strain at the same point. Rather, one must adopt the more general concept of a nonlocal continuum, defined as a continuum in which the stress at a point depends also on the strains in the neighborhood of that point or some type of average strain of the neighborhood. The reasons for introducing the nonlocal concept are both mathematical and physical:

1. The mathematical reason is that, as we discussed in Chapter 8 (Bažant 1986c), a local strain softening continuum exhibits spurious damage localization instabilities, in which all damage is localized into a zone of measure zero. This leads to spurious mesh sensitivity. The energy that is consumed by cracking damage during structural failure depends on the mesh size and tends to zero as the mesh size is refined to zero. The reason is that the energy dissipation, as described by the local stress–strain relation per unit volume, and thus also the total dissipation, converges to zero if all damage is localized into a band of single element width as the element size tends to zero. Such spurious localization on a set of measure zero is prevented by the nonlocal concept.

2. The physical reason is that microcracks interact (Bažant 1994b; Bažant and Jirásek 1994a,b)). The formation or growth of one microcrack either promotes or inhibits the formation or growth of adjacent microcracks. Continuum smearing of such interactions inevitably leads to some kind of a nonlocal continuum. The interaction of microcracks is the physical reason why a continuum model for distributed strain softening damage ought to be nonlocal. A secondary physical reason is that a crack has a macroscopically nonnegligible dimension, causing the crack growth to depend on the macroscopic stress field in a zone larger than the crack (Bažant 1987c, 1991b).

Spatial averaging integrals and interaction integrals are not the only way to describe a nonlocal continuum. If the strain field in the neighborhood of a point is expanded into a Taylor series, the strains in the neighboring points are approximately characterized by the spatial partial derivatives (gradients) of the strain tensor at the given point. Thus, the nonlocal continuum may alternatively be defined as a continuum in which the stress at a point depends not only on the strain at that point but also on the successive gradients of the strain tensor at that point (Bažant 1984b, Triantafyllidis and Aifantis 1986). This approach may be regarded as a generalization of Cosserat's couple stress continuum or Eringen's micropolar elasticity. Cosserat's continuum was considered as an alternative approach to achieve regularization of the strain-localization problem in softening materials, but they have been superseded by fully nonlocal or high-gradient models, and will not be presented here. The interested reader may refer, among others, to the works by de Borst, R. (1990) Vardoulakis (1989), de Borst (1991), de Borst and Sluys (1991), and Dietsche and Willam (1992).

In Chapter 8 we have already seen the crudest but simplest type of nonlocal approach—the crack band model, in which the dependence of stress on the average deformation of a certain representative volume

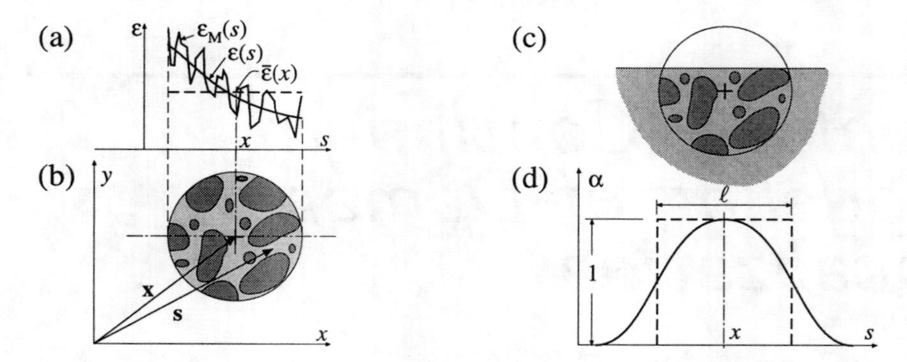

Figure 13.1.1 Spatial averaging. (a) Profiles of micro strain and average strain along a segment centered at point **x** in the center of a representative volume. (b) Sketch of the representative volume centered at **x**. (c) Representative volume near the surface of the body. (d) Uniform vs. smoothly decaying averaging functions. (Adapted from Bažant 1990c.)

of the material is enforced by prescribing the minimum crack band width, coinciding with a minimum element size (the reason is that, in a constant strain finite element, the strain approximates the average strain of the material within the element area). The first part of this chapter will describe the nonlocal models based on the idea of averaging, approached in a phenomenological manner. The second part of this chapter will present a recent development in which the nonlocal concept is derived from micromechanical analysis of crack interactions. The mathematical aspects of localization instabilities and bifurcations will not be discussed in detail and the reader is referred to the book by Bažant and Cedolin (1991, Ch. 13).

13.1 Basic Concepts in Nonlocal Approaches

13.1.1 The Early Approaches

The concept of nonlocal continuum for materials with a randomly heterogeneous microstructure was originally conceived and extensively studied for elastic materials (Eringen 1965, 1966; Kröner 1967; Levin 1971; Kunin 1968; Eringen and Edelen (1972); Eringen and Ari 1983). For such materials, the constitutive relation is considered as a relation between the average continuum stress tensor $\overline{\sigma}(\mathbf{x})$ and strain tensor $\overline{\varepsilon}(\mathbf{x})$, which are defined as the statistical averages of the randomly scattered microstresses over a suitable representative volume of the material centered at point $\mathbf{x}$ (Fig. 13.1.1a–b).

Intuitively, the justification for nonlocal averaging may be explained by Fig. 13.1.1b, showing a representative volume of the material with an aggregate microstructure. (The representative volume is, in the statistical theory, defined as the smallest volume for which the statistics of the microstructure are not changed by shifting the volume.) The formation of a crack in the center of this element obviously does not depend only on the continuum strain at the center of the crack, but on the overall deformation of this representative volume, which determines the strain energy content and thus the energy release from this volume.

The simplest way to introduce a nonlocal strain measure is to define the average strain tensor as

$$\overline{\varepsilon}(\mathbf{x}) = \frac{1}{V_r(\mathbf{x})}\int_V \alpha(|\mathbf{x}-\mathbf{s}|)\varepsilon(\mathbf{s})dV(\mathbf{s}) = \int_V \hat{\alpha}(\mathbf{x},\mathbf{s})\varepsilon(\mathbf{s})dV(\mathbf{s}) \qquad (13.1.1)$$

in which $\varepsilon(\mathbf{x})$ is the usual (local) strain tensor at point $\mathbf{x}$, V the volume of the structure, and $\alpha(r)$ is a scalar function of the distance $r = |\mathbf{x}-\mathbf{s}|$ between the point at which the average is taken and the point contributing to that average; V_r is a normalizing factor introduced so that, for uniform strain, the average is also uniform and coincides with the local value. It is a simple matter to find the required relationships between V_r, α, and $\hat{\alpha}$

$$V_r(\mathbf{x}) = \int_V \alpha(|\mathbf{x}-\mathbf{s}|)dV(\mathbf{s}) \quad \text{and} \quad \hat{\alpha}(\mathbf{x},\mathbf{s}) = \frac{\alpha(|\mathbf{x}-\mathbf{s}|)}{V_r(\mathbf{x})} \tag{13.1.2}$$

The function $\alpha(r)$ decays with the distance from point $\mathbf{x}$ and is zero or nearly zero at points sufficiently remote from $\mathbf{x}$. The simplest is the uniform averaging function, for which $\alpha = 1$ in a sphere of diameter ℓ and 0 outside. For points in the interior of the body whose distance to any boundary is larger than $\ell/2$, V_r is the volume of the averaging sphere $V_r = \pi\ell^3/6$. However, for points closer to the surface, the part of the sphere that protrudes outside the body does not contribute to the integral in (13.1.2) and V_r must be considered a variable (Fig. 13.1.1c).

We may note that according to (13.1.1) and (13.1.2), α may be multiplied by an arbitrary factor without introducing any change in the nonlocal variable, because V_r also gets multiplied by the same factor. This means that we can rescale α at will. In the following, we scale α so that the value of α at the origin is 1, i.e.,

$$\alpha(0) = 1 \tag{13.1.3}$$

as depicted in Fig. 13.1.1d.

The convergence of numerical solutions is slightly better if α is a smooth bell-shaped function (Fig. 13.1.1d, full line) rather than rectangular (Fig. 13.1.1d, dashed line). According to Bažant (1990c), an effective choice is the function

$$\alpha = \left[1-(r/\rho_0\ell)^2\right]^2 \quad \text{if} \quad |r| < \rho_0\ell, \qquad \alpha = 0 \quad \text{if} \quad |r| \geq \rho_0\ell \tag{13.1.4}$$

where $r = |\mathbf{x}-\mathbf{s}|$ is the distance from point $\mathbf{x}$, ℓ is the characteristic length (a material property, Fig. 13.1.1), and ρ_0 is a coefficient chosen in such a manner that the volume under function α given by Eq. (13.1.4) is equal to the volume of the uniform distribution in Fig. 13.1.1d. From this requirement, one may calculate that $\rho_0 = \sqrt[3]{35}/4 = 0.8178$. In the earlier works, the normal (Gaussian) distribution function has also been used instead of Eq. (13.1.4) and was found to work well, although its values are nowhere exactly zero. Note that the limit of nonlocal continuum for $\ell \to 0$ is the local continuum (because $\bar{\varepsilon} \to \varepsilon$).

The foregoing approximation deals with three-dimensional averaging. In many cases, however, two- or one-dimensional approximations are required. In those cases, similar definitions can be written for the averaging operator. For the two-dimensional case V_r must be replaced by A_r, a representative area, and the integrals become surface integrals. For the uniaxial case the integrals reduce to simple integrals and V_r is replaced by a reference length L_r. It is an easy matter to see that if the size of the representative zone is ℓ in either dimension, and the averaging is uniform, then, for interior points $A_r = \pi\ell^2/4$ and $L_r = \ell$. For the bell-shaped function (13.1.4), the values of ρ_0 to be used for two and one dimensions are adjusted so that they give the same values for A_r and L_r as the uniform distribution (see the exercises at the end of this section).

Now that we have introduced the concept of nonlocal averaging, formulating the equations of a nonlocal continuum seems to be a simple matter: Just replace some or all of the classical local variables by their nonlocal averages. However, this is not easy because, in general, some physically problematic features appear and the model does not work at all. In the remainder of this section we focus on uniaxial models to illustrate some of the problems that may arise and the approaches devised to overcome them, leading to various useful models.

13.1.2 Models with Nonlocal Strain

The simplest model imaginable is the nonlocal version of the classical linear elastic model. Its uniaxial version simply reads

$$\sigma(x) = E\bar{\varepsilon}\,, \qquad \bar{\varepsilon} = \frac{1}{L_r}\int_L \alpha(|s-x|)\,\varepsilon(s)\,ds \tag{13.1.5}$$

where σ and ε are the uniaxial stress and strain. Consider, for the sake of simplicity, an infinitely long bar subjected to uniaxial stress σ, and assume that the averaging rule is rectangular, such that $\alpha(r) = 1$

for $|r| < \ell/2$ and zero otherwise. The equilibrium of the bar requires σ to be constant, and so we need to solve the equation

$$\frac{E}{\ell}\int_{x-\ell/2}^{x+\ell/2} \varepsilon(s)\, ds = \sigma \qquad (= \text{const.}) \tag{13.1.6}$$

This is an integral equation that accepts as a trivial solution $\varepsilon = \sigma/E$. However, the solution is not unique. Indeed, we can write the general solution as $\varepsilon = \sigma/E + \varepsilon^*$ and substitute into the foregoing equation to find that the condition to be satisfied by the unknown function ε^* is

$$\int_{x-\ell/2}^{x+\ell/2} \varepsilon^*(s)\, ds = 0 \tag{13.1.7}$$

This equation simply states that the mean of the function over any segment of length ℓ is zero. There are infinitely many solutions of this equation since any harmonic function whose wavelength is a submultiple of ℓ satisfies this condition, i.e., any function of the type

$$\varepsilon^*(s) = A \cos\frac{2\pi n\, s}{\ell} + B \cos\frac{2\pi n\, s}{\ell} \tag{13.1.8}$$

is a solution whatever the constants A and B and the nonzero integer n.

It may be shown that many bell-shaped curves also lead to multiple solutions. To avoid this problem, the averaging function α must have the property that its Fourier transform is positive for any wave number (see Bažant and Cedolin 1991, Sec. 13.10). One particular possibility is to take a weight function with a Dirac δ-spike at its center. Then the nonlocal elastic equation (13.1.5) can be rewritten as

$$\sigma = \gamma E \varepsilon + (1-\gamma) E \overline{\varepsilon}\,, \qquad \overline{\varepsilon} = \frac{1}{L_r}\int_L \alpha_s(|s-x|)\, \varepsilon(s)\, ds \tag{13.1.9}$$

in which the first term comes from the spike, and $0 < \gamma < 1$; γ is a constant that measures the relative weight of the spike, and α_s is the smooth part of the weight function. It is obvious that for $\gamma = 1$ the response is purely local (in which case the elastic solution is unique), while for $\gamma = 0$ the response is purely nonlocal and displays the aforementioned multiple solutions. If γ is selected large enough, then the multiple solutions can be avoided.

This kind of approach, with intermediate values of γ, can be interpreted as a parallel coupling of a local elastic model with a nonlocal model, in which γ has the meaning of the volume fraction of local medium. Such a model may also be regarded as a nonlocal continuum model overlaid by a local elastic continuum. The overlay by an ordinary elastic continuum (called the imbricate continuum) was introduced to stabilize the solutions for softening nonlocal continua (Bažant, Belytschko and Chang 1984; Bažant 1986c). However, such an overlay prevents strain-softening from reducing the stress to zero. Other later formulations were proposed, such that this artificial expedient could be avoided.

Before proceeding to other models, we may note that in the foregoing simple analysis the multiplicity of solutions arises because the strain can accept *alternating* solutions. This is so because the strain can, in principle, take any value, positive or negative. However, if nonlinear ever-increasing variables (such as cumulated plastic work or damage) were to appear in nonlocal equations similar to (13.1.7), then no arbitrary solution could exist, because the average of a nonnegative variable cannot be zero unless the variable vanishes everywhere. This is at the root of the most recent nonlocal models in which the stress and strain are considered to be local, while some nondecreasing internal variable is taken as nonlocal. We will examine a number of models of this kind in the sequel. But before doing so, it is useful to generalize the nonlocal idea of averaging to other kinds of operators, in particular, differential operators which lead to the so called *gradient models*.

13.1.3 Gradient Models

Consider an uniaxial model in which a certain scalar variable —for example, the uniaxial strain— is assumed to be nonlocal, as given by the second of (13.1.5), and assume further that the function α is a rectangular or bell-shaped function as sketched in Fig. 13.1.1d. If we further assume that the bar is very long compared to ℓ, so that the averaging integral would extend from $-\infty$ to $+\infty$, then, setting $u = s - x$, we may rewrite the expression for the nonlocal variable as

$$\bar{\varepsilon} = \frac{1}{\ell} \int_{-\infty}^{+\infty} \alpha(|u|)\, \varepsilon(x+u)\, du \tag{13.1.10}$$

If the local variable ε is assumed to be smooth and varying slowly over a segment centered at x in which α is different from zero, we can approximate $\varepsilon(x+u)$ by a truncated Taylor power series expansion about point x. Thus, we get the following expansion:

$$\bar{\varepsilon}(x) = \varepsilon(x) + \frac{\partial \varepsilon}{\partial x}(x)\ell\mu_1 + \frac{\partial^2 \varepsilon}{2\partial x^2}(x)\ell^2\mu_2 + \cdots + \frac{\partial^n \varepsilon}{n!\partial x^n}(x)\ell^n\mu_n \tag{13.1.11}$$

in which μ_i are the dimensionless moments of the weight function, defined as

$$\mu_n = \int_{-\infty}^{+\infty} \alpha(|s|) \frac{s^n\, ds}{\ell^{n+1}} \tag{13.1.12}$$

Since α is even, the odd moments are equal to zero, and only the even moments need to be retained in the foregoing expansion. In cases in which the local variable (ε in this example) varies slowly over the length ℓ (and thus can be approximated by an arc of second-degree parabola), a two-term expansion is a good approximation of the nonlocal variable. Therefore, setting $\mu_2\ell^2/2 = (\lambda/2\pi)^2$, we get

$$\bar{\varepsilon} \approx \varepsilon(x) + \ell^2\mu_2 \frac{\partial^2 \varepsilon}{2\partial x^2}(x) = \varepsilon(x) + \left(\frac{\lambda}{2\pi}\right)^2 \frac{\partial^2 \varepsilon}{\partial x^2}(x) \tag{13.1.13}$$

as proposed by Bažant (1984b, Eqs. 44, 55, 64,70 and 73).

We thus see that, under certain smoothness conditions, the nonlocal integral operator can be approximated by a differential operator involving even-order gradients. For the second-order case, the differential operator reduces to the *harmonic* operator in (13.1.13).

The harmonic operator as well as fourth-order differential operators have been proposed to describe materials with softening. They have the advantage of leading to differential equations which are easier to treat analytically and numerically than the integral equations posed by the full nonlocal approach (following Bažant 1984b, such models, called second gradient models, have also been introduced by Mühlhaus and Aifantis 1991, de Borst and Mühlhaus 1992, and others).

It is easy to show that the gradient approaches also display multiple solutions if the nonlocal variable can take arbitrary positive or negative values. We thus turn attention to the nonlocal models (including their gradient approximations) based on assuming that the nonlocal variable is one irreversible (nondecreasing) internal variable.

13.1.4 A Simple Family of Nonlocal Models

A set of nonlocal models with a common underlying local formulation can be formulated with relative ease, as done by Planas, Elices and Guinea (1993, 1994). These models have the advantage of decoupling the material nonlinearity, involved in the softening curve, from the nonlinearity introduced by the localization. To be specific, we select the uniaxial softening model with strength degradation described in Section 8.4.2, for which

$$\varepsilon = \frac{\sigma}{E} + \varepsilon^f \tag{13.1.14}$$

$$\tilde{\varepsilon}^f = \max(\varepsilon^f) \tag{13.1.15}$$

$$\sigma \leq \phi(\tilde{\varepsilon}^f) \tag{13.1.16}$$

where in the last condition the equality holds whenever ε^f and $\tilde{\varepsilon}^f$ are both increasing.

We know from the analysis in Section 8.3 that this model leads to strain localization into a zone of measure zero. To avoid this, various nonlocal modifications are possible. The simplest is probably to modify only the last equation (13.1.16) and let it depend on a nonlocal variable Ω. Since the last equation defines the evolution of strength, we can call this type of model a nonlocal strength model.

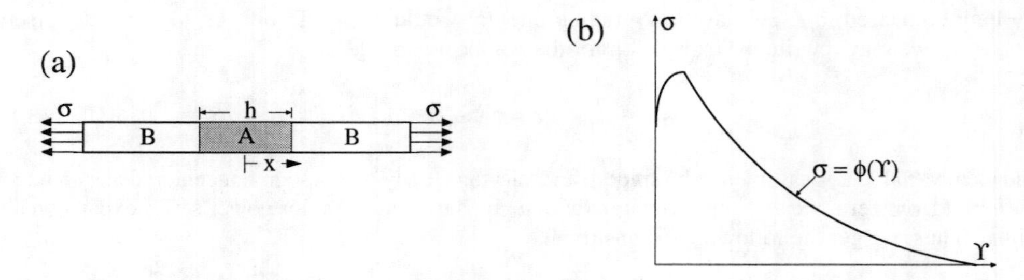

Figure 13.1.2 (a) Bar subjected to tension, with a localized zone A and unloading zones B. (b) Postpeak softening curve of stress vs. nonlocal fracturing strain Υ. (Adapted from Planas, Elices and Guinea 1993.)

There are various ways to include nonlocality in the strength equation. A simple one is to modify (13.1.16) to read

$$\sigma \leq \phi(\Upsilon) \tag{13.1.17}$$

Υ is a nonlocal variable defined from the fracturing strain distribution as

$$\Upsilon(x) = \mathcal{F}\left[\varepsilon^f(s); x\right] \tag{13.1.18}$$

in which $\mathcal{F}\left[\epsilon^f(s); x\right]$ denotes a spatial operator relating the distribution of inelastic strains to the nonlocal variable. This operator can, in principle, be of the differential or integral type, or of other types.

To see how the general equations are obtained, consider a very long bar (i.e., neglect the end effects) and assume that, upon reaching the peak, strain localization occurs within zone A while over the remainder of the bar unloading takes place (Fig. 13.1.2a). This means that $\varepsilon^f > 0$, and $\sigma = \phi(\Upsilon)$ for $x \in A$, and $\varepsilon^f = 0$ and $\sigma \leq \phi(\Upsilon)$ for $x \in B$. Equilibrium further requires that $\sigma =$ constant along the bar. Therefore, if we assume, as usual, that after peak the function $\phi(\Upsilon)$ is monotonically decreasing as depicted in Fig. 13.1.2b, the foregoing conditions can be rewritten as

$$\varepsilon^f > 0 \text{ and } \Upsilon = \Upsilon_A \quad \text{for } x \in A \tag{13.1.19}$$

$$\varepsilon^f = 0 \text{ and } \Upsilon \leq \Upsilon_A \quad \text{for } x \in B \tag{13.1.20}$$

in which Υ_A is the constant value that the nonlocal variable assumes in the softening zone. Given Υ_A, the stress is obviously obtained as $\sigma = \phi(\Upsilon_A)$.

Substituting now Υ from (13.1.18) in the two last equations, the problem is reduced to the functional equation

$$\mathcal{F}\left[\epsilon^f(s); x\right] = \Upsilon_A \qquad \text{for} \qquad x \text{ and } s \in A \tag{13.1.21}$$

subjected to the restriction

$$\mathcal{F}\left[\epsilon^f(s); x\right] \leq \Upsilon_A \qquad \text{for} \qquad x \in B \text{ and } s \in A \tag{13.1.22}$$

The solution of this equation yields the distribution of ε^f for each Υ_A, which is the basic problem to solve. Note that appropriate jump conditions at the interface between zones A and B may be necessary to complete the solution. They depend on the type of operator envisaged. Note also that the zone A over which localization takes place is not known *a priori,* and so it must ensue as a part of the solution. This means that, even if the nonlocal operator $\mathcal{F}$ is linear, the overall problem is not.

We turn next to the analysis of three types of operators and their properties. They are all linear operators, and so the localization problem in Eqs. (13.1.21)–(13.1.22) is quasi-linear. In this way, the material nonlinearity, included in the softening function, is decoupled from the localization problem, which sheds light on the mathematical aspects of the problem.

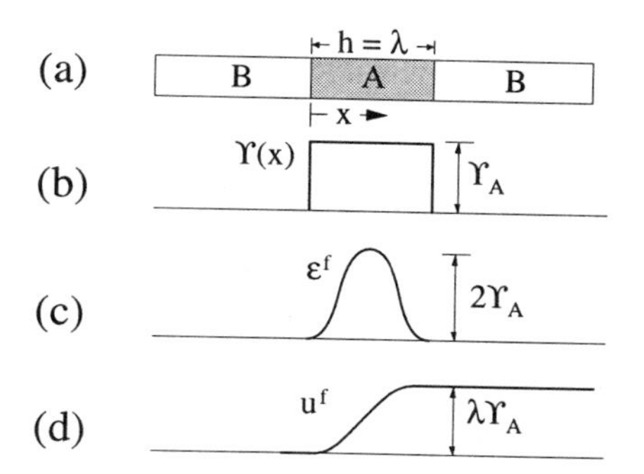

Figure 13.1.3 Nonlocal gradient model with harmonic operator. (a) Bar subjected to tension, with a localized zone A and unloading zones B. Distributions for Υ (b), ε^f (c) and inelastic displacement u^f (d). (Adapted from Planas, Elices and Guinea 1993.)

13.1.5 A Second-Order Differential Model

Consider the differential harmonic operator in Eq. (13.1.13). Then (13.1.21) reduces to the equation

$$\varepsilon^f + \left(\frac{\lambda}{2\pi}\right)^2 \frac{\partial^2 \varepsilon^f}{\partial x^2} = \Upsilon_A \qquad \text{for} \qquad x \in A \tag{13.1.23}$$

whose general solution is

$$\varepsilon^f = \Upsilon_A + C\cos\left(\frac{2\pi\, x}{\lambda} + \psi\right) \qquad \text{for} \qquad x \in A \tag{13.1.24}$$

where C and ψ are arbitrary constants. To determine these constants and the possible size of the localization zone, the jump conditions between the regions A and B must be determined. These conditions are obtained easily if the solution is analyzed in the sense of the theory of distributions. Then, since $\varepsilon^f = \Upsilon = 0$ at the interior points of B and $\Upsilon = \Upsilon_A$ at the interior points of A, the solution for Υ has C^{-1} continuity, therefore, the solution for for ε^f must have C^{n-1} continuity, where n is the differential order of the operator (a jump in the n-th derivative exists). In our case $n = 2$, and so ε^f must have C^1 continuity, i.e., it must be continuous, with a continuous first derivative.

Taking the x origin to lie at the left interface between parts A and B, as shown in Fig. 13.1.3a, and requiring that at this point both ε^f and its first derivative must vanish, we get $C = -\Upsilon_A$ and $\psi = 0$, from which the possible solution takes the form

$$\varepsilon^f = \Upsilon_A\left(1 - \cos\frac{2\pi\, x}{\lambda}\right) = 2\Upsilon_A \sin^2\frac{\pi\, x}{\lambda} \tag{13.1.25}$$

Writing now the continuity conditions at the right interface between A and B, we find that the size h of the localization zone can take only discrete values $h = m\lambda$, with $m = 1, 2, \cdots$. Thus, a periodic solution with an integer number of wavelengths is possible. However, we immediately see that the inelastic displacement and energy requirements are minimum for the smallest possible size, i.e., for a single wavelength. Figs. 13.1.3b–d show the resulting distributions for Υ, ε^f and u^f, where $u^f = \int_{-\infty}^{x} \varepsilon^f\, dx$ is the displacement associated with the fracturing strain.

Note that in solving this problem we assume that there is a region B in which the material unloads, and that the softening region A is continuous. Obviously, there also exist solutions in which (a) the strain is uniform along the bar and $\varepsilon^f = \Upsilon_A$ everywhere, or (b) there exist various nonoverlapping distributions identical in shape to that in Fig. 13.1.3c. However, it is easy to see that the single wavelength solution in this figure is energetically preferred.

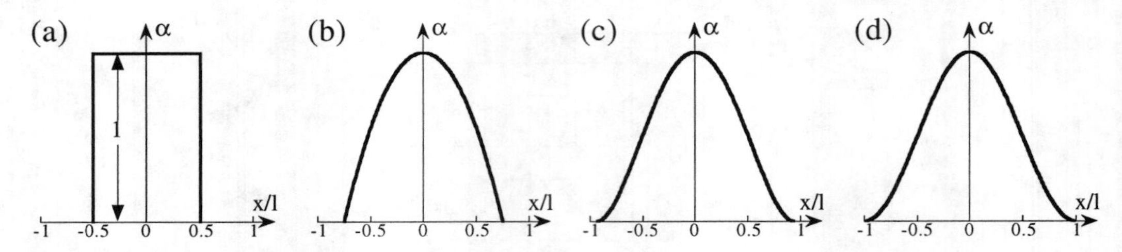

Figure 13.1.4 Examples of weight functions: (a) Rectangular; (b) second-degree parabola, $\alpha(x) = 1 - (x/0.75\ell)^2$; (c) fourth-degree parabola, Eq. (13.1.4) with $\rho_0 = 15/16$; (d) cosine, $\alpha = 1 + \cos(\pi x/\ell)$. (Adapted from Planas, Elices and Guinea 1993.)

13.1.6 An Integral-Type Model of the First Kind

A simple integral functional was investigated by Planas, Elices and Guinea (1993, 1994). It leads to an integral equation of the first kind. Its solution, surprisingly, can be obtained in a closed form, and it turns out to be a cohesive crack.

In this model, the expression for the nonlocal variable is

$$\Upsilon(x) = \mathcal{F}\left[\epsilon^f(s); x\right] \equiv \frac{1}{L_r} \int_L \alpha(|s - x|)\, \varepsilon^f(s)\, ds \tag{13.1.26}$$

in which the weight function α is assumed to be smooth and to have a maximum only at the center. This function is normalized so that $\alpha(0) = 1$, and L_r is given by the uniaxial version of (13.1.2), which ensues by replacing the volume integral by a simple integral and V_r by L_r. For very long bars (L extending from $-\infty$ to $+\infty$) $L_r = \ell =$ characteristic length. Examples of such weight functions are given in Fig. 13.1.4.

Planas, Elices and Guinea (1993, 1994) showed that when a very long bar is considered and the foregoing expression for the functional is substituted into Eqs. (13.1.21)–(13.1.22), the resulting problem accepts a solution consisting of a Dirac's δ-function:

$$\varepsilon^f = w\delta(x)\,, \qquad \text{with} \qquad w = \Upsilon_A \ell \tag{13.1.27}$$

where we assume the origin of coordinates to coincide with the spike location; w is the displacement jump associated with the δ-function, i.e., the crack opening. Since $\sigma = \phi(\Upsilon_A)$, the foregoing result indicates that the solution of this nonlocal model is physically equivalent to a cohesive crack model with a softening function

$$\sigma = f(w) = \phi\left(\frac{w}{\ell}\right) \tag{13.1.28}$$

Note also the remarkable similitude of the foregoing result and Eq. (8.3.2) for the crack band model.

That the foregoing expression is indeed a solution is easily shown by substituting (13.1.27) into (13.1.26) and performing the integration; the result is

$$\Upsilon(x) = \Upsilon_A\, \alpha(x) \tag{13.1.29}$$

which shows that, since $\alpha(0) = 1$, $\Upsilon = \Upsilon_A$ at the origin where $\varepsilon^f > 0$, and $\Upsilon < \Upsilon_A$ everywhere else, as required. Fig. 13.1.5b shows the distribution for the nonlocal variable Υ; Figs. 13.1.5c–d display the distributions for the fracturing strain and displacement.

Certainly, however, this is not the only solution, at least on pure mathematical grounds. First, the location of the δ-spike is arbitrary. Second, an array of any number of δ-functions is also possible, which is equivalent to having multiple cohesive cracks. However, the principle of minimum second-order work indicates, similar to the localization analysis in Chapter 8, that only one crack will occur in reality. Planas, Elices and Guinea (1994) further showed that if the weight function satisfies very mild conditions, solutions with bounded strains distributed over a finite support are not possible. Therefore, it appears that the single δ-spike is *the* solution of this simple nonlocal model. This provides theoretical support for the cohesive crack models.

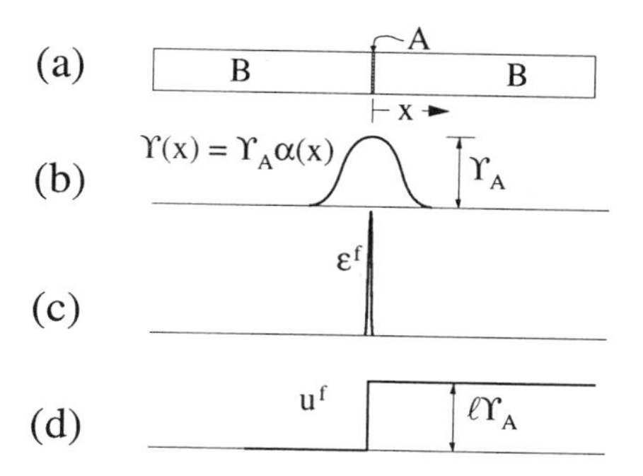

Figure 13.1.5 Nonlocal integral model of the first kind. (a) Bar subjected to tension, with a localized zone reduced to point A and unloading zones B. Distributions for Υ (b), ε^f (c), and inelastic displacement u^f (d). (Adapted from Planas, Elices and Guinea 1993.)

13.1.7 An Integral-Type Model of the Second Kind

Although the foregoing integral model involves a localization limiter in the sense that the solution for the inelastic strain has a finite measure (i.e., the model gives a finite inelastic displacement, w, and a finite energy dissipation), the localization still occurs over a segment of vanishing size. Planas, Guinea and Elices (1996) have extended the analysis to include a linear term along with the integral term in (13.1.26). They take the integral operator as

$$\Upsilon(x) = \mathcal{F}\left[\epsilon^f(s); x\right] \equiv -\gamma\,\varepsilon^f(x) + \frac{1+\gamma}{L_r}\int_L \alpha(|s-x|)\,\varepsilon^f(s)\,ds \qquad (13.1.30)$$

in which γ is constant. Obviously, for $\gamma = 0$ we recover the previous model. Considering again a very long bar in which localization takes place in region A far from both ends as sketched in Fig. 13.1.2a, we have $L_r = \ell$. Taking the x-origin to lie at the center of the localization zone, as depicted in Fig. 13.1.2a, we can reduce Eqs. (13.1.21)–(13.1.22) to the following Fredholm integral equation of the second kind:

$$-\gamma\,\varepsilon^f(x) + \frac{1+\gamma}{\ell}\int_{-h/2}^{h/2} \alpha(|s-x|)\,\varepsilon^f(s)\,ds = \Upsilon_A \qquad \text{for } x \in [-h/2, h/2] \qquad (13.1.31)$$

subjected to the restrictions

$$\varepsilon^f(x) \geq 0 \qquad \text{for } x \in [-h/2, h/2] \qquad (13.1.32)$$

$$\frac{1+\gamma}{\ell}\int_{-h/2}^{h/2} \alpha(|s-x|)\,\varepsilon^f(s)\,ds \leq \Upsilon_A \qquad \text{for } x \notin [-h/2, h/2] \qquad (13.1.33)$$

Here it is understood that $\varepsilon^f(x) = 0$ for $x \notin [-h/2, h/2]$. The integral equation (13.1.31) is a Fredholm equation of the second kind that can be solved for a given h by any of several known methods (see, e.g., Mikhlin 1964; Press et al. 1992). The key point here is that h is not known *a priori*, but that it has to be obtained as part of the solution, because if h is picked at random the solution will fail to satisfy (13.1.32) or (13.1.33), or both.

Planas, Guinea and Elices (1996) investigated the behavior of the problem both theoretically and numerically. On the theoretical side they showed that, for the solutions with a zero-measure to be excluded, γ must be positive. They also showed that the solution for ε^f must be continuous across the interfaces between the softening and unloading regions. On the numerical side, they investigated symmetric modes of localization by discretizing the bar in equal elements of constant ε^f and using point collocation at the center of the elements. The integral was evaluated using a single integration point in the center of each element. A certain value of h was initially assumed and the resulting linear system was solved using standard methods (LU decomposition). It was found that if h was too small, condition (13.1.33) was violated, as shown for one particular case in Fig. 13.1.6a by the full lines, while if h was too large, (13.1.32) could not be fulfilled, as shown in the same figure by the dashed line. The solution was

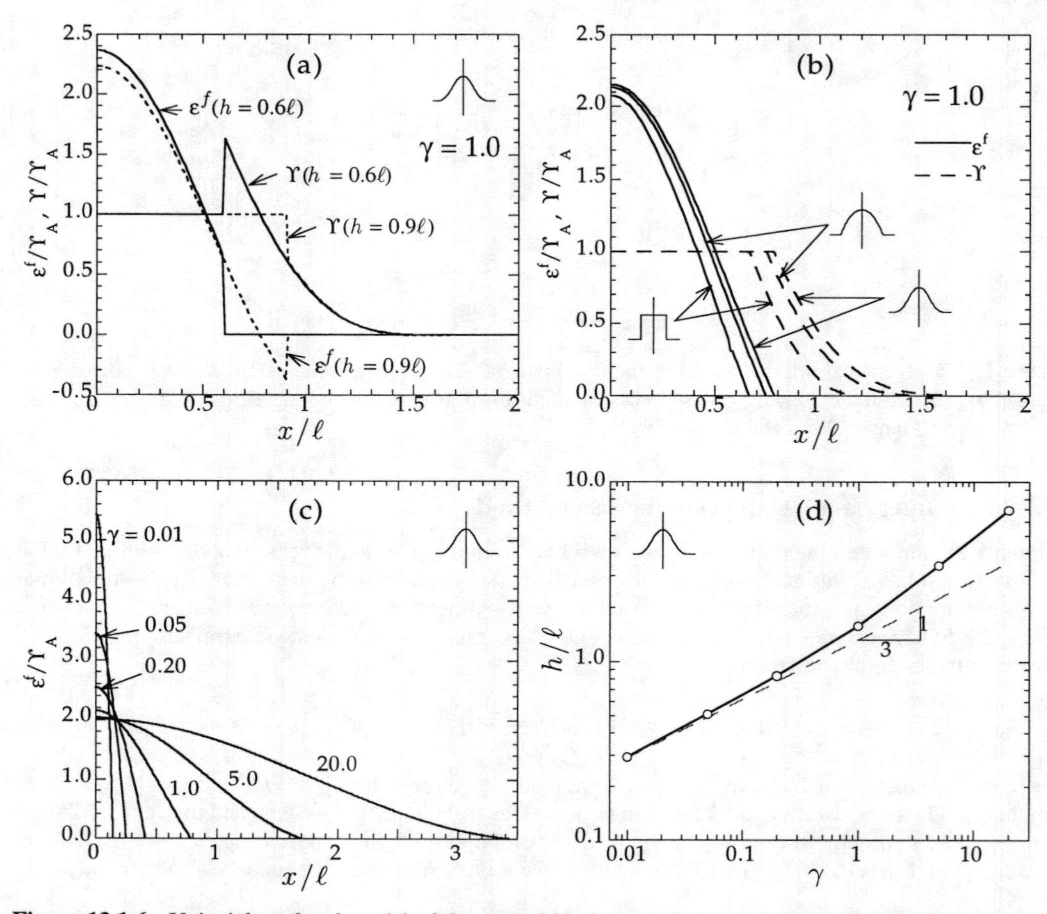

Figure 13.1.6 Uniaxial nonlocal model of the second kind. (a) Solutions of the integral equation (13.1.31) for too small a value of h (full lines) and for too large a value of h (dashed lines). (b) Complete solutions of the problem for various weight functions. (c) Distributions of fracturing strains for various γ. (d) Influence of the factor γ on the width h of the localization zone. (After Planas, Guinea and Elices 1996.)

found iteratively, first with relatively large elements ($\ell/12$ in size) and then for a refined mesh (element size $\ell/100$ to $\ell/1000$ depending on the cases). The results can be summarized as follows:

1. The distribution of ε^f is parabolic in shape and is not very sensitive to the shape of the weight function α, as shown in Fig. 13.1.6b, in which the distributions for the α-functions in Figs. 13.1.4a, b, and d are compared for $\gamma = 1$. We see that h varies only between 1.3ℓ and 1.6ℓ, approximately.

2. The width h of the softening zone is very much influenced by the value of γ, as shown in Fig. 13.1.6c. Indeed, since the exact solution for $\gamma = 0$ is known to be the Dirac δ-function for which $h = 0$, we must have $h \to 0$ for $\gamma \to 0$. For the cases investigated by Planas, Guinea and Elices (1996) the asymptotic relationship is of the power-type: $h \propto \gamma^m \ell$ where m is of the order of 1/3, as shown in Fig. 13.1.6d.

13.1.8 Nonlocal Damage Model

In a series of papers, Bažant and Pijaudier-Cabot developed an isotropic nonlocal damage model, whose uniaxial version was thoroughly investigated (Bažant and Pijaudier-Cabot 1988; Pijaudier-Cabot and Bažant 1987, 1988). The underlying local damage model is similar to the damage models analyzed in

Sections 8.4.1 and 8.4.4, which can be rewritten as

$$\sigma = (1 - \Omega) E\, \varepsilon \tag{13.1.34}$$

Ω is the damage, a nondecreasing variable that is made nonlocal using adequate flow rules. Bažant and Pijaudier-Cabot used two different sets of flow rules which are equivalent in the local version but lead to slightly different models in the nonlocal version. For the underlying local model they assume, based on energy considerations, that the driving force for the growth of damage is the damage energy release rate Y, defined as

$$Y = -\frac{\partial \overline{\mathcal{U}}}{\partial \Omega} = \frac{1}{2} E \varepsilon^2 \tag{13.1.35}$$

in which $\overline{\mathcal{U}} = (1/2)\sigma\varepsilon = (1/2)(1-\Omega)\varepsilon^2$ is the elastic energy density. Once the driving force is defined, the evolution of Ω is assumed to be described by a unique function of the maximum driving force ever experienced by the material:

$$\Omega = F(\tilde{Y}) \qquad \text{with} \qquad \tilde{Y} = \max(Y) \tag{13.1.36}$$

where $F(\tilde{Y})$ is a monotonically increasing function of $\tilde{Y}$. Because $F(\tilde{Y})$ is monotonic, it turns out that $F[\max(Y)] = \max[F(Y)]$, and thus, on purely nonlocal grounds, the foregoing growth rule is strictly equivalent to writing

$$\Omega = \max(\omega) \qquad \text{with} \qquad \omega = F(Y) \tag{13.1.37}$$

Although these two formulations are equivalent in the local framework, they lead to two different nonlocal models according to whether the nonlocal averaging is applied to Y in (13.1.36) or to ω in (13.1.37). In the first case Pijaudier-Cabot and Bažant (1987) introduced the nonlocal variable $\overline{Y}$ as

$$\overline{Y}(x) = \frac{1}{L_r} \int_L \alpha(|s - x|)\, Y(s)\, ds \tag{13.1.38}$$

and then modified (13.1.36) to read

$$\Omega = F(\tilde{\overline{Y}}) \qquad \text{with} \qquad \tilde{\overline{Y}} = \max(\overline{Y}) \tag{13.1.39}$$

They called this the *energy averaging* approach because of the meaning of Y. In their second formulation (Bažant and Pijaudier-Cabot 1988) they averaged the intermediate variable ω in (13.1.37) (which they called the *damage averaging* approach). The new nonlocal variable $\overline{\omega}$ is defined as

$$\overline{\omega}(x) = \frac{1}{L_r} \int_L \alpha(|s - x|)\, \omega(s)\, ds \qquad \text{with} \qquad \omega = F(Y) \tag{13.1.40}$$

and then the evolution of Ω is defined as

$$\Omega = \tilde{\overline{\omega}} \qquad \text{with} \qquad \tilde{\overline{\omega}} = \max(\overline{\omega}) \tag{13.1.41}$$

Recently, Jirásek (1996) showed that averaging of different variables yields models with very different postpeak responses, and suggested that averaging of the inelastic strain or damage seems to be most realistic.

Pijaudier-Cabot and Bažant (1987, 1988) used the energy average model to investigate dynamic strain localization in a bar subjected to two shock waves traveling from both ends and converging in the center of the bar. The analyses confirmed that the nonlocal formulation does prevent zero measure fracture modes. Furthermore, the computations were shown to be mesh-objective. In a further work, Bažant and Pijaudier-Cabot (1988) analyzed the static localization in a bar subjected to tension. Although the complete analysis is globally nonlinear, it is incrementally linear, and the incremental formulation for the initiation of localization takes a form similar to that described in the preceding paragraph, namely, that of a Fredholm equation of the second kind subjected to certain restrictions. To see this, consider the damage average formulation (13.1.41), and assume that the bar is homogeneously deformed up to a point

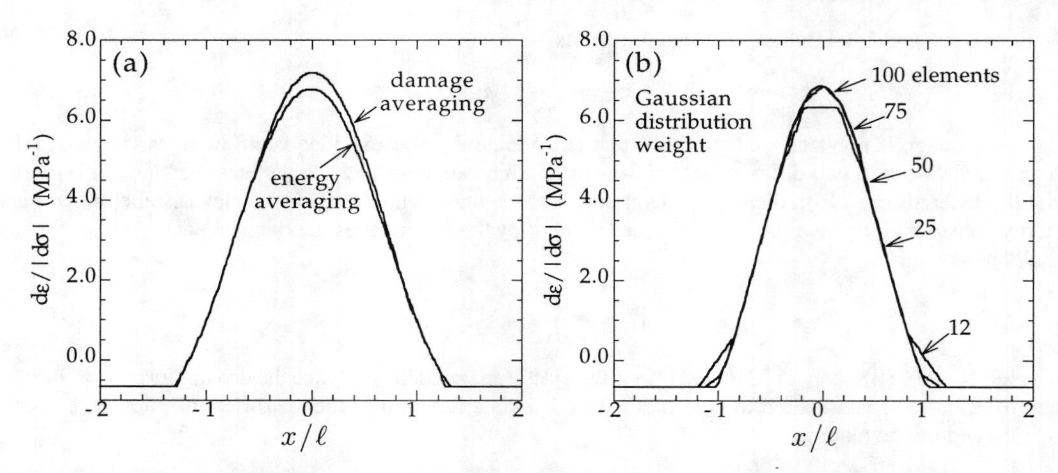

Figure 13.1.7 (a) Incremental strain distribution at the onset of localization in the nonlocal damage model. (b) Convergence to the solution for successive mesh refinement. (After Bažant and Pijaudier-Cabot 1988.)

on the softening branch. We want to analyze the initiation of localization, and so we consider the rate (or incremental) equation derived by differentiating (13.1.34) with respect to time:

$$E(1-\Omega_0)\dot{\varepsilon}_0 - E\varepsilon_0\dot{\Omega} = \dot{\sigma} \qquad (= \text{const.}) \tag{13.1.42}$$

Here we have set $\Omega = \Omega_0$ and $\varepsilon = \varepsilon_0$. These are the values reached prior to localization, which are, by hypothesis, uniform. Note also that equilibrium requires $\dot{\sigma}$ to be uniform. Inserting (13.1.40) into (13.1.41) and differentiating, we get

$$\dot{\Omega} = \left\langle \frac{F'(Y_0)E\varepsilon_0}{L_r}\int_L \alpha(|s-x|)\,\dot{\varepsilon}(s)\,ds \right\rangle = \frac{F'(Y_0)E\varepsilon_0}{L_r}\left\langle \int_L \alpha(|s-x|)\,\dot{\varepsilon}(s)\,ds \right\rangle \tag{13.1.43}$$

in which $F'(Y) = dF(Y)/dY$ and $\langle\cdot\rangle$ are the Macauley brackets, equal to its argument if it is positive, or zero if it is negative; the second equality holds because $F'(Y)$ and ε are positive (remember that $F(Y)$ is monotonically increasing). Therefore, (13.1.42) can be rewritten as follows:

$$E(1-\Omega_0)\dot{\varepsilon}(x) - 2EF'(Y_0)Y_0\left\langle \frac{1}{L_r}\int_L \alpha(|s-x|)\,\dot{\varepsilon}(s)\,ds \right\rangle = \dot{\sigma} \qquad (= \text{const.}) \tag{13.1.44}$$

This is an integral equation of the second kind in $\dot{\varepsilon}$ which, however, is not linear because of the presence of the Macauley brackets. Bažant and Pijaudier-Cabot solved this integral equation numerically for various weight functions and characteristic lengths. They found the typical strain-rate profiles shown in Fig. 13.1.7a. They also found that the size h of the localized zone was proportional to ℓ, and that the convergence with mesh refinement was fully satisfactory, as shown in Fig. 13.1.7b.

The results for the strain-rate profiles are remarkably similar to those found in the previous paragraph for the so-called integral model of the second kind, and Planas, Guinea and Elices (1996) examined whether the two problems were related. It turned out that they were: it suffices to write the problem in terms of an inelastic strain rate $\dot{\varepsilon}^f$ defined as

$$\dot{\varepsilon}^f = \dot{\varepsilon} - \frac{\dot{\sigma}}{E(1-\Omega_0)} \tag{13.1.45}$$

Solving this equation for $\dot{\varepsilon}$ and inserting the result into (13.1.44), the integral equation is transformed into

$$\frac{(1-\Omega_0)}{2F'(Y_0)Y_0}\dot{\varepsilon}^f(x) - \left\langle \frac{\dot{\sigma}}{E(1-\Omega_0)} + \frac{1}{L_r}\int_L \alpha(|s-x|)\,\dot{\varepsilon}^f(s)\,ds \right\rangle = 0 \tag{13.1.46}$$

Now, to analyze this problem we can split the bar, as before, into region A in which localization occurs (and hence the expression into angle brackets is positive) and region B in which unloading occurs and thus $\dot{\varepsilon}^f = 0$ and the expression in brackets is negative. Taking further into account that the stress evolves along the softening branch for which $\dot{\sigma} = -|\dot{\sigma}| < 0$, the foregoing equation can be split into two:

$$-\frac{(1-\Omega_0)}{2F'(Y_0)Y_0}\dot{\varepsilon}^f(x) + \frac{1}{L_r}\int_L \alpha(|s-x|)\,\dot{\varepsilon}^f(s)\,ds = \frac{|\dot{\sigma}|}{E(1-\Omega_0)} \qquad \text{for } x \in A \quad (13.1.47)$$

$$\frac{1}{L_r}\int_L \alpha(|s-x|)\,\dot{\varepsilon}(s)\,ds \leq \frac{|\dot{\sigma}|}{E(1-\Omega_0)} \qquad \text{for } x \in B \quad (13.1.48)$$

For very long bars, this system reduces to (13.1.31)–(13.1.33) if we introduce the following correspondences

$$\dot{\varepsilon}^f \to \varepsilon^f\,, \qquad \frac{|\dot{\sigma}|/E}{2F'(Y_0)Y_0-(1-\Omega_0)} \to \Upsilon_A\,, \qquad \frac{(1-\Omega_0)}{2F'(Y_0)Y_0-(1-\Omega_0)} \to \gamma \qquad (13.1.49)$$

This result, combined with the analysis in the previous paragraph, shows that the relationship between the characteristic length ℓ and the extent of the localization zone h depends on the characteristics of the softening function at the point where the localization occurs. This was pointed out by Bažant and Pijaudier-Cabot (1988), and can now be quantitatively assessed using the plot in Fig. 13.1.6d and the expression for γ from the preceding formula.

Exercises

13.1 Consider the bell-shaped averaging function defined in (13.1.4), restricted to two dimensions. Determine ρ_0 so that the value of $A_r = \int_A \alpha(|\mathbf{s}-\mathbf{x}|)dA(\mathbf{s})$ coincide with that for a uniform distribution over a circle of diameter ℓ. [Hint: use polar coordinates to carry out the integral and get $\rho_0 = \sqrt{3}/2$.]

13.2 Consider the bell-shaped averaging function defined in (13.1.4), restricted to one dimension. Determine ρ_0 so that the value of $L_r = \int_L \alpha(|s-x|)ds$ coincide with that for a uniform distribution over a segment of length ℓ.

13.3 Consider a nonlocal model with a uniform weight function and its high gradient harmonic approximation. Determine the relationship between ℓ and λ.

13.4 Consider a nonlocal model with the parabolic weight function defined in Fig. 13.1.4b, and its high gradient harmonic approximation. Determine the relationship between ℓ and λ.

13.5 Show that the energy and damage averaging in (13.1.39) and (13.1.41) are exactly equivalent if $F(Y)$ is linear in Y.

13.6 Show that along the softening branch $2F'(Y)Y - (1-\Omega) > 0$ if no localization occurs, and that the denominators in (13.1.49) are always positive.

13.7 Bažant and Pijaudier-Cabot use a (local) damage function given by

$$F(Y) = 1 - \frac{1}{1 + b\langle Y - Y_t\rangle} \qquad (13.1.50)$$

where $b = 20.5\ (\mathrm{MPa})^{-1}$ and $Y_t = 8.54$ MPa. With $E = 32$ GPa, they analyzed the initiation of localization at various uniform strains. For the particular strain $\varepsilon_0 = 0.003$, they found that the size of the localization zone was $h \approx 2\ell$ for a bell-shaped weight function. Verify that this is consistent with the results predicted by the nonlocal strength theory of the second kind. To this end, compute first Y_0 and Ω_0; then compute γ from (13.1.49) and use the curve in Fig. 13.1.6d to find h/ℓ.

13.2 Triaxial Nonlocal Models and Applications

In the preceding section we discussed the most basic issues of the nonlocal models based on simple uniaxial cases. Now we address various other possible phenomenological formulations with features that are essentially three-dimensional, and show the general aspects of some practical applications. We also address the most basic problem of making an experimental determination of the characteristic length ℓ.

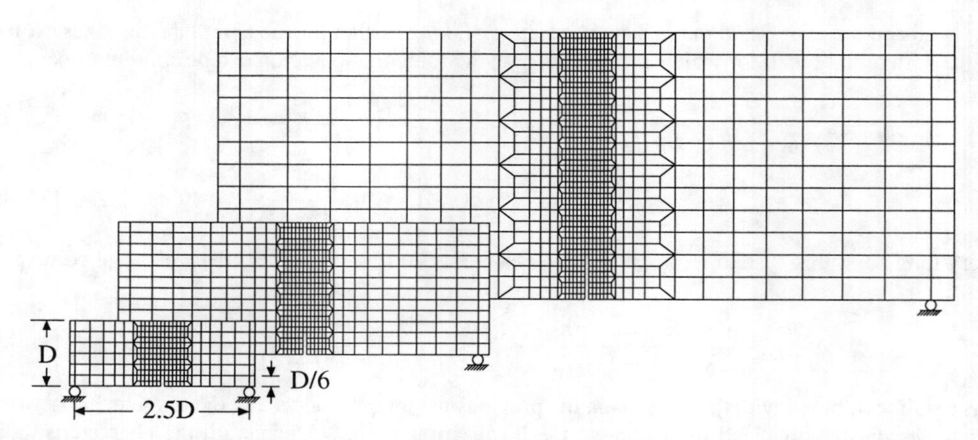

Figure 13.2.1 Finite element meshes used by Bažant and Lin (1988a) to analyze three-point bending of notched specimens of three sizes in the ratios 1:2:4 (from Bažant and Lin 1988a).

13.2.1 Triaxial Nonlocal Smeared Cracking Models

The nonlocal concept can, in principle, be applied to any inelastic constitutive model. It has been applied to the smeared cracking model described in Chapter 8. There are two variants of this model, both of which have been studied. One variant is the cracking of fixed direction (Section 8.5.3) in which the damage ω, which is used to modify the compliance matrix, is considered to be a function of the normal strain ε_{nn} in the direction normal to the cracks. The nonlocal generalization is obtained by considering the nonlocal damage $\overline{\omega}$ to be the same function of the averaged strain $\overline{\varepsilon}_{nn}$ in the direction normal to the cracks (for details, see Bažant and Lin 1988a).

Another variant is the rotating crack model, for which the local formulation was presented in Section 8.5.6. Again, the nonlocal generalization is obtained by replacing the dependence of the normal compliance C_N on the local principal strain by an identical dependence calculated as a function of the nonlocal principal strain $\overline{\varepsilon}_{\text{I}}$. Of course, when the cracks do not rotate, the first and second variant coincide. When they rotate, the second variant seems to be closer to reality.

The model was used by Bažant and Lin (1988a) to simulate three-point-bend fracture specimens, and particularly the size effect. Fig. 13.2.1 shows the finite element meshes for three specimens sizes in the proportions 1:2:4. Fig. 13.2.2 shows a comparison of the nonlocal finite element analysis with test results. The strain-softening law has been considered in two forms: exponential (dashed) and linear (dash-dot). The calculations are compared to the test results of Bažant and Pfeiffer (1987) and to the optimum fit of these results with the size effect law, Eq (1.4.10). The results demonstrate that the nonlocal model eliminates mesh sensitivity (because the ratio of the element size to specimen size is very different for the three specimens). They also demonstrate that the transitional size effect is well described by the nonlocal model. The width of the fracture process zone is, in these calculations, found to be roughly 2.7 ℓ, where $\ell =$ characteristic length, in agreement with the calculations of Bažant and Pijaudier-Cabot (1988).

Fig. 13.2.3 shows finite element calculations on unnotched beams with deliberately slanted meshes. These calculations show that the nonlocal model in which the characteristic length is sufficiently larger than the element size is free of directional mesh bias. The cracking band can propagate in any direction, without bias, to the mesh lines or the diagonal directions.

13.2.2 Triaxial Nonlocal Models with Yield Limit Degradation

The plasticity models can also be adapted to nonlocal analysis of distributed damage. To this end, plastic hardening is replaced by softening, which means, for example, that the plastic hardening modulus H becomes negative, as illustrated by the negative slope in Fig. 13.2.4 for a Mohr-Coulomb yield surface model. If this is done, of course, Drucker's stability postulate for plasticity ceases to be satisfied, but this is not fundamentally incorrect (see Chapters 10 and 13 in Bažant and Cedolin 1991) because this postulate cannot be expected to apply in the case of damage. The nonlocal concept is introduced into the model

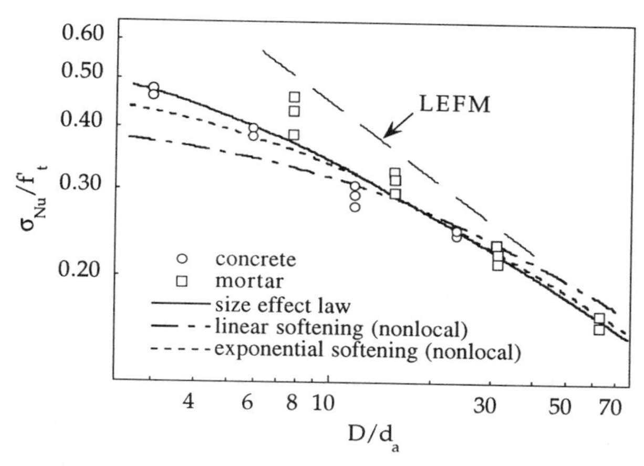

Figure 13.2.2 Size effect plot comparing the test results of Bažant and Pfeiffer (1987) to the size effect law as well as to finite element results of the nonlocal smeared cracking model for linear and exponential softenings (after Bažant and Lin 1988a).

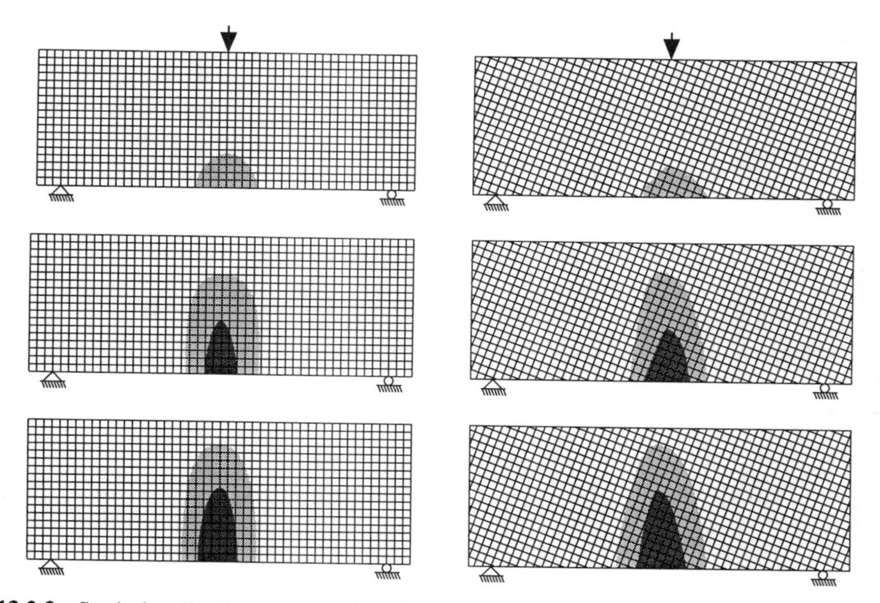

Figure 13.2.3 Strain localization zones at three loading stages for a mesh aligned with the crack path (left) and for a skew mesh (right) (from Bažant and Lin 1988a).

by replacing the plastic strain increment, as soon as it is calculated, by its spatial average and using this average in the constitutive relation.

A debatable feature of this formulation is the fact that Prager's continuity condition of plasticity (consistency condition) is satisfied by the local rather than the nonlocal plastic strain increments, which means that the constitutive law is local and the nonlocality is introduced as separate adaptation. This approach appears to be in line with the conclusions of the analysis of crack interactions (Bažant 1994b) which will be explained later. Some theorists (e.g., de Borst) have insisted that the continuity relation must be satisfied by the nonlocal strains, which, however, would cause a tremendous complication of the model because the continuity condition would become an integral equation over the entire structure. Such a complexity would defeat the advantages of the nonlocal approach. It is true, however, that if Prager's continuity condition is not satisfied by nonlocal strains, there is no precisely defined nonlocal constitutive law. Theoretically, this is a weak point of this type of formulation.

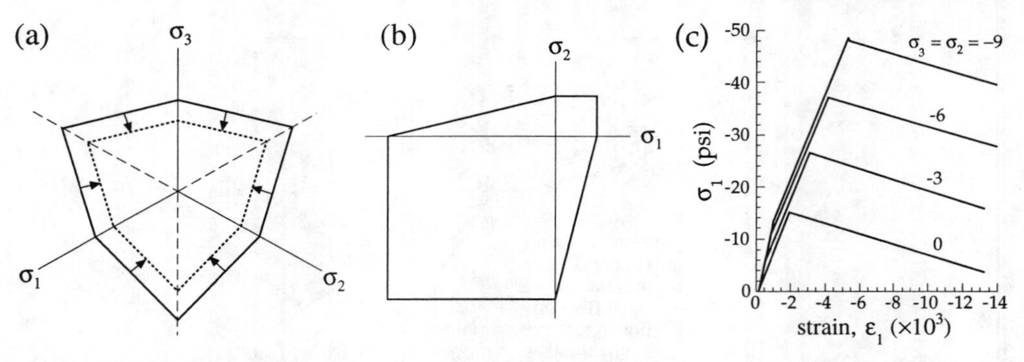

Figure 13.2.4 Mohr-Coulomb yield criterion with strain-softening due to yield limit degradation: (a) yield locus in the deviatoric stress space; (b) yield locus in the principal stress plane; (c) triaxial stress-strain curves with softening for various confining stresses. (Adapted from Bažant and Lin 1988b.)

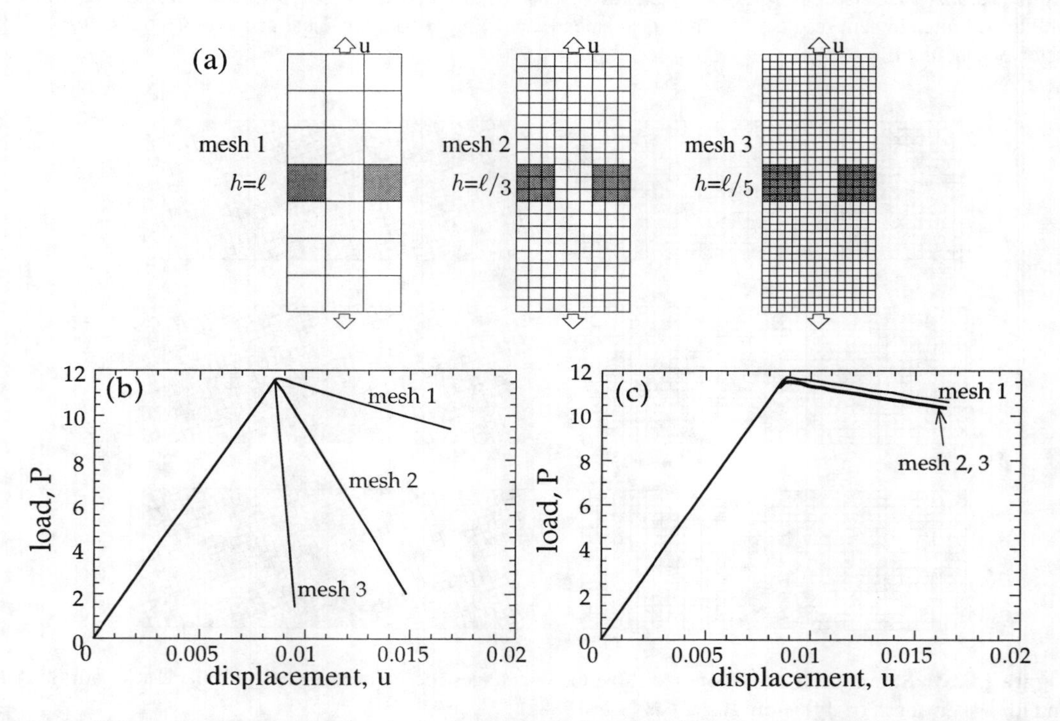

Figure 13.2.5 Analysis of mesh sensitivity. (a) Rectangular panel with various mesh subdivisions; and the corresponding load-displacement curves for (b) local modeling, and (c) nonlocal modeling . (Adapted from Bažant and Lin 1988b.)

Fig. 13.2.5 shows an example (Bažant and Lin 1988b) of a rectangular panel solved by meshes of three different refinements. The local plasticity solution with a degrading yield limit gives the response in Fig. 13.2.5b and the nonlocal model gives the responses shown in Fig. 13.2.5c.

This model has also been applied to the analysis of failure of a tunnel excavation in grouting soil; see Fig. 13.2.6, which shows meshes of four different refinements and the boundaries of the strain softening zones obtained by the four meshes. Note again that the nonlocal approach is basically free of mesh sensitivity.

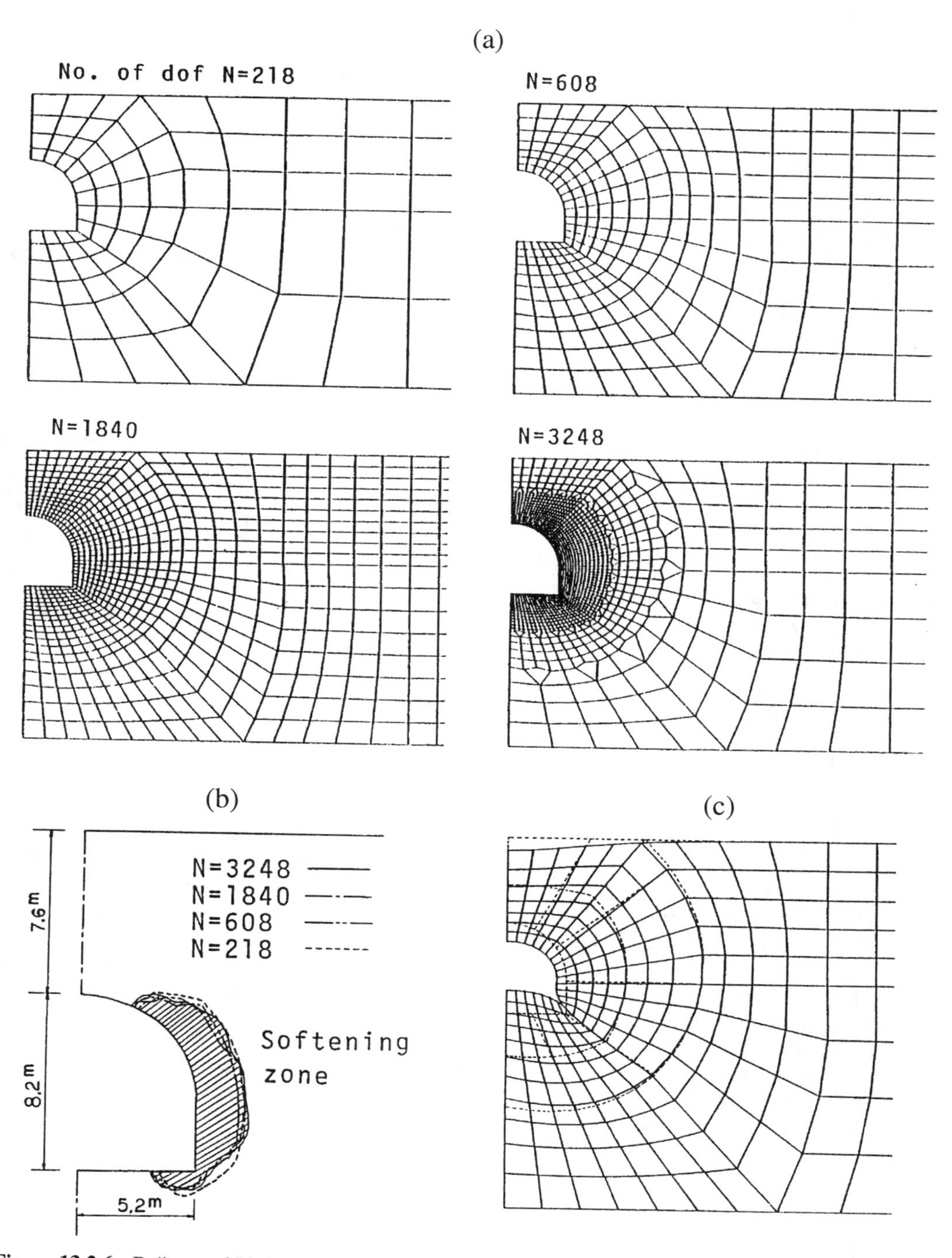

Figure 13.2.6 Bažant and Lin's (1988b) finite element analysis of a tunnel excavation in a grouted soil with a degrading yield limit: (a) finite element meshes; (b) boundaries of the softening zone at full tunnel excavation obtained for the four meshes shown in (a); (c) exaggerated deformation at full excavation. (Adapted from Bažant and Lin 1988b.)

13.2.3 Nonlocal Microplane Model

The most powerful and versatile approach to complex constitutive modeling, including strain softening, appears to be the microplane model which will be explained in the next chapter. Suffice to say at this point that the nonlocal microplane approach has proven very effective and provided, so far, the best finite element results in the modeling of damage and failure of concrete structures, including the size effect.

13.2.4 Determination of Characteristic Length

The characteristic length is a parameter that controls the spread of the nonlocal weight function. It may be defined as the diameter of an averaging region (line segment, circle, or sphere in one-, two-, or three-dimensions) with a uniform weight function that has the same volume as the actual weight function used. The characteristic length ℓ cannot be directly measured but must be inferred indirectly from test of suitable types. There are two types of tests suitable for this purpose: (1) the use of size effect, and (2) the use of elastically restrained tests. Let us examine each of the two possibilities.

(a) Use of size effect. The size effect is the most blatant and most important manifestation of nonlocality. It is necessary to carry out tests of geometrically similar notched specimens of sufficiently different sizes and determine the size effect plot (Chapter 6). Then the characteristic length of the nonlocal model needs to be varied until the finite element calculations match the experimentally determined size effect curve in the optimum way. Generally, it is observed that the transitional size D_0 of the size effect plot (intersection of the horizontal and inclined asymptotes) is approximately (but not exactly) proportional to the value of characteristic length ℓ. Therefore, an effective strategy is to assume characteristic length ℓ', calculate by a nonlocal finite element code the nominal strength of specimens of different sizes, and trace the size effect curve. Optimum fitting of this curve with the size effect law makes it possible to obtain the horizontal and vertical asymptotes and determine their intersection D_0. Then the best estimate of the corrected characteristic length is

$$\ell = \ell' D_0 / D_0' \qquad (13.2.1)$$

The process is then repeated and the value of ℓ corrected iteratively. Normally no more than two corrections are required for convergence.

(b) Elastically restrained tensile test. Another approximate way of determining ℓ was proposed by Bažant and Pijaudier-Cabot (1989). A long prismatic specimen of concrete, with a thickness of only a few aggregate sizes, is cast and many longitudinal thin steel rods are glued to its surfaces by epoxy as shown in Fig. 13.2.7. It is assumed that the glued steel bars are sufficient to force the strain in the specimen to be uniformly distributed, and for this reason the specimen must be as thin as possible. If that is the case, the tensile load-deflection diagram directly yields the stress-strain curve for the fracture process zone of concrete. This is illustrated in Fig. 13.2.7c, where the inclined straight line of slope K_s gives the stress carried by steel bars and epoxy alone, and the shaded zone represents the additional contribution due to concrete. If the slope of the load-deflection curve is always positive, localization should not happen according to uniaxial localization analysis. Thus, plotting the results in terms of the average stress and average strain, the shaded area in Fig. 13.2.7c gives the energy W_s dissipated per unit volume of the fracture process zone, on the average. Hence, the average width of the softening zone $\hat{h}$ should approximately be given by

$$\hat{h} = G_f / W_s \qquad (13.2.2)$$

which has the dimension of length because $G_f \sim$ J/m^2 and $W_s \sim$ J/m^3. The fracture energy G_f is determined by any of the previously discussed methods. A particular nonlocal model is then needed to correlate $\hat{h}$ and ℓ, although it may be assumed that $h \approx \ell$ (Bažant and Pijaudier-Cabot 1989).

In practice, however, it turned out that this method gives only a crude estimate of the characteristic length because the specimen with tensile restraining elastic bars does not behave uniaxially. The deformation becomes nonuniform transversely and there is some degree, although not a large degree, of localization, as transpired from a thorough investigation by Berthaud, Ringot and Schmitt (1991). Further development would be required before ℓ can be accurately determined by this method.

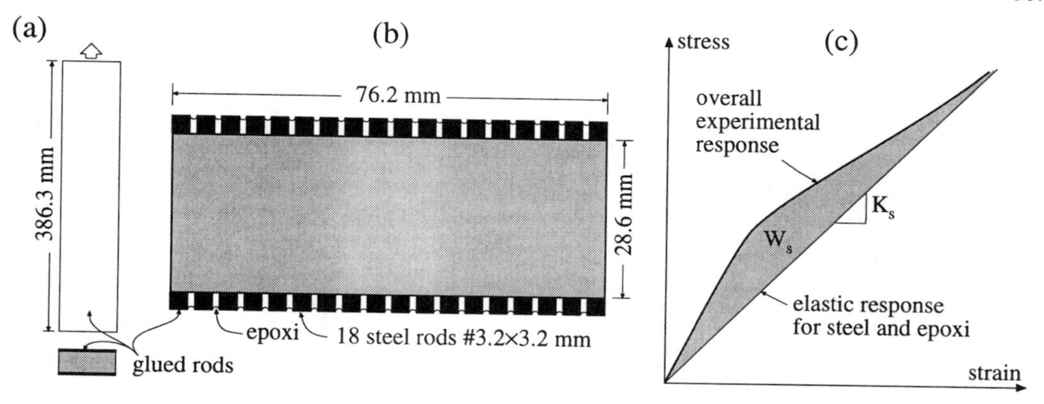

Figure 13.2.7 Bažant and Pijaudier-Cabot's (1989) method to determine the characteristic length for concrete: (a) sketch of restrained specimen; (b) cross-section with the arrangement of the steel bars at the surface of concrete; (c) sketch of the stress-strain curves for the specimen with the glued rods. (Adapted from Bažant and Pijaudier-Cabot 1989.)

13.3 Nonlocal Model Based on Micromechanics of Crack Interactions

13.3.1 Nonlocality Caused by Interaction of Growing Microcracks

The local constitutive law may be written in the incremental form

$$\Delta\boldsymbol{\sigma} = \boldsymbol{E}(\Delta\boldsymbol{\varepsilon} - \Delta\boldsymbol{\varepsilon}'') = \boldsymbol{E}\Delta\boldsymbol{\varepsilon} - \Delta\mathbf{S} \tag{13.3.1}$$

where $\Delta\boldsymbol{\sigma}$ and $\Delta\boldsymbol{\varepsilon}$ are the increments of the stress and strain tensors, $\boldsymbol{E}$ is the fourth-rank tensor of elastic moduli of uncracked material, $\Delta\boldsymbol{\varepsilon}''$ the inelastic strain increment tensor, and $\Delta\mathbf{S}$ the inelastic stress increment tensor.

In a nonlocal continuum formulation, this equation is replaced by

$$\Delta\boldsymbol{\sigma} = \boldsymbol{E}\Delta\boldsymbol{\varepsilon} - \Delta\tilde{\mathbf{S}} \tag{13.3.2}$$

where $\Delta\tilde{\mathbf{S}}$ is the nonlocal inelastic stress increment tensor. In the phenomenological approach discussed in the previous sections, this tensor is directly obtained by a spatial averaging integral

$$\Delta\tilde{\mathbf{S}}(\mathbf{x}) = \Delta\overline{\mathbf{S}} = \int_V \hat{\alpha}(\mathbf{x}, \boldsymbol{\xi})\Delta\mathbf{S}(\boldsymbol{\xi})\, dV(\boldsymbol{\xi}) \tag{13.3.3}$$

completely analogous to (13.1.1), in which we remember that the weight function α is to be postulated. Following Bažant (1994b), we now describe how the equation governing the evolution of $\Delta\tilde{\mathbf{S}}$ can be developed from the mechanics of crack interactions.

Consider an elastic solid that contains, at the beginning of the load step, many microcracks numbered as $\mu = 1, \ldots N$ (Fig. 13.3.1). On the macroscale, the microcracks are considered to be smeared, as required by a continuum model. Exploiting the principle of superposition, we may decompose the loading step of prescribed load or displacement increment into two substeps:

I In the first substep, the cracks (already opened) are imagined temporarily "frozen" (or "filled with a glue"), that is, they can neither grow and open wider nor close and shorten. Also, no new cracks can nucleate. The stress increments, caused by strain increments $\Delta\boldsymbol{\varepsilon}$ and transmitted across the temporarily frozen (or glued) cracks (I in Fig. 13.3.1), are then simply given by $\boldsymbol{E}\Delta\boldsymbol{\varepsilon}$. This is represented by the line segment $\overline{13}$ (Fig. 13.3.2) having the slope of the initial elastic modulus E.

II In the second substep, the prescribed boundary displacements and loads are held constant, the cracks are "unfrozen" (or "unglued"), and the stresses transmitted across the cracks are relaxed. This is equivalent to applying pressures (surface tractions) on the crack faces (II in Fig. 13.3.1). In

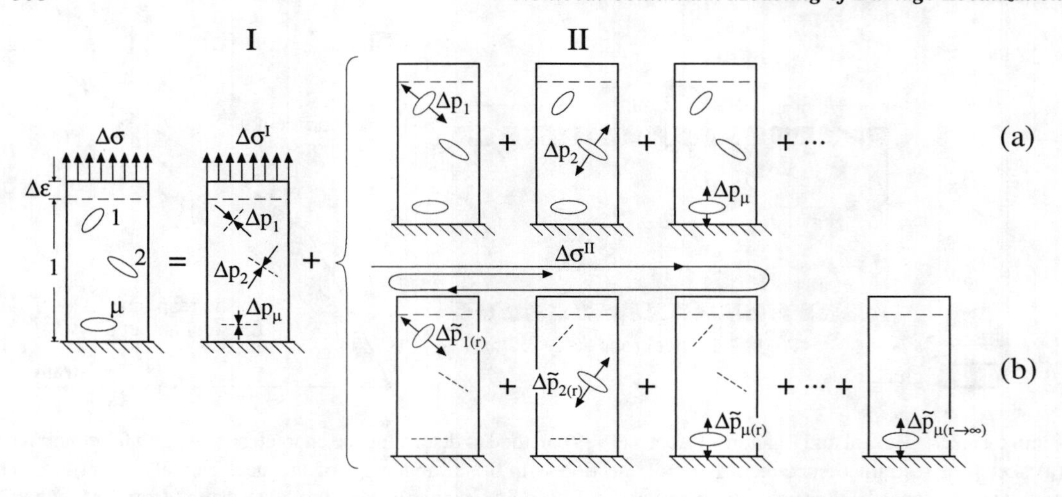

Figure 13.3.1 Superposition method for solid with many cracks. In part I, the cracks are closed and $\Delta\boldsymbol{\sigma}^{\mathrm{I}} = \boldsymbol{E}\Delta\boldsymbol{\varepsilon}$. In part II, the stresses Δp_i on the crack faces generated in part I are released, either simultaneously (alternative **a**) or iteratively keeping all the cracks closed but one (alternative **b**); adapted from Bažant 1994b.

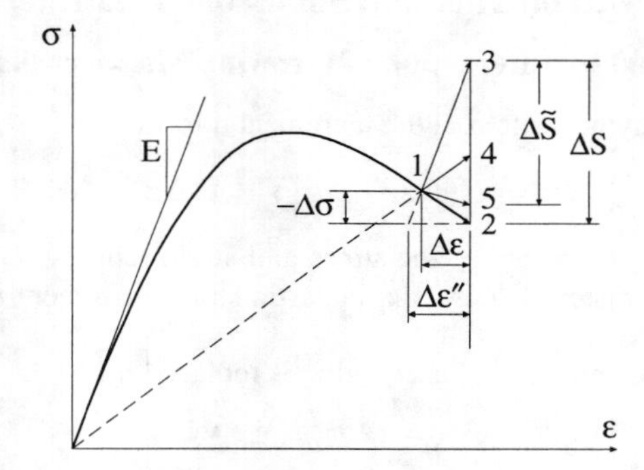

Figure 13.3.2 Local and nonlocal inelastic stress increments during loading step (adapted from Bažant 1994b).

response to these pressures, the cracks are now allowed to open wider and grow (remaining critical according to the crack propagation criterion), or to close and shorten. Also, new cracks are now allowed to nucleate.

If cracks neither grew nor closed (nor new cracks nucleated), the unfreezing (or unglueing) at prescribed increments of loads or boundary displacements that cause macro-strain increment $\Delta\varepsilon$ would engender the stress drop $\overline{34}$ down to point 4 on the secant line $\overline{01}$ (Fig. 13.3.2). The change of state of the solid would then be calculated by applying the opposite of this stress drop onto the crack surfaces. However, when the cracks propagate (and new cracks nucleate), a larger stress drop defined by the local strain-softening constitutive law and represented by the segment $\Delta\mathbf{S} = \overline{32}$ in Fig. 13.3.2 takes place. Thus, the normal surface tractions

$$\Delta p_\mu = \mathbf{n}_\mu \cdot \Delta\mathbf{S}_\mu \mathbf{n}_\mu \tag{13.3.4}$$

representing the normal component of tensor $\Delta\mathbf{S}_\mu$, must be considered in the second substep as loads Δp_μ that are applied onto the crack surfaces (Fig. 13.3.1), the unit normals of which are denoted as $\mathbf{n}_\mu$. (Note that for mode II or III cracks, a similar equation could, in general, be written for the tangential tractions on the crack faces.)

Let us now introduce two simplifying hypotheses:

(a)

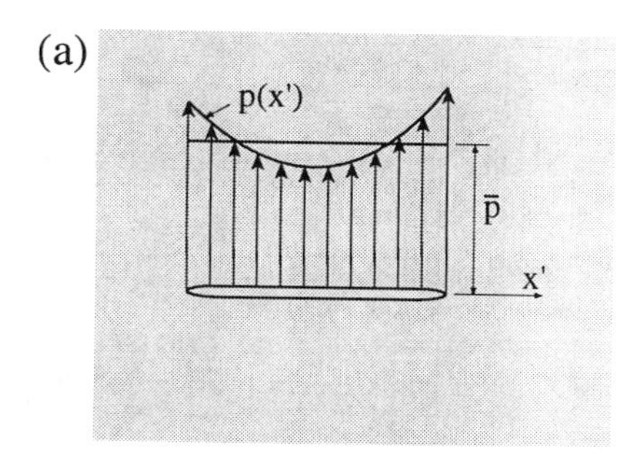

(b)

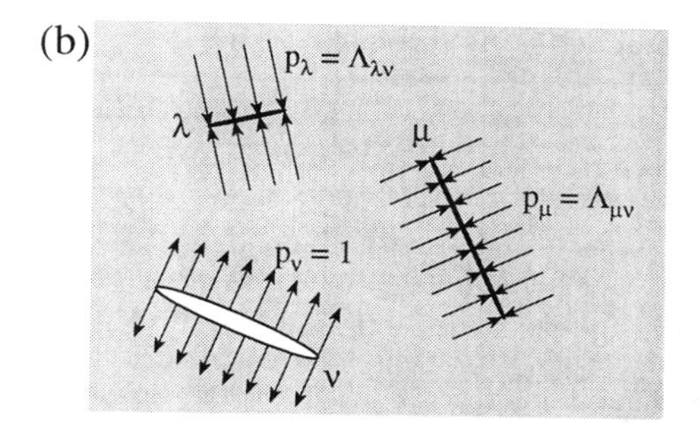

Figure 13.3.3 Details of crack interactions: (a) Actual crack pressure distribution and mean pressure; (b) mean pressure distributions generated at cracks number λ and μ by a unit uniform pressure at crack ν, all other cracks being frozen. (Adapted from Bažant 1994b.)

1. Although the stress transmitted across each temporarily frozen crack varies along the crack, we consider only its average, i.e., Δp_μ is constant along each crack (Fig. 13.3.3a). This approximation, which is crucial for our formulation, was introduced by Kachanov (1985, 1987a). He discovered by numerical calculations that the error is negligible except for the rare case when the distance between two crack tips is at least an order of magnitude less than their size.

2. We consider only mode I crack openings, i.e., neglect the shear modes (modes II and III). This is often justified, for instance in materials such as concrete, by a high surface roughness which prevents any significant relative slip of the microcrack faces (the mode II or III relative displacements that can occur on a macroscopic crack are mainly the result of Mode I openings of microcracks that are inclined with respect to the macrocrack).

A simple-minded kind of superposition method would be to unfreeze all the cracks, load by pressure only one crack at a time, and then superpose all the cases (Fig. 13.3.1a). In this approach, the pressure on each crack, Δp_μ, would be known. But one would still have to solve a body with many cracks.

A better kind of superposition method is that adopted by Kachanov (1985, 1987a, which was also used by Datsyshin and Savruk (1973), Chudnovsky and Kachanov (1983), Chudnovsky, Dolgopolski and Kachanov (1987), and Horii and Nemat-Nasser (1985), and, in a displacement version, was introduced by Collins (1963). In this kind of superposition, all that is needed is the solution of the given body for the case of only *one* crack, with all the other cracks considered frozen (Fig. 13.3.1b). The cost to pay for this advantage is that the pressures to be applied at the cracks are unknown in advance and must be solved. By virtue of Kachanov's approximation, we apply this kind of superposition only to the average crack pressures. The opening and the stress intensity factor of crack μ are approximately characterized by the *uniform* crack pressure $\Delta\tilde{p}_\mu$ that acts on a *single* crack within the given solid that has elastic moduli $\boldsymbol{E}$ and contains no other crack. This pressure is solved from the superposition relation:

$$\Delta\tilde{p}_\mu = \overline{\Delta p_\mu} + \sum_{\nu=1}^{N} \Lambda_{\mu\nu}\Delta\tilde{p}_\nu \qquad \mu = 1, ...N \qquad (13.3.5)$$

where the superimposed bar indicates averaging over the crack length; $\Lambda_{\mu\nu}$ are the crack influence coefficients representing the average pressure (Fig. 13.3.3b) at the frozen crack μ caused by a unit uniform pressure applied on unfrozen crack ν, with all the other cracks being frozen (Fig. 13.3.1b); and $\Lambda_{\mu\mu} = 0$ because the summation in (13.3.5) must skip $\nu = \mu$. The reason for the notation for $\Delta\tilde{p}_\mu$ with a tilde instead of an overbar is that the unknown crack pressure is uniform by definition and thus its distribution over the crack area never needs to be calculated and no averaging of pressure actually needs to be carried out.

Note that the exact solution requires considering pressures $\Delta p_\mu(x')$ and $\Delta\tilde{p}_\mu(x')$ that vary with coordinate x' along each crack. In numerical analysis, the crack must then be subdivided into many intervals. This could hardly be reflected on the macroscopic continuum level, but is doubtless unimportant at that level.

Substituting (13.3.4) into (13.3.5), we obtain

$$\Delta(\mathbf{n}_\mu \cdot \tilde{\mathbf{S}}_\mu \mathbf{n}_\mu) = \overline{\Delta(\mathbf{n}_\mu \cdot \mathbf{S}_\mu \mathbf{n}_\mu)} + \sum_{\nu=1}^{N} \Lambda_{\mu\nu} \Delta(\mathbf{n}_\nu \cdot \tilde{\mathbf{S}}_\nu \mathbf{n}_\nu) \tag{13.3.6}$$

Now we adopt a third simplifying hypothesis. In each loading step, the influence of the microcracks at macro-continuum point of coordinate vector $\boldsymbol{\xi}$ upon the microcracks at macro-continuum point of coordinate vector $\mathbf{x}$ is determined only by the dominant microcrack orientation. This orientation is normal to the unit vector $\mathbf{n}_\mu$ of the maximum principal inelastic macro-stress tensor $\Delta\tilde{S}^{(1)}$ at the location of the center of microcrack μ. We use the definition:

$$\Delta\tilde{S}^{(1)}_\mu = \Delta(\mathbf{n}_\mu \cdot \tilde{\mathbf{S}}_\mu \mathbf{n}_\mu) = [\mathbf{n}_\mu \cdot \tilde{\mathbf{S}}_\mu \mathbf{n}_\mu]_{\text{new}} - [\mathbf{n}_\mu \cdot \tilde{\mathbf{S}}_\mu \mathbf{n}_\mu]_{\text{old}} \tag{13.3.7}$$

The subscripts 'new' and 'old' denote the values at the beginning and end of the loading step, respectively. According to this hypothesis, the dominant crack orientation generally rotates from one loading step to the next. Eq. (13.3.6) may now be written as:

$$\Delta\tilde{S}^{(1)}_\mu - \sum_{\nu=1}^{N} \Lambda_{\mu\nu} \Delta\tilde{S}^{(1)}_\nu = \overline{\Delta S^{(1)}_\mu} \tag{13.3.8}$$

The values of $\Delta\tilde{\mathbf{S}}_\mu$ are graphically represented in Fig. 13.3.2 by the segment $\Delta\tilde{S} = \overline{35}$. This segment can be smaller or larger than segment $\overline{32}$.

Alternatively, one might assume $\mathbf{n}_\mu$ to approximately coincide with the direction of the maximum principal strain. Such an approximation is simpler to use in finite element programs, and it might be realistic enough, especially when the elastic strains are relatively small.

When the principal directions of the inelastic stress tensor $\mathbf{S}$ do not rotate, the increment operators Δ can of course be moved inside each product in (13.3.6), i.e., $\Delta(\mathbf{n}_\mu \cdot \tilde{\mathbf{S}}_\mu \mathbf{n}_\mu) = \mathbf{n}_\mu \cdot \Delta\tilde{\mathbf{S}}_\mu \mathbf{n}_\mu$, etc. One might wonder whether this should not be done even when these directions rotate (i.e., when $\mathbf{n}_\mu$ varies), which would correspond to crack orientations being fixed when the cracks begin to form. But according to the experience with the so-called rotating crack model, empirically verified for concrete, it is more realistic to assume that the orientation of the dominant cracks rotates with the principal direction of $\mathbf{S}$.

It might seem in the foregoing equations we should have taken only the positive part of tensor $\Delta\mathbf{S}_\mu$. But this is not necessary since the unloading criterion prevents $\Delta\mathbf{S}_\mu$ from being negative.

13.3.2 Field Equation for Nonlocal Continuum

Now comes the most difficult step. We need to determine the nonlocal macroscopic field equation which represents the continuum counterpart of (13.3.8). The homogenization theories as known are inapplicable, because they apply only to macroscopically uniform fields while the nonuniformity of the macroscopic field is the most important aspect for handling localization problems. The following simple concept has been proposed (Bažant 1994b):

The continuum field equation we seek is the equation whose discrete approximation can be written in the form of the matrix crack interaction relation (13.3.8).

This concept leads to the following field equation for the continuum approximation of microcrack interactions (Bažant 1994b):

$$\Delta\tilde{S}^{(1)}(\mathbf{x}) - \int_V \Lambda(\mathbf{x}, \boldsymbol{\xi}) \Delta\tilde{S}^{(1)}(\boldsymbol{\xi}) dV(\boldsymbol{\xi}) = \overline{\Delta S^{(1)}(\mathbf{x})} \tag{13.3.9}$$

Indeed, an approximation of the integral by a sum over the continuum variable values at the crack centers yields (13.3.8). Here we denoted $\Lambda(\mathbf{x}_\mu, \boldsymbol{\xi}_\nu) = \mathcal{E}(\Lambda_{\mu\nu})/V_c$ = crack influence function, V_c is a constant that may be interpreted roughly as the material volume per crack, and $\mathcal{E}$ is a statistical averaging operator which yields the average (moving average) over a certain appropriate neighborhood of point $\mathbf{x}$ or $\boldsymbol{\xi}$. Such statistical averaging is implied in the macro-continuum smoothing and is inevitable because, in a random crack array, the characteristics of the individual cracks must be expected to exhibit enormous random scatter.

It must be admitted that the sum in (13.3.8) is an unorthodox approximation of the integral from (13.3.9) because the values of the continuum variable are not sampled at certain predetermined points such as the chosen mesh nodes but are distributed at random, that is, at the microcrack centers. Another point to note is that (13.3.8) is only one of various possible discrete approximations of (13.3.9). Since this approximation is not unique, the uniqueness of (13.3.9) as a continuum approximation is not proven. Therefore, acceptability of (13.3.9) will also depend on computational experience (which has so far been favorable; see Ožbolt and Bažant 1996).

When (13.3.9) is approximated by finite elements, it is again converted to a matrix form similar to (13.3.8). However, the subscripts for the sum then runs over the integration points of the finite elements. This means the crack pressures (or openings) that are translated into the inelastic stress increments are only sampled at these integration points, in the sense of their density, instead of being represented individually as in (13.3.8). Obviously, such a sampling can preserve only the long-range interactions of the cracks and the averaging. The individual short-range crack interactions will be lost, but they are so random and vast in number that aspiring to represent them in any detail would be futile.

For macroscopic continuum smearing, the averaging operator $\overline{\cdots}$ over the crack length now needs reinterpretation. Because of the randomness of the microcrack distribution, the macro-continuum variable at point $\mathbf{x}$ should represent the spatial average of the effects of all the possible microcrack realizations within a neighborhood of point $\mathbf{x}$ whose size is roughly equal to the spacing ℓ of the dominant microcracks (which is, in concrete, approximately determined by the spacing of the largest aggregates, which is in turn proportional to the maximum aggregate size); hence,

$$\overline{\Delta S^{(1)}(\mathbf{x})} = \int_V \Delta S^{(1)}(\boldsymbol{\xi})\alpha(\mathbf{x}, \boldsymbol{\xi})dV(\boldsymbol{\xi}) \tag{13.3.10}$$

The weight function $\alpha(\mathbf{x}, \boldsymbol{\xi})$ is analogous to that in Eq. (13.1.1). It should vanish everywhere outside the domain of a diameter roughly equal to ℓ. For computational reasons, it seems preferable that α have a smooth bell shape. Because of randomness of the microcrack distribution, function $\alpha(\mathbf{x}, \boldsymbol{\xi})$ may be considered as rotationally symmetric (i.e., same in all directions, or isotropic).

Strictly speaking, the macroscopic averaging domain could be a line segment in the direction of the dominant microcrack (that is, normal to $\Delta S^{(1)}(\mathbf{x})$), or an elongated, roughly elliptical domain. However, averaging only along a line segment seems insufficient for preventing damage from localizing into a line, in the case of a homogeneous uniaxial tension field, and it would also be at variance with the energy release argument for nonlocality of damage presented in Bažant (1987c, 1991b).

Equation (13.3.9) represents a Fredholm integral equation (i.e., an integral equation of the second kind with a square-integrable kernel) for the unknown $\Delta\tilde{S}^{(1)}(\mathbf{x})$, which corresponds in Fig. 13.3.2 to the segment $\overline{35}$. The inelastic strain increment tensors $\Delta S^{(1)}(\mathbf{x})$ on the right-hand side, which correspond in Fig. 13.3.2 to the segment $\overline{32}$, are calculated from the strain increments using the given local constitutive law (for example, the microplane model, continuum damage theory, plastic-fracturing theory, or plasticity with yield limit degradation).

13.3.3 Some Alternative Forms and Properties of the Nonlocal Model

The solution of (13.3.9) can be written as:

$$\Delta\tilde{S}^{(1)}(\mathbf{x}) = \overline{\Delta S^{(1)}(\mathbf{x})} - \int_V K(\mathbf{x}, \boldsymbol{\xi})\overline{\Delta S^{(1)}(\boldsymbol{\xi})}dV(\boldsymbol{\xi}) \tag{13.3.11}$$

in which function $K(\mathbf{x}, \boldsymbol{\xi})$ is the resolvent of the kernel $\Lambda(\mathbf{x}, \boldsymbol{\xi})$. (This resolvent could be calculated numerically in advance of the nonlocal finite element analysis, but it would not allow a simple physical interpretation and a closed-form expression.) With the notation

$$\Psi_{\mu\nu} = \delta_{\mu\nu} - \Lambda_{\mu\nu} \tag{13.3.12}$$

where $\delta_{\mu\nu}$ = Kronecker delta, Eq. (13.3.8) can be transformed to

$$\sum_\nu \Psi_{\mu\nu}\Delta\tilde{S}_\nu^{(1)} = \overline{\Delta S_\mu^{(1)}} \tag{13.3.13}$$

The macro-continuum counterpart of this discrete matrix relation is

$$\int_V \Psi(\mathbf{x},\boldsymbol{\xi})\Delta\tilde{S}^{(1)}(\boldsymbol{\xi})dV(\boldsymbol{\xi}) = \overline{\Delta S^{(1)}(\mathbf{x})} = \int_V \Delta S^{(1)}(\boldsymbol{\xi})\alpha(\mathbf{x},\boldsymbol{\xi})dV(\boldsymbol{\xi}) \tag{13.3.14}$$

which represents an integral equation of the first kind for the unknown function $\Delta\tilde{S}^{(1)}(\boldsymbol{\xi})$. Obviously,

$$\Psi(\mathbf{x},\boldsymbol{\xi}) = \delta(\mathbf{x}-\boldsymbol{\xi}) - \Lambda(\mathbf{x},\boldsymbol{\xi}) \tag{13.3.15}$$

where $\delta(\mathbf{x}-\boldsymbol{\xi})$ = Dirac delta function in two or three dimensions; indeed, substitution of this expression into Eq. (13.3.14) yields Eq. (13.3.9).

Defining the inverse square matrix:

$$[B_{\mu\nu}] = [\Psi_{\mu\nu}]^{-1} \tag{13.3.16}$$

we may write the solution of the equation system (13.3.13) as

$$\Delta\tilde{S}_\mu^{(1)} = \sum_\nu B_{\mu\nu}\overline{\Delta S_\nu^{(1)}} = \sum_\lambda C_{\mu\lambda}\Delta S_\lambda^{(1)}, \quad C_{\mu\lambda} = \sum_\nu B_{\mu\nu}\alpha_{\nu\lambda}. \tag{13.3.17}$$

with $\alpha_{\nu\lambda} = \alpha(\mathbf{x}_\nu, \boldsymbol{\xi}_\lambda)$. The macro-continuum counterpart of the last equation is

$$\Delta\tilde{S}^{(1)}(\mathbf{x}) = \int_V B(\mathbf{x},\boldsymbol{\xi})\overline{\Delta S^{(1)}(\boldsymbol{\xi})}dV(\boldsymbol{\xi}) = \int_V C(\mathbf{x},\boldsymbol{\xi})\Delta S^{(1)}(\boldsymbol{\xi})dV(\boldsymbol{\xi}) \tag{13.3.18}$$

where $B(\mathbf{x}_\mu, \boldsymbol{\xi}_\nu) = \mathcal{E}(B_{\mu\nu})/V_c$ and $C(\mathbf{x},\boldsymbol{\xi}) = \int_V B(\mathbf{x},\boldsymbol{\xi})\alpha(\boldsymbol{\xi},\mathbf{x})dV(\boldsymbol{\xi})$. The kernel $B(\mathbf{x},\boldsymbol{\xi})$ represents the resolvent of the kernel $\Psi(\mathbf{x},\boldsymbol{\xi})$ of (13.3.14). Furthermore,

$$B(\mathbf{x},\boldsymbol{\xi}) = \delta(\mathbf{x}-\boldsymbol{\xi}) - K(\mathbf{x},\boldsymbol{\xi}) \tag{13.3.19}$$

because substitution of this equation into Eq. (13.3.18) furnishes Eq. (13.3.11). With (13.3.18) we have reduced the nonlocal formulation to a similar form as (13.3.3) for the previous nonlocal damage formulations (Pijaudier-Cabot and Bažant 1987; Bažant and Pijaudier-Cabot 1989; Bažant and Ožbolt 1990, 1992). However, the presence of the Dirac delta function in the last equation makes Eq. (13.3.18) inconvenient for computations. Aside from that, it seems inconvenient to calculate in finite element codes function $B(\mathbf{x},\boldsymbol{\xi})$. Another difference is that the weight function (i.e., the kernel) is anisotropic (and, in the present simplification, associated solely with the principal inelastic stresses).

Note also that if we set $\Lambda(\mathbf{x},\boldsymbol{\xi}) = 0$, the present model would become identical to the aforementioned previous nonlocal damage model. But this would not be realistic. The directional and tensorial interactions characterized by $\Lambda(\mathbf{x},\boldsymbol{\xi})$ appear to be essential.

Because the nonlocal integral in (13.3.21) is additive to the local stress $\Delta\mathbf{S}$, the present nonlocal model can be imagined as an overlay of two solids that are forced to have equal displacements at all points: (i) the given solid with all the damage due to cracks, but local behavior (no crack interactions); and (ii) an overlaid solid that describes only crack interactions. The nonlocal stress $\Delta\tilde{\mathbf{S}}$ represents the sum of the stresses from both solids. This is the stress that is to be used in formulating the differential equilibrium equations for the solid.

For the sake of simplicity, we have so far assumed that the influence of point $\boldsymbol{\xi}$ on point $\mathbf{x}$ depends only on the orientation of the maximum principal inelastic stress at $\boldsymbol{\xi}$. Since at $\boldsymbol{\xi}$ there might be cracks normal to all the three principal stresses (denoted now by superscripts $i = 1, 2, 3$ in parentheses), it might be more realistic to consider that each of them separately influences point $\mathbf{x}$. In that case, Eqs. (13.3.8) and (13.3.9) can be generalized as follows:

$$\Delta\tilde{S}_\mu^{(i)} - \sum_{\nu=1}^{N}\sum_{j=1}^{3}\Lambda_{\mu\nu}^{(ij)}\Delta\tilde{S}_\nu^{(j)} = \overline{\Delta S_\mu^{(i)}} \tag{13.3.20}$$

$$\Delta\tilde{S}^{(i)}(\mathbf{x}) - \int_V \sum_{j=1}^{3} \Lambda^{(ij)}(\mathbf{x},\boldsymbol{\xi})\Delta\tilde{S}^{(j)}(\boldsymbol{\xi})dV(\boldsymbol{\xi}) = \overline{\Delta S^{(i)}(\mathbf{x})} \qquad (i = 1, 2, 3) \quad (13.3.21)$$

Similar generalizations can be made in the subsequent equations, too. Note that when the body is infinite, all the present summations or integrations are assumed to follow a special path labeled by $\odot$, which will be defined in the next section.

The heterogeneity of the material, such as the aggregate in concrete, is not specifically taken into account in our equations. Although the heterogeneity obviously must influence the nonlocal properties (e.g., Pijaudier-Cabot and Bažant 1991), this influence is probably secondary to that of microcracking. The reason is that the prepeak (hardening) inelastic behavior, in which microcracking is much less pronounced than after the peak while the heterogeneity is the same, can be adequately described by a local continuum. The main effect of heterogeneity (such as the aggregates in concrete, or grains in ceramics) is indirect; it determines the spacing, orientations, and configurations of the microcracks.

13.3.4 Admissibility of Uniform Inelastic Stress Fields

In the previous nonlocal formulations, the requirement that a field of uniform inelastic stress and damage must represent at least one possible solution led to the aforementioned normalizing condition for the weight function α. Similarly, we must now require that the homogeneous stress field $\Delta\tilde{S}^{(1)} = \overline{\Delta S^{(1)}}$ satisfy (13.3.8) and (13.3.9) identically. This yields the conditions that the integral of $\Lambda(\mathbf{x},\boldsymbol{\xi})$ or the sum of $\Lambda_{\mu\nu}$ over an infinite body vanish. However, the asymptotic behavior of $\Lambda(\mathbf{x},\boldsymbol{\xi})$ for $r \to \infty$ which will be discussed later causes this integral or sum to be divergent. Therefore, the conditions must be imposed in a special form—the integral in polar coordinates is required to vanish only for a special path, labeled by $\odot$, in which the angular integration is completed before the limit $r \to \infty$ is calculated, that is,

$$\int_V^{\odot} \Lambda(\mathbf{x},\boldsymbol{\xi})dV(\boldsymbol{\xi}) = \lim_{R\to\infty} \int_0^R \left(\int_0^{2\pi} \Lambda(\mathbf{x},\boldsymbol{\xi})r d\phi \right) dr = 0 \text{ (for 2D)}$$

$$\int_V^{\odot} \Lambda(\mathbf{x},\boldsymbol{\xi})dV(\boldsymbol{\xi}) = \lim_{R\to\infty} \int_0^R \left(\int_0^{2\pi} \int_0^{\pi} \Lambda(\mathbf{x},\boldsymbol{\xi})r^2 \sin\theta d\theta d\phi \right) dr = 0 \text{ (for 3D)} \quad (13.3.22)$$

r, ϕ are polar coordinates; r, θ, ϕ are spherical coordinates. Furthermore, again labeling by $\odot$ a similar summation path (or sequence) over all the cracks ν in an infinite body, the following discrete condition needs to also be imposed:

$$\sum_{\nu}^{\odot} \Lambda_{\mu\nu} = 0 \quad (13.3.23)$$

This condition applies only to an array of infinitely many microcracks that are, on the macroscale, perfectly random and distributed statistically uniformly over an infinite body (or are periodic). By the same reasoning, for an infinite body we must also have

$$\int_V^{\odot} K(\mathbf{x},\boldsymbol{\xi})dV(\boldsymbol{\xi}) = 0 \quad (13.3.24)$$

$$\int_V^{\odot} \Psi(\mathbf{x},\boldsymbol{\xi})dV(\boldsymbol{\xi}) = \int_V^{\odot} B(\mathbf{x},\boldsymbol{\xi})dV(\boldsymbol{\xi}) = \int_V^{\odot} C(\mathbf{x},\boldsymbol{\xi})dV(\boldsymbol{\xi}) = 1; \quad (13.3.25)$$

and in the discrete form

$$\sum_{\nu}^{\odot} \Psi_{\mu\nu} = \sum_{\nu}^{\odot} \alpha_{\mu\nu} = \sum_{\nu}^{\odot} B_{\mu\nu} = \sum_{\nu}^{\odot} C_{\mu\nu} = 1 \quad (13.3.26)$$

For integration paths in which the radial integration up to $r \to \infty$ is carried out before the angular integration, the foregoing integrals and sums are divergent.

13.3.5 Gauss-Seidel Iteration Applied to Nonlocal Averaging

For the purpose of finite element analysis, we will now assume that subscripts μ and ν label the numerical integration points of finite elements, rather than the individual microcracks. This means that the microcracks are represented by their mean statistical characteristics sampled only at the numerical integration points.

In finite element programs, nonlinearity is typically handled by iterations of the loading steps. Let us, therefore, examine the iterative solution of (13.3.8) or (13.3.13), which represents a system of N linear algebraic equations for N unknowns $\Delta\tilde{S}_\mu^{(1)}$ if $\Delta S_\mu^{(1)}$ are given. The matrix of $\Psi_{\mu\nu}$ is, in general, nonsymmetric (because the influence of a large crack on a small crack is not the same as the influence of a small crack on a large crack). This nonsymmetry seems disturbing until one realizes that this is so only because of our choice of variables $\Delta\tilde{S}_\nu^{(1)}$ and $\overline{\Delta S_\mu^{(1)}}$, which do not represent thermodynamically conjugate pairs of generalized forces and generalized displacements. If $\overline{\Delta S_\mu^{(1)}}$ were expressed in terms of the crack opening volumes, then the matrix of the equation system resulting from (13.3.8) or (13.3.13) would have to be symmetric (because of Betti's theorem) and also positive definite (if the body is stable). These are the attributes mathematically required for convergence of the iterative solution by Gauss-Seidel method (e.g., Rektorys 1969; Collatz 1960; Korn and Korn 1968; Varga 1962; Fox 1965; Strang 1980). Aside from that, convergence of the iterative solution of (13.3.8) or (13.3.13) must also be expected on physical grounds (because it is mechanically equivalent to the relaxation method, which always converges for stable elastic systems).

In the r-th iteration, the new, improved values of the unknowns, labeled by superscripts $[r+1]$, are calculated from the previous values, labeled by superscript $[r]$, either according to the recursive relations:

$$\Delta\tilde{p}_\mu^{[r+1]} = \overline{\Delta p_\mu} + \sum_{\nu=1}^{N} \Lambda_{\mu\nu}\Delta\tilde{p}_\nu^{[r]} \tag{13.3.27}$$

$$\Delta\tilde{S}_\mu^{(1)[r+1]} = \overline{\Delta S_\mu^{(1)}} + \sum_{\nu=1}^{N} \Lambda_{\mu\nu}\Delta\tilde{S}_\nu^{(1)[r]} \qquad (\mu = 1, ..., N) \tag{13.3.28}$$

or according to the recursive relations:

$$\Delta\tilde{p}_\mu^{[r+1]} = \overline{\Delta p_\mu} + \sum_{\nu=1}^{\mu-1} \Lambda_{\mu\nu}\Delta\tilde{p}_\nu^{[r+1]} + \sum_{\nu=\mu+1}^{N} \Lambda_{\mu\nu}\Delta\tilde{p}_\nu^{[r]} \qquad (\mu = 1, ..., N) \tag{13.3.29}$$

$$\Delta\tilde{S}_\mu^{(1)[r+1]} = \overline{\Delta S_\mu^{(1)}} + \sum_{\nu=1}^{\mu-1} \Lambda_{\mu\nu}\Delta\tilde{S}_\nu^{(1)[r+1]} + \sum_{\nu=\mu+1}^{N} \Lambda_{\mu\nu}\Delta\tilde{S}_\nu^{(1)[r]} \quad (\mu = 1, ..., N) \tag{13.3.30}$$

Equation (13.3.28), also known as the Gauss method or Jacobi method, is less, but normally only slightly less efficient than (13.3.30), in which the latest approximations are always used. The values of $\Delta S_\mu^{(1)}$ may be used as the initial values of $\Delta\tilde{S}_\mu^{(1)[r]}$ in the first iteration.

It is possible to derive Eqs. (13.3.27) and (13.3.28) more directly, rather than from (13.3.5). To this end, we note that the sequence of iterations is identical to a solution by the relaxation method in which one crack after another is relaxed (i.e., its pressure reduced to zero) while all the other cracks are frozen (so in each relaxation step, one has a problem with one crack only), as illustrated in Fig. 13.3.1b. Each relaxation produces pressure on the previously relaxed cracks. After relaxing all the cracks one by one, the cycle through all the cracks is repeated again and again. This kind of relaxation is known in mechanics to converge in general (this was numerically demonstrated for a system of cracks and inclusions by Pijaudier-Cabot and Bažant 1991). The solution to which the relaxation process converges is obviously that defined by Eq. (13.3.8). (Note also that this relaxation argument in fact represents a simple way to prove the superposition equation (13.3.5).)

For structural engineers, it is interesting to note the similarity with the Cross method (moment distribution method) for elastic frames. Relaxing the pressure at one crack while all the other microcracks are frozen (glued) is analogous to relaxing one joint in a frame while all the other joints are held fixed.

Repeating this for each joint, and then repeating the cycles of such relaxations of all the joints, eventually converges to the exact solution of the frame.

The macro-continuum counterpart of the Gauss-Seidel iterative method, which converges to the solution of the Fredholm integral equation (13.3.9), is analogous to (13.3.28) and is given by the following relation for successive approximations (iterations):

$$\Delta\tilde{S}^{(1)[r+1]}(\mathbf{x}) = \overline{\Delta S^{(1)}(\mathbf{x})} + \int_V \Lambda(\mathbf{x},\boldsymbol{\xi})\Delta\tilde{S}^{(1)[r]}(\boldsymbol{\xi})\, dV(\boldsymbol{\xi}) \tag{13.3.31}$$

The discrete approximation of the last relation is the equation that ought to be used in finite element programs with iterations in each step. We see that the form of averaging is different from that assumed in the phenomenological models we described. There are now two additive spatial integrals: one for close-range averaging of the inelastic stresses from the local constitutive relation, and one for long-range crack interactions based on the latest iterates of the inelastic stresses.

In programming, the old iterates need not be stored in the computer memory. So the superscripts $[r]$ and $[r+1]$ may be dropped and equations (13.3.9) and (13.3.31) may be replaced by the following assignment statements:

$$\Delta\tilde{S}^{(1)}_{\mu} \leftarrow \overline{\Delta S^{(1)}_{\mu}} + \sum_{\nu=1}^{N} \Lambda_{\mu\nu}\Delta\tilde{S}^{(1)}_{\nu} \qquad (\mu = 1, 2, ...N) \tag{13.3.32}$$

$$\Delta\tilde{S}^{(1)}(\mathbf{x}) \leftarrow \overline{\Delta S^{(1)}(\mathbf{x})} + \int_V \Lambda(\mathbf{x},\boldsymbol{\xi})\Delta\tilde{S}^{(1)}(\boldsymbol{\xi})\, dV(\boldsymbol{\xi}) \tag{13.3.33}$$

A strict implementation of Gauss-Seidel iterations suggests programming each iteration loop for Eq. (13.3.32) to be contained within another loop for the iterations of the loading step in which the displacement and strain increments in the structure are solved. However, it is computationally more efficient to use one common iteration loop serving both purposes. Then, of course, the iteration solution is not exactly the Gauss-Seidel method because the strains are also being updated during each iteration. There is already some computational experience (Ožbolt and Bažant 1996) showing that convergence is still achieved.

The common iteration loop has the advantage that it permits using the explicit load-step algorithm for structural analysis. In each loading step of this algorithm, one evaluates in each iteration at each integration point the elastic stress increments $\boldsymbol{E}\Delta\boldsymbol{\varepsilon}$ and the local inelastic stress increments $\Delta\mathbf{S}$ from fixed strains $\Delta\boldsymbol{\varepsilon}$; then one uses (13.3.32) to calculate from $\Delta\mathbf{S}$ the nonlocal inelastic stress increments $\Delta\tilde{\mathbf{S}}$ for all the integration points, solves new nodal displacements by elastic structural analysis, and, finally, updates the strains.

13.3.6 Statistical Determination of Crack Influence Function

The basic characteristic of the new formulation is the crack influence function Λ, whose rate of decay is determined by a certain characteristic length ℓ. This function represents the stress field due to pressurizing a single crack in the given elastic structure, all other cracks being absent. In practice, the structure is always finite, and thus the values of $\Lambda_{\mu\nu}$ should, in principle, be calculated taking into account the geometry of the structure. However, the crack is often very small compared to the dimensions of the structure. Then the present formulation has the advantage that one can use, as a very good approximation, the stress field for a single crack in an infinite body, which is well known and calculated easily. This is, of course, not possible for cracks very near the boundary of the structure.

The cracks in structures are distributed randomly and their number is vast. Thus, on the macro-continuum level, function $\Lambda_{\mu\nu}$ cannot characterize the stress fields of the individual cracks. Rather, it should characterize the stress field of a representative crack obtained by a suitable statistical averaging of the random situation on the microstructure level.

A method of rigorous mathematical formulation of the macroscopic continuum crack influence function Λ was briefly proposed in the addendum to Bažant (1994b) and was developed in detail in Bažant and Jirásek (1994a). This method will now be described.

The crack that is pressurized by unit pressure, as specified in the definition of Λ, will be called the source crack, and the frozen crack in the structure on which the influence is to be found will be called

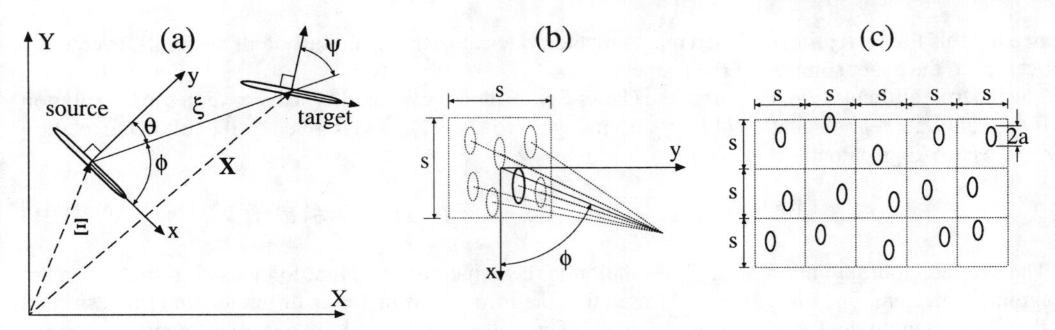

Figure 13.3.4 Definition of relative crack locations: (a) general position of the source crack and target crack; (b) various possible random locations of the source crack influencing a target crack; (c) dominant cracks appearing in regions of size s that determine their typical spacing. (Adapted from Bažant and Jirásek 1994a.)

the target crack (Fig. 13.3.4a). For the purpose of calculations, the target crack is, of course, closed and glued, as if it did not exist, and the stresses transmitted across the target crack are calculated assuming the body to be continuous. In the following, the global axes will be denoted with capital letters and the position vectors of the target and source cracks by $\mathbf{X}$ and Ξ (Fig. 13.3.4a). We take axes (x, y) to be, respectively, parallel and perpendicular to the source crack with origin at its center, and call ξ the vector from the source to the target crack (Fig. 13.3.4a). Then function $\Lambda(\mathbf{0}, \boldsymbol{\xi})$ represents the influence of a source crack centered at $\mathbf{x} = \mathbf{0}$ on a target crack centered at $\boldsymbol{\xi}$ (Fig. 13.3.4a).

At the given macro-continuum point, there may or may not be a crack in the microstructure. Function Λ corresponding to that point must reflect the smeared statistical properties of all the possible microcracks occurring near that point. To do this, we must idealize the random crack arrangements in some suitable manner.

We will suppose that the center of the source crack can occur randomly anywhere within a square of size s centered at point $\mathbf{x} = \mathbf{0}$; see Fig. 13.3.4b,c, where various possible cracks are shown by the dashed curves, but only one of these, the crack showed by the solid lines, is actually realized. The value of s is imagined to represent the typical spacing of the dominant cracks. In a material such as concrete, approximately $s \approx m d_a$ where d_a = spacing of the largest aggregate pieces and m = coefficient larger than 1 but close to 1 (m would equal d_a if the aggregates were arranged at the densest ideal packing and if there were no mortar layers within the contact zones).

To simplify the statistical structure of the system of dominant cracks, one may imagine the material to be subdivided by a square mesh of size s as shown in Fig. 13.3.4c, with one and only one crack center occurring within each square of the mesh. This is, of course, a simplification of reality because the underlying square mesh introduces a certain directional bias (as is well known from finite element analysis of fracture). It would be more realistic to assume that the possible zone of occurrence of the center of each crack is not a square but has a random shape and area about $s \times s$, and that all these areas are randomly arranged. But this would be too difficult for statistical purposes, and probably unimportant with respect to the other simplifications of the model.

Let us now center coordinates x and y in the center of the square $s \times s$, as shown in Fig. 13.3.4b, and consider the influence of a source crack within this square on a target crack at coordinates $\boldsymbol{\xi} \equiv (\xi, \eta)$. The macroscopic crack influence function should describe the influence of any possible source crack within the given square in the average, smeared macroscopic sense. Therefore, $\Lambda(\mathbf{0}, \boldsymbol{\xi})$ is defined as the mathematical expectation $\mathcal{E}$ with regard to all the possible random realizations of the source crack center within the given square $s \times s$, that is

$$\Lambda(0, \boldsymbol{\xi}) = \mathcal{E}\left[\sigma^{(1)}(\xi - x, \eta - y)\right] \tag{13.3.34}$$

The vector $(\xi - x, \eta - y) = \mathbf{r}$ = vector from the center $\mathbf{x} \equiv (x, y)$ of a source crack to the center

$\boldsymbol{\xi} \equiv (\xi, \eta)$ of the target crack. In detail,

$$\Lambda(0, \boldsymbol{\xi}) = \frac{1}{s^2} \int_{-s/2}^{s/2} \int_{-s/2}^{s/2} w(x, y) \sigma^{(1)} (\xi - x, \eta - y) \, dx dy \tag{13.3.35}$$

Here $\sigma^{(1)}$ is the field of the maximum principal stresses caused by applying a unit pressure on the faces of the source crack, and the integrals represent the statistical averaging over the square $s \times s$ (Fig. 13.3.4b). Certain specified weights $w(x, y)$ have been introduced for this averaging. At first, one might think that uniform weights w might be appropriate, but that would not be realistic near the boundaries of the square because a crack cannot intersect a crack centered in the adjacent square, and, in practice, cannot even lie too close. Rigorously, one would have to consider the joint probability of the occurrences of the crack center locations in the adjacent squares, but this would be too complicated. We prefer to simply reduce the probability of occurrence of the source crack as the boundary of the square is approached. For numerical computations we choose a bell-shaped function in both the x and y directions, given as

$$w(x, y) = w_0 \left[1 - \left(\frac{2x}{s} \right)^2 \right]^2 \left[1 - \left(\frac{2y}{s} \right)^2 \right]^2, \qquad w_0 = \frac{225}{64} \tag{13.3.36}$$

for $x \leq s/2, y \leq s/2$, and $w(x, y) = 0$ otherwise; constant w_0 is selected so the integral of $w(x, y)$ over the square $s \times s$ be equal to 1. It may be added that there is also a practical reason for introducing this weight function. If the weights were uniform over the square, function Λ would not have a smooth shape, which would be inconvenient and probably also unrealistic for a continuum model.

The stress field $\sigma^{(1)}$ to be substituted into (13.3.36) is given for two dimensions by the well-known Westergaard's solution (see Chapter 4). However, the integral in Eq. (13.3.35) is difficult to evaluate analytically, and it is better to use numerical integration to obtain Λ.

The asymptotic properties of function Λ for large r can nevertheless be determined easily (Bažant 1992b, 1994b) by considering the lines of influence from various possible source cracks to the given target crack as shown in Fig. 13.3.4b. If the target crack is very far from the square in which the source crack is centered, all the possible rays of influence are nearly equally long and come from nearly the same direction. Therefore, the integral in Eq. (13.3.35) should exactly preserve the long range asymptotic field $\sigma^{(1)}$.

13.3.7 Crack Influence Function in Two Dimensions

Consider now a crack in an infinite solid, subjected to uniform pressure σ (Fig. 13.3.3b). According to Westergaard's solution (Chapter 4) the stress distribution can be written as

$$\sigma_{xx} = \mathrm{Re} Z - y \, \mathrm{Im} \, Z' - \sigma, \qquad \sigma_{yy} = \mathrm{Re} Z + y \, \mathrm{Im} \, Z' - \sigma, \qquad \tau_{xy} = -y \, \mathrm{Re} \, Z' \tag{13.3.37}$$

in which σ_{xx} and σ_{yy} are the normal stresses, τ_{xy} is the shear stress, and

$$Z = \sigma z \, (z^2 - a^2)^{-1/2}; \qquad z = r \, \mathrm{e}^{\mathrm{i}\phi} \tag{13.3.38}$$

Here $2a$ = crack length, $\mathrm{i}^2 = -1$, $Z' = dZ/dz$, and r, ϕ = polar coordinates with origin at the crack center and angle ϕ measured from the crack direction x. For $r \gg a$ we have the approximation:

$$Z = \sigma \left(1 - \frac{a^2}{r^2 \, \mathrm{e}^{2\mathrm{i}\phi}} \right)^{-1/2} = \sigma \left(1 + \frac{a^2}{2r^2} \mathrm{e}^{-2\mathrm{i}\phi} + ... \right) = \sigma \left(1 + \frac{a^2}{2z^2} + ... \right) \tag{13.3.39}$$

From this, we calculate

$$\mathrm{Re} \, Z = \sigma \left(1 + \frac{a^2}{2r^2} \cos 2\phi + ... \right), \qquad Z' = \sigma(-a^2 z^{-3} + ...)$$

$$y \, \mathrm{Im} \, Z' = \sigma a^2 r \, \sin\phi \, \mathrm{Im} \, (-r^{-3} \mathrm{e}^{-3\mathrm{i}\phi}) = -\sigma a^2 r^{-2} \sin\phi \, (-\sin 3\phi) \tag{13.3.40}$$

Substituting this into (13.3.37) and using the formulas for products of trigonometric functions, we get the following simple result for the long-range ($r \gg a$) asymptotic field (Bažant 1992b, 1994b):

$$\sigma_{xx} = \sigma k(r)\,\frac{\cos 4\phi}{2}, \qquad \sigma_{yy} = \sigma k(r)\left(\cos 2\phi - \frac{\cos 4\phi}{2}\right)$$

$$\tau_{xy} = \sigma k(r)\,\frac{\sin 4\phi - \sin 2\phi}{2} \tag{13.3.41}$$

where $k(r) = a^2/r^2$. Subscripts x, y refer to cartesian coordinates with origin at point $\boldsymbol{\xi}$ coinciding with the crack center and axis y normal to the crack; σ_{xx} and σ_{yy} are the normal stresses, τ_{xy} is the shear stress; and ϕ are polar coordinates with origin at the crack center, with the polar angle ϕ measured from axis x. The principal stresses $\sigma^{(1)}$ and $\sigma^{(2)}$ and the first principal stress direction $\phi^{(1)}$ are given by:

$$\sigma^{(1)} = \sigma k(r)\left(\frac{\cos 2\phi}{2} + \sin\phi\right), \qquad \sigma^{(2)} = \sigma k(r)\left(\frac{\cos 2\phi}{2} - \sin\phi\right) \tag{13.3.42}$$

$$\tan 2\phi^{(1)} = -\cot 3\phi$$

The foregoing expressions describe the long-range form of function $\Lambda(\mathbf{x}, \boldsymbol{\xi})$. It does not matter that they have a r^{-2} singularity at the crack center, because they are invalid for not too large r. Note that the average of each expression over the circle r = constant is zero, which is, in fact, a necessary property.

By virtue of considering only principal stress directions, $\Lambda(\mathbf{x}, \boldsymbol{\xi})$ is a scalar. All the information on the relative crack orientations is embedded in the values of this function. The principal stress direction at point $\boldsymbol{\xi}$, which can be regarded as the dominant crack direction at that location (Fig. 13.3.4a), is all the directional information needed to calculate the stress components at point $\mathbf{x}$; see (40), in which $r = \|\mathbf{x}-\boldsymbol{\xi}\|$ = distance between points $\mathbf{x}$ and $\boldsymbol{\xi}$. The value of $\Lambda(\mathbf{x}, \boldsymbol{\xi})$, needed for (13.3.31) or (9), may be determined as the projection of the stress tensor at point $\mathbf{x}$ onto the principal inelastic stress direction at that point. According to Mohr circle: $2\Lambda(\mathbf{x}, \boldsymbol{\xi}) = (\sigma_{xx}+\sigma_{yy})+(\sigma_{xx}-\sigma_{yy})\cos 2(\psi-\theta)-2\tau_{xy}\sin 2(\psi-\theta)$ in which $\theta, \psi =$ angles of the principal inelastic stress directions at points $\boldsymbol{\xi}$, $\mathbf{x}$, respectively, with the line connecting these two points (i.e., with the vector $\mathbf{x} - \boldsymbol{\xi}$). Substituting here for σ_{xx}, etc., the expressions from (13.3.41), one obtains a trigonometric expression which can be brought by trigonometric transformations (Planas 1992) to the form:

$$\Lambda(\mathbf{x}, \boldsymbol{\xi}) = -\frac{k(r)}{2\ell^2}\left[\cos 2\theta + \cos 2\psi + \cos 2(\theta + \psi)\right] \tag{13.3.43}$$

where $\theta = 90° - \phi$. Note that the function $\Lambda(\mathbf{x}, \boldsymbol{\xi})$ is symmetric. This is, of course, a necessary consequence of the fact that the body is elastic.

Two properties contrasting with the classical nonlocal formulations explained before should be noted: (1) the crack influence function is not isotropic but depends on the polar angle (i.e., is anisotropic), and (2) it exhibits a shielding sector and an amplification sector. We may define the amplification sector as the sector in which σ_{yy} (the stress component normal to the crack plane) is positive, and the shielding sector as the sector in which σ_{yy} is negative. The amplification sector $\sigma_{yy} \geq 0$, according to (13.3.41), is the sector $\phi \leq \phi_b$ where

$$\phi_b = 55.740° \tag{13.3.44}$$

The sector in which the volumetric stress $\sigma_{xx} + \sigma_{yy}$ (first stress invariant) is positive is $\phi \leq 45°$. The sector in which $\sigma_{xx} \geq 0$ is $\phi \leq 22.5°$ and $\phi \geq 67.5°$. The sector in which $2\,\tau_{\max} = \sigma_{xx} - \sigma_{yy} \geq 0$ is $\phi \leq 45°$. The maximum principal stress $\sigma^{(1)}$ is positive for all angles ϕ, and the minimum principal stress $\sigma^{(2)}$ is positive for $\phi < 21.471°$.

The consequence of the anisotropic nature of the crack influence function is that interactions between adjacent cracks depend on the direction of damage propagation with respect to the orientation of the maximum principal inelastic stress. In a cracking band that is macroscopically of mode I (Fig. 13.3.5a), propagating in the dominant direction of the microcracks, the microcracks assist each other in growing because they lie in each other's amplification sectors. In a cracking band that is macroscopically of mode II (Fig. 13.3.5b), the microcracks are mutually in the transition between their amplification and shielding sectors, and thus interact little. Under compression, a band of axial splitting cracks may propagate

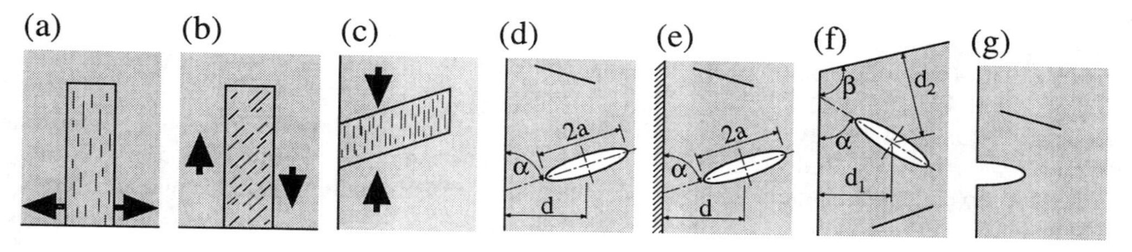

Figure 13.3.5 Crack bands and cracks near boundary (from Bažant 1994b).

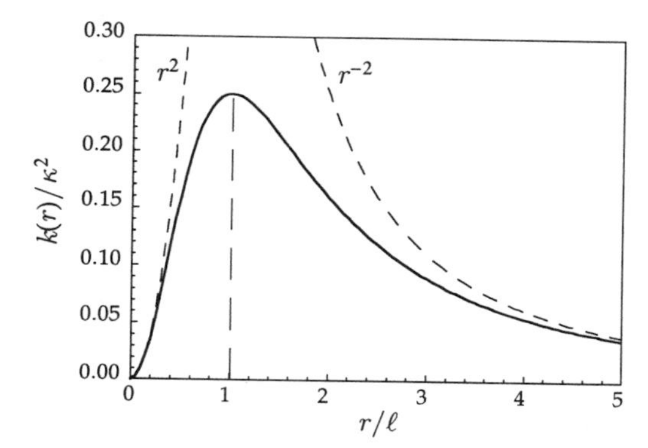

Figure 13.3.6 Radial dependence of Λ.

sideways (Fig. 13.3.5c), and in that case, the microcracks inhibit each other's growth because they lie in each other's shielding sectors. Differences in the kind of interaction may explain why good fitting of test data with the previous nonlocal microplane model required using a different material characteristic length for different types of problems (e.g., mode I fracture specimens vs. diagonal shear failure of reinforced beam).

For small r, function $\Lambda(\mathbf{x}, \boldsymbol{\xi})$ is a result of interactions in all directions. As the first approximation, these interactions may be assumed to cancel each other. Accordingly, we replace function $k(r) = a^2/r^2$ by a simple function of the same asymptotic properties for $r \to \infty$ which does not have a singularity at $r = 0$ and for $r \to 0$ approaches 0 with a horizontal tangent:

$$k(r) = \left(\frac{\kappa \ell r}{r^2 + \ell^2} \right)^2 \tag{13.3.45}$$

Here κ is an empirical constant such that $\kappa\ell$ roughly represents the average or effective crack size a for the macro-continuum; ℓ is a certain constant representing what may be called the characteristic distance of crack interactions (it represents the radial distance to the peak in Fig. 13.3.6). This length may be identified with what has been called the characteristic length of the nonlocal continuum. It reflects the dominant spacing of the microcracks, which in turn is determined by the size and spacing of the dominant inhomogeneities such as aggregates in concrete.

The foregoing expressions give the crack influence function Λ_∞ which is exact asymptotically for $r \to \infty$ but is only a crude approximation for small r. It is now convenient to represent the complete crack influence function Λ in the form:

$$\Lambda(\mathbf{0}, \boldsymbol{\xi}) = \Lambda_\infty(\xi, \eta) + \Lambda_1(\xi, \eta) \tag{13.3.46}$$

where Λ_1 represents a difference that is decaying to infinity faster (i.e., as a higher power of r) than Λ_∞ and can, therefore, be neglected for sufficient distances r from the center of the source crack.

The complete function Λ was determined by numerical integration of Eq. (13.3.35) using a dense square mesh; see 13.3.7a (Bažant and Jirásek 1994a). The target crack was considered parallel to the source crack, and $a/s = 0.25$. The asymptotically correct analytical expression for the crack influence function from Eq. (13.3.45) is plotted in Fig. 13.3.7b. After its subtraction from the Λ_∞ values, one obtains the plot of the difference Λ_1 shown in Fig. 13.3.7c. A table of numerical values of Λ_1 was reported in Bažant and Jirásek (1994a)).

Function $\Lambda_1(x, y)$ obviously depends on the relative crack size a/s. However, it has been found that it depends on a/s only very little when $a/s \geq 0.25$. For smaller a/s, the crack interactions are probably unimportant. So perhaps a single crack influence function expression could be used for all the cases.

A statistical definition of Λ in three dimensions that is analogous to Eq. (13.3.35) can obviously be written, too.

13.3.8 Crack Influence Function in Three Dimensions

The case of three dimensions (3D) is not difficult when the cracks are penny-shaped (i.e., circular) and the boundary is remote. The stresses around such cracks have traditionally been expressed as integrals of Bessel functions (Sneddon and Lowengrub 1969; Kassir and Sih 1975), which are however cumbersome for calculations. Recently, though, Fabrikant (1990) ingeniously derived the following closed-form expressions:

$$\sigma_{xx} = \frac{\sigma_1 + \mathrm{Re}\,\sigma_2}{2}, \qquad \sigma_{yy} = \frac{\sigma_1 - \mathrm{Re}\,\sigma_2}{2}, \qquad \tau_{xy} = \frac{\mathrm{Im}\,\sigma_z}{2}$$
$$\tau_{xz} = \mathrm{Re}\,\tau_z, \qquad \tau_{yz} = \mathrm{Im}\,\tau_z \tag{13.3.47}$$

in which

$$\sigma_z = \frac{2\sigma}{\pi}(B - D), \qquad \sigma_1 = \frac{2\sigma}{\pi}[(1 + 2\nu)B + D]$$
$$\sigma_2 = \mathrm{e}^{2\mathrm{i}\phi}\frac{2\sigma}{\pi}\frac{a l_1^2 l_3}{l_2^2 l_4^2}\left(1 - 2\nu + \frac{z^2[a^2(6l_2^2 - 2l_1^2 + \rho^2) - 5l_2^4]}{l_4^4 l_3^2}\right)$$
$$\tau_z = -\mathrm{e}^{\mathrm{i}\phi}\frac{2\sigma}{\pi}\frac{z l_1[a^2(4l_2^2 - 5\rho^2) + l_1^4]l_3}{l_2 l_4^6} \tag{13.3.48}$$
$$B = \frac{a l_3}{l_4^2} - \arcsin\frac{a}{l_2}, \qquad D = \frac{a z^2[l_1^4 + a^2(2a^2 + 2z^2 - 3\rho^2)]}{l_4^6 l_3}$$
$$l_1 = \frac{L_2 - L_1}{2}, \qquad l_2 = \frac{L_2 + L_1}{2}, \qquad l_3 = \sqrt{l_2^2 - a^2}, \qquad l_4 = \sqrt{l_2^2 - l_1^2}$$
$$L_1 = \sqrt{(a - \rho)^2 + z^2}, \qquad L_2 = \sqrt{(a + \rho)^2 + z^2}$$

in which a = crack radius; r, θ, ϕ are the spherical coordinates (Fig. 13.3.8) attached to cartesian coordinates x, y, z at point $\boldsymbol{\xi}$, with angle θ measured from axis z which is normal to the crack at point $\boldsymbol{\xi}$; r = distance between points $\mathbf{x}$ and $\boldsymbol{\xi}$; ρ, ϕ, z are the cylindrical coordinates with origin at the crack center; and ρ, ϕ are polar coordinates in the crack plane, angle ϕ being measured from axis x.

The long-range asymptotic form of the foregoing stress field has been derived (Bažant 1994b). The derivation is easy if one notes that, for large r, $L_1 \approx r - a\sin\theta, L_2 \approx r + a\sin\theta$ (see the meaning of L_1 and L_2 in Fig. 13.3.8a), $l_1 \approx a\sin\theta, l_2 \approx r$ and, for $r \gg a$, $\arcsin(a/l_2) \approx [1 + (a^2/6l_2^2)]a/l_2$, $\sqrt{l_2^2 - a^2} \approx r[1 - (a^2/2r^2)]$. The result is the following long-range asymptotic field:

$$\sigma_{\rho\rho} = \sigma k(r)\left[(1 + 2\nu)\left(\sin^2\theta - \frac{2}{3}\right) + (1 - 2\nu - 5\cos^2\theta)\sin^2\theta\right]$$
$$\sigma_{\phi\phi} = \sigma k(r)\left[(1 + 2\nu)\left(\sin^2\theta - \frac{2}{3}\right) - (1 - 2\nu - 5\cos^2\theta)\sin^2\theta\right]$$
$$\sigma_{zz} = \sigma k(r)\left(\sin^2\theta - \frac{2}{3}\right) \tag{13.3.49}$$
$$\sigma_{\rho z} = -\,\sigma k(r)\sin 2\theta\,(4 - 5\sin^2\theta), \qquad \sigma_{\rho\phi} = \sigma_{\phi z} = 0$$

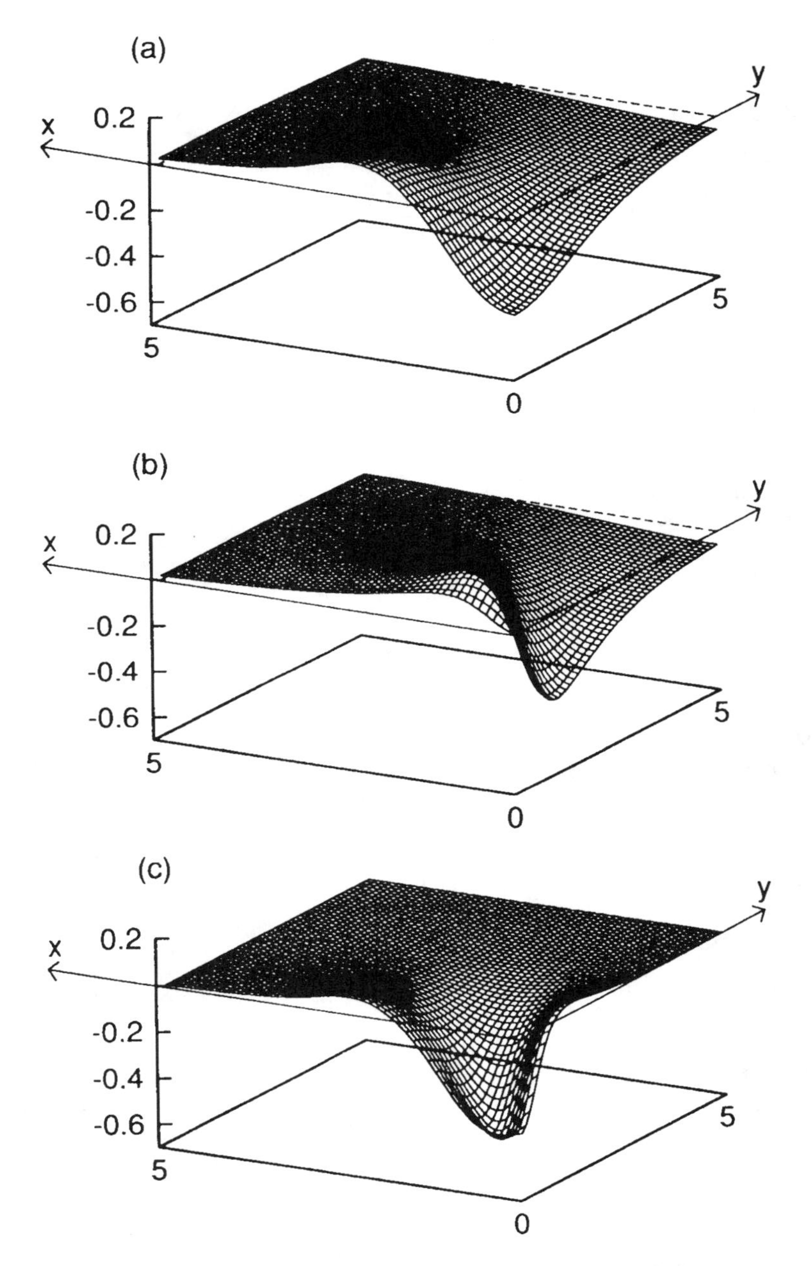

Figure 13.3.7 Crack influence function determined by Bažant and Jirásek (1994a)): (a) total crack influence function for the case of parallel source and target cracks, (b) analytical expression having the correct long-range asymptotic field, and (c) difference of the crack influence functions in (a) and (b). (From Bažant and Jirásek 1994a.)

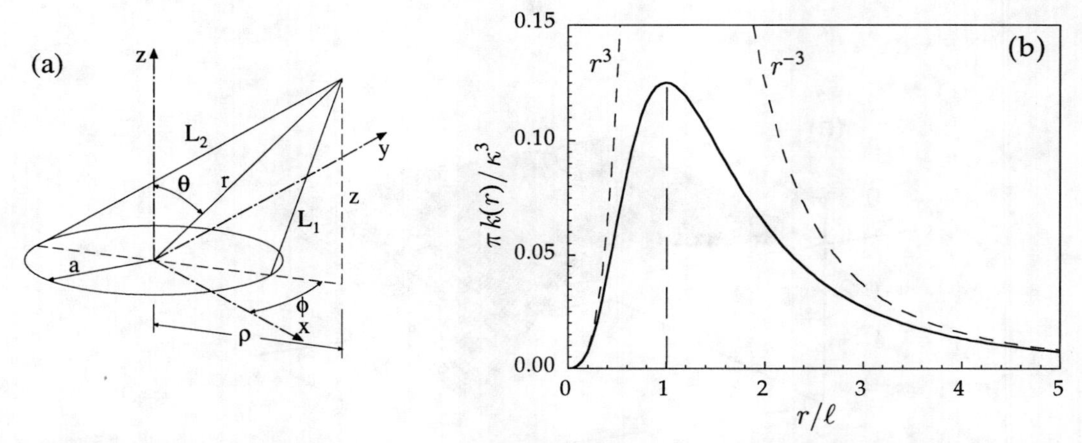

Figure 13.3.8 Crack influence function in three dimensions: (a) definition of coordinates for a penny-shaped crack (adapted from Bažant 1994b); (b) radial dependence of the influence function.

in which, for three dimensions, $k(r) = a^3/(\pi r^3)$. For the same reasons as those that led to Eq. (13.3.45), this expression may be replaced by

$$k(r) = \frac{1}{\pi}\left(\frac{\kappa \ell r}{r^2 + \ell^2}\right)^3 \tag{13.3.50}$$

(Fig. 13.3.8b) which is asymptotically correct for $r \to \infty$ and nonsingular at $r = 0$. The crack influence function based on (13.3.49) satisfies again the condition that its spatial average over every surface $r =$ constant be zero.

For large distances r, the crack influence function in three dimensions asymptotically decays as r^{-3}, whereas in two dimensions, it decays as r^{-2}. Again, in contrast to the phenomenological models we expounded before, the weight function (crack influence function) is not axisymmetric (isotropic) but depends on the polar or spherical angles (i.e., is anisotropic).

Further note that one can again distinguish a shielding sector and an amplification sector. According to the change of sign of σ_{zz} in Eq. (13.3.49), the boundary of these sectors is given by the angle

$$\theta_b = \arcsin\sqrt{2/3} = 54.736° \tag{13.3.51}$$

or $90° - \theta_b = 35.264°$. Thus, the amplification sector $\theta \geq \theta_b$ is significantly narrower in three than in two dimensions.

In the case of a field that is translationally symmetric in z, one might wonder whether integration over z might yield the two-dimensional crack influence function. However, this is not so because the two-dimensional crack influence function represents in three dimensions the effect of an infinite strip (of thickness dx) at coordinate x of pressurized cracks aligned in the z direction on the stresses in a strip of glued cracks at coordinate ξ. This cannot yield the same properties as the field of one penny-shaped crack.

13.3.9 Cracks Near Boundary

When the boundary is near, the crack influence function should be obtained by solving the stress field of a pressurized crack located at a certain distance d from the boundary; Fig. 13.3.5d–g. Obviously, the function will depend on d as a parameter, i.e., $\Lambda(\mathbf{x}, \boldsymbol{\xi}, d)$. Functions Λ will be different for a free boundary, fixed boundary, sliding boundary, and elastically supported boundary or interface with another solid (Fig. 13.3.5d–g). When the crack is near a boundary corner (Fig. 13.3.5f), Λ represents the solution of the stress field of a pressurized crack in the wedge, and will depend on the distances from both boundary planes of the wedge. These solutions will be much more complicated than for a crack in infinite body, and simplifications will be needed. On the other hand, because of the statistical nature of the crack system, exact solutions of these problems are not needed. Only their essential features are.

A crude but simple approach to the boundary effect is to consider the same weight function as for an infinite solid, protruding outside the given finite body. In the previous nonlocal formulations, based on the idea of spatial averaging, the same weight function as for the infinite solid has been used in the spatial integral and the weight function has simply been scaled up (renormalized), so that the integral of the weight function over the reduced domain would remain 1. In the present formulation, such scaling would have to be applied to all the weight functions whose integral should be 1, i.e., α, ψ, B, C. For those weight functions whose integral should vanish, a different scaling would be needed to take the proximity of the boundary into account; for example, the values at the boundary should be scaled up so that the spatial integral would always vanish, as indicated in (13.3.22). As a reasonable simplification, this might perhaps be done by replacing the $\Lambda_{\mu\nu}$ values for the integration points $\boldsymbol{\xi}_\nu$ of the boundary finite elements by $k_b \Lambda_{\mu\nu}$ where the multiplicative factor k_b is determined from the condition that $\sum_{\nu=1}^{N} \Lambda_{\mu\nu} = 0$ (with the summation carried over all the points in the given finite body);

$$k_b = - \sum_{\text{interior } \nu} \Lambda_{\mu\nu} \Big/ \sum_{\text{boundary } \nu} \Lambda_{\mu\nu} \tag{13.3.52}$$

13.3.10 Long-Range Decay and Integrability

Consider now an infinite two-dimensional elastic solid in which the stress, strain, and cracking are macroscopically uniform. All the microcracks are of the same size a, and the area per crack is s^2. The stress σ applied on each microcrack is the same. From (13.3.41) we calculate the contribution to the nonlocal integral from domain V_1 outside a circle of radius R_1 that is sufficiently large for permitting the approximation $k(r) \approx a^2/r^2$;

$$\int_{V_1}^{\odot} (\sigma_{xx} + \sigma_{yy}) dV = \lim_{R \to \infty} \int_{r=R_1}^{R} \int_{\phi=0}^{2\pi} \frac{\sigma a^2 \cos 2\phi \; r d\phi dr}{2r^2 \qquad s^2} = \frac{\sigma a^2}{2s^2} \int_{r=R_1}^{\infty} \int_{\phi=0}^{2\pi} \frac{\cos 2\phi}{r} d\phi dr \tag{13.3.53}$$

Now an important observation, to which we already alluded: the last expression is an improper integral which is divergent (because it is divergent when the integrand is replaced by its absolute value; see e.g., Rektorys 1969). This also means that the value of the integral depends on the integration path. For some path the integral may be convergent, and that path, shown in (13.3.53), has been labeled by $\odot$. So we must conclude that a homogeneous $\Delta \mathrm{S}$ field, that is, a field with a uniform length increment of all the cracks in an infinite body that is initially in a statistically uniform state, is impossible.

But this is not all that surprising. As known from the analysis of bifurcation and stable equilibrium path, strain-softening damage (which is due to microcrack growth) must localize (e.g., Bažant and Cedolin 1991). So, in practice, the two-dimensional domain of the integrals such as the last one must not be infinite in two directions. It can be infinite in only one direction, as is the case for a localization band. The basic reason for this situation is that the asymptotic decay r^{-2}, which we have obtained, is relatively weak—much weaker than the exponential decay assumed in previous works (for an exponential decay, the integration domain could be infinite in all directions without causing this kind of problem).

A similar analysis of uniform damage can be carried out for an infinite three-dimensional solid, and the conclusion is that the integration domain, that is, the zone of growing microcracks, can be infinite in only two directions (a localization layer), but not in three.

A similar divergence of the integral over infinite space has been known to occur in other problems of physics, for example, in calculation of the stresses from periodically distributed inclusions, or the light received from infinitely many statistically uniformly distributed stars. For a perspicacious mathematical study of this type of problem, see Furuhashi, Kinoshita and Mura (1981).

13.3.11 General Formulation: Tensorial Crack Influence Function

In Eq. 13.3.9, the principal stress orientations at points $\mathbf{x}$ and $\boldsymbol{\xi}$ are reflected in the values of the scalar function $\Lambda(\mathbf{x}, \boldsymbol{\xi})$. For the purpose of general analysis, however, it seems more convenient to use a tensorial crack influence function referred to common structural cartesian coordinates $X \equiv X_1, Y \equiv X_2, Z \equiv X_3$, and transform all the inelastic stress tensor components to X, Y, Z. The local cartesian coordinates $x \equiv x_1, y \equiv x_2, z \equiv x_3$ at point $\boldsymbol{\xi}$ are chosen so that axis y coincides with the direction of the maximum principal value of the inelastic stress tensor $\overline{\mathbf{S}}(\boldsymbol{\xi})$, and axes x and z coincide with the other two principal directions.

Equations (13.3.32) and (13.3.33) may be rewritten in common structural coordinates as follows:

$$\Delta\tilde{S}_{\mu IJ} \leftarrow \overline{\Delta S_{\mu IJ}} + \sum_{\nu=1}^{N}\sum_{i=1}^{3} R^{(i)}_{\nu IJkl}\Lambda^{(i)}_{\mu\nu kl}\Delta\tilde{S}^{(i)}_{\nu} \qquad (\mu = 1, 2, ...N) \tag{13.3.54}$$

$$\Delta\tilde{S}_{IJ}(\mathbf{x}) \leftarrow \overline{\Delta S_{IJ}(\mathbf{x})} + \int_V \sum_{i=1}^{3} R^{(i)}_{IJkl}(\boldsymbol{\xi})\Lambda^{(i)}_{kl}(\mathbf{x}, \boldsymbol{\xi})\Delta\tilde{S}^{(i)}(\boldsymbol{\xi})dV(\boldsymbol{\xi}) \tag{13.3.55}$$

in which, similar to (13.3.21), we included the influence of the dominant cracks normal to the principal stress direction at each point; $R^{(i)}_{IJkl}(\boldsymbol{\xi})$ or $R^{(i)}_{\nu IJkl} = c_{kI}c_{lJ}$ = fourth-rank coordinate rotation tensor (programmed as a square matrix when the stress tensors are programmed as column matrices) at point $\boldsymbol{\xi}$ or $\boldsymbol{\xi}_\mu$; c_{kI}, c_{lJ} = coefficients of rotation transformation of coordinate axes (direction cosines of new axes) from local coordinates x_i at point $\boldsymbol{\xi}$ (having, in general, a different orientation at each $\boldsymbol{\xi}$) to common structural coordinates X_I ($c_{kI} = \cos(x_k, X_I), X_I = c_{kI}x_k, \sigma_{IJ} = c_{kI}c_{lJ}\sigma_{kl}$); subscripts I, J, or k, l refer to cartesian components in the common structural coordinates or in the local coordinates at $\boldsymbol{\xi}$; and $\Lambda^{(i)}_{\mu\nu kl}$ or $\Lambda^{(i)}_{kl}(\mathbf{x}, \boldsymbol{\xi})$ = components of a tensorial discrete or continuous nonlocal weight function (crack influence function, replacing the scalar function Λ), which are equal to ℓ^{-2} times the cartesian stress components σ_{kl} for $\sigma = 1$ as defined by (13.3.41) for two dimensions, or ℓ^{-3} times such cartesian components as defined by (13.3.49) for three dimensions (with $r = \|\mathbf{x} - \boldsymbol{\xi}\|$).

13.3.12 Constitutive Relation and Gradient Approximation

As is clear from the foregoing exposition, the constitutive relation is defined only locally. It yields the inelastic stress increment $\overline{\Delta S^{(1)}(\mathbf{x})}$, illustrated by segment $\overline{32}$ shown in Fig. 13.3.2. In the previous nonlocal formulations, by contrast, the nonlocal inelastic strain, stress, or damage was part of the constitutive relation. This caused conceptual difficulties as well as continuity problems with formulating the unloading criterion. Furthermore, in the case of nonlocal plasticity, this may also cause difficulties with the consistency condition for the subsequent loading surfaces.

Here these difficulties do not arise, because the nonlocal spatial integral is separate from the constitutive relation. Thus the unloading criterion can, and must, be defined strictly locally. If plasticity is used to define the local stress-strain relation, the consistency condition of plasticity is also local.

In principle, the nonlocal model based on crack interaction can be applied to any constitutive model for strain-softening, for example, parallel smeared cracking, isotropic damage theory, plasticity with yield limit degradation, plastic-fracturing theory, and endochronic theory. But to fully realize the potential of this approach, a more realistic model, such as the microplane model, appears more appropriate and has already been applied by Ožbolt and Bažant (1996). This will be discussed and documented in the next chapter.

Recently there has been much interest in limiting localization of cracking by means of the so-called gradient models. These models can be looked at as approximations of the nonlocal integral-type models, and can be obtained by expanding the nonlocal integral in Taylor series (Bažant 1984b); see Section 13.1.3. Unlike the present model, there have been only scant and vague attempts at physical justifications for the gradient models, especially for aggregate-matrix composites such as concrete. It seems that the physical justification for the gradient models of such materials must come indirectly, through the integral-type model. However, if that is the case, the present results signal a problem. If the spatial integral in (13.3.9) were expanded into Taylor series and truncated, the long-range decay of the type r^{-2} or r^{-3} could not be preserved. Yet it seems that this decay is important for microcrack systems. If so, then the gradient approximations are physically unjustified.

13.3.13 Localization of Oriented Cracking into a Band

The nonlocal model based on crack interactions has been applied to the problem of localization of unidirectional cracking into an infinite planar band parallel to the cracks (Bažant and Jirásek 1994b)). The body either is infinite or is an infinite planar layer parallel to the cracks, of thickness L. This represents the most fundamental localization problem, which is one dimensional, with the coordinate x normal to the cracks as the only coordinate. Due to translational symmetry in directions x and z parallel to the layer, the constitutive relation given by Eq. (13.3.9) with (13.3.43) can be integrated in the direction y parallel to the layer (the original problem is considered two-dimensional, although generalization to three dimensions would be possible). For the approximate crack interaction function (Eq. (13.3.43) with (13.3.45)) with $\theta = \psi = 0$), which is asymptotically correct at infinity, the integral can be evaluated analytically. This yields the following one-dimensional field equations for the increments of stresses and strains

$$\Delta\sigma = C(x)\Delta\varepsilon(x) - \Delta\tilde{S}(x) \tag{13.3.56}$$

$$\Delta\tilde{S} = \int_{-\infty}^{\infty} \hat{\Phi}(x,\xi)\Delta S(\xi)d\xi + \int_{-\infty}^{\infty} \hat{\Lambda}(x,\xi)\Delta\tilde{S}(\xi)d\xi \tag{13.3.57}$$

$$\hat{\Phi}(x,\xi) = \int_{-\infty}^{\infty} \Phi(x,0;\xi,\eta)d\eta, \quad \hat{\Lambda}(x,\xi) = \int_{-\infty}^{\infty} \Lambda(x,0;\xi,\eta)d\eta \tag{13.3.58}$$

$$\hat{\Lambda}(x,\xi) = \frac{\pi\kappa^2}{\ell}\left[\frac{16\zeta^6 + 24\zeta^4 + 6\zeta^2 + 1}{4(1+\zeta^2)^{3/2}} - 4|\zeta|^3\right] \tag{13.3.59}$$

$\zeta = (x-\xi)/\ell$; C = elastic material stiffness (modulus); $\hat{\Phi}$ is the one-dimensional weight function for spatial averaging, corresponding to the averaging over crack surface in Kachanov's method; and $\hat{\Lambda}$ is the one-dimensional crack influence function. Note that function $\hat{\Lambda}(x,\xi)$ is always positive, in contrast to the two-dimensional function Λ.

The solutions of Eqs. (13.3.56)–(13.3.59) have been studied numerically, by introducing a discrete subdivision in coordinate x and reducing the equations to a matrix form. As the boundary condition, the layer of thickness L was considered fixed at both surfaces. The localization profiles of strain increment $\Delta\varepsilon$, beginning from a state of uniform strain of various magnitudes, have been calculated and the evolution of the strain profile during loading has been followed. Fig. 13.3.9a shows the evolution of the strain profile across the layer, obtained for a local stress-strain relation that is linear up to the peak and then decays exponentially. Fig. 13.3.9b shows the stress-displacement diagram obtained for various ratios L/ℓ of the thickness of the layer to the nonlocal characteristic length. It is clear that the formulation prevents localization into a layer of zero thickness and enforces a smooth strain profile through the localization band. It is also seen that the size effect on the postpeak softening slope is obtained realistically. An interesting point is that localizations according to this formulation can happen even before attaining the maximum load. For further details, see Bažant and Jirásek (1994b).

13.3.14 Summary

The inelastic stress increments in a microcracking material are equal to the stresses that the load increment would produce on the cracks if they were temporarily "frozen" (or "glued"), i.e., prevented from opening and growing. The nonlocality arises from two sources: (1) *crack interactions*, which means that application of the pressure on the crack surfaces that corresponds to the "unfreezing" (or "ungluing") of one crack produces stresses on all the other frozen cracks; and (2) averaging of the stresses due to unfreezing over the crack surface, which is needed because crack interactions depend primarily on the stress average over the crack surface (or the stress resultant) rather than the macroscopic stress corresponding to the microcrack center. The crack interactions (source 1) can be solved by Kachanov's (1987a) simplified version of the superposition method, in which only the average crack pressures are considered.

The resulting nonlocal continuum model involves two spatial integrals. One integral, which corresponds to source (1) and has been absent from previous nonlocal models, is long-range and has a weight function whose spatial integral is 0; it represents interactions with remote cracks and is based on the long-range

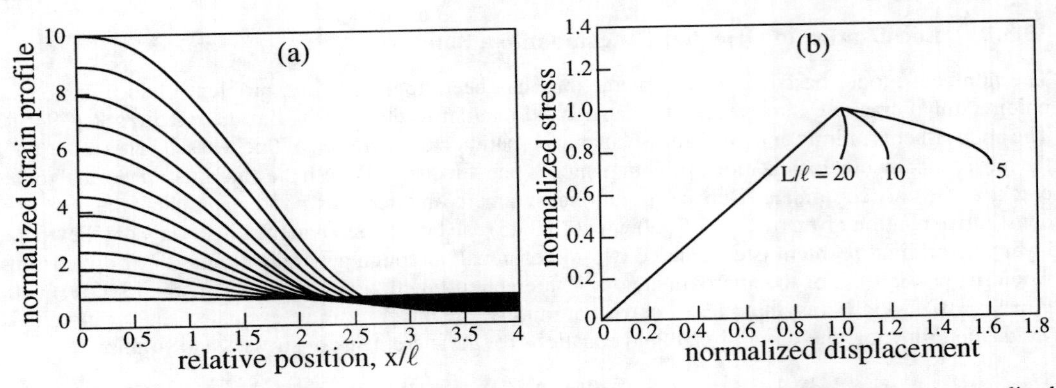

Figure 13.3.9 Localization in a bar predicted by nonlocal model based on crack interactions: (a) normalized strain profiles along a bar; the profiles are symmetric with respect to the origin, and an exponential softening law was used; (b) load-displacement diagrams for various bar lengths L. (From Jirásek and Bažant 1994.)

asymptotic form of the stress field caused by pressurizing one crack while all the other cracks are frozen. Another integral, corresponding to source (2), is short-range, involves a weight function whose spatial integral is 1, and represents spatial averaging of the local inelastic stresses over a domain whose diameter is roughly equal to the spacing of major microcracks (which is roughly equal to the spacing of large aggregates in concrete).

As an approach to continuum smoothing when the macroscopic field is nonuniform, one may seek a continuum field equation whose possible discrete approximation coincides with the matrix equation governing a system of interacting microcracks.

The long-range asymptotic weight function of the nonlocal integral representing crack interactions (source 1) has a separated form which is calculated as the remote stress field of a crack in infinite body. It decays with distance r from the crack as r^{-2} in two dimensions and r^{-3} in three dimensions. This long-range decay is much weaker than assumed in previous nonlocal models. In consequence, the long-range integral diverges when the damage growth in an infinite body is assumed to be uniform. This means that only the localized growth of damage zones can be modeled.

In contrast to the previous nonlocal formulations, the weight function (crack influence function) in the long-range integral is a tensor and is not axisymmetric (isotropic). Rather, it depends on the polar or spherical angle (i.e., is anisotropic), exhibiting sectors of shielding and amplification. The weight function is defined statistically and can be obtained by evaluating a certain averaging integral in which the integrand is the stress field of one pressurized crack in the given structure.

When an iterative solution of crack interactions according to the Gauss-Seidel iterative method is considered, the long-range nonlocal integral based on the crack influence function yields the nonlocal inelastic stress increments explicitly. This explicit form is suitable for iterative solutions of the loading steps in nonlinear finite element programs. The nonlocal inelastic stress increments represent a solution of a tensorial Fredholm integral equation in space, to which the iterations converge.

The constitutive law, in this new formulation, is strictly local. This is a major advantage. It eliminates difficulties with formulating the unloading criterion and the continuity condition, experienced in the previous nonlocal models in which nonlocal inelastic stresses or strains have been part of the constitutive relation.

14

Material Models for Damage and Failure

Computer analysis of concrete structures requires a general and robust material model for distributed cracking and other types of strain-softening damage such as softening plastic-frictional slip. The material model must perform realistically under a wide range of circumstances. The problem can be approached through two types of models: (1) the continuum approach, in which case the structure is usually solved by finite element discretization (although boundary elements and other methods are possible), and (2) the discrete (or lattice) approach, taking the form of discrete element method or its variants—the random particle model or lattice model. In the former approach, the material is characterized by a general nonlinear triaxial stress-strain relation coupled with a nonlocal formulation. In the latter approach, the material is represented by a lattice of particles and connecting bars for which simple rules of deformation and breakage must be devised.

At present, the continuum approach is more general, more widely applicable to structural analysis under general types of loading. The discrete approach provides some valuable insight into the micromechanics of failure and the role of heterogeneity, but only when the failure is due principally to tensile cracking and fracture. The computational demands of the discrete approach are still prohibitive for large structures and three-dimensional analysis, and attempts to develop the discrete approach for compression or compression-shear failures have so far been unsuccessful. In this chapter, we will first discuss the continuum approach and later briefly review the lattice approach.

The preceding chapter, dealing with nonlocal formulations, already presented one of two necessary components of the continuum approach to general analysis of softening damage due to microcracking and frictional-plastic slip. The other necessary component—the general triaxial stress-strain relation—will be described in this chapter. We have already touched this subject in chapter 8 while describing the triaxial stress-strain relation for the crack-band model, such as the fixed-crack and rotating crack models. But these simple models are not sufficiently general to deal with compression splitting and compression shear, cracking combined with plastic-frictional slip and softening slip.

Formulation of a general constitutive relation for such phenomena is a rather difficult problem, to which numerous studies have been devoted during the last two decades. Although many valuable advances have been made, this chapter will present in detail only one approach—the microplane model, which currently appears most realistic, powerful, and versatile. Other approaches, which use classical types of constitutive relations based on the invariants of the stress and strain tensors and include models such as plasticity, continuum damage mechanics, fracturing theory, plastic-fracturing theory, and endochronic theory, will not be treated.

All the constitutive models describing fracture exhibit properties such as post-peak strain softening and deviations from the normality rule (or Drucker's postulate). As discussed in Chapter 8, these properties, which are inevitable if the constitutive relation should describe cracking, friction, and loss of cohesion realistically, cause well-known mathematical difficulties such as ill-posedness of the boundary value problem, spurious localization instabilities, and spurious mesh sensitivity. To avoid these difficulties, the constitutive relations presented in this chapter must be combined with some kind of localization limiter. The nonlocal approach described in the preceding chapter is an effective method of solving these problems.

It is often thought that the continuum approach cannot be applied to the final stages of failure, in which damage localizes into large continuous cracks. However, the continuum approach can provide a relatively good (albeit not perfect) model for the propagation of such cracks. The reasons have already been explained in Chapter 8, in connection with the crack band model, which may be regarded as the simplest version of the nonlocal approach. The width of the localized damage band has, in most cases, negligible influence on the results of structural analysis. A zero width, that is, a distinct crack, and a finite width (not excessively large, of course) often yields about the same results. Forcing, through the nonlocal concept, the distinct crack to spread over a width of several finite element sizes, or forcing a narrow damage band to be wider than the real width, is usually admissible, provided that the energy dissipation per unit length of advance of the band is adjusted to remain the same. It should be noted that such spreading of damage over a width of several element sizes is also a convenient way to avoid the directional bias of finite element mesh.

14.1 Microplane Model

The microplane model (Bažant 1984c) trades simplicity of concept for increased numerical work left to the computer. This model represents a generalization of the basic idea of G.I. Taylor (1938), who proposed that the constitutive behavior of polycrystalline metals may be characterized by relations between the stress and strain vectors acting on planes of all possible orientations within the material, and that the macroscopic strain or stress tensor may then be obtained as a summation (or resultant) of all these vectors under the assumption of a static or kinematic micro-macro constraint.

Taylor's idea was soon recognized as the most realistic way to describe the plasticity of metals, but the lack of computers prevented practical application in the early times. Batdorf and Budianski (1949) were first to describe hardening plasticity of polycrystalline metals by a model of this type, and many other researchers subsequently refined or modified this approach to metals (Kröner 1961; Budianski and Wu 1962; Lin and Ito 1965, 1966; Hill 1965, 1966; Rice 1970). Taylor's idea was also developed for the hardening inelastic response of soils and rocks (Zienkiewicz and Pande 1977; Pande and Sharma 1981, 1982; Pande and Xiong 1982).

In all the aforementioned approaches, it was assumed that the stress acting on various planes in the material, called the slip planes, was the projection of the macroscopic stress tensor. This is a static constraint. As shown later, the static constraint prevents such models from being generalized to postpeak strain softening behavior or damage. In an effort to model concrete, it was realized that the extension to damage requires replacing the static constraint by a kinematic constraint, in which the strain vector on any inclined plane in the material is the projection of the macroscopic strain tensor (Bažant 1984c). The kinematic constraint makes it possible to avoid spurious localization among orientations in which all the strain softening localizes preferentially into a plane of only one orientation.

In all applications to metals, the formulations based on Taylor's idea were called the slip theory of plasticity, and in applications to rock, the multi-laminate model. These terms, however, became unsuitable for the description of damage in quasibrittle materials. For example, the salient inelastic behavior of concrete does not physically represent plastic slip (except under extremely high confining stresses), but microcracking. For this reason, the neutral term "microplane model", applicable to any physical type of inelastic behavior, was coined (Bažant 1984c) (although a nondescriptive term such as "Taylor-Batdorf-Budianski model", possibly with the names of further key contributors, could also be used). The term "microplane" reflects the basic feature that the material properties are characterized by relations between the stress and strain components independently for planes of various orientation within the microstructure of the material. This term also avoids confusion with the type of micro-macro constraint, which has always been static in the slip theory of plasticity but must be kinematic for strain-softening of concrete. Also, as introduced for the microplane model (Bažant 1984c), the tensorially invariant macroscopic constitutive relations are obtained from the responses on the microplanes of all orientations in a more general manner than in the slip theory of plasticity—by means of a variational principle (or the principle of virtual work).

The microplane model of concrete was developed in detail first for the tensile fracturing (Bažant and Oh 1983b, 1985; Bažant and Gambarova 1984), and later for nonlinear triaxial behavior in compression with shear (Bažant and Prat 1988b). The reason that these new models used the kinematic rather than static constraint for the microplanes was to avoid spurious instability of the constitutive model due to strain softening (which always occurs for the static constraint). Because the tangential material stiffness matrix loses positive definiteness (due to postpeak strain softening as well as lack or normality), the nonlocal approach, which prevents spurious excessive localization of damage in structures and spurious mesh sensitivity, was combined with the microplane model (Bažant and Ožbolt 1990, 1992; Ožbolt and Bažant 1991, 1992). An explicit formulation and an efficient numerical algorithm for the microplane model of

Bažant and Prat (1988b) was recently presented by Carol, Prat and Bažant (1992). It was also shown that the microplane model with a kinematic constraint can be cast in the form of continuum damage mechanics in which the damage, understood as a reduction of the stress-resisting cross section area fraction in the material, represents a fourth-order tensor independent of the microplane material characteristics (Carol, Bažant and Prat 1991; Carol and Bažant 1997).

Although the microplane model of Bažant and Prat (1988b) was initially thought to perform well for postpeak softening damage in both compression and tension, Jirásek (1993) found that, in postpeak uniaxial tension, large positive lateral strains develop at large tensile strains. He showed that this was caused by localization of tensile strain softening into the volumetric strain while the deviatoric strains on the strain softening microplanes exhibited unloading. It was recognized that this localization of tensile softening damage into one of the two normal strain components in tension (that is the volumetric one), was an inevitable consequence of separating the normal strains into the volumetric and deviatoric parts. However, this separation was previously shown necessary (Bažant and Prat 1988b) for correct modeling of triaxial behavior in compression as well as for achieving the correct elastic Poisson ratio. The problem was overcome by introducing a new concept—the stress-strain boundaries (Bažant 1993c; Bažant, Jirásek et al. (1994); Bažant, Xiang and Prat 1996), which will be described in detail. This concept allows an explicit algorithm and is computationally efficient.

The basic philosophy of microplane model blends well with the philosophy of finite elements. Finite elements represent a discretization with respect to space (or distance), while the microplane model represents a discretization with respect to orientations. In both, the principle of virtual works is used, as will be seen, in analogous ways—to establish the equilibrium relations and stiffness for the postulated kinematic constraint, which is given by the shape (or interpolation) functions for finite elements or by the kinematic constraint between orientations. This analogous structure is suitable for explicit programs (Fig. 14.1.1).

In another sense, the microplane model can be regarded as complementary to the nonlocal concept. Whereas the nonlocal concept handles interactions at distance, the microplane model handles interactions between orientations (Fig. 14.1.2). The nonlocality prevents spurious localization in space, whereas the kinematic constraint of the microplane model prevents localizations between orientations, as will be pointed out.

14.1.1 Macro-Micro Relations

In the classical approach, the constitutive relation is defined by algebraic or differential relations between the stress tensor $\boldsymbol{\sigma}$ and the strain tensor $\boldsymbol{\varepsilon}$, based on the theory of tensorial invariants. In the microplane approach, the constitutive relation is defined as a relation between the stress and strain *vectors* acting on a plane of arbitrary orientation in the material. The orientation of this plane, called the microplane, is characterized by the unit normal $\vec{n}$. The basic hypothesis, which makes it possible to describe strain softening (Bažant 1984c), is that the strain vector $\vec{\varepsilon}_N$ on the microplane (Fig. 14.1.3a) is the projection of the macroscopic strain tensor $\boldsymbol{\varepsilon}$, that is,

$$\vec{\varepsilon}_N = \boldsymbol{\varepsilon}\vec{n} \tag{14.1.1}$$

The stress vector $\vec{\sigma}_N$ on the microplane cannot be exactly equal to the projections of the macroscopic stress tensors $\boldsymbol{\sigma}$ if the strains represent the projections of $\boldsymbol{\varepsilon}$. Thus, static equivalence or equilibrium between the macro and micro levels must be enforced only approximately, by other means. The way to enforce it is to use a variational principle, that is, the principle of virtual work. For equilibrium, it suffices that, for any variation $\delta\boldsymbol{\varepsilon}$, the virtual work of the macrostresses within a unit sphere be equal to the virtual work of the microstresses on the surface elements of the sphere (Bažant 1984c). This condition is written as:

$$\frac{2\pi}{3}\,\boldsymbol{\sigma}\cdot\delta\boldsymbol{\varepsilon} = \int_{\Omega} \vec{\sigma}_N \cdot \delta\vec{\varepsilon}_N \, d\Omega \tag{14.1.2}$$

where the dot represents scalar product of two vectors or two second-order tensors.

Remark: A more detailed justification of this relation may be given as follows. We consider a small representative volume of the material, given by a small cube of side Δh. A pair of two parallel sides corresponds

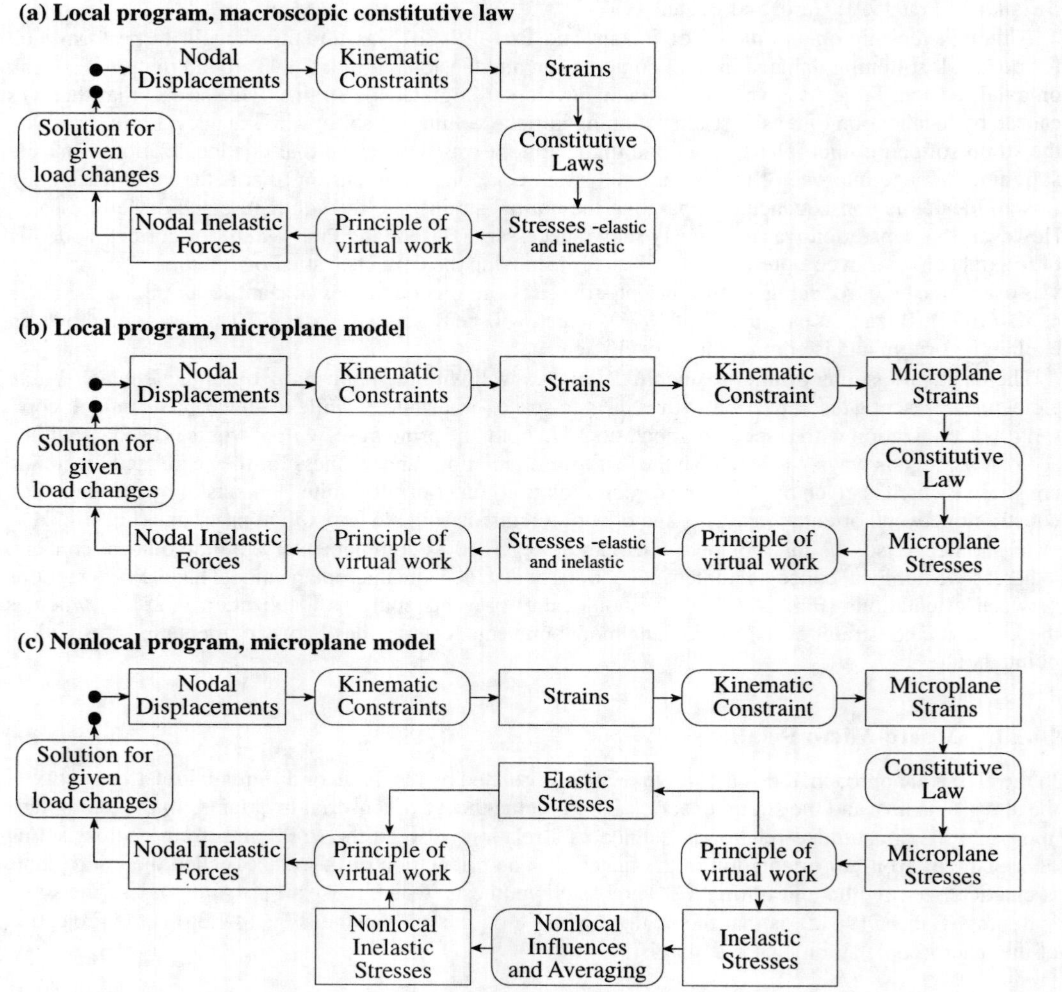

(a) Local program, macroscopic constitutive law

Nodal Displacements → Kinematic Constraints → Strains → Constitutive Laws → Stresses -elastic and inelastic → Principle of virtual work → Nodal Inelastic Forces → Solution for given load changes → Nodal Displacements

(b) Local program, microplane model

Nodal Displacements → Kinematic Constraints → Strains → Kinematic Constraint → Microplane Strains → Constitutive Law → Microplane Stresses → Principle of virtual work → Stresses -elastic and inelastic → Principle of virtual work → Nodal Inelastic Forces → Solution for given load changes → Nodal Displacements

(c) Nonlocal program, microplane model

Nodal Displacements → Kinematic Constraints → Strains → Kinematic Constraint → Microplane Strains → Constitutive Law → Microplane Stresses → Principle of virtual work → Elastic Stresses; Inelastic Stresses → Nonlocal Influences and Averaging → Nonlocal Inelastic Stresses → Principle of virtual work → Nodal Inelastic Forces → Solution for given load changes → Nodal Displacements

Figure 14.1.1 General flow charts of iteration cycles in load steps of explicit finite element programs (using initial elastic stiffness matrix).

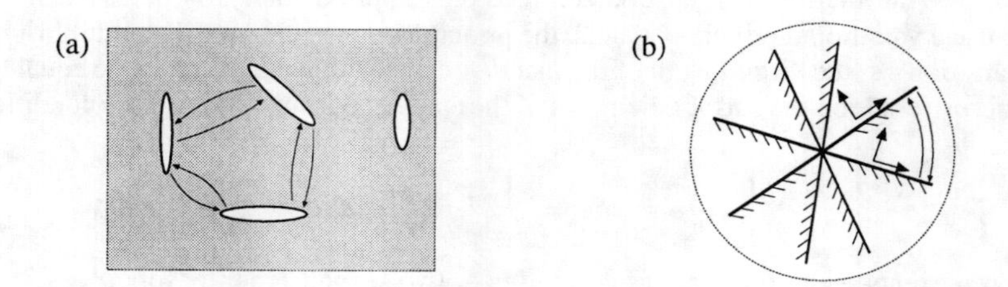

Figure 14.1.2 (a) Interaction at distance (nonlocality) and (b) interaction between orientations (microplane concept).

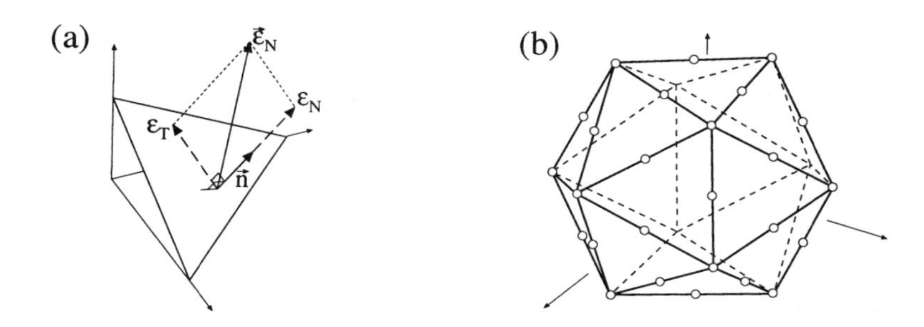

Figure 14.1.3 (a) Microplane normal and microstrain vectors, and normal and shear components of the microstrain vector. (b) Directions of microplane normals (circles) for a system of 21-microplanes per hemisphere (after Bažant and Oh 1986, adapted from Bažant, Xiang and Prat 1996).

to microplane labeled by subscript N, and the other two pairs of sides correspond to orthogonal microplanes labeled by subscripts P, and Q. The strain vectors on these microplanes may be assumed to have the meaning defined by $\Delta\vec{u}_N/\Delta h = \vec{\varepsilon}_N$, $\Delta\vec{u}_P/\Delta h = \vec{\varepsilon}_P$, $\Delta\vec{u}_Q/\Delta h = \Delta\vec{\varepsilon}_Q$ in which $\Delta\vec{u}_N, \Delta\vec{u}_P$, and $\Delta\vec{u}_Q$ are the differences in the displacement vector between the opposite sides of the cube in the directions by labeled by N, P, and Q. The equality of the incremental virtual work of stresses within the representative volume on the macrolevel and the work of stresses on the three microplanes representing the sides of the cube implies that $\Delta h^3\boldsymbol{\sigma}\cdot\delta\boldsymbol{\varepsilon} = \Delta h^2(\vec{\sigma}_N\cdot\delta\Delta\vec{u}_N + \vec{\sigma}_P\cdot\delta\Delta\vec{u}_P + \vec{\sigma}_Q\cdot\delta\Delta\vec{u}_Q)$, where δ denotes the variations and Δh^2 = area of the sides of the elementary cube. The strain vectors $\vec{\varepsilon}_N, \vec{\varepsilon}_P$ and $\vec{\varepsilon}_Q$ include the contributions of elastic deformations as well as displacements due to cracking (and possibly also to plastic slip). The cracking or other inelastic deformation happens randomly on planes of various orientations within the material, and the macroscopic continuum must represent these strains statistically, in the average sense. Therefore,

$$\Delta h^3\boldsymbol{\sigma}\cdot\delta\boldsymbol{\varepsilon} = \Delta h^2\frac{1}{\Omega_0}\int_\Omega\left(\vec{\sigma}_N\cdot\delta\Delta\vec{u}_N + \vec{\sigma}_P\cdot\delta\Delta\vec{u}_P + \vec{\sigma}_Q\cdot\delta\Delta\vec{u}_Q\right)d\Omega \tag{14.1.3}$$

in which the integral represents averaging over all spatial orientations; $d\Omega = \sin\theta d\theta d\phi$ where θ, ϕ = spherical angles, Ω = surface of a unit hemisphere, and $\Omega_0 = 2\pi$ = its surface area. Now, obviously, $\int_\Omega \vec{\sigma}_N\cdot\delta\Delta\vec{u}_N d\Omega = \int_\Omega \vec{\sigma}_P\cdot\delta\Delta\vec{u}_P d\Omega = \int_\Omega \vec{\sigma}_Q\cdot\Delta\vec{u}_Q d\Omega$. Thus the variational equation (14.1.3) yields (14.1.2). $\triangle$

Substituting (14.1.1) into the integral in (14.1.2) and factorizing $\delta\boldsymbol{\varepsilon}$, we obtain

$$\left[\frac{2\pi}{3}\boldsymbol{\sigma} - \int_\Omega(\vec{\sigma}_N\otimes\vec{n})^S d\Omega\right]\cdot\delta\boldsymbol{\varepsilon} = 0 \tag{14.1.4}$$

where $\otimes$ indicates tensorial product and superscript S for a tensor denotes the symmetric part of such tensor, i.e., $\mathbf{T}^S = (\mathbf{T}+\mathbf{T}^T)/2$, in which $\mathbf{T}$ is an arbitrary second-order tensor and $\mathbf{T}^T$ its transpose. Since the variational equation (14.1.4) must be satisfied for any variation $\delta\boldsymbol{\varepsilon}$, it is not only sufficient but also necessary that the expression in parentheses vanish. This yields the following fundamental relation from which the macroscopic stress tensor is calculated:

$$\boldsymbol{\sigma} = \frac{3}{2\pi}\int_\Omega(\vec{\sigma}_N\otimes\vec{n})^S d\Omega \tag{14.1.5}$$

To compute the integral over the unit sphere, Gaussian integration can be used, and so the cartesian stress components σ_{ij} are computed as

$$\sigma_{ij} = \frac{3}{2\pi}\int_\Omega s_{ij}d\Omega \approx \sum_{\mu=1}^{N_m} w_\mu s_{ij}^\mu\ , \tag{14.1.6}$$

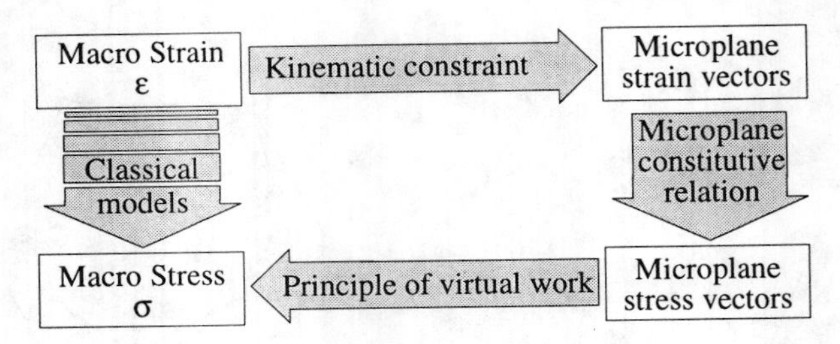

Figure 14.1.4 Flow of calculation between micro- and macro-levels (adapted from Bažant, Xiang and Prat 1996).

in which

$$s_{ij} = \left[(\vec{\sigma}_N \otimes \vec{n})^S\right]_{ij} = \frac{1}{2}(\sigma_{Ni}n_j + \sigma_{Nj}n_i) \tag{14.1.7}$$

and the last expression represents an approximate numerical evaluation of the integral over the hemisphere; subscripts μ represent a chosen set of integration points representing orientations of discrete microplanes defined by unit vectors n_i^μ (shown by the circled points in Fig. 14.1.3b); w_μ are the integration weights associated with these microplanes, normalized so that $\sum_\mu w_\mu = 1$; and superscript μ labels the values corresponding to these directions. While the integral over Ω represents integration over infinitely many microplanes, the numerical approximation represents summation over a finite number of suitably chosen discrete microplanes. The flow of calculation between the macro- and micro-levels is explained by Fig. 14.1.4

Formulation of an optimal numerical integration formula over the surface of a hemisphere is not a trivial matter. The problem has been studied extensively by mathematicians, and Gaussian integration formulas of various degrees of approximation have been developed. The simplest integration formulas, for which all the weights are equal, are obtained by taking the discrete microplanes identical to the faces of a regular polyhedron (Platonic solid). But the regular polyhedron of the largest number of sides is the icosahedron, with 20 faces, which yields 10 microplanes per hemisphere. It has been shown that the accuracy of the corresponding integration formula is insufficient for representing the postpeak stress-strain curve of concrete (this was demonstrated by the fact that rigid-body rotations of the set of discrete microplanes can yield unacceptably large differences in stresses); see Bažant and Oh (1985, 1986). Thus, formulas based on a regular polyhedron cannot be used, which means that the discrete microplanes cannot have equal weights. Determination of the optimum weights is not a trivial matter. The weights must be determined so that the formula would exactly integrate polynomials up to the highest possible degree and that the integration error due to the next higher-degree term of the polynomial be minimized (Bažant and Oh 1986).

One sufficiently accurate formula, which consists of 28 microplanes (i.e., 28 integration points) over a hemisphere, is given by Stroud (1971). A more efficient and only slightly less accurate formula, involving 21 microplanes, was derived by Bažant and Oh (1986) (and was used in the nonlocal finite element microplane program by Ožbolt, and in the program EPIC by Adley at WES). This 21-point formula exactly integrates polynomials up to the 9th degree. The normals to the microplanes of this formula represent the radial directions to the vertices and to the centers of the edges of a regular icosahedron (as shown in Fig. 14.1.3b). Fewer than 21 microplanes cannot give sufficient accuracy (Bažant and Oh 1985).

14.1.2 Volumetric-Deviatoric Split of the Microstrain and Microstress Vectors

It is well known in continuum mechanics that for many purposes it is useful to decompose the strain tensor into its hydrostatic and deviatoric parts, by writing $\varepsilon = (1/3)\mathrm{tr}\,\varepsilon\,\mathbf{1} + \varepsilon'$, where $\mathbf{1}$ is the unit tensor and ε' the deviatoric strain tensor. When applied to (14.1.1), the following decomposition of the microstrain vector follows

$$\vec{\varepsilon}_N = \varepsilon_V\,\vec{n} + \vec{\varepsilon}_D \tag{14.1.8}$$

in which ε_V is called the volumetric strain and $\vec{\varepsilon}_D$ the deviatoric strain vector acting on the microplane; they are defined as

$$\varepsilon_V = \frac{1}{3}\text{tr}\,\boldsymbol{\varepsilon} \qquad \text{and} \qquad \vec{\varepsilon}_D = \boldsymbol{\varepsilon}'\vec{n} \tag{14.1.9}$$

The deviatoric strain vector is further decomposed into its normal component ε_D that we call deviatoric strain for short, and its component tangential to the microplane that we call the shear strain vector $\vec{\varepsilon}_T$:

$$\varepsilon_D = \vec{\varepsilon}_D \cdot \vec{n} = \boldsymbol{\varepsilon}'\vec{n} \cdot \vec{n} \qquad \text{and} \qquad \vec{\varepsilon}_T = \vec{\varepsilon}_D - \varepsilon_D\,\vec{n} \tag{14.1.10}$$

The microplane strain vector can, thus, be written as

$$\vec{\varepsilon}_N = \varepsilon_V\vec{n} + \varepsilon_D\vec{n} + \vec{\varepsilon}_T \tag{14.1.11}$$

Analogous components σ_V, σ_D, and $\vec{\sigma}_T$ are defined for the microstress vector, and so we write

$$\vec{\sigma}_N = \sigma_V\vec{n} + \sigma_D\vec{n} + \vec{\sigma}_T \tag{14.1.12}$$

Note that both the volumetric and deviatoric components contribute to the normal component at the microplane. We can thus define the total normal microstrain and microstress ε_N and σ_N as

$$\varepsilon_N = \varepsilon_V + \varepsilon_D\,, \qquad \sigma_N = \sigma_V + \sigma_D \tag{14.1.13}$$

Based on the foregoing definitions, a particular microplane constitutive law consists in a set of rules specifying how the microstress components σ_V, σ_D, and $\vec{\sigma}_T$ change as $\varepsilon_V, \varepsilon_D$, and $\vec{\varepsilon}_T$ evolve. The simplest case to be solved is the linear elastic case that we analyze next.

14.1.3 Elastic Response

In the elastic regime we must have a linear relationship between $\vec{\sigma}_N$ and $\boldsymbol{\varepsilon}$ for every $\vec{n}$; therefore, we must seek a relationship of the form

$$\vec{\sigma}_N = \vec{L}(\boldsymbol{\varepsilon}, \vec{n}) \tag{14.1.14}$$

where the function $\vec{L}(\boldsymbol{\varepsilon}, \vec{n})$ is linear in $\boldsymbol{\varepsilon}$. Moreover, isotropy requires that if the microplane (and its normal vector) and the macrostrain tensor are both rotated through any orthogonal tensor $\mathbf{Q}$, the resulting microstress must be correspondingly rotated, i.e.,

$$\vec{L}(\mathbf{Q}\boldsymbol{\varepsilon}\mathbf{Q}^T, \mathbf{Q}\vec{n}) = \mathbf{Q}\vec{\sigma}_N = \mathbf{Q}\vec{L}(\boldsymbol{\varepsilon}, \vec{n}) \tag{14.1.15}$$

which indicates that the function $\vec{L}(\vec{\varepsilon}_N, \vec{n})$ is an isotropic vector-valued function of a second-order tensor and a vector. The most general function of this type that is linear in $\boldsymbol{\varepsilon}$ can be written as

$$\vec{L}(\vec{\varepsilon}_N, \vec{n}) = a_1\text{tr}\,\boldsymbol{\varepsilon}\,\vec{n} + b_1(\boldsymbol{\varepsilon}\vec{n}\cdot\vec{n})\vec{n} + c_1\boldsymbol{\varepsilon}\vec{n} \tag{14.1.16}$$

where a_1, b_1, and c_1 are scalar constants. This can readily be rewritten in terms of the volumetric, deviatoric, and shear components of the microstrain:

$$\vec{\sigma}_N = E_V\varepsilon_V\,\vec{n} + E_D\varepsilon_D\,\vec{n} + E_T\vec{\varepsilon}_T \tag{14.1.17}$$

where E_V, E_D, and E_T are microplane elastic moduli corresponding to volumetric, deviatoric, and shear straining. In view of (14.1.12), the foregoing expression can be split into the following three relations:

$$\sigma_V = E_V\varepsilon_V\,, \qquad \sigma_D = E_D\varepsilon_D \qquad \text{and} \qquad \vec{\sigma}_T = E_T\vec{\varepsilon}_T \tag{14.1.18}$$

The microplane elastic moduli can be determined in terms of the macroscopic elastic moduli by identifying the macroscopic stress-strain response predicted by the microplane model with the classical Hooke equations. The macroscopic response is obtained by substituting $\vec{\sigma}_N$ from (14.1.17) into (14.1.5) and then into the expressions for the microplane strain components in terms of the macrostrain tensor. The resulting macroscopic relationship is

$$\boldsymbol{\sigma} = \frac{E_V}{3}\text{tr}\;\boldsymbol{\varepsilon}\;\mathbf{A} + \frac{E_T}{2}\left(\mathbf{A}\boldsymbol{\varepsilon}' + \boldsymbol{\varepsilon}'\mathbf{A}\right) + (E_D - E_T)\boldsymbol{B}\boldsymbol{\varepsilon}' \tag{14.1.19}$$

where $\mathbf{A}$ and $\boldsymbol{B}$ are, respectively, the following second- and fourth-order tensors:

$$\mathbf{A} = \frac{3}{2\pi}\int_{\Omega} \vec{n}\otimes\vec{n}\;d\Omega \qquad \text{and} \qquad \boldsymbol{B} = \frac{3}{2\pi}\int_{\Omega} \vec{n}\otimes\vec{n}\otimes\vec{n}\otimes\vec{n}\;d\Omega \tag{14.1.20}$$

These two tensors can be computed with relative ease using various methods (see the exercises at the end of this section). The result is simple:

$$\mathbf{A} = \mathbf{1} \qquad \text{and} \qquad 5B_{ijkl} = \delta_{ij}\delta_{kl} + \delta_{ik}\delta_{jl} + \delta_{il}\delta_{jk} \tag{14.1.21}$$

where B_{ijkl} are the rectangular cartesian components of $\boldsymbol{B}$. Substituting these expressions into (14.1.19) we get the final expression for the macrostress tensor as:

$$\boldsymbol{\sigma} = \frac{E_V}{3}\text{tr}\;\boldsymbol{\varepsilon}\;\mathbf{1} + \frac{2E_D + 3E_T}{5}\boldsymbol{\varepsilon}' \tag{14.1.22}$$

Comparing now this expression with the classical expression of isotropic elasticity

$$\boldsymbol{\sigma} = \frac{E}{3(1-2\nu)}\text{tr}\;\boldsymbol{\varepsilon}\;\mathbf{1} + \frac{E}{1+\nu}\boldsymbol{\varepsilon} \tag{14.1.23}$$

where E is the elastic modulus and ν the Poisson ratio, we easily find that

$$E_V = \frac{E}{1-2\nu}\,, \qquad E_D = \frac{5E}{(2+3\mu)(1+\nu)}\,, \qquad E_T = \mu E_D \tag{14.1.24}$$

where $\mu = E_T/E_D$ is a free parameter which may be chosen.

Parameter μ can be optimized so as to best match the given test data. Bažant and Prat (1988b), who gave relations equivalent to (14.1.24) but in terms of parameter $\eta = E_D/E_V$ instead of μ, found the range of η-values giving the optimum fits of test data for concrete. This range corresponds to μ-values close to 1. Therefore, the value μ= 1 has subsequently been used in all the data fitting that we cite later in this section. Note also that the inverse of (14.1.24) yields E and ν in terms of E_V, E_D, and μ.

As revealed by the study of Carol, Bažant and Prat (1991), the value $\mu = 1$ is also conceptually advantageous because it makes it possible to characterize damage, in the sense of continuum damage mechanics, by a fourth-rank tensor that is independent of the material stiffness properties. This will be discussed in more depth later.

It is interesting to note that for the choice $\mu = (1-4\nu)/(1+\nu)$, one has $E_V = E_D$. Then one can set $\sigma_N = E_N\varepsilon_N$, where $E_N = E_V = E_D$. So, in that case there is no volumetric-deviatoric split. But that would not be realistic for concrete.

One reason that the normal strain on the microplane must be split into the volumetric and deviatoric normal components is that a general model ought to be capable of giving (for any μ) any thermodynamically admissible value of Poisson's ratio, that is, $-1 < \nu < 0.5$. That this is indeed so can be checked by eliminating μ from (14.1.24) and solving for ν, which yields $\nu = (5E_V - 2E_D - 3E_T)/(10E_V + 2E_D + 3E_T)$. This relation also shows that, for the case of no split (which corresponds to the case $E_V = E_D = E_N$), one would have $\nu = (E_N - E_T)/(4E_N + E_T)$, and so the Poisson ratio would be restricted to the range $-1 < \nu < 0.25$. Although this range would suffice for concrete, the microplane model, in principle, could not be fully realistic if it were restricted to Poisson's ratios less than 0.25.

It may also be noted that if the shear stiffness were neglected ($E_T = 0$ or $\mu = 0$), then any Poisson ratio between -1 and 0.5 could still be obtained, provided that the volumetric and deviatoric normal microplane strains would be split. However, if they were not (i.e., $\sigma_N = E_N\varepsilon_N$), which was implied in the initial model of Bažant and Oh (1983b, 1985) for tensile fracturing only, then Poisson's ratio would be restricted to the value $\nu = 0.25$. Such a restriction is not realistic, and besides, the shear stiffness on the microplane level appears to be important for correct modeling of the effect of confining pressure on compression failure.

The main reason for the volumetric-deviatoric split with independent moduli E_V and E_D (Bažant and Prat 1988b) is the absence of a peak and of postpeak strain softening for hydrostatic compression test and uniaxial strain compression test (see the tests of Bažant, Bishop and Chang 1986), while at the same time the loading by uniaxial compressive stress or other compressive loadings with uninhibited volume expansion exhibits stress peak followed by postpeak strain softening. Without the aforementioned split, compressive loading with restricted volume expansion (hydrostatic compression and uniaxial strain) would also, incorrectly, exhibit a peak stress and postpeak strain softening.

In the initial proposal of microplane model with strain softening (Bažant 1984c), the stress-strain relation for the normal and shear components of stresses and strains of the microplanes had the form of incremental plasticity, based on subsequent yield surfaces and loading potentials for the microplane. However, subsequent studies have shown that this was unnecessarily complicated. As it turned out (Bažant and Oh 1985; Bažant and Prat 1988b), one can assume a total algebraic stress-strain relation for these components for the case of virgin loading, that is, σ_V, σ_D, and $\vec{\sigma}_T$ can be assumed to be functions of $\varepsilon_V, \varepsilon_D$, and $\vec{\varepsilon}_T$. Further it turned out that each stress component can be considered to depend only on the associated strain component, with the exception of shear stress $\vec{\sigma}_T$, which is considered to depend on σ_N to express internal friction (and, at high pressures, plasticity). Without the frictional aspect, it is not possible to model standard triaxial tests at high confining pressures.

14.1.4 Nonlinear Microplane Behavior and the Concept of Stress-Strain Boundaries

In the original microplane model for compressive failure (Bažant and Prat 1988b), the stress-strain relations for the microplanes were smooth curves. However, difficulties arose in the handling of the transition from reloading to virgin inelastic loading in the quadrants of negative stress-strain ratio, and complicated rules had to be devised (Hasegawa and Bažant 1993; and Ožbolt and Bažant 1992). Also, the modeling of cyclic loading was difficult. These difficulties can be circumvented with the concept of stress-strain boundaries. However, the main reason for introducing this concept is the modeling of triaxial behavior in tension.

The condition that the response must not exceed a specified boundary curve $\sigma_X = F_X(\varepsilon_X)$ —where X indicates the appropriate microplane component— makes it easy to ensure continuity at the transition from elastic behavior, which is defined separately for volumetric and deviatoric components, to the strain-softening damage behavior in tension, which is defined without the volumetric-deviatoric split (Bažant 1993c). It seems next to impossible to devise explicit algebraic stress-strain relations that would describe such transitions without any discontinuity.

The stress-strain boundaries, shown in Fig. 14.1.5, are defined as (Bažant, Xiang and Prat 1996):

$$\sigma_N = F_N(\varepsilon_N), \sigma_V = -F_V(-\varepsilon_V), \sigma_D = -F_D(-\varepsilon_D), \sigma_D = F_D^+(\varepsilon_D), \sigma_T = F_T(\sigma_N) \quad (14.1.25)$$

in which σ_T stand for either σ_M or σ_L (the components of shear stress vector on two arbitrarily assigned orthogonal axes M and L within the microplane). It might seem that, from the viewpoint of rotational invariance in the microplane, the shear stress vector $\vec{\sigma}_T = (\sigma_M, \sigma_L)$ should be considered parallel to $\vec{\varepsilon}_T$, i.e., $\vec{\sigma}_T/|\vec{\sigma}_T| = \vec{\varepsilon}_T/|\vec{\varepsilon}_T|$. Such a formulation (Bažant and Prat 1988b), however, did not perform very well for complex loading paths. It appeared preferable and simpler to consider that σ_T in (14.1.25) stands either for σ_M or σ_L, i.e., $\sigma_M = F_T(\sigma_N)$ and $\sigma_L = F_T(\sigma_N)$, thus allowing $\vec{\sigma}_T$ and $\vec{\sigma}_T$ to be, in general, nonparallel. Of course, this implies a directional bias for the chosen orientations of axes M and L on each microplane. However, due to averaging on the macroscale, such a bias becomes negligible on the macroscale if the orientations of M and L on various microplanes are chosen with nearly equal probability (or frequency) for various possible orientations, and if there are many integration points of finite elements within the representative volume of material.

Function F_T defines only the boundary for positive stresses. The other, for negative stresses, is symmetric. The reason for writing the minus signs in (14.1.25) is that functions F_N, F_V, F_D are defined as positive-valued functions of positive arguments. Function F_T defines only the boundary for the magnitudes of the shear stresses (Fig. 14.1.5d). The dependence of σ_T on σ_N characterizes friction on the microplane, as well as the fact that a widely opened rough crack offers less resistance to shear than a narrow rough crack.

The response anywhere within the boundaries may be simply assumed to be elastic, as given in the rate

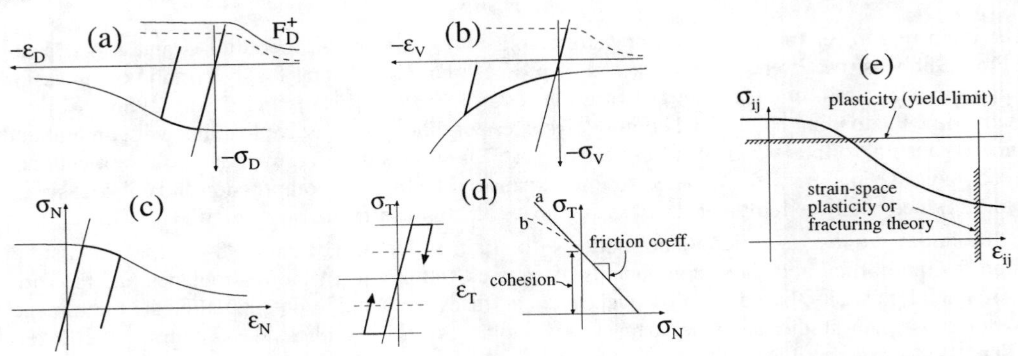

Figure 14.1.5 Stress-strain boundaries. General form for the deviatoric (a), volumetric (b), normal (c), and shear (d) components. For a classical macroscopic formulation, the boundary would be an arbitrary surface in the 12-dimensional σ_{ij}–ε_{ij} as indicated by the thick curve in (e). (Adapted from Bažant, Xiang and Prat 1996.)

form by equations similar to (14.1.18):

$$\dot{\sigma}_V = E_V \dot{\varepsilon}_V \;, \qquad \dot{\sigma}_D = E_D \dot{\varepsilon}_D \;, \qquad \text{and} \qquad \dot{\vec{\sigma}}_T = E_T \dot{\vec{\varepsilon}}_T \tag{14.1.26}$$

This simple assumption, of course, implies the stress-strain path for the microplane to exhibit a sudden change of slope when the elastic response arrives to the boundary curve. However, such changes of slope on the macroscale are not so abrupt because different microplanes reach the boundary at different times. Nevertheless, the response can be made smoother by the formulation in the following remark.

Remark: The response on the microscale can be made smooth by introducing a transition curve between the elastic straight line and the boundary curve. The transition curve, however, cannot be defined as a simple function of strains because the elastic lines and boundary curves are functions of different components. A helpful idea is to define the transition implicitly, in terms of (i) the elastic stress value σ^e and (ii) the boundary curve value σ^b, both of them corresponding to the same strain ε. When $\sigma^b \gg \sigma^e > 0$, the transition curve must nearly coincide with σ^e, and when $\sigma^e = \sigma^b$, it must lie farthest below both curves. These required properties can be achieved by the following formula for the transition curve (Bažant, Xiang and Prat 1996):

$$T(\sigma^e, \sigma^b) = \frac{\sigma^b + \sigma^e + \delta_1}{2} - \delta_0 \ln\left(2\cosh \frac{\sigma^b - \sigma^e - \delta_1}{2\delta_0}\right) \tag{14.1.27}$$

where $\sigma^e = \sigma^e_D, \sigma^e_N$, or σ^e_T; $\sigma^b = \sigma^b_D, \sigma^b_N$, or σ^b_T; and δ_1, δ_0 are constants, which have been chosen as $\delta_1 = 0.10\, f_0 \,\mathrm{sign}(\sigma^b)$ and $\delta_0 = 0.24\, f_0 \,\mathrm{sign}(\sigma^b)$ with $f_0 = f^0_D, f^0_N$, or f^0_T. For the volumetric boundary, no transition curve is introduced because the slope change is mild.

For $\delta_1 = 0$ the transition curve would approach the elastic curve and the boundary curve asymptotically at $\pm\infty$ (this may be easily checked by noting that, for large $|x|$, $2\cosh x = \exp|x|$). But the response near the origin of stress-strain space must be exactly elastic. Therefore, the left-side asymptote of the transition curve is shifted up by distance δ_1. This causes the transition to intersect the elastic curve. By choosing a small enough δ_1, the slope change at the intersection is small and acceptable.

The transition curve (14.1.27) with $\delta_1 = 0$ approaches the elastic line and the boundary curve exponentially, i.e., very rapidly. Another formula of similar properties was also explored: $T(\sigma^e, \sigma^b) = \{\sigma^b + \sigma^e + \delta_1 - [(\sigma^b - \sigma^e - \delta_1)^2 + \delta_0^2]^{1/2}\}/2$. This formula would be faster to execute computationally (which matters somewhat because it is evaluated a great many times). However, for $\delta_1 = 0$, it approaches the elastic and boundary curves too slowly, much slower than (14.1.27), which is therefore preferable. △

The stress-strain boundary may be regarded as a strain-dependent yield limit. Such an idea could hardly be introduced in the classical macroscopic invariant approach to plasticity, because the boundary would be a surface in a 12-dimensional space of all σ_{ij} and ε_{ij} components. The microplane concept makes the idea of strain-dependent yield limit feasible, in fact simple, because there are only a few components on the microplane level. The strain-dependent yield limit may be illustrated by the curve in Fig. 14.1.5e. The classical (stress space) plasticity is in this figure represented by the horizontal line for the yield limit. Now

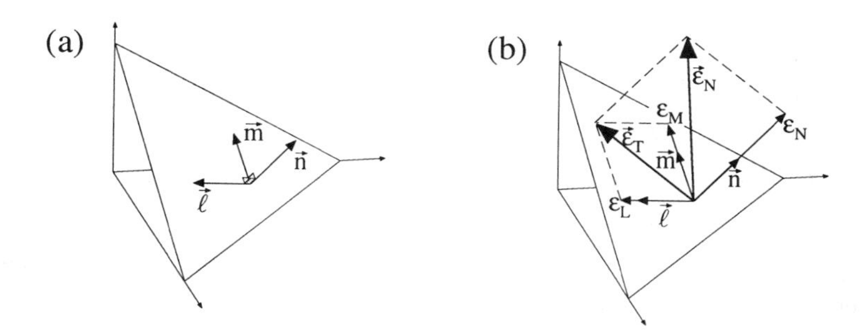

Figure 14.1.6 (a) Orthonormal base associated to a microplane. (b) Microstrain components.

note that plastic metals and fracturing materials (Dougill 1976) have also been satisfactorily described by strain-space plasticity, which corresponds to the vertical line in this figure. Obviously, a general curve should allow a better description because it is a combination of stress-space and strain-space plasticity theories.

14.1.5 Numerical Aspects

In finite element programs, a system of at least 21 microplanes must be associated with each integration point of each finite element (Bažant and Oh 1985, 1986; used in Bažant and Ožbolt 1990). Their number, however, can be reduced for the symmetries of plane stress, plane strain, axisymmetric behavior, and uniaxial stress.

For a given microplane, the normal and shear components of the microstrain vector are conveniently handled by defining an orthonormal base $\{\vec{n}, \vec{m}, \vec{\ell}\}$ (Fig. 14.1.6). Since the selection of $\vec{m}$ and $\vec{\ell}$ is arbitrary, we may, for example, choose vector m_i to be normal to the global axis x_3, in which case the cartesian components of $\vec{m}$ in the global coordinate system are $m_1 = n_2(n_1^2 + n_2^2)^{-1/2}, m_2 = -n_1(n_1^2+n_2^2)^{-1/2}, m_3 = 0$ but $m_1 = 1$ and $m_2 = m_3 = 0$ if $n_1 = n_2 = 0$. To get a vector m_i normal to axis x_1 or axis x_2, we carry out permutations $123 \rightarrow 231 \rightarrow 312$ of the indices in the preceding equations. (To minimize directional bias, the procedure of generating vectors m_i should be such that if for one microplane m_i is normal to x_1, for the next numbered microplane it is normal to x_2, for the next to x_3, for the next again to x_1, etc.) The other coordinate vector ℓ_i within the microplane is obtained as vector product, $\vec{\ell} = \vec{m} \times \vec{n}$.

Once the components of the base vectors for the microplane are obtained, the determination of the components of the microstrain vector given the macrostrain tensor immediately follow as:

$$\varepsilon_N = N_{ij}\varepsilon_{ij}\ , \qquad \varepsilon_M = M_{ij}\varepsilon_{ij}\ , \qquad \varepsilon_L = L_{ij}\varepsilon_{ij} \tag{14.1.28}$$

where ε_M and ε_L are the components of the shear microstrain vector (i.e., $\vec{\varepsilon}_T = \varepsilon_M\vec{m} + \varepsilon_L\vec{\ell}$), and the projection tensors $\mathbf{N}, \mathbf{M}$, and $\mathbf{L}$ are given in component form by

$$N_{ij} = n_i n_j\ , \qquad M_{ij} = \frac{1}{2}(m_i n_j + m_j n_i)\ , \qquad L_{ij} = \frac{1}{2}(\ell_i n_j + \ell_j n_i) \tag{14.1.29}$$

To write an efficient finite element program, the values of $N_{ij}^\mu, M_{ij}^\mu, L_{ij}^\mu$ should be calculated, for all the discrete microplanes (labeled here by superscript μ), in advance of finite element analysis and stored in memory. The values of n_i^μ and of the weights w_μ in the integration formula (14.1.6) must also be stored in advance.

In each loading step, an explicit computational algorithm can be formulated as follows. First, the new values of macro-strains ε_{ij} are calculated at each integration point from the new (incremented) values of nodal displacements. Then, for each integration point, the new values of $\varepsilon_N, \varepsilon_M$, and ε_L are calculated for all the microplanes from (14.1.28) and the volumetric and deviatoric components ε_V and ε_D are determined from the first of (14.1.9) and (14.1.13). Using these values, the following new stress values are calculated for each microplane:

$$\sigma_V^e = \sigma_V^i + E_V(\varepsilon_V - \varepsilon_V^i)\,,\ \sigma_D^e = \sigma_D^i + E_D(\varepsilon_D - \varepsilon_D^i)\,,\ \sigma_N^e = \sigma_V^e + \sigma_D^e \tag{14.1.30}$$

$$\sigma_M^e = \sigma_M^i + E_T(\varepsilon_M - \varepsilon_M^i)\,,\ \sigma_L^e = \sigma_L^i + E_T(\varepsilon_L - \varepsilon_L^i)\,,\ \sigma_V' = \mathrm{Max}[\sigma_V^e, -F_V(\varepsilon_V)] \tag{14.1.31}$$

$$\sigma_N' = \sigma_V' + \mathrm{Min}\{\mathrm{Max}[\sigma_D^e, -F_D(\langle -\varepsilon_D\rangle)], F_D^+(\langle\varepsilon_D\rangle)\} \tag{14.1.32}$$

$$\sigma_N = \mathrm{Min}[\sigma_N', F_N(\langle\varepsilon_N\rangle, \hat{\sigma}_V), \hat{\sigma}_V] \tag{14.1.33}$$

Superscripts i denote the previously calculated initial values at the beginning of the loading step, and superscripts e denote the new stress values based on elastically calculated increments; $\langle x\rangle = \mathrm{Max}(x, 0) =$ positive part of x (this symbol, called the Macauley bracket, is used so that functions $F_N, \ldots F_T$ could be defined for only the positive values of strain arguments), and $\hat{\sigma}_V = \sigma_V^i$, but if the load step is iterated, it helps accuracy to take $\hat{\sigma}_V$ as the value of σ_V obtained in the previous iteration. After sweeping through all the microplanes $\mu = 1, \ldots N_m$, one must calculate

$$\overline{\sigma}_V = \sum_{\mu=1}^{N_m} w_\mu \sigma_N^\mu \tag{14.1.34}$$

Then, for each microplane one can calculate

$$\sigma_V = \mathrm{Min}(\sigma_V', \overline{\sigma}_V) \tag{14.1.35}$$

for $\varepsilon_V - \varepsilon_V^i > 0$:

$$\sigma_T' = \mathrm{Min}\{\sigma_T^e, T[\sigma_T^e, F_T(\langle\varepsilon_T\rangle, \sigma_V)]\},\quad \sigma_T = \mathrm{Max}[\sigma_T', -F_T(\langle -\varepsilon_T\rangle, \sigma_V)] \tag{14.1.36}$$

for $\varepsilon_V - \varepsilon_V^i \leq 0$:

$$\sigma_T' = \mathrm{Max}\{\sigma_T^e, T[\sigma_T^e, -F_T(\langle -\varepsilon_T\rangle, \sigma_V)]\},\quad \sigma_T = \mathrm{Min}[\sigma_T', F_T(\langle\varepsilon_T\rangle, \sigma_V)] \tag{14.1.37}$$

After sweeping again through all the microplanes, all the new values of the microplane stresses at the end of the loading step are known, and the macrostresses can then be calculated from (14.1.6) where the expression for the components of s_{ij} in terms of the components of the microstress vector is easily seen to be:

$$s_{ij} = \sigma_N N_{ij} + \sigma_M M_{ij} + \sigma_L L_{ij} \tag{14.1.38}$$

The inelastic parts of the new macrostresses must subsequently be modified according to a suitable nonlocal formulation. This subject is discussed later in Section 14.3.

Note that the foregoing algorithm gives the new stresses as explicit functions of the new strains. No equations need to be solved. This is important for computational efficiency.

14.1.6 Constitutive Characterization of Material on Microplane Level

By fitting of various types of test data for concrete, the following functions, characterizing the constitutive properties of the material, have been identified (Bažant, Xiang and Prat 1996):

$$F_V(-\varepsilon_V) = f_V^0 \exp\left(-\frac{\varepsilon_V}{k_1 k_5}\right),\quad f_V^0 = E k_1 k_4,\quad (\text{any } \varepsilon_V) \tag{14.1.39}$$

$$F_D(-\varepsilon_D) = f_D^0\left(1 - \frac{\varepsilon_D}{k_1 c_2}\right)^{-1},\quad f_D^0 = E k_1 c_4,\quad (\varepsilon_D \leq 0) \tag{14.1.40}$$

$$F_D^+(\varepsilon_D) = c_5 f_D^0\left(1 + \frac{\varepsilon_D}{k_1 c_2 c_5}\right)^{-1},\quad (\varepsilon_D \geq 0) \tag{14.1.41}$$

$$F_N(\varepsilon_N, \sigma_V) = f_N^0\left[1 + \left(\frac{\varepsilon_N}{c}\right)^2\right]^{-1},\quad c = c_1 k_1 + \left\langle\frac{-c_3\hat{\sigma}_V}{\varepsilon_V}\right\rangle,\quad f_N^0 = E k_1,\quad (\varepsilon_N \geq 0) \tag{14.1.42}$$

$$F_T(\sigma_N) = \langle E k_1 k_2 - k3\sigma_N\rangle \tag{14.1.43}$$

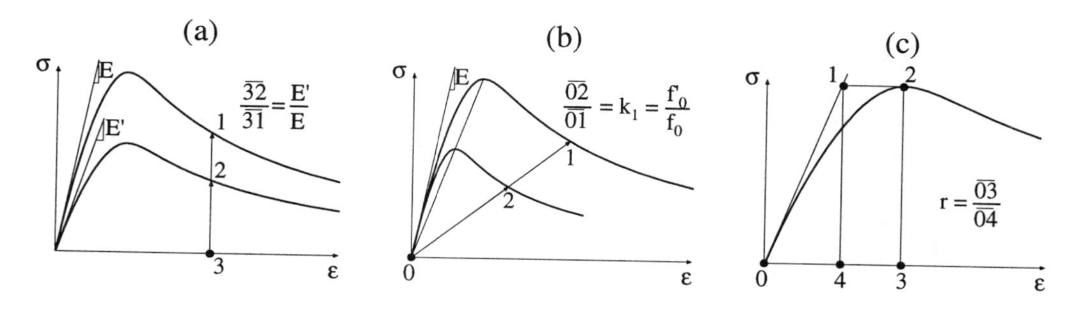

Figure 14.1.7 (a-b) Vertical and radial scaling (affinity transformations) of the stress-strain curves, and (c) concept of ductility as a ratio $\overline{03}/\overline{04}$ (from Bažant, Xiang and Prat 1996).

in which $k_1, \ldots, k_5$ are adjustable empirical constants, which take different values for different types of concretes, while $c_1, \ldots, c_5$ are fixed empirical constants that can be kept the same for all normal concretes. They have the values $c_1 = 5, c_2 = 6, c_3 = 50, c_4 = 130$, and $c_5 = 6$ (parameter c_5 affects almost only the standard triaxial tests at very high pressures). It has been recommended that, in absence of sufficient test data, the adjustable parameters may be taken with the following reference values $k_1 = 72 \times 10^{-6}, k_2 = 0.1, k_3 = 0.05, k_4 = 15$, and $k_5 = 150$. The value of Poisson's ratio may be considered as $\nu = 0.18$. Except for E, all the parameters are dimensionless.

The macroscopic Young's modulus is a parameter whose change causes a vertical scaling transformation (affinity transformation) of all the response stress-strain curves. If this parameter is changed from E to some other value E', all the stresses are multiplied by the ratio E'/E at no change of strains (Fig. 14.1.7a). Parameter k_1 describes radial scaling (affinity transformation) with respect to the origin. If this parameter is changed from k_1 to some other value k_1', all the stresses and all the strains are multiplied by the ratio k_1'/k_1 (Fig. 14.1.7b).

The aforementioned reference values of material parameters along with E =58000 MPa yield the uniaxial compression strength f_c' =42.4 MPa, as calculated by simulating the uniaxial compression test by incremental loading. The strain corresponding to the stress peak has been found to be $\varepsilon_p = 0.0022$. Now, if the user needs a microplane model that yields the uniaxial compressive strength f_c^* and the corresponding strain at peak ε_p^*, one needs to modify the reference values of only two parameters as follows:

$$k_1^* = k_1 \frac{\varepsilon_p^*}{\varepsilon_p}, \qquad E^* = E \frac{f_c^*}{f_c'} \frac{\varepsilon_p}{\varepsilon_p^*} \tag{14.1.44}$$

Table 14.1.1 shows the values of f_c', f_t' for some typical values of material parameters ("R" in Table 14.1.1 refers to the reference values stated above). It also gives the corresponding ductility $r = \varepsilon_p E/f'$, representing the ratio $\overline{03}/\overline{04}$ in Fig. 14.1.7c. The smaller r, the steeper the postpeak softening. The transformations according to (14.1.44) do not change the ratio r.

The aforementioned reference values of material parameters have been selected so that the ratio of tensile to compressive uniaxial strengths be approximately $f_t'/f_c' = 0.082$; the ratio of equitriaxal to uniaxial compression strength $f_{bc}'/f_c' = 1.17$; the ratio of the strength in pure shear to the uniaxial compressive strength approximately $f_c^s/f_c' = 0.069$; the ratio of residual stress for very large uniaxial compressive strain to the uniaxial compression strength approximately $\sigma_r/f_c' = 0.07$; and the ratio of residual stress for very large shear strain to the shear strength approximately $\tau_r/f_c^s = 0.3$. The transformations according to (14.1.44) do not change these ratios. These ratios can be changed only by adjusting material parameters other than E and k_1.

Parameters k_4 and k_5 can be determined exclusively from the data on hydrostatic compression tests. Taking the logarithm of (14.1.43), the equation can be reduced to a linear regression plot, and thus parameters k_4 and k_5 can be obtained by fitting the data on the hydrostatic compression test, separate from all other parameters (because the value of ε_V for hydrostatic compression is the same for all microplanes and $\varepsilon_D = \varepsilon_M = \varepsilon_L = 0$ for all microplanes). The softening tail in uniaxial compression can be lengthened by increasing c_2 while reducing k_1 a little, and for tension by reducing k_3 while reducing k_1

Table 14.1.1 Strength, ductility, and typical material parameters

Tests	E	k_1	k_2	k_3	k_4	bf k_5	$k_6(10^{-6})$	f'_c	r_c	f'_t	r_t
Hognestad	3900	120	R	R	R	R	R	5.18	1.96	0.46	1.78
van Mier	29000	R	R	R	R	R	R	40.0	1.98	3.75	1.80
Petersson	26000	R	R	R	0.4	R	R	32.1	2.43	3.62	1.65
Bažant	6000	112	R	R	R	12	175	7.55	1.91	0.69	1.73
Green	5100	R	R	R	R	R	125	7.39	1.93	0.66	1.78
Balmer	3500	90	20	R	0.5	18	R	3.05	2.29	0.33	1.71
Bresler	5100	R	R	0.2	R	R	R	6.49	2.36	0.40	2.44
Kupfer	5100	R	40	0.3	.06	R	R	4.86	4.41	0.62	2.22
Launay	5100	R	R	R	0.3	R	R	4.87	2.93	0.77	1.73
Sinha	3200	113	R	R	R	R	R	4.00	2.07	0.36	1.79

"R" means reference values; $r_c = \varepsilon_c^p E/f'_c$, $r_t = \varepsilon_t^p E/f'_t$

a little. The ratio of the tensile-to-compressive strength can be increased by reducing c_4 or k_3. The ratio of the strength in pure shear to the uniaxial compressive strength can be increased by increasing k_2 while reducing c_4 or k_3 a little.

14.1.7 Microplane Model for Finite Strain

In some applications of the microplane model, for example, the impact of missiles into hardened concrete structures or nuclear reactor containments, or the analysis of energy absorption of a highly confined column in an earthquake, very large strains, ranging from 10% to 200%, and shear angles up to 40°, have been encountered in calculations. For such situations, the microplane model must be generalized to finite strain. However, a thorough exposition of the finite strain generalization would require introducing advanced mathematical apparatus that has not yet appeared in this book. Therefore, only a summary of the main results will be given here. The interested reader can find the details in Bažant, Xiang and Prat (1996) and Bažant, Xiang et al. (1996) for the case of moderately large strains (up to about 10%), and in Bažant (1997b) for the case of very large strains (100% or more, with shear angles up to 40°).

The simplest finite strain tensor to use is Green's Lagrangian strain tensor $\mathbf{E} = (\mathbf{F}^T\mathbf{F} - \mathbf{1})/2$ where $\mathbf{F}$ is the deformation gradient and $\mathbf{1}$ the unit tensor (see, e.g., Bažant and Cedolin 1991, Chapter 11). Its conjugate stress tensor, that is the tensor for which Green's Lagrangian strain tensor gives a correct work expression $dW = \mathbf{T} \cdot d\mathbf{E}$, is a tensor called the second Piola-Kirchhoff stress tensor, $\mathbf{T} = \mathbf{F}^{-1}J\mathbf{S}\mathbf{F}^{-T}$ where $J = \det \mathbf{F}$ is the Jacobian of the transformation (giving the relative volume change); $\mathbf{F}^{-T} = (\mathbf{F}^{-1})^T = (\mathbf{F}^T)^{-1}$.

Difficult problems arise in the modeling of very large strains. In finite-strain generalization of the microplane model, a definite physical meaning needs to be attached to the normal and shear strain components on the microplanes. In this regard, the following two conditions must be met:

Condition I. The normal and shear components of the stress tensor used in the constitutive relation must uniquely characterize the normal and shear components of the tensor of true stress $\mathbf{S}$ in the deformed material, called the Cauchy stress tensor.

Condition II. The normal strain component e_N, characterizing the stretch λ_N of a material line segment in the direction $\vec{n}$ initially normal to the microplane, must be independent of the stretches of material line segments in other initial directions. Furthermore, the shear strain component e_{NM} (or e_{NL}), characterizing the change of angle θ_{NM} or θ_{NL} between two initially orthogonal material line segments with initial unit vectors $\vec{n}$ and $\vec{m}$ (or $\vec{n}$ and $\vec{\ell}$), must be independent of the stretches and angle changes in planes other than $(\vec{n}, \vec{m})$ or $(\vec{n}, \vec{\ell})$.

Consider first only condition I. It turns out that, in this regard, the use of the second Piola-Kirchhoff stress tensor is possible only if the largest magnitude of the principal strains is less than about 7% to 10%, i.e., if the strain is only moderately large. It has been shown by numerical examples (Bažant 1997b) that, for large isochoric deformations, the shear components of the second Piola-Kirchhoff stress tensor $\mathbf{T}$ strongly depend on the volumetric component of the Cauchy (true) stress tensor $\mathbf{S}$ (i.e., the true hydrostatic pressure), and the volumetric component of the second Piola-Kirchhoff stress tensor strongly depends on the shear components of the Cauchy stress tensor. This indicates that the projections of the second Piola-Kirchhoff stress tensor on the microplanes have no physical meaning. They cannot be used to characterize the strength, yield limit and damage on the microplane, nor the phenomenon of friction.

The stress tensor must be referred to the initial configuration of the material (as required for the modeling of a solid remembering the initial state). The only such tensor whose microplane components have a physical meaning is the rotated Kirchhoff stress tensor $\boldsymbol{\tau} = \mathbf{R}^T J\mathbf{S}\mathbf{R}$, where $J\mathbf{S}$ represents the Kirchhoff stress tensor, and $\mathbf{R}$ is the material rotation tensor defined in the polar decomposition of the deformation gradient $\mathbf{F} = \mathbf{R}\mathbf{U} = \mathbf{V}\mathbf{R}$. Here, $\mathbf{U}$ and $\mathbf{V}$ are the right and left stretch tensors. When the principal stress axes do not rotate against the material, the rotated Kirchhoff stress tensor is equal to the Cauchy (true) stress tensor scaled by a scalar factor, J. Only this tensor is free of the aforementioned problems revealed by numerical examples.

A variational procedure can be used to obtain an expression for the finite strain tensor γ that is conjugated by work with the rotated Kirchhoff stress tensor $\boldsymbol{\tau}$. If the principal strain axes do not rotate against the material, this tensor is found to be identical to Hencky's (logarithmic) strain tensor. However, when the principal strain axes rotate, one obtains an incremental expression for $d\gamma$ that cannot be integrated. This means that the strain tensor conjugate to the rotated Kirchhoff stress tensor is nonunique, path-dependent (nonholomonic).

The aforementioned path-dependence is strong and unacceptable, except for moderately large strains less than about 7% to 10%. For such strains, and for larger strains for which the rotations of principal strain axes is small, the use of Hencky's strain tensor is advantageous. However, there is also the problem of the efficient calculation of the Hencky tensor. This tensor is defined by the spectral representation, which is computationally demanding for large finite element programs in which this tensor may have to be calculated up to a billion times. Nevertheless, an easy-to-compute very close approximation of the Hencky strain tensor has recently been found (Bažant 1997c).

Consider now condition II. The relative length change of a segment normal to the microplane from length dS (in the initial configuration) to length ds (in the deformed configuration) is characterized, e.g., by $e_N = (ds - dS)/dS$ (called Biot strain or engineering strain). The change of angle between the microplane normal vector $\vec{n}$ and vector $\vec{m}$ in the microplane represents the shear angle θ_{NM}. When Green's Lagrangian strain tensor $\mathbf{E}$ is used, e_N can be expressed (exactly) in terms of the normal component E_N, and θ_{NM} can be expressed (exactly) in terms of the shear component E_{NM} and the normal components E_N, E_M (see, e.g., Malvern 1969, pp.165–166). In other words, the exact change of length in normal direction and of shear angle for a microplane can be expressed solely in terms of the strain tensor components on the same microplane. This is not true, however, for all the other strain tensors, including Hencky's (logarithmic) strain tensor and Biot's strain tensor. For them, the exact e_N and θ_{NM} depend also on the ratio of the principal strains (which seems an inconvenient feature for the programming of microplane model and would increase demands on computer time). This dependence can be neglected only when the maximum principal strain is less than about 25% (Bažant 1997d).

It thus appears that, for large strains (i.e., when the maximum principal strain exceeds 7% to 10%) the only suitable strain and stress tensors are Green's Lagrangian strain tensor and the rotated Kirchhoff stress tensor. These two tensors are not conjugate.

It has normally been considered a taboo to use nonconjugate stresses and strains. However, due to the special character of the present microplane model, the use of nonconjugate stresses and strains in formulating a constitutive relation is admissible if certain precautions are taken (see Bažant 1995d and Bažant, Adley and Xiang 1996). One point to note in this regard is that the constitutive relation in terms of the aforementioned nonconjugate stress and strain tensors is a transformation of the constitutive relation in terms of conjugate stress and strain tensors such that the transformation depends only Green's Lagrangian strain tensor (or the stretch tensor) but, importantly, is independent of the material rotation tensor. Such a transformation is perfectly admissible. The second point to note is that nonnegativeness of energy dissipation is ensured for two reasons: (1) The elastic parts of strains are always small (which ensures that the elastic part of the nonconjugate stress-strain relation preserves energy), and (2) the drop of stress to the boundary surface is carried out in each load step at constant strain and cannot cause negative energy dissipation.

A further precaution that must be taken is that the work done by the stresses (or by the nodal forces on displacements) cannot be directly calculated from the stresses and strains used in the constitutive law, because the areas under the stress-strain curves for the nonconjugate constitutive law do not correctly characterize energy dissipation. If the work needs to be calculated, one can easily obtain the second Piola-Kirchhoff stress tensor from the rotated Kirchhoff stress tensor and evaluate the work that way. Also, elastic response cannot be described as a functional relation between nonconjugate stresses and strains.

For moderately large strains, of course, the conjugate pair or Green's Lagrangian strain tensor and second Piola-Kirchhoff strain tensor can be used, and has been used by Bažant, Xiang and Prat (1996) in a finite strain generalization of the microplane model. There is, however, a gap in the experimental data for very large strains of concrete. To fill this gap, one must get reconciled with the fact that it is next to impossible to keep the specimen deformation uniform when triaxial deformations become large. Triaxial test data on concrete at strains up to shear angle of 35° at very high hydrostatic pressures (several times the uniaxial compression strength) have recently been obtained by Bažant and Kim (1996b) using a novel type of test, called the 'tube-squash' test. In this test, a thick-walled tube of very ductile steel is filled with concrete, and after curing, it is compressed axially to about half the initial length. The concrete undergoes shear angles over 30°. Due to high confining pressure (which exceeds 1000 MPa), the concrete in the tube retains integrity and small cores can be drilled out from the concrete. These cores show uniaxial compression strength between 20% and 50% of the virgin concrete, both for normal and high strength concretes. In the evaluation of the 'tube-squash test' one must fit the measured load-displacement curves with a finite element program incorporating finite strain constitutive models for both the concrete and the steel.

Finite strain tests need to be also carried out at small hydrostatic (confining) pressures, at which concrete turns into rubble when large deformations occur. A constitutive relation for such rubbelized concrete at finite strain needs to be developed.

Another problem that needs to be resolved for the microplane model is the split of total normal strain into deviatoric and volumetric components. The decomposition of large deformations into their volumetric and deviatoric (strictly speaking, isochoric) parts is, in general, multiplicative. Specifically, it has the form $\mathbf{U} = \mathbf{F}_D\mathbf{U}_V$ (Flory 1961; Sidoroff 1974; Simo 1988; Simo and Ortiz 1985; Lubliner 1986; Bell 1985) where $\mathbf{U}$ is the right stretch tensor, $\mathbf{U}_V$ the volumetric right-stretch tensor, and $\mathbf{F}_D$ = the deviatoric transformation tensor. An additive volumetric-deviatoric decomposition exists only for the Hencky (logarithmic) strain tensor $\mathbf{H}$.

For any type of finite strain tensor, however, an approximate additive decomposition in terms of volumetric strain tensor $\mathbf{E}_V = \varepsilon_V \mathbf{1}$ and deviatoric strain tensor $\mathbf{E}_D$ is possible for materials that can exhibit only large deviatoric strains but not large volumetric strains (Bažant 1996c), as is the case for concrete. Unlike $\mathbf{F}_D$, the components of $\mathbf{E}_D$ depend on J, i.e., the relative volume change (unless the Hencky strain tensor is used). However, their dependence on J is, in the case of Green's Lagrangian strain tensor, negligible if the volume change is less than about 3% in magnitude (Bažant, Xiang and Prat 1996; Bažant, Xiang et al. 1996). For Biot strain tensor $\mathbf{E}^b = \mathbf{U} - \mathbf{1}$ (Biot 1965; Ogden 1984; Bažant and Cedolin 1991), the limit is about 8% (for concrete, the volume change is -3% at highest pressure tested so far, which is 300000 psi or 2069 MPa; Bažant, Bishop and Chang 1986). Thus, the classical multiplicative decomposition, which is not as convenient for calculations as the additive decomposition, seems to be inevitable only for materials exhibiting very large volume changes, such as stiff foams. An additive decomposition of the aforementioned kind, developed in Bažant (1996c), was used by Bažant, Xiang and Prat (1996) in the generalization of the microplane model for moderately large finite strains of concrete.

The multiplicative decomposition could nevertheless be implemented in the microplane model by decomposing each loading step into two substeps, pure volumetric deformation followed by pure isochoric deformation, but that would greatly complicate the analysis, especially if the solution is not explicit.

14.1.8 Summary of Main Points

This section has explained the basic concept and the latest formulation of the general microplane model for concrete—a constitutive model in which the nonlinear triaxial behavior is characterized by relations between the stress and strain components on a microplane of any orientation under the constraint that the strains on the microplane are the projections of the macroscopic stress tensor. The microplane model simplifies constitutive modeling because the stress-strain relation on the microplane level involves only a few stress and strain components that have a clear physical meaning. The passage from elastic response to softening damage defined in terms of different variables is effectively handled by the concept of boundaries in the stress-strain space. The advantage of this recently proposed concept is that various boundaries and the elastic behavior can be defined as a function of different variables (strain components). While the stress-strain boundaries for compression are defined separately for volumetric and deviatoric components, the boundary for tension is defined in terms of the total normal strains. This is necessary to achieve a realistic triaxial response at large tensile strains. A smooth transition from the elastic behavior to the boundary curve has also been formulated. The formulation is fully explicit, that is, the stress can be explicitly calculated from given strains.

Exercises

14.1 Based on symmetry properties, show that the rectangular cartesian components of the tensor $\mathbf{A}$ in (14.1.20) satisfy the following properties: (a) the off-diagonal components are zero; (b) the diagonal elements are equal. (c) Demonstrate that $\mathbf{A} = A\mathbf{1}$, where A is a scalar, and (d) compute A. (Hint: compute tr $\mathbf{A} = 3A = (3/2\pi)\int d\Omega$ —Why?)

14.2 Show that $\mathbf{A}$ in (14.1.20) can be written as $\mathbf{A} = (3/4\pi)\int_\Omega \vec{r}\otimes\vec{n}\,d\Omega$ where the integral is now extended to the surface of the whole unit sphere and $\vec{r}$ is the position vector relative to the center of this sphere. Apply the divergence theorem to this surface integral and show that $\mathbf{A} = (3/4\pi)\int_V \text{grad}\,\vec{r}dV$ where V is the region defined by the unit sphere. Use this expression to determine $\mathbf{A}$.

14.3 Let B_{ijkl} be the rectangular cartesian components of the fourth-order tensor $\boldsymbol{B}$ in (14.1.20). Show that they satisfy the relations (a) $B_{ijkl}\delta_{kl} = A_{ij}$ and $B_{ijkl}\delta_{jk} = A_{il}$. A basic property of linear elasticity is that the most general cartesian form of an isotropic fourth-order tensor of elastic moduli, say $\boldsymbol{B}$, such that $\boldsymbol{\sigma} = \boldsymbol{B\varepsilon}$ is isotropic, is $B_{ijkl} = B_0\delta_{ij}\delta_{kl} + B_1\delta_{ik}\delta_{jl}$, where B_0 and B_1 are constants. (b) Use the results in (a) to show that for $\boldsymbol{B}$ in (14.1.20) $B_0 = 0$ and $3B_1 =$ tr $\mathbf{A}$. (c) Use the result of the previous exercise to determine B_1.

14.4 Show that $\boldsymbol{B}$ in (14.1.20) can be written as $\boldsymbol{B} = (3/4\pi)\int_\Omega \vec{r}\otimes\vec{r}\otimes\vec{r}\otimes\vec{n}\,d\Omega$ where now the integral is extended to the surface of the whole unit sphere and $\vec{r}$ is the position vector relative to the center of this sphere. Apply the divergence theorem to this surface integral and show that $\boldsymbol{B} = (3/4\pi)\int_V \text{grad}\,(\vec{r}\otimes\vec{r}\otimes\vec{r})\,dV$ where V is the region defined by the unit sphere. Show that the component form of this integral can be reduced to $B_{ijkl} = \delta_{il}J_{jk} + \delta_{jl}J_{ik} + \delta_{kl}J_{ij}$ where $\mathbf{J}$ is the Euler tensor of inertia products for a sphere of unit radius and unit density with respect to its center:

$$\mathbf{J} = \int_V \vec{r}\otimes\vec{r}\,dV \qquad \text{or} \qquad J_{ij} = \int_V x_i x_j\,dV \tag{14.1.45}$$

Use the well-known result that the inertia moment of a homogeneous sphere relative to any diameter is $2mR^2/5$, with m = mass and R = radius of the sphere, to prove that $\mathbf{J} = (1/5)\mathbf{1}$. Finally, determine the general expression for the components B_{ijkl}.

14.2 Calibration by Test Data, Verification and Properties of Microplane Model

Following the general theoretical formulation in the preceding section, we will now demonstrate calibration and verification of the microplane model by fitting of the relevant test data from the literature. We will also show how the data afflicted by localization of damage within the gage length can be decontaminated.

14.2.1 Procedure for Delocalization of Test Data and Material Identification

Until very recently it has been general practice to identify the postpeak stress-strain relation from test data ignoring the fact that the deformation of the specimen within the gage length often becomes nonuniform, due to localization of cracking damage. The fact that damage must localize, except in the smallest possible specimens, was shown in detail in Chapter 8. The correct analysis of localization in strain softening materials led first to the development of the crack band model (Chapter 8), and later to the more sophisticated models described in the preceding chapter. The localization phenomena were already documented in the early eighties. However, because the general problem of identification of material parameters in presence of strain-softening localization (Ortiz 1987) is tremendously complex, the contamination of test

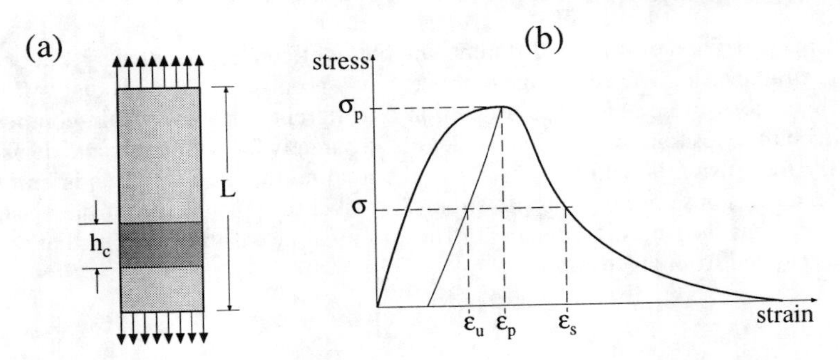

Figure 14.2.1 Underlying crack-band model (series coupling model) for filtering of strain softening localization from laboratory test data.

data by localization has typically been ignored. At the present state of knowledge, however, this is no longer acceptable. The data must be decontaminated, delocalized. An approximate procedure to do that, applicable to any type of constitutive model, was recently proposed by Bažant, Xiang et al. (1996).

The delocalization cannot, and need not, be done with a high degree of accuracy and sophistication. In the identification of the microplane model by Bažant, Xiang et al. (1996), the test data from laboratory specimens have been analyzed taking into account the strain localization in an approximate manner. The idea is to exploit two simple approximate concepts: (1) localization in the series coupling model described in Sections 8.1–8.3, and (2) the effect that energy release due to localization within the cross section of specimen has on the maximum load, as described by Bažant's size effect law (Section 1.4 and Chapter 6).

The strain as commonly observed is the average strain ε_m on a gage length L. According to the series coupling model and the crack band model (Sections 8.1–8.3), the strain may be assumed to localize after the peak into a band of width h_c, as depicted in Fig. 14.2.1a, while the remainder of the gauge length unloads. In this way, the strain of the material inside the localized zone is ε_s —corresponding to the softening branch— while in the remaining part the strain is ε_u, as given by the unloading curve from the peak (Fig. 14.2.1b).

The strain that the constitutive model for damage should predict is the strain ε_s in the localization zone. But this strain is difficult to measure, for three reasons: (1) the size of the localization zone is small, which reduces the accuracy of strain measurements; (2) the location of the localization zone is uncertain, and so one does not know where to place the gage; and (3) the deformation of the localization zone is quite random while the constitutive model predicts the statistical mean of many random realizations (determining this mean requires taking measurements on many specimens). Therefore, a simplified method is desirable based on measuring only the average strain ε_m.

To find the simplified formula, we note that the total increment of the gauge length ΔL (equal to $L\varepsilon_m$ by definition) is obtained by adding the contributions of the softening and unloading regions, i.e.,

$$L\varepsilon_m = h_c\varepsilon_s + (L - h_c)\varepsilon_u \tag{14.2.1}$$

If we further assume that the unloading proceeds parallel to the initial elastic loading (i.e., stiffness degradation up to the peak is negligible), then the unloading strain is $\varepsilon_u = \varepsilon_p - (\sigma_p - \sigma)/E$, where E is the elastic modulus and ε_p and σ_p are the strain and stress at the peak of the stress-strain curve for the given type of loading (Fig. 14.2.1b). So we finally get

$$\text{for } \varepsilon_m > \varepsilon_p : \qquad \varepsilon_s = \frac{L}{h_c}\varepsilon_m - \frac{L - h_c}{h_c}\left(\varepsilon_p - \frac{\sigma_p - \sigma}{E}\right) \tag{14.2.2}$$

To correct the given test data according to (14.2.2), one must obviously know the value of the localization length h_c. It is impossible to determine this length from the reports on the uniaxial, biaxial, and triaxial tests of concrete found in the literature. However, a reasonable estimate can be made by experience from other studies; $\ell \approx 3d_a$ where d_a = maximum size of the aggregate in concrete (for high-strength concretes, ℓ is likely smaller, perhaps as small as $\ell = d_a$).

Bažant, Xiang et al. 1996 further proposed an approximate procedure to filter out of the given tensile test data the size effect on the maximum tensile stress. This procedure was based on the size effect law. According to the size effect on maximum load, they scaled the measured response curve by affinity transformation with respect to the strain axis and in the direction parallel to the elastic slope. Thus, they obtained the response curve with the peak tensile stress corresponding to specimen of size h_c.

14.2.2 Calibration of Microplane Model and Comparison with Test Data

The microplane model we described has been calibrated and compared to the typical test data available in the literature (Bažant, Xiang et al. 1996). They included: (**1**) uniaxial compression tests by van Mier (1984, 1986; Fig. 14.2.2a), for different specimen lengths and with lateral strains and volume changes measured, and by Hognestad, Hanson and McHenry (1955; Fig. 14.2.2b); (**2**) uniaxial direct tension tests by Petersson (1981; Fig. 14.2.2c); (**3**) uniaxial strain compression tests of Bažant, Bishop and Chang (1986; Fig. 14.2.2d); (**4**) hydrostatic compression tests by Green and Swanson (1973; Fig. 14.2.2e); (**5**) standard triaxial compression tests (hydrostatic loading followed by increase of one principal stress) by Balmer (1949; Fig. 14.2.2f); (**6**) uniaxial cyclic compression tests of Sinha, Gerstle and Tulin (1964; Fig. 14.2.2g). (**7**) tests of shear-compression failure envelopes under torsion by Bresler and Pister (1958) and Goode and Helmy (1967; Fig. 14.2.3a); (**8**) tests of biaxial failure envelope by Kupfer, Hilsdorf and Rüsch (1969; Fig. 14.2.3b); and (**9**) failure envelopes from triaxial tests in octahedral plane (π-projection) by Launay and Gachon (1971; Fig. 14.2.3c).

As seen from the figures, good fits of test data can be achieved with the microplane model. In Fig. 14.2.2a it should be noted that the uniaxial compression stress-strain diagrams are well represented for three specimens lengths, $\ell = 5, 10$, and 20 cm (it was already shown that the series coupling describes well the length effect in these tests; see Bažant and Cedolin 1991, Sec. 13.2). Fig. 14.2.2d serves as the basis for calibrating the volumetric stress-strain boundary, and a good fit is seen to be achieved for these enormous compressive stresses (up to 300 ksi or 2 GPa). Fig. 14.2.2f shows that the large effect of the confining pressure in standard triaxial tests can also be captured.

In Fig. 14.2.2g, note that the subsequent stress peaks in cycles reaching into the softening range are modeled quite correctly, and so are the initial unloading slopes. Significant differences, however, appear at the bottom of the cyclic loops, which is due to the fact that the unloading modulus is, in the present model, kept constant (a refinement would be possible by changing the constant unloading slope on the microplane level to a gradually decreasing slope, of course, with some loss of simplicity). It should also be noted that the loading in these tests was quite slow and much of the curvature may have been due to relaxation caused by creep.

In Fig. 14.2.3c note that the model predicts well the shape of the failure envelopes, which is noncircular and nonhexagonal, corresponding to rounded irregular hexagons squashed from three sides. Fig. 14.2.3b shows that the ratio of uniaxial and biaxial compression strengths found in these tests can be modeled.

It must be emphasized that all the solid curves plotted in the figures are the curves that are predicted by the microplane model. The dotted curves in Fig. 14.2.2 are those after correction according to the series coupling model. The dashed curves in Fig. 14.2.2c are those after correction according to the size effect law, and the dotted curves are those after a further correction according to the series coupling model.

Note that only six parameters need to be adjusted if a complete set of uniaxial, biaxial, and triaxial test data is available, and two of them can be determined separately in advance from the volumetric compression curve. If the data are limited, fewer parameters need to be adjusted. The parameters are formulated in such a manner that two of them represent scaling by affinity transformation. Normally only these two parameters need to be adjusted, which can be done by simple closed-form formulas. Thus, we can conclude that the model may be efficiently used to describe concrete behavior in uniaxial, biaxial, and triaxial situations.

14.2.3 Vertex Effects

There is another important property that is exhibited by the microplane model, and not, for example, by macroscopic plasticity models. For a nonproportional path with an abrupt change of direction such that the load increment in the σ_{ij} space is directed parallel to the yield surface, the response of a plasticity model is perfectly elastic, unless this change of direction happens at a corner of the yield surface. But

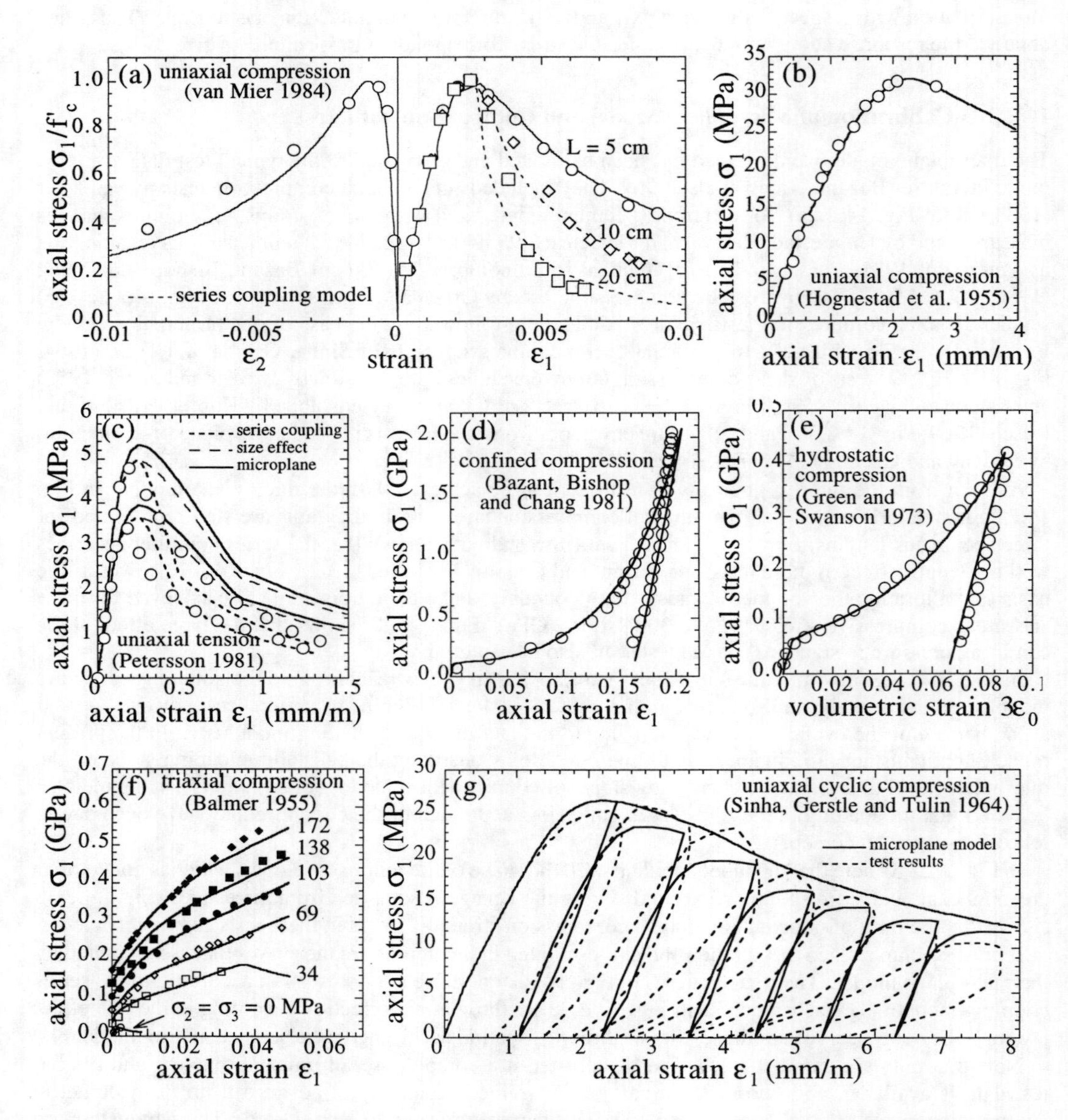

Figure 14.2.2 Experimental results from various sources and best fits with the microplane model (after Bažant, Xiang et al. 1996): (a) uniaxial compression tests by van Mier (1984); (b) uniaxial compression tests by Hognestad, Hanson and McHenry (1955; (c) uniaxial tension tests by Petersson (1981); (d) confined compression test (uniaxial strain) of Bažant, Bishop and Chang (1986); (e) hydrostatic compression test by Green and Swanson (1973); (f) triaxial test data (increasing axial compression at constant lateral confining pressure) by Balmer (1949); (g) uniaxial cyclic compression tests of Sinha, Gerstle and Tulin (1964). (Adapted from Bažant, Xiang et al. 1996.)

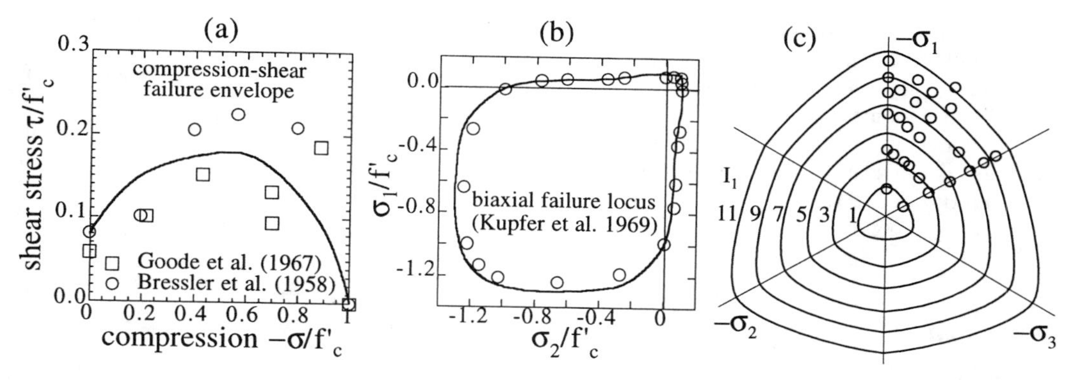

Figure 14.2.3 Fitting of (a) shear compression failure envelope (in torsion) measured by Bresler and Pister (1958), (b) biaxial failure envelope measured by Kupfer, Hilsdorf and Rüsch (1969), and (c) failure envelopes in hydrostatic planes at various pressures measured by Launay and Gachon (1971). (Adapted from Bažant, Xiang et al. 1996.)

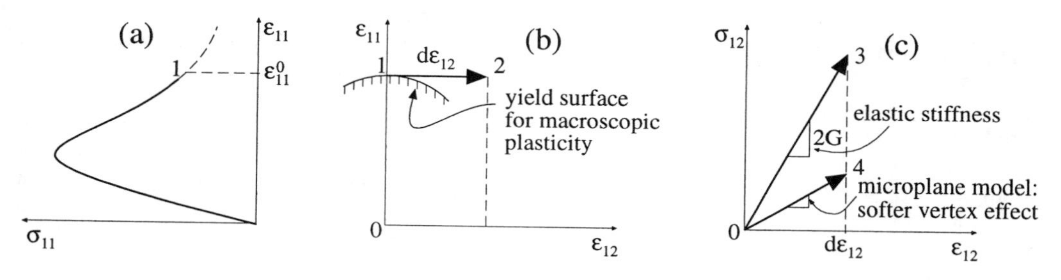

Figure 14.2.4 Vertex effect: (a) preloading in the σ_{11}-ε_{11} space at increasing ε_{11} and zero shear strain; (b) in the ε_{11}-ε_{12} space preloading corresponds to segment $\overline{01}$ and further tangent loading to segment $\overline{12}$; (c) the further tangent loading in the σ_{12}-ε_{12} diagram corresponds to segments $\overline{03}$ in classical plasticity models (fully elastic loading) and to segment $\overline{04}$ when vertex effect is present (after Bažant, Xiang et al. 1996).

in reality, for all materials, this response is softer, in fact much softer, than elastic. It is as if a corner or vertex of the yield surface traveled with the state point along the path.

This effect, called the vertex effect (see Sec. 10.7 in Bažant and Cedolin 1991), is automatically described by the microplane model, but is very hard to model with the usual plastic or plastic-fracturing models. It can be described only by models with many simultaneous yield surfaces, which are prohibitively difficult in the σ_{ij} space. The microplane model is, in effect, equivalent to a set of many simultaneous yield surfaces, one for each microplane component (although these surfaces are described in the space of microplane stress components rather than in the σ_{ij} space).

This is one important advantage of the microplane approach. It is, for example, important for obtaining the correct incremental stiffness for the case when a $d\varepsilon_{12}$-increment (segment $\overline{12}$ in Fig. 14.2.4b) is superimposed on a large strain ε_{11}^0 (segment $\overline{01}$) in the inelastic range. Segment $\overline{03}$ in Fig. 14.2.4c is the predicted response according to all classical macroscopic models with yield surfaces, which is elastic, and segment $\overline{04}$ is the prediction of microplane model, which is much softer than elastic (i.e., $d\sigma_{12}/d\varepsilon_{12} < 2G$ where G = elastic shear modulus). Fig. 14.2.4c shows the incremental stiffness $\overline{04}$ calculated for the case of the present reference parameters and $\varepsilon_{11}^0 = 0.005$. Indeed, the slope $\overline{04}$ is almost $1/5$ of the slope $\overline{03}$ which would be predicted by plasticity with a simple yield surface.

14.2.4 Other Aspects

To check for the limit of stability and for bifurcations of the response path, the tangential stiffness matrix is needed. The microplane model does not provide it directly, but it can always be computed by incrementing the strain components (or the displacements) one by one and solving for the corresponding stress changes with the microplane model.

A greater insight into the microplane model, which may be useful for data fitting, can be achieved by separating the geometric aspect of damage (i.e., the effect of reduction of the stress-resisting cross section of the material) from other inelastic phenomena. This separation has been achieved in Carol, Bažant and Prat (1991). Correlation to plasticity models and continuum damage mechanics has been elucidated in Carol and Bažant (1997).

14.3 Nonlocal Adaptation of Microplane Model or Other Constitutive Models

In unconfined straining, the microplane model displays softening. Therefore, localization limiters of some kind must be used to avoid spurious localization and mesh sensitivity, as for all other models with strain softening. This can be easily implemented using a nonlocal adaptation of the microplane model in which the inelastic stress increment is made nonlocal following the theory of microcrack interactions presented in the previous chapter (Bažant 1994b; see §13.3). This approach affects the flow of calculation only partially and a general finite element scheme can be used. Fig. 14.1.1c shows the basic calculation flowchart for this approach in which the nonlocal adaptation is implemented just after the microplane stresses get computed; the flow bifurcates and the inelastic incremental stress is computed following the nonlocal theory with microcrack interactions.

The microplane model as presented, or for that matter any constitutive model for damage, gives a prescription to calculate the stress tensor $\boldsymbol{\sigma}$ as some tensor-valued function $\mathbf{R}$ of the strain tensor ε (and of some further parameters depending on the loading history, e.g., on whether there is loading or unloading). So, $\boldsymbol{\sigma} = \mathbf{R}(\varepsilon)$. The most robust (although not always the most accurate) method of structural analysis is to base the solution of a loading step or time step on the incremental elastic stress-strain relation with inelastic strain involving the initial elastic moduli tensor $\boldsymbol{E}$, as explained in Section 13.3.1. Then, for a local formulation, the inelastic stress increment tensor $\Delta\mathbf{S}$ defined in (13.3.1) can be computed as

$$\Delta\mathbf{S} = \boldsymbol{E}(\varepsilon_{\text{new}} - \varepsilon_{\text{old}}) - \mathbf{R}(\varepsilon_{\text{new}}) + \mathbf{R}(\varepsilon_{\text{old}}) \tag{14.3.1}$$

in which subscripts old and new label the old and new value of the variables at the beginning and end of the loading step (or time step); and $\mathbf{S}$ is the inelastic stress tensor due to nonlinear behavior. This stress-strain relation is used for both dynamic explicit analysis and static implicit analysis (as the iterative initial stiffness method).

A possible simple approach to introduce nonlocal effects is similar to the isotropic scalar nonlocal approach (Pijaudier-Cabot and Bažant 1987), which was applied to the microplane model by Bažant and Ožbolt (1990, 1992) and Ožbolt and Bažant (1992). In this approach, the elastic parts of stress increments are calculated locally. The inelastic parts of the increments of $\mathbf{S}$ must be calculated nonlocally. This is accomplished by first determining, at each integration point of each finite element, the average (or nonlocal) strains $\bar{\varepsilon}$, and then calculating nonlocal $\overline{\Delta\mathbf{S}}$ from these, i.e.,

$$\Delta\boldsymbol{\sigma} = \boldsymbol{E}\Delta\varepsilon - \overline{\Delta\mathbf{S}}; \qquad \overline{\Delta\mathbf{S}} = \boldsymbol{E}(\bar{\varepsilon}_{\text{old}} - \bar{\varepsilon}_{\text{new}}) - \mathbf{R}(\bar{\varepsilon}_{\text{new}}) + \mathbf{R}(\bar{\varepsilon}_{\text{old}}) \tag{14.3.2}$$

The only modification required in a local finite element program is to insert the spatial averaging subroutine just before the calculation of $\overline{\Delta\mathbf{S}}$.

A better approach is to introduce the crack interaction concept explained in Section 13.3 and write the incremental elastic stress-strain relation as

$$\Delta\boldsymbol{\sigma} = \boldsymbol{E}\Delta\varepsilon - \Delta\tilde{\mathbf{S}} \tag{14.3.3}$$

in which $\Delta\tilde{\mathbf{S}}$ is given by Eq. (13.3.9). The spatially averaged strains are not calculated in this approach. The nonlocal part of the analysis proceeds in the following steps:

1. First, $\Delta\mathbf{S}$ is calculated (in the local form) from (14.3.1) according to the microplane model. Then one calculates at each integration point of each finite element the maximum principal direction vectors $\mathbf{n}^{(i)}$ $(i = 1, 2, 3)$ of strain tensor ε, for which the value of ε_{old} may be used as an approximation.

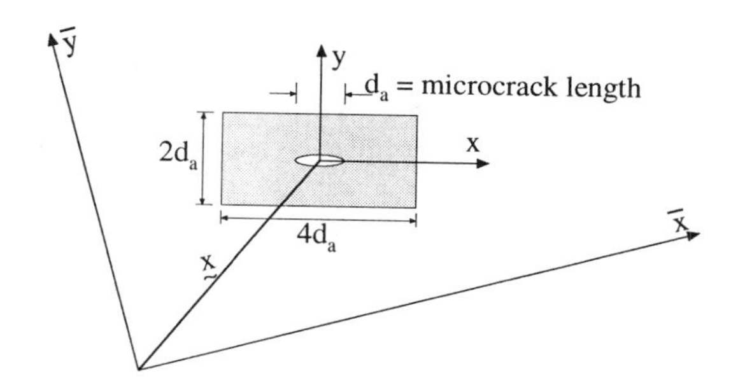

Figure 14.3.1 Local representative volume: orientation and size (adapted from Ožbolt and Bažant 1996).

2. Then one starts a loop on principal strain directions $\mathbf{n}^i$ ($i = 1, 2, 3$) of tensor $\boldsymbol{\varepsilon}$ and evaluates the inelastic stress changes in the directions $\mathbf{n}^{(i)}$, that is, $\Delta S^{(i)} = \mathbf{n}^{(i)} \cdot \Delta \mathbf{S} \mathbf{n}^{(i)}$ or $\Delta S^{(i)} = n_i^{(i)} \Delta S_{ij} n_j^{(i)}$. For those principal directions $\mathbf{n}^{(i)}$ for which $\Delta S^{(i)} \leq 0$, the nonlocal calculations are skipped because the inelastic strain is not due to cracking, i.e., one jumps directly to the end of this loop; here we make the assumption that, on the microscale (but not on the macroscale), there is no softening in compression, which is true for the microplane model.
3. The values of $\Delta S^{(i)}$ for the integration points of finite elements are then spatially averaged:

$$\overline{\Delta S}_{\mu}^{(i)} = \frac{1}{V_{\mu}} \sum_{\nu=1}^{n} \Delta S_{\nu}^{(i)} \alpha_{\mu\nu} \Delta V_{\nu} \tag{14.3.4}$$

where $V_{\mu} = \sum_{\nu=1}^{n} \alpha_{\mu\nu} \Delta V_{\nu}$ = normalizing factor, n = number of all the integration points inside the averaging volume, and $\alpha_{\mu\nu}$ = given weight coefficients, whose distribution is suitably chosen with a bell shape in both x and y directions, described by a polynomial of the fourth degree. The bell shape, which is similar to that in the nonlocal damage approach (Chapter 13, Eq. (13.1.5)) is reasonable in that it gives larger contributions to the sum from points that lie closer. Because the spacing of major cracks in concrete is approximately the same as the spacing of the largest aggregate pieces, the size of the averaging volume may be assumed to be approximately proportional to the maximum aggregate size, d_a. For two-dimensional analysis, the region of averaging should probably be taken as a rectangle with its longer side in the direction normal to $\mathbf{n}^{(1)}$ (Fig. 14.3.1)
4. The values of the nonlocal principal inelastic stress increments $\Delta \tilde{S}^{(i)}$ must then be solved from the system of linear equations (13.3.31) based on the crack influence function $\Lambda_{\mu\nu}$. However, as discussed before, exact solution is normally not needed. Depending on the type of program, one of two approximate methods can be used:

   (a) In programs in which the loading step is iterated, these equations may be solved iteratively within the same iteration loop as that used to solve the nonlinear constitutive relation, using the following equation:

$$\Delta \tilde{S}_{\mu}^{(i)\text{new}} = \overline{\Delta S}_{\mu}^{(i)} + \frac{1}{V_{\mu}} \sum_{\nu=1}^{N} \Delta V_{\nu} \Lambda'_{\mu\nu} \Delta \tilde{S}_{\nu}^{(i)\text{old}} \qquad (\mu = 1, 2, ...N) \tag{14.3.5}$$

   in which $\Lambda'_{\mu\nu}$ = crack influence matrix defined in Eq. (13.3.43), which must, however, be adjusted with factor k_b for finite elements close to the boundary of concrete (Section 13.3.9).

   (b) In explicit finite element programs without iteration, one may calculate from (14.3.5) only the first iterate ($r = 1$), which represents one explicit calculation, requiring only the values of $\Delta S_{\mu}^{(i)}$ and $\langle \Delta S_{\mu}^{(i)} \rangle$. The premise of this approximation is that the repetitions of similar calculations (for $r = 1$) in the next loading step (or time step) effectively serve as the subsequent iterations (for $r = 2, 3, 4, ...$) because the loading steps in the explicit programs are very small. This of course means that the correct value of $\overline{\Delta S}_{\mu}^{(i)}$ gets established with a delay of several steps or time intervals (in other words, the computer program is using nonlocal inelastic stress increments that are several steps old; the nonlocal interactions expressed by the crack influence function are delayed by several steps).

We recall from Section 13.3.9 that the adjustment by factor k_b must ensure that, even if part of the influencing volume protrudes beyond the boundary, the condition $\sum_{\nu=1}^{N} \Lambda'_{\mu\nu} = 0$ be met. Because this condition may be written as $\sum_{\text{interior}} \Lambda_{\mu\nu} + k_b \sum_{\text{boundary}} \Lambda_{\mu\nu} = 0$, the following adjustment is needed for the integration points of the elements adjoining the boundary:

$$\Lambda'_{\mu\nu} = k_b \Lambda_{\mu\nu}; \qquad k_b = -\sum_{\text{interior } \nu} \Lambda_{\mu\nu} \Big/ \sum_{\text{boundary } \nu} \Lambda_{\mu\nu} \tag{14.3.6}$$

For the remaining integration points in the interior, no adjustment is done, i.e., $\Lambda'^{\mu\nu} = \Lambda^{\mu\nu}$.

5. At each integration point of each finite element, the nonlocal inelastic stress increment tensor is then constituted from its principal values according to the following equation:

$$\Delta\tilde{S}_{kl} = \sum_{i=1}^{3} \Delta\tilde{S}^{(i)} n_k^{(i)} n_l^{(i)} \qquad \text{or} \qquad \Delta\tilde{\mathbf{S}} = \sum_{i=1}^{3} \tilde{S}^{(i)} \mathbf{n}^{(i)} \otimes \mathbf{n}^{(i)} \tag{14.3.7}$$

based on the spectral decomposition theorem of a tensor.

Note that if, at some integration point, all the principal values of tensor $\Delta\mathbf{S}$ are nonpositive, then the foregoing nonlocal procedure may be skipped for that point.

14.4 Particle and Lattice Models

A large amount of research, propitiated by the advent of powerful computers, has been devoted to the simulation of material behavior based directly on a realistic but simplified modeling of the microstructure —its particles, phases, and the bonds between them. A spectrum of diverse approaches can be found in the literature spanning an almost continuous transition from the finite element simulations, with the classical hypothesis of continuum mechanics, to discrete particle models and lattice models in which the continuum is approximated *a priori* by a system of discrete elements: particles, trusses, or frames.

An extreme example of the continuum approach —in view of the fineness of material subdivision— is the *numerical concrete* of Roelfstra, Sadouki and Wittmann (1985), Wittmann, Roelfstra and Kamp (1988) and Roelfstra (1988), in which the mortar, the aggregates, and their interfaces are independently modeled by finite elements. This requires generation of the geometry of the material (random placement of aggregates within the mortar) and the detailed discretization of the elements to adequately reproduce the geometry of the interfaces. With a completely different purpose, but with the same kind of analysis, Rossi and Richer (1987) and Rossi and Wu (1992) developed a random finite element model in which the microstructure is not directly modeled, but is taken into account by assigning random properties to the element interfaces. The common feature of these approaches is that, before cracking starts, the displacement field is approximated by a continuous function.

The particle and lattice models do not model the material continuously, but substitute the continuum by an array of discrete elements in the form of particles in contact, trusses, or frames, in such a manner that the displacements are defined only at the centers of the particles, or at the nodes of the truss or frame.

The origin of the particle approach can generally be traced to the development of the so-called distinct element method by Cundall (1971, 1978), Serrano and Rodriguez-Ortiz (1973), Rodriguez-Ortiz (1974), Kawai (1980), and Cundall and Strack (1979) in which the behavior of particulate materials (originally just cohesionless soils and rock blocks) was analyzed simulating the interactions of the particles in contact. This kind of analysis, which deals with a genuine problem of discrete particle systems, used highly simplified contact interaction laws permitted by the fact that the overall response is controlled mainly by kinematic restrictions (grain interlock) rather than by the details of the force-deformation relation at the contacts. However, although the kinematics of the simulations appeared very realistic, the quantitative stress-strain (averaged) response was not quite close to the actual behavior. This shortcoming, which still

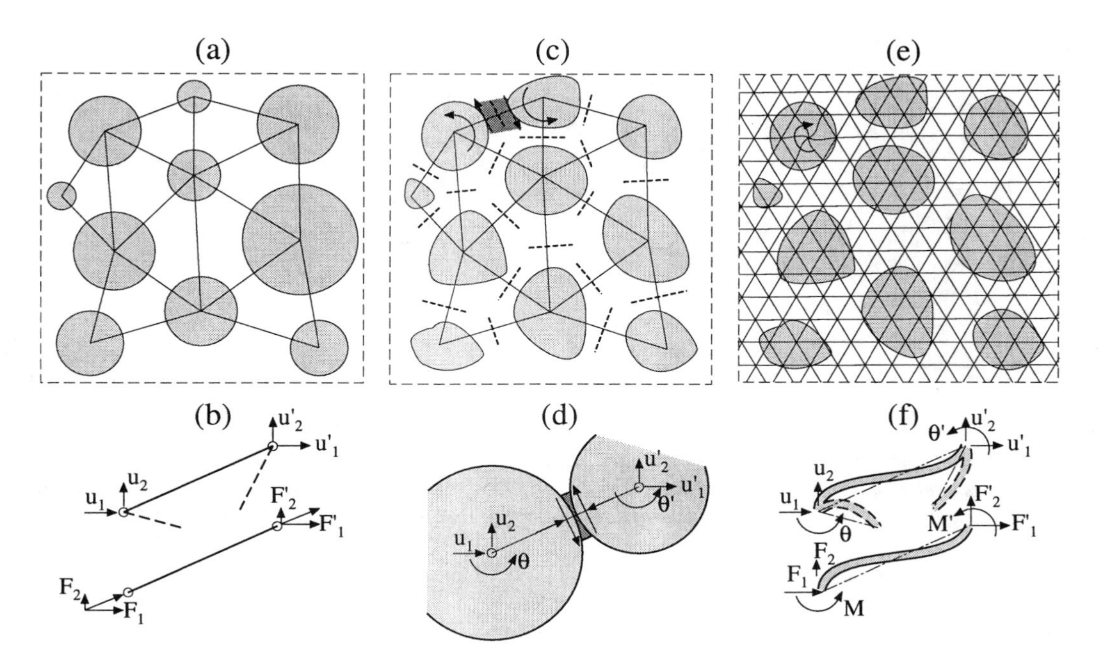

Figure 14.4.1 Various types of lattice models. (a) Pin-joined truss with (b) corresponding displacements and forces at the nodes. (c) Rigid particles in a deformable matrix with (d) the displacements and rotations are transmitted through a deformable layer by normal and shear forces. (e) Triangular regular lattice (originally used by van Mier and Schlangen) formed by (f) beams that stretch and bend.

persists in many modern particle and lattice models, is largely caused by the fact that the simulations are usually two-dimensional while a realistic simulation ought to be three-dimensional.

The basic idea of the particle model can be extended to simulate the particular structure of composite materials, for example, the configuration of the large aggregate pieces of concrete, as done by Zubelewicz (1980, 1983), Zubelewicz and Mróz (1983), and Zubelewicz and Bažant (1987), or the grains in a rock (Plesha and Aifantis 1983). In these cases the model requires defining the force interaction between particles (aggregates or grains) which are caused mainly by the relative displacements and rotations of neighboring particles. Although, for computational purposes, the problem is reduced to a truss (Fig. 14.4.1a–b) or to a frame (Fig. 14.4.1c–d), the basic ingredient of such models is that the geometry (size) of the truss or frame elements and their properties (stiffness, strength, etc.) are dictated by the geometry of the physical structure of the material (stiffness, size, shape, and relative position of aggregates or grains).

In contrast to this, the pure lattice models replace the actual material by a truss or frame whose geometry and element sizes are not related to the actual internal geometry of the material, but are selected freely by the analyst. The truss approach to elasticity, elementary atomistic representations of the physics of elasticity (i.e., arrays of atoms linked by springs shown in textbooks of solid state physics), was already proposed as early as 1941 by Hrennikoff. The lattice models have been championed by theoretical physicists for the simulation of fracturing in disordered materials (Herrmann, Hansen and Roux 1989; Charmet, Roux and Guyon 1990; Herrmann and Roux 1990; Herrmann 1991) and have been developed to analyze concrete fracture by Schlangen and van Mier at Delft University of Technology (Schlangen and van Mier 1992; Schlangen 1993, 1995; van Mier, Vervuurt and Schlangen 1994). In their approach, a regular triangular frame of side length less than the dimensions of the smallest aggregates, is laid over the actual material structure (Fig. 14.4.1e–f) and the properties of each beam are assigned according to the material the beam lies over: mortar, aggregate or interface. However, to eliminate directional bias of fracture, the lattice must be random (see Section 14.4.2).

This section presents a brief overview of the main concepts and results of the particle and lattice models as far as concrete fracture is concerned.

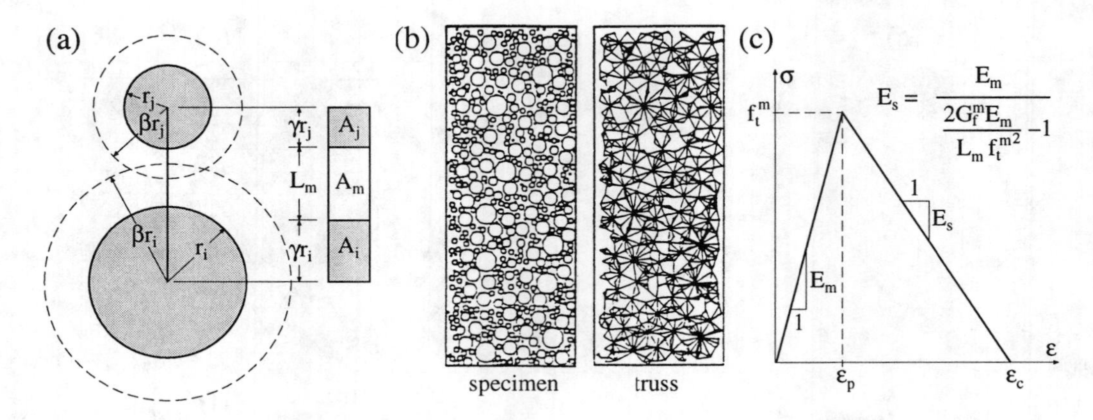

Figure 14.4.2 Random particle model of Bažant, Jirásek et al. (1994): (a) two adjacent circular particles with radii r_i and r_j and corresponding truss member ij; (b) typical randomly generated specimen and its corresponding mesh of truss elements; (c) constitutive law for matrix. (Adapted from Bažant, Tabbara et al. 1990.)

14.4.1 Truss, Frame, and Lattice Models

The simplest model is a pin-jointed truss, in which only the center-to-center forces between the particles are considered (Fig. 14.4.1a–b, Bažant, Tabbara et al. 1990). A more refined model is that of Zubelewicz and Bažant (1987), which imagines rigid particles separated by deformable thin contact layers of matrix that respond primarily by thickness extension-contraction and shear (Fig. 14.4.1c–d). Since the internodal links also transmit shear, moment equilibrium of the nodes needs to be considered, while for the pin-jointed truss it need not. Therefore, this model has three degrees-of-freedom per node (two displacements and one rotation, with corresponding two force components and one moment) for planar lattices, while the pin-jointed model has only two degrees-of-freedom per node. In the spatial case, the model of Zubelewicz and Bažant requires six degrees-of-freedom per node, i.e., three displacements and three rotations, while the pin-jointed The simplest model truss requires only three degrees-of-freedom per node. There is an additional important advantage of shear transmissionThe simplest model —it makes it possible to obtain with the lattice any Poisson ratio, while a random or regular pin-jointed lattice (truss) has Poisson ratio always 1/3 in two dimensions and 1/4 in three dimensions.

The simplest model In the model of Bažant, Tabbara et al. (1990) and Zubelewicz and Bažant (1987), the major particles in the material (large aggregate pieces) are imagined as circular and interacting through links as shown in Fig. 14.4.2. In the initial work of Zubelewicz and Bažant, the link between particles was assumed to transmit both axial forces and shear forces, the latter based on the rotations of particles. In the subsequent model by Bažant, Tabbara et al. (1990), the particle rotations and transmission of shear were neglected and only axial forces were assumed to be transmitted through the links. In such a case, the system of particle links is equivalent to a truss. As pointed out before, the penalty to pay for this simplification is that the Poisson ratio of a random planar truss is always 1/3 (and for a spatial truss 1/4). Another consequence of ignoring particle rotations and interparticle shears is that the fracture process zone obtained becomes narrower. But this can be counteracted by assuming a smaller postpeak softening slope for the interparticle stress displacement law, and also by introducing a greater random scatter in the link properties, both of which tend to widen the fracture process zone.

A random particle configuration must be statistically homogeneous and isotropic on the macroscale. In the simulation of concrete, the configuration must meet the required granulometric distribution of the particles of various sizes, as prescribed for the mix of concrete. The problem of generation of random configurations of particles in contact under such constrains involves some difficult and sophisticated aspects (see, e.g., Plesha and Aifantis 1983).

However, the problem becomes much simpler when the particles do not have to be in contact, as is the case for aggregate pieces in concrete. In that case, a rather simple procedure (Bažant, Tabbara et al. 1990) can proceed as follows: **(1)** using a random number generator, coordinate pairs of particle centers (nodes) are generated one after another, assuming a uniform probability distribution of the coordinates within the area of the specimens; (**2**) for each generated pair a check for possible overlaps of the particles is made, and if the generated particle overlaps with some previously generated one, it is rejected; (**3**) the random generation of coordinate pairs proceeds until the last particle of the largest size has been placed within the specimen, (**4**) then the entire random placement process is repeated for the particles of the next smallest size, and then again for the next smallest size, etc. (The number of particles of each size is determined in advance according to the prescribed mix ratio and granulometry.)

To determine which particles interact, a circle of radius βr_i is drawn around each particle i (with $\beta \approx 5/3$) as shown by the dashed lines in Fig. 14.4.2a: Two particles interact if their dashed circles intersect each other. See Bažant, Tabbara et al. (1990) for the details of the assignment of the dimensions of the truss element, particularly the cross-section A_m and length L_m of the deformable portion (labeled with subscript m for matrix). In a later study by Jirásek and Bažant (1995a), a uniform stiffness of all the links was assumed.

Fig. 14.4.2b shows a typical computer-generated random particle arrangement resembling concrete, and the corresponding truss (random lattice). Fig. 14.4.2c shows the stress-strain relation for the interparticle links, characterized by the elastic modulus E_m, tensile strength limit f_t^m, and the postpeak softening slope E_s (or alternatively by strain ε_f at complete failure, or by G_f^m, each of which is related to the foregoing three parameters). The microscopic fracture energy of the material, G_f^m, is represented by the area under the stress-strain curve in Fig. 14.4.2c, multiplied by the length of the link. The ratio of ε_f to the strain ε_p at the peak stress may be regarded as the microductility of the material.

The lattices in Fig. 14.4.1a–d attempt to directly simulate the major inhomogeneities in the microstructure of concrete. By contrast, the model introduced by Schlangen and van Mier (1992) takes a lattice (in the early versions regular, but later randomized) that is much finer than the major inhomogeneities. Its nodal locations and links are not really reflections of the actual microstructure (Fig. 14.4.1e–f). Rather, the microstructure is simulated by giving various links different properties, which is done according to the match of the lattice to a picture of a typical aggregate arrangement.

Van Mier and Schlangen take advantage of the available simple computer programs for frames and assume the lattice to consist of beams which resist not only axial forces but also bending. Due to bending, the internodal links (beams), of course, also transmit shear, same as in the model of Zubelewicz and Bažant (Figs. 14.4.1c–d and e–f). This feature is useful, because shears are indeed transmitted between adjacent aggregate pieces and across weak interfaces in concrete, and because arbitrary control of the Poisson ratio is possible. However, the idea of bending of beams is a far-fetched idealization that has nothing to do with reality. No clear instances of bending in the microstructures of concrete can be identified.

The idealization of the links as beams subject to bending implies that a bending moment applied at one node is transmitted to the adjacent node with the carry-over factor 0.5, as is well known from the theory of frames. This value of the carry-over factor is arbitrary and cannot not have anything in common with real behavior. In the model of Zubelewicz and Bažant (Fig. 14.4.1c-d), the shear resistance also causes a transmission of moments from node to node, however, the carry-over factor is not 0.5 and can have different values. The transmission of moments is, in that model, due to shear in contact layers between particles, which is a clearly identifiable mechanism. In consequence of this analysis, it would seem better to consider the carry-over factor in the lattices of van Mier and Schlangen to be an arbitrary number, determined either empirically or by some microstructural analysis. This means that the 6×6 stiffness matrix for the element of the lattice, relating the 6 generalized displacement and force components of a beam sketched in Fig. 14.4.1f, should be considered to have general values in its off diagonal members, not based on the bending solutions for a beam but on other considerations. In fact, the use of such a stiffness matrix would require only an elementary change in the computer program for a frame (or lattice with bending). Of course, if the need for such a modification is recognized, the model and van Mier and Schlangen becomes essentially equivalent to that of Zubelewicz and Bažant, except that the nodes do not represent actual particles and the lattice is much finer than the particles.

The beams in the lattice model of Schlangen and van Mier are assumed to be elastic-brittle, and so, when the failure criterion is met at one of the beams, the link may be removed. This means that at each step the computation is purely elastic. This is computationally efficient, but makes the model predict a far too brittle behavior, even for three-dimensional lattices (van Mier, Vervuurt and Schlangen 1994).

Another important aspect in lattice models is the size of the links. Unlike the lattice of Zubelewicz and Bažant, which directly reflects the particle configurations and thus cannot (and should not) be refined, the

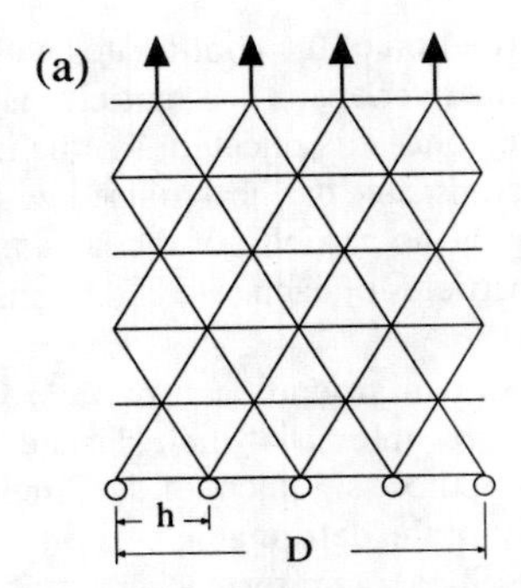

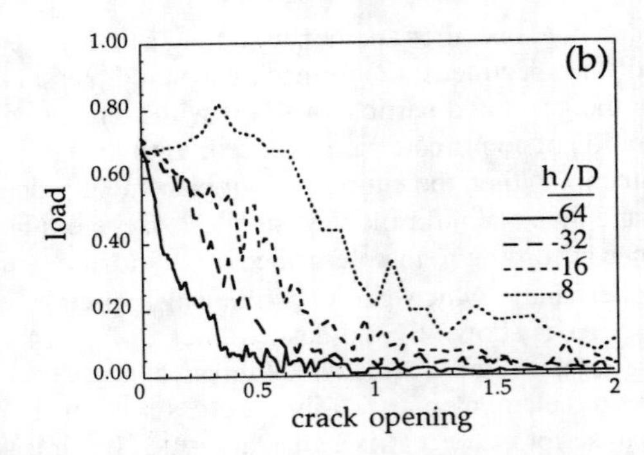

Figure 14.4.3 Dependence of the load-displacement curve on the size of the lattice links: (a) basic element geometry, (b) load-displacement curves for various lattice spacings (adapted from Schlangen 1995).

lattice of van Mier and Schlangen has an undetermined nodal spacing, which raises additional questions. First, as is well known, frames or lattices with bending are on a large scale asymptotically equivalent to the so-called micro-polar (or Cosserat) continuum (e.g., Bažant and Cedolin 1991, Sec. 2.10-2.11). Pin-jointed trusses, on the other hand, asymptotically approach a regular continuum on a large scale. The micropolar continuum is a continuum with nonsymmetric shear stresses and with couple stresses. It possesses a characteristic length, which is essentially proportional to the typical nodal spacing of the lattice approximated by the micro-polar continuum. While, in principle, the presence of a characteristic length is a correct property for a model of concrete, the characteristic length should not be arbitrary but should be of the order of the spacing of the major aggregate pieces. In this regard, the lattice of van Mier and Schlangen appears to be too refined. Moreover, as transpired from recent researches and the previous chapter on nonlocal concepts, the micro-polar character or the presence of characteristic length should refer only to the fracturing behavior and not to the elastic part of its bonds. The model of Schlangen and van Mier goes against this conclusion, since even the elastic response of the lattice is asymptotically approximated by a micro-polar rather than regular continuum.

Furthermore, a question arises about the dependence of the response on the lattice spacing. A recent study of Schlangen (1995) shows that the crack pattern is not strongly affected by the size of the beams, but the load-displacement is affected in much the same way as mesh refinement in local strain-softening models: the finer the lattice, the less the inelastic displacement and the dissipated energy, as illustrated in the load-crack opening curves in Fig. 14.4.3 for a square specimen subjected to pure tension. Indeed, it is easy to imagine that upon infinite refinement the stresses in a beam close to a crack tip must tend to infinity and thus a precracked specimen must fail for a vanishingly small load (roughly proportional to the square root of the beam size). Also, the shorter the beams, the smaller the dissipated energy, because the volume of material affected by the crack is smaller the smaller the elements. Note that the lattice analyzed by Schlangen in Fig. 14.4.3 has random strength in all the cases with identical probabilistic distribution. Therefore, randomness does not relieve mesh sensitivity as sometimes claimed.

The mesh-sensitivity of Schlangen and van Mier's model can probably be artificially alleviated or eliminated by taking a beam strength inversely proportional to the square root of the beam size, similar to the equivalent strength method described in Section 8.6.4 for crack band analysis. However, this is purely speculative and a more sound basis should be built for the lattice models before they can be confidently used as predictive (rather than just descriptive) models. A nonlocal fracture criterion may serve as an alternative solution to the problem, but this would break the computational efficiency of the elastic-brittle beam lattice model. Note that nonlocality (i.e., interaction at finite distance) is automatically implemented in the particle models because the particle distances are finite and fixed, so the lattice size should also be fixed as dictated by the microstructure.

14.4.2 Directional Bias

An important aspect of the model is the generation of the lattice configuration. In many works regular lattices have been used. However, recently Jirásek and Bažant (1995a), and also Schlangen (1995)

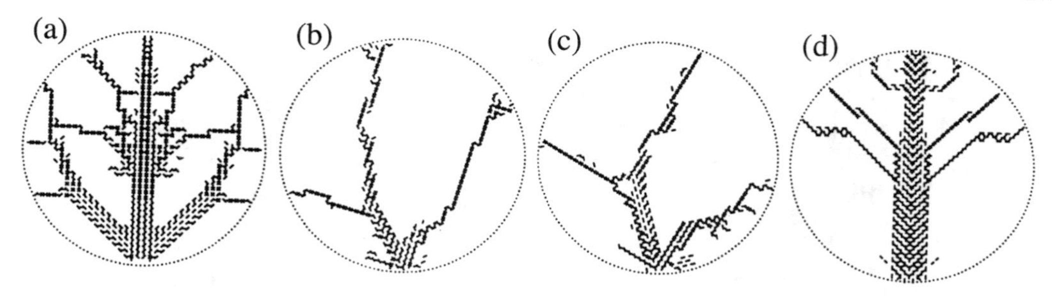

Figure 14.4.4 Failure patterns for various values of α: (a) 0°, (b) 15°, (c) 30°, (d) 45° using a regular lattice with deterministic properties of the links (from Jirásek and Bažant 1995a).

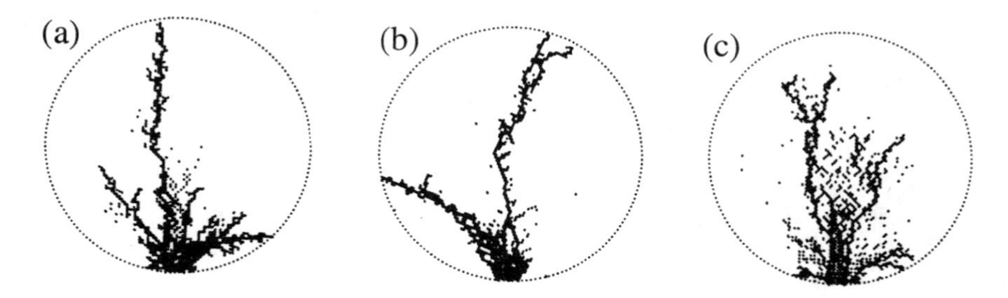

Figure 14.4.5 Failure patterns for various values of α: (a) 0°, (b) 22.5°, (c) 45° using a regular lattice with random strength, elastic stiffness and microfracture energy of the links (from Jirásek and Bažant 1995a).

demonstrated that a regular lattice always impresses a strong bias on the direction of fracture propagation.

For the square lattices with diagonals analyzed by Jirásek and Bažant, it is, of course, possible to choose the elastic stiffnesses of the links in the main directions of the square mesh and the diagonal directions, the corresponding strength limits of the links and the corresponding microfracture energies in such ratios that the lattice is isotropic in terms of elastic properties, strength along straight line cuts, and fracture energies dissipated on such cuts for any orientation of the cut. However, even in that case, the fracture tends to run preferentially among the mesh lines. This has been blatantly demonstrated by simulations of fracture of a circular specimen on which a regular square mesh with diagonals was overlaid; see Fig. 14.4.4. In this particle simulation the fracture was caused by an impact at the bottom of the circle in upward direction. In Fig. 14.4.4a the impact was in the direction of the square mesh lines, in Fig. 14.4.4d in the direction of the diagonals, and in Fig. 14.4.4b and 14.4.4c in two intermediate directions. Note the enormous differences in fracture patterns, which were also manifested by great differences in peak loads and energies dissipated. When all the properties of the links of a regular lattice were randomized, strong directional bias of fracture still remained; see Fig. 14.4.5.

Only when a geometrically random lattice was used in Jirásek and Bažant's (1995a) study, the directional bias was eliminated, except for small random differences between meshes. Similar results were found by Schlangen (1995) for a double-edge notched specimen subjected to shear. These results indicate that random (unstructured) lattices must be used to avoid directional bias.

14.4.3 Examples of Results of Particle and Lattice Models

Bažant, Tabbara et al. (1990) used the random particle system described before (Figs. 14.4.1a–b and 14.4.2) to simulate tensile tests and bending tests on notched specimens. A similar model was used by Jirásek and Bažant (1995a,b) to relate the microscopic features of the model (such as the softening curve and the statistics of strength distribution) to the macroscopic properties, particularly size effect and fracture energy.

Fig. 14.4.6 shows direct tension specimens of various sizes studied computationally by Bažant, Tabbara et al. (1990), with the results displayed in Fig. 14.4.7 as the calculated curves of load (axial force resultant)

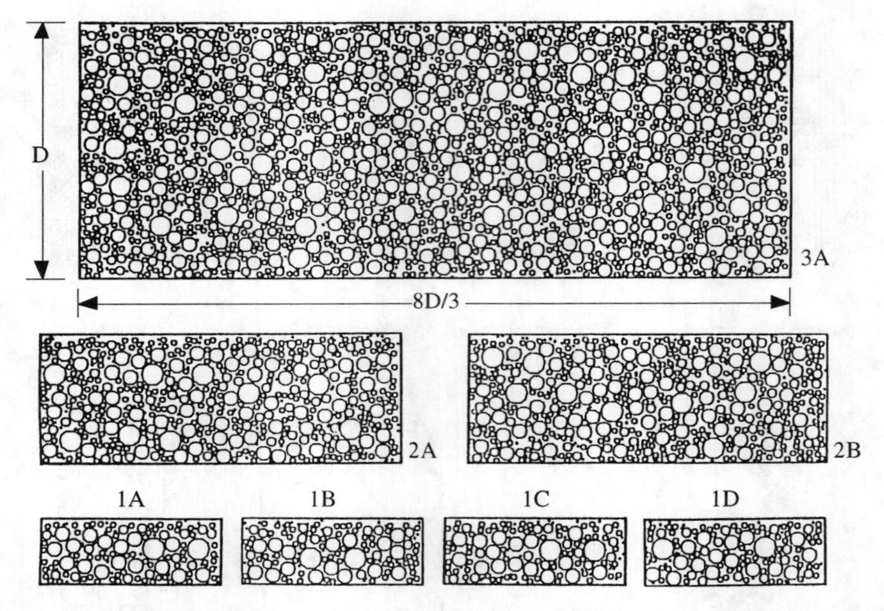

Figure 14.4.6 Geometrically similar specimens of various sizes with randomly generated particles (adapted from Bažant, Tabbara et al. 1990).

vs. relative displacement between the ends.

Fig. 14.4.7b gives the curves for several specimens of the smallest size from Fig. 14.4.6. Fig. 14.4.7c shows the curves for the medium size specimens and Fig. 14.4.7d the curve for one large size specimen. Fig. 14.4.7e shows, in relative coordinates, the average response curves calculated for the small, medium, and large specimens. Note that while the prepeak shape of the load displacement curve is size independent, the postpeak response curve is getting steeper with increasing size.

Fig. 14.4.8 shows the progressive spread of cracking in one of the smallest specimens from Fig. 14.4.6. The cracking patterns are shown for four different points on the load displacement diagram, as seen in Fig. 14.4.8a, the first point corresponding to the peak load. The dashed black lines are the normals to the links that undergo softening and correspond to partially formed cracks. The solid lines are normal to completely broken links and represent fully formed cracks. The gray dashed lines represent normals to the links that partly softened and then unloaded, and correspond to partially formed cracks that are closing. Note from Fig. 14.4.8 that the cracking is at first widely distributed, but then it progressively localizes.

Fig. 14.4.9 shows the calculated peak loads for the specimens from Fig. 14.4.6 in the usual size effect plot of the logarithm of nominal strength vs. logarithm of the size. Bažant, Tabbara et al. (1990) interpreted the results in terms of the classical size effect law Eq. (1.4.10) with relatively good results. The recent results of Bažant explained in Section 9.1, particularly the size effect formula for failures at crack initiation from a smooth surface (Section 9.1.6) suggest that these results must be interpreted using Eq. 9.1.42. Thus the results of Bažant, Tabbara et al. (1990) have been fitted here by the simplest version of this curve (for which $\gamma = 0$). Fig. 14.4.9 shows the resulting fit, which is excellent for the mean values of the data.

Three-point-bend fracture specimens of three sizes in the ratio 1:2:4 were simulated in the manner illustrated in Fig. 14.4.10a. Fig. 14.4.10b shows the size effect plot obtained from the three sizes of the specimens in Fig. 14.4.10a for three different materials. As can be seen, the calculated maximum loads can be well approximated by Bažant's size effect law, Eq. (1.4.10).

By fitting the size effect law to the maximum load obtained by the lattice or particle model for similar specimens of different sizes, one can determine the macroscopic fracture energy G_f of the particle system and the effective length c_f of the fracture process zone (see Chapter 6). It thus appears that fracture simulations with the lattice model or random particle system provide a further verification of the general applicability of the size effect law.

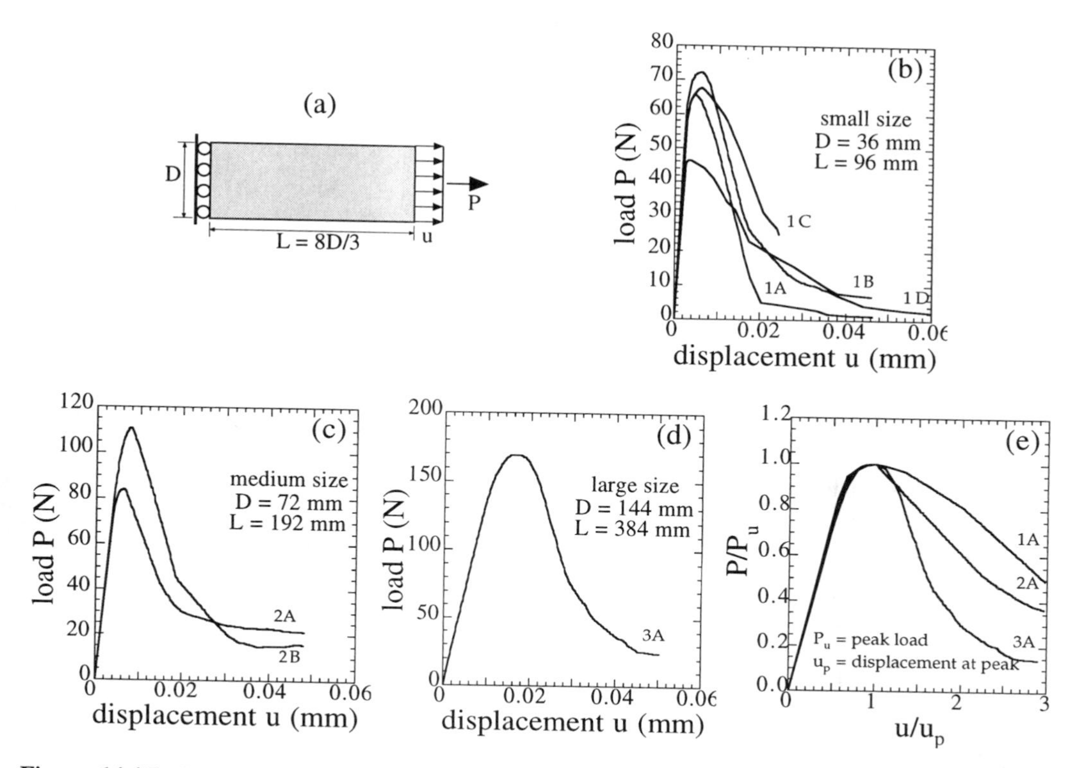

Figure 14.4.7 Results of Bažant, Tabbara et al. (1990) for direct tension of random particle specimens: (a) specimen with $D = 36, 72$, and 144 mm; (b) load-displacement curve for small specimens; (c) load-displacement curve for medium specimens; (d) load-displacement curve for large specimens; (e) normalized load-displacement curves for specimens 1A, 2A, and 3A. (Adapted from Bažant, Tabbara et al. 1990.)

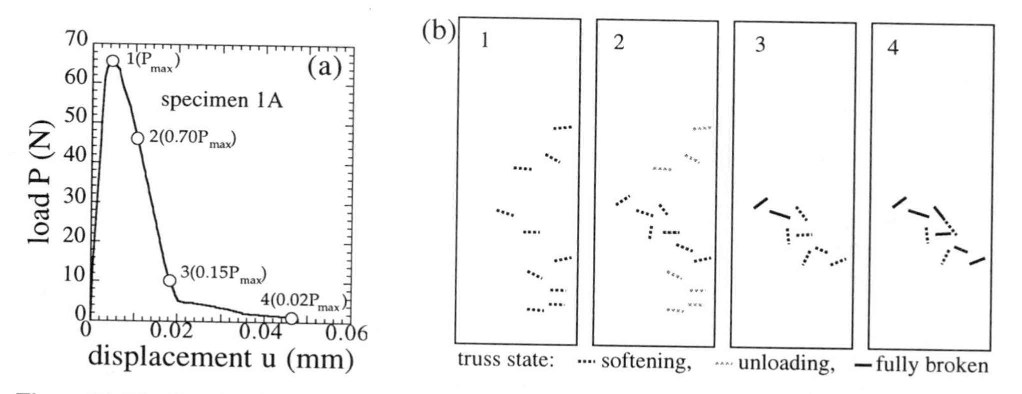

Figure 14.4.8 Results of Bažant, Tabbara et al. (1990) for the evolution of cracking in direct tension of random particle specimens: (a) load-displacement curve for specimen 1A; (b) evolution of cracking with loading and localization. (Adapted from Bažant, Tabbara et al. 1990.)

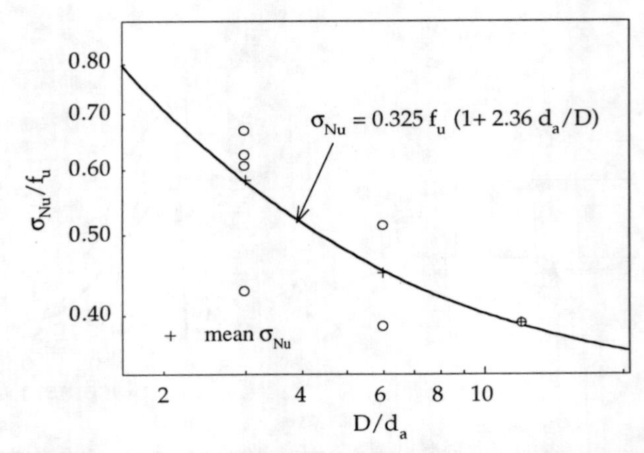

Figure 14.4.9 Size effect plot constructed form maximum load values calculated for direct tension specimens of various sizes by Bažant, Xiang et al. (1996). The solid line is the fit by Eq. (9.1.42) with $\gamma = 0$. f_u is a reference strength taken to be equal to the matrix tensile strength f_t^m (see Fig. 14.4.2c).

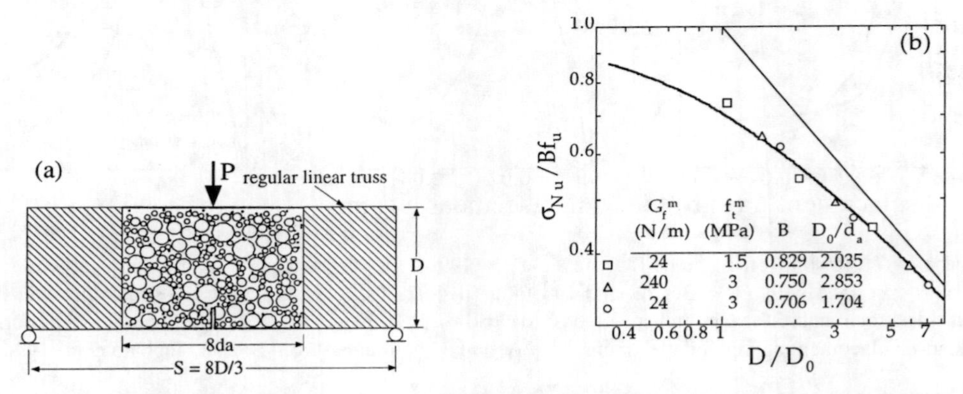

Figure 14.4.10 Simulation of three three-point-bend tests by random particle model. (a) Three-point-bend specimens with $d = 36, 72$, and 144 mm and (b) corresponding size effect plot. (Adapted from Bažant, Tabbara et al. 1990.)

At the same time, the size effect law is seen to be an effective approach for studying the relationship between the microscopic characteristics of the particle system, simulating the microscopic properties of the material, and the macroscopic fracture characteristics.

Such studies have been undertaken by Jirásek and Bažant (1995b). Fig. 14.4.11 show the results of a large number of such simulations, dealing with two dimensional three-point-bend fracture specimens of different sizes. In these specimens, the microductility number, representing the ratio $\varepsilon_c/\varepsilon_p$, was varied (see Fig. 14.4.2b). The coefficient of variation of the microstrength of the particle length, used in random generation of the properties of the links, was also varied (the microstrength was assumed to have a normal distribution).

It was found that both the microductility and the coefficient of the microstrength of the links have a significant effect on the macroscopic fracture energy G_f and on the effective length c_f of the process zone; see Fig. 14.4.11a,c. Randomness of these plots is largely due to the fact that the number of simulations was not very large (the values in the plot are the averages of the values obtained in individual sets of simulations of specimens of different sizes). The plots in Fig. 14.4.11a,c have been smoothed as shown in Fig. 14.4.11b,d by the following bilinear polynomials which provide optimum fits:

$$\bar{G}_f = 2.16 - 1.08\omega_f + 0.48\gamma_f - 0.71\omega_f\gamma_f, \tag{14.4.1}$$

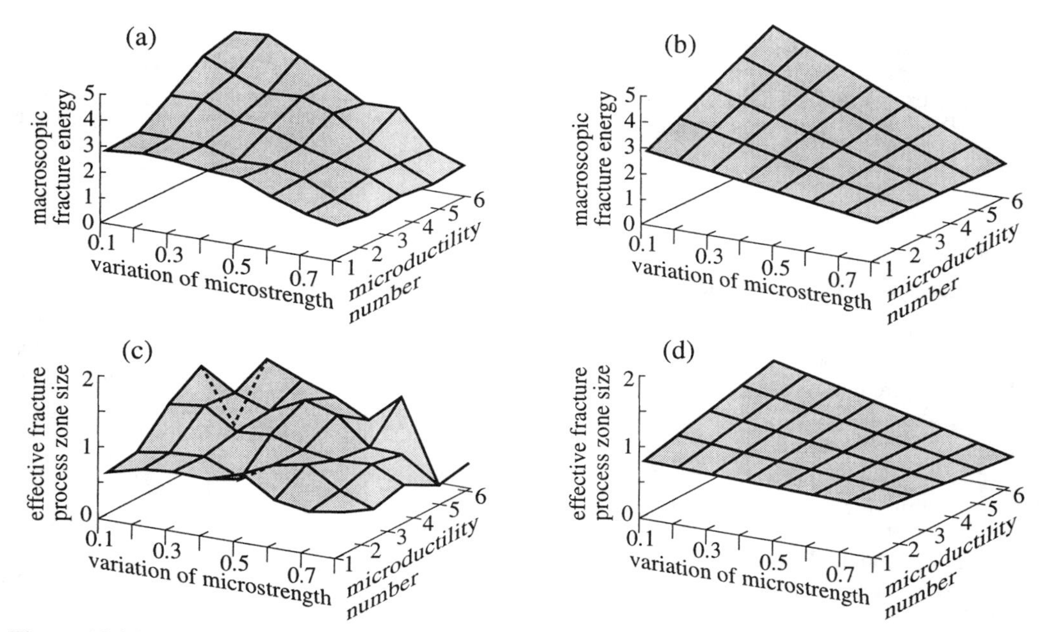

Figure 14.4.11 Normalized fracture energy and normalized effective process zone size as a function of two parameters: (a) and (c) computed, (b) and (d) fitted by bilinear functions (from Jirásek and Bažant 1995b).

$$\bar{c}_f = 0.64 + 0.08\omega_f + 0.09\gamma_f - 0.19\omega_f\gamma_f, \tag{14.4.2}$$

in which the superimposed bars refer to average values, and ω_f and γ_f are the coefficient of variation of microstrength and the microductility number. Obviously, the effect of various other microscopic characteristics of particle systems on their macroscopic fracture properties could also be studied in this manner, exploiting the size effect law.

14.4.4 Summary and Limitations

The lattice models or particle systems are computationally very demanding and a very efficient computational algorithm must be used. A highly efficient algorithm, which was applied to simulation of sea ice fracture, is presented in Jirásek and Bažant (1995a). It is an explicit algorithm for fracture dynamics, but it can also be used for static analysis in the sense of dynamic relaxation method. (Fracturing in particle systems with more than 120,000 degrees-of-freedom was simulated with this algorithm on a desktop 1992 work station.) Computational effectiveness will be particularly important for three-dimensional lattices, the use of which is inevitable for obtaining a fully realistic, predictive model.

As it now stands, the lattice or particle models can provide a realistic picture of tensile cracking in concrete in two-dimensional situations. However, solution of significant three-dimensional problems or nonlinear triaxial behavior as well as simulation of behavior in which compression and shear fracturing is important is still beyond reach. Thus, the lattice or particle models have not attained the degree of generality already available with the finite element approach.

Although computer programs for lattices are attractive by their simplicity, it must nevertheless be recognized that a lattice modeling of a continuum is far less powerful than the finite element method because the stress and strain tensors cannot be simulated by the elements of the lattice and, thus, nonlinear tensorial behavior cannot be directly described. From this viewpoint, the numerical concrete of Roelfstra and Wittmann, seems preferable to the lattice model of van Mier and Schlangen because the particles and mortar are discretized by a much finer mesh of finite elements.

To sum up, lattice models or particle systems have proven to be a useful tool for understanding fracture process and clarifying some relationship between the micro- and macro- characteristics of quasibrittle heterogeneous materials. These models appear to be particularly suited for failures due to tensile fracturing and capture well the distributed nature of such fracturing and its localization. However, one must keep in mind that these models, in their present form, cannot simulate three-dimensional situations, larger structures (even two-dimensional), compressive and shear fractures, and nonlinear triaxial stress-strain relations. Overall, these models are still far inferior to finite element models and do not really have a predictive capability. No doubt significant improvements may be expected in the near future.

14.5 Tangential Stiffness Tensor Via Solution of a Body with Many Growing Cracks

The power of the microplane model is limited by the assumption of a kinematic (or static) constraint, which is a simplification of reality. To avoid this simplification, one needs to tackle the boundary value problem of the growth of many statistically uniformly distributed cracks in an infinite elastic body. This problem is not as difficult as it seems. It appears possible that an approximate solution of this problem might once supersede the microplane model as the most realistic predictive approach to cracking damage.

The problem of calculating the macroscopic stiffness tensor of elastic materials intersected by various types of random or periodic crack systems has been systematically explored during the last two decades and effective methods such as the self-consistent scheme (Budianski and O'Conell 1976; Hoenig 1978), the differential scheme (Roscoe 1952; Hashin 1988), or the Mori-Tanaka method (Mori and Tanaka 1973) have been developed. A serious limitation of these studies was that they dealt with cracks that neither propagate nor shorten (Fig. 14.5.1a). This means that, in the context of response of a material with growing damage illustrated by the curve in Fig. 14.5.1a, these formulations predict only the secant elastic moduli (such as E_s in Fig. 14.5.1a). Such information does not suffice for calculating the response of a body with progressing damage due to cracking.

To calculate the response of a material with cracks that can grow or shorten, it is also necessary to determine the tangential moduli, exemplified by E_t in Fig. 14.5.1b. Knowledge of such moduli makes it possible, for a given strain increment, to determine the inelastic stress drop $\delta\sigma_{cr}$ (Fig. 14.5.1b). This problem has recently been studied by Bažant and Prat (1995, 1997) and Prat and Bažant (1997). Its solution will now be briefly reviewed.

We consider a representative volume V of an elastic material containing on the microscale many cracks (microcracks). On the macroscale, we imagine the cracks to be smeared and the material to be represented by an approximately equivalent homogeneous continuum whose local deformation within the representative volume can be considered approximately homogeneous over the distance of several average crack sizes. Let $\boldsymbol{\varepsilon}$ and $\boldsymbol{\sigma}$ be the average strain tensor and average stress tensor within this representative volume. To obtain a simple formulation, we consider only circular cracks of effective radius a.

Consider the material to be intersected by N families of random cracks, labeled by subscripts $\mu = 1, 2, \ldots, N$. Each crack family may be characterized by its spatial orientation $\vec{n}_\mu$, its effective crack radius a_μ, and the number n_μ of cracks in family μ per unit volume of the material. The compliance tensor may be considered as the function $\boldsymbol{C} = \boldsymbol{C}(a_1, a_2, \ldots, a_N;\, n_1, n_2, \ldots, n_N)$. Approximate estimation of this function has been reviewed by Kachanov and co-workers (Kachanov 1992, 1993; Sayers and Kachanov 1991; Kachanov, Tsukrov and Shafiro 1994).

To obtain the tangential compliance tensor of the material on the macroscale, the cracks must be allowed to grow during the prescribed strain increment $\delta\varepsilon$. This means that the energy release rate per unit length

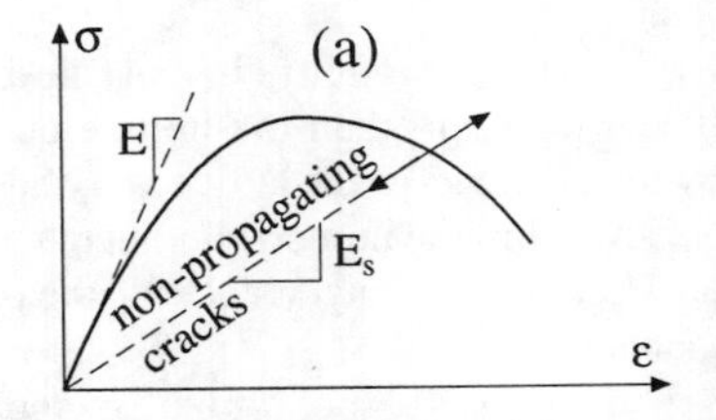

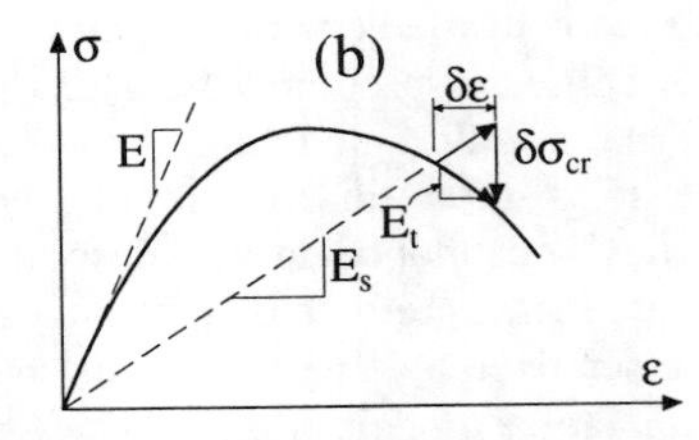

Figure 14.5.1 Stress-strain curves and moduli: (a) effective secant moduli; (b) tangential moduli and inelastic stress decrement due to crack growth.

of the front edge of one crack must be equal to the given critical value $R(a_\mu)$ (or to the fracture energy G_f, in the case of LEFM). For the sake of simplicity, we will enforce the condition of criticality of cracks only in an overall (weak) sense, by assuming that the average overall energy release rate of all the cracks of one family within the representative volume equals their combined energy dissipation rate.

Our analysis will be restricted to the case when the number of cracks in each family is not allowed to change ($\delta n_\mu = 0$), i.e., no new cracks are created and no existing cracks are allowed to close. This does not seem an overly restrictive assumption because a small enough crack has a negligible effect on the response and can always be assumed at the outset. Besides, concrete is full of microscopic cracks (or flaws) to begin with, and no macroscopic crack nucleates from a homogeneous material.

The incremental constitutive relation can be obtained by differentiation of Hooke's law. It reads $\delta\boldsymbol{\varepsilon} = \boldsymbol{C}\delta\boldsymbol{\sigma} + \sum_{\mu=1}^{N}(\partial\boldsymbol{C}/\partial a_\mu)\boldsymbol{\sigma}\,\delta a_\mu$ from which

$$\delta\boldsymbol{\sigma} = \boldsymbol{E}\left(\delta\boldsymbol{\varepsilon} - \sum_{\mu=1}^{N}\frac{\partial\boldsymbol{C}}{\partial a_\mu}\boldsymbol{\sigma}\,\delta a_\mu\right) \tag{14.5.1}$$

where δ denotes infinitesimal variations and $\boldsymbol{C}(a_\mu)$ is the fourth-order macroscopic secant compliance tensor of the material with the cracks, and $\boldsymbol{E}(a_\mu)$ is the fourth-order secant stiffness tensor, whose 6×6 matrix is inverse of the matrix of $\boldsymbol{C}(a_\mu)$.

The surface area of one circular crack of radius a_μ is $A = \pi a_\mu^2$, and the area change when the crack radius increases by δa is $\delta A_\mu = 2\pi a_\mu \delta a_\mu$. We assume we can replace the actual crack radii by their effective radius a_μ.

The crack radius increments δa_μ must be determined in conformity to the laws of fracture mechanics. Let us assume that the cracks (actually microcracks) can be described by equivalent linear elastic fracture mechanics (LEFM) with an R-curve $R(a_\mu)$. This means that the energy release rates must be equal to R_μ (or G_f, in the case of LEFM). For the special case $R(a_\mu) = G_f$ = fracture energy of the matrix of the material, the cracks follow LEFM.

The complementary energy density of the material is $\overline{\mathcal{U}}^* = \overline{\mathcal{U}}^*(\boldsymbol{\sigma}, a_\mu) = \frac{1}{2}\boldsymbol{\sigma}\cdot\boldsymbol{C}(a_\mu)\boldsymbol{\sigma}$, where the dot indicates scalar product of two second-order tensors. To make the problem tractable, we impose the energy criterion of fracture mechanics (energy balance condition) only in the overall, weak sense. This leads to the following N conditions of crack growth (Bažant and Prat 1995; Prat and Bažant 1997):

$$\left[\frac{\partial\overline{\mathcal{U}}^*}{\partial a_\mu}\right]_{\boldsymbol{\sigma}=\text{const.}} = \frac{1}{2}\boldsymbol{\sigma}\cdot\frac{\partial\boldsymbol{C}}{\partial a_\mu}\boldsymbol{\sigma} = 2\pi a_\mu n_\mu \kappa_\mu R(a_\mu) \qquad (\mu = 1, 2, ...N) \tag{14.5.2}$$

(repetition of subscript μ in this and subsequent equations does not imply summation); κ_μ = crack growth indicator which is equal to 1 if the crack is growing ($\delta a_\mu > 0$) and 0 if it is closing ($\delta a_\mu < 0$), while any value $0 < \kappa_\mu < 1$ can be used if $\delta a_\mu = 0$.

Differentiation of (14.5.2) provides the incremental energy balance conditions:

$$\boldsymbol{\sigma}\cdot\frac{\partial\boldsymbol{C}}{\partial a_\mu}\delta\boldsymbol{\sigma} + \sum_{\nu=1}^{N}\left(\frac{1}{2}\boldsymbol{\sigma}\cdot\frac{\partial^2\boldsymbol{C}}{\partial a_\mu\partial a_\nu}\boldsymbol{\sigma} - 2\pi n_\mu\kappa_\mu G_f\delta_{\mu\nu}\right)\delta a_\nu = 2\pi n_\mu a_\mu G_f \delta\kappa_\mu \qquad (\mu = 1, 2, ...N) \tag{14.5.3}$$

where $\delta\kappa_\mu = 0$ except when the crack growth changes to no growth or to closing, or vice versa. The handling of the large jump in κ_μ is exact if $\kappa_\mu = \kappa_\mu^{\text{new}}$ and $a_\mu = a_\mu^{\text{old}}$ because $\delta(\kappa_\mu a_\mu) = (\kappa_\mu a_\mu)^{\text{new}} - (\kappa_\mu a_\mu)^{\text{old}} = \kappa_\mu^{\text{new}}\delta a_\mu + \delta\kappa_\mu a_\mu^{\text{old}}$, exactly.

Substitution of (14.5.1) into (14.5.3) leads to a system of N equations for N unknowns $\delta a_1, ...\delta a_N$:

$$\sum_{\nu=1}^{N} A_{\mu\nu}\delta a_\nu = B_\mu \qquad (\mu = 1, ...N) \tag{14.5.4}$$

where

$$A_{\mu\nu} = \boldsymbol{\sigma}\cdot\left(\frac{\partial\boldsymbol{C}}{\partial a_\mu}\boldsymbol{E}\frac{\partial\boldsymbol{C}}{\partial a_\nu} - \frac{1}{2}\frac{\partial^2\boldsymbol{C}}{\partial a_\mu\partial a_\nu}\right)\boldsymbol{\sigma} + 2\pi G_f n_\mu \kappa_\mu^{\text{new}}\delta_{\mu\nu} \tag{14.5.5}$$

$$B_\mu = \boldsymbol{\sigma} \cdot \frac{\partial \boldsymbol{C}}{\partial a_\mu} \boldsymbol{E} \delta \boldsymbol{\varepsilon} - 2\pi n_\mu a_\mu^{\text{old}} G_f \delta \kappa_\mu \tag{14.5.6}$$

After solving (14.5.4), evaluating $\delta\boldsymbol{\sigma}$ from (14.5.1) and incrementing $\boldsymbol{\sigma}$, one must check whether the case $\delta\bar{a}_\mu > 0$ and $\sigma_{N\mu} < 0$ (or $\epsilon^{cr}_{N\mu} < 0$) occurs for any μ. If it does, the corresponding equation in the system (14.5.4) must be replaced by the equation $\delta\bar{a}_\mu = 0$. The solution of such a modified equation system must be iterated until the case $\delta\bar{a}_\mu > 0$ and $\sigma_{N\mu} < 0$ (or $\epsilon^{cr}_{N\mu} < 0$) would no longer occur for any μ.

The formulation needs to be further supplemented by a check for compression. The reason is that the energy expression is quadratic, which means that (14.5.2) is invariant when $\boldsymbol{\sigma}$ is replaced by $-\boldsymbol{\sigma}$. Thus, a negative stress intensity factor $K_{I\mu}$ can occur even when (14.5.4) is satisfied, and so the sign of $K_{I\mu}$ must be checked for each crack family μ. The case $K_{I\mu} < 0$ is inadmissible.

Since the present formulation yields only the values of $(K_{I\mu})^2 = E\mathcal{R}(a_\mu)$ but not the values of $K_{I\mu}$, the sign of $K_{I\mu}$ must be inferred approximately. It can be considered the same as the sign of the stress $\sigma_{N\mu} = \vec{n}_\mu \cdot \boldsymbol{\sigma}\vec{n}_\mu$ in the direction normal to the cracks of μ-th family ($\vec{n}_\mu$ = unit vector normal to the cracks). [Alternatively, the sign of $K_{I\mu}$ could be inferred from the sign of the normal component $\epsilon^{cr}_{N\mu} = \vec{n}_\mu \cdot \epsilon^{cr}_N \vec{n}_\mu$ of the cracking strain tensor $\boldsymbol{\varepsilon}^{cr} = \boldsymbol{C}^{cr}\boldsymbol{\sigma}$.] Approximate though such estimation surely is on the microscale of an individual crack, it nevertheless is fully consistent with the macroscopic approximation of $\boldsymbol{C}$ implied in this model. The reason is that all the composite material models for cracked solids are based on the solution of one crack in an infinite solid, for which the sign of $K_{I\mu}$ coincides with the sign of $\sigma_{N\mu}$ (or $\epsilon^{cr}_{N\mu}$).

Usually the six independent components of $\delta\boldsymbol{\varepsilon}$ are known or assumed, and then (14.5.4) represents a separate system of only N equations for $\delta a_1, ...\delta a_N$ (this is a simplification compared to the formulation in the paper, which led to a system of $N + 6$ equations). In each iteration of each loading step, the values of κ_μ are set according to the sign of δa_μ in the preceding step or preceding iteration.

If $\delta a_\mu = 0$, or if (due to numerical error) $|\delta a_\mu|$ is nonzero but less than a certain chosen very small positive number δ, κ_μ is arbitrary and can be anywhere between 0 and 1, which makes equation (14.5.3) superfluous. The condition (14.5.2) of energy balance in the of constant crack length becomes meaningless and must be dropped. It needs to be replaced by an equation giving κ_μ (or $K_{I\mu}$) as a function of $\boldsymbol{\sigma}$), which must be used to check whether κ_μ indeed remains within the interval (0,1). However, it seems that for proportionally increasing loads the case $\delta a_\mu = 0$ is not important and its programming could be skipped, using conveniently the value $\kappa_\mu = 0.5$.

The foregoing solution was outlined in Bažant and Prat (1995) and worked out in detail in Prat and Bažant (1997) (with Addendum in a later issue).

To be able to use (14.5.5), we must have the means to evaluate the effective secant stiffness $\boldsymbol{C}$ as a function of a_μ. Bažant and Prat (1995) and Prat and Bažant (1997) adopted the approximate approach developed by Sayers and Kachanov (1991) using the symmetric second-order crack density tensor

$$\boldsymbol{\alpha} = \sum_{\mu=1}^{N} n_\mu a_\mu^3 \, \vec{n}_\mu \otimes \vec{n}_\mu \tag{14.5.7}$$

(Vakulenko and Kachanov 1971; Kachanov 1980, 1987b). In this approach, the effective secant compliance $\boldsymbol{C}$ is derived from an elastic potential F which is considered as a function of the crack density tensor $\boldsymbol{\alpha}$ (in addition of the stress tensor $\boldsymbol{\sigma}$):

$$\boldsymbol{\varepsilon} = \frac{\partial F(\boldsymbol{\sigma}, \boldsymbol{\alpha})}{\partial \boldsymbol{\sigma}} = \boldsymbol{C}\boldsymbol{\sigma} \tag{14.5.8}$$

The elastic potential $F(\boldsymbol{\sigma}, \boldsymbol{\alpha})$ can be expanded into a tensorial power series. Sayers and Kachanov (1991) proposed to approximate potential F by a tensor polynomial that is quadratic in $\boldsymbol{\sigma}$ and linear in $\boldsymbol{\alpha}$:

$$F(\boldsymbol{\sigma}, \boldsymbol{\alpha}) = \frac{1}{2} \boldsymbol{\sigma} \cdot \boldsymbol{C}^\circ \boldsymbol{\sigma} + \eta_1 \, (\boldsymbol{\sigma} \cdot \boldsymbol{\alpha}) \text{tr}\, \boldsymbol{\sigma} + \eta_2 \, \boldsymbol{\sigma}^2 \cdot \boldsymbol{\alpha} \tag{14.5.9}$$

in which η_1 and η_2 are assumed to depend only on $\alpha = \text{tr}\, \boldsymbol{\alpha} = \sum_\mu n_\mu a_\mu^3$, the first invariant of $\boldsymbol{\alpha}$ (Sayers and Kachanov 1991). The strain tensor follows from (14.5.9):

$$\boldsymbol{\varepsilon} = \boldsymbol{C}^\circ \boldsymbol{\sigma} + \eta_1 \, \text{tr}\, \boldsymbol{\sigma} \, \boldsymbol{\alpha} + \eta_1 \, \text{tr}\, \boldsymbol{\alpha} \, \boldsymbol{\sigma} + \eta_2 \, (\boldsymbol{\alpha}\boldsymbol{\sigma} + \boldsymbol{\sigma}\boldsymbol{\alpha}) \tag{14.5.10}$$

The functions $\eta_1(\alpha)$ and $\eta_2(\alpha)$ can be obtained by taking the particular form of the preceding formulation for the case of random isotropically distributed cracks and equating the results to those obtained using, e.g., the differential scheme (Hashin 1988). To this end, we note that if the orientation, density, and size of cracks is isotropically distributed, then $\boldsymbol{\alpha}$ must be spherical, and since its trace is equal to α, it must be $\boldsymbol{\alpha} = (\alpha/3)\mathbf{1}$. Substituting this into the preceding equation we get for this case:

$$\boldsymbol{\varepsilon} = \boldsymbol{C}^\circ \boldsymbol{\sigma} + \frac{2}{3}\eta_1 \operatorname{tr} \boldsymbol{\sigma}\, \mathbf{1} + \frac{2}{3}\eta_2 \boldsymbol{\sigma} \tag{14.5.11}$$

Noting now that the resulting behavior in (14.5.11) must be elastic with effective elastic modulus E_{eff} and effective Poisson's ratio ν_{eff}, we get the following relationship between the functions η_i, the effective elastic constants and the α:

$$\eta_1(\alpha) = -\frac{3}{2\alpha}\left(\frac{\nu_{\text{eff}}(\alpha)}{E_{\text{eff}}(\alpha)} - \frac{1+\nu}{E}\right), \qquad \eta_2(\alpha) = \frac{3}{2\alpha}\left[\frac{1+\nu_{\text{eff}}(\alpha)}{E_{\text{eff}}(\alpha)} - \frac{1+\nu}{E}\right] \tag{14.5.12}$$

where E and ν are the elastic constants of the uncracked material (included in $\boldsymbol{C}^\circ$); here we indicate explicitly that the effective elastic constants depend on α. This dependence can be obtained, for example, by using the differential scheme which, for quasibrittle materials, gives better predictions than the self-consistent scheme (Bažant and Prat 1997). The resulting relationships (see e.g., Hashin 1988 for the details of the derivation) are as follows:

$$\alpha = \frac{5}{8}\ln\frac{\nu}{\nu_{\text{eff}}} + \frac{15}{64}\ln\frac{1-\nu_{\text{eff}}}{1-\nu} + \frac{45}{128}\ln\frac{1+\nu_{\text{eff}}}{1+\nu} + \frac{5}{128}\ln\frac{3-\nu_{\text{eff}}}{3-\nu} \tag{14.5.13}$$

$$\frac{E_{\text{eff}}}{E} = \left(\frac{\nu_{\text{eff}}}{\nu}\right)^{10/9}\left(\frac{3-\nu}{3-\nu_{\text{eff}}}\right)^{1/9} \tag{14.5.14}$$

From Eqs. (14.5.7), (14.5.10), and (14.5.12)–(14.5.14) the effective compliance tensor is obtained as a function of a_μ. Then δa_μ is calculated from Eq. 14.5.4 for a given $\delta\varepsilon$ as indicated before.

The crack density may be characterized as a continuous function n_μ of spherical angles ϕ and θ (Prat and Bažant 1997). Function n_μ is then sampled at spherical angles ϕ_μ and θ_μ corresponding to the orientations of the microplanes in the microplane model. For isotropic materials such as concrete, the distribution of n_μ is initially uniform, and a very small but nonzero value must be assigned to every n_μ as the initial condition because no new cracks are allowed to nucleate. For an initially anisotropic material such a rock with joints, function n_μ is assumed to peak at a few specified discrete orientations ϕ_i^*, θ_i^*.

In on-going studies, the R-curves are used by Bažant and co-workers as a means to approximate the effect of plastic strains in the matrix of the material occurring simultaneously with the crack growth. Unlike classical plasticity, the plastic strain cannot be considered here to be smeared in a continuum manner because the cracks cause stress concentrations. Therefore, plasticity of the matrix will get manifested by the formation of a plastic zone at the front edge of each microcrack. As is well known, the effect of such a plastic zone can be approximately described by an R-curve.

References

Achenbach, J. D. and Bažant, Z. P. (1975) "Elastodynamic near-tip stress and displacement fields for rapidly propagating cracks in orthotropic materials." *J. Appl. Mech.-T. ASME,* **42**, 183–189.

Achenbach, J. D., Bažant, Z. P. and Khetan, R. P. (1976a) "Elastodynamic near-tip fields for a rapidly propagating interface crack." *Int. J. Eng. Sci.,* **14**, 797–809.

Achenbach, J. D., Bažant, Z. P. and Khetan, R. P. (1976b) "Elastodynamic near-tip fields for a crack propagating along the interface of two orthotropic solids." *Int. J. Eng. Sci.,* **14**, 811–818.

ACI Committee 318 (1989) *Building Code Requirements for Reinforced Concrete (ACI 318-89) and Commentary (ACI 318R-89).* U.S. Standard, American Concrete Institute, Detroit.

ACI Committee 318 (1992) *Building Code Requirements for Reinforced Concrete (ACI 318-89 Revised 1992).* U.S. Standard, American Concrete Institute, Detroit.

ACI Committee 349 (1989) "Code requirements for nuclear safety structures (ACI 349.1R), Appendix B - Steel embedments." In *Manual of Concrete Practice, Part IV*, American Concrete Institute, Detroit.

ACI Committee 408 (1979) "Suggested development, splice and standard hook provisions for deformed bars in tension." *Concrete Int.,* **1**(7), 44–46. (ACI 408.1R-79.)

ACI Committee 446 (1992) "Fracture mechanics of concrete: Concepts, models and determination of material properties." In *Fracture Mechanics of Concrete Structures*, Z. P. Bažant, ed., Elsevier Applied Science, London, pp. 1–140. (State of Art Report.)

ACI–ASCE Committee 426 (1973) "The shear strength of reinforced concrete members: Chapters 1 to 4." *J. Struct. Div.-ASCE,* **99**, 1091–1187.

ACI–ASCE Committee 426 (1974) "The shear strength of reinforced concrete members: Chapter 5." *J. Struct. Div.-ASCE,* **100**, 1543–1591.

ACI–ASCE Committee 426 (1977) "Suggested revisions to shear provisions for building codes." *ACI J.,* **74**(9), 458–469.

Adamson, R. M., Dempsey, J. P., DeFranco, S. J. and Xie, Y. (1995) "Large-scale *in-situ* ice fracture experiments. Part I: Experimental aspects." In *Ice Mechanics 1995*, J. P. Dempsey and Y. D. S. Rajapakse, eds., The American Society of Mechanical Engineers, New York, pp. 107–128. (AMD-Vol. 207, ASME Summer Meeting, Los Angeles, CA.)

Alexander, M. G. (1987) Test data in Shah and Ouyang (1994).

Alfrey, T. (1944) "Nonhomogeneous stress in viscoelastic media." *Quart. Appl. Math.,* **2**(2), 113–119.

Aliabadi, M. H. and Rooke, D. P. (1991) *Numerical Fracture Mechanics*, Computational Mechanics Publications, Southampton.

Alvaredo, A. M. and Torrent, R. J. (1987) "The effect of the shape of the strain-softening diagram on the bearing capacity of concrete beams." *Mater. Struct.,* **20**, 448–454.

Argon, A. S. (1972) "Fracture of composites." In *Treatise of Materials Science and Technology*, Vol. 1, Academic Press, New York, pp. 79.

Ashby, M. F. and Hallam, S. D. (1986) "The failure of brittle solids containing small cracks under compressive stress states." *Acta Metall.,* **34**(3), 497–510.

ASTM (1983) "Standard test method for plane-strain fracture toughness of metallic materials." In *Annual Book of ASTM Standards*, Vol. 03.01, ASTM, Philadelphia, pp. 519–554. (Standard E399-83.)

ASTM (1991) "Standard test methods for plane-strain fracture toughness and strain energy release rate of plastic materials." In *Annual Book of ASTM Standards*, Vol. 08.03, ASTM, Philadelphia. (Standard D 5045-91.)

Atkins, A.G. and Mai. Y.W. (1985) *Elastic and Plastic Fracture*, Ellis Horwood Ltd., John Wiley & Sons, Chichester, New York.

Ballarini, R., Shah, S. P. and Keer, L. M. (1985) "Crack growth in cement based composites." *Eng. Fract. Mech.,* **20**(3), 433–445.

Balmer, G. G. (1949) *Shearing Strength of Concrete under High Triaxial Stress—Computation of Mohr's Envelope as a Curve.* Report No. SP-23, Struct. Res. Lab., Denver, CO.

Baluch, M. H., Azad, A. K. and Ashmawi, W. (1992) "Fracture mechanics application to reinforced concrete members in flexure." In *Application of Fracture Mechanics to Reinforced Concrete*, A. Carpinteri, ed., Elsevier Applied Science, London, pp. 413–436.

Baluch, M. H, Qureshi, A. B. and Azad, A. K. (1989) "Fatigue crack propagation in plain concrete." In *Fracture of Concrete and Rock*, S.P. Shah and S.E. Swartz, eds., Springer-Verlag, New York, pp. 80–85.

Barenblatt, G. I. (1959) "The formation of equilibrium cracks during brittle fracture, general ideas and hypothesis, axially symmetric cracks." *Prikl. Mat. Mech.,* **23**(3), 434–444.

Barenblatt, G. I. (1962) "The mathematical theory of equilibrium of cracks in brittle fracture." *Adv. Appl. Mech.,* **7**, 55–129.

Barenblatt, G. I. (1979) *Similarity, Self-Similarity and Intermediate Asymptotics*, Plenum Press, New York.

Barsoum, R. S. (1975) "Further application of quadratic isoparametric finite elements to linear fracture mechanics of plate bending and general shells." *Int. J. Fracture,* **11**, 167–169.

Barsoum, R. S. (1976) "On the use of isoparametric finite elements in linear fracture mechanics." *Int. J. Numer. Meth. Eng.,* **10**, 25–37.

Batdorf, S. B. and Budianski, B. (1949) *A Mathematical Theory of Plasticity Based on the Concept of Slip*. Technical Note No. 1871, Nat. Advisory Committee for Aeronautics, Washington, D.C.

Batto, R. A. and Schulson, E. M. (1993) "On the ductile-to-brittle transition in ice under compression." *Acta Metall. Mater.,* **41**(7), 2219–2225.

Bažant, Z. P. (1967) "Stability of continuum and compression strength." *Bulletin RILEM,* **39**, 99–112. (In French.)

Bažant, Z. P. (1968) "Effect of folding of reinforcing fibers on the elastic moduli and strength of composite materials." *Mekhanika Polimerov,* **4**, 314–321. (In Russian.)

Bažant, Z. P. (1972a) "Thermodynamics of hindered adsorption with application to cement paste and concrete." *Cement Concrete Res.,* **2**, 1–16.

Bažant, Z. P. (1972b) "Thermodynamics of interacting continua with surfaces and creep analysis of concrete structures." *Nucl. Eng. Des.,* **20**, 477–505.

Bažant, Z. P. (1974) "Three-dimensional harmonic functions near termination or intersection of singularity lines: A general numerical method." *Int. J. Eng. Sci.,* **12**, 221–243.

Bažant, Z.P. (1975) "Theory of creep and shrinkage in concrete structures: A precis of recent developments." In *Mechanics Today*, Vol. 2, S. Nemat-Nasser, ed., Pergamon Press, Oxford, pp. 1–93.

Bažant, Z. P. (1976) "Instability, ductility and size effect in strain-softening concrete." *J. Eng. Mech. Div.-ASCE,* **102**, 331–344. (Discussion **103**, 357-358, 775-777 and **103**, 501-502.)

Bažant, Z. P. (1982) "Crack band model for fracture of geomaterials." In *Proc. 4th Int. Conf. on Numerical Methods in Geomechanics*, Vol. 3, Z. Eisenstein, ed., pp. 1137–1152.

Bažant, Z. P. (1983) "Fracture in concrete and reinforced concrete." In *Preprints IUTAM Prager Symposium on Mechanics of Geomaterials: Rocks, Concretes, Soils*, Z. P. Bažant, ed., pp. 281–316. (See also Bažant 1985c.)

Bažant, Z. P. (1984a) "Size effect in blunt fracture: Concrete, rock, metal." *J. Eng. Mech.-ASCE,* **110**, 518–535.

Bažant, Z. P. (1984b) "Imbricate continuum and its variational derivation." *J. Eng. Mech.-ASCE,* **110**, 1693–1712.

Bažant, Z. P. (1984c) "Microplane model for strain controlled inelastic behavior." In *Mechanics of Engineering Materials*, C. S. Desai and R. H. Gallagher, eds., J. Wiley, London, pp. 45–59.

Bažant, Z. P. (1985a) "Mechanics of fracture and progressive cracking in concrete structures." In *Fracture Mechanics of Concrete: Structural Application and Numerical Calculation*, G. C. Sih and A. DiTommaso, eds., Martinus Nijhoff, Dordrecht, pp. 1–94.

Bažant, Z. P. (1985b) "Fracture mechanics and strain-softening in concrete." In *Preprints U. S.- Japan Seminar on Finite Element Analysis of Reinforced Concrete Structures*, Vol. 1, pp. 47–69.

Bažant, Z. P. (1985c) "Fracture in concrete and reinforced concrete." In *Mechanics of Geomaterials: Rocks, Concretes, Soils*, Z. P. Bažant, ed., John Wiley & Sons, Chichester, New York, pp. 259–303.

Bažant, Z.P. (1985d) "Comment on Hillerborg's size effect law and fictitious crack model." In *Dei Poli Anniversary Volume*, L. Cedolin et al., eds., Politecnico di Milano, Italy, pp. 335–338.

Bažant, Z. P. (1986a) "Fracture mechanics and strain-softening of concrete." In *Finite Element Analysis of Reinforced Concrete Structures*, C. Meyer and H. Okamura, eds., ASCE, New York, pp. 121–150.

Bažant, Z. P. (1986b) "Distributed cracking and nonlocal continuum." In *Finite Element Methods for Nonlinear Problems*, P. Bergan et al., eds., Springer, Berlin, pp. 77–102.

Bažant, Z. P. (1986c) "Mechanics of distributed cracking." *Appl. Mech. Rev.*, **39**, 675–705.

Bažant, Z. P. (1987a) "Fracture energy of heterogeneous materials and similitude." In *Preprints SEM-RILEM Int. Conf. on Fracture of Concrete and Rock*, S. P. Shah and S. E. Swartz, eds., Society for Experimental Mechanics (SEM), Bethel, pp. 390–402. (See also Bažant 1989a.)

Bažant, Z. P. (1987b) "Snapback instability at crack ligament tearing and its implication for fracture micromechanics." *Cement Concrete Res.*, **17**, 951–967.

Bažant, Z. P. (1987c) "Why continuum damage is nonlocal: Justification by quasi-periodic microcrack array." *Mech. Res. Commun.*, **14**(5/6), 407–419.

Bažant, Z. P., ed. (1988a) *Mathematical Modeling of Creep and Shrinkage of Concrete*, John Wiley & Sons, Chichester, New York.

Bažant, Z. P. (1988b) *Fracture of Concrete*. Lecture Notes for Course 720-D30, Northwestern University, Evanston, Illinois 60208, U.S.A.

Bažant, Z. P. (1989a) "Fracture energy of heterogeneous materials and similitude." In *Fracture of Concrete and Rock*, S. P. Shah and S. E. Swartz, eds., Springer Verlag, New York, pp. 229–241.

Bažant, Z. P. (1989c) "Advances in material modeling of concrete." In *Tenth International Conference on Structural Mechanics in Reactor Technology (SMiRT10)*, Vol. A, A. H. Hadjian, ed., pp. 301–330.

Bažant, Z. P. (1990a) "Justification and improvement of Kienzler and Herrmann's estimate of stress intensity factors of cracked beam." *Eng. Fract. Mech.*, **36**(3), 523–525.

Bažant, Z. P. (1990b) "A critical appraisal of 'no-tension' dam design: A fracture mechanics viewpoint." *Dam Eng.*, **1**(4), 237–247.

Bažant, Z. P. (1990c) "Recent advances in failure localization and nonlocal models." In *Micromechanics of Failure of Quasi-Brittle Materials*, S. P. Shah, S. E. Swartz and M. L. Wang, eds., Elsevier, London, pp. 12–32.

Bažant, Z. P. (1990d) "Smeared-tip superposition method for nonlinear and time-dependent fracture." *Mech. Res. Commun.*, **17**(5), 343–351.

Bažant, Z. P. (1991a) "Rate effect, size effect and nonlocal concepts for fracture of concrete and other quasi-brittle materials." In *Toughening Mechanisms in Quasi-Brittle Materials*, S.P. Shah, ed., Kluwer Academic Publishers, Dordrecht, The Netherlands, pp. 131–154.

Bažant, Z. P. (1991b) "Why continuum damage is nonlocal: micromechanics arguments." *J. Eng. Mech.-ASCE*, **117**(5), 1070–1087.

Bažant, Z. P. (1992a) "Large-scale thermal bending fracture of sea ice plates." *J. Geophys. Res.*, **97**(C11), 17739–17751.

Bažant, Z. P. (1992b) "New concept of nonlocal continuum damage: Crack influence function." In *Macroscopic Behavior of Heterogeneous Materials from Microstructure*, S. Torquato and D. Krajcinovic, eds., The American Society of Mechanical Engineers, New York, pp. 153–160. (AMD–Vol.147, ASME Winter Annual Meeting, Anaheim.)

Bažant, Z. P. (1992c) "Large-scale fracture of sea ice plates." In *Proc., 11th IAHR International Ice Symposium*, Vol. 2, T. M. Hrudey, ed., Univ. of Alberta, Edmonton, Canada, pp. 991–1005.

Bažant, Z. P. (1993a) "Scaling laws in mechanics of failure." *J. Eng. Mech.-ASCE*, **119**(9), 1828–1844.

Bažant, Z. P. (1993b) "Current status and advances in the theory of creep and interaction with fracture." In *Creep and Shrinkage of Concrete*, Z. P. Bažant and I. Carol, eds., E & FN Spon, London, pp. 291–307.

Bažant, Z. P. (1993c) *Concept of Boundaries for Microplane Model*. Internal Research Note, Dept. of Civil Engrg., Northwestern University, Evanston, Illinois 60208, U.S.A.

Bažant, Z. P. (1993d) "Discussion of 'Fracture mechanics and size effect of concrete in tension', by Tang et al. (1992)." *J. Struct. Eng.-ASCE*, **119**(8), 2555–2558.

Bažant, Z. P. (1994a) "Size effect in tensile and compressive quasibrittle failures." In *Size Effect in Concrete Structures*, H. Mihashi, H. Okamura and Z. P. Bažant, eds., E & FN Spon, London, pp. 161–180.

Bažant, Z. P. (1994b) "Nonlocal damage theory based on micromechanics of crack interactions." *J. Eng. Mech.-ASCE*, **120**(3), 593–617. (Addendum and Errata **120**, 1401–1402.)

Bažant, Z. P. (1994c) *Is size effect caused by fractal nature of crack surfaces?*. Report No. 94-10/402i, Department of Civil Engineering, Northwestern University, Evanston, Illinois.

Bažant, Z. P. (1995a) "Scaling theories for quasibrittle fracture: Recent advances and new directions." In *Fracture Mechanics of Concrete Structures*, F. H. Wittmann, ed., Aedificatio Publishers, Freiburg, Germany, pp. 515–534.

Bažant, Z. P. (1995b) *Scaling of Quasibrittle Fracture: I. Asymptotic Analysis Based on Laws of Thermodynamics. II. The Fractal Hypothesis, its Critique and Weibull Connection.* Report No. 95-7/C402s, Dep. of Civil Engineering, Northwestern University, Evanston, Illinois 60208, U.S.A.

Bažant, Z. P. (1995c) "Creep and Damage in Concrete." In *Materials Science of Concrete IV*, J. Skalny and S. Mindess, eds., Am. Cer. Soc., Westerville, Ohio, pp. 355–389.

Bažant, Z. P. (1995d) *Microplane model for concrete. I. Stress-strain boundaries and finite strain.* Internal Report to WES, Vicksburg, Northwestern University.

Bažant, Z. P. (1995e) "Scaling of quasibrittle fracture and the fractal question." *J. Eng. Mater. Technol.-T. ASME,* **117**, 361–367.

Bažant, Z. P. (1995f) "Scaling theories for quasibrittle fracture: Recent advances and new directions." In *Fracture Mechanics of Concrete Structures*, Vol. 1, F.H. Wittmann, ed., Aedificatio Publishers, Freiburg, Germany, pp. 515–534.

Bažant, Z. P. (1996a) "Is no-tension design of concrete or rock structures always safe?—Fracture analysis." *J. Struct. Eng.-ASCE,* **122**(1), 2–10.

Bažant, Z. P. (1996b) *Fracturing Truss Model: Explanation of Size Effect Mechanism in Shear Failure of Reinforced Concrete.* Report No. 96-3/603f, Dept. of Civil Engrg., Northwestern University, Evanston, Illinois; also *J. of Engrg. Mechanics ASCE* 123 (12), in press.

Bažant, Z. P. (1996c) "Finite strain generalization of small-strain constitutive relations for any finite strain tensor and additive volumetric-deviatoric split." *Int. J. Solids Struct.,* **33**(20–22), 2887–2897. (Special issue in memory of Juan Simo.)

Bažant, Z. P. (1996d) "Untitled." Personally communicated research note to J. Planas, November 1996.

Bažant, Z. P. (1996e) "Can scaling of structural failure be explained by fractal nature of cohesive fracture?" In *Size-Scale Effects in the Failure Mechanisms of Materials and Structures*, A. Carpinteri, ed., E & FN Spon, London, pp. 284–289. (Appendix to a paper by Li and Bažant.)

Bažant, Z. P. (1997a) "Scaling of quasibrittle fracture: Asymptotic analysis." *Int. J. Fracture,* **83**(1), 19–40.

Bažant, Z. P. (1997b) "Recent advances in brittle-plastic compression failure: damage localization, scaling and finite strain." In *Computational Plasticity*, D. R. J. Owen, E. Oñate and E. Hinton, eds., Int. Center for Num. Meth. in Eng. (CIMNE), Barcelona, Spain, pp. 3–19.

Bažant, Z. P. (1997c) *Easy-to-compute finite strain tensor with symmetric inverse, approximating Hencky strain tensor.* Report No. 96-9/425e, Northwestern University, Evanston, Illinois. (Submitted to *J. of Engrg. Materials and Technology* ASME.)

Bažant, Z. P. (1997d) "Scaling of quasibrittle fracture: hypotheses of invasive and lacunar fractality, their critique and Weibull connection." *Int. J. Fracture,* **83**(1), 41–65.

Bažant, Z. P. (1997e) "Scaling of structural failure." *Appl. Mech. Rev.,* **50**(10), 593–627.

Bažant, Z. P. and Baweja, S. (1995a) "Justification and refinement of Model B3 for concrete creep and shrinkage. 1. Statistics and sensitivity." *Mater. Struct.,* **28**, 415–430.

Bažant, Z. P. and Baweja, S. (1995b) "Justification and refinement of Model B3 for concrete creep and shrinkage. 2. Updating and theoretical basis." *Mater. Struct.,* **28**, 488–495.

Bažant, Z. P. and Baweja, S. (1995c) "Creep and shrinkage prediction model for analysis and design of concrete structures – model B3." *Mater. Struct.,* **28**, 357–365. (RILEM Recommendation, in collaboration with RILEM Committee TC 107-GCS. Erratum **29**, 126.)

Bažant, Z. P. and Beissel, S. (1994) "Smeared-tip superposition method for cohesive fracture with rate effect and creep." *Int. J. Fracture,* **65**, 277–290.

Bažant, Z. P. and Belytschko, T. B. (1985) "Wave propagation in strain-softening bar: Exact solution." *J. Eng. Mech.-ASCE,* **111**, 381–389.

Bažant, Z. P. and Cao, Z. (1986) "Size effect in brittle failure of unreinforced pipes." *ACI J.,* **83**(3), 369–373.

Bažant, Z. P. and Cao, Z. (1987) "Size effect in punching shear failure of slabs." *ACI Struct. J.,* **84**, 44–53.

Bažant, Z. P. and Cedolin, L. (1979) "Blunt crack band propagation in finite element analysis." *J. Eng. Mech. Div.-ASCE,* **105**, 297–315.

Bažant, Z. P. and Cedolin, L. (1980) "Fracture mechanics of reinforced concrete." *J. Eng. Mech. Div.-ASCE,* **106**, 1257–1306.

Bažant, Z. P. and Cedolin, L. (1983) "Finite element modeling of crack band propagation." *J. Struct Eng.-ASCE,* **109**, 69–92.

Bažant, Z. P. and Cedolin, L. (1991) *Stability of Structures: Elastic, Inelastic, Fracture and Damage Theories*, Oxford University Press, New York.

Bažant, Z. P. and Cedolin, L. (1993) "Why direct tension specimens break flexing to the side." *J. Struct. Eng.-ASCE,* **119**(4), 1101–1113.

Bažant, Z. P. and Chern, J.-C. (1985a) "Strain-softening with creep and exponential algorithm." *J. Eng. Mech.-ASCE,* **111**, 391–415.

Bažant, Z. P. and Chern, J.-C. (1985b) "Concrete creep at variable humidity: Constitutive law and mechanism." *Mater. Struct.,* **18**, 1–20.

Bažant, Z. P. and Desmorat, R. (1994) "Size effect in fiber of bar pullout with interface softening slip." *J. Eng. Mech.-ASCE,* **120**(9), 1945–1962.

Bažant, Z. P. and Estenssoro, L. F. (1979) "Surface singularity and crack propagation." *Int. J. Solids Struct.,* **15**, 405–426. (Addendum **16**, 479–481.)

Bažant, Z. P. and Gambarova, P. (1984) "Crack shear in concrete: Crack band microplane model." *J. Struct. Eng.-ASCE,* **110**, 2015–2035.

Bažant, Z. P. and Gettu, R. (1989) "Determination of nonlinear fracture characteristics and time dependence from size effect." In *Fracture of Concrete and Rock: Recent Developments*, S. P. Shah, S. E. Swartz and B. Barr, eds., Elsevier Applied Science, London, pp. 549–565.

Bažant, Z. P. and Gettu, R. (1990) "Size effect in concrete structures and influence of loading rates." In *Serviceability and Durability of Construction Materials*, B. A. Suprenant, ed., American Society of Civil Engineers (ASCE), New York, pp. 1113–1123.

Bažant, Z. P. and Gettu, R. (1992) "Rate effects and load relaxation: Static fracture of concrete." *ACI Mater. J.,* **89**(5), 456–468.

Bažant, Z. P. and Jirásek, M. (1993) "R-curve modeling of rate and size effects in quasibrittle fracture." *Int. J. Fracture,* **62**, 355–373.

Bažant, Z. P. and Jirásek, M. (1994a) "Damage nonlocality due to microcrack interactions: Statistical determination of crack influence function." In *Fracture and Damage in Quasibrittle Structures: Experiment, Modelling and Computer Analysis*, Z. P. Bažant, Z. Bittnar, M. Jirásek and J. Mazars, eds., E&FN Spon, London, pp. pp.3–17.

Bažant, Z. P. and Jirásek, M. (1994b) "Nonlocal model based on crack interactions: A localization study." *J. Eng. Mater. Technol.-T. ASME,* **116**, 256–259.

Bažant, Z. P. and Kazemi, M. T. (1990a) "Determination of fracture energy, process zone length and brittleness number from size effect, with application to rock and concrete." *Int. J. Fracture,* **44**, 111–131.

Bažant, Z. P. and Kazemi, M. T. (1990b) "Size effect in fracture of ceramics and its use to determine fracture energy and effective process zone length." *J. Am. Ceram. Soc.,* **73**(7), 1841–1853.

Bažant, Z. P. and Kazemi, M. T. (1991) "Size effect on diagonal shear failure of beams without stirrups." *ACI Struct. J.,* **88**(3), 268–276.

Bažant, Z. P. and Keer, L. M. (1974) "Singularity of elastic stresses and of harmonic functions at conical notches or inclusions." *Int. J. Solids Struct.,* **10**, 957–964.

Bažant, Z. P. and Kim, J.-J. (1996a) *Penetration fracture and size effect in sea ice plates with part-through bending cracks.* Report No. 96-10/402p, Dep. of Civil Engineering, Northwestern University, Evanston, Illinois. (Also *J. of Eng. Mech.,* in press.)

Bažant, Z. P. and Kim, J.-J. (1996b) *Tube-squash test and large strains of normal and high-strength concretes with shear angle over 30°.* Report, Northwestern University, Evanston, Illinois. (Submitted to ACI Materials Journal.)

Bažant, Z. P. and Kim, J.-K. (1984) "Size effect in shear failure of longitudinally reinforced beams." *ACI J.,* **81**, 456–468. (Discussion and Closure **82**, 579-583.)

Bažant, Z. P. and Kwon, Y. W. (1994) "Failure of slender and stocky reinforced concrete columns: Tests of size effect." *Mater. Struct.,* **27**, 79–90.

Bažant, Z. P. and Li, Y.-N. (1994a) "Cohesive crack model for geomaterials: Stability analysis and rate effect." *Appl. Mech. Rev.,* **47**(6), S91–S96. (Part of *Mechanics U.S.A. 1994.* ed. by A.S. Kobayashi, Proc. 12th U.S. Nat. Congress of Appl. Mechanics, Seattle, WA, June.)

Bažant Z. P. and Li Y.-N. (1994b) "Penetration fracture of sea ice plate: Simplified analysis and size effect." *J. Eng. Mech.-ASCE,* **120**(6), 1304–1321.

Bažant, Z. P. and Li, Y.-N. (1995a) "Stability of cohesive crack model: Part I—Energy principles." *J. Appl. Mech.-T. ASME,* **62**, 959–964.

Bažant, Z. P. and Li, Y.-N. (1995b) "Stability of cohesive crack model: Part II—Eigenvalue analysis of size effect on strength and ductility of structures." *J. Appl. Mech.-T. ASME,* **62**, 965–969.

Bažant, Z. P. and Li Y.-N. (1995d) "Penetration fracture of sea ice plate." *Int. J. Solids Struct.,* **32**(3/4), 303–313.

Bažant, Z. P. and Li, Y.-N. (1997) "Cohesive crack model with rate-dependent crack opening and viscoelasticity: I. Mathematical model and scaling." *Int. J. Fracture,* in press.

Bažant, Z. P. and Li, Z. (1995c) "Modulus of rupture: Size effect due to fracture initiation in boundary layer." *J. Struct. Eng.-ASCE,* **121**(4), 739–746.

Bažant, Z. P. and Li, Z. (1996) "Zero-brittleness size-effect method for one-size fracture test of concrete." *J. Eng. Mech.-ASCE,* **122**(5), 458–468.

Bažant, Z. P. and Lin, F.-B. (1988a) "Nonlocal smeared cracking model for concrete fracture." *J. Struct. Eng.-ASCE,* **114**(11), 2493–2510.

Bažant, Z. P. and Lin, F.-B. (1988b) "Nonlocal yield limit degradation." *Int. J. Numer. Meth. Eng.,* **26**, 1805–1823.

Bažant, Z. P. and Oh, B. H. (1982) "Strain rate effect in rapid triaxial loading of concrete." *J. Eng. Mech.-ASCE,* **108**(5), 767–782.

Bažant, Z. P. and Oh, B.-H. (1983a) "Crack band theory for fracture of concrete." *Mater. Struct.,* **16**, 155–177.

Bažant, Z. P. and Oh, B.-H. (1983b) "Microplane model for fracture analysis of concrete structures." In *Proc. Symp. on the Interaction of Non-Nuclear Munitions with Structures*, pp. 49–53.

Bažant, Z. P. and Oh, B.-H. (1985) "Microplane model for progressive fracture of concrete and rock." *J. Eng. Mech.-ASCE,* **111**, 559–582.

Bažant, Z. P. and Oh, B.-H. (1986) "Efficient numerical integration on the surface of a sphere." *Z. Angew. Math. Mech.,* **66**(1), 37–49.

Bažant, Z. P. and Ohtsubo, H. (1977) "Stability conditions for propagation of a system of cracks in a brittle solid." *Mech. Res. Commun.,* **4**(5), 353–366.

Bažant, Z. P. and Osman, E. (1976) "Double power law for basic creep of concrete." *Mater. Struct.,* **9**, 3–11.

Bažant, Z. P. and Ožbolt, J. (1990) "Nonlocal microplane model for fracture, damage, and size effect in structures." *J. Eng. Mech.-ASCE,* **116**(11), 2484–2504.

Bažant, Z. P. and Ožbolt, J. (1992) "Compression failure of quasi-brittle material: Nonlocal microplane model." *J. Eng. Mech.-ASCE,* **118**(3), 540–556.

Bažant, Z. P. and Panula, L. (1978) "Practical prediction of time-dependent deformations of concrete. Part II – Basic creep." *Mater. Struct.,* **11**, 317–328.

Bažant, Z. P. and Pfeiffer, P. A. (1986) "Shear fracture tests of concrete." *Mater. Struct.,* **19**, 111–121.

Bažant, Z. P. and Pfeiffer, P. A. (1987) "Determination of fracture energy from size effect and brittleness number." *ACI Mater. J.,* **84**(6), 463–480.

Bažant, Z. P. and Pijaudier-Cabot, G. (1988) "Nonlocal continuum damage, localization instability and convergence." *J. Appl. Mech.-T. ASME,* **55**, 287–293.

Bažant, Z. P. and Pijaudier-Cabot, G. (1989) "Measurement of characteristic length of nonlocal continuum." *J. Eng. Mech.-ASCE,* **115**(4), 755–767.

Bažant, Z. P. and Prasannan, S. (1989) "Solidification theory for concrete creep: I. Formulation." *J. Eng. Mech.-ASCE,* **115**(8), 1691–1703.

Bažant, Z. P. and Prat, P. C. (1988a) "Effect of temperature and humidity on fracture energy of concrete." *ACI Mater. J.,* **85**, 262–271.

Bažant, Z. P. and Prat, P. C. (1988b) "Microplane model for brittle plastic material: I. Theory; and II. Verification." *J. Eng. Mech.-ASCE,* **114**, 1672–1702.

Bažant, Z. P. and Prat, P. C. (1995) "Elastic material with systems of growing or closing cracks: Tangential Stiffness." In *Contemporary Research in Engineering Science*, R. Batra, ed., Springer Verlag, New York, pp. 55–65.

Bažant, Z. P. and Prat, P. C. (1997) "Tangential stiffness tensor of material with growing cracks." *Mech. Res. Commun.,* submitted.

Bažant, Z. P. and Schell, W. F. (1993) "Fatigue fracture of high-strength concrete and size effect." *ACI Mater. J.,* **90**(5), 472–478.

Bažant, Z. P. and Şener, S. (1987) "Size effect in torsional failure of longitudinally reinforced concrete beams." *J. Struct Eng.-ASCE,* **113**(10), 2125–2136.

Bažant, Z. P. and Şener, S. (1988) "Size effect in pullout tests." *ACI Mater. J.,* **85**, 347–351.

Bažant, Z. P. and Sun, H.-H. (1987) "Size effect in diagonal shear failure: Influence of aggregate size and stirrups." *ACI Mater. J.,* **84**, 259–272.

Bažant, Z. P. and Wahab, A. B. (1979) "Instability and spacing of cooling or shrinkage cracks." *J. Eng. Mech. Div.-ASCE,* **105**, 873–889.

Bažant, Z. P. and Wahab, A. B. (1980) "Stability of parallel cracks in solids reinforced by bars." *Int. J. Solids Struct.,* **16**, 97–105.

Bažant, Z. P. and Xi, Y. (1991) "Statistical size effect in quasi-brittle structures: II. Nonlocal theory." *J. Eng. Mech.-ASCE,* **117**(11), 2623–2640.

Bažant, Z. P. and Xi, Y. (1994) "Fracture of random quasibrittle materials: Markov process and Weibull-type models." In *Structural Safety and Reliability*, G. I. Schuëller, M. Shinozuka and J. T. P. Yao, eds., A. A. Balkema, Rotterdam–Brookfield, pp. 609–614. (Proc. of ICOSSAR'93—6th Intern. Conf. on Struct. Safety and Reliability, Innsbruck, Austria, Aug. 9–13, 1993.)

Bažant, Z. P. and Xiang, Y. (1994) "Compression failure of quasibrittle materials and size effect." In *Damage Mechanics in Composites*, D. H. Allen and J. W. Ju, eds., The American Society of Mechanical Engineers, New York, pp. 143–148. (AMD–Vol. 185, Winter Annual Meeting, Chicago.)

Bažant, Z. P. and Xiang, Y. (1997) "Size effect in compression fracture: splitting crack band propagation." *J. Eng. Mech.-ASCE,* **123**(2), 162–172.

Bažant, Z. P. and Xu, K. (1991) "Size effect in fatigue fracture of concrete." *ACI Mater. J.,* **88**(4), 390–399.

Bažant, Z. P., Adley, M. D. and Xiang, Y. (1996) "Finite strain analysis of deformations of quasibrittle material during missile impact and penetration." In *Advances in Failure Mechanisms in Brittle Materials*, R. J. Clifton and H. D. Espinosa, eds., The American Society of Mechanical Engineers, New York. (MD–Vol 75, AMD–Vol. 219.)

Bažant, Z. P., Bai, S.-P. and Gettu, R. (1993) "Fracture of rock: Effect of loading rate." *Eng. Fract. Mech.,* **45**(3), 393–398.

Bažant, Z. P., Belytschko, T. B. and Chang, T.-P. (1984) "Continuum model for strain softening." *J. Eng. Mech.-ASCE,* **110**, 1666–1692.

Bažant, Z. P., Bishop, F. C. and Chang, T.-P. (1986) "Confined compression tests of cement paste and concrete up to 300 ksi." *ACI J.,* **83**(4), 553–560.

Bažant, Z. P., Daniel, I. M. and Li, Z. (1996) "Size effect and fracture characteristics of composite laminates." *J. Eng. Mater. Technol.-T. ASME,* **118**, 317–324.

Bažant, Z. P., Gettu, R. and Kazemi, M. T. (1991) "Identification of nonlinear fracture properties from size-effect tests and structural analysis based on geometry-dependent R-curves." *Int. J. Rock Mech. Min. Sci.,* **28**(1), 43–51.

Bažant, Z. P., Glazik, J. L. and Achenbach, J. D. (1976) "Finite element analysis of wave diffraction by a crack." *J. Eng. Mech. Div.-ASCE,* **102**, 479–496. (Discussion **103**, 226–228, 497–499, 1181–1185.)

Bažant, Z. P., Glazik, J. L. and Achenbach, J. D. (1978) "Elastodynamic fields near running cracks by finite elements." *Comput. Struct.,* **26**, 567–574.

Bažant, Z. P., Gu, W.-H. and Faber, K. T. (1995) "Softening reversal and other effects of a change in loading rate on fracture of concrete." *ACI Mater. J.,* **92**, 3–9.

Bažant, Z. P., He, S., Plesha, M. E. and Rowlands, R. E. (1991) "Rate and size effect in concrete fracture: Implications for dams." In *Proc. Int. Conf. on Dam Fracture*, V. Saouma, R. Dungar, and D. Morris, eds., University of Colorado, Boulder, CO, pp. 413–425.

Bažant, Z. P., Jirásek, M., Xiang, Y. and Prat, P. C. (1994) "Microplane model with stress-strain boundaries and its identification from tests with localized damage." In *Computational Modeling of Concrete Structures*, H. Mang et al., eds., Pineridge Press, Swansea, pp. 255–261.

Bažant, Z. P., Kazemi, M. T. and Gettu, R. (1989) "Recent studies of size effect in concrete structures." In *Transactions of the Tenth International Conference on Structural Mechanics in Reactor Technology.*, Vol. H, A. H. Hadjian, ed., pp. 85–93.

Bažant, Z. P., Kazemi, M. T., Hasegawa, T. and Mazars, J. (1991) "Size effect in Brazilian split-cylinder tests: Measurement and fracture analysis." *ACI Mater. J.,* **88**(3), 325–332.

Bažant, Z. P., Kim, J.-J. and Li, Y.-N. (1995) "Part-through bending cracks in sea ice plates: Mathematical modeling." In *Ice Mechanics 1995*, J. P. Dempsey and Y. Rajapakse, eds., The American Society of Mechanical Engineers, New York, pp. 97–105. (AMD-Vol. 207, ASME Summer Meeting, Los Angeles, CA.)

Bažant, Z. P., Kim, J-K. and Pfeiffer, P. A. (1986) "Nonlinear fracture properties from size effect tests." *J. Struct. Eng.-ASCE,* **112**, 289–307.

Bažant, Z. P., Lee, S-G. and Pfeiffer, P. A. (1987) "Size effect tests and fracture characteristics of aluminum." *Eng. Fract. Mech.,* **26**(1), 45–57.

Bažant, Z. P., Li, Z. and Thoma, M. (1995) "Identification of stress-slip law for bar or fiber pullout by size effect tests." *J. Eng. Mech.-ASCE,* **121**(5), 620–625.

Bažant, Z. P., Lin, F.-B. and Lippmann, H. (1993) "Fracture energy release and size effect in borehole breakout." *Int. J. Numer. Anal. Meth. Geomech.,* **17**, 1–14.

Bažant, Z. P., Ohtsubo, R. and Aoh, K. (1979) "Stability and post-critical growth of a system of cooling and shrinkage cracks." *Int. J. Fracture,* **15**, 443–456.

Bažant, Z. P., Ožbolt, J. and Eligehausen, R. (1994) "Fracture size effect: Review of evidence for concrete structures." *J. Struct Eng.-ASCE,* **120**, 2377–2398.

Bažant, Z. P., Şener, S. and Prat, P. C. (1988) "Size effect tests of torsional failure of plain and reinforced concrete beams." *Mater. Struct.,* **21**, 425–430.

Bažant, Z. P., Tabbara, M. R., Kazemi, M. T. and Pijaudier-Cabot, G. (1990) "Random particle model for fracture of aggregate or fiber composites." *J. Eng. Mech.-ASCE,* **116**, 1686–1705.

Bažant, Z. P., Xi, Y. and Baweja, S. (1993) "Improved prediction model for time-dependent deformations of concrete: Part 7 — Short form of BP-KX model, statistics and extrapolation of short-time data." *Mater. Struct.,* **26**, 567–574.

Bažant, Z. P., Xi, Y. and Reid, S. G. (1991) "Statistical size effect in quasi-brittle structures: I. Is Weibull theory applicable?" *J. Eng. Mech.-ASCE,* **117**(11), 2609–2622.

Bažant, Z. P., Xiang, Y., Adley, M. D., Prat, P. C. and Akers, S. A. (1996) "Microplane model for concrete. II. Data delocalization and verification." *J. Eng. Mech.-ASCE,* **122**(3), 255–262.

Bažant, Z. P., Xiang, Y. and Prat, P. C. (1996) "Microplane model for concrete. I. Stress-strain boundaries and finite strain." *J. Eng. Mech.-ASCE,* **122**(3), 245–254. (Erratum in **123**.)

Bell, J. F. (1985) "Contemporary perspectives in finite strain plasticity." *Int. J. Plasticity,* **1**, 3–27.

Belytschko, T. B., Bažant, Z. P., Hyun, Y. W. and Chang, T.-P. (1986) "Strain-softening materials and finite element solutions." *Comput. Struct.,* **23**(2), 163–180.

Belytschko, T., Fish, J. and Englemann, B. E. (1988) "A finite element with embedded localization zones." *Comput. Meth. Appl. Mech. Eng.,* **70**, 59–89.

Bender, M. C. and Orszag, S. A. (1978) *Advanced Mathematical Methods for Scientists and Engineers,* McGraw Hill, New York.

Beremin, F. M. "A local criterion for cleavage fracture of a nuclear pressure vesse steel." *Metall. Trans. A,* **14**, 2277–2287.

Berthaud, Y., Ringot, E. and Schmitt N. (1991) "Experimental measurements of localization for tensile tests on concrete." In *Fracture Processes in Concrete, Rock and Ceramics*, J. G. M. van Mier, J. G. Rots and A. Bakker, eds., E & FN Spon, London, pp. 41– 50.

Bieniawski, Z. T. (1974) "Estimating the strength of rock materials." *J. S. Afr. Inst. Min. Metall.,* **74**, 312–320.

Biot, M. A. (1955) "Variational principles of irreversible thermodynamics with application to viscoelasticity." *Phys. Rev.,* **97**, 1163–1169.

Biot, M. A. (1965) *Mechanics of Incremental Deformations*, John Wiley & Sons, New York.

Bittencourt, T. N., Ingraffea, A. R. and Llorca, J. (1992) "Simulation of arbitrary, cohesive crack propagation." In *Fracture Mechanics of Concrete Structures*, Z. P. Bažant, ed., Elsevier Applied Science, London, pp. 339–350.

Blanks, R. F. and McNamara, C. C. (1935) "Mass concrete tests in large cylinders." *ACI J.,* **31**, 280–303.

Bocca, P., Carpinteri, A. and Valente, S. (1990) "Size effects in the mixed crack propagation: Softening and snap-back analysis." *Eng. Fract. Mech.,* **35**, 159–170.

Bocca, P., Carpinteri, A. and Valente, S. (1991) "Mixed mode fracture of concrete." *Int. J. Solids Struct.,* **27**, 1139–1153.

Bocca, P., Carpinteri, A. and Valente, S. (1992) "Fracture mechanics evaluation of anchorage bearing capacity in concrete." In *Application of Fracture Mechanics to Reinforced Concrete*, A. Carpinteri, ed., Elsevier Applied Science, London, pp. 231–265.

Bogdanoff, J. L. and Kozin, F. (1985) *Probabilistic Models of Cumulative Damage*, John Wiley & Sons, New York.

Borodich, F. M. (1992) "Fracture energy in a fractal crack propagating in concrete or rock." *Doklady Akademii Nauk.,* **325**(6), 1138–1141. (In Russian. Transl. in *Trans. Russian Ac. Sci.,* Earth Sci. Sec., **327**(8), 36–40.)

de Borst, R. (1986) *Non-Linear Analysis of Frictional Materials.* Doctoral thesis. Delft University of Technology, Delft, The Netherlands.

de Borst, R. (1990) "Simulation of localization using Cosserat theory." In *Proc., Int. Conf. on Computer-Aided Analysis and Design of Concrete Structures*, N. Bićanić and H.A. Mang, eds., Pineridge Press, Swansea, pp. 931–944.

de Borst, R. (1991) "Simulation of strain localization: A reappraisal of the Cosserat continuum." *Eng. Comput.,* **8**, 317–332.

de Borst, R. and Mühlhaus, H.-B. (1991) "Continuum models for discontinuous media." In *Fracture Processes in Concrete, Rock and Ceramics*, Vol. 2, J. G. M. van Mier, J. G. Rots and A. Bakker, eds., E & FN Spon, London, pp. 601–618.

de Borst, R. and Mühlhaus, H.-B. (1992) "Gradient-dependent plasticity: Formulation and algorithmic aspects." *Int. J. Numer. Meth. Eng.*, **35**, 521–539.

de Borst, R. and Sluys, L. J. (1991) "Localization in a Cosserat continuum under static and dynamic loading conditions." *Comput. Meth. Appl. Mech. Eng.*, **90**, 805–827.

Bosco, C. and Carpinteri, A. (1992) "Fracture mechanics evaluation of minimum reinforcement in concrete structures." In *Applications of Fracture Mechanics to Reinforced Concrete*, A. Carpinteri, ed., Elsevier Applied Science, London, pp. 347–377.

Bosco, C., Carpinteri, A. and DeBernardi, P. G. (1990a) "Fracture of reinforced concrete: Scale effect and snap-back instability." *Eng. Fract. Mech.*, **35**, 665–677.

Bosco, C., Carpinteri, A. and DeBernardi, P. G. (1990b) "Minimum reinforcement in reinforced concrete beams." *J. Struct Eng.-ASCE*, **116**, 427–437.

Brennan, G. (1978) *A Test to Determine the Bending Moment Resistance of Rigid Pipes.* TRRL Supplementary Report No. SR 348, Transport and Road Research Laboratory, Crowthorne, Berkshire.

Bresler, B. and Pister, K. S. (1958) "Strength of concrete under combined stresses." *ACI J.*, **55**(9), 321–345.

Broek, D. (1986) *Elementary Engineering Fracture Mechanics*, 4th edition, Martinus Nijhoff Publishers, Dordrecht.

Broms, C. E. (1990) "Punching of flat plates — A question of concrete properties in biaxial compression and size effect." *ACI Struct. J.*, **87**(3), 292–304.

Brown, J. H. (1972) "Measuring the fracture toughness of cement paste and mortar." *Mag. Concrete Res.*, **24**, 185–196.

Brown, W. F. and Srawley, J. E. (1986) *Plane Strain Crack Toughness Testing of High Strength Metallic Materials*, ASTM Special Technical Publication, No. 410.

Bruckner, A. and Munz, D. (1984) "Scatter of fracture toughness in the brittle-ductile transition region of a ferritic steel." In *Advances in Probabilistic Fracture Mechanics*, C. Sundararajan, ed., The American Society of Mechanical Engineers, New York, pp. 105–111.

Brühwiler, E. (1988) *Fracture Mechanics of Dam Concrete Subjected to Quasi-static and Seismic Loading Conditions.* Doctoral thesis. Laboratory for Building Materials, Swiss Federal Institute of Technology, Lausanne. (Thesis No. 739. In German.)

Brühwiler, E. and Wittmann, F. H. (1990) "The wedge splitting test, a new method of performing stable fracture mechanics tests." *Eng. Fract. Mech.*, **35**, 117–125.

Budianski, B. (1983) "Micromechanics." *Computers and structures,* **16**(1–4), 3–12.

Budianski, B. and Fleck, N. A. (1994) "Compressive kinking of fiber composites: A topical review." *Appl. Mech. Rev.*, **47**(No. 6, Part 2–Supplement, Proc. of 12th U.S. Nat. Congress of Applied Mechanics, Seattle), pp. S246–S255.

Budianski, B. and O'Conell, R. J. (1976) "Elastic moduli of a cracked solid." *Int. J. Solids Struct.*, **12**, 81–97.

Budianski, B. and Wu, T. T. (1962) "Theoretical prediction of plastic strains of polycrystals." In *Proc. Fourth U.S. National Congress of Applied Mechanics*, The American Society of Mechanical Engineers, New York, pp. 1175–1185.

Budianski, B., Fleck, N. A. and Amazigo, J. C. (1997) *On compression kink band propagation.* Report MECH No. 305, Harvard University, Cambrigde, Massachussets. (Submitted to *J. Mech. Phys. Solids.*)

Bueckner, H. F. (1970) "A novel principle for the computation of stress intensity factors." *Z. Angew. Math. Mech.*, **50**, 529–546.

Bui, H. D. (1978) *Mécanique de la Rupture Fragile*, Masson, Paris.

Buyukozturk, O. and Lee, K.M. (1992a) "Implication of mixed-mode fracture concepts." In *Concrete Design Based on Fracture Mechanics*, W. Gerstle and Z. P. Bažant, eds., American Concrete Institute, Detroit, pp. 47–62. (ACI Special Publication SP-134.)

Buyukozturk, O. and Lee, K.M. (1992b) "Interface fracture mechanics of concrete composites." In *Fracture Mechanics of Concrete Structures*, Bažant, Z. P., ed., Elsevier Applied Science, London, pp. 163–168.

Buyukozturk, O., Bakhoum, M. M. and Beattie, S. M. (1990) "Shear behavior of joints in precast concrete segmental bridges." *J. Struct Eng.-ASCE,* **116**(12), 3380–3401.

Carneiro, F. L. L. and Barcellos, A. (1953) "Tensile strength of concrete." *RILEM Bulletin,* **13**, 97–123.

Carol, I. and Bažant, Z. P. (1993) "Solidification theory: A rational and effective framework for constitutive modeling of aging viscoelasticity." In *Creep and Shrinkage of Concrete*, Z. P. Bažant and I. Carol, eds., E & FN Spon, London, pp. 177–188.

Carol, I. and Bažant, Z. P. (1997) "Damage and plasticity in microplane theory." *Int. J. Solids Struct.*, **34**. (In press.)

Carol, I. and Prat, P. C. (1990) "A statically constrained microplane model for the smeared analysis of concrete cracking." In *Computer Aided Analysis and Design of Concrete Structures*, Vol. 2, N. Bićanić and H. Mang, eds., Pineridge Press, Swansea, pp. 919–930.

Carol, I. and Prat, P. C. (1991) "Smeared analysis of concrete fracture using a microplane based multicrack model with static constraint." In *Fracture Processes in Concrete, Rock and Ceramics*, J. G. M. van Mier, J. G. Rots and A. Bakker, eds., E & FN Spon, London, pp. 619–628.

Carol, I., Bažant, Z. P. and Prat, P. C. (1991) "Geometric damage tensor based on microplane model." *J. Eng. Mech.-ASCE,* **117**(10), 2429–2448.

Carol, I., Bažant, Z. P. and Prat, P. C. (1992) "Microplane-type constitutive models for distributed damage and localized cracking in concrete structures." In *Fracture Mechanics of Concrete Structures*, Bažant, Z. P., ed., Elsevier Applied Science, London, pp. 299–304.

Carol, I., Prat, P. C. and Bažant, Z. P. (1992) "New explicit microplane model for concrete: Theoretical aspects and numerical implementation." *Int. J. Solids Struct.,* **29**(9), 1173–1191.

Carpinteri, A. (1980) *Static and Energetic Fracture Parameters for Rocks and Concretes.* Report, Istituto di Scienza delle Costruzioni-Ingegneria, University of Bologna, Italy.

Carpinteri, A. (1981) "A fracture mechanics model for reinforced concrete collapse." In *Proc. IABSE Colloquium on Advanced Mechanics of Reinforced Concrete*, pp. 17–30.

Carpinteri, A. (1982) "Notch sensitivity in fracture testing of aggregative materials." *Eng. Fract. Mech.,* **16**, 467–481.

Carpinteri, A. (1984) "Stability of fracturing process in RC beams." *J. Struct Eng.-ASCE,* **110**, 2073–2084.

Carpinteri, A. (1986) *Mechanical Damage and Crack Growth in Concrete*, Martinus Nijhoff, Dordrecht.

Carpinteri, A. (1989) "Decrease of apparent tensile and bending strength with specimen size: Two different explanations based on fracture mechanics." *Int. J. Solids Struct.,* **25**(4), 407–429.

Carpinteri, A. (1994a) "Fractal nature of material microstructure and size effects on apparent mechanical properties." *Mech. Mater.,* **18**, 89–101.

Carpinteri, A. (1994b) "Scaling laws and renormalization groups for strength and toughness of disordered materials." *Int. J. Solids Struct.,* **31**, 291–302.

Carpinteri, A. (1996) "Strength and toughness in disordered materials: Complete and incomplete similarity." In *Size-Scale Effects in the Failure Mechanisms of Materials and Structures*, A. Carpinteri, ed., E & FN Spon, London, pp. 3–26.

Carpinteri, A. and Chiaia, B. (1995) "Multifractal scaling Law for the fracture energy variation of concrete structures." In *Fracture Mechanics of Concrete Structures*, Vol. 1, F. H. Wittmann, ed., Aedificatio Publishers, Freiburg, Germany, pp. 581–596.

Carpinteri, A. and Chiaia, B. (1996) "A multifractal approach to the strength and toughness scaling of concrete structures." In *Fracture Mechanics of Concrete Structures*, Vol. 3, F. H. Wittmann, ed., Aedificatio Publishers, Freiburg, Germany, pp. 1773–1792.

Carpinteri, A. and Ferro, G. (1993) "Apparent tensile strength and fictitious fracture energy of concrete: A fractal geometry approach to related size effects." In *Fracture and Damage of Concrete and Rock*, H. P. Rossmanith, ed., E & FN Spon, London, pp. 86–94.

Carpinteri, A. and Ferro, G. (1994) "Size effects on tensile fracture properties: A unified explanation based on disorder and fractality of concrete microstructure." *Mater. Struct.,* **27**, 563–571.

Carpinteri, A. and Valente, S. (1989) "Size-scale transition from ductile to brittle failure: A dimensional analysis approach." In *Cracking and Damage, Strain Localization and Size Effect*, J. Mazars and Z. P. Bažant, eds., Elsevier Applied Science, London, pp. 477–490.

Carpinteri, A., Chiaia, B. and Ferro G. (1994) "Multifractal scaling law for the nominal strength variation of concrete structures." In *Size Effect in Concrete Structures*, H. Mihashi, H. Okamura and Z. P. Bažant, eds., E & FN Spon, London, pp. 193–206.

Carpinteri, A., Chiaia, B. and Ferro, G. (1995a) "Multifractal nature of material microstructure and size effects on nominal tensile strength." In *Fracture of Brittle Disordered materials: Concrete, Rock and Ceramics*, G. Baker and B.L. Karihaloo, eds., E & FN Spon, London, pp. 21–50.

Carpinteri, A., Chiaia, B. and Ferro, G. (1995b) "Size effects on nominal tensile strength of concrete structures: Multifractality of material ligaments and dimensional transition from order to disorder." *Mater. Struct.,* **28**, 311–317.

Carpinteri, A., Chiaia, B. and Ferro, G. (1995c) *Multifractal scaling law: An extensive application to nominal strength size effect of concrete structures.* Atti del Dipartimento di Ingegneria Strutturale No. 50, Politecnico di Torino, Italy.

Carter, B. C. (1992) "Size and stress gradient effects on fracture around cavities." *Rock Mech. Rock Eng.*, **25**(3), 167–186.

Carter, B. C., Lajtai, E. Z. and Yuan, Y. (1992) "Tensile fracture from circular cavities loaded in compression." *Int. J. Fracture,* **57**, 221–236.

Castillo, E. (1987) *Extreme Value Theory in Engineering*, Academic Press, Inc., San Diego.

CEB (1991) "CEB-FIP Model Code 1990, final draft." *Bulletin d'Information du Comité Euro-International du Béton.,* **203–205**.

Cedolin, L. and Bažant, Z. P. (1980) "Effect of finite element choice in blunt crack band analysis." *Comput. Meth. Appl. Mech. Eng.,* **24**, 305–316.

Cedolin, L., Dei Poli, S. and Iori, I. (1983) "Experimental determination of the fracture process zone in concrete." *Cement Concrete Res.*, **13**, 557–567.

Cedolin, L., Dei Poli, S. and Iori, I. (1987) "Tensile behavior of concrete." *J. Eng. Mech.-ASCE,* **113**(3), 431–449.

Červenka, J. (1994) *Discrete Crack Modeling in Concrete Structures.* Doctoral thesis. University of Colorado, Boulder, CO.

Červenka, V. and Pukl, R. (1994) "SBETA analysis of size effect in concrete structures." In *Size Effect in Concrete Structures*, H. Mihashi, H. Okamura and Z. P. Bažant, eds., E & FN Spon, London, pp. 323–333.

Červenka, J. and Saouma, V. E. (1995) "Discrete crack modeling in concrete structures." In *Fracture Mechanics of Concrete Structures*, F. H. Wittmann, ed., Aedificatio Publishers, Freiburg, Germany, pp. 1285–1300.

Červenka, V., Pukl, R., Ožbolt, J. and Eligehausen, R. (1995) "Mesh sensitivity in smeared finite element analysis of concrete fracture." In *Fracture Mechanics of Concrete Structures*, F. H. Wittmann, ed., Aedificatio Publishers, Freiburg, Germany, pp. 1387–1396.

Charmet, J. C., Roux, S. and Guyon, E., eds. (1990) *Disorder and Fracture*, Plenum Press, New York.

Chen, W.F. (1982) *Plasticity in Reinforced Concrete*, McGraw-Hill, New York.

Chen, W-F. and Yuan, R. L. (1980) "Tensile strength of concrete: Double-punch test." *J. Struct. Div.-ASCE,* **106**, 1673–1693.

Cho, K. Z., Kobayashi, A. S., Hawkins, N. M., Barker, D. B. and Jeang, F. L. (1984) "Fracture process zone of concrete cracks." *J. Eng. Mech.-ASCE,* **110**(8), 1174–1184.

Choi, O.C., Darwin, D. and McCabe, S.L. (1990) *Bond Strength of Epoxy-Coated Reinforcement to Concrete.* S. M. Report No. 25, University of Kansas Center for Research, Lawrence, KS.

Christensen, R. M. (1971) *Theory of Viscoelasticity*, Academic Press, New York.

Christensen, R. M. and DeTeresa, S. J. (1997) "The kink band mechanism for compressive failure of fiber composite materials." *J. Appl. Mech.-T. ASME,* **64**, 1–6.

Chudnovsky, A. and Kachanov, M. (1983) "Interaction of a crack with a field of microcracks." *Int. J. Eng. Sci.,* **21**(8), 1009–1018.

Chudnovsky, A. and Kunin, B. (1987) "A probabilistic model of crack formation." *J. Appl. Phys.,* **62**(10), 4124–4129.

Chudnovsky, A., Dolgopolsky, A. and Kachanov, M. (1987) "Elastic interaction of a crack with a microcrack array (parts I and II)." *Int. J. Solids Struct.,* **23**, 1–21.

Collatz, L. (1960) *The Numerical Treatment of Differential Equations*, Springer, Berlin.

Collins, M. P. (1978) "Towards a rational theory for RC members in shear." *J. Struct. Div.-ASCE,* **104**, 396–408.

Collins, M. P. and Mitchell, D. (1980) "Shear and torsion design of prestressed and non-prestressed concrete beams." *J. Prestressed Concrete Inst.,* **25**(5), 32–100. (Discussion **26**(6), 96–118.)

Collins, M. P., Mitchell, D., Adebar, P. and Vecchio, F. J. (1996) "General shear design method." *ACI Struct. J.,* **93**(1), 36–45.

Collins, W. D. (1963) "Some coplanar punch and crack problems in three-dimensional elastostatics." *Philos. T. Roy. Soc. A,* **274**(1359), 507–528.

Commission of European Communities (1984) *Eurocode No. 2.* Commission of European Communities.

Cope, R. J., Rao, P. V., Clark, L. A. and Norris, P. (1980) "Modelling of reinforced concrete behaviour for finite element analysis of bridge slabs." In *Numerical Methods for Nonlinear Problems*, Vol. 1, C. Taylor et al., eds., Pineridge Press, Swansea, pp. 457–470.

Cornelissen, H. A. W., Hordijk, D. A. and Reinhardt, H. W. (1986a) "Experimental determination of crack softening characteristics of normal weight and lightweight concrete." *Heron,* **31**(2), 45–56.

Cornelissen, H. A. W., Hordijk, D. A. and Reinhardt, H. W. (1986b) "Experiments and theory for the application of fracture mechanics to normal and lightweight concrete." In *Fracture Mechanics and Fracture Energy of Concrete*, F. H. Wittmann, ed., Elsevier, Amsterdam, pp. 565–575.

Corres, H., Elices, M. and Planas, J. (1986) "Thermal deformation of loaded concrete at low temperatures. 3: Lightweight concrete." *Cement Concrete Res.*, **16**, 845–852.

Costin, D. M. (1991) "Damage mechanics in the post-failure regime." *Mech. Mater.*, **4**, 149–160.

Cotterell, B. (1972) "Brittle fracture in compression." *Int. J. Fract. Mech.*, **8**(2), 195–208.

Cotterell, B. and Rice, J. R. (1980) "Slightly curved or kinked cracks." *Int. J. Fracture*, **16**, 155–169.

Cottrell, A.H. (1961) ISI Special Report No. 69, Iron and Steel Institute.

Cox, J. V. (1994) *Development of a Plasticity Bond Model for Reinforced Concrete — Theory and Validation for Monotonic Applications*. Technical Report No. TR-2036-SHR, Naval Facilities Engineering Service Center, Port Hueneme, CA 93043-4328.

Crisfield, M. A. and Wills, J. (1987) "Numerical comparisons involving different 'concrete-models'." In *IABSE Reports 54, Colloquium on Computational Mechanics of Reinforced Concrete*, Delft University Press, pp. 177–187.

Cundall, P. A. (1971) "A computer model for simulating progressive large scale movements in blocky rock systems." In *Proc. Int. Symp. Rock Fracture*.

Cundall, P. A. (1978) *BALL – A Program to Model Granular Media Using the Distinct Element Method.* Technical Note, Advanced Tech. Group, Dames and Moore, London.

Cundall, P. A. and Strack, O. D. L. (1979) "A discrete numerical model for granular assemblies." *Geotechnique*, **29**, 47–65.

Darwin, D. (1985) "Concrete crack propagation — Study of model parameters." In *Proc. Finite Element Analysis of Reinforced Concrete Structures*, Meyer, C. and Okamura, H., eds., ASCE, New York, pp. 184–203.

Darwin, D. and Attiogbe, E. K. (1986) "Effect of loading rate on cracking of cement paste in compression." In *Proc. Mat. Res. Soc. Symp. No. 64 on Cement Based Composites: Strain Rate Effects on Fracture*, S. Mindess and S. P. Shah, eds., pp. 167–180.

Datsyshin, A. P. and Savruk, M. P. (1973) "A system of of arbitrarily oriented cracks in elastic solids." *J. Appl. Math. and Mech.*, **37**(2), 326–332.

Dauskardt, R. H., Marshall, D. B. and Ritchie, R. O. (1990) "Cyclic fatigue-crack propagation in in magnesia-partially-stabilized zirconia." *J. Am. Ceram. Soc.*, **73**, 893–903

Davies, J. (1992) "Macroscopic study of crack bridging phenomenon in mixed-mode loading." In *Fracture Mechanics of Concrete Structures*, Z. P. Bažant, ed., Elsevier Applied Science, pp. 713–718.

Davies, J. (1995) "Study of shear fracture in mortar specimens." *Cement and Concrete Research*, **25**(5), 1031–1042.

Dietsche, A. and Willam, L. J. (1992) "Localization analysis of elasto-plastic Cosserat continua." In *Damage and Localization*, J. W. Ju and K. C. Valanis, eds., The American Society of Mechanical Engineers, New York, pp. 25–40. (AMD–Vol.142, Winter Annual Meeting, Anaheim.)

Dougill, J. W. (1976) "On stable progressively fracturing solids." *J. Appl. Math. Phys.*, **27**, 423–436.

Droz, P. and Bažant, Z. P. (1989) "Nonlocal analysis of stable states and stable paths of propagation of damage shear bands." In *Cracking and Damage, Strain Localization and Size Effect*, J. Mazars and Z. P. Bažant, eds., Elsevier Applied Science, London, pp. 183–207.

Du, J., Kobayashi, A. S. and Hawkins, N. M. (1989) "FEM dynamic fracture analysis of concrete beams." *J. Eng. Mech.-ASCE*, **115**(10), 2136–2149.

Dugdale, D. S. (1960) "Yielding of steel sheets containing slits." *J. Mech. Phys. Solids*, **8**, 100–108.

Dvorkin, E. N., Cuitiño, A. M. and Gioia, G. (1990) "Finite elements with displacement interpolated embedded localization lines insensitive to mesh size and distortions." *Int. J. Numer. Meth. Eng.*, **30**, 541–564.

Elfgren, L., ed. (1989) *Fracture Mechanics of Concrete Structures*, Chapman and Hall, London.

Elfgren, L. (1990) "Round robin analysis and tests of anchor bolts – Invitation." *Mater. Struct.*, **23**, 78.

Elfgren, L. and Swartz, S.E. (1992) "Fracture mechanics approach to modeling the pull-out of anchor bolts." In *Design Based on Fracture Mechanics*, W. Gerstle and Z. P. Bažant, eds., American Concrete Institute, Detroit, pp. 63–78. (ACI SP-134.)

Elfgren, L., Ohlsson, U. and Gylltoft, K. (1989) "Anchor bolts analyzed with fracture mechanics." In *Fracture of Concrete and Rock*, S.P. Shah and S.E. Swartz, eds., Springer-Verlag, New York, pp. 269–275.

Elices, M. (1987) *Mecánica de la Fractura Aplicada a Sólidos Elásticos Bidimensionales*, Dep. de Publicaciones de Alumnos, ETS de Ingenieros de Caminos, Canales y Puertos, Ciudad Universitaria, 28040, Madrid, Spain. (Fracture Mechanics Applied to Two-Dimensional Elastic Solids.)

Elices, M. and Planas, J. (1989) "Material models." In *Fracture Mechanics of Concrete Structures*, L. Elfgren, ed., Chapman and Hall, London, pp. 16–66.

Elices, M. and Planas, J. (1991) "Size effect and experimental validation of fracture models." In *Analysis of Concrete Structures by Fracture Mechanics*, L. Elfgren, ed., Chapman and Hall, London, pp. 99–127.

Elices, M. and Planas, J. (1992) "Size Effect in Concrete Structures: an R-Curve Approach." In *Application of Fracture Mechanics to Reinforced Concrete*, A. Carpinteri, ed., Elsevier Applied Science, London, pp. 169–200.

Elices, M. and Planas, J. (1993) "The equivalent elastic crack: 1. Load-Y equivalences." *Int. J. Fracture,* **61**, 159–172.

Elices, M. and Planas, J. (1996) "Fracture mechanics parameters of concrete: An overview." *Adv. Cem. Bas. Mat.,* **4**, 116–127.

Elices, M., Guinea, G. V. and Planas, J. (1992) "Measurement of the fracture energy using three-point bend tests: 3. Influence of cutting the P-δ tail." *Mater. Struct.,* **25**, 327–334.

Elices, M., Guinea, G. V. and Planas, J. (1995) "Prediction of size effect based on cohesive crack models." In *Size-Scale Effect in the Failure Mechanisms of Materials and Structures*, A. Carpinteri, ed., E & FN Spon, London, pp. 309–324.

Elices, M., Guinea, G. V. and Planas, J. (1997) "On the measurement of concrete fracture energy using three-point bend tests." *Mater. Struct.,* **30**, 375–376.

Elices, M., Planas, J. and Corres, H. (1986) "Thermal deformation of loaded concrete at low temperatures, 2: Transverse deformation." *Cement Concrete Res.,* **16**, 741–748.

Elices, M., Planas, J. and Guinea, G. V. (1993) "Modeling cracking in rocks and cementitious materials." In *Fracture and Damage of Concrete and Rock*, Rossmanith, H.P, ed., E & FN Spon, London, pp. 3–33.

Eligehausen, R. and Ožbolt, J. (1990) "Size effect in anchorage behavior." In *Fracture Behavior and Design of Materials and Structures*, Vol. 2, D. Firrao, ed., Engineering Materials Advisory Services Ltd. (EMAS), Warley, West Mindlands, U.K., pp. 721–727.

Eligehausen, R. and Sawade, G. (1989) "A fracture mechanics based description of the pull-out behavior of headed studs embedded in concrete." In *Fracture Mechanics of Concrete Structures*, L. Elfgren, ed., Chapman and Hall, London, pp. 281–299.

Eligehausen, R., Fusch, W., Ick, U., Mallée, R., Reuter, M., Schimmelphenning, K. and Schmal, B. (1991) *Tragverhalten von KopfbolzenVerankerung bei Zentrischer Zugbeansprunchung*. Report, Stuttgart University.

England A.H. (1971) *Complex Variable Methods in Elasticity*, Wiley-Interscience.

Entov, V. M. and Yagust, V. I. (1975) "Experimental investigation of laws governing quasi-static development of macrocracks in concrete." *Mech. Solids,* **10**(4), 87–95. (Translation from Russian.)

Eo, S. H., Hawkins, N. M. and Kono S. (1994) "Fracture characteristics and size effect for high-strength concrete beams." In *Size Effect in Concrete Structures*, H. Mihashi, H. Okamura and Z. P. Bažant, eds., E & FN Spon, London, pp. 245–254.

Erdogan, F. (1963) "Stress distribution in a nonhomogeneous elastic plane with cracks." *J. Appl. Mech.-T. ASME,* **30**(2), 232–236.

Erdogan, F. and Sih, G. C. (1963) "On the crack extension in plates under plane loading and transverse shear." *J. Basic Eng.,* **85**, 519–527.

Eringen, A. C. (1965) "Theory of micropolar continuum." In *Proc. Ninth Midwestern Mechanics Conference*, pp. 23–40.

Eringen, A. C. (1966) "A unified theory of thermomechanical materials." *Int. J. Eng. Sci.,* **4**, 179–202.

Eringen, A. C. and Ari, N. (1983) "Nonlocal stress field at Griffith crack." *Cryst. Latt. Def. Amorph. Mat.,* **10**, 33–38.

Eringen, A. C. and Edelen D. G. B. (1972) "On nonlocal elasticity." *Int. J. Eng. Sci.,* **10**, 233–248.

Eshelby, J. D. (1956) "The continuum theory of lattice defects." In *Solid State Physics*, Vol. 3, F. Seitz and D. Turnbull, eds., Academic Press, New York, pp. 79–141.

ESIS Technical Committee 8 (1991) "Recommendation for use of FEM in fracture mechanics." *ESIS Newsletter,* (15), 3–7.

Evans, A. G. and Fu, Y. (1984) "The mechanical behavior of alumina." In *Fracture in Ceramic Materials*, Noyes Publications, Park Ridge, NJ, pp. 56–88.

Evans, R. H. and Marathe M. S. (1968) "Microcracking and stress-strain curves for concrete in tension." *Mater. Struct.*, **1**(1), 61–64.

Fabrikant, V. I. (1990) "Complete solutions to some mixed boundary value problems in elasticity." *Adv. Appl. Mech.*, **27**, 153–223.

Fairhurst, C. and Cornet, F. (1981) "Rock fracture and fragmentation." In *Proc. 22nd U.S. Symp. on Rock Mechanics*, pp. 21–46.

Fathy, A. M. (1992) *Application of Fracture Mechanics to Rocks and Rocky Materials.* Doctoral thesis. Universidad Politécnica de Madrid, Departamento de Ciencia de Materiales, ETS de Ingenieros de Caminos, Ciudad Universitaria, 28040 Madrid, Spain. (In English.)

Feddersen, C. E. (1966) In *Plane Strain Crack Toughness Testing of High Strength Metallic Materials*, W. F. Brown and J. E. Srawley, eds., American Society for Testing and Materials, Philadelphia, pp. 77–79. (Contribution to Discussion, ASTM Special Technical Publication No. 410.)

Fenwick, R. C. and Paulay, T. (1968) "Mechanics of shear resistance of concrete beams." *J. Struct Eng.-ASCE*, **94**, 2235–2350.

Ferguson, P. M. and Thompson, J. N. (1962) "Development length of high strength reinforcing bars in bond." *ACI J.*, **59**, 887–922.

Ferguson, P. M. and Thompson, J. N. (1965) "Development length for large high strength reinforcing bars." *ACI J.*, **62**, 71–93.

Fischer, R. A. and Tippett L. H. C. (1928) "Limiting forms of the frequency distribution of the largest and smallest member of a sample." *Proc., Cambridge Philosophical Society*, **24**, 180–190.

Flory, T. J. (1961) "Thermodynamic relations for high elastic materials." *T. Faraday Soc.*, **57**, 829–838.

Forman, R. G., Kearney, V. E. and Engle, R. M. (1967) "Numerical analysis of crack propagation in cyclic-loaded structures." *J. Basic Eng.*, **89**, 459–464.

Fox, L. (1965) *An Introduction to Numerical Linear Algebra*, Oxford University Press, New York.

Fréchet, M. (1927) "Sur la loi de probabilité de l'écart maximum." *Ann. Soc. Polon. Math*, **6**, 93.

Freudenthal, A. M. (1968) "Statistical approach to brittle fracture." In *Fracture – An Advanced Treatise*, Vol. 2, H. Liebowitz, ed., Academic Press, New York, pp. 591–619.

Freudenthal, A. M. (1981) *Selected Papers by Alfred M. Freudenthal*, Am. Soc. of Civil Engrs., New York.

Freund, L. B. (1990) *Dynamic Fracture Mechanics*, Cambridge University Press, Cambridge and New York.

Furuhashi, R., Kinoshita, N. and Mura, T. (1981) "Periodic distributions of inclusions." *Int. J. Eng. Sci.*, **19**, 231–236.

Galileo Galilei Linceo (1638) *Discorsi i Demostrazioni Matematiche intorno à due Nuove Scienze*, Elsevirii, Leiden. (English transl. by T. Weston, London (1730), pp. 178–181.)

Gálvez, J, Llorca, J. and Elices, M. (1996) "Fracture mechanics analysis of crack stability in concrete gravity dams." *Dam Eng.*, **7**(1), 35–63.

Gdoutos, E. E. (1989) *Problems of Mixed Mode Crack Propagation*, Martinus Nijhoff Publishers, The Hague.

Gerstle, W. H., Partha, P. D., Prasad, N. N. V., Rahulkumar, P. and Ming, X. (1992) "Crack growth in flexural members — A fracture mechanics approach." *ACI Struct. J.*, **89**(6), 617–625.

Gettu, R., Bažant, Z. P. and Karr, M. E. (1990) "Fracture properties and brittleness of high-strength concrete." *ACI Mater. J.*, **87**, 608–618.

Gioia, G, Bažant, Z. P. and Pohl, B. P. (1992) "Is no-tension dam design always safe? – a numerical study." *Dam Eng.*, **3**(1), 23–34.

Gjørv, O. E., Sorensen, S. I. and Arnesen, A. (1977) "Notch sensitivity and fracture toughness of concrete." *Cement Concrete Res.*, **7**, 333–344.

Go, C. G. and Swartz, S. E. (1986) "Energy methods for fracture-toughness determination in concrete." *Exp. Mech.*, **26**(3), 292–296.

Gonnermann, H. F. (1925) "Effect of size and shape of test specimen on compressive strength of concrete." *Proc. ASTM*, **25**, 237–250.

Goode, C. D. and Helmy, M. A. (1967) "The strength of concrete under combined shear and direct stress." *Mag. Concrete Res.*, **19**(59), 105–112.

Gopalaratnam, V. S. and Shah S. P. (1985) "Softening response of plain concrete in direct tension." *ACI J.*, **82**(3), 310–323.

Graham, G. A. C. (1968) "The correspondence principle of linear viscoelasticity theory for mixed boundary value problems involving time-dependent boundary regions." *Q. Appl. Math.*, **26.**, 167–174.

Green, S. J. and Swanson, S. R. (1973) *Static Constitutive Relations for Concrete*. Report No. AFWL-TR-72-2, Air Force Weapons Lab., Kirkland Air Force Base.

Griffith, A. A. (1921) "The phenomena of rupture and flow in solids." *Philos. T. Roy. Soc. A*, **221**, 163–197.

Griffith, A. A. (1924) "The theory of rupture." In *Proceedings of the First International Conference of Applied Mechanics*, pp. 55–63.

Gross, D. (1982) "Spannungsintensitätsfaktoren von rißsystemen." *Ingenieur-Archiv*, **51**, 301–310.

Guinea, G. V. (1990) *Medida de la Energía de Fractura del Hormigón*. Doctoral thesis. Dep. Ciencia de Materiales, Universidad Politecnica de Madrid, ETS de Ingenieros de Caminos, Ciudad Universitaria, 28040 Madrid, Spain. ('Measurement of the Fracture Energy of Concrete', in Spanish.)

Guinea, G. V., Planas, J. and Elices, M. (1990) "On the influence of bulk dissipation on the average specific fracture energy of concrete." In *Fracture Behaviour and Design of Materials and Structures*, Vol. 2, D. Firrao, ed., Engineering Materials Advisory Services Ltd. (EMAS), Warley, West Mindlands, U.K., pp. 715–720.

Guinea, G. V., Planas, J. and Elices, M. (1992) "Measurement of the fracture energy using three-point bend tests: 1. Influence of experimental procedures." *Mater. Struct.*, **25**, 212–218.

Guinea, G. V., Planas, J. and Elices, M. (1994a) "Correlation between the softening and the size effect curves." In *Size Effect in Concrete Structures*, H. Mihashi, H. Okamura and Z. P. Bažant, eds., E & FN Spon, London, pp. 233–244.

Guinea, G. V., Planas, J. and Elices, M. (1994b) "A general bilinear fit for the softening curve of concrete." *Mater. Struct.*, **27**, 99–105.

Guo, Z. and Zhang X. (1987) "Investigation of complete stress-deformation curves for concrete in tension." *ACI Mater. J.*, **84**, 278–285.

Gupta, A. K. and Akbar, H. (1984) "Cracking in reinforced concrete analysis." *J. Struct. Eng.-ASCE*, **110**(8), 1735–1746.

Gustafsson, P. J. (1985) *Fracture Mechanics Studies of Non-Yielding Materials Like Concrete: Modeling of Tensile Fracture and Applied Strength Analyses*. Report No. TVBM-1007, Division of Building Materials, Lund Institute of Technology, Lund, Sweden.

Gustafsson, P. J. and Hillerborg, A. (1985) "Improvements in concrete design achieved through the application of fracture mechanics." In *Application of Fracture Mechanics to Cementitious Composites*, S. P. Shah, ed., Martinus Nijhoff, Dordrecht, pp. 667–680.

Gustafsson, P. J. and Hillerborg, A. (1988) "Sensitivity in the shear strength of longitudinally reinforced beams to fracture energy of concrete." *ACI Struct. J.*, **85**(3), 286–294.

Gylltoft, K. (1983) *Fracture Mechanics Models for Fatigue in Concrete Structures*. Doctoral thesis. Luleå University of Technology, Luleå, Sweden.

Gylltoft, K. (1984) "A fracture mechanics model for fatigue in concrete." *Mater. Struct.*, **17**(97), 55–58.

Haimson, B. C. and Herrick, C. G. (1989) "*In-situ* stress calculation from borehole breakout experimental studies." In *Proc., 26th U.S. Symp. on Rock Mechanics*, pp. 1207–1218.

Hasegawa, T. and Bažant, Z. P. (1993) "Nonlocal microplane concrete model with rate effect and load cycles. I. General formulation. II. Application and verification." *J. Mater. Civil Eng.*, **5**(3), 372–417.

Hasegawa, T., Shioya, T. and Okada, T. (1985) "Size effect on splitting tensile strength of concrete." In *Proc. Japan Concrete Inst. 7th Conf.*, pp. 309–312.

Hashin, Z. (1988) "The differential scheme and its application to cracked materials." *J. Mech. Phys. Solids*, **36**(6), 719–734.

Hassanzadeh, M. (1992) *Behaviour of Fracture Process Zones in Concrete Influenced by Simultaneously Applied Normal and Shear Displacements*. Report No. TVBM-1010, Division of Building Materials, Lund Institute of Technology, Lund, Sweden.

Hawkes, I. and Mellor, M. (1970) "Uniaxial testing in rock mechanics laboratories." *Eng. Geol.*, **4**, 177–285.

Hawkins, N. (1985) "The role for fracture mechanics in conventional reinforced design." In *Application of Fracture Mechanics to Cementitious Composites*, S. P. Shah, ed., Martinus Nijhoff, Dordrecht, pp. 639–666.

Hawkins, N.M. and Hjorteset, K. (1991) "Minimum reinforcement requirements for concrete flexural members." In *Application of Fracture Mechanics to Reinforced Concrete*, A. Carpinteri, ed., Elsevier Applied Science, London, pp. 379–412.

He, M.-Y. and Hutchinson, J. W. (1989) "Crack deflection at an interface between dissimilar elastic materials." *Int. J. Solids Struct.*, **25**, 1053–1067.

He, S., Plesha, M. E., Rowlands, R. E. and Bažant, Z. P. (1992) "Fracture energy tests of dam concrete with rate and size effects." *Dam Eng.*, **3**(2), 139–159.

Hededal, O. and Kroon, I. B. (1991) *Lightly Reinforced High Strength Concrete.* Master thesis. University of Åalborg, Denmark.

Heilmann, H. G., Hilsdorf H. and Finsterwalder, K. (1969) "Festigkeit und Verformung von Beton unter Zugspannungen." *Deustcher Ausshuss fur Stahlbeton,* (Heft 203).

Henshell, R. D. and Shaw, K. G. (1975) "Crack tip finite elements are unnecessary." *Int. J. Numer. Meth. Eng.,* **9**, 495–507.

Herrmann, G. and Sosa, H. (1986) "On bars with cracks." *Eng. Fract. Mech.,* **24**, 889–894.

Herrmann, H. J. (1991) "Patterns and scaling in fracture." In *Fracture Processes in Concrete, Rock and Ceramics*, J. G. M. van Mier, J. G. Rots and A. Bakker, eds., E & FN Spon, London, pp. 195–211.

Herrmann, H. J. and Roux, S., eds. (1990) *Statistical Models for the Fracture of Disordered Media*, North-Holland, New York.

Herrmann, H. J., Hansen, H. and Roux, S. (1989) "Fracture of disordered, elastic lattices in two dimensions." *Phys. Rev. B,* **39**, 637–648.

Hetényi, M. (1946). *Beams on Elastic Foundation*, The University of Michigan Press, Ann Arbor.

Higgins, D. D. and Bailey, J. E. (1976) "Fracture measurements on cement paste." *J. Mater. Sci.,* **11**, 1995–2003.

Hill, R. (1965) "Continuum micromechanics of elastoplastic polycrystals." *J. Mech. Phys. Solids,* **13**, 89–101.

Hill, R. (1966) "Generalized constitutive relations for incremental deformations of metal crystals by multi-slip." *J. Mech. Phys. Solids,* **14**, 95–102.

Hillerborg, A. (1984) *Additional Concrete Fracture Energy Tests Performed by 6 Laboratories According to a Draft RILEM Recommendation.* Report No. TVBM-3017, Division of Building Materials, Lund Institute of Technology, Lund, Sweden.

Hillerborg, A. (1985a) "The theoretical basis of a method to determine the fracture energy G_F of concrete." *Mater. Struct.,* **18**, 291–296.

Hillerborg, A. (1985b) "Numerical methods to simulate softening and fracture of concrete." In *Fracture Mechanics of Concrete: Structural Application and Numerical Calculation*, G. C. Sih and A. DiTomasso, eds., Martinus Nijhoff, Dordrecht, pp. 141–170.

Hillerborg, A. (1989) "Fracture mechanics and the concrete codes." In *Fracture Mechanics: Applications to Concrete*, V. C. Li and Z. P. Bažant, eds., American Concrete Institute, Detroit, pp. 157–169. (ACI Special Publication SP-118.)

Hillerborg, A. (1990) "Fracture mechanics concepts applied to moment capacity and rotational capacity of reinforced concrete beams." *Eng. Fract. Mech.,* **35**, 233–240.

Hillerborg, A. (1991) "Reliance upon concrete tensile strength." In *IABSE Colloquium StuttGart 91: Structural Concrete*, IABSE, Zürich., pp. 589–604.

Hillerborg, A., Modéer, M. and Petersson, P. E. (1976) "Analysis of crack formation and crack growth in concrete by means of fracture mechanics and finite elements." *Cement Concrete Res.,* **6**, 773–782.

Hinch, E. J. (1991) *Perturbation Methods*, Cambridge University Press, Cambridge.

Hodge, P. G. (1959) *Plastic Analysis of Structures*, McGraw Hill, New York.

Hoek, E. and Bieniawski, Z. J. (1965) "Brittle fracture propagation in rock under compression." *Int. J. Fract. Mech.,* **1**, 137–155.

Hoenig, A. (1978) "The behavior of a flat elliptical crack in an anisotropic solid." *Int. J. Solids Struct.,* **14**, 925–934.

Hognestad, E., Hanson, N. W. and McHenry, D. (1955) "Concrete stress distribution in ultimate strength design." *ACI J.,* **52**(4), 455–477.

Hondros, G. (1959) "Evaluation of Poisson ratio and the modulus of materials of low tensile resistance by the Brazilian (indirect tensile) test with particular references to concrete." *Aust. J. Appl. Sci.,* **10**, 243–268.

Hong, A.-P., Li, Y.-N. and Bažant, Z. P. (1997) "Theory of crack spacing in concrete pavements." *J. Eng. Mech.-ASCE,* **123**(3), 267–275.

Hordijk, D. A. (1991) *Local Approach to Fatigue of Concrete.* Doctoral thesis. Delft University of Technology, Delft, The Netherlands.

Hordijk, D. A. and Reinhardt, H. W. (1991) "Growth of discrete cracks in concrete under fatigue loading." In *Toughening Mechanisms in Quasi-Brittle Materials*, S. P. Shah, ed., Kluwer Academic Publishers, Dordrecht, The Netherlands, pp. 541–554.

Hordijk, D. A. and Reinhardt, H. W. (1992) "A fracture mechanics approach to fracture of plain concrete." In *Fracture Mechanics of Concrete Structures*, Z. P. Bažant, ed., Elsevier Applied Science, London, pp. 924–929.

Horii, H. (1989) "Models of fracture process zone and a system of fracture mechanics for concrete and rock." In *Fracture Toughness and Fracture Energy: Test Methods for Concrete and Rock*, H. Mihashi, H. Takahashi and F. H. Wittmann, eds., Balkema, Rotterdam, pp. 409–422.

Horii, H. (1991) "Mechanisms of fracture in brittle disordered materials." In *Fracture Processes in Concrete, Rock and Ceramics*, J. G. M. van Mier, J. G. Rots and A. Bakker, eds., E. & FN Spon, London, pp. 95–110.

Horii, H. and Nemat-Nasser, S. (1982) "Compression-induced non planar crack extension with application to splitting, exfoliation and rockburst." *J. Geophys. Res.*, **87**, 6806–6821.

Horii, H. and Nemat-Nasser, S. (1985) "Elastic fields of interacting inhomogeneities." *Int. J. Solids Struct.*, **21**, 731–745.

Horii, H. and Nemat-Nasser, S. (1986) "Brittle failure in compression, splitting, faulting and brittle-ductile transition." *Philos. T. Roy. Soc.*, **319**(1549), 337–334.

Horii, H., Hasegawa, A. and Nishino, F. (1989) "Fracture process and bridging zone model and influencing factors in fracture of concrete." In *Fracture of Concrete and Rock*, S. P. Shah and S. E. Swartz, eds., Springer-Verlag, New York, pp. 205–214.

Horii, H., Shin, H. C. and Pallewatta, T. M. (1990) "An analytical model of fatigue crack growth in concrete." *Proc. of the Japan Concrete Institute*, **12**, 835–840.

Horii, H., Zihai, S. and Gong, S.-X. (1989) "Models of fracture process zone in concrete, rock, and ceramics." In *Cracking and Damage, Strain Localization and Size Effect*, J. Mazars and Z. P. Bažant, eds., Elsevier Applied Science, London, pp. 104–115.

Hrennikoff, A. (1941) "Solution of problems of elasticity by the framework method." *J. Appl. Mech.-T. ASME*, **12**, 169–175.

Hsu, T. T. C. (1968) "Torsion of structural concrete – Plain concrete rectangular sections." In *Torsion of Structural Concrete*, American Concrete Institute, Detroit, pp. 203–238. (ACI Special Publication SP-18.)

Hsu, T. T. C. (1988) "Softened truss model theory for shear and torsion." *ACI Struct. J.*, **85**(6), 624–635.

Hsu, T. T. C. (1993) *Unified Theory of Reinforced Concrete*, CRC Press, Boca Raton, FL.

Huang, C. M. J. (1981) *Finite Element and Experimental Studies of Stress Intensity Factors for Concrete by Means of Fracture Mechanics and Finite Elements.* Doctoral thesis. Kansas State University, Kansas.

Hughes, B. P. and Chapman, G. P. (1966) "The complete stress-strain for concrete in direct tension." *RILEM Bulletin*, **30**, 95–97.

Humphrey, R. (1957) "Torsional properties of prestressed concrete." *Structural Eng.*, **35**(6), 213–224.

Hutchinson, J. W. (1968) "Singular behaviour at the end of a tensile crack in a hardening material." *J. Mech. Phys. Solids*, **16**, 13–31.

Hutchinson, J. W. (1990) "Mixed mode fracture mechanics of interfaces." In *Metal-Ceramic Interfaces*, M. Ruhle et al., eds., Pergamon Press, New York, pp. 295–306.

Inglis, C. E. (1913) "Stresses in a plate due to the presence of cracks and sharp corners." *T. Inst. Naval Architects*, **55**, 219–241.

Ingraffea, A. R. (1977) *Discrete Fracture Propagation in Rock: Laboratory Tests and Finite Element Analysis.* Doctoral thesis. University of Colorado, Boulder.

Ingraffea, A. R. and Gerstle, W. H. (1985) "Nonlinear fracture models for discrete crack propagation." In *Application of Fracture Mechanics to Cementitious Composites*, S. P. Shah, ed., Martinus Nijhoff, Dordrecht, pp. 247–285.

Ingraffea, A. R. and Heuzé, F. E. (1980) "Finite element models for rock fracture mechanics." *Int. J. Numer. Anal. Meth. Geomech.*, **4**, 25–43.

Ingraffea, A. R. and Saouma, V. (1984) "Numerical modeling of fracture propagation in reinforced and plain concrete." In *Fracture Mechanics of Concrete: Structural Application and Numerical Calculation*, G. Sih and A. DiTommasso, eds., Martinus Nijhoff, Dordrecht, pp. 171–225.

Ingraffea, A. R., Gerstle, W. H., Gergely, P. and Saouma, V. (1984) "Fracture mechanics of bond in reinforced concrete." *J. Struct Eng.-ASCE*, **110**(4), 871–890.

Ingraffea, A. R., Linsbauer, H. and Rossmanith, H. (1989) "Computer simulation of cracking in large arch dam — Downstream side cracking." In *Fracture of Concrete and Rock*, S.P. Shah and S.E. Swartz, eds., Springer-Verlag, New York, pp. 334–342.

Irwin, G. R. (1957) "Analysis of stresses and strains near the end of a crack traversing a plate." *J. Appl. Mech.-T. ASME*, **24**, 361–364.

Irwin, G. R. (1958) "Fracture." In *Handbuch der Physik*, Vol. 6, Flügge, ed., Springer-Verlag, Berlin, pp. 551–590.

Irwin, G. R. (1960) *Structural Mechanics*, Pergamon Press, London.

Irwin, G. R., Kies, J. A. and Smith, H. L. (1958) "Fracture strengths relative to the onset and arrest of crack propagation." *Proc ASTM,* **58**, 640–657.

Isida, M. (1973) "Analysis of stress intensity factors for the tension of a centrally cracked strip with stiffened edges." *Eng. Fract. Mech.,* **5**, 647–655.

Janssen, J. G. (1990) *Mode I Fracture of Plain Concrete under Monotonic and Cyclic Loading: Implementation and Evaluation of a Constitutive Model in DIANA*. Graduate thesis. Delft University of Technology, Delft, The Netherlands.

Jenq, Y. S. and Shah, S. P. (1985a) "A fracture toughness criterion for concrete." *Eng. Fract. Mech.,* **21**(5), 1055–1069.

Jenq, Y. S. and Shah, S. P. (1985b) "Two parameter fracture model for concrete." *J. Eng. Mech.-ASCE,* **111**(10), 1227–1241.

Jenq, Y. S. and Shah, S. P. (1988a) *Geometrical Effects on Mode I Fracture Parameters.* Report to RILEM Committee 89-FMT.

Jenq, Y. S. and Shah, S. P. (1988b) "On the concrete fracture testing methods." In *Fracture Toughness and Fracture Energy: Test Methods for Concrete and Rock*, H. Mihashi, H. Takahashi and F. H. Wittmann, eds., Balkema, Rotterdam, pp. 443–463.

Jenq, Y.S. and Shah, S.P. (1989) "Shear resistance of reinforced concrete beams — A fracture mechanics approach." In *Fracture Mechanics: Applications to Concrete*, V. Li and Z. P. Bažant, eds., American Concrete Institute, Detroit, pp. 237–258. (ACI Special Publication SP-118.)

Jirásek, M. (1993) *Modeling of Fracture and Damage in Quasi-Brittle Materials.* Doctoral thesis. Northwestern University, Evanston, IL.

Jirásek, M. (1996) "Nonlocal models for concrete cracking." Oral presentation at 38th Annual Technical Meeting of Society of Eng. Science in Tempe, Arizona; to appear in *Int. J. Solids Struct.*

Jirásek, M. and Bažant, Z. P. (1994) "Localization analysis of nonlocal model based on crack interactions." *J. Eng. Mech.-ASCE,* **120**(7), 1521–1542.

Jirásek, M. and Bažant, Z. P. (1995a) "Particle model for quasibrittle fracture and application to sea ice." *J. Eng. Mech.-ASCE,* **121**(9), 1016–1025.

Jirásek, M. and Bažant, Z. P. (1995b) "Macroscopic fracture characteristics of random particle systems." *Int. J. Fracture,* **69**(3), 201–228.

Jirásek, M. and Zimmermann, T. (1997) "Nonlocal rotating crack model with transition to scalar damage." In *Computational Plasticity*, Vol. 2, D. R. J. Owen, E. Oñate and E. Hinton, eds., Int. Center for Numer. Meth. in Eng. (CIMNE), pp. 1514–1521.

Jishan, X. and Xixi, H. (1990) "Size effect on the strength of a concrete member." *Eng. Fract. Mech.,* **35**, 687–696.

John, R. and Shah, S. P. (1986) "Fracture of concrete subjected to impact loading." *Cement, Concrete and Aggregates,* **8**(1), 24–32.

John, R. and Shah, S. P., (1990) "Mixed mode fracture of concrete subjected to impact loading." *J. Struct. Eng.-ASCE,* **116**(3), 585–602.

Kachanov, M. (1958) "Time of rupture process under creep conditions." *Izv. Akad. Nauk. SSR, Otd. Tekh. Nauk.,* **No. 8**, 26–31.

Kachanov, M. (1980) "A continuum model of medium with cracks." *J. Eng. Mech.-ASCE,* **106**, 1039–1051.

Kachanov, M. (1982) "A microcrack model of rock inelasticity—Part I. Frictional sliding on microcracks." *Mech. Mater.,* **1**, 19–41.

Kachanov, M. (1985) "A simple technique of stress analysis in elastic solids with many cracks." *Int. J. Fracture,* **28**, R11–R19.

Kachanov, M. (1987a) "Elastic solids with many cracks: A simple method of analysis." *Int. J. Solids Struct.,* **23**, 23–43.

Kachanov, M. (1987b) "On modelling of anisotropic damage in elastic-brittle materials—a brief review." In *Damage Mechanics in Composites*, A. Wang and G. Haritos, eds., The American Society of Mechanical Engineers, New York, pp. 99–105.

Kachanov, M. (1992) "Effective elastic properties of cracked solids: Critical review of some basic concepts." *Appl. Mech. Rev.,* **45**(8), 304–335.

Kachanov, M. (1993) "Elastic solids with many cracks and related problems." In *Advances in Applied Mechanics*, Vol. 30, J. Hutchinson and T. Wu, eds., Academic Press, New York, pp. 259–445.

Kachanov, M. and Laures, J.-P. (1989) "Three-dimensional problems of strongly interacting arbitrarily located penny-shaped cracks." *Int. J. Fracture,* **41**, 289–313.

Kachanov, M., Tsukrov, I., and Shafiro, B. (1994) "Effective moduli of solids with cavities of various shapes." *Appl. Mech. Rev.,* **47**(1), S151–S174.

Kani, G. N. J. (1966) "Basic facts concerning shear failure." *ACI J.,* **63**(6), 675–692.

Kani, G. N. J. (1967) "How safe are our large reinforced concrete beams?." *ACI J.,* **64**(3), 128–141.

Kanninen, M. F. and Popelar, C. H. (1985) *Advanced Fracture Mechanics*, Oxford University Press, New York.

Kaplan, M. F. (1961) "Crack propagation and the fracture of concrete." *ACI J.,* **58**(5), 591–610.

Karihaloo, B. L. (1992) "Failure modes of longitudinally reinforced beams." In *Application of Fracture Mechanics to Reinforced Concrete*, A. Carpinteri, ed., Elsevier Applied Science, London, pp. 523–546.

Karihaloo, B. L. (1995) "Approximate fracture mechanical approach to the prediction of ultimate shear strength of RC beams." In *Fracture Mechanics of Concrete Structures*, F. H. Wittmann, ed., Aedificatio Publishers, Freiburg, Germany, pp. 1111–1123.

Karihaloo, B. L. and Nallathambi, P. (1991) "Notched beam test: Mode I fracture toughness." In *Fracture Mechanics Test Methods for Concrete*, S. P. Shah and A. Carpinteri, eds., Chapman and Hall, London, pp. 1–86.

Karp, S. N. and Karal, F. C. (1962) "The elastic field behavior in the neighbourhood of a crack of arbitrary angle." *Commun. Pur. Appl. Math.,* **15**, 413–421.

Kassir, M. K. and Sih, G. C. (1975) *Three-Dimensional Crack Problems*, Noordhoff International Publishing, Leyden, The Netherlands.

Kawai, T. (1980) "Some considerations on the finite element method." *Int. J. Numer. Meth. Eng.,* **16**, 81–120.

Kemeny, J. M. and Cook, N. G. W. (1987) "Crack models for the failure of rock under compression." In *Proc. 2nd Int. Conf. on Constitutive Laws for Eng. Mat.*, Vol. 2, C. S. Desai et al., eds., Elsevier Science Publisher, New York, pp. 879–887.

Kemeny, J. M. and Cook, N. G. W. (1991) "Micromechanics of deformation in rock." In *Toughening Mechanisms in Quasibrittle Materials*, S. P. Shah, ed., Kluwer, Dordrecht, The Netherlands, pp. 155–188.

Kendall, K. (1978) "Complexities of compression failure." *Philos. T. Roy. Soc. A,* **361**, 254–263.

Kesler, C. E., Naus, D. J. and Lott, J. L. (1972) "Fracture mechanics — Its applicability to concrete." In *Proc. Int. Conf. on the Mechanical Behavior of Materials*, Vol. 4, The Soc. of Mater. Sci., pp. 113–124.

Kienzler, R. and Herrmann G. (1986) "An elementary theory of defective beams." *Acta Mech.,* **62**, 37–46.

Kim, J.-K. and Eo, S.-H. (1990) "Size effect in concrete specimens with dissimilar initial cracks." *Mag. Concrete Res.,* **42**, 233–238.

Kim, J.-K. et al. (1989) *Size Effect on the Splitting Tensile Strength of Concrete and Mortar.* Report No. CM 89-3, Korea Advanced Institute of Science and Technology, Seoul. (Data reported by Kim and Eo 1990.)

Kittl, P. and Díaz, G. (1988) "Weibull's fracture statistics, or probabilistic strength of materials: State of the art." *Res. Mechanica,* **24**, 99–207.

Kittl, P. and Díaz, G. (1989) "Some engineering applications of the probabilistic strength of materials." *Appl. Mech. Rev.,* **42**(11), 108–112.

Kittl, P. and Díaz, G. (1990) "Size effect on fracture strength in the probabilistic strength of materials." *Reliab. Eng. Syst. Safe.,* **28**, 9–21.

Klisinski, M., Olofsson, T. and Tano, R. (1995) "Mixed mode cracking of concrete modelled by inner softening band." In *Computational Plasticity*, D. R. J. Owen et al., eds., Pineridge Press, Swansea, U.K., 1595–1606

Klisinski, M., Runesson, K. and Sture, S. (1991) "Finite element with inner softening band." *J. Eng. Mech.-ASCE,* **117**(3), 575–587.

Knauss, W. G. (1970) "Delayed Failure – The Griffith problem for linearly viscoelastic materials." *Int. J. Fracture,* **6**, 7–20.

Knauss, W. G. (1973) "The mechanics of polymer fracture." *Appl. Mech. Rev.,* **26**, 1–17.

Knauss, W. G. (1974) "On the steady propagation of a crack in a viscoelastic sheet: Experiments and analysis." In *The Mechanics of Fracture*, F. Erdogan, ed., The American Society of Mechanical Engineers, New York, pp. 69–103. (AMD-19.)

Knauss, W. G. (1976) "Fracture of solids possessing deformation rate sensitive material properties." In *Deformation and Fracture of High Polymers*, H. H. Kausch et al., eds., Plenum Press, New York, pp. 501–541.

Knauss, W. G. (1989) "Time dependent fracture of polymers." In *Advances in Fracture Research*, Vol. 4, K. Salama, K. Ravi-Chandar, D. M. R. Taplin and P. Rama Rao, eds., Pergamon Press, Oxford, pp. 2683–2711.

Knein, M. (1927) "Zur theorie des druckversuchs." *Abhandlungen aus dem Aerodynamischen Institut an der Technische Hochschule Aachen,* **7**, 43–62.

Knott, J. F. (1973) *Fundamentals of Fracture Mechanics*, Butterworths, London.

Knowles, J. K. and Sternberg, E. (1972) "On a class of conservation laws in linearized and finite elastostatics." *Arch. Ration. Mech. An.,* **44**, 187–211.

Kobayashi, A. S., Hawkins, M. N., Barker, D. B. and Liaw, B. M. (1985) "Fracture process zone of concrete." In *Application of Fracture Mechanics to Cementitious Composites*, S. P. Shah, ed., Martinus Nijhoff, Dordrecht, pp. 25–50.

Korn, G. A. and Korn, T. M. (1968) *Mathematical Handbook for Scientists and Engineers*, 2nd edition, McGraw Hill, New York.

Krafft, J. M., Sullivan, A. M. and Boyle, R.W. (1961) "Effect of dimensions on fast fracture instability of notched sheets." In *Proc. of the Crack-Propagation Symposium*, Vol. 1, pp. 8–28.

Krausz, A. S. and Eyring, H. (1975) *Deformation Kinetics*, Wiley-Interscience.

Krausz, A. S and Krausz, K. (1988) *Fracture Kinetics of Crack Growth*, Kluwer Academic Publishers, Dordrecht.

Kröner, E. (1961) "Zur plastischen verformung des vielkristalls." *Acta Metall.,* **9**, 155–161.

Kröner, E. (1967) "Elasticity theory of materials with long-range cohesive forces." *Int. J. Solids Struct.,* **3**, 731–742.

Kunin, I. A. (1968) "The theory of elastic media with microstructure and the theory of dislocations." In *Mechanics of Generalized Continua*, E. Kröner, ed., Springer Verlag, Berlin, pp. 321–328.

Kupfer, H., Hilsdorf, H. K. and Rüsch, H. (1969) "Behavior of concrete under biaxial stresses." *ACI J.,* **66**, 656–666.

Kyriakides, S., Ascerulatne, R., Perry, E. J. and Liechti, K. M. (1995) "On the compressive failure of fiber reinforced composites." *Int. J. Solids Struct.,* **32**(6–7), 689–738.

Labuz, J. F., Shah, S. P. and Dowding, C. H. (1985) "Experimental analysis of crack propagation in granite." *Int. J. Rock Mech. Min. Sci. & Geomech. Abstr.,* **22**(2), 85–98.

Larsson, R. and Runesson, K. (1995) "Cohesive crack models for semi-brittle materials derived from localization of damage coupled to plasticity." *Int. J. Fracture,* **69**, 101–122.

Larsson, R., Runesson, K. and Åkesson, M. (1995) "Embedded cohesive crack models based on regularized discontinuous displacements." In *Fracture Mechanics of Concrete Structures*, F. H. Wittmann, ed., Aedificatio Publishers, Freiburg, Germany, pp. 899–911.

Launay, P. and Gachon, H. (1971) "Strain and ultimate strength of concrete under triaxial stress." In *Proc. First Int. Conference on Struct. Mechanics in Reactor Technology*, T. Jaeger, ed., paper H1/3, 12 pp.

Lehner, F. and Kachanov, M. (1996) "On modeling of "winged" cracks forming under compression." *Int. J. Fracture,* **77**, R65–R75.

Leibengood, L. D., Darwin, D. and Dodds, R. H. (1986) "Parameters affecting FE analysis of concrete structures." *J. Struct. Eng.-ASCE,* **112**(2), 326–341.

Lemaitre, J. and Chaboche, J.-L. (1985) *Mécanique des Matériaux Solides*, Dunod, Paris.

Leonhardt, F. (1977) "Schub bei stahlbeton und spannbeton—Grundlagen der neueren schubbemessung." *Beton und Stahlbetonbau,* **72**(11–12), 270–277 and 295–392.

Levin, V. M. (1971) "The relation between the mathematical expectation of stress and strain tensors in elastic microheterogeneous media." *Prikl. Mat. Mekh.,* **35**, 694–701. (In Russian.)

Li, V. C., Chan, C. M. and Leung, C. K. Y. (1987) "Experimental determination of the tension-softening relations for cementitious composites." *Cement Concrete Res.,* **17**, 441–452.

Li, Y.-N. and Bažant, Z. P. (1994a) "Eigenvalue analysis of size effect for cohesive crack model." *Int. J. Fracture,* **66**, 213–226.

Li Y.-N. and Bažant Z. P. (1994b) "Penetration fracture of sea ice plate: 2D analysis and size effect." *J. Eng. Mech.-ASCE,* **120**(7), 1481–1498.

Li, Y.-N. and Bažant, Z. P. (1996) "Scaling of cohesive fracture (with ramification to fractal cracks)." In *Size-Scale Effects in the Failure Mechanisms of Materials and Structures*, A. Carpinteri, ed., E & FN Spon, London, pp. 274–299.

Li, Y.-N and Bažant, Z. P. (1997) "Cohesive crack model with rate-dependent crack opening and viscoelasticity: Numerical algorithm, behavior and size effect." *Int. J. Fracture,* in press.

Li, Y.-N., Hong, A. N. and Bažant, Z. P. (1995) "Initiation of parallel cracks from surface of elastic half-plane." *Int. J. Fracture,* **69**, 357–369.

Liaw, B. M., Jeang, F. L., Du, J. J., Hawkins, N. M. and Kobayashi, A. S. (1990) "Improved non-linear model for concrete fracture." *J. Eng. Mech.-ASCE,* **1106**(2), 429–445.

Lin, C. S. and Scordelis, A. (1975) "Nonlinear analysis of RC shells of general forms." *J. Struct Eng.-ASCE,* **101**, 523–538.

Lin, T. H. and Ito, M. (1965) "Theoretical plastic distortion of a polycrystalline aggregate under combined and reversed stresses." *J. Mech. Phys. Solids,* **13**, 103–115.

Lin, T. H. and Ito, M. (1966) "Theoretical plastic stress-strain relationship of a polycrystal." *Int. J. Eng. Sci.,* **4**, 543–561.

Lindner, C. P. and Sprague, I. C. (1955) "Effect of depth of beams upon the modulus of rupture of plain concrete." *ASTM Proc.,* **55**, 1062–1083.

Linsbauer, H. and Tschegg, E. K. (1986) "Fracture energy determination of concrete with cube-shaped specimens." *Zement und Beton,* **31**, 38–40. (In German.)

Linsbauer, H., Ingraffea, A. R., Rossmanith, H. and Wawrzynek, P. A. (1988a) "Simulation of cracking in large arch dam: Part I." *J. Struct Eng.-ASCE,* **115**(7), 1599–1615.

Linsbauer, H., Ingraffea, A. R., Rossmanith, H. and Wawrzynek, P. A. (1988b) "Simulation of cracking in large arch dam: Part II." *J. Struct Eng.-ASCE,* **115**(7), 1615–1630.

Llorca, J., Planas, J. and Elices, M. (1989) "On the use of maximum load to validate or disprove models for concrete fracture behaviour." In *Fracture of Concrete and Rock, Recent Developments*, S. P. Shah, S. E. Swartz and B. Barr, eds., Elsevier Applied Science, London, pp. 357–368.

Lofti, H. R. and Shing, P. B. (1994) "Analysis of concrete fracture with an embedded crack approach." In *Computational Modeling of Concrete Structures*, H. Mang, N. Bicanic and R. de Borst, eds., Pineridge Press, Swansea, pp. 343–352.

Lubliner, J. (1986) "Normality rules in large-deformation plasticity." *Mech. Mater.,* **5**, 29–34.

Lundborg, N. (1967) "Strength-size relation of granite." *Int. J. Rock Mech. Min. Sci.,* **4**, 269–272.

MacGregor, J. G. and Gergely, P. (1977) "Suggested revisions to ACI code – Clauses dealing with shear in beams." *ACI J.,* **74**(10), 493–500.

Mai, Y.-W. (1991) "Fracture and fatigue of non-transformable ceramics: The role of crack-interface bridging." In *Fracture Processes in Concrete, Rock and Ceramics*, J. G. M. van Mier, J. G. Rots and A. Bakker, eds., E & FN Spon, London, pp. 3–26.

Maji, A. K. and Shah, S. P. (1988) "Process zone and acoustic emission measurement in concrete." *Exp. Mech.,* **28**, 27–33.

Malvern, L. E. (1969) *Introduction to the Mechanics of a Continuous Medium*, Prentice-Hall, Englewood Cliffs, New Jersey.

Mandelbrot, B. B., Passoja, D.E. and Paullay, A. (1984) "Fractal character of fracture surfaces of metals." *Nature,* **308**, 721–722.

Mariotte, E. (1686) *Traité du mouvement des eaux*, Posthumously edited by M. de la Hire, Engl. transl. by J.T. Desvaguliers, London (1718), p. 249. (Also *Mariotte's collected works*, 2nd ed., The Hague (1740).)

Martha, L. F., Llorca, J., Ingraffea, A. R. and Elices, M. (1991) "Numerical simulation of crack initiation and propagation in an arch dam." *Dam Eng.,* **2**(3), 193–213.

Marti, P. (1980) *Zur Plastischen Berechnung von Stahlbeton*. Bericht No. 104, Institute für Baustatik und Konstruktion, ETH, Zürich.

Marti, P. (1985) "Basic tools of reinforced concrete beam design." *ACI J.,* **82**(1), 46–56. (Discussion **82**(6), 933–935.)

Marti, P. (1989) "Size effect in double-punch tests on concrete cylinders." *ACI Mater. J.,* **86**(6), 597–601.

Massabò, R. (1994) *Mechanismi di Rottura nei Materiali Fibrorinforzati.* Doctoral thesis. Dottorato di Ricerca in Ingegneria Strutturale, Politecnico di Torino, Torino, Italia.

Maturana, P., Planas, J. and Elices, M. (1990) "Evolution of fracture behaviour of saturated concrete in the low temperature range." *Eng. Fract. Mech.,* **35**(4-5), 827–834.

Mazars, J. (1981) "Mechanical damage and fracture of concrete structures." In *Advances in Fracture Research, Preprints 5th. Int. Conf. Fracture*, Vol. 4, D. François, ed., Pergamon Press, Oxford, pp. 1499–1506.

Mazars, J. (1984) *Application de la Mécanique de l'Endomagement au Comportement Non-Linéaire et à la Rupture du Béton de Structures.* Doctoral thesis. Université de Paris 6.

Mazars, J. (1986) "A model for a unilateral elastic damageable material and its application to concrete." In *Fracture Toughness and Fracture Energy of Concrete*, F. H. Wittmann, ed., Elsevier, Amsterdam, pp. 61–71.

McHenry, D. (1943) "A new aspect of creep in concrete and its application to design." *Proc. ASTM*, **43**, 1069–1086.

McKinney, K. R. and Rice, R. W. (1981) "Specimen size effects in fracture toughness testing of heterogeneous ceramics by the notch beam method." In *Fracture Mechanics Methods for Ceramics, Rocks, and Concrete*, S. W. Freiman and E. R. Fuller Jr., eds., American Society for Testing and Materials, Philadelphia, pp. 118–126. (ASTM Special Technical Publication No. 745.)

McMullen, A. E. and Daniel, H. R. (1975) "Torsional strength of longitudinally reinforced beams containing an opening." *ACI J.*, **72**(8), 415–420.

Melan (1932) "Der spannungzustand der durch eine einzelkraft im innern beanspruchten halbschiebe." 2 *Angew. Math. Mech*, **12**(6).

Meyer, C. and Okamura, H., eds. (1986) *Finite Element Analysis of Reinforced Concrete Structures*, ASCE, New York.

van Mier, J. G. M. (1984) *Strain-Softening of Concrete Under Multiaxial Loading Conditions.* Doctoral thesis. De Technische Hogeschool Eindhoven, The Netherlands.

van Mier, J. G. M. (1986) "Multiaxial strain-softening of concrete; Part I: Fracture; Part II: Load histories." *Mater. Struct.*, **19**(111), 179–200.

van Mier, J. G. M. and Vervuurt, A. (1995) "Micromechanical analysis and experimental verification of boundary rotation effects in uniaxial tension tests on concrete." In *Fracture of Brittle Disordered Materials: Concrete, Rock, Ceramics*, G. Baker and B. L. Karihaloo, eds., E & FN Spon, London, pp. 406–420.

van Mier, J. G. M., Nooru-Mohamed, M. B. and Schlangen, E. (1991) "Experimental analysis of mixed mode I and II behavior of concrete." In *Analysis of Concrete Structures by Fracture Mechanics*, L. Elfgren and S. P. Shah, eds., Chapman and Hall, London, pp. 32–43.

van Mier, J. G. M., Schlangen, E. and Vervuurt, A. (1996) "Tensile cracking in concrete and sandstone: Part 2 - Effect of boundary rotations." *Mater. Struct.*, **29**, 87–96.

van Mier, J. G. M., Vervuurt, A. and Schlangen, E. (1994) "Boundary and size effects in uniaxial tensile tests: A numerical and experimental study." In *Fracture and Damage in Quasibrittle Structures*, Z. P. Bažant, Z. Bittnar, M. Jirásek and J. Mazars, eds., E & FN Spon, London, pp. 289–302.

Mihashi, H. (1983) "A stochastic theory for fracture of concrete." In *Fracture Mechanics of Concrete*, F. H. Wittmann, ed., Elsevier Science Publishers, Amsterdam, pp. 301–339.

Mihashi, H. (1992) "Material structure and tension softening properties of concrete." In *Fracture Mechanics of Concrete Structures*, Z. P. Bažant, ed., Elsevier, London, pp. 239–250.

Mihashi, H. and Izumi, M. (1977) "Stochastic theory for concrete fracture." *Cement Concrete Res.*, **7**, 411–422.

Mihashi, H. and Wittmann, F. H. (1980) "Stochastic approach to study the influence of rate of loading on strength of concrete." *Heron*, **25**(3).

Mihashi, H. and Zaitsev, J. W. (1981) *Statistical Nature of Crack Propagation.* Report to RILEM TC 50-FMC.

Mikhlin, S. G. (1964) *Integral Equations*, Pergamon Press, Oxford.

Miller, R. A., Shah, S. P. and Bjelkhagen, H. (1988) "Measurement of crack profiles in mortar using laser holographic interferometry." *Exp. Mech.*, **28**(4), 388–394.

Mindess, S. (1983) "The application of fracture mechanics to cement and concrete: A historical review." In *Fracture Mechanics of Concrete*, F. H. Wittmann, ed., Elsevier Science Publishers, Amsterdam, The Netherlands, pp. 1–30.

Mindess, S. and Shah, S. P., eds. (1986) *Proc. MRS Symp. No. 64 on Cement Based Composites: Strain Rate Effect on Fracture*, Materials Research Society.

Mindess, S., Lawrence, F. V. and Kesler, C. E. (1977) "The J-integral as a fracture criterion for fiber reinforced concrete." *Cement Concrete Res.*, **7**, 731–742.

von Mises, R . (1936) "La distribution de la plus grande de n valeurs." *Rev. Math. Union Interbalcanique*, **1**, 1.

Mitchell, D. and Collins, M. P. (1974) "Diagonal compression field theory – a rational model for structural concrete in pure torsion." *ACI J.*, **71**(8), 346–408.

Modéer, M. (1979) *A Fracture Mechanics Approach to Failure Analyses of Concrete Materials.* Report No. TVBM-1001, Division of Building Materials, Lund Institute of Technology, Lund, Sweden.

Mori, T. and Tanaka, K. (1973) "Average stress in matrix and average elastic energy of materials with misfit inclusions." *Acta Metall.*, **21**, 571–574.

Mörsch, E. (1922) *Der Eisenbetonbau — Seine Theorie und Anwendung*, Vol. 1, 5th edition, Wittwer, Stuttgart. (Reinforced Concrete Construction—Theory and Application.)

Mosolov, A. B. and Borodich, F. M. (1992) "Fractal fracture of brittle bodies under compression." *Doklady Akademii Nauk.*, **324**(3), 546–549. (In Russian.)

Mueller, H. K. and Knauss, W. G. (1971) "Crack propagation in a linearly viscoelastic strip." *J. Appl. Mech.-T. ASME,* **38**, 483–488.

Mühlhaus, H.-B. and Aifantis, E. C. (1991) "A variational principle for gradient plasticity." *Int. J. Solids Struct.,* **28**, 845–858.

Mulmule, S. V., Dempsey, J. P. and Adamson, R. M. (1995) "Large-scale *in-situ* ice fracture experiments. Part II: Modeling aspects." In *Ice Mechanics 1995*, J. P. Dempsey and Y. D. S. Rajapakse, eds., The American Society of Mechanical Engineers, New York, pp. 129–146. (AMD-Vol. 207, ASME Summer Meeting, Los Angeles, CA.)

Mura, T. (1987) *Micromechanics of Defects in Solids*, 2nd edition, Martinus Nijhoff Publishers, Dordrecht.

Murakami, Y. (1987) *Stress Intensity Factors Handbook*, Pergamon Press, Oxford.

Nallathambi, P. and Karihaloo, B. L. (1986a) "Determination of specimen- size independent fracture toughness of plain concrete." *Mag. Concrete Res.,* **38**(135), 67–76.

Nallathambi, P. and Karihaloo, B. L. (1986b) "Stress intensity factor and energy release rate for three-point bend specimen." *Eng. Fract. Mech.,* **25**(3), 315–321.

Naus, D. J. (1971) *Applicability of Linear-Plastic Fracture Mechanics to Portland Cement Concretes.* Doctoral thesis. University of Illinois at Urbana-Champaign.

Naus, D. J. and Lott, J. L. (1969) "Fracture toughness of portland cement concretes." *ACI J.,* **66**, 481–498.

Nemat-Nasser, S. and Obata, M. (1988) "A microcrack model of dilatancy in brittle material." *J. Appl. Mech.-T. ASME,* **55**, 24–35.

Nesetova, V. and Lajtai, E. Z. (1973) "Fracture from compressive stress concentration around elastic flaws." *Int. J. Rock Mech. Min. Sci.,* **10**, 265–284.

Newman Jr., J. C. (1971) *An Improved Method of Collocation for the Stress Analysis of Cracked Plates with Various Shaped Boundaries.* Technical Note No. TN D-6376, NASA.

Nielsen, K. E. C. (1954) "Effect of various factors on the flexural strength of concrete test beams." *Mag. Concrete Res.,* **15**, 105–114.

Nielsen, M. P. and Braestrup, N. W. (1975) *Plastic Shear Strength of Reinforced Concrete Beams.* Techn. Report No. 3, Bygningsstatiske Meddelesler (Vol. 46).

Nixon, W. F. (1996) "Wing crack models of the brittle compressive failure of ice." *Cold Reg. Sci. Technol.,* **24**, 41–55.

Noghabai, K. (1995a) *Splitting in Concrete in the Anchoring Zone of Deformed Bars.* Graduate thesis. Division of Structural Engineering, Luleå University of Technology, Luleå, Sweden.

Noghabai, K. (1995b) "Splitting of concrete covers – a fracture mechanics approach." In *Fracture Mechanics of Concrete Structures,* F. H. Wittmann, ed., Aedificatio Publishers, Freiburg, Germany, pp. 1575–1584.

Nuismer, R. J. (1975) "An energy release rate criterion for mixed mode fracture." *Int. J. Fracture,* **11**, 245–250.

Ogden, R. W. (1984) *Non-linear elastic deformations*, Ellis Horwood, Ltd. and John Wiley & Sons, Chichester, U.K.

Oglesby, J. J. and Lamackey, O. (1972) *An Evaluation of Finite Element Methods for the Computation of Elastic Stress Intensity Factors.* Report No. No. 3751, NSRDC.

Ohgishi, S., Ono, H., Takatsu, M. and Tanahashi, I. (1986) "Influence of test conditions on fracture toughness of cement paste and mortar." In *Fracture Toughness and Fracture Energy of Concrete*, F. H. Wittmann, ed., Elsevier Science, Amsterdam, The Netherlands, pp. 281–290.

Ohtsu, M. and Chahrour, A. H. (1995) "Fracture analysis of concrete based on the discrete crack model by the boundary element method." In *Fracture of Brittle Disordered Materials: Concrete, Rock, Ceramics*, G. Baker and B. L. Karihaloo, eds., E & FN Spon, London, pp. 335–347.

Oliver, J. (1989) "A consistent characteristic length for smeared cracking models." *Int. J. Numer. Meth. Eng.,* **28**, 461–474.

Oliver, J. (1995) "Modeling strong discontinuities in solid mechanics via strain softening constitutive equations." *Monograph CIMNE,* **28**.

Olsen, P. C. (1994) "Some comments on the bending strength of concrete beams." *Mag. Concrete Res.,* **46**, 209–214.

Ortiz, M. (1985) "A constitutive theory for the inelastic behaviour of concrete." *Mech. Mater.,* **4**, 67–93.

Ortiz, M. (1987) "An analytical study of the localized failure modes in concrete." *Mech. Mater.*, **6**, 159–174.

Ottosen, N. S. (1977) "A failure criterion for concrete." *J. Eng. Mech. Div.-ASCE,* **103**(4), 527–535.

Ouchterlony, F. (1975) *Concentrated Loads Applied to the Tips of a Symmetrically Cracked Wedge.* Report No. DS-1975:3, Swedish Dectonic Research Foundation.

Owen, D. R. J. and Hinton, E. (1980) *Finite Elements in Plasticity: Theory and Practice*, Pineridge Press, Swansea, U.K.

Ožbolt, J. and Bažant, Z. P. (1991) "Cyclic microplane model for concrete." In *Fracture Processes in Concrete, Rock and Ceramics*, J. G. M. van Mier, J. G. Rots and A. Bakker, eds., E & FN Spon, London, pp. 639–650.

Ožbolt, J. and Bažant, Z. P. (1992) "Microplane model for cyclic triaxial behavior of concrete and rock." *J. Eng. Mech.-ASCE,* **118**(7), 1365–1386.

Ožbolt, J. and Bažant, Z. P. (1996) "Numerical smeared fracture analysis: Nonlocal microcrack interaction approach." *Int. J. Numer. Meth. Eng.,* **39**, 635–661.

Ožbolt, J. and Eligehausen, R. (1995) "Size effect in concrete and reinforced concrete structures." In *Fracture Mechanics of Concrete Structures*, Vol. 1, F. H. Wittmann, ed., Aedificatio Publishers, Freiburg, Germany, pp. 665–674.

Palmer, A. C. and Sanderson, T. J. O. (1991) "Fractal crushing of ice and brittle solids." *Philos. T. Roy. Soc. A,* **443**, 469–477.

Pan, Y. C., Marchertas, A. H. and Kennedy, J. M. (1983) "Finite element analysis of blunt crack propagation, a modified J-integral approach." In *Transactions of the Seventh International Conference on Structural Mechanics in Reactor Technology*, North-Holland, New York, pp. 235–292.

Pande, G. N. and Sharma, K. G. (1981) "Implementation of computer procedures and stress-strain laws in geotechnical engineering." In *Proc. Symp. on Implementation of Computer Procedures and Stress-Strain Laws in Geotechnical Engineering.*, C. S. Desai and S. K. Saxena, eds., Acorn Press, Durham, N.C., pp. 575–590.

Pande, G. N. and Sharma, K. G. (1982) *Multi-Laminate Model of Clays—A Numerical Evaluation of the Influence of Rotation of the Principal Stress Axis.* Report, Dept. of Civil Engrg., University College of Swansea, Swansea, U.K.

Pande, G. N. and Xiong, W. (1982) "An improved multi-laminate model of jointed rock masses." In *Proc. Int. Sym. on Numerical Models in Geomechanics*, R. Dungar, G. N. Pande and G. A. Studder, eds., Balkema, Rotterdam, pp. 218–226.

Paris, P. C. and Erdogan, F. (1963) "Critical analysis of propagation laws." *J. Basic Eng.,* **85**, 528–534.

Paris, P. C., Gomez, M. P. and Anderson, W. E. (1961) "Rational analytic theory of fatigue." *Trends Eng.,* **13**(1).

Park, R. and Paulay, T. (1975) *Reinforced Concrete Structures*, John Wiley & Sons, New York.

Pastor, J. Y. (1993) *Fractura de Materiales Cerámicos Estructurales Avanzados.* Doctoral thesis. Facultad de Ciencias Físicas, Dep. de Ciencia de Materiales, Univeridad Complutense de Madrid, Ciudad Universitaria, 28040 Madrid, Spain. ('Fracture of Advanced Structural Ceramics', in Spanish.)

Pastor, J. Y., Guinea, G., Planas, J. and Elices, M. (1995) "Nueva expresión del factor de intensidad de tensiones para la probeta de flexión en tres puntos." *Anales de Mecánica de la Fractura,* **12**, 85–90. ('A new expression for the stress intensity factor of a three-point bend specimen', in Spanish.)

Paul, B. (1968) "Macroscopic criteria for plastic flow and brittle fracture." In *Fracture – An Advanced Treatise*, Vol. 2, H. Liebowitz, ed., Academic Press, New York, pp. 313–496. (Chapter 4.)

Peirce, F. T. (1926) "Tensile strength of cotton yarns. V.—The weakest link theorems on the strength of long and composite specimens." *J. Textile Inst.,* **17**, T355–368.

Perdikaris, P. C. and Calomino, A. M. (1989) "Kinetics of crack growth in plain concrete." In *Fracture of Concrete and Rock*, S. P. Shah and S. E. Swartz, eds., Springer-Verlag, New York, pp. 64–69.

Petersson, P.-E. (1981) *Crack Growth and Development of Fracture Zone in Plain Concrete and Similar Materials.* Report No. TVBM-1006, Division of Building Materials, Lund Institute of Technology, Lund, Sweden.

Petrovic, J. J. (1987) "Weibull statistical fracture theory for the fracture of ceramics." *Metall. Trans. A,* **18**, 1829–1834.

Phillips, D. V. and Binsheng, Z. (1993) "Direct tension tests on notched and un-notched plain concrete specimens." *Mag. Concrete Res.,* **45**, 25–35.

Pietruszczak, S. and Mróz, Z. (1981) "Finite element analysis of deformation of strain-softening materials." *Int. J. Numer. Meth. Eng.,* **17**, 327–334.

Pijaudier-Cabot, G. and Bažant, Z. P. (1987) "Nonlocal damage theory." *J. Eng. Mech.-ASCE,* **113**(10), 1512–1533.

Pijaudier-Cabot, G. and Bažant, Z. P. (1988) "Dynamic stability analysis with nonlocal damage." *Comput. Struct.,* **29**(3), 503–507.

Pijaudier-Cabot, G. and Bažant, Z. P. (1991) "Cracks interacting with particles or fibers in composite materials." *J. Eng. Mech.-ASCE,* **117**(7), 1611–1630.

Planas, J. (1992) Untitled letter. Privately communicated comment to Z. P. Bažant, Northwestern University, July 13.

Planas, J. (1993) *A Note on the Effect of Specimen Self Weight on the Effective Crack Extension Measurement by the Compliance Method.* Report No. 93-jp02, Departamento de Ciencia de Materiales, ETS de Ingenieros de Caminos, Universidad Politécnica de Madrid, Ciudad Universitaria sn. 28040 Madrid, Spain.

Planas, J. (1995) *Crack growth in a n elastic medium with random crack growth resistance.* Report No. 95-jp03, Departamento de Ciencia de Materiales, ETS de Ingenieros de Caminos, Universidad Politécnica de Madrid, Ciudad Universitaria sn. 28040 Madrid, Spain.

Planas, J. and Bažant, Z. P. (1997) *Statistics of crack growth based on random R-curves.* Report No. 97-jp02, Departamento de Ciencia de Materiales, ETS de Ingenieros de Caminos, Universidad Politécnica de Madrid, Ciudad Universitaria sn. 28040 Madrid, Spain.

Planas, J. and Elices, M. (1986a) "Towards a measure of G_F: An analysis of experimental results." In *Fracture Toughness and Fracture Energy of Concrete,* F. H. Wittmann, ed., Elsevier, Amsterdam, pp. 381–390.

Planas, J. and Elices, M. (1986b) "Un nuevo método de análisis del comportamiento de una fisura cohesiva en Modo I." *Anales de Mecánica de la Fractura,* **3**, 219–227.

Planas, J. and Elices, M. (1989a) "Size effect in concrete structures: Mathematical approximations and experimental validation." In *Cracking and Damage, Strain Localization and Size Effect,* J. Mazars and Z. P. Bažant, eds., Elsevier Applied Science, London, pp. 462–476.

Planas, J. and Elices, M. (1989b) "Conceptual and experimental problems in the determination of the fracture energy of concrete." In *Fracture Toughness and Fracture Energy: Test Methods for Concrete and Rock,* H. Mihashi, H. Takahashi and F. H. Wittmann, eds., Balkema, Rotterdam, pp. 165–181.

Planas, J. and Elices, M. (1990a) "Fracture criteria for concrete: Mathematical approximations and experimental validation." *Eng. Fract. Mech.,* **35**, 87–94.

Planas, J. and Elices, M. (1990b) "Anomalous structural size effect in cohesive materials like concrete." In *Serviceability and Durability of Construction Materials,* Vol. 2, B. A. Suprenant, ed., American Society of Civil Engineers (ASCE), New York, pp. 1345–1356.

Planas, J. and Elices, M. (1990c) "The approximation of a cohesive crack by effective elastic cracks." In *Fracture Behaviour and Design of Materials and Structures,* Vol. 2, D. Firrao, ed., Engineering Materials Advisory Services Ltd. (EMAS), Warley, West Midlands, U.K., pp. 605–611.

Planas, J. and Elices, M. (1991a) "Nonlinear fracture of cohesive materials." *Int. J. Fracture,* **51**, 139–157.

Planas, J. and Elices, M. (1991b) "Asymptotic analysis of cohesive cracks and its relation with effective elastic cracks." In *Toughening Mechanisms in Quasi-Brittle Materials,* S. P. Shah, ed., Kluwer Academic Publishers, Dordrecht, pp. 189–202.

Planas, J. and Elices, M. (1991c) "The influence of specimen size and material characteristic size on the applicability of effective crack models." In *Fracture Processes in Concrete, Rock and Ceramics,* J. G. M. van Mier, J. G. Rots and A. Bakker, eds., E & FN Spon, London, pp. 375–385.

Planas, J. and Elices, M. (1991d) *On the Bazant-Herrmann Approximation to the Stress Intensity Factor in Single Edge Notched Beams.* Report No. 91-jp03, Departamento de Ciencia de Materiales, ETS de Ingenieros de Caminos, Universidad Politécnica de Madrid, Ciudad Universitaria sn. 28040 Madrid, Spain.

Planas, J. and Elices, M. (1992a) "Asymptotic analysis of a cohesive crack: 1. Theoretical background." *Int. J. Fracture,* **55**, 153–177.

Planas, J. and Elices, M. (1992b) "Shrinkage eigenstresses and structural size-effect." In *Fracture Mechanics of Concrete Structures,* Z. P. Bažant, ed., Elsevier Applied Science, London, pp. 939–950.

Planas, J. and Elices, M. (1993a) "Asymptotic analysis of a cohesive crack: 2. Influence of the softening curve." *Int. J. Fracture,* **64**, 221–237.

Planas, J. and Elices, M. (1993b) "Drying shrinkage effect on the modulus of rupture." In *Creep and Shrinkage of Concrete,* Z. P. Bažant and I. Carol, eds., E & FN Spon, London, pp. 357–368.

Planas, J., Corres, H., Elices, M. and Chueca, R. (1984) "Thermal deformation of loaded concrete during thermal cycles from 20 C to –165 C." *Cement Concrete Res.,* **14**, 639–644.

Planas, J., Elices, M. and Guinea G. V. (1992) "Measurement of the fracture energy using three-point bend tests: 2. Influence of bulk energy dissipation." *Mater. Struct.*, **25**, 305–312.

Planas, J., Elices, M. and Guinea, G. V. (1993) "Cohesive cracks versus nonlocal models: Closing the gap." *Int. J. Fracture,* **63**, 173–187.

Planas, J., Elices, M. and Guinea, G. V. (1994) "Cohesive cracks as a solution of a class of nonlocal models." In *Fracture and Damage of Quasibrittle Structures*, Z. P. Bazant, Z. Bittnar, M. Jirásek and J. Mazars, eds., E & FN Spon, London, pp. 131–144.

Planas, J., Elices, M. and Guinea, G. V. (1995) "The extended cohesive crack." In *Fracture of Brittle Disordered Materials: Concrete, Rock and Ceramics*, G. Bakker and B. L. Karihaloo, eds., E & FN Spon, London, pp. 51–65.

Planas, J., Elices, M. and Ruiz, G. (1993) "The equivalent elastic crack: 2. X-Y equivalences and asymptotic analysis." *Int. J. Fracture,* **61**, 231–246.

Planas, J., Elices, M. and Toribio, J. (1989) "Approximation of cohesive crack models by R-CTOD curves." In *Fracture of Concrete and Rock, Recent Developments*, S. P. Shah, S. E. Swartz and B. Barr, eds., Elsevier Applied Science, London, pp. 203–212.

Planas, J., Guinea, G. V. and Elices, M. (1994a) *SF-1. Draft Test Method for Flexural Strength and Elastic Modulus of Notched Concrete Beams Tested in Three-Point Bending*. Report No. 94-jp02, Departamento de Ciencia de Materiales, ETS de Ingenieros de Caminos, Universidad Politécnica de Madrid, Ciudad Universitaria sn. 28040 Madrid, Spain. (Contribution to the Japan Concrete Institute International Collaboration Project on Size Effect in Concrete Structures.)

Planas, J., Guinea, G. V. and Elices, M. (1994b) *SF-2. Draft Test Method for Linear Initial Portion of the Softening Curve of Concrete*. Report No. 94-jp03, Departamento de Ciencia de Materiales, ETS de Ingenieros de Caminos, Universidad Politécnica de Madrid, Ciudad Universitaria sn. 28040 Madrid, Spain. (Contribution to the Japan Concrete Institute International Collaboration Project on Size Effect in Concrete Structures.)

Planas, J., Guinea, G. V. and Elices, M. (1995) "Rupture modulus and fracture properties of concrete." In *Fracture Mechanics of Concrete Structures*, Vol. 1, F. H. Wittmann, ed., Aedificatio Publishers, Freiburg, Germany, pp. 95–110.

Planas, J., Guinea, G. V. and Elices, M. (1996) *Basic Issues on Nonlocal Models: Uniaxial Modeling*. Report No. 96-jp03, Departamento de Ciencia de Materiales, ETS de Ingenieros de Caminos, Universidad Politécnica de Madrid, Ciudad Universitaria sn. 28040 Madrid, Spain.

Planas, J., Guinea, G. V. and Elices, M. (1997) "Generalized size effect equation for quasibrittle materials." *Fatigue Fract. Eng. Mater. Struct.*, **20**(5), 671–687.

Planas, J., Ruiz, G. and Elices, M. (1995) "Fracture of lightly reinforced concrete beams: Theory and experiments." In *Fracture Mechanics of Concrete Structures,* Vol. 2, F. H. Wittmann, ed., Aedificatio Publishers, Freiburg, Germany, pp. 1179–1188.

Plesha, M. E. and Aifantis, E. C. (1983) "On the modeling of rocks with microstructure." In *Proc. 24th U.S. Symp. Rock Mech.*.

Post-Tensioning Institute (1988) *Design and Construction Specifications for Segmental Concrete Bridges*. Final Report, Post-Tensioning Institute, Phoenix, Arizona.

Priddle, E. K. (1976) "High cycle fatigue crack propagation under random and constant amplitude loadings." *Int. J. Pres. Ves. Pip.*, **4**, 89–117.

Prandtl, L. (1904) "Uber die Flüssigkeitsbewebung bei sehr kleiner Reibung." In *Verhandlungen, III. Int. Math.-Kongr.*, Heidelberg, Germany.

Prat, P. C. and Bažant, Z. P. (1997) "Tangential stiffness of elastic materials with systems of growing or closing cracks." *J. Mech. Phys. Solids,* **45**(4), 611–636; wth Addendum and Errata **45**(8), 1419–1420.

Press, W. H., Teukolsky, S. A., Vetterling, W. T. and Flannery, B. P. (1992) *Numerical Recipes in C*, Cambridge University Press, New York.

Primas, R. J. and Gstrein, R (1994) *ESIS TC Round Robin on Fracture Toughness*. EMPA Report No. 155'088, Swiss Federal Laboratories for Material Testing And Research (EMPA). (Draft Nov. 1994.)

Pugh, E. M. and Winslow, G. H. (1966) *The Analysis of Physical Measurements*, Addison-Wesley, Reading, MA.

Rashid, Y. R. (1968) "Analysis of prestressed concrete pressure vessels." *Nucl. Eng. Des.*, **7**(4), 334–355.

Reagel, F. V. and Willis, T. F. (1931) "The effect of dimensions of test specimens on the flexural strength of concrete." *Public Roads,* **12**, 37–46.

Reece, M. J., Guiu, F. and Sammur M. F. R. (1989) "Cyclic fatigue crack propagation in alumina under direct tension-compression loading." *J. Am. Cera. Soc.,* **72**, 348–352.

Reich, R. Červenka, J. and Saouma, V. (1994) "Merlin: A computational environment for 2D/3D discrete fracture analysis." In *Computational Modeling of Concrete Structures*, H. Mang, N. Bicanic and R. de Borst, eds., Pineridge Press, Swansea, pp. 999–1008.

Reineck, K.-H. (1991) "Model for structural concrete members without transverse reinforcement." In *Proc. IABSE Colloquium on Structural Concrete*, Stuttgart, pp. 643–648. (IABSE Report Vol. 62.)

Reinhardt, H. W. (1981a) "Masstabeinfluss bei schubversuchen im licht der bruchmechanik." *Beton und Stahlbetonbau*, **7**, 19–21. ("Size Effect in Shear Tests in the Light of Fracture Mechanics". In German.)

Reinhardt, H. W. (1981b) "Similitude of brittle fracture of structural concrete." In *Proc. IABSE Colloquium on Advances in Mechanics of Reinforced Concrete*, pp. 201–210.

Reinhardt, H. W. (1982) "Length influence on bond shear strength of joints in composite concrete slabs." *Int. J. Cement Compos. Lightweight Concrete*, **4**(3), 139–143.

Reinhardt, H. W. (1984) "Fracture mechanics of fictitious crack propagation in concrete." *Heron*, **29**(2), 3–42.

Reinhardt, H. W. (1992) "Bond of steel to strain-softening concrete taking account of loading rate." In *Fracture Mechanics of Concrete Structures*, Z. P. Bažant, ed., Elsevier Applied Science, London, pp. 809–820.

Reinhardt, H. W. and Cornelissen, H. A. W. (1984) "Post-peak cyclic behavior of concrete in uniaxial and alternating tensile and compressive loading." *Cement Concrete Res.*, **14**(2), 263–270.

Reinhardt, H. W. and van der Veen, C. (1992) "Splitting failure of a strain-softening material due to bond stresses." In *Application of Fracture Mechanics to Reinforced Concrete*, A. Carpinteri, ed., Elsevier Applied Science, London, pp. 333–346.

Rektorys, K. (1969) *Survey of Applicable Mathematics*, Kluwer Acad. Publ., Dordrecht. (Also Iliffe Books Ltd., London, 1969.)

Rice, J. R. (1968a) "A path independent integral and the approximate analysis of strain concentrations by notches and cracks." *J. Appl. Mech.-T. ASME*, **35**, 379–386.

Rice, J. R. (1968b) "Mathematical analysis in the mechanics of fracture." In *Fracture – An Advanced Treatise*, Vol. 2, H. Liebowitz, ed., Academic Press, New York, pp. 191–308.

Rice, J. R. (1970) "On the structure of stress-strain relations for time-dependent plastic deformation of metals." *J. Appl. Mech.-T. ASME*, **37**, 728–737.

Rice, J. R. (1988) "Elastic fracture concepts for interfacial cracks." *J. Appl. Mech.-T. ASME*, **55**, 98–103.

Rice, J. R. and Levy, N. (1972) "The part-through surface crack in an elastic plate." *J. Appl. Mech.-T. ASME*, **39**, 185–194.

Rice, J. R. and Rosengren, G. F. (1968) "Plane strain deformation near a crack tip in a power law hardening material." *J. Mech. Phys. Solids*, **16**, 1–12.

Rice, J. R. and Sih, G. C. (1965) "Plane problem of cracks in dissimilar media." *J. Appl. Mech.-T. ASME*, **32**(2), 418–425.

Riedel, H. (1989) "Recent advances in modelling creep crack growth." In *Advances in Fracture Research*, Vol. 2, K. Salama, K. Ravi-Chandar, D. M. R. Taplin and P. Rama Rao, eds., Pergamon Press, Oxford, pp. 1495–1523.

RILEM (1985) "Determination of the fracture energy of mortar and concrete by means of three-point bend tests on notched beams." *Mater. Struct.*, **18**, 285–290. (RILEM Draft Recommendation, TC 50-FMC Fracture Mechanics of Concrete.)

RILEM (1990a) "Determination of fracture parameters (K_{Ic}^s and CTOD$_c$) of plain concrete using three-point bend tests." *Mater. Struct.*, **23**, 457–460. (RILEM Draft Recommendation, TC 89-FMT Fracture Mechanics of Concrete–Test methods.)

RILEM (1990b) "Size-effect method for determining fracture energy and process zone size of concrete." *Mater. Struct.*, **23**, 461–465. (RILEM Draft Recommendation, TC 89-FMT Fracture Mechanics of Concrete–Test methods.)

Ritter, W. (1899) "Die bauweise hennebique." *Schweiz. Bauzeitung Zürich*, **33**(7), 59–61.

Rocco, C. G. (1996) *Influencia del Tamaño y Mecanismos de Rotura del Ensayo de Compresión Diametral.* Doctoral thesis. Dep. Ciencia de Materiales, Universidad Politecnica de Madrid, ETS de Ingenieros de Caminos, Ciudad Universitaria, 28040 Madrid, Spain. ('Size-Dependence and Fracture Mechanisms in the Diagonal Compression Splitting Test', in Spanish.)

Rocco, C., Guinea, G. V., Planas, J. and Elices, M. (1995) "The effect of the boundary conditions on the cylinder splitting strength." In *Fracture Mechanics of Concrete Structures*, F. H. Wittmann, ed., Aedificatio Publishers, Freiburg, Germany, pp. 75–84.

Rodriguez-Ortiz, J. M. (1974) *Study of Behavior of Granular Heterogeneous Media by Means of Analogical and Mathematical Discontinuous Models.* Doctoral thesis. Universidad Politécnica de Madrid, 28040-Madrid, Spain.

Roelfstra, P. E. (1988) *Numerical Concrete.* Doctoral thesis. Laboratoire de Matériaux de Construction, Ecole Polytéchnique Féderale de Lausanne, Lausanne, Suisse.

Roelfstra, P. E. and Wittmann, F. H. (1986) "Numerical method to link strain softening with failure of concrete." In *Fracture Toughness and Fracture Energy of Concrete*, F. H. Wittmann, ed., Elsevier Science, Amsterdam, pp. 163–175.

Roelfstra, P. E., Sadouki, H. and Wittmann, F. H. (1985) "Le béton numérique." *Mater. Struct.,* **18**, 327–335

Rokugo, K., Iwasa, M., Suzuki, T. and Koyanagi, W. (1989) "Testing methods to determine tensile strain softening curve and fracture energy of concrete." In *Fracture Toughness and Fracture Energy: Test Methods for Concrete and Rock*, H. Mihashi, H. Takahashi and F. H. Wittmann, eds., Balkema, Rotterdam, pp. 153–163.

Rolfe, S. T. and Barsom, J. M. (1987) *Fracture and Fatigue Control in Structures*, 2nd edition, Prentice-Hall, Englewood Cliffs, NJ.

Rooke, D. P. and Cartwright, D. J. (1976) *Compendium of Stress Intensity Factors*, Her Majesty's Stationary Office, London.

Rosati, G. and Schumm, C. (1992) "Modelling of local bar-to-concrete bond in reinforced concrete beams." In *Proc. Int. Conf. on Bond in Concrete – From Research to Practice*, Vol. 3, A. Skudra and R. Tepfers, eds., pp. 34–43.

Roscoe, R. A. (1952) "The viscosity of suspensions of rigid spheres." *Brit. J. Appl. Phys.,* **3**, 267–269.

Rosen, B. W. (1965) "Mechanics of composite strengthening." In *Fiber Composite Materials*, Chapter 3, Am. Soc. for Metals Seminar.

Ross, C. A. and Kuennen, S. T. (1989) "Fracture of concrete at high strain-rates." In *Fracture of Concrete and Rock: Recent Developments*, S. P. Shah, S. E. Swartz and B. Barr, eds., Elsevier Applied Science, London, pp. 152–161.

Ross, C. A., Thompson, P. Y. and Tedesco, J. W. (1989) "Split-Hopkinson pressure-bar tests on concrete and mortar in tension and compression." *ACI Mater. J.,* **86**(5), 475–481.

Rossi, P. and Richer, S. (1987) "Numerical modeling of concrete cracking based on a stochastic approach." *Mater. Struct.,* **21**, 3–12.

Rossi, P. and Wu, X. (1992) "Probabilistic model for material behavior analysis and appraisement of concrete structures." *Mag. Concrete Res.,* **44**, 271–280.

Rossmanith, H. P., ed. (1993) *Fracture and Damage of Concrete and Rock*, E & FN Spon, London.

Rots J. G. (1988) *Computational Modeling of Concrete Fracture.* Doctoral thesis. Delft University of Technology, Delft, The Netherlands.

Rots, J. G. (1989) "Stress rotation and stress locking in smeared analysis of separation." In *Fracture Toughness and Fracture Energy, Test Methods for Concrete and Rock*, H. Mihashi, H. Takahashi and F. H. Wittmann, eds., Balkema, Rotterdam, pp. 367–382.

Rots, J. G. (1992) "Simulation of bond and anchorage: Usefulness of softening fracture mechanics." In *Application of Fracture Mechanics to Reinforced Concrete*, A. Carpinteri, ed., Elsevier Applied Science, London, pp. 285–306.

Rots, J. G., Nauta, P., Kusters, G. M. A. and Blauwendraad, J. (1985) "Smeared crack approach and fracture localization in concrete." *Heron,* **30**(1), 1–48.

Ruiz, G. (1996) *El Efecto de Escala en Vigas de Hormigón Débilmente Armadas y su Repercusión en los Criterios de Proyecto.* Doctoral thesis. Dep. Ciencia de Materiales, Universidad Politecnica de Madrid, ETS de Ingenieros de Caminos, Ciudad Universitaria, 28040 Madrid, Spain. ('Size Effect in Lightly Reinforced Concrete Beams and its Repercussion on Design Criteria', in Spanish.)

Ruiz, G. and Planas, J. (1994) "Propagación de una fisura cohesiva en una vigas de hormigón debilmente armadas: modelo de la longitud efectiva de anclaje." *Anales de Mecánica de la Fractura,* **11**, 506–513.

Ruiz, G. and Planas, J. (1995) "Estudio experimental del efecto de escala en vigas debilmente armadas." *Anales de Mecánica de la Fractura,* **12**, 446–451.

Ruiz, G., Planas, J. and Elices, M. (1993) "Propagación de una fisura cohesiva en vigas de hormigón debilmente armadas." *Anales de Mecánica de la Fractura,* **10**, 141–146.

Ruiz, G., Planas, J. and Elices, M. (1996) "Cuantía mínima en flexión: Teoría y normativa." *Anales de Mecánica de la Fractura,* **13**, 386–391.

Rüsch, H. (1960) "Researches toward a general flexural theory for structural concrete." *ACI J.,* **57**(1), 1–28.

Russ, J. C. (1994) *Fractal Surfaces*, Plenum Press, New York.

Sabnis, G. M. and Mirza, S. M. (1979) "Size effects in model concretes?" *J. Struct. Div.-ASCE,* **106**, 1007–1020.

Saenz. L. P. (1964) "Discussion of 'Equation for stress-strain curve of concrete' by P. Desay and S. Krishnan." *ACI J.*, **61**, 1229–1235.

Saleh, A. L. and Aliabadi, M. H. (1995) "Crack growth analysis in concrete using boundary element method." *Eng. Fract. Mech.*, **51**, 533–545.

Sallam, S. and Simitses, G. J. (1985) "Delamination buckling and growth of flat, cross-ply laminates." *Compos. Struct.*, **4**, 361–381.

Sallam, S. and Simitses, G. J. (1987) "Delamination buckling of cylindrical shells under axial compression." *Compos. Struct.*, **8**, 83–101.

Sammis, C. G. and Ashby, M. F. (1986) "The failure of brittle porous solids under compressive stress state." *Acta Metall.*, **34**(3), 511–526.

Sanderson, T. J. O. (1988) *Ice Mechanics Risks to Offshore Structures*, Graham and Trotman, Boston.

Saouma, V. E., Ayari, M. L. and Boggs, H. (1989) "Fracture mechanics of concrete gravity dams." In *Fracture of Concrete and Rock*, S. P. Shah and S. E. Swartz, eds., Springer-Verlag, New York, pp. 311–333.

Saouma, V. E., Barton, C. C. and Gamaleldin, N. E. (1990) "Fractal characterization of fracture surfaces in concrete." *Eng. Fract. Mech.*, **35**, 47–53.

Saouma, V. E., Broz, J. J., Brühwiler, E. and Ayari, M. L. (1990) *Fracture Mechanics of Concrete Dams, Vols. I, II, and III.* Reports submitted to the Electric Power Research Institute, Department of Civil Engineering, University of Colorado, Boulder, CO.

Saouma, V. E., Broz, J. J., Brühwiler, E. and Boggs, H. L. (1991) "Effect of aggregate and specimen size on fracture properties of dam concrete." *J. Mater. Civil Eng.*, **3**(3), 204–218.

Sayers, C. M. and Kachanov, M. (1991) "A simple technique for finding effective elastic constants of cracked solids for arbitrary crack orientation statistics." *Int. J. Solids Struct.*, **27**(6), 671–680.

Scanlon, A. (1971) *Time Dependent Deflections of Reinforced Concrete Slabs.* Doctoral thesis. Univ. of Alberta, Edmonton, Canada.

Schapery, R. A. (1975a) "A theory of crack initiation and growth in viscoelastic media I. Theoretical development." *Int. J. Fracture,* **11**, 141–159.

Schapery, R. A. (1975b) "A theory of crack initiation and growth in viscoelastic media II. Approximate methods of analysis." *Int. J. Fracture,* **11**, 369–388.

Schapery, R. A. (1975c) "A theory of crack initiation and growth in viscoelastic media III. Approximate methods of analysis." *Int. J. Fracture,* **11**, 549–562.

Schijve, J. (1979) "Four lectures on fatigue crack growth." *Eng. Fract. Mech.*, **11**, 167–221.

Schlaich, J., Schafer, K. and Jannewein, M. (1987) "Toward a consistent design for structural concrete." *J. Prestressed Concrete Inst.*, **32**(3), 75–150.

Schlangen, E. (1993) *Experimental and Numerical Analysis of Fracture Processes in Concrete.* Doctoral thesis. Delft University of Technology, Delft, The Netherlands.

Schlangen, E. (1995) "Computational aspects of fracture simulations with lattice models." In *Fracture Mechanics of Concrete Structures*, F. H. Wittmann, ed., Aedificatio Publishers, Freiburg, Germany, pp. 913–928.

Schlangen, E. and van Mier, J. G. M. (1992) "Shear fracture in cementitious composites, Part II: Numerical simulations." In *Fracture Mechanics of Concrete Structures*, Z. P. Bažant, ed., Elsevier Applied Science, New York, pp. 671–676.

Schulson, E. M. (1990) "The brittle compressive fracture of ice." *Acta Metall. Mater.*, **38**, 1963–1976.

Schulson, E. M. and Nickolayev, O. Y. (1995) "Failure of columnar saline ice under biaxial compression: failure envelopes and the brittle-to-ductile transition." *J. Geophys. Res.*, **100**(B11), 22383–22400.

Şener, S. (1992) "Bond splice tests." In *FIP'92 Symposium*, G. Tassi, ed., Hungarian Scientific Society of Building, Budapest, pp. 357–362.

Şener, S., Bažant, Z. P. and Becq-Giraudon, E. (1997) *Sizew effect in failure of lap splice of reinforcing bars in concrete beams.* Report, Northwestern University, Evanston, IL.

Serrano, A. A. and Rodriguez-Ortiz, J. M. (1973) "A contribution to the mechanics of heterogeneous granular media." In *Proc. Symp. Plasticity and Soil Mech.*.

Shah, S. P. and John, R. (1986) "Strain rate effects on mode I crack propagation in concrete." In *Fracture Mechanics and Fracture Energy of Concrete*, F. H. Wittmann, ed., Elsevier, Amsterdam, pp. 453–465.

Shah, S. P. and McGarry, F. J. (1971) "Griffith fracture criterion and concrete." *J. Eng. Mech. Div.-ASCE,* **97**, 1663–1676.

Shetty, D. K., Rosenfield, A. R. and Duckworth, W. H. (1986) "Mixed mode fracture of ceramics in diametrical compression." *J. Am. Ceram. Soc.*, **69**(6), 437–443.

Sidoroff, F. (1974) "Un modéle viscoélastique non linéaire avec configuration intermédiaire." *J. de Mécanique,* **13**, 679–713.

Sih, G. C. (1973) *Handbook of Stress Intensity Factors for Researchers and Engineers*, Lehigh University, Bethlehem, PA.

Sih, G. C. (1974) "Strain energy density factor applied to mixed mode crack problems." *Int. J. Fracture,* **10**, 305–322.

Simo, J. C. (1988) "A framework for finite strain elastoplasticity based on maximum plastic dissipation and the multiplicative decomposition." *Comput. Meth. Appl. Mech. Eng.,* **66**, 199–219. (Also **68**, 1–31.)

Simo, J. C. and Oliver, J. (1994) "A new approach to the analysis and simulation of strong discontinuities." In *Fracture and Damage in Quasibrittle Structures*, Z. P. Bažant, Z. Bittnar, M. Jirásec and J. Mazars, eds., E & FN Spon, London, pp. 25–39.

Simo, J. C. and Ortiz, M. (1985) "A unified approach to finite deformation elasto-plasticity based on the use of hyperelastic constitutive equations." *Comput. Meth. Appl. Mech. Eng.,* **49**, 177–208.

Simo, J. C. and Rifai, S. (1990) "A class of mixed assumed strain methods and the method of incompatible nodes." *Int. J. Numer. Meth. Eng.,* **29**, 1595–1638.

Simo, J. C., Oliver, J. and Armero, F. (1993) "An analysis of strong discontinuities induced by strain-softening in rate-independent inelastic solids." *Comput. Mech.,* **12**, 227–296.

Sinha, B. P., Gerstle, K. H. and Tulin, L. G. (1964) "Stress-strain relations for concrete under cyclic loading." *ACI J.,* **62**(2), 195–210.

Slepyan, L. I. (1990) "Modeling of fracture of sheet ice." *Izvestia AN SSSR, Mekhanika Tverdogo Tela,* **25**(2), 151–157.

Smith, C. W. and Kobayashi, A. S. (1993) "Experimental fracture mechanics." In *Handbook of Experimental Mechanics*, 2nd edition, A. S. Kobayashi, ed., VCH Publishers and Society for Experimental Mechanics (SEM), pp. 905–968.

Smith, E. (1995) "Recent research on the cohesive zone description of an elastic softening material." In *Fracture of Brittle Disordered Materials: Concrete, Rock, Ceramics*, G. Baker and B. L. Karihaloo, eds., E & FN Spon, London, pp. 450–463.

Sneddon, I. N. (1946) "The distribution of stress in the neighbourhood of a crack in an elastic solid." *Philos. T. Roy. Soc. A,* **187**, 229–260.

Sneddon, I.N. and Lowengrub, M. (1969) *Crack Problems in the Classical Theory of Elasticity*, John Wiley & Sons, New York.

So, K. O. and Karihaloo, B. L. (1993) "Shear capacity of longitudinally reinforced beams – A fracture mechanics approach." *ACI Struct. J.,* **90**, 591–600.

Sok, C., Baron, J. and François, D. (1979) "Mécanique de la rupture appliquée au béton hydraulique." *Cement Concrete Res.,* **9**, 641–648.

Sozen, M. A., Zwoyer, E. M. and Siess, C. P. (1958) "Strength in shear of beams without web reinforcement." *Bulletin Eng. Experiment Station,* **452**, 1–69.

Spencer, A. J. M. (1971) "Theory of invariants." In *Continuum Physics, 1 –Mathematics*, A. Cemal Eringen, ed., Academic Press, New York, pp. 239–353.

Srawley, J. E. (1976) "Wide range stress intensity factor expressions for ASTM E-399 standard fracture toughness specimens." *Int. J. Fracture,* **12**, 475–476.

Steif, P. S. (1984) "Crack extension under compressive loading." *Eng. Fract. Mech.,* **20**, 463–473.

Strang, G. (1980) *Linear Algebra and Its Applications*, 2nd edition, Academic Press, New York.

Stroh, A. N. (1957) "A theory of the fracture of metals." *Adv. Phys., Philos. Mag. Suppl.,* **6**, 418–65.

Stroud, A. H. (1971) *Approximate Calculation of Multiple Integrals*, Prentice-Hall, Englewood Cliffs, NJ.

Suidan, M. and Schnobrich, W. C. (1973) "Finite element analysis of reinforced concrete." *J. Struct. Div.-ASCE,* **99**(10), 2109–2122.

Suo, Z. and Hutchinson, J. W. (1989) "Sandwich test specimens for measuring interface crack toughness." *Mater. Sci. Eng. A,* **107**, 135–143.

Suresh, S. (1991) *Fatigue of Materials*, Cambridge University Press, Cambridge.

Swartz, S. E. and Go, C. G. (1984) "Validity of compliance calibration to cracked concrete beams in bending." *Exp. Mech.,* **24**(2), 129–134.

Swartz, S. E. and Refai, T. M. E. (1989) "Influence of size effects on opening mode fracture parameters for precracked concrete beams in bending." In *Fracture of Concrete and Rock*, S. P. Shah and S. E. Swartz, eds., Springer-Verlag, New York, pp. 242–254.

Swartz, S. E. and Taha, N. M. (1990) "Mixed-mode crack propagation and fracture in concrete." *Eng. Fract. Mech.,* **35**(1-3), 137–144.

Swartz, S. E. and Taha, N. M. (1991) "Crack propagation and fracture of plain concrete beams subjected to shear and compression." *ACI Struct. J.,* **88**(2), 177–196.

Swartz, S. E., Hu, K. K. and Jones, G. L. (1978) "Compliance monitoring of crack growth in concrete." *J. Eng. Mech. Div.-ASCE,* **104**, 789–800.

Swartz, S. E., Hu, K. K., Fartash, M. and Huang, C. M. J. (1982) "Stress intensity factors for plain concrete in bending – Prenotched versus precracked beams." *Exp. Mech.,* **22**(11), 412–417.

Swenson, D. V. and Ingraffea, A. R. (1991) "The collapse of the Schoharie Creek bridge: A case study in concrete fracture mechanics." *Int. J. Fracture,* **51**(1), 73–92.

Tada, H., Paris, P. C. and Irwin, G. R. (1973) *The Stress Analysis of Cracks Handbook*, Del Research Corporation, Hellertown, PA.

Tada, H., Paris, P. C. and Irwin, G. R. (1985) *The Stress Analysis of Cracks Handbook*, Paris Productions, Saint Louis, MO.

Tandon, S., Faber, K. T., Bažant, Z. P. and Li, Y.-N. (1995) "Cohesive crack modeling of influence of sudden changes in loading rate on concrete fracture." *Eng. Fract. Mech.,* **52**(6), 987–997.

Tang, T., Shah, S. P. and Ouyang, C. (1992) "Fracture mechanics and size effect of concrete in tension." *J. Struct. Eng.-ASCE,* **118**(11), 3169–3185.

Taylor, G. I. (1938) "Plastic strain in metals." *J. Inst. Metals,* **62**, 307–324.

Tepfers, R. (1973) *A Theory of Bond Applied to Overlapped Tensile Reinforcement Splices for Deformed Bars.* Doctoral thesis. Division of Concrete Structures, Chalmers University of Technology, Göteborg, Sweden.

Tepfers, R. (1979) "Cracking of concrete cover along anchored deformed reinforcing bars." *Mag. Concrete Res.,* **31**(106), 3–12.

Thouless, M. D., Hsueh, C. H. and Evans, A. G. (1983) "A damage model of creep crack growth in polycrystals." *Acta Metall.,* **31**(10), 1675–1687.

Thürlimann, B. (1976) *Shear Strength of Reinforced and Prestressed Concrete Beams, CEB Approach.* Technical Report, E. T. H. Zürich.

Timoshenko, S. (1956) *Strength of Materials*, Van Nostrand, New York.

Timoshenko, S. and Goodier, J. N. (1951) *Theory of Elasticity*, 2nd edition, McGraw-Hill, New York.

Tippett, L. H. C. (1925) "On the extreme individuals and the range of samples taken from a normal population." *Biometrika,* **17**, 364–387.

Triantafyllidis, N. and Aifantis, E (1986) "A gradient approach to localization of deformation. I. Hyperelastic materials." *J. Elasticity,* **16**, 225–237.

Tschegg, E. Kreuzer, H. and Zelezny, M. (1992) "Fracture of concrete under biaxial loading — Numerical evaluation of wedge splitting test results." In *Fracture Mechanics of Concrete Structures*, Z. P. Bažant, ed., Elsevier Applied Science, London, pp. 455–460.

Uchida, Y., Kurihara, N., Rokugo, K. and Koyanagi, W. (1995) "Determination of tension softening diagrams of various kinds of concrete by means of numerical analysis." In *Fracture Mechanics of Concrete Structures*, Vol. 1, F. H. Wittmann, ed., Aedificatio Publishers, Freiburg, Germany, pp. 17–30.

Uchida, Y., Rokugo, K. and Koyanagi, W. (1992) "Application of fracture mechanics to size effect on flexural strength of concrete." *Proceedings of JSCE, Concrete Engineering and Pavements,* (442), 101–107.

Ulfkjær, J. P. and Brincker, R. (1993) "Indirect determination of the σ–w relation of HSC through three-point bending." In *Fracture and Damage of Concrete and Rock*, H. P. Rossmanith, ed., E & FN Spon, London, pp. 135–144.

Ulfkjær, J. P., Brincker, R. and Krenk, S. (1990) "Analytical model for complete moment-rotation curves of concrete beams in bending." In *Fracture Behavior and Design of Materials and Structures*, Vol. 2, D. Firrao, ed., Engineering Materials Advisory Services Ltd. (EMAS), Warley, West Midlands, U.K., pp. 612–617.

Ulfkjær, J. P., Hededal, O., Kroon, I. and Brincker, R. (1994) "Simple application of fictitious crack model in reinforced concrete beams — analysis and experiments." In *Size Effect in Concrete Structures*, H. Mihashi, H. Okamura and Z. P. Bažant, eds., E & FN Spon, London, pp. 281–292.

Vakulenko, A. A. and Kachanov, M. (1971) "Continuum theory of medium with cracks." *Mech. Solids,* **6**, 145–151.

Valente, S. (1995) "On the cohesive crack model in mixed-mode conditions." In *Fracture of Brittle Disordered Materials: Concrete, Rock and Ceramics*, G. Bakker and B. L. Karihaloo, eds., E & FN Spon, London, pp. 66–80.

Vardoulakis, I. (1989) "Shear banding and liquefaction in granular materials on the basis of Cosserat continuum theory." *Ingenieur-Archiv*, **59**, 106–113.

Varga, R. S. (1962) *Matrix Iterative Analysis*, Prentice-Hall, Englewood Cliffs, NJ.

van der Veen, C. (1991) "Splitting failure of reinforced concrete at various temperatures." In *Fracture Processes in Concrete, Rock and Ceramics*, J. G. M. van Mier, J. G. Rots and A. Bakker, eds., E & FN Spon, London, pp. 629–638.

da Vinci, L. (1500s) —*see* The Notebooks of Leonardo da Vinci *(1945), Edward McCurdy, London (p. 546); and* Les Manuscrits de Léonard de Vinci, *transl. in French by C. Ravaisson-Mollien, Institut de France (1881-91), Vol. 3.*.

Wagner, H. (1929) "Ebene blechwandträger mit sehr dünnem stegblech." *Z. Flugtechnik Motorluftschifffahr,* **20**, 8–12.

Walker, S. and Bloen, D. L. (1957) "Studies of flexural strength of concrete–Part 3: Effects of variations in testing procedures." *ASTM Proc.*, **57**, 1122–1139.

Walraven, J. C. (1978) *The Influence of Depth on the Shear Strength of Lightweight Concrete Beams Without Shear Reinforcement.* Stevin Laboratory Report No. 5-78-4, Delft University of Technology, Delft, The Netherlands.

Walsh, P. F. (1972) "Fracture of plain concrete." *Indian Concrete J.*, **46**(11), 469–470 and 476.

Walsh, P. F. (1976) "Crack initiation in plain concrete." *Mag. Concrete Res.*, **28**, 37–41.

Wawrzynek, P. A. and Ingraffea, A. R. (1987) "Interactive finite element analysis of fracture processes: An integrated approach." *Theor. Appl. Fract. Mech.*, **8**, 137–150.

Wecharatana, M. (1986) "Specimen size effect on non-linear fracture parameters in concrete." In *Fracture Toughness and Fracture Energy of Concrete*, F. H. Wittmann, ed., Elsevier Science, Amsterdam, pp. 437–440.

Wecharatana, M. and Shah, S. P. (1980) *Resistance to Crack Growth in Portland Cement Composites.* Report, Department of Material Engineering, University of Illinois at Chicago Circle, Chicago, IL.

Weibull, W. (1939) "A statistical theory of the strength of materials." *Proc. Royal Swedish Academy of Eng. Sci.*, **151**, 1–45.

Weibull W. (1949) "A statistical representation of fatigue failures in solids." *Proc. Roy. Inst. of Techn.*, (27).

Weibull, W. (1951) "A statistical distribution function of wide applicability." *J. Appl. Mech.-T. ASME,* **18**, 293–297.

Weibull W. (1956) "Basic aspects of fatigue." In *Proc. Colloquium on Fatigue*, Springer–Verlag.

Weihe, S. and Kröplin, B. (1995) "Fictitious crack models: A classification approach." In *Fracture Mechanics of Concrete Structures*, F. H. Wittmann, ed., Aedificatio Publishers, Freiburg, Germany, pp. 825–840.

Wells, A. A. (1963) "Application of fracture mechanics at and beyond general yielding." *Brit. Weld. J.*, **10**, 563–570.

Willam, K., Bićanić, N. and Sture, S. (1986) "Composite fracture model for strain-softening and localised failure of concrete." In *Computational Modelling of Reinforced Concrete Structures*, E. Hinton and D. R. J. Owen, eds., Pineridge Press, Swansea, pp. 122–153.

Willam, K., Pramono, E. and Sture, S. (1989) "Fundamental issues of smeared crack models." In *Fracture of Concrete and Rock*, S. P. Shah and S. E. Swartz, eds., Springer Verlag, New York, pp. 142–157.

Williams, M. L. (1952) "Stress singularities resulting from various boundary conditions in angular corners of plates in extension." *J. Appl. Mech.-T. ASME,* **19**, 526–528.

Wilson, W. K. (1966) Untitled. In *Plane Strain Crack Toughness Testing*, Brown, W. F. and J. E Srawley, eds., American Society for Testing and Materials, Philadelphia, pp. 75–76. (Contribution to Discussion, ASTM Special Technical Publication No. 410.)

Wilson, W. K. (1971) *Crack Tip Finite Elements for Plane Elasticity.* Report No. 71-1E7-FM-PWR-P2, Westinghouse.

Wiss, A.N. (1971) *Application of Fracture Mechanics to Cracking in Concrete Beams.* Doctoral thesis. University of Washington.

Wiss, Janney, Eltsner Associates, Inc. and Mueser Rutledge Consulting Engineers (1987) *Collapse of the Thruway Bridge at Schoharie Creek.* Final Report submitted to the New York State Thruway Authority, Wiss, Janney, Eltsner Associates, Inc. and Mueser Rutledge Consulting Engineers, Albany, New York.

Wittmann, F. H., ed. (1983) *Fracture Mechanics of Concrete*, Elsevier, Amsterdam.

Wittmann, F. H. and Zaitsev, Y. V. (1981) "Crack propagation and fracture of composite materials such as concrete." In *Advances in Fracture Research, Preprints 5th. Int. Conf. Fracture*, Vol. 5, D. François, ed., Pergamon Press, Oxford, pp. 2261–2274.

Wittmann, F. H., Roelfstra, P. E. and Kamp, C. L. (1988) "Drying of concrete: an application of the 3L-approach." *Nucl. Eng. Des.*, **105**, 185–198.

Wittmann, F. H., Roelfstra, P. E., Mihashi, H., Huang, Y. Y., Zhang, X. H. and Nomura, N. (1987) "Influence of age of loading, water-cement ratio, and rate of loading on fracture energy of concrete." *Mater. Struct.*, **20**, 103–110.

Wittmann, F. H., Rokugo, K., Brühwiler, E., Mihashi, H. and Simonin, P. (1988) "Fracture energy and strain softening of concrete as determined by means of compact tension specimens." *Mater. Struct.*, **21**, 21–32.

Wright, P. J. F. (1952) "The effect of the method of test on the flexural strength of concrete." *Mag. Concrete Res.*, **11**, 67–76.

Wu, X-R. and Carlsson A. J. (1991) *Weight Functions and Stress Intensity Factor Solutions*, Pergamon Press, Oxford.

Wu, Z. S. and Bažant, Z. P. (1993) "Finite element modeling of rate effect in concrete fracture with influence of creep." In *Creep and Shrinkage of Concrete*, Z. P. Bažant and I. Carol, eds., E & FN Spon, London, pp. 427–432.

Xi, Y. and Bažant, Z. P. (1992) "Markov process model for random growth of crack with R-curve." In *Fracture Mechanics of Concrete Structures*, Z. P. Bažant, ed., Elsevier Applied Science, London, pp. 179–182.

Xi, Y. and Bažant, Z. P. (1993) "Continuous retardation spectrum for solidification theory of concrete creep." In *Creep and Shrinkage of Concrete*, Z. P. Bažant and I. Carol, eds., E & FN Spon, London, pp. 225–230.

Yankelevsky, D. Z. and Reinhardt H. W. (1989) "Uniaxial behaviour of concrete in cyclic tension." *J. Struct. Eng.-ASCE,* **115**(1), 166–182.

Yin, W.-L., Sallam, S., and Simitses, G. J. (1986) "Ultimate axial capacity of a delaminated beam-plate." *AIAA J.,* **24**(1), 123–128.

Young , T. (1807) *A course of lectures on natural philosophy and the mechanical arts*, Vol. 1, , London. (p. 144.)

Yuan, Y. Y., Lajtai, E. Z. and Ayari, M. L. (1993) "Fracture nucleation from a compression-parallel finite-width elliptical flaw." *Int. J. Rock Mech. Min. Sci.*, **30**(7), 873–876.

Yuzugullu, O. and Schnobrich, W. C. (1973) "A numerical procedure for the determination of the behavior of a shear wall frame system." *ACI J.*, **70**(7), 474–479.

Zaitsev, Y. V. (1985) "Inelastic properties of solids with random cracks." In *Mechanics of Geomaterials: Rocks, Concretes, Soils*, Z. P. Bažant,, ed., John Wiley & Sons, Chichester, New York, pp. 89–128.

Zaitsev, Y. V. and Wittmann, F. H. (1974) "Verformung und Bruchvorgang poröser Baustoffe unter kurzzeitiger Belastung und unter Dauerlast." *Deutscher Ausschuss für Stahlbeton,* (232), 65–145.

Zaitsev, Y. V. and Wittmann, F. H. (1981) "Simulation of crack propagation and failure of concrete." *Mater. Struct.,* **14**, 357–365.

Zech, B. and Wittmann, F. H. (1977) "A complex study on the reliability assessment of the containment of a PWR, Part II. Probabilistic approach to describe the behavior of materials." In *Trans. 4th Int. Conf. on Structural Mechanics in Reactor Technology*, H, T. A. Jaeger and B. A. Boley, eds., European Communities, Brussels, pp. 1–14.

Zhang, C.-Y. and Karihaloo, B.L. (1992) "Stability of a crack in a linear viscoelastic tension-softening material." In *Fracture Mechanics of Concrete Structures*, Z. P. Bažant, ed., Elsevier Applied Science, London, pp. 155–162.

Zhou, F. P. (1992) *Time-Dependent Crack Growth and Fracture in Concrete*. Report No. TVBM-1011, Division of Building Materials, Lund University, Lund, Sweden.

Zhou, F. P. (1993) "Cracking analysis and size effect in creep rupture of concrete." In *Creep and Shrinkage of Concrete*, Z. P. Bažant and I. Carol, eds., E & FN Spon, London, pp. 407–412.

Zhou, F. P. and Hillerborg, A. (1992) "Time-dependent fracture of concrete: Testing and modelling." In *Fracture Mechanics of Concrete Structures*, Z. P. Bažant, ed., Elsevier Applied Science, London, pp. 906–911.

Zienkiewicz, O. C. and Pande, G. N. (1977) "Time-dependent multi-laminate model of rocks—A numerical study of deformation and failure of rock masses." *Int. J. Numer. Anal. Meth. Geomech.*, **1**, 219–247.

Zubelewicz, A. (1980) *Contact Element Method.* Doctoral thesis. Technical University of Warsaw, Warsaw, Poland. (In Polish.)

Zubelewicz, A. (1983) "Proposal of a new structural model for concrete." *Archiwum Inzynierii Ladowej,* **29**, 417–429. (In Polish.)

Zubelewicz, A. and Bažant, Z. P. (1987) "Interface element modeling of fracture in aggregate composites." *J. Eng. Mech.-ASCE,* **113**, 1619–1630.

Zubelewicz, A. and Mróz, Z. (1983) "Numerical simulation of rockburst processes treated as problems of dynamic instability." *Rock Mech. Rock Eng.*, **16**, 253–274.

Reference Citation Index

Index